Handbook on Thermal Hydraulics in Water-Cooled Nuclear Reactors

Woodhead publishing series in energy

Handbook on Thermal Hydraulics in Water-Cooled Nuclear Reactors

Volume 1: Foundations and Principles

Second Edition

Edited by

Francesco D'Auria
University of Pisa, Pisa, Italy

Yassin A. Hassan
Texas A&M University, College Station, TX, United States

WP
WOODHEAD PUBLISHING
An imprint of Elsevier

ELSEVIER

Woodhead Publishing is an imprint of Elsevier
50 Hampshire Street, 5th Floor, Cambridge, MA 02139, United States
125 London Wall, London EC2Y 5AS, United Kingdom

ISBN: 978-0-323-85606-5

For information on all Woodhead Publishing publications visit our website at https://www.elsevier.com/books-and-journals

Publisher: Megan Ball
Editorial Project Manager: Joshua Mearns
Production Project Manager: Surya Narayanan Jayachandran
Cover Designer: Vicky Pearson Esser

Typeset by TNQ Technologies

Dedication

We recognize that nuclear thermal hydraulics technology exists because of the contributions of tens of thousands of researchers since 1950, as well as a few hundred other contributions before that time. Regrettably and unavoidably, only 4000+ researchers belong to the list of reference authors, i.e., a tiny fraction of the overall number of contributors. They include theoreticians of thermodynamics, experimentalists, model developers, code developers, nuclear reactor safety specialists, proponents of procedures (e.g. scaling, validation, uncertainty), and others and form the group of "unreferenced researchers" (as mentioned in the Acknowledgments). The book's editors and authors had the privilege of summarizing the contributions of those researchers.

The number who have been leviathans in nuclear thermal hydraulics science and application technology during the last few decades unavoidably drops because of aging and because technology-relevant targets are achieved, and consequently, incentives to inject life into the area (i.e., not only financial incentives) diminish. Recognizing their roles, we dedicate the book to

- The unreferenced researchers
- Those among the leviathans who passed away during the period between the issuance of the first and second editions of the book.

For obvious reasons, we cannot list 100,000+ unreferenced researchers in the former category. From the latter category, however, we mention (regrettably again, not all of them) G. Bandini, T. Bajs, B. Boyack, A. Bucalossi, I. Catton, M. Cumo, M. Dusic, G. F. Hewitt, M. Jankowski, P. Kirillov (chapter author in the first edition), Wilson Lam, B. R. Sehgal, W. T. Sha, H. Staedtke, N. Tricot, E. Uspuras, Gary Wilson, K. Wolfert and G. Yadigaroglu. Their teachings of and dedication to nuclear thermal hydraulics will be a model for future generations of researchers.

Francesco D'Auria and Yassin Hassan, on behalf of the authors

Contents

For additional information on the topics covered in the book, visit the companion site: https://www.elsevier.com/books-and-journals/book-companion/9780323856065

List of contributors

Nusret Aksan Paul Scherrer Institut, Villingen, Switzerland

Dominique Bestion CEA, Grenoble, France

Francesco D'Auria University of Pisa, Pisa, Italy

Jordi Freixa Polytechnic University of Catalonia, Barcelona, Spain

Horst Glaeser GRS, Garching, Germany

Yassin Hassan Texas A&M University, College Station, TX, United States

Jae Jun Jeong Pusan National University, Pusan, Korea

Pavel L. Kirillov IPPE, Obninsk, Russia

Christophe Morel CEA, Grenoble, France

Hisashi Ninokata Tokyo Institute of Technology, Tokyo, Japan; Polytechnic University of Milan, Italy

Francesc Reventos Polytechnic University of Catalonia, Barcelona, Spain

Richard R. Schultz Idaho State University, Pocatello, ID, United States

Klaus Umminger Framatome, Erlangen, Germany

Contributors for volumes 1, 2 and 3

Nusret Aksan was born on Apr. 13, 1945, in Bursa, Turkey. He holds a BSc degree (1967) and an MSc degree (1969) in mechanical engineering from Middle East Technical University, Ankara, and an MSc degree (1971) in nuclear engineering from the University of London. He has more than 45 years of experience in the nuclear engineering field. After working in the commissioning of the Karachi Nuclear Power Plant (CANDU), he joined the Paul Scherrer Institute PSI (formerly the Swiss Federal Institute for Reactor Research, EIR) and conducted two-phase flow and heat transfer research for light water reactor (LWR) safety (1973–2008). His main area of interest is thermal-hydraulic safety analysis of nuclear reactor systems, LWRs, and advanced LWRs. His five other areas of interest are (1) experimental data analysis, (2) modeling of phenomena, (3) computer code modeling and analysis, (4) licensing of nuclear power plants, and (5) active participation and leading roles in the work of various international organizations. Presently he is an independent science and engineering consultant.

[2]**Andrea Bersano** was born on Oct. 17, 1992, in Asti, Italy. He holds an MSc degree in energy and nuclear engineering from Politecnico di Torino (2016) and in mechanical engineering from the University of Illinois at Chicago (2017). He received a Ph.D. in energetics from Politecnico di Torino in 2020 with a thesis on passive systems experiments and modeling. He was named a Siebel Scholar in the class of 2019. He joined the Italian National Agency for New Technologies, Energy and Sustainable Economic Development (ENEA) in 2021. His five main areas of interest are deterministic safety analysis, TH system codes, experimental analysis, BEPU, and severe accidents.

[1] He contributed to the first Edition of the book
[2] He contributed to the second Edition of the book

Dominique Bestion was born on Aug. 9, 1955. He graduated from an engineering school in 1978 and received a Ph.D. in fluid mechanics in 1982. He has been working at the Commissariat à l'Energie Atomique since 1983 in the Thermal-Hydraulics Department at CEA-Grenoble. He is currently a research director at CEA, "Professeur Charge de Cours" at the Ecole Polytechnique, and an editor at *Nuclear Engineering and Design Review.* He has worked extensively in modeling two-phase flow and has been the project manager of the CATHARE code. He coordinated two-phase flow modeling activities in the NEPTUNE multiscale thermal hydraulic project and the European NURESIM, NURISP, and NURESAFE projects for a multidisciplinary and multiscale software platform. He coordinated a writing group at the OECD-CSNI on the application of CFD to nuclear reactor safety issues. His five areas of interest are (1) interfacial friction, (2) flashing and condensation, (3) two-phase modeling for SYS TH codes, (4) modeling for two-phase CFD, and (5) multiscale analysis of safety issues.

Francesco D'Auria was born on Aug. 28, 1954 in Benevento, Italy. He holds a master's degree in nuclear engineering (1978) and Ph.D. in nuclear reactor safety (1980) from the University of Pisa. He has been an assistant professor (1981), the University of Pisa); an associate and chair professor (1990), the University of Roma "La Sapienza," and 2000, the University of Pisa); and an adjunct professor (1993–2004, the University of Genova). His primary research activity is concerned with nuclear thermal hydraulics and reactor safety. His other professional positions and awards include Foreign Member of the Argentinean Academy of Science in Buenos Aires (since 2013); the "N.A. Dollezhal Gold Medal" by NIKIET in Moscow (2006); founder and editor in chief (2006–2011) of *J STNI*; member of the J NED editorial board since 1990 and editor since 2019; recipient of TAA from ANS in 2021; GC or TPC Chair of Int. Confs., NURETH (2013), TopSafe (2008, 2012, 2017); NEA SM (2022); author and editor of books; NURETH Fellow (2017); NEA SOAR Editor-Author (1980–2022); CSNI member since 2016; life recognition for activities in BEPU (2018); and ANS member since 1981.

[2]**Juan Carlos Ferreri** was born on Feb. 11, 1944. He graduated as an aeronautical engineer from La Plata University in 1967 and has dedicated his professional career to the field of computational fluid mechanics and heat and mass transfer. He has devoted his work to numerical modeling in nuclear safety and engineering at the Nuclear Regulatory Authority of Argentina. He is currently (a) a fellow member at the National Academy of Sciences of Buenos Aires (since 2009) and was its president from 2017 to 2021 and (b) a retired researcher at the National Research Council (CONICET). He received the Senior 2004 Award for Research, Professional and Teaching Achievements in Argentina from the Argentinean Association for Computational Mechanics. He has been a member and president of the Argentine Committee for Heat and Mass Transfer (CONICET). He has lectured on numerical methods applied to natural circulation in nuclear installations and heat transfer effects in soils for archaeometry purposes in Argentina and abroad (Italy, USA, China, France, and Perú). In the period 1995–2010, he developed research activities in collaboration with researchers at the University of Pisa. He has published more than 110 papers.

[2]**Jordi Freixa** was born on Jun. 12, 1978. He graduated with a degree in physics from the University of Barcelona (2001) and received a Ph.D. in nuclear engineering (2007) from the Universitat Politècnica de Catalunya. Since 2001 he has consistently worked in the field of thermal hydraulics, combining R&D tasks and engineering-oriented activities. He has worked in research institutions (Paul Scherrer Institute, 2007–2012) and cooperated in international projects in connection with regulatory bodies and utilities. In 2013 he joined the Advanced Nuclear Technologies group at UPC and in July 2020 became a professor with a Serra-Húnter contract at UPC, where he leads research projects related to thermal hydraulics and nuclear safety in combination with teaching activities.

Giorgio (Maria) Galassi was born on Oct. 31, 1952, in Riccione, Italy. He received his master's degree in nuclear engineering and Ph.D. in nuclear reactor safety (1987) with a thesis dealing with the assessment of DBA and post-DBA computer codes. From 1982 to 2015 he held the position of senior engineer at the Tuscany Region and from 1999 to 2016 was a member of the "Consiglio del Programma di Dottorato di Ricerca in Ingegneria Nucleare e Sicurezza Industriale" of the University of Pisa. His five main topics of interest in nuclear thermal hydraulics are (1) accuracy quantification and code validation, (2) counterpart tests and scaling, (3) BEPU, (4) ITF experiments, and (5) development and qualification of nodalization.

[2]**Jean-Luc Gandrille** was born on Mar. 14, 1953, in Paris, France. He received engineering degrees (1975) from both the École Nationale Supérieure de Techniques Avancées, Paris, and the Institut National Supérieur des Sciences et Techniques Nucléaires, Saclay. After 1 year of military service as a nuclear intelligence analyst, he joined the Framatome company at the end of 1976 as an analyst engineer of large-break LOCA. He took an active part in post-TMI actions, new safety injection system designs (including coauthoring one patent), and a new symptom-oriented approach to accident management. In 1986 he became section manager in charge of codes and methods development for small-break LOCA and non-LOCA transients. In 1998 he was put in charge of LOCA analyses. In 2006 he was appointed department manager of codes and methods for neutronics and thermal hydraulics, encompassing core physics, core TH, LOCA and non-LOCA transients, CFD, and severe accidents. In 2010 he became the technical advisor to the Process and Safety Division manager, responsible for R&D budgets. He has been retired since 2016.

Horst Glaeser was born on Jan. 11, 1947, in Berlin, Germany. He received his Ph.D. in process engineering in 1976 from the Technical University Berlin, Germany. Later he worked in the design and construction of chemical plants and nuclear-reprocessing plants at UHDE GmbH in Dortmund, Germany. He joined the Gesellschaft fuer Anlagen und Reaktorsicherheit (GRS) in Garching near Munich, Germany, in 1980. From 2002 onward, he was the head of the Cooling Systems Department at GRS until his retirement in July 2014. He is now a senior consultant in nuclear reactor safety analysis. His main interests are (1) the development of thermal-hydraulic computer codes and their verification and validation; (2) the design, scaling, and evaluation of thermal-hydraulic experiments; (3) the development and application of the GRS statistical uncertainty analysis method; (4) deterministic safety analyses and assessment for licensing of nuclear power plants; and (5) safety requirements and guides.

Yassin Hassan is a University Distinguished Professor, a Regents Professor, and the L.F. Peterson ‘36 Chair II in the Department of Nuclear Engineering and the J. Mike Walker ’66 Department of Mechanical Engineering at Texas A&M University. He is the Director of the Center for Advanced Small Modular and Microreactors.

Prior to joining Texas A&M in September 1986, he worked for seven years at the Nuclear Power Division, Babcock & Wilcox Company, Lynchburg, Virginia. His research is in computational and experimental thermal hydraulics and advanced nuclear reactors.

He is a member of the United States National Academy of Engineering, the Academy of Medicine, Engineering & Science of Texas, and the Slovenian Academy of Engineering. He is a fellow of the American Association for the Advancement of Science, American Nuclear Society (ANS), and American Society of Mechanical Engineers (ASME). His awards include the 2008 ANS Seaborg Medal, the 2003 George Westinghouse Gold Medal Award, 2004 Thermal Hydraulics Technical Achievement Award, 2003 Arthur Holly Compton Award of the ANS, 2001 Glenn Murphy Award of the American Association for Engineering Education, 2017 James N. Landis Medal

of the ASME, Distinguished Achievement Award in Research given by the Association of Former Students and Texas A&M University, and 2022 Fluid Engineering Award from the American Society of Mechanical Engineers.
He is the editor of the ANS journal *Nuclear Technology* and an honorary editor of the journal *Nuclear Engineering and Design*. He is a recipient of the 2020 Alumni Award for Distinguished Service from the Grainger College of Engineering at the University of Illinois.

Jae Jun Jeong was born on Jan. 16, 1962, in Yeongcheon, Korea. He joined Seoul National University (1980), where he majored in nuclear engineering. He received his master's and Ph.D. degrees in nuclear engineering at the Korea Advanced Institute of Science and Technology. Thereafter, he joined the Korea Atomic Energy Research Institute (1990), where he was responsible for the safety analysis of PWRs, the development and assessment of a TH system code, and a component-scale two-phase flow code. In Mar. 2011, he moved to the School of Mechanical Engineering at Pusan National University. His five main areas of interest are (1) safety analysis, (2) TH system code, (3) component-scale code, (4) two-phase flow modeling, and (5) numerical methods for two-phase flow.

[2]**Kyung Doo Kim** was born on Apr. 11, 1959, in Seoul, Korea. He graduated from Inha University in engineering in 1981. He received a Ph.D. in nuclear engineering from North Carolina State University in 1991. Since then, he has been working in the Thermal-Hydraulic Department at the Korea Atomic Energy Research Institute and is a professor at the University of Science and Technology. He worked as a project manager for the nuclear reactor safety analysis code development project for 25 years until 2021, playing a leading role in the development and validation of the nuclear reactor safety analysis code, SPACE, obtaining a license for PWR design from the Nuclear Safety and Security Commission in Korea. He serves as chair of the nuclear thermal hydraulics division of the Korea Nuclear Society. His five areas of interest are SYS-TH code, two-phase flow modeling, safetyanalysis for PWR and SMR, integration and application of multidisciplinary codes, and nuclear reactor simulator development.

[1]**Pavel Leonidovich Kirillov** was born on Aug. 20, 1927. He earned his Ph.D. and became a professor in 1971. He graduated from the Moscow Power Engineering Institute, specializing in thermal physics (1950). After graduation, he worked at the Institute of Physics and Power Engineering (IPPE, Obninsk) in various positions from research assistant to director (1950–95). He was a director advisor at the Institute of Thermal Physics and Nuclear Science in the State Scientific Centre of the Russian Federation, IPPE, named after A. I. Leypunsky. He taught at the Obninsk branch of the Moscow Engineering Physics Institute (1956–96) and headed the Departments of Thermal Physics (1972–85) and Nuclear Power Plants (1985–92). He is a member of the International and Russian Engineering Academies, an Honored Worker of Science and Technology (1988), an Honorable Doctor of Obninsk State Technical University for Nuclear Energy, National Research Nuclear University, Moscow Engineering Physics Institute (2003), and a member of the editorial board of *Atomic Energy*. He has more than 350 publications, including monographs, textbooks, handbooks, and seven patents. He is an ASME member. Pavel passed away in Nov. 2021.

[2]**Tomoaki Kunugi** was born on Sep. 15, 1953, in Kofu, Japan. He holds a BE and an ME in chemical engineering from Keio University (1979) and a PhDPh.D. in quantum engineering and system science from the University of Tokyo (1994). He has also been a researcher, senior researcher, and principal researcher at JAERI (1979–1997), a professor at Tokai University, Japan (1998), an associate professor at Kyoto University, Japan (1999–2006), a professor at Kyoto University (2007–2019), a professor emeritus of Kyoto University (2019), and finally, a professor in the College of Energy Engineering, Zhejiang University, P.R. China (2019 to present). His research activities are in thermal hydraulics and reactor safety for nuclear fission and fusion reactors and fundamental studies on nucleate boiling and droplet impact phenomena. His other activities and distinguished achievements have included GC of the 10th Japan-Korea Symp. on Nuclear Thermal Hydraulics & Safety (2016) and GC of the 13th Int. Symp. on Fusion Nuclear Technology (2017). He was awarded a fellow of JSME (2010) and a fellow of AESJ (2009). He received Heat Transfer Society of Japan Academic Awards (2004) and an Int. SFT Award from La Société Française de Thermique at the 12th IHTC in Grenoble, France (2002). He has published more than 400 publications (journals, conference papers, books).

[2]**Fulvio Mascari** was born on Jan. 20, 1978. He has a master's degree in nuclear engineering (2006) and his Ph.D. in nuclear, chemical and safety technology (2010) from the University of Palermo, Italy. Since September 2013 he has worked as a researcher at the ENEA. His technical field is nuclear reactor thermal-hydraulics in reactor coolant systems/ containment, and their coupling and the analyses of severe accident phenomena. In relation to that, he is an expert on the use of best estimate thermal-hydraulic system code (RELAP5 and TRACE) and severe accident code (ASTEC and MELCOR). His current interests are (1) the analyses of the capability of best estimate thermal-hydraulic system code to simulate the main thermal-hydraulic phenomena typical of light water small modular reactor and of fusion reactor; (2) the analyses of the scaling issues and how to address it; (3) investigation of severe accident issues in current and advanced designs; (4) the analyses of the capability of severe accident code to simulate degradation phenomena; and (5) the application of the probabilistic method to propagate input uncertainty in deterministic safety analyses. He is involved in several international activities (e.g. H2020 EURATOM Projects, OECD/NEA/CSNI/WGAMA activities, IAEA activities, etc.).

[1]**Christophe Morel** was born on Mar. 31, 1969. He graduated from an aerospace engineering school in 1993 and received a Ph.D. in fluid mechanics and energetics in 1997. He has been working at the Commissariat à l'Energie Atomique for 20 years in the Thermal-Hydraulics Department at CEA-Grenoble. He currently works on numerical simulation for new energies and materials studies. He has been teaching fluid mechanics and energetics in several engineering schools and universities and published a book on two-phase flow modeling for CFD application in 2015. His interests are numerical simulations in thermal, fluid mechanics (single phase or multiphase), solid mechanics applications, and thermoelectricity.

[2]**Hideo Nakamura** was born in 1956 in Nagoya, Japan, and received a BA (1979) in nuclear engineering, an MS (1981) in crystalline material engineering, and a Ph.D. (1992) in nuclear engineering, all from Nagoya University. He joined the former Japan Atomic Energy Research Institute (JAERI) at Tokai, Japan, in 1981 to work for the ROSA (Rig-of-Safety Assessment) program to study reactor accidents and thermal-hydraulic phenomena for both BWR & PWR with prototypical IETs and SETs as well as code/model developments. He became head of the Thermo-hydraulic Safety Research Group in 2001 involving BDBAs and SAs, and head of the operating agent for the OECD/NEA ROSA and ROSA-2 joint projects with LSTF experiments since 2005, across from the Fukushima-Daiichi accident in 2011. He was an executive editor of *Nuclear Engineering and Technology* for 6 years, from 2015. He has been a technical associate at the Japan Atomic Energy Agency (JAEA) since 2018, and a chair of OECD/NEA WGAMA since 2022.

Hisashi Ninokata was born on Sep. 15, 1946, in Kagoshima Prefecture, Japan. He received his BA in physics and MS and Ph.D. in nuclear engineering from the University of Tokyo. He started working at Tokyo Electric Power Co. in 1977, was engaged in R&D of Japanese prototype LMFBR development from 1980, became a professor at Tokyo Institute of Technology in 1993, and a professor emeritus in 2012. Since 2012 he has been a professor at Dipartimento di Energia, Politecnico di Milano, Italy. His areas of interest include (1) physical modeling of microscopic phenomena, (2) liquid metal boiling, (3) BWR subchannel analysis, (4) natural convection phenomena, and (5) uncertainty identification and quantification. He is a fellow of the American Nuclear Society (ANS) and a recipient of the ANS Technical Achievement Award in Thermal Hydraulics.

[2]**Shripad Revankar** is a professor and director of the Multiphase and Fuel Cell Research Laboratory in the School of Nuclear Engineering at Purdue University, West Lafayette, IN, USA. He also serves as BK21 Plus Visiting Professor in the Division of Advanced Nuclear Engineering at Pohang University of Science and Technology, South Korea. He has more than 40 years of research experience in advanced reactor systems, reactor safety, reactor thermal hydraulics, composite fuel for advanced nuclear reactors, instrumentation, multiphase flow, and heat transfer, microgravity multiphase flow, direct energy conversion, hybrid power systems, nuclear hydrogen generation, solar energy storage, packed bed reactor, renewable energy, and fuel cell technology. He has published more than 400 peer-reviewed technical articles in archival scientific journals and conference proceedings and is the author/coauthor of six books. He is a Fellow of ASME, ANS, and AIChE. He received the Technical Achievement Award from the American Nuclear Society and the Long Service Award from the Nuclear Engineering Division of ASME.

Francesc Reventos was born on May 27, 1951, in Barcelona, Spain. He received his Doctor-Engineer (1982) at the "Institut National Polytechnique de Grenoble" (France). He became an attached professor in the Department of Physics at the Technical University of Catalonia (UPC), also collaborating at the Institute of Energy Technologies of the same university. From 1985 to 2001, he was responsible for plant dynamic analysis in the utility organization operating Asco' NPP and Vandello's NPP, leading the preparation and qualification of corresponding thermal-hydraulic models as well as its use in fields like operation support, probabilistic safety assessment, training, licensing, and fuel management. His teaching and research responsibilities encompass the areas of nuclear safety, reactor physics, plant dynamic analysis, thermal-hydraulic analysis, management of nuclear power plants and energy technology.

[2]**José N. Reyes, Jr** is co-founder and chief technology officer of NuScale Power. He is a professor emeritus and the former head of the Oregon State University School of Nuclear Science and Engineering. He holds a BE degree in nuclear engineering from the University of Florida (1978) and MS and Ph.D. degrees in Nuclear Engineering from the University of Maryland, (1984, 1986). His areas of expertise and publications include systems scaling analysis, reactor thermal hydraulics, passive safety systems, and small modular reactors. He is a coinventor on 116 patents granted or pending in 20 countries, a recipient of several national awards, a NURETH Fellow, a fellow of the American Nuclear Society, and a member of the National Academy of Engineering.

[2]**Jovica Riznic** is a Technical Specialist at the Canadian Nuclear Safety Commission (CNSC), working on regulatory analysis and fitness-for-service assessment of major components of operating nuclear power plants (NPP). After receiving his BSc in 1975, he joined the Vinca Institute of Nuclear Sciences and continued to pursue his graduate studies at the University of Belgrade where he earned his MSc and Sc.D. degrees in 1983 and 1989, respectively. He conducted research at Argonne National Laboratory, Purdue University, and the University of Wisconsin-Milwaukee in the areas of heat and mass transfer, nuclear reactor thermal hydraulics, multi-phase thermo-fluid systems, nuclear reactor safety and reliability, and engineering management. He is an adjunct professor/thesis advisor at the University of Waterloo and Purdue Universities and a faculty at Algonquin College in Ottawa in the School of Business and the Centre for Continuing and Online Education.

Upendra Singh Rohatgi was born in 1949 in Kanpur, India. He received his bachelor of technology degree in mechanical engineering from the Indian Institute of Technology, Kanpur. He then received his master's and Ph.D. degrees in fluids and thermal sciences from Case Western Reserve University, Cleveland, OH, in 1972 and 1975, respectively. He joined the scientific staff of Brookhaven National Laboratory in 1975 in the Department of Applied Sciences. He is currently a senior scientist. He has contributed to US Nuclear Regulatory programs since 1975 in the areas of development and validation of reactor simulation codes, specifically the TRAC, RELAP5, and RAMONA series, and participated in the development of code scaling, applicability, and uncertainty methodology, the first method of estimating uncertainty in code prediction. His expertise is in two-phase flows, nuclear simulation codes, LOCA and instabilities, scaling, and turbomachinery.

Richard R. Schultz received his Ph.D. in nuclear science & engineering (2010) from Idaho State University, Pocatello, ID; MSc in mechanical engineering (1971) from Rensselaer Polytechnic Institute, Troy, NY; and BSc in mechanical engineering (1967) from the University of Florida, Gainesville, FL. He has served in various positions including professor of practice—Texas A&M University; distinguished research engineer (retired)—Idaho National Laboratory, Idaho Falls, Idaho; and research professor—Idaho State University, Pocatello, ID. He has received several recognitions including the Outstanding Engineer Award, 2013, School of Engineering, Idaho State University, Pocatello, ID; Fellow, ASME; Chair, ASME V&V 30 Standards Committee: V&V in Computational Simulation of Nuclear Systems Thermal Fluids Behavior. He has more than 40 years of professional experience in (1) modeling and analyzing the steady-state and transient behavior of power plants and steam supply systems (PWRs, BWRs, gas-cooled reactors, and commercial steam supply systems); (2) V&V of advanced thermal-hydraulic (TH) engineering numerical models; (3) design, scaling, specification, and conduct of thermal-hydraulic experiments; and (4) evaluation of user effect on application of advanced thermal-hydraulic software and BEPU techniques applied to advanced thermal-hydraulic software.

[2]**Mario A. Storti** was born on Aug. 10, 1960. He is a senior researcher of CONICET at CIMEC (Center for Computational Methods Research, Santa Fe, Argentina). He graduated in physics from the Balseiro Institute, National University of Cuyo, Bariloche, in 1983 and in 1990 obtained his Ph.D. at UN Litoral (Santa Fe, Arg), where he is currently a professor of undergraduate and graduate programs. He received the 2012 Senior Scientist Award from the Argentine Association of Computational Mechanics (AMCA). He has published more than 80 articles in indexed journals, has been an advisor to seven doctoral theses, organized international conferences, and edited seven volumes of the Computational Mechanics (AMCA) series. He has conducted numerous transfer activities in the field of computational fluid dynamics for industrial processes. Prof. Storti currently participates in the BIOTRAFO project of the European Union.

[1]**Klaus Umminger** was born on Jan. 13, 1956, in Oberlauda, Germany. After graduating from the Technical University of Munich (diploma in mechanical engineering), he joined AREVA (formerly Siemens/KWU) in 1982, where he has been engaged in experimental projects in the field of reactor safety research and in the design of fluid instrumentation. Since 1995 he has been in charge of the large experimental program PKL. The PKL facility (operated at AREVA Germany) is used to investigate the T/H system behavior of PWRs during postulated accidents. Since 2001 the PKL project has continued as an international OECD project. His main areas of work include the design, performance, and interpretation of experiments for application in code validation, and experimental verification of AM measures for PWRs.

[2]**Bao-Wen Yang** was born in 1956 in Taiwan, received his Ph.D. from Columbia University and joined the Heat Transfer Research Facility (CU-HTRF) (1988), where he was appointed acting director (1993) and then director (1995–2003). He served as the exclusive advisor for Westinghouse/Toshiba/ABB (2003——2012), chief scientist for CGN (2012–2018), and dean/professor for XJTU's School of Nuclear Science and Technology (2012–2017). He founded XJTU's Science and Technology Center for Advanced Fuel Research (2013), was appointed CICHME Chair Professor (2017), established DEQD (2018), and has served as the chief scientist/director at the DEQD Institute for Advanced Research in Multiphase Flow and Energy Transfer. He specializes in reactor thermal hydraulics, nuclear safety, multiphase flow, mixing vane spacer design, rod bundle CHF, subchannel analysis, and CFD simulation. He was awarded the National Outstanding Scholars Award (2011) and NURETH Fellow (2019), has over 250 publications and 23 patents, and has served as chief guest editor for several journals (NED, NSE, NT, etc.) and keynote speaker for numerous conferences/seminars. He organized the first NURETH (NURETH-17, 2017) in China with more than 800 papers and 1000 participants. As the lead organizer and the TPC chair, he helped organize ISACC-I&-II (2013, 2015) and cochair IS-ReCTHA (2018).

Foreword

The latest progress in nuclear thermal hydraulics may not seem to justify the second edition of the book *Thermal Hydraulics in Water-Cooled Nuclear Reactors*, or (*The Handbook*) *of Nuclear Thermal Hydraulics*, just 6 years after the publication of the first edition (2017). Therefore, the main motivations for the present edition are:

- To expand the coverage domain of topics in nuclear thermal hydraulics that were part of the first edition of the book
- To consider findings and directions from a brainstorming activity dealing with nuclear thermal hydraulics at an international level (see Chapter 24), an undertaking in which the authors of the book were deeply involved.
- To make the book interactive for students and interested learners by introducing an appendix with more than 300 solutions to questions and exercises

A proliferation of new books on nuclear thermal hydraulics (i.e., having the words "nuclear thermal hydraulics" in the title) during the last few years has been noted, as discussed in Chapter 1. Two distinct strategies characterize those books:

(1) An innovative two-dimensional (2-D) vision for the discipline focusing on the description of individual, detailed topics, e.g., those found in the list of subjects in the first edition (here, 2-D means topic versus description)
(2) Information transfer and a learning process implying the presentation of bases and background, climbing through details, or a three-dimensional vision for the discipline (i.e., topic versus description versus background)

The former strategy may result in content that is fascinating and up to date; readers may observe the characteristic topics coming together (as in a 2-D space). This observation starts from a single, narrow topic that may connect horizontally to the rest of the disciplines. In this case, an understanding of the origins and motivations for the topic may be missing, and systematic repetitions of concepts are thus unavoidable. A fashion encyclopedia or a structured database of information is the outcome, consistent with current web features and related search mechanisms.

The latter strategy pursues the tedious pileup of knowledge. One may imagine virtual stairs leading from the basement of principles implying supportive and appropriate fundamentals to the boundaries of knowledge. The latter strategy is pursued in the second edition of the book in continuity with the first, namely by

➢ Bringing the discussion of fundamentals up to date, e.g., more systematic consideration of thermal-hydraulic phenomena, specific attention to reflood, etc.

- ➢ Introducing science and technology areas inherent to the domain of nuclear thermal hydraulics, like numerical solution methods and instrumentation
- ➢ More deeply connecting the fundamentals of thermal hydraulics with applications in reactor technology, e.g., nuclear containment issues, core subchannels, and advances in scaling and validation
- ➢ Characterizing the thermal-hydraulic design of nuclear reactor systems and components

Three volumes make up the second edition. This improves the readability of the paper format of the book and helps distinguish the three topical areas. The following titles reflect the bulk of the content in each volume:

- ⇒ Vol I, Foundations and Principles (Chapters 1–9)
- ⇒ Vol II, Modelling (Chapters 10–17)
- ⇒ Vol III, Procedures and Applications (Chapters 18–24 plus nomenclature, including abbreviations, an appendix with Q&A, a subject index, a list of general references, and a list of reference authors)

The ordering of Chapters 1–16 of the first edition is unchanged, avoiding a dismantling of the original structure of the book and minimizing editorial imperfections and modifications that could distort the proposed vision of nuclear thermal hydraulics. In a few cases, this resulted in a nonoptimal number of chapters within a volume, as clarified in Chapter 1.

To improve the technological impact of the book, specialist authors revised existing chapters, performed consistency checks, and added new chapters and sections. Namely, each chapter of the first edition (i.e., Chapters 1–16) has been expanded and improved, with eight chapters (i.e., Chapters 17–24) added. Each chapter now includes at least 10 questions or exercises for which we provide answers or solutions in the Appendix.

The editors and a dozen committed lead authors tried to produce a monograph (handbook) on nuclear thermal hydraulics rather than simply an edited book as the formal outcome of these efforts. Readers may evaluate the success of this attempt.

A key question arose when planning the first edition of the book and is still valid for the second edition: Do the contents of the book monograph (second edition) make all key applications of nuclear thermal hydraulics to nuclear reactor technology possible? The answer is no.

To fulfill industrial needs and requirements for design, safety, and operations, proper coupling is required between nuclear thermal hydraulics and other disciplines. A broad spectrum of a dozen disciplines must be considered for coupling: neutron physics, probabilistic safety assessment, and structural mechanics are key areas, as discussed in Chapter 1. Books exist for any of those disciplines, and any valuable summary implies expertise and resources beyond the current framework (i.e., the issuance of the second edition). A possible third edition might focus on coupling nuclear thermal hydraulics and eventually constitute Volume IV.

A simplified "List of Contents" (identifying three levels of sectional hierarchy) is provided at the beginning of each chapter; this substitutes for highlights and keywords if missing.

Notwithstanding the efforts made, repeated discussions of individual topics were unavoidable. To minimize the inconvenience to the reader, a section has been added to the last chapter of the book (Section 24.10) that identifies duplicated discussions; if applicable, the motivations for those repetitions are provided.

Finally, the preface to the first edition of the book is also provided in this second edition.

Glossary

Glossary applies for the whole book. It constitutes a support and a complement of the Introduction to the Book (Chapter 1). Therefore, Glossary is part of Chapter 1.

This page is here to stress the importance of Glossary.

Preface to the first edition of the book

The technology created after the first self-sustained nuclear chain reaction took place in Chicago, USA, at Chicago Pile-1, in 1942 under the leadership of Enrico Fermi (also known as the "architect of the nuclear age") constitutes the general framework for nuclear reactor thermal-hydraulics and the present book. That event initiated the nuclear power plant (NPP) technology. Related cornerstone events and achievements are summarized as follows:

- Production of electricity by the heat released from the nuclear fission of uranium dioxide
- Water used as a coolant (and as a moderator in the majority of designs) for the region where the chain reaction is sustained, i.e., the core of the reactor
- Construction of large (high power) NPP units
- Safety and design advances for NPPs as well as the advent of computational science

Thermal hydraulics is a pillar discipline that makes it possible to exploit nuclear fission to produce electrical power. The complexity and relevance of the discipline are described by three characteristics:

(A) Turbulence: The existence of turbulence has been known for several hundred years in single-phase flow conditions; at least, it was noticed by Leonardo da Vinci in the 16th century. Three questions can be formulated: (1) Does turbulence exist? (2) What is turbulence? and (3) Can we model turbulence? "Yes" is the easy answer to the first question. The second question can hardly be answered, i.e., not in a way that covers all the details of turbulence. "No" is the generic answer to the third question if all relevant appearances of turbulence in nature or human-developed systems are considered. The intricacy is amplified in nuclear thermal hydraulics by the presence of two phases that interact by turbulence-controlled mass, momentum, and energy-transfer processes.

(B) Bubble motion: I was visiting an institute of the Russian Academy of Sciences in Siberia during the time of the Soviet Union. During my presentation about the capabilities of thermal-hydraulic system codes, a senior researcher who had a "personal" underground lab in the building asked, "How is it possible to predict transient evolutions of two-phase mixtures in complex systems when I cannot calculate the motion of a single bubble?" The problem is more challenging if it concerns thc production, growth coalescence, and collapse of a multitude of bubbles. Averaging is the answer to the question, but it introduces unpredictable errors.

(C) Void fraction and neutron flux: Massive boiling occurs in the core under the design conditions of some reactors: a two-phase mixture is responsible for generating thermal neutrons by a moderation process other than removing fission energy. One key affecting parameter is the water mass in the region. Under high-velocity two-phase flow conditions, the void fraction associated with flow regimes determines the fluid mass. Expertise in thermal hydraulics is required to determine the void fraction with acceptable error including transient conditions; this is directly connected with various flow configurations (flow regimes).

The book attempts to provide a vision of nuclear thermal hydraulics as it appears following more than half a century of development by covering fundamentals to applications. This implies that a universe of topics are considered, as discussed in Chapter 1, that may require an encyclopedia rather than simply a book. So a window is opened to describe selected elements with an emphasis on phenomena and principle equations on the side of fundamentals and nuclear reactor accident scenarios on the side of applications.

Currently, system thermal-hydraulic codes are used to predict accident scenarios in nuclear reactors. At some possible future date, computational fluid-dynamics codes (or sets of equations ready to be solved by computers) may predict similar accident scenarios. In this respect, computer codes constitute a central element of the book.

The processes of identifying and characterizing thermal-hydraulic phenomena, developing and validating code, demonstrating scale, and predicting errors in code calculations, i.e., evaluating uncertainty, may appear to be matters for gurus—rigorous procedures are not always applied or simply do not exist. This is also the consequence of inadequate modeling of some aspects of nature, e.g., behaviors like turbulence, bubble motion, and coalescence, as already noted.

The vision provided in the book is expected to help demonstrate the ability of thermal hydraulics to address the need for design and safety evaluation of existing nuclear reactors, although this is achieved by complex processes involving averaging techniques, validation, scaling, uncertainty, etc. Consequently, the book may resemble the description of a dynamic target rather than an archival product. However, the inspiration to further displace barriers to knowledge can be attained from various discussed topics, of which thermal-hydraulic phenomena and accident scenarios are especially relevant.

The enormous research invested during the period 1970−1990 or 1970−2000, i.e., the golden period for research in nuclear thermal hydraulics, provided the major source of knowledge to justify the book. An understanding of the phenomena sufficient to satisfy technological purposes and a worldwide decline in interest in nuclear fission caused a sharp decrease in the financing and support of research on thermal hydraulics in the subsequent decades.

The established expertise, sometimes of the guru type, will be lost with the retirement of top scientists who were prime actors during the golden period. The knowledge built resembles a small town, one now at risk of becoming a "Pompeii of the year 2020s," buried by the dust of oblivion. Accidents like Chernobyl and, more powerfully, Fukushima mined the trustworthiness of nuclear technology: they produced the dust responsible for clouding the minds of decision-makers. The book contributes to the hope for knowledge preservation, although coordinated efforts by all nuclear technology stakeholders are needed in this regard.

F. D'Auria
September 2016
Pisa, Italy

Preface to the second edition of the book

This second edition is organized like a monograph; however, several authors contributed to provide views and insights that are as complementary as possible; this is true for both the first and the second editions. More than a dozen (i.e., 15) authors and a couple of dozen scientists contributed to the first and second editions, respectively.

The table below includes details to cross-connect editors, authors, and contributors (to sections), with chapters and sections in the first and second editions of the book.

Table of Editors, Authors (of Chapters), and Contributors (to Sections)

Name	Chapters	Sections (2nd Ed.)
Aksan N.	1, 6	
Bersano A.[b]		21.2
Bestion D.	5, 11, 22, 24	
D'Auria F. (Ed.)	1, 2, 8, 14, 15, 16, 18, 19, 21, 22, 23, 24, App. 1	5.12.1, 7.12, 12.1, 11.13, 11.14, 17.9
Ferreri J.C.[b]	20	
Freixa J.[b]	3, 21, App. 1	
Galassi G.M.	15, 16	
Gandrille J-L.[b]	14	
Glaeser H.	4, 10, 13, App. 1	
Hassan Y. (Ed.)	1, 12, 24	
Jeong J.J.	9, 22, App. 1	
Kim K-D.[b]	24	
Kirillov P.[a]	7	
Kunugi T.[b]	17, App. 1	
Mascari F.[b]	21	
Morel C. [a]	5	
Nakamua H.[b]	21	
Ninokata H.	7, 18	
Revankar S.[b]		23.7
Reventos F.	3, 21, App. 1	
Reyes J.N. Jr[b]	21	
Riznic J.[b]		23.7
Rohatgi U.S.	10, 19, App. 1	
Schultz R.R.[a]	4	
Storti M.A.[b]	20	
Umminger K.[a]	8	
Yang B–W.[b]	18	

[a]Contributed to the first edition of the book.
[b]Contributed to the second edition of the book.

A single bound volume was printed for the first edition (Chapters 1–16), although distinguishable as four parts. Three volumes constitute the second edition (Chapters 1–24). Rather than rearranging the chapters, the contents of the first edition were expanded in planning the second edition, with each chapter from the first edition revised. The second edition consists of three volumes of similar size with sequential pagination across all three.

- Volume I: Foundations and Principles. This volume consists of Chapters 1–9; it includes topics ranging from definitions in thermal hydraulics and related applications to balance equations to phenomena identified in accident scenarios.
- Volume II: Modelling. This volume consists of Chapters 10–17; the focus is on modeling the features (Chapters 10–12), procedures, and solution methods needed to apply the models (Chapters 13, 14, and 17) and cross-link between models and phenomena (Chapters 15 and 16).
- Volume III: Procedures and Applications. This volume consists of Chapters 18–24; extensions (e.g. numerical solution methods and scaling), insights (e.g. subchannel thermal hydraulics), and outcomes from the applications (e.g. elements of water-cooled nuclear reactor design) of topics covered in previous chapters make up this volume. The final chapter of the second edition (i.e. Chapter 24) is devoted to critically evaluating the present capabilities of thermal hydraulics and research envisaged for the future.

As a companion to these three volumes, a website has been developed and hosts the complete set of questions and answers as well as a list of reference authors.

Acknowledgments (for the past) and wishes (for the future)

The editors undertook the endeavor of putting out this book after lifetimes dedicated to research in nuclear thermal hydraulics where fundamentals and technology applications intimately merge. The complexity of the topic, over a time span of around 7 decades, results from and promotes deep international cooperation. All this presents a challenge in creating a representative and convincing collection of individuals to acknowledge. Regardless, the following is our best attempt at doing so.

In the first stage of acknowledgments, we recognize that the work of several thousand scientists and technologists has contributed to the topics of concern here. We are profoundly indebted to each of them and we have proposed an alphabetic list of authors derived from the cited references, as every author created or inspired at least one piece of the mosaic constituting the book.

However, and unfortunately, the list of authors outlined above is incomplete. We regret that many more scientists and researchers than those we cite (i.e., 3000+) provided decisive contributions and ideas to move the boundaries of knowledge within nuclear thermal hydraulics. Thus, in the second stage of acknowledgments, we recognize "the unreferenced researchers" and apologize to them for any oversight on our part.

In the final stage, we envision a comprehensive and encompassing but virtually iceberg-shaped collection of individual scientists and groups of scientists to acknowledge for their contributions to the present book on nuclear thermal hydraulics, where only the tip of the iceberg [or item (a) below] has a recognizable identification:

(a) The authors and contributors of chapters who have deep specific expertise in concerned areas, who accepted the duty to depict new visions for those areas
(b) The authors of reference works and the "unreferenced researchers"
(c) Scientists and technologists who participated in international groups, notably at the premises of the Nuclear Energy Agency of the Organisation for Economic Co-operation and Development in Paris and of the International Atomic Energy Agency in Vienna, who provided innumerable ideas and research outcomes
(d) Secretaries, managers, and directors of a variety of institutions, who ensured, sometimes by hidden effort, research sustainability or knowledge financing
(e) Members of acting groups that we had the privilege to lead, who have been functional in solving deep issues, training experts, providing data, and identifying new directions for research in nuclear thermal-hydraulics
(f) Scientists contributing within established international contexts, e.g.,

- The American and European Nuclear Societies and the American Society of Mechanical Engineers
- Conference frameworks like Nuclear Reactor Thermal-Hydraulics, the International Conference on Nuclear Engineering, the International Congress on Advances in Nuclear Power Plants, and the International Topical Meeting of Nuclear Reactor Thermal Hydraulics, Safety and Operation
- Journals like *Nuclear Engineering and Design*, *Science and Technology of Nuclear Installations*, and *Nuclear Engineering and Technology*

(g) Colleagues at universities where we have been teaching for more than 40 years and coworkers including master's and Ph.D. students from many countries

(h) Teachers who opened the doors of knowledge and made our professional careers possible

(i) Parents, relatives, and friends who created or accepted the conditions to keep in touch for a few hours, a few weeks, several years, or a lifetime. International friendships with those who shared a common interest in nuclear thermal hydraulics and applications provided remarkable, unrepeatable, and inexpressible feelings representing signal events in our lives.

Acknowledgments imply considerations of the past. Now what can we extract (i.e., from acknowledgments) by looking into the future? In addition, other than acknowledgments, shall we consider disacknowledgments, or persons who put up obstacles to research? Alternatively, have we learned anything that could help young generations? Obviously, these are uncommon questions that may be answered easily by philosophers but not so easily by researchers.

An indirect address to those questions may derive from the following notes:

- As discussed in the book (e.g., Chapter 2) the issuing of the Interim Acceptance Criteria for the design of emergency cooling systems by the United States Nuclear Regulatory Commission triggered the massive research that has been brought to thermal hydraulics today. The editors had the opportunity to spend the key part of their working lives in the wake of that fluorescent environment. Therefore, only positive acknowledgments are appropriate.
- Ironically, once accidents occurred, such as Three Mile Island and Chernobyl, the massive research that followed confirmed the safety of existing reactors; nonetheless, a decline in nuclear technology started around the year 2000. This notably occurred in countries that contributed to the development of the technology, including nuclear thermal hydraulics. National policies and economic interests have overtaken the technological aspects. A proliferation of research managers and a decrease in attention to reactor safety demonstrations were the consequences. Fashion, more than industrial needs, drove research.

These acknowledgments remain valid for work from the last few decades. However, owing to changes in nuclear technology strategies (not in all countries!) and a lack of realistic objectives, young readers may be unaware of the danger of entering dead-end trails for research and interacting with those who create roadblocks and attempt to slow down or stop valuable initiatives. Young researchers should accept undue challenges but be ready to react. This book may help in connection with those responses.

Francesco D'Auria and Yassin Hassan

Introduction

1

Francesco D'Auria[1], *Yassin Hassan*[2] *and Nusret Aksan*[3]
[1]University of Pisa, Pisa, Italy; [2]Texas A&M University, College Station, TX, United States; [3]Paul Scherrer Institut, Villingen, Switzerland

Foreword

Top-level researchers involved with applying and developing thermal hydraulics worldwide cooperated to write this book (either the first or the second edition).

Any scientist involved with nuclear thermal hydraulics may benefit from reading the book.

Reading this chapter is a prerequisite (although not wholly sufficient) for becoming acquainted with the book.

Referring to the names of the authors of the references cited in Chapter 1 and topics related to thermal hydraulics but not discussed in the book (although listed in Chapter 1) will greatly help in understanding the contents of the book.

1.1 Introduction

The term thermal hydraulics, here preferred to thermohydraulics, thermal-hydraulics, and thermalhydraulics, is derived from two Greek words used by the Romans to indicate heat and water, respectively. Thus, thermal hydraulics is the discipline that studies the transient evolution of engineering systems where (hot heated) water is the coolant fluid and heating and cooling or boiling and condensing processes typically occur.

Nuclear thermal hydraulics addresses the performance of water-cooled nuclear reactors (WCNRs), which are typically water-moderated—i.e., they represent the majority of research reactors (RRs) and power plants in operation and under design today. Selected features for the discipline are as follows:

- The design and safety features of nuclear reactors greatly depend on thermal hydraulics; this has been established since the beginning of the nuclear technology era.
- The following can be noted about the numerical simulation of phenomena:
  - With fluid dynamics and gas dynamics, it shares the difficulty associated with turbulence modeling—the problem becomes even more complex because of the presence of two phases.
  - Time and space averaging are unavoidable when modeling the phenomena; discontinuities, including the interface between the phases, are solved by continuum equations.
- Approximating results from numerical models and needing to bound the errors of calculations, i.e., uncertainty, are consequences of the preceding factors.

Handbook on Thermal Hydraulics in Water-Cooled Nuclear Reactors
https://doi.org/10.1016/B978-0-323-85606-5.00019-2

- Experiments are necessary and highly dependent on geometric configurations; they are not conclusive for suitable knowledge attainment because they are performed at a reduced scale in the case of power reactors.
- Controversy is not surprising when evaluating the quality, applicability, and acceptability of experimental and numerical data for situations of technological interest.

D'Auria et al. (2019), put forth the question "What is nuclear thermal hydraulics?" This is not a science like water chemistry, the theory of elasticity, or neutron physics. Rather, it is an ensemble of topics including procedures, equations, models, interpretation of experimental data, definitions, and assumptions for designing systems, developing numerical codes, or writing papers or entire books. Therefore,

(i) the term nuclear thermal hydraulics shall be associated with water and WCNRs
(ii) experiments are essential for expanding the boundaries of knowledge;
(iii) basic, system, and computational fluid dynamics (CFD) of nuclear thermal hydraulics are distinguishable, although interactions do exist;
(iv) suitable quality for applications requires procedures, e.g., verification and validation (V&V), scaling, uncertainty, and best estimate plus uncertainty (BEPU); and
(v) rethinking the fundamentals of turbulence and closely connecting experimentalists and theoreticians may be ways toward progress.

Thus nuclear thermal hydraulics can be considered a sliding and lapsing technology—expertize based on decades of activity may prove insufficient to address the targets and challenges in the design and safety of various nuclear installations, and a book built on weak foundations may not be justified. The former issue remains open; the present endeavor constitutes an attempt to address the latter—technologists and scientists deeply involved with nuclear thermal hydraulics are invited to be coauthors. The plan is to present, as much as possible, a comprehensive vision of modern nuclear thermal hydraulics, assuming that rudimentary, established knowledge about reactor technology is available to readers.

That the word nuclear is associated with thermal hydraulics testifies to how much knowledge has progressed as nuclear technology has been developed. Providing further detail, the term system thermal hydraulics (SYS TH), or technically nuclear SYS TH, was proposed during the planning of the Specialist Meeting on Transient Two-Phase Flow—Current Issues in System Thermal Hydraulics organized by the Organisation for Economic Co-operation and Development/Nuclear Energy Agency/ Committee on the Safety of Nuclear Installations (OECD/NEA/CSNI), held in Aix-en-Provence (OECD/NEA/CSNI, 1992). The SYS TH acronym reflects the connection between thermal hydraulics and the nuclear system, namely, the size of the concerned system and interactions among phenomena that occur or are expected in complex geometrical configurations and transient conditions during postulated accidents within nuclear reactors.

SYS TH is considered an evolution of fundamental thermal hydraulics and constitutes the main topic and framework of the book.

The computational tools, called SYS TH codes, represent baskets of area expertize and are the direct or indirect targets of the individual chapters, emphasizing the development, qualification, and application of those codes to demonstrate the safety and acceptability of nuclear installations. The bases, applicability of CFD, and role of nuclear reactor safety and design are described to complement the vision provided for nuclear thermal hydraulics.

1.1.1 Scope and framework

A comprehensive list of topics is introduced hereafter to define the framework of thermal hydraulics. The scope and framework of the book are discussed in the last paragraph of the section.

1.1.1.1 Origin of nuclear thermal hydraulics

The words *thermal* and *hydraulics* are etymologically associated with heat and water, respectively. Heat is embedded with concepts like material thermal properties, (thermal) energy balance, thermodynamic efficiency, and energy dissipation. At the same time, the water constituent in the term has limited the scope of the book to WCNRs. Owing to this and based on the historical use of the term thermal hydraulic, any process connected with heat generation and transport is considered part of the discipline called thermal hydraulics. Notably, thermal energy transport in sodium and gas-cooled reactors, natural circulation, and thermal mixing of lead–bismuth systems are important fragments in nuclear thermal hydraulics. The book by Ciofalo (2022) provides a cross-connection between thermotechnics (a fundamental discipline in engineering courses) and existing nuclear thermal hydraulics.

Here, the words' etymological origins are retained while formally establishing their connection to the nuclear world. To that aim, we refer to the facility constructed by Fermi in Chicago to show the sustainability of the fission chain reaction. Nuclear thermal hydraulics was born on Dec. 2, 1942, when Fermi triggered a fission reaction within a multiton graphite brick pile. This ensured, in addition to neutron energy moderation capability and radiation protection, suitable thermal capacity with a heat sink to accommodate the fission power.

In Jul. 1951 the US Congress authorized the construction of a nuclear submarine that entered the sea on January 17, 1955. In the meantime, according to the World Nuclear Association (WNA, 2014), the first nuclear reactor to produce electricity, albeit only a few hundred watts, was the liquid metal-cooled EBR-1 reactor in Idaho on December 20, 1951 (Michal, 2001), with a description indicating that "the liquid metal flowed into water-jacketed pipes to form steam that turned a turbine." The first substantial production of nuclear electricity, i.e., 5 MW, occurred in the Soviet Union by way of a water-cooled and graphite-moderated reactor called AM-1—it commenced operation as the world's first nuclear-powered electricity generator in Obninsk on June 27, 1954.

Thus nuclear thermal hydraulics is an unavoidable part of nuclear technology. Fission power heats a material, which is typically nuclear fuel, and processes are needed to transport generated energy to their application purposes, e.g., toward heat exchangers for thermal power and turbines for electricity generation or (marine) propulsion (Marcus and Mirsky, 2021; Guimaraes, 2020).

1.1.1.2 Single- and two-phase flows

Liquid water and gas water are commonly referred to as liquid and steam or vapor, with the latter term using the word gas to indicate superheated conditions. The thermodynamic evolutions of individual liquid and steam fluids (one-phase or 1Φ), as well as their mixtures or two-phase flows (two-phase or 2Φ), are topics of interest in nuclear thermal hydraulics (e.g., Ishii, 1975; Yadigaroglu and Lahey, 1976). Furthermore, steam flows are associated with noncondensable gas species of substances other than water, and liquid flows may transport solid particles like impurities, including solid fission products or high-temperature-boiling boric acid (typically measured in parts per million, or ppm).

Thus nuclear thermal hydraulics deals with 1Φ and 2Φ water flows plus noncondensable gas plus solid impurities without treating the chemical interactions. However, the metal−water chemical reaction is a notable exception that creates hydrogen at high temperatures, which typically come from a heating surface within the boundary of interest.

1.1.1.3 Prominent scientists, origins, and textbooks

Technology needs associated with single- and two-phase flows arose well before the nuclear era. The transition from horsepower on land and wind power at sea to steam-moved machinery imposed the need to obtain expertize in thermal hydraulics during the 19th and late 18th centuries.

The creation of modern physics by G. Galilei (1564−1642) and I. Newton (1642−1727) and mathematics by R. Descartes (1596−1650) and G. W. von Leibnitz (1646−1716) before the advent of A. Einstein (whose theories have little or no connection to current nuclear thermal hydraulics) was followed by the development of thermodynamics and fluid dynamics with the formulation of the mechanics of continuum by L. Euler (1707−1783), which is the basis of modern thermal hydraulics. Prominent scientists contributed, notably including mathematicians D. Bernoulli (1700−1782), J.L. Lagrange (1736−1813), J. Watt (1736−1819), P.S. Laplace (1749−1827), J.B.J. Fourier (1768−1830), C.L. Navier (1785−1836), N.L.S. Carnot (1796−1832), J.P. Joule (1818−1889), G.G. Stokes (1819−1903), J.A. Froude (1818−1894), H. von Helmholtz (1821−1894), R. Clausius (1822−1888), W. Thomson (Lord of) Kelvin (1824−1907), F. Grashof (1826−1893), J.C. Maxwell (1831−1879; he introduced the use of statistics in thermodynamics in addition to contributing to electromagnetism), W. Gibbs (1839−1903), O. Reynolds (1842−1912; e.g., see Reynolds, 1896), L. Boltzmann (1844−1906), and M. Planck (1858−1947; see, e.g., Planck, 1923).

During the 18th century, the seeds were planted for the birth and growth (during the 20th century) of disciplines such as thermal hydraulics. The idea of linking mathematics and physics was introduced (e.g., Laplace), the concept of deterministic analysis (i.e., one can predict the future state of any system such as the universe by knowing the initial state) already present in Cicero's writing was restored, and the use of statistics to average the behavior of an infinite number of parts was highlighted. Complex technology systems (e.g., nuclear reactors) came later, although implementing those developments was straightforward. Within such a context, L. Prandtl (1875–1953; e.g., see Prandtl, 1956), T. von Karman (1881–1963), K.E.W. Nusselt (1882–1957; e.g., see Nusselt, 1943), S. Nukiyama (1896–1983; e.g., see Nukiyama, 1966), A. Kolmogorov (1903–1987), L.D. Landau (1908–1968), A.V. Lykov (1910–1974; see, e.g., Lykov, 1958), S.S. Kutateladze (1914–1986; see, e.g., Kutateladze, 1963), and C.A. Truesdell (1919–2000; he investigated partial differential equations and outlined Euler's role) directly contributed to the bases of thermal hydraulics without taking leadership in applying it to nuclear technology.

Parallel to nuclear engineering and having limited interactions with it, thermal and fluid engineering/physics evolved during the first half of the 20th century. Fundamentals in those areas are fixed in textbooks by Grober and Erk (1933), Reichardt (1941), Jakob (1949), Landau and Lifchitz (1954), Lykov and Mikhailov (1961), Bird et al. (1960), Spalding (1963), Sparrow and Cess (1966), and Patankar and Spalding (1967).

Rudimentary thermal hydraulics, or elements and concepts of thermal hydraulics, can be found in pioneering nuclear engineering textbooks by Glasstone and Edlund (1952), Schultz (1955), Glasstone and Sesonske (1955), Stephenson (1958), and Etherington (1958) and later by Lamarsh (1966), as well as by Margoulova (1969) in the former Soviet Union.

Nuclear thermal hydraulics was inspired by the rudiments of nuclear engineering textbooks rather than textbooks on thermal and fluid engineering. Professor El-Wakil made the first effort to gather the concepts of nuclear thermal hydraulics into a textbook (El-Wakil, 1962; 1971a,b).

Modern nuclear thermal hydraulics is associated with a couple dozen internationally distributed and recognized textbooks. A representative and nonexhaustive list of those textbooks (ordered by publication date) includes the names of key scientists in that area: Tong (1965), Wallis (1970), Tong and Weisman (1970), Collier (1975), Ishii (1975) [see also Ishii and Hibiki (2006)], Incropera and Dewitt (1981), Delhaye et al. (1981), Fenech (1981), Cumo (1982), Hsu and Graham (1986), Todreas and Kazimi (1990), Hewitt et al. (1992), Lahey and Moody (1993), Levy (1999), and Kirillov et al. (2010, 2013, 2014).

To best obtain a complete and current picture of nuclear thermal hydraulics reference books and scientists, the following should be considered:

- Reports issued by OECD/NEA and the International Atomic Energy Agency (IAEA) dealing with applications and needs in nuclear thermal hydraulics (e.g., Aksan et al., 1987 and Lewis et al., 1989) from the OECD/NEA/CSNI side and Allison et al. (2002) from the IAEA side
- Reports (code manuals) and support papers describing SYS TH computer codes, e.g., from RELAP2 (covered by Moore and Rettig, 1968, to Berry et al., 2015) to RELAP7 (addressed by Anders et al., 2012), to the closure laws of the CATHARE code (dealt with by Bestion, 1990)
- CFD technology and related CFD and direct numerical simulation (DNS) codes (e.g., Hassan, 2005)
- Textbooks dealing with topics other than SYS TH for nuclear reactors where thermal hydraulics plays a key role, like nuclear plants and nuclear energy (e.g., Knief, 2008; Cacuci, 2010; Oka, 2014; Murray and Holbert, 2014; Sjoden, 2015; Chao, 2022); nuclear fuel subchannel analysis (e.g., Ninokata and Aritomi, 1992); industrial plant safety (e.g., Banerjee, 2003);

investigations into selected phenomena (e.g., Yadigaroglu, 1975); fluid–structure interactions (FSIs) (e.g., Païdoussis et al., 2014); thermal-hydraulic instrumentation (e.g., Mayinger, 1994; Hassan, 2019); and the design of key components for nuclear and thermal plants (e.g., Cumo and Naviglio, 1990), including those focused on centrifugal pumps (e.g., Gülich, 2010), steam generators (e.g., Annaratone, 2008), and turbines (e.g., Bloch and Singh, 2009)

- Edited and issued open-access textbooks (see later) focusing on CFD applications
- Some scientists have published outstanding papers over a period exceeding a half-century, like Novak Zuber (see e.g., Zuber, 1958; Zuber, 2010), others have published single papers that remained important for a half-century, like Yamanouchi's work on the reflood modeling idea (see, e.g., Yamanouchi, 1968), and others still have devoted their half-century working lives to a single thermal-hydraulic phenomenon, like D.C. Groeneveld's work on critical heat flux (see, e.g., Groeneveld, 1996).
- An integrated vision was proposed by N. Kolev through a series of five volumes dealing with multiphase flow dynamics (Kolev, 2011–2025). That series is the outcome of that author's lifetime effort. Basilar physical phenomena are considered in an attempt to reconstruct complex system performance from the knowledge and modeling of local interactions among substances. Both water and molten material interactions are concerned with handling severe accident conditions at nuclear power plants (NPPs).

Recent textbooks containing the words nuclear thermal hydraulics in the title deserve specific attention. Seven books are considered (in alphabetic order by first author or editor) by Akimoto et al. (2016), Guillen (2013), Lopez De Bertodano et al. (2017), Masterson (2019); Saha (2017), Wang et al. (Eds.) (2018), and Zohuri and Fathi (2015). The focuses of those books differ quite a bit (as pointed out by D'Auria et al. (2019), who elaborated on a cross-link of contents among those books, except Masterson's and the first edition of the present book), testifying that the common understanding of the term nuclear thermal hydraulics has diverged rather than converged during the last few years. Notably, (1) the book by Akimoto is a summary of teaching experience, i.e., it does not address the progress of science; (2) the book by Guillen discusses the framework of a few CFD applications for nuclear and rocket technologies; (3) numerics and instability in two-phase flows is the key topic of the book by Lopez de Bertodano; (4) a two-dimensional vision of nuclear thermal hydraulics, already mentioned in the foreword of the present book, characterizes the book by Masterson. A systematic connection between fundamentals and applications is missing in that work; rather, all topics, e.g., from a suitable list of subjects, are interconnected like stations in a metro network. Such a book idea is undoubtedly fascinating (innovative in the present area) and reflects the website structure of competences; whether this is valuable for progress will be seen in future decades; (5) the book by Saha constitutes a coherent industrial view of nuclear thermal hydraulics bounded by the lifetime expertize of the author; (6) a collection of a dozen randomly selected papers on nuclear thermal hydraulics constitutes the book by Wang and others; and (7) a broad range of information covering economics, reactor description, fuel cycle, including waste, risk assessment, and other topics, is covered in the nuclear thermal hydraulics book by Zohuri.

The textbooks referenced in this section provide a snapshot, still narrow and incomplete, of the complexity of nuclear thermal hydraulics. Taken together, they constitute a challenge for the present book (first and second editions).

1.1.1.4 *Journals, conferences, the web, and selected international institutions*

Nuclear thermal hydraulics, like many disciplines, has space and an interest in devoted journals, national and international conferences, and the web. Namely, the web not only is a mirror for journal and conference papers but may also include much more, like master's theses, PhD dissertations, documents issued by institutions and companies, and lectures by professors and experts.

Listing journals and conferences covering nuclear thermal hydraulics is not the purpose here. Granted, a couple dozen journals regularly publish papers dealing with nuclear thermal hydraulics, and on average, more than one large (more than 500 papers presented) conference deals with the topic. The purpose here, rather, is to give an idea of the area's overall number of papers and its rate of growth.

Starting from the 1950s and considering the exponential growth in publications after the advent of computers and the Internet, roughly 10,000–20,000 papers have thus far been published within the nuclear thermal-hydraulic area. The current growth rate is around 1000 papers/year when both journals and conferences are considered. This may also provide an idea of how much effort is needed to follow or interpret developments in the field.

The importance of a few international institutions (one is actually a national institution) in the development of nuclear thermal hydraulics is mentioned. The concerned institutions are the already cited OECD/NEA/CSNI, IAEA, and US Nuclear Regulatory Commission (USNRC). The role of each is discussed in more detail in Chapter 2 and can be summarized as follows:

- CSNI has had, for half a century, a leading role in achieving a common understanding of SYS TH, identifying and characterizing phenomena, see e.g., Aksan et al. (2018a), supporting experimental research, and developing procedures for applications to nuclear reactor safety and design. The following reports provide deep insights into this role: OECD/NEA/CSNI (1992), OECD/NEA/CSNI and USNRC (1997), and OECD/NEA/CSNI (2022).
- The role of the IAEA is twofold. On the one hand, the needs of thermal hydraulics for safety analysis of nuclear reactors are streamlined, e.g., IAEA (2002). On the other hand, deep investigations into specific topics requested by member countries are supported, .e.g., IAEA (2002a).
- USNRC imposed directions on industry and research organizations for the development of thermal hydraulics, e.g., USAEC (1971). At the same time, it supported finalized investigations for applying thermal hydraulics, USNRC (1988), and fixed requirements for connecting thermal hydraulics and licensing, e.g., USNRC (2013a).

1.1.1.5 *Education and training*

Nuclear thermal hydraulics is an educational topic in university undergraduate, graduate, and PhD courses. Several dozen universities around the world offer nuclear thermal hydraulics courses; one might estimate that a few thousand students each year attend those courses, although the country-specific numbers change each year.

Specialized training courses exist and cover thermal-hydraulic fundamentals and application topics (see the course founded in Zurich by Prof. G. Yadigaroglu) and the technology for using SYS TH codes (e.g., see Petruzzi et al., 2008). Remarkably, international institutions like IAEA in Vienna and NEA in Paris organize workshops and seminars devoted to transferring competence in nuclear thermal hydraulics; see, e.g., D'Auria and Bousbia Salah (2004). The material in the present book can be further used for training and education in nuclear thermal hydraulics.

1.1.1.6 Target nuclear power plant and research reactor types and structures, systems, and components of nuclear installations

The construction and operation of NPPs have resulted from one of the most complex endeavors of human minds—discoveries and technological applications from various technological sectors are necessary to design, construct, operate, and evaluate the safety of NPPs.

Exploiting fission reactions for thermal energy and, after proper transformation, mechanical or electrical power is not exclusively connected to water. Various options are available, including gas or liquid metal as a coolant, graphite as a moderator, or even dissolving nuclear fuel into the coolant, i.e., homogeneous reactors. Nuclear power reactors are categorized roughly according to their "generation" as follows, https://fr.wikipedia.org/wiki/Générations-de-réacteurs-nucléaires:

- Gen I: reactors designed and constructed before the end of the 1970s (in France, UNGG natural uranium graphite gas)
- Gen II: current reactors constructed from the 1970s to the end of the 1990s and in operation, including pressurized water reactors (PWRs), boiling water reactors (BWRs), AGR, CANDU, RBMK, and VVER
- Gen III: advanced reactors, including CE system80+, CANDU6, and AP600
- Gen III+: evolutionary reactors, including EPR, AP1000, APR1400, ACR1000, ABWR, and ESBWR
- Gen IV: reactors of the future, including sodium fast neutron reactor, gas fast neutron reactor, lead fast neutron reactor, very high-temperature gas-cooled reactor, molten salt reactor, and supercritical water reactor

The current (2021) situation arose from strategic, financial, marketing, and technological considerations, which resulted in 99% of operated NPPs being cooled by water and more than 95% being cooled and moderated by water, whether light or heavy. Using water as the coolant brought high pressure to the cooling loop encasing the nuclear core, e.g., 7–15 MPa, to ensure suitable thermal efficiency. The high pressure, high operational power (up to 4500 MWt), and power density (up to 100 kW/dm^3 in the core region, or 50 kW/m linear power for the fuel pin) impose (1) a large geometry (up to around 300 m^3) for the overall cooling system and (2) an equivalent diameter of flow channels in the range of around 0.01 m in the core region. All of this, combined with:

- possible neutron flux (reactivity) excursions translating into thermal power pulses;
- unavoidable production of decay power (which cannot be stopped); and
- unavoidable production of radioactive products (dangerous for humans and the environment) retained in the fuel material part of the core;

Forms the basis of the needs or requirements of nuclear thermal hydraulics at the design and safety demonstration levels in nuclear installations.

The operational conditions and system configuration of water-cooled RRs compared with NPPs are characterized by (1) much lower power (usually in the range of a few kW and less than about 30 MW in any case), (2) low-pressure values (in most cases, close to atmospheric values), and (3) much smaller geometric dimensions. Even so, challenges for nuclear thermal hydraulics are not therefore smaller than for NPPs, e.g., the predictions of margins to subcooled nucleate boiling and the natural circulation mass flow rate under flow reversal conditions.

Systems, structures, and components (SSCs) include any hardware part in NPPs or RRs that is functional for operations or to guarantee safety. SSCs may have features, either geometric or ranges of key behavioral variables, that require specific thermal-hydraulic modeling capabilities. Examples are centrifugal pumps, steam separators, and containment.

Key SSCs and related thermal-hydraulic performance are discussed in the book. A specific paragraph that follows deals with containment (additional details about Gen IV and "Gen" classifications can be found in Section 1.1.1.16).

1.1.1.7 Experiments and instrumentation

Any progress in 2Φ nuclear thermal hydraulics has been made possible by experiments. Experiments can be basic, like pressure drops in a cylindrical pipe and heat transfer between a wire and surrounding fluid, or make reference to NPPs and RRs, including SSCs. Nonbasic experiments are generated from separate-effect test facilities (SETFs) or integral test facilities (ITFs) (e.g., Umminger and Del Nevo, 2012). Key issues for experiments are measurement errors, time or space averaging, perturbation introduced by sensors, difficulty dealing with heat losses, the role of radiation heat transfer when solid-surface temperatures surpass a threshold of about 500°C, the reproducibility mainly of transient experiments, the variance of measured data, and the suitability of the needed scaling processes to demonstrate similarity or correspondence between gathered data and data expected in prototype systems.

Adopted instrumentation can be quite simple and not expensive, like a thermocouple for 1 USD, or complex and expensive, like gamma-densitometers and laser-doppler anemometry costing up to 10^6 USD per single instrument.

A few textbooks are aimed at or restricted to experiments and instrumentation in nuclear thermal hydraulics (e.g., Cumo, 1982). However, most of the previously mentioned textbooks reference and properly consider experimental data.

Recently, following the OECD/CSNI Specialists Meeting on Advanced Instrumentation and Measurement Techniques held in Santa Barbara, California, USA, OECD/NEA/CSNI (1997), the Specialists Workshop on Advanced Instrumentation and Measurement Techniques for Nuclear Reactor Thermal Hydraulics (SWINTH) was organized in Livorno, Italy, in June 2016 and October 2019. In the former case (June 2016 conference), a special issue of the *Journal of Nuclear Engineering and Design* was published (see Aksan et al., 2018). In the latter case (October 2019 conference), the workshop was supported by NEA/CSNI (see, e.g., Moretti, 2016). Issues related to experiments are marginally and occasionally considered in various chapters of the book (e.g., Chapter 8, Chapter 10, and Chapter 15). A specific chapter (Chapter 17) has been added to the second edition of the book that deals with advancements in thermal-hydraulic instrumentation.

1.1.1.8 Numerical methods and computer science

Nuclear thermal-hydraulic models and equations are typically nonlinear and require numerical methods to achieve solutions (e.g., Patankar, 1980). Programming equations and achieving solutions requires proper competence in computer science (e.g., Ferziger and Peric, 1999). Numerical solution methods may substantially affect the simulated performance of thermal-hydraulic systems (e.g., Ferreri and Ambrosini, 2002).

The design and use of SYS TH codes are discussed in Chapter 11, and related V&V and applications are presented in subsequent chapters: all of this requires consideration and competence in numerical methods and computer science. Therefore, a specific chapter (Chapter 20) has been added to the second edition of the book dealing with numerical methods in nuclear thermal hydraulics.

1.1.1.9 Nuclear safety, licensing process, and design basis accident (moving) boundaries

Nuclear reactor safety constitutes the essential technology for exploiting nuclear energy (e.g., Lewis, 1978). Furthermore, licensing can be seen as the legal part of nuclear reactor safety; already in 1981, Prof. Okrent published a related book (Okrent, 1981). The key objective is to protect humans and the environment from radiation or contact with radioactive nuclei generated by fission. Principles established for this aim and supported by regulatory requirements have become mandatory within the licensing process. Safety barriers and safety functions connect those requirements with the hardware of nuclear installations within a virtual environment known as defense in depth. Thermal-hydraulic parameters like pressure, fluid temperatures, and velocities must be controlled to protect barriers and safety functions.

Notably, fuel cladding and containment constitute barriers that can be broken by high temperatures and/or high pressures. High temperatures in core materials may be a consequence of inadequate cooling of the concerned fission power level. High pressure inside cladding and containment may derive from gas released by nuclear fuel or a loss of pressurized coolant from the primary circuit, respectively. Ensuring the protection of the barriers implies the design of so-called engineered safety features (ESFs), including emergency core cooling systems (ECCSs) and demonstrated related capabilities.

A key issue in this context is the determination of safety margins or a space that separates safe operations from potentially dangerous conditions during an accident. This is done by introducing postulated accidents and, shortly, related design basis accidents (DBAs; the term design basis condition, DBC, was recently introduced and is preferred to DBA; e.g., IAEA, 2019).

The ensemble of all hypothetical situations, accidents, or transients against which a nuclear installation (typically an NPP) must be protected is called the DBA envelope.

The probability of any event occurring plays a role in determining the envelope (see later). The DBA envelope is important because it is associated with regulatory requirements—whatever is inside the envelope must be known with the highest quality available at the time. This sets the boundary for thermal hydraulics and the present book. Thus the book covers situations in the design and nominal operating conditions of NPP and RR, as well as accident situations until the loss of geometric integrity for the core. This implies a space of knowledge outside the DBA envelope that is covered by accident management (AM). This space is now called design extension conditions (DECs) and is distinguished into two regions, DEC-A and DEC-B. The boundary between the two regions is constituted by core structural integrity that is retained or reduced (see also Sections 1.1.1.10 and 1.1.1.15).

Modern thermal hydraulics received decisive input from the regulatory requirements posed by authorities in each country (e.g., Reocreux, 1994; see also Chapter 2 of the book). Namely, regulations required demonstrated acceptability of ECCSs under transient conditions; system and geometry complexities prevented the possibility of human mind solutions without the support of experiments and computational science; this opened the way to complex SYS TH codes; the issues that then came about related to scaling, scaled experiments, V&V for computational tools, and the uncertainty characterizing predictions by computational tools.

SYS TH may be seen as a key sector for nuclear safety and the topics discussed in the book.

1.1.1.10 Severe accidents

Severe accidents have occurred and may occur in NPPs and RRs. Those events may damage the core, primary circuit, and eventually, containment (e.g., DEC-B). Severe accidents may cause serious damage to humans and the environment; see the fundamental book by Prof. Sehgal (2012).

Other than proper competences and efforts, specific regulatory requirements exist and are necessary to prevent severe accident evolution and mitigate consequences. Furthermore, nuclear designers must demonstrate the licensing acceptability of NPPs and RRs related to severe accidents, and operators must be in a condition to make decisions if severe accidents occur.

Thermal hydraulics has a key role concerning the severe accident topic overall. Namely, a "pure" thermal-hydraulic analysis is needed to analyze any transient from its beginning till the moment when core degradation occurs. However, severe accident conditions identified by DEC-B and related evolutions are beyond the domain of interest for the book.

1.1.1.11 Containment and reactor coolant system

The reactor coolant system (RCS) of any NPP is embedded into a protective structure called containment that also constitutes a safety barrier, as already discussed. Unavoidably, RCS and containment interact should an accident occur. In case of a break in the RCS, counterpressure develops in the containment from the release of steam water mixtures from RCS that affects the flow rate at the break (e.g., in the late period of a typical accident scenario). During an SBLOCA event and the initial part of an LBLOCA event (blowdown phase), the break flow is critical, i.e., independent of the downstream backpressure in the containment. Only the reflooding phase during an LBLOCA event is influenced by containment backpressure after RCS is depressurized down to near equilibrium with containment pressure. The higher both pressures—RCS and containment—the faster the reflooding and the better the core cooling conditions will be, since steam production in the core produces less of a loop pressure drop when steam density is higher, thus reducing the adverse "steam binding" effect. The fluid temperature and liquid level in the sump affect the injection of ECCS in the long-term cooling period. In the case of BWRs, the role of the sump is taken by the pool of the wet-well region. Furthermore, hydrogen diffusion and transport may occur inside the containment and are also affected by temperature stratification and steam condensation at the walls. The last two phenomena are relevant for pressure buildup, containment cooling, and possible controlled releases to the environment.

The role of containment is essential should an accident occur. In the case of some advanced reactor designs like advanced PWRs or the AP-1000, core cooling is ensured by gravity and affected by condensation phenomena (see also passive systems, Section 1.1.1.12) occurring in the containment. Thermal-hydraulic capabilities are needed to predict containment performance; these are discussed primarily in Chapter 15. A specific chapter (Chapter 19) dealing with containment pressure has been added to the second edition of the book.

1.1.1.12 Passive systems, reliability, and stability issues

Following the events of Three-Mile Island (TMI) in 1979 and more notably Chernobyl in 1986, so-called passive systems have gained importance in NPP design and, recently, construction and operation. Further motivation to design new passive systems came from the Fukushima event in 2011. Components like accumulators and containments have been passive parts of NPP system design since the beginning of the nuclear era. Passive systems here primarily mean systems introduced in the aftermath of the Chernobyl accident.

Notably, the original idea was that natural forces, gravity in particular, could improve the reliability of core cooling (beyond reducing construction and maintenance costs). Research performed since the year 2000 has shown that this is not correct (e.g., D'Auria and Galassi, 2000; Jafari et al., 2003; Marques et al., 2005). Reasons for this are discussed in the recently completed status report, OECD/NEA/CSNI (2020), see also D'Auria (2020), and are summarized here:

- Passive systems, namely based on natural circulation involving a heat sink and a heat source at different elevations, never operate under (perfect) steady-state conditions: temperature and/or pressure and/or level and/or exchanged/removed power changes with time. Difficulties arise in demonstrating the reliability of passive systems because of the wide variations in governing parameters to be investigated, and the combination of selected parameter values, even within narrow ranges, may cause malfunctions.
- Boiling and condensation processes cannot be avoided in most passive systems: density changes associated with both processes make those systems prone to instability.

- Minor changes in boundary conditions that are uninfluential on system performance when pumps are installed, like a 1-degree inclination of horizontal pipes (specifically carrying two-phase mixtures), small leakages (unavoidable, if large valves are present to isolate a large pressurized pipe), and heat losses (e.g., in the region where steam or two-phase flows are present), challenge system performance and may greatly reduce system reliability.
- A stable natural circulation system characterized by a fixed-power heat source may become unstable when heat source power is generated by another natural circulation system, i.e., when the heat sink of a natural circulation system is also the heat source for a downstream natural circulation system. This is the case when a passive cooling system is associated with the steam generator of a PWR. In this situation, oscillations in the passive system, i.e., downstream of the steam generator, may reflect oscillations in the core mass flow rate, thus endangering core cooling.

Nuclear thermal-hydraulic investigations and reliability analyses are essential for the quality design of passive systems and the assessment of related and expected ranges of operational parameters.

1.1.1.13 Accident phenomenology

Knowledge of thermal hydraulics has evolved according to the notes given in Chapter 2. Complex transient two-phase flow phenomena occurring in case of accidents in water-cooled reactors (i.e., as they are understood today) did not play a major role at the beginning of the nuclear era up to around 1970. However, the logical path for nuclear thermal hydraulics became (starting from around 1970) [NPP-geometry-and-operating-conditions-&- design-requirements] + [DBA envelope] ➔ [phenomena] ➔ [range-of-variation-of-parameters- of-interest] ➔ [requirements-&-development-for-thermal-hydraulics]. This (virtual) view is pursued in the book, with the phenomena discussed in Chapter 6 and Chapter 15 playing the central role.

Scaling constitutes one key issue here: the derivation of phenomena, their evolutions, and related variation ranges of important parameters based on data measured in small-sized facilities compared with NPPs, with a few exceptions. Ensuring a reasonable understanding, including predictive capabilities, of phenomena within the DBA envelope in the NPP constitutes a challenge for specialists in thermal hydraulics. This is the starting point for defining the objective of the book in Section 1.1.2.

1.1.1.14 Role of startup and shutdown phenomena

Toward the end of the 1980s it was realized that the risks to NPPs during startup and shutdown were not negligible compared with risks during power operation. This triggered research and development in various areas including thermal hydraulics.

A distinction is made hereafter between nuclear thermal-hydraulic issues after a power operation postulated initiating event (PIE) and after a shutdown or startup PIE. Both types may be related to the DBA envelope.

- In the former case the problem to be addressed is the demonstration of the adequacy of ECCS in satisfying licensing criteria for core cooling. Components like centrifugal pumps that inject ECCS water are assumed to function properly even though failures are possible, i.e., according to outcomes from reliability analyses. Applications of thermal hydraulics are essential to determine accident evolutions and provide a solution to safety and licensing issues.

- In the latter case the problem is to determine what happens following the failure of residual heat removal (RHR) pumps, typically starting from zero fission power operation. In this case, thermal-hydraulic calculations focus on determining the time available to operators before irreversible damage to the core. The (only possible) solution is an improved design for pumps or related operating conditions, rather than the outcome of a thermal-hydraulic calculation.

Failure of the RHR pump is one possible PIE, but not the only one. The unit 4 Civaux plant (in France) experienced a small break in RHR piping during RHR operation due to thermal fatigue in a region connecting two pipes of the RHR system, with mixing of flows at different temperatures resulting in fluid thermal fluctuations. Thermal-hydraulic conditions challenging RHR operation through long-term accidents and NPP shutdowns are similar in some cases to the range of variations in pressure, temperatures, and mass flows. This is also true for RR operating conditions.

1.1.1.15 Accident management and its procedures

The terms AM and AMP were introduced in the 1990s. They indicate not only the possibility but also the need and capability for NPP operators to manage or drive the evolution of any accident. This is true for situations outside the DBA envelope (currently characterized as DBCs), i.e., beyond design basis accident (BDBA) conditions (currently characterized by DEC, including DEC-A and DEC-B, where the loss of geometrical integrity of the core distinguishes between the two), while for DBA conditions, emergency operating procedures are planned to support the operator actions.

Two categories are distinguishable for AMP within the concerned (BDBA) domain (partly consistent with DEC-A and DEC-B introduced earlier), respectively, aimed at preventing massive core degradation and mitigating the consequences of core melt including recovery of the plant. The phenomena expected to be relevant for the first category, i.e., including (recent nomenclature) DBC and DEC-A conditions, are considered in Chapter 6 and further discussed in Chapter 15.

1.1.1.16 Generation IV and small modular reactor thermal hydraulics

The following mentions Generation IV (Gen IV) and small modular reactor (SMR) nuclear reactor thermal hydraulics without pretending to define the terms Gen IV and SMR. What is stated may be seen as a complement to the discussion in Section 1.1.1.6.

Exploiting fission reactions for thermal energy and, after proper transformation to get mechanical or electrical power, is not exclusively connected to water. Various options are available, including gas or liquid metal as a coolant, graphite as a moderator, or even dissolving nuclear fuel into the coolant, i.e., homogeneous reactors. In the following definitions (from https://fr.wikipedia.org/wiki/Générations-de-réacteurs-nucléaires), nuclear power reactors are categorized roughly according to their "generation":

- Gen I: reactors designed and constructed before the end of the 1970s (in France, UNGG natural uranium graphite gas)
- Gen II: current reactors constructed from the 1970s to the end of the 1990s and under operation, including pressurized water reactors (PWRs), boiling water reactors (BWRs), AGR, CANDU, RBMK, and VVER

- Gen III: advanced reactors, including CE system80+, CANDU6, and AP600
- Gen III+: evolutionary reactors, including EPR, AP1000, APR1400, ACR1000, ABWR, and ESBWR
- Gen IV: reactors of the future, including sodium fast neutron reactor, gas fast neutron reactor, lead fast neutron reactor, very high-temperature gas-cooled reactor, molten salt reactor, and supercritical water reactor

Following the parallel growth of WCNR technology and thermal hydraulics, around the year 2000, the words and the associated research framework "innovative reactors", SMR, and Gen IV appeared. This also occurred with words or acronyms and technologies, other than research, like evolutionary and/or advanced reactors and Gen III+ that have been in use since the 1990s of the last century.

The thermal hydraulics of the latter category of reactors (i.e., evolutionary and/or advanced reactors, as they are designed, built, and operated so far) is considered in the book, e.g., Aksan et al. (2018a).

Non-water-cooled reactors primarily belong to the categories SMR and Gen IV. The thermal hydraulics of those reactors fall outside the scope of the book.

Water-cooled SMRs, e.g., Reyes and Young (2011); Ricotti and Fomin (2020), could benefit from nuclear thermal-hydraulic investigations of WCNRs. Therefore the background research and several focused findings like the phenomena classification, procedures, validation activities, and selected correlations discussed in this book apply to those reactors. Nevertheless, the following warning/caution statement is needed. Each specific SMR design is characterized by geometries (e.g., helicoid heat exchanger, in-vessel pressurizer, short core) and operating conditions (e.g., gravity-driven circulation) different from those in a WCNR. In general, new research is needed to characterize possible new phenomena and combinations of parameter ranges, including addressing the scaling issue. Furthermore, if passive systems are used in SMRs for either normal operations or accident conditions, reliability analyses that consider thermal-hydraulic phenomena are needed (OECD/NEA/CSNI, 2020).

Only one (supercritical) water-cooled reactor is nowadays classified as a Gen IV reactor. Related thermal-hydraulic phenomena, see e.g., Pioro (2019); Liu et al. (2019), are marginally considered in the present book. Research, including research on thermal hydraulics, is still required to finalize the design.

1.1.1.17 Connections to neutron physics, probabilistic safety assessment, radioprotection, chemistry, mechanics, nuclear fuel, and electronics, including instrumentation and control

The application of thermal hydraulics within NPP technology and nuclear reactor safety is not possible without an interaction (or coupling) with different disciplines: exceptions (i.e., thermal-hydraulic applications alone) also exist but are not universal and do not allow overall safety assessment. In addition to textbooks dealing with other disciplines already cited in this chapter, examples are given in the following of connections between thermal hydraulics and selected disciplines. The descriptions that follow are not exhaustive or systematic: rather, they provide an idea of the complexity of applying thermal hydraulics to characterize needs in WCNS and RR technologies; more details can be found in D'Auria et al. (2019).

Neutron physics. Thermal power is generated by neutrons and induced fission. The neutron flux depends, among the other parameters, on the temperature of the moderator and fuel and notably on the density of the moderator coolant. Thus, fuel temperature and moderator density and temperature are calculated by thermal hydraulics and affect instantaneous fission power. So any NPP or RR core-related analysis implies coupled neutron physics thermal-hydraulic calculations. A zero-dimensional neutron kinetics model was embedded into the primitive SYS TH codes since the end of the 1960s (e.g., Lamarsh and Baratta, 2011). Toward the end of the 1990s, three-dimensional core physics models were coupled with thermal-hydraulic codes (e.g., Ivanov and Baratta, 1999; see also D'Auria et al., 2004). This opened the way to the analysis of core transients involving local reactivity changes like those associated with boron mixing (e.g., Pla et al., 2010) and control rod ejection (e.g., Avvakumov et al., 2007).

Probabilistic safety assessment (PSA). PSA constitutes a developed technology by itself (e.g., see Apostolakis and Moieni, 1987; Fulwood, 2000; Arkadov et al., 2012). Identification of all interactions between thermal hydraulics and PSA is, again, beyond the scope here. To give an idea of the tight connection between PSA and thermal hydraulics it is sufficient to recall that (1) the transients part of the DBA envelope, i.e., those analyzed by thermal hydraulics as given in Chapter 15 of the book, need specifications for probability of occurrence; this also determines the acceptability criteria (e.g., IAEA, 2010a); (2) a transient thermal-hydraulic analysis is needed to determine the consequences of any accident relevant to PSA (e.g., IAEA, 2010b).

Radioprotection. Radioactivity and doses (e.g., Knoll, 2010) are concerned with the application of thermal hydraulics. The amount of radioactivity released from the core (or the source term calculation, e.g., Herviou, 2005), should an accident occur, and the possible paths to doses to humans or radioactivity impact for the environment require thermal-hydraulic calculations. One may also state that atmospheric dispersion calculations for radioactive products (e.g., Thielen et al., 2007) are part of a nuclear reactor safety analysis based on thermal hydraulics.

Chemistry. The chemistry of the coolant and notably the control and transport of boron are required for long-term reactivity control or fuel burnup compensation in any NPP (e.g., Benedict et al., 1981). A devoted system called a chemical and volume control system is designed, installed, and operated in each NPP. Time delays for boron arriving in the core since the injection time and mixing in regions like the downcomer or lower plenum of the vessel require a thermal-hydraulic calculation (e.g., Theofanous and Shabana, 1991; Cheng et al., 2014). This is particularly true in the case of anticipated transient without scram (ATWS). The calculation of the amount of hydrogen produced by the metal—water reaction (e.g., Baker and Just, 1962) and the transport of the H_2 gas into the containment (e.g., Camp et al., 1983; Royl et al., 2009; Xiao and Travis, 2013) require specific thermal-hydraulic analyses even in relation to accidents within the DBA envelope.

Mechanics. Eight examples are given hereafter of the interaction between mechanics and thermal hydraulics (without having the target to provide a comprehensive view): (1) main coolant pump (MCP) transient operation, including the consideration of two-phase flows, cavitation and vibrations, and net positive suction head (e.g., Tuzson, 2000)—various coupled sets of mechanics and thermal-hydraulic equations are needed to address the MCP simulation; (2) flap valve operation with changing pressure differences at the boundaries (e.g., Abu-Zeid, 2013)—the inertia of the flap body (mechanics) is correlated with the pressure difference calculated by thermal hydraulics; (3) pressurized thermal shock (PTS) stresses are caused by imperfect mixing of cold liquid from ECCSs injected in the cold leg and challenge reactor pressure vessel (RPV) integrity. PTS is an example of a multidisciplinary topic also involving, beyond local thermal stresses, long-term damage to the vessel by neutron radiation (also affected by water density), material aging and embrittlement, welding technologies, crack growth and propagation, or fracture mechanics (e.g., IAEA, 2010c; see also Araneo and D'Auria, 2012); (4) condensation-induced water hammer—fast condensation causes the movement of liquid slug that may determine dynamic load on piping and restraints (e.g., Schultz et al., 2001; Barna et al., 2010); (5) thermal fatigue, which constitutes a long-term degradation mechanism important in NPP aging caused by fluid temperature stratification and turbulent mixing effects; those effects induce continuous (e.g., during the life of the system) local mass flows of fluid having different temperatures that are at the origin of thermal stresses in structures (see, e.g., Courtin, 2013); (6) loads on RPV internals originated by pressure wave propagation from postulated break (e.g., Sokolowski and Koszela, 2012); (7) fluid–structure interaction—FSI is a scientific and technological area dealing with loads on RPV internals caused by pressure wave (previous topic) as well as vibrations (e.g., Bungartz and Schäfer, 2006; see also the textbook by Païdoussis et al., 2014) already mentioned; (8) jet impingement (and jet thrust) and pipe whip—the break of a pipe causes the release of the elastic energy stored in the pressurized structures (e.g., piping) that adds to the fluid dynamic loads generated by the flow motion exiting the break. This generates (see, e.g., Vigni and D'Auria, 1979; D'Auria and Vigni, 1981) (1) loads on restraints; (2) mechanical and thermal loads by the jet impinging on close surfaces; (3) potential missiles, including parts of components like the rotor of a pump, blade of a turbine, and pieces of broken material; and (4) pipe whip involving the displacement of piping.

Nuclear fuel. Any analytical simulation of NPP or RR requires the modeling of nuclear fuel. Issues and challenges are connected to fuel, geometric, mechanical, and physical or material properties that continuously change during the fuel life in the core. This is due to burnup. Furthermore, the values of those properties depend on thermal-hydraulic parameters, notably pressure and the temperatures of the coolant and moderator (e.g., OECD/NEA/CSNI, 2010; see also Lisovvy et al., 2015).

Electronics, including instrumentation and control (I&C). Many parameters are continuously monitored during the operation of NPPs, and several systems and components are called into operation during either nominal or accident conditions. The hardware and software to control those components and systems are identified as part of the I&C: they have a role in determining the evolution of accidents or the performance of WCNR. I&C technology is beyond the scope of the present book. However, the realistic (BEPU) simulation of WCNR and RR requires the modeling of I&C: see, e.g., Pakonen et al., 2021; D'Auria et al., 2012b for the impact of I&C in accident evolution and simulation of accident scenarios, respectively.

1.1.1.18 Summary of scope and framework of the textbook

The enormous amount of interest, scope, and framework for nuclear thermal hydraulics and SYS TH derives from the nonsystematic summary in the previous sections. A textbook covering with sufficient detail each of the mentioned topics is well beyond the planning of the present effort. Nevertheless, any of those topics may have a connection with what is discussed in the book. An attempt is made hereafter to summarize what topics are considered or not considered in parts of the book.

- Origin of nuclear thermal hydraulics: Chapter 2 deals with a vision of the history of SYS TH.
- Single- and two-phase flows: these are of main concern in Chapters 3−11. Fundamentals in fluid dynamics and gas dynamics are not discussed. However, subjects like definitions, balance equations, pressure drops, and heat transfer, e.g., Chapters 3, 5, 7, and 8, are part of the present book.
- International reports issued by OECD/NEA and IAEA as well as reports documenting the features of SYS TH and CFD codes constitute key references for the present book with related details in Chapters 9 and 11−15.
- Prominent scientists and textbooks: the idea is to not replicate any historical bibliography; the reader could guess from cited names (and references in some cases) the contributions to the knowledge in thermal hydraulics by prominent scientists.
- Journals, conferences, and the web: a systematic review of publications (journals, conferences, and web) is beyond the scope of this textbook. However, the latest developments in selected areas are considered (till around the year 2022 in the current second edition of the book).
- Education and training: considering existing education and training courses is beyond the scope of the book. However, the book itself may become a reference for education and training courses.
- The target NPP and the RR types, the SSC for concerned nuclear installations: the reader of the book is expected to be familiar with the features of concerned NPP, RR, and SSC including containment. No devoted descriptions are part of the book.
- Experiments and instrumentation: experiments are unavoidable for the progress of nuclear thermal hydraulics as already mentioned. The phenomena covered in Chapter 6 are derived from SETF and ITF experiments, e.g., Chapter 21. Instrumentation and experimental techniques are discussed in Chapter 17 of the book.
- Numerical methods and computer science are needed for solutions of thermal-hydraulic models as for applications of SYS TH codes, as discussed in Chapter 20.
- Nuclear Safety, the licensing process, and the DBA/DBC moving boundaries: the reader of the book is expected to have a fundamental knowledge of nuclear reactor safety technology; otherwise, an idea of the licensing process can be derived from the cited references. In Chapter 4 the need for nuclear safety is considered, and the results of applying thermal hydraulics to nuclear safety are discussed in Chapters 14−16. The scope of the book in this context is clarified by the statements in the following item.
- Severe accidents: the book does not cover severe accidents, i.e., the boundary of interest is set to the time of any accident or the reactor configuration when the core starts to be degraded. The two key reasons for this are
  - keeping the book as small as possible (i.e., to avoid an encyclopedia), and
  - differing licensing requirements, capabilities, computational tool rigor and quality, and numbers of experimental investigations between the areas of thermal hydraulics and severe accidents.
- Containment and RCS: Containment is part of NPP, and containment response is part of accident analysis (AA). Thus thermal hydraulics of containment is part of the book, e.g., Chapter 4

in relation to the needs for thermal hydraulics, Chapter 6 concerning occurring phenomena, Chapter 7 concerning condensation heat transfer, Chapter 15 concerning containment performance in case of accidents, and Chapter 19 specifically related to containment design and performance in case of an accident.

- Passive systems, reliability, and stability issues: thermal-hydraulic phenomena expected due to the operation of passive systems are presented in Chapter 6. Furthermore, SYS TH codes, Chapter 11, can predict the transient performance of passive systems, Chapter 15. The design of passive systems requiring reliability investigation is considered in Chapter 23.
- The accident phenomenology: for the current vision, phenomena expected in NPP following off-normal conditions are at the bottom of the pyramid of knowledge for nuclear thermal hydraulics. Phenomena are presented in Chapter 6.
- Role of startup and shutdown phenomena: phenomena during RHR operation including startup and shutdown in WCNR are considered in Chapters 6 and 15. The reliability of RHR-related pumps (i.e., reliability of mechanical components and systems) is outside the scope of the book.
- The AM and the AM procedures (AMPs): AM phenomena occurring to prevent severe core degradation are part of the scope of the book and discussed in Chapter 15.
- Connection with neutron physics, PSA, radioprotection, chemistry, mechanics, nuclear fuel, and electronics including I&C: the application of nuclear thermal hydraulics to the safety and design of NPP and RR is not possible without the support of other disciplines, as it also derives from the cited references. Related technological and scientific aspects are beyond the scope of the book. Nevertheless, results of accident analyses and occurred accidents in WCNR, Chapters 15 and 16 necessarily involve considering "other disciplines."
- Procedures: Chapter 14 deals with the BEPU approach to AA: procedures that are part of BEPU are discussed in Chapter 13, and recent achievements in the areas of scaling and V&V are considered in Chapters 21 and 22.
- Thermal-hydraulic design of WCNR: the application of thermal hydraulics within the design of WCNR is considered in Chapter 23; insights and/or background technology for core design can be derived from Chapter 18 dealing with subchannel analysis.
- Critical issues in thermal hydraulics: these are considered in Chapter 24 together with perspectives for developments.

Computer codes in SYS TH (e.g., Miro et al., 1990) and CFD are the focus of the book even though they are not explicitly in the given list of topics. The computational tools, mainly the SYS TH codes, constitute the end product of knowledge and expertize in nuclear thermal hydraulics; the related technology includes code development (e.g., Bestion, 1990; Jeong et al., 2010); code V&V (e.g., D'Auria et al., 2012a); code scaling (e.g., Glaeser and Karwat, 1993; D'Auria and Galassi, 2010; Rohatgi, 2015); and code application (e.g., Reventos et al., 2010). In this case, the evaluation of uncertainty in code predictions (e.g., D'Auria et al., 2008) is a requisite for properly considering applications in nuclear reactor safety. Therefore, any concept or information reported in the book may be considered relevant to the SYS TH code development and application processes (see also Section 1.1.3).

1.1.2 Objectives, innovations, and target audience

Nuclear thermal hydraulics appears like a universe of matters and themes: galaxies and planetary systems of topics and skills can be distinguished and are needed for proper comprehension of the discipline. Within such a universal vision of thermal hydraulics, Dr. Zuber introduced the concept of quanta to deal with scaling (Zuber, 2010), and G. Yadigaroglu wrote about a trip to the moon to describe the computational multiphase fluid dynamics (CMFD) approach to calculate critical heat flux, specifying that "*we have not landed yet*" (Yadigaroglu, 2014). Here we may add that black holes in the knowledge of thermal hydraulics do exist and may be associated, e.g., with the modeling of turbulence at either the micro or macro scale, including the occurrence of boiling and condensation in multifluid systems and the need for averaging at cross-sectional, volumetric, and temporal levels.

A systematic and comprehensive treatment of all the matters and themes is not realistic or possible, and an encyclopedia or a series of books is required to ensure the reasonable coverage of each matter. Rather, one may note that various textbooks (some included in the list of references) address one or more individual topics in nuclear thermal hydraulics. In this case, it is difficult to exceed their quality and novelty.

Here, items that do not represent the objectives of the book can be stated: comprehensively covering or mentioning each important matter or theme in nuclear thermal hydraulics; moving the boundary of knowledge by considering quanta and black holes as mentioned earlier; replicating any existing textbook rather than referencing those books.

Several requirements are identified hereafter. The objective of the book is fixed on fulfilling those requirements:

- A novel approach or new vision for nuclear thermal hydraulics is envisaged or attempted.
- Topics are discussed covering fundamentals in thermal hydraulics, i.e., mostly by referencing appropriate literature (e.g., textbooks), to applications in NPP design and safety technologies, including RRs to the extent possible.
- The impulse to acquire knowledge from reactor design and operation and nuclear safety, with the latter unavoidably connected to licensing rules, shall be recognized. A simplified chain is considered: regulatory goals and reactor design needs ➔ reactor configuration and accident scenarios ➔ operation and AA ➔ phenomena ➔ thermal-hydraulic features.
- From a logical viewpoint and not necessarily connected to the order of chapters, a top-down approach is pursued: then, needs come first (basically Chapter 4), and applications are expected at the end (e.g., Chapters 15 and 23); notably, phenomena of interest come after needs (e.g., Chapters 6 and 4, respectively).
- Technology impacts are considered in each chapter (e.g., at least in introductory or closing remarks).
- SYS TH code is seen throughout the book as the main/key repository of expertize in the area. This implies the consideration of code development (namely thermal-hydraulic bases for development, e.g., Chapters 9−11 and Chapter 18).
- Procedures are the intermediary elements between models and applications. V&V, scaling, and uncertainty (e.g., Chapters 13, 21 and 22) are prerequisites for applications (e.g., Chapters 15 and 19). The BEPU framework (Chapter 14) is the infrastructure of procedures.

The innovations introduced in the book include (1) shedding light on the connection between fundamentals of thermal hydraulics and application in nuclear technology; (2) streamlining the essential role in thermal-hydraulic applications of procedures including V&V, scaling, and uncertainty evaluation; and (3) identifying the BEPU approach as the infrastructure of models, codes, procedures, and (licensing) rules. This includes an integrated and nonconflicting role for deterministic safety assessment

(DSA) and PSA. The book has been written with targets in mind to promote the application of SYS TH codes within the licensing process of existing and new nuclear reactors and to disseminate various steps of the BEPU approach to software technologies different from SYS TH.

1.1.2.1 Reformulation of textbook objective

No numerical tool can predict the movement, growth, and collapse of a single bubble moving inside a liquid when condensation and evaporation processes take place. Similarly, there is no capability to simulate all the effects of turbulence even in single-phase flows, e.g., in predicting pressure drops at geometric discontinuity with an error that is acceptable for the design-operation of thermal-hydraulic systems. Prediction capabilities are even poorer or nonexistent when a few bubbles move together and interact or when two-phase turbulence is considered, respectively. Decades of development until now, e.g., in the area of two-phase CFD, have not been sufficient to achieve a solution to the computational challenges. So the obvious question that arises is, How can one predict transient two-phase mixture evolutions in complex geometrical systems like a WCNR or an RR? The objective of this book is to address such a question.

The instant, simplified answer is as follows and may be seen as the reformulated objective of the book.

Time and space averaging is adopted in SYS TH codes, or approximate equations are solved by approximate numerical methods—the impacts of approximations on results or errors are unknown. However, V&V procedures supported by experimental data and suitable scaling analyses allow the quantification of those errors. Qualified (as much as possible) uncertainty methods are established to achieve this aim. In other words, one could state that expertize, including knowledge of the data measured in transient experiments, is necessary to achieve qualified calculation results from the application of thermal hydraulics.

The definite side objectives of the book are, on the one hand, to recognize the irreplaceable role of SYS TH codes in establishing the safety of nuclear installations (i.e., determined on transients that involve two-phase fluid mixture time evolutions) and, on the other hand, to clarify the need for rigorous V&V and support by an uncertainty evaluation of any valuable set of SYS TH calculation results.

1.1.2.2 To whom this book is addressed

Any scientist involved with nuclear thermal hydraulics may benefit from the book (as already mentioned in the "Chapter foreword"). This is specifically true for code developers, nuclear thermal-hydraulic analysts, AA specialists including their supervisors, and more generally, nuclear safety technologists.

Students in the last years of engineering courses and master's or PhD students/researchers are expected to become acquainted with the vision of nuclear thermal hydraulics from the reading of the book.

In all cases, knowledge of thermodynamics and nuclear thermal hydraulics fundamentals is desirable, although not essential, for comprehending the concepts in the book. The questions-problems in each chapter with answers provided in Appendix 1 have been introduced in the second edition of the book: this is expected to find favor with readers of the book.

1.1.3 Structure and content

The textbook is organized like a monograph; however, several authors contributed to provide views and insights that are complementary as far as possible: this is true either for the first or the second edition. More than a dozen (i.e., 15 authors) and a couple dozen scientists contributed to the first and the second edition, respectively.

A single bounded volume is printed for the first edition (Chapters 1–16), although four parts are distinguished. Three volumes constitute the second edition (Chapters 1–24) of the book. Rather than rearranging, preference was given to expanding the list of contents from the first edition when planning the second edition. Each of the chapters in the first edition was revised. The second edition consists of three volumes of similar size.

- Volume I: Background and Bases. This is made up of Chapters 1–9; it includes topics ranging from definitions in thermal hydraulics and related applications, to balance equations to phenomena identified in accident scenarios.
- Volume II: Modeling. This is made up of Chapters 10–17; the focus is on modeling features (Chapters 10–12), procedures and solution methods needed for the application of the models (Chapters 13, 14 and 17), and cross-links between models and phenomena (Chapters 15 and 16).
- Volume III: Procedures and applications. This is made up of Chapters 18–24; extensions (e.g., numerical solution methods and scaling), insights (e.g., subchannel thermal hydraulics) and outcomes from applications (e.g., elements of design of WCNR) of the topics covered in the previous chapters make up this volume. The final chapter of the second edition book (i.e., Chapters 24) is devoted to a critical evaluation of present capabilities in thermal hydraulics and to research envisaged for the future. Related to each Chapter (from 1 to 24), problems proposed to readers with solutions are added. The list of authors in the references, providing an idea of all those more than 5000 scientists who have been identified as contributors of current thermal hydraulics is available in electronic format.

Thermal Hydraulics in Water-Cooled Nuclear Reactors is the title (and the subject-content) of the book. Having recognized that it is not possible even to mention all important topics, a selection of themes and matters has been assembled to form the content. This shall be seen as one of the (infinite and possible) break-through views of the universe of matters and themes mentioned in Section 1.1.2. Then, an attempt is made to link bases and applications in nuclear thermal hydraulics; however, connections with other disciplines, which are recognized as essential within the framework of applications, are not discussed in the book.

The sketch in Fig. 1.1 is built to facilitate the reader, thus providing a synthesis for the rationale of the book and suggesting a reading strategy. The top row (the bottom blue one) and the column on the left side distinguish, for each chapter, the focus and the book volume, respectively. Colors are used to differentiate each chapter according to five characterizing topics (these are listed on the right side of Fig. 1.1 and are

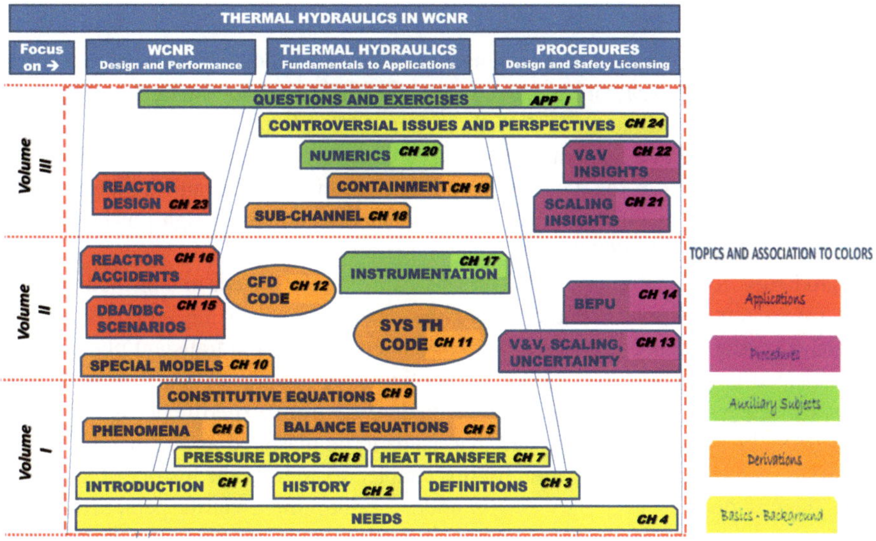

Figure 1.1 Log frame for the structure and contents of the book.

associated with colors). The title of each topic is expected to be self-explanatory. However a few words are added hereafter in the case of dark-yellow and green topics: "derivations" imply that the content of the concerned chapter needs a basis background (e.g., the yellow topics), and "auxiliary subjects" are not in the main flow of information although they are required to arrive at "applications" (i.e., including the answer to questions and the solution of exercises in Appendix I), typically passing through procedures. Furthermore, arrows connecting the chapters are avoided for simplicity (too many arrows would be needed). Rectangular boxes are reported for each chapter except for Chapters 11 and 12 characterized by an ellipse: the computer codes (i.e., the contents of those chapters) are interpreted as the basket-repository of information in all other chapters. The target is to deeply investigate the topics brought to partial duplication in the titles of Chapter 13 and Chapters 21 and 22: an established and coherent vision for V&V, scaling, and uncertainty is part of the former chapter, otherwise selected insights (related to scaling and V&V) are considered in the latter chapters. In addition, BEPU (Chapter 14) may be seen as the framework of procedures and competences such that nuclear thermal hydraulics may impact the WCNR technology: in different terms, in the absence of rigorous and quality recognized procedures, any deep knowledge of bases in thermal hydraulics is not applicable to demonstrate safety of WCNR or to optimize their design.

Each chapter is introduced in the short summaries that follow.

1.1.3.1 Chapter 1: Introduction

The framework of nuclear thermal hydraulics considered for the planning and the issuing of the book (second edition) is discussed in the present chapter.

1.1.3.2 Chapter 2: Historical remarks

The nuclear thermal hydraulics discipline was developed following the need to demonstrate the safety of the reactors and, to a more limited extent, to optimize the design. As in all other fields where analytical methods are involved, nuclear thermal hydraulics benefitted from the development of computers. Thermodynamics, rather than fluid dynamics, is at the basis of the development of nuclear thermal hydraulics together with experiments in complex two-phase situations, namely geometry, high power density, and pressure typical of WCNR.

Mandatory licensing requirements originated from safety needs and pushed tremendous investments in research that drove the developments in the area. All of this created a unique technology where highly geometry-dependent transient phenomena, sophisticated two-phase flow modeling, continuously advancing numeric and computer science, and unavoidable empiricism derived from experiments are merged. This forms the iceberg of knowledge that needs to be mastered and managed.

The objective and the motivation for this chapter are to trace the history of the discipline. This is deemed necessary for its understanding and creating perspectives for development.

1.1.3.3 Chapter 3: Definitions

Definitions may appear not necessary for an established discipline like nuclear thermal hydraulics: all terms should have a consolidated meaning within the myriad and the variety of publications mentioned in Section 1.1.1. Unfortunately, this is not the case, one reason being the complexity of the discipline that reflects in the difficulty of conjugating basic concepts like the principles of thermodynamics and complex displays like subcooled void appearance during depressurization that impact heat transfer and fission-generated power.

Then, objective and motivation for the chapter is to show the complexity of the nuclear thermal hydraulics discipline from the point of view of what are called "definitions" (see following): then the chapter is complementary to the discussion in Section 1.1.2.

"Definitions" here imply in most cases providing a message that shows the difficulty associated with the concerned term. For instance, in the case of convection heat transfer coefficient (HTC), the huge number of variables that affect HTC are balanced by the simplicity of its correlation with heat flux and temperatures; in the case of turbulence, the lack of any direct predictive capability, or the current status of the DNS technology, is stressed; in the case of pressure drops at geometric discontinuity, difficulties in measuring a free-from-(large or not-acceptable)-error value for the local coefficient arise, e.g., because of need to distinguish between reversible and irreversible contributions to the total pressure change: the problem becomes more complex when properly considering the condition of fully developed flow.

Three categories of definitions are envisaged.

(A) Variables
(B) Phenomena
(C) Concepts

The variables include the fundamental quantities that are also part of balance equations, like temperature, pressure, velocity, and material properties. Errors characterizing or expected from the measurement and the prediction by analytical tools of those quantities are outlined. The void fraction is introduced to characterize thermal hydraulics. Furthermore, nondimensional quantities like Re, Pr, Nu, and Fr are also part of this class. The derivation of key nondimensional quantities in thermal hydraulics and their physical meaning are discussed. Difficulties to apply those quantities in scaling analysis are mentioned: e.g., during a transient, at a given time, different Re values apply to different locations of an assigned loop: a comprehensive mapping of Re values in the entire loop may not be possible e.g., by adopting a scaling method alone not supported by a system code calculation.

Examples of discussed phenomena or concepts in thermal hydraulics include natural circulation, two-phase critical flow (TPCF), critical heat flux, blowdown, reflood and rewet, friction pressure drop (empirical) with main reference to two-phase flow multiplier, conduction heat transfer (theoretical, but utilizing the empirical/approximate Fourier law), convection heat transfer (empirical), and radiation heat transfer (supported by theory, but needing empirical parameters). Aspects connected with WCNR and RR technology are emphasized.

The concepts considered encompass a variety of subjects and themes of the nuclear thermal-hydraulic universe, ranging from phenomena-modeling terms like drift flux formulation to the characterization of equations like balance or conservation equations, constitutive or closure equations, state equations, to approaches to perform a calculation like conservative or best estimate (BE), to modeling approaches like lumped parameter model or UVUT (unequal velocities unequal temperatures) model dealing with balance equations. Widely adopted terms like PIE, DBA/DBC, BDBA (or DEC-A and DEC-B), and AM are clarified.

Within the class of concepts, a connection is created between nuclear thermal hydraulics and fluid dynamics and gas dynamics, i.e., by including suitable definitions. Then, the concept of turbulence can be used to connect thermal hydraulics with fluid dynamics and gas dynamics emphasizing the added complexity, as far as turbulence is concerned, associated with a void fraction: flow regimes (bubbly, slug, churn, annular, etc.) can be seen as the connection between turbulence and two-phase flows.

1.1.3.4 Chapter 4: Needs

The development of nuclear thermal hydraulics has several origins and motivations including thermodynamics and fluid dynamics sciences: nuclear technology had an important role. In this chapter, a reverse approach is pursued: starting points are the existing configurations, operating conditions, and safety aspects of WCNR and RR; then an attempt is made to establish what capabilities are required for nuclear thermal hydraulics.

To accomplish this goal, the following sectors, not independent of each other, are considered and reported hereafter together with examples of the origins of needs.

- Reactor coolant system for WCNR: all water-cooled power reactor designs are included. Needs are identified that are associated with the design of core and other RPV internals, steam generators, separators, dryers (in the last two cases with main reference to BWRs), main coolant pumps, pressurizer, and recirculation legs.

- Nonnuclear part of the WCNR also known as balance of plant (BoP). The BoP has similar features for NPP and oil or coal-fueled plants for electricity generation. However, in the case of NPP the size of components, like turbines and condensers, may be larger than in conventional plants and the attention originally devoted (e.g., in the 1950 and 1960s) to maximizing the thermal efficiency might have been lower (because of the smaller impact of the fuel on the cost of the produced electricity). So needs may derive from the modernization or revamping projects for existing WCNR like decreasing pressure drops in the steam lines.
- Containment: the designs of tight (or zero leakage target after DBA) containment for almost all NPP and of vented containments in "not-the-latest-generation" of Russian reactors need thermal-hydraulic capabilities. Interesting to note that controlled venting became a desirable feature after the Fukushima event in 2011. Containments equipped with a pressure suppression pool of liquid need deep thermal hydraulics competence (e.g., estimation of vibrations and loads on compartment walls).
- RR designs: a variety of designs characterize the few hundred RRs now in operation. RRs are equipped with highly enriched uranium cores or, the larger number, a low enriched uranium core. Pool type or forced circulation RRs are distinguished. Challenging topics for nuclear thermal hydraulics are the evaluation of margins available to the onset of nucleate boiling even during possible flow reversal consequences of the main coolant pump trip, i.e., the pathway from downward forced circulation to upward natural circulation. The narrow passage for fluid across flat plate fuel elements also puts challenges to nuclear thermal-hydraulic investigations.
- Safety analysis of water-cooled reactors: the need for thermal hydraulics primarily comes from the licensing requirements, e.g., the ECCS design criteria. Updating requirements and creating the link with computational capabilities constitute challenges for applications. Needs and challenges for nuclear thermal hydraulics are (unavoidably) defined outside the subject boundaries for the book: namely, the connection between nuclear thermal hydraulics and neutron physics, nuclear fuel (modeling), chemistry (e.g., H_2 production), mechanics (e.g., FSIs, modeling of flap or inertia valves, modeling of the inertia of pump rotor, etc.), radioprotection (e.g., transport of radioactivity from core to containment) are at the origin of needs.

Addressing the listed topics constitutes motivation and objectives for the chapter.

1.1.3.5 Chapter 5: Balance equations

Balance equations constitute the seed around which the predictive capabilities of nuclear thermal hydraulics have been built with the main reference to transient two-phase flow evolutions. This is the motivation for the chapter: predictive capabilities in thermal hydraulics do not exist without these equations. Balance equations, synonymous with conservation equations, are established from the principles of science and technology: mass and energy balance are derived from mass conservation and the first principle of thermodynamics, and momentum balance is derived from the first principle of mechanics. Nuclear reactor thermal hydraulics may be studied using a large variety of models that simplify more or less the physical reality.

The first problem of an engineer when investigating a reactor issue is to select the right degree of simplification depending on the situation and the required accuracy and reliability of the solution. Simplifications may result from the space and time resolution and simulation tools include several scales from lumped parameter models to

CFD and CMFD. Zero-, one-, two-, and three-dimensional models are available for single- and two-phase flow that use either time averaging or combined space-time averaging. Each approach uses a set of balance equations for mass, momentum, and energy, and the derivation of these equations may need complex mathematical developments. Attention is drawn to the physical assumptions and simplifications used at some steps of the derivation of equations and to the resulting limitations of the predictive capabilities.

The objective of this chapter is then to give a picture of the available models with an evaluation of the predictive capabilities and limitations considering the degree of accuracy and the maturity of each model. More information is given on 1-D models since they are the most currently used in design and safety studies using system codes. However, 3-D models are also increasingly used, and the balance equations are also presented to introduce CFD and CMFD.

1.1.3.6 Chapter 6: Phenomena

The transient performance of NPP following plausible initiating events or accidents and changes in operating conditions (e.g., startup and shutdown) brings to the identification of phenomenological windows and phenomena, see Aksan et al. (2018a). These are characterized by significant parameters and related ranges of variations and may involve steady-state or transient conditions.

Thermal-hydraulic phenomena are important owing (at least) to the following reasons.

(1) Experiments in thermal hydraulics either "integral" or "separate-effect" type are designed and executed to characterize the phenomena.
(2) A variety of empirical or non-first-principle-derived formulations needs to be developed making reference to phenomena, e.g., including special models discussed in Chapter 10.
(3) A comprehensive list of phenomena properly characterized by significant parameters is needed to ensure suitable validation for SYS TH codes as discussed in Chapter 13.

International activities performed at OECD/NEA/CSNI and by IAEA brought to the identification of phenomena in the following conditions (sample list, more details in Chapter 15).

- LBLOCA in PWR and VVER
- SBLOCA in PWR and VVER
- LOCA in BWR
- Separate-effect phenomena in BWR and PWR
- Shutdown conditions in PWR
- AM in PWR
- Containment response in PWR and BWR
- Performance of passive systems characterizing the coming reactor design
- Phenomena associated with RR

The motivation for the chapter derives from the needs (Chapter 4), and the objective is to create a comprehensive list of phenomena that are described in a (so much as possible) homogeneous way.

1.1.3.7 Chapter 7: Heat transfer

Heat transfer and pressure drops are the main targets within the design of thermal-hydraulic systems: a wide variety of investigation approaches and precision needs in simulation have been established and are available.

Chapters 7 and 8 are planned to constitute a bridge between the past and the future for nuclear thermal hydraulics: the evaluation of convection HTC has been among the most important research targets since the beginning of the nuclear era and is among the most important targets for the future with needs for increased calculation precision.

Selected important issues in the area that also constitute the motivation for the chapter are outlined hereafter. Those issues may imply the need for further development, to achieve a common understanding and/or fix precision goals for calculations.

- The convection HTC is defined based on an empirical equation also known as Newton's law. This makes possible the quantitative evaluation from experiments but implies the existence of a multidimensional surface (actually not derived so far) where HTC is a function of dozens of variables.
- Reflood heat transfer implies the definition of an HTC that assumes (very) high values for a few millimeters around the quench front; the reflood occurrence is associated with the quench front propagation and may be associated with a surface temperature also called minimum film boiling; in some situations, e.g., when a short (a few decimeters length) pieces of fuel rod remains to be quenched, a sort of homogeneous-instantaneous reflood takes place (i.e., sudden formation of the liquid film in contact with the solid wall).
- Heat transfer from fuel-clad surface at high temperature in the conditions of interest for nuclear thermal hydraulics on the one hand includes radiation and on the other hand may occur toward different fields of the same phase, i.e., liquid film and droplets. Modeling of both radiation heat transfer and liquid fields heat transfer poses challenges: the presence of spacer grids causes additional complexity to the modeling.
- Wall condensation occurs in containment during postulated accident conditions. Condensation is affected by the presence of noncondensable gases including H_2 and determines the global convection motion of the fluid that at its time governs H_2 concentration, again putting a challenge to nuclear thermal hydraulics.
- Conjugate heat transfer, in rough words connecting convection and conduction heat transfer, has already been known for decades; however related analytical formulations are still based on empirical formulations (e.g., HTC and "Newton law").

The objective of the chapter is to present the current status and the vision for the future in this (selected) area of thermal-hydraulic science.

1.1.3.8 Chapter 8: Pressure drops

Pressure drops and heat transfers are the main targets within the design of thermal-hydraulic systems and a wide variety of approaches and precision needs have been established and are available (as already mentioned).

Chapters 7 and 8 are planned in the book also to constitute a bridge between the past and the future for nuclear thermal hydraulics: the evaluation of pressure drops, at geometric discontinuities and in two-phase conditions, has been among the first targets since the beginning of the nuclear era and is among the most important targets for the future with needs for increased calculation precision.

Selected important issues in the area that also constitute the motivation for the chapter are outlined in the following. Those issues may imply the need for further development, to achieve a common understanding and/or fix precision goals for calculations.

- Modeling of friction pressure drop in single-phase flows is based on an empirical approach widely validated by experimental and theoretical data. This is not the case for two-phase flows where the ancient approach characterized by a two-phase flow multiplier is adopted: this approach may have validity when fully developed flow conditions are established, i.e., a situation never verified when accident conditions in NPP are calculated.
- Pressure drops at geometric discontinuities in either single- or two-phase conditions are currently modeled by the application of empirically based coefficients. Assumptions are needed to determine those coefficients from experiments and, usually, no dependence is available from Reynolds number and void fraction: the impact of assumptions and the lack of specification of dependencies (e.g., Re and void fraction) on the results of prediction is unknown. This may largely contribute to the uncertainty of SYS TH code predictions.
- Evaluation of pressure drops in case of sonic and supersonic velocity conditions implying the formation of shock waves.
- Evaluation of acceleration pressure drops in a (not necessarily) closed boiling and condensing system where irreversibility of the thermodynamic process may add up.

The objective of the chapter is to present the current status and the vision for the future in this (selected) area of thermal-hydraulic science.

1.1.3.9 Chapter 9: Constitutive equations

Principles of thermodynamics and mechanics imply the conservation of mass energy and momentum that give the origin to the balance equations (Chapter 5). However, the same principles do not allow the characterization of the interactions between the phases, including the appearance of geometry- and pressure-dependent flow patterns.

Thus, constitutive equations are needed to model the interactions between the phases and between each phase or field and the environment. As a target, the constitutive equations must account for all possible conditions and phenomena expected in WCNR (and RR) within the DBA/DBC boundaries.

Selected important issues in the area that also constitute the motivation for the chapter are outlined in the following. Those issues may imply the need for further development, to achieve a common understanding and/or fixing precision goals for calculations.

- The words "closure equations" are also used and may embed the constitutive equations or may identify a class of equations having an ordinary numerical formulation that is focused on achieving hyperbolic solutions from balance equations (as discussed in Section 1.1.3.5).
- Steady state relations among water physical properties including derivatives are not usually classified as constitutive equations and are rather called state equations: nevertheless, tight interactions are established between water properties and constitutive equations.
- Constitutive and/or closure equations may have an empirical nature where measurement errors and scatter of data are unavoidable: characterizing the impact of those errors and scatters on the results obtained from solving balance equations coupled with constitutive equations may not have a commonly agreed answer.
- Assumptions are needed to create solvable sets of closure equations. As an example, reference can be made to the repartition of wall heat flux, i.e., a unique value derived from

experiments, into contribution terms to liquid, steam, and in some cases, steam-liquid interface. Arbitrariness in the repartition is mitigated by consistency checks and partially validated "a-posteriori" making use of experimental data independent from those used to generate the original repartition.
- An additional level of arbitrariness and complexity, following up on the preceding item (assumptions to create solvable sets of constitutive equations), occurs when either three fields or three dimensions are part of the concerned thermal-hydraulic model.

The objective of the chapter is to provide the current status in the areas of development and applications of constitutive equations and to provide evaluations or discussion of the preceding list of topics.

1.1.3.10 Chapter 10: Special models

Reference is made here to the prediction of transient scenarios occurring in the concerned water-cooled reactors. The physical systems, the physical processes, or the phenomena that are not (or cannot be) modeled by what are called constitutive equations embedded into balance equations are considered in the category of special models. Namely, two broad categories of special models are distinguished and related to

(A) components and
(B) physical processes or phenomena

Examples in the former category are centrifugal pumps, jet pumps, steam separators, steam dryers, valves (various configurations), and turbines.

Examples in the latter category are pressure drop coefficients at geometric discontinuities (also discussed in Chapter 8), flooding and countercurrent flow limitation (CCFL), TPCF when not derived from the set of balance and constitutive equations that are standard for a given SYS TH code, and direct contact condensation, typically at the ECC port.

The motivation for the chapter is the importance of concerned components and physical processes or phenomena in the frame of the prediction of transient scenarios.

The objective of the chapter is the description of models that constitute essential parts of the SYS TH codes.

1.1.3.11 Chapter 11: System thermal-hydraulic codes

The envisaged role of the SYS TH codes within nuclear thermal hydraulics and the pivot role for the same codes in the present book are discussed in this chapter (see also Fig. 1.1).

As it may be derived from the notes in Section 1.1.2, the development of an SYS TH code needs a variety of expertise, or matter and themes of the thermal-hydraulic universe, including numerical methods, neutron physics, computer science, etc. Only thermal-hydraulic topics are discussed in Chapter 11.

Important aspects for BE SYS TH codes and related applications are (the last item in the list that follows is considered a consequence of those that come before):

- Demonstration of the qualification level
- User effect on the results of the calculation

- Justification of the reason why approximate modeling may produce acceptable results
- Need and way to evaluate the uncertainty
- Acceptability by regulatory authorities

The motivation for Chapter 11 is the importance of the SYS TH code. The objectives are a contribution to establishing the capabilities of the codes based on the thermal-hydraulic modeling and, as a consequence, making the applications robust to WCNR technology.

1.1.3.12 Chapter 12: Computational fluid dynamics codes

The CFD matter is apparently outside the main path across the nuclear thermal-hydraulic universe selected for the book. However, at least two reasons are suggested considering the CFD codes in Chapter 12:

(1) CFD technology has the potential for further development in the future also connected with more powerful computing machines.
(2) Currently the application of CFD codes provides an irreplaceable role to support SYS TH code calculations in a large number of situations; examples of those situations are PTS, boron mixing and transport, temperature stratification, and H_2 transport in the containment.

CFD technology was born well before the advent of computers at the end of the 19th century; however, only the availability of powerful computers after the 1970s (i.e., during the 20th century) allowed the application of CFD codes to problems of industrial interest. Key differences between SYS TH codes and CFD codes are connected with the turbulence modeling that is direct in the case of CFD and the averaging process (e.g., at cross-sectional or volumetric levels) that in principle is not needed in the case of CFD codes. Namely, convergence of results is expected with a reduction in mesh sizes in CFD codes, while the same goal cannot be demonstrated in a general situation in the case of SYS TH codes. Moreover, in the case of CFD codes, attempts are made to reduce the dependence on the empiricism that characterizes the modeling of SYS TH codes.

Currently debated aspects within CFD technology are:

- Results are still mesh dependent: i.e., the convergence of results cannot be established for each application, and changes in the mesh topology cause different results.
- Wall functions are needed, e.g., to account for friction or HTC: this implies the reintroduction of empiricism, contrary to the no-empiricism target that characterizes the development of CFD codes.
- Maturity and/or robustness of solutions for single-phase applications in nuclear reactor safety are claimed by code developers. However, this is not the case for two-phase applications.
- V&V methods including qualitative and quantitative acceptability targets and Uncertainty Quantification (UQ) are under discussion within the scientific community.
- Precision targets, as in the case of the SYS TH codes technology area, e.g., for punctual velocity vectors, pressure drop scalar between two positions, punctual mixing scalar, etc., are not established.

All the foregoing shall be seen as a framework for the motivation of the chapter. The objective of Chapter 12 is to present the current status of the development of CFD codes with a focus on modeling, capabilities, and perspectives.

1.1.3.13 Chapter 13: Verification, validation, scaling, and uncertainty

This chapter can be characterized as "putting the right perspectives for exploiting the current capabilities of SYS TH codes." This means addressing the areas of V&V, scaling, and uncertainty. With some simplification, scaling (and scaling needs; see also "Glossary") is an outcome of V&V needs, as well as UQ becomes a need following the evaluation of V&V results.

If one uses the human body to create a parallel with nuclear thermal hydraulics where the body movement constitutes the application of the discipline, then the SYS TH code is the heart of the body, and V&V, including scaling, and uncertainty constitute the blood.

V&V, scaling, and uncertainty are also the bridges between SYS TH code development and the application in nuclear technology. As a result of decades of R&D (as already discussed) qualified numerical tools are available including the capability to evaluate the uncertainty; however, current applications within the licensing process do not exploit those capabilities, and simplified (so-called conservative) methods and approaches are used. This situation may appear controversial.

Then, the objectives and motivations for the chapter are to show the robustness of the processes and the methodologies adopted for V&V, scaling, and uncertainty.

1.1.3.14 Chapter 14: Best estimate plus uncertainty approach

Several definitions are suitable for BEPU, e.g., "BEPU constitutes the link between SYS TH code development and V&V, Scaling, Uncertainty on the one side and licensing process on the other side," and "BEPU is the application framework for SYS TH codes."

Thus, the BEPU process shall be associated with the licensing process: this notably includes the interaction of DSA and PSA and the consideration of several related methods (those methods are not discussed in the present book: however, an overview of the interaction between DSA and PSA with focus on the reliability of passive systems can be found in OECD/NEA/CSNI, 2020). AA is a key part of the licensing process: thus, looking at AA, one may deduct features of BEPU and the requirements.

The first attribute of AA is its compliance with established regulatory requirements.

The second attribute deals with the adequacy and completeness of the selected spectrum of events: this is the result of the combined applications of deterministic and probabilistic methods.

The third attribute is the knowledge base including that captured by qualified computational tools and analytical procedures and by properly scaled experiments.

The complexity of a WCNR and/or of the expected accident scenarios prove to be challenging for a conservative analysis, thus justifying the choice for a BEPU approach. This conveys two main needs for nuclear thermal hydraulics, i.e., to develop (1) suitable computational tools, proving (to the regulatory authority) adequate quality via V&V and scaling; (2) qualified uncertainty methods, possibly estimating errors in code calculation results with algorithms not affected by the inadequacies of those codes (i.e., in relation to which the uncertainty is estimated).

The motivation for Chapter 14 is to establish a role for BEPU within science and technology, breaking the isolation that restricts now BEPU knowledge to a narrow group of specialists. The idea is to consider BEPU an essential piece of nuclear thermal hydraulics thus facilitating and spreading the application of SYS TH codes.

The objective of the chapter is to provide a suitable overview of the techniques and procedures part of BEPU, properly emphasizing the benefits expected from its application.

1.1.3.15 Chapter 15: Design basis accident/condition calculations

AA constitutes one key branch of nuclear reactor safety and is needed for individual WCNR and RR safety demonstrations, as already mentioned. PIEs are at the origin of AA. Among those, the internal PIEs need transient calculations: the most powerful tool to perform transient calculations is the SYS TH code. Thermal-hydraulic phenomena, which are needed to validate SYS TH codes, derive (primarily) from scaled experiments. Demonstration of consistency between phenomena derived from experiments and resulting from AA scenarios constitutes the motivation and the objective for Chapter 15.

An answer to the question "What calculations must be performed?" is needed in advance. This leads to the DBA/DBC envelope for categories of WCNR. The following list of AA scenarios (i.e., the category of events), part of the DBA/DBC envelope, is created by considering the principal cause for PIE:

- Reactivity anomalies due to control rod malfunctions
- Reactivity anomalies due to boron dilution or cold-water injection
- Coast-down of the main circulation pumps
- Loss of primary system integrity (LOCA)
- Interfacing systems LOCA
- Loss of integrity of secondary system
- Loss of power supply
- Malfunctions in the primary systems
- Malfunctions in the secondary systems
- ATWS
- Accidents in fuel handling
- Accidents in auxiliary systems
- Accidents due to external events

Sample results from WCNR transient analyses performed by SYS TH codes are discussed in Chapter 15, and a connection is established with the phenomena presented in Chapter 6.

1.1.3.16 Chapter 16: Thermal hydraulics of nuclear power plant accident occurrences

Accidents have occurred in WCNR and RR and are part of the history of nuclear technology. Those accidents had a notable impact on the development of the technology and, unfortunately, a negative impact on the exploitation of the technology for electricity production. This constitutes the motivation for Chapter 16 although this is an unexpected chapter for those who planned the development of nuclear thermal hydraulics till the late 1950s of the previous century.

The nuclear events considered are hereafter listed together with selected main implications (or lessons learned) in nuclear thermal hydraulics; the low role that thermal-hydraulic competences and expertize had in the occurrence of the event or in mitigating their consequences (where applicable) is recognized in advance:

- TMI, Unit 2 (TMI-2), LOFW-SBLOCA combined event, March 1979. As a consequence of the event, the focus of investigation in nuclear thermal hydraulics was shifted from LBLOCA to SBLOCA: this means moving the attention from homogeneous conditions in the vessel to separated fluids or stratified conditions. This can also be translated as a shift from HEM or EVET modeling to UVUT modeling. The TMI-2 event also triggered huge investments for experiments in large-scale ITF.
- LaSalle Unit 2 and Oskarshamn Unit 2 BWR stability events, respectively occurred in March 1988 and February 1999. As a consequence of the first event, a systematic reevaluation of the capabilities to predict density wave instability (DWI) phenomena took place. This brought on the understanding that DWI depends on a large number of factors including their combined effect; furthermore, (very) small changes in the value of each factor that are decisively affecting DWI and are difficult to predict. Nevertheless, DWI events are found to have an incubation period that is detectable making mild or negligible the safety implications of possible modeling inadequacies.
- Chernobyl Unit 4, neutron flux and thermal power excursion event, April 1986. As a consequence of the event, operator-independent systems became a fashion in the new design. Thus, the event triggered the appearance of passive systems; both new-generation PWR and BWR were equipped with passive systems, notably AP-600 (later on AP-1000) and SBWR (later on ESBWR).
- Fukushima Units 1−3, LOOSP event, March 2011. As a consequence of the event, controlled containment venting, already introduced in Germany (a concept already part of the design of the early Soviet Union reactors to avoid the construction of containment walls resistant to the full pressure caused, e.g., by a LOCA) became of interest. The event also solicited more attention to low-probability initiating events (e.g., high-intensity earthquake in the deep water of the ocean) that, once occurred, may reveal precursors with a high probability of highly-safety-impacting accident scenarios (e.g., loss of ECCS availability following the tsunami).

The objective of the chapter is to describe the transient thermal-hydraulic scenarios that occurred during the accidents.

1.1.3.17 Chapter 17: Instrumentation and (basic) experiments

Inadequate modeling capabilities impose relevance of phenomena (e.g., Aksan et al., 2018a) in nuclear thermal hydraulics. Phenomena that constitute the target for computer code design and calculation, need characterization primarily throughout experiments. Then, instrumentation is needed to detect and portray measured phenomena. A broad technological field is involved with instrumentation: transducers, electronic chains to arrive at digitized values from each signal, and, in most cases, elaboration algorithms with embedded assumptions, are part of the field. Specific textbooks have been devoted to nuclear thermal-hydraulic instrumentation since the last century, e.g., Cumo (1982), and conferences have been held (e.g., SWINTH, Aksan et al., 2018). Simple instruments applicable to the direct detection of parameters like temperature and pressure shall be distinguished from measurement techniques like Particle Image Velocimetry to investigate turbulence that may require laboratory infrastructure and procedure/methods for the elaboration of signals, Hassan et al. (2014); Arlit et al. (2018); Hassan (2019).

The importance of instrumentation in nuclear thermal hydraulics constitutes the main motivation for the chapter; the objective is to depict the current capabilities and to streamline research in the future.

1.1.3.18 Chapter 18: Subchannel thermal hydraulics

Subchannel constitutes a branch or an attribute for nuclear thermal hydraulics. Its key feature is the focus given to the solution domain comprising the subchannels of a WCNR core. The subchannel is characterized by a cross-sectional shape, orthogonal to the flow direction, far away from a circle, and by the unavoidable interaction with neighboring channels. It was early discovered during the last century that correlations and solution techniques valid for a single cylindrical pipe may barely apply to core subchannels. Furthermore, the aim of subchannel analysis is the core design (nominal operating conditions) rather than the accident (transient conditions), Ninokata and Aritomi (1992). The in-principle capability of SYS TH codes to handle subchannel conditions was demonstrated early in this century, e.g., Petruzzi et al. (2004). However, research and technology applications in the areas of subchannel TH and SYS TH continued so far in two parallel trails with limited interactions. The objective and the motivation for the chapter are to shed light on subchannel thermal hydraulics and to peculiarities that keep it different from SYS TH.

1.1.3.19 Chapter 19: Containment thermal hydraulics

The synthesis in Section 1.1.3.18 related to the relationship between subchannel TH and SYS TH can be repeated here for the containment of TH. Therefore, containment constitutes a branch or an attribute of nuclear thermal hydraulics. Its key feature is the focus given to the solution domain that is the containment of the WCNR core, where full pressure design typical for PWR and pressure suppression design typical for BWR are distinguished. For both the cases of PWR and BWR, the subchannel is characterized by flow areas and (integration) volumes that are 10–1000 times bigger than the equivalent case for an RCS. Correlations and solution techniques valid for a containment region may be different from those adopted for a pipe, this is specifically true in the case of condensation, Yadav et al. (2016). The in-principle capability of SYS TH codes to handle containment conditions was demonstrated at the end of last century or early in this century, e.g., Bieder and Barré (1997); Souza Dutre et al. (2002). However, research and technology applications in the areas of containment TH and SYS TH continued so far in two parallel trails with limited interactions. The objective and the motivation for the chapter are to shed light on containment thermal hydraulics and to peculiarities that keep it different from SYS TH. The scope is limited to DBA/DBC, e.g., issues like H_2 burning or aerosol diffusion are out of the scope of the chapter.

1.1.3.20 Chapter 20: Numerics in nuclear thermal hydraulics

The balance equations for mass momentum and energy of fluids are at the center of attention here. Numerics and/or numerical solution methods and approaches constitute the key element to derive results from those equations, i.e., to calculate the values of thermal-hydraulic parameters that characterize the WCNR performance during the time of an accident scenario. Typical activities performed by a (group of) specialist(s) in numerical methods are (1) accepting the equations in differential form as derived from specialists in fluid dynamics or even contributing to derive those equations: the geometric dimensions of the problem (e.g., 1D, 2D, or 3D) are relevant, as discussed in Chapter 5; (2) transforming the equations in algebraic form: this implies selecting the solution variables (i.e., the unknowns) and imposing averaging domains typically at the levels of the cross section for the stream of fluid and the integration region-volume; (3) solving the equations, e.g., pursuing an implicit or an explicit pathway where all unknowns are obtained all together at each time step (implicit method), or guessing the solution for some unknowns and then iteratively derive the solution (explicit method); (4) checking/imposing the hyperbolicity of the solution, Ahn et al. (2017); (5) establishing the consistency between the numerical methods and the design features of the computer(s) where the concerned numerical method is expected to produce results; (6) contributing to the verification process (part of V&V, as mentioned in Chapters 13 and 22). For instance, the method of characteristics and hyperbolicity of the solution are considered by Rakhmatulin (1956); Lyczkowski et al. (1978); Lyczkowski (2018).

Dozens of textbooks are available on the subject of numerical methods, e.g., Chaskalovic (2014), and specifically for fluid dynamics problems, Durran (2010). The objective of the chapter is not to replicate those books; rather the objective is to provide insights into technology and research issues associated with the development and the application of numerical methods adopted to solve Partial Differential Equations (PDEs) for fluid dynamics and considered by developers of current SYS TH and CFD computer codes. The motivation for the chapter is to complement the information given in Chapter 5 (balance equations) and Chapters 11 and 12 (codes) by discussing an unavoidable component of the technological puzzle constituted by nuclear thermal hydraulics.

1.1.3.21 Chapter 21: Scaling insights

Scaling in nuclear thermal hydraulics is part of Chapter 13 and constitutes a key element of BEPU discussed in Chapter 14 for both the first and the second editions of the book. The reason why a new chapter on the topic of scaling is added in the second edition of the book is clarified hereafter.

Scaling and addressing the scaling issue are well-known subjects in fluid dynamics established since the XIX century, D'Auria and Galassi (2010). Science and technology aspects ranging from deriving dimensionless parameters from balance equations, to fixing requirements for experiments (design of facilities and tests), accepting results of experiments, proving the scaling capabilities of codes, to demonstrating that the scaling issue is sufficiently considered within the safety evaluation of a WCNR, are part of scaling. Scaling requirements and the interconnections of scaling with V&V

and BEPU are considered in Chapters 13 and 14. Otherwise, scaling methods and experimental databases, namely from ITFs, suitable for scaling investigations are among the scaling insights in Chapter 21. The outcomes from a recent international project, i.e., see OECD/NEA/CSNI (2017), aimed at achieving a consensus concerning scaling are also summarized in the last chapter. Therefore it has been preferred to keep the self-standing contents with suitable modifications of the existing chapters in the first edition and to add insights in the present chapter of the second edition. The relevance of scaling in nuclear thermal hydraulics justifies the new chapter.

1.1.3.22 Chapter 22: New perspectives for verification and validation

A similar situation to the case of scaling occurs for V&V: e.g., V&V in nuclear thermal hydraulics is part of Chapter 13 and constitutes a key element for BEPU discussed in Chapter 14, for both the first and the second edition of the book. Then, the reason why a new chapter on the topic of V&V is added in the second edition of the book is clarified hereafter.

The need for V&V in nuclear thermal hydraulics was clear once SYS TH codes started to produce results applicable to the WCNR, i.e., toward the end of the 1970s in the last century, Oberkampf and Roy (2010) (see also recent, Oberkampf and Balch, 2020; and older literature, D'Auria and Galassi, 1998). Two main motivations are at the basis of Chapter 22 in the second edition of the book.

- The lack of publication (and then of dissemination for the related contents) of the report IAEA (2015) already cited in the first edition of the book;
- The proposal for V&V&C (where "C" stands for Consistency), D'Auria and Lanfredini (2019).

In the former case, a systematic vision is provided for V&V, and additional requirements are proposed in its application. In the latter case, the starting observation is that validation (primarily) contributes little to the improvement of modeling where V&V is applied, although this is not the purpose of V&V; therefore, the V&V domain is extended outside the parameter ranges that characterize the experimental database adopted for validation.

As in the case of scaling, e.g., Section 1.1.3.21, it has been preferred to keep the self-standing contents with suitable modifications of the chapters existing in the first edition and to add the discussion of new perspectives for V&V in the present chapter of the second edition. The relevance of V&V in nuclear thermal hydraulics justifies the new chapter.

1.1.3.23 Chapter 23: Thermal-hydraulic design of water-cooled nuclear reactors

The expectation of being able to derive the geometric configuration and the operational characteristics of WCNR from elaborations of balance equations in thermal hydraulics is immaterial. Nevertheless, several design features of the RCS, BoP, ECCSs, and containment derive from basic thermal-hydraulic principles; incidentally, SYS TH and CFD codes are adopted to improve the design of those systems and to confirm their safety. Example of questions or topics considered in the chapter are (1) what is/are the key motivation(s) at the basis of the layout of RCS in PWR, i.e., the mutual position between the RPV and steam generator? (2) What is the role of the MCPs in PWR and BWR? (3) Calculate the mechanical/electrical power needed for an MCP in a PWR. (4) How can one determine the height of a jet pump in a BWR? (5) What is the design principle for the operation of a pressurizer (PRZ) spray line in PWR? (6) Why in PWR do we have low-pressure coolant injection, whereas in BWR we need low-pressure core spraying water at the top of the core? (7) Why does BWR operate at 7 MPa ($\approx$1000 psia)? (8) Why are full pressure and pressure suppression containments considered for BWR and PWR (in most designs)? (9) Why are MCPs in BWR installed at the bottom of the drywell? (10) Why is the condenser inlet at the bottom exit of (low-pressure) turbines? (11) What is a "loop seal" in the RCS of a PWR? (12) What is the reason for the mutual gravity elevation of the cold leg (CL) and core in a PWR? (13) What is the limiting size for the CL pipe diameter? (14) Why is the iPilot-operated relief valve installed at the top and not at the bottom of a PRZ in a PWR? (15) Why is the PRZ in a PWR connected to the hot leg and not the CL?

The objective and motivation of the chapter are to connect nuclear thermal hydraulics and reactor design, making clear the reason why for the value of several reactor designs.

1.1.3.24 Chapter 24: Controversial issues and perspectives

Controversial issues, challenges, and stagnation in development characterize, among other things, nuclear thermal hydraulics during the last couple of decades, D'Auria and Hassan (2021). A textbook shall not be the house of open issues or unsolved questions, rather consolidated topics and concepts should be part of the contents; however, in the case of nuclear thermal hydraulics, controversial issues are part of the consolidated state of the art. Therefore, the discussion around selected controversial issues constitutes the motivation for Chapter 24; the objective is to establish a background of information suitable for identifying valid research lines. In this connection, turbulence modeling and averaging techniques for two-phase flows are among the key aspects that need improvement, while the attention of decision-makers and researchers is oriented toward fashion topics or even research objectives that can be accomplished within an assigned framework of resources (time and budget). It is recognized, D'Auria and Hassan (2021), that breakthrough ideas in modeling are needed to exit the stagnation.

1.1.3.25 Chapter interconnections

Fig. 1.1 and the discussion in Sections 1.1.3.1–1.1.3.24 are relevant to establishing connections among individual chapters of the book. These can be of help to those who access only parts of the book.

Chapters 1–4 are directly or indirectly connected (namely Chapter 1 is connected to Chapters 2–4 and connections exist between Chapters 3 and 4) and to all remaining chapters.

Chapters 5–10 are directly connected to Chapter 11 (SYS TH code).

Chapter 11 (SYS TH code) is at the origin of Chapters 13–16. Thus, Chapter 11 plays a pivotal role in the present book as already emphasized.

Chapter 13 should be seen at the origin (i.e., discussing the background) of Chapters 21 and 22.

Chapters 18 and 19 should be seen as "in parallel" and as "complementary" to Chapter 11.

Chapter 20 constitutes an integration into Chapter 11 (primarily) and Chapter 12.

Chapter 24 provides comments on all previous chapters.

Additional connections among chapters are:

- Chapters 4 and 6: the knowledge of thermal-hydraulic phenomena (Chapter 6) constitutes a need that is clarified in Chapter 4.
- Chapter 5 (balance equations) constitutes the framework and the basis for Chapters 9 and 10: namely, in the case of Chapter 10 (special models), the phenomena of the "physical situations" that cannot be calculated by balance equations shall be dealt with by special models. Constitutive equations (Chapter 9) must be integrated into balance equations (Chapter 5). Altogether, Chapters 5, 9, and 10 are an input to Chapter 11 (SYS TH code). Chapter 5 shall also be seen as providing (partial) input to Chapter 12 (CFD code).
- Chapter 6 (phenomena) has a direct connection and interrelation with Chapters 11, 13, 15, and 17: the development of SYS TH codes (Chapter 11) needs the knowledge of phenomena; the validation of SYS TH codes (Chapter 13) must be based on the phenomena; finally, phenomena must be derived from the envelope of DBA/DBC (Chapter 15).
- The knowledge of subjects from Chapter 7 (heat transfer) and Chapter 8 (pressure drop) is embedded into the SYS TH code, thus creating a direct connection between those chapters and Chapter 11.
- Chapter 9 (constitutive equations) shall be seen as a necessary support for (the solution of) balance equations discussed in Chapter 5. Both Chapters 5 and 9 provide input to the development of the SYS TH code discussed in Chapter 11. Duplication of concepts appears unavoidable between Chapters 5 and 9 from the one side and Chapter 11 from the other (however, the authors made an effort to present complementary information and different viewpoints). The balance equations in Chapter 5 also constitute the background for Chapters 18 and 19.
- The knowledge and the subjects from Chapter 10 (special models) are embedded into the SYS TH code thus creating a direct connection between this chapter and Chapter 11. Furthermore, pressure drops at geometric discontinuities are also discussed in Chapter 8.
- Chapter 11 (SYS TH code) is the end product of scientific competence and knowledge from all previous chapters.
- Chapters 12, 18, and 19 (CFD, subchannel, and containment codes, respectively) have a connection in the book to Chapter 5 (balance equations) and Chapter 11 (SYS TH codes). In this last case, the connection is not only to provide support to the SYS TH code calculations but also to provide differences in modeling approaches that may be closer to experimental reality.
- Chapter 13 (V&V, scaling, and uncertainty) and the related insights in Chapters 21 and 22, constitute the articulated joint between Chapter 11 (SYS TH codes) and Chapter 15 (codes application to NPP), see also role of BEPU approach (Chapter 14).
- Chapter 14 (BEPU approach) shall be seen as having the upstream interface with Chapter 13 (V&V, scaling, and uncertainty), therefore with Chapters 21 and 22, and the downstream interface with Chapter 15 (DBA/DBC calculations), namely when the licensing process is concerned.

- Chapter 15 (DBA/DBC calculations) is possible because of SYS TH (Chapter 11) code applications; in addition, the results from the concerned code calculations establish a feedback type of link with phenomena in Chapter 6.
- Chapter 16, i.e., occurrences of WCNR accidents, has a weak connection to nuclear thermal hydraulics. Rather, the nuclear thermal hydraulics discipline and, notably, the SYS TH codes allowed or facilitated the understanding of the progression of the events.
- Chapter 23, i.e., selected features of thermal-hydraulic design of WCNR, may be seen as having limited (direct) connections with previous chapters; however, competence in nuclear thermal hydraulics is essential for finalizing the design and proving its compliance with safety requirements.

Glossary

The book is devoted to researchers, scientists, and professionals involved in safety and design analyses of WCNR on both the industry side and the regulatory side as well as to students working for master's or PhD theses. Therefore, readers are expected to have expertize in or suitable knowledge of nuclear technology. Thus the Glossary only includes terms that may generate misunderstanding or not have wide diffusion. Furthermore, streamlining concepts rather than providing formal definitions (readers should consult terminology proposed by recognized international organizations) is the objective for the terms considered hereafter.

Accident Management Procedures Procedures for the operation of ESFs and of any available system outside the DBA/DBC boundaries when core geometry is still coolable (so-called DEC-A conditions). AMP in the DEC-B conditions is not of interest in the present framework.

Accident scenario The thermal-hydraulic time evolution of a single- or two-phase mixture inside the WCNR following a postulated accident created an accident scenario.

Accuracy The word accuracy is used as a measure and a figure of merit (or even as a part of the metrics) to characterize the (known) error resulting from the comparison between measured data (typically from ITF and SETF experiments) and calculation results (typically of SYS TH and CFD codes). The comparison refers to parameters that characterize the phenomena. The evaluation of accuracy can be performed at qualitative and quantitative levels. Acceptability thresholds to demonstrate the quality of results should be adopted.

Adequacy of Evaluation Model (EM) The definition of EM implies the definition of related adequacy. The adequacy of the EM implies a series of actions (or a "process"; see later definition) that are requested from the code user, dealing with the code, concerned transient scenario, nodalization, and initial conditions at various levels or steps within the framework using system thermal-hydraulic codes in nuclear reactor safety. This is summarized in suitable detail in the USNRC RG 1.203, where the "Adequacy of EM" is reported as the evaluation model development and assessment process (EMDAP).

Assessment See code assessment.

Balance Equations See also field equations.

Balance of Plant (BoP) The part of a WCNR external to the containment and including (among the others: turbine, condenser, preheaters, and pumps) is called BoP.

Basic Test A basic test is a rather simple test to check the capability of a numerical tool to correctly simulate a simple physical problem. A basic test can be a simple test with an analytical solution to check some capability of the numerical scheme. It can also be a simple physical test with a trivial solution or an experimental test in some prototypical condition with measurements to compare with predictions.

Benchmark A benchmark problem is a test problem for code comparison (either numerical scheme comparison or physical model comparison) that does not necessarily have experimental results but has a generally agreed numerical solution.

Best Estimate (BE) The terms "best estimate" should be intended as the best simulation (for expected accident scenarios) attainable within the current knowledge boundaries. Based on the validation process (that is the analysis of the comparison between measured data and calculated results), the target of the best simulation can only be obtained introducing the uncertainty. Therefore the BEPU acronym and framework are created. See Chapter 14.

Best Estimate Code A code that: (1) is free of conservative assumptions in the modeling (deliberate pessimism regarding selected acceptance criteria), and (2) contains a sufficiently detailed model to describe the relevant processes that require modeling.

Best Estimate Plus Uncertainty (BEPU) See Best Estimate.

Bias of Calculation Result A bias is a measure of a systematic difference between an actual or true value and a predicted or measured value. Bias is the tendency of a model to systematically over- or underpredict data that are representative of an assigned phenomenon.

Blind International Standard Problem This is an ISP where concerned experimental data are not known to the code user before starting the processes of developing the nodalization and performing the calculation to simulate the assigned test

Cliff edge effect

When there are discontinuities in physical models, a small change in a parameter may lead to a great change in the evolution of whole physical processes. This is called the cliff edge effect. For example, some phenomena, such as loop seal clearing, flow regime, heat transfer near critical heat flux condition, are very sensitive to the values of certain parameters. Thus, these phenomena are characterized by "cliff edge effect." This effect evokes some challenges in the numerical simulation of those phenomena

Closure Relations

After averaging of local instantaneous fluid balance equations, many terms of the resulting equations are new unknowns or contain new unknowns. At this step, the number of unknowns is more than the number of equations and the system of equations is not closed. In particular, averaged equations exhibit terms for transfers of mass, momentum and energy at the walls and at the interfaces (in two-phase or multi-fluid flow conditions). Constitutive relations are mathematical expressions for these phase transfer terms as functions of the selected averaged principal variables. These are part of closure relations. Other relations are also necessary to express the average of nonlinear terms as functions of averaged main variables. The terminology "closure relations" derives from considering that they are necessary to close the system of equations. Some closure relations are simplifying assumptions such as $\langle \rho(P,T) \rangle = \rho(\langle P \rangle, \langle T \rangle)$

Code Assessment

Code assessment is the process of validation and verification that proves the capability of the physical models and of the numerical scheme of a code to simulate with sufficient accuracy and confidence the physical behavior in a given domain of application

Code Manual

Code manuals are a set of documents for the code users that are written by the code developers. The code manual usually consists of a theory manual, a models and correlation manual, user guidelines, an assessment report, an installation manual, a programmer's manual, etc.

Code Qualification

Code qualification is the set of procedures that are used to prove that a code is actually able to do what it is designed for and to meet all the associated requirements. Code qualification includes QA, V&V, full documentation, and validated tools for uncertainty evaluation for BE codes

Computational Fluid Dynamics (CFD) code Software aimed at the solution of Navier–Stokes (balance) equations based on interpretative (typically empirical) models of turbulence. CFD is the activity to simulate fluid dynamics with a numerical tool that solves multidimensional fluid dynamic equations, either local instantaneous equations or averaged equations, or filtered equations, mainly for single fluid models. Recently, CMFD (computational multifluid dynamics) was also introduced to extend the capability of CFD to multifluid models. CFD or CMFD tools may use either the open medium approach or the porous medium approach. In the common understanding, CFD and CMFD refer to a fine resolution. Therefore, the terminologies within CFD and CMFD are not applicable to SYS TH codes because SYS TH codes use 0-D and 1-D models and they may also use 3-D models with the porous body approach, but with a rather coarse nodalization. See Chapter 12.

Confidence in Validation In statistics "the confidence level" refers to the likelihood that the true parameter value lies within the range specified by the confidence interval. The confidence level is usually expressed as a percentage. Thus a 95% confidence level implies that the probability that the true parameter value lies within the confidence interval is 0.95. Here the confidence level implies a probability 0.95. In the area of V&V for system thermal-hydraulic codes, the confidence in validation indicates the level of satisfaction on the results of the validation process by an observer. Therefore, the confidence in validation should be based on the quantitative evaluation of accuracy where proper thresholds of acceptability for the calculations (i.e., comparison between measured and predicted values) are fixed and fulfilled

Conservatism See conservative.

Conservative Calculation A conservative calculation is a calculation carried out to produce pessimistic results relative to specified acceptance criterion/criteria by using a conservative code and an unfavorable input data set including conservative initial and boundary conditions

Conservative Code A conservative code is a safety analysis code that was designed to produce pessimistic results relative to specified acceptance criterion/criteria. Some phenomena are modeled in a way to overestimate the severity of the accident sequence (e.g., a wall-to-fluid heat transfer coefficient is minimized to overestimate a rod-clad temperature)

Conservative: A set of input parameters (i.e., for s SYS TH code), or a calculation, or a model can be conservative; "conservative" set of input parameters, calculation and model means different-from-reality and implies the target of producing results that confirm (or, more in general establish) the safety of the system under assigned conditions. Conservatism is associated with lack of knowledge (in some cases, ignorance) of phenomena and of boundary-initial conditions for the integration of PDE. In general, the achievement of the "target" just mentioned cannot be demonstrated. Whatever discussed in the book aims at BE simulation (therefore the word conservative is irrelevant for the book)

Constitutive Equations The constitutive equations provide a supplement to the PDE typically derived from the analysis of experimental phenomena. Averaged fluid balance equations exhibit terms for transfers of mass, momentum and energy at the walls and at the interfaces. Constitutive relations are expressions for these transfers as functions of principal variables. These are also called closure relations since they are necessary to close the system of equations. See Chapter 9.

Convergence See mesh convergence

Courant Limit See material Courant limit

Design Basis Accident (DBA) The acronym is used interchangeably with the more recent DBCs (design basis conditions). The DBA/DBC envelope of postulated accidents is at the origin of phenomena that need investigation in nuclear thermal hydraulics

Design Basis Conditions See DBA

Design Extension Condition (DEC) Outside of DBC; cab be DEC-A and DEC-B, see also Accident Management Procedures

Deterministic Safety Assessment (DSA) A branch of nuclear reactor safety suitable for demonstrating the compliance between the safety objective and the safety requirements on the one side and the individual WCNR design and operation features on the other side. Nuclear thermal hydraulics is the key instrument for DSA

Developmental Assessment This is the assessment that is performed by code developers during the development phase of the code to select appropriate physical models and numerical algorithms and to check that they have a good behavior and that they are properly implemented. See also Validation (Chapter 22).

EMDAP (Evaluation Model Development and Assessment Process)	See Adequacy of EM
Emergency Core Cooling Systems (ECCS)	Set of systems to protect safety functions in WCNR and to minimize radioactive releases to containment and to environment following DBA/DBC
Emergency Operating Procedures (EOPs)	Procedures for the operation of ESF (ECCS, primarily) within the DBA/DBC boundaries (core geometry is intact or coolable: ballooning of a limited number of rods is possible)
Evaluation Model (EM)	EM is the calculation framework for evaluating the behavior of the reactor system during a postulated transient or DBA. The EM may include one or more computer programs, special models, and all other information needed to apply the calculation framework to a specific event
Explicit Numerical Scheme	An explicit numerical scheme is a numerical scheme for which unknown quantities (i.e., the quantities at the next time step) are evaluated in terms of known quantities at the current time step only. In the fully explicit scheme, all unknowns of all terms are taken at the beginning of the time step except in the time derivative term
Field Equations	Field equations are the equations that are solved to determine the transport of mass, energy and momentum throughout the system
Frozen	The condition whereby the analytical tools and associated facility input decks remain unchanged and under configuration control throughout a safety analysis, thereby ensuring traceability and consistency in the final results
Graded Approach	The terms "graded approach" are used in nuclear reactor safety in relation to the requirements for safety analyses in RRs. Those requirements shall be commensurate with the risk, including the potential damage, associated with any failure. The risk can be lower than in the case of NPPS; thus requirements shall be milder according to the "graded approach." In the present context the terms "graded approach" are related to the Evaluation Model Development and Assessment Process (EMDAP), namely to the "additional process" performed when passing from one evaluation model (EM) to another. In this situation, the graded approach implies the systematic consideration of the differences between the two EMs: these include the amount of innovation (novelty) introduced in the new EM, the complexities of the events under concern, the possible conservatism introduced in one or both EMs and the

	importance of the plant design or operational changes that would require reanalysis by the new EM
IET (Integral Effect Test)	Integral Effect Tests (or experiments performed in ITFs) are experimental tests that intend to simulate the behavior of a complex system with all interactions between various flows and heat transfer processes occurring in various system components. IETs relative to reactor accidental thermal hydraulics simulate the whole primary cooling circuit and simulate the accidental scenario through initial and boundary conditions
Ill-posed Equations	Ill-posed equations are equations that do not satisfy the conditions of the well-posedness (see well-posed equations)
Implicit Numerical Scheme	An implicit numerical scheme is a numerical scheme for which unknown quantities (i.e., the quantities at the next time step) are evaluated in terms of unknown quantities. The implicit formulation requires the simultaneous solution of several coupled equations involving the unknowns. In the fully implicit scheme, all unknowns of all terms are taken at the end of time-step except in the time derivative term
Independent Assessment	Verification and validation by an organization that is both technically and managerially separate from the organization responsible for developing the software or the code. See Validation (Chapter 22).
Interim Acceptance Criteria	Requirement set by US AEC in 1971 for the design of ECCS: these constituted a cornerstone for the development of nuclear thermal hydraulics
International Standard Problem	International standard problems (ISPs) are selected problems in the framework of the independent assessment of codes, namely system codes, used in AA. The ISPs are selected by relevant committees inside the OECD/NEA/CSNI and are based on experiments performed in facilities available within OECD member countries. One Host Institution manages the ISP and is responsible for issuing related documents. Participants perform calculations and contribute to the preparation of the document. ISP-issued documents are approved by the CSNI following proper review processes.
Integral test Facility (ITF)	see IET
Licensing Analysis	Licensing analysis is an analysis that may be performed by system codes following legal regulations by Regulatory Authority
Lumped Model (or Lumped Parameter Model)	A zero-dimensional or point model aimed at the simulation of steady state or transient performance

of a finite space (typically, dimensions are consistent with the entire NPP or with a significant part of it) is known as lumped model or lumped parameter model. A lumped model is also constituted by the coupling of lumped models as just defined. For instance, a point neutron kinetics model coupled with a point hydraulic model for the core and a point model for conducting heat transfer of fuel rods still constitutes a lumped model.

Manufactured Solution (Method of Manufactured Solution)

The method of manufactured solution is a method to test the capabilities and the accuracy of a numerical solution algorithm for a given set of equations. The first step is to select a manufactured solution as a closed analytic form for the solution that we want for the final test problem. Then the analytic solution is substituted into the base PDEs (partial differential equations) to generate new or modified source terms in the equations. Finally, initial and boundary conditions for the test problem are obtained by evaluating the selected solution form at zero time and at spatial boundaries of the problem

Material Courant Limit (or Courant, Friedrichs, and Lewy limit)

The material Courant limit is a time step limit for numerical stability imposed by the explicit treatment of the convective terms. When the convective quantities are treated explicitly, the time step should be equal to or less than Min (dz/U) for numerical stability, where dz and U are the mesh size and fluid velocity, respectively. This is also called the convective limit. The origin of the term CFL is a famous paper written by Courant, Friedrichs, and Lewy (CFL), Courant et al. (1928). See also the revisiting by De Moura and Kubrusly (2013) that addressed the stability requirement for the numerical solution of hyperbolic PDE.

Mesh Convergence

Mesh convergence is the process to decrease the mesh size of a nodalization up to the degree where the solution no longer changes. When the numerical scheme is consistent and convergent, the convergence is obtained when the exact solution of the PDE is reached. In practice, convergence tests are made up to a given tolerance limit by defining a convergence criterion. In system thermal-hydraulic codes, mesh convergence is possible in 1-D modules that use a well-posed system of equations

Module

A module of a thermal-hydraulic code is a part of the code that models a component of a reactor or of a test facility with a given set of physical assumptions and physical models. A module of a system code can be

	0-D, 1-D, or multi-D. A system thermal-hydraulic code generally includes several modules and submodules to be able to model all components of a reactor. Submodules represent parts of a code that model systems of a reactor such as pumps, sources, sinks, or valves that are coupled to a module
Natural circulation	Natural circulation implies the presence of a loop where a heat sink and a heat source can be identified at two different gravity elevations (heat source "below" the heat sink). No mechanical pumps are in operation in a natural circulation loop.
Natural convection	Natural convection is a heat transfer regime: it is expected that the motion of fluid is generated (not all situations of natural convection) by heat transfer
Neutron physics	Neutron moderation occurs within the coolant fluid in the majority of WCNR (e.g., notable exceptions are the CANDU, RBMK, and Atucha reactors). Doppler effect calculation is affected by HTC and HTC changes at the surface of fuel pins. Thus, there is a tight connection between neutron physics (e.g., calculation of neutron flux) and calculation, by nuclear thermal hydraulics, of fluid status
Nodalization	The nodalization of a thermal-hydraulic system is the schematization of the system so that it may be reduced to a finite number of nodes or control volumes, to which the governing equations are applied and solved. In practice, it consists in defining a simplified geometry of the system and in defining nodes or control volumes for modeling fluid flow and heat conduction in solids. The nodalization is the interface between the code and the reality constituted by the NPP or a test facility. The main function of the nodalization is to transfer the peculiarities of the NPP into a format suitable for the system thermal-hydraulic code. Thus, the choice of the nodalization must take into account the real system characteristics (e.g., the NPP and the boundary conditions), the code features, the objective of the analysis, and the available computational resources
NPP Data	The set of data needed to perform an assigned safety or design analysis related to an NPP is known as NPP data. The data that constitute the NPP data, including the related level of detail, depend on the requirements and objectives of the analysis. Typical NPP data categories include (all or part of the following) fuel, neutron kinetics, topology, geometry, thermal-hydraulic initial conditions at steady state, components, I & C, BoP, and boundary

conditions (e.g., heat losses, leakages of valves, leakages of containment, etc.).

Nuclear installation The ensemble of NPP and RR (including WCNR) is characterized by the phrase nuclear installation.

Numerical Consistency Numerical consistency is the extent to which discretized equations approximate PDEs. A discretized representation of a PDE is said to be consistent if we can show that the difference between the exact solution of a PDE and its discretized representation (i.e., truncation error) vanishes as the mesh size tends to zero.

Numerical Convergence This is a refinement of the numerical feature that the solution to the discretized equation by a numerical scheme approaches the true solution to a PDE with the same initial and boundary conditions as the mesh. A consistent, stable scheme is convergent.

Numerical Diffusion Numerical diffusion is the numerical feature that the discretized equation of a PDE possesses an artificial diffusion term when a first-order approximation is used for the discretized equation, i.e., the leading term in the truncation error of a discretized equation for a PDE that has second-order derivatives when a first-order approximation is used. The truncation error then behaves like an artificial diffusion term and is thus called numerical diffusion, false diffusion, or numerical dissipation.

Numerical Stability Numerical stability is the numerical feature that perturbations, physical or otherwise, from any source are not permitted to grow during the sequence of numerical procedures as the calculation proceeds from one marching step to the next. This concept applies in a strict sense only to transient problems.

Open International Standard Problem This is an ISP where the experimental data of concern are known to the code user before starting the process of developing the nodalization and simulating the assigned test.

Open Medium (Media) Approach Open medium (media) approach in CFD is applied when local instantaneous equations are either not averaged (DNS) or simply time-averaged or ensemble-averaged (RANS: Reynolds-Averaged Navier–Stokes equations), or space averaged (or filtered) over a space scale much smaller than the dimensions of the fluid domain to be simulated. Equations are written only in the fluid domain and solid walls may exist only at the boundaries of the simulation domain. The space resolution of the fluid flow simulation is much finer than the hydraulic diameter or the dimensions of the flow domain. (See also Porous Media Approach)

Partial Differential (or Derivative) Equations (PDE) Mass momentum and energy balance equations written in terms of derivative of key unknowns and derived from the principles of thermodynamics and of solid mechanics

Passive system Passive systems are defined in devoted IAEA reports. Here thermal-hydraulic phenomena (notably natural circulation) are expected to occur in passive systems (... and thermal-hydraulic evolutions of those phenomena may be different than expected: hence the words reliability of passive systems)

Phenomena Identification Phenomena identification is the process of analyzing and subdividing a complex system thermal-hydraulic scenario (depending on a large number of thermal-hydraulic quantities) into several simpler processes or phenomena that depend mainly on a limited number of thermal-hydraulic quantities

Phenomenon and phenomena Phenomena are identified to characterize accident scenarios. Phenomena are the connection between modeling (and modeling capabilities) of SYS TH codes and Accident Scenarios. Phenomena are relevant for the validation process. See Chapters 6 and 15

Platform independence (of a code) See portability

Porous Body See Porous Media Approach

Porous Media Approach Porous medium (media) approach is applied when three-dimensional equations are written in a space domain containing both fluid and solid structures. Equations include a porosity factor representing the ratio of fluid volume to total volume and transfers with walls are modeled at every calculation node or in every mesh through source terms. The space resolution of the fluid flow simulation is at least equal to or larger than the hydraulic diameter. (See also Open Media Approach)

Portability (of a Code) or platform independence The portability of a code is the capability of a code to be run without failure and to predict the same results on various platforms, including various computers, various compilers, and various operating systems. Since predicted results cannot be the same up to the last digit, portability is often defined with a tolerance limit, and acceptance criteria are defined based on some target variables (e.g., relative difference lower than 10−5 on pressure and velocity)

Postulated accidents Thermal-hydraulic transient scenarios expected in nuclear installations following a Postulated Initiating Event (PIE). Postulated accidents are used to

	evaluate the safety of individual NPP and RR (WCNR in the present book). SYS TH codes supported by uncertainty evaluation are needed to calculate nuclear thermal-hydraulic scenarios during postulated accidents
Postulated Initiating Events (PIE)	See postulated accidents
Probabilistic Safety Assessment (PSA)	A branch of nuclear reactor safety suitable for demonstrating the compliance between the safety objective and the safety requirements on the one side and the individual WCNR design and operation features on the other side. The probability and the consequences of any PIE are concerned. Event tree and fault tree methods are concerned: nuclear thermal hydraulics is used as a tool
Quench (or Quench Front or Quench Front Propagation)	see reflood
Reactor Coolant (or Cooling) System (RCS)	The primary loop (case of BWR) and the primary and secondary loop till the isolation valves (case of PWR) is indicated as RCS in WCNR
Reflood	The word reflood is a phenomenon expected following LOCA in WCNR (as such, among 116 thermal-hydraulic phenomena, it is exceptionally considered in present Glossary). Namely, the word reflood is defined in Chapters 5, 7, 11 and 15. Here we note that the word reflooding (e.g., used in Chapter 11) is a synonymous of reflood and the words rewet, quench front and quench front propagation are used for aspects connected with reflood. Sometimes, those words are used instead of reflood (the reader should not be confused)
Reliability	The word reliability is used to characterize the capabilities of (typically) thermal-hydraulic passive systems to accomplish an assigned target mission. It may be noted that accuracy and uncertainty refer to a code calculation, while reliability refers to a system; however, uncertainty may have a role in the (mathematical) evaluation of reliability
Rewet	see reflood
Robustness	A code or a numerical scheme is said to be robust if it is capable of coping well with various physical situations without numerical failure. It includes the quality of being able to withstand numerical perturbations due to sudden changes in physical processes
Safety Analysis	Safety analysis is an analytical study usually performed by means of computer codes by which it is demonstrated to what extent safety requirements are met

Scaling or Scaling Issue	The issue is originated by the difference in size, pressure, and power (including combinations), between any prototype WCNR and any model facility where phenomena are investigated. As pointed out in the OECD/NEA/CSNI, 2017, report, two main scaling activities can be distinguished: (1) aimed at design of experiments; (2) aimed at demonstrating the scaling capabilities of SYS TH codes
Semi-implicit Numerical Scheme	A semi-implicit numerical scheme is a numerical scheme for which unknown quantities (i.e., the quantities at the next time-step) are evaluated in terms of both known and unknown quantities. Sometimes this term indicates a specific numerical scheme, such as the RELAP5 numerical scheme, in which the "convected" quantities are explicitly treated and the sonic quantities are implicitly treated so that the time step is limited by the material Courant limit but not limited by the sonic Courant limit
Sensitivity	Sensitivity or sensitivity analysis is a quantitative examination of how the behavior of a system varies with the changes of related parameters, usually in the values of the governing parameters. Systematic variation in code input variables or modeling parameters is needed to determine the influence of physical phenomena and/or of code input variables on the overall analysis results. In this report, sensitivity analysis is defined by varying individually one input parameter; then sensitivity is the effect on the overall analysis results of the variation of this input parameter. Other more sophisticated measures exist such as Pearson's correlation coefficients or standardized regression coefficients; in these cases, all the input parameters are simultaneously varied. Such sensitivity measures are outside the scope of this report. It is worth noting that performing a sensitivity analysis has nothing to do with performing an uncertainty analysis; the sensitivity analysis neither addresses the weakness of the code and the nodalization nor the sources of uncertainty for an assigned calculation
Separate Effects Test (SET)	Separate effects tests are experimental tests that intend to investigate a single physical process either in the absence of other processes or in conditions that allow for the measuring of the effects of the process of interest. SET may be used to validate a constitutive relation independently from the others
Software Quality Engineering	Software quality engineering is the activity related to the management of the quality of software. It

	includes QA procedures for code development, code testing including portability tests, code assessment including V&V, code delivery, code documentation, configuration management, management of the life cycle of the code, code debugging, code maintenance, correction of defects, and delivery of versions and updates
Special models	Special models, like constitutive equations, provide a supplement to PDE. However, as a difference from constitutive equations, special models refer to components (e.g., centrifugal pumps) and phenomena (e.g., TPCF and CCFL) that are "external" to PDE in an SYS TH code. See Chapter 10
Subsystem	A subsystem is a part of a system
System code [from SRS No. 23, IAEA, 2002]	System code is a computer model that is capable of simulating the transient performance of a complex system like an NPP. A system code typically includes equations for thermal hydraulics, neutronics and heat transfer and must be equipped with special models to simulate the performance of components such as pumps and separators. The code should typically also simulate the control logic implemented in the plant and be able to predict the accident evolution
System Modeling	System modeling is the process including input development for a system thermal-hydraulic code, and building a nodalization including the needed assumptions
System Thermal-hydraulic Code	A system thermal-hydraulic code is a thermal-hydraulic simulation tool that can model a whole thermal-hydraulic system made of several subsystems and components and that can simulate DBA scenarios. The system is modeled by modules and submodules assembled to describe the circuits. System code main modules are 0-D lumped, 1-D and 3-D in the porous media approach. Submodules model heat conduction in solids, heat exchangers, sources, sinks, breaks, pumps, turbines, etc., see Chapter 11
System	A system is an ensemble of material elements coupled together to perform a given functionality (e.g., a reactor cooling circuit is a system made of all the components of the reactor that allow the cooling fluid to cool the reactor core)
Thermal-Hydraulic Phenomenon (THP or T-HP, or TP)	See Phenomenon earlier and refer to the 116 phenomena listed in Chapters 6 and 15
Uncertainty (see also "Uncertainty Analysis")	Uncertainty is a measure of the expected error range in experimental data or in calculated values by a code. It may ideally be defined by a PDF (probability density function) of the considered output parameter.

It is often more simply expressed by two percentiles (for example the 2.5% and 97.5% percentiles) of the output parameter, from which a variation interval is deduced. The uncertainty can also be described by a bias and a standard deviation

Uncertainty Analysis (see also "Uncertainty")
Uncertainty analysis is an analysis to estimate the uncertainties (expected error range) of the quantities involved in and the results from the solution of a problem. If the applied method is based on propagation of the uncertainties of input parameters, uncertainty analysis includes estimation of individual modeling or overall code uncertainties, representation uncertainties, numerical inadequacies, user effects, computer/compiler effects and plant data uncertainties for the analysis of an individual event

User or Code-User
A code user is a person or a group of persons or an organization that will use the software product under concern (here the system thermal-hydraulic code). The roles of the users are the development of the nodalization for an experimental facility or an NPP, the independent assessment, and the code application. Normally, the code-user is also the analyst who interprets the data obtained from a calculation

User Effect
A user effect is what makes the difference between two sets of calculation results obtained by two code users (or two groups of code users) who use the same code to simulate the same problem (here a thermal-hydraulic system with initial and boundary conditions), and have access to the same information for setting up the nodalization and for determining the needed input and boundary condition values

User Guidelines
User Guidelines are provided to code users to give them the necessary recommendations for building an adequate nodalization and input deck to perform an appropriate simulation of a thermal-hydraulic problem. User guidelines of a system code primarily address the following:

- the choice of modules (0-D, 1-D, 3-D) and sub modules (pumps, sources, sinks, valves, breaks, heat exchanger, neutron kinetics models, etc.) to model the components of a system (test facility or reactor),
- the space resolution by giving recommendations on the mesh size for hydraulic meshes and for heat conduction in solids,
- the time resolution by giving recommendations on the time step and time step control.

User Guidelines are necessary to eliminate or at least to reduce as far as possible the user effect.

User Qualification
The process aiming at reducing the user effect constitutes the User Qualification. Extensive training and performing a suitable number of code calculations involving comparison with experimental data is at the basis of User Qualification. A well-qualified user may require experience in the area for at least 5 years (continuous-documented application of codes) or even 10 years in the case that he/she has the responsibility for interpreting the results of system analyses relevant to the licensing process

Validation
Validation is a process for accessing the accuracy of a physical model in an SYS TH code based on comparisons between computational simulations and experimental data. In a broad sense, validation is performed to provide confidence in the ability of a code to predict the values of the safety parameter or parameters of interest realistically or conservatively. It may also quantify the accuracy. The results of a validation may be used to determine the uncertainty of some constitutive laws of the code. The validation can be conducted by the code developers and/or by the code users. The former is called developmental assessment, and the latter is called independent assessment

Validation Matrix (code, user, specific system, and application)
A validation matrix is a set of selected experimental data for the purpose of extensive and systematic validation of a code. The validation matrix usually includes (1) basic tests, (2) separate effects tests, (3) integral effects tests, and (4) NPP data. Various validation matrices can be established by code developers and/or code users for their own purposes.

Verification
Verification is a process for accessing the software correctness and the numerical accuracy of the solution to a given physical model defined by a set of equations. In a broad sense, the verification is performed to demonstrate that the design of the code numerical algorithms conforms to the design requirements, that the source code conforms to programming standards and language standards, and that its logic is consistent with the design specification. The verification is usually conducted by the code developers and, sometimes, independent verification is performed by the code users

Versatility [of a Numerical Scheme]
Versatility is the ability to simulate all of the physical conditions that need to be modeled to meet the requirements of the code. The versatility of system codes should cover the domain of safety analyses of an NPP.

Water-Cooled Nuclear Reactors (WCNRs) See nuclear installations. The list of WCNRs that are the target for the book can be found in Chapter 15

Water-Packing Water-packing is related to predicting pressure spikes that are not related to known physical phenomena. These fictitious pressure spikes are sometimes calculated when steam or even gas is disappearing from, and water is about to fill, a control volume

Well-posed Problem According to Adaamar (1978), a problem is well-posed when a solution exists, is unique, and depends continuously on the initial condition values. In thermal hydraulics a problem is defined by a set of PDE plus initial and boundary conditions. Heat conduction equations with specified initial conditions to predict a future temperature field is well-posed. But the inverse problem of predicting a past temperature field from final temperatures is ill-posed since the solution is very sensitive to small changes in the final data. For fluid dynamic equations hyperbolicity is a condition of the well-posedness. A system of equations is hyperbolic when all eigenvalues (solutions of the characteristic equation) are real

Exercises and questions

Exercises and questions are proposed hereafter with solutions given in Appendix 1 (Volume III). Answers and solution are expected to be provided by readers before reading Appendix 1. Please provide extended answers.

1.1 Provide a distinction between thermal hydraulics and thermodynamics.
1.2 Provide a distinction between thermal hydraulics and fluid dynamics.
1.3 Identify the key features of nuclear thermal hydraulics based on your knowledge and the reading of Chapter 1 of the book.
1.4 Why validation has a key role in nuclear thermal hydraulics?
1.5 Who (which institution) is fixing the DBA/DBC envelope?
1.6 What are the challenging parameters for modeling convective heat transfer?
1.7 What are the challenging parameters for modeling pressure drops?
1.8 What are the key differences between an SYS TH code and a CFD code?
1.9 What are the main similarities between an SYS TH code and a CFD code?
1.10 Discuss the role of phenomena in nuclear thermal hydraulics.
1.11 Why turbulence and averaging constitute current challenges in nuclear thermal hydraulics?
1.12 Can you distinguish the Lagrangian approach and Eulerian approach in nuclear thermal hydraulics?
1.13 Summarize, possibly based on a web search, the role of Euler in the progression of fluid models.

1.14 Provide examples of the use of nuclear thermal hydraulics in PSA.
1.15 Discuss the connection between nuclear thermal hydraulics and neutron physics (e.g., not only moderation).

Acknowledgments

The authors wish to acknowledge comments to the present chapter in the first edition of the book received by D. Bestion, H. Glaeser, and P. Kirillov: in particular, D. Bestion provided suggestions for the structure of the chapter, H. Glaeser performed a review of the original manuscript, and P. Kirillov supplied various references. T. Kunugi and J.L Gandrille provided comments for the second edition.

A historical perspective of nuclear thermal hydraulics

2

Francesco D'Auria
University of Pisa, Pisa, Italy

Foreword

The discipline of nuclear thermal hydraulics (NTH) developed from the needs of nuclear power plants (NPPs) and, to a limited extent, research reactor (RR) design and safety. As in all other fields involving analytical methods, NTH benefited from the development of computers. Thermodynamics, rather than fluid dynamics, is the basis for its development together with experiments in transient two-phase situations, namely, complex geometry, high thermal power density, and pressure.

Mandatory licensing requirements originating from safety needs pushed tremendous investments starting in the late 1960s that drove developments in the area. All of this created a unique technology that merged highly geometry-dependent transient phenomena, sophisticated two-phase flow modeling, continuously advancing numerics science, and unavoidable empiricism derived from experiments. This forms the knowledge that new generations of scientists must master and manage.

The objective and motivation for this chapter is to trace the history of the discipline, which is necessary for understanding it and creating perspectives for its development.

Selected perspectives, or examples of directions for improvement, constitute the content of this chapter. However, a caution or warning statement is this—a single scientist or a group of scientists (or researchers, e.g., the authors of this book) has little or no possibility, nor the capability, of predicting forthcoming investments or determining a road map for future activities. Rather, the global market in science and technology—see, e.g., D'Auria (2019)—without the breakthrough idea of a genius, will inexorably remain slow and flat at the front of development.

A paper by the author, D'Auria (2012), was the inspiration for this chapter in the first edition of the book. Substantiation for the current chapter in this second edition of the book comes from D'Auria and Hassan (2019), Cheung (2021), D'Auria et al. (2021), and D'Auria and Hassan (2021), all issued in the period between the publication of the first edition and the submittal for publication of the second edition. A vision deeper and wider than what is provided in this chapter can be derived from the mentioned references. Nevertheless, the synthesis that follows (from the first edition) is consistent with the new documents, and only a few notes have been added. Moreover, the newly added Chapter 24 deals with perspectives in thermal hydraulics and is based on "elements for reflection", item (3) of the Abstract.

Handbook on Thermal Hydraulics in Water-Cooled Nuclear Reactors
https://doi.org/10.1016/B978-0-323-85606-5.00002-7

2.1 Introduction

Thermal hydraulics and system thermal hydraulics (SYS TH) make up one of the five key disciplines essential for progress in nuclear science and refer to NPP design and nuclear reactor safety (NRS) technologies; see Chapter 1 of the book. The other disciplines are neutron physics, structural mechanics, radioprotection, and reliability including statistics.

SYS TH deserves and achieves proper attention from national and international institutions engaged with NPP design and NRS. This is also true from the standpoints of research and development (R&D) and innovation.

Therefore, motivations are proposed hereafter for planning one's work through this chapter of the book. Reference is made to developments in complex technology, i.e., SYS TH in the present case, that may require multimillion-dollar investment to achieve concrete improvements. A few even contradictory concepts are discussed in the next paragraphs.

The epoch that started 2 or 3 decades ago is characterized by a fast increase in easiness-to-travel and communication power. Under these conditions, political and market-driven decisions, large institutions, or industry agreements may have contributed to the average but also overshadowed the ideas of individuals. In science and technology, including planning for the future, joint decisions constitute a democratic process. This also minimizes the resources needed for development and, in generic terms, is functional in obtaining peace for humankind. Bejan (2021), provides views on this connection, also emphasizing the difficulties of attributing research findings to originators. However, the ideas of individuals in research and innovation remain the engine for commercial success and advancements, specifically when noncomplex technologies or fragments of complex technologies are involved. All of this explains the status of NTH in the new millennium: many involved technologists and researchers, many ideas for innovations, and (very) slow perceptible progress. Developments in this area are like the expansion of a marsh rather than the flow of a river in one direction.

The political and strategic importance of NPP design, construction, and operation and NRS, primarily the acceptance criteria set by regulators, e.g., 10 CFR 50.46, USNRC (2015), results in the following needs: (1) to construct and operate scaled facilities having "suitable-achievable" power, pressure, and geometric dimensions as part of the design and (2) to develop and qualify sophisticated computer codes (or computational tools). Endeavors associated with both SYS TH facilities and computational tools involve research groups formed by a few dozen or several dozen researchers and corresponding financial investments.

However, one may decide the following for NTH (and not only for that discipline):

- Averaging inputs from several decision-makers is a democratic process, and "fashion-of-the-period" activities are privileged for funding.
- Targets set by individuals may become part of a dictatorial process if approved for development and may experience difficulties in fundraising; therefore, targets set by individuals may remain hidden.
- (Brilliant) ideas of individuals may be overshadowed where important projects are concerned with complex technologies.

2.1.1 Key actors and stakeholders in nuclear thermal hydraulics

It seems that establishing a framework in advance for key actors and stakeholders in nuclear reactor thermal hydraulics (NURETH) would be worthwhile. The list that follows, derived from Chapter 1 (where references can be found), provides an idea:

(1) Authors of the textbook, spread worldwide: El-Wakil, Tong-Weisman, Hsu, Collier, Lahey-Moody, Ishii, Todreas-Kazimi, and Levy are eminent scientists who contributed to establishing the NURETH technology.
(2) US Nuclear Regulatory Commission (USNRC): its role was to establish targets for the acceptability of NPP, consequently requesting a suitable level of knowledge and understanding of SYS TH.
(3) International institutions, namely, the Organisation for Economic Co-operation and Development/Nuclear Energy Agency (OECD/NEA) and the International Atomic Energy Agency (IAEA), an organization of the United Nations with the role of fixing the status of knowledge, spreading that same knowledge, and identifying perspectives for development.
(4) National Institutions, namely, the European Commission (EC) and the US Department of Energy (DOE: its key role was to finance research consistent with R&D investment strategies).
(5) Nuclear industry, including vendors-designers (primarily but not only Westinghouse, General Electric, AREVA-Siemens-Framatome, and Babcock & Wilcox) and utilities or associations of utilities (primarily but not only EPRI and EdF) financed R&D in various sectors of thermal hydraulics with the objectives of addressing regulatory needs and creating profits through better design.
(6) The NURETH series of conferences within the framework of the American Nuclear Society (ANS), started in 1980 (NURETH-01), e.g., Cheung (2021), can be used to synthesize the state of the art and progress in the area—around 5000 papers coauthored by more than 6000 scientists were collected on the occasion of NURETH-15, D'Auria et al. (2013).
(7) A dozen journals currently publish NURETH papers. A few journals in the previous century were leaders in nuclear SYS TH information: NED, NT, ANE, and PNE. More recently, other journals have been founded and publish papers in the area, for instance, STNI and NET. The fundamentals of thermal hydraulics are a reference topic for the journals ETFS, IJMF, and JHT.

A recent effort completed by D'Auria et al. (2021), embedding the paper PI-02 from D'Auria and Hassan (2019), showed a continuous enlargement in the scope of NTH and related topics. Seven recent books have been published with the words nuclear thermal hydraulics appearing in their titles, i.e., they were published after those mentioned under item (1), notably excluding Todreas et al. (2022), and were not available at the time of the present writing. The difficulty in providing a synthetic (and/or straight and commonly agreed) definition for NTH is confirmed.

2.1.2 Objective

The objective of the chapter is to provide a historical perspective of SYS TH, distinguishing between basic and SYS TH. No consideration goes (as far as possible) to research budgets, time schedules, fashion-research, and road maps of established and ongoing projects and strategies of major stakeholders in the sector.

The given list of actors and stakeholders is useful for estimating the complexity of the endeavor for shaping the future of NURETH.

A side or additional objective for the chapter is to provide insights for understanding NTH.

The current and envisaged roles of NTH inside NRS technology are relevant here. This implies establishing a link with Chapter 14, which deals with the best estimate plus uncertainty (BEPU) approach to exploit the capabilities of existing and qualified SYS TH codes discussed in Chapter 11. In the second edition of the book, Chapter 24 includes a more comprehensive and systematic discussion about perspectives for development than what is provided in Section 2.3.

2.2 System thermal hydraulics history and trends

This section aims to provide an outline of the history of NTH development, i.e., the first objective for the chapter (although we recognize that the history of science is a matter for specialists and we do not have the competences to invade that field). Nonetheless, we discuss various topics of concern in different parts of the book and make an effort to minimize the duplication of concepts from Chapters 1, 3, and 24. Moreover, the outline of history allows (or makes easier) an interpretation of current trends in SYS TH.

The following is a streamlined synthesis of the pioneering history of thermodynamics taken from Gavezzotti (2020):

- Avogadro A. (1776–1856) introduced the term "molecule".
- Gay-Lussac J. L. (1778–1850) performed fundamental research related to gas.
- Carnot S. (1796–1832) discovered the importance of temperature differences.
- Clapeyron E. (1799–1864) emphasized the role of Carnot from the viewpoint of analytical derivations.
- Joule J. (1818–89) found the mechanical equivalence of heat.
- Clausius R. (1822–89) adopted Carnot documents to arrive at the second principle of thermodynamics.
- Thomson W (Lord Kelvin, 1824–1907) emphasized the role of absolute temperature.
- Maxwell J. C. (1831–79) established the microscopic gas theory.
- Gibbs J. W. (1939–03) established chemical thermodynamics.
- Boltzmann L. (1844–1906) introduced statistical thermodynamics.

At this point, the reader may not want to miss the history of thermodynamics from the ideas of Greek scientists until the modern attempts to connect entropy, gravity, and relativity, as given by Dreyer et al. (2000). At some point in their paper, those authors state "*Nowhere else in science is there, or has there ever been, such an extreme and sometimes absurd struggle over models, concepts, interpretations, and fictitious problems as in thermodynamics.*"

2.2.1 Role of nuclear thermal hydraulics in nuclear reactor safety

The wide role of thermal hydraulics in nuclear technology is outlined in Chapter 1 and discussed by Gu (2018). So why is NRS of special concern here (i.e., relative to NTH)? The answer is that a key feature of NTH that partly distinguishes it from other key disciplines relevant to nuclear technology (mentioned earlier) is the importance and complexity of dealing with time-changing or transient situations—this feature is a need originating from NRS.

NRS, like any technology, entails two main parts, the fundamentals and the application plus the assessment. The sketch in Fig. 2.1 (e.g., see D'Auria et al., 2015) illustrates this. An idea of the complexity of NRS, at least the part of it dealing with accident analysis, can also be derived from IAEA (2000), IAEA (2006), and IAEA (2009a).

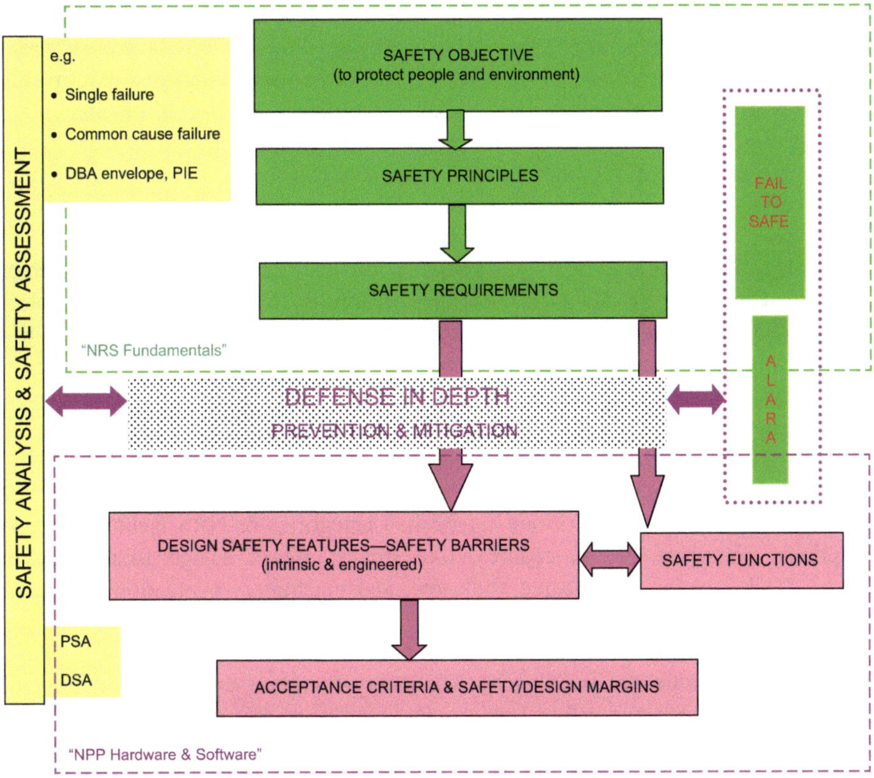

Figure 2.1 Simplified sketch of nuclear reactor safety technology showing the role of nuclear thermal hydraulics.

Fundamentals (e.g., IAEA, 2000) include the key safety objective, i.e., to protect people and the environment from ionizing radiation, related safety principles, and safety requirements according to established IAEA nomenclature. Application (e.g., IAEA, 2006) refers to whatever is done for the design, licensing, construction, displacement, operation, and decommissioning of a nuclear installation involving the presence of radioactive material (NPP equipped with water-cooled reactors are of concern here). Assessment (e.g., IAEA, 2009a) implies establishing how to demonstrate that the safety features introduced in the design are consistent with the safety objective.

The bases and procedures that make up the defense-in-depth (DiD) framework form the link between NRS fundamentals and NRS applications. Prevention and mitigation are distinguished in this connection, and DiD concepts apply to both.

NRS technology implies the existence of a road map originating with the established safety objective and becoming concrete with the design, construction, and operation (including plant dismantling and the fuel cycle) of an NPP or an RR. For acceptable safety and/or design margins, a demonstration is necessary for each step of the process in compliance with the safety fundamentals. Safety and/or design margins imply the acceptance of criteria established by devoted institutions, typically regulatory authorities in the case of safety margins. Furthermore, concepts like fail-safe and as-low-as-reasonably-achievable (ALARA) are part of the overall picture.

Safety fundamentals in an NPP design are accomplished through suitable safety analysis and assessment. A comprehensive safety analysis report, also known as a final safety analysis report (FSAR), for individual NPPs demonstrates that the safety objective has been met and notably that acceptable safety margins exist.

Here it is worthwhile to distinguish between analysis and assessment: the former relates to work done by designers, vendors, or utilities to implement fundamentals into the NPP (hardware and software), whereas the latter relates to confirmation activities typically performed under the oversight of a regulatory authority. Tools and procedures can be similar or even the same for analysis and assessment; however, independent specialist groups from the designer and regulator side that perform the same activity should be independent of each other.

Probabilistic safety assessments (or analyses) (PSAs) and deterministic safety assessments (or analyses) (DSAs) are established categories of NRS technology. Assumptions, concepts, and procedures, like single failure, design basis accident (DBA), DBA envelope, event-tree, fault-tree, and uncertainty evaluations, are used in PSA, DSA, or both.

A path is established in Fig. 2.1 between the safety objective (top of the sketch) and safety margins (bottom of the sketch). The safety functions ensure the integrity of design-safety features and safety barriers. Parameters characterizing the pink block parts of the NPP hardware and software are objects of calculations performed within the context of DSA and PSA. The concepts of prevention and mitigation are relevant to the overall road map and are considered key DiD elements.

A comprehensive description of individual elements in the diagram is beyond the scope of the chapter. Our discussion focuses on the roles of PSA and DSA and the crosscutting issue, which is addressed in a later diagram on licensing.

2.2.1.1 Roles of probabilistic and deterministic safety assessment

DSA (IAEA, 2009c) is associated with the availability of qualified best estimate (BE) computational tools or codes and has been in use since the 1990s. Conservative DSA is a key practice for the design and safety confirmation of existing reactors, i.e., activities performed since the 1950s. On the other hand, uncertainty is the keyword in the application of BE codes.

PSA (IAEA, 2010b) has been used in NRS since the issuance of WASH-700 (subsequently WASH-1400) by N. Rasmussen in the early 1970s. Three PSA levels are distinguished to estimate risk; those levels cover the probability and consequences (i.e., the radiological impact) of faulting events at any time of NPP life. Notably, DSA tools can only calculate consequences (see later).

Both DSA and PSA are necessary for the issuance of a consistent safety analysis report (i.e., as discussed primarily in Chapters 15 and 19 for the generally accepted FSAR structure). Furthermore, various interactions are envisaged and exist between the two NRS technology categories.

2.2.1.2 Role of nuclear thermal hydraulics

The role of NTH in NRS illustrated in the (simplified) sketch of Fig. 2.1 can be seen in the left part of the diagram—other than allowing the design of safety features and safety barriers (pink region of the diagram), NTH is essential to link the fundamentals (green region in the diagram) to the application (pink region of the diagram). This brings us to the demonstration of safety (and design) margins for nuclear installations. For this, NTH is embedded into PSA and DSA procedures and processes (dark white column in the diagram).

2.2.2 Definitions

The distinction between basic thermal hydraulics (simply, thermal hydraulics), SYS TH, and computational fluid dynamics (CFD) in single- and two-phase flows is a topic of concern from this point forward, consequent to the discussion in Chapter 1. Commonly accepted (i.e., by the international scientific community) boundaries among the three sectors of knowledge do not exist, and we do not wish to create unnecessary new definitions. Furthermore, D'Auria et al. (2021) provide a broader view of what NTH consists of today:

- Thermal hydraulics refers to single- and two-phase mixture performance under steady and transient conditions for an individual fluid or multiple fluids. Each phase can involve different fields. The numerical domain for solutions to equations is assumed to be bounded by a solid surface at "one-point" (giving rise to zero-dimensional (0D) or lumped-parameter models) within a one-dimensional (1D), two-dimensional (2D), or three-dimensional (3D) domain. The solid-surface-bounding hypothesis is straightforward for 0D and 1D situations, and a virtual solid surface may be considered in 2D and 3D situations. Pursuing a numerical solution within solid-bounded regions is known (sometimes and not necessarily) as a porous media approach or finite space averaged.

- SYS TH refers to the geometries, materials, and boundary conditions characterizing an NPP. Experimental (or test) facilities (TFs) and computational tools are the pillars of SYS TH. Within the broader category of TFs, differences exist between integral test facilities (ITFs) and separate effects test facilities (SETFs). ITFs aim to "reproduce" (typically by scaling down) overall reactor coolant system (RCS) performance, whereas SETFs aim to "reproduce" limited space regions or components of the NPP during limited periods following an accident. In the case of computational tools, the code and the input deck (or nodalization) are distinguished. Furthermore, SYS TH code includes neutron kinetics and instrumentation and control (I&C) system modeling capability. Qualification activities involving the so-called scaling issue are addressed concerning the design of ITFs, SETFs, related experiments, and associated computational tools (e.g., IAEA, 2002; D'Auria and Galassi, 2010). We distinguish the following key branches in SYS TH, where "code" implies modeling and numerics developments, and "experiment" implies a suitable database (see D'Auria et al., 2021):
  - Code1: SYS TH (codes)
  - Code2: subchannel (codes)
  - Code3: containment (codes)
  - Code4: specific phenomena, components, steady-state core, and critical heat flux (CHF) (codes)
  - Experiment1: data from NPP measurements
  - Experiment2: data from properly scaled ITF
  - Experiment3: data from properly scaled SETF and selected "basic" experiments
- CFD refers, primarily to single-phase fluid dynamics and heat transfer. Two-phase capability is available for CFD, but it is far from suitable for industrial needs across a broad range of applications, as discussed in Chapter 12. The numerical domain for the solution of equations is unbounded; 3D geometry is adopted for the integration of equations; the turbulence phenomenon has a direct connection with the solved equations and the computed unknowns; the linear dimension of the integration domain can be as small as 1/10 mm; variation profiles for quantities like velocity can be calculated inside a pipe for a cross section orthogonal to the flow axis. Pursuing the numerical solution in the unbounded fluid regions is known as the open media approach.

The key difference between basic thermal hydraulics and CFD is the modeling approach for two-phase mixtures and the related capability. When addressing single-phase flow, CFD does not require the geometry-averaging assumptions adopted in thermal hydraulics. Turbulence is calculated directly only for CFD. Furthermore, the mesh (or calculation region) dimension reduction is expected to generate convergence only in the case of CFD.

The difference between SYS TH and basic thermal hydraulics mostly applies when experiments and computational tool validation processes are planned. SYS TH implies processes for identifying and characterizing phenomena that are relevant to the conditions of the NPP (e.g., OECD/NEA/CSNI, 1987, 1993, 1996).

2.2.3 History

The history of NTH can be covered in several ways, including (1) a chronology of related events; (2) a list of topics of interest for different periods; (3) a history of nuclear technology or NRS connected to accidents and major (nuclear) catastrophes that have occurred; and (4) the consideration of recognized achievements, such as milestone reports issued by international institutions, publications of widely distributed textbooks, findings from certain experiments, and the availability of certain (also widely available) computational tools.

Keeping in mind the motivation for introducing the history of thermal hydraulics, i.e., to clarify the role of the discipline in NPP design and NRS technologies and to streamline possible future developments, an attempt will be made to consider (1) to (4) together.

The history shall start with the E. Fermi pile in 1942 (Dec. 2nd), as given in Fig. 2.2—several tons of graphite were assembled in the pile having a cross section $>20\ m^2$ to ensure suitable thermal capacity for dissipating the thermal power possibly produced by the chain fission reaction.

Afterward, the design, construction, and operation of energy systems in the range from a few kW to thousands of MW were possible with a parallel and consistent development of the thermal hydraulics discipline.

2.2.3.1 Regulatory history in the United States

A snapshot of the history of the Regulatory Authority in the United States, USNRC, appears worthwhile at this point (e.g., Bajorek, 2015).

In 1946, the Atomic Energy Commission (AEC) was established to foster and control the peacetime development of atomic science. The congressional act that created the AEC transferred from military to civilian control of many nuclear laboratories that were instrumental in the initial understanding of nuclear science and safety. This also

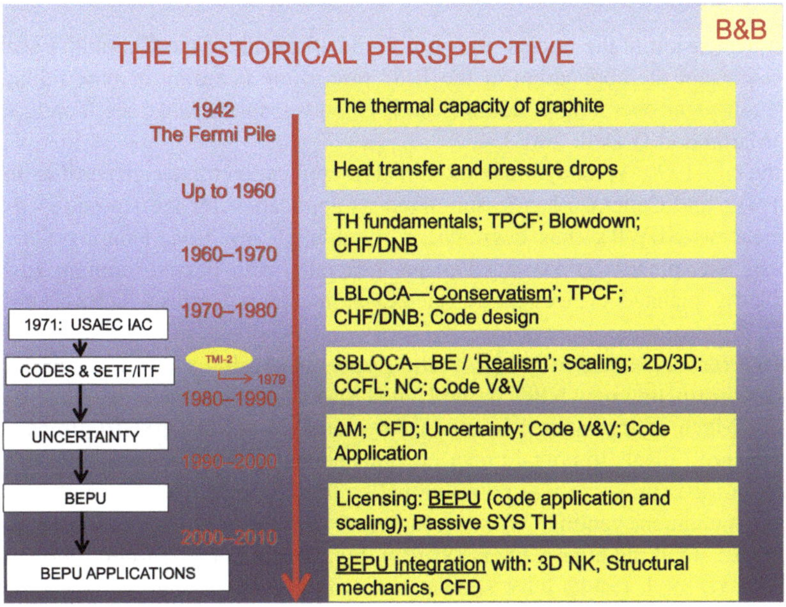

Figure 2.2 Simplified sketch of the history of nuclear thermal hydraulics.

puts the AEC in the dual position of promoting the nuclear power industry and ensuring the safety of the public.

In 1974, the Energy Research and Development Administration (which later originated the present DOE) and the Nuclear Regulatory Commission, USNRC, replaced the AEC. At this moment, the independent role (e.g., independent from industry and promoters of nuclear technology) of USNRC became clear.

Following a long-lasting process (see also in the succeeding text) 10 CFR 50.46 and related Appendix K were formally published in Jan. 1974. Because technology was limited and there were few experimental studies available the rule and Appendix K restrictions were conservative. From the regulator's perspective, the state-of-the-art, namely concerning NTH, was poor, and restrictions associated with conservatism on the application of analysis methods reflected that perception.

The demonstration of acceptable conservatism suddenly became an issue (Lewis et al., 1975; Bajorek, 2015):

> *... If the degree of conservatism of the ECCS codes is to be established (assuming they are indeed conservative) then codes which predict the LOCA phenomena realistically must also be available for comparison.*

The recommendations of the APS study group helped establish advanced code development programs and the experimental programs provided data for model development and assessment. USNRC initiated a broad series of research including large-scale system tests and support for the development and validation of codes.

All of this enabled a rule change in 1988, which allowed "best estimate" loss-of-coolant accident (LOCA) calculations as an alternative to the models stipulated by Appendix K.

The depression in the nuclear industry in the aftermath of the Chernobyl accident did not prevent the application of the 1988 rule to the licensing of core reload and changes in plant operating conditions such as power uprate (e.g., see Young et al., 1998; Martin and O'Dell, 2005).

In the mid-1990s Westinghouse submitted its application for design certification for the AP600, and General Electric submitted its application for the simplified boiling water reactor (SBWR), thus beginning a roughly 15-year period during which an emphasis was placed on advanced plants generally with passive cooling systems. The USNRC planned and executed related experimental programs which also covered the AP1000 design.

The current era started in 2011 with the events at Fukushima, as interest has diminished in constructing new NPPs in the United States and worldwide, with notable exceptions. Much of the interest in thermal hydraulics from the USNRC transitioned from advanced plants to containment phenomena and severe accident analysis. This included a focus on spent fuel pool thermal hydraulics under LOCA conditions and filtered containment venting. Work on in-vessel thermal hydraulics has been limited primarily to small modular reactors (SMRs). Furthermore, high burnup fuel received attention—research results have shown that hydrogen uptake into the cladding can affect integrity at high burnup (Billone et al., 2008) and it may be necessary to revise the existing LOCA ECCS rule. The proposed new limits would restrict the maximum equivalent cladding reacted to lower than current values for high burnup fuel (USNRC, 2014). In the area of modeling nuclear fuel, greater consideration should be given to oxide and crud.

2.2.3.2 Before 1960

The fission reaction offered (it still offers) innumerable possibilities to exploit nuclear power—a wide variety of reactor designs have been conceived, with breeding and several fuel cycles made and still possible. Water became the thermal power vector in commercial reactors and nuclear system complexity became unavoidable—the high pressure associated with suitable thermal efficiency opened the way to the possibility of LOCA. The large-break LOCA (LBLOCA) became of concern.

Accident initiators and subsequent scenarios in NPP were identified and characterized to demonstrate the safety of NPP since the 1950s, i.e., when computers and computational tools did not exist. Experiments, pioneering (thermal hydraulics) models, and engineering evaluations were the basis of the reactor safety analyses at the time.

NTH and reactor physics (or neutron kinetics) as well as nuclear fuel materials were subjects of integrated studies.

2.2.3.3 During 1960–70

Systematic thermal hydraulic studies and experiments were conducted in the 1960s, notably concerning individual "physical" phenomena like two-phase critical flow (TPCF), critical heat flux, depressurization, and blowdown. Thermal hydraulics became a "self-standing" discipline. The works by Zuber and Findlay (1965), Forslund and Rohsenow (1968), and Wallis (1969) give an idea of the interests at the time.

Several small-scale fundamental programs were launched and completed. New findings from that research were considered in reactor safety and licensing documents. In the meantime, already in 1968, the seed of what would have become the reference computational tool in NRS adopted by several generations of researchers and analysts, i.e., the RELAP code, became available (Moore and Rettig, 1968).

2.2.3.4 During 1970–80

Massive use of computers for NRS started in the 1970s. The accident analysis could also benefit from primitive (SYS TH) numerical codes and measurements taken in integral-system experiments. The nuclear regulatory point-of-view was established with the publication of the "Interim Acceptance Criteria for ECCS" (1971) (USAEC, 1971); this triggered a wide variety of research aiming to evaluate safety margins and focusing on the estimation of the maximum temperature on the surface of fuel rods following LBLOCA. The key role of LOFT experiments (see various citations of LOFT in the list of references) in clarifying Large Break LOCA phenomenology shall be noted. Appendix K to paragraph 10 CFR 50.46 of the Code of Federal Regulation followed in 1974. The issue of the "Interim Acceptance Criteria for ECCS" shall be taken as the starting date for "modern" SYS TH—updated competences in thermal hydraulics are necessary to comply with those criteria.

The technological community and the industry reacted to the request by regulators—comprehensive research projects were started in experimental and code development areas. Large experimental facilities were designed and operated, and SYS TH codes were made available for transient analyses of accidents relevant to NRS. Large-scale experimental ITF implied international cooperation projects, and "relevant" measured data were gathered to understand transient SYS TH performance.

TPCF (see, e.g., D'Auria and Vigni, 1980) and CHF/departure from nucleate boiling) (see e.g., Groeneveld, 1973) can be identified as the key thermal-hydraulic phenomena of interest during the decade associated with the LBLOCA scenario.

"Conservatism" (as already mentioned) is the keyword that characterizes the application of Appendix K (to 10 CFR 50.46) in licensing analyses. During this decade the WASH-1400 or the "Rasmussen Report" was issued, becoming the basis for the application of PSA in NRS; significant results from the execution of probabilistic analyses were produced (USNRC, 1975).

At the end of the decade, in 1979, the Three Mile Unit 2 accident happened (e.g., Henry, 2011). In the area of SYS TH, this shifted the attention from LBLOCA to small-break LOCA (SBLOCA) phenomena.

2.2.3.5 During 1980–90

Within the framework of SYS TH code use, verification and validation (V&V) was soon recognized (e.g., D'Auria and Galassi, 1998) as a mandatory process to be completed before the application of those computational tools for safety and licensing. In this context, the bases were set for addressing the scaling issue (e.g., D'Auria and Galassi, 2010) (see also Section 2.3 later). The reference SYS TH codes are APROS**, ATHLET*, CATHARE*, KORSAR*, MARS**, RELAP*, SPACE**, TRAC*, TRACE** (where * precursor code; ** lately developed code).

International activities were conducted at CSNI (Committee on the Safety of Nuclear Installations of OECD/NEA, Organization for Economic Cooperation and Development/Nuclear Energy Agency) proposing viable ways for V&V (OECD/NEA/CSNI, 1987, 1993, 1996). The importance of user effect upon the predictions was recognized (Aksan et al., 1993; Ashley et al., 1998; D'Auria, 1998) as well as the roles of the input deck (or nodalization) development and the related qualification (see, e.g., Bonuccelli et al., 1993).

The contribution to the understanding of NRS important phenomena from ITF experimental programs, conducted or initiated during this decade, shall be realized. Key acronyms for ITFs or large-scale SETFs within boiling water reactor (BWR), pressurized water reactor (PWR), and CANDU technologies are (related research programs may have developed during different decades than the current one in cases identified by an asterisk; however, for the sake of synthesis, all major research programs in SYS TH are listed hereafter in alphabetic order) APEX*, ATLAS*, BETHSY, CCTF, FIST, FIX-II, GIRAFFE, HDR, ISB, LOBI, LOFT, LSTF, MARVIKEN, MIST, PACTEL, PANDA, PIPER-ONE, PKL, PMK, PSB*, PUMA, RD-14M, ROSA, SCTF,

SEMISCALE, SPES, UM, THTF, UPTF. In this framework, the 2D/3D international cooperation program (Damerell and Simons, 1992) (involving UPTF, SCTF, and CCTF) provided key information to address the scaling issue, i.e., connecting the measured data with expected NPP conditions, from the experimental viewpoint. Enormous benefits were derived in the area of demonstrating SYS TH code capabilities.

Countercurrent flow limitation (CCFL; e.g., Glaeser, 1989) and natural circulation (NC; e.g., D'Auria and Vigni, 1985; D'Auria et al., 1991) can be identified as the key thermal-hydraulic phenomena of interest during the decade associated with the SBLOCA event, together with reflood. The State of the Art Report on thermal hydraulics of ECCS (OECD/NEA/CSNI, 1989) shall be mentioned in this context.

The words SYS TH were proposed (later on, those words had a spread use) inside the scientific community on the occasion of a specialist conference held in Aix-en-Provence (OECD/NEA/CSNI, 1992).

The need for uncertainty methods (UMs) suitable for predicting unavoidable errors to be added to the results of calculations performed by SYS TH codes became clear at the beginning of the 1990s (or even at the end of the 1980s). Working approaches were proposed; notably, the pioneering effort by USNRC shall be mentioned (USNRC, 1989a) which brought to the formulation of code scaling and applicability and uncertainty (CSAU). The phenomena identification and ranking table (PIRT) process was proposed.

In the middle of the decade, in 1986, the Chernobyl Unit 4 accident happened (e.g., Khan, 1990). In the area of SYS TH, this moved increased attention toward passive Systems; the processes for the design of AP-600 and SBWR had a strong impulse.

2.2.3.6 During 1990–2000

Addressing the uncertainty in SYS TH as a follow-up of V&V was at the center of attention in the period. Following and considering the CSAU, the Wilks formulation and the uncertainty methodology based on accuracy extrapolation (UMAE) were proposed (Hofer, 1990; D'Auria et al., 1995). The Uncertainty Method Study (UMS) project was launched by the CSNI in 1993 and completed in 1998 (OECD/NEA/CSNI, 1998). The fundamental features of UMs were described in detail, and a suitable demonstration was achieved concerning their robustness and qualification level.

The USNRC issued Regulatory Guide (RG) 1.157 (USNRC, 1989b); the application of SYS TH codes was envisaged despite recommending the use of selected conservative models. Those models are concerned with phenomenological areas where the knowledge was not considered satisfactory. Requirements in RG 1.157 did allow a few attempts of practical applications. However, Appendix K to 10 CFR 50.46 continued to be used during the decade for licensing purposes. The acronym BEPU started to circulate.

A breakthrough type of workshop for planning the future of SYS TH was held in OECD/NEA/CSNI and USNRC (1997) under the combined effort by OECD/NEA and USNRC. The development of a new SYS TH code was launched (current name is TRACE), following the identification of inadequacies of existing codes at the time. Notably, the keywords "Internal Assessment of Uncertainty" were proposed during the workshop. At the end of the decade, the Code with the Capability of Internal Assessment of Uncertainty (CIAU) method was developed (D'Auria and Giannotti, 2000) and ready for practical applications. CIAU used UMAE (mentioned before) as the "engine" and the qualification tool for the process of code application.

The development of CFD technology, mostly connected with single-phase flows, had a strong impulse possible with the increase in computational power (Yadigaroglu et al., 2003).

SYS TH coupling with 3D neutron physics was possible by the availability of more powerful computers and numerical techniques (OECD/NEA, 2004).

2.2.3.7 During 2000–10

Applications of BEPU approaches in licensing processes, implying the exploitation of the capabilities of SYS TH codes and UM, definitely started in the 2000s. The following key events, not an exhaustive list, not in the order of importance, not in the order of time, see also Petruzzi et al. (2005), give an idea of the technology development in the area.

(a) The KWU (NPP designer later on incorporated by Siemens, and lately became AREVA) on behalf of the ETN (Brazilian utility owner for the nuclear plant) proposed a BEPU methodology to analyze the LBLOCA for the licensing of Angra-2 NPP in Brazil (KWU-Siemens, 1997). The submission was analyzed by the regulatory authority of the Country, which also requested the application of a different UM by an assessor independent from AREVA.

(b) USNRC issued the RG 1.203, USNRC (2005), which provided clarification of the regulatory expectation for transient and accident analysis including the application of BEPU approaches.

(c) CSNI launched and completed the 6-year project BEMUSE. The aim was to demonstrate the maturity of UM and approaches with the main concern being LBLOCA applications. The objective was achieved. However, differences among the results by participants (mainly in predicting reflood time) caused the need for a careful interpretation of related findings. The difficulty in harmonizing, from the side of applicants of UM, the choice of input uncertainty parameters, and the related ranges of variations was an outcome of the project.

(d) Three important BEPU-concerned documents were issued by the IAEA, two Safety Report Series, SRS 23 and SRS 52 (IAEA, 2002, 2008a), and one Specific Safety Guide, SSG-2 (IAEA, 2009c). SRS 52 deals with the description of workable uncertainty approaches and methods. The SSG-2, dealing with DSA in general, proposes the BEPU approach in licensing as consistent with the technological state of the art in the area of accident analysis.

(e) BE conferences, BE-2000 and BE-2004, ANS (2000), and ANS (2004), respectively, were held under the auspices of the ANS. This series of conferences was continued by V&V Workshops in the United States held in Idaho Falls (Idaho), Myrtle Beach (North Carolina), and Las Vegas (Nevada), in 2008, 2010, and, up to now, till 2016, respectively, with the cooperation of the nuclear division of the ASME.

(f) The BEAU (application of the best estimate analysis and uncertainty) method was proposed by Canadian experts (Abdul-Razzak, 2009).

(g) A variety of BEPU (it shall be clear that the BEPU acronym is not always adopted) applications all over the world during the concerned decade, mostly within the license renewal framework, are summarized in the paper by Glaeser (2008).

The first decade of the current millennium is characterized by the application of NRS technology of expertize in thermal hydraulics—the BEPU approach is the key part of this connection.

2.2.3.8 During 2010–20

The 2010s decade started with the submission of Chapter 15 of the FSAR of the Atucha-II NPP to the regulatory authority in Argentina, by the NA_SA utility. The entire Chapter 15 of the FSAR is based on the application of BEPU. The approach itself was endorsed by the regulatory body, UNIPI-GRNSPG (2008); D'Auria et al. (2012a,b). Two complementary books, D'Auria et al. (2021a); Riznic (2021), have been issued that describe the details of the effort: keywords for the books are the motivation for BEPU, reconstruction of reactor design technology (e.g., by BEPU), code-coupling, and results from BEPU application.

At the international level within CSNI, projects dealing with inverse uncertainty evaluation (IUQ), have been completed, i.e., PREMIUM, Skorek et al. (2019), and SAPIUM, Baccou et al. (2019). Difficulties of the IUQ idea or deriving ranges of variation for (code) input uncertainty parameters (IUPs) are confirmed. Those difficulties derive from the arbitrariness in associating individual IUP and experiments, although sophisticated analytical techniques exist, which are applicable once the association is established.

Within the IAEA, the DSA guidance SSG-2 has been updated, IAEA (2019). Notably, the terms DBA and beyond DBA (BDBA) were changed to design basis conditions and design extension conditions (DECs). Furthermore, two categories for DEC, i.e., DEC-A and DEC-B were introduced; the distinction deals with the level of nuclear reactor core degradation and affects the strategies for minimizing radiation releases.

USNRC issued draft documents dealing with damaging mechanisms for fuel rods, USNRC (2018a,b); these triggered investigations aiming to demonstrate compliance between LBLOCA scenarios and new possible licensing requirements, e.g., D'Auria et al. (2019a); D'Auria (2021).

In the area of thermal hydraulics, international FONESYS and SILENCE networks were launched and operated for nearly the entire decade, e.g., Ahn et al. (2017) and Aksan et al. (2018), in the area of SYS TH code development and experimentation, respectively.

An interpretation of trends and perspectives for development in nuclear thermal hydraulic research established for the first edition of the book can be found in Sections 2.2.4 and 2.3. Chapter 24, second edition of the book, deals with a view for future research resulting from a specific CSNI activity completed in 2021.

2.2.3.9 Historical list of topics

A list of topics can be identified from each nuclear thermal hydraulic conference held during the last 4 or 5 decades. A recent summary provided by Yadigaroglu and Lakehal (2015) and related to the NURETH series of conferences brought to the following list of topics of interest applicable to years 2013 and 2015 [similar, noncounterfeiting lists can be found in more recent papers, e.g., Cheung (2021), and Riznic (2021).

- Multifield two-phase flow modeling (several sessions)
- Two-phase flow and heat transfer fundamentals (several sessions)
- Boiling and condensation fundamentals (several sessions)
- CHF and post-CHF heat transfer, flooding, and CCFL
- Advances in enhancement, understanding, and prediction of CHF and quenching (several sessions)
- Investigation of reflood phenomena (now in partially blocked core with fuel relocation)
- Operation and safety of existing reactors (several sessions)
- Safety systems and related phenomena
- NPP transient and accident analysis (several sessions)
- Plant system code development and validation (several sessions)
- Realistic BWR LOCA evaluation: methodology development and application
- Containment analysis (with V&V)
- Natural convection and mixing phenomena, modeling, and experiments
- Critical heat flux in a fuel bundle: modeling, prediction, and experimental measurements (several sessions)
- Core thermal hydraulics and subchannel analysis, fluid dynamics, and heat transfer
- Interfacial area (IA) transport
- Severe accident phenomenology (several sessions)
- Modeling and experiments of severe accidents (several sessions)
- Debris bed cooling

One may note the presence of "historical" words like CHF, CCFL, and reflood, mixed with established concepts like V&V, modeling and experiments, and containment analysis, and "newcomer" topics like IA transport and debris bed cooling. Furthermore, topics like severe accident phenomenology, modeling, and experiments of severe accidents and debris bed cooling although part of the conference in NTH are not within the domain of interest for the present book, as already discussed in Chapter 1 of the book.

2.2.3.10 Summary history of nuclear thermal hydraulics

A summary history of NTH is depicted in Fig. 2.3 or can be based on the keywords in Table 2.1.

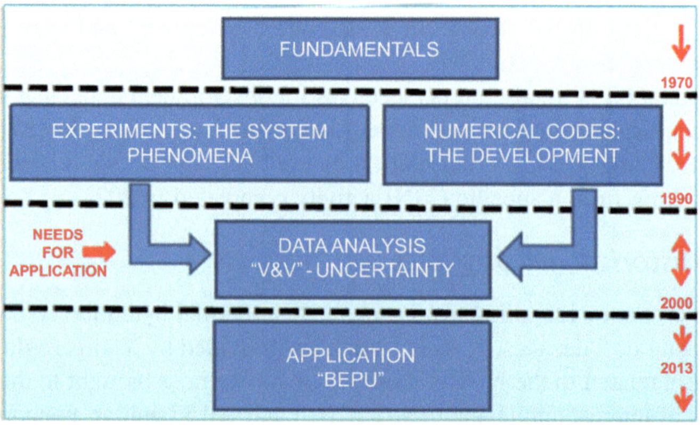

Figure 2.3 Key elements in the history of nuclear thermal hydraulics.

Table 2.1 Keywords for a spot historical outline of system thermal hydraulics.

Actors and stakeholders[a] → authors of textbooks, USNRC, international and national institutions, industry, NURETH (conferences), journals		
Period or event	**Keywords/phenomena**	**Key document/activity**
Fermi fission reaction (1942)	Thermal capacity (of graphite)	
Up to 1960	Heat transfer and pressure drops	E.g., Dittus-Boelter equation for HTC, multiplier approach for TPPD
1960–70	TH fundamentals, TPCF, blowdown, CHF/DNB	E.g.: Moody and H–F models for TPCF, LUT for CHF
1970–80	LBLOCA—conservatism, TPCF, CHF/DNB, code design	USNRC IAC for ECCS, Appendix K to 10 CFR 50.46
1980–90	SBLOCA—BE, V&V and scaling, 2D/3D, CCFL, NC, code validation	CSAU, USNRC compendium, CSNI SOAR on TECC, CCVM-ITF
1990–2000	AM, CFD, UM, code validation, and application	CCVM-SETF, UMS[b], USNRC RG 1.157, UMAE, GRS-method
2000–10	Licensing: BEPU (code application) and scaling, passive system thermal hydraulics	USNRC RG 1.203, IAEA SRS 23 and 52, IAEA SSG-2, BEMUSE[b], NURESIM[b], NURISP[b], CASL[b],
2015		NURESAFE[b], CASL, PREMIUM[b]
After 2015	Consolidation in the above areas[c]	

[a]See Section 2.1 in this chapter.
[b]Acronym for an International Project.
[c]See Section 2.3 in this chapter.

One may see from Fig. 2.3 how fundamentals evolved into system experiments and numerical code development. This gave origin to the data analysis and V&V procedures and activities. The maturity of the NTH discipline is testified by the applications to NPP and NRS technologies; in the case of NRS, BEPU has a key role (as mentioned in Section 2.2.4 and with additional detail in Chapter 14 of the book).

Keywords for the TH spot historical outline are shown in Table 2.1. In the first row, the key "actors" and the "stakeholders" for the NTH discipline are listed as derived by items (1)–(7) in Section 2.1 in this chapter. The cross-correlation between phenomena and keywords representative of topics that form the NTH discipline (second column of the table) and selected reference documents or activities (third column of the table) provides a sample vision for the development of the discipline. Namely, starting from the second row of the table a few examples of cross-correlations are provided.

(a) Convection heat transfer and friction pressure drop phenomena are addressed by numerical equations like the Dittus-Boelter one and by approaches like the two-phase multiplier—the second concerned historical period.
(b) The evaluation of TPCF and blowdown phenomena is possible by models like the one proposed by Moody—the third concerned historical period.
(c) The LBLOCA analysis (key topic in NTH) shall be performed in accordance with the Appendix K rule associated with the Interim Acceptance Criteria—the fourth concerned historical period.
(d) Validation and scaling activities shall be performed based on the knowledge synthesized in the USNRC Compendium—the fifth concerned historical period.
(e) Accident management and UMs are addressed in documents like those describing CSNI Code Validation Matrices (CCVM and UMS)—the sixth concerned historical period.
(f) Reference documents for applying BEPU consist of the USNRC RG 1.203 and IAEA SRS 52—the seventh concerned historical period.

Table 2.1 provides a synthesis vision of NTH connecting with historical periods.

2.2.4 An interpretation of current trends

There are important research projects and initiatives in progress (reference year is 2015, see Chapter 24 for updated and additional information) in the area of SYS TH that also imply future plans and perspectives. A nonsystematic and (maybe) incomplete list of relevant programs, projects, or activity frameworks includes (each of the given acronyms corresponds to a website where the reader can find detailed information; for the few notable exceptions, references are provided).

- Recent or ongoing projects on the side of EC: EUROFASTNET, NURESIM, NURISP, and NURESAFE 1
- Ongoing project on the side of US DOE: CASL 1
- OECD/NEA (CSNI & NSC) activities or projects: UAM, PREMIUM, W-GAMA initiatives including ISP, "VVER-1000 Kalinin addressing NPP 3D measurements," and "Oskarshamn-II addressing the BWR stability issue"
- IAEA planning and carrying out of CRP and TM, namely, in the areas of BEPU, connection DSA-PSA, CFD for NRS, and thermal hydraulics of innovative systems including SMRs
- USNRC supporting the development of SYS TH codes (e.g., TRACE) and running the related CAMP program, other than continuously evaluating new reactors and updating the licensing process (in this last case with several connections to TH)
- ATLAS, MARS, and SPACE projects in Korea, altogether constituting an integrated approach to SYS TH that involves experiments, modeling, and numerical development—e.g., see the recent comprehensive paper by Song (2015)
- China and India designing, building, or even operating large-scale thermal-hydraulic facilities, and China primarily developing new system codes (e.g., AERB, 2010; Chang et al., 2011)
- FONESYS and SILENCE networks, established and recently launched, in the areas of code development and experimental programs, respectively.

Typical subjects for activities in progress or envisaged include (the list that follows shall be integrated with subjects mentioned in Section 2.2.3.9).

- Improvement in the CHF database and understanding
- Improvement in reflood understanding
- Improving in understanding the role of spacer grids
- Finalizing, as far as possible, the development of IA transport models
- Coupling of SYS TH with other branches of nuclear technology: 3D NK, structural mechanics, nuclear fuel, etc.
- Improvement in CFD modeling including two-phase
- Addressing the V&V of CFD
- Modeling of CRUD (Chalk River unidentified deposit) deposition
- Connection DSA-PSA
- Use of BE techniques in PSA
- Operation of experimental ITF or large-scale facilities, e.g., in Finland (PACTEL), Germany (PKL-III), Japan (LSTF), Korea (ATLAS), Switzerland (PANDA), Russia (PSB)

BEPU is an additional element for interpreting current trends, as mentioned in Section 2.3.16 and discussed in more detail in Chapter 14.

2.3 Perspectives for system thermal hydraulics

This chapter aims to provide an outline of the current perspectives (time of the first edition of the book; see Chapter 24 for an updated discussion) for the development of NTH, i.e., the second objective for this chapter.

The historical outline in the previous section and the expected incomplete list of key actors and stakeholders provide another picture of the "thermal hydraulics universe" and its complexity (Section 2.2.2 and Table 2.1). A broad set of activities is already ongoing (Section 2.2.3) in the same framework. Thus, the destiny of any individual or personal position for development may be impractical and unusable, as also discussed in the "Chapter foreword" of this chapter.

However, a list of envisaged ideas for the future is prepared looking from an optimistic viewpoint and in an attempt to invite the reader to a (further) reflection. The concerned domains for investigation include modeling features of computational tools, validation and application of those tools, and planning of experiments. The list unavoidably reflects personal experience and shall not be seen as in competition with any ongoing international projects or with any similar list supported by proper contexts.

The envisaged needs in SYS TH, or better, elements for reflection may need elaborations and details beyond what is possible in one chapter of an edited book. Furthermore, what are called here envisaged needs or, better, elements for reflection shall be seen as complementary (or in addition) to the needs discussed in Chapter 4 of the book.

No attempt has been made to provide an order of importance for the issues that follow and priority classification is left to the reader if deemed necessary[1].

Sixteen elements for reflection (or categories/targets for improvement) are considered. Interactions in virtual activities related to the elements are unavoidable although an attempt is made to select independent research topics. A description of each "element" is provided and rough proposals are formulated dealing with ideas to address the concerned open issues.

2.3.1 "Local form loss" coefficients (also reported as "K-factors")

The K-factor is the multiplicative term to the product ρw^2 (where "ρ" is the "reference" density and "w" is the "reference" velocity of the concerned fluid) to calculate pressure drops at geometric discontinuities.

The origin of the issue lies in deriving the balance equations at the heart of SYS TH codes—balance equations are derived by referring to a geometric space without discontinuities (some advancement is proposed for the integration domain by TRACE code, but this is insufficient for practical applications). Therefore, the missing information about form loss coefficients at geometric discontinuities, either in the forward fluid motion direction, typically nominal conditions for the steady-state operation of NPP and ITF, or in the reverse direction (expected during a transient), must be fixed by the analyst or code-user. In a few situations tabulated (precalculated) or experimental data exist. Under most conditions, the user must consider analogies with those data or "invent" a value based on his expertise.

The user has almost no possibility to estimate the effect of the not-fully developed flow upon the value of the "K-factor." Additional origins of misunderstandings and consequently of errors in the values of "K-factors" derive from inherent difficulties in gathering qualified experimental data: (1) irreversible pressure drops shall be distinguished from reversible pressure drops; (2) practical difficulties arise when measuring the pressure in narrow spaces and complex geometries; (3) the influence of local velocities and void fraction, other than the non−fully developed flow condition and the transient situation upon the adopted values of the coefficients, can be large and may require a broad range of experimental conditions. All of this is confirmed in a recent activity by Benhamadouce (2015), where referring to spacer grids and high Re numbers, errors in a comparison of experimental and CFD-calculated data were in the range of 15%−30%.

The impact of the issue in NRS can be derived from the following: (1) many discrepancies among results of the application of SYS TH codes to target cases (either

[1] Commonly, this has been done in the past by adopting PIRT or related techniques. Then, the occasion is taken to state a viewpoint concerning PIRT. The present author expressed his doubts concerning the use of the "ranking" process in SYS TH, since its presentation in 1989. Recently, the sentences by the key proponent of the PIRT process, "... the PIRT process as presently implemented produces absolutely useless documents ... ranking is based purely on subjective opinions ..." (Zuber, 2010), put those doubts in the right perspective.

of the same code or different codes) originate from the process of "fixing" the "K-factors"; (2) in some cases, based on the comparison between experimental data and calculated results involving unacceptable discrepancies, code models other than the formulas for pressure drops are modified to minimize those discrepancies without accounting for the effect of the arbitrarily fixed "K-factors": this brings to never-ending efforts for SYS TH code improvement.

The proposals to address the issue are:

- an effort shall be made to develop traceable procedures for calculating local form loss coefficients in cases of interest to the nuclear technology and/or supporting those procedures with a comprehensive data bank. Experiments and experimental conditions all over the world should be collected and archived. Examples of this could be the libraries for lookup tables for CHF and the microscopic XS (cross section) for neutrons.
- CFD computational tools, primarily in single-phase conditions, should be systematically used to generate proper values of K-factors in the situations of interest for NPP. Related results represent traceable support for applying SYS TH codes. Specific attention shall be devoted to rounded or sharp edges.

2.3.2 Multidimensional heat transfer coefficient surface

The convection heat transfer, i.e., one mechanism for the transfer of thermal power between fluid and wall, is modeled by the well-established empirical relationships between the heat transfer coefficient (HTC) and the heat flux. This gives rise to the current need to use around two dozen correlations to cover the range of interest in NPP design and NRS, notably including the consideration of phenomena like CHF, reflood (also characterized by the acronym RNB and the MFB temperature), and condensation (also in the presence of noncondensable gases).

At least three fundamental issues of concern shall be identified relating to the current status for predicting HTC.

(a) The continuity among HTC values calculated from the various correlations, namely at the boundaries of the respective validity ranges.
(b) The repartition of heat flux from the wall to the fluid, distinguishing contributions to the gas and to the liquid phases and even directly (from the wall) to the interface liquid-steam.
(c) Heat flux by radiation heat transfer adds to total heat flux and in some experimental situations (notably, film boiling in the presence of fluid droplets) separation between convection and radiation becomes difficult to be quantified.

The proposal here, related to issue (a), is to construct, as much as possible, a continuous multidimensional (multi-D) surface covering all the heat transfer regimes (examples of HT (heat transfer) regimes are nucleate boiling, saturated boiling, film boiling) and all the flow regimes (example of flow regimes are bubbly, slug, and annular flow). Reflood and condensation should be included in this picture. No proposal is formulated related to issues (b) and (c), which also deserve proper consideration.

It should also be noted that the problem (i.e., of constructing a multi-D HTC surface) is quite well known in NTH and applicable to other (nonnuclear) technologies (e.g., Divo et al., 2000; see also Tanaka and Dulikravich, 2000).

2.3.3 Energy and entropy balance following reactor coolant system blowdown and containment pressurization

The system of interest is a pressure vessel surrounded by full-pressure containment. The vessel is filled with high pressure and high temperature (two-phase fluid), and the containment is ideally isolated from the environment and filled by air at atmospheric conditions. The opening of a break in the pressure vessel triggers the "irreversible phenomenon" called blowdown or pressurized two-phase mixture discharge into the containment. As containment pressure increases, energy exchanges with the environment are (ideally) prevented. In this situation, the isolated system consisting of the vessel and containment is expected to experience increased entropy following the start of blowdown, i.e., an irreversible thermodynamic process for an isolated system. The results of code calculations are not consistent with expectations—this is the issue.

In the case of SYS TH codes, difficulty may arise in estimating the energy flowing from the break (TPCF model results can be qualified against experimental data and are more complex relative to energy flow from the break).

Additional issues include:

(a) Supersonic conditions. Supersonic conditions may occur in discharge pipes (e.g., the line connecting the PORV with the sump or the ADS 1−3 to the IRWST in the case of AP1000 design). Pressure drops in the discharge lines and mechanical loads upon constraints are affected by the occurrence of supersonic conditions.
(b) TPCF on-off. This is the situation that occurs when the pressure difference between upstream (reactor pressure vessel side) and downstream (containment side) is not enough to provoke critical flow at the assigned throat and cycles of TPCF and "Bernoulli-flow" occur.
(c) TPCF location jumps. It has been known for decades that TPCF conditions in a nozzle may migrate (or jump) from one section to another. The same situation can be more complex inside a valve. TPCF location jumps may originate by sharp edge cavitation—steam generated at any geometric discontinuity upstream of an assigned critical section may cause the shift of the critical section at the upstream geometric discontinuity.

The capability to predict the discussed phenomena is not demonstrated. The recommendation here is to perform specific experiments and eventually modify SYS TH codes. Again, as in the previous case, the problem is well known in NTH (e.g., D'Auria et al., 2003; IAEA, 2014; Qiu et al., 2015)—this is an issue given the importance ascribed to the problem.

2.3.4 Precision targets

The word precision has not commonly been used within SYS TH. Common words, other than "error," are "accuracy" and "uncertainty," the former being the known error resulting from the comparison between measured and calculated data, and the latter being the unknown error in NPP predictions. The word precision encompasses both the words and the concepts of accuracy and uncertainty. No values for precision or precision targets are fixed by any institution (or by stakeholders identified in Table 2.1). So the common practice is to produce new codes or improved code versions without knowing the target for acceptability.

The recommendation is to solicit the issuance of precision targets in SYS TH which also should be moving targets associated with the advancements in technology and the understanding. In this case, the reluctance of all actors in NTH to fix the precision target is an issue. An attempt to fix precision targets for a narrow range of conditions can be found in D'Auria et al. (2004).

2.3.5 Applying computational fluid dynamics—like approaches to nuclear power plant design and nuclear reactor safety technologies

CFD was mentioned in Chapter 1 (under an eponymous heading and in Section 1.1.1) and, more importantly, is the topic of Chapter 12. CFD applications in NRS are the subject of many publications (e.g., see Bestion, 2014; Smith et al., 2008). Some notes consistent with those key documents are discussed hereafter to make this a standalone chapter of the book.

CFD technology impacted the area of NPP technology and NRS since 1990, as already mentioned. The CFD code design targeting steady-state conditions, the lack of traceability of embedded equations (due to commercial reasons), the common assumption to neglect the pressure influence upon density changes, and the difficulty of modeling an entire NPP loop in two-phase conditions may represent obstacles for the application of CFD in accident analysis or may delay its full exploitation. Otherwise, the quality assurance and the needs of the DiD (see Fig. 2.1) impose the use of the best techniques (i.e., the least restrictive assumptions) and the possibility of "independent" checks.

CFD codes for single-phase flows form a mature technology ready for applications even for NPP design and NRS situations. The issue for full application to NPP design and mainly to NRS is the qualification.

(a) Best practice guidelines are available and may need enforcement and improvement
(b) the convergence of results, when the mesh sizes are decreased, shall be demonstrated in any application
(c) the demonstration of independence of results from mesh generation might be required

A further issue is related to the coupling between CFD and SYS TH codes. The coupled computational tool may generate inconsistent results (e.g., in a closed NC loop partly modeled by SYS TH and CFD codes). CFD code results are (strongly) affected by boundary and initial condition values at the boundaries of the CFD domain of integration which are necessarily approximate (otherwise, CFD modeling of the entire loop is needed).

A recommendation is to use CFD for the evaluation of local pressure drops including derivation of direct and reverse K-factors (Section 2.3.1)—properly qualified CFD codes and approaches may be exploited to generate a databank for K-factors suitable for qualified SYS TH code applications.

Finally, CFD codes for two-phase flows are going to be developed. The full exploitation (the useful application) of the capabilities for those codes in NPP design and NRS may take decades or even never happen. Thus, two-phase CFD codes may not be strictly considered (within the present context) as part of the needs in SYS TH.

2.3.6 Thermal hydraulics of passive systems

Passive systems and related thermal hydraulics are part of the history for NPP design and NRS (e.g., IAEA, 2005; 2009b)—the configuration of the RCS for both BWR and PWR (for instance) is determined considering the possibility of cooling the core via NC, at least removing decay heat (e.g., D'Auria et al., 1991, 1997). Passive ECC systems and components like accumulators are installed in most of the NPPs. The Chernobyl event in 1986 triggered new attention toward passive systems—new (i.e., other than accumulators) passive ECCS are part of NPP design, e.g., AP1000 and passive phenomena are at the basis of the RCS operation, e.g., the economic simplified BWR (ESBWR). Passive systems are also used in the category of "small" NPP, the so-called SMR (small and medium reactors or small modular reactors).

The following is a not-comprehensive list of issues connected with the design and the operation of passive systems (again, not in order of importance):

(a) Stability of flow of two-phase mixture in a dozen meter high, large diameter riser of ESBWR.
(b) Triggering of NC flow rate, in some cases a consequence of the opening of a valve: high values of flow rate may cause too fast cooling and large dynamic loads on the structures.
(c) Stability issue: all NC systems are characterized by "small" driving forces (i.e., small related to a possible similar system equipped by a centrifugal pump). "Small" perturbations may challenge stability. The stability issue is even more complicated for the category of NC systems where boiling and condensation occur.
(d) The stability issue for "composite" systems: Either different NC systems are connected among each other by a heat exchanger (series connection, e.g., passive cooling of SG when PS is working in NC) or they are hydraulically connected (parallel connection, e.g., PRHR (pressurized residual heat removal), CMT (core makeup tank), and IRWST loops in AP1000). The issue here is that the instability of one loop may propagate to connected loops, and the stability issue must be addressed for the entire system.
(e) Local pressure drop coefficients. The issue here, see also item (1) in the present list, is the precision needed for the "K-factor" values: owing to the impact on stability (e.g., a "cliff-edge" type of thermal-hydraulic phenomenon) and the "small" driving forces available, a higher precision is needed for the "K-factor" values than needed for a system equipped with energized pumps. Then, qualification results of calculations, without having access to suitable values for "K-factors," may reveal questionable.
(f) The reliability of passive systems is a thermal-hydraulic issue that is not well established among designers of passive systems (e.g., Jafari et al., 2003; D'Auria, 2020).

2.3.7 Scaling issue and experiments

Notwithstanding the large number of papers and reports related to scaling, "scaling" is still a source of controversy in SYS TH and applications, e.g., see the paper by D'Auria and Galassi (2010). The scaling issue in licensing is considered by D'Auria and Galassi (2010), who introduced the scaling pyramid to explain the current situation (Fig. 2.4) and proposed a road map for addressing the issue (Fig. 2.5). Further discussion about scaling can be found in Chapter 21 (second edition of the book), including the consideration of the OECD/NEA/CSNI Scaling State of the Art Report, Bestion et al. (2017).

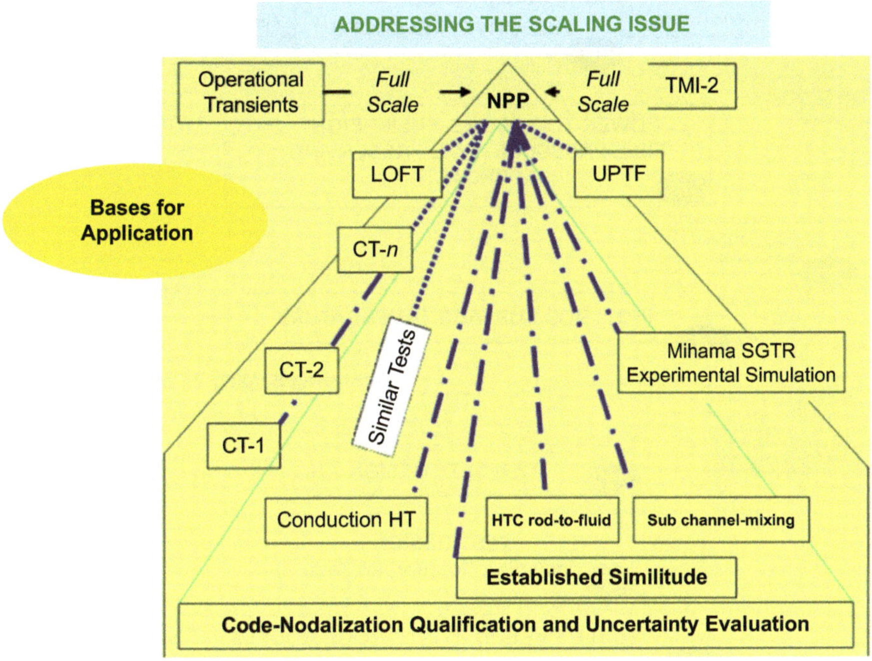

Figure 2.4 The scaling pyramid proposed by D'Auria and Galassi (2010).

Even though there is no common acceptance of the proposed road map, the reader can refer to the last-mentioned paper for details. The scaling issue for designing experiments is considered hereafter.

SETF and ITF shall be distinguished when addressing the scaling issue for experiments.

The SETF design as well as the design of SETF experiments may or may not be connected with scaling; rather, the need to characterize a TH phenomenon, coming from the process of code validation, shall be at the basis of the SETF design. So the connection between scaling and SETF may be weak and is not further discussed here.

In the case of ITF, a suitable design process shall include, namely, when the planned experiments are relevant for NRS, the consideration of principles like ALARA or a framework like DiD. ALARA imposes that the design effort shall be consistent with the safety relevance. In other words, the inadequacy of budget may not justify the construction of a "very" (too) small facility or a "short" (reduced height) facility. The framework of DiD imposes providing proper consideration to (in this case "the simulation of") parameters affecting the safety functions and the integrity of safety barriers. Then, "the simulation of" PCT shall be considered a key target when designing ITF. This requirement may become more stringent when NC phenomena (or passive systems; see earlier section) are involved in the simulation. The scaling freedom in relation to the (design of a possible reduced) height for ITF offered by the so-called Ishii-scaling may reveal inconsistency with the DiD framework and the

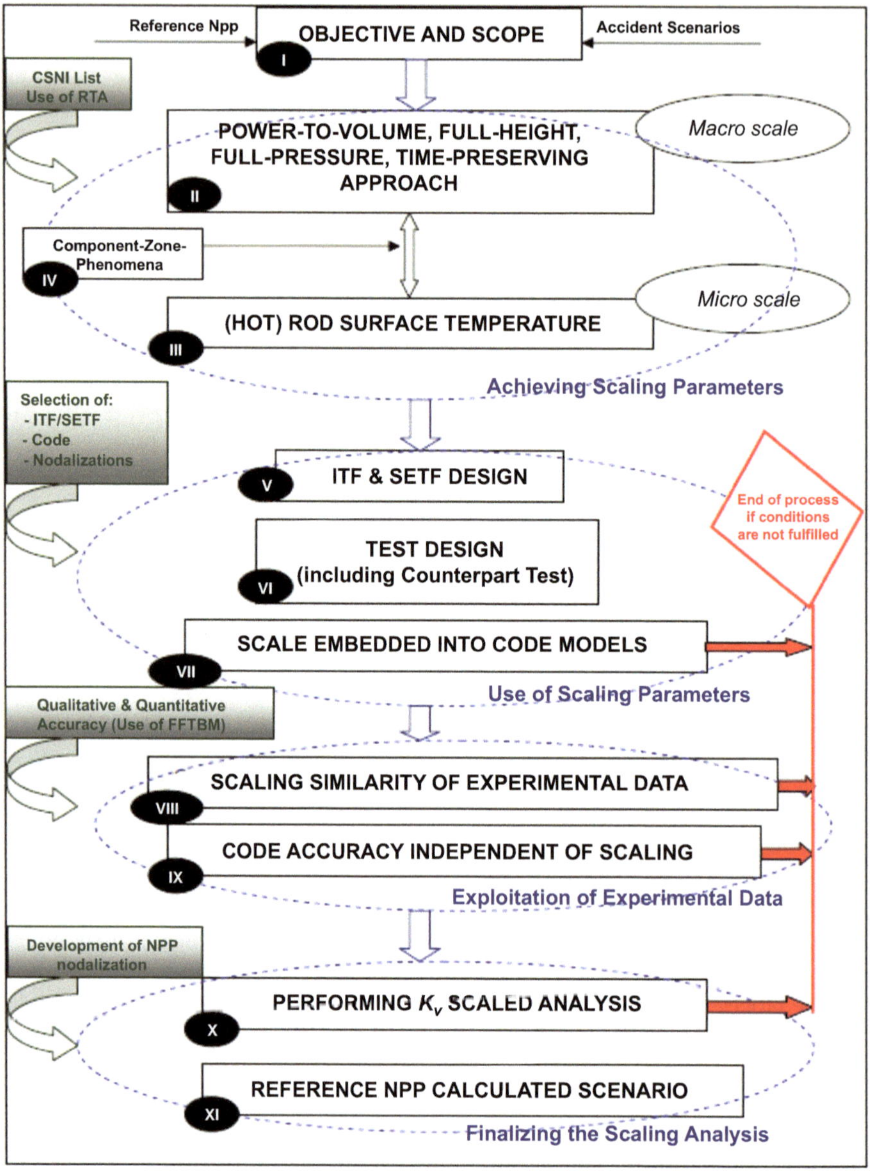

Figure 2.5 The scaling road map proposed by D'Auria and Galassi (2010).

ALARA principle. Thus, the basic limitations of the Ishii-scaling (actually reduced height is an interpretation of the Ishii papers) are:

- No criteria can determine an acceptable "shorter" length; i.e., to simulate 4 m for the active length of the reference core (prototype) height, one can choose 3.9 m or even 0.5 m in the model.

- The rod surface temperature cannot be optimally simulated when a shorter length is adopted because of the unknown relationships (this typically represents the objective of any experimental campaign) between linear power, HTC, and fluid dynamic conditions like local velocities and void fraction.

Scaling is closely connected with experiments. The issue here is the preservation of the experimental database (EDB). Preservation of the EDB implies the characterization of what must be preserved and the gathering of the data; indications for the process can be found in the paper by Petruzzi et al. (2012).

2.3.8 Verification and validation of system thermal hydraulics codes

As in the case of CFD (Section 2.3.5) the V&V subject has already been mentioned in Chapter 1 of the book (under the headings and in Section 1.1.1) and, more important, is the topic of Chapter 13 of the book. Furthermore, a recent IAEA report deals with a comprehensive vision for V&V (IAEA, 2015), as discussed in Chapter 22 (second edition of the book). A synthesis picture of the V&V for an SYS TH code derived from that report can be found in Fig. 2.6. Hereafter some notes supporting those key documents are discussed to make self-standing the present chapter.

The V&V applies to SYS TH codes and, as such, shall be considered a technological branch established since the beginning of 1980. Precision targets, covered in Section 2.3.4, should be considered a prerequisite for V&V and part of the present V&V issue. Additional aspects deal with the procedures for V&V and with the comparison between calculated and experimental data.

An attempt to synthesize current V&V understanding is too long for the present chapter (see Chapters 13 and 22 for details). Some selected elements for reflection are the following:

(a) Matrices for V&V should be available as a complement to code development (i.e., developmental assessment matrices) and should be endorsed by the scientific community based on the fulfillment of acceptability conditions (e.g., see OECD/NEA/CSNI, 1993, 1996, and the recent collection of 116 thermal-hydraulic phenomena by Aksan et al., 2018a).
(b) Independent assessment matrices (i.e., matrices at the basis of code application), mostly dealing with Validation and not with verification, should also be available based on requirements—the exploitation of those matrices should be the basis and the prerequisite for any code application to NRS problems.
(c) V&V&C, where "C" stands for consistency (and adds requirements to V&V), was proposed by D'Auria and Lanfredini (2019) and is outlined in Chapter 22.

The recommendation here is to pursue the development and the application of those matrices and introduce acceptability requirements in connection with either the coverage range of phenomena (see also Section 2.3.12 in this chapter), and the evaluation of comparison results coming from the use of the experiments part of the matrices (see also Section 2.3.4 in this chapter).

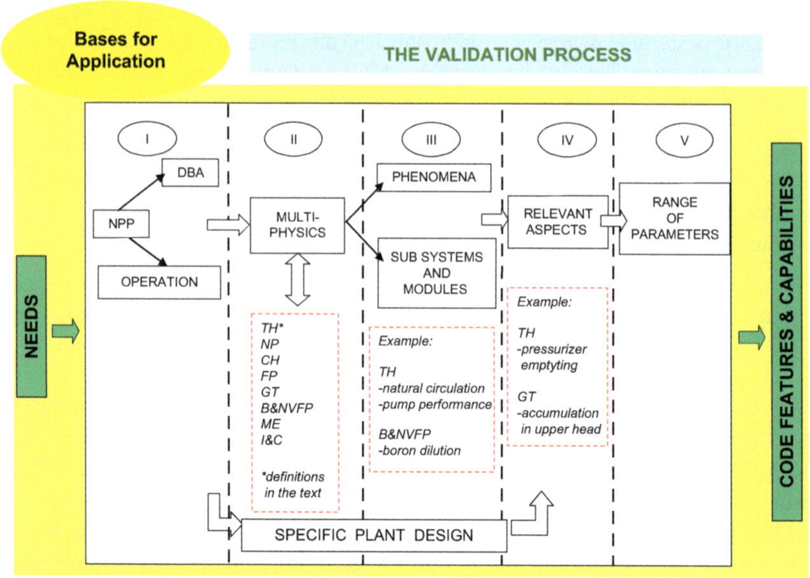

Nomenclature associated with Fig. 6

B	= born including mixing and separation
CH	= chemistry restricted to the zirconium-water reaction
FP	= fuel performance
GT	= gas transport (N_2, H_2 and fission products) including mixing and separation
I&C	= instrumentation and control including the logics
ME	= mechanics, e.g., dealing with the inertial behaviour of rotors
NP	= neutron physics
NVFP	= non volatile fission products transport including mixing and separation
TH	= thermal-hydraulics including conduction and radiation heat transfer

Figure 2.6 A snapshot of the V&V for SYS TH codes (IAEA, 2015).

2.3.9 *Uncertainty analysis*

The uncertainty needs to come from the V&V process. Evaluation of uncertainty must be connected with the application of BE SYS TH codes and must be supported by suitable UM. In other words, one can state that the application of SYS TH code to the analysis of experimental data (e.g., obtained in ITF or SETF within the framework of [code-developers independent] validation activities) shows unavoidable errors, typically referred to as accuracy; thus, errors shall be expected in the prediction of an NPP transient scenario and are referred to as the uncertainty of code calculations. A recent IAEA report deals with the current status of UM (IAEA, 2008a). Uncertainty is the topic of Chapter 13 and is considered in Chapter 14 of the book. A synthesis picture providing an idea of UM can be found in Fig. 2.7. Hereafter some notes supporting those key documents are discussed to make self-standing this chapter of the book.

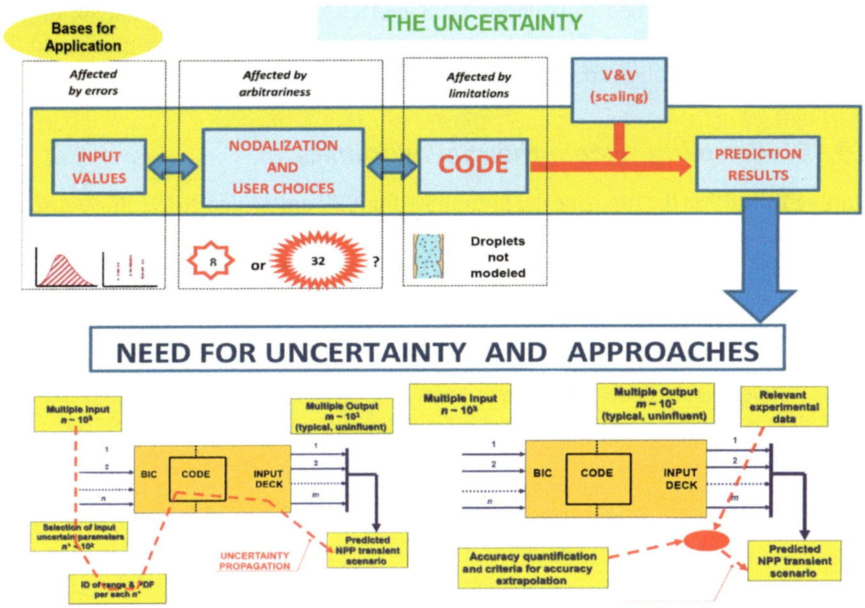

Figure 2.7 A snapshot of the motivation and the processes to estimate the uncertainty of SYS TH code predictions.

International projects like UMS, BEMUSE, and PREMIUM are worthwhile to spread the expertise and shall be pursued. However, from the outside, e.g., from the view of scientists not on the working teams, it definitely may appear that those initiatives generate more questions than answers or more doubts than solutions.

Bifurcation analysis, or identification of cliff edges, is possible with current UM (e.g., see D'Auria et al., 2000).

The elements for reflection, or the issues for possible development, are:

(a) The correct interpretation of results from international projects
(b) The setup and the acceptability procedure and criteria for V&V of UM
(c) The integration of CIAU and GRS UM is desirable

The elements (a) and (b) apply to different areas of NTH and other disciplines.

International projects like those just mentioned are important to achieve a common understanding and even to open the way for further development in the concerned areas. However, in the general case because of (the lack of) proper resources, (the presence of) proprietary information, and (the need for) consensus among several participants from different countries, written conclusions from the activities are smooth and may not reflect the current status. Therefore, item (a), expert participant evaluation may be needed to supplement the information from the final report of the activities.

Any method or set of equations in NTH should be supported by a validation framework, item (b). In other words, because of a lack of knowledge, e.g., associated with turbulence modeling or the evaluation of convection heat transfer, suitable proofs of validation and related documentation are strictly necessary for any method including UM.

An idea was launched to combine GRS and CIAU methods (Kovtonyuk et al., 2008) item (c). The completion of the activity, which never started, may provide additional proof of validation for both methods.

2.3.10 Coupling system thermal hydraulics

Starting from 1960 the fundamental disciplines for NPP design and NRS followed separate paths of development, and related expertize was built in "separate compartments," as discussed in Chapter 1. To some extent, this was also true for SYS TH if one considers computational tools in the analysis sectors of transient RCS performance, core design (primarily aimed at nominal conditions), subchannel analysis, and CO.

During the decade 1990–2000 the increase in computer power allowed the numerical coupling among codes in different areas like structural mechanics, neutron physics, CFD, and nuclear fuel, other than different branches in SYS TH. SYS TH codes played a pivotal role in this connection.

The coupling among codes implies addressing "numerical and informatics" interfaces between the codes and the computer compilers. Those interfaces are not necessarily homogeneous at the design level for different codes and generate issues connected with the transfer of information among codes and with the selection of time steps in transient calculations. The elements for reflection and the recommendations are:

(a) To merge into one SYS TH code the capabilities of modeling RCS, core, subchannel, and CO. This confers the advantage of avoiding troublesome code coupling in some situations.
(b) To propose systematic V&V procedures for coupling, also involving acceptance criteria for the coupled results (e.g., see D'Auria et al., 2004, concerning the coupling between SYS TH and 3D NK codes).

An example of cross-correlation between codes, technological areas in NRS, and modeling of physical barriers, materials, and NPP components/systems can be seen in Fig. 2.8 (D'Auria et al., 2012a); see also CFD coupling in Section 2.3.5 and D'Auria et al. (2021a), where results from the application of coupled codes are discussed.

2.3.11 Modeling and structuring of computational tools

The structuring and the modeling of existing TH SYS codes (or computational tools) are based on the application of the "six-partial-derivative-balance-equations" to simulate the transient performance of the two-phase mixtures. Those equations shall be seen as the result of the application of the principles of thermodynamics and mechanics. However, approximations or even inadequacies (so far unavoidable) are part of the structuring and modeling, like (more details can be found in the paper by D'Auria et al., 1998):

- No direct accounting for two-phase flow regimes
- Averaging the cross-sectional flow area, integration volume, and time, all of which are associated with the concept of control volume
- Considering two (or three) fluid fields as solution domains (this may be a consequence of the two preceding items)

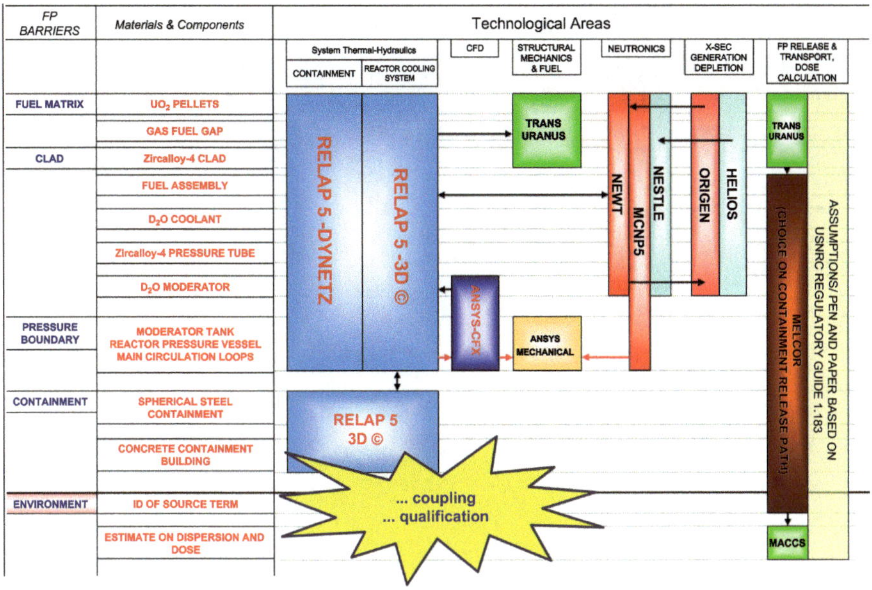

Figure 2.8 Coupling between codes: cross-correlation with technological areas in NRS and modeling of physical barriers, materials, and NPP components/systems (D'Auria et al., 2012a).

- Needing various constitutive and/or closure equations that are unavoidably used outside their range of applicability; this also implies the need for "virtual correlations" (or numerical terms) to obtain solutions within the real domain

The concept of control volume implies the need to reconstruct the continuous reality (i.e., the NPP configuration) by more or less large pieces, thus introducing undue discontinuities at the connection faces. The well-known issue of "water-packing" is a consequence.

V&V with the main reference to the validation is the only way to qualify and, eventually, to make acceptable those inadequacies.

Nevertheless, SYS TH codes are the basket for the expertise gathered in the areas of modeling and experiments. Validation has been, and is, the key to overcome the modeling approximations. From pursuing the validation process one shall recognize that errors exist in the results from code application and shall be capable of quantifying those errors—this imposes, among other things, the need to estimate uncertainty.

The recommendation here is a systematic effort to characterize the listed deficiencies, including a detailed evaluation of their impact on the results. This also represents the target recently established by the international network for code developers, FONESYS (e.g., see Ahn et al., 2015).

2.3.12 Licensing needs

Licensing can be seen as the legal part of NRS and as the exit edge for NRS findings and requirements. A picture of the licensing framework in the area of accident analysis

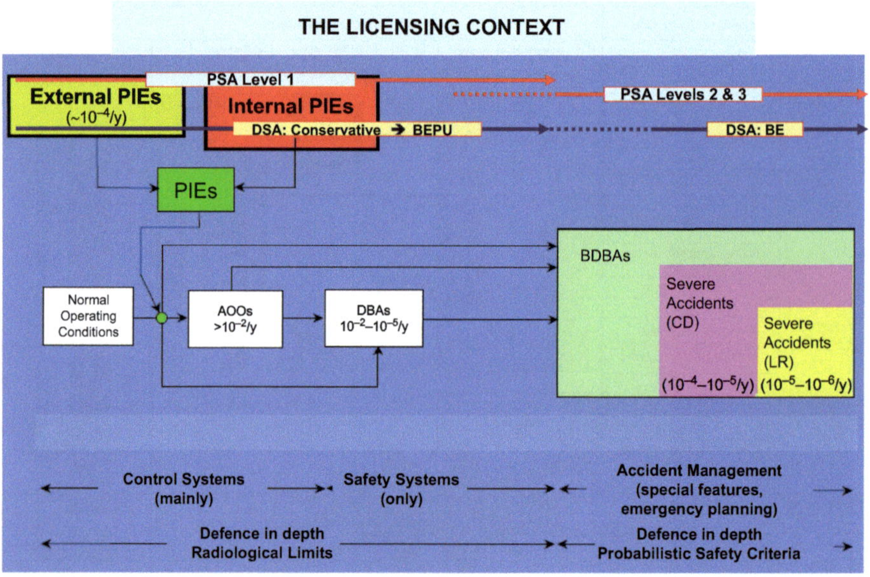

Figure 2.9 A picture of the licensing framework.
Based on a proposal by J. Misak, later modified by M. Dusic and F. D'Auria.

(or Chapter 15 of the FSAR) can be derived from Fig. 2.9. An example of the hierarchy of documents applicable for licensing analysis can be found in Fig. 2.10. Recent proposals for requirements (i.e., connected with possible weaknesses for nuclear fuel identified by USNRC) are discussed in Section 2.2.3.8.

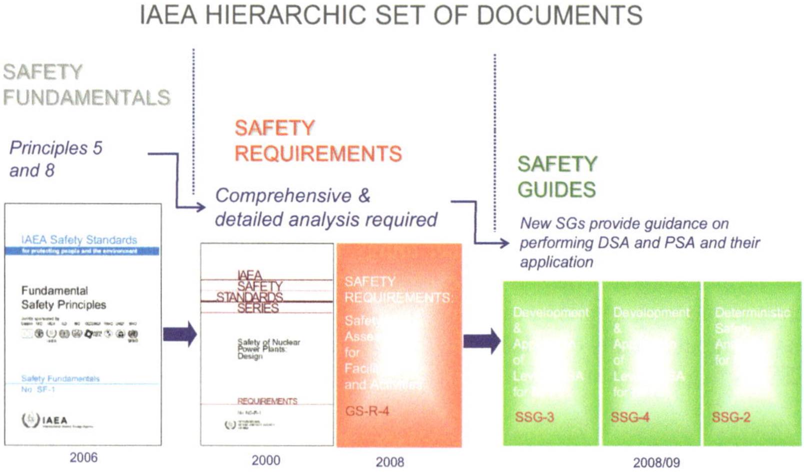

Figure 2.10 An example of the hierarchy of documents from IAEA, applicable for licensing analysis.

Accident analysis is an important component of licensing. DBA anticipated operational occurrence, and BDBA, as well as emergency operating procedures (EOPs) and accident management guidelines, are established concepts in accident analysis. Moreover, following the start of any accident (e.g., a PIE (postulated initiating event) part of the DBA envelope) I&C has the potential to transition the NPP far from its initial condition (e.g., D'Auria et al., 2012a).

The first element for reflection on accident analysis deals with the use of conservatism. The conservatism in input data, also known as "Option 2" in IAEA documents (e.g., see IAEA, 2008a, 2009c) does not ensure conservatism in the output (or code calculation results). Conservatism in the output can only be demonstrated when BEPU analyses are performed; otherwise, adequacy of conservatism on the output can only be inferred by deep expertize. Thus, the IAEA "Option 2" should be considered an "interim" option in the history of the application of SYS YH codes to licensing.

The key issue in the case of accident analysis and licensing is the application of the BEPU approach also identified as "Option 3" in the IAEA documents (e.g., see IAEA, 2008a, 2009c). Pioneering efforts have been completed in this direction as discussed under the historical outline in Section 2.3. However, developments are needed to make traceable and robust the regulatory process, by addressing subissues like (not a comprehensive not a systematic list):

- use of 3D NK to determine the maximum linear power of individual fuel rods
- use of 3D NK to perform the transient analysis of events where nonhomogeneous core behavior is expected
- consideration of single failure
- consideration of the availability of systems and components part of the NPP (notably including engineered safety features, ESF)
- BEPU analysis of transients or situations identified as ATWS, SBO, and PTS (pressurized thermal shock)
- establish procedures for determining the worst break position and, at a higher level of detail, the break opening time, which affects the calculation of mechanical load on internal and external components and structures
- establish procedures to account for the impact of burnup upon the transient analysis
- establish a procedure to calculate the quantities affecting the acceptance criteria other than the PCT and the H_2 production, i.e., the demonstration of the core remaining intact (mostly, mechanical loads at the transient start in case of LBLOCA) and the long-term core cooling involving the specific subissues of sump-recirculation, debris-effect, and, where applicable, containment leakages and long-term condensation.

Further elements for reflections are:

(a) A procedure shall be established connecting the accident analysis with the design and the applicability domain of system codes: DBA envelope → TH phenomena → significant physical quantities (e.g., pressure, heat flux, flow rate) → range of variations for those quantities and related combinations → design domain and V&V domain for SYS TH codes (e.g., see Fig. 2.6).

(b) I&C simulation shall be part of the analysis of accidents (e.g., see D'Auria et al., 2012b).

The first recommendation also applies in relation to V&V for SYS TH codes (Section 2.3.8 in this chapter). The second recommendation also applies in relation to PSA and SYS TH (Section 2.3.13 in this chapter).

2.3.13 Probabilistic safety assessment and system thermal hydraulics

PSA process at any of Levels 1, 2, or 3 requires deterministic calculations to establish the success of any sequence, decide about the failure modes of systems structures and components, and estimate the "source term." Historically, approaches and computational tools have been adopted for deterministic calculations which are simplified in relation to the methods and the numerical codes available within the SYS TH domain (the reason for this being the enormous number of requested calculations). A currently controversial issue is the quality and qualification levels of deterministic calculations used in PSA (e.g., see IRSN, 2013).

The role of I&C in PSA is an additional concern. I&C systems and components are designed with the target of being uninfluential or providing benefits after any PIE. However, the complexity of the entire NPP system and the interaction among various logics and components following any PIE has the potential to bring the NPP conditions far from the configuration associated with the originating PIE.

The issues or the elements for reflection are:

(a) The design of complex EOPs and severe accident management guidelines (SAMGs) on the one hand and the availability of UM on the other, together with the advent of more powerful computers, impose reconsidering the use of simplified approaches when estimating the success of any sequence or when calculating radiological releases. This may imply the need for extraordinary computational resources.
(b) The role of "deterministic" uncertainty shall be included in probabilistic results.
(c) I&C simulation shall be considered for PSA and DSA.

Finally, Duffey and D'Auria (2020), provide a discussion about the interconnection between DSA and PSA when low probability events are concerned.

2.3.14 Severe accidents and system thermal hydraulics

Referring to the predicted evolution of any accidents, the intact (or nearly intact) core geometry is normally taken as the boundary between SYS TH and SA analyses. Furthermore, severe accidents are outside the domain of interest for this book as stated in Chapter 1.

The quality of DBA analyses (or of analyses performed within the DBA envelope) shall be distinguished from the quality of SA analyses or analyses involving core degradation. Enormous progress has been made during the last 2 decades in the understanding of SA phenomena; however, the qualification level of results from SA analyses shall be considered far lower than the qualification level expected or achieved in the area of SYS TH analyses. A picture of the SA area can be obtained from documents (USNRC, 1995; OECD/NEA/CSNI, 2007, 2009; IAEA, 2008b).

The design impact from SA analyses shall be avoided or minimized, rather SA analyses shall be used to ensure thorough bases for nuclear technology understanding and supporting the decision of NPP staff in the unluckily SA situation. New systems to mitigate SA conditions may reveal improper conditions not directed toward facilitating operator actions and, even worse, at the origin of new accident sequences (e.g., the undue flooding of the reactor cavity during nominal operation).

Recommendations or elements for reflection are:

(a) The analysis until the prediction of important core degradation shall be performed by the BE SYS TH code. This is specifically true in relation to the calculation of the time of A historical perspective of NTH 77 occurrence of degraded core conditions. In other words, the use of simplified approaches might produce misleading results far from what can be reliably obtained by the BEPU approach.

(b) SA code results properly supported by SYS TH code results can be relevant for planning operator actions: optimized SAMG and connected procedures can be designed.

(c) Impact on the NPP hardware design from SA code results, even supported by SYS TH codes, should be avoided owing to the low qualification level, as already mentioned. Furthermore, design changes derived from SA analyses should imply changes in the licensing approach that currently distinguishes between DBA and BDBA—the DBA envelope shall be changed according to the design, and a proper quality level should be ensured for SA computational tools.

2.3.15 User effects and training

The current user effects in applying SYS TH codes and coupled codes undermine the applicability of those codes, confidence in the derived calculation results, and the possible benefits of code applications (e.g., see Aksan et al., 1993; Ashley et al., 1998; D'Auria, 1998; Petruzzi et al., 2005).

The effect of code-users upon "poor" code calculation results shall be seen as similar to the effect of inexperienced drivers bringing top-quality bus or trucks against a wall or out-of-the-road. In these cases (truck or bus off-road), little or no blame is put on the truck makers or designers, and discipline and training are invoked for drivers. In the case of SYS TH code, poor code calculation results are preferably attributed to poor code capabilities; codes appear unreliable and code development processes become repetitive and never-ending.

More focus should be on the qualification of researchers in the SYS TH area, and additional emphasis should be given to the requested (mainly from the side of regulators) qualification level of code users. International programs are ongoing in this direction (e.g., Petruzzi et al., 2005), but they appear still weak and not properly supported by regulatory authorities. Targets and accomplishments for code users shall be connected with the number of years of experience and the modalities of performing the analysis (e.g., see IAEA, 2015). The role of the input deck, or the interface between the NPP (as built and operated) and the code, should receive proper attention. Specific recommendations are:

– The input deck or the nodalization developed by group 1 of code users should be "independently checked" by group 2 of code users.

- Proper acceptance criteria shall be used to endorse the quality of any input deck.
- The results of any safety or design-relevant calculation should be "independently checked" and approved by a "supervisor" having at least 10 years of documented expertize in SYS TH analyses.
- In addition to uncertainty analysis needed in conjunction with any BE code calculation, properly planned sensitivity analyses are necessary to minimize the risk of neglecting possible bifurcation or cliff-edge effects.

2.3.16 Best estimate plus uncertainty approach

Only the definition of the BEPU approach and its key elements are given here to complete the list of perspectives for NTH in the future. BEPU is discussed in papers by D'Auria et al. (2012a) and D'Auria et al. (2012b). Details related to BEPU can be found in Chapter 14.

The following statements may help provide an understanding of the BEPU approach:

(a) BEPU is composed of the words best estimate ($^{1}/_{4}$ the best one can do, within current technology— i.e., a moving target—in predicting/calculating an assigned, typically complex, transient scenario) and Uncertainty ($^{1}/_{4}$ expected, or unavoidable, error associated to the calculation process).

(b) The process of combined application of an SYS TH code and a UM, to the analysis of accidents in NPP, is called BEPU.

(c) BEPU can be seen as the connection between NTH and licensing (needs) (see also Fig. 2.11).

(d) A metaphor has been used to describe BEPU (D'Auria, 2013). The knowledge acquired in NTH resembles a city suddenly abandoned by inhabitants. Everything there is at rest, and no information is available to trigger life in the city. Wild-experienced travelers arriving there feel lost and unable to use their competence and existing, visible knowledge. In the allegory, the wild-experienced traveler is the (expert) thermal-hydraulic specialist, the city and its components and systems are nuclear thermal-hydraulic knowledge, and BEPU represents the civilization needed to make the city alive.

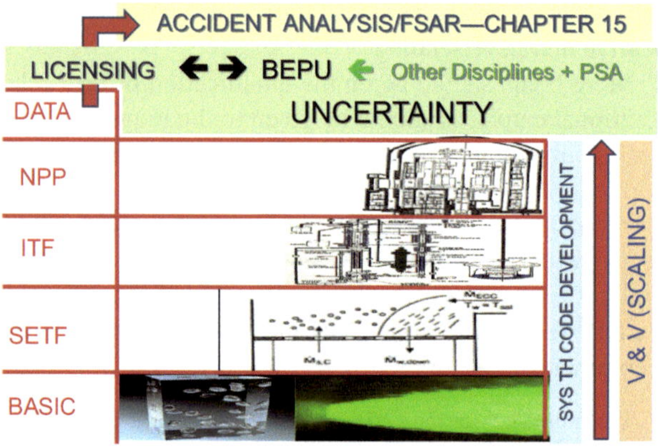

Figure 2.11 The BEPU process focusing on accident analysis and Chapter 15 of FSAR (D'Auria, 2015).

(e) BEPU is a process that implies the widest exploitation of data and information in NTH, which can be derived from Fig. 2.11 (D'Auria, 2015).

The element for reflection is as follows: the application of BEPU is strongly recommended; benefits from the application can be listed and partly quantified (see also D'Auria et al., 2015); however, given those circumstances, neither Regulatory Authorities nor Nuclear Industry (more reasonable) is pushing for the application of BEPU.

2.3.17 Summary remarks

A synthesis of the items discussed earlier is provided in Table 2.2. The following modalities or areas are proposed to plan possible improvements (third column in Table 2.2).

- The SYS TH code: this implies that improvements are expected for the codes and includes the (fundamental) modeling.
- Modeling: this implies that fundamental processes, basic research, or experimental data are needed.

Table 2.2 Synthesis of author's views of perspectives of system thermal hydraulics.

Topic no.	Category/target for improvement	Modality for improvement	Proposals/notes
1	Local "form loss" coefficients—*K*-factors	Modeling	Libraries shall be created with similar complexity as the ones for neutron microscopic XS
2	Multidimensional (multi-D) HTC surface	Modeling	Either a new approach for calculating wall-to-fluid HT, or a suitable multi-D surface for HTC
3	RCS depressurization and CO pressurization	SYS TH code	New experiments may be needed
4	Precision targets	Application	Precision targets shall be established (also part of V&V)
5	CFD role for NPP design and NRS	Application	Specific V&V needed, including improved best practice guidelines
6	Passive systems	Design and SYS TH code	Specific experiments needed to address stability issue
7	Scaling and experiments	ITF design and EDB	Scaling issue in licensing is addressed in OECD/NEA/CSNI (2016), see also D'Auria and Galassi (2010). Scaling for experiments is discussed here. "Full" height ITF may be needed. EDB to be properly stored

Continued

Table 2.2 Continued

Topic no.	Category/target for improvement	Modality for improvement	Proposals/notes
8	V&V	Application	Establishing developmental and independent assessment matrices. Connected with items (4) and (12)
9	Uncertainty methods	Application	Feedback from international projects. Specific V&V needed. Integration OF CIAU and GRS uncertainty method desirable
10	Coupling	Application	Numeric and informatics issues. Specific V&V needed
11	Modeling and structure	Modeling and SYS TH code	Inadequacies of SYS TH codes. FONESYS network established
12	Licensing	SYS TH code and application	Comprehensive BEPU approach. Licensing at the origin of the applicability range for SYS TH code (also part of V&V). Consideration of I&C
13	PSA connection	Application	Consideration of complexities in SYS TH including uncertainty. Consideration of I&C
14	Severe accident	Design and application	Current qualification level far below SYS TH. Design of optimized SAMG expected
15	User effect	Application	Code-user role. Training and qualification needed
16	BEPU	Application	BEPU application strongly recommended for a "new vision of NRS"

- Application: this deals with modalities and requirements for the processes and the procedures that foresee the application of computational tools and data to the NPP design, safety, or licensing. The word Application is adopted here to signify "exploitation of achieved knowledge and understanding in the technological practice."
- Design: this implies that improvements are expected in the design of thermal-hydraulic systems.
- ITF design: this implies that improvements (or acceptable strategies) are expected in the design of the new ITF.
- EDB: this should be taken as synonymous with "EDB preservation" and implies that investments are expected in preserving and storing experimental data.

It shall be noted how the SYS TH code is both a target for improvement and a way to pursue envisaged improvements.

2.4 Conclusions

A noncomprehensive and nonsystematic review of perspectives and proposals for future activities in SYS TH has been performed and applies to NPP design and NRS frameworks. "Elements for reflection" are also presented in the chapter.

Numerical codes, experimental facilities, associated research programs, and related data and applications constitute the heart of SYS TH—the identification of numerical codes and experimental programs is provided in Section 2.2.2 and repeated in the "Nomenclature."

A spot historical review emphasized the publication of the Interim Acceptance Criteria for the design of ECCS by the USNRC in 1971 as the triggering event for the development of modern NTH (or SYS TH). Other milestones are the qualification of SYS TH codes and their use within the BEPU approach together with the development and validation of UM.

Sixteen topics were discussed in the present chapter and are viewed as the "elements for reflection." An alphabetic list of those topics follows and refers to the keywords in Table 2.2:

- BEPU
- CFD role for NPP design and NRS
- Coupling (with SYS TH codes)
- Licensing (implying application of SYS TH codes)
- Local form loss coefficients, "K-factors"
- Modeling and structure (of SYS TH codes)
- Multi-D HTC surface
- Passive systems
- Precision targets
- PSA (connection with SYS TH code application)
- RCS depressurization and containment pressurization
- Severe accident (connection to SYS TH code application)
- Scaling and experiments
- Uncertainty methods
- User effect (and user training-qualification)
- Verification and validation (V&V)

Arbitrariness in the use of direct and reverse local pressure drop coefficients (K-factors) and inadequate qualification of code users have thus far undermined confidence in the application of currently available SYS TH codes.

Procedures shall be established in several areas (identified as an application in Table 2.2) which are consistent with the present level of knowledge, to make robust and reproducible the results from SYS TH code calculations. The BEPU approach notably falls within this category. "Closing the issue" in areas like Validation (rather than V&V), Scaling, and application of UM is also deemed possible based on the current understanding—spreading available information within the scientific community may be worthwhile for reaching consensus.

Improvements are considered necessary in areas like selected phenomena modeling, coupling among codes in different disciplines, the design of passive systems, the evaluation of related reliability, the interface between probabilistic and

DSA, and the application of CFD codes. A peculiar area that deserves proper attention is the code-user training and qualification.

From a global perspective viewpoint, the results from envisaged activities should converge for a more efficient licensing process (and a better NPP design) minimizing the unavoidable lag between scientific and technological achievements and practical applications.

The implementation of the thoughts and ideas, necessarily incomplete and not systematic, shaped in this chapter is expected to be difficult, and the practical relevance of connected proposals may be questionable. Nevertheless, the provided discussion is expected to contribute to a deeper understanding of the NTH discipline. Chapter 24 (second edition of the book) deals with suggestions for future developments in the area.

Exercises and questions

Exercises and questions are proposed hereafter with solutions (or hints to get solutions) given in Appendix 1 (Volume III of the book). Answers and solutions are expected to be provided by readers before reading Appendix 1. Please provide extended answers.

2.1 Discuss the importance of USNRC Interim Acceptance Criteria for the ECCS design (issued in 1971 at a time when NRC was AEC) for NTH.
2.2 In which international context the process code Validation for SYS TH was established?
2.3 Please provide suitable definitions for "scaling," the "scaling issue," and "addressing the scaling issue."
2.4 Can you outline the history of scaling in NTH?
2.5 Is V&V part of scaling or scaling part of V&V?
2.6 Why Uncertainty and Uncertainty Evaluation are relevant (essential) for the applications of NTH to the WCNR technology?
2.7 What is the connection between Probabilistic Safety Assessment and NTH?
2.8 What are the key elements of the BEPU approach and how those elements are connected with NTH?
2.9 Who are the stakeholders for NTH?
2.10 Explain the cross-connection between NTH and the design of passive systems.

Acknowledgments

J-L. Gandrille revised the original version (Edition 1) of this chapter; he is gratefully acknowledged. R.B. Duffey provided key concepts considered in the chapter and suitable references dealing with the history of thermodynamics.

Parameters and concepts in nuclear thermal hydraulics

3

Francesc Reventos and Jordi Freixa
Polytechnic University of Catalonia, Barcelona, Spain

Foreword

This chapter uses a wide variety of parameters and concepts to provide a view of the nuclear thermal-hydraulic universe. The content covers everything from bases to applications within the discipline. The views expressed "parallel" those elsewhere in the book and aim to provide help in understanding its content. After reading this chapter, readers should be able to confirm their understanding of thermal hydraulics.

The reader should expect (or try) to obtain answers to the following questions from the present chapter:

- What is nuclear thermal hydraulics (or what are its key elements)?
- How is nuclear thermal hydraulics connected to other disciplines within nuclear technology?
- What is the application domain for nuclear thermal hydraulics?
- Why is an understanding of nuclear thermal hydraulics essential for nuclear engineering?

Thus, this chapter provides the background for most other chapters, as noted in Chapter 1. For simplicity, we avoid mentioning the interconnections between topics here and in the other chapters of the book.

Because of the fundamental nature of the topics considered here, the list of references in the second edition of the book is nearly the same as that in the first edition.

3.1 General remarks

3.1.1 General remarks on parameters and concepts

This chapter introduces the parameters and concepts required for developments in and an understanding of nuclear thermal hydraulics.

Since the book includes fundamental topics ranging from nuclear thermal hydraulics (or system thermal hydraulics) to advanced applications of thermal hydraulics in nuclear power plant (NPP) safety and design, the chapter provides discussions for understanding the boundaries of the overall text. It includes concepts for depicting connections between fluid dynamics and gas dynamics while emphasizing their added complexity. We introduce the concept of turbulence to clarify modeling boundaries and complexities.

Handbook on Thermal Hydraulics in Water-Cooled Nuclear Reactors
https://doi.org/10.1016/B978-0-323-85606-5.00007-6

The chapter consists of several sections, with the Section 3.1 devoted to general remarks. Since two-phase flow and heat transfer are the most relevant subjects to appear in subsequent development, a full section is devoted to each.

Section 3.2 deals with parameters and concepts of two-phase flow for deeper insights in other chapters. We specifically introduce void fraction and quality together with their relationship. We describe parameters involved in the mass, energy, and momentum balance equations of two-phase flows.

Section 3.3, devoted to heat transfer, introduces concepts and variables related to conduction, convection, and radiation. The attention is on characterizing heat transfer, but the discussion also includes the empirical and theoretical aspects required for understanding expected developments.

Section 3.4 develops descriptions of selected phenomena to complete the concepts required in subsequent treatments (among them are natural circulation, critical flow, blowdown, and countercurrent flow limit). Those descriptions cover the essentials of expected occurrences.

Section 3.5 is devoted to analytical tools. It includes brief explanations of simple approaches—homogeneous, separate, drift-flux, and lumped parameter models—to illustrate the complexity of the involved physics. It introduces parameters and concepts referred to in system codes and concludes with some additional concepts. Concepts related to verification and validation (V&V) are presented in Section 3.6, with an emphasis on experimental thermal hydraulics and the issue of scaling.

A block of three sections (i.e., Sections 3.7–3.9) connects thermal hydraulics with design basis (DB), general safety concepts, and concepts in deterministic safety analysis (DSA).

The last block of the chapter comprises two sections (i.e., Sections 3.10 and 3.11) related to accidents and other uses of thermal-hydraulic analysis.

When a section includes a topic of particular interest, the context of that topic is noted in a brief introduction.

3.1.2 Importance of two-phase flow

The analysis of water reactors' thermal hydraulics must consider two-phase flow and related phenomena. Phenomena that occur in normal operations in boiling water reactors (BWRs) and the specific components of pressurized water reactors (PWRs)—mainly steam generators (SGs)—can also occur under both PWR and BWR accident scenarios. Most safety scenarios involve depressurization; among them, loss-of-coolant accidents (LOCAs) and malfunctions are related to pressure control or pressure relief valves.

When depressurization occurs in a PWR, system saturation temperature may decrease, and two-phase flow could appear. It also appears in the advanced phases of most scenarios due to decay heat and the presence of water. Vapor generation occurs in the core, and any mixture of water and vapor could flow, driven by a break, an injection, or natural circulation. The importance of two-phase flow developments in the thermal hydraulics of nuclear systems has been stressed in conferences (e.g., the OECD/NEA/CSNI, 1992) as well as in internationally recognized textbooks (e.g., Tong, 1965; Wallis, 1969; and Collier, 1975).

3.1.3 Multiscale and multiphysics

The current scientific approach describes involved scenarios that include multiscale calculations. Thermal-hydraulic analysis devoted to nuclear systems must cover the entire range of modeling scales. By now, it is widely accepted that those scales are the following: system, subchannel, and computational fluid dynamics (CFD).

The aim of calculations at the system scale is to predict the overall behavior of the plant, including the reactor and components such as the primary circuits, SG, and pumps. System codes are the appropriate tool for addressing this subject. We identify phenomena at such scales as system effects. System codes usually solve the problem with a one-dimensional (1D) formulation for most modeled systems and three-dimensional (3D) options for the relevant parts of reactor vessels. System codes are widely used in the thermal hydraulics of nuclear systems (Freixa et al., 2015a, 2015b; IAEA, 2014).

The subchannel scale is also firmly established. It deals with modeling the reactor core or fuel bundles. More strictly, at the definitional level, phenomena occur on the component scale given that, beyond the core, other components have more detailed codes and models. Since the core is the most often studied component, it seems appropriate to nominate that scale to be the subchannel scale. At such a scale, 3D aspects enter into play, and fuel bundle models require sufficient detail to predict local cladding temperatures or expected DNB values. Analogously, SGs and other heat exchanger tube bundles are modeled using specific codes at similar levels.

The third group is the so-called CFD scale. CFD produces simulations at an extremely fine scale. Investigations at such a scale deal with modeling on a level closer to physics, e.g., adopting nodalizations (or equation integration domains) for a massive number of nodes and applying more updated numerical methods. Accordingly, the required computing power is also enormous and must address a rather complex problem definition; see Smith et al. (2008). CFD codes are numerical tools that primarily clarify the behavior of phenomena in small (local) areas or critical components during specific scenarios.

Many initiatives of the international research community have dealt with and deal with coupling thermal-hydraulic models at various scales. Similarly, many initiatives have connected thermal-hydraulic codes with codes that describe other physical aspects of the simulated scenario. Among them, one can find neutron physics (e.g., Ivanov and Baratta, 1999; D'Auria et al., 2004; Pericas et al., 2013, 2016; Pericas, 2015) or mechanical or severe accident codes (e.g., Pla, 2004; Espinosa-Paredes et al., 2012; IAEA, 2008b). Such activities are now encouraged because computers are becoming faster every day. Multiscale and multiphysics platforms and projects are ongoing.

Neither the scale nor required coupling with other physical formulations affects the thermal-hydraulic concepts introduced in the sections that follow.

3.1.4 Turbulence and two-phase flow

The essential characteristic of turbulent flow is its chaotic property variation. In turbulent flow, pressure and velocity tend to change in space and time rapidly due to unsteady vortices that appear on many scales and interact among them. The Reynolds number provides a first approach to limiting the question—flows with large Reynolds numbers (more than 5000) are typically turbulent. This is only a first approach, and turbulence, despite the important analytical effort currently devoted to it, is considered an unresolved problem.

The analysis of turbulent two-phase flow has added difficulty. The abovementioned chaotic property variations could change or suppress any two-phase flow regime. At this point, it seems that obtaining an accurate picture of two-phase flow requires turbulence to be previously mastered.

Once again, experimentation is crucial to overcoming the problem. An important number of tests related to two-phase flow have been performed, and observations of their results provide some confirmed outcomes. Two-phase flow regimes have distinctly steady properties and transitions among them at definite key parameter values. All this has uncertainty but opens up the possibility of proceeding with a two-phase flow study. More detail on flow regimes and key parameters is provided below.

3.1.5 Empirical database and instrumentation

An empirical database supports the use of equations, correlations, and codes. This is an important point since the validation of theories and practices depends on the right correspondence between what is defined and what is measured. The reliability of the database is supported by its systematic documentation and the qualification process of facility instrumentation.

An enormous database has been produced for many years and is now available for use at the various levels where it is needed; see also Addabbo et al. (2002), Pla et al. (2015, 2016b). It consists of separate effects and integral tests designed to support developments devoted to predicting the behavior of a reactor system under normal, abnormal, or accident conditions. After following a Committee on the Safety of Nuclear Installations (CSNI) initiative, the results of the experiments are organized into two different matrices (separate effects and integral tests) to establish their suitability for each phase of the validation process of best estimate (BE) thermal-hydraulic computer codes; see Aksan et al. (1987) and OECD/NEA/CSNI (1992, 1993, 1996).

Instruments for measuring relevant parameters are required in the complete qualification process used to certify reliability. The design of such instruments must ensure the right correspondence between the signal produced and the physical parameter that the analyst is following. The complexity of thermal-hydraulic processes adds difficulty to instrumentation design and setup. Another difficulty is designing and qualifying instruments that are suitable for measuring wide-ranging parameters. For instance, certifying the validity of thermocouple temperature measurements in accident scenarios is rather complex.

3.2 Concepts involved in two-phase flow development

3.2.1 General aspects

The purpose of this section is to define the most relevant parameters related to two-phase flow and discussed in subsequent chapters. Again, some recognized textbooks provide interesting details on such parameters and concepts, among them (El-Wakil, 1962; El-Wakil, 1971a; El-Wakil, 1971b; Tong and Weisman, 1970; Fenech (Ed.), 1981; Todreas and Kazimi, 1990; and Anglart, 2010). Parameters, usually known as water properties, are only mentioned when the added comments are sufficiently significant for the specific development.

Two comments are presented in this general section related to intensive and extensive properties and time- and volume-averaging techniques. By definition, intensive properties are independent of the system's extent or size, whereas extensive properties are dependent. Temperature is a good example of an intensive property, whereas mass and volume are examples of extensive properties.

Despite extensive mass and volume, density and specific volume are intensive and can be easily written down in a table. Water property tables are quite useful, and most codes have versions of them.

Since thermal-hydraulic analysis deals with the dynamic behavior of two-phase mixtures, discretization of time and space is an important issue. To accomplish this, control volumes or space nodes on the one hand and time steps on the other hand are defined accordingly. Such definitions may change depending on the problem and are significant in distinguishing local from volume-average values and instantaneous from time-averaged values.

Mass and energy conservation usually provide a criterion for averaging quantities. The criterion is applied somewhat differently depending on whether the properties are intensive or extensive. For extensive quantities like mass, for example, the application is straightforward. The application for intensive properties like liquid water temperature proceeds as follows—the volume-averaged temperature is that of liquid water where the mass and energy of the node are conserved.

Such concepts become important when designing measurement techniques and at the numerical method level when considering instantaneous and time-averaged values.

3.2.2 Parameters and concepts

3.2.2.1 Void fraction/quality

In the first and most intuitive approach, the void fraction (α) of a two-phase mixture is the volumetric fraction of a given control volume occupied by vapor.

Since different measurement techniques identify the cross-sectional area (A) of a given channel and the areas occupied by the gas (A_g) and the liquid (A_l), the void fraction is

$$\alpha = \frac{A_g}{A_g + A_l} \tag{3.1}$$

$$1 - \alpha = \frac{A_l}{A_g + A_l} \tag{3.2}$$

The void fraction equals 1 when all volume is vapor and 0 when it is liquid.

Similarly, the quality (x) of a two-phase mixture is the mass fraction of vapor at a given control volume. If W_g is vapor mass flow and W_l is liquid mass flow, quality is written as

$$x = \frac{W_g}{W_g + W_l} \tag{3.3}$$

$$1 - x = \frac{W_l}{W_g + W_l} \tag{3.4}$$

Quality equals 1 when all volume is vapor and 0 when it is liquid.

Interfacial area/interfacial area concentration

Interfacial area for a given control volume is the area of contact between the two phases, liquid and vapor. It is usually described using the parameter defined as the interfacial area concentration, which is the area of contact between the two phases per unit of volume; see Bestion and Serre (2013). Such a parameter is strongly connected with the discussion of flow regimes that follows. Characterizing such a connection is an ongoing research activity with interesting results thus far.

3.2.2.2 Mass velocity

The total mass flow (W) per unit of the cross-flow area is the so-called mass velocity (G):

$$W = W_g + W_l \tag{3.5}$$

$$G = \frac{W}{A} \tag{3.6}$$

Mass velocity is especially interesting in subchannel analysis since it directly correlates with the remaining capability to extract energy from a fuel bundle.

3.2.2.3 Equilibrium/subcooling/superheating

In general, two substances are in thermal equilibrium when no energy transfer occurs between them when they are at the same temperature. For a given pressure, saturation temperature is the temperature at which vapor and liquid coexist without exchanging energy. At the saturation temperature, vapor and liquid are in thermal equilibrium. For further developments, saturation temperature is useful as a reference.

Thus, liquid with a temperature (T) below the saturation temperature is called subcooled, and $\Delta T_{SUB} = T_{sat} - T$ is called subcooling.

Similarly, vapor at a temperature (T) higher than the saturation temperature is called superheated, and $\Delta T_{SAT} = T - T_{sat}$ is called superheating.

3.2.2.4 Pressure drop

The difference in pressure between two points of a fluid system is known as the pressure drop. That difference can result from static head, acceleration, friction, or local resistance.

Static head occurs between two points with different heights. The acceleration term for a given fluid comes from changes in cross-areas. Forces exerted between channel walls and the fluid occur because wall roughness causes friction. Local resistance is due to changes in geometry like those produced by grids and elbows.

The proper evaluation of pressure drop is crucial in thermal-hydraulic analysis and becomes more complex when dealing with two-phase flow; see Martinelli and Nelson (1948), Lockhart and Martinelli (1949). The start-up of natural circulation in any dynamic situation depends on the correct evaluation of pressure drops. In addition, correctly predicting water mass distribution in LOCA scenarios depends on knowledge about the pressure drops in the involved system. Once the mass distribution is dynamically established versus time, core dryout can also be properly predicted.

3.2.2.5 Friction

The concept of friction already introduced and used requires some additional explanation since it could appear related to wall-to-fluid friction or interfacial friction. In both cases, the friction force is exerted on the considered fluid in an opposite sense to its motion. Wall-to-fluid friction is calculated as a function of fluid velocity and density. Simple correlations are available.

For one-phase flow,

$$\Delta P = 2 \times f \times (L / D_h) \times \rho \times v^2 \tag{3.7}$$

For two-phase flow,

$$\Delta P = 2 \times f \times (L / D_h) \times \rho \times v^2 \times R(P, \chi) \tag{3.8}$$

where ΔP is the pressure drop, f is the roughness coefficient (function of fluid viscosity), L is the length, D_h is the hydraulic diameter; ρ is the fluid density; and v is the fluid velocity and Nelson-Martinelli multiplier correlated to P (pressure) and χ (quality of two-phase mixture).

Interfacial friction is a function of relative velocity between the two phases, and its key parameter is the interfacial area. Flow regimes with a high interfacial area value also have high interfacial friction, which is a crucial phenomenon for determining phase separation.

3.2.2.6 Vaporization/evaporation/boiling

Vaporization is the phase transition from liquid to vapor and can be a surface or bulk phenomenon. In the former case, it is called evaporation, whereas in the latter, it is known as boiling. Heat balance and pressure changes are the most relevant parameters driving this phenomenon. Vaporization could be due to either depressurization or heat input. Despite this similarity, it is important to point out that differences will appear.

Near a heated surface, evaporation could occur following one of two mechanisms: bubble-wise, commonly known as "nucleate boiling," or film-wise. Heat transfer improves depending on the evaporation mechanism (further comments follow). Depressurization can also result in boiling. Since saturation temperature decreases with pressure, a depressurization event can cause the system to experience bulk boiling. LOCAs are examples of such a behavior. Other accident scenarios could show the combined vaporization effect of depressurization and heating.

3.2.2.7 Condensation

Condensation is the transition from vapor to liquid, so the opposite phenomenon of vaporization. Condensation occurs when vapor cool down occurs sufficiently below the saturation temperature. Similarly, it could be due either to pressurization or to heat removal. In addition, it could be droplet or film-wise.

3.2.2.8 Phase separation/mixture levels

Vaporization, condensation, and the dynamic behavior of phases in a water system will establish its physical configuration. In this sense, vapor and liquid could appear in different forms. The stagnant liquid at the bottom of a vertical component with vapor in the upper part is a clear situation of phase separation with a defined level. If the liquid temperature in this case is close to saturation and heat is added, bubbles will be produced, and the mixture level will increase. Alternatively, one could find a homogeneous mixture of liquid and vapor having a common velocity in a pipe. Between the two depicted physical configurations, various phase separations could occur. To deal with such phenomena, all forces acting on the fluid must be considered; among them, gravity, inertia, and friction are the most relevant. Balance equations and flow-regime definitions, discussed in the following sections, are crucial in assessing phase separation.

3.2.2.9 Parameters of balance equations

The balance equations typically considered for two-phase flow developments are conservation of mass, conservation of momentum, and conservation of energy.

The mass equation is usually postulated for a defined control volume. The most relevant parameters and concepts involved are commented on below.

Each phase has a volume, and velocities and densities are usually considered at their average values. If u_g and u_l are the velocities of the vapor and liquid phases, respectively, and ρ_g and ρ_l are the densities of the vapor and liquid phases, respectively, then

$$W_g = A_g \rho_g u_g = W\alpha \tag{3.9}$$

$$W_l = A_l \rho_l u_l = W(1-\alpha) \tag{3.10}$$

The equation of conservation of mass is usually solved for each phase by considering its existing mass for a given control volume and adding or subtracting any addition or removal of mass at the boundaries of the volume.

Vapor generation (by boiling) and liquid generation (by condensation) are also terms of such a balance. The accumulation of mass in a control volume equals the inlet/outlet of mass across its boundaries and the mass generated within it. Indeed, boiling and condensation correspond to this generation term, with a corresponding sign for each phase.

The equation of conservation of momentum is usually formulated for each phase by considering the forces acting on it at each control volume and equating it to the change in momentum of the phase. Under these conditions, the most relevant parameters and concepts require the following additional discussion.

The forces exerted by each phase against the frictional resistance of the channel wall in contact with it will be in the direction of the fluid and the opposite sense.

The forces exerted by each phase against the other phase at the interfacial area are in the direction of the fluid and the opposite sense. Their magnitude is proportional to the relative velocity.

The forces due to static pressure against each phase can be calculated as $P \times A_g$ for vapor and $P \times A_l$ for liquid.

Since the weight of each phase is vertical, in a pipe with an angle θ with respect to the horizontal direction, the force due to weight (per unit of volume) is computed as $\rho_g\ g\ A_g \sin\theta$ for vapor and $\rho_l\ g\ A_l \sin\theta$ for liquid.

A complete momentum equation for two-phase flow should include a virtual mass term and momentum production due to phase change.

Solving the resulting momentum equations has additional complications since frictional resistance is difficult to handle. Pressure drop at geometric discontinuities is usually measured by assuming that a fully developed flow is present and becomes an important source of uncertainty.

The energy balance equation is usually established for a defined control volume. The most relevant parameters and concepts involved are commented on below.

General parameters such as flow, velocity, and density are considered for each phase in a similar way as in the previous equations. Other relevant specific parameters are change in enthalpy, heat absorbed, and work done. Variations in enthalpy are not commented on since they are state variables.

The heat absorbed by each phase could be provided by the other phase or by specific heat sources. Such a source could be internal (chemical or nuclear reaction) or external (hot or cold wall).

The term "work done" (positive or negative) is only relevant for components producing or converting mechanical energy, like turbines and compressors.

3.2.2.10 Flow-regime definition

Many experiments have been performed with heated and nonheated channels with various sizes and inclinations (mainly horizontal and vertical). By observing the results of such experiments, important conclusions can be extracted.

This observation of results leads to the so-called flow-regime definition that is crucial to the thermal hydraulics of nuclear systems. This development is highly motivated by the difficulties encountered in detailed calculations. As noted in a previous section, full 3D calculations with advanced physical models and detailed geometries are still difficult to carry out. The process of defining and improving flow-regime definition methods using the existing test database leads the analyst to something feasible, which induces the thermal-hydraulic community to follow this approach. Some examples will help us understand some situations that have already been tested.

In nonheated channels with a two-phase flow at their inlet, boiling is not powered because no energy is added. Some energy exchange could take place depending on the subcooling of liquid and the superheating of vapor.

In heated channels with a two-phase flow at their inlet (see, e.g., Lahey and Moody, 1993) additional boiling could occur but could be higher or lower, again depending on the subcooling of liquid and the superheating of vapor.

In horizontal channels with a two-phase flow at their inlet, gravity can produce phase separation depending on the void fraction and velocity. Separated phases could easily obtain different velocities for each phase. Friction between phases could appear and cause them to exchange momentum and again alter their relative velocities.

In vertical channels with a two-phase flow at their inlet, gravity can only produce phase separation if velocity is low and boiling is considerable.

Experimental results fulfill two purposes: characterizing flow regimes and establishing flow-regime maps.

A given group of experiments dealing with horizontal channels provides useful information for characterizing the following flow regimes listed at increasing void fractions: bubbly, slug, annular mist, and dispersed, as shown in Fig. 3.1. All these flow regimes are considered for this group at high values of mass velocity that avoid horizontal stratification in this first approach. Combinations of effects will come later along with a discussion of other relevant parameters. Characterizations produced by this first group of tests are the following:

- Bubbly flow (very low void fraction): liquid appears as a continuous phase, and vapor is formed by bubbles that are usually spherical and small.

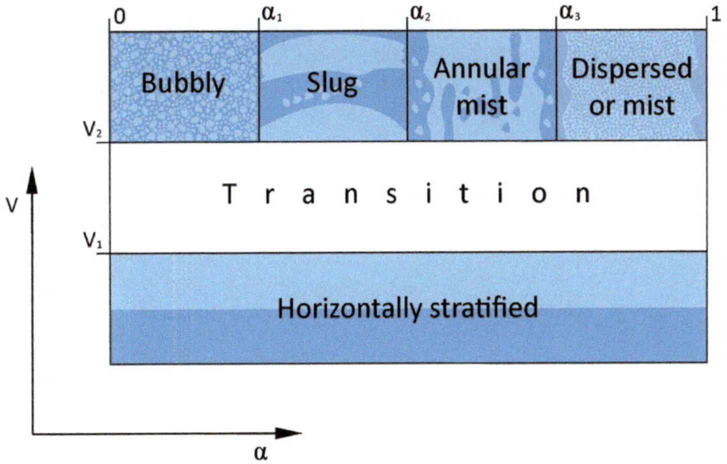

Figure 3.1 Flow regimes in horizontal control volumes. Notes from the lecture "Thermal Hydraulic Concepts," F. Reventos. UPC Master of Nuclear Engineering.

- Slug flow (low void fraction): vapor bubbles are neither spherical nor small. Their diameter is close to that of the channel.
- Annular-mist flow (medium void fraction): liquid appears as a thick film in the vicinity of the wall, and vapor is in the central part of the annulus.
- Dispersed flow (high void fraction): vapor appears as a continuous phase, and liquid is formed as droplets or mist.

The considered group of experiments provides information for deriving transitions from one regime to another. The void fraction is the most relevant parameter for deriving transitions specific to horizontal channels. Despite being the most significant parameter, the void fraction is not the only one. This means that Fig. 3.1 is quite a good approach for understanding the approximate transition between various flow patterns. This approach is quite good because some system codes use similar flow-regime maps. Some authors suggest more complex maps, and research on smooth flow pattern transitions is still an open issue.

Other specific aspects of horizontal channels appear at lower mass velocity values. Horizontally stratified flow occurs in situations where gravity is significant compared with inertia force and thus produces phase separation. Once in a stratified pattern, if vapor velocity is increased, friction between phases produces a wavy flow. Such a flow does not appear in the Fig. 3.1 map because it appears after evaluating an additional parameter. Mass velocity and vapor-liquid relative velocity are the parameters driving specific transitions between flow regimes in horizontal channels.

Another group of experiments provides the same information but for vertical channels, as shown in Fig. 3.2. At a mean velocity sufficient to avoid vertical stratification before critical heat flux (CHF), the basic flow regimes are the same as those in horizontal channels.

Combining effects for vertical channels are more complex except for vertically stratified flow, which appears for a sufficiently low mixture velocity. In this case, transitions depend mainly on one parameter.

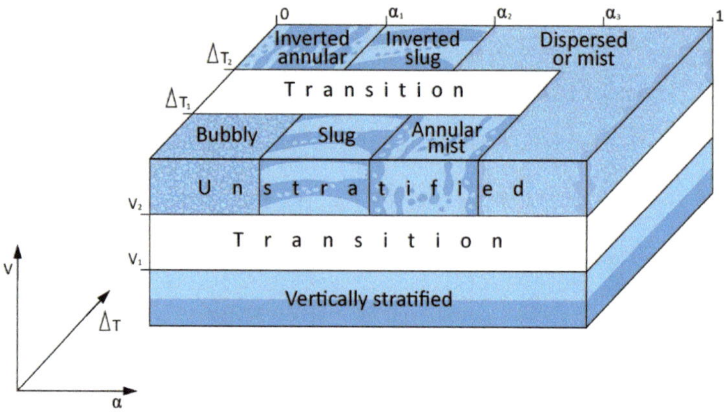

Figure 3.2 Flow regimes in vertical control volumes. Notes from the lecture "Thermal Hydraulic Concepts" F. Reventos. UPC Master of Nuclear Engineering.

The other specific aspect is connected to heat transfer (see section 3.3) and a post-CHF situation. After CHF, an important number of bubbles are formed and coalesced in the vicinity of the wall, so each pattern accordingly moves to a new flow regime. Bubbly flow will move to inverted annular flow, slug flow to inverted slug flow, and annular-mist flow to dispersed flow. In this case, transitions depend not only on usual parameters like the void fraction but also on the conditions of the heat transfer mode established at the wall.

At this point, an important consistency check can be performed. Let us remember that ΔT is the difference between the wall temperature and that of the bulk of the fluid. The next section shows that once steam bubbles coalesce in the vicinity of the hot wall, the heat transfer coefficient decreases and therefore ΔT increases. Inverted annular flow appears as a thick vapor film in the vicinity of the wall, and liquid is in the central part of the annulus, which is consistent with having a high ΔT and a rather low void fraction.

3.3 Concepts involved in heat transfer developments

3.3.1 General aspects

3.3.1.1 Importance of heat transfer developments

An analysis of water reactor thermal hydraulics must consider heat transfer mainly in fuel elements to produce the required results for safety analysis; see Kutateladze (1963). Predictions of temperature distributions in fuel pins and reactor structures must be accurate since the acceptance criteria of safety analysis are defined using the limiting values of the thermal parameters.

3.3.1.2 Scope

Most of the results needed to analyze safety and deal with heat transfer are related to fuel. Therefore, most efforts to improve prediction capabilities are devoted to fuel. Heat transfer in fuel includes heat conduction, convection, and radiation.

Other areas and components of power plants are also needed in heat transfer calculations, including passive heat structures, electrical heaters, heat losses to the environment, and heat exchangers.

Passive heat structures are usually metallic structures like the internals of reactor vessels or vessel walls or others in contact with the fluid of the analyzed system. Despite being passive, which means they do not generate heat, they must be accounted for because they affect certain scenarios for storing thermal energy during certain transient periods and releasing it later.

Electrical heaters, like pressurizer heaters, are important for energy balance and pressure control.

Heat losses to the environment are sometimes negligible, for instance in a full-size plant. This is not the case, however, in experimental facilities.

Finally, heat exchangers, mainly SGs, are crucial in many scenarios. Sometimes they are essential components for performing an emergency function, and sometimes they are components wherein the initiator takes place, as with an SG tube rupture.

3.3.2 Parameters and concepts

3.3.2.1 Related to power generation

Three parameters help in dealing with power generated or transferred: volumetric power, heat flux, and linear power.

Volumetric power (q''') is the thermal energy generation rate per unit of volume. Usually defined for the fuel pin, it is a key parameter in the general equation of heat conduction.

Heat flux (q'') is the rate of thermal energy transferred through a surface per unit of area. Usually defined for cladding surface, it is a key parameter to evaluate heat transfer by convection.

Linear power (q') is the thermal energy generation rate per unit of length of fuel rod. It is relevant in nuclear reactor safety statements.

3.3.2.2 Related to heat conduction

The general equation of heat conduction is derived by combining Fourier's law.

Heat flux $= -K(\text{temperature gradient})$, or

$$q'' = -K\nabla T \tag{3.11}$$

with an energy balance equation:

$$q'''(r,t) - \nabla q'' = \delta/\delta t(\rho c T) \tag{3.12}$$

$$q'''(r,t) - \nabla(-K\nabla(T(r,t))) = \delta/\delta t(\rho c T) \tag{3.13}$$

where K is the thermal conductivity and c is the specific heat of the solid where generation and conduction take place.

The medium is usually isotropic, so K is a scalar and in many cases only depends on temperature. If the medium is not isotropic, K becomes a tensor, and more complex options must be considered.

If the equation is applied to fuel, $q''' \neq 0$ and the general equation will be integrated after calculating, as follows.

At power, it will be supplied by the neutron physics formulation:

$$q'''(r,z) = |K_f \Sigma_f \Phi(r,z) \tag{3.14}$$

where $\Sigma_f \Phi$ is the fission rate at the r, z location, and K_f is the factor converting fission/s in W after a scram, either by a devoted table (like ANS79):

$$q''' / q_0''' = f(t_f, t_a) \tag{3.15}$$

where t_f = irradiation time and t_a = cooling time, or by a specific code to deal with decay heat.

If the equation is applied to the gap, $q''' = 0$ and the general equation will be integrated by considering it a region with a given thermal resistance. The evaluation of that resistance is rather complex and usually not performed online since it is based on results from a thermal-mechanical analysis of fuel rods as a function of fuel burn-up.

If the equation is applied to cladding, $q''' = 0$ and the general equation are integrated accordingly.

3.3.2.3 Related to heat convection—general aspects

Heat convection is a mechanism of heat extraction from a hot wall through a fluid with a certain movement. The mechanism also involves molecular diffusion and fluid movement that can be forced (forced convection) or free (natural convection). Heat flux extracted by convection follows Newton's law:

$$q'' = h_s(T_{\text{wall}} - T_{\text{bulk}}) \tag{3.16}$$

where h_s = convection heat transfer coefficient, T_{wall} = temperature of the hot wall (cladding), and T_{bulk} = average temperature of the fluid in the considered area.

Evaluating the convection heat transfer coefficient is rather complex and the object of important experimental activities. On the one hand, it depends on the convection mode (in some code manuals, it is ambiguously called the heat transfer mode), and on the other hand, its range of variation is very wide. Its units are W/(K m^2), and approximate values are typically between 35,000 and 55,000 in a PWR nominal steady state and up to 5000 in one-phase natural circulation or 50 or less in a dryout situation.

The convection mode depends on the mechanism of heat extraction. Besides the already mentioned conditions of movement and diffusion, another relevant aspect generates an additional difficulty for coefficient evaluation—the vicinity of saturation. Convection modes resulting from this vicinity have a certain relation or proximity to flow regimes but are essentially something different. The goal of the former is to calculate the heat transfer coefficient, while that of the latter is to allow detailed two-phase flow studies.

Establishing a correlation for each heat transfer coefficient from data produced by specific experiments is the culmination of the activities devoted to evaluating them.

3.3.2.4 Related to heat convection—pre–critical heat flux convection modes

Convection to a single-phase liquid is the basic convection mode that occurs when the bulk temperature is far from saturation. It takes place in PWRs at a nominal steady state for most fuel rods, primarily near core inlets for pins with medium or low linear power, as shown in Fig. 3.3.

If the bulk temperature is close to saturation (if ΔT_{SUB} is a few degrees K) and that of the wall is near saturation, bubbles could build up in contact with the hot wall and

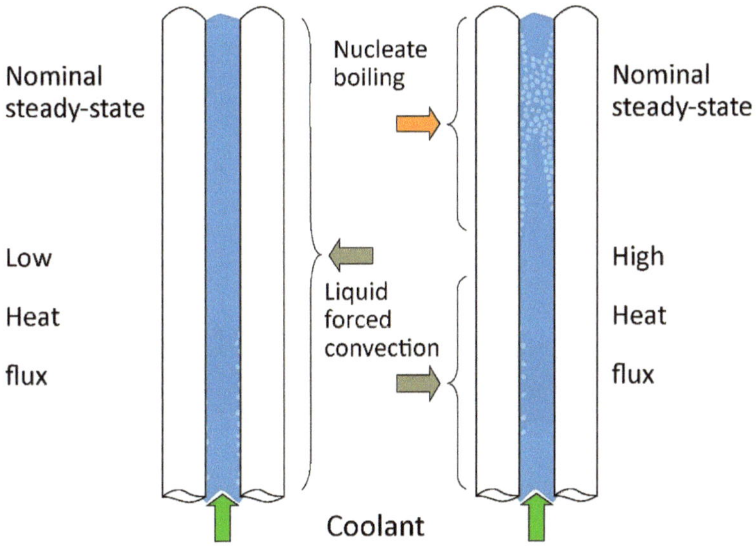

Figure 3.3 Convection heat transfer modes in a pressurized water reactor at nominal steady state. Notes from the lecture “First Approach to Reactor Case” F. Reventos. UPC Master of Nuclear Engineering.

extract necessary latent heat from it. Since the fluid is subcooled, the built-up bubble could easily mix with liquid water and aid general heat extraction. This mechanism occurs under certain conditions and is extremely positive since it helps heat transfer. Consequently, the wall temperature could stabilize at a lower value. We call this mechanism nucleate boiling (or subcooled nucleate boiling), with a heat transfer coefficient that is greater than in the previous situation. This convection mode takes place in PWRs for nominal steady-state fuel rods with higher linear powers, as shown in Fig. 3.4.

If the bulk temperature is at saturation (ΔT_{SUB} is 0) and the wall temperature is slightly above saturation, bubbles in contact will easily build up. Similarly, bubbles could easily mix with liquid water and aid general heat extraction. The mechanism is then called saturated nucleate boiling. This convection mode takes place in most fuel rods of BWRs at a nominal steady state and in accident scenario progressions.

3.3.2.5 Related to heat convection—critical heat flux/boiling crisis/post–critical heat flux convection modes

The progress of built-up bubbles at the hot wall and its subsequent coalescence produce a partial film covering its surface, resulting in a strong reduction in the heat transfer coefficient; see Arias (2009), Arias et al. (2009), Arias and Reventos (2010). Many authors define this phenomenon as a boiling crisis, with the value of the heat flux at which it appears being the CHF.

The phenomenon can take place under three circumstances. When the previous heat transfer mode is nucleate boiling at a subcooled bulk temperature, it is called a

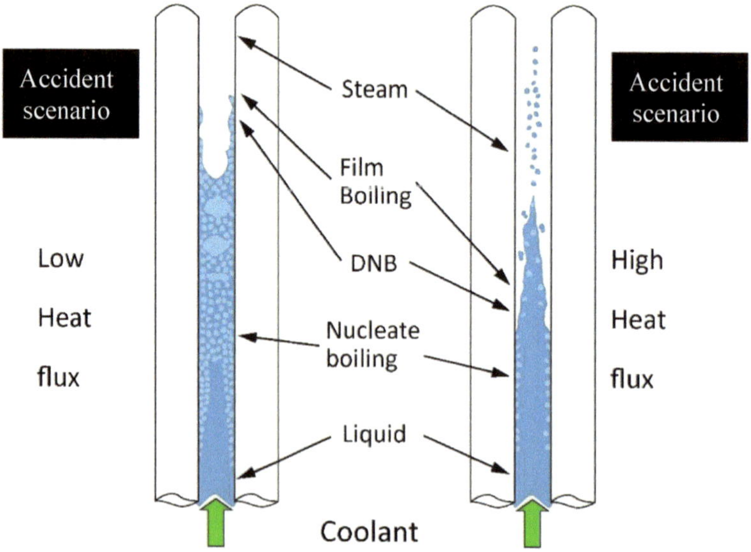

Figure 3.4 Convection heat transfer modes in an accident scenario. Notes of lecture "First Approach to Reactor Case" F. Reventos. UPC Master of Nuclear Engineering.

"departure from subcooled nucleate boiling" (DNB subcooled); when the previous heat transfer mode is nucleate boiling at a saturated bulk temperature, it is called a "departure from saturated nucleate boiling" (DNB saturated); and finally, at high liquid enthalpy, it is called "dryout."

This phenomenon has high relevance for safety. When it takes place, the sudden reduction in heat transfer could produce so-called burnout or physical destruction of the hot surface.

A tremendous number of experiments have clarified the physics of the phenomenon and characterized post-CHF heat transfer modes. Again, such experiments have been useful in establishing correlations and lookup tables that are valid for corresponding predictions.

3.3.2.6 Related to heat radiation

Interest in thermal hydraulics of nuclear systems in radiation heat transfer is limited to the need to compute the rate of radiation exchanged between hot surfaces in an advanced phase of an accident scenario assuming no participating media. Such interest is limited because disregarding the effect of heat radiation is usually slightly conservative and traditionally has been ignored. The surfaces that could cool down by heat radiation must have a line-of-sight or a reflection path through which they can communicate. The central problem is calculating the view factors.

3.4 Concepts involved in target phenomena

3.4.1 General aspects

In the progression of scientific thermal hydraulics devoted to nuclear safety, many phenomena have been identified as significant to understanding system behavior in accident scenarios. Each target phenomenon has been studied through analytical and experimental activities and is presented in corresponding chapters. A few have been selected and are summarized below, while a complete list can be found in chapter 6. The purpose of such a selection is connected to the basic goal of the current chapter, which is to cover relevant concepts.

3.4.2 Selected phenomena

3.4.2.1 Natural circulation

Any closed circuit heated at its lower part and cooled at its upper part could start natural circulation. The driving force of that natural circulation is the density difference between its cold and hot legs. Depending on the magnitude of the density difference and the pressure drop along the circuit, equilibrium can be reached among the involved parameters: power generated, natural circulation mass flow, and power extraction at the heat sink.

This well-known phenomenon, which is widely used in heat transfer technology in conventional heat exchangers and evaporators, appears in NPPs in two different fields. The first is related to safety scenarios analyzed for current water reactors—see D'Auria et al. (1991), IAEA (2005)—and the second is related to third- and fourth-generation reactors; see IAEA (2009). A thermal-hydraulic analyst of nuclear plants must be able to address both. In a nuclear system such as the primary loop of a PWR, the reactor is supposed to act as a thermal power-generating device and the SG as a heat sink.

3.4.2.2 Critical and choked flows

Increasing ΔP between the upstream and downstream of a valve or junction results in an increase in the mass flow. Once ΔP becomes higher, a limiting value is observed after which the mass flow rate no longer increases. That value is called critical flow, and the phenomenon occurs when fluid velocity reaches the speed of sound, which is physical because mechanical perturbations propagate at the speed of sound. Theoretical and empirical approaches are available to overcome the difficulty of trying to solve the momentum equation in such a situation.

This phenomenon is extremely significant in nuclear safety since water reactors operate at quite high pressures, so any break could produce a high ΔP between the reactor coolant system (RCS) (upstream of the break) and the reactor building space (downstream of the break). Accurately evaluating mass flow through the break is crucial for matching the correct phenomenology of the scenario.

The mass flow rate when ΔP takes on values falling within such a critical interval is also called choked flow (since it is limited or choked), and special models are needed to evaluate it (see Section 3.4.6).

3.4.2.3 Blowdown/reflood/rewet

Blowdown, reflood, and rewet are the three time phases expected with a large-break loss-of-coolant accident (LBLOCA); see USNRC (1987).

Blowdown is the first phase and takes place just after the break. In this period, a very high ΔP occurs between the RCS and the containment. For the entire time window, the break produces the driving force, and an important amount of coolant mass is lost through the break at a time while pressure and saturation temperature rapidly decrease and vapor is formed. During the blowdown phase, a quick dryout is produced, and heat transfer decreases quite a bit. Energy stored in the fuel results in a high peak temperature.

Reflood is the second phase in an LOCA scenario. It starts when emergency core coolant injection brings water to the core. During this period, coolant in the core is vertically stratified, the water level in the reactor vessel increases progressively, and heat transfer becomes a rather complex situation. On the one hand, heat transfer phenomena that appear are very different depending on their location, and on the other hand, axial conduction in fuel becomes rather important due to temperature gradient. Water level in the core, in this situation is called quench front, increases at a rate that is a function of injected mass flow. Under the quench front, coolant is dominantly liquid (low void fraction), and bubbles are formed since cladding is still hot. In this situation flow regime is probably bubbly and convection mode nucleate boiling either subcooled or saturated. Above the quench front, coolant is dominantly vapor (high void fraction), and droplets are dragged by vapor produced under the front. In that other situation flow regime is probably dispersed or mist and convection mode is post-CHF.

Finally, rewet is the last phase of an LOCA scenario. It starts when quench front reaches the upper part of the core. In this period, coolant is dominantly in liquid phase and heat transfer does not have any particular situation.

3.4.2.4 Reflux condenser mode

Reflux condenser mode is a cooling mechanism that could occur in a PWR usually in defined phases of small-break LOCA (SBLOCA) (Pretel, 1997). At a reduced inventory vapor is produced in the core and flows through the hot leg to SG U-tubes where it can condense, fall back to the vessel, and help cool the core. The secondary side of the SG must be active (or at least full of water) to provide the condensing capability. In addition, during such cooling mode countercurrent flow occurs in the hot leg since vapor flows forward to the SG and liquid flows backward to the vessel.

3.4.2.5 Loop seal clearing

In the progression of an SBLOCA scenario, vapor production is due to depressurization and heat addition from core decay power (Pretel, 1997). Such vapor could fill the upper parts of the primary side. More in detail, two different regions comprise such upper parts: on the one hand upper plenum, hot legs, pressurizer, and SG U-tubes and, on the other hand, cold legs and downcomer upper ring. Liquid water accumulates in the lower parts of the primary system especially reactor vessel and intermediate leg or loop seal region. While vapor pressure in the upper head remains smaller than the compensating water column, loop seal water blocks its path and avoids the merging of water regions. During this interval, core uncovering could occur. In the event of a pressure increase (due to additional vaporization), loop seal water could be cleared or displaced to the cold leg and from there to the vessel. We call this phenomenon "loop seal clearing." The above description is valid for scenarios where reactor coolant pumps (RCPs) are stopped (due to loss of electric power supply, or operator action). TMI-2 event (see Section 16.3.1.1) illustrated the influence and phenomena associated with RCPs running for long periods during an SBLOCA event. There is less or no stratification in the RCS compared with the situation of a stopped RCP, no loop seal formation/clearing, impact on break flow quality, and RCS residual inventory in case of late RCP stopping.

3.4.2.6 Steam binding

In the progression of an LOCA scenario, secondary pressure could become higher than primary pressure. This can happen after a loop seal clearing when vapor is the dominant phase of the fluid that fills the upper part of the primary side including SG U-tubes. In this situation, such primary fluid could receive heat from the secondary side and from hot passive heat structures. Consequently, void fraction and primary pressure could increase, and this could produce difficulties in existing natural circulation or the core level recover. We call this phenomenon "steam binding."

The above mention of loop seal clearing (see Section 3.4.2.5) rather refers to an SBLOCA scenario; however, steam binding is more often associated with the LBLOCA scenario. Namely, during the reflooding phase of LBLOCA, when RCS pressure is nearly equal to containment pressure, i.e., a few bars, with T_{sat} of 120°C−140°C, steam exiting the quench front in the core progressively superheats up to the top of the core reaching some hundreds of centigrade, while droplets of entrained liquid leave the quench front at T_{sat}. During their transit throughout the RCS toward the break, these droplets may partly evaporate due to contact with superheated steam, but more important is their evaporation due to important heat transfer area in the SG tubes merged in secondary side water still at high temperature, close to the initial SG secondary side temperature. The SG then becomes a heat source for this droplet liquid flow in the RCS. The additional steam production due to their evaporation dramatically increases the steam binding effect, and more pressure drops in the RCS loop due to higher steam flow, which reduces the reflooding progression, given the driving head provided by the difference of elevation between downcomer liquid level and core mixture elevation.

3.4.2.7 Boron dilution accidents

Boric acid is used as a soluble neutron absorber in the primary coolant of PWRs. In normal reactor operation, the required boron concentration is maintained by the volume control system. The event by which the boron concentration in the core is accidentally reduced and initiates a reactivity excursion is called a boron dilution accident; see Freixa (2007), Freixa et al. (2007, 2009a):

- The most useful classification of boron dilution transients considers external boron dilution. In such transients, unborated water from outside the primary cooling system is injected.
- In inherent boron dilution transients, the primary coolant is separated into highly borated water and nearly boron-free water. This process occurs through evaporation and condensation.

Thermal-hydraulic analysis is greatly involved with boron dilution transients. Both the buildup of a nonborated slug in the primary system and its transport to the core are connected to thermal-hydraulic phenomena. When the level in the reactor pressure vessel is sufficiently low and the core temperature is sufficient to boil the coolant in that region, reflux-condensation conditions are established. Under such conditions, produced steam condenses in SG tubes, and nearly boron-free water collects within a region of the primary system (usually in the loop seal) to form a nonborated slug. If natural circulation is later established or an RCP is restarted, the slug can be transported to the core and produce a reactivity excursion.

Experiments related to boron dilution scenarios have been carried out over the last decade using the framework of two Organisation for Economic Co-operation and Development (OECD) projects, OECD-SETH (2001−04), OECD/NEA/CSNI

(2004), and OECD-PKL (2004–07) (e.g., Mull et al., 2007; Umminger et al., 2012). Tests were performed at the PKL facility and covered the most relevant scenarios. Analytical activities were also included in those projects; see Freixa et al. (2009b), Reventós et al. (2008b).

3.4.2.8 Boron dilution/deboration

Boron dilution is the process where a local decrease in boron concentration occurs within the primary system. Deboration is the process where a net loss of boron from the primary system occurs, which implies a decrease in average boron concentration. In the same way, the following definitions are also accepted: boration (the opposite of deboration), boron concentration recovery (the opposite of boron dilution), deboration rate (the rate of deboration), and dilution rate (the rate of boron dilution). These definitions were proposed by D'Auria et al. (2005), and are accepted nowadays within the international community.

3.4.2.9 Countercurrent flow limitation

Gas and liquid phases could flow in a pipe or an orifice, and in some specific conditions, in opposite senses. This is called countercurrent flow. The most intuitive situation where this occurs is a vertical pipe with an upward low flow of gas concurrent with a downward and low flow of liquid. The flow of gas is driven by the required delta-p, while the downward flow of liquid is gravity-driven. It is intuitive that the upward flow of gas increases in the described situation and may prevent liquid fallback. This phenomenon is called the countercurrent flow limitation (CCFL); it also takes place in horizontal pipes with stratified two-phase flow.

CCFL is relevant for nuclear safety thermal hydraulics since it can occur at various phases under several accident scenarios and various locations within plant circuitsj, like hot and cold legs, downcomers, pressurizer surge lines, or orifices, such as those within the upper tie plate or channel inlet of a BWR.

This phenomenon has been studied with experiments at various scales and in a variety of facilities. Special models and correlations are available to address CCFL.

3.5 Analytical tools

3.5.1 General aspects

This Section 3.3 introduces and defines various concepts that are analytical tools originating from pioneering investigations, e.g., by Sha et al. (1976). The goal of the first subsection is to present available approximations. The second subsection centers on system codes as the most used option. Finally, the third subsection introduces relevant parameters and dimensionless numbers that are useful in current analyses.

3.5.2 Models and approximations

3.5.2.1 Homogeneous model

Solving two-phase flow equations is complex. Various simplified approximations are now available, includinghomogeneous, separated flow, and drift-flux models. For the homogeneous model, the key assumptions are the following:

- Liquid and vapor have equal velocities.
- Liquid and vapor are in thermodynamic equilibrium.
- A single friction factor suitable for the mixture is used.

The model can produce suitable results both when differences between the properties of the liquid and vapor are unimportant (in the vicinity of the critical point) and when the mass velocity of the two-phase flow is very high and the flow regime has a flow that is either bubbly or misty.

3.5.2.2 Separated flow model

The key assumptions of the separated flow model are the following:

- Liquid and vapor velocities are not necessarily equal.
- Liquid and vapor are in thermodynamic equilibrium.
- An empirically correlated friction factor multiplier is used.

The separated flow model has been successfully used for calculating the pressure drop in flow channels. As a general statement, this model fits better at low mass velocity.

3.5.2.3 Drift-flux model

The drift-flux model is built on the concepts of drift velocity (the relative velocity between phases) and drift flux (the volumetric rate at which vapor passes forward or backward through the unit of area normal to the traveling flow).

The drift-flux model is currently used for various codes in the thermal hydraulics of nuclear systems and other industries. It helps analyze system behavior for distinct transient conditions. Simplicity and robustness are its most significant advantages.

3.5.2.4 Lumped parameter models

Generally, a lumped parameter model is built by replacing the behavior of a specific parameter from a more detailed model with a tested correlation or logic. Its use is justified when running the detailed code is time-consuming. The validity of the approach depends on the correlation or logic involved.

3.5.2.5 One-dimensional/three-dimensional

The decision to use a 1D or 3D option for a given calculation is rarely straightforward. Three-dimensional options are generally better because they have more detail, but sometimes this is not the case because they do not add specific relevant output to the established engineering goals or their validation process is incomplete and produces uncertain results.

Heat transfer in nuclear fuel is a case where the 1D option works properly in many significant scenarios. Given the geometry and dimensions of a fuel rod, it is quite correct to assume that heat is transferred in a radial direction and that almost no axial transfer occurs. This statement is valid for most involved scenarios. Reflood situations are different, however. During reflood, cold water fills only the lower part of the vessel, while most of the core is noncooled and at a high temperature. The axial temperature gradient is high, and axial heat conduction is considerable, so a 3D option is required.

Whether to use a 1D or 3D option is determined case by case.

3.5.2.6 Two-fluid model

State-of-the-art knowledge about thermal-hydraulic two-phase flow predictions currently resides on a level between the approximations depicted above and the detailed simulations of CFD codes.

On the one hand, basic models (like homogeneous, separated flow, drift-flux, and even some lumped parameter models) are good tools for helping us understand phenomena and calculate specific scenarios, but they provide insufficient detail for many situations. On the other hand, CFD studies based on more advanced physical models, nodalizations, and numerical methods are appropriate for certain aspects but have two main drawbacks: high computing effort is required, and their validation process is not fully complete for typical nuclear safety scenarios.

Nowadays, the most commonly used option is the so-called two-fluid model, which other authors sometimes refer to as the "flow pattern model" or even the *UVUT* model (where $U=$ unequal, $V=$ velocity, $U=$ unequal, and $T=$ temperature). This approach allows unequal velocities and temperatures for the two phases and a more accurate picture of the scenario's thermal-hydraulic behavior to be built. This approach is followed in most system codes used today.

To summarize, since so-called system codes are the main repository of expertise for thermal hydraulics in nuclear systems, they provide a reliable picture of current options in the area. Such codes usually employ quite complex techniques represented by nonhomogeneous and nonequilibrium two-phase flow models. Their two-fluid equations are formulated using volume- and time-averaged flow parameters. Friction and heat transfer are formulated from bulk properties and empirical heat transfer coefficients. Flow regime information is supported by empirical results and accurate implementation.

3.5.3 Concepts related to system codes

3.5.3.1 General aspects

System codes can solve the six balance equations of mass, momentum, and energy (for liquid and steam in each case). Such a resolution is supplemented by a suitable set of closure laws. Equations are coupled with heat transfer models and sometimes with neutron kinetics coding. The two-phase flow field is organized into several control volumes connected by junctions. Thermal-hydraulic components such as separators, valves, accumulators, or pumps are coded to complete the simulation of the described system.

Typically, system codes follow a finite difference method based on staggered grids to avoid the checkerboard problem (artificial pressure oscillations). Depending on the code, many options provide temporal discretization. Options are chosen to satisfy an acceptable compromise between a relatively large time step and computational cost savings. Spatial discretization is typically defined as a combination of the central difference and the upwind (donor-like) numerical schemes to guarantee the maximum stability of the simulation.

Information describing the plant or facility is read by the code in a file called the "input deck." The initial and boundary conditions of the scenario to be analyzed, as well as nodalization characteristics, are also included in the input deck.

The results produced by code runs can be written to an ASCII or binary file. The former can be directly accessed, while the latter can be used to produce time trends for key parameters or calculation results.

RELAP5 (developed by the USNRC and USDOE), CATHARE (developed by the French CEA), Trace (developed by NRC), and ATHLET (developed by the German GRS) are the most used system codes.

3.5.3.2 Closure laws or constitutive equations

Closure laws or constitutive equations are the set of equations that handle mass, momentum, and energy exchanges between phases and between each phase and the walls of the involved control volumes. Such laws, which are rigorously necessary for solving the whole set of equations, are usually obtained empirically and based on separate effects experiments, which are typically designed specifically for that purpose; see Bestion (1990).

Heat and mass transfer, along with friction at the interface and from (or to) the wall, are the most relevant closure laws.

3.5.3.3 Heat transfer correlations/lookup tables

Empirical information, usually about heat transfer, is provided to the code through correlations or lookup tables. Correlations exist for many important parameters in physical models. Technical literature supports and validates this information. This is the case for work by Dittus and Boelter (1930) on correlations for heat transfer under forced convection and for correlations like those discussed by Kays (1975), by Chen et al. (1977), or Chen-Sundaram-Ozkaynak, and by Bromley (1950), and Nusselt, Shah, and Colburn-Hougen. Lookup tables are used with system codes to supply the values of empirical information. Their fit is better for the more complex functions of multiple arguments. Such is the case for the Groeneveld lookup table in CHF evaluation; see Groeneveld (1973), Groeneveld et al. (1986, 1996).

3.5.3.4 Special models

System codes are used when incorporating special models. Such models are called by the main code when needed and interact with the equations to be solved to complete the description of the simulated scenario. Due to their relevance, the following are briefly mentioned or depicted: choked flow, user-supplied hydrodynamic losses, abrupt area change, and CCFL.

Most system codes include a choked flow model as a special model. Such a model predicts whether choked flow occurs at the considered junction (usually a break or a nozzle) and allows for the calculation of expected mass flow.

The issue of form hydrodynamic losses in junctions is related to what was noted in the section devoted to pressure drop and is needed for some flexibility. To make such flexibility available, most system codes provide the option of a general user-specified form loss coefficient. User-specified coefficients can be introduced and used by the code. This special model provides the user with the possibility of applying form coefficients taken directly from experiments, derived from user calculations, or adapted from detailed codes.

Another special model, the so-called abrupt area change model, is related to the previous one. Quite often, the code user can connect two nodes of different cross-areas in a junction with a sharp or abrupt edge. In such sudden enlargement or contraction of the equivalent pipe, the known vena contracta effect will occur. This special model is often implemented to avoid repetitive calculations of the corresponding coefficients.

CCFL, as briefly introduced above, is a phenomenon that occurs in junctions with a partially closed structure under conditions of an ascending developed vapor flow and a countercurrent fallback liquid. Such structures could include orifices, the upper core tie plate (Glaeser, 1989) SG tube support plates, or the entrance to the tube sheet in the SG at the inlet plenum. At low vapor mass flow rate values, both countercurrent flows are perfectly compatible. Increases in vapor mass flow, however, could limit or even stop the gravity drainage of liquid.

The special models typically implemented with system codes provide acceptable results if the user supplies empirically determined coefficients that are specific for the involved detailed geometry.

3.5.3.5 Special components

System codes are used for the incorporation of special components. Such components include control volumes or junctions with specific additional conditions that interact with the equations to be solved. Such specific conditions depend on the component itself. Examples are provided below.

An SG separator works by providing centrifugal separation of liquid and vapor. Some system codes include a nonmechanistic model that consists of a special volume with junction flows determined as a function of void fraction. At a high void fraction, the vapor junction ejects only vapor; at a medium or low void fraction, an empirical curve provides the results.

Valves are special junctions. They can be classified as various types, including those that open or close in a time step or that open or close following the value of a given variable. Both kinds are driven by selected variables that connect hydrodynamics with protection and controls.

Accumulators are modeled in system codes as lumped-parameter components. Usually, a simple ideal gas equation of state simulates gas behavior during the discharge.

Pump components are also a special volume. The pump model connects with the two-fluid hydrodynamic model through the head term appearing in the momentum equation. Homologous curves provide (empirical) information about pump behavior.

3.5.3.6 Nodalization/nodalization diagram

Spatial discretization of the two-phase flow field in which equations are solved is organized in control volumes connected by junctions and is called nodalization. Sometimes the meaning of nodalization is extended to the entire set of parameters and user-supplied information.

A nodalization diagram is a drawing or illustration that shows control volumes and their connecting junctions. Such diagrams, when they provide sufficient detail and are set properly, are quite useful for understanding, at first glance, the involved calculation.

3.5.3.7 User effect/good practices

Aksan et al. (1993) proposed the characterization of user effect, following an early investigation by D'Auria and Galassi (1990c).

Later, in 1998, it was defined by OECD/Nuclear Energy Agency (NEA) as follows: "User effects are any differences in calculations that use the same code version and the same specifications for a given plant or facility." Ashley et al. (1998) in a document providing such a definition, analyzed the primary reasons behind the differing appearances and established a series of "good practices" devoted to avoiding or minimizing the user effect. Good practices are based on user training, as noted by D'Auria (1998), Petruzzi et al. (2008) user discipline, quality assurance, and in a wider context, organizational requirements and safety culture.

3.5.4 Additional concepts involved in analytical thermal hydraulics

3.5.4.1 Hydraulic diameter/Reynolds and Froude numbers

Hydraulic diameter (D_h) is defined as

$$D_h = 4A/p \tag{3.17}$$

where A is the cross-sectional area and p is the wetted perimeter of the cross section.

This equation helps handle flow in noncylindrical tubes and channels. Using it, one can calculate flow and heat transfer conditions in the same way as for a cylindrical tube.

The most used dimensionless numbers in two-phase flow are the Reynolds number (*Re*) and the Froude number (*Fr*).

The Reynolds number (*Re*) is defined as

$$Re = \rho \times v \times D_h / \mu \tag{3.18}$$

The Reynolds number is the ratio between inertia forces and viscous forces. When viscous forces are dominant, *Re* has low values and flow becomes laminar; when inertia forces are dominant, *Re* has higher values and flow becomes turbulent.

The Froude number (*Fr*) is defined as

$$Fr = v / \sqrt{(g \times L)} \tag{3.19}$$

The Froude number is the ratio between inertia forces and gravity forces. The Froude number is crucial when determining the transition to stratified flows. When gravity forces are dominant, the flow becomes stratified.

3.5.4.2 Prandtl and Nüsselt numbers

Besides the Reynolds and Froude numbers just introduced, the most used dimensionless numbers in heat transfer developments are the Prandtl number (*Pr*) and the Nüsselt number (*Nu*).

The Prandtl number (*Pr*) is defined as

$$Pr = \mu \times C_e \,/\, k \tag{3.20}$$

It is the ratio between momentum diffusion and heat diffusion.

The Nüsselt number (*Nü*) is defined as

$$N\ddot{u} = h_s \times D_h / k \tag{3.21}$$

It is the ratio between convective and conduction heat transfer.

3.5.4.3 Nuclear factors

Nuclear factors are ratios between parameters related to power generation. They are useful for understanding radial and axial power distortion and helpful in producing a quick estimation of heat transfer in different areas of the reactor.

The radial nuclear hot channel factor (F_R^N) is defined as the ratio between the average heat flux of the hot channel and the average heat flux of the whole core. A typical value for such a factor at the beginning of the cycle in a PWR is about 1.5. At the end of the cycle, the value decreases. Similarly, the radial nuclear channel (i) factor is the ratio between the average heat flux of the channel (i) and the average heat flux of the whole core.

The axial nuclear hot channel factor (F_Z^N) is defined as the ratio between the maximum heat flux of the hot channel and the average heat flux of the hot channel. A typical value for such a factor at the beginning of the cycle in a PWR is about 1.5. At the end of the cycle, the value decreases.

Similarly, the axial nuclear channel (i) factor is the ratio between the maximum heat flux of the channel (i) and the average heat flux of the channel (i).

The total nuclear heat flux factor is defined as the ratio between the maximum heat flux of the hot channel and the average heat flux of the whole core. A typical value of such a factor at the beginning of the cycle in a PWR can be about 2.25 (obtained as 1.5 × 1.5). At the end of the cycle, its value decreases.

3.5.4.4 Engineering factors

Engineering factors are ratios between parameters related to two significant engineering concepts, tolerances and cross-flow paths. They are useful for quickly estimating the impact of such phenomena. Engineering tolerances are a source of uncertainty with an impact on heat transfer; the "engineering tolerances factor" (F_Q^E) is a simple figure that accounts for this. A typical value for F_Q^E is 1.03. Such a factor helps compute the so-called total power peaking factor, which has a typical value of 2.32. Based on these definitions,

$$F_Q^E = (\text{Engineering tolerances factor}) \text{ and}$$

$$F_Q = (\text{Total power peaking factor}), \text{ we have}$$

$$F_Q = F_Q^N \times F_Q^E \tag{3.22}$$

The "enthalpy-rise hot channel factor" accounts for the effect of cross-flow paths in a PWR core. The mass interchange between parallel channels results usually in smoothing enthalpy rise. In an ideal situation assuming that coolant would be strictly associated with its fuel rod, $F_{\Delta H}$ and F_R^N are identical. In practice, cross-flows produce a mixture among channels leading to moderate values of ΔH, and then, $F_{\Delta H}$ is slightly lower than F_R^N. Typical values are 1.53 ($F_{\Delta H}$) and 1.56 (F_R^N), where the following definitions apply:

$$F_{\Delta H} = \Delta\text{Enthalpy}_{\text{Max}} / \Delta\text{Enthalpy}_{\text{Average}} \quad \text{and}$$

$$F_{\Delta H_i} = \Delta\text{Enthalpy}_i / \Delta\text{Enthalpy}_{\text{Average}}$$

3.6 Verification and validation

3.6.1 Definitions

V&V concepts are used in many engineering fields. As a first approach, such concepts can be applied to any product, service, or system. In this context, to verify is to check that the product, service, or system is built as intended, and to validate is to check whether it works.

In the context of thermal hydraulics (D'Auria et al., 2012; IAEA, 2015) model verification is the process of confirming that the model is correctly implemented with respect to its theoretical principles and equations. Similarly, model validation is the process of checking the model's accuracy in representing the real system.

In the context of thermal hydraulics, V&V can be applied to nodalizations (models or input decks); see Bonuccelli et al. (1993), Reventós et al. (2007a).

The verification of a nodalization is the process of confirming that the nodalization is implemented using documentation that properly describes the facility or plant. Similarly, validation of a nodalization is the process of checking its accuracy in representing the dynamic behavior of the real plant. Methodologies exist to guide such processes.

3.6.2 Concepts involved in experimental thermal hydraulics

3.6.2.1 Separate effects tests

Each separate effects test (SET) facility has been designed to investigate the local occurrence of a particular phenomenon. Usually, such facilities physically represent either relevant zones of the NPP or geometry in which the studied phenomenon can be identified and studied. The results of performed experiments are used to validate the models created to describe the studied phenomenon. The final goal of the process is to include the model in the system code.

The SET Validation Matrix is a document by OECD/NEA/CSNI (1993) established under the auspices of the OECD to identify phenomena covered by the matrix and provide an outline of existing facilities and test types. The document shows a cross-reference matrix considering the relationship between phenomena, the test type, and the facility.

3.6.2.2 Integral effects tests/integral test facilities

Each integral effects test (IET) performed in an integral test facility (ITF) has been designed to investigate the behavior of its reference plant under accident conditions. All the significant components and systems of the involved NPP are usually reproduced at a lower scale. For this reason, scaling considerations must be considered to ensure the suitability of results.

The expected results are the responses of the plant to simulated initiating events. The results are also used to improve models or check code capabilities.

The IET Validation Matrix is a document established under the auspices of the OECD to identify scenarios covered by the matrix and provide an outline of existing facilities and test types.

The PKL facility in Erlangen (Germany), LSTF in Tokai-Mura (Japan), and ATLAS in Daejeon (Korea) are three relevant ITFs that are currently in operation.

The LOBI facility in Ispra (Italy), LOFT in Idaho Falls (Idaho, USA), and BETHSY in Grenoble (France) were ITFs operating in the past that are now dismantled.

3.6.2.3 Analytical support to experimental thermal hydraulics

Analytical support for experimental thermal hydraulics (Umminger et al., 2012) is based on calculations simulating the tests. Pretest calculations are usually performed before running the test to complete test preparation or clarify uncertain aspects of what is expected.

Posttest calculations are performed following the availability of experimental data (e.g., Reventòs et al., 2012). The results obtained are compared with experimental data, and the comparison is useful in code and nodalization validation. A posttest can be open or blind. Both are helpful for validation, and blind calculations have an additional outcome since they show the actual capabilities of predictive analysis.

3.6.2.4 Databases of experimental thermal hydraulics

The importance of experimental data production for code validation is clearly accepted. Test result databases must be preserved and properly organized. Data must be accessed and easily retrieved by the analysts who use them; see Addabbo et al. (2002).

The information stored in such databases includes files containing experimental data for various timeframes (ASCII or binary formats) to facilitate comparison between predictions and test results. In addition, typical documents related to experiments, among them the well-known Quick-Look Report and Experimental Data Report, complete the information associated with each test.

The Storage of Thermal Reactor Safety Analysis Data (STRESA) web-based informatics platform is a good example of how such a database can be implemented (Pla et al., 2015, 2016b). In its current version, STRESA forms a network of nodes and stores data from an important number of relevant thermal-hydraulic tests.

3.6.2.5 International standard problems

International standard problems are comparative exercises in which the predictions or recalculations of a given physical problem with different best-estimate computer code are compared with each other and above all with the results of a carefully specified experimental study.

Such exercises have been organized within the OECD/NEA framework since 1975 to increase confidence in the validity and use of the different tools used in assessing the safety of nuclear installations. Extensive information is available. It includes reports on particular exercises (Kimber et al., 1998), a compendium of groups of exercises (OECD/NEA/CSNI, 1997a), and articles on the results of revisiting any exercise with updated codes and models; see Freixa et al. (2015a).

3.6.3 Concepts involved with scaling

3.6.3.1 Scaling/scaling issue/addressing the scaling issue

Scaling in nuclear thermal hydraulics is the topic of an important document completed in 2016 by a Specialist Scaling Group (SSG) formed in 2013 by the Working Group on Analysis and Management of Accidents; see OECD/NEA/CSNI (2016). The document testifies not only to the importance of scaling in nuclear technology but also to controversial evaluations of scaling-related findings by the scientific community.

"Scaling," "scaling issue," and "addressing the scaling issue" indicate actions, methods, and techniques aimed at connecting the parameter values related to experiments with NPP conditions. Scaling is the process of demonstrating the applicability of any parameter value to the reactor conditions. Addressing the scaling issue refers to the process of demonstrating the applicability of data from models to prototypes:

- A general guideline for addressing the scaling issue can be found in the USNRC code scaling, applicability, and uncertainty method (CSAU) (USNRC, 1989a). It is an extensive work aimed at providing a suitable framework for uncertainty and accuracy analyses using best-estimate codes. This methodology was presented as a response to the revision of the 10 CFR part 50 emergency core cooling system (ECCS) rule (USNRC, 2015). Under this approach, acceptance criteria can be evaluated using realistic codes under certain conditions. Literally, uncertainties in the analysis method and inputs must be identified and assessed so that uncertainty in calculated results can be estimated.
- There is a high level of probability that the criteria will not be exceeded.
- The document faces the scaling issue, emphasizing the need to consider two main aspects to evaluate the capability of BE codes to scale up processes from reduced-scale test facilities to full-scale NPPs.
- To quantify the effects of scale distortion or a limited database on code uncertainty to calculate a safety parameter of interest, for example, peak clad temperature.

Related to these concepts, several scaling criteria and scaling analysis methodologies can be found in the technical literature. Regarding scaling criteria and the design of experimental facilities, one can find terms like linear scaling, power-to-volume scaling, and modified time-preserving scaling. Regarding scaling analysis and experimental data extrapolation, hierarchical two-tiered scaling, Ishii three-level scaling, and fractional scaling analysis (FSA) are the most common methodologies. Some techniques are described below.

3.6.3.2 Power-to-volume scaling

The power-to-volume scaling (or volume scaling) method conserves time, length (or height), velocity, and heat flux equivalent to those of the prototype. A test facility designed following power-to-volume scaling has a full-height scale, while the area and volume of the facility are reduced to the same scale. The experiments involved also conserve pressure, temperature, and time.

3.6.3.3 Hierarchical two-tiered scaling

Hierarchical two-tiered scaling (H2TS) is a comprehensive and systematic yet practical, auditable, and traceable methodology for evaluating the applicability of experimental data from scaled-down facilities to NPP conditions, and hence, implicitly, the capability of the system codes and their special models (because they are qualified just at the facility level) to scale up TH data to full-scale plant conditions. It is based on the premise that each TH process occurring within a system can be characterized by three parameters—volumetric concentrations, transfer areas, and timescales—and all of them can be included just in the timescale parameter dependencies. Hence, TH process relevance can be determined with just one parameter. H2TS is divided into two levels of analysis: the "top-down system scaling analysis" and the "bottom-up process scaling analysis."

3.6.3.4 Ishii three-level scaling

The Ishii three-level scaling approach is a systematic methodology for designing IET facilities and analyzing their generated experimental data relative to prototype conditions. The bases of the methodology are conservation principles and constitutive laws. This approach is divided into three levels of analysis: integral system scaling, control volume scaling, and local phenomena scaling. The first two levels correspond to top-down scaling, and the third level represents bottom-up scaling.

3.6.3.5 Fractional scaling analysis

FSA is a quantitative methodology based on the H2TS approach that has been developed to facilitate two aspects:

- Scaling up of time-dependent evolution processes involving an aggregate of interacting modules and processes (such as an NPP)
- Integration and organization of information and data of interest to NPP design and safety analyses

3.6.3.6 Scaling distortion

Scaling distortion refers to any discrepancy between the scaled and the referenced plant parameters. In a perfectly scaled experiment, all the scaled parameters are equally reproduced at the intended scaled time. Only in very limited cases, like very specific SETs, is perfect scaling perfectly feasible. Usually, tests are rather complex and involve phenomena that require diverse scaling criteria for the design of the scaled-down facility.

Following the Buckingham π theorem, equations and laws can be written in nondimensional form. A long list of nondimensional groups can be generated to define similarity conditions. Nevertheless, not all of them can be matched simultaneously in the design of scaled-down facilities, and distortions will appear once the analyst or designer decides which are the most relevant to preserve. Analysis and justification of such distortions are some of the most significant purposes of current scaling developments.

3.6.3.7 Counterpart tests/similar/special counterparts

Counterpart (CP) tests are tests performed in IET facilities at different scales but with the same initial and boundary conditions. Their purpose is to study the effects of scale on the behavior of the plant. If tests are properly designed and facilities have close configurations, similarities in the time trends of important parameters will appear, and the sources of distortions can be identified and explained. Such distortions come primarily from differences in configuration; see Martínez-Quiroga (2014), Martinez-Quiroga et al. (2014), Martinez-Quiroga and Reventos (2014).

Similar tests refer to tests that, although not exactly meeting the conditions of a CP test, are greatly similar in their initial and boundary conditions. Similar tests are at least useful to corroborate the similarities expected in the time trends of the main parameters.

A special CP test is a scaled-down SET designed with full-scale test data as a reference and has the goal of clarifying the same phenomena and processes. A special CP test may record significant phenomena in greater detail.

3.6.3.8 Using system codes in scaling analysis

System codes can help at every step in the analysis of complex transients. As discussed in OECD/NEA/CSNI (2016): "*System codes can reduce the cost of manpower by quickly doing evaluations that are done 'by hand' in H2TS and FSA. System codes are not an alternative to scaling analysis, but they are a tool for assisting scaling analysis and solving problems. For example, the identification of the phases of a transient may be easier after some preliminary simulations of the transient using system code, even before having simulated the transient with an appropriate IET.*"

System codes can also identify (or help identify) dominant processes and predict how the relative importance of each process may change all along the transient. They open the possibility of investigating phenomena that may be of second-order importance but require some attention.

3.6.3.9 K_V-scaled calculations

A K_V-scaled calculation is a system code simulation in which defined IET conditions are scaled up to an NPP nodalization to reproduce the same scenario. It allows the behavior of the NPP and ITF nodalizations to be compared under the same conditions. It may then be used to check the validity of the NPP nodalization and improve it if needed.

In the framework of qualifying the NPP nodalization and meeting its quality assurance procedures, K_V-scaled calculations, introduced by D'Auria et al. (1995), are of great value because they allow testing of their capabilities and plant responses under accident conditions. One of the important points of this activity relies on calculating the NPP scaling factor (the K_v-factor), which is commonly computed as the ratio between the volume of the primary liquid of the NPP and the ITF. In addition, K_v-scaled calculations are also of great value for extrapolating code accuracy, as described in the UMAE methodology. Many examples are available (e.g., D'Auria and Galassi, 2010).

Finally, the UPC scaling-up methodology (Martinez-Quiroga and Reventos, 2014; Martinez-Quiroga et al., 2014; Martínez-Quiroga, 2014) further established the development of K_v-scaled calculations. This method is a systematic approach for qualifying NPP nodalizations by extrapolating ITF posttest simulations. It includes defining and using concepts like "hybrid nodalization" and "scaled-up nodalization."

Both concepts are necessary for justifying the qualitative discrepancies related to K_v-scaled calculations. "Scaled-up nodalizations" allow for the determination of scaling effects related to the criterion applied in ITF design, and "hybrid nodalizations" help the user establish how design differences change experimental results. To carry out these calculations, specific software has been developed; see Martinez-Quiroga (2014).

3.7 Concepts connected to design and licensing bases

3.7.1 Structures, systems, and components

"Structures," "systems," and "components" (SSCs) are well-known concepts that are widely used in current engineering and design terminology. In this sense, no definition is needed for each concept, but often design or safety requirements are specified that apply to all of them. This is the case with the IAEA Specific Safety Guide No. SSG-30 (IAEA, 2014c) which establishes a set of requirements for all SSCs that are important for safety.

3.7.2 Acceptance criteria/design criteria

Both wordings refer to the same essential functional and safety requirements of the design.

The words acceptance criterion and acceptance criteria are preferred when pointing out a statement that establishes the conditions that the designed SSC must fulfill. It is used and defined in the DB.

The words design criterion and design criteria are mostly used in legislation, standards, and the licensing basis (see below). The NRC General Design Criteria established in Appendix A of 10 CFR 50 are a good example of general design criteria, whereas the velocities of fluids in a pipe as established in an engineering handbook are a good example of detailed design criteria.

3.7.3 Design basis

The DB is defined as the set of conditions, needs, and requirements (including legal requirements) considered when designing a facility or product. The concept can be applied to a particular SSC or an entire plant. The DB establishes the essentials of the engineering procedures needed to design the plant.

The DB of SSC is an engineering document (or part of a document) used to correlate the involved SSC with the scenario/s used in its design along with the criterion or criteria that will be considered in the corresponding acceptance.

Such scenarios and acceptance criteria are usually supported by standards, regulations, or engineering handbooks. For brand-new or very specific SSCs, the supporting information could reflect other particular data.

In design development that follows the DB, a calculation or engineering judgment must demonstrate that the acceptance criteria are fulfilled in each scenario for a given dimensioning of the involved SSCs. The DB is or must be an adequate and self-standing guide for design activity. The DB must be the central guideline in design. For this reason, it must be a concise document.

Additionally, the DB must classify SSCs and establish whether they are safety-related. It also facilitates design verification.

3.7.4 Licensing basis

A licensing basis is a complete and comprehensive set of regulations in force from which DB statements are retrieved. It is complemented by the set of terms and conditions imposed on the license and regulatory commitments adopted. It provides general criteria, scenarios, detailed criteria, standards, and codes.

The licensing basis is generic to an important extent and only plant-specific in the results of commitments and complementary terms. Differently, DB is plant-specific but also includes more general statements defined on a licensing basis. Both terms are often mixed in technical literature, but such an error is not a strong inconvenience.

The licensing basis is strongly connected with safety regulations, and the involved scenarios are design basis accidents (DBAs). If all considered safety-related SSCs during assigned accident scenarios are fulfilled, the reactor can be licensed.

3.7.5 Quality attributes/availability/reliability

Modern system engineering usually includes rigorous treatment of quality requirements. A list of quality attributes can be long and vary depending on the engineering field involved. In the context of nuclear safety, two such attributes are especially significant: availability and reliability.

A system or function is available when it is ready to operate. Availability provides information about the interval of time during which the system or function can actuate.

A system or function is reliable when few failures are expected. Reliability provides information about the failure-free interval.

Both availability and reliability are highly appreciated in the design of safety systems. Probabilistic safety assessment (PSA) treats both in detail and considers similarities and differences. Many provisions of system design are devoted to improving both at the same time. Such provisions are usually derived from general criteria.

3.7.6 Design basis accidents

DBAs are the scenarios considered for the licensing basis. They are the accidents against which an NPP is designed according to established design criteria. They are called "reference accidents" and connect with the licensing basis and DB. Their definition comes from a process called "grouping and bounding" (see below). The DB classifies them according to their estimated frequency of occurrence. The Standard Review Plan (USNRC, 2000) classifies them as anticipated operational occurrences and postulated accidents.

Regulatory Guide 1.206 (USNRC, 1997) classifies DBAs as Category 2 (moderate frequency), Category 3 (not frequent accidents), or Category 4.

In an appropriate design, DBAs are events for which fuel damage and the release of radioactive material are kept within authorized limits.

3.7.7 Design basis accident phenomenology

The most relevant DBAs are the following: LBLOCA, intermediate-break loss-of-coolant accident (IBLOCA), SBLOCA, loss of feedwater (LOFW), station blackout (SBO), loss of off-site power (LOOP), steam generator tube rupture (SGTR), and main steam line break (MSLB).

The initiating event in LOCA scenarios is usually a pipe break but could also result from the failure of an isolation valve connected to the primary system. The size of the pipe break determines whether the scenario is large (LBLOCA), intermediate (IBLOCA), or small (SBLOCA). The expected phenomenology is specific for each kind of LOCA, and in fact, this is the most relevant aspect of classification. Despite this, approximate break sizes are provided as an approach to such classification.

In an LBLOCA scenario, the primary pipe is supposed to break into a double guillotine; this means an equivalent break area of 200%. Therefore, quick depressurization occurs, and well-known two-phase flow phenomena take place: blowdown, refill, and reflood (see Section 3.2.3). The ECCS is designed to avoid out-of-range increases in the peak cladding temperature.

In an SBLOCA, the broken pipe has an area roughly in the range of 1%–6%. Both limits are arbitrary and questionable. The lower limit is related to high-pressure injection system (HPIS) performance. Any "break" that can be compensated by HPIS is no longer a break but instead a leakage. The other end of the interval is more ambiguous and related to phenomenology. The most relevant aspect of such phenomenology is that depressurization due to the break takes time, and while the primary pressure remains at a considerable value, the only injection available is HPIS. Under these conditions and depending on the circumstances, different events could occur, including dryout and temperature excursion. The vapor produced in the core could remain in the core or move to hot legs and SG U-tubes. A loop seal can be formed by liquid in the intermediate leg and can be cleared depending on pressure evolution (see Section 3.2.3). The vapor produced can also condensate in SG U-tubes, fall back, and produce the so-called reflux-condenser cooling mode as explained previously (see Section 3.2.3). DSA usually sweeps the SBLOCA spectrum, but most of its concern relates to emergency operations since the surveillance of critical parameters and manual actuation is crucial in addressing these scenarios. Among the manual actuations, cooling and depressurizing through the secondary side are the most significant. This is done manually by feeding the SG with auxiliary feedwater (AFW) and relieving vapor to the atmosphere or the condenser. Reducing the primary pressure also reduces the break flow. Further reductions could reduce the primary side pressure to the set point required for accumulator injection.

In an IBLOCA, the broken pipe has an intermediate area. Phenomenology is neither that of the LBLOCA nor of the SBLOCA. The peak cladding temperature is expected to be solved by accumulators' injections.

LOFW is normally detected as a decrease in the affected SG level. An automatic turbine trip will occur, and as a consequence, the scram. AFW will automatically start up, and plant recovery will be completed. Additional failures can be considered. If LOFW is in coincidence with LOOP, RCP trips will take place, and AFW will undergo additional difficulties—either a delay due to diesel generator start-up or a flow reduction because its flow will be limited to that of the turbopump (assuming that electrical pumps are not available). In this situation, one SG (at least one of 3) could ensure a heat sink sufficient to maintain core cooling by natural circulation on the primary side. The vapor produced is sent to the atmosphere through SG relieve valves.

Under a LOOP scenario, all electrical equipment becomes unavailable. Later, an emergency diesel generator (EDG) starts up, and in about 60 s, safety systems are recovered by following a defined sequence. Among the unavailable components, the most relevant are RCPs, pressurizer heaters, and steam dumps. LOOP produces the RCP trip, and directly or indirectly, the scram. From this point forward, the expected phenomena are the same as those in the LOFW scenario (see previous paragraph).

Under an SBO scenario, all electrical equipment becomes unavailable and EDG does not start up. In other words, SBO can be considered a LOOP without an available safety system. Additional failures must also be accounted for. Since injection to RCP seals is also unavailable, leakage through such seals must be assumed. Since seals could be further damaged during accident progression, such leakages must be estimated accordingly. The AFW turbopump could remain available while vapor is still produced on the secondary side, but that availability is also limited. For a certain time, the plant, despite being in an SBLOCA situation (due to seal leakage), can be cooled by natural circulation. This is done by feeding the secondary side using an AFW turbopump. Since conditions that allow vapor production are also deteriorating, this situation is only guaranteed for a limited time, which is called the SBO coping time. SBO coping time is specific for each plant and is in the range of a few hours. To calculate it, information on times to recover the availability of systems must be taken into account, and thermal-hydraulic calculations must be performed. Recovering the AFW, DC for instruments and valves, and EDG are the crucial steps.

An SGTR originates from a break in one or more tubes of an SG. In an SGTR, the break brings water from the primary to the secondary side. The most important point of this scenario is that primary water, which could have radioactivity, is flowing to a secondary system that is not prepared to handle it. Studies of such scenarios are of two kinds: DBA and emergency. The DBA scenario makes some conservative assumptions, and a calculation is performed to estimate the maximum (or pessimistic) release of radioactivity to the atmosphere through a broken SG relief valve. If such a release falls below the established limit, the DBA is positively evaluated. The general strategy for the emergency scenario is to minimize Δp from the primary to the secondary side to minimize leakage flow. Once the rupture takes place, scram will occur due to overtemperature or low pressure, and safety injection will automatically actuate. The first steps of the corresponding procedure identify and isolate the broken SG. After doing so, radioactivity is confined to the primary circuit and broken SG. The next actuations are devoted to reducing primary pressure and cooling the plant using intact SGs. Intact SGs are now fed with AFW, and the vapor produced is relieved. Pressurizer spray is also used to reduce primary pressure and limit break flow. Subcooling and core exit temperature are the main parameters surveyed.

An MSLB accident in a PWR consists of a rupture of one steam line upstream of the steam header. Immediately after the break, high differential pressure between steam lines causes actuation of the HPIS, followed by no delay in reactor and turbine trips and steam isolation. Because of the break, the broken secondary loop mass flow will greatly increase and produce important heat extraction in the corresponding SG. The primary side of such an SG will quickly cool to a rather cold temperature. When cold water reaches the core, asymmetric cooling occurs in the core with sudden positive reactivity feedback. A major concern for the MSLB accident is returning to power and criticality. In conservative calculations, the MSLB scenario uses assumptions that maximize the conditions for returning to power. One such condition could be assuming a stuck-out control rod after the reactor trip.

3.7.8 Beyond design basis accidents/severe accidents

Beyond design basis accidents (BDBAs) are scenarios with boundary conditions that are more strict or rigorous than DBAs.

Severe accidents are scenarios involving significant core degradation.

As can be seen, severe accidents are a part of BDBAs. Accidents belonging to the latter category and not to the former include those in which damage is avoided by applying accident management (AM) strategies or actions.

3.8 General safety concepts

3.8.1 Safety objectives/safety culture

Fundamentals of the safety of NPPs are explained in available documents of the International Atomic Energy Agency (IAEA); see IAEA (2000, 2002, 2006).

The fundamental safety objective is to protect people and the environment from the harmful effects of ionizing radiation. This means guaranteeing that the risk added by technology is affordable and ensuring the avoidance of fission product release. To do so, it is necessary to establish and execute engineering procedures that accomplish the mentioned objective.

Safety culture is a type of organizational culture where safety is a value of utmost priority and is considered essential for the long-term success of the organization. The issue is to make the safety culture strong and sustainable so that safety becomes a primary responsibility and focus in all types of activities. Nuclear safety must be highly integrated into the nuclear business.

3.8.2 Accident management/emergency operating procedures/severe accident management guidelines

AM is the systematic treatment or handling of accident-related conditions to prevent it, avoid its escalation to more serious events, and mitigate its consequences. To fulfill its goals, AM establishes policies and plans that are identified as strategies, guidelines, and procedures. Procedures describe actions intended to be as concise as they are feasible. They establish a sequence of actions to be followed step by step. Guidelines are less strict since the actions to be taken need some evaluation. Strategies establish the general aspects of safety function recovery.

Emergency operating procedures (EOPs) are designed to avoid or delay the progression of accidents to severe accident conditions. Bringing the plant up to an optimal plant state is the general goal of EOPs.

Severe accident management guidelines (SAMGs) are designed to mitigate severe accident consequences by maintaining containment functions and minimizing radiation relief (inside and outside). Attaining a secure and stable plant condition in the long term is their main goal.

3.8.3 Heat extraction safety relevance

Correct heat extraction from fuel maintains the integrity of the first safety barrier and avoids the accidental release of radioactivity. Generally, avoiding damage prevents an accidental release of radioactivity from progressing.

3.8.4 Energy sources

Since damage can be produced by the failure of heat extraction, to predict accident progression, complete studies of energy sources are needed. Various energy sources must be accounted for; among them are fission energy, stored energy, residual energy, and energy due to core degradation.

Fission energy is produced by a fission reaction. A neutron physics model usually allows its calculation. Such a model can be rather simple, such as the so-called point kinetics (or zero-D) model, or quite complex, such as those based on neutron diffusion, neutron transport, or Monte Carlo simulation. The thermal-hydraulic analyst must decide on the need to couple the involved code with any neutron-kinetics model and how detailed the related neutron physics model must be.

Energy stored in fuel and structures is nonnegligible in some transient analyses. The amount of stored energy depends on the materials' heat capacity and temperature. As a general statement, stored energy is the most significant heat source at the early stages of most accidents. Blowdown dryout in an LBLOCA is an example of core heat-up produced by energy stored in the fuel. Another situation related to stored energy is the injection of rather cold water in a way that allows it to contact the hot structures of pipes and vessels. The energy stored in those structures is partially absorbed by the injected water, which increases its temperature and hence decreases the cooling capability of the safety system.

Residual energy or decay heat is the energy produced by fission products within fuel rods. When the fission chain is stopped due to reactor scram, decay heat starts to decrease but never becomes zero. Despite being a decreasing function, decay heat can produce damage by itself. An accurate evaluation of decay heat can be performed by analyzing the buildup and decay of fission products. Such a calculation is connected with fuel burn-up, historical power, and time after scram. In practice, lookup tables are available and helpful for producing results that are sufficient for safety and design purposes.

During core degradation, many chemical reactions can occur among the fuel, structures, and degradation materials. Some reactions could be exoenergetic and become significant in powering core degradation. Among those reactions, oxidation is the most relevant. Oxidation can combine oxygen or water with zirconium, stainless steel, or other materials. Zirconium oxidation is exoenergetic at high temperatures (more than 1000 K) and produces hydrogen that can produce explosions when combined with air. Safety limits are set to avoid such interactions.

3.8.5 Damage

Besides its essential goal of fulfilling acceptance criteria, thermal-hydraulic analysis of DBA can ensure that damage is not involved in the progression of such scenarios. Although the definitions are not fully established, authors distinguish between failure and damage when addressing fuel behavior in an accident scenario. In most technical literature, core damage occurs when strong degradation starts. Such degradation may include mechanical and chemical transformation (including strong cladding oxidation), component melting, and fission product release. All of these occur at high or very high temperatures. The cladding temperature is typically 1200°C (or 2200°F), which is traditionally related to a well-known acceptance criterion on a licensing basis and is indirectly used as the threshold of damage. In this sense, it is accepted that other incipient degradation phenomena, for example, cladding oxidation at lower temperatures, are usually not pointed out as core damage.

Clad ballooning, burst (with the release of gaseous fission products to the coolant), and fuel relocation (notably originated by a loss of fuel ductility) may occur. Fuel relocations imply the possible release of heavy fission products into the coolant and a change in the linear power profile for individual rods, typically during the reflood period.

3.8.6 Safety functions

Safety functions are specific purposes that must be accomplished to ensure the general safety objective. They must be achieved in operational states during and following DBA, and to the extent practicable, during and following the considered plant conditions beyond DBA. The fulfillment of such functions is continuously checked in emergency operations. They constitute an important guideline for DBAs and emergency operating guides and procedures. General safety functions are the following: subcriticality, heat extraction, and radioactivity confinement.

Subcriticality is the safety function related to reactivity control. It is used and developed in the DB of systems dealing with reactivity control, like the control rod system, reactor protection systems, and safety injection. Emergency operations require that the ability to scram the reactor can be checked.

Heat extraction is the safety function related to removing heat from the core. It is used and developed in the DB of systems established to operate in any operational situation, such as the ECCS. Emergency operations require that their capability to operate accordingly can be checked.

Radioactivity confinement is the safety function related to radioactivity control. It is used and developed in the DB of systems dealing with containment integrity and radioactivity extraction like containment building systems. Emergency operations require that the availability of such systems can be checked.

3.8.7 Defense in depth

Following the IAEA definition, defense in depth is a comprehensive approach to safety that has been developed by nuclear power experts to ensure, with high confidence, that the public and the environment are protected from hazards posed by the use of nuclear power to generate electricity. The concepts of defense in depth and safety culture have served the nuclear power industry well as a basic philosophy for the safe design and operation of NPPs.

In such an approach, the general safety objective is accomplished by making provisions at various levels and considering various physical barriers to confine radioactivity.

3.8.8 Defense-in-depth levels

The defense-in-depth approach uses five levels of actuation:

Level 1 Prevention of abnormal operations and failures (provisions are made to avoid abnormal behavior)
Level 2 Control of abnormal operations and detection of failures (provisions are made to recover normal behavior)
Level 3 Control of accidents within the DB (provisions are made to avoid damage)
Level 4 Control of severe plant conditions, including prevention of accident progression and mitigation of the consequences of severe accidents (provisions are made to limit damage)

Level 5 Mitigation of radiological consequences of significant releases of radioactive materials (provisions are made to limit radioactivity release)

3.8.9 Safety barriers

Safety barriers are physical obstructions or walls to confine radioactive material at specified locations. For water reactors, the typical barriers are the fuel cladding, RCS pressure boundary, and containment building. Some authors consider the fuel matrix an additional safety barrier.

3.8.10 Safety systems/safety-related systems

Two kinds of systems should be noted as important for safety: safety systems and safety-related systems.

Safety systems are directly necessary to ensure the performance of safety functions. They are designed by following specific codes and include protections and safeguards.

Protection systems monitor the operation of a component (usually the reactor) using the values of relevant parameters. If an abnormal condition is detected through a critical value for a surveyed parameter, the system automatically initiates actions to prevent an unsafe situation.

The most significant preventive action is usually to stop the equipment (to scram the reactor).

Engineered safeguards are SSC-designed to maintain safety functions. The safeguards have various natures, including containment systems, ECCSs, secondary side safeguards, systems related to control room habitability, and complementary systems.

Despite control systems (or controls) contributing to levels 1 and 2 of defense in depth, they are not considered safety systems.

Safety-related systems are those whose failure has a safety impact. They are classified into three safety classes: 1, 2, and 3.

Safety Class 1 systems are those whose failure could produce an LOCA (category 3 or 4).
Safety Class 2 systems are the remaining SSCs belonging to the RCS pressure boundary.
Safety Class 3 systems include those whose failure could produce a release to the environment, those that support other safety systems, and those controlling radioactivity.

3.8.11 Inherent safety/passive safety/active safety

Safety is highly integrated in design. Safety targets are properly achieved because of many different implementations (e.g., redundancy).

Among the different ways to perform safety in a plant, inherent safety is maybe the most effective. Inherent safety features are those that bring the plant to safe situation based on the initiator itself without the need of designing any additional safety system. Fail-safe designs are in this category. Current plants have only a few inherent safety features. Inherent safety is in the conceptual design of some future plants.

Passive safety is the one that brings the plant to safe situation-based passive systems actuation. A passive system performs its function without any manual action and without the need for electrical power. Second-generation power plants have some passive safety systems. Some third-generation power plants are fully passive (Batet, 2004; IAEA, 2009b).

So safety in current plants, with a few inherent safety features and some passive safety systems, is essentially active. This means that safety performance in current plants requires short-term electrical power and midterm manual actions. Some engineered features are needed to help with safety system availability and thus avoid plant vulnerability. IAEA identifies and provides the definition (see below) of three such features: redundancy, diversity, and independence. Design robustness is an additional feature connected with usual design practices.

Redundancy is the use of more than the minimum number of sets of equipment to fulfill a given safety function. It is an important design principle for achieving high reliability in systems important to safety and for meeting the single failure criterion of safety systems. Redundancy enables the failure or unavailability of at least one set of equipment to be tolerated without a loss of function.

Diversity is applied to redundant systems or components that perform the same safety function by incorporating different attributes into systems or components. Such attributes could be different principles of operation, different physical variables, different conditions of operation, or production by different manufacturers. Diversity helps minimize the risk of common-cause failure.

- Independence is accomplished in systems design through functional isolation and physical separation.
- Functional isolation should be used to reduce the likelihood of adverse interaction between equipment and components of redundant or connected systems resulting from normal or abnormal operations or the failure of any system component.
- Physical separation and layout of plant components should use physical separation as much as practicable to further ensure that independence will be achieved, particularly for certain common-cause failures.

Design practices consider robustness an important additional feature to account for, especially when system reliability is a priority. The adequate size of safeguards is an example of design robustness.

3.8.12 International institutions

The NEA is an OECD institution. Its strategic mission can be derived from the following statements:

- It serves as a forum for sharing and analyzing information and experience among member countries.
- It fosters international cooperation in the nuclear field.
- It helps member countries to pool and maintain their technical expertise and human infrastructure and support their nuclear activities.
- It provides member countries with nuclear policy analyses.

Activities and projects, conferences, and specialist meetings (e.g., OECD/NEA/CSNI, 1992, 2000, 2011), and state-of-the-art reports (e.g., OECD/NEA/CSNI, 1989, 2016a), are typical initiatives used to develop strategic areas.

The EU EURATOM's tasks definition includes statements related to research and safety, such as promoting research, ensuring the dissemination of technical information, establishing uniform safety standards to protect the health of workers and the general public, and ensuring that they are applied.

Additional statements relate to the following (not an exhaustive list):

a Ensuring an equitable supply of nuclear fuels (making certain, by appropriate supervision, that nuclear materials are not diverted to purposes other than those for which they are intended)
b Exercising the right of ownership conferred upon it concerning special fissile materials
c Establishing relations with countries and international organizations that will foster progress in the peaceful uses of nuclear energy

The former group resulted in a network of joint research centers throughout Europe: Petten (Netherlands), Geel (Belgium), Karlsruhe (Germany), Ispra (Italy), and Sevilla (Spain). The latter resulted in activities related to nuclear materials and peaceful uses of nuclear energy.

The IAEA's tasks definition includes statements dealing with research, science, and technical information for specific purposes, such as the following (not an exhaustive list):

- Encouraging and assisting research
- Making provisions, e.g., for materials, services, equipment, and facilities, to meet safety requirements
- Fostering the exchange of scientific and technical information on the peaceful uses of atomic energy and encouraging the exchange and training of scientists and experts in the field on peaceful uses of atomic energy

Tasks with a greater impact on development may be those related to establishing or adopting standards of safety for protection in consultation or collaboration with competent organs of the United Nations and specialized agencies.

3.9 Concepts related to deterministic safety assessment

3.9.1 Safety analyses

The IAEA Specific Safety Guide SSG-2 (IAEA, 2009c) defines safety analyses as the following:

- Analytical evaluations of physical phenomena occurring at NPP, made for the purpose of demonstrating that safety requirements, such as the requirement for ensuring the integrity of barriers against the release of radioactive material and various other acceptance criteria, are met. This hall be demonstrated for all postulated initiating events that could occur over a broad range of operational states, including different levels of availability of the safety systems.

There are two basic types of safety analysis: DSA and PSA.

3.9.2 Deterministic safety analysis

DSA is a safety analysis devoted to studying all initiators independently of frequency (IAEA, 2009c). In principle, this kind of analysis must address all the mechanisms that lead to barrier failure for each operating mode and the entire operating range.

In practice, for each initiator, only a few limiting cases are analyzed. This is due to an important engineering activity called "grouping and bounding" that configures the practical scope and practice of DSA. Accidents are grouped. In each group, a bounding representative is chosen or depicted in a way that the acceptance of the results of its analysis ensures the acceptance of those of the other accidents of the group. To do so, the results of previous studies are taken into account, failure mechanisms are identified and compared, transient categorization is developed, and conservative assumptions are confirmed for all group members. The primary result of "grouping and bounding" is a list of bounding scenarios (also called DBAs or reference accidents), and the ultimate result is the feasibility of DSA. DBAs or reference accidents have already been cited in a previous section that stressed their connection with DB. The present section, meanwhile, emphasizes the way they are defined.

At this point, it can be established that DSA provides design verification. For each scenario considered, DSA demonstrates the success of the involved SSC as stated in the DB.

It is important to understand the scope and objective of DSA.

It must also be pointed out that DSA does not provide any information on overall plant risk or the suitability of the DB. Another safety analysis will take care of these points.

3.9.3 Safety margin/safety limit

The acceptance criterion has been defined as a statement that establishes the condition/s that must be fulfilled by the designed SSC. Most acceptance criteria defined in the licensing basis are set by selecting a key parameter and establishing a limiting value for it in a defined scenario. The limiting value is called the safety limit, and its difference from the calculated worse value is called the safety margin.

3.9.4 Deterministic safety assessment codes and methodologies

Four different options are currently available to address DSA, according to IAEA (2009c):

- Conservative
- BE with conservative boundary conditions
- Best estimate plus uncertainty (BEPU) evaluation
- BE with systems availability supplied through probabilistic assessment

Each methodology uses specific codes and procedures.

3.9.5 Conservative methodologies

A conservative approach is carried out by replacing one or more parameters with values that will unfavorably affect specific acceptance criteria. The plant conditions, physical models, and initial and boundary conditions used are conservatively set.

In conservative analyses, the single failure criterion is applied when determining the availability of systems and components. This establishes that safety systems should be able to perform their specified functions when any single failure occurs. Besides that criterion, factors related to plant status are also considered—credit is not given to operator action unless it limits conservatism, and controls are not operated unless their actuation limits conservatism.

Finally, additional conservatism is considered related to calculating tools. Conservative codes are used. Conservative codes are simpler than BE codes, and their application has proved useful from the start of NPP technology.

The main drawback of conservative codes and methodologies is that they do not produce a realistic picture of accident phenomenology.

3.9.6 Methodologies using best estimate codes with conservative boundary conditions

This methodology tries to overcome the main drawback of the previous one. Using BE codes with conservative boundary conditions, a realistic picture of accident phenomenology is recovered, and safety margins become clearer. Methodologies usually establish the main parameters that must be considered and the conservative values at which they must be set.

3.9.7 Uncertainty methodologies part of best estimate plus uncertainty approach

BEPU evaluation methodologies analyze DBA on a realistic basis; see IAEA (2008a). Once a base case calculation is performed using a BE and a qualified nodalization, uncertainties must be evaluated to establish error bands for the key parameters to allow checking of the acceptance criterion. Such an uncertainty evaluation can be performed by following various procedures. Depending on the procedures used, one can find statistical or accuracy extrapolation methodologies.

Statistical methodologies, as in the work of Glaeser (2008), can propagate input uncertainties and obtain output error bands for key parameters. Most are based on Wilks' theorem; see Wilks (1941, 1942).

Methodologies based on accuracy extrapolation (e.g., D'Auria et al., 1995; D'Auria and Giannotti, 2000) apply in deriving error bands from an accuracy database obtained from posttest calculations performed with experiments for different facilities. D'Auria et al. (1988) proposed the pioneering paper dealing with accuracy extrapolation, i.e., before the CSAU publication (1989, Section 3.6.3.1).

Within the framework of OECD/NEA/CSNI activities, various comparative exercises have been carried out with the general objective of improving the capabilities of such techniques, among them. UMS (OECD/NEA/CSNI, 1998; Wickett et al., 1998) and BEMUSE (OECD/NEA/CSNI, 2006, 2007a, 2008, 2009a, 2011).

The debate on lessons learned from BEMUSE activity has produced an important number of review articles (e.g., de Crécy et al., 2008; Pérez et al., 2010; Perez et al., 2011), conference papers (e.g., Glaeser et al., 2011; Reventós et al., 2011), and further developments (e.g., Perez et al., 2009; Pérez, 2011).

BEPU methods are now available and used by some organizations; (e.g., see the following publications: D'Auria et al., 2012a; Martin and O'Dell, 2008), and the discussions in Chapter 14. Current applications of BEPU approaches do not eliminate all conservative assumptions (e.g., single failure).

3.9.8 Inverse methods

In modeling and simulation, the inverse method consists of a technique where model input parameters are estimated (with uncertainty) from a comparison of model output magnitudes with experimental data. The solution to the inverse problem can include model calibration, model uncertainty quantification, or both. In the specific field of BEPU methodologies, such techniques are helpful in deriving the uncertainties of selected input parameters from specific SET (or intermediate test) measurements. OECD/PREMIUM is devoted to the quantification of model uncertainty. The objective of the PREMIUM (Post-BEMUSE Reflood Model Input Uncertainty Methods) benchmark is progression on the issue of quantifying the uncertainty of physical models in system thermal-hydraulic codes by considering a concrete case: physical models involved in the prediction of core reflooding. Two OECD/CSNI reports (OECD/NEA/CSNI, 2016a,b) describe the project. The former establishes the objectives and a description of the methods used, while the latter provides the final results and identifies open items. Freixa et al. (2016) provide a comparison of two key methods used in PREMIUM.

3.9.9 Final safety analysis report

The final safety analysis report (FSAR) is an enormous technical document certifying that the safety requirements of a plant have been accomplished. It is plant-specific. It puts together the licensing basis (DB of all SSCs important for safety) and the calculation results and engineering judgments that confirm its fulfillment. FSAR contains a verification of safety design.

3.10 Concepts connected to actual accidents

3.10.1 Context

An accident is defined as an undesired event that produces distortion in a preestablished process. Some authors add that such an event results in damage and that such damage could be human or material.

Advanced societies devote important effort to minimizing the number and severity of accidents.

Research, regulations, and accident investigation are actions taken to fulfill such purposes. Accidents are investigated for legal reasons or as part of a preventive strategy.

In the nuclear field, since the release of radioactivity into the environment is an event to avoid, all these investigations and preventive actions are more significant. Plant operators and safety authorities conduct investigations. Operators' associations share experiences about actual accidents.

Since the behavior of plants involved in actual accidents is often a topic in discussions addressing the preparation of new regulations or launching new research activities, it is important to bear in mind some specific background of actual accident concepts and phenomenology. To do so, the next section highlights accident classification and some relevant notes on the three most significant accidents that have occurred to date: Three Mile Island, Chernobyl, and Fukushima.

3.10.2 Concepts

3.10.2.1 Classification and treatment

NPP accidents are classified or treated using the following scales and groups:

- International Nuclear and Radiological Event Scale (INES)
- Estimated core damage frequency (CDF)
- Recurrent accidents

The INES is used most. It is properly a classification system prepared by the IAEA. It classifies events on eight levels according to their relevance from a safety point of view. INES classification criteria define the level of an event as the highest of the values corresponding to three factors: off-site impact, on-site impact, and the degradation of defense in depth. Its highest level is 7, defined as "events with large impact off-site having effects on health and environment."

CDF is a PSA concept. For event classification, CDF is calculated using the established PSA of the plant, quantifying the probability of bringing the plant to a damaged situation provided that known failures have occurred. It is a good indicator of the severity of the event and relies on the existing PSA of the plant.

This treatment is not properly a classification. Recurrent accidents are studied by following the frequency of their initiators. The most relevant accidents in western water reactors are the loss of the residual heat removal system during midloop operation, errors in valve operation, BWR instabilities, and SG tube ruptures.

3.10.2.2 Relevant notes on Three-Mile Island TMI-2 accident

The Three Mile Island accident occurred on March 28, 1979 (see also Chapter 16). It was rated as level 5 on the INES scale. A stuck-open pressurizer relief valve and the subsequent loss of primary coolant were relevant causes of the accident, but the most significant failures were in procedures and human factors. Operators failed to handle the situation after misunderstandings about the state of the plant and the actions to be taken. The reactor core partially melted down. Among the main lessons learned from the accident was the clear need to improve EOPs.

3.10.2.3 Relevant notes on Chernobyl accident

The Chernobyl accident occurred on April 26, 1986 (e.g., Martinez Val et al., 1990; Khan, 1990) (see also Chapter 16). It was rated as level 7 on the INES scale. The reactor suffered a catastrophic power increase. Explosions took place. Radioactive materials were dispersed into the environment, and the graphite moderator was burned. The accident occurred during an experiment planned to test additional emergency core cooling capabilities. Some causes of the accident were design deficiencies, like the positive void reactivity coefficient and the fact that the reactor did not have a hard containment vessel. Other causes were related to operating procedures, like disregarding the operating reactivity margin and switching off reactor protections after initiating the test. Most specialists agree that the lack of a safety culture was the most relevant cause of the accident.

3.10.2.4 Relevant notes on Fukushima accident

The Fukushima site underwent an earthquake and a tsunami on March 11, 2011; see OECD/NEA/CSNI (2015) (see also Chapter 16). Consequently, the plants lost their ability to cool some operating reactors and spent fuel pools. Two units degraded severely and released substantial amounts of radioactivity into the environment. Several explosions of H_2 resulted in the severe deterioration of the involved buildings. As stated by L.E. Herranz "Fukushima: A multiple severe accident within a natural disaster." The root cause of the accident was Tsunami "sweep" away of the EDG. Some lessons were learned for the technical community to reconsider:

- Maximum historical values of external causes
- Sites with more than one NPP
- Spent fuel storage and the availability of portable equipment to face SBO scenarios

3.11 Concepts related to other uses of thermal-hydraulic analysis

3.11.1 Context

An important amount of analytical thermal hydraulics is performed in a framework slightly different from deterministic analysis. Three areas are identified:

- Probabilistic safety assessment
- Emergency operations
- Support plant operations

Many engineering organizations work in these three areas and perform thermal-hydraulic calculations, so some concepts must be introduced.

3.11.2 Concepts

3.11.2.1 Probabilistic safety assessment sequences/success criteria

PSA (Apostolakis and Moieni, 1987) sequences are the scenarios involved in the construction of event trees. Thermal-hydraulic analysis of such sequences facilitates evaluating success criteria (IAEA, 2010a).

3.11.2.2 Emergency operations/strategies, guidelines, and procedures

The goal of emergency operations is to keep general and specific safety functions and to recover the plant back to normal procedures or to an optimal plant end state.

Optimal plant end state is a key concept in emergency operations in which radiation release is minimized, equipment release is minimized, and plant conditions are stable with equipment operating in long-term alignments.

Strategies, guidelines, and procedures are documents establishing the plans for emergency operations. They range from more generic to more detailed. Strategies are general processes and maneuvers such as secondary side cool down, primary feed and bleed, equalizing pressure, or facilitating natural circulation. Guidelines are recommendations to be followed during emergency operations. Procedures are detailed practices that can be followed step by step.

3.11.2.3 Severe accident management guidelines

SAMGs (IAEA, 2008b) are plant-specific documents providing recommendations to meet safety requirement during a severe accident.

3.11.2.4 Supporting plant operations: analytical tasks

The responsibilities of analysts involved in supporting plant operations are somewhat different from those of other analysts that currently produce usual thermal-hydraulic studies (Reventós et al., 2007a, 2008a). There could be some coincidences but usually their tasks are quite specific.

Support tasks for commercial plants are alive. Their guidelines strongly depend on organizational requirements, plant status, and the availability of external help, including the following:

- supporting dialogues with regulatory bodies and fuel designers;
- analyzing actual transient occurrences in the plant (operating events) (Reventós et al., 2010) NPP start-up test analysis (Llopis et al., 2007);
- analyzing hypothetical transients (Reventós et al., 2007b; Freixa et al., 2015c; Pla et al., 2016a);
- analyzing transients for training support, design modifications, control system design support (Llopis, 2006) or plant availability improvements.

Exercises and questions

Exercises and questions are proposed hereafter with solutions given in Appendix 1 (Volume III). The readers should answer the questions or provide solution to exercises before reading Appendix 1. Please provide extended answers.

3.1 Consider a vertical pipe with an internal diameter of D = 0.6 m and bubbly flow conditions. If the linear interfacial drag area is 89.6 m and the mean bubble diameter is 1: 5 mm, what is the void fraction?

3.2 Consider a vertical pipe with an internal diameter of D = 0.5 m where the flow conditions are inverted annular flow, and the void fraction is 0.06. What is the linear interfacial heat transfer area in a cut section?

3.3 Consider a vertical pipe with an internal diameter of D = 0.7 m where the flow conditions are annular flow with a film thickness $\delta = 0.005$ m. Calculate the void fraction.

3.4 A flow has a void fraction of 0.3. Use the Fauske void correlation model to obtain the quality for a density ratio of $\rho_l/\rho_g = 900$.

3.5 A flow has a void fraction of 0.3. Use the HEM void correlation model to obtain the quality for a density ratio of $\rho_l/\rho_g = 900$.

3.6 Calculate the convection heat transfer coefficient in the core during normal operation (liquid forced convection). Consider the following information:

- The thermal conductivity of saturated water at 300°C is $k_{H2O} = 0.55$ W/mK.
- The dynamic viscosity of saturated water at 300°C is 8.6×10^{-5} N s/m^2.
- The fluid density is 716 kg/m^3.
- The specific heat is 5.7 kJ/kg K.
- The hydraulic diameter is 0.013 m.
- The core flow velocity is constant and equal to 5 m/s.

3.7 For each of the following processes, what (nondimensional) number best represents/explains the phenomenon?

- CCFL
- stratification
- convection heat transfer
- laminar flow
- inverted annular film boiling

3.8 For each of the following processes, what (nondimensional) number best represents/explains the phenomenon?

- choking flow
- offtake model
- end of natural circulation
- turbulence
- annular film heat transfer

3.9 Consider the geometry of the flooding experiment with blocked arrays shown in Fig. 3.3a, with values expressed in millimeters.

(a) Calculate the hydraulic diameter for the flow path.

(b) Calculate the approximate hydraulic diameter of a PWR reactor with 157 fuel assemblies and a distribution of fuel rods similar to that in Fig. 3.3a.

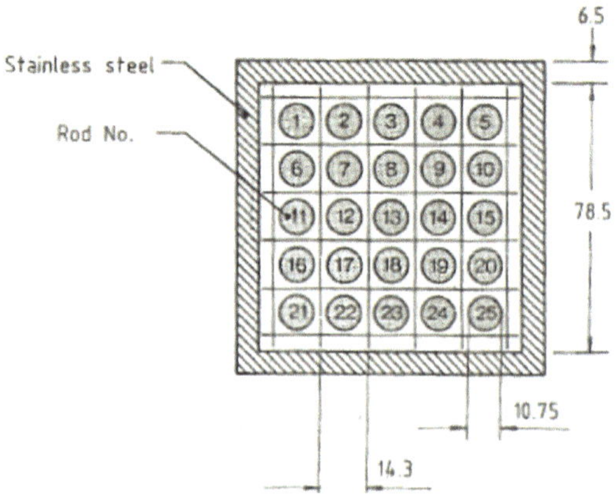

Figure 3.3a Cross-sectional view of fuel assembly setup for flooding experiment with blocked arrays.

3.10 Consider the geometry of the PWR subchannel and bundle test (PSBT) experiment shown in Fig. 3.3b, with values expressed in millimeters.

(a) Calculate the wetted hydraulic diameter for the flow path.

(b) Calculate the heated hydraulic diameter for the flow path.

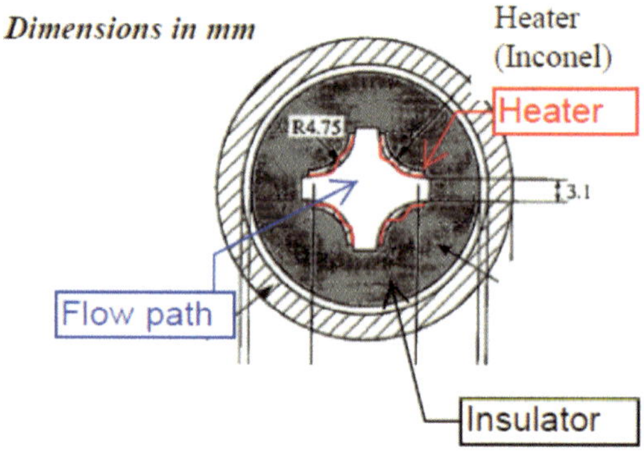

Figure 3.3b Cross-sectional view of the pressurized water reactor subchannel and bundle test experimental setup.

3.11 Consider the geometry of the PSBT experiment shown in Fig. 3.3c, with values expressed in millimeters.

(a) Calculate the wetted hydraulic diameter for the flow path.

(b) Calculate the heated hydraulic diameter for the flow path.

(c) This experiment represents a particular subchannel in a fuel assembly. Draw a portion of the fuel assembly where this particular subchannel appears and identify it.

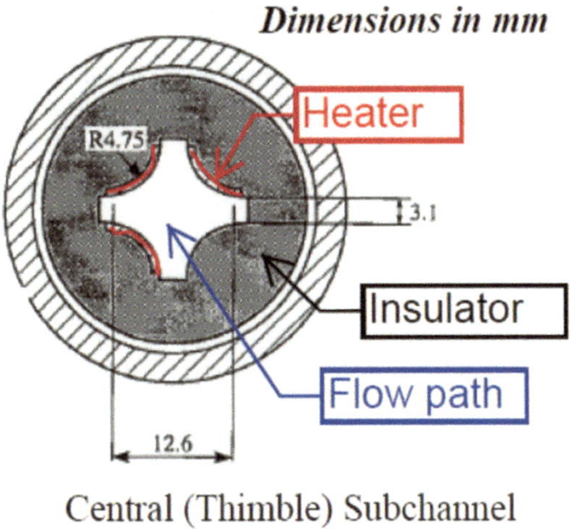

Figure 3.3c Cross-sectional view of the PSBT subchannel experimental setup (note the differences related to Fig. 3.3b).

3.12 Consider the geometry of the PSBT experiment shown in Fig. 3.3d, with values expressed in millimeters.

(a) Calculate the wetted hydraulic diameter for the flow path.

(b) Calculate the heated hydraulic diameter for the flow path.

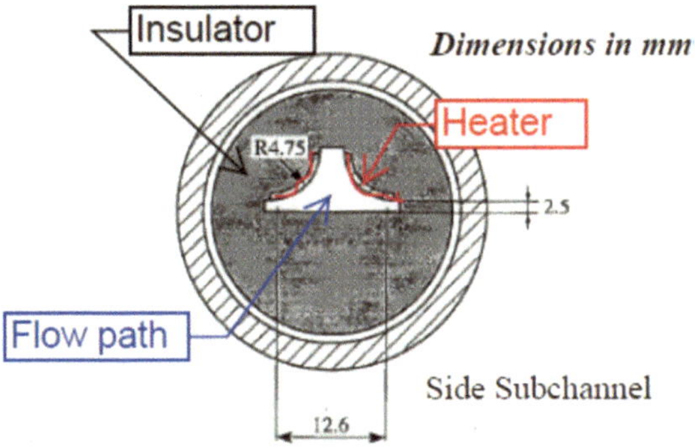

Figure 3.3d Cross-sectional view of the PSBT subchannel experimental setup (note the differences related to Figs. 3.3b and 3.3c).

3.13 Consider the geometry of the PSBT experiment shown in Fig. 3.3e, with values expressed in millimeters.

(a) Calculate the wetted hydraulic diameter for the flow path.

(b) Calculate the heated hydraulic diameter for the flow path.

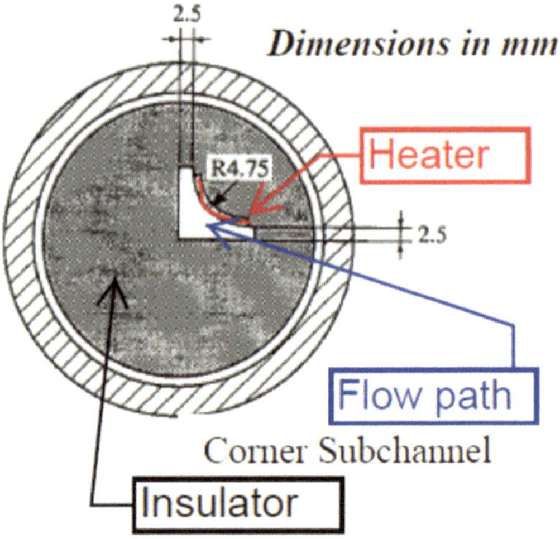

Figure 3.3e Cross-sectional view of the PSBT subchannel experimental setup (note the differences related to Figs. 3.3b, 3.3c and 3.3d).

3.14 Fig. 3.3f displays a U-tube steam generator section (related to a test facility simulating a PWR) at midelevation. The numbered small circles represent the U-tubes; there are a total of 30, each with an inner diameter of 19 mm. The crossed circles represent fillers (metallic structures). The inner diameter of the steam generator secondary side section is 0.38 m.

(a) Calculate the hydraulic diameter of the flow path through the primary side of the U-tube bundle.

(b) Estimate the hydraulic diameter of the secondary side flow path. Explain your hypotheses.

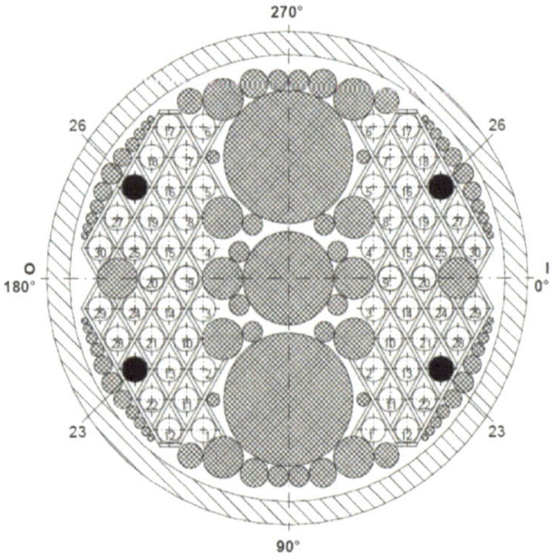

Figure 3.3f Cross-sectional view of a U-tube SG (related to an experimental facility).

Acknowledgments

The authors wish to acknowledge J.L. Gandrille, who provided comments for the second edition of the book.

Role of thermal hydraulics in nuclear power plant design and safety

4

Richard R. Schultz[1] *and Horst Glaeser*[2]
[1]Idaho State University, Pocatello, ID, United States; [2]GRS, Garching, Germany

Foreword

Following the definitions in nuclear thermal hydraulics (Chapter 3), the present chapter attempts to establish the need for the (nuclear thermal hydraulics) discipline. The need comes from the design/operation of nuclear power plants, namely water-cooled nuclear reactors, as well as from safety evaluations. This chapter focuses on the need for safety, namely coming from the licensing process.

Needs are translated into the features of computational tools and the procedures used to prove their validity, including scaling capabilities. So ideally, the reader may jump from Chapter 1 (a general overview of nuclear thermal hydraulics), to Chapter 4 (the present chapter), to Chapter 11 (which discusses the features of computational tools), to Chapter 13 (which describes validation procedures and methods, scaling issues, and the evaluation of the uncertainty of code results) and maybe Chapter 22 (for more details on verification and validation), to Chapter 14 (accident analysis focused on applying the best estimate approach to licensing), to Chapter 15 (which discusses selected application results from computational tools), and finally to Chapter 23 (which discusses the thermal-hydraulic design of water-cooled reactors).

Some chapter information is specifically related to gas-cooled reactors and is also applicable to the technology of water-cooled reactors.

4.1 Introduction

Thermal hydraulics is the physics of fluid flow, energy (e.g., heat) transfer, and interactions among fluid flow, energy, and surrounding structures. In nuclear systems, material interactions with nuclear radiation are also sometimes an important consideration. Thermal hydraulics in nuclear reactors is a wide-ranging subject and includes such diverse topics as gravity-dominated stratified flow in pipes with a free surface—including hydraulic jumps, momentum-driven critical flow through passages and restrictions, fluid–structure interactions such as fluid jet impingement on structures, flow boiling and condensation, flow and heat transfer regimes for flow in various geometries at various

Handbook on Thermal Hydraulics in Water-Cooled Nuclear Reactors
https://doi.org/10.1016/B978-0-323-85606-5.00021-0

orientations, energy generation and transfer to fluids and structures from nuclear sources, and a multitude of other phenomena. The study of these topics may range from one-dimensional to three-dimensional flow fields that interact with structures, which may be represented by bodies ranging from points to three-dimensional structures with heat sources or sinks. For many of the nuclear reactor scenarios that must be considered, important variables and phenomena change as a function of time.

Thermal-hydraulic analyses are governed not only by their boundary and initial conditions but also by the limiting values of important figures of merit that define maximum or minimum acceptable limits stemming from design/operational or safety requirements. Therefore, thermal-hydraulic analysis objectives may often be separated into two broad categories:

- Design: limits defined to ensure that plant design is robust and thus that the plant can operate economically with minimal shutdowns and operational costs and
- Safety: limits defined by laws or regulatory rules specified to protect public health or living standards.

4.1.1 Design

Thermal-hydraulic design analyses focus on creating nuclear systems and components that are robust and reliable, have a long and trouble-free lifetime, are efficient and economical to operate, do not tend to stray into unwanted transient scenarios leading to shutdowns that challenge equipment design limits, and are easy to inspect and maintain. A well-designed and operated nuclear system should consistently achieve capacity factors exceeding 90%. nuclear power plants (NPPs) were originally designed for a minimum lifetime of 40 years. In addition, many NPPs have been granted life-extension licenses, sometimes up to 60 years, after providing a plant life extension file demonstrating continuous and acceptable safe operation to their regulatory organization.

Because design analyses generally deal with normal operational conditions and the transition from one normal operational condition to another, including power-up and shutdown, they generally center on non-safety-grade equipment performance and phenomena known to affect plant efficiency, reliability, and long-term operations. Examples include the following:

1. *Efficiency*: steam separator and dryer performance for Rankine cycle plants; core power distributions (axial and radial) that match specifications; and uniform core inlet conditions. Optimization of such parameters enhances power output by enabling the plant to function at or near technical specification limits.
2. *Reliability and long-term operations*: unwanted fluid-structure interactions, such as fretting, throughout the system but particularly in the core at spacers and steam generator U-tubes (in the case of a pressurized water reactor [PWR]) and factors known to cause crud formation within the core. By minimizing phenomena known to be adverse in the long term, plant reliability is enhanced, and long-term operations are assured.

Thermal-hydraulic analyses centered on system design include numerical analysis tools, including those for safety analyses, such as (1) system analysis codes, fuel behavior codes, and containment performance evaluation tools, and some not typical in safety analyses, such as (2) operational core power distribution analysis tools, computational fluid dynamics (CFD), heat transfer numerical models, and many specialized thermal-hydraulic tools tuned to analyze specialized operational conditions.

4.1.2 Safety

Thermal-hydraulic safety analyses focus on ensuring that nuclear systems and components always operate with operational margins consistent with the "defense-in-depth" approach required by regulatory agencies. In essence, the principle of defense in depth requires the presence of four levels of protection for public health and safety, as follows (Sorensen et al., 1999):

1. NPPs are designed to not fail within the range of normal and abnormal conditions they might reasonably experience during their operating lives, including extreme natural phenomena and human-caused hazards.
2. Redundant and diverse support systems are provided to detect initiating events or incipient failures and shut the reactor down before fuel damage can occur following events that are within the design basis.
3. Redundant and diverse emergency systems, including ECCSs and containment systems, are provided to control reactivity, cool the core, limit the release of radioactivity into the environment, and thereby mitigate accident consequences if the first two levels of defense fail.
4. Reactors are sited remotely, and emergency plans are prepared and practiced in advance with local and regional authorities to limit the public consequences of a catastrophic accident involving a large release of radioactivity.

Thermal-hydraulic analyses are important ingredients of all four defense-in-depth protection levels. Adherence to the defense-in-depth principle has resulted in less than 0.1 significant safety events annually per plant, with no discernible radiological health effects from operations and incidents.

Acceptance criteria are set by regulatory authorities for the entire range of operational states and accident conditions. Acceptance criteria may be related to the frequency of an event. Events that occur frequently, such as anticipated operational occurrences, should have more restrictive acceptance criteria than those for less frequent events such as design basis accidents (IAEA, 2009c). For example, an operational parameter known as the maximum linear heat generation rate (MLHR) is established to prevent critical nucleate boiling at the cladding of fuel rods during anticipated operational occurrences. This is a departure from nucleate boiling (DNB) for PWRs and dryout for boiling water reactors. However, the MLHR may have to be reduced to satisfy a safety requirement, such as a loss-of-coolant accident (LOCA) criterion, to stay below the less restrictive 1200°C maximum cladding temperature of the fuel rods. During an accident, critical boiling occurs, and criteria are set for the design of ECCSs and parameters to stay below a specific threshold, e.g., the maximum fuel cladding temperature, to limit failure of the cladding.

Because thermal-hydraulic safety analysis tools are typically used to assess the margin between calculated key figures of merit and limits for those figures of merit that are imposed on plant operations by national law, analysis tools are assessed[1] using a rigorous protocol that has been closely examined and is monitored by licensing authorities. The thermal-hydraulic tools used in safety analyses—usually system analysis, containment analysis, core performance, and fuel performance tools—are all created to study the behavior of nuclear systems under especially adverse transient conditions that may lead to fission product release.

4.1.3 The role of thermal hydraulics

Whether the objective is to design a nuclear system or to demonstrate a system's safety by meeting the necessary licensing requirements, that objective is achieved by:

1. Identifying the nuclear reactor operational and accident envelopes
2. Isolating the key phenomena in each of the plant scenarios that comprise the operational and accident envelopes
3. Developing numerical models that accurately calculate the behavior of the key phenomena for the scenarios of interest
4. Assessing the numeric models *to demonstrate their adequacy* for calculating the scenarios and thus calculate the nuclear reactor behavior for these scenarios in a manner that can be trusted by the plant designers and regulatory authorities
5. Performing the necessary calculations showing the subject nuclear reactor meets the necessary design and licensing requirements.

Thus the role of thermal-hydraulic analyses is to provide evidence showing that the subject nuclear reactor meets design and licensing requirements by achieving the goals outlined in steps 1 to 5 just mentioned.

4.1.4 Organization of Chapter 4

The role of thermal hydraulics, as a key ingredient in the analysis of nuclear reactor design and licensing, is directly correlated with the types of analyses that must be performed, that is, the thermal-hydraulic analysis envelope. This topic is discussed in Section 4.2. Finally, Section 4.3 outlines how analysis tools, that is, numeric models, have demonstrated their adequacy for performing analyses on key scenarios that must be analyzed.

4.2 The thermal-hydraulic analysis envelope

The design and safety requirements that underlie NPP design and operation combine to form the operational and accident domain of an NPP, including conditions (structural loads, temperatures, pressures, life expectancy, radiation levels, etc.) representative of:

- Normal operations (e.g., power production, shutdown, refueling), Anticipated operational occurrences (e.g., reactor trip, water hammer)
- Infrequent events or transients (e.g., loss of offsite power, loss of feedwater)

[1] The term "assessed" includes verification and validation of numerical models using a top-down and bottom-up protocol that has been rigorously examined and is approved by the licensing authorities.

- Design basis events (e.g., floods, earthquakes, external human-caused hazards)
- Design basis accidents (e.g., loss of coolant, large reactivity changes).

These design basis conditions are used in the design of structures, systems, and components. In addition, design extension conditions are defined as follows:

- DEC-A: Domain of extended conditions corresponding to multiple failures (examples: a total loss of steam generator feedwater—e.g., normal, auxiliary, or emergency; a small-break LOCA with a loss of part of the emergency core cooling system (ECCS), the high or medium head safety injection system, or the low head safety injection system)
- DEC-B: Domain of extended conditions corresponding to a core meltdown.

The primary safety requirements governing NPP design in the United States are codified in Appendix A of 10 CFR 50, "General Design Criteria for Nuclear Power Plants." Internationally, other nations have similar requirements that must be followed by both the designers and the operators of NPPs in those countries.

The design basis for each NPP is described in its safety analysis report (SAR), which must be produced in preliminary form by the plant designer for review by regulatory authorities as well as the plant owner and other stakeholders before obtaining a license. The technical information contained in such reports includes the seismic, meteorological, hydrologic, and geologic characteristics of the plant site, "with appropriate consideration of the most severe of the natural phenomena that have been historically reported for the site and surrounding area and with sufficient margin for the limited accuracy, quantity, and time in which the historical data have been accumulated." The SAR also contains safety assessments of plant performance under a specified set of conditions that define a design envelope. All the safety assessments must show that the reactor core is maintained in a coolable geometry and that there would be limited releases of fission products to the reactor coolant system or the reactor containment. In addition, the combination of site characteristics, demography, and plant design must be shown to meet certain radiation exposure criteria for offsite individuals in the event of an arbitrarily large release of radioactive material from the reactor coolant system into the reactor containment.

The complete operational and accident domain of an NPP, which makes up the plant's design basis, is represented by the yellow region of the Venn diagram shown in Fig. 4.1. The owner of an NPP will only qualify for an operating license if all the operational conditions, infrequent events/transients, and design basis events/accidents are shown to result in conditions that do not violate national standards, laws, and operating requirements—for example, 10 CFR 50 in the United States (USNRC, 2015).

The design basis includes accident prevention and accident mitigation requirements for each NPP for a defined scope of events. The requirements for all events within the design basis are that the fission process can be shut down, the reactor remains subcritical, the core can be cooled and its geometry maintained, coolant makeup can be supplied as needed, and in the event of an accident involving core damage, any fission products released from the core can be controlled within a containment structure that ensures limited radioactivity leakage into the environment. Information contained within the design basis for a particular piece of equipment or building includes, for example, the specific functions to be performed and specific values or ranges of values

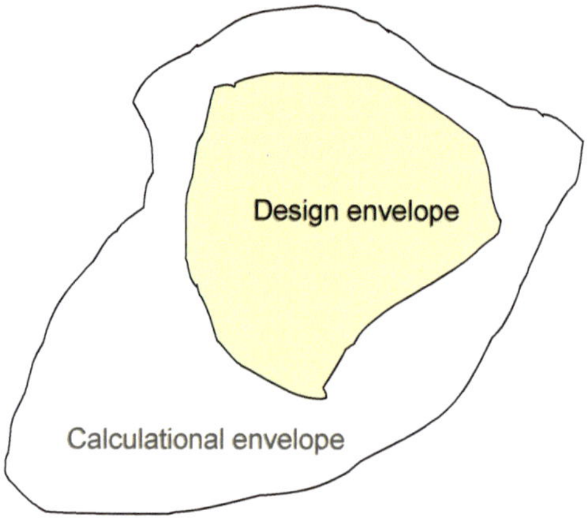

Figure 4.1 Venn diagram showing relationship between nuclear power plant system envelope and thermal-hydraulic analysis calculational envelope (Venn, 1880, 1881).

for the conditions that serve as the bounds for its design. These conditions are derived from an engineering analysis or experiments to predict the consequences of events postulated for design, including those in the preceding list. An important lesson learned from decades of operating NPPs is that rigorous configuration control must be applied throughout a plant's operating life to ensure that the design basis is maintained.

A plant designer and plant operator demonstrate that their nuclear reactor system does not violate national laws and national NPP operational requirements and standards using thermal-hydraulic analysis tools, that is, numeric models, that are adequate for performing trustworthy predictions of the behavior of the NPP for all the scenarios, conditions, and transients within the design envelope. The thermal-hydraulic tools must be capable of calculating trustworthy analyses demonstrating that a domain either equals or exceeds the design envelope. The domain for which thermal-hydraulic analyses may be used constitutes the "calculational envelope," as shown in Fig. 4.1.

Beyond-design-basis accidents, including severe accidents, are usually treated separately in deterministic safety analyses. Some initiating events may be the same as those for design basis accidents. The results help determine the necessary measures to prevent severe accidents and mitigate their radiological consequences if they occur. The analyses should be carried out using specialized computer codes that are available to model relevant physical phenomena. For light-water reactors, these phenomena include thermal-hydraulic effects, heating and melting of the reactor core, retention of the molten core in the lower plenum, interactions between molten core and concrete, steam explosions, hydrogen generation and combustion, and fission product behavior.

Preventive measures are recovery strategies to prevent core damage. They should be analyzed to investigate possible actions to inhibit or delay the onset of core damage. Examples are various manual restorations of systems, primary and secondary feed and bleed, depressurization of the primary or secondary system, and restarting of the reactor coolant pumps. Conditions for initiation of the actions should be specified, and criteria for when to stop the actions or to change to another action.

Mitigative measures are strategies for managing severe accidents to mitigate the consequences of core melt. Such strategies include coolant injection into the degraded core, depressurization of the primary circuit, operation of containment sprays, and the use of fan coolers, hydrogen recombiners, and filtered venting, if available, in the reactors of different types in operation or being constructed. Possible adverse effects as a consequence of taking mitigative measures should be accounted for, like pressure spikes, hydrogen generation, return to criticality, steam explosion, thermal shock, or hydrogen deflagration or detonation.

The US safety authority requires, through Section 50.34 of 10 CFR Part 50, (USNRC, 2015), that each application for a construction permit for a nuclear reactor facility includes a preliminary SAR and that each application for a license to operate such a facility includes a final SAR. Section 50.34 specifies in general terms the information to be supplied in SARs. Further information is provided in "*Standard Format and Content of Safety Analysis Reports for Nuclear Power Plants*" (Revision 3), Regulatory Guide 1.70 (USNRC, 1978). The release of Reg. Guide 1.206 included a plan to incorporate this content from Reg. Guide 1.70 while superseding it.

The *Standard Review Plan* [SRP] *for the Review of Safety Analysis Reports for Nuclear Power Plants*, NUREG 0800 (USNRC, 2013a), is continuously updated and is provided to guide staff reviewers in the Office of Nuclear Reactor Regulation in performing safety reviews of applications to construct or operate NPPs. The principal purpose of the SRP is to ensure the quality and uniformity of staff reviews and to present a well-defined base from which to evaluate proposed changes in the scope and requirements of reviews. The SRP also makes information about regulatory matters widely available and improves communication and understanding of the staff review process by interested members of the public and the nuclear power industry. Some information is also provided here in Chapter 14.

4.3 Demonstration of adequacy of thermal-hydraulic analysis tool

The process of qualifying software, for example, system analysis codes, such that the software is considered adequate for calculating the behavior of NPPs, is in concept relatively straightforward, even though there are many questions to be answered and the process itself is iterative. From a high-level perspective, the process consists of five stages (see Fig. 4.2) that must be exercised in their entirety for each NPP design. The five stages consist of

(1) identifying scenarios of importance,
(2) identifying the key phenomena for the scenarios of importance using a phenomena identification and ranking table (PIRT) process,
(3) verification and validation (V&V) of the analysis tools/software, that is, determining whether the tools for analyzing the scenario progressions can calculate the important phenomena and figures of merit for the validation matrix,
(4) correcting or enhancing existing software or possibly developing new software, and finally,
(5) performing an adequacy evaluation.

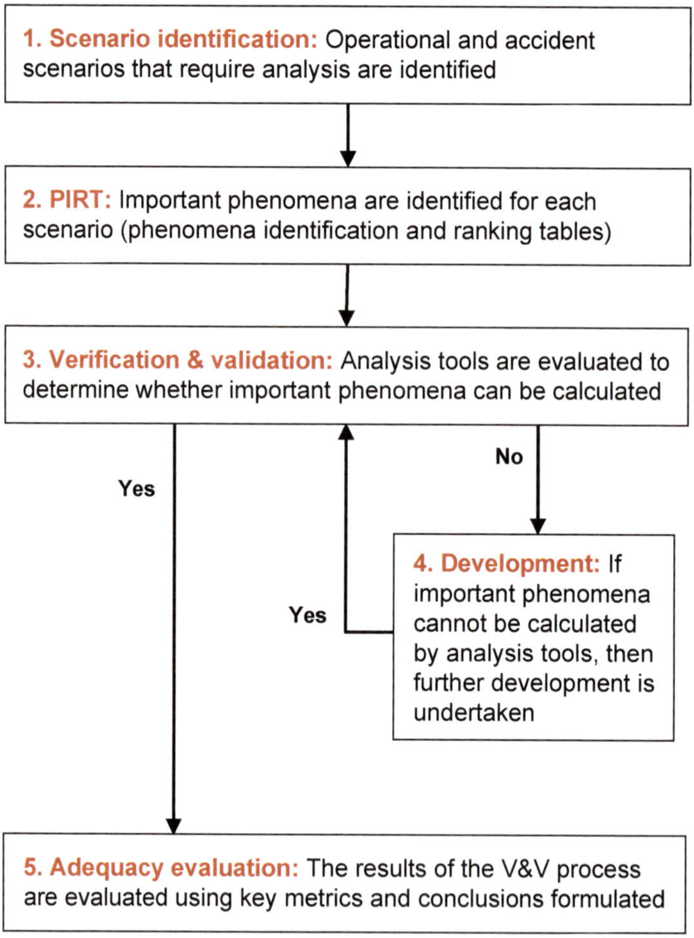

Figure 4.2 Phenomena identification and ranking table–informed R&D process.

The final stage, the adequacy evaluation, functions to bring together all the data—both experimental and calculational—obtained in earlier steps—for a thorough review and the formulation of conclusions regarding whether the software can make trustworthy predictions about nuclear system behavior for the scenarios of interest. Therefore, software deemed adequate in calculating a particular plant scenario is capable of calculating the dominant phenomena and physics with an acceptable calculational uncertainty that governs the behavior of the plant during the scenario in question.

Each stage consists of several important substages that are discussed individually in the subsequent sections. It is important to note that a positive outcome from the adequacy evaluation sets the stage for obtaining uncertainty quantifications on the figures of merit required by stakeholders—either for licensing or design purposes.

Online NPPs can only operate within limits defined by the capability of the licensee to demonstrate that the magnitude of all important figures of merit for plant safety parameters will not exceed acceptable limits. Thus, a sufficient margin must be maintained so these figures of merit are not in danger of being violated. A figure of merit is a key parameter indicating whether a safety limit or an equipment failure limit is in danger of being violated. Important figures of merit include fuel temperature, the quantity of water inventory in a light-water reactor pressure vessel, and reactor vessel wall temperature. Although a number of figures of merit are crucial to the operation of a power plant, for the sake of demonstrating the role of thermal-hydraulic tools/software, the fuel cladding temperature is perhaps the most common, particularly when considering safety limits during LOCAs in online light-water reactors.

In some cases, the limits translate to operating the plant at power levels below the designed power level. Another significant operational characteristic is the MLHR, or core power peaking factor in nondimensional form (Fq), which may have to be reduced to meet the acceptance criteria of LOCAs (max PCT of 1200°C). Such a reduction affects plant operational flexibility, which is an evident drawback for the utility but is not as detrimental as plant power reduction.

4.3.1 Software adequacy: A more detailed look at the process

To ensure that plant design analyses adequately cover the operational and accident domains of concern, several methodologies have been developed. In the United States, the Nuclear Regulatory Commission (NRC) issued Regulatory Guide 1.203 (USNRC, 2005), which describes historically acceptable processes for the development and assessment of numeric models used to evaluate transient and accident behaviors within the design basis of an NPP. The process is known as the "evaluation model development and assessment process" (EMDAP; see Fig. 4.3).

The term "evaluation model (EM)" is commonly used in the licensing arena to include all numerical models used to calculate nuclear system behavior, including system analysis and CFD software. The NRC's EMDAP flowchart comprises four major elements followed by a final "adequacy decision" step.

4.3.1.1 Element 1

Element 1 establishes the plant scenario types and phenomena ranges that define the operational and accident envelopes for the plant of interest. It is common practice to combine these two envelopes to form a single operational/accident envelope for the system that may be represented by validation matrices specific to each scenario type. In subsequent discussions, this envelope will be called the design envelope, as shown in Fig. 4.1. The envelope for the system is the domain within which the numeric models, which will be used to calculate the plant behavior, must be shown to be applicable and capable of calculating the key phenomena and figures of merit with acceptable calculational uncertainty. Once the domain for adequacy evaluation is defined, each scenario must be dissected to isolate regions with different controlling phenomena. Perhaps the most widely known example of a scenario being divided into regions with different controlling phenomena is the division of the light-water reactor double-ended guillotine break, also known as the large-break LOCA scenario, into blowdown,

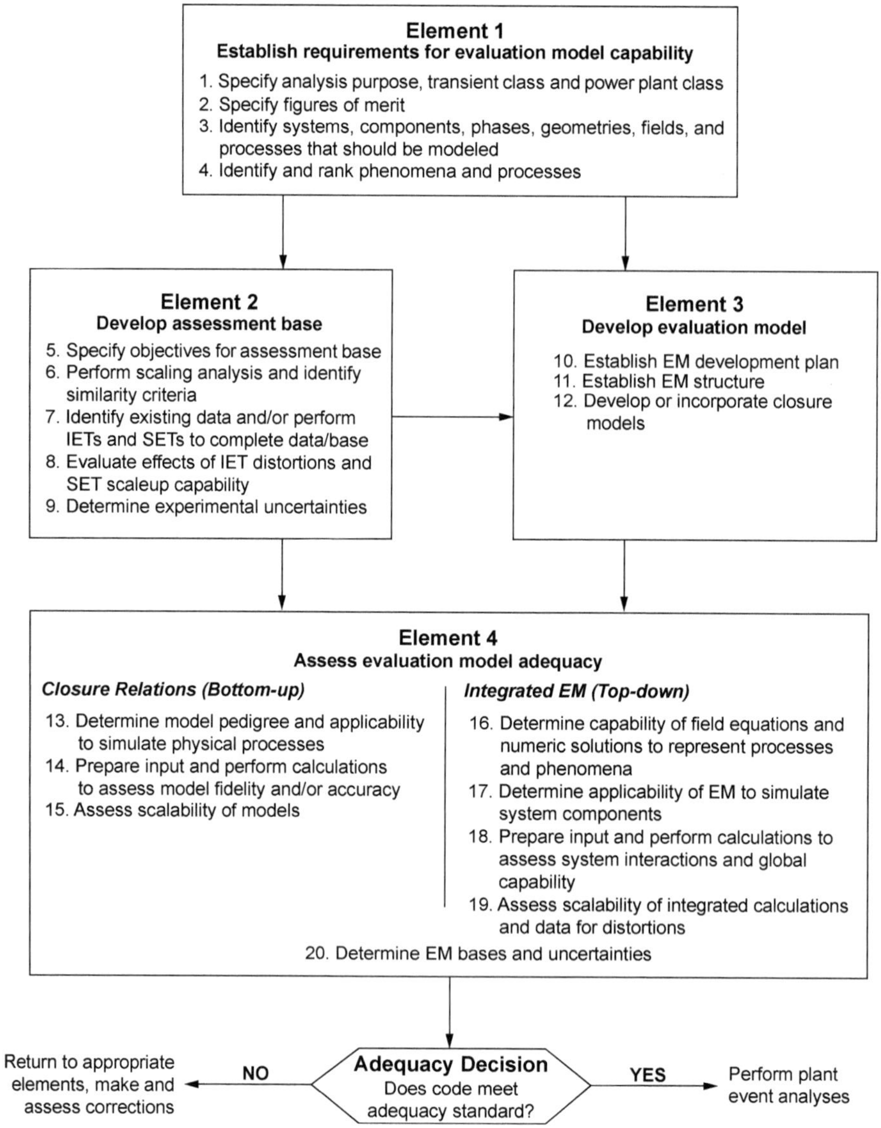

Figure 4.3 Elements of evaluation model development and assessment process (USNRC, 2005).

refill, and reflood regions (USNRC, 1988). For each region, different phenomena dictate the scenario progression—and the dominant phenomena together with less important phenomena that must still be considered are identified using a PIRT. The PIRT is an important ingredient in adequacy evaluation because the identification of key phenomena points to the physics that must be adequately represented numerically within each region (Wilson and Boyack, 1998).

4.3.1.2 Element 2

Element 2 is focused on developing the "assessment base," which is equivalent to developing the validation matrices used as the validation calculation benchmarks. The validation matrices are comprised of the data from experiments that have a rigorous, scaled relationship to the prototype as well as known measurement uncertainties that are sufficiently low to enable high-fidelity validation calculations to be performed.

The set of experiments designed to generate validation data is based on nuclear reactor operational and accident envelopes. The envelopes, in turn, are composed of the various scenarios that may occur. For example, the operational envelope may be composed of a set of scenarios that includes plant operations at various power levels—such as scenarios A–E shown in Fig. 4.4. The system of equations that describes the plant operational conditions for any region within a given scenario (e.g., the blue region within the scenario A domain) will change when a phenomenon or phenomena, which has negligible influence in the blue region, becomes influential, for example, the presence of subcooled boiling within the core denoted by the gold region within the scenario A domain. The boundary between the blue and gold regions denotes the change in the systems of equations that govern the thermal-hydraulic behavior in the two regions, respectively.

Such boundaries may signify changes to the systems of equations that, although significant enough to change the trajectory of the scenario progression, will not lead to a reduction in the plant safety margin. Conversely, some boundaries represent the crossover to a region where a "cliff-edge" effect may occur, for example, critical heat flux in the core, which if encountered will lead to significant temperature increases in the core fuel rods.

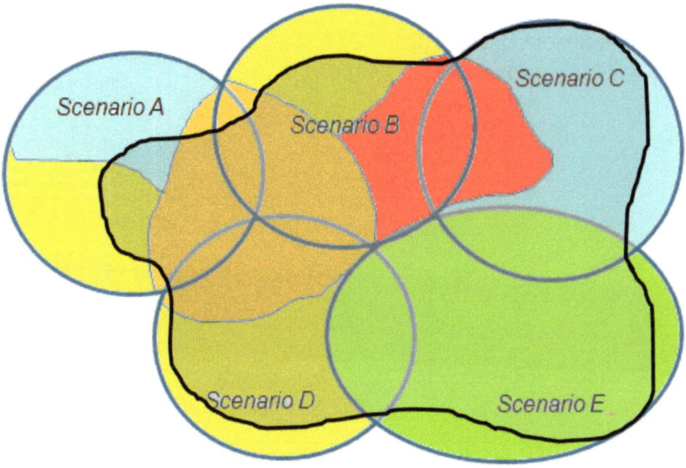

Figure 4.4 Plant operating envelope (irregular curve) overlaid by plant operational scenarios. Regions within each scenario, denoted by different colors, denote domains where a specific system of equations applies.

This concept is important from several perspectives: (1) an important role of thermal-hydraulic analysis is to define the boundaries that separate the various regions within scenarios that reflect important changes in key characteristics of the governing equations[2] and (2) scaled experiments designed to generate validation data focus on characterizing the fluid/heat transfer behavior within each region as well as defining the boundaries between regions.

If the validation matrix used to qualify the numeric models is rigorously defined, the experiments properly scaled such that experimental distortions are both minimized and quantified, and the data have sufficiently low uncertainties, then the groundwork for completing the Element 4 numeric model validation and numeric models adequacy assessment is in place.

The design envelope for any NPP is the domain of conditions and phenomena that the numeric models must be capable of calculating (see Figs. 4.1 and 4.5). The definition of the design envelope is addressed in Element 1 of the EMDAP. The region where the numeric models are valid, as confirmed by performing validation studies, is the calculational envelope and is the domain where the numeric models are shown to have the desired calculational fidelity. The numeric model's calculational envelope must encompass the design envelope, as shown in Fig. 4.5, if the numeric models will be used to calculate the projected behavior of the system for licensing studies.

The range of applicability for a given set of numeric models (the software) is in part determined by the physics and the numeric models contained within the software developed in Element 3. Therefore, the assessment or validation process, as implemented in Element 4, and the decision regarding the adequacy of the numerical models, must confirm that the software physics models properly calculate the key phenomena over the entire range of conditions encompassed by the calculation envelope. Success is only achieved when an adequate, high-fidelity data matrix and/or exact

Figure 4.5 Relationship of validation data to calculational and system envelopes.

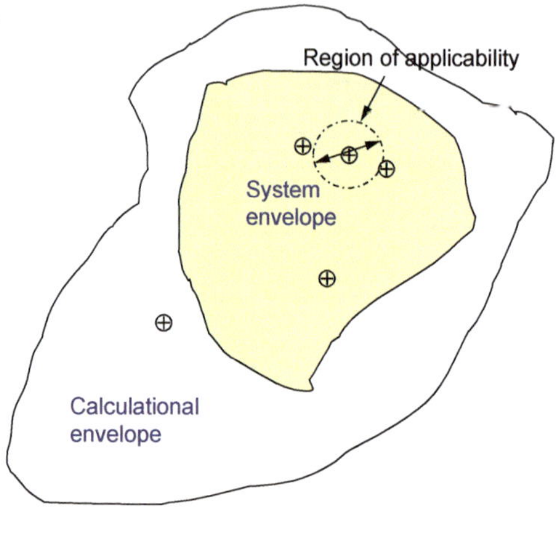

[2] Such changes may occur when fluid behavior moves into the region of applicability of various correlations.

analytical solution set are available to benchmark the calculation results over the range of conditions that encompass the entire design envelope.

The data included in the validation matrix are indicated in Fig. 4.5 by ⊕, where + indicates the data coordinates and the circle around the + indicates data uncertainty. An important point to be addressed is the applicability of the individual data points and the validations performed based on the selected data to the region between the data points. The "region of applicability" for a particular data point within the validation matrix is indicated by the subdomain defined by the dotted line surrounding an individual data point (see Fig. 4.5).

The validation matrix is constructed using specially designed experiments that are linked to the prototype by the scaling laws. Historically experiments such as LOFT, BETHSY, and ROSA were scaled using power-to-volume scaling with accommodations to achieve density-gradient scaling important for small break LOCA-related phenomena by using 1:1 height scaling (Ybarrondo et al., 1974; Levy, 1999). Over the last 20 years, more advanced scaling techniques have been employed, including hierarchical two-tiered scaling as first proposed by Zuber (1991).

Acceptably using the scaling laws, such as outlined in Zuber (1991), is crucial because when the numerical models are validated using data from correctly scaled facilities, as noted in Zuber et al. (1990), "Implicit to applications of computer codes to analyses of postulated accident events in nuclear plants is the premise that these codes have the capability to scale-up phenomena and processes from test facilities to full-scale plant conditions."

These scaling methods typically combine top-down and bottom-up scaling analyses. The top-down scaling approach is most applicable to the validation of system analysis codes and is usually performed for scaled integral experiments designed to simulate the integral behavior of the complete prototype plant. These scaled integral experimental facilities include the simulation of most of the major components. On the other hand, the bottom-up scaling approach is useful not only for system analysis software but especially for the validation of CFD software used to evaluate localized phenomena. The bottom-up scaling approach typically looks at phenomena occurring in individual components (local heat transfer, countercurrent flow, etc.) to confirm that scaling distortion does not occur at the local level or to quantify whatever distortion is present.

In the top-down scaling approach, the system to be addressed is defined, and the transient scenarios within the operational and accident domain are divided into several phases, and subphases if necessary. The system responses of interest in each phase are represented by the governing conservation equations that determine the behavior of the systems and/or components. These equations are then nondimensionalized and the nondimensional groups are evaluated to determine the relative importance of each group. A set of nondimensionalized coefficients characterizing the system response is then defined. The nondimensional groups from the calculated response of the prototype plant can then be compared with the measured data from the experimental facilities to quantitatively evaluate the scalability of the test facilities to the prototype plant for the parameters and scenarios of interest.

As indicated earlier, the bottom-up scaling approach is primarily concerned with localized phenomena, so it is primarily used to assess the applicability of individual models and correlations implemented into the system analysis and/or CFD software. This approach addresses the details lost in the averaging at the component level in the top-down scaling, thereby providing insights into qualitatively different responses between the test facility and plant. The bottom-up scaling approach is also useful in understanding apparent distortions between the test facility and plant that are observed but not easily explained by the nondimensional groups obtained from the top-down scaling approach. Identification and quantification of the distortions inherent to the experimental facility design and conditions are a key element of the process. This activity is identified as item eight of the EMDAP Element 2 (see Fig. 4.3).

The data needed for software validation depends on the particular software being considered. System analysis software is used to analyze the overall behavior of the prototype reactor or experimental facility for selected scenarios in the operational or accident domain, while the CFD software is used to study more detailed flow behavior and the potential for hot spots and/or unacceptably large thermal gradients in localized areas of the reactor system. Depending on the particular application, the system analysis and CFD software may be used separately or coupled together to perform the analyses.

When new experimental facilities are required the design of these new facilities must be rigorous to ensure compliance with NRC requirements for licensing a new reactor plant, particularly an advanced reactor design with a limited history of regulatory review. Important features to be considered in the experiment design include (1) an assurance that the proposed experiment facility captures key phenomena being investigated; (2) scaling of the experiment to provide a direct link between the scaled facility and prototype plant; (3) the inherent distortions of the scaled experiments are identified and quantified; (4) adequate high-quality measurements are available to ensure that experimental data uncertainties are quantifiable and acceptable low; and (5) experimental results that can be decomposed to the lowest level modeled by the software to ensure that system behavior at the component level is properly being calculated by the governing software physics.

All the data from the experimental facilities, taken together, form a data matrix. The data used in the validation of the numeric models form the validation matrix. Ideally, the validation matrix data are defined in a sufficiently ordered manner - for example, as depicted in Fig. 4.6 - such that the datasets are spaced at intervals sufficiently small that all of the system envelope domain is covered by either the data points or the combined subdomains of the individual data points as defined by the regions of applicability as shown in Fig. 4.5.

In general, the data comprising the validation matrix may include NPP, integral facility, separate effects, and basic (sometimes called fundamental) experimental data. The relationship between these types of data is shown in a "validation pyramid" as illustrated in Fig. 4.7. The experiments performed to investigate and characterize the behavior of individual phenomena are basic experiments and the data obtained from these experiments form the foundation of the validation pyramid. Also forming the foundation of the pyramid are analytical solutions to the phenomena. Separate-

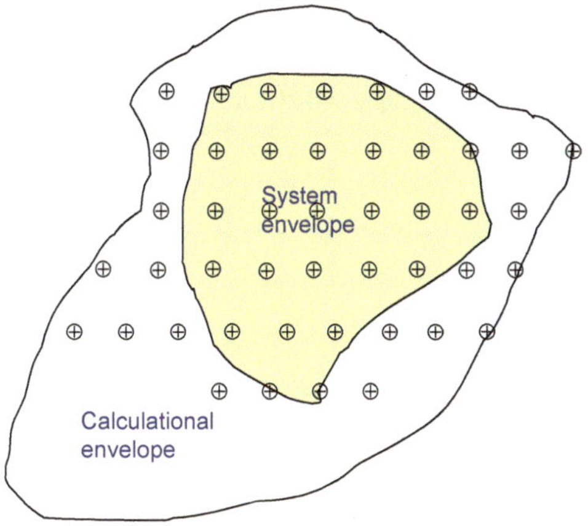

Figure 4.6 Ideal coverage of the system envelope using an ordered validation matrix approach.

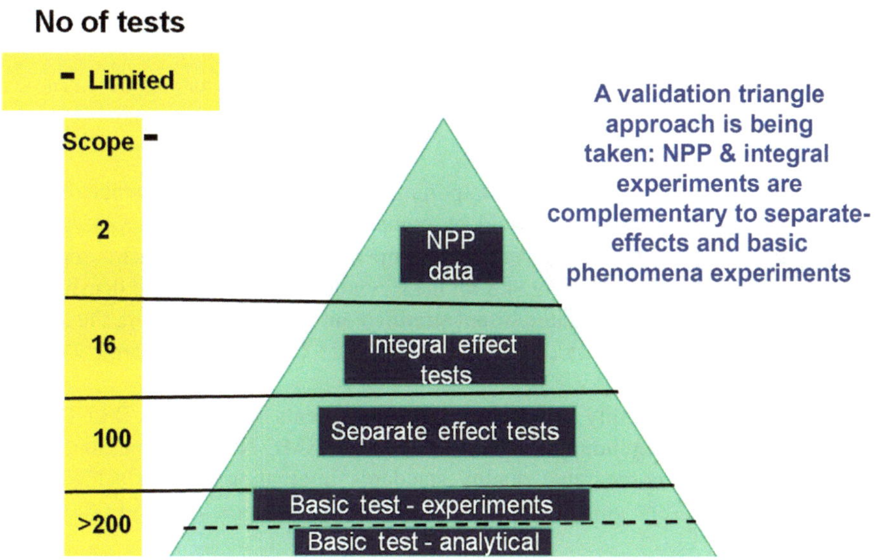

Figure 4.7 Concept of a validation pyramid.

effects experiments are designed to investigate and characterize the behavior of individual or a limited number of phenomena in a well-defined flow geometry, for example, turbulent flow in an inlet plenum. The separate-effects data obtained from such experiments form the next tier. The upper tiers of the validation pyramid are composed of the data obtained from integral-effects experiments and operational plants. The overall sum of the data forms the individual elements of the validation matrices.

In principle, each experiment's design relates directly to the system envelope of the prototype for the key phenomena or figures of merit. Thus the ranges of the data

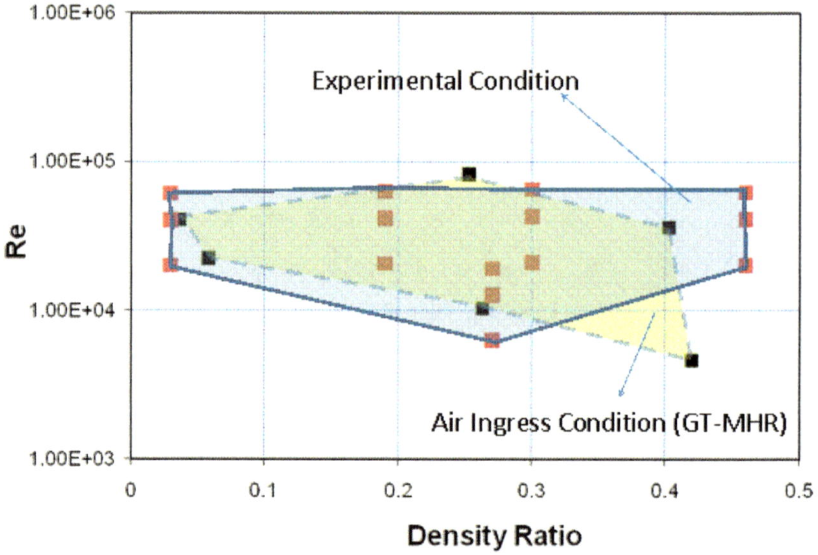

Figure 4.8 Comparison of scaled experimental and prototype envelopes for key phenomena in a design basis scenario.

measured in the experiments are quite important. The experimental data range of interest for the subject phenomena should be comparable to that of the prototype. An example comparison done for a very high-temperature gas-cooled reactor separate-effects experiment is shown in Fig. 4.8 for the Reynolds number (Re) and density ratio of density-gradient driven countercurrent stratified air-helium flow, where the system envelope for a gas-turbine modular high-temperature reactor is compared with the experimental envelope.

Experimental data quality must satisfy the requirements of Appendix B of Title 10 of Code of Federal Regulations Part 50.46 and the ASME NQA-1 standard—in the United States—and comparable quality standards internationally. To achieve this goal, the data should be generated in conformance with the practices and procedures described in the

i. International Organization for Standardization: "Guide to the expression of uncertainty in measurement"
ii. ASME PTC 19.1−2005: "Test Uncertainty" standard.

4.3.1.3 Element 3

Element 3 activities are centered on the creation of the numeric models that make up the EM and thus comprise the physics and mathematical models, including mathematical approximations, for example, finite difference models, finite element models, and correlation approximations. Many widely used numeric models describing thermal-hydraulic phenomena are already available. Examples of commonly used system analysis software include ATHLET, RELAP5, RETRAN, CATHARE, and TRACE. Examples of commonly used CFD software include Fluent, STAR-CCM +, and CFX. However, the existence of such software does not eliminate the need for V&V of the software.

Stemming from the presence of readily available software, such as the important and commonly known examples given in the previous paragraph, verification of the numeric models and developmental validation of the existing software packages presently usually focuses on revisions and additions that are specific to advanced NPP designs that have design features that are not present in online NPPs that were licensed and for which the above software are adequate and thus have been used to perform NPP behavioral calculations for licensing submittals to the applicable regulatory agencies.

4.3.1.4 Element 4

Element 4 encompasses the creation of input for the numeric models and initial and boundary conditions for the numeric models. Subsequently, the input models are verified, and finally, the validation calculations are performed as a part of the Element 4 activities. The degree to which the numeric model validation calculations compare to the experimental data is a primary ingredient in determining the adequacy of the numeric models for calculating the behavior of the system within the design envelope. After completion of the Element 4 activities, a decision regarding the adequacy of the EM to calculate the scenarios, including consideration of the calculational uncertainty, within the system envelope may be made. To summarize Element 4 activities, the V&V processes are discussed.

Verification

Practices and procedures are divided into several categories to indicate the goal and intent of each. These categories include code verification, code and calculation documentation, reduction of numerical error, quantification of numerical uncertainty, and calculation validation. Quantification of numerical uncertainty is discussed in some detail in Johnson et al. (2006).

Code Verification: Code verification of system analysis codes is generally done via traditional techniques centered on first performing detailed reviews of the coding by parties who are independent of the code development group followed by a rigorous quality assurance protocol designed to track the code results obtained using verification benchmark problems with respect to known high-fidelity results.

Code verification of CFD codes involves the determination of coding correctness, (Roache, 1998) a process separate from calculation verification (the quantification of numerical uncertainty). Users demand that the analysis software that will be applied to reactor safety analyses will already have been subjected to a variety of code verification tests. What will be required is documentation of these tests.

Only tests that exercise the options used in the particular computations need to be documented. The tests as a suite must be designed to exercise all the terms in the governing partial differential equations. For example, it is not adequate to only test the code on linearly varying solutions such as planar Couette flow, since this solution does not exercise vertical convection terms and others. The most complete and convincing type of code verification test uses the method of manufactured solutions (MMS), (Roache, 1998) but this will not be required. If MMS is not used, it will probably be necessary to use a suite of test problems to demonstrate code correctness. For all of these problems, the observed rate of discretization error convergence should be documented and compared to a theoretical value for the discretization algorithms employed. If it is not, then more stringent requirements will be enforced during calculation certification (Johnson et al., 2006). The code verification must also include some data on the effect of iterative convergence criteria on numerical results (see Chapters 11–13 for details).

Code and Calculation Documentation: Software that is used for nuclear reactor safety analysis must be described in detail in code documentation. Such documentation should include describing equations used and their discretization as well as the basic methods used to obtain a solution. The truncation error and its formal order or accuracy should be given. The code documentation must include all the details for implementing the turbulence models used in calculations, for example, turbulence models for CFD software. The code documentation should be available for reference to reviewers who must review the associated calculational results.

For each calculation performed and submitted as a safety analysis, assumptions must be listed along with the details of the methods and models used. Other details, including boundary and initial conditions, model constants (parameters), and other relevant information, must also be provided. Options not used in the calculation need not be documented.

Reduction of Numerical Error: The reduction of numerical error is a desirable objective for numerical calculations. Lessons have been learned about what not to do when using computational techniques for numerical analysis. These lessons learned have been canonized in the requirements for manuscripts submitted to well-known journals such as the ASME Journal of Fluids Engineering. It therefore seems prudent to apply them to relevant software for reactor safety analysis. Examples of such requirements are those given in the ASME's *Journal of Fluids Engineering* "Statement on Numerical Accuracy." Details regarding the philosophy and meaning of the various key points are discussed in Johnson et al. (2006). Examples of the content include requirements that (1) methods must be at least second-order accurate in space, (2) grid independence or convergence must be established, and (3) in transient calculations phase error must be assessed and minimized.

Grid independence is the process of refining the grid from the starting point until numerical results stop changing or change by negligible amounts. Theoretically, the results will continue to change until the grid spacing approaches zero. The precision of the machine, however, will halt this process at a finite grid spacing. This is sometimes referred to as achieving machine zero (of the residuals). Not only is the process of obtaining grid independence is not only important to reducing numerical error but also a good way to obtain estimates of numerical uncertainty (see "Quantification of Numerical Uncertainty" in Johnson et al. (2006)).

Iterative convergence relates to the number of iterations required to obtain residuals that are sufficiently close to zero, either for a steady-state problem or for each time step in an unsteady problem. This error is in addition to the numerical error associated with the truncation error terms. Because of the well-known and unacceptable sensitivity of some commercial codes to the iteration tolerance and too-lax default tolerance, the final calculations must determine this effect. At least two levels of iteration tolerance must be shown, and the sensitivity presented. For example, if results for a solution functional f are presented using a default iteration tolerance of (say) 10^{-3} reduction in residual from the initial condition, as required in Freitas et al. (2003), then another calculation with 10^{-4} will be required, and the sensitivity f_{10}' will be stated as the normalized % change in f per decade of change in iteration tolerance—as given in Eq. (4.1).

$$f_{10}' = \Delta f\left[f\left(10^{-4}\right) - f\left(10^{-3}\right)\right] / f_{\text{norm}} \times 100\% \tag{4.1}$$

The normalization can be based on $f_{\text{norm}} = f(10^{-3})$ when divides by near zero are not a problem, otherwise by another appropriate normalization. The final test of sufficient tightness of the iterative tolerance will be the acceptability of the final results based on the estimation of numerical uncertainty and validation metrics.

For transient calculations, the same convergence criterion should apply to spatial convergence (grid independence). The time step should be refined until a negligible change is obtained. Additionally, though not required by the *Journal of Fluids Engineering*, it is recommended that the time-wise discretization scheme should have second-order accuracy or better. While there are other practices to reduce numerical error, the above will constitute the required practices for reactor safety analysis at present. Other practices that reduce numerical error are certainly allowed and even encouraged.

Validation

Whether the software is adequate for performing best-estimate very high-temperature reactor (VHTR) analyses is determined using both "top-down" and "bottom-up" evaluations, as summarized in Fig. 4.9 and described in this section. Although an illustration of this process is given in the case of these VHTR analyses, it is perfectly generic, and the same remark stands for Fig. 4.9 as well.

"Bottom-Up" Code Adequacy: Bottom-up evaluation of code adequacy entails examination of four features: the pedigree, applicability, fidelity, and scalability of the code under consideration. The pedigree of a systems code consists of its history, its development procedures, and the basis for each correlation that is used in the code. Any correlations, data sources, and approximations used in the code must be documented in textbooks, laboratory reports, papers, etc. The uncertainty data used to bound the correlation(s), data, and approximations such as instrumentation and data system uncertainties must be included in the documentation. The basis for the uncertainties should be traceable and reproducible. The assumptions and limitations of the models must be known and documented.

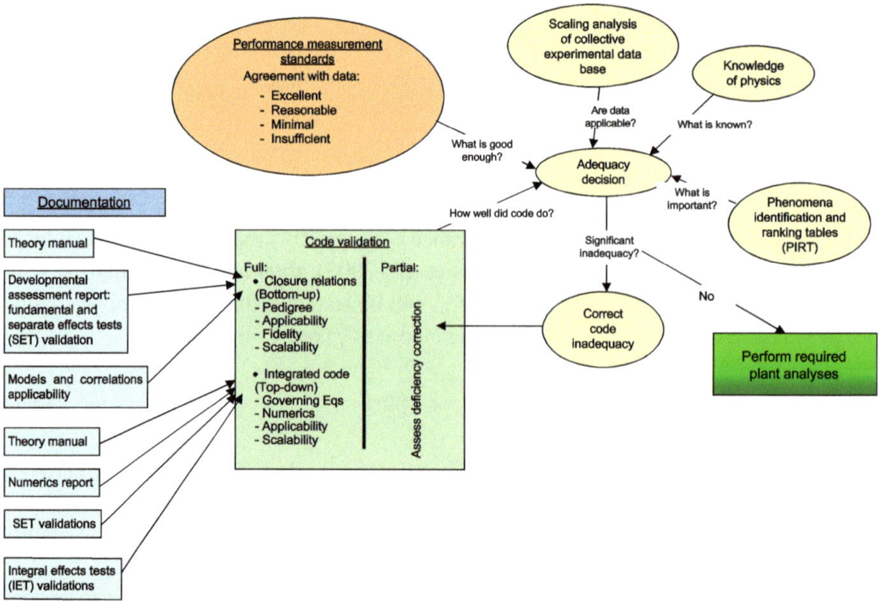

Figure 4.9 Very high-temperature reactor system design software—elements of adequacy evaluation and acceptance testing practices.

The applicability of a systems code depends on the range of use of each of its correlations, data, and approximations. Those ranges must be documented and referenced. Finally, the range of applicability claimed in the code manual should be consistent with the pedigree, or, if a greater range is claimed, the justification for the increase in range must be reported.

The fidelity of a systems code means the degree to which the code's predictions agree with physical reality. High fidelity requires that the mathematical models and correlations used in the code are not altered in an ad hoc manner from their documented formulation. A code is validated when it is shown that the code's predictions of key parameters agree within allowable tolerances with experimental data. The validation effort should be complete for all the key phenomena in the events of interest. Finally, benchmarking studies may either supplement the validation effort or make up the validation effort if appropriate standards are available, for example, comparisons of code calculation with a closed-form solution.

"Bottom-up" scaling stems from the need to.

- Build experimental facilities that model the desired full-scale system.
- Closely match the expected behavior of the most important transient phenomena in the scenarios of interest.
- Demonstrate the applicability of data from a scaled facility to a full-scale system and defend the use of data from a scaled facility in a code used to calculate the behavior of a full-scale system.
- Relate a calculation of a scaled facility to a calculation of a full-scale system.

Scalability studies are usually performed to scale key parameters for a portion of the system behavior, not to correlate global system behavior. Therefore, scalability analyses consist of four steps: (1) isolate the "first-order" phenomena, (2) characterize the "first-order" phenomena, (3) convert the defining equations into nondimensional form, and (4) adjust the experimental facility conditions to give equivalent behavior with the full-scale system within the limitation of the facility (or nearly equivalent, based on nondimensional numbers that follow from Step 3).

As implied in the earlier discussion, "bottom-up" code adequacy techniques focus principally on closure relationships. Thus, the field equations used in the code must be correctly formulated and programmed. In addition, the field equations must be reviewed by the scientific community, and its agreement on the correct formulation and insertion of the governing equations in the code must be obtained.

"Top-Down" Code Adequacy: The "top-down" approach for ensuring code adequacy focuses on the capabilities and performance of the integrated code. The top-down approach consists of four parts: numerics, fidelity, applicability, and scalability.

Numerics. Evaluation of the numerical solution considers convergence, stability, and property conservation.[3] Again, agreement by the scientific community on acceptable convergence, stability, and property conservation must be obtained.

Fidelity. The fidelity of the code is demonstrated by performing thorough code assessments based on applicable integral-effects and separate-effects data. The data are part of an agreed-upon code assessment matrix constructed based on the transients of importance and the key phenomena for each phase of the transients.

Applicability. The code must be shown to be capable of modeling the key phenomena in the system components and subsystems by conducting thorough validation studies. The key phenomena are identified in the PIRT.

The method to determine whether the code is capable of modeling key phenomena is to compare the calculation produced by the code to data that have known uncertainties. For example, "excellent" agreement between the code calculation and data is achieved if the calculated value is at all times within the data uncertainty band.

The degree of agreement between the code calculation and the data is generally divided into four categories as given in Table 4.1. A more rigorous definition is given by Schultz (1993). A code is considered adequate in applicability when it shows either excellent or reasonable agreement with the highly ranked phenomena (sometimes identified as the dominant phenomena) for a transient of interest. If the code gives minimal or unacceptable agreement, additional work must be performed, which may range from additional code development to additional analyses if needed to better understand the phenomena.

Scalability. Experimental scaling distortions are identified and isolated, for example, inappropriate environmental heat losses that stem from the larger surface-to-volume ratios that are inherent to scaled facilities. Finally, an effort to isolate all code scaling distortions is performed through code assessment calculations. Scaling distortions may arise from an inappropriate use of a correlation developed in a small-scale system when applied to a full-scale system.

[3] Property conservation issues arise when two calculations for the same property are performed by a system code using two different algorithms or methods. This practice may result from an attempt to enhance the accuracy of the code result. Because the two methods are likely to calculate slightly different values for the same property, such as pressure, property conservation must be considered.

Table 4.1 Code adequacy identifiers.

Classifier	Description
Excellent	The calculation lies within or near the data uncertainty band at all times during the phase of interest.
Reasonable	The calculation sometimes lies within the data uncertainty band and shows the same trends as the data. Code deficiencies are minor.
Minimal	Significant code deficiencies exist. Some major trends and phenomena are not predicted. Incorrect conclusions may be drawn based on the calculation without the benefit of data.
Unacceptable	A significant difference between the calculation and the data is present—and the difference is not understood. Such a difference could follow from errors in either the calculation or the portrayal of the data—or an inadequate code model of the phenomenon.

4.3.2 Software tool selection and software development

When confronted with the need to calculate key phenomena for an advanced NPP, analysts will inevitably be required to choose one software tool over another. This will be particularly true of system analysis software and CFD software. To assist the analyst in formally choosing software, a methodology is given in Fig. 4.10 where a flowchart summarizes key factors and questions such as.

- Has the software ever been used to analyze the phenomena or scenario? By answering this question, the analyst may be introduced to references and other experts who have applied the software to similar phenomena or scenarios. Hence a body of useful information may be available.
- Are the phenomena modeled properly? And does the model region of applicability correspond to the system phenomena or scenario envelope? These questions may be most easily answered by using the required manuals and documentation identified in Fig. 4.8, for example, models and correlations, theory manual, scaling relationships and applications, developmental assessment reports, and validations.
- Have validation studies been completed for the phenomena or scenario? Were the validation results reasonable or excellent (as defined in Table 4.1)—or were the results minimal or unacceptable? If a body of validation results are not available, or if the validation results were not "reasonable" as a minimum, then either the software should not be used, or it should be validated to ensure that the calculated results are reliable rather than misleading.

Only when acceptable answers are obtained for the questions listed earlier can the software under consideration be used for the required analysis with confidence.

4.4 Final remarks and connections with other chapters of the book

The chapter provides a summary picture of the deep intersection of thermal hydraulics into the legal process, or licensing, to demonstrate the safety of water-cooled nuclear reactors. The process constitutes the basis of requirements for transient thermal

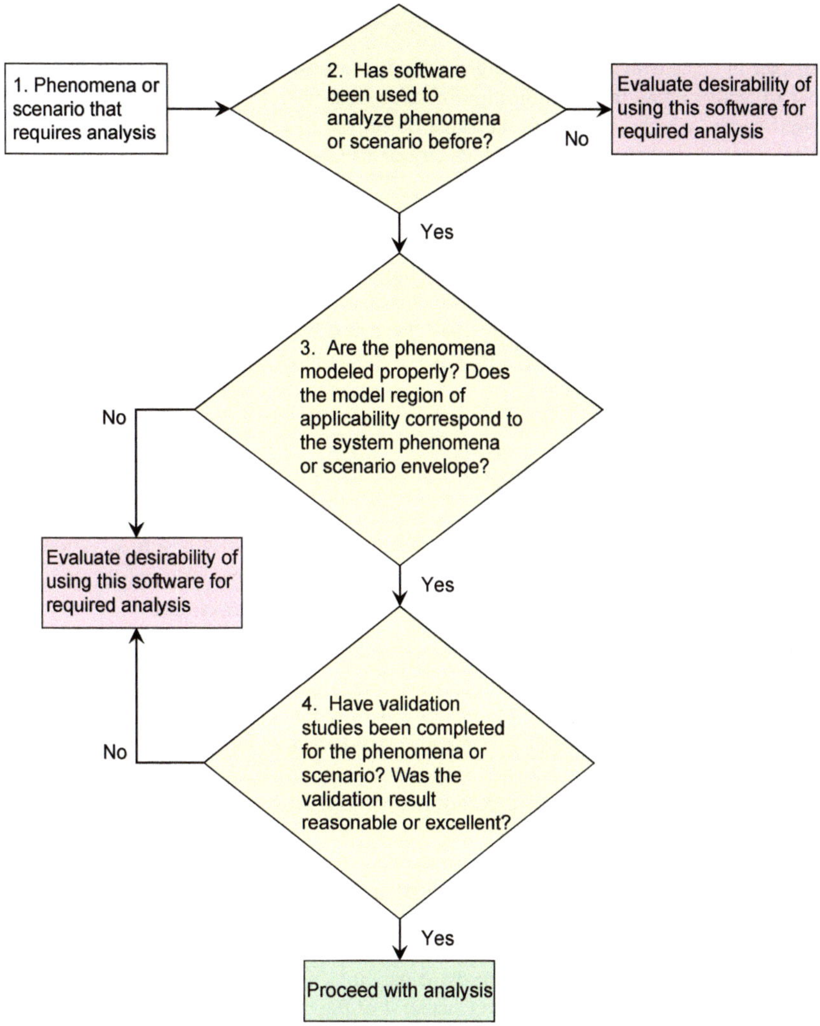

Figure 4.10 Flowchart to evaluate applicability of analysis software.

hydraulics. This, among other things, establishes the continuous moving boundaries for the acceptability of current knowledge and understanding in the area.

The chapter shows that thermal hydraulics and related knowledge are not sufficient for applications without procedures digging deeply into nuclear reactor safety technology. In different terms, procedures and methods constitute the additional leg, i.e., in addition to modeling and experimentation, to sustain the applicability of nuclear thermal hydraulics.

The concepts in the present chapter, as well as the questions posed by regulators, are considered with possible answers provided, in the following chapters of the book, primarily.

- Chapter 6 deals with phenomena descriptions, which have safety relevance and substantiate the EMDAP requirements.
- Chapter 11 deals with the design features of codes, i.e., system thermal hydraulics (SYS TH), that (directly) allow the quantification of safety margins. Additional thermal hydraulics–based codes, which accomplish the same objectives in combination with SYS TH or as stand-alone computational tools are discussed in Chapters 12 (CFD), 18 (subchannels), and 19 (containment).
- Chapters 13, 21 and 22 deal with the key procedures, or V&V, uncertainty, and scaling, essential for the safety-related application of SYS TH codes.
- Chapter 14 presents features of the general interactive framework called best estimate plus uncertainty; interactive, at least in the provided vision, implies the combined consideration of phenomena, modeling, procedures, and licensing requirements, including probabilistic safety assessment (PSA), although PSA is not the topic of the book.

Exercises and questions

Exercises and questions are proposed hereafter with solutions given in Appendix 1 (Volume III). The readers should answer the questions or provide solutions to exercises before reading Appendix 1. Please provide extended answers.

4.1 Please explain what a PIRT is and its connection to this book (e.g., Chapter 6). Separately, discuss the words "phenomena," "phenomena identification," and "phenomena ranking."
4.2 Discuss the challenges of a PIRT process, possibly including examples.
4.3 What are the elements of software adequacy? Where are these elements discussed in the book?
4.4 What is EMDAP? Identify connections between EMDAP and the topics in different chapters of the book.
4.5 Discuss the connections between thermal hydraulics and nuclear reactor safety.
4.6 What are the bases for arriving at the "thermal-hydraulic analysis envelope"?
4.7 Define, in your own words, CFR, RG, SRP, and finally, SAR. Please discuss the connections among the key objectives embedded in each acronym.
4.8 What are the connections between the capabilities and key features of system thermal-hydraulic codes and the content in Chapter 4 of the book?
4.9 Discuss key differences and similarities between Interim Acceptance Criteria (IAC) for the design of ECCS (1971), RG 1.70 (1978), and RG 1.203 (2005). *Please note that the concerned documents are not homogeneous (in relation to their content and objectives).*
4.10 What is the role of experiments within the context of Chapter 4?

Acknowledgments

The authors wish to acknowledge the comments on the present chapter provided by J.L. Gandrille.

Balance equations

5

Dominique Bestion and Christophe Morel
CEA, Grenoble, France

Foreword

It is mandatory for any scientist working with nuclear thermal hydraulics, namely those who use computer codes, to know the balance equations, their starting points, and the unavoidable approximations needed to achieve solutions.

The balance equations constitute the bases for the modern computational tools used in thermal hydraulics. An effort was made (by the authors of the chapter) to present the subject in steps of increasing difficulty until the description of current trends and foreseeable developments. One key issue considered is the distinctions and the similarities between balance equations as the bases of system thermal hydraulics and those of computational fluid dynamics (CFD).

The complexity of the topic should not discourage readers from obtaining insights into the equations; rather, this should encourage readers to find a path toward understanding nuclear thermal hydraulics.

Although the equations in this chapter are general and valid for any fluid where continuity assumptions and the stress-velocity relationship for Newtonian fluid are valid, the technology of water-cooled nuclear reactors (WCNRs) is the primary field for the derivation and application of these equations.

The subject of the chapter is the focal point in nuclear thermal hydraulics, as discussed in Chapter 1. The definitions and framework outlined in Chapter 3 are the basis for the balance equations. In addition, constitutive equations, Chapter 9, are embedded into balance equations that require special models, Chapter 10, to derive solutions applicable for modeling WCNRs. System codes, CFD, subchannels, and containment codes, Chapters, 11, 12, 18, and 19, respectively, use balance equations, with different formats and hypotheses typically used in derivation.

5.1 Introduction to balance equations

The objective of this chapter is to provide the minimum information to engineers on the most up-to-date systems of equations used to describe and predict single- and two-phase flow for thermal-hydraulic studies in nuclear reactors. Various modeling approaches are possible, and a few reference books exist on the mathematical derivation of the systems of equations (Delhaye et al., 1981; Lahey and Moody, 1993; Ishii and Hibiki, 2011; Morel, 2015). Fundamental research in the area was also done by Boure (see, e.g., Boure, 1975). Details of all the existing approaches will not be repeated here.

Handbook on Thermal Hydraulics in Water-Cooled Nuclear Reactors
https://doi.org/10.1016/B978-0-323-85606-5.00009-X

Rather, the following sections focus on important aspects of the most common sets of equations used in system codes and some CFD codes. Attention will be drawn to the physical assumptions and simplifications used to derive the equations and to the resulting limitations of their predictive capabilities.

After recalling the basic principles of thermodynamics and fluid mechanics, the equations for local instantaneous mass, momentum, and energy are provided. Then, the various approaches to single- and two-phase flow modeling are presented and classified. Time-averaged, single-phase equations are derived to introduce the CFD approach and turbulence modeling. Space-averaged 1-D equations for single-phase flow are derived, as they may provide reasonable approximations in many situations. Then, two-phase modeling is considered within the frame of the two-fluid approach. Time-averaged two-fluid equations are given, and the derivation of space-averaged 1-D equations is summarized to identify the most important simplifications before presenting the time-space-averaged 1-D two-fluid system of equations as it is used in current system codes. Some simple two-phase models, such as the homogeneous equilibrium model (HEM) and the drift flux model, although previously developed, are presented here as simplifications of the two-fluid model to better identify the degree of approximation as a function of the flow regime. Recent advances in two-phase flow modeling include the transport equation for the interfacial area (TIA) and the three-field model. The equations used in these new approaches are also given.

5.2 The principles of thermodynamics and mechanics

All fluids considered here in reactor thermal hydraulics are considered Newton–Stokes fluids. They may be present as a single-component (species) or multicomponent fluid. They may be present as a single-phase liquid, a single-phase gas, or even as a two-phase gas-liquid. A dominant species (e.g., water) may be present in both liquid and vapor phases, with other species present as noncondensable gas in a gas mixture or diluted in liquid (boron, nitrogen, ...).

5.2.1 Local instantaneous equations

The three basic equations result from these basic principles:

- mass conservation
- Newton's second law of motion applied to fluid flow
- energy conservation (first principle of thermodynamics)

Let us consider a fluid volume $V(t)$ limited by a surface $A(t)$, as in Fig. 5.1. One may consider the variation of a specific quantity ψ using the general conservation equation:

$$\frac{d}{dt}\int_{V(t)} \rho\, \psi dv = \int_{V(t)} \rho\phi\, dv - \int_{A(t)} \rho\psi\left(\vec{v} - \vec{v_A}\right)\cdot \vec{n}\, da - \int_{A(t)} \left(J \cdot \vec{n}\right) da \tag{5.1}$$

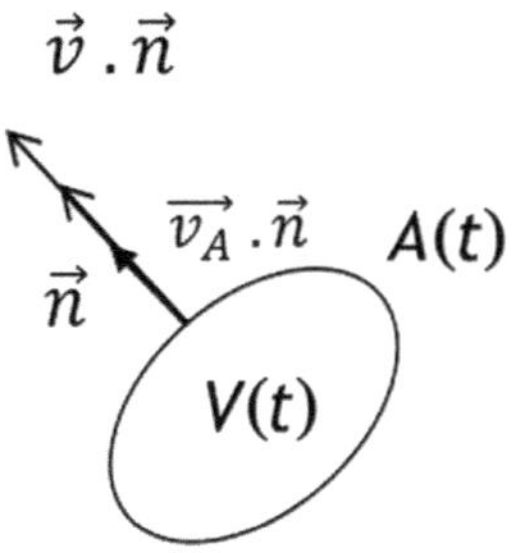

Figure 5.1 Volume *V(t)* limited by a surface *A(t)*.

where $-\int_{A(t)}\rho\psi\left(\vec{v}-\vec{v_A}\right)\cdot\vec{n}\,da$ is the flowrate of ψ entering through *A*(*t*), and

$\vec{n}$ unit vector directed outside *V*(*t*)

ρ is the fluid density (kg/m3)

$\vec{v}$ is the velocity vector (m/s)

$\vec{v_A}$ surface velocity vector (m/s)

ϕ volumetric source of ψ within *V*(*t*)

J diffusive flux of ψ through *A*

Mass momentum and total energy balance equations can be written considering the respective ψ, ϕ and $\vec{J}$ given in Table 5.1.

$\vec{F}$ is the specific external force (J/kg or m/s^2) usually limited to gravity force ($\vec{F}=\vec{g}$)

T is the stress tensor (Pa or N/m^2)

$\vec{q}$ heat flux vector (W/m^2)

q_{ext} external volumetric source of energy (usually radiation)

The Leibniz rule applied to a sufficiently regular function *f*(*x*, *y*, *z*, *t*) can be written as

$$\frac{d}{dt}\int_{V(t)}f\,dv=\int_{V(t)}\frac{\partial f}{\partial t}\,dv+\int_{A(t)}f\vec{v_A}\cdot\vec{n}\,da \tag{5.2}$$

Applied to $f=1$, this shows that

$$\int_{A(t)}\vec{v_A}\cdot\vec{n}\,da=\frac{dV(t)}{dt} \tag{5.3}$$

Table 5.1 Volumetric source and surface source in mass momentum and total energy equations.

	ψ	ϕ	J
Mass	1	0	0
Momentum	$\vec{v}$	$\vec{F}$	$-T$
Energy	$u+\frac{v^2}{2}$	$q_{ext}+\vec{F}\cdot\vec{v}$	$\vec{q}-T\cdot\vec{v}$

The Gauss theorem applied to a vector $\vec{a}$ and a tensor B can be written as

$$\begin{aligned}\int_{V(t)} \nabla\cdot \vec{a}\, dv &= \int_{A(t)} \vec{a}\cdot \vec{n}\, da \\ \int_{V(t)} \nabla\cdot B\, dv &= \int_{A(t)} B\cdot \vec{n}\, da\end{aligned} \tag{5.4}$$

Using the Leibniz rule and Gauss theorem, one can write all the terms of balance Eq. (5.1) using only volume-averaged terms. Applied to an elementary volume of fluid, this results in local instantaneous mass, momentum, and energy equations (see Delhaye et al., 1981).

The mass balance equation or continuity equation is

$$\frac{\partial \rho}{\partial t} + \nabla\cdot\left(\rho\, \vec{v}\right) = 0 \tag{5.5}$$

The momentum balance equation is

$$\frac{\partial \rho\, \vec{v}}{\partial t} + \nabla\cdot\left(\rho\, \vec{v}\, \vec{v}\right) - \rho\vec{F} - \nabla\cdot\ T = 0 \tag{5.6}$$

The total specific energy (internal energy + kinetic energy) balance equation is

$$\begin{aligned}&\frac{\partial}{\partial t}\left[\rho\left(u + \frac{v^2}{2}\right)\right] + \nabla\cdot\left[\rho\, \vec{v}\left(u + \frac{v^2}{2}\right)\right] - \rho\vec{F}\cdot \vec{v} \\ &\quad - \nabla\cdot\left(T\cdot\ \vec{v}\right) + \nabla\cdot\ \vec{q} = q_{ext}\end{aligned} \tag{5.7}$$

where u is the specific internal energy

$v^2 = \vec{v}\cdot\vec{v}$ is the specific kinetic energy (m^2/s^2)

T is split into a pressure tensor and a deviator stress tensor $\mathcal{T}$:

$$\mathcal{T} \triangleq T + p\, I$$

where τ_{ij} are the components of tensor $\mathcal{T}$

p is the pressure (Pa)

I is the identity tensor

Separating the contributions of pressure and $\mathcal{T}$,

$$\begin{aligned}&\frac{\partial}{\partial t}\left[\rho\left(u + \frac{v^2}{2}\right)\right] + \nabla\cdot\left[\rho\, \vec{v}\left(u + \frac{v^2}{2}\right)\right] - \rho\vec{F}\cdot \vec{v} + \nabla\cdot\left(p\, \vec{v}\right) \\ &\quad - \nabla\cdot\left(\mathcal{T}\cdot\ \vec{v}\right) + \nabla\cdot\vec{q} = q_{ext}\end{aligned} \tag{5.8}$$

A complete specific enthalpy equation can be written as

$$\frac{\partial}{\partial t}\left[\rho\left(h+\frac{v^2}{2}\right)-p\right]+\nabla\cdot\left[\rho\,\overrightarrow{v}\left(h+\frac{v^2}{2}\right)\right]-\rho\overrightarrow{F}\cdot\overrightarrow{v}-\nabla\cdot\left(\mathcal{T}\cdot\overrightarrow{v}\right)+\nabla\cdot\overrightarrow{q}=q_{ext} \tag{5.9}$$

where h is the specific enthalpy:

$$h=u+\frac{p}{\rho}$$

The momentum equation can then be written as

$$\frac{\partial\rho\,\overrightarrow{v}}{\partial t}+\nabla\cdot\left(\rho\,\overrightarrow{v}\,\overrightarrow{v}\right)+\nabla p-\rho\overrightarrow{F}-\nabla\cdot\mathcal{T}=0 \tag{5.10}$$

and projected along each coordinate in a Cartesian frame using summation on the repeated indices:

$$\frac{\partial\rho\,v_j}{\partial t}+\frac{\partial\rho v_i v_j}{\partial x_i}+\frac{\partial p}{\partial x_j}-\rho F_j-\frac{\partial\tau_{ij}}{\partial x_i}=0 \tag{5.11}$$

One can also write an equation for specific kinetic energy $\frac{v^2}{2}$ by multiplying the momentum energy by the velocity vector $\overrightarrow{v}$:

$$\frac{\partial\rho\,v^2/2}{\partial t}+\nabla\cdot\left(\rho\,\overrightarrow{v}\frac{v^2}{2}\right)-\rho\,\overrightarrow{F\cdot}\,\overrightarrow{v}+\overrightarrow{v}\cdot\nabla p-\overrightarrow{v}\cdot\nabla\cdot\mathcal{T}=0 \tag{5.12}$$

which can be subtracted from total energy or total enthalpy equations to obtain internal energy and enthalpy equations:

$$\frac{\partial\rho u}{\partial t}+\nabla\cdot\left(\rho\,\overrightarrow{v}\,u\right)+p\nabla\cdot\overrightarrow{v}-\mathcal{T}:\nabla\overrightarrow{v}+\nabla\cdot\overrightarrow{q}=q_{ext}$$

$$\frac{\partial\rho h}{\partial t}+\nabla\cdot\left(\rho\,\overrightarrow{v}\,h\right)-\frac{\partial p}{\partial t}-\overrightarrow{v}\cdot\nabla p-\mathcal{T}:\nabla\overrightarrow{v}+\nabla\cdot\overrightarrow{q}=q_{ext} \tag{5.13}$$

The flow can be solved using a system of equations for mass balance, momentum balance, and energy or entropy (one of the various equations available) complemented by equations of state (EoSs) and expressions of $\overrightarrow{q}$ and $\mathcal{T}$.

The Fourier law is used for $\overrightarrow{q}$:

$$\overrightarrow{q}=-\lambda\nabla T \tag{5.14}$$

where λ is the heat conductivity (W/m/K).

In Newton–Stokes fluid, $\mathcal{T}$ can be written as

$$\mathcal{T} = -\frac{2}{3}\mu\nabla\cdot\vec{v}\,I + \mu\left(\nabla\vec{v} + {}^{t}\nabla\vec{v}\right)$$

where μ is the dynamic viscosity (kg/m/s)

The thermodynamic state of a single-phase single-component fluid is characterized by the knowledge of two independent state variables, such as pressure and temperature (p,T) or (p,h), (p,u).

Let us consider that (p,h) are chosen as the independent variables; the necessary EoSs are then

$$\rho(p,h); T(p,h); \mu(p,h); \lambda(p,h)$$

Combining the above equations results in three equations with three independent unknowns p, h, and $\vec{v}$ such that

$$\frac{\partial\rho}{\partial t} + \nabla\cdot\left(\rho\,\vec{v}\right) = 0 \tag{5.15}$$

$$\frac{\partial\rho\,\vec{v}}{\partial t} + \nabla\cdot\left(\rho\,\vec{v}\,\vec{v}\right) + \nabla p - \rho\vec{F} - \nabla\cdot\left[-\frac{2}{3}\mu\nabla\cdot\vec{v}\,I + \mu\left(\nabla\vec{v} + {}^{t}\nabla\vec{v}\right)\right] = 0 \tag{5.16}$$

Often, μ is considered nearly homogeneous, so

$$\frac{\partial\rho\,\vec{v}}{\partial t} + \nabla\cdot\left(\rho\,\vec{v}\,\vec{v}\right) + \nabla p - \rho\vec{F} - \mu\left(\frac{1}{3}\nabla\nabla\cdot\vec{v} + \nabla^{2}\vec{v}\right) = 0$$

$$\frac{\partial}{\partial t}\left[\rho\left(h + \frac{v^{2}}{2}\right) - p\right] + \nabla\cdot\left[\rho\,\vec{v}\left(h + \frac{v^{2}}{2}\right)\right] - \rho\vec{F}\cdot\vec{v} - \nabla\cdot\left(\mathcal{T}\cdot\vec{v}\right) - \nabla\cdot\left(\lambda\nabla T\right) = q_{ext} \tag{5.17}$$

The momentum equation for each component is

$$\frac{\partial\rho v_j}{\partial t} + \frac{\partial\rho v_i v_j}{\partial x_i} + \frac{\partial p}{\partial x_j} = -\rho g_j + \mu\left[\frac{\partial^2 v_j}{\partial {x_i}^2} + \frac{1}{3}\frac{\partial^2 v_i}{\partial x_j \partial x_i}\right] \tag{5.18}$$

The entropy equation can be written as

$$\rho\frac{ds}{dt} = \frac{q_{ext}}{T} - \nabla\cdot\left(\frac{\vec{q}}{T}\right) + \frac{\mathcal{T}:\mathcal{T}}{2\mu T} + \frac{\lambda\left(\nabla T\right)^2}{T^2} \tag{5.19}$$

The last two terms of the entropy equation are the irreversible entropy sources associated with viscous dissipation and heat conduction.

The second principle of thermodynamics indicates that $\mu > 0$ and $\lambda > 0$.

The single-component single-phase flow equations of Newton–Stokes fluid can be summarized as in the following using the total enthalpy equation. Note that the principal variables here are p, h, and $\vec{v}$:

Basic equations: one-component one-phase flow equations for a Newton–Stokes fluid

Principal variables: p, $\vec{v}$, h

$$\frac{\partial \rho}{\partial t} + \nabla \cdot \left(\rho \vec{v}\right) = 0$$

$$\frac{\partial \rho \vec{v}}{\partial t} + \nabla \cdot \left(\rho \vec{v}\,\vec{v}\right) + \nabla p - \rho \vec{F} - \mu\left(\frac{1}{3}\nabla\nabla \cdot \vec{v} + \nabla^2 \vec{v}\right) = 0$$

$$\frac{\partial}{\partial t}\left[\rho\left(h + \frac{v^2}{2}\right) - p\right] + \nabla \cdot \left[\rho \vec{v}\left(h + \frac{v^2}{2}\right)\right] - \rho \vec{F} \cdot \vec{v} - \nabla \cdot \left(\mathcal{T} \cdot \vec{v}\right)$$
$$- \nabla \cdot \lambda \nabla T = q_{ext}$$

$$\mathcal{T} = -\frac{2}{3}\mu \nabla \cdot \vec{v}\, I + \mu\left(\nabla \vec{v} + {}^{t}\nabla \vec{v}\right);$$

$$\rho = \rho(p, h); T = T(p, h); \mu = \mu(p, h); \lambda = \lambda(p, h)$$

In a multicomponent fluid flow with n species, n-1 equations are added to predict the mass concentration of each species:

$$\frac{\partial \rho Y_k}{\partial t} + \nabla \cdot \left(\rho Y_k \vec{v}\right) = \nabla \cdot \left(\overrightarrow{\dot{m}_{Y_k}}\right) + S_{Y_k} \tag{5.20}$$

$$\overrightarrow{\dot{m}_{Y_k}} = D_{Y_k} \nabla Y_k$$

which can be written as

$$\frac{\partial \rho Y_k}{\partial t} + \frac{\partial \rho v_j Y_k}{\partial x_j} = \frac{\partial}{\partial x_j}\left(D_{Y_k}\frac{\partial Y_k}{\partial x_j}\right) + S_{Y_k} \tag{5.21}$$

where Y_k is the mass concentration of species k, and ρ is the density of the mixture:

$$\sum\nolimits_{k=1,n} Y_k = 1$$

where $\overrightarrow{\dot{m}_{Y_k}}$ is the diffusion mass flux vector of component k within the mixture (kg/m^2/s).

One may choose (p,h, $Y_{k=2,n}$) as independent variables, and the necessary EoSs are then functions of $2+n-1$ state variables:

$$\rho(p, h, Y_{k=2,n}); T(p, h, Y_{k=2,n}); \mu(p, h, Y_{k=2,n}); \lambda(p, h, Y_{k=2,n})$$

5.2.2 Some properties of local instantaneous equations

This set of equations has not yet proven itself to be a well-posed problem as defined by Jacques Hadamard, who requires that three conditions are met:

1. A solution exists.
2. The solution is unique.
3. The solution's behavior changes continuously with the initial conditions.

The third condition is directly linked to the predictability of fluid flows since it relates to sensitivity to initial conditions. Exponential sensitivity would result in an ill-posed problem, and high sensitivity may, at minimum, provide an ill-conditioned problem.

It is simply "believed" that fluid equations are exact and correspond to a well-posed problem—even if the demonstration is still considered open by mathematicians—but turbulence has been observed to result in rather high sensitivity to initial conditions. In addition, macroscopic quantities (time-averaged, space-averaged, or space-time-averaged) have shown lower sensitivity to initial conditions. This opens the possibility of predicting macroscopic fluid flow parameters using averaged balance equations.

Fluid flow equations are strongly nonlinear, which may explain the rather high sensitivity to initial conditions. The main source of nonlinearity is the advection term in momentum equations.

The turbulent nature of many flows induces a chaotic or stochastic character and a large spectrum of flow fluctuations, including very small scales. This makes it practically impossible to solve the local instantaneous equations. Analytically exact solutions are not known in (almost) all cases of interest, and numerical solutions are and for a long time will remain beyond the capabilities of current computers.

Since exact equations cannot be solved, simplified or averaged equations must be used to predict fluid flows. The main difficulty in industrial thermal-hydraulic studies is applying the right simplification to exact equations to provide an approximate solution with a sufficient degree of confidence. The next sections present some simplified sets of equations used in nuclear thermal hydraulics.

5.3 Various approaches to single-phase modeling

One can classify single-phase modeling approaches (Bestion et al., 2011) first according to the space dimension of their models, which considers 0-D, 1-D, 2-D, and 3-D models. For greater specificity, one may consider the time or space averaging used.

Three-dimensional models are often called CFD models. They can be first split into 3-D for a porous body and 3-D CFD for an open medium approach.

Reynolds-averaged Navier–Stokes (RANS)-type, large eddy simulation (LES)-type, and direct numerical solution (DNS) are the three main categories of single-phase CFD models for an open medium; see Table 5.2. RANS models are the ones most commonly used in nuclear reactor applications and include a large variety of turbulence models. Recommendations were made by the Writing Group of OECD-WGAMA for selecting models by application in reactor safety analysis (Mahaffy 2007).

5.3.1 Reynolds-averaged Navier–Stokes-type models

Time or ensemble averaging of local instantaneous equations (mass momentum and energy) is a common way to derive equations for the so-called RANS approach. Although time averaging and ensemble averaging are not the same, they can be considered reasonably equivalent (ergodicity) in the steady or quasi-steady flows where the RANS approach is applied. Since closure laws (e.g., for Reynolds stress) are almost always systematically validated against time-averaged measured flow parameters, one may consider the practical validated RANS model a time-averaged model. Time-averaged equations are supposed to filter all turbulent eddies and predict only a mean velocity field. Time averaging affects both time and space resolution in a Eulerian approach since eddies are transported by mean flow. The most popular RANS model ($k - \varepsilon$) uses a two-equation turbulence model with a Boussinesq approximation and turbulent viscosity. Many variations of two-equation turbulence modeling exist, such as k-l, k-ω, SST, RNG-k-ε, k-ε-V_2, and nonlinear k-ε. RANS was initially devoted to work on steady flow but may also be applied to unsteady or transient flow (U-RANS or T-RANS) if the time scale of the mean flow is larger than the time scale of the largest eddies. A RANS approach has been shown to predict large-scale Von Karman alleys in the wake of the cylinder provided that the numerical scheme does not induce artificial numerical viscosity.

RANS-type models require closure models for the Reynolds stress tensor and turbulent diffusion of heat. Wall functions are required for both momentum and energy to avoid too-fine meshes close to walls. The modeling effort is rather heavy because models for Reynolds stress require additional equations with several closure laws.

5.3.2 Scale/eddy-resolving-type models

When the computational grid is not sufficient to resolve the entire spectrum of scales (from integral to Kolmogorov), a space filter is applied to the basic balance equations. This allows the prediction of large eddies (sometimes called supergrid scale), whereas the effects of smaller eddies must be modeled statistically. This is the basis of LES. Detached eddy simulation and very large eddy simulation (VLES) belong to the same class of scale/eddy-resolving models. Some hybrid methods between U-RANS and LES exist, such as scale-adaptive simulation. LES requires a transient calculation with rather fine 3-D space nodalization; a high-order space discretization scheme and a

Table 5.2 Characteristics and modeling needs of various types of single-phase computational fluid dynamics.

	3-D			3-D			
	Open medium			Porous	2-D	1-D	0-D
		LES	RANS				
Type of model	**DNS**	**(DES, VLES, SAS, …)**	**(U-RANS, T-RANS)**				
Time or ensemble averaging	No	LES (No) DES, VLES, SAS (partially)	Yes	Yes	Yes	Yes	Yes
Space filtering Averaging	No	Yes	No	Filtering or averaging	Averaging	Averaging	Averaging
Treatment of eddies	All eddies simulated	Large eddies simulated Effects of small eddies modeled	No eddy simulated (Largest scale fluctuations—e.g., vortex shedding-may be simulated in U-RANS)	Effects of all eddies modeled	Effects of all eddies modeled	Effects of all eddies modeled	Effects of all eddies modeled
Required closure models	No model	Subgrid turbulent diffusion of momentum and scalars Wall functions	Reynolds stress tensor Turbulent diffusion of scalars Wall functions	3-D-diffusion-dispersion Wall friction Wall HT	2-D-diffusion-dispersion Wall friction Wall HT	Axial-diffusion-dispersion Wall friction Wall HT	Wall HT
Required modeling effort	0	X	XX	XXX	XXX	XXX	XX
Locality and universality of the modeling	*****[a]	****	***	*	**	*	–

DES, detached eddy simulation; *SAS*, scale-adaptive simulation. X is the unit modeling effort.
[a]See the last paragraph in Section 5.3.4.

high-order time integration method are needed, which makes the method much more CPU time-demanding than RANS methods. Statistical models must mimic the diffusive effects of the unresolved subgrid scales (called simply "subscale" in the VLES context) on momentum and energy transfers. Very often, turbulent viscosity and diffusivity models are used. Wall functions are also required for both momentum and energy to avoid fine meshes near walls. However, the modeling effort is much lighter than for RANS.

5.3.3 Direct numerical solution

DNS solves exact local instantaneous equations without any averaging or filtering. In turbulent flow, this requires that the nodalization is extremely fine to be able to capture the smallest eddies at the Kolmogorov scale. This approach is extremely CPU-costly and limited to some investigations of simple problems to create reference solutions or "numerical experiments" to assess other approaches.

5.3.4 Locality and universality of modeling

Both RANS- and LES-type methods require some closure laws for additional terms of equations coming from the averaging of the filtering procedure. The modeling of these terms most often assumes a kind of "principle of locality" where the physics of the process to model depends only on local flow variables (i.e., velocity, density, internal energy) and their time and space local derivatives. This is the case for the treatment of Reynolds stresses and turbulent heat fluxes in both RANS- and LES-type methods. The only exceptions to this principle of locality are the so-called wall functions where some closure terms depend on the distance to a wall. This allows for rather coarse nodalization in high-shear regions close to a wall.

It must be stressed that the principle of locality is much more relevant in a LES-type method than in a RANS approach. The subgrid local turbulent viscosity at a given point $P(x, y, z, t)$ of a flow in a LES approach models only effects of small eddies that are close to point P. In the RANS approach, the effects of the largest eddies on local physics have also to be modeled, and such large eddies depend on flow conditions far from point P. This is why different RANS model options are recommended for different flows like boundary layer, wake, mixing layer, or jet, whereas the best LES models tend to be more universal. In the RANS approach, local models are much more geometry-dependent than in LES. One may also consider that only DNS provides a full application of the principle of locality since the only transfers that are present are molecular transfers, which are expressed as functions of the first derivatives (Fourier law, Newtonian fluid law).

One may define a figure of merit to quantify the degree of locality and then the degree of universality of a model that would give five stars to the DNS *****, four stars **** to the LES, three stars *** to the RANS, two stars ** to the 2-D, only one star* to porous-3-D and 1-D, and no stars to 0-D.

5.4 Time-averaged single-phase equations

5.4.1 Reynolds averaging—time averaging

Time-averaged single-phase equations are used in single-phase RANS CFD models. CFD is presented in more detail in Chapter 12. Thus, these equations are briefly presented here for use in the double-space time-averaging of 0-D, 1-D, 2-D, and 3-D models for porous bodies.

When the boundary conditions of a flow are steady—not considering turbulent fluctuations—time-averaged equations are used. Reynolds averaging is applied to steady or quasi-steady flows where the time scales of mean flow variations are significantly larger than those of turbulent fluctuations. Time averaging over period T is

$$\overline{f} = \frac{1}{T}\int_{T} f\, dt \tag{5.22}$$

$$f \triangleq \overline{f} + f'$$

If variations in $\overline{f}$ over a time scale T are negligible,

$$\overline{f'} = 0$$

The properties of Reynolds averaging are such that

$$\overline{cf} = c\overline{f} \;\; if \; c = cte$$

$$\overline{\overline{f}} = \overline{f}$$

$$\overline{\overline{f}g} = \overline{f}\,\overline{g}$$

$$\overline{\frac{\partial f}{\partial x}} = \frac{\partial \overline{f}}{\partial x}$$

$$\overline{\frac{\partial f}{\partial t}} = \frac{\partial \overline{f}}{\partial t}$$

One can consider three levels of fluid compressibility:

1. Incompressible flow for low-Mach flow without dilatation:

$$\rho = \rho_0 = cte$$

2. Dilatable flow for low-Mach flow with dilatation:

$$\rho = \rho_0[1 - \beta(T - T_0)]$$

3. Compressible flow:

$$\rho = (P, T)$$

In the most general case, one may average local instantaneous equations and obtain the following mass, momentum, and total enthalpy equations:

$$\frac{\partial \overline{\rho}}{\partial x} + \frac{\partial \overline{\rho v_j}}{\partial x_j} = 0 \tag{5.23}$$

$$\frac{\partial \overline{\rho v_j}}{\partial t} + \frac{\partial \overline{\rho v_i v_j}}{\partial x_i} + \frac{\partial \overline{p}}{\partial x_j} = -\overline{\rho} g_j + \mu \left[\frac{\partial^2 \overline{v_j}}{\partial {x_i}^2} + \frac{1}{3} \frac{\partial^2 \overline{v}_i}{\partial x_j \partial x_i} \right] \tag{5.24}$$

$$\frac{\partial}{\partial t} \left[\overline{\rho u} + \frac{\overline{\rho v^2}}{2} \right] + \nabla \cdot \left[\overline{\rho u \vec{v}} + \frac{\overline{\rho \vec{v} v^2}}{2} \right] - \vec{F} \cdot \overline{\rho \vec{v}} - \nabla \cdot \left(\overline{\mathcal{T} \cdot \vec{v}} \right) + \nabla \cdot \overline{\vec{q}} = \overline{q_{ext}} \tag{5.25}$$

$$\frac{\partial}{\partial t} \left[\overline{\rho h} + \frac{\overline{\rho v^2}}{2} - \overline{p} \right] + \nabla \cdot \left[\overline{\rho h \vec{v}} + \frac{\overline{\rho \vec{v} v^2}}{2} \right] - \vec{F} \cdot \overline{\rho \vec{v}}$$
$$- \nabla \cdot \left(\overline{\mathcal{T} \cdot \vec{v}} \right) + \nabla \cdot \overline{\vec{q}} = \overline{q_{ext}} \tag{5.26}$$

It is usual to define mass-weighted average quantities of velocity, internal energy, or enthalpy:

$$\widetilde{v_j} = \overline{\rho v_j} / \overline{\rho}$$

$$\widetilde{u} = \overline{\rho u} / \overline{\rho} \tag{5.27}$$

$$\widetilde{h} = \overline{\rho h} / \overline{\rho}$$

The mass balance equation thus simplifies to

$$\frac{\partial \overline{\rho}}{\partial t} + \frac{\partial \overline{\rho} \widetilde{v_j}}{\partial x_j} = 0 \tag{5.28}$$

Averaging the nonlinear momentum balance advection term induces an additional tensor $\mathcal{T}^t$ of components R_{ij}:

$$\mathcal{T}^t \triangleq \overline{\rho \vec{v} \vec{v}} - \overline{\rho} \widetilde{\vec{v}} \widetilde{\vec{v}}; \ R_{ij} \triangleq \overline{\rho v_i v_j} - \overline{\rho} \widetilde{v_i} \widetilde{v_j} \tag{5.29}$$

where R_{ij} is the Reynolds stress tensor.

The averaging of the nonlinear enthalpy balance advection term induces an additional vector term $\overrightarrow{q}^t$, which is turbulent heat transfer or

$$\overrightarrow{q}^t \triangleq \overline{\rho h \overrightarrow{v}} - \overline{\rho}\widetilde{h}\widetilde{\overrightarrow{v}};\ q_j^t \triangleq \overline{\rho h v_j} - \overline{\rho}\widetilde{h}\widetilde{v}_j \tag{5.30}$$

5.4.2 Incompressible flow

In the case of an incompressible flow,

$$\frac{\partial \overline{v_j}}{\partial x_j} = 0 \tag{5.31}$$

$$\overline{\rho v_i v_j} \triangleq \rho \overline{v}_i \overline{v}_j + \rho \overline{v_i' v_j'};\ R_{ij} = \rho \overline{v_i' v_j'} \tag{5.32}$$

$$\overline{\rho h v_j} \triangleq \rho \overline{h}\overline{v}_j + \rho\, \overline{h' v_j'}\ ;\ q_j^t = \rho \overline{h' v_j'}$$

$$\frac{\partial \overline{v_j}}{\partial t} + \frac{\partial \overline{v_i v_j}}{\partial x_i} + \frac{1}{\rho}\frac{\partial \overline{p}}{\partial x_j} = -g_j - \frac{\partial R_{ij}}{\partial x_i} + \nu\left[\frac{\partial^2 \overline{v_j}}{\partial {x_i}^2}\right] \tag{5.33}$$

$$\frac{\partial}{\partial t}\left[\rho\left(\overline{h} + \frac{\overline{v}^2}{2} + k\right) - \overline{p}\right] + \nabla\cdot\left[\rho\left(\overline{h\overrightarrow{v}} + \frac{\overline{\overrightarrow{v}v^2}}{2}\right)\right] - \rho\overrightarrow{F}\cdot\overline{\overrightarrow{v}}$$
$$-\nabla\cdot\left(\overline{\mathcal{T}\cdot\overrightarrow{v}}\right) + \nabla\cdot\left(\overline{\overrightarrow{q}} + \overrightarrow{q}^t\right) = \overline{q_{ext}} \tag{5.34}$$

where $k = \frac{1}{2}\overline{v_i' v_i'}$; is turbulent kinetic energy. When $h = C_p T$.

$$\overline{\rho h v_j} \triangleq \rho \overline{h}\overline{v}_j + \rho C_p \overline{T' v_j'}\ ;\ q_j^t = \rho C_p \overline{T' v_j'} \tag{5.35}$$

This shows that R_{ij} and q_j^t represent, respectively, the transport of momentum and energy by velocity fluctuations. By analogy with transport by molecular agitation, they were first modeled as diffusive terms.

Usual modeling consists of using a turbulent viscosity and turbulent conductivity:

$$\mathcal{T}^t \triangleq R_{ij} = -\rho\overline{v_i' v_j'} = \rho\left[-\frac{2}{3}k\delta_{ij} + \nu_t\left(\frac{\partial \overline{v_i}}{\partial x_j} + \frac{\partial \overline{v_j}}{\partial x_i}\right)\right] \tag{5.36}$$

$$q_j^t = \rho C_p \overline{u_j' T'} = -\lambda_t \frac{\partial \overline{T}}{\partial x_j} \qquad \lambda_t = \frac{C_p \mu_t}{Pr_t} \tag{5.37}$$

$$\overline{u'_j Y_k'} = -D_{Y_k t}\frac{\partial \overline{Y_k}}{\partial x_j} \quad D_{Y_k t} = \frac{\nu_t}{Sc_t} \tag{5.38}$$

In addition, one can model isotropic turbulent viscosity as the product of turbulent length scale l with turbulent velocity scale v':

$$\nu_t \sim l\,v' \tag{5.39}$$

Following Kolmogorov, in high-Reynolds-number flows, the dissipation ε is controlled by inertial processes and does not depend on viscosity. Then from dimensional analysis one can write

$$\varepsilon \sim \frac{v'^3}{l} \tag{5.40}$$

Using the turbulent kinetic energy k as the square of the velocity scale:

$$\nu_t \sim \frac{v'^4}{\varepsilon} = C_\mu \frac{k^2}{\varepsilon} \tag{5.41}$$

This relation opens the way to the very popular k-ε turbulence model, which requires transport equations for k and ε. It belongs to the first-order RANS models.

Many other turbulence models exist for R_{ij} and q_j^t . More advanced second-order RANS models write a separate equation for each component of the R_{ij} tensor and $\overrightarrow{q}^t$ vector, which may be complemented by equations for ε, temperature variance $k_\theta = \overline{T'^2}$, and its associated dissipation ε_θ.

5.4.3 Closure of time-averaged equations

To close the system of time-averaged equations, some simplifications and models are necessary:

- A choice is first made to use only time averaging of all variables or use mass-weighted averaging of velocity and enthalpy.
- Expression for R_{ij}: models developed first for incompressible flows were extrapolated to compressible flow with some adaptation.
- Expression for q_j^t: models developed first for incompressible flows were extrapolated to compressible flow with some adaptation.
- Expression for $\overline{\frac{\rho v^2}{2}}$:

  a It may be approximated by $\overline{\rho\frac{v^2}{2}} \cong \overline{\rho}\frac{\overline{v}^2}{2} + \overline{\rho}k$, and k may be neglected considering that $\frac{\overline{v}^2}{2} \gg k$.

  b It may also be approximated by $\overline{\rho\frac{v^2}{2}} \cong \overline{\rho}\frac{\tilde{v}^2}{2} + \overline{\rho}k$, and k may be neglected considering that $\frac{\tilde{v}^2}{2} \gg k$.

- Expression for $\nabla\cdot\left(\overline{\mathcal{T}\cdot\overrightarrow{v}}\right)$ may require simplification of the average of products:

c $\nabla\cdot\left(\overline{\mathcal{T}\cdot\overrightarrow{v}}\right)\cong\nabla\cdot\left(\overline{\mathcal{T}}\cdot\overline{\overrightarrow{v}}\right)$

d or $\nabla\cdot\left(\overline{\mathcal{T}\cdot\overrightarrow{v}}\right)\cong\nabla\cdot\left(\overline{\mathcal{T}}\cdot\widetilde{v}\right)$

A possible set of time-averaged equations for a single-component single-phase flow equation of Newton−Stokes fluid can be summarized as follows if we use the total enthalpy equation and mass-weighted velocity and enthalpy. Note that the principal variables here are $\overline{p}$, $\widetilde{h}$, and $\widetilde{v_j}$:

Time-averaged equations: one-component one-phase flow equations for a Newton−Stokes fluid

Principal variables: $\overline{p}$, $\widetilde{\overrightarrow{v}}$, $\widetilde{h}$

$$\frac{\partial\overline{\rho}}{\partial t}+\nabla\overline{\rho}\,\widetilde{\overrightarrow{v}}=0$$

$$\frac{\partial\overline{\rho}\,\widetilde{\overrightarrow{v}}}{\partial t}+\nabla\overline{\rho}\,\widetilde{\overrightarrow{v}}\,\widetilde{\overrightarrow{v}}+\nabla\overline{p}=+\overline{\rho}\,\overrightarrow{g}+\nabla\cdot\left(\overline{\mathcal{T}}+\mathcal{T}^t\right)$$

$$\frac{\partial}{\partial t}\left[\overline{\rho}\left(\widetilde{h}+\frac{\widetilde{v}^2}{2}\right)-\overline{p}\right]+\nabla\cdot\left[\overline{\rho}\,\widetilde{\overrightarrow{v}}\left(\widetilde{h}+\frac{\widetilde{v}^2}{2}\right)\right]$$

$$=\overrightarrow{F}\cdot\overline{\rho}\,\widetilde{\overrightarrow{v}}+\nabla\cdot\left[\left(\overline{\mathcal{T}}+\mathcal{T}^t\right)\cdot\widetilde{\overrightarrow{v}}\right]+\nabla\cdot\left(\overline{\overrightarrow{q}}+\overrightarrow{q}^t\right)+\overline{q_{ext}}$$

$$\overline{\mathcal{T}}=-\frac{2}{3}\mu\nabla\cdot\widetilde{\overrightarrow{v}}\,I+\mu\left(\nabla\widetilde{\overrightarrow{v}}+{}^t\nabla\widetilde{\overrightarrow{v}}\right);\quad\overline{\overrightarrow{q}}=-\lambda\nabla\overline{T}$$

$$\overline{\rho}=\rho\left(\overline{p},\widetilde{h}\right);\overline{T}=T\left(\overline{p},\widetilde{h}\right);\mu=\mu\left(\overline{p},\widetilde{h}\right);\lambda=\lambda\left(\overline{p},\widetilde{h}\right)$$

+ *expressions or equations for* $\mathcal{T}^t$ and $\overrightarrow{q}^t$

One may note that time averaging only adds $\mathcal{T}^t$ and $\overrightarrow{q}^t$ to the momentum and energy equations.

5.5 Space-time averaged 1-D equations for single-phase flows

One-dimensional equations commonly describe flows and heat transfers within circuits. They can result from space averaging over the cross section of the duct and produce a set of three partial derivative equations as a function of time and curvilinear abscissa. It applies to ducts with a continuous cross-sectional area and is not adapted to abrupt area changes. Another approach is the use of successive "control volumes," where equations are integrated over a volume between two sections of the duct. This produces interconnected 0-D differential equations as a function of time only rather than real 1-D equations. This allows abrupt area changes and may be equivalent to 1-D equations when the distance between the sections tends to zero.

5.5.1 One-dimensional equations for single-phase flow

The derivation of space-time-averaged equations in 1-D is a very common approach for flows in pipes and ducts. As shown by Delhaye et al. (1981), one can proceed in one of two ways: first space averaging and then time averaging or time averaging and then space averaging. In the case of single-phase flow, the second approach is summarized here.

One will here use the set of time-averaged equations given in section 4, where $\overline{p}$, $\widetilde{h}$, and $\widetilde{v_j}$ are the principal variables, which can also be written (with summation on repeated indices) as

$$\frac{\partial \overline{\rho}}{\partial t}+\frac{\partial \overline{\rho}\,\widetilde{v_j}}{\partial x_j}=0$$

$$\frac{\partial \overline{\rho}\,\widetilde{v_j}}{\partial t}+\frac{\partial \overline{\rho}\widetilde{v_i}\,\widetilde{v_j}}{\partial x_i}+\frac{\partial \overline{p}}{\partial x_j}= -\overline{\rho}g_j+\frac{\partial}{\partial x_j}\left(\overline{\tau_{ij}}-R_{ij}\right)$$

$$\frac{\partial\left[\overline{\rho}\left(\widetilde{h}+\frac{\widetilde{v^2}}{2}\right)-\overline{p}\right]}{\partial t}+\frac{\partial\left[\overline{\rho}\,\widetilde{v_j}\left(\widetilde{h}+\frac{\widetilde{v^2}}{2}\right)\right]}{\partial x_j}-\overline{\rho}\,g_i\,\widetilde{v_i}$$
$$-\frac{\partial\left[\left(\overline{\tau_{ij}}-R_{ij}\right)\widetilde{v_i}\right]}{\partial x_j}+\frac{\partial\left(\overline{\overrightarrow{q_j}}+\overrightarrow{q}^{\,t}_j\right)}{\partial x_j}=\overline{q_{ext}} \tag{5.42}$$

Now one must average these equations for a duct cross section using the following notation:

$$<f>^S=\frac{1}{A}\int_A f\,ds \tag{5.43}$$

Fig. 5.2 shows a portion of a duct with the following notations:
z is the curvilinear abscissa of the duct.
$A(z)$ is the cross-sectional area.
C is the contour of A.
$\overrightarrow{n}$ is the vector normal to the duct wall directed to the exterior of the duct.

Figure 5.2 Portion of a duct.

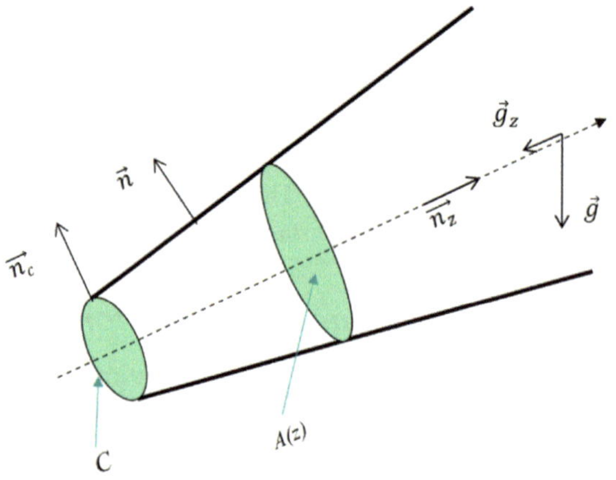

$\overrightarrow{n_c}$ is the vector normal to the contour C directed to the exterior of the duct.
$\overrightarrow{g_z}$ is the gravity acceleration vector projected along z axis.
One will use a particular form of the Leibniz rule:

$$\frac{d}{dt}\int_A f(x,y,z,t)\,da = \int_A \frac{\partial f}{\partial t}\,da + \int_C f\,\vec{v}\cdot\vec{n}\,\frac{dC}{\vec{n}\cdot\overrightarrow{n_C}} \tag{5.44}$$

Since there is no flow through the wall, $\vec{v}\cdot\vec{n} = 0$ and

$$\left\langle\frac{\partial f}{\partial t}\right\rangle^S = \frac{\partial\langle f\rangle^S}{\partial t} \tag{5.45}$$

Notation: $\hat{f} = <\bar{f}>^S$

One will use a particular form of the Gauss theorem applied to a vector $\vec{B}$ or a tensor T:

$$\int_A \nabla\cdot\vec{B}\,da = \frac{\partial}{\partial z}\int_A B_z\,da + \int_C \vec{n}\cdot\vec{B}\,\frac{dC}{\vec{n}\cdot\overrightarrow{n_C}} \tag{5.46}$$

$$\vec{B} = \overrightarrow{n_z} \Rightarrow \frac{\partial A}{\partial z} = -\int_C \vec{n}\cdot\overrightarrow{n_z}\,\frac{dC}{\vec{n}\cdot\overrightarrow{n_C}} \tag{5.47}$$

$$\int_A \nabla\cdot T\,da = \frac{\partial}{\partial z}\int_A \overrightarrow{n_z}\cdot T\,da + \int_C \vec{n}\cdot T\,\frac{dC}{\vec{n}\cdot\overrightarrow{n_C}} \tag{5.48}$$

5.5.1.1 Mass balance equation

$$\left\langle \frac{\partial \overline{\rho}}{\partial t} + \nabla \cdot \left(\overline{\rho}\, \widetilde{\vec{v}} \right) \right\rangle^S = \frac{\partial \langle \overline{\rho} \rangle^S}{\partial t} + \left\langle \nabla \left(\overline{\rho}\, \widetilde{\vec{v}} \right) \right\rangle^S = \frac{\partial \langle \overline{\rho} \rangle^S}{\partial t} + \frac{1}{A} \frac{\partial}{\partial z} \int_A \overline{\rho}\, \widetilde{v_z}\, da$$
$$= \frac{\partial \langle \overline{\rho} \rangle^S}{\partial t} + \frac{1}{A} \frac{\partial A \langle \overline{\rho}\, \widetilde{v_z} \rangle^S}{\partial z} = 0 \tag{5.49}$$

One may define the space-time-averaged velocity as

$$V \triangleq \frac{\langle \overline{\rho}\, \widetilde{v_z} \rangle^S}{\langle \overline{\rho} \rangle^S} = \frac{\langle \overline{\rho}\, \widetilde{v_z} \rangle^S}{\widehat{\rho}} \tag{5.50}$$

$$A \frac{\partial \widehat{\rho}}{\partial t} + \frac{\partial A \widehat{\rho} V}{\partial z} = 0 \tag{5.51}$$

5.5.1.2 Energy equation

$$\frac{\partial \overline{\rho} \left(\widetilde{h} + \frac{\widetilde{v^2}}{2} \right) - \overline{p}}{\partial t} + \frac{\partial \overline{\rho}\, \widetilde{v_j} \left(\widetilde{h} + \frac{\widetilde{v^2}}{2} \right)}{\partial x_j} - \overline{\rho}\, g_i\, \widetilde{v_i}$$
$$- \frac{\partial \left(\overline{\tau_{ij}} + R_{ij} \right) \widetilde{v_i}}{\partial x_j} + \frac{\partial\, \overline{\vec{q_j}} + \vec{q}^{\,t}_j}{\partial x_j} = \overline{q_{ext}} \tag{5.52}$$

$$\left\langle \frac{\partial \overline{\rho} \left(\widetilde{h} + \frac{\widetilde{v^2}}{2} \right) - \overline{p}}{\partial t} \right\rangle = \frac{\partial \langle \overline{\rho}\, \widetilde{h} \rangle^S}{\partial t} + \frac{\partial \left\langle \overline{\rho}\, \frac{\widetilde{v^2}}{2} \right\rangle^S}{\partial t} - \frac{\partial \langle \overline{p} \rangle^S}{\partial t} \tag{5.53}$$

Let us define the mass-weighted space-time average enthalpy H and the space-time average pressure P:

$$H \triangleq \frac{\langle \overline{\rho}\, \widetilde{h} \rangle^S}{\langle \overline{\rho} \rangle^S};$$

$$P \triangleq \langle \overline{p} \rangle^S = \widehat{p} \tag{5.54}$$

An approximation is necessary for the second term:

$$\left\langle \overline{\rho}\frac{\widetilde{v}^2}{2} \right\rangle \cong \langle \overline{\rho} \rangle^S \frac{V^2}{2} \tag{5.55}$$

$$\left\langle \frac{\partial \overline{\rho}\left(\widetilde{h}+\frac{\widetilde{v}^2}{2}\right)-\overline{p}}{\partial t} \right\rangle = \frac{\partial \widehat{\rho} H}{\partial t} + \frac{\partial \widehat{\rho}\frac{V^2}{2}}{\partial t} - \frac{\partial P}{\partial t} \tag{5.56}$$

$$\left\langle \frac{\partial \overline{\rho}\ \widetilde{v}_j\left(\widetilde{h}+\frac{\widetilde{v}^2}{2}\right)}{\partial x_j} \right\rangle^s = \frac{1}{A}\frac{\partial A\left\langle \overline{\rho}\ \widetilde{v}_z\left(\widetilde{h}+\frac{\widetilde{v}^2}{2}\right)\right\rangle^S}{\partial z} = \frac{1}{A}\frac{\partial A\langle \overline{\rho}\ \widetilde{v}_z\, \widetilde{h}\rangle^S}{\partial z} + \frac{1}{A}\frac{\partial A\left\langle \overline{\rho}\ \widetilde{v}_z\ \frac{\widetilde{v}^2}{2}\right\rangle^S}{\partial z} \tag{5.57}$$

Again, the averaging of the advection term induces an additional term q_z^d:

$$q_z^d \triangleq \langle \overline{\rho}\ \widetilde{v}_z\, \widetilde{h}\rangle^S - \widehat{\rho} V H \tag{5.58}$$

where q_z^d is called a "dispersion" term and has an axial diffusive effect on H. An approximation is necessary for the second term:

$$\left\langle \overline{\rho}\ \widetilde{v}_z\ \frac{\widetilde{v}^2}{2} \right\rangle^S \cong \langle \overline{\rho} \rangle^S V \frac{V^2}{2} \tag{5.59}$$

$$\begin{aligned}\left\langle \frac{\partial \overline{\rho}\ \widetilde{v}_j\left(\widetilde{h}+\frac{\widetilde{v}^2}{2}\right)}{\partial x_j} \right\rangle^s &= \frac{1}{A}\frac{\partial}{\partial z}\int_A \overline{\rho}\ \widetilde{v}_z\left(\widetilde{h}+\frac{\widetilde{v}^2}{2}\right)da = \frac{1}{A}\frac{\partial A\left\langle \overline{\rho}\ \widetilde{v}_z\left(\widetilde{h}+\frac{\widetilde{v}^2}{2}\right)\right\rangle^S}{\partial z} \\ &= \frac{1}{A}\left[\frac{\partial A\widehat{\rho} V H}{\partial z} - \frac{\partial A q_z^d}{\partial z} + \frac{\partial A\widehat{\rho} V\frac{V^2}{2}}{\partial z}\right]\end{aligned} \tag{5.60}$$

$$\langle \overline{\rho}\, g_i\, \widetilde{v}_i \rangle^S \cong \widehat{\rho}\, g_z V$$

$$\left\langle \frac{\partial\left(\overline{\tau_{ij}} - R_{ij}\right)\widetilde{v_i}}{\partial x_j} \right\rangle^S = \frac{1}{A}\frac{\partial}{\partial z}\left[A< \left(\left(\overline{\mathcal{T}} + \mathcal{T}^t\right)\cdot \overrightarrow{\widetilde{v}}\right)\overrightarrow{n_z} >^S\right]$$

$$\left\langle \frac{\partial\, \overrightarrow{\overline{q_j}} + \overrightarrow{q}_j^t}{\partial x_j} \right\rangle^S = \frac{1}{A}\frac{\partial}{\partial z} A< \overline{q_z} + q_z^t >^S + \frac{1}{A}\int_C \overrightarrow{n}\cdot\left(\overrightarrow{\overline{q_j}} + \overrightarrow{q}_j^t\right)\frac{dC}{\overrightarrow{n}\cdot\overrightarrow{n_C}}$$

$$= \frac{1}{A}\frac{\partial A\left(\widehat{q_z} + \langle q_z^t\rangle^S\right)}{\partial z} + \frac{1}{A}Cq_w$$

Cq_w is the wall heat transfer term:

$$A\frac{\partial}{\partial t}\left[\widehat{\rho}\left(H + \frac{V^2}{2}\right) - P\right] + \frac{\partial}{\partial z}\left[A\widehat{\rho}\, V\left(H + \frac{V^2}{2}\right)\right] - A\widehat{\rho}g_z\, V =$$

$$\frac{\partial}{\partial z}A< \left(\left(\overline{\mathcal{T}} + \mathcal{T}^t\right)\cdot \overrightarrow{\widetilde{v}}\right)\overrightarrow{n_z} >^S - \frac{\partial A q_z^{eff}}{\partial z} + Cq_w + A\widehat{q_{ext}}$$

$$q_z^{eff} \triangleq \widehat{q_z} + \langle q_z^t\rangle^S + q_z^d = -\left(\lambda + \lambda_t + \lambda_d\right)\frac{\partial T}{\partial z}$$

where q_z^{eff} is the effective axial diffusion-dispersion term.

The axial contribution of the stress tensor $\overline{\mathcal{T}} + \mathcal{T}^t$ is usually neglected so that the equation can be written as

$$A\frac{\partial}{\partial t}\left[\widehat{\rho}\left(H + \frac{V^2}{2}\right) - P\right] + \frac{\partial}{\partial z}\left[A\widehat{\rho}\, V\left(H + \frac{V^2}{2}\right)\right] - A\widehat{\rho}g_z\, V = Cq_w + A\widehat{q_{ext}} - \frac{\partial A q_z^{eff}}{\partial z} \quad (5.61)$$

The effective axial diffusion-dispersion term may be neglected when heat sources at the wall have a much higher impact than axial diffusion.

5.5.1.3 Momentum equation

One can apply the particular form of the Leibniz rule and Gauss theorem to the vectorial equation and then project along z axis or apply the particular form of the Leibniz rule and Gauss theorem to the equation projected along the z-axis:

$$\int_A \frac{\partial \overline{\rho}\widetilde{v}_z}{\partial t}\, ds = A\left\langle \frac{\partial \overline{\rho}\widetilde{v}_z}{\partial t}\right\rangle^S = A\frac{\partial \langle \overline{\rho}\widetilde{v}_z\rangle^S}{\partial t} = A\frac{\partial \widehat{\rho} V}{\partial t} \quad (5.62)$$

$$A\left\langle \nabla\cdot\left(\overline{\rho}\widetilde{v}_z\,\overrightarrow{\widetilde{v}}\right)\right\rangle^S=\frac{\partial A\langle\overline{\rho}\widetilde{v}_z\widetilde{v}_z\rangle^S}{\partial z} \tag{5.63}$$

One must define a momentum axial dispersion term T_z^d:

$$T_z^d\triangleq\langle\overline{\rho}\ \widetilde{v}_z\ \widetilde{v}_z\rangle^S-\widehat{\rho}VV \tag{5.64}$$

$$A\left\langle \nabla\cdot\left(\overline{\rho}\widetilde{v}_z\,\overrightarrow{\widetilde{v}}\right)\right\rangle^S=\frac{\partial A\langle\overline{\rho}\widetilde{v}_z\widetilde{v}_z\rangle^S}{\partial z}=\frac{\partial A\widehat{\rho}VV}{\partial z}+\frac{\partial AT_z^d}{\partial z} \tag{5.65}$$

The contribution of pressure term is split into an axial gradient and a contribution Fp along the duct wall contour C:

$$A\left\langle\frac{\partial\overline{p}}{\partial z}\right\rangle^S=A\left\langle\nabla\overline{p}\cdot\overrightarrow{n_z}\right\rangle^S=\frac{\partial A<\overline{p}>^S}{\partial z}+A\int_C\overline{p}\ \overrightarrow{n}\cdot\overrightarrow{n_z}\frac{dC}{\overrightarrow{n}_c\cdot\overrightarrow{n}}=\frac{\partial AP}{\partial z}+Fp \tag{5.66}$$

with

$$Fp=\int_C\overline{p}\ \overrightarrow{n}\cdot\overrightarrow{n_z}\frac{dC}{\overrightarrow{n}_c\cdot\overrightarrow{n}} \tag{5.67}$$

The local time-averaged pressure at a wall is split into the space-time-averaged pressure P and a local time-averaged pressure difference $\delta\overline{p}$:

$$\overline{p}=<\overline{p}>^S+\delta\overline{p}=P+\delta\overline{p} \tag{5.68}$$

$$Fp=-P\frac{dA}{dz}+\int_C\delta\overline{p}\ \overrightarrow{n}\cdot\overrightarrow{n_z}\frac{dC}{\overrightarrow{n}_c\cdot\overrightarrow{n}}=-P\frac{dA}{dz}+F'p$$

with

$$F'p=\int_C\delta\overline{p}\ \overrightarrow{n}\cdot\overrightarrow{n_z}\frac{dC}{\overrightarrow{n}_c\cdot\overrightarrow{n}}$$

then

$$A\left\langle\frac{\partial\overline{p}}{\partial z}\right\rangle^S=A\frac{\partial P}{\partial z}+F'p$$

No difficulty with the external force (gravity) term:

$$\langle\overline{\rho}g_z\rangle^S=\widehat{\rho}g_z$$

Two contributions of the deviator stress tensor:

$$A\left\langle \nabla\cdot\left(\overline{T}+T^{t}\right)\cdot\overrightarrow{n}_{z}\right\rangle^{S} = -\frac{\partial}{\partial z}A < \overrightarrow{n_z}\cdot\left(\overline{T}+T^{t}\right)\cdot\overrightarrow{n_z}) > + \int_{C}\overrightarrow{n}\cdot\left(\overline{T}+T^{t}\right)\cdot\overrightarrow{n_z})\frac{dC}{\overrightarrow{n}_{c}\cdot\overrightarrow{n}}$$

The first RHS term is usually small and neglected. The second RHS term is the wall friction term:

$$-C\tau_w \triangleq \int_{C}\overrightarrow{n}\cdot\left(\overline{T}+T^{t}\right)\cdot\overrightarrow{n_z})\frac{dC}{\overrightarrow{n}_{c}\cdot\overrightarrow{n}}$$

Finally:

$$A\frac{\partial\widehat{\rho}V}{\partial t}+\frac{\partial A\widehat{\rho}VV}{\partial z}+A\frac{\partial P}{\partial z}=\widehat{\rho}g_z - C\tau_w - F'p - \frac{\partial AT_z^d}{\partial z} \tag{5.69}$$

where $F'p$ is nonzero when the duct shape or cross section is not constant $\left(\overrightarrow{n}\cdot\overrightarrow{n_z}\neq 0\right)$.

It considers geometric effects on pressure forces and is usually modeled as a singular pressure loss:

$$-F'p = A\left(\frac{\partial P}{\partial z}\right)_{sing} = -A\frac{K_{sing}}{\delta z}\rho\frac{V^2}{2}$$

It can play an important role in singular geometry such as convergent, divergent, bend, and any other complex geometry that can influence the flow. The form-loss coefficient K_{sing} depends on the geometry and may be a function of the Reynolds number $Re=\frac{\rho VD_h}{\mu}$, with D_h being the hydraulic diameter $D_h=\frac{4A}{C}$:

$$-F'p = +\left(p_p - <p>\right)\frac{dA}{dz}$$

The last dispersion term, $\frac{\partial AT_z^d}{\partial z}=\frac{\partial A\left(\langle\overline{\rho}\ \widetilde{v_z}\ \widetilde{v_z}\rangle^S-\widehat{\rho}VV\right)}{\partial z}$, is zero in established flow (transverse profiles of $\widetilde{v_z}$ do not depend on z) in a constant-cross-section duct ($A=Ct$). In nonestablished flow within a constant-cross-section duct, it is usually neglected. In a variable cross section or the presence of complex geometry, its effect may be modeled together with $F'p$ as a geometric effect.

These equations must be complemented by EoSs that are necessarily approximated due to the averaging of nonlinear functions, such as

$$\widehat{\rho}=\left\langle\overline{\rho(p,h)}\right\rangle^S \cong \rho(P,H)=\rho\left(\langle\overline{p}\rangle^S,\frac{\left\langle\overline{\rho}\,\widetilde{h}\right\rangle^S}{\langle\overline{\rho}\rangle^S}\right)$$

These equations must be complemented by closure laws for τ_w, K_{sing}, q_w, and q_z^{eff}. The usual expressions for τ_w and K_{sing} are in the following form:

$$\tau_w = Cf(Re)\rho\frac{V^2}{2}; Cf = aRe^{-b};\ K_{sing} = cRe^{-d} \tag{5.70}$$

The usual expressions for q_w and q_z^{eff} are

$$q_w = h(T_w - T) = \frac{\lambda Nu(Re, Pr)}{D_h}(T_w - T); Nu = C\ Re^p\ Pr^q$$

$$q_z^{eff} = -\lambda^{eff}\frac{\partial \overline{T}}{\partial z} \tag{5.71}$$

The resulting set of 1-D equations is then:

One-phase equations: one-component one-phase flow space-time-averaged equations

Principal variables: $P,\ V, H$

$$A\frac{\partial \widehat{\rho}}{\partial t} + \frac{\partial A\widehat{\rho}V}{\partial z} = 0$$

$$A\frac{\partial \widehat{\rho}V}{\partial t} + \frac{\partial A\widehat{\rho}VV}{\partial z} + A\frac{\partial P}{\partial z} = \widehat{\rho}g_z - C\tau_w - A\frac{K_{sing}}{\delta z}\rho\frac{V^2}{2}$$

$$\mathrm{A}\frac{\partial}{\partial t}\left[\widehat{\rho}\left(H + \frac{V^2}{2}\right) - P\right] + \frac{\partial}{\partial z}\left[A\widehat{\rho}\ V\left(H + \frac{V^2}{2}\right)\right] - A\widehat{\rho}g_z\ V = Cq_w$$
$$+ A\widehat{q_{ext}} - \frac{\partial A q_z^{eff}}{\partial z}$$

$$\tau_w = Cf(Re)\rho\frac{V^2}{2}; K_{sing} = K_{sing}(Re); Re = \frac{\widehat{\rho}VD_h}{\mu}$$

$$q_w = h(T_w - T) = \frac{\lambda Nu(Re, \mathrm{Pr})}{D_h}(T_w - T); q_z^{eff} = -\lambda^{eff}\frac{\partial \mathrm{T}}{\partial z}; Pr = \frac{\mu Cp}{\lambda}$$

$$P = \langle\overline{p}\rangle^S; H = \frac{\langle\overline{\rho}\,\widetilde{h}\rangle^S}{\langle\overline{\rho}\rangle^S};\ \widehat{\rho} = \rho(P, H); \mu = \mu(P, H); \lambda = \lambda(P, H);$$

$$T = T(P, H)$$

+ *expressions for* $Cf,\ K_{sing}, Nu$ and λ^{eff}

5.5.2 Predictive capabilities of the one-dimensional set of equations

One may note that the usual expressions for wall friction and wall heat transfers use the "principle of locality" in the modeling, where the physics of the process to model depends only on local flow variables (i.e., velocity, pressure, enthalpy) and their time and space local derivatives. It assumes that the knowledge of local instantaneous principal variables (*P*, *V*, *H*) is sufficient to model the process. Moreover, algebraic expressions are most often used without partial derivatives. This corresponds to the assumption of a quasi-steady quasi-established flow. The quasi-steady approximation is a reasonable approach when the time scales of the process are smaller than the time scales for variations in principal variables. The quasi-established flow is a reasonable approach in the case of regular geometry after some distance from entry (say $Z/D > 50$) when radial profiles of velocity and temperature are similar (not dependent on the abscissa). It also requires that the time scales of the process are smaller than those of transit time scales V/L.

The set of equations can give very good predictions of many one-phase flows with heat transfers. The main limitations are the following:

- If the flow is not truly one-dimensional, i.e., if the direction of the flow is not unique in a section of the duct, for example, when buoyancy forces induce natural circulation, cells in the duct, pressure, and temperature field cannot be precisely described by 1-D equations.
- Closure laws for τ_w and q_w are available for a simple geometry like a circular pipe, a square or rectangular duct, or an annulus, but predictive capabilities in more complex geometry (rod bundle) may require that specific closure laws are established based on prototypical experiments.
- In the case of local complex geometry, no universal model can predict singular pressure losses. Specific form loss coefficients must be used depending on the geometry, but the 1-D model cannot predict it by itself. The geometric effects on wall heat transfer may also be very important. Usually, any perturbation of the flow that increases turbulence may increase the heat transfer coefficient. The 1-D model cannot predict this effect by itself.
- Rapid transient conditions may require more sophisticated wall momentum and heat transfers depending on partial derivatives of *V*, *H*, and T_w.
- Nonestablished flow exists in a significant part of any industrial system, and using classical closure laws for established flow may involve some inaccuracy of predictions.

5.6 Various approaches to two-phase flow modeling

One can classify two-phase modeling approaches (Bestion et al., 2011) in a similar manner to how it was done for a single phase in Section 5.3. Classification is made of the space dimension of the model considering 0-D, 1-D, 2-D, and 3-D models. Then, to be more specific one may consider the type of time and/or space averaging used. Three-dimensional models of two-phase or multiphase flow are often called computational multifluid dynamics (CMFD) models. They are split first into 3-D for a porous body and 3-D CFD for an open medium approach.

RANS-type time-averaged models, LES-type space-filtered models, and DNS-type models are the three main categories of two-phase CFD models in open medium (see Table 5.3). Because two-phase flow is substantially more complex than single-phase flow, 1-D models are the most mature and common models used in nuclear reactor applications. Three-dimensional RANS models are also progressively used for some reactor issues, but there is no mature 3-D CMFD model in an open medium covering all flow regimes. Modeling covering all flow regimes is still a challenge (see Bestion, 2014).

5.6.1 Time-space and phase averaging in two-phase flow

Phase averaging means that a multiplication of local instantaneous equations by a phase indicator function (or phase characteristic function) is made before time or space averaging.

Let χ_k be the phase in the indicator function (see Fig. 5.3) of phase k:

$\chi_k(x, y, z, t) = 1$ if point (x,y,z) is occupied by phase k at time t

$\chi_k(x, y, z, t) = 0$ if point (x,y,z) is not occupied by phase k at time t

In a two-phase flow with liquid and gas, $\boldsymbol{\chi_g + \chi_l = 1}$.

A two-phase flow can be split into several fields including, for example, continuous liquid, continuous gas, one or Nb bubble, and one or Nd droplet fields. In this case, field averaging means that local instantaneous equations are multiplied by field indicator (or characteristic) functions before time or space averaging.

Let us define some useful quantities and operators:

1-D space averaging:	$\langle f\rangle^L \triangleq \frac{1}{L}\int_L f\, dl$
Area averaging:	$\langle f\rangle^S \triangleq \frac{1}{A}\int_A f\, ds$
Volume averaging:	$\langle f\rangle^V \triangleq \frac{1}{V}\int_V f\, dv$
Volume filtering:	$\langle f(x,y,z,t)\rangle^{\delta v} \triangleq \iiint_{R^3} g(x' - x, y' - y, z' - z, t)\cdot f(x', y', z', t)dx'dy'dz'$

with

$$\iiint_{R^3} g(x, y, z)dxdydz = 1$$

The filter function g may be a uniform function over a distance δ_f

Time averaging:	$\overline{f} \triangleq \frac{1}{T}\int_T f\, dt$
Time fraction of phase or field k:	$\varepsilon_k \triangleq \overline{\chi_k} = \frac{1}{T}\int_T \chi_k\, dt = \frac{T_k}{T}$
Phase or field space (surface) averaging:	$\breve{f}_k^{\,k} \triangleq \frac{\int_A \chi_k f_k ds}{AR_k}$

Table 5.3 Classification of Eulerian fluid dynamic simulation approaches for two-phase computational fluid dynamics.

Treatment of equations	Pseudo DNS	Scale/Eddy-resolving approaches			RANS U-RANS
Time or ensemble averaging	No	No	No	No	**Yes**
Space filtering	No	**Yes**	**Yes**	**Yes**	No
Multiplication by phase (field) characteristic functions	No	No	**Yes**	**Yes**	**Yes**
Nb of phases or fields	One-fluid	One-fluid	Two-fluid n-field	Two-fluid n-field	Two-fluid n-field
TM or interface recognition/ reconstruction	**ITM**	**ITM**	**IRM**	No	No (IRT)
Statistical interfaces/ Deterministic interfaces/ Filtered interfaces	Deterministic interfaces	Deterministic interfaces	Filtered and statistical interfaces	Statistical interfaces	Statistical and filtered interfaces
Type of approach	**Pseudo DNS**	**LES with deterministic interfaces**	**Hybrid LES with filtered and statistical interfaces**	**LES with statistical interfaces**	**RANS U-RANS statistical and filtered interfaces**
Required modeling effort	ε	**X**	**XXXXX**	**XX**	**XXXX**
Locality–universality of the modeling	*****[a]	****	***	***	*

IRT, interface reconstruction technique; *ITM*, interface tracking method. X is the unit modeling effort.
[a]See the last paragraph in Section 5.3.4.

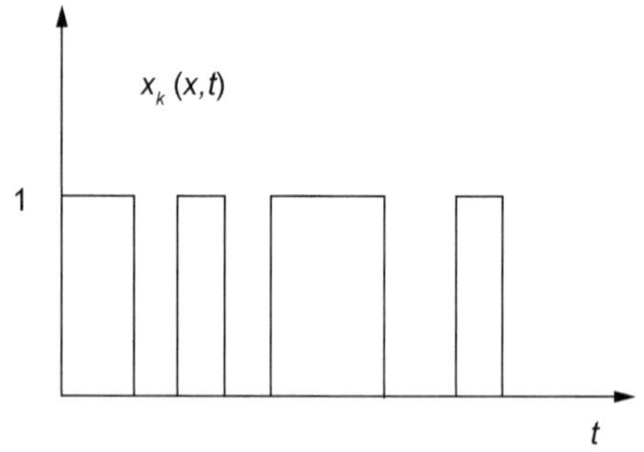

Figure 5.3 Definition of time fractions for phase k.

One should note that $R_k \breve{\overline{f_k}}^{k} = \langle \chi_k f_k \rangle^S$:

Phase or field time averaging:	$\overline{f}^k \triangleq \dfrac{\int_T \chi_k f_k dt}{\int_T \chi_k dt}$
Space-time averaging (time-area averaging)	$\widehat{\overline{f}} \triangleq \langle \overline{f} \rangle^S$
Mass-weighted space-time averaging	$\breve{\overline{f}} \triangleq \dfrac{\langle \rho \overline{f} \rangle^S}{\langle \overline{\rho} \rangle^S}$
Phase or field space-time averaging	$\widehat{\overline{f}_k}^k \triangleq \dfrac{\langle \overline{\chi_k f_k} \rangle^S}{\alpha_k}$
Mass-weighted phase space-time averaging	$\breve{\overline{f_k}}^k \triangleq \dfrac{\langle \overline{\chi_k \rho_k f_k} \rangle^S}{\langle \overline{\chi_k \rho_k} \rangle^S} = \dfrac{\widehat{\overline{\rho_k f_k}}^k}{\widehat{\overline{\rho_k}}^k}$

5.6.2 Computational multifluid dynamics models

Five types of Eulerian two-phase CFD approaches for the open medium are distinguished in Table 5.3. They differ in the treatment applied to basic mass, momentum, and energy balance equations. The four choices can be made successively:

1. Multiplication or not of mass momentum and energy equations by the phase indicator functions (for each phase in the two-fluid approach) or multiplication by the field indicator functions (for each field for a multifield approach)
2. Time averaging or not
3. Space filtering or not

Addition of an interface tracking method (ITM) or interface reconstruction technique (IRT) for some or all interfaces or not.

These choices allow the distinguishing of various methods according to the following aspects:

- Phase averaging or field averaging:
  - Mixture model under homogeneous and algebraic slip
  - Two-fluid model
  - Multifield models

- Filtering turbulent scales and two-phase intermittency scales:
  - All turbulent and two-phase scales are filtered (RANS-type models).
  - Only some scales are filtered (LES-type models).
  - All scales are predicted or simulated (DNS-type models).
- Treatment of interfaces
  - Use of ITM
  - Use of IRT
  - Use of a pure statistical treatment of interfaces

As a result of the various choices, three kinds of interface treatments are possible:

- *Deterministic interface* or simulated interface: an interface is said to be "deterministically treated" when its space and time position is simulated or actually predicted without any simplification. This requires the use of an ITM.
- *Statistical interface*: an interface is said to be "statistically treated" when a space or time averaging or filtering procedure does not allow the predicting of its space and time position. Only statistical or averaged information on several interfaces may be predicted through quantities such as a void fraction or an interfacial area density. This is the case for the small bubble and droplet interfaces of Figs. 5.3 and 5.4.
- *Filtered interface*: an interface is considered a "filtered interface" when its space and time position is predicted with some filtering of the smaller scale deformations. This filtering may result either from space filtering or time averaging.

The modeling of wall transfers, interfacial transfers, interfield transfers, and turbulent transfers depends on the approach. Each approach is described briefly below.

5.6.2.1 The pseudo-direct numerical solution method

Pseudo-DNS does not apply any space or time averaging to the equations. However, an ITM is required to track the interface. A one-fluid approach is used. Additional models are often required, for example, for implementing a film-splitting criterion when two bubbles coalesce or for the contact angles at a triple line of solid–liquid–gas. Such additional models are why, in two-phase flow, the word DNS must be replaced by pseudo-DNS—some very small-scale physics is not solved but only modeled.

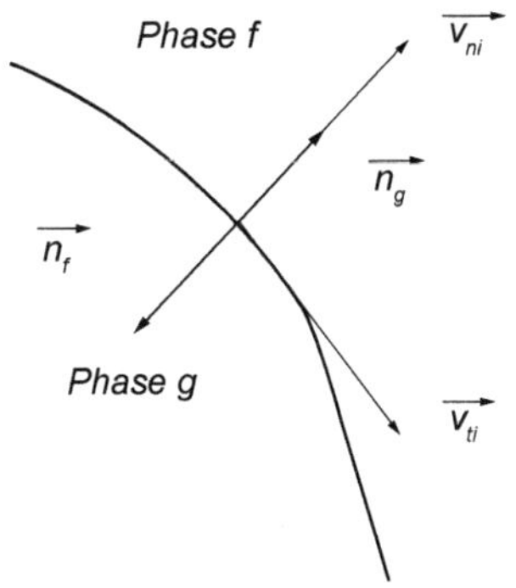

Figure 5.4 Interface between phases *f* and *g*.

Since no averaging or filtering is used, pseudo-DNS should be able to simulate all interfaces including smaller-scale interface deformations and predictions of smaller-scale eddies up to the Kolmogorov dissipative scale. Since even pseudo-DNS requires extremely expensive CPU time, its use is restricted to the investigation of very small-scale flow processes to complement experimental investigations and as support for the modeling and validation of more macroscopic approaches.

The pseudo-DNS model can in principle simulate all flow regimes but in practice is limited by the extremely expensive CPU time required.

5.6.2.2 The Reynolds-averaged Navier–Stokes method

The RANS (or U-RANS) method is the simplest, cheapest (in terms of CPU), most advanced, and most used two-phase CFD method. All turbulent scales and all two-phase intermittency scales are filtered. All interfaces are statistically treated. The method is applied to steady flows or quasi-steady (U-RANS) flows when the time scales of variation of mean variables are larger than the largest time scales of turbulence and two-phase intermittency (time between passage of two interfaces at a given point).

In the case of stratified flow or annular flow, large interfaces (free surface or film interface) can be treated as "filtered interfaces." RANS is in principle compatible with all two-phase flow regimes provided that they are steady or quasi-steady. For dispersed bubbly flow or dispersed droplet flow, the condition is easy to satisfy. For separate-phase flows (stratified flow, annular flow), the averaging filters interfacial waves in a way that is not fully clear: although the RANS modeling of turbulent diffusion prevents simulation of large eddies it does not prevent irrotational waves to be predicted resulting from Kelvin-Helmholtz instability. Therefore RANS and U-RANS are compatible with statistical interfaces and some filtered interfaces.

For slug and churn flow regimes with large bubbles (either Taylor bubbles in slug flow or distorted large bubbles in churn flow), intermittency due to the passage of these bubbles corresponds to rather large time scales. Since RANS filters even these large scales, it cannot predict this intermittency, although it can predict average behavior.

RANS is compatible with all possible choices for the number of fields, including single-fluid, two-fluid, and all kinds of multifield models.

5.6.2.3 Large eddy simulation with deterministic interface

LES with a deterministic interface combines filtering of turbulent fluctuations with an ITM for all interfaces. This method was first proposed by Toutant et al. (2009a,b), Bois et al. (2010), Magdeleine et al. (2010), Lakehal (2008), and Lakehal et al. (2008) for free surface flow. The concept has since been better reformulated by Lakehal and Labois (2011), who refer to it as LEIS, short for large-eddy and interface simulation.

5.6.2.4 Large eddy simulation with statistical interfaces

When the largest interfacial scale is rather small and significantly smaller than the largest turbulent eddies, the LES method applies to a filter scale smaller than larger eddies but larger than the two-phase scale to allow a statistical treatment. The idea was first developed by Lakehal (2002), who put filtered two-fluid equations on firmer ground and proposed a bubble-induced diffusion model incorporated within the dynamic procedure of Lilly and Germano (see Lilly, 1991). It has since then been applied with some success to turbulent dispersed flow by Dhotre et al. (2007), Dhotre et al. (2008), Niceno et al. (2008), and Ničeno et al. (2009).

Such a method clearly uses fewer CPU resources than pseudo-DNS and LES with a deterministic interface, but it is restricted to certain flow situations, typically the dispersed bubble or dispersed drops in which large eddies are much larger than the largest bubbles or eddies. In slug or churn flow where the largest bubbles and largest eddies are of the size of the geometric dimensions of the flow (such as a hydraulic diameter), this method is clearly not applicable.

This method is compatible with the two-fluid model and multifield models where dispersed bubbles or droplets may be treated by size groups.

5.6.2.5 Hybrid large eddy simulation method with both filtered and statistical interfaces

This hybrid method filters smaller eddies and treats statistically the small droplets or small bubbles, while the other interfaces are simply filtered and treated as large interfaces. The space filter eliminates the smaller bubbles that are treated statistically, and it thickens and filters the interface of the large bubble, which may be reconstructed with a simplification of the shape.

Lakehal (2010) discussed the difficulties of this method compared with LES with simulated interfaces.

It may be a promising way of modeling the most complex two-phase flow (such as churn or slug flow) at a reasonable CPU cost without filtering the large two-phase structures as RANS does. However, the closure issue is rather complex, and the present state of the art is not very advanced.

This method is compatible with the two-fluid approach and with multifield models. A four-field model with a continuous gas field, a continuous liquid field, a dispersed bubble field, and a dispersed droplet field may help reconstruct large filtered interfaces and model mass transfers between continuous and dispersed fields.

This method includes many more closure terms to model than LES with deterministic interfaces. In particular, the effects due to the subgrid waves of large interfaces must be modeled.

It has been observed that many applications that use an ITM pretend to be of the LES type with deterministic interfaces despite actually being hybrid methods. Interface tracking by the ITM is "underresolved," and simulated interfaces are "filtered interfaces" instead of "deterministic interfaces." All effects due to the subgrid waves of large interfaces should be added to the modeling.

5.6.2.6 Locality and universality of the modeling

The "principle of locality of the modeling" is used in two-phase CFD as it is in single-phase CFD since closure laws are expressed only as functions of local flow variables (i.e., velocity, density, internal energy) and their time and space derivatives. Here the exceptions to this principle of locality are the so-called wall functions, where some closure terms depend on the distance to a wall, and their equivalents for large interfaces, which may be called "interfacial functions." This allows rather coarse nodalization in high-shear regions close to a wall or a large interface.

It has been stressed that the principle of locality is more relevant when the processes to model at point $P(x,t)$ depend on flow conditions close to point P. In single-phase CFD the principle of locality is better satisfied by a LES-type method than a RANS approach since the subfilter local turbulent viscosity in a LES approach models only the effects of small eddies close to point P. Two-phase models include small interfaces (small bubbles and droplets) and large-scale interfaces, such as large, distorted bubbles in churn flow, the free surface in stratified flow, and the film surface in annular flow. Therefore, in flow regimes with large interfaces like slug and churn flow, the principle of locality may only be reasonably applied in LES with deterministic or filtered interfaces.

5.6.2.7 *Applicability and degree of maturity of various computational multifluid dynamics approaches*

Table 5.4 summarizes the applicability and degree of maturity of various two-phase CFD approaches to every flow regime. As shown, only two modeling approaches can be applied to simulate physical situations where all two-phase flow regimes may occur: RANS (or U-RANS) and LES with filtered and statistical interfaces. However, the former is more mature than the latter. Pseudo-DNS and LES with deterministic interfaces can address all flow regimes but are practically limited by required CPU time, which is prohibitive for the most complex flows. LES with statistical interfaces cannot treat "large interfaces" and is intrinsically limited to dispersed flows.

5.7 Time-averaged two-fluid equations

Phase time averaging or field time averaging is the most natural approach to extend CFD RANS models to two-phase flow, Fig. 5.4. It consists of multiplying basic equations by the phase indicator function χ_k of phase k.

The derivation of equations is presented in detail by Ishii and Hibiki (2011), and various methods are also presented for dispersed flow by Morel (2015).

Following Ishii and Hibiki (2011), a general form of the transport equation for a specific quantity integrated over period Δt is (see Fig. 5.3)

$$\frac{\partial \varepsilon_k \overline{\rho_k}^k \widetilde{\psi_k}^k}{\partial t} + \nabla \cdot \left(\varepsilon_k \overline{\rho_k}^k \widetilde{\overrightarrow{v_k}}^k \widetilde{\psi_k}^k \right) + \nabla \cdot \left[\varepsilon_k \left(\overline{J_k}^k + J_k^t \right) \right]$$

$$= \varepsilon_k \overline{\rho_k}^k \widetilde{\phi_k}^k + \frac{1}{\Delta t} \sum_j \left\{ \frac{1}{v_{ni}} \left[\overrightarrow{n}_k \cdot \rho_k \left(\overrightarrow{v_k} - \overrightarrow{v_i} \right) \psi_k \right] + \overrightarrow{n}_k \cdot J_k \right\} \quad (5.72)$$

Table 5.4 Applicability and degree of maturity of the various two-phase computational fluid dynamics approaches to every flow regime including CPU cost.

		LES	**LES-VLES**	**LES-VLES**	**RANS / U-RANS**
	Pseudo-DNS	**Simulated interfaces**	**Statistical and filtered interfaces**	**Statistical interfaces**	**Statistical and filtered interfaces**
Bubbly	Applied	Applied	Applied	Applied	Applied
Slug-churn	Possible Too expensive	Possible Too expensive	Possible	Not possible	Possible
Annular	Possibly expensive	Possibly expensive	Possible	Not possible	Applied
Annular-mist	Possible Too expensive	Possible Too expensive	Possible	Not possible	Possible
Mist flow	Applied	Applied	Applied	Applied	Applied
Stratified	Applied	Applied	Applied	Not possible	Applied
Stratified-mist	Possible Too expensive	Possible Too expensive	Possible	Not possible	Possible
All flow regimes	**Too expensive**	**Too expensive**	**Possible**	**Not possible**	**Possible**
Degree of maturity	**Average in some flow regimes** **Zero for "all flow regimes"**	**Average in some flow regimes** **Zero for "all flow regimes"**	**Average in some flow regimes** **Very low for "all flow regimes"**	**Average in dispersed flow**	**Average in some flow regimes** **Low for "all flow regimes"**

Applied: is applicable and has been applied; *Not possible*: cannot be applied due to intrinsic limitations; *Possible*: can be applied but not very mature; *Expensive*: requires a very high CPU time; *Too expensive*: unaffordable with current computer power.

Table 5.5 Volumetric source and surface source in mass momentum and total energy equations.

	ψ_k	ϕ_k	J_k
Mass	1	0	0
Momentum	$\overrightarrow{v_k}$	$\overrightarrow{F_k}$	$-T_k$
Energy	$u_k + \frac{\overrightarrow{v_k}^2}{2}$	$q_{kext} + \overrightarrow{F_k} \cdot \overrightarrow{v_k}$	$\overrightarrow{q_k} - T_k \cdot \overrightarrow{v_k}$

where $\_\phi_k$ is the volumetric source of ψ_k in phase k

J_k diffusive flux of ψ_k in phase k

Mass momentum and total energy balance equations can be written considering the respective ψ_k, ϕ_k and J_k, as given in Table 5.5.

$\overrightarrow{F_k}$ is the specific external force on phase k(N/kg or m/s^2) usually limited to gravity force ($\overrightarrow{F} = \overrightarrow{g}$)

T_k is the stress tensor in phase k(Pa or N/m^2)

$\overrightarrow{q}_k$ heat flux vector in phase k (W/m^2)

q_{ext} external volumetric source of energy (usually radiation)

$\overrightarrow{n}_k$ unit vector normal to the interface oriented outside of phase k

$\overrightarrow{v}_{ni}$ component of velocity of interface normal to the interface

$\overrightarrow{v}_{ti}$ component of velocity of interface tangential to the interface

J_k^t turbulent flux of ψ_k

$$J_k^t \triangleq \frac{\overline{\chi_k \rho_k \overrightarrow{v_k} \psi_k}}{\varepsilon_k} - \overline{\rho_k}^k \overrightarrow{\widetilde{v_k}}^k \widetilde{\psi_k}^k$$

This term results from averaging a product for convection terms.

The second RHS term accounts for convective and diffusive fluxes through all interfaces passed at point x during period Δt .

One obtains the following set of phase mass, momentum, and energy equations;

$$\frac{\partial \varepsilon_k \overline{\rho_k}^k}{\partial t} + \nabla \cdot \left(\varepsilon_k \overline{\rho_k}^k \overrightarrow{\widetilde{v_k}}^k \right) = \frac{1}{\Delta t} \sum_j \left\{ \frac{1}{v_{ni}} \left[\overrightarrow{n}_k \rho_k \left(\overrightarrow{v_k} - \overrightarrow{v_i} \right) \right] \right\} \triangleq \gamma_{ik} \tag{5.73}$$

where γ_{ik} is the time-averaged interfacial mass transfer per unit volume ($kg/m^3/s$):

$$\frac{\partial \varepsilon_k \overline{\rho_k}^k \overrightarrow{\widetilde{v_k}}^k}{\partial t} + \nabla \cdot \left(\varepsilon_k \overline{\rho_k}^k \overrightarrow{\widetilde{v_k}}^k \overrightarrow{\widetilde{v_k}}^k \right) + \varepsilon_k \nabla \overline{p_k}^k$$

$$= \varepsilon_k \overline{\rho_k}^k \overrightarrow{g} + \nabla \cdot \left[\varepsilon_k \left(\overline{\mathcal{T}_k} + \mathcal{T}_k^t \right) \right] + \overrightarrow{M}_{ik}$$

$$\vec{M}_{ik} = \frac{1}{\Delta t}\sum_j \left\{ \frac{1}{v_{ni}} \left[\vec{n}_k \cdot \rho_k \left(\vec{v_k} - \vec{v_i} \right) \vec{v_k} \right] - \vec{n}_k \cdot T_k \right\}$$

A common approximation is to consider a unique pressure for both phases is

$$\overline{p_f}^f = \overline{p_g}^g = \overline{p}$$

where M_{ik} is the time-averaged interfacial momentum transfer per unit volume (N/m^3). The first term is usually modeled as

$$\frac{1}{\Delta t}\sum_j \left\{ \frac{1}{v_{ni}} \left[\vec{n}_k \rho_k \left(\vec{v_k} - \vec{v_i} \right) \vec{v_k} \right] \right\} \cong \gamma_{ik} \tilde{\vec{v_k}}^k$$

The second term is the most complex contribution since it contains pressure and viscous forces applied to a moving deformable interface, which may have very diverse forms depending on the flow regime.

A very common approximation is to consider a part equal to $\overline{p}\nabla\varepsilon_k$ and to model the rest as several forces:

$$\vec{M}_{ik} = \overline{p}\nabla\varepsilon_k + \vec{M}_{ik}^{D} + \vec{M}_{ik}^{L} + \vec{M}_{ik}^{AM} + \vec{M}_{ik}^{TD}$$

The drag force $\vec{M}_{ik}^{D}$ is a coupling between average phase velocities. For a bubble, this is written as

$$\vec{M}_{ig}^{D} = -\vec{M}_{if}^{D} = -\frac{a_i C_d}{8} \rho_f \left| \tilde{\vec{v_g}}^g - \tilde{\vec{v_f}}^f \right| \left(\tilde{\vec{v_g}}^g - \tilde{\vec{v_f}}^f \right) \tag{5.74}$$

The lift force $\vec{M}_{ik}^{L}$ is a force applied to a particle (bubble or drop) of a dispersed phase d. It is perpendicular to the velocity difference between phases and is proportional to the rotational of the continuous phase c:

$$\vec{M}_{id}^{L} = -\vec{M}_{ic}^{L} = -C_L \varepsilon_d \rho_c \left(\tilde{\vec{v_d}}^d - \tilde{\vec{v_c}}^c \right) \wedge \left(\nabla \wedge \tilde{\vec{v_c}}^c \right) \tag{5.75}$$

The added mass force $\vec{M}_{ik}^{AM}$ is a coupling between phase accelerations. A particle (bubble or drop) that accelerates must accelerate the fluid around it, and this increases its inertia. It was established first for a sphere accelerating in an infinite medium by solving inviscid flow equations. This force is particularly important for dispersed gas since the fluid around the bubble has a high density and may be neglected for drops since the gas around the drop has a low density. It is also not considered in separate-phase flows such as annular or stratified flow.

A possible expression for a bubbly flow is

$$\overrightarrow{M}_{ig}^{AM} = -\overrightarrow{M}_{if}^{AM}$$

$$= C^{AM}\frac{1+2\varepsilon_g}{1-\varepsilon_g}\varepsilon_g\rho_f\left[\left(\frac{\partial\widetilde{\overrightarrow{v_g}}^d}{\partial t}+\widetilde{\overrightarrow{v_g}}^g\cdot\nabla\widetilde{\overrightarrow{v_g}}^g\right)-\left(\frac{\partial\widetilde{\overrightarrow{v_f}}^f}{\partial t}+\widetilde{\overrightarrow{v_f}}^f\cdot\nabla\widetilde{\overrightarrow{v_f}}^f\right)\right] \tag{5.76}$$

The turbulent dispersion force $\overrightarrow{M}_{ik}^{TD}$ is associated with the fluctuating part of all forces, which has been identified as an important contributor to dispersed flow (see Morel, 2015). It includes the mixing effect of dispersed particles by turbulent eddies. One of the simplest models has the following form, which uses the turbulent kinetic energy k_c of the continuous phase:

$$\overrightarrow{M}_{id}^{TD} = -\overrightarrow{M}_{ic}^{TD} = C^{TD}\rho_c k_c\nabla\varepsilon_d$$

The total energy equation can be written as

$$\frac{\partial}{\partial t}\left[\varepsilon_k\overline{\rho_k}^k\left(\widetilde{u_k}^k+\frac{\widetilde{\overrightarrow{v_k}}^{k2}}{2}\right)\right]+\nabla\cdot\left[\varepsilon_k\overline{\rho_k}^k\widetilde{\overrightarrow{v_k}}^k\left(\widetilde{u_k}^k+\frac{\widetilde{\overrightarrow{v_k}}^{k2}}{2}\right)\right]$$

$$= \varepsilon_k\overline{\rho_k}^k\overrightarrow{g}\cdot\widetilde{\overrightarrow{v_k}}^k-\nabla\cdot\left[\varepsilon_k\left(\overline{\overrightarrow{q}_k}^k+\overrightarrow{q}_k^t\right)\right]+q_{rk}+E_{ik}+\nabla\cdot\left[\varepsilon_k\left(\overline{T}_k+T_k^t\right)\cdot\widetilde{\overrightarrow{v_k}}^k\right] \tag{5.77}$$

$$E_{ik} = -\frac{1}{\Delta t}\sum_j\left\{\frac{1}{v_{ni}}\left[\overrightarrow{n}_k\rho_k\cdot\left(\overrightarrow{v_k}-\overrightarrow{v_i}\right)\left(u_k+\frac{{v_k}^2}{2}\right)\right]-\overrightarrow{n}_k\cdot T_k\cdot\overrightarrow{v_k}+\overrightarrow{n}_k\overrightarrow{q}_k\right\} \tag{5.78}$$

where E_{ik} is time-averaged interfacial energy transfer per unit volume ($W/\,m^3$).

The contribution of the stress tensor is often neglected, and E_{ik} is usually modeled as a function of interfacial mass transfer γ_{ik} and interfacial heat transfer q_{ik}:

$$E_{ik}\cong q_{ik}+\gamma_{ik}\left(\widetilde{h_k}^k+\frac{\widetilde{\overrightarrow{v_k}}^{k2}}{2}\right) \tag{5.79}$$

A RANS two-fluid model used for adiabatic bubbly flow by Krepper et al. (2011) and boiling bubbly flow by Koncar et al. (2011) is summarized hereafter.

RANS two-fluid equations for computational multifluid dynamics: 3-D two-phase flow equations

Principal variables: $\overline{p}$, ε_g , $\overrightarrow{\widetilde{v_g}}^{g}$, $\overrightarrow{\widetilde{v_f}}^{f}$, $\widetilde{h_g}^{g}$, $\widetilde{h_f}^{f}$

$$\frac{\partial \varepsilon_k \overline{\rho_k}^k}{\partial t} + \nabla \cdot \left(\varepsilon_k \overline{\rho_k}^k \overrightarrow{\widetilde{v_k}}^k \right) = \gamma_{ik}$$

$$\frac{\partial \varepsilon_k \overline{\rho_k}^k \overrightarrow{\widetilde{v_k}}^k}{\partial t} + \nabla \cdot \left(\varepsilon_k \overline{\rho_k}^k \overrightarrow{\widetilde{v_k}}^k \overrightarrow{\widetilde{v_k}}^k \right) + \varepsilon_k \nabla \overline{p}$$

$$= \varepsilon_k \overline{\rho_k}^k \overrightarrow{g} - M_{ik} + \gamma_{ik} \overrightarrow{\widetilde{v_k}}^k + \nabla \cdot \left[\varepsilon_k \left(\overline{\mathcal{T}}_k + \mathcal{T}_k^t \right) \right]$$

$$\frac{\partial}{\partial t}\left[\varepsilon_k \overline{\rho_k}^k \left(\widetilde{h_k}^k + \frac{\overrightarrow{\widetilde{v_k}}^{k2}}{2} \right) \right] + \nabla \cdot \left[\varepsilon_k \overline{\rho_k}^k \overrightarrow{\widetilde{v_k}}^k \left(\widetilde{h_k}^k + \frac{\overrightarrow{\widetilde{v_k}}^{k2}}{2} \right) \right] - \varepsilon_k \frac{\partial \overline{p}}{\partial t}$$

$$= \varepsilon_k \overline{\rho_k}^k \overrightarrow{g} \cdot \overrightarrow{\widetilde{v_k}}^k + \nabla \cdot \left[\varepsilon_k \left(\overrightarrow{\overline{q}}_k + \overrightarrow{q}_k^t \right) \right] + q_{ik} + \gamma_{ik} \left(\widetilde{h_k}^k + \frac{\overrightarrow{\widetilde{v_k}}^{k2}}{2} \right) + q_{rk}$$

$$\sum_k \varepsilon_k = 1; \sum_k \gamma_{ik} = 0; \sum_k M_{ik} = 0; \sum_k \left[q_{ik} + \gamma_{ik} \left(\widetilde{h_k}^k + \frac{\overrightarrow{\widetilde{v_k}}^{k2}}{2} \right) \right] = 0;$$

$$M_{ik} = M_{ik}^D + M_{ik}^L + M_{ik}^{AM} + M_{ik}^{TD}$$

$$\overline{\mathcal{T}_k} = -\frac{2}{3} \mu_k \nabla \cdot \overrightarrow{\widetilde{v_k}}^k I + \mu_k \left(\nabla \overrightarrow{\widetilde{v_k}}^k + {}^t \nabla \overrightarrow{\widetilde{v_k}}^k \right); \overrightarrow{\overline{q}}_k = -\lambda_k \nabla \overline{T_k}$$

$$\overline{\rho_k}^k = \rho_k \left(\overline{p}, \widetilde{h_k}^k \right); \overline{T_k} = T_k \left(\overline{p}, \widetilde{h_k}^k \right); \mu_k = \mu_k \left(\overline{p}, \widetilde{h_k}^k \right); \lambda_k = \lambda_k \left(\overline{p}, \widetilde{h_k}^k \right)$$

+ *expressions or equations for* $\mathcal{T}_k^{\,t}$, $\overrightarrow{q}_k^{\,t}$, $M_{ik}^D, M_{ik}^L, M_{ik}^{AM}, M_{ik}^{TD}$

5.7.1 Predictive capabilities of the time-averaged two-fluid model

Much information is lost when averaging a two-phase flow due to the complex and moving geometries of interfaces. The main source of difficulty is associated with the contributions of pressure and viscous forces on the interfaces that require the main modeling efforts. Another difficulty comes from the turbulence in a two-phase flow, which is far less understood than that of a single phase. Modeling efforts so far have provided reasonable predictions of dispersed flow for bubbles and droplets, but accuracy remains significantly degraded compared with single-phase turbulent flow.

The time-averaged two-fluid model has also been used with some success in separate-phase flow with large interfaces (Lucas et al., 2011). Specific effort was spent on the identification of large interfaces using an IRT or ITM and mass momentum and energy transfers at large interfaces.

5.8 Space averaged one-dimensional equations for two-phase flow

The 1-D two-fluid six-equation model of two-phase flow is the most common one for describing two-phase flow with heat transfers in circuits. It results from space-time phase averaging (Delhaye et al., 1981, Ishii and Hibiki, 2006; Morel, 2015). Phase averaging means that a multiplication of local instantaneous equations by a phase indication function is made before time or space averaging.

One can obtain the 1-D two-fluid six-equation model of two-phase flow in two ways:

a multiplying local instantaneous equations by χ_k, then time averaging, and then space averaging

b multiplying local instantaneous equations by χ_k, then space averaging, and then time averaging

Here, the decision was made to show steps from the second method since it allows identifying and justifying some of the most important of the necessary simplifications and approximations.

5.8.1 *Space-phase averaging*

Let us define some useful quantities and operators (see also Fig. 5.5).

Space averaging:	$\langle f\rangle^S \triangleq \frac{1}{A}\int_A f\, ds$
Time averaging:	$\overline{f} \triangleq \frac{1}{T}\int_T f\, dt$
Time fraction of phase k:	$\varepsilon_k \triangleq \overline{\chi_k} = \frac{1}{T}\int_T \chi_k\, dt = \frac{T_k}{T}$
Space fraction of phase k:	$R_k \triangleq \langle \chi_k\rangle^S = \frac{1}{A}\int_A \chi_k\, ds$
Phase space (surface) averaging:	$\widehat{f_k}^k \triangleq \frac{\int_A \chi_k f_k ds}{AR_k}$
Phase mass weighted area averaging	$\breve{f_k}^k \triangleq \frac{\langle \chi_k \rho_k f_k\rangle^S}{\langle \chi_k \rho_k\rangle^S}$
One should note:	$R_k \widehat{f_k}^k = \langle \chi_k f_k\rangle^S$
Phase time averaging:	$\overline{f}^k \triangleq \frac{\int_T \chi_k f_k dt}{\int_T \chi_k\, dt}$
One should note:	$\varepsilon_k \overline{f}^k = \overline{\chi_k f_k}$

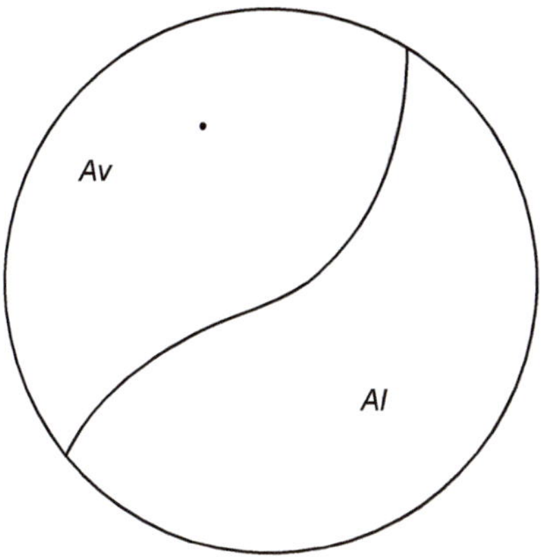

Figure 5.5 Definition of space fractions for phase k.

Considering commutativity of space and time integration:

$$\int_A \int_T \chi_k f_k \, dtds = \int_T \int_A \chi_k f_k \, dsdt \tag{5.80}$$

$$\overline{R_k \widehat{f_k}^k} = \left\langle \varepsilon_k \overline{f_k}^k \right\rangle \tag{5.81}$$

A space-time average phase fraction α_k is defined, i.e., according to Fig. 5.6:

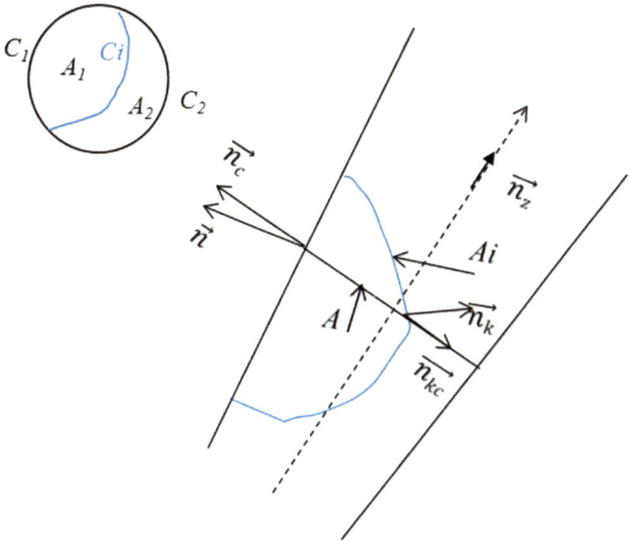

Figure 5.6 Portion of a duct with a two-phase flow.

$$\alpha_k \triangleq \overline{R_k} = \langle \varepsilon_k \rangle \tag{5.82}$$

where A is the cross section of the duct
C_k is the contour of A occupied by phase k
Ai is the interfacial surface
C_i is the contour of phase k in A (intersection of A and Ai)
$\overrightarrow{n}$ is the vector normal to the duct wall directed to the exterior of the duct
$\overrightarrow{n_c}$ is the vector normal to the contour C directed to the exterior of the duct
$\overrightarrow{n_{kc}}$ is the vector normal to the contour C_i directed to the exterior of phase k
$\overrightarrow{n_k}$ is the vector normal to the interface A_i directed to the exterior of phase k
$\overrightarrow{n_z}$ is the unit vector along the duct axis

5.8.2 Space-phase-averaged equations

Mass balance of phase k

$$\frac{\partial AR_k \langle \rho_k \rangle_k^S}{\partial t} + \frac{\partial AR_k \left\langle \rho_k \overrightarrow{v_{kz}} \right\rangle_k^S}{\partial z} = -\int_{Ci} \rho_k \left(\overrightarrow{v_k} - \overrightarrow{v_i} \right) \cdot \overrightarrow{n_k} \frac{dCi}{\overrightarrow{n_k} \cdot \overrightarrow{n_{kC}}} \triangleq A\Gamma_{ik} \tag{5.83}$$

The interface has no volume and no mass, so $\Gamma_{iv} = -\Gamma_{il}$.
Momentum balance of phase k
Time derivative term:

$$\int_{A_k} \left(\frac{\partial \rho v_z}{\partial t} \right) dA = \frac{\partial AR_k \langle \rho_k v_{kz} \rangle_k^S}{\partial t} = A \frac{\partial R_k \widehat{\rho_k}^k \breve{V}_k^k}{\partial t} \tag{5.84}$$

Advection term:

$$\int_{A_k} \left(\frac{\partial y}{\partial z} \rho v_z \, v_z \right) dA = \frac{\partial AR_k \langle \rho_k v_{kz} v_{kz} \rangle_k^S}{\partial z} + M_{vik} \tag{5.85}$$

$$M_{vik} = \int_{Ci} \rho_k v_{kz} \left(\overrightarrow{v_k} - \overrightarrow{v_i} \right) \cdot \overrightarrow{n_k} \frac{dCi}{\overrightarrow{n_k} \cdot \overrightarrow{n_{kC}}} \tag{5.86}$$

where M_{vik} is the volumetric axial momentum transfer through the interface. Considering that both phases have the same axial velocity V_i at the interface, $M_{vik} = A \, \Gamma_{ik} V_i$.

V_i must be modeled as a function of R_k, $\breve{V}_l$, $\breve{V}_g$ depending on the flow regime.

One must define a phase momentum axial dispersion term T_{kz}^d:

$$T_{kz}^d \triangleq \langle \rho_k v_{kz} v_{kz} \rangle_k^S - \widehat{\rho_k}^k \breve{V}_k^k \breve{V}_k^k \tag{5.87}$$

$$\int_{A_k}\left(\frac{\partial y}{\partial z}\rho v_z\, v_z\right)dA=\frac{\partial AR_k\widehat{\rho_k}^k\breve{V}_k{}^k\breve{V}_k{}^k}{\partial z}+\frac{\partial AR_kT^d_{kz}}{\partial z}+M_{vik}$$

Stress tensor term (deviator part):

$$\int_{A_k}-\nabla\cdot\mathcal{T}\cdot\overrightarrow{n}_z\,dA=-\frac{\partial AR_k\left\langle\overrightarrow{n}_z\cdot\mathcal{T}\cdot\overrightarrow{n}_z\right\rangle_k^S}{\partial z}-F_{frik}-F_{frwk} \tag{5.88}$$

The first RHS term is usually small and neglected.

F_{frik} is the axial component of the friction of phase k along the interface.

F_{frwk} is the axial component of friction term of phase k along a wall.

$$F_{frik}\triangleq\int_{Ci}\overrightarrow{n}_z\cdot\mathcal{T}_k\cdot\overrightarrow{n}_k\frac{dCi}{\overrightarrow{n}_{kc}\cdot\overrightarrow{n}_k} \tag{5.89}$$

$$F_{frk}\triangleq\int_{C_k}\overrightarrow{n_z}\cdot\mathcal{T}_k\cdot\overrightarrow{n_k}\frac{dC_k}{\overrightarrow{n}_c\cdot\overrightarrow{n}} \tag{5.90}$$

External force (gravity) term:

$$\int_{A_k}\rho g_z\,dA=AR_k\widehat{\rho_k}^k g_z \tag{5.91}$$

Pressure term:

$$\int_{A_k}\frac{\partial p}{\partial z}\,dA=\int_{A_k}\nabla p\cdot\overrightarrow{n}_z\,dA=\frac{\partial AR_k\breve{p}_k{}^k}{\partial z}-F_{pwk}-F_{pik} \tag{5.92}$$

$$F_{pwk}\triangleq\int_{Ck}\overrightarrow{n_k}\cdot p\,\overrightarrow{n_z}\frac{dC}{\overrightarrow{n}_c\cdot\overrightarrow{n}} \tag{5.93}$$

$$F_{pik}\triangleq\int_{Ci}\overrightarrow{n}_k p\cdot\overrightarrow{n_z}\frac{dC}{\overrightarrow{n}_{kc}\cdot\overrightarrow{n}_k} \tag{5.94}$$

where F_{pwk} is the axial contribution of pressure force on the part of the duct wall occupied by phase k.

F_{pik} is the axial contribution of pressure force along the interface.

Let us define an average mixture pressure P and a local pressure difference p'_k in phase k:

$$P\triangleq R_v\widehat{p_v}^v+R_l\widehat{p_l}^l \tag{5.95}$$

$$p'_k \triangleq p_k - P$$

$$F_{pwk} + F_{pik} = -P\frac{\partial A_k}{\partial z} + F'_{pwk} + F'_{pik} = -P\frac{\partial AR_k}{\partial z} + F'_{pwk} + F'_{pik}$$

$$F'_{pwk} = \int_{Ck} \vec{n}_k \cdot p'_k \, \overrightarrow{n_z} \frac{dC}{\vec{n}_c \vec{n}_k}; F'_{pik} = \int_{Ck} \vec{n}_k p'_k \cdot \overrightarrow{n_z} \frac{dC}{\vec{n}_{kc} \cdot \vec{n}_k} \tag{5.96}$$

Note that

$$\widehat{p_v}^v = P - R_l\left(\widehat{p_l}^l - \widehat{p_v}^v\right) \triangleq P - \delta p_v \tag{5.97}$$

$$\widehat{p_l}^l = P - R_v\left(\widehat{p_v}^v - \widehat{p_l}^l\right) \triangleq P - \delta p_l \tag{5.98}$$

Then $\delta p_{vl} = R_v \delta p_v = -R_l \delta p_l$:

$$\frac{\partial AR_v \widehat{p_v}^v}{\partial z} = \frac{\partial AR_v P}{\partial z} - \frac{\partial AR_v \delta p_v}{\partial z} = \frac{\partial AR_v P}{\partial z} - \frac{\partial A \delta p_{vl}}{\partial z}$$

$$\frac{\partial AR_l \widehat{p_l}^l}{\partial z} = \frac{\partial AR_l P}{\partial z} - \frac{\partial AR_l \delta p_l}{\partial z} = \frac{\partial AR_l P}{\partial z} + \frac{\partial A \delta p_{vl}}{\partial z}$$

Within the framework of a one-pressure model, one can eliminate the phase pressures $\check{p}_k^{\,k}$ from the momentum equation to keep only the mixture pressure P:

$$\int_{A_k} \frac{\partial p}{\partial z} dA = \frac{\partial AR_k \widehat{p_k}^k}{\partial z} - F_{pwk} - F_{pik} = \frac{\partial AR_k \widehat{p_k}^k}{\partial z} - P\frac{\partial AR_k}{\partial z} + F'_{pwk} + F'_{pik}$$

$$= \frac{\partial AR_k P}{\partial z} - P\frac{\partial AR_k}{\partial z} + F'_{pwk} + F'_{pik} - \frac{\partial AR_k \delta p_k}{\partial z}$$

$$= AR_k \frac{\partial P}{\partial z} + F'_{pwk} + F'_{pik} - \frac{\partial AR_k \delta p_k}{\partial z} \tag{5.99}$$

Let us define $F_{iv}{}^*$ and $F_{il}{}^*$:

$$F_{iv}{}^* \triangleq F'_{piv} + F_{friv} + \frac{\partial A \delta p_{vl}}{\partial z}$$

$$F_{il}{}^* \triangleq F'_{pil} + F_{fril} - \frac{\partial A \delta p_{vl}}{\partial z}$$

One must model the three components of $F_{iv}{}^{*}$ and $F_{il}{}^{*}$ that depend on the phase repartition within each flow regime:

$$A\frac{\partial R_k\widehat{\rho_k}^k\breve{V}_k^{\,k}}{\partial t}+\frac{\partial AR_k\widehat{\rho_k}^k\breve{V}_k^{\,k}\breve{V}_k^{\,k}}{\partial z}+AR_k\frac{\partial P}{\partial z}$$
$$=\mathrm{A}\,R_k\widehat{\rho_k}^k g_z+F'_{pwk}+F_{frwk}+F'_{pik}+F_{frik}+\mathrm{A}\,\Gamma_{ik}V_i\mp\frac{\partial A\delta p_{vl}}{\partial z}+\frac{\partial AR_kT^d_{kz}}{\partial z} \tag{5.100}$$

The law of action and reaction can be written as

$$\sum\nolimits_k\left(F'_{pik}+F_{frik}+M_{vik}\right)=0 \tag{5.101}$$

It is extremely difficult to model $F'_{pwk}, F_{frwk}, F'_{pik}, F_{frik}, \delta p_{vl}$and T^d_{kz} due to the complexity and diversity of the interface shape. In particular, pressure force integration on complex and moving boundaries is a real challenge.

Simplifications and approximations are necessary.

One common simplification is used in dispersed flow (bubbly flow and droplet flow) that considers that $\delta p_{vl}=0$ so that $\widehat{p_v}^{\,v}=\widehat{p_l}^{\,l}=\mathrm{P}$

However, it may be found that the average pressure at the interface is usually lower than the average mixture pressure. From F'_{pik} one may identify a part corresponding to an average interface pressure and a part due to differences or interface pressure along the interface.

If the interface pressure shown in Fig. 5.7 is higher than the average pressure at the stagnation point at the leading edge, it decreases due to the curvature of streamlines and never recovers the stagnation pressure at the trailing edge due to flow detachment in the wake. The average interface pressure is lower than the average pressure. The part of F'_{pik} due to average interface pressure may be modeled as a term of the form $Af_i\frac{\partial R_k}{\partial z}$ with $f_i\sim\rho\left(V_g-V_l\right)^2$ (see Stuhmiller, 1977).

The other part of F'_{pik} corresponds to the difference between the local and the average interface pressure and is responsible for a part of the drag on the particle (higher pressure on the leading edge than on the trailing edge). It is then combined into a part of F_{frik} in the drag force $A\ \tau_{ik}$ that couples the velocities of the phases.

In accelerating flow, another part of F'_{pik} and F_{frik} is known as the added mass force that couples the acceleration of the phases.

A possible expression is

$$A\ F_v^{AM}=A\ \beta R_k(1-R_k)\rho_m\left[\frac{\partial\breve{V}_v^{\,v}}{\partial t}+\breve{V}_v^{\,v}\frac{\partial\breve{V}_v^{\,v}}{\partial z}-\frac{\partial\breve{V}_l^{\,l}}{\partial t}+\breve{V}_l^{\,l}\frac{\partial\breve{V}_l^{\,l}}{\partial z}\right];F_l^{AM}=-F_v^{AM} \tag{5.102}$$

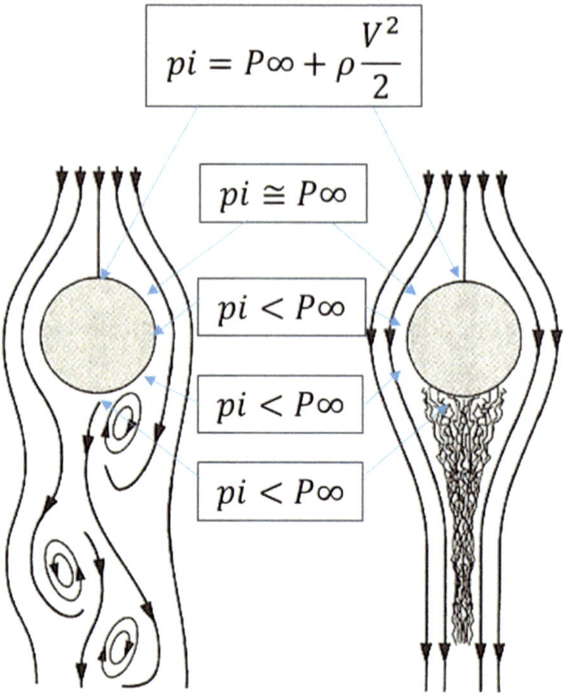

Figure 5.7 Interface pressure along a spherical particle.

where ρ_m is a mixture density and β is an added mass coefficient.

In the same way, in a duct with a varying cross section, one may model F'_{pwk} as a term of the form $g_k\frac{\partial A}{\partial z}$.

It may also be integrated into a singular pressure loss term as in a single-phase flow:

$$-AR_k\frac{\delta p_{sing,k}}{\delta z}=AR_k\frac{K_{sing,k}}{\delta z}\rho_k\frac{\breve{V}_k^{\,2}}{2}$$

F_{frwk} is usually modeled as a wall friction term:

$$F_{frwk}=-C_k\tau_{wk}$$

The momentum equation can then be written as

$$A\frac{\partial R_k\widehat{\rho_k}^k\breve{V}_k^{\,k}}{\partial t}+\frac{\partial AR_k\widehat{\rho_k}^k\breve{V}_k^{\,k}\breve{V}_k^{\,k}}{\partial z}+AR_k\frac{\partial P}{\partial z}+Af_i\frac{\partial R_k}{\partial z}+g_k\frac{\partial A}{\partial z}$$

$$=AR_k\widehat{\rho_k}^k g_z-\frac{\partial AR_kT_{kz}^d}{\partial z}-A\tau_{ik}-C_k\tau_{wk}-AR_k\frac{\delta p_{sing,k}}{\delta z}+AF_k^{AM}+A\Gamma_{ik}V_i$$

The terms $g_k\frac{\partial A}{\partial z}$ and $\frac{\partial AR_k T^d_{kz}}{\partial z}$, which are influenced by geometric effects, may be modeled together with the singular pressure loss $\delta p_{sing,k}$ depending on the flow regime.

Energy balance of phase k

Time derivative term

Let us define the mass-weighted phase space average enthalpy, $\breve{H_k}^k = \frac{\langle \rho_k h_k \rangle^S_k}{\widehat{\rho_k}^k}$:

$$\int_{A_k} \frac{\partial \rho\left(h+\frac{v^2}{2}\right)}{\partial t} dA = \frac{\partial A_k < \rho_k\left(h_k+\frac{v_k{}^2}{2}\right)>_k}{\partial t} = \frac{\partial A\, R_k \widehat{\rho_k}^k\left(\breve{H_k}^k + \frac{\breve{V_k}^{k2}}{2}\right)}{\partial t} \tag{5.103}$$

An approximation was necessary for the kinetic energy term:

$$\left\langle \rho_k \frac{v_k{}^2}{2}\right\rangle_k \cong \widehat{\rho_k}^k \frac{\breve{V_k}^{k2}}{2}$$

Advection term:

$$\int_{A_k} \nabla\cdot\rho\,\vec{v}\left(h+\frac{v^2}{2}\right) dA = \frac{\partial A R_k \langle \rho_k v_{kz} h_k\rangle^S_k}{\partial z} + \frac{\partial A R_k \left\langle \rho_k v_{kz}\frac{v_k{}^2}{2}\right\rangle^S_k}{\partial z} + MH_{ik} + EC_{ik} \tag{5.104}$$

$$\cong \frac{\partial A\, R_k \widehat{\rho_k}^k \breve{V_k}^k \left(\breve{H_k}^k + \frac{\breve{V_k}^{k2}}{2}\right)}{\partial z} + \frac{\partial A\, R_k q^d_{kz}}{\partial z} + MH_{ik} + EC_{ik} \tag{5.105}$$

MH_{ik} and EC_{ik} are the volumetric energy transfer through the interface:

$$MH_{ik} = -\int_{Ci} \rho_k\left(\vec{v_k} - \vec{v_i}\right) h_k \cdot \vec{n_k}\, \frac{dCi}{\vec{n_k}\cdot\vec{n_{kC}}} \tag{5.106}$$

$$EC_{ik} = -\int_{Ci} \rho_k\left(\vec{v_k} - \vec{v_i}\right) \frac{v_k{}^2}{2} \cdot \vec{n_k}\, \frac{dCi}{\vec{n_k}\cdot\vec{n_{kC}}} \tag{5.107}$$

MH_{ik} and EC_{ik} are usually modeled as

$$MH_{ik} = A\Gamma_{ik}\breve{H_k}^k$$

$$EC_{ik} = \Gamma_{ik}\frac{V_i^2}{2}$$

Again, the averaging of the advection term induces an additional dispersion term q_{kz}^d:

$$q_{kz}^d \triangleq \langle \rho_k v_{kz} h_k \rangle_k^S - \widehat{\rho_k}^k \breve{V}_k^k \breve{H}_k^k$$

q_{kz}^d can be modeled as

$$q_{kz}^d = \lambda_{kz}^d \frac{\partial \breve{T}_k^k}{\partial z}$$

Pressure term:

$$\int_{A_k} \frac{\partial p}{\partial t}\, dA = A\frac{\partial R_k \langle p_k \rangle_k^S}{\partial z} = A\frac{\partial R_k \widehat{p_k}^k}{\partial t} \tag{5.108}$$

Contribution of the deviator stress tensor:

$$\int_{A_k} \nabla \cdot \mathcal{T} \cdot \vec{v}\, dA = \frac{\partial}{\partial z} A_k \vec{v_k} \cdot \left(\mathcal{T}_k \cdot \vec{n_z}\right)_k^S + EF_{ik} + EP_{ik} \tag{5.109}$$

This first term is often neglected.
The second term EF_{ik} is a work of interfacial friction and can be modeled as

$$EF_{ik} = \tau_{ik} V_i$$

The third term EP_{ik} is the work of interfacial pressure and can modeled as

$$EP_{ik} = EP'_{ik} - P\frac{\partial R_k}{\partial t} \cong -P\frac{\partial R_k}{\partial t}$$

Gravity term:

$$\int_{A_k} \rho \vec{g}_z \cdot \vec{v}\, dA = A_k \langle \rho_k g_z u_{kz} \rangle_k^S = A R_k \widehat{\rho_k}^k \breve{V}_k^k g_z \tag{5.110}$$

Heat flux term:

$$\begin{aligned} -\int_{A_k} \nabla \cdot \vec{q}\, dA = & -\frac{\partial}{\partial z} A_k \left\langle \vec{q_k} \cdot \vec{n_z} \right\rangle_k^S - \int_{C_k} \vec{n}_k q_k \frac{dC}{\vec{n}_{kc} \cdot \vec{n}_k} \\ & - \int_{C_k} \vec{n} \cdot \vec{q}_k \frac{dC}{\vec{n} \cdot \vec{n_C}} \end{aligned} \tag{5.111}$$

$$= -\frac{\partial A\, R_k q_{kz}}{\partial z} + C_k q_{wk} + A\, q_{ik}$$

with

$$C_k q_{wk} \triangleq -\int_{C_k} \overrightarrow{n_k} \cdot \overrightarrow{q}_k \frac{dC}{\overrightarrow{n_k} \cdot \overrightarrow{n_{C_k}}};\ A q_{ik} \triangleq -\int_{Ci} \overrightarrow{n}_k q_k \frac{dC}{\overrightarrow{n}_{kc} \cdot \overrightarrow{n}_k}$$

$$\sum\nolimits_k q_{ik} + \Gamma_{ik} \breve{H}_k{}^k = 0 \tag{5.112}$$

$$A\frac{\partial\, R_k \widehat{\rho_k}^k \left(\breve{H}_k{}^k + \frac{\breve{V}_k{}^{k2}}{2} \right)}{\partial t} + \frac{\partial A\, R_k \widehat{\rho_k}^k \breve{V}_k \left(\breve{H}_k{}^k + \frac{\breve{V}_k{}^{k2}}{2} \right)}{\partial z}$$
$$- A R_k \frac{\partial P}{\partial t} + \frac{\partial A\, R_k \left(q_{kz} + q_{kz}^d \right)}{\partial z}$$

$$= +A R_k \widehat{\rho_k}^k \breve{V}_k{}^k g_z + C_k q_{wk} + A\, q_{ik} + A \Gamma_{ik} \left(\breve{H}_k{}^k + \frac{V_i{}^2}{2} \right) + A\, \tau_{ik} V_i + A \widehat{q_{kext}} \tag{5.113}$$

Considering orders of magnitude, the diffusion-dispersion and $\tau_{ik} V_i$ terms are often smaller than others and neglected.

5.8.3 Space-phase-averaged equations for stratified flow

The case of a stratified flow in a horizontal or inclined duct is particular since the interface has a simple shape, and the pressure field in a cross section may be reasonably approximated. Therefore, the integration of pressure forces is easier than it is in more complex flow regimes.

Keeping in mind that the main difficulty in horizontal legs is to correctly predict the stratification occurrence and the characteristics of stratified flow, attention was paid in CATHARE code development to derive the best possible system of equations.

The CATHARE code two-fluid six-equation model to describe stratified flows was derived by de Crecy (1986), who also analyzed its properties, Fig. 5.8, assuming the following:

- A hydrostatic transverse pressure gradient
- Vapor and liquid densities uniform over the cross section
- Continuity of pressure across the interface (no surface tension effect and no pressure jump induced by mass transfer)

Under such conditions, pressure terms can be calculated and momentum equations written as

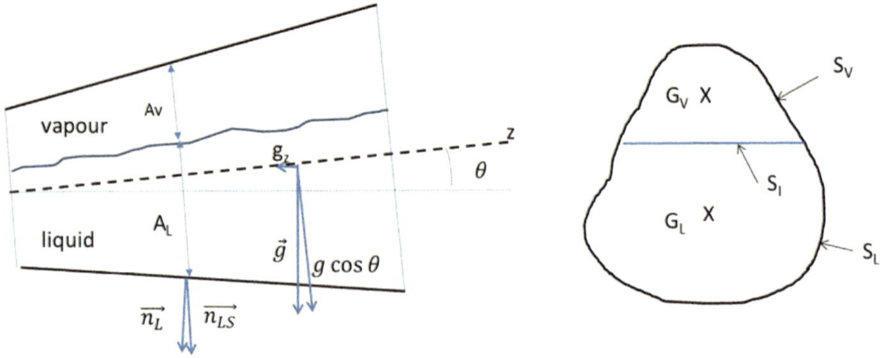

Figure 5.8 Stratified flow.

$$A\frac{\partial R_k \widehat{\rho_k}^k \breve{V_k}^k}{\partial t}+\frac{\partial A R_k \widehat{\rho_k}^k \breve{V_k}^k \breve{V_k}^k}{\partial z}+AR_k\frac{\partial P}{\partial z}+Af_i\frac{\partial R_k}{\partial z}\Delta_k\frac{\partial A}{\partial z}+A\epsilon_k G_z-D_k$$

$$=AR_k\widehat{\rho_k}^k g_z-A\tau_{ik}-C_k\tau_{wk}+A\Gamma_{ik}V_i \qquad (5.114)$$

with

$$g_z=-g\sin\theta$$

$$\Delta_k=-\rho_k g\sin\theta(X_k-X_i)$$

Δ_k is the pressure difference between the average pressure p_k in phase k and pressure at the interface pi:

$$f_i=R_g\Delta_l-\left(1-R_g\right)\Delta_v+R_g\left(1-R_g\right)\frac{\partial(\Delta_v-\Delta_l)}{\partial\alpha}$$

$$G_z=R_g\left(1-R_g\right)\frac{\partial(\Delta_v-\Delta_l)}{\partial z}$$

$$D_k=\int_{S_k}\overrightarrow{n_z}\cdot\overrightarrow{n_k}[p_k(x,y,z)-p_i]\frac{ds}{\overrightarrow{n_k}\cdot\overrightarrow{n_{ks}}}$$

$$\epsilon_v=1;\epsilon_l=-1$$

For each phase k, D_k considers the pressure difference due to gravity between the interface and the wall perimeter in contact with phase k.

D_k and Δ_k are terms dependent on the pipe shape. In the particular case of a square duct, the following expressions are found:

$$G_z = A^{3/2}\varepsilon_k \alpha(1-\alpha)\frac{\rho_m}{2}\frac{\partial g\cos\theta}{\partial z}$$

$$f_i = R_g(1-R_g)(\rho_L-\rho_V)g\cos\theta A^{1/2}$$

$$\Delta_k = \frac{1-\alpha_k}{4}f_i$$

$$\rho_m = R_g\rho_V + (1-R_g)\rho_L$$

The momentum equation is

$$A\frac{\partial R_k\widehat{\rho_k}^k\breve{V}_k^{\,k}}{\partial t} + \frac{\partial AR_k\widehat{\rho_k}^k\breve{V}_k^{\,k}\breve{V}_k^{\,k}}{\partial z} + AR_k\frac{\partial P}{\partial z} + \Delta_k\frac{\partial A}{\partial z} + A\epsilon_k G_z - D_k$$

$$= AR_k\widehat{\rho_k}^k g_z - A\tau_{ik} - C_k\tau_{wk} + A\Gamma_{ik}V_i - Af_i\frac{\partial R_k}{\partial z} + \frac{1-R_k}{4}f_i\frac{\partial A}{\partial z}$$

$$+ A^{3/2}\varepsilon_k R_k(1-R_k)\frac{\rho_m}{2}\frac{\partial g\cos\theta}{\partial z} \tag{5.115}$$

The expression for f_i in a circular pipe should include a shape function of the void fraction, which is often forgotten and replaced with 1.

In the CATHARE code 1-D module (de Crecy, 1986; Bestion and Serre, 2012), f_i is written as

$$f_i = R_g(1-R_g)(\rho_L-\rho_V)g\cos\theta D \tag{5.116}$$

The $f_i\frac{\partial R_k}{\partial z}$ term controls the characteristic velocity corresponding to void fraction waves that are shallow water waves at the free surface of the stratified flow (see Bestion and Serre, 2012).

The properties of the momentum equations in stratified flow are illustrated in Fig. 5.9 below where water is injected very slowly at the entrance of a pipe with area restriction and enlargement followed by a bend. Gas velocity is zero at the entrance and pressure is imposed at the outlet. It has been shown that the water level plotted from the predicted void fraction respects a horizontal free surface even in varying diameter zones and in the bend. This is the result of the three last terms of the momentum equations.

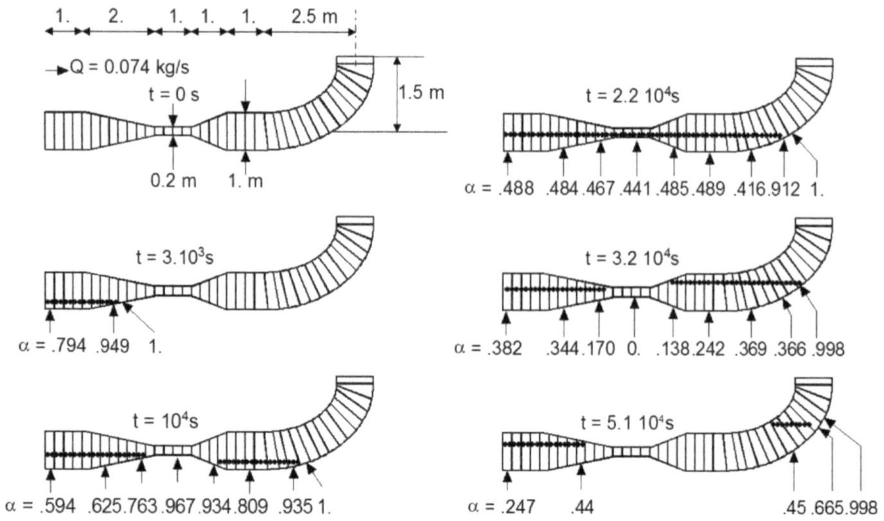

Figure 5.9 Liquid level at rest in a pipe with varying diameter and inclination.

5.9 Time-space-averaged one-dimensional equations for two-phase flow

Let us define some simple notations for space-time phase-averaged parameters.

$\varrho_k \triangleq \widehat{\overline{\rho_k}}^k = \frac{\langle \overline{\chi_k \rho_k} \rangle^S}{\alpha_k}$	Space-time phase average density
$\varrho_m \triangleq \alpha_v \varrho_v + \alpha_l \varrho_l$	Space-time mixture average density
$P_k \triangleq \widehat{\overline{p_k}}^k \triangleq \frac{\langle \overline{\chi_k p_k} \rangle^S}{\alpha_k}$	Space-time phase average pressure
$P \triangleq \alpha_v P_v + \alpha_l P_l$	Space-time mixture average pressure
$V_k \triangleq \breve{\overline{v_{zk}}}^k \triangleq \frac{\langle \overline{\chi_k \rho_k v_{zk}} \rangle^S}{\langle \overline{\chi_k \rho_k} \rangle^S} = \frac{\langle \overline{\chi_k \rho_k v_{zk}} \rangle^S}{\alpha_k \, \widehat{\overline{\rho_k}}^k}$	Mass weighted space-time phase average velocity
$H_k \triangleq \breve{\overline{h_k}}^k = \frac{\langle \overline{\chi_k \rho_k h_k} \rangle^S}{\langle \overline{\chi_k \rho_k} \rangle^S} = \frac{\widehat{\overline{\rho_k h_k}}^k}{\overline{\rho_k}^k}$	Mass weighted space-time phase average enthalpy

In momentum equations, a turbulent stress contribution q_{kz}^t resulting from time averaging of convection term is added to the dispersion term q_{kz}^d.

In energy equations, a turbulent heat flux contribution T_{kz}^t resulting from time averaging of convection term is added to the dispersion term T_{kz}^d.

Some minor approximations are still necessary to derive a time-space-averaged 1-D equation.

Simplified notations and approximations:

$\overline{\Gamma_{ik}}$ is simply noted as Γ_{ik}

$\overline{V_i}$ is simply noted as V_i

with the following approximation: $\overline{\Gamma_{ik} V_i} \cong \Gamma_{ik} V_i$

$\overline{F_k^{AM}}$ is simply noted as F_k^{AM}
$\overline{\tau_{ik}}$ is simply noted as τ_{ik}
$\overline{C_k}$ is simply noted as C_k
$\widehat{\tau_{wk}}$, defined as $\overline{C_k\tau_{wk}} \triangleq \overline{C_k}\,\widehat{\tau_{wk}}$, is simply noted as τ_{wk}
$\overline{f_i\frac{\partial R_k}{\partial z}}$ is modeled as a term in the form $\widehat{f_i}\,\frac{\partial \alpha_k}{\partial z}$, and $\widehat{f_i}$ is simply noted as f_i
$\widehat{q_{wk}}$, defined as $\overline{C_k q_{wk}} \triangleq \overline{C_k}\,\widehat{q_{wk}}$, is simply noted as q_{wk}
$\overline{q_{ik}}$ is simply noted as q_{ik}
$\overline{q_{kext}}$ is simply noted as q_{kext}
This results in the following system of equations.

One-dimentional two-fluid equations

Principal variables: α_g, P, V_g, V_f, H_g, H_l

$$A\frac{\partial \alpha_k \varrho_k}{\partial t} + \frac{\partial A\alpha_k \varrho_k V_k}{\partial z} = A\Gamma_{ik}$$

$$A\frac{\partial \alpha_k \varrho_k V_k}{\partial t} + \frac{\partial A\alpha_k \varrho_k V_k V_k}{\partial z} + A\alpha_k\frac{\partial P}{\partial z} + Af_i\frac{\partial \alpha_k}{\partial z} + g_k\frac{\partial A}{\partial z}$$

$$= A\,\alpha_k\varrho_k g_z + \frac{\partial AR_k\left(T_{kz}^t + T_{kz}^d\right)}{\partial z} - A\tau_{ik} - C_k\tau_{wk} - A\alpha_k\frac{\delta P_{ksing}}{\delta z} + AF_k^{AM}$$
$$+ A\Gamma_{ik}V_i$$

$$A\frac{\partial \alpha_k \varrho_k\left(H_k + \frac{V_k{}^2}{2}\right)}{\partial t} + \frac{\partial A\alpha_k \varrho_k V_k\left(H_k + \frac{V_k{}^2}{2}\right)}{\partial z} - A\alpha_k\frac{\partial P}{\partial t}$$

$$= -\frac{\partial A\,\alpha_k\left(\overline{q_{kz}} + q_{kz}^t + q_{kz}^d\right)}{\partial z} + \mathrm{A}\alpha_k\varrho_k g_z V_k$$

$$+ C_k q_{wk} + Aq_{ik} + A\Gamma_{ik}\left(H_K + \frac{V_i^2}{2}\right) + A\,\tau_{ik}V_i + Aq_{kext}$$

$$\sum_k \alpha_k = 1; \sum_k \Gamma_{ik} = 0; \sum_k \tau_{ik} = 0; \sum_k F_k^{AM} = 0; \sum_k \left[q_{ik} + \Gamma_{ik}H_k\right] = 0;$$

$$q_{kz} + q_{kz}^t + q_{kz}^d = -\lambda_k^{eff}\nabla T_k$$

$$\varrho_k = \rho_k(P, H_k); T_k = T_k(P, H_k); \mu_k = \mu_k(P, H_k); \lambda_k = \lambda_k(P, H_k)$$

$+$*expressions or equations for* $\tau_{ik}, T_{kz}^t, T_{kz}^d, q_{ik}, \overrightarrow{q}_k^t, C_k, \tau_{wk}, q_{wk}, \delta P_{ksing}, F_k^{AM}, \lambda_k^{eff}$

5.10 One-dimensional two-fluid equations in system codes

System codes such as CATHARE (Bestion, 1990), RELAP-5 (RELAP5/MOD3, 1998), and TRACE (TRACE V5.0, 2007) use the 1-D two-fluid equations as in section 9 below with some differences. TRACE writes volume-averaged equations on a part of a duct between two sections. Equations are somewhat similar to the discretized form of 1-D equations when a staggered grid is used.

Axial diffusion and dispersion terms of momentum are never modeled and may be partly modeled through the singular pressure loss term.

Momentum equation may be written using the secondary form (CATHARE, RELAP-5) instead of the conservative form:

$$A\alpha_k\varrho_k\frac{\partial V_k}{\partial t}+A\alpha_k\varrho_k V_k\frac{\partial V_k}{\partial z}+A\alpha_k\frac{\partial P}{\partial z}+Af_i\frac{\partial \alpha_k}{\partial z}+g_k\frac{\partial A}{\partial z}$$

$$=A\alpha_k\varrho_k g_z-A\tau_{ik}-C_k\tau_{wk}-A\alpha_k\frac{\delta P_{ksing}}{\delta z}+AF_k^{AM}+A\Gamma_{ik}(V_i-V_k) \qquad (5.117)$$

The term $g_k\frac{\partial A}{\partial z}$ is not present in all codes and may be present (in the CATHARE code) only for stratified flow for which an expression was derived.

The term $f_i\frac{\partial \alpha_k}{\partial z}$ is present in all codes for stratified flow and may be present (in the CATHARE code) for all flow regimes.

The axial diffusion and dispersion terms of energy are not included in standard models, but some options may exist and be actuated for specific transients. Axial diffusion and dispersion terms of mass concentrations of transported species like boron or radio-active components, which are treated as passive scalars, may also be added as options.

Energy equation may be written as a total enthalpy equation (CATHARE), an internal energy equation (RELAP-5), or a total energy equation (TRACE):

$$A\frac{\partial \alpha_k\varrho_k U_k}{\partial t}+\frac{\partial A\alpha_k\varrho_k V_k U_k}{\partial z}=-AP\frac{\partial \alpha_k}{\partial t}-P\frac{\partial A\alpha_k V_k}{\partial z}$$

$$+C_k q_{wk}+Aq_{ik}+A\Gamma_{ik}H_K+Aq_{kext}+D_k \qquad (5.118)$$

where D_k is viscous dissipation in phase k:

$$A\frac{\partial \alpha_k\varrho_k\left(U_k+\frac{V_k^2}{2}\right)}{\partial t}+\frac{\partial A\alpha_k\varrho_k V_k\left(U_k+\frac{V_k^2}{2}\right)}{\partial z}+\frac{\partial A\alpha_k V_k P}{\partial z}=\mathrm{A}\ \alpha_k\varrho_k g_z V_k$$

$$+C_k q_{wk}+Aq_{ik}+A\Gamma_{ik}H_{ki}+A\ \tau_{ik}V_k+A\ \tau_{wk}V_k+Aq_{kext}$$

5.11 Simplified two-phase models: homogeneous equilibrium drift-flux model

5.11.1 One- to two-fluid models

Historically, two-phase steam-water flows were described as one fluid forming a homogeneous mixture following approximately a single-phase set of equations with some minor modifications. A 1-D one-fluid three-equation two-phase flow model was the first set of equations used to describe steam-water flow in a reactor. It can use both mechanical and thermodynamic equilibrium in the HEM or may accept some velocity difference (slip between phases) expressed by an algebraic relation. Before going to the 1-D six-equation two-fluid model, some four-equation or five-equation models were used in some codes for steam-water flow to describe the thermodynamic nonequilibrium of one phase or both phases.

In two-phase steam-water flow, the interface between phases is at saturation temperature. A phase can be at thermodynamic equilibrium in a two-phase system when it is at saturation temperature ($T_k = Tsat(P)$). A phase can be out of thermodynamic equilibrium in a two-phase system when it is either subcooled ($T_k < Tsat(P)$) or superheated ($T_k > Tsat(P)$).

Phases may have different velocities when different forces are applied to them. For example, the gravity force being proportional to the phase density creates some differences between phases. In addition, the same friction force along walls is not applied to both phases, which may affect the two velocities. Mechanical interactions between phases (drag force, added mass force) tend to couple phases and reduce the slip between phases. As a result of all forces, there may be an equilibrium slip between phases. Due to the different inertia of phases—corresponding to different densities—there may be acceleration of phases before reaching an equilibrium slip that may be expressed as an algebraic relation between gravity, drag, and wall friction forces. The simplest assumption considers an equilibrium slip with no velocity differences. This corresponds to an infinite drag coefficient. A more realistic assumption considers that phases are always at equilibrium slip; only an algebraic relation between phase velocities is sufficient. The most realistic treatment is to consider the different inertia of phases, which requires separate momentum equations for each phase. Table 5.6 summarizes the existing models from one-fluid to two-fluid system of equations.

Let us define some quantities.

$\varrho_k \triangleq \widehat{\overline{\rho_k}}^k \triangleq \frac{\langle \overline{\chi_k \rho_k} \rangle^S}{\alpha_k}$	Space-time phase average density
$\varrho_m \triangleq \alpha_v \varrho_v + \alpha_l \varrho_l$	Space-time mixture average density
$P_k \triangleq \widehat{\overline{p_k}}^k \triangleq \frac{\langle \overline{\chi_k p_k} \rangle^S}{\alpha_k}$	Space-time phase average pressure
$P \triangleq \alpha_v P_v + \alpha_l P_l$	Space-time mixture average pressure
$V_k \triangleq \breve{\overline{v_{zk}}}^k \triangleq \frac{\langle \overline{\chi_k \rho_k v_{zk}} \rangle^S}{\langle \overline{\chi_k \rho_k} \rangle^S} = \frac{\langle \overline{\chi_k \rho_k v_{zk}} \rangle^S}{\alpha_k \hat{\overline{\rho_k}}^k}$	Mass weighted space-time phase average velocity
$G_k \triangleq \alpha_k \varrho_k V_k = \langle \overline{\chi_k \rho_k v_{zk}} \rangle^S$	Phase mass flux

Continued

$G \triangleq G_v + G_l$	Mixture mass flux
$H_k \triangleq \overline{\breve{H}_k}^k \triangleq \frac{\langle \overline{\chi_k \rho_k h_k} \rangle^S}{\langle \overline{\chi_k \rho_k} \rangle^S} = \frac{\overline{\widehat{\rho_k h_k}}^k}{\overline{\widehat{\rho_k}}^k}$	Mass weighted space-time phase average enthalpy
$P_k \triangleq \widehat{\overline{p_k}}^k \triangleq \frac{\langle \overline{\chi_k p_k} \rangle^S}{\alpha_k}$	Space-time phase average pressure
$P \triangleq \alpha_v P_v + \alpha_l P_l$	Mixture space-time average pressure
$H_m \triangleq \alpha_v \frac{\varrho_v H_v + \alpha_l \varrho_l H_l}{\varrho_m}$	Mass weighted space-time mixture average enthalpy
$\widehat{H}_k \triangleq \frac{\langle \overline{\chi_k \rho_k v_{zk} h_k} \rangle^S}{\langle \overline{\chi_k \rho_k v_{zk}} \rangle^S} = \frac{\alpha_k \langle \overline{\rho_k v_{zk} h_k} \rangle^S}{G_k}$	Mass flux weighted space-time phase average enthalpy
$\widehat{H}_m \triangleq \frac{G_v \widehat{H}_v + G_l \widehat{H}_l}{G}$	Mass flux weighted space-time mixture average enthalpy
$x \triangleq \frac{\alpha_v \varrho_v}{\varrho_m}$	Static quality
$X \triangleq \frac{G_v}{G}$	Dynamic quality
$X_{th} \triangleq \frac{\widehat{H}_m - h_{lsat}}{h_{vl}}$	Thermodynamic quality (or equilibrium quality)

5.11.2 Homogeneous equilibrium model

The HEM is the simplest model to describe a steam-water flow. It assumes that both phases are tightly coupled mechanically and thermally so that they have the same mean velocity $(V_v = V_l)$ and same temperature. Since the two-phase mixture is supposed to be at thermodynamic equilibrium, it results that both phases are at saturation temperature when they coexist.

The 1-D HEM model is here reestablished starting from the 1-D two-fluid formulation.

Adding the two phasic mass balance equations provides the mixture mass balance equation. Adding the two phasic momentum balance equations provides the mixture momentum balance equation. Adding the two phasic total enthalpy balance equations provides the mixture total enthalpy balance equation. Variations in kinetic energy and the work of gravity are neglected. The axial diffusion and dispersion terms are also neglected.

5.11.2.1 Single set of equations from single-phase liquid to single-phase steam

The one-fluid HEM model can continuously treat a flow of subcooled water that receives heat and vaporizes progressively up to a full superheated vapor flow that can then be cooled and condensate to come back to a pure liquid flow.

Depending on the value of the mean mixture enthalpy, one can consider three cases, given the following set of equations with P, G, and H as the principal variables.

Table 5.6 The various steam water models from one-fluid to two-fluid system of equations.

	Mass balance	Momentum Balance	Energy Balance	Thermodynamic Equilibrium Assumptions	Mechanical Equilibrium Assumptions
Three-Equation HEM	One mixture equation	One mixture equation	One mixture equation	Both phases at saturation	No slip Between phases
Three-Equation + Drift flux	One mixture equation	One mixture equation	One mixture equation	Both phases at saturation	Algebraic equilibrium-slip Between phases
Four-Equation + Drift flux	Two equations	One mixture equation	One mixture equation	One phase at saturation One phase out of equilibrium	Algebraic equilibrium-slip Between phases
Five-Equation + Drift flux	Two equations	One mixture equation	Two equations	Both phases out of equilibrium	Algebraic equilibrium-slip Between phases
Six-Equation	Two equations	Two equations	Two equations	Both phases out of equilibrium	No mechanical equilibrium Between phases

1-D HEM equations

Main variables: $P, G, \widehat{H}_m$

$$A\frac{\partial \rho_m}{\partial t}+\frac{\partial AG}{\partial z}=0$$

$$A\frac{\partial G}{\partial t}+\frac{\partial A\,G^2/\rho_m}{\partial z}+A\frac{\partial P}{\partial z}=A\rho_m g_z-C\tau_w-\frac{A}{\delta z}\delta p_{sing}$$

$$A\rho_m\frac{\partial H_m}{\partial t}+AG\frac{\partial \widehat{H}_m}{\partial z}-A\frac{\partial P}{\partial t}=Cq_w$$

$$\widehat{H}_m<h_{lsat}\,;X_{th}\triangleq\frac{H_l-h_{lsat}}{h_{lv}}<0;\ \alpha_v=x=X=0$$

$$H_m=\widehat{H}_m=H_l;\ \rho_m=\rho_l(P,H_l);\ T=T_l(P,H_l)$$

$$\widehat{H}_m>h_{vsat}\,;\ X_{th}\triangleq\frac{H_v-h_{lsat}}{h_{lv}}>1;\alpha_v=x=X=1$$

$$H_m=\widehat{H}_m=H_v;\ \rho_m=\rho_v(P,H_v);\ T=T_v(P,H_v)$$

$$h_{lsat}\le\widehat{H}_m\le h_{vsat}\ ;\ 0<X_{th}<1;X_{th}=X=x=\frac{\widehat{H}_m-h_{lsat}}{h_{lv}}$$

$$H_m=\widehat{H}_m=xh_{vsat}+(1-x)h_{lsat}$$

$$\alpha_v=\frac{\left(\widehat{H}_m-h_{lsat}\right)\rho_{lsat}}{\left(h_{vsat}-\widehat{H}_m\right)\rho_{vsat}+\left(\widehat{H}_m-h_{lsat}\right)\rho_{lsat}}$$

$$\rho_m=\frac{\rho_{lsat}\rho_{vsat}h_{lv}}{\left(h_{vsat}-\widehat{H}_m\right)\rho_{vsat}+\left(\widehat{H}_m-h_{lsat}\right)\rho_{lsat}}$$

$+$ *expressions for* $\tau_w, \delta p_{sing}, q_w$

5.11.2.2 Predictive capabilities of the one-dimensional homogeneous equilibrium model set of equations

Although the 1-D HEM set of equations looks much like the single-phase 1-D set of equations, its predictive capabilities are much worse and can be very poor depending on the two-phase flow situation. The main limitations are:

- The no-slip assumption is almost always far from reality and is the main source of error. It may be a reasonable approximation only in the case of a bubbly flow with very small bubbles. It is completely wrong in separate flows such as annular flow and stratified flow and is also a bad approximation for droplet flow.
- The thermodynamic equilibrium of both phases does not allow good prediction of condensing flow, and under post-CHF conditions, with inverse-annular, inverse-slug, and dispersed flow.
- Closure laws for τ_w and q_w are available but are much more complex and must account for a large number of flow and heat transfer regimes. They are generally much less accurate than the closure laws of single-phase flow.

5.11.3 The drift flux model

The drift flux model was the first approach that could better model mechanical interactions between phases and describe the slip between phases.

It can easily be obtained from a simplification of the two-fluid model only if phase density variations within the time-space averaging domain are neglected.

Let us define some quantities:

Local phase time-averaged velocity:	$\overline{v}_{zk}^{k} \triangleq \frac{\overline{\chi_k v_{zk}}}{\varepsilon_k}$
Phase k local superficial velocity:	$j_k \triangleq \overline{\chi_k v_{zk}} = \varepsilon_k \overline{v}_{zk}^{k}$
Local mixture superficial velocity:	$j \triangleq j_v + j_l$
Phase k superficial velocity:	$J_k \triangleq \langle j_k \rangle^S = \left\langle \varepsilon_k \overline{v}_{zk}^{k} \right\rangle^S$
Mixture superficial velocity:	$J_m \triangleq J_v + J_l$
Space-time phase-averaged velocity:	$\mathcal{V}_k \triangleq \frac{J_k}{\alpha_k} = \frac{\langle \overline{\chi_k v_{zk}} \rangle^S}{\langle \overline{\chi_k} \rangle^S}$
Distribution parameter	$Co \triangleq \frac{\langle \varepsilon_v j \rangle^S}{\langle \varepsilon_v \rangle^S \langle j \rangle^S} = \frac{\langle \varepsilon_v j \rangle^S}{\alpha_v J}$
Drift velocity	$V_{gj} \triangleq \frac{\left\langle \varepsilon_v \left(\overline{v}_{zv}^{v} - j \right) \right\rangle^S}{\alpha_v}$
From the identity:	$j_v = \varepsilon_v \overline{v}_{zv}^{v} = \varepsilon_v j + \varepsilon_v \left(\overline{v}_{zv}^{v} - j \right)$
One can write:	$\langle j_v \rangle^S = \langle \varepsilon_v j \rangle^S + \left\langle \varepsilon_v \left(\overline{v}_{zv}^{v} - j \right) \right\rangle^S$

and dividing by α_v, obtain the drift flux identity:

$$\mathcal{V}_v = CoJ + V_{gj}$$

where the difference between the average gas velocity $\mathcal{V}_v$ and the mixture superficial velocity J is due to two effects:

- void and velocity profiles effects when $Co \neq 1$
- average local slip between phases through $V_{gj} \neq 0$

An approximation neglecting the density variations within the time-space averaging domain simplifies expressions of accumulation and advection terms of the equations.

One can then consider that

$$V_k \triangleq \frac{\left\langle \overline{\chi_k \rho_k v_{zk}} \right\rangle^S}{\left\langle \overline{\chi_k \rho_k} \right\rangle^S} \cong V_k = \frac{\left\langle \overline{\chi_k v_{zk}} \right\rangle^S}{\left\langle \overline{\chi_k} \right\rangle^S}$$

$$H_k \triangleq \overline{\check{h}_k}^k \triangleq \frac{\left\langle \overline{\chi_k \rho_k h_k} \right\rangle^S}{\left\langle \overline{\chi_k \rho_k} \right\rangle^S} \cong \widehat{\overline{h_k}}^k = \frac{\left\langle \overline{\chi_k h_k} \right\rangle^S}{\left\langle \overline{\chi_k} \right\rangle^S}$$

Provided that good models are available to express Co and V_{gj}, the simple algebraic relation between the phase velocities and the void fraction can eliminate one partial differential equation of the 1-D six-equation two-fluid model:

$$V_v(1 - Co\alpha_v) = Co(1 - \alpha_v)V_l + V_{gj} \quad (5.119)$$

It can also be used with a 1-D three-equation model (using thermodynamic equilibrium) to remove the main drawback of the HEM model.

5.11.4 Drift-flux model versus two-fluid model

Let us consider the momentum equations of the two-fluid model and the drift flux approach. In the two-fluid model, one can write the two-phase momentum equations as follows:

$$A\frac{\partial \alpha_k \varrho_k V_k}{\partial t} + \frac{\partial A \alpha_k \varrho_k V_k V_k}{\partial z} + A\alpha_k \frac{\partial P}{\partial z} + Af_i \frac{\partial \alpha_k}{\partial z} + g_k \frac{\partial A}{\partial z}$$

$$= \mathrm{A}\ \alpha_k \varrho_k g_z - A\tau_{ik} - C_k \tau_{wk} - A\alpha_k \frac{\delta P_{ksing}}{\delta z} + AF_k^{AM} + A\Gamma_{ik} V_i \quad (5.120)$$

One can write the mixture momentum equation by adding them:

$$A\frac{\partial \alpha_v \varrho_v V_v + \alpha_l \varrho_l V_l}{\partial t} + \frac{\partial A\left[\alpha_v \varrho_v V_v V_v + \alpha_l \varrho_l V_l V_l\right]}{\partial z} + A\frac{\partial P}{\partial z} + (g_v + g_l)\frac{\partial A}{\partial z}$$

$$= A\ \varrho_m g_z - C_v \tau_{wv} - C_l \tau_{wl} - A\left(\frac{\alpha_v \delta P_{vsing} + \alpha_l \delta P_{lsing}}{\delta z}\right) \quad (5.121)$$

and one can obtain a "cross-momentum equation" by eliminating the pressure gradient term between the two phase equations:

$$A\left[\alpha_l\frac{\partial\alpha_v\varrho_v V_v}{\partial t}-\alpha_v\frac{\partial\alpha_l\varrho_l V_l}{\partial t}\right]+\alpha_l\frac{\partial A\alpha_v\varrho_v V_v V_v}{\partial z}-\alpha_v\frac{\partial A\alpha_l\rho\varrho_l V_l V_l}{\partial z}$$

$$=-Af_i\frac{\partial\alpha_v}{\partial z}-[g_v\alpha_l-\alpha_v g_l]\frac{\partial A}{\partial z}-A\,\alpha_v\alpha_l[\varrho_l-\varrho_v]g_z$$

$$-A\tau_{iv}-C_v\alpha_l\tau_{wv}+C_l\alpha_v\tau_{wl}-A\alpha_v\alpha_l\left(\frac{\delta P_{vsing}-\delta P_{lsing}}{\delta z}\right)+AF_v^{AM}+A\Gamma_{iv}V_i$$

(5.122)

Let us consider a quasi-steady flow $\left(\frac{\partial}{\partial t}\cong 0\right)$ and quasi-established flow $\frac{\partial\alpha_v}{\partial z}\cong 0$; $\frac{\partial V_k}{\partial z}\cong 0$ without strong geometric complexity $\left(g_v\cong 0;\ \delta P_{ksing}\cong 0\right)$ and without large interfacial mass transfer ($\Gamma_{iv}V_i$ term is negligible) so that the cross-momentum equation is simplified:

$$\text{A}\ \tau_{iv}\cong\ -A\ \alpha_v\alpha_l[\varrho_l-\varrho_v]g_z-C_v\alpha_l\tau_{wv}+C_l\alpha_v\tau_{wl} \tag{5.123}$$

Under such conditions, the force balance between phases is reduced to equilibrium between drag force gravity and wall friction forces. Considering a dispersed gas phase and a continuous liquid as encountered in bubbly, slug, and churn flow, in a vertical channel at moderate velocity, wall friction terms may be significantly smaller than interfacial friction due to a much higher interfacial area than wall friction area. Then a rough approximation in vertical upward-oriented flow leads to

$$\tau_{iv}\cong\alpha_v\alpha_l[\varrho_l-\varrho_v]g \tag{5.124}$$

This simple balance between buoyancy and drag forces has a clear similarity to drift flux models, where drift velocity models are of the following form:

$$V_{gj}=C\sqrt{\frac{g\Delta\rho\delta}{\rho_l}} \tag{5.125}$$

where δ is a length scale characteristic of the bubble size. This expression can be obtained from a balance between buoyancy B and drag force D applied to a bubble of size

$$B\sim g\Delta\rho\delta^3;D\sim C_d\rho_l\delta^2V_{gj}^2$$

If the drag coefficient is assumed constant and δ is proportional to the Laplace scale, one obtains the form of the Zuber and Findlay (1965) model for churn-turbulent bubbly flow:

$$V_{gj} \sim \left[\frac{g\Delta\rho\sigma}{\rho_l^2}\right]^{0.25} \tag{5.126}$$

If the drag coefficient is assumed constant and δ is bounded by the duct diameter, one obtains the form of the Zuber and Findlay model for slug flow:

$$V_{gj} \sim \sqrt{\frac{g\Delta\rho D}{\rho_l}} \tag{5.127}$$

From the drift flux equation, V_{gj} general expression, and simplified cross-momentum equation, one can obtain an expression for interfacial friction consistent with a drift flux model:

$$\tau_{iv} = \alpha_v\alpha_l\varrho_l\frac{V_{gj}^2}{C^2\delta} = \alpha_v\alpha_l\varrho_l(1 - Co\alpha_v)^2\frac{(V_v - C_\alpha V_l)^2}{C^2\delta} \tag{5.128}$$

$$C_\alpha = \frac{Co(1 - \alpha_v)}{(1 - Co\alpha_v)} \tag{5.129}$$

This simply shows that a drift flux approach can be equivalent to a full two-fluid six-equation model when the approximations made in the simplified cross-momentum equation are valid. When some simplifications are not justified, a full two-fluid six-equation model may have better capabilities.

In situations where inertial forces are not negligible compared with buoyancy, drag, and friction forces, the drift flux may be significantly worse than the full six-equation model. For example:

- Droplet flow: a water droplet created at a quench front during a core reflooding process is entrained by steam but it does not reach an equilibrium slip within the core.
- Stratified flow: in a stratified flow, due to a weak coupling between phases, many situations cannot be predicted precisely by an algebraic slip since the $Af_i\frac{\partial\alpha_v}{\partial z}$ term is not negligible and some acceleration terms may play a significant role in the phase slip.
- Transient flow: when $\frac{\partial}{\partial t}\cong 0$ is no more valid in cross-momentum equation, the algebraic slip relation is not valid.
- Nonestablished flow: when $\frac{\partial\alpha_v}{\partial z}\cong 0$; $\frac{\partial V_k}{\partial z}\cong 0$ is no more valid in cross-momentum equation, the algebraic slip relation is not valid.
- High condensation and flashing flows: such flows are usually accelerating flows and the term $\Gamma_{iv}V_i$ is no more negligible.

5.12 Characteristic velocities and hyperbolicity

The notion of well-posedness was introduced by Hadamard in his analysis of the nature of mathematical modeling. According to Hadamard's classification, well-posedness requires the existence and uniqueness of the solution of the model and that the solution depends continuously on the initial and boundary data. This requisite is induced by the fact that it is necessary to ensure that a small perturbation in the data should give rise to a small variation in the solution at any point of the domain at a finite distance from the boundaries.

A first-order conservative set of PDEs is well posed if and only if it is hyperbolic. As shown in sections 4 and 7, 1-D systems of equations for single-phase and two-phase flows are not first-order since there are diffusion terms. However, they are often neglected, and the system becomes a first-order conservative set of PDEs.

A set of PDEs as a function of time and space is said to be hyperbolic when all the roots of the characteristic equation are real.

More precisely, let a 1-D system of equations be written in the following form:

$$A\frac{\partial X}{\partial t}+\mathrm{B}\frac{\partial X}{\partial z}=C \tag{5.130}$$

where X is the vector of state variables, A and B are matrices, and C is the vector of sources/sinks.

The characteristic equation is

$$Det(B-\lambda A)=0 \tag{5.131}$$

In the 1-D one-phase three-equation model, this is a third-order polynomial function with three roots:

$\lambda_1 = V$ is the transport velocity of enthalpy.

$\lambda_{2,3} = V \mp C_s$ are the propagation velocities of pressure (sound) waves.

$\frac{1}{C_s^2}=\left(\frac{\partial\rho}{\partial P}\right)_S$ The celerity is related to the compressibility of the fluid and is usually considered an isentropic compressibility since the time scales of pressure fluctuations are smaller than those of heat conduction.

In the 1-D two-fluid six-equation model this is a sixth-order polynomial function with six roots.

When solving this equation for a 1-D two-fluid six-equation model, one obtains the following:

$\lambda_1 = V_g$ is the transport velocity of gas enthalpy,

$\lambda_2 = V_l$ is the transport velocity of liquid enthalpy,

$\lambda_{3,4} = W_\alpha \mp C_\alpha$ are the propagation velocities of void waves,

$\lambda_{5,6} = W_s \mp C_s$ are the propagation velocities of pressure (sound) waves.

A problem was raised when the two-fluid single pressure set of equations was found not hyperbolic when no closure law uses partial derivatives. In particular, the propagation velocities of void waves may be imaginary. This may induce an unstable void wave that increases in amplitude very rapidly. Such a system could not be used in any numerical tool.

Then it was observed that terms like the added mass force or $f_i \frac{\partial \alpha_v}{\partial z}$ term can change the nature of the system and make it hyperbolic. It was also found that a nonhyperbolic system of equations could be solved in some codes without being unstable. From the initial problem, two new problems arise:

1. Should a nonhyperbolic system of equations be prohibited in two-fluid modeling?
2. What way should be followed to make the two-fluid model hyperbolic?

Different answers have been given to both questions.

The second-order diffusive terms that have been neglected in the establishment of the six-equation model would have made the system parabolic and thus probably well-posed. At least they can stabilize the highest-frequency waves. In addition, many transfer terms of the equations, like wall friction and interfacial friction, can induce dissipation that may stabilize lower-frequency waves, but it has not been demonstrated that this is sufficient. A complex stability analysis would be necessary and is not currently available. Two answers to question 1 were given:

- A very pragmatic point of view has been developed: the second-order diffusive terms that have been neglected in the six-equation model are small compared with numerical diffusion, and the cell size used is such that these second-order terms are not needed. This point of view requires cells large enough to prevent instability through numerical diffusion. However, when decreasing the mesh size, numerical diffusion is decreased and the possible unstable nature of the equation is revealed. In this approach, no convergence in cell size is possible.
- Although diffusion terms could stabilize high-frequency waves, it is more convenient to use a system of equations that is hyperbolic without diffusion.

As with question 1, there are also two answers to question 2.

- Using added mass force and $f_i \frac{\partial \alpha_v}{\partial z}$ to guarantee hyperbolicity is artificial, and one should use an unconditionally hyperbolic system of equations without such closure differential terms. For example, a seven-equation model with two phasic pressures is unconditionally hyperbolic. A seventh equation is added (usually a void fraction equation) leading to a new and very difficult closure problem that can be understood as finding a model for relaxation time to the single pressure model, with this relaxation time being very difficult to understand and measure.
- Using the added mass force and the $f_i \frac{\partial \alpha_v}{\partial z}$ term is justified by the physics of the two-phase flow. However, the expression for f_i is not known for all flow regimes in two-phase flow and can only be approximated after some simplifying assumptions are made. Adding a condition that these terms make the system hyperbolic may be justified if the condition remains compatible with the limited understanding of the complex physics. This position is illustrated below.

The characteristic equation of the CATHARE 1-D six-equation model was calculated by making some simplifying assumptions.

Both the added mass force (with a coefficient β) and the $f_i \frac{\partial \alpha_v}{\partial z}$ terms are used in momentum equations. When calculating the two characteristic velocities relative to void fraction waves, it was assumed that the two densities are constant (infinite sound speed), and only mass and momentum equations are used. The following velocities were found:

$$\lambda_{3,4} = W \mp C_\alpha$$

$$W_\alpha = \frac{\alpha_l \rho_g^{am} V_G + \alpha_G \rho_l^{am} V_l}{\alpha_l \rho_g^{am} + \alpha_G \rho_l^{am}}$$

$$C_\alpha{}^2 = \frac{f_i \rho_m^{am} - \alpha(1-\alpha)\rho_l^{am}\rho_g^{am}(V_G - V_l)^2}{\rho_m^{am2}}$$

with

$$\rho_k^{am} = \rho_K + \beta \rho_m$$

$$\rho_m = \alpha_G \rho_G + \alpha_L \rho_L$$

$$\rho_m^{am} = \alpha_L \rho_g^{am} + \alpha_G \rho_l^{am}$$

One can see that void wave velocities can only be real if $C_\alpha{}^2 > 0$, which gives the condition

$$f_i > f_{ihyp} = \frac{\alpha(1-\alpha)\rho_l^{am}\rho_g^{am}(V_G - V_l)^2}{\rho_m^{am}} \tag{5.132}$$

Since this form is close to the formulation suggested by Stuhmiller (1977), it was accepted and chosen to satisfy this condition in CATHARE. One can see that the added mass force does not play any role here in hyperbolicity, and one may obtain a hyperbolic system of equations without added mass force provided that

$$f_i > f_{ihyp} = \frac{\alpha(1-\alpha)\rho_g \rho_l (V_G - V_l)^2}{\rho_m} \tag{5.133}$$

The ATHLET code (Lerchl and Autregesilo, 2013) uses a slightly different form of f_i:

$$f_i = C\alpha\rho(V_G - V_l)^2 \tag{5.134}$$

where the C value can also provide hyperbolicity and stability.

5.12.1 Insights into selected topics

Considerations dictated by Dan Hughes, Hughes (2020), provide insights into the method of characteristics (further details are given in Chapter 20) and the hyperbolicity condition, discussed by Ahn et al. (2017). These are concerned with attempts to progress, rather than with established achievements in the area.

5.12.1.1 Characteristics of two-pressure modeling

A large number of investigations into the characteristics of two-phase flow model equation systems were carried out in the early 1970s. At that time, it was determined that the generally accepted model equations for two-phase flows do not have real characteristics over all of phase space. The complex characteristics of the equation system implied that the system is not hyperbolic, which additionally implied that the problem is not well posed in a mathematical sense. To a certain extent, investigations continue to this day. Hundreds, if not thousands, of publications have been written on the subject. Publications cited as an example of the initial finding of the problem are Rakhmatulin (1956) and Lyczkowski et al. (1978), and see also Lyczkowski (2018). The paper by Rakhmatulin (1969) deals with the speed of sound for two- and multiphase mixtures. The author understood some of the consequences of the lack of hyperbolical equation systems. The equation system of interest is the ubiquitous one-pressure model, as summarized in the following simplified notes. The mechanical case without mass and energy exchanges is used. The EoS gives density as a function of pressure alone. This approach does not introduce any limitations to the analyses and greatly simplifies equation processing. So we take

$$\rho_l = \rho_l(p) \text{ and } \rho_g = \rho_g(p) \tag{5.135}$$

The derivatives of these are

$$\frac{1}{c_l^2} = \left(\frac{\partial \rho_l}{\partial p}\right) \text{ and } \frac{1}{c_g^2} = \left(\frac{\partial \rho_g}{\partial p}\right) \tag{5.136}$$

Note that if the mixture consists of the liquid and vapor phases of a pure substance at the same constant temperature, the thermodynamic state of each phase is constrained to the coexistence line for the substance. Neglecting this does not affect the results of the analyses in these notes. One then obtains, for the mass conservation of liquid and steam:

$$\frac{\partial \alpha_l \rho_l}{\partial t} + \frac{\partial \alpha_l \rho_l u_l}{\partial z} = 0 \text{ and } \frac{\partial \alpha_g \rho_g}{\partial t} + \frac{\partial \alpha_g \rho_g u_g}{\partial z} = 0 \tag{5.137}$$

Expanding the derivatives of products and considering Eq. (5.136) for the derivatives of the density gives

$$\frac{\alpha_l}{c_l^2}\frac{\partial p}{\partial t} - \rho_l \frac{\partial \alpha_g}{\partial t} + \alpha_l \rho_l \frac{\partial u_l}{\partial z} + \frac{\alpha_l u_l}{c_l^2}\frac{\partial p}{\partial z} - \rho_l u_l \frac{\partial \alpha_g}{\partial z} = 0 \tag{5.138}$$

and

$$\frac{\alpha_g}{c_g^2}\frac{\partial p}{\partial t}+\rho_g\frac{\partial \alpha_g}{\partial t}+\alpha_g\rho_g\frac{\partial u_g}{\partial z}+\frac{\alpha_g\, u_g}{c_g^2}\frac{\partial p}{\partial z}+\rho_g u_g\frac{\partial \alpha_g}{\partial z}=0 \tag{5.139}$$

The momentum equations for liquid and vapor are

$$\frac{\partial \alpha_l\rho_l\, u_l}{\partial t}+\frac{\partial \alpha_l\rho_l\, u_l\, u_l}{\partial z}+\frac{\partial \alpha_l p}{\partial z}=0 \tag{5.140}$$

and

$$\frac{\partial \alpha_g\rho_g\, u_g}{\partial t}+\frac{\partial \alpha_g\rho_g\, u_g\, u_g}{\partial z}+\frac{\partial \alpha_g p}{\partial z}=0 \tag{5.141}$$

Carefully note how the pressure gradient contribution has been written in the last terms in the momentum balance equations; this is a formulation that has appeared in the literature. This formulation will be labeled as the void fraction inside the pressure gradient term, or alpha inside. This form is used here because it is related to the situation with RELAP-7 model equations. The formulation (generally rejected for one-pressure models) was explicitly rejected in Lyczkowski et al. (1978). Expanding the derivatives of the products in the momentum equations and substituting the basic form of the mass conservation equation into those returns a zero value, which gives

$$\alpha_l\rho_l\frac{\partial u_l}{\partial t}+\alpha_l\rho_l u_l\frac{\partial u_l}{\partial z}+\alpha_l\frac{\partial p}{\partial z}+\mathrm{p}\frac{\partial \alpha_l}{\partial z}=0 \tag{5.142}$$

$$\alpha_g\rho_g\frac{\partial u_g}{\partial t}+\alpha_g\rho_g u_g\frac{\partial u_g}{\partial z}+\alpha_g\frac{\partial p}{\partial z}+\mathrm{p}\frac{\partial \alpha_g}{\partial z}=0 \tag{5.143}$$

Note that there is no explicit equation for the void fraction. The void is obtained through the solution of the equation system. For the HEM, various drift flux models, and the one-pressure, six-equation model, the void fraction is basically determined by the EoS by use of the solutions of the conservation and balance equations. Equations (5.138), (5.139), (5.142), and (5.143) are four equations for the four unknowns:

$$\boldsymbol{W}=\left(\alpha_g, p, u_l, u_g\right)^T \tag{5.144}$$

that can be written in the (matrix) form

$$A(\boldsymbol{W})\frac{\partial \boldsymbol{W}}{\partial t}+B(\boldsymbol{W})\frac{\partial \boldsymbol{W}}{\partial z}=0 \tag{5.145}$$

The characteristic polynomial for the system is given by the evaluation of

$$\det=|A(\boldsymbol{W})-\lambda B(\boldsymbol{W})|=0 \tag{5.146}$$

where

$$A(\mathbf{W}) = \begin{vmatrix} -\rho_l & \dfrac{\alpha_l}{c_l^2} & 0 & 0 \\ \rho_g & \dfrac{\alpha_l}{c_l^2} & 0 & 0 \\ 0 & 0 & \alpha_l\,\rho_l & 0 \\ 0 & 0 & 0 & \alpha_g\,\rho_g \end{vmatrix} \tag{5.147}$$

and

$$B(\mathbf{W}) = \begin{vmatrix} -\rho_l u_l & \dfrac{\alpha_l u_l}{c_l^2} & \alpha_l\,\rho_l & 0 \\ \rho_g u_g & \dfrac{\alpha_g u_g}{c_g^2} & 0 & \alpha_g u_g \\ -p & \alpha_l & \alpha_l\,\rho_l\,u_l & 0 \\ p & \alpha_g & 0 & \alpha_g \rho_g\,u_g \end{vmatrix} \tag{5.148}$$

The characteristic polynomial is quite complex and not shown here. Other forms of Eq. (5.146) are equally valid and appear in the literature. Generally, the characteristic equation polynomial has not yet been successfully solved analytically for the roots. Instead, numerical solutions are obtained by sweeping over the ranges of interest for all dependent variables in phase space. This approach is required for the cases of alpha inside or outside the pressure gradient.

Typical results are given by Lyczkowski et al. (1978) for mixtures of liquid and vapor phases of water. Generally, the formulation with alpha inside gives significantly larger regions of real roots and thus a hyperbolic system.

The primary interest for these notes is the entries in the first column of the third and fourth rows of Eq. (5.148). Those entries come from the alpha inside the pressure gradient. As mentioned above, that modeling is universally rejected. The basis for the rejection can be ascertained by looking at the momentum balance of Eqs. (5.92) and (5.93). The first two terms in these equations are the acceleration of the fluid, and the last two terms are the forces that produce the acceleration. The last term in those equations shows that the modeling approach results in a force that can induce mean motions in each phase, and the force is determined by the void gradient. This force does not arise in the physical domain. The model equations will produce non-physical results.

The discovery of the lack of hyperbolicity for the alpha outside model occurred from the search for a stable numerical solution method. No stable method could be devised. All numerical methods, including fully implicit, and dispersion analyses of the linearized continuous equations, indicated unbounded growth of the highest-frequency perturbations, which is not in accord with the physical domain. Someone decided to look into the characteristics, and that led to the finding of complex characteristic roots.

5.12.1.2 From early history of two-pressure model to RELAP-7 development

Initial work on two-pressure seven-equation model equation systems started in the late 1970s and early 1980s. Over the decades since, many efforts have been made to patch the one-pressure model by accounting for various forces in the physical domain acting at the interface between the liquid and vapor. These are not of interest here, and in general, none have proven decisive for simulating each detail of physical phenomena.

Let us first consider a short digression; we looked at the following system. We simply postulated that a void-fraction equation existed, and blindly added a second pressure. The void fraction equation was taken to be

$$\frac{\partial \alpha_g}{\partial t} + U_{\alpha g} \frac{\partial \alpha_g}{\partial z} = 0 \tag{5.149}$$

where $U_{\alpha g}$ is the propagation velocity (m/s) of void fraction perturbation. The EoS now becomes

$$\rho_l = \rho_l(p_l) \text{ and } \rho_g = \rho_g\left(p_g\right) \tag{5.150}$$

The derivatives are

$$\frac{1}{c_l^2} = \left(\frac{\partial \rho_l}{\partial p_l}\right) \text{ and } \frac{1}{c_g^2} = \left(\frac{\partial \rho_g}{\partial p_g}\right) \tag{5.151}$$

The mass conservation Eqs. (5.138) and (5.139) become, for liquid and vapor phases respectively,

$$\frac{\alpha_l}{c_l^2} \frac{\partial p_l}{\partial t} - \rho_l \frac{\partial \alpha_g}{\partial t} + \alpha_l \rho_l \frac{\partial u_l}{\partial z} + \frac{\alpha_l\, u_l}{c_l^2} \frac{\partial p_l}{\partial z} - \rho_l u_l \frac{\partial \alpha_g}{\partial z} = 0 \tag{5.152}$$

and

$$\frac{\alpha_g}{c_g^2} \frac{\partial p_g}{\partial t} + \rho_g \frac{\partial \alpha_g}{\partial t} + \alpha_g \rho_g \frac{\partial u_g}{\partial z} + \frac{\alpha_g\, u_g}{c_g^2} \frac{\partial p_g}{\partial z} + \rho_g u_g \frac{\partial \alpha_g}{\partial z} = 0 \tag{5.153}$$

The momentum Eqs. (5.142) and (5.143), or "alpha inside" equations, become

$$\alpha_l \rho_l \frac{\partial u_l}{\partial t} + \alpha_l \rho_l u_l \frac{\partial u_l}{\partial z} + \alpha_l \frac{\partial p_l}{\partial z} + p_l \frac{\partial \alpha_l}{\partial z} = 0 \tag{5.154}$$

$$\alpha_g \rho_g \frac{\partial u_g}{\partial t} + \alpha_g \rho_g u_g \frac{\partial u_g}{\partial z} + \alpha_g \frac{\partial p_g}{\partial z} + p_g \frac{\partial \alpha_g}{\partial z} = 0 \tag{5.155}$$

where the unknowns are

$$\mathbf{W} = \left(\alpha_g, p_l, p_g, u_l, u_g\right)^T \tag{5.156}$$

The matrices needed for Eq. (5.146) are

$$A(\mathbf{W}) = \begin{vmatrix} 1 & 0 & 0 & 0 & 0 \\ -\rho_l & \alpha_l/c_l^2 & 0 & 0 & 0 \\ \rho_g & 0 & \alpha_g/c_g^2 & 0 & 0 \\ 0 & 0 & 0 & \alpha_l\,\rho_l & 0 \\ 0 & 0 & 0 & 0 & \alpha_g\,\rho_g \end{vmatrix} \tag{5.157}$$

and

$$B(\mathbf{W}) = \begin{vmatrix} U_{\alpha g} & 0 & 0 & 0 & 0 \\ -\rho_l u_l & \alpha_l u_l/c_l^2 & 0 & \alpha_l\,\rho_l & 0 \\ \rho_g u_g & 0 & \alpha_g u_g/c_g^2 & 0 & \alpha_g\,\rho_g \\ -p_l & \alpha_l & 0 & \alpha_l\,\rho_l u_l & 0 \\ p_g & 0 & \alpha_g & 0 & \alpha_g\,\rho_g u_g \end{vmatrix} \tag{5.158}$$

Thus the characteristic polynomial and the roots do not contain p_l or p_g (equivalent to the case of a single pressure with alpha inside the pressure gradient). The roots are

$$\lambda_1 = u_g + c_g; \lambda_2 = u_g - c_g; \lambda_3 = u_l + c_l; \lambda_4 = u_l - c_l; \lambda_5 = U_{\alpha g} \tag{5.159}$$

These roots are also the solutions for the case with the void outside the pressure gradient, for which the entries in the first column and fourth and fifth rows are zero. Therefore

(1) so long as two pressures are used along with an additional equation, the roots will be real, and they will be, intuitively, consistent with the single flow case. It makes no difference if the void fraction is inside or outside the pressure gradient;
(2) the void fraction inside the pressure gradient introduces, as discussed above, properties that are not consistent with the physical domain;
(3) real, and unique, characteristic roots are necessary for a well-posed mathematical problem, but that requirement is not sufficient to ensure reality with the physical domain.

Specific mathematical analyses are necessary to determine the consistency of the continuous equations with the physical domain and validation (implying a comparison between model prediction and experimental data) is the ultimate test of the model.

While the characteristic roots look familiar from the single-phase flow case, this superficial property in no way whatsoever is related to the reality of the roots in the physical domain. In fact, it is difficult to envision a propagation of pressure waves in a heterogeneous mixture of a continuous–discrete flow regime like bubbly flow, for example. Note that if $U_{\alpha g}$ happens to have the same value as any of the other four roots, the system is no longer strictly hyperbolic.

The seven-equation, two-pressure, two-temperature, two-velocity (or UVUTUP), hyperbolic, well-posed models for two-phase flows, i.e., the RELAP7 model, is described in the code theory manual Berry et al. (2015) with details provided by Berry (2003, 2005 and 2005a) and Berry et al. (2008). The authors of the last document cite Saurell and Abgrall (1999 and 1999a) and Saurel and Lemetayer (2001) as primary sources for the complete, closed continuous system. Further, the model by Baer and Nunziato (1986), is a fundamental template for the equation systems investigated by the RELAP7 team.

Without reporting the equations hereafter [to avoid copying from the cited code manual and the difficulty of considering the equations "as-coded"], the RELAP7 UVUTUP model is formally similar to the UVUT model (i.e., the mass, momentum, and energy balance of each phase), namely related to the 1-D standard UVUT model; see also Delchini et al., (2017):

- p_l and p_g substitute p in the momentum and energy equations.
- A void equation similar to Eq. (5.149) is added: constitutive terms are considered in the RHS of the equation and instead of $U_{\alpha g}$, u_{int} is adopted; u_{int} is the interface velocity and is the velocity of the phase with the smallest acoustic impedance; therefore $u_{int} \equiv u_g$.
- Specific constitutive terms are added: these include "numerical" terms, p_{int} (pressure at the interface) and u_{int} (defined above), which appear in the determinant (equivalent to Eqs. 5.157 and 5.158) to fix characteristics polynomial and the hyperbolicity of the set of equations.
- Relaxation terms are extensively adopted for pressure, velocity, etc., according to Leveque (2002).

The seven resulting unknowns are $\alpha_g, \rho_g, \rho_l, u_g, u_l, p_g, p_l$, including the condition $\alpha_g + \alpha_l = 1$.

5.12.1.3 Implications of the (RELAP-7) two-pressure, seven-equation model

Following the discussion in Sections 5.12.1.1 and 5.12.1.2, an evaluation of the RELAP7 author viewpoints may contribute to making clear the current understanding of two-pressure modeling. Therefore, the RELAP7 author viewpoints and comments to those viewpoints are given hereafter.

Let us start with statements by the RELAP7 team quoted below:

> *"This seven-equation two-phase model allows both phases to be compressible. Because p_g is not, in many practical problems, very different from p_l (with the exception of surface tension effects), most traditional two-phase models assume $p_g = p_l$ which allows the elimination of one dependent variable and serves as a substitute for the volume fraction evolution equation. However, $p_g \approx p_l$ does not entail the same property for their partial derivatives, Rousseau and Ferch (1979). Therefore, the assumption of $p_g = p_l$ is very restrictive when derivatives are involved. As pointed out by Boure and Delhaye (1982), it requires that pressure disturbances have the same average effect on the two phases and, in particular, that they propagate at the same velocity within the phases. While the assumption $p_g = p_l$ has proved useful in many cases, it is definitely too restrictive when propagation phenomena are important.[1] The RELAP-7 approach forgoes this assumption and retains the seven-equation model as its basis. More importantly, the seven-equation model allows for complete mechanical and thermodynamic nonequilibrium between the phases and it is hyperbolic and well-posed in the sense of Hadamard,[2] Hadamard (1902). The hyperbolicity (and thus well-posedness) of this model is a direct result of incorporation of both phases' compressibilities, and not of a manipulation of interfacial variables as is done in CATHARE, Bestion (1990). The system has symmetrically occurring eigenvalues and eigenvectors with respect to the two-phases; its wave speeds (eigenvalues) are $(u \pm c)_l$ and $(u \pm c)_g$ for the genuinely nonlinear fields, and u_l, u_g and u_{int} (multiplicity 2) for the linear degenerate fields. This seven-equation two-phase model is a relaxation model and it has the very desirable feature of naturally devolving to simpler, even classical, models upon mechanical and/or thermodynamical relaxation, Berry et al. (2013). Thus, this model can readily couple to simpler models via a natural transition from the seven-equation model to a classical six-equation (ill-posed) model, a five-equation Kapila model, Kapila (2001), and Murrone and*

[1] With the complex characteristics that can occur with the classical six-equation model, it is not clear how to set the boundary conditions, and high wave number instabilities occur during convergence testing. It has been argued that equation sets with complex characteristics may still model a range of phenomena quite adequately if the numerical method introduces sufficient dissipation to damp the high frequency instabilities. There are obviously real physical effects that do this but are left out of the equations. As pointed out by Banerjee and Hancox (1978), one does not always know whether these effects are important and under what conditions they are important

[2] The mathematician Jacques Hadamard espoused that a "well-posed" mathematical model of physical phenomena should have the properties that (1) a solution exists, (2) the solution is unique, and (3) the solution's behavior depends continuously on the initial conditions. Problems that are not well-posed are said to be "ill-posed." Early researchers in two-phase flow knew that, if due diligence was not exercised, an ill-posed formulation could result; and they understood the need for a well-posed model, as summarized in Hughes et al. (1976).

Guillard (2005), a four-equation homogeneous relaxation model (HRM), or a three-equation homogeneous equilibrium model (HEM). It is noted that, because of this feature, experience shows that some physically and mathematically realistic solutions may, upon first examination, appear counter-intuitive to the inexperienced modeler."

After 5 decades since the idea of two-pressure modeling was considered and 2 decades devoted to the development of RELAP7, at the time of writing the second edition of the current book, "*there appears to be a cone of absolute silence around the code*," Hughes (2020). Furthermore, no systematic proof of validation in situations of industrial interest for nuclear thermal hydraulics, e.g., nuclear reactor accidents, is available, if one excludes the demonstration to predict pressure wave propagation, e.g., Delchini et al. (2017).

However strong the theoretical bases, constitutive laws and assumptions require comprehensive validation backed by experimental evidence. The overall two-pressure model may require deep motivation and support from experiments. Furthermore, no experiment, except for stratification (e.g., in situations where gravity is relevant), is available to provide measurement of different liquid and vapor pressures.

The above two statements reveal (yet so far) a dead-end attempt at two-pressure modeling; nonetheless, its description is useful for readers to decide on future directions of research. Additional comments are in the following.

The hyperbolicity resulting from Eq. (5.159) is determined by the decision to introduce two pressures and the void fraction equation. That is, it does not make any difference that the void is inside the pressure gradient and it does not make any difference what dependent variable multiplies the void gradient on the right-hand side: so long as the equation still meets the requirements for calculating the characteristic roots, the system will be hyperbolic. In different terms,

a hyperbolicity is an important, though not decisive, advantage for the two-pressure model (over the UVUT model), but
  - two-pressure modeling has no support from experimental evidence, and
  - the void fraction equation, essential for obtaining hyperbolicity, requires additional specific validation effort for its justification.

Related to the convergence issue (i.e., raised in footnote "[1]" above by the RELAP7 developers), six-equation UVUT models have a large and quite significant proven track record, including comparisons with experimental data. The need to strictly show spatial and temporal convergence has not been identified as a practical limitation for these models in any industrial applications, although this is certainly not a completely satisfying "solution" to the theoretical problem (i.e., with one-pressure six-equation models). Additionally, it is well known that fooling around with various modifications to the basic forms, primarily adding terms for different physical phenomena and processes into the equations, has not yet provided a universal "solution" to the complex characteristics problem. However, it is not clear how and if the approach taken for RELAP7 provides a universal "solution," including addressing the convergence issue.

The relaxation terms represent sources and sinks of mass, momentum, and energy and affect the calculated responses of the bulk of fluids/phases: while the micro scale is invoked for their introduction, the terms modify the response at the macro scale. Spurious non-physical energy sources and sinks are also potential sources of numerical instability induced by relaxation terms.

The possible local computation of two different sonic velocities for liquid and vapor implies the need for a specific procedure and assumptions to ensure the convergence of a calculation at the level of the system.

5.13 Transport of interfacial area and polydispersion models

Two-phase flow has a complex nature with a large diversity of length scales associated with bubble size, droplet size, distance between bubbles or droplets, film thickness, turbulent eddy sizes, and wavelength at a free surface or film surface. Similarly, a large variety of time scales can be identified. All basic processes, such as mass momentum and energy transfers at walls and interfaces, to be modeled depend on length scales, which depend on the flow regime. Transport equations may be added to predict the evolution of some scales, such as turbulent scales. The first approach in two-phase flow modeling was to write an equation for interfacial area concentration a_i that may be defined as the ratio of interface area (IA) to the volume of a two-phase mixture. The TIA was developed in both system codes in a 1-D approach and 3-D CMFD, with TIA focused on bubbly and droplet flows. Three factors promote the use of the TIA in system codes:

Nonestablished flows: it was observed that dispersed bubbly flow and droplet flows have a mean bubble and drop size that depends on several mechanisms such as breakup, coalescence, nucleation, collapse, and interfacial mass transfer. Some of these phenomena have relatively long relaxation times that are larger than convective time scales so that the "average size" is not an "equilibrium size" that can be expressed based only on the local principal variable of the two-fluid six-equation model used in the current system codes. TIA is capable of better modeling such nonestablished flows by accounting for relaxation times provided that source and sink terms are modeled and validated.

Flow regime transitions: it was observed that many flow regime transitions correspond to a change in IA: bubbly-to-slug flow or bubbly-to-churn flow results from a coalescence process that decreases IA, and bubbly-to-stratified flow transition also results from bubble sedimentation and coalescence of bubbles. Therefore, TIA should be used for a prediction of some flow regime transitions and could also model intermediate states between two flow regimes. Other flow regime transitions, such as the onset of droplet entrainment, dramatically increase IA, but the governing phenomenon is interfacial instability. One may distinguish transitions that result from a change in IA from transitions that induce a change in IA.

Interfacial transfers: interfacial transfers of mass, momentum, and energy depend on IA (or on drop/bubble size, for example). Therefore a dynamic prediction of IA would also improve the prediction of interfacial friction and interfacial heat transfers that control void fraction and slip between phases and phase temperatures and mass transfers by vapourization and condensation.

Usual expressions of interfacial friction and interfacial heat transfers in system codes are of the form

$$\tau_i = -a_i C_i \frac{\rho}{2} |V_v - V_l|(V_v - V_l) \tag{5.160}$$

$$q_{ik} = a_i h_{ik}(T_{sat} - T_k) \tag{5.161}$$

Let us consider a dispersed flow with a continuous phase c and a dispersed phase d (bubbles or drops) with a particle number density n_d and a Sauter mean diameter δ_{32}.

5.13.1 Monodispersed approximation

The simplest assumption is to consider a population of uniform-sized spherical particles. In this case, simple relations exist between a_i, the dispersed phase fraction α_d, δ_{32}, and n_d:

$$a_i = \frac{6\alpha_d}{\delta_{32}}; n_d = \frac{6\alpha_d}{\pi\delta_{32}^3} \tag{5.162}$$

The usual expressions for interfacial friction and heat transfer coefficients h_{id} in dispersed flow are

$$\tau_{id} = -C_d(Re_d)\frac{\pi\delta_{32}^2}{4} n_d \frac{\rho_c}{2} |V_d - V_c|(V_d - V_c) \tag{5.163}$$

$$= -\frac{3\alpha_d}{4\delta_{32}} C_d \rho_c |V_d - V_c|(V_d - V_c)$$

$$q_{id} = \frac{6\alpha_d}{\delta_{32}} h_{ik}(T_{sat} - T_d) = \frac{6\alpha_d}{\delta_{32}} \frac{\lambda_c Nu_d}{\delta_{32}}(T_{sat} - T_d) \tag{5.164}$$

with

$$Nu_d = a\, Re_d{}^p Pr_c{}^q; Re_d = \frac{\rho_c(V_d - V_c)\delta_{32}}{\mu_c}; Pr_c = \frac{\mu_c Cp_c}{\lambda_c} \tag{5.165}$$

Both models require expressions for δ_{32} or a_i or n_d.

5.13.2 Monodispersed approximation with algebraic interfacial area

In standard industrial fully validated versions of system codes, only algebraic expressions of δ_{32} or a_i or n_d are available. It is often assumed that a droplet or bubble diameter is equal to the limiting value corresponding to a critical Weber number:

$$We_{crit} = \frac{\rho_c (V_d - V_c)^2 \delta_{32}}{\sigma} \tag{5.166}$$

This assumes that all droplets or bubbles have reached the maximum diameter above which they would be broken. It is clear that this is a very strong assumption that cannot be valid except in some particular situations.

5.13.3 Monodispersed approximation with dynamic interfacial area modeling

A more general approach within the monodispersed approximation consists in writing a transport equation for a_i or n_d. The derivation of such equations can be found in Ishii and Hibiki (2011), or in Morel (2015), see also Shen and Hibiki (2018), and Schlegel et al. (2015):

$$\frac{\partial n_d}{\partial t} + \nabla \cdot n_d v_n = \Phi_n^{Nuc} + \Phi_n^{Col} + \Phi_n^{Coa} + \Phi_n^{BU} \tag{5.167}$$

$$\frac{\partial a_i}{\partial t} + \nabla \cdot a_i v_i = \frac{2a_i}{3\alpha_d \rho_d}\left(\Gamma_{id} - \alpha_d \frac{\partial \rho_d}{\partial t} - \alpha_d V_d \nabla \rho_d\right) + \Phi_{a_i}^{Nuc} + \Phi_{a_i}^{Col} + \Phi_{a_i}^{Coa} + \Phi_{a_i}^{BU} \tag{5.168}$$

v_n or v_i must be modeled, and the dispersed phase velocity v_d is most often used. The following source and sink terms must also be modeled.

Φ_n^{Nuc} source term of n_d due to nucleation (creation of bubble by flashing or boiling or drop creation by condensation)

Φ_n^{Col} sink term of n_d due to collapse (vapourization of a drop or condensation of a bubble)

Φ_n^{Coa} sink term of n_d due to coalescence

Φ_n^{BU} source terms of n_d due to breakup

$\Phi_{a_i}^{Nuc}$ source term of a_i due to nucleation

$\Phi_{a_i}^{Col}$ sink term of a_i due to collapse

$\Phi_{a_i}^{Coa}$ sink term of a_i due to coalescence

$\Phi_{a_i}^{BU}$ source terms of n_d due to breakup

with the relations:

$$\Phi_{a_i}^{Coa} = 12\pi \left(\frac{\alpha_d}{a_i}\right)^2 \Phi_n^{Coa}; \Phi_{a_i}^{BU} = 12\pi \left(\frac{\alpha_d}{a_i}\right)^2 \Phi_n^{BU} \tag{5.169}$$

5.13.4 Polydispersed dynamic modeling with the method of statistical moments

A more complex modeling consists of predicting bubble or drop size distribution along the flow.

The bubble diameter distribution function $f(\delta; x, t)$ is defined such that $f(\delta; x, t)d\delta d^3x$ represents the probable number of bubbles in a volume element d^3x at time t, having their diameters in the interval $[\delta, \delta + d\delta]$

The jth-order moment of the bubble diameter distribution function is defined by

$$S_j(x,t) \triangleq \int_0^{\infty} \delta^j f(\delta; x, t) d\delta \tag{5.170}$$

The four first-order moments give the bubble number density, the product of the bubble number density n_d by the mean bubble diameter δ_{10}, the IA a_i, and the fraction of dispersed phase α_d:

$$S_0 = n_d; S_1 \triangleq n_d \delta_{10}; S_2 = \frac{a_i}{\pi}; S_3 = \frac{6\alpha_d}{\pi} \tag{5.171}$$

The Sauter mean diameter is the ratio

$$\delta_{32} = \frac{S_3}{S_2} \tag{5.172}$$

A transport equation for the moment S_j can be written as

$$\frac{\partial S_j}{\partial t} + \nabla \cdot (S_j v_j) - j\dot{D}_{j-1} S_{j-1} = P_j \tag{5.173}$$

where closure relations are necessary for the transporting velocity v_j, the mean bubble or drop growth rate $\dot{D}_j$, and the source term P_j.

Selecting a law of distribution function and writing transport equations for some moments, one can predict the evolution with x and t of the distribution. This was tested with log-normal and quadratic laws by Zaepffel et al. (2012), since these are simple laws that can be defined with knowledge of only two statistical moments.

When a bubble size distribution function is known, one can express liquid-to-interface heat transfers as

$$q_{il} = \left[n_d \pi \int_{\delta_{min}}^{\delta_{max}} \delta f(\delta) Nu(\delta) d\delta \right] \lambda_c (T_{sat} - T_c) \tag{5.174}$$

which is likely to be more precise than the monodispersed expression.

5.13.5 Polydispersed dynamic modeling with a multifield approach

Another approach for modeling polydispersion is the multisize group (MUSIG) method, where mass balance equations are written for tje different size groups of bubbles or drops, and all transfers between the groups are modeled. This provides a discrete function *f*. The MUSIG model was developed and validated (Krepper et al., 2011) to describe bubble size polydispersion, where the disperse phase is divided into several groups, each characterized by its volume fraction and velocity field by writing several mass and momentum equations. This approach has already shown good capability for capturing the different behavior of small and large bubbles and their influence on the formation of a certain flow pattern.

5.13.6 Predictive capabilities of the monodispersed and polydispersed approaches

Using a simple algebraic model for the interfacial area provides an average bubble size—the Sauter mean diameter—that cannot take all effects into account. Coalescence and breakup may have rather long relaxation times so that equilibrium between them is not reached in the situation of interest. Moreover, boiling and condensation may affect size and depart from equilibrium.

Using a transport equation for either the bubble number density or the IA may improve the prediction of coalescence and breakup. However, there are common limitations of both algebraic and dynamic "monodispersed" bubble size models since almost all dispersed flows of interest in reactor thermal hydraulics have a rather large bubble and drop size distribution.

To go beyond these limitations, polydispersed models offer much better inherent capabilities but require an enormous effort for modeling coalescence and fragmentation.

A state of the art in using monodispersed or polydispersed approaches can be found in Krepper et al. (2011) for CMFD, and status and perspective for TIA in system codes were presented by Bestion and Serre (2013). Some conclusions are:

- The TIA models available today cover only a very small fraction of flow situations of interest.
- The analysis of some flow regimes with new measurements giving access to IA and bubble size distribution has confirmed that almost all flows are nonestablished. Under such a condition, algebraic IA modeling has many limitations, and a TIA equation adds some value.
- Polydispersion modeling would be required to achieve significant progress.

5.14 Three-field models in system codes

When a phase is present in both a dispersed and a continuous field, the two fields may have very different behaviors. Using a simple mass, momentum, and energy equation for both fields may not predict important phenomena. This was identified in nuclear reactor thermal hydraulics, particularly for annular-mist flow, where the liquid is present with droplets in the core steam flow and films along the walls. A similar situation may exist in horizontal pipes when there is a free surface and a steam flow carrying droplets above. This may be encountered in a boiling water reactor (BWR) core under normal and abnormal situations and in some accidental situations in a pressurized water reactor (PWR). Three-field models were developed to better capture the physics of flow. Film velocity is usually small (on the order of a few cm/s or mm/s), whereas droplets may be entrained at higher velocities (several m/s) in the gas phase. Both liquid fields may be countercurrent with falling films and upward-flowing droplets. The interest in a three-field model for large-break loss-of-coolant accident simulation was discussed by Valette et al. (2009) and Skorek et al. (2022).

One-dimensional three-field equations are derived following the same method as for the two-fluid model by multiplying by field characteristic functions of vapor χ_v, continuous liquid χ_{cl}, and dispersed liquid χ_{dl} before applying time and space averaging.

New transfers are defined:

$\Gamma_{vcl} = -\Gamma_{clv}$ interfacial mass transfer from continuous liquid to vapor $(kg/m^3/s)$

$\Gamma_{vdl} = -\Gamma_{dlv}$ interfacial mass transfer from dispersed liquid to vapor $(kg/m^3/s)$

Γ_d mass transfer from dispersed to continuous liquid by droplet deposition $(kg/m^3/s)$

Γ_e mass transfer from continuous to dispersed liquid by droplet entrainment $(kg/m^3/s)$

q_{idl} heat flux per unit volume from interface to droplet (W/m^3)

q_{ivdl} heat flux per unit volume from vapor to interface of droplet (W/m^3)

q_{icl} heat flux per unit volume from interface to continuous liquid (W/m^3)

q_{ivcl} heat flux per unit volume from vapor to interface of continuous liquid (W/m^3)

V_{idl} interfacial axial velocity at dispersed liquid interface (m/s)

V_{icl} interfacial axial velocity at continuous liquid interface (m/s)

The term $Af_{ivcl}\frac{\partial \alpha_{cl}}{\partial z}$ may be modeled as the equivalent term in two-fluid for stratified flow

The term $Af_{ivdl}\frac{\partial \alpha_{dl}}{\partial z}$ may be modeled as the equivalent term in two-fluid for dispersed flow

An added mass term AF_v^{AM} is used for interactions between gas in bubbly flow and continuous liquid but such a term is unnecessary for interactions between droplets and gas would have only a minor effect on droplet behavior.

It is assumed that there is no direct contact between dispersed liquid and walls since drops would deposit in pre-CHF conditions and would bounce in post-CHF conditions. Then there is no wall friction for droplets, no wall direct heat transfer for droplets, no $g_{dl}\frac{\partial A}{\partial z}$ term and no $A\alpha_{dl}\frac{\delta P_{dlsing}}{\delta z}$ term.

The set of equations is given below for the case where axial diffusion-dispersion terms are neglected.

One-dimensional three-field equations for gas-continuous liquid-droplets

Principal variables: $\alpha_v, \alpha_{cl}, P,\ V_v,\ V_{cl},\ V_{dl},\ H_v,\ H_{cl},\ H_{dl}$

$$A\frac{\partial \alpha_v \varrho_v}{\partial t} + \frac{\partial A \alpha_v \varrho_v V_v}{\partial z} = A\Gamma_{vcl} + A\Gamma_{vdl}$$

$$A\frac{\partial \alpha_{cl} \varrho_{cl}}{\partial t} + \frac{\partial A \alpha_{cl} \varrho_{cl} V_{cl}}{\partial z} = A\Gamma_{clv} + A\Gamma_d - A\Gamma_e$$

$$A\frac{\partial \alpha_{dl} \varrho_{dl}}{\partial t} + \frac{\partial A \alpha_{dl} \varrho_{dl} V_{dl}}{\partial z} = A\Gamma_{dlv} - A\Gamma_d + A\Gamma_e$$

$$A\frac{\partial \alpha_v \varrho_v V_v}{\partial t} + \frac{\partial A \alpha_v \varrho_v V_v V_v}{\partial z} + A\alpha_v \frac{\partial P}{\partial z} - Af_{ivcl}\frac{\partial \alpha_{cl}}{\partial z} - Af_{ivdl}\frac{\partial \alpha_{dl}}{\partial z} + g_v \frac{\partial A}{\partial z}$$

$$= \mathrm{A}\ \alpha_v \varrho_v g_z - A\tau_{ivcl} - A\tau_{ivdl} - C_v \tau_{wv} - A\alpha_v \frac{\delta P_{vsing}}{\delta z} + AF_v^{AM} + A\Gamma_{vcl} V_{icl}$$

$$+ A\Gamma_{vdl} V_{idl}$$

$$A\frac{\partial \alpha_{cl} \varrho_{cl} V_{cl}}{\partial t} + \frac{\partial A \alpha_{cl} \varrho_{cl} V_{cl} V_{cl}}{\partial z} + A\alpha_{cl} \frac{\partial P}{\partial z} + Af_{ivcl}\frac{\partial \alpha_{cl}}{\partial z} + g_{cl} \frac{\partial A}{\partial z}$$

$$= \mathrm{A}\ \alpha_{cl} \varrho_{cl} g_z - A\tau_{iclv} - C_{cl} \tau_{wcl} - A\alpha_{cl} \frac{\delta P_{clsing}}{\delta z} + AF_{cl}^{AM} + A\Gamma_{clv} V_{icl}$$

$$+ A\Gamma_d V_{dl} - A\Gamma_e V_{cl}$$

$$A\frac{\partial \alpha_{dl} \varrho_{dl} V_{dl}}{\partial t} + \frac{\partial A \alpha_{dl} \varrho_{dl} V_{dl} V_{dl}}{\partial z} + A\alpha_{dl} \frac{\partial P}{\partial z} + Af_{ivdl}\frac{\partial \alpha_{dl}}{\partial z}$$

$$= \mathrm{A}\ \alpha_{dl} \varrho_{dl} g_z - A\tau_{idlv} + A\Gamma_{dlv} V_{idl} - A\Gamma_d V_{dl} + A\Gamma_e V_{cl}$$

$$A\frac{\partial \alpha_v \varrho_v \left(H_v + \frac{V_v^2}{2}\right)}{\partial t} + \frac{\partial A \alpha_v \varrho_v V_v \left(H_v + \frac{V_v^2}{2}\right)}{\partial z}$$

$$- A\alpha_v \frac{\partial P}{\partial t} = \mathrm{A}\ \alpha_v \varrho_v g_z V_v + C_v q_{wv} + Aq_{idl} + Aq_{ivdl}$$

$$+ A\Gamma_{vdl}\left(H_v + \frac{V_{idl}^2}{2}\right) + \Gamma_{vcl}\left(H_v + \frac{V_{icl}^2}{2}\right) + Aq_{extv}$$

One-dimensional three-field equations for gas-continuous liquid-droplets—cont'd

$$A\frac{\partial \alpha_{cl}\varrho_{cl}\left(H_{cl}+\frac{V_{cl}^2}{2}\right)}{\partial t}+\frac{\partial A\alpha_{cl}\varrho_{cl}V_{cl}\left(H_{cl}+\frac{V_{cl}^2}{2}\right)}{\partial z}$$

$$-A\alpha_{cl}\frac{\partial P}{\partial t}=\mathrm{A}\ \alpha_{cl}\varrho_{cl}g_zV_{lc}+C_{cl}q_{wcl}+Aq_{icl}+A\Gamma_{vcl}\left(H_{cl}+\frac{V_{icl}^2}{2}\right)$$

$$-\Gamma_e\left(H_{cl}+\frac{V_{cl}^2}{2}\right)+\Gamma_d\left(H_{dl}+\frac{V_{dl}^2}{2}\right)+Aq_{extcl}$$

$$A\frac{\partial \alpha_{dl}\varrho_{dl}\left(H_{dl}+\frac{V_{dl}^2}{2}\right)}{\partial t}+\frac{\partial A\alpha_{dl}\varrho_{dl}V_{dl}\left(H_{dl}+\frac{V_{dl}^2}{2}\right)}{\partial z}$$

$$-A\alpha_{dl}\frac{\partial P}{\partial t}=\mathrm{A}\ \alpha_{dl}\varrho_{dl}g_zV_{lc}+C_{dl}q_{wdl}+Aq_{idl}+A\Gamma_{vdl}\left(H_{dl}+\frac{V_{idl}^2}{2}\right)$$

$$+\Gamma_e\left(H_{cl}+\frac{V_{cl}^2}{2}\right)-\Gamma_d\left(H_{dl}+\frac{V_{dl}^2}{2}\right)+Aq_{extdl}$$

$$\Gamma_{vdl}=-\Gamma_{dlv}=\frac{q_{idl}+q_{ivdl}}{h_{vl}};\ \Gamma_{vcl}=-\Gamma_{clv}=\frac{q_{icl}+q_{ivcl}}{h_{vl}};\ \tau_{ivcl}=-\tau_{iclv};$$

$$\tau_{idlv}=-\tau_{ivdl};\ F_v^{AM}=-F_{cl}^{AM}$$

$$\alpha_v+\alpha_{cl}+\alpha_{dl}=1;\ \varrho_k=\rho_k(P,H_k);T_k=T_k(P,H_k);$$

$$\mu_k=\mu_k(P,H_k);\ \lambda_k=\lambda_k(P,H_k)$$

$$+exp\ ressions\ for\ \Gamma_e,\Gamma_d,f_{ivdl},f_{ivcl},g_v,g_{cl}\ \tau_{idlv},\tau_{ivcl},q_{icl},q_{ivcl},q_{idl},q_{ivdl}\ C_k,$$

$$\tau_{wk},q_{wk},\delta P_{ksing},F_v^{AM},V_{icl},V_{idl}$$

5.15 Summary remarks

The equations forming the basis of computational tools adopted for transient analysis of WCNRs have been presented. Those equations have a broad validity that derives from their origin, i.e., the principles of thermodynamics and mechanics. The continuum body assumption is invoked to translate the principles into workable equations for fluids, either single-phase or multiphase.

The process for the development of the equations has been outlined in this chapter, starting from the fundamentals and involving steps of increasing complexity. The set of six PDEs dealing with steam and water mixtures is also known as unequal velocity unequal temperature (UVUT)—qualified sets of UVUT-PDE, including the constitutive/closure equations (Chapter 9) implemented in current system thermal-hydraulic codes, are considered the key outcomes of decades of research in nuclear thermal hydraulics (Chapter 2); the same equations also constitute a synthesis of knowledge and the state of the art in the area.

Time and space averaging constitute at the same time unavoidable processes for achieving a solution and sources of approximation in the results of numerical integration that are difficult to quantify.

Nevertheless, the translation of equations into the algebraic forms necessary for numerical computation and the solutions of algebraic equations have been marginally considered in this chapter (more information is given in Chapter 11).

The distinction between open and porous media modeling approaches is discussed—the former deals with CFD-like codes and includes DNS codes; the latter deals with system thermal hydraulics and may be characterized by the number of closure models and types (i.e., including constitutive equations) needed for achieving a solution. Only in the case of single-phase DNS, closure models can be avoided for predicting fluid transient evolutions at the expense of computational power that is impractical for NPP applications. More information about current CFD models can be found in Chapter 12.

Special topics related to the nature and the structure of the balance equations are also considered in the present chapter. These are listed below together with summary notes.

(a) *The HEM*, also called equal velocity equal temperature (EVET): steam and liquid form a compressible fluid.

(b) *The drift flux model*: this can be considered an attempt to establish a relationship between liquid and steam velocities in a gravity environment; the drift flux model, as conceived in the 1960s, can be derived by introducing approximations to the UVUT momentum equation.

(c) *The hyperbolicity of the solution*: the issue of hyperbolicity is connected with the numerical solution of the UVUT set of equations and has been thoroughly investigated for the case of 1-D models. Nonhyperbolicity or lack of hyperbolicity of a set of equations implies nonreal roots for the characteristic equation and may cause abortion of a calculation process.

(d) *The CFD modeling*: this is considered by the paragraph above and included in this list for the sake of completeness (more details in Chapter 12).

(e) *The Transport of Interfacial Area*: efforts in the last 2 decades attempting to introduce an additional equation (and additional unknown) to the existing set of balance equations did not demonstrate yet to be able to catch the complex two-phase phenomena even in a limited range of conditions, notably droplet and bubbly flows, considered in modeling. The poly-dispersed modeling approach may reveal more promising in this area.

(f) *The 3-D modeling*: all major system thermal-hydraulic codes have a three-dimensional modeling capability; however, systematic and comprehensive validation demonstrations are not achieved. Differences in relation to 3-D open media and porous media approaches are important, as discussed in the chapter.

(g) *The multifield modeling*: the complex nature of two-phase flow imposes a mulfield consideration on modeling. The capability appears important for the modeling of reflood, with main reference to the flow configuration from the quench front where droplets are entrained up to the upper plenum, hot legs, and steam generators where they can deposit and vaporize. This complex behavior has an important impact on the overall reflooding process.

It may be noted that the last three topics constitute areas for current research for which industrial applications are not fully mature at the time of this writing. Furthermore, a problem common to the TIA approach, 3-D modeling (at least in porous media), and multifield modeling derives from the difficulty in utilizing the enormous and expensive experimental database available in nuclear thermal hydraulics (see Chapter 6): the same level of information from experiments that is suited for 1-D modeling needs to be extended to more complex modeling approach. Alternative and innovative modeling approaches are proposed by Hu (2015); Roth and Aydogan (2015); Liu et al. (2025); Silverman (2022), and Komen et al. (2022). New experimental programs with advanced local measurement techniques are then necessary.

This chapter constitutes a hinge, or has a pivot role, for all the other chapters. Namely, the development of equations is the technology response to needs established in Chapter 4 within the framework in nuclear thermal hydraulics set by Chapters 1–3. The equations in this chapter are supplemented by constitutive/closure equations in Chapter 9 and constitute the basis for the code discussed in Chapter 11. The code must have the capability to predict phenomena in Chapter 6 and shall undergo the process of validation discussed in Chapter 13. The envisaged framework for code applications and the results from the calculation of transient scenarios are discussed in Chapters 14 and 15, respectively.

Exercises and questions

Exercises and questions are proposed hereafter with solutions given in Appendix 1 (Volume III). The readers should answer the questions or provide solutions to exercises before reading Appendix 1. Please provide extended answers.

5.1 Please provide the distinctions between HEM, EVUT, UVET, and UVUT equations for system thermal-hydraulic codes.

5.2 Please discuss the role of the drift flux model in UVET and UVUT models.

5.3 Please discuss the features and challenges of the UVUTUP model.

5.4 Please discuss the features and challenges of the TIA model.

5.5 Please discuss the challenges of developing three-field or multifield models in nuclear thermal hydraulics. Please refer to thc droplet field model primarily.

5.6 What are the difficulties in developing and qualifying 3-D models for two-phase flow? Please refer to 3-D equation models suitable for modeling selected components of reactors. For instance, explain why a 3-D model exists for the core and not necessarily for the hot leg; consider the case of steam generator 3-D modeling (primary and secondary side).

5.7 Discuss the issue of heat flux partitioning for a (1) UVUT 1-D model; (2) UVUT 3-D model; and (3) three-field 1-D model.

5.8 Discuss why balance and conservation equations currently suitable for modeling single- and two-phase flow do not include directly (another equation dealing with) the conservation of angular momentum.

5.9 Discuss the connection between balance equations (refer to a specific set of equations with which you are familiar) and entropy increases, should a nonreversible situation (e.g., blowdown) be the objective of the solution.

5.10 Explain why geometric discontinuities are not solved within balance equations (e.g., the momentum equation may lack a term for pressure drop at a geometric discontinuity). Please distinguish the case of abrupt discontinuity (e.g., an orifice in a pipe or the core inlet plate) and smooth discontinuities (e.g., a bend in the hot leg or the connection between the downcomer and lower plenum in the reactor pressure vessel of a BWR or PWR). In the case of a BWR, you have jet pumps—if possible, consider them in your answer.

5.11 Explain why 1-D equations can simulate (reproduce) experimental data with sufficient accuracy in some cases (if any), 3-D situations that occur in the core, and imply the consideration of transversal flows (e.g., mixing in a fuel bundle).

Acknowledgments

The present chapter takes benefit from notes provided by E.D. Hughes (section 5.1.1) which have been edited and commented by one chapter author (DB) and one book editor (FDA).

Target phenomena in nuclear thermal hydraulics

6

Nusret Aksan
Paul Scherrer Institut, Villingen, Switzerland

Dedication

Dedicated to my parents, Meliha and Hüseyin Aksan, and my family, Berrin Aksan, Elif Aksan Akkaya, Ercan Akkaya, Defne Akkaya, Celine Le Roch, and Gökhan Aksan. Thank you for your patience, love, and support through the years.

Foreword

Nuclear power produces 15% of the world's electricity. Many countries plan to either introduce nuclear energy or expand their nuclear-generating capacity. Design organizations are incorporating both proven means and new approaches to reduce the capital costs of their advanced designs. In the future the newest nuclear plant designs will be evolutionary, often pursuing economies of scale. In the longer term, innovative designs could help promote a new era of nuclear power.

The introduction of this chapter provides background history on identifying and characterizing phenomena at the Organisation for Economic Co-operation and Development (OECD), Nuclear Energy Agency (NEA; part of the OECD), and Committee on the Safety of Nuclear Installations (CSNI; part of the NEA) and thereafter outlines the complementary activities of the International Atomic Energy Agency (IAEA; part of the United Nations).

Since the 1980s, passive safety systems (i.e., those whose operations take advantage of natural forces such as convection gravity) have been recognized for contributing to simplification with the potential to improve the economics of new nuclear power plant (NPP) designs. The IAEA Conference on The Safety of Nuclear Power: Strategy for the Future, which was convened in 1991, noted for new plants that "the use of passive safety features is a desirable method of achieving simplification increasing the reliability of the performance of essential safety functions, should be used wherever appropriate." Some new designs use natural circulation (NC) to remove core power during normal operations. The use of passive systems can eliminate costs associated with the installation, maintenance, and operation of active systems that require multiple pumps and independent, redundant electric power supply. However, considering the weak forces driving passive systems based on NC, careful design analysis methods must be employed to ensure that systems perform their intended functions.

Handbook on Thermal Hydraulics in Water-Cooled Nuclear Reactors
https://doi.org/10.1016/B978-0-323-85606-5.00008-8

To support the development of advanced water-cooled reactor (AWCR) designs with passive systems, investigations of NC are conducted in several IAEA Member States with advanced reactor development programs. To foster international collaboration on the enabling technology of passive systems that utilize NC, the IAEA initiated a Coordinated Research Project (CRP) on Natural Circulation Phenomena, Modeling and Reliability of Passive Systems that Utilize Natural Circulation in 2004. The published output of that CRP (see the general list of references herein), describes passive safety systems for a wide range of advanced water-cooled NPP designs to gain insights into system design, operation, and reliability.

The identification and characterization of the thermal-hydraulic phenomena (THPs) to expect during postulated accident scenarios in light-water reactors (LWRs), including AWCRs and supercritical water-cooled reactors (SCWRs), are the topic of this chapter. Knowledge about THPs is an essential component of development in the nuclear thermal hydraulics subject area.

After the publication of the first edition of this book and before the second edition was issued, the author of this chapter together with two colleagues made an effort to gather THPs identified in OECD/NEA/CSNI and IAEA activities supported by (a couple of dozen) publications. The related process and its main outcome, the "116 thermal hydraulic phenomena," are outlined by Aksan et al. (2018a), who cite the originating publications for THPs, and D'Auria et al. (2021b); see also Table 15.2. In the former paper the THPs are derived and listed in alphabetic order including an identifying label for each phenomenon (PH); the latter paper focuses on the predictive capabilities of codes and models and emphasizes four THPs. Hereafter (the second edition of the book), identifying labels are introduced (in Chapter 6), as derived from Table 15.2 and Aksan et al. (2018a), and connected to a discussion in D'Auria et al. (2021a) whenever appropriate.

6.1 Introduction

Thermal-hydraulic computer codes have been developed over the last 40 years to analyze transients and loss-of-coolant accidents (LOCAs) in LWRs. The relatively simple computer codes from the early 1970s have been continuously developed to provide more realistic THP descriptions and more detailed system representations.

At the beginning of the 1970s, codes were requested for analyzing large-break loss-of-coolant accidents (LBLOCAs). The codes were based on the homogeneous equilibrium model (HEM), assuming equal velocities and temperatures for the vapor and liquid phases. The next effort in code development was directed by the demand for simulations of transients and small-break accidents. The implementation of new models allowed for the separation of vapor and liquid by gravity. The representations of the primary and secondary sides with control systems and balance of plant models were extended.

In the mid-1970s, the development of a new generation of thermal-hydraulic codes was initiated to provide analytical tools for a more realistic simulation of LWR behavior under transient and accident conditions with thermal and mechanical nonequilibrium phenomena taken into account. The effects of noncondensable and boron tracking were also considered. These codes allow the simulation of transients and the entire range of break sizes as well as beyond-design-basis accidents (DBA), including accident management (AM) procedures with operator interventions.

Parallel to the development of analytical tools, a large variety of experimental programs have been executed to improve the understanding of THPs, study system behavior, and provide the required databases for code development and code validation. Quite a number of separate effects tests (SETs) have been performed for the development and validation of code models. SETs investigate individual phenomena under clear boundary conditions. Whereas experiments during the 1970s were conducted mainly at small-scale test facilities, the 1980s saw more attention directed toward integral effects tests that simulated system behavior and emphasized scaling. For example, in 1986 the first tests were performed at the UPTF test facility, represented by a four-loop 1300 MWe pressurized water reactor (PWR) with an upper plenum (UP), a downcomer, and main coolant pipes in a full-scale reactor geometry (US NRC, 1987).

The overall results of code calculations are validated primarily with data from integral test facilities (ITFs) representing the primary and secondary coolant systems. While experiments in the early 1970s focused on large-break issues, in the follow-up parallel to the advancement of code development, integral effects tests have been carried out to investigate LWR system behavior during transients, small breaks, transients under shutdown (SHD) conditions, and beyond DBA. In addition to the results of integral effects tests, LWR plant data of transients or accidents are used to validate the predictive capability of codes.

The construction of code validation matrices is an attempt to collect the best sets of test data for code validation and improvement from the wide range of experiments that have been conducted worldwide in the field of thermal hydraulics. The first formulation of a validation matrix was proposed by Wolfert and Frisch (GRS) (1983). This activity was taken by a CSNI subgroup to establish matrices for PWRs and boiling water reactors (BWRs). During this activity, many related works were performed by experts in the field, and reports were published. These activities also included the identification and characterization of THPs for the current generation of LWRs.

By 1987, the OECD/NEA/CSNI had published a document that systematically identified a set of physical phenomena and the tests considered best for assessing the performance of thermal-hydraulic codes, the "CSNI Code Validation Matrix of Thermohydraulic Codes for LWR LOCA and Transients" (OECD/NEA/CSNI, 1987). The report included all the typical phenomena expected to occur in plant transients and LOCA analyses. As a follow-up, state-of-the-art report (SOAR) on the thermohydraulics of emergency core cooling (ECC) in LWRs, Lewis et al. (1989) emphasized the importance of phenomena identification in transient analysis for LWRs. After publishing these two reports, it turned out that a continued comparison of calculations with additional SET data is also necessary to consider particular code applications, especially where a quantitative evaluation of prediction accuracy is required for best estimate computer codes, as well as for code model improvement.

Based on these needs, the OECD/NEA-CSNI issued an independent and separate two-volume report on the SET validation matrix (Aksan et al., 1994). An updated integral effects test validation matrix report was also prepared and issued (Annunziato et al., 1996). In addition, to set up validation matrices for Russian pressurized water-cooled and water-moderated energy reactor (VVER or WWER) analyses, an International Working Group was formed on the initiative of the Federal Ministry for Research and Technology (BMFT) of the Federal Republic of Germany. A further evaluation of the VVER matrices was performed by a specific technical support group, OECD/NEA/CSNI (2001). One of the few AWCR studies within OECD/NEA-CSNI was directed to some characteristics of advanced reactor designs (ARDs), for example, AP600, SBWR, advanced boiling water reactor (ABWR), EPR, PIUS, etc. In this status report, specific needs for thermal-hydraulic codes together with the the identified relevant and important THPs for ARDs are summarized to guide the development of research plans to consider code development and assessment needs and plan experimental programs (Aksan and D'Auria, 1996). These identified phenomena for ARDs are in addition to the relevant THPs identified for current-generation LWRs.

As part of the IAEA's overall effort to foster international collaborations to improve the economics and safety of future water-cooled NPPs, an IAEA CRP was started in early 2004. This CRP, titled Natural Circulation Phenomena, Modeling and Reliability of Passive Safety Systems that Utilize Natural Circulation, focused on the use of passive safety systems to help meet the safety and economic goals of a new generation of advanced NPPs. This CRP has been organized within the framework of the IAEA Department of Nuclear Energy's Technical Working Groups for Advanced Technologies for LWR and Heavy Water Reactors (the TWG-LWR and TWG-HWR) and provided international cooperation on research work underway at the national level in several IAEA Member States.

The use of passive safety systems such as accumulators, condensation and evaporative heat exchangers, and gravity-driven safety injection systems eliminates the costs associated with the installation, maintenance, and operation of active safety systems that require multiple pumps with independent and redundant electric power supplies. As a result, passive safety systems are being considered for numerous reactor concepts (including Generation III and III + concepts) and are expected to find applications in Generation IV reactor concepts as identified by the Generation IV International Forum (GIF). Another motivation for the use of passive safety systems is the potential for enhanced safety through increased safety system reliability.

The CRP on NC and passive safety systems benefits from earlier IAEA activities that include developing databases on physical processes of significant importance to water-cooled reactor operations and safety. For example, in the area of THPs in AWCRs, recent IAEA activities have assimilated data internationally on heat transfer (HT) coefficients and pressure drop (PD) (IAEA, 2001) and have shared information on NC data and analytical methods (IAEA, 2001) and on experimental tests and the qualification of analytical methods (IAEA, 2000a). To establish the progress of work in this CRP, an integrated research plan was established describing the tasks for addressing CRP objectives. These tasks were:

- Establish the state of the art in NC.
- Identify and describe reference systems.
- Identify and characterize phenomena that influence NC.
- Examine the application of data and codes to design and safety.
- Examine the reliability of passive systems that utilize NC.

The activity under the first task aims to summarize the current understanding of NC system phenomena and the experimental methods used to investigate and model such phenomena. In November 2005, the IAEA issued a technical document (IAEA, 2005) developed by the collaborative effort of CRP participants and major contributions from selected experts in the CRP to document existing knowledge in six specific areas: advantages and challenges of NC systems in advanced designs, local transport phenomena and models, integral system phenomena and models, NC experiments, advanced computation methods, and reliability assessment methodology.

The activity for the third task aims to identify and categorize NC phenomena of importance to advanced reactors and passive safety system operations and reliability. This task is the major link between the second and fourth tasks. During the phase to perform activities for the mentioned tasks, one of the unique features was to collect and include plant design descriptions with a strong emphasis on the passive safety systems of the specific design. These descriptions of passive safety systems were considered during phenomena identification and characterization by the group of experts. As a basis for PH identification in AWCRs, earlier works performed within the OECD/NEA framework during 1983–1997 were considered in detail (OECD/NEA/CSNI, 1987; Lewis et al., 1989; Aksan et al., 1994; Annunziato et al., 1996; Aksan and D'Auria, 1996). The list of relevant phenomena established in Aksan and D'Auria (1996) has been taken as the basis for AWCR work and has been modified according to the reactor types and passive safety systems considered. It should also be noted that in identifying the relevant THPs in the list provided in Section 6.5, expert judgment is the main contributor.

IAEA-TECDOC-626 provides definitions for safety-related terms as applied to advanced reactors (IAEA, 1991). In that document, the concepts of passive and active safety systems are defined and discussed. The definition of a passive safety system is as follows: *Either a system which is composed entirely of passive components and structures or a system which uses active components in a very limited way to initiate subsequent passive operation*. Four categories were established to distinguish the different degrees of passivity.

This chapter provides details on important and relevant THPs for current-generation LWRs (e.g., PWRs, BWRs, and VVERs) including separate effects and integral effects phenomena, AWCRs (AP1000, ESBWR, EPR, VVER-1200, etc.), and SCWRs. These identified phenomena are the backbone of nuclear reactor thermal hydraulics and help define research and development activities in the experimental and analytical field.

6.2 Classification of thermal-hydraulic phenomena for separate effects tests

The basic conservation equations used in thermal-hydraulic two-phase flow models for analyzing LWR systems under normal and off-normal operating conditions describe the transport of mass, momentum, and energy of steam and water (as discussed in Chapters 5 and 11). These local instantaneous partial differential equations are generally integrated over a control volume of the fluid to result in a set of finite difference equations (algebraic equations) that can be numerically solved using known algorithms. When making the integration a kind of averaging procedure is applied, which implies that the resulting equations will lose their ability to resolve phenomena in a microscale sense. Thus phenomena such as turbulent fluctuations will either be ignored or be integrated into correlations. Only the average three-dimensional, time-varying characteristics of flow over a scale dictated by the size of mesh cells can be retained. The most commonly used averaging procedure includes double averages in time and space. The procedure generates distribution coefficients for various transport phenomena. These coefficients appear as the ratio of the averages of products divided by the products of the averages. Due to a lack of information about these coefficients, they are usually taken to be unity (Yadigaroglu et al., 1990).

In earlier two-phase models the fluid was treated as a homogeneous mixture of liquid and vapor and consequently, only three conservation equations were needed to describe two-phase flow. In the HEM the phases are assumed to move with the same velocity and exist at the same temperature. An extension to the mixture model is the drift flux model in which the relation between phasic velocities is described through an algebraic equation, thus allowing for slip between phases (Sonnenburg, 1990). In further developed models, the conservation equations are set up for each phase separately, and six equations are used (two-fluid two-phase flow models). Additional equations can be incorporated into the models to account for liquid solutes and noncondensable gases. The significant property of the two-fluid model is that it allows the two phases to have different velocities and temperatures. Consequently, this type of model has the potential to calculate the effects of phase separation and thermal nonequilibrium.

The two-phase flow model is in most cases simplified to a one-dimensional model regardless of whether it is a mixture or two-fluid model. Possible three-dimensional effects cannot then be taken into account but must be included through special process models (e.g., special models for simulating the phenomena associated with subcooled ECC injection into a voided pipe) and by simulation of the actual flow path by the appropriate connections of several one-dimensional flow paths.

Some two-phase models retain three-dimensional characteristics, although simplifications are usually introduced into the equations to make system calculations practical. Despite the simplifications, these models have the potential to simulate the three-dimensional effects that may be most important, for example, in some time windows of an LBLOCA where flow redistributions are supposed to occur within the reactor vessel.

Solving the set of finite difference equations requires closure or constitutive equations along with relations describing steam–water properties under different conditions. In this context, steam–water properties are assumed as well known for most application ranges, and associated tables and functions are supposed to be adequately programmed in any codes used. The same is assumed true for the needed derivatives of the properties.

The constitutive equations (see also Chapters 7–10) describe the HT between wall structures and the two-phase fluid, the interphase mass, momentum, HT, and wall-to-fluid drag. Wall HT and wall drag to each phase are dependent on the extent of the wall contact of the phase and the associated transfer process, whereas the interphase exchange of mass, momentum, and energy depends on the interfacial area along with associated interphase transport processes. Given the averaged nature of the conservation equations mentioned above, the wall contact of the phases and the interfacial area must be expressed as volume-based or averaged quantities. The adequate description of the interfacial area density, and thus the prevailing flow regimes, and the wall-to-fluid and interphase transport processes is one of the most important requirements for accurate modeling of the two-phase flow behavior.

Most constitutive equations are based on correlations with different levels of empiricism, from pure empirical models to those based on physical principles with possibly only a few constants that must be evaluated from experimental data. Pump degradation curves under two-phase conditions are examples of the former, whereas models for frictional PD are examples of the latter. In some cases, the inherent capabilities of the two-fluid model are not fully exploited, and a separate special process model or correlation is utilized. An example is the modeling of critical flow (Yadigaroglu et al., 1990). Integration of two-fluid conservation equations can predict a choked condition. Choking calculations in a component with a smooth area change could be peformed using a sufficiently fine axial mesh and noting the occurrence of critical conditions. However, this requires a very fine mesh to adequately predict critical conditions.

The interactions of correlations and conservation equations constitute a code's ability to simulate important phenomena during an LOCA or a transient. Identifying each relevant PH and ascertaining whether the code adequately simulates the specific PH is essential. This is usually done by comparing test data with code results from the simulations of well-defined tests conducted in properly designed test facilities. In this context a distinction is made between integral effects tests and SETs.

Integral effects tests are conducted in scaled test facilities and provide data on the overall behavior of a simulated reactor system during LOCAs or transients. These tests are used for code assessment purposes related to overall reactor behavior. A definition of sets of such tests (matrices) is provided in OECD/NEA/CSNI (1987). These facilities have complicated configurations and are expensive to operate. Compromises, for instance scaling to real plants, are inevitable. In addition, the instrumentation for measuring parameters governing different two-phase flow phenomena is limited. Thus, integral effects tests are less suited to detailed investigations of specific two-phase flow phenomena. As a consequence, assessment analyses are required to determine whether the code user is confident about which phenomena prevail during a transient to avoid "good results but for the wrong reasons" (compensating errors).

SETs are employed not only to develop correlations of specific processes but also to investigate individual or localized two-phase phenomena, which in most cases are dependent on several specific processes. These kinds of tests can also characterize the behavior of single components such as pumps or steam separators.

Interactions with other phenomena or components, as in a full-scale LWR, are either imposed by the external boundary conditions of the test or purposely eliminated. In fact, an important consideration in the design of SETs is having well-defined boundary conditions. In some cases, the test section in an SET is locally at or very near full scale. This minimizes concerns about scale effects. On the other hand, it is important in all SETs to evaluate the influence of the chosen boundary conditions on measured parameters.

In SETs the instrumentation for measuring two-phase flow parameters can be quite extensive. Thus data for assessing details about the models that codes use to simulate localized phenomena can be provided and can help in understanding the interactions of processes that combine to produce the overall phenomena of interest. These data have also proved valuable when evaluating prediction uncertainties for specific phenomena at or near full scale.

As part of defining integral effects test matrices in OECD/NEA/CSNI, (1987), various phenomena were identified. In particular, phenomenological windows were distinguished in the case of:

- LBLOCAs in PWRs
- SBLOCAs and intermediate-break LOCAs and transients in PWRs
- LOCAs in BWRs
- transients in BWRs

Typical scenarios simulated in ITFs formed the basis for this analysis. The presumed evolution paths of accidents in the reference plants (PWR and BWR) were also considered and played a role in this connection. The phenomena were identified essentially based on the experiences of individual members of the writing group formed to investigate the simulated scenarios. Those identified phenomena formed the first basis for a more detailed list of relevant phenomena.

An expansion of that list covering distinct and important LOCA phenomena was presented for both BWRs and PWRs in Lewis et al. (1989). A weighting and appraisal of each PH were also included. The main support for these phenomena was provided by several integral effects tests and SETs. That list of phenomena is the basis for the one given in Table 6.1. Only minor modifications were introduced to the list. Apart from some rearrangement, a new group named "Basic phenomena" has been added. Each PH in this group is independent of other phenomena and is constitutive of two-phase flow.

In this section, the identification of relevant two-phase flow phenomena is made, and a short description of each PH is provided. A total of 67 phenomena are identified, and the resulting phenomena list is given in Table 6.1. As stated above, the identification is based on the phenomena list in OECD/NEA/CSNI (1987). All representative

Table 6.1 List of thermal-hydraulic phenomena.

0	Basic phenomena	1	Evaporation due to depressurization
		2	Evaporation due to heat input
		3	Condensation due to pressurization
		4	Condensation due to heat removal
		5	Interfacial friction in vertical flow
		6	Interfacial friction in horizontal flow
		7	Wall-to-fluid friction
		8	Pressure drops at geometric discontinuities
		9	Pressure wave propagation
1	Critical flow	1	Breaks
		2	Valves
		3	Pipes
2	Phase separation/vertical flow with and without mixture level	1	Pipes/plena
		2	Core
		3	Downcomer
3	Stratification in horizontal flow	1	Pipes
4	Phase separation at branches	1	Branches
5	Entrainment/deentrainment	1	Core
		2	Upper plenum
		3	Downcomer
		4	Steam generator tube
		5	Steam generator mixing chamber (PWR)
		6	Hot leg with ECCI (PWR)

Continued

Table 6.1 Continued

6	Liquid-vapor mixing with condensation	1	Core
		2	Downcomer
		3	Upper plenum
		4	Lower plenum
		5	Steam generator mixing chamber (PWR)
		6	ECCI in hot and cold leg (PWR)
7	Condensation in stratified conditions	1	Pressurizer (PWR)
		2	Steam generator primary side (PWR)
		3	Steam generator secondary side (PWR)
		4	Horizontal pipes
8	Spray effects	1	Core (BWR)
		2	Pressurizer (PWR)
		3	Once-through steam generator secondary side (PWR)
9	Countercurrent flow/countercurrent flow limitation	1	Upper tie plate
		2	Channel inlet orifices (BWR)
		3	Hot and cold leg
		4	Steam generator tube (PWR)
		5	Downcomer
		6	Surge line (PWR)

10	Global multidimensional fluid temperature, void, and flow distribution	1	Upper plenum	
		2	Core	
		3	Downcomer	
		4	Steam generator secondary side	
11	Heat transfer:	Natural or forced convection	1	Core, steam generator, structures
		Subcooled/nucleate boiling	2	Core, steam generator, structures
		DNB/dryout	3	Core, steam generator, structures
		Post-critical heat flux	4	Core, steam generator, structures
		Radiation	5	Core
		Condensation	6	Steam generator, structures
12	Quench front propagation/rewet	1	Fuel rods	
		2	Channel walls and water rods (BWR)	
13	Lower plenum flashing			
14	Guide tube flashing (BWR)			
15	One and two-phase impeller-pump behavior			
16	One and two-phase jet-pump behavior (BWR)			
17	Separator behavior			
18	Steam dryer behavior			
19	Accumulator behavior			
20	Loop seal filling and clearance (PWR)			
21	ECC bypass/downcomer penetration			
22	Parallel channel instabilities (BWR)			
23	Boron mixing and transport			
24	Noncondensable gas effect (PWR)			
25	Lower plenum entrainment			

phenomena occurring during an LOCA or a transient are included, although in some cases several phenomena are combined under a general heading, such as for various instances of countercurrent flow limitation (CCFL) and critical flow. It should also be realized that some phenomena are dependent on each other, for instance, spray effects and condensation. It will be seen that several types of phenomena exist, varying from interphase friction, which is a very basic attribute of two-phase flow, to loop seal clearing, which is essentially a system PH localized in its occurrence but very dependent on events and conditions elsewhere in the loop. In such cases, influences from the loop must be provided as boundary conditions.

6.3 Characterization of thermal-hydraulic phenomena for separate effects tests

A brief description of each of the 67 phenomena listed in Table 6.1 is provided in this section.

This indicates the interpretation made of the PH to assess suitable SET data. Thus, for example, the term horizontal stratified flow is used to include the THP of thermal stratification as well as aspects of two-phase stratified flow associated with the presence of a surface. The description also indicates where and under what circumstances a THP is likely to occur within a reactor system. Any THP discussion should indicate features such as the degree of scale or geometry dependence that affect the range of data likely to be required in code model validation.

Some groups of phenomena are linked. In this case an introductory description prefaces the group to avoid the repetition of material. For instance, CCFL can occur at varioius locations in an LWR. CCFL is quite geometry-dependent and encompasses a wide range of situations, requiring data from various SETFs. Nevertheless, some aspects of CCFL are common to all these circumstances.

In some cases, there is an apparent overlap of the phenomena description, since the same THP occurs in combination with other phenomena. The details and specifics of overlaps and clarifications are provided in detail for each specific THP.

6.3.1 Basic phenomena

Basic phenomena are the transfer processes at the interface between different fluids (liquid and steam or a noncondensable gas) and the solid surfaces facing the fluid. They comprise momentum, heat, and mass transfer, which are divided into evaporation due to depressurization (Section 6.3.1.1), evaporation due to heat input (Section 6.3.1.2), condensation due to pressurization (Section 6.3.1.3), condensation due to heat removal (Section 6.3.1.4), interfacial friction in vertical flow (Section 6.3.1.5), interfacial friction in horizontal flow (Section 6.3.1.6), wall-to-fluid friction (Section 6.3.1.7), PD at geometry discontinuities (Section 6.3.1.8), and pressure wave propagation (Section 6.3.1.9).

In the six equation models used in thermal-hydraulic safety computer codes, the constitutive laws are intended to model these basic transfers between phases. The method of discretization for fluid dynamics and the numerical solution significantly affect the calculation of pressure wave propagation.

Basic phenomena govern all fluid behavior during an accident sequence, so they are inevitably important for nuclear safety.

In most cases of practical interest, there are no direct measurements of basic phenomena for two-phase flow. They can be only evaluated based on global measurements (e.g., *DP*, *P*, *T*).

Models are necessary to evaluate basic transfers from global measurements. From these models, correlations are set up. However, it is important to realize that the use of these correlations is not independent of the chosen model. Normally correlations should be used in a way consistent with their derivation. Usually, basic phenomena (Sections 6.3.1.1–6.3.1.7) are treated as independent of modeling.

With the present state of knowledge, models employed in codes give a large range of differences in the evaluation of basic phenomena, Forge et al. (1988). But global thermal-hydraulic behavior is nevertheless predicted quite similarly by different approaches; Ambrosini et al. (1988).

6.3.1.1 Evaporation due to depressurization

A good illustration of a PH is the case of water and saturated steam in an adiabatic cylinder with a piston. By increasing the volume with the piston, one can observe a sudden pressure decrease followed by a pressure increase due to the vaporization of the liquid. The common understanding is that there is a delay before vaporization begins and then a gradual return to thermodynamic equilibrium. The delay depends on water purity, which means that the presence and number of microscopic nucleation sites play a role.

In a closed cylinder of fixed dimensions, the pressure variation can be obtained by extracting a certain amount of liquid at the bottom or steam at the top.

Another way to depressurize two-phase fluid is to use a steady-state flow in a Venturi nozzle. In this case, fluid acceleration is the main source of depressurization and induces steam evaporation. A strong gradient of pressure can be obtained in a steady state with this process.

In real experiments, the perfect adiabatic situation is not achieved. There is always heat exchange with walls, as well as heat losses.

In blowdown experiments, the flow can enhance transfer phenomena between phases.

This PH makes a major contribution to general behavior during an accident, for example, pressure evolution, upstream flow conditions at the breach, and the stagnation point for an LBLOCA.

This PH is easily predicted under the thermal and mechanical equilibrium conditions that occur in slow transients (in which mass and energy conservation are important but kinematics are not). It is the prediction of departures from the equilibrium behavior that is particularly interesting in the safety context. The prediction of a PH is challenging when it occurs at geometric discontinuities (i.e., originated by singular pressure drop), as discussed in Section 8.3.2.1.

The PH is identified as B-3-EV1 (or No 38 in the list of 116 THPs) by Aksan et al. (2018a), and D'Auria et al. (2021b).

6.3.1.2 Evaporation due to heat input

Evaporation of liquid due to heat input is one of the most essential phenomena occurring both in normal and off-normal operation of LWR. The evaporation or boiling process on a heated surface is a very effective HT mode with more than one order of magnitude higher HT coefficient than for single-phase convection. On the other hand, at high heat flux on the heated surface, or when a high vapor content has accumulated under forced flow conditions, a sudden deterioration of the effective boiling HT mode can occur which can be hazardous to the heated surface.

Because of its contribution to determining boiling HT and dry-out phenomena, a relatively detailed description of the nucleation process is given below. HT from a heated surface is described in Section 6.3.12.

The evaporation process implies that vapor bubbles can be created and then are able to grow within the liquid. This requires that bubble nuclei are present and that the surrounding liquid is somewhat superheated. The nucleation process is described in terms of the difference of free energy between existing small vapor clusters in a liquid and the surrounding liquid (Hsu and Graham, 1986).

The free energy will reach a maximum as a function of increasing vapor cluster radius and then decrease. When the cluster radius is less than this critical radius, the natural tendency is for the vapor cluster to collapse since this will reduce free energy. On the other hand, if the cluster radius is greater than the critical value, an increase in the cluster size will also reduce free energy, and thus a spontaneous growth of the cluster will occur. The vapor clusters with sizes greater than the critical value will serve as nuclei for vapor bubbles.

In the case when a solid heated surface is immersed into a liquid, the free energy difference required for nucleation is usually smaller than for the case of a homogeneous liquid. The reason is that on the solid surface small cavities exist which will act as nucleation sites. The liquid temperature gradient across the thermal boundary layer adjacent to the heated surface will determine the bubble growth.

When a bubble leaves the heated surface (or collapses) relatively cold liquid will replace the space earlier occupied by the bubble. This liquid is gradually heated by transient conduction until the temperature is sufficiently high for a new bubble to grow. The period until the appearance of the second bubble is called the waiting period. Following the appearance of the bubble, the growth period starts. During this period the bubble will grow because of evaporation taking place at the vapor-liquid interface and to some extent from the microlayer of liquid at the bubble base. The adjacent liquid is subjected to enhanced convection because of bubble agitation. The growth period terminates with bubble departure or collapse. Then the process repeats itself and a new bubble is created.

When the thermal boundary layer temperature is increased, more nucleation sites are activated. The bubble frequency at each site is increased, while the waiting period is decreased. A multibubble region will be established in which the creation of one bubble is strongly influenced by existing adjacent bubbles. Additionally, coalescence of bubbles in the thermal boundary layer will occur, and this will be increased with heat flux. As a result, liquid access to the heated surface will be restricted and momentary dry patches will begin to form. When this situation is reached the heat flux is approaching the critical heat flux (CHF).

The evaporation process due to heat input is a basic PH that contributes substantially to determining the overall course of an accident or transient event. The evaporation and coalescence of bubbles occurring at small scales give rise to different macroscopic two-phase flow regimes. The associated HT in the different regimes determines the maximum fuel temperature encountered during subsequent phases of a transient and, for example, the quenching velocity of the fuel during the reflood phase of an LOCA. The evaporation is also a basic process when a three-dimensional distribution of vapor and liquid is developed in pressure vessel plena after an accident and in a steam generator (SG) secondary side.

The PH is identified as B-4-EV2 (or No 39 in the list of 116 THPs) by Aksan et al. (2018a), and D'Auria et al. (2021b).

6.3.1.3 Condensation due to pressurization

In the case of the experiment described in Section 6.3.1.1, by reducing the volume with the piston, one can observe a sudden increase in pressure followed by a small decrease due to condensation of steam. In a closed cylinder of fixed geometry, the pressure increase can be obtained by injection of saturated water at the bottom or saturated steam at the top. This effect is also very sensitive to the film condensation process which occurs as walls become subcooled when pressure increases. It is possible that this condensation process is more important than the condensation at the interface.

In real experiments the same remarks as in Section 6.3.1.1 apply. The pressurization is, generally, due to injected subcooled water (e.g., safety injection).

The PH contributes to the general behavior of the pressurizer (PRZ) which governs the pressure evolution of transients such as anticipated transients without scram (ATWS), and feed and bleed procedures for transients such as a loss of feedwater (LOFW). Pressure recovery in the upper head at the end of some transients and SG secondary side behavior in the case of tube rupture are related to this PH, and void collapse in the upper head during refill and reflood of the core at high pressure may delay core recovery.

The PH is identified as B-2-COP (or No 21 in the list of 116 THPs) by Aksan et al. (2018a), and D'Auria et al. (2021b).

6.3.1.4 Condensation due to heat removal

Condensation of vapor due to heat removal is an essential PH occurring in certain phases of normal and off-normal operation of LWR and in various types of condenser equipment. The mode of condensation can be classified as dropwise or film-wise condensation on a solid surface, or as direct contact condensation on the surface of a subcooled liquid.

Dropwise and film-wise condensations are obtained on surfaces with temperatures below the saturation temperature in a superheated or saturated vapor environment. The exact mechanism of initiation of condensation is not yet fully understood. However, the consensus view is that the initiation of condensation starts within small surface cavities that contain liquid with sufficient subcooling. The magnitude of the subcooling depends on the curvature of the liquid surface that faces the vapor.

If the condensed liquid does not wet the surface the condensate will form droplets that run off the surface depending on the surrounding flow velocity, surface conditions, and surface inclination. This is known as dropwise condensation.

If the liquid wets the surface, or the dropwise condensation is sufficiently intense, a continuous liquid film will be created and will cover the surface. This mode of condensation has a much lower HT coefficient than the dropwise mode. However, film-wise condensation is the predominant mode and is usually assumed when specifying correlations for condensation HT coefficients on a surface. These correlations are almost entirely based on Nusselt's theory.

Direct contact condensation is obtained when existing subcooled liquid droplets are passing through a saturated or superheated vapor environment. An example of this kind of condensation will be the spraying in a PWR PRZ. Relatively cold liquid will be provided as spray by means of spray nozzles in the PRZ top. The condensation will take place at the surface of the droplets and the temperature of the droplets will gradually increase as the droplets pass through the vapor phase. The spray flow rate is normally rather high compared with the amount of vapor being condensed and hence the droplet radius will increase only by a small amount.

The amount of condensed vapor will depend on the temperature difference between the vapor and liquid droplets, the droplet size, the droplet transit time in the vapor phase, and vapor and liquid properties.

The condensation processes due to heat removal are important during some parts of the course of the evolution of an accident in LWR and generally in condensers.

In condensers, the condensation of vapor takes place on the shell side while the cooling liquid passes on the tube side in the usually horizontal tubes. The tubes are often arranged in a triangular pattern in such a way that the condensate can drain as freely as possible due to gravity and at the same time allow for high vapor velocities at low PD within the condenser.

During a small-break loss-of-coolant accident (SBLOCA) in a PWR, a situation can occur where vapor from the more or less uncovered core flows through the hot leg and condenses on the hot leg side of the SG tubes. The condensed liquid flows back due to gravity along the tube wall and back through the hot leg to the core. This mode of cooling the core is quite effective and is characterized as the reflux condensation mode. The condensation in the SG tubes will basically be of the film-wise type.

The spray process is important in PRZ control and for the effectiveness of ECC in BWR.

The PH is identified as B-1-COH (or No 20 in the list of 116 THPs) by Aksan et al. (2018a), and D'Auria et al. (2021b).

6.3.1.5 Interfacial friction in vertical flow

The momentum transfer between phases is principally a function of relative velocity between steam and water, of fluid densities, and of a so-called interfacial friction coefficient which is highly dependent on the flow regime. For vertical flow, the basic flow regimes are bubbly flow, slug flow, chum flow, annular flow, and inverted annular flow (Govier and Aziz, 1976). The interfacial friction in general is highest in those flow regimes that result in high interfacial areas. In relating the interphase friction to the difference in average phase velocities, the effect of nonuniform velocity (velocity profile) needs to be considered, as does the effect of averaging flow structures (e.g., use of a single droplet diameter). Although interfacial friction is a basic PH, it is still a highly complex one, in which much simplification must be done to reduce it to a one-dimensional two-fluid description.

The interfacial friction has a direct influence on carryover and carryunder: entrainment and deentrainment of water by steam. During core uncover, the swell level is related to interfacial drag. If the friction is high, the swell level is higher. This PH, coupled with wall friction, also influences CCFL.

To study the fundamentals of this PH air-water flow is useful (no heat and mass transfer).

Interphase friction is a key PH affecting the overall fluid behavior in a system under two-phase conditions, and its correct modeling is crucial for the adequate prediction of a number of other phenomena. This PH, together with wall-to-fluid friction (see Section 6.3.1.7), contributes to determining PD for two-phase flow in vertical situations. It governs phase separation and swell level, particularly in the core, and so it is of high relevance for reactor safety.

The PH is identified as B-6-IF2 (or No 53 in the list of 116 THPs) by Aksan et al. (2018a), and D'Auria et al. (2021b).

6.3.1.6 Interfacial friction in horizontal flow

On the same physical basis as described in Section 6.3.1.5, the interfacial friction in horizontal flow is highly influenced by the flow regime. Possible flow regimes are bubbly flow, plug flow, stratified flow, wavy flow, slug flow, and annular flow (Delhaye et al., 1981). It also influences countercurrent flow (CCF) in stratified conditions, and CCFL (Hewitt, 1989). At high flow rates horizontal and vertical flows behave similarly. There are major differences, resulting from the effects of phase separation, at low flows.

This PH, together with wall-to-fluid friction (see Section 6.3.1.7), contributes to PD for two-phase flow in horizontal conditions. It plays a role in the reflux condenser heat removal mode and in the rate at which water can drain from the upper parts of the circuit to the core.

The PH is identified as B-5-IF1 (or No 52 in the list of 116 THPs) by Aksan et al. (2018a), and D'Auria et al. (2021b).

6.3.1.7 *Wall-to-fluid friction*

Wall-to-fluid friction causes, as is well known, a force opposite to the direction of the motion of the fluid. In single-phase fluid dynamics, the force is known to depend on the cross-sectional average squared velocity and fluid density, as well as external parameters. These parameters are the roughness of the walls and the equivalent diameter of the flow duct. The fluid viscosity also plays a role in the determination of the friction coefficient.

In two-phase flows, the wall-to-fluid interaction may depend greatly on the flow regime and local void fraction.

The PH makes a significant contribution to the overall PD in a plant during normal and off-normal conditions. Essentially, three kinds of geometric situations can be distinguished in a nuclear reactor system concerning wall-to-fluid friction.

(a) large plena,
(b) piping,
(c) core and SGs.

The last item (c) is characterized by small values of equivalent diameter.

The PH is identified as B-9-FR (or No 116 in the list of 116 THPs) by Aksan et al. (2018a), and D'Auria et al. (2021b).

6.3.1.8 *Pressure drops at geometric discontinuities*

Important pressure variations occur due to abrupt changes of geometry in any thermal-hydraulic system including variations of curvature. The void fractions, the cross-sectional profile of the fluid velocities, and geometric peculiarities of the system (e.g., the shapes of edges that constitute the discontinuity) are among the main parameters that affect the PH.

Strong three-dimensional flow paths may occur in the proximity of abrupt geometry variations making difficult analytical modeling and a complete experimental characterization of the PH (see also section 8.3.2.1).

The prediction of this PH is essential to evaluate scenarios in the plant. The following main zones are relevant as far as local PD are concerned:

(a) Inlet and outlet of the core
(b) Inlet and outlet of the vessel
(c) Inlet and outlet of the U-tube bundle on the primary and secondary sides of PWR SGs
(d) Bypass flow paths, mostly inside the vessel
(e) Special components like pumps, steam separators, and valves

The PH is identified as B-7-PD-c (or No 84 in the list of 116 THPs) by Aksan et al. (2018a), and D'Auria et al. (2021b).

6.3.1.9 Pressure wave propagation

Because the primary system's coolant is subcooled everywhere except in the PRZ, extremely rapid decompression occurs following an almost instantaneous rupture of the primary system piping. The pressure can fall well below the hot-leg saturation pressure (12 MPa), to around 5 MPa. The subsequent motion of decompression waves through the piping and vessels induces severe dynamic loads on their internal structures. The larger the breach, the larger the instantaneous differences of pressure applied to structures. The decompression waves have a smaller effect for smaller breaches. On the secondary side in the case of a steam line break, pressure waves can propagate and induce dynamic loads on elbows.

Later in the transient in the case of a large amount of steam in the primary or secondary side, the injection of the cold water can cause a condensation shock and a related water hammer, for example, on dead ends or bends.

The BWR vessel has the same features as the secondary side of SGs in PWRs as far as the propagation of pressure waves is concerned. Sudden condensation and evaporation are a consequence of the transit of pressure wave attenuation of the amplitude being strongly affected by the geometry and the average value of the void fraction (including the degree of subcooling) for a given volume.

Core internals must be designed to withstand these dynamic loads without severe deformation to ensure that a coolable geometry is retained during an LBLOCA. Severe deformation could also prevent the control rods from entering the core. For most plants, control rods are not essential during an LBLOCA in a PWR because coolant voids and boron injected with the ECC water will make the reactor subcritical anyway. It is nevertheless good engineering practice and essential for small breaches to ensure that the reactor is scrammed.

Condensation shock and related water hammer can damage pipes and some problems have been experienced on the secondary side for some plants.

Positive pressure propagation causes void collapse in the BWR core and may be the origin of a significant neutron flux excursion. The interaction between neutronics and thermal hydraulics in the case of instability may also be affected by the propagation of small pressure waves (e.g., initial perturbation).

Furthermore, especially in the case of LBLOCA (steam line or recirculation line break), pressure wave propagation from the break toward the vessel may cause significant loads on internal structures and damage to the systems relevant to safety.

The PH is identified as B-8-PW (or No 85 in the list of 116 THPs) by Aksan et al. (2018a), and D'Auria et al. (2021b).

6.3.2 Critical flow

When a flow discharges from a reservoir at higher pressure to a reservoir at lower pressure through a joining pipe, providing the pressure difference is not too large, the rate of flow will be determined by the size of the pressure difference and (inter alia) the wall frictional losses at the pipe walls, entrance, and exit. If the ratio of the upstream to downstream pressure is sufficiently large, however, the flow is independent of the downstream reservoir conditions and is said to be critical or choked. In single-phase flow this condition is simply interpreted: the speed of the flow at some point along the flow path, the choke plane, is then equal to the maximum speed at which pressure disturbances can propagate within the flow. Such a situation is conceptually simpler in the case of flow through a pipe, where the flow is essentially one-dimensional, than for flow through a short constriction such as a break in the side of a pipe, or a valve, where multidimensional effects can be more important (Brittain et al., 1982).

The three cases of breaks, valves, and pipes are treated separately below. For discussion, the term "pipe" is taken to apply to a duct with a constant cross section and a length-to-diameter ratio (L/D) greater than 5. In practice, there will be a gradual change in behavior as the length of a duct with critical flow at exit is increased from $L/D < 1$ to $L/D \gg 1$, and the choice of five is rather arbitrary. The length of the flow path affects velocity profile development, the degree of thermal disequilibrium between phases, and the importance of wall friction. Two possible sites for choking in an LBLOCA, in addition to the break itself, are the PRZ surge line and the pump. No specific SETs relating to either of these two cases have been found. The PRZ surge line should in principle be covered by the database for critical flow in pipes. The pump rotor and diffuser have a more complex geometry, and present problems akin to those encountered in valves. In addition, the rotor may be spinning or locked.

The CSNI published a State of the Art Report on critical flow modeling in OECD/NEA/CSNI (1982) which covered both critical flow models and experimental data. An extended review of the models is reported in the technical addendum to this report (D'Auria and Vigni, 1980). Current predictive capabilities are discussed by D'Auria et al. (2021b).

6.3.2.1 Critical flow in breaks

When a break occurs in a separating wall structure between high-pressure and low-pressure systems, flow through the break depends on conditions upstream of the break and the break's area and shape. Critical flow through a break is similar to critical flow through a nozzle, but because the geometry of the break can encompass any shape, location, and size from a small crack to a complete 200% guillotine break in a flow pipe, multidimensional effects and losses are important but also difficult to account for accurately.

For a guillotine break in a large pipe, the flow will pass across the entire cross-sectional break area rather homogeneously, but for a small break, the geometry of the break and the upstream flow pattern close to it are critical. At low flow rates, stratification effects upstream of the break are crucial to the flow rate. Under such conditions, the orientation of the break relative to the liquid-vapor interface governs the actual flow rate and this is also likely to change during drainage through the break. Two-phase flow processes influencing the flow rate through a break in a horizontal pipe include upstream flow regime transitions, liquid entrainment in the break flow, and vapor pull-through.

The critical flow through breaks is a very important element in analyzing plant behavior during the complete spectrum of LOCA. The break flow determines the depressurization rate of the system and the time to core uncovery which in turn are of major concern for when and how different mitigation auxiliary systems will be initiated and function. Because of the empirical nature of critical flow models as well as the arbitrariness in location, orientation, shape, and size of a break, the licensing requirements of a plant include that a spectrum of postulated breaks is analyzed. It is important that in such analyses the envelope of the influences from the break conditions can be found.

The PH is identified as S-59-CF1 (or No 110 in the list of 116 THPs) by Aksan et al. (2018a), and D'Auria et al. (2021b).

6.3.2.2 *Critical flow in valves*

Critical flow through a valve is essentially no different from critical flow through a nozzle, but because the geometry of the flow path through a valve is variable, and usually more complicated than a nozzle, multidimensional effects and losses are likely to be more important and more difficult to account for. Several studies have shown that the choking position and the effective minimum area in a valve are not always easy to determine. They may even vary with downstream conditions. There are many different types of valves, and because of the uncertainties inherent in using 1-D models, data from at least a scaled version of the relevant geometry is highly desirable if reasonably accurate critical flow predictions are required.

Critical flow through valves is an important element in many analyses of plant behavior. In safety assessments, it is central to the stuck-open PRZ valve LOCA, ATWS, LOFW, and accident recovery procedures based on dumping primary side coolant.

The PH is identified as S-61-CF3 (or No 112 in the list of 116 THPs) by Aksan et al. (2018a), and D'Auria et al. (2021b).

6.3.2.3 *Critical flow in pipe*

In pipes with a constant cross section, choking is expected to occur at the pipe exit. In this case the flow rate is still influenced by the entrance losses and wall friction in the pipe since these produce a fall in pressure before the choke point is reached. Thus *L*/*D* is an important parameter in determining critical flow from pipes. Pipe length is also important in determining the degree of thermal and mechanical disequilibrium in the critical flow. The experimental evidence suggests that unless the pipe is very short, even in the absence of significant wall friction the phases will approach thermal equilibrium at the choke plane. The relative importance of length and length-to-diameter ratio are discussed in Holmes and Allen (1990). It is also assumed that for critical flow in pipes, slip ratios are close to unity.

The subject of critical flow through cracks is closely related to that occurring in pipes, being dominated by friction and the attainment of thermal equilibrium.

A pipe break is a prime mechanism for initiating an LOCA. The pipe involved could be a main coolant pipe, in which case complete severance leads to an LBLOCA or a pipe connected to the main coolant loop—e.g., an emergency core cooling system (ECCS) line—that could lead to an intermediate or SBLOCA. The flow rate from the pipe is an important factor in determining the course of the resulting transient. How the flow evolves as a function of time can be different in the case of a small broken pipe from that corresponding to a small hole in a large pipe even when the initial break flow is the same in both cases. In many licensing applications the actual flow rate through a break of a given size is not required because what is of interest is the behavior of a plant for a range of break sizes. Exceptions are the determination of the maximum flow from particular types of break (for instance from an instrument penetration in the pressure vessel), and the likely flow from a broken SG tube.

The PH is identified as S-60-CF2 (or No 111 in the list of 116 THPs) by Aksan et al. (2018a), and D'Auria et al. (2021b).

6.3.3 *Phase separation/vertical flow with and without mixture level*

Flow regimes in two-phase conditions are included under the title "phase separation/ vertical flow with and without mixture level"; this comprises the limit case of a stagnant or nearly stagnant two-phase mixture when all the liquid has collapsed into the bottom of the vertical component. The consideration of the so-called mixture level is also part of this section.

Three subsections have been distinguished owing to their relevance in nuclear reactor thermal hydraulics and to the differences in the flow regimes that are a consequence of the geometry and of the boundary conditions (essentially, heat input in the case of the core):

- pipes and plena
- core
- downcomer of the vessel

The first item essentially includes the SG tubes, both primary and secondary sides, and the vertical parts of the PRZ surge line and the hot and cold legs of PWR main loops. The subitem plena include all the zones of the primary and secondary sides where the fluid velocities are substantially lower than the neighboring zones. Examples of this are the lower and UP of the vessel (in both BWR and PWR), the steam dome of the SGs, the inlet and outlet chamber of the SG primary side, and the PRZ. Stratified conditions are typical of plena, particularly the lower plenum (LP) of both PWR and BWR, while several slow regimes may occur and are important in vertical pipes and in other types of plena (e.g., SG chambers, see also Section 6.3.7.5).

All flow and HT regimes occur in the core region of BWR under nominal conditions and are expected for PWR during off-normal conditions (second item). The difference between the two kinds of reactors is the possibility of cross-flow in PWR that is inhibited in BWR owing to the presence of the fuel boxes.

The third item has been introduced mainly to take account of the complex three-dimensional situations occurring in a PWR vessel downcomer during several kinds of accidents: the ECC bypass/downcomer penetration, specifically considered as item 21 in Tables 6.1, is an example of this.

6.3.3.1 Phase separation/vertical flow with and without mixture level—pipes and plena

Steam and liquid phases remaining at rest or flowing cocurrently in vertical channels tend to separate owing to either the gravity force or the inertia forces, both essentially driven by the different density of the two phases. With the phases at rest or moving at very low velocity, gravity dominates and the PH is very simple with the lighter phase stratifying in the upper part of the available volume. Essentially, the only physical aspect to be considered in modeling this situation is the time required to reach a stable condition starting from mechanical nonequilibrium (e.g., liquid in the upper part of a voided system).

The situation is much more complex in a flowing system: various flow configurations may occur as a function of system parameters (e.g., equivalent diameter, flow area, geometric flow path, wall HT, etc.), of the fluid thermodynamic parameters (e.g., pressure, liquid and steam velocities, void fraction, etc.) and of the fluid properties. The well-known flow regime zones are the result of the interaction of the parameters mentioned. The formation of a mixture level is also a consequence of the interaction between the phases in flowing and/or wall-heated two-phase fluid systems. Finally, special flow regimes may occur in the case of strong mechanical and/or thermodynamic nonequilibrium such as in the case of abrupt area change or reflood.

The PH is relevant to nuclear reactor safety because the values of PD and HT depend significantly on the flow regimes. The prediction of flow regime may be important in the case of BWR (nominal conditions) or BWR and PWR (conditions of accidents without scram) in the evaluation of local void fraction, and as a consequence, in predicting neutron flux.

The PH is identified as S-48-PS3 (or No 82 in the list of 116 THPs) by Aksan et al. (2018a), and D'Auria et al. (2021b).

6.3.3.2 Phase separation vertical flow with and without mixture level—core

Phase separation in the core will develop during some phases of an LOCA in PWR and under normal and off-normal conditions in BWR. The separation effect is caused by gravity, inertia, or drag-dominated forces acting on the phases along with the density difference between the phases.

At low flow velocities and accelerations, gravity will cause the phases to stratify. If the transient is sufficiently slow, thermal nonequilibrium may have enough time to develop. Vapor and droplets can be driven out of the mixture creating a mixture level. This type of separation can be found, for example, toward the end of the blowdown phase of an LOCA. The rate of phase separation will depend on parameters controlling the velocity of rising bubbles in the liquid and the stability of the separated flow interface, such as bubble size, geometry, and different types of drag (form, interfacial, and wall) and pressure gradient. HT from the fuel cladding to the fluid depends greatly on the location of the mixture level. Below the level, the HT coefficient can be quite high, resulting in moderate fuel temperatures, whereas the conditions above the level can be the opposite.

At higher flow velocities the vapor—liquid interface becomes distributed within the flow, creating complex two-phase flow patterns that depend on various parameters. Governing system parameters are, for example, core flow area and geometry, possibly spacer vanes characteristics, etc., while fluid thermodynamic parameters are pressure, phase velocities, void fraction, etc. This flow behavior is typical in BWR at normal and off-normal conditions and in PWR at LOCA conditions.

Another form of phase separation that can occur in the core is droplet entrainment and deentrainment. In this case liquid droplets can be entrained from for instance a mixture surface through the vapor flow. The liquid droplets carried along in the vapor flow will enhance the downstream cooling of the core.

The phase separation in the core is crucial for the fuel-coolant HT during some phases in LOCA in both PWR and BWR and normal operation in a BWR. This process determines for instance the quenching of the core during an LOCA and consequently has a great impact on the fuel maximum temperature.

The PH is identified as S-46-PS1 (or No 80 in the list of 116 THPs) by Aksan et al. (2018a), and D'Auria et al. (2021b).

6.3.3.3 Phase separation vertical flow with and without mixture level—downcomer

Phase separation with and without mixture level in vertical flow is mainly determined by gravity or inertia forces, both essentially influenced by the density of the gas and liquid phases. If the phases are at rest or moving at very low velocity the gravity force dominates and the lighter phase separates in the upper part of the downcomer. One aspect is the time required to obtain a stable condition starting from well-mixed two-phase distribution situations.

In a flowing system, flow regimes depend on the void fraction, mass fluxes of steam and water, and dimensions of the downcomer flow path. Mixture level formation also depends on the evaporation of water at the hot downcomer walls, if the walls are still at higher temperatures than that of water during or after blowdown. If the flow is directed from the downcomer to the cold leg(s) the flow conditions in the cold leg are affected (e.g., cold-leg break flow rate, pressure decrease in the system, etc.).

The downcomer collapsed water level is the driving force for core reflood and consequently core cooling conditions. The phase separation in the downcomer influences the water inventory loss during blowdown via a cold-leg break due to its influence on the pressure loss mainly in the broken loop and on the critical flow rate at the break.

The PH is identified as S-47-PS2 (or No 81 in the list of 116 THPs) by Aksan et al. (2018a), and D'Auria et al. (2021b).

6.3.4 Stratified flow in horizontal pipes

At high velocities, the flow of a two-phase mixture in a horizontal pipe can be fairly homogeneous. As the velocity falls, the gravitational forces tending to cause phase separation across the direction of the flow become more important than the inertial forces. The ratio of these forces is expressed by the Froude number, a key number in describing stratified flows. The occurrence of stratified flow has a significant effect on a number of two-phase flow processes. Interfacial friction and the interfacial area for HT are much reduced over other flow regimes, and in branching flows the occurrence of stratified flow can significantly alter the flow quality within a branch. Stratified flow conditions are a prerequisite for CCF in a horizontal pipe.

Adjacent flow regimes range from bubbly (at low gas flow velocity and low void fraction) through plug flow and slug flow to wavy-annular flow (at high gas velocity and high void fraction). The transition between stratified and nonstratified flow is often not a sharp one. For instance, the boundary between stratified and slug flow can be difficult to define. When the slug frequency is low the flow retains many of the characteristics of purely stratified flow.

The prediction of behavior within the stratified regime requires knowledge of appropriate wall friction, interphase friction, and HT relationships. These are less difficult to determine than for other more complex flow regimes, although these relationships depend on whether the stratified surface is smooth or wavy. However, inhomogeneous effects within the liquid phase, such as thermal stratification, are more noticeable in this low-flow regime. The extent to which inlet effects influence the flow regime (when the flow is not fully developed) is still not fully understood.

It should be noted that stratified flow is not confined to horizontal pipes, but can occur in quite steeply inclined pipes.

The main sites for stratified pipe flow are the hot legs, the cold legs, and the loop seals of a PWR. Stratified flow can also occur in auxiliary piping such as feedwater lines. The development of stratified conditions in the hot legs is closely bound up with the transition from NC to refluxing, and the draining of the SG tubes in an LOCA. Its occurrence in the cold legs affects the condensation rates of steam on the ECCS liquid in SBLOCA. Where a break in the primary circuit occurs in a horizontal pipe, break flow is much more strongly dependent on the orientation of the break once the flow in that pipe has become stratified, when the break flow quality becomes very level dependent. Similar considerations apply to a stuck-open PRZ valve during LOCA. Flow stratification was found to occur in LOFT at reduced primary inventory even while the primary pumps were still running.

Thermal stratification within the liquid phase is an important consideration in the determination of thermal stresses associated with ECC injection.

The PH is identified as S-56-STR (or No 102 in the list of 116 THPs) by Aksan et al. (2018a), and D'Auria et al. (2021b).

6.3.5 Phase separation at branches

At low velocities, where the flow tends to stratify, phase separation at branches is dominated by gravity effects. In this case the flow through the branch will be liquid-rich if the branch is at the bottom of the pipe and gas-rich if it is at the top. In general, branch flow quality will greatly depend on the liquid level within the main pipe. Liquid entrainment and vapor pull-through, which are primarily inertial effects, both exert a powerful influence on this relationship.

At high velocities the effect of inertial forces on branching flow becomes dominant: the heavier phase (liquid) tends to flow in the straight direction, while the lighter one (steam) is more easily diverted by the low pressure in the side junction of the branch. This causes in most cases a larger void fraction in the side pipe with respect to the value at the branch inlet. The mechanism is multidimensional and is significantly affected by all the geometric peculiarities of the system.

The flow regime also plays an important role in this connection; as an example, in bubbly or slug flow the steam phase can prefer to flow in the side branch as mentioned above, but in annular flow, the liquid on the walls may be diverted more easily to the side branch.

Branching situations can occur in LOCA scenarios at break locations. The prediction of the phase separation at branches is important for evaluating the energy and mass lost from the system. Vapor pull-through and liquid entrainment are examples of phenomena expected.

Other typical branching configurations include the connections of the main piping with the vessel and the surge line with the hot leg in the PWR.

The PH is identified as S-45-PSB (or No 79 in the list of 116 THPs) by Aksan et al. (2018a), and D'Auria et al. (2021b).

6.3.6 Entrainment/deentrainment

Droplet entrainment and deentrainment are also forms of phase separation and are related to interface friction dependent on the flow regime. Water can, for example, be entrained by sufficiently high steam velocities from the bottom of the reactor vessel and be carried out of the vessel. This water can be lost from the liquid inventory in the pressure vessel.

Compared with phase separation in vertical flow with and without mixture level (Section 6.3.3), entrainment and deentrainment processes relate more to high void fraction flows, including void fraction discontinuities.

The main source of entrainment, in the case of core uncovery, is the core, where steam is produced. Another place closely connected to the core is the UP where both entrainment and deentrainment can occur. These phenomena also take place in the downcomer, mainly in the case of reverse steam flow for LBLOCA. SG mixing chambers, SG tubes, and the hot legs, in the case of ECCS injection, are also affected by these phenomena.

For a given mass inventory, entrainment and deentrainment influence the mass distribution in the primary circuit and as a consequence the available mass inventory in the core. So they have consequences on core cooling and are relevant for nuclear safety.

The important parameters related to entrainment and deentrainment to be measured are the droplet sizes and distribution, the interfacial area, and the field of velocities. Only limited global measurements are available based mainly on DP measurements. Catch tanks are also used, and carryover tanks are available.

The present state of knowledge is limited to allow a detailed description of these phenomena. Only global correlations are available. The geometry of internals and geometric discontinuities also play an important role in relation to this PH. This limits the degree of extrapolation of particular models.

6.3.6.1 Entrainment/deentrainment in the core

Droplet entrainment and deentrainment are forms of phase separation. Water can, for example, be entrained by sufficiently high steam velocities from the bottom of the reactor vessel and be carried out of the vessel. Thus water can be lost from the liquid inventory in the vessel. Droplets carried out of the reactor core by rising steam during reflooding can be deentrained by impinging on the various structures situated in the UP and, consequently, may fall back into the core.

Entrainment of liquid droplets by the flow of steam has a considerable influence on the effectiveness of ECC. The importance of the entrainment has been especially recognized as associated with heat and mass transfer processes during LOCA, in particular, during the recovery phase of these accidents through reflooding of a core. In this case entrainment improves HT since droplets act as a heat sink through droplet evaporation. This will lead to lower vapor superheating and improved cooling of fuel rods. Indeed, the hot portion of the fuel rods in the core is cooled by the mixture of steam and entrained droplets. Liquid entrainment affects the liquid core inventory and is also important in the determination of the CHF location. Another indirect effect of the entrainment is that the vaporization of the droplets carried to SGs pressurizes the UP and prevents the core water level from rising (i.e., steam binding effect). Liquid deentrainment occurs when water entrained in the steam generated in the core is removed from the steam flow at other places in the core, UP, or beyond. Deentrainment occurs from gravitational and inertial forces. It is enhanced when the flow slows because of a flow area increase or when it changes direction to pass around flow obstructions or structures or to turn out through a nozzle. Deentrainment removes entrained water from the two-phase flow mixture. Deentrained water that accumulates in the core and UP provides a source of water for core cooling supplementing the cooling by bottom reflood. This enhances core cooling near fuel assembly grid spacers and is also important for the upper regions of the core. Clearly, such entrainment and deentrainment phenomena are controlled by interfacial shear. Correct modeling of the interfacial shear force is crucial for the adequate prediction of several phenomena and interfacial shear models also vary according to the flow regimes.

The importance of the entrainment and deentrainment in a core has been recognized as associated with heat and mass transfer processes during LOCA, in particular, during the recovery phase of these accidents through reflooding of a core and under boil-off situations, see e.g., Arai et al. (2015). Since these phenomena will lead to improved cooling of the fuel rods, the knowledge of the extent of entrainment and deentrainment is necessary to predict the hydrodynamic and thermal response in the core.

The PH is identified as S-15-ED1 (or No 32 in the list of 116 THPs) by Aksan et al. (2018a), and D'Auria et al. (2021b).

6.3.6.2 Entrainment/deentrainment in the upper plenum

These phenomena are very much related to the entrainment and deentrainment in the core (Section 6.3.6.1). Some of the carryover out of the core will be captured in the water pool, which may form on the upper core support plate, and by UP structures.

Pool formation refers to the collection of water in the UP during reflood. The source of this water is entrained water carried up from the core which is then deentrained in the UP. The UP contains a great deal of internal structures that act as a steam separator to deentrain liquid in the two-phase flow. In addition, the drop in flow velocities from the core to the UP allows gravitational separation of liquid from steam. The flow behavior of a collected pool may be highly three-dimensional because of the tendency of the water pool to collect in low-flow areas and regions near walls and structures. UP pool formation is important to ECCS performance during an LOCA. The reasons can be summarized as: the pool, particularly due to its multidimensional nature, provides a source of additional cooling water for the core; deentrainment and pool formation reduce carryover to the SG; the pool can reduce the flooding rate by creating a static head PD in the UP.

The extent of liquid carryover to the UP is expressed by the liquid carryover rate fraction, that is, the ratio of the total liquid mass flow rate out of the core to the liquid mass entering the core. Liquid carryover reduces the rate of accumulation of water within the core and provides a source of water for deentrainment in the UP or beyond. Carryover has a major effect, because the rate of advance of the quench front (QF) is tied directly to the increase in core mass inventory, and is therefore highly relevant to reactor safety.

The PH is identified as S-20-ED6 (or No 37 in the list of 116 THPs) by Aksan et al. (2018a), and D'Auria et al. (2021b).

6.3.6.3 Entrainment/deentrainment in the downcomer

In a large cold-leg break LOCA, the beginning of the "reflood" phase occurs when the bottom of the core. This creates a seal between the core and the downcomer, and further ECC water injection tends to fill the downcomer to near the cold-leg elevation. The difference in water level between the downcomer and the core provides the driving head for core flooding. This driving head also creates a pressure difference from the top of the core to the top of the downcomer which tends to cause core steam generation to flow out of the core and through the loops to the downcomer. In combined injection PWR, essentially all of the steam flow is condensed by hot-leg and/or cold-leg ECC water and there is no steam flow into the downcomer. In cold-leg injection or downcomer injection PWR, steam flows to the downcomer via intact loops or vent valves. Some of the steam is condensed by ECC injected in the cold legs or the downcomer. Any steam not condensed, along with steam generated in the downcomer due to superheated walls, flows circumferentially around the downcomer and out of the broken cold leg, potentially entraining and carrying away a portion of the ECC water (Damerell and Simons, 1993b). Some deentrainment may occur due to the higher inertia of water droplets compared with the steam phase where the flow changes its direction from the downcomer to the broken leg.

The PH is mainly important if a large cold-leg break occurs and the ECC water is injected only at the cold side of the reactor (into the cold legs or the downcomer). The circumferential flow of steam around the downcomer and the generation of steam on superheated downcomer walls tend to entrain ECC in the downcomer region. The interaction of steam flow, wall boiling, and ECC entrainment is important since it affects the water level in the downcomer. Reduction of the downcomer liquid level below the level (which is at the bottom of the cold-leg nozzles) reduces the available driving head and tends to reduce the core flooding rate. This prolongs quench times and potentially allows higher clad temperatures in the core.

The PH is not important if subcooled ECC water is injected at the hot side of the reactor vessel and condenses a considerable steam flow from the core to the hot legs. The PH is not important for intermediate and small breaks (rupture area less than 25% of maximum pipe area connected with the pressure vessel) since the circumferential steam flow in the downcomer is not high enough to cause significant entrainment.

The PH is identified as S-16-ED2 (or No 33 in the list of 116 THPs) by Aksan et al. (2018a), and D'Auria et al. (2021b).

6.3.6.4 Entrainment/deentrainment in steam generator tubes

In the case of a high steam flow rate droplets can be entrained into the SG tubes, from the mixing chambers. In the case where the SG is a heat source (such as a large-break accident), the droplets may vaporize creating additional steam binding. This entrainment process is strongly connected with the behavior of fluid in the mixing chambers.

This PH contributes to the so-called steam binding for LBLOCA.

For a smaller break size, entrainment contributes to reducing the two-phase NC, producing reverse flow in some SG tubes, and reducing mass inventory in the core.

The PH is identified as S-19-ED5 (or No 36 in the list of 116 THPs) by Aksan et al. (2018a), and D'Auria et al. (2021b).

6.3.6.5 Entrainment/deentrainment in steam generator mixing chambers

Mixing chambers, by reducing the steam velocity, have a large impact on the phase separation process and can reduce droplet entrainment in SG tubes. But when a mixture level appears within a mixing chamber a sufficiently high steam flow can also generate droplets with consequences for fluid behavior in the SG tubes (see Section 6.3.6.4).

As in Section 6.3.6.4, this PH contributes to steam binding for large breaks and to flow distribution in U-tubes (3-D effects) for smaller sizes of break and NC conditions. It also influences the mass distribution in the primary circuit.

The PH is identified as S-18-ED4 (or No 35 in the list of 116 THPs) by Aksan et al. (2018a), and D'Auria et al. (2021b).

6.3.6.6 Entrainment/deentrainment in the hot leg with emergency core cooling injection (pressurized water reactor)

A pressure difference exists from the UP to the downcomer through the loops, if a large break occurs at the cold-leg side of one of the primary coolant loops, or if condensation occurs in the SG primary side in a small-break scenario. The pressure difference causes a steam flow from the UP to the SGs. ECC water injected into the hot legs via a special nozzle (Hütze in KWU PWR) at the bottom of the hot legs, for instance, may be entrained toward the SG by sufficiently high steam velocities. The entrained water may reach the SG tubes where it may evaporate in the large-break case. The evaporated water is no longer available for core cooling. On the other hand, the pressure in the SG primary side increases due to the evaporation, reducing the steam velocity toward the SG, or even causing a steam flow from the SG to the UP. Consequently, the total ECC water injected into the hot legs will be delivered into the UP for a certain time span, causing an oscillatory flow in the hot legs.

In the case of condensation in the SG, the condensed water flows back toward the UP. Due to the decreasing steam flow in the SG, the entrained water also flows back to the hot leg toward the UP.

Since the hot-leg ECC injection water is subcooled (injection temperature 30–35°C) the condensation may cause a water plug to form in the hot legs where ECC water is injected. Plug formation is dependent on the steam flow rate in the hot leg and the condensation potential of the ECC water. The hot-leg ECC injection entrainment behavior is mainly determined by condensation (see Section 6.3.7.6). An oscillatory plug flow establishes itself or a stable stratified flow occurs depending on these parameters. The whole process is closely connected with CCFL including condensation and momentum effects of ECC water injection.

The amount of ECC water injected into the hot legs (i.e., via the special nozzles at the bottom of the hot legs in some reactors) reaching the core is important for ECC effectiveness. The water entrained in the steam flow is lost for core cooling at first. The pressure increase due to the evaporation of entrained water in the SG may support the ECC water delivery subsequently, resulting in an oscillating ECC water delivery to the UP. SCTF test results show that core quench in an LBLOCA is delayed by approximately 15 s if ECC water is injected into the UP intermittently instead of continuously.

The PH is identified as S-17-ED3 (or No 34 in the list of 116 THPs) by Aksan et al. (2018a), and D'Auria et al. (2021b).

6.3.7 Liquid-vapor mixing with condensation

Direct contact condensation of steam on cold water is a very efficient heat removal mechanism that often takes place at very rapid heat and mass transfer rates. Violent pressure oscillations due to the rapid condensation of steam and the resulting volume reductions have been observed in several situations when a subcooled liquid is brought into intimate contact with steam. For example, such oscillations can take place at the ECC injection points. The mixing of liquid and vapor affects the magnitude of the interfacial area and consequently, in conjunction with temperature differences between steam and water, it determines the condensation efficiency. The mixing is mainly influenced by the void fraction and the mass fluxes of steam and water as well as steam and water temperatures and other properties.

Condensation processes are relevant to reactor safety in that they

- reduce the reactor primary side pressure below the set points for low-pressure ECCSs by SG secondary side heat removal and high-pressure ECCS injection,
- reduce steam flows which otherwise reduce water delivery to the core due to CCFL, as shown by experiments in UPTF (Damerell and Simons, 1993a,b; Riegel, 1990a,b); however, a higher steam upflow to the condensation location preventing water delivery has occurred in small-scale facilities like LOBI,
- reduce pressure in the UP region (by means of hot-leg ECC injection) to support a faster reflood of the core.

Condensation may be reduced if noncondensable gases, for example, nitrogen from the accumulators, are present, see Section 6.3.25.

The measurement of condensed steam is mainly performed by measurements of mass flow, water inventories, temperatures, pressures, and void fractions (gamma-densitometer, pressure difference). The interfacial area has not yet been measured successfully.

Computer code models describe mixing and condensation in the way that interphase heat and mass transfer rates are empirically correlated to the flow regimes that are assumed to occur (Forge et al., 1988). An interfacial area then needs to be specified, which is difficult for all flows except for well-defined, gravity-separated flow conditions (see Section 6.3.8). The flow map is not used explicitly in every computer code, but implicitly in the transfer laws. It is not clear if these flow regime maps are scale-independent (Lewis et al., 1989; Yadigaroglu et al., 1990).

Interfacial heat and mass transfer laws and the resulting condensation models are mainly determined by the interfacial area models that the computer codes use (Fletcher and Schultz, 1992). In the three-dimensional, highly unsteady two-phase flow situation, interfacial area and surface temperature can change rapidly. Two fluid codes employ an average interfacial area over a chosen numerical cell size which, in practice, is orders of magnitude larger than the characteristic turbulent mixing dimensions or the molecular transfer dimensions of the flow in or perpendicular to the main flow directions (e.g., turbulent mixing length or thermal boundary layer thickness). Thus predictions of mixing and condensation often depend on the discretization chosen for a problem (Yadigaroglu et al., 1990; Forge et al., 1988; OECD/NEA/CSNI, 1989). The stability of the calculations is particularly sensitive to pressure spikes from artificial water packing.

Liquid-vapor mixing with condensation is of special interest in the core, downcomer, UP, LP, and SG mixture chamber (PWR) and in the ECC water injection regions in the hot and cold legs (PWR) (Hafner and Fischer, 1990). The differences in geometry and boundary conditions cause differences in the flow regimes of the different locations. Spray cooling effects for BWR are covered by Section 6.3.9.1.

6.3.7.1 Liquid-vapor mixing with condensation—core

Liquid-vapor mixing in the core occurs during ECC water injection to prevent overheating of the core. Subcooled ECC water may enter the core from the cold injection location via the downcomer and LP or from the hot-leg injection location via the UP down to the core.

Condensation on subcooled water entering the core removes a part of the heat in the core. Local reduction of steam flow and consequently higher water delivery under CCF conditions may occur due to condensation in the core. Fast water delivery to the core is essential for ECC. The PH is relevant for plants with combined, UP, or upper head injection.

The PH is identified as S-33-LVM1 (or No 59 in the list of 116 THPs) by Aksan et al. (2018a), and D'Auria et al. (2021b).

6.3.7.2 Liquid-vapor mixing with condensation—downcomer

Subcooled ECC water is injected into the cold legs or into the downcomer directly depending on the reactor design. The injected water flows down the downcomer, and, on its way down, the water is warmed up mainly due to steam condensing on the water interface.

Direct contact condensation on the ECC water which is injected into the cold legs or directly into the downcomer has an influence on the CCF in the downcomer region. The upward-flowing steam is reduced by the condensation rate. Consequently, the water downflow rate toward the core can increase. The condensation rate again depends on the downward water mass flow rate. The main water downflow was observed in UPTF tests below the cold legs which are farthest from the broken cold leg. In small-scale facilities like LOBI, however, an opposite effect was observed. Condensation at the accumulator ECC water injection location caused a high steam upflow in the downcomer which prevented water downflow for some time.

Condensation in the cold legs and the downcomer is ranked as very important for LBLOCA refill in a four-loop PWR (Wilson et al., 1990).

Steam condensation causes a warm-up of the subcooled ECC water which may reduce pressurized thermal shock (PTS) in the reactor vessel wall, in the case of high-pressure ECC water injection.

The PH is identified as S-34-LVM2 (or No 60 in the list of 116 THPs) by Aksan et al. (2018a), and D'Auria et al. (2021b).

6.3.7.3 Liquid-vapor mixing with condensation—upper plenum

Liquid-vapor mixing with condensation is important in the case of UP and hot-leg ECC injection in PWR and BWR. It is highly dependent on the injection device, for example, whether injection is via spray systems or nozzles in the UP or hot leg.

Condensation in the UP may decrease the steam binding effect substantially and consequently may allow faster reflood of the core.

Direct contact condensation on the ECC water in the UP influences the CCF in the UP, especially in the upper tie plate (UTP) region at the top of the core. The upward-flowing steam is reduced by the condensation rate. Consequently, the water downflow rate into the core can increase. The condensation rate is dependent on the downward water mass flow rate.

The PH is identified as S-38-LVM6 (or No 64 in the list of 116 THPs) by Aksan et al. (2018a), and D'Auria et al. (2021b).

6.3.7.4 Liquid-vapor mixing with condensation—lower plenum

At the end of the blowdown period of an LBLOCA, the depressurization of the reactor coolant system (RCS) allows the ECCS to inject subcooled water into the system. When this cold water comes into intimate contact with the superheated steam or saturated two-phase mixture intense condensation occurs with associated pressure oscillations. The ECCS water will penetrate the downcomer as described in Section 6.3.7.2 and reach the LP of the vessel. At this instant, the refill period starts. Refill will be terminated when the LP water level reaches the bottom end of the core. Some water could remain in the LP after the blowdown and the amount of water has a high impact on the fuel cladding temperature. Evaporation and the entrainment of liquid droplets, which to some extent pass through the core, limit the fuel cladding temperature. Additionally, some subcooled water could enter the LP from leakage in the control rod drives (BWR) and contribute to the water contents of the plenum.

When the ECCS water reaches the LP, the water will partly be evaporated on the superheated mechanical structures but will also contribute to the condensation of the existing saturated or superheated vapor in a complicated manner. The ECCS water has a strongly three-dimensional random distribution when it enters the LP because of the CCF in the downcomer. This will become even more pronounced in the LP because of the complex evaporation on irregularly shaped and distributed structures and the simultaneous condensation of existing vapor.

Direct-contact condensation often takes place at very rapid heat and mass transfer rates. Thus the condensation of vapor can lead to the occurrence of violent pressure fluctuations.

Liquid-vapor mixing in the LP is an important PH, as it influences the fuel temperature through related entrainment and vaporization processes. The entrained liquid and vapor will partly pass through the core and enhance the core HT, and the final cooling of the core.

The PH is identified as S-36-LVM4 (or No 62 in the list of 116 THPs) by Aksan et al. (2018a), and D'Auria et al. (2021b).

6.3.7.5 Liquid-vapor mixing with condensation—steam generator mixing chambers (inlet/outlet plenum)

The SG mixing chambers connect the RCS piping with the SG tube bundle. Because of the geometric shape of these chambers high turbulence and effective mixing can be expected even at rather low flow conditions. Thus the inlet conditions to the tube bundle can in most cases be assumed fairly uniform across the tube bundle flow area.

However, during SBLOCA, situations occur when the NC in the loops can no longer be sustained because the liquid inventory is insufficient to allow flow over the top of the U tubes. Complete phase separation occurs and at further loss of liquid inventory, CCF develops in the hot legs. Thus saturated or slightly superheated vapor flows from the core through the hot leg into the SG mixing chamber and up into the tube bundle. If the secondary side of the SG has a temperature lower than the saturation temperature of the primary side, condensation of vapor in the tubes will be initiated.

The condensate will fall back down the tubes due to gravity and CCF will prevail in those parts of the U tubes. The liquid will enter the mixing chamber in which stratification occurs. Some condensation of the vapor in the chamber can be expected. The liquid flows "over the edge" of the SG nozzle into the hot leg and back to the core where it evaporates. The process is known as reflux condensation and can contribute significantly to the cooling of the core.

The condensation in the SG tubes is not likely to be uniform and is also oscillatory because of the behavior of the condensation process. As a result, the condensate flow into the mixing chamber will be quite nonuniform, and this along with the stratification will create multidimensional effects that will influence, for instance, the reflux condensation process. It is also quite possible that conditions in each SG are different, which may contribute to the asymmetric behavior of the loops.

Under low-flow or almost stagnant conditions, liquid–vapor mixing within the SG mixing chambers will influence SG efficiency as a heat sink. In the reflux condensation mode, the detailed processes in SG tubes will be governed by, for instance, multidimensional conditions in the mixing chamber. However, the exact details of the conditions in the mixing chamber seem to be of second-order importance in most situations because the large HT area of SGs (sized to remove full power) is more than adequate for decay heat removal. Exceptions can occur when HT efficiency in SGs is greatly reduced by the presence of noncondensables or dryout of the secondary side.

The PH is identified as S-37-LVM5 (or No 63 in the list of 116 THPs) by Aksan et al. (2018a), and D'Auria et al. (2021b).

6.3.7.6 Emergency core cooling injection in hot and cold legs

Depending on the ratio of the steam flow rate in the cold or hot leg to the condensation potential of the ECC water, a stratified flow pattern or a water plug will be established in these legs. Conditions favoring plugging are most likely to be found in an LBLOCA scenario. While the mechanisms of plug onset are the same during condensation and adiabatic flows, an additional mechanism takes place during condensation due to the local decrease in pressure in the pipe as a result of reduced steam volume. Plug formation causes an intermittent water delivery to the downcomer or the UP, respectively. Plug formation delays the water delivery to the core region.

Water plugs formed in hot legs (despite nozzle momentum toward the UP) enter the SG tubes and steam is generated. The increased pressure in the SG pushes the plug toward the UP. A new plug is formed after the old plug is delivered to the UP and causes intermittent water delivery to the UP and subsequently into the core.

The temperature of plug end faces in the cold legs increases up to saturation temperature due to steam condensation. Consequently, the condensation decreases, and the plug is delivered to the downcomer. A new plug is formed after delivery of the old one. This process causes intermittent water delivery to the downcomer.

ECC water delivery from hot- and cold-leg injection locations toward the core is essential for the effective performance of the ECCS. Small delays in ECC water delivery are produced due to plug formation in the legs.

The PH is identified as S-35-LVM3 (or No 61 in the list of 116 THPs) by Aksan et al. (2018a), and D'Auria et al. (2021b).

6.3.8 Condensation under stratified conditions

When the flow is stratified with subcooled liquid, experiments have shown that due to liquid temperature stratification, the condensation rate is relatively low because the liquid *interface* is close to the saturation temperature of steam.

Condensation in stratified conditions plays an important role in the behavior of the PRZ (Section 6.3.8.1), the SG primary side (Section 6.3.8.2), and the SG secondary side (Section 6.3.8.3). In these components a so-called piston effect can occur, that is, an increase in pressure, sometimes limited by safety valves, following cold water injection. In horizontal pipes condensation in stratified conditions can occur during emergency cooling water injection and can cause water hammer (Section 6.3.1.9).

Condensation in stratified conditions governs the pressure behavior during recovery procedures and the mass distribution in the circuits, and so is relevant to nuclear safety.

The important parameters are the vertical temperature profile, local heat and mass transfer, local liquid and steam velocities, and void fraction. Often only *P* and *DP* measurements, and limited temperature measurements, are available—densitometer measurements are sometimes available in pipes. The database is generally poor.

This PH is not well predicted by codes mainly due to 1-D assumptions (1 liquid temperature). Codes generally predict too large a condensation rate.

Generally, fine mesh nodding can improve code calculation results for vertical components because that gives a better simulation of the water temperature profile. Justified user guidelines are strongly recommended.

6.3.8.1 Condensation under stratified conditions—pressurizer (pressurized water reactor)

During recovery procedures the safety injection system refills the primary side. When the water enters the surge line, and then the PRZ, condensation occurs. But with stratified conditions in the PRZ heat and mass transfer are limited. The so-called piston effect can occur and the pressure increases, which induces some condensation.

Condensation in stratified conditions governs the pressure behavior at the end of recovery procedures and can make it difficult to reach the set point for the normal residual heat removal system.

The PH is identified as S-10-CO2 (or No 23 in the list of 116 THPs) by Aksan et al. (2018a), and D'Auria et al. (2021b).

6.3.8.2 Condensation in stratified conditions—steam generator primary side

During an accident, in the case of refilling of the primary side with stopped pumps the water level there rises into the SG primary side. Steam is compressed by subcooled water and condensation occurs. However, the PH is complicated by axial heat conduction in the substantial number of tubes and depends on secondary side conditions. This can be complicated by the presence of noncondensable gas (Section 6.3.25). Under these conditions, reestablishing NC is difficult.

This PH is related to the pressure behavior of the primary side during recovery procedures after an accident. The pressure influences the injected water flow rate, so there is a feedback influence on condensation. Finally, the total amount of injected water depends on condensation.

The PH is identified as S-11-CO3 (or No 24 in the list of 116 THPs) by Aksan et al. (2018a), and D'Auria et al. (2021b).

6.3.8.3 Condensation in stratified conditions—steam generator secondary side

In the case of stratification in the SG secondary side, different behaviors can be expected depending on the water level, the recirculation flow, and the geometry. This problem is encountered in the particular case of a steam generator tube rupture (SGTR), where intact SGs are used to cool down the primary side. After a certain time, the riser becomes cooler than the downcomer and the recirculation can reverse.

If the upper recirculation flow path is close to the water level, significant condensation occurs. If this flow path is below the water level the condensation is strongly reduced and the so-called piston effect becomes important.

This PH governs the pressure behavior in the SG in the case of SGTR, so it has significant effects on the amount of primary fluid flowing to the secondary side and on fission product release by SG safety relief valves. (Fission product transport and distribution themselves are not considered in this work.)

The PH is identified as S-12-CO4-c (or No 25 in the list of 116 THPs) by Aksan et al. (2018a), and D'Auria et al. (2021b). To minimize the number of phenomena in the above list, condensation occurring in a BWR containment wet well (i.e., in a pressure suppression pool) is assumed to be part of this PH; see section 6.5.7.

6.3.8.4 Condensation in stratified conditions—horizontal pipe

The typical case is associated with high pressure injection system (HPIS) flow in the cold leg during an SBLOCA. For example, in the ISP18 test in LOBI, a large temperature difference (100 K) was observed in a 50 mm cold-leg pipe for at least 1000 s.

This behavior (stable buildup of a large steam/liquid temperature difference) is due to liquid temperature stratification. The liquid interface reaches saturation temperature and the condensation rate becomes lower.

The mass distribution in the primary circuit depends on the condensation PH in connection with bypass flows. The pressure behavior is also governed by condensation.

The PH is identified as S-9-CO1 (or No 22 in the list of 116 THPs) by Aksan et al. (2018a), and D'Auria et al. (2021b).

6.3.9 Spray effects

The phenomena that occur when a more or less subcooled liquid is sprayed into a zone containing superheated steam or a two-phase mixture (usually at a high void fraction), are considered in this section.

Three subitems have been distinguished essentially to emphasize the influence that system geometry may have on the evolution of phenomena.

- core in the case of a BWR,
- PRZ in the case of a PWR,
- secondary side in the case of once-through steam generators (OTSGs).

Spraying of liquid in the UP zone of BWR is essential to quench core rods from the top in the case of inadequate effectiveness of the low-pressure injection system. Flooding situations typically occur at the top of the fuel elements and obstruct the downward liquid flow. The distance of the sprayers from the core upper plate, the subcooling of the sprayed liquid, and the distribution of droplet diameters are parameters affecting the PH.

Spray in the upper part of PRZ (steam zone) is used to control the pressure during nominal conditions and minor transients in PWR plants. The liquid is directed from the cold leg downstream of the main circulation pumps and arrives at the PRZ top with a subcooling of about 60 K. The sprayed droplets may interact with the PRZ walls and decrease the stored heat of the structures. Spraying of liquid in the PRZ may prevent the so-called piston effect.

Injection of a liquid spray into the secondary side of a voided OTSG occurs as a consequence of auxiliary feedwater intervention. The effects of this are essentially twofold. First, there is direct HT to the superheated steam (decrease in stored heat and condensation), or to the saturated steam (condensation). Second, there is direct HT with the primary side. In this last case, HT depends on the geometric configuration of the SG tubes and the thermal-hydraulic conditions of the primary side fluid.

6.3.9.1 Spray effects—core

The ECCS of a BWR is designed to spray liquid into the steam-filled UP region of a BWR, downstream of the core. The low-pressure core spray (LPCS) starts to inject water when the pressure reaches values below about 2 MPa. In BWR-6-type reactors, a high-pressure system (high-pressure core spray—HPCS) working similarly to the LPCS is also installed. The injection set point pressure is higher than the nominal pressure of the reactor.

The spray of the liquid into the UP creates condensation of steam and, owing to flooding, a liquid level may form on the top of the upper core plate. Three-dimensional phenomena occur and make possible a breakdown of the CCFL in some bundles, leading to liquid falling to the LP. This can facilitate the bottom-up quench of the fuel bundles.

In the case of liquid formation in the UP, the liquid level may reach the elevation of the spray nozzles that are located a few dozen centimeters above the top of the core. At this time the liquid mass may quickly become subcooled, making more effective the condensation of steam coming from the core. The final result is a CCFL breakdown and penetration of the liquid into the core.

At a basic level, the heating of the sprayed droplets and the consequent condensation of steam, constitute the main PH occurring in the case of spray. The size of the droplets (distribution of diameter), the spray angle, and the degree of subcooling are the main parameters affecting the PH apart from the total sprayed flow rate. The distance between the spray nozzles and the upper core plate and interactions among the various jets can also play a role in the evolution of the PH.

The actuation of spray systems (especially of the LPCS) is required in the case of major accidents for the plant such as LBLOCA or SBLOCA. The ability to maintain the system within safety limits depends strongly on the capability of the LPCS or HPCS to quench the core.

The PH is identified as S-52-SPR1 (or No 96 in the list of 116 THPs) by Aksan et al. (2018a), and D'Auria et al. (2021b).

6.3.9.2 *Spray effects in the pressurizer*

Liquid spray into the top of the PRZ is used to control the PRZ pressure by reducing it. (Pressure increase can be obtained by heaters.) Spray water is taken from the cold leg downstream of the main circulation pump, so the subcooling in normal operating conditions is about 50 K. In most of the reactors, a small flow (of the order of 1 kg/s) flows continuously through the line connecting the cold leg with the PRZ to prevent thermal shock problems in the case of sudden operation.

In the case of undesired PRZ pressure increase, the spray valve is open to control the pressure. Two main phenomena occur.

(a) direct condensation between spray droplets and steam,
(b) heating of the sprayed droplets due to direct HT from the PRZ wall.

Both the above effects contribute to lowering the pressure. The size of the droplets and the spray angle are the main parameters affecting the PH (droplet temperature is fixed and droplet velocity and PRZ pressure can also play a role). The amount of space occupied by steam, including the distance between the spray nozzle and the actual PRZ level, is also important.

Finally, in some cases a shield is installed inside the PRZ to protect the PRZ wall from thermal shock.

Actuation of the PRZ spray system will occur during the normal operation of the plant. In the case of off-normal transients like steam line closure, turbine trip, LOFW, etc., opening of the valve may occur but, usually, the delivered flow rate is not sufficient to stop the pressure increase (opening of the PRZ relief valve is necessary).

Considering the above the spray effect in the PRZ cannot be considered highly relevant from a safety point of view. This consideration does not exclude the possibility that spurious actuation of the spray valve may be the origin of a more complex transient for the plant.

The PH is identified as S-54-SPR3 (or No 98 in the list of 116 THPs) by Aksan et al. (2018a), and D'Auria et al. (2021b).

6.3.9.3 Spray effects in once-through steam generator secondary sides (pressurized water reactor)

Auxiliary feedwater is injected into the SG secondary side via a spray system. The water is heated up and evaporated by the hot water on the primary side flowing through the SG tubes.

The feedwater inlet into the OTSG secondary side downcomer of US OTSG PWR is performed via a ring pipe with holes. No spray system is designed for feedwater inlet into the German OTSG plant. A simple pipe connection to the OTSG secondary side downcomer is installed. The German OTSG auxiliary feedwater inlet is constructed as a ring with holes to inject radially directly into the secondary side. The US OTSG auxiliary feedwater is injected radially by a nozzle directly into the secondary side. The US OTSG nozzle is located close to the top of the secondary side whereas the German OTSG plant, auxiliary feedwater ring is located close to the bottom. The primary side water flow in normal operation is downward within the SG tubes. A design goal of the auxiliary feedwater spray system is to prevent thermal stress as much as possible. The spray effect is not considered highly relevant from a safety point of view but it may determine the number of SG tubes wetted by the auxiliary feedwater.

The PH is identified as S-53-SPR2 (or No 97 in the list of 116 THPs) by Aksan et al. (2018a), and D'Auria et al. (2021b).

6.3.10 Countercurrent flow/countercurrent flow limitation

At high flow rates, the gas and liquid phases in a two-phase flow move down the pressure gradient in the same direction. At lower flow rates, however, the existence of a gravitational field can result in significantly different velocities for the two phases, and liquid can flow downward while the lighter gas phase flows upward. This is CCF, and it can occur in vertical pipes, inclined pipes, and in horizontal pipes in stratified flow (Bankoff and Lee, 1985, 1986).

There is a limit for the rate at which liquid can flow down for a given gas upflow, which is known as the CCF limit. The term "flooding" is sometimes used to mean the CCF limit, but it can also be used for the special case of zero liquid phase penetration down the duct. In this document, the former interpretation is used.

If the liquid flow rate is held constant, while the gas upflow is slowly increased, the onset of flooding often occurs abruptly, accompanied by a change of the flow regime from a relatively separated flow to a churn flow pattern.

Flooding is a local PH, and often takes place at the entrance or exit of a piping section. The flooding characteristics are thus fairly sensitive to the local geometry. Flooding in some geometries, such as the hot leg/SG inlet plenum, has been found to display significant hysteresis with respect to gas flow rate.

CCFL is also of importance in calculating flow behavior at a perforated plate (such as the UTP). Here the question is how fast can water, which may have formed a pool on top of the plate, drain through (some of) the perforations against a given steam flow from below the plate. This is more like flow through a set of parallel tubes than CCFL in a single tube and is likely to be quite asymmetric.

If the liquid phase is subcooled then condensation will reduce the volumetric flow of steam as the steam proceeds along the duct. This can modify the flooding behavior appreciably.

In the following discussion, the CCFLs associated with specific regions within an LWR are described. These are

- upper tie plate (UTP),
- channel inlet orifices (BWR),
- hot and cold leg,
- SG tube (PWR),
- downcomer,
- surge line (PWR).

The current predictive capabilities for this PH are discussed by D'Auria et al. (2021b).

6.3.10.1 Countercurrent flow/countercurrent flow limitation at the upper tie plate

CCF of steam and water through the UTP is a multidimensional PH. It can occur due to water being deentrained in the UP flowing back to the core region, or because ECC water is injected into the UP or the hot legs.

CCFs of steam (or steam/water mixtures) and water play an important role regarding the distribution of coolant in the primary system both in PWR and BWR. In CCF a limiting condition appears first when the downward flow of liquid is influenced by the upward flow of vapor. Any increase in vapor flux then leads to a reduction in the liquid counterflow until, eventually, this becomes zero and the incoming liquid may be carried upward with the gas flow. CCFL tends to occur at flow cross-section restrictions, like the tie plate, where the gas velocity is maximum (Siebe and Stumpf, 1990).

Many CCFL data have been obtained with air and water. For steam/water the CCFL situation may be different due to steam condensation on subcooled ECC water resulting in a reduced steam upflow rate. Steam condensation reduces the vapor flux, allowing a greater liquid flux which in turn increases condensation. This process may lead to a complete water "break-through."

If the up flow consists of a two-phase gas/droplet flow at the CCFL location the momentum of the entrained droplets opposes the liquid downflow leading to more limitation of the liquid downflow rate.

In some PWR designs the ECC injection location is not only on the cold side of the primary system but also on the hot side. The injection locations are, for example.

- into the hot legs via a special "Hütze" nozzle at the bottom of the hot legs directed toward the UP, and into the cold legs,
- into the UP via two injection nozzles (low-pressure coolant injection system), and into the cold legs (high-pressure coolant injection system and accumulators).

Either deentrained water from the UP or ECC water is injected into the hot legs and the UP must flow down through the UTP countercurrently to steam generated within the reactor core or to steam generated by evaporation due to depressurization or to steam expansion due to depressurization. The steam flows toward the broken loop hot leg. The injected water is needed in the core region to prevent overheating. The limited water delivery into the core can delay the core reflood. UTP CCFL has a high ranking of importance for UP or hot-leg ECC injection during LBLOCA refill and reflood phases.

The PH is identified as S-8-CCF6 (or No 17 in the list of 116 THPs) by Aksan et al. (2018a), and D'Auria et al. (2021b).

6.3.10.2 Countercurrent flow/countercurrent flow limitation at channel inlet orifices (boiling water reactor)

Fuel rod bundles of BWR are equipped with side-inlet orifices at the bottom. These side orifices have different flow cross sections to control the radial power profile of the core during normal reactor operation. CCFL can occur in the side-inlet orifices under accident conditions. Water evaporation in the LP causes a steam upflow (see Section 6.3.14) which can limit the draining of the fuel rod bundles (Sun, 1979).

BWR core bundles form parallel channels with different heating powers, and in accident situations there would be nonuniform liquid holdup and steam flow distribution among the bundles.

The first limiting condition appears in a CCF situation when the downward flow of liquid in the channels is influenced by the upward flow of vapor. Any increase in vapor flux leads to a reduction in the liquid counterflow until, eventually, this becomes zero. CCFL tends to occur at flow cross section restrictions like the channel side-inlet orifices, where the gas velocity is maximum. Marked differences in the behavior of individual channels can be expected.

Limiting draining of the fuel rod channels due to CCFL in the side-inlet orifices at the bottom of the fuel channels is important for cooling the fuel rods during blowdown, refill, reflood, and spray periods.

The PH is identified as S-3-CCF1 (or No 12 in the list of 116 THPs) by Aksan et al. (2018a), and D'Auria et al. (2021b).

6.3.10.3 Countercurrent flow/countercurrent flow limitation in hot and cold legs

Under certain circumstances, the high- and low-density phases of a two-phase mixture can flow in opposite directions within a pipe. The driving force for such flows is gravity. For CCF to occur in a horizontal pipe the flow must be stratified, and the driving force corresponds to changes of liquid level along the pipe. For a given liquid depth at the pipe inlet, increasing the countercurrent gas velocity will tend to reduce the liquid flow rate. Complete CCFL, often known as "flooding," when the liquid phase can no longer flow against the gas, will occur when the stratified flow regime breaks down. A typical mechanism for this to occur is for waves on the liquid surface to develop to the extent of bridging the pipe and producing slug flow, but other mechanisms are possible. CCFL behavior can be substantially modified when significant subcooling is present in the liquid phase since condensation of the steam phase changes the steam flow along the pipe (Macken, 1990).

In inclined pipes the driving force for CCF is larger than in horizontal pipes when the liquid is in downflow. An inclined pipe of significant length is, however, unable to sustain liquid upflow, and at a transition from the horizontal to an upwardly inclined direction churn flow can develop.

CCFL in the hot legs is associated with behavior in the inclined pipe sections leading to the SGs. CCFL in this part of the circuit could affect the rate of liquid drain back from the SG tubes to the core in a small LOCA. This in turn would affect the depth to which the core became uncovered. Furthermore, for reflux condensing to occur, the steam flow produced by core decay heat and flashing in the vessel must be less than that required to cause flooding in the hot legs. CCFL is also of interest in the case of injection of ECCS into the hot legs, where for the injected water to flow directly to the vessel it must flow against steam exiting the vessel.

CCFL may also occur in the horizontal part of the cold legs and is effectively involved in the PH of loop seal clearing, in the intermediate leg.

The PH is identified as S-5-CCF3 (or No 14 in the list of 116 THPs) by Aksan et al. (2018a), and D'Auria et al. (2021b).

6.3.10.4 Countercurrent flow/countercurrent flow limitation in steam generator tubes

CCFL can occur within the vertical tubes of the SG at the tube plate. As the hot legs and SG inlet plena empty during an SBLOCA, liquid can be held up within the SG tubes by steam produced within the reactor vessel that flows into the SGs. If the direction of heat flow is from primary to secondary some of this steam is condensed within the tubes. CCFL in SG U tubes is characterized by nonuniform parallel channel behavior, that is, nonuniform liquid holdup and steam flow rate. Some of the tubes may be flooded, while others are cleared (D'Auria and Galassi, 1990).

CCFL at the inlet to the SG tubes will affect the rate of liquid drain back from the SG tubes to the core in a small LOCA. This in turn affects the depth to which the core uncovers. Significant drainback occurs when NC is terminated by the buildup of steam in the downside of the SG tubes. Furthermore, the steam flow produced by core decay heat and flashing in the vessel must be less than that required to cause flooding at the tube plate to prevent a buildup of liquid in the SG tube upsides later on in an LOCA if reflux cooling is to be sustained. The buildup of liquid would again have the effect of depressing the core mixture level.

CCFL is less likely to occur in the downward side of the SG tubes.

The PH is identified as S-6-CCF4 (or No 15 in the list of 116 THPs) by Aksan et al. (2018a), and D'Auria et al. (2021b).

6.3.10.5 Countercurrent flow/countercurrent flow limitation in the downcomer

CCFs of steam (or steam/water mixtures) and water play an important role regarding the distribution of coolant in the primary system both in PWR and

BWR. In CCF a limiting condition appears first when the downward flow of liquid is influenced by the upward flow of vapor. Any increase in vapor flux then leads to a reduction in the liquid counterflow until, eventually, this becomes zero. The liquid which cannot flow down may be carried upward with the gas flow. CCFL tends to occur at flow cross-section restrictions like the downcomer, where the gas velocity is maximum.

ECC water is injected into the cold legs or directly into the downcomer. The injected water is needed to refill the LP and reflood the core in an LOCA situation. This water has to flow countercurrent to steam flowing upward toward the broken loop cold leg. Downcomer CCF may limit the water delivery into the LP. Additional steam production in the downcomer may occur if some of the injected water evaporates due to the hot downcomer walls. This hot wall effect is more pronounced in small-scale test facilities than in large-scale facilities due to the different ratios of structure surface to downcomer volume.

CCF in the downcomer is a multidimensional PH, and the flows of steam and liquid are far from evenly distributed around the downcomer annulus (Mullen et al., 1992).

Many CCFL data have been obtained with air and water. For steam/water the CCFL situation may be different due to steam condensation on subcooled ECC water resulting in a reduced steam upflow rate. Steam condensation reduces the vapor flux, allowing a greater liquid flux which in turn increases condensation. This process may lead to a complete water "break-through." In small-scale facilities like LOBI, however, an opposite effect was observed. Condensation at the accumulator ECC water injection location caused a high steam upflow in the downcomer that prevented water downflow for some time.

If the upflow consists of a two-phase gas/droplet flow at the CCFL location, the momentum of the entrained droplets opposes the liquid downflow, leading to additional limitations on the liquid downflow rate.

ECC water injected into the downcomer or the cold legs must flow downward through the downcomer to refill the LP and subsequently the total core to prevent overheating. This water flow can be limited by steam flowing upward through the downcomer. Consequently, core reflood can be delayed. Downcomer CCFL is one of the most important phenomena during the refill phase of an LBLOCA in PWR (OECD/NEA/CSNI, 1989; Wilson et al., 1990).

The PH is identified as S-4-CCF2 (or No 13 in the list of 116 THPs) by Aksan et al. (2018a), and D'Auria et al. (2021b).

6.3.10.6 Countercurrent flow/countercurrent flow limitation in the pressurizer surge line

The PRZ surge line contains both horizontal and vertical sections of large diameter (around 0.3 m ID) piping. If fluid is discharged from the top of the PRZ, by opening the PRZ valve, for instance, there is the possibility that under two-phase primary circuit conditions, steam flowing to the PRZ will delay or prevent liquid draining back from the PRZ to the primary circuit. CCFL could occur in the horizontal sections or at the upward elbows in the system.

CCFL in the surge line has a direct bearing on faults such as the stuck-open power-operated relief valve (PORV). It results in the liquid level within the PRZ being decoupled from that within the rest of the primary circuit, as in the TMI-2 accident. It also has implications for the use of the PRZ PORV in AM procedures; for instance, if the PORV is discharging to depressurize the primary side, liquid held within the PRZ may not be available to supplement the inventory of the primary loop.

The PH is identified as S-7-CCF5 (or No 16 in the list of 116 THPs) by Aksan et al. (2018a), and D'Auria et al. (2021b).

6.3.11 Global multidimensional fluid temperature, void, and flow distribution

The phenomena in this section primarily address three-dimensional flow behavior in selected components compared with the phenomena described in Section 6.3.3. Global distribution means large-scale behavior of the fluid temperature, void, and flow distribution within special components such as the UP, core, downcomer, or SG secondary side. Nearly all two-phase flow phenomena are inherently three-dimensional and unsteady on a fluid-element scale, for example, in a droplet and the gas field around a droplet. In addition, turbulent fluctuations are ignored within all practical problems, and space-time averaging of conservation equations filters them out. Only the average three-dimensional, time-varying characteristics of the flow over the size of the numerical mesh cells can be retained. Within this average framework, the terms for the production and convection of fluxes of mass, momentum, and energy are all specified in a mathematically rigorous form (Yadigaroglu et al., 1990; OECD/NEA/CSNI, 1989). However, the averaging process generates unknown distribution coefficients for the convective terms and several unknown coefficients for the wall and interface transport terms. Distribution coefficients are usually taken to be unity. Transport or constitutive relationships are often assumed to be algebraic relationships involving parameters such as friction factors, drag coefficients, and HT coefficients, derived from the fully developed situation and related to one-dimensional variables such as the cross-sectional average velocity—all these have to be established empirically.

On the other hand, three-dimensional phenomena are sometimes very evident, even dominating the local thermal hydraulics, in large-size flow passages of the primary system, such as the UP in the reactor vessel, the core, and the downcomer in a PWR. Mixing, condensation, and CCF of liquid and vapor during ECC injection are typical examples that give rise to three-dimensional velocity, temperature, and void distributions in these regions. One-dimensional mixture and one-dimensional two-fluid models can only average these three-dimensional effects into their one-dimensional framework, which may penalize the scaling and generality of their prediction capabilities. Thus, uncertainties introduced by the one-dimensional approximation have to be assessed from near full-scale experiments.

UPTF results indicate that the larger the size of a flow passage (such as the reactor vessel downcomer, or the UP and core), combined with the nonuniform geometric distribution of water inlet and flow outlet, the more multidimensional the flow distribution becomes (Glaeser, 1992). Fluids of different densities and different directions of flow find it much easier to establish their own preferred flow paths in which the flow is predominantly cocurrent rather than countercurrent.

The two-fluid code TRAC allows a three-dimensional representation of the pressure vessel. However, interface transfer like shear stress, heat, and mass transfer between the fluid in the neighboring fluid is included only where there is bulk flow exchange between the cells (Glaeser, 1990a,b). The averaging process within the cells is described above.

One of the major sources of uncertainty in current thermal-hydraulic computer code simulation is the fact that it is not adequately known for all flow situations, how to describe the local interfacial area density and the momentum, mass, and energy transfer rates at those interfaces, irrespective of the computational cell size.

All the above considerations specifically address the geometric configuration of the primary circuit of PWR. Nevertheless, these are essentially valid for BWR, for which case the following aspects can be added.

- The multidimensional behavior in the lower plenum occurs as in the case of PWR with the additional complexity arising from the presence of the control rod guide tubes (CRGTs) with related CCFL and delayed flashing phenomena.
- The multidimensional behavior in the core originates from the interaction of single channels, where the flow can be locally two-dimensional, from channel to channel, and involve multidimensional interchange with the upper and lower plena. Additional details can be found in Section 6.3.11.2.
- The multidimensional behavior in the UP is less complex than in the PWR case because of the geometric configuration that allows only vertical flows from the core to separators; no horizontal connections corresponding to PWR hot legs are present. The injection of spray systems inside this zone creates thermodynamic situations that are similar to those created in the case of hot-leg injection in a PWR (see also Section 6.3.9.1). Multidimensional phenomena may take place in the steam dome in the case of a main steam line break (MSLB) or during the period of actuation of the fast depressurization system (ADS, automatic depressurization system), also causing eccentric loading of the internal structures. These are not further considered here because of their low relevance in terms of safety and the global behavior of the system.
- The geometric dimensions of the vessel downcomer (thickness and flow area) are much greater in a BWR than in a PWR. This makes the 3-D behavior of the BWR vessel downcomer less important than it is in the PWR. The relatively small ratio between the feedwater flow area and the downcomer cross-sectional area (compared with the ratio of cold leg area to downcomer cross-sectional area in a PWR) also reduces three-dimensionality. Three-dimensional effects can occur in the case of the break of one recirculation line (for additional details, see Section 6.3.11.3) while limited or no (in the majority of BWR) 3-D effects are expected following the injection of emergency cooling system water.

The phenomena in this section also incorporate the PH of temperature stratification. Temperature stratification in the water in the LP and the U-tubes of the SGs are not specifically addressed in this section but may influence system behavior if this water is pushed to other locations (e.g., into the core). Its relevance to nuclear safety is considered low.

6.3.11.1 Global multidimensional fluid temperature, void, and flow distribution—upper plenum

The flow in the UP is three-dimensional. During the blowdown period of an LBLOCA, for example, water is delivered from the PRZ and the SG primary side into the UP via the hot legs.

In the case of hot-leg ECC water injection, water is injected toward the UP via a special nozzle at the bottom of each hot leg. The water jet is directed out of the hot legs horizontally into the UP. The ECC water impinges on the columns (guide tubes) in the UP and is directed downward.

The water downflow location in the UP determines the downflow location through the upper-end box, including the UTP, into the core region. This observation has been made in UPTF and other large-scale test facilities like SCTF and CCTF (Damerell et al., 1993a,b; MPR Associates, 1990). The magnitude of cross-sectional ECC water downflow in the UP and the core is important for partially cooling the core during the end of the blowdown and refill period of a postulated LOCA. Multidimensional effects are also of prime importance in estimating fuel rod damage during the heat-up period. The UP multidimensional flow is one of the most important phenomena during refill and reflood with hot leg or UP ECC injection.

The liquid inventory in the UP is relevant for steam binding if no subcooled ECC is injected into the UP (only cold leg or downcomer injection). The UP liquid inventory for all injection designs may be important for the washout of fission products released from defective fuel rods. (Fission product transport and distribution themselves are not considered in this chapter.)

The PH is identified as S-24-GM4 (or No 44 in the list of 116 THPs) by Aksan et al. (2018a), and D'Auria et al. (2021b).

6.3.11.2 Global multidimensional fluid temperature, void, and flow distribution—core

During normal operation of a PWR, the core is completely covered with water whereas in a BWR the steam is produced in the core. Even when the core is liquid-covered, flow patterns, which are essentially three-dimensional rather than 1-D, can be set up. In the case of an accident, the water level may drop and the cooling of the core has to be reestablished by means of the ECC system.

Cold-leg injection leads to a core refill from the bottom upward. The combined cold- and hot-leg ECC injection or hot-leg-only injection leads to local top-down cooling of the core in addition to refilling from the bottom. The hot-leg ECC injection water flows down within preferred flow paths. The precursory cooling in these water downflow locations is higher than in the steam-dominated core regions. The precursory cooling may reduce the possible fuel rod damage during the heat-up period.

A three-dimensional distribution of water downflow and steam upflow establishes itself above the swell level in the core. After the start of the bottom reflood, the void fraction and flow distribution determine the further core cooling.

There is also some local top-down water flow in the core from the PRZ and the SGs during blowdown.

In the case of BWR the different parallel channels bounded by the fuel boxes may cause different flow regimes (Dix, 1983): three of these may occur at the same time during a typical LOCA transient. In the peripheral region of the core, where the coolant is sprayed, and in the bypass region, where there is almost no power production, the subcooled liquid may flow downward from UP to LP (CCFL breakdown condition), thus condensing almost all the steam eventually produced. In other zones of the core, a few fuel channels vent the flashing vapor from the LP which results in the cocurrent upflow condition of steam and entrained droplets in those channels. At the same time, the remaining parts of the fuel channels may experience typical CCFL conditions at the bottom, with level formation inside the fuel bundles (see Section 6.3.10.2).

Looking at the overall scenario, a typical horizontal "sandwich" situation is expected with level formation in the LP (the steam is produced due to boil-off caused by the depressurization), core (the steam is produced by the evaporation of the coolant and by depressurization), and UP (the liquid is produced by the deentrainment of droplets from the core and by the injection of the spray systems).

The cooling of the core using available remaining water or by injecting subcooled ECC water is the most essential safety aspect. The increase in water inventory through ECC injection determines the HT from the fuel rods.

Three-dimensional flow is important for combined injection (into hot and cold legs during refill) and is ranked important for cold-leg injection only during reflood in LBLOCA (OECD/NEA/CSNI, 1989; Wilson et al., 1990).

The 3-D flow behavior is important in the case of BWR considering that the flow patterns discussed above constitute a typical configuration for the core.

The PH is identified as S-21-GM1 (or No 41 in the list of 116 THPs) by Aksan et al. (2018a), and D'Auria et al. (2021b).

6.3.11.3 Global multidimensional fluid temperature, void, and flow distribution—downcomer

During the blowdown and LP refill of an LOCA, the ECC water injected into the cold legs flows down the downcomer below the injection cold legs, if the steam upflow is sufficiently low. With increasing steam upflow the water downflow may be limited (see Section 6.3.10.5).

If the collapsed water level is higher than the bottom of the downcomer (above the lower edge of the downcomer skirt) the steam upflow from the core is stopped.

During core reflood, the water level in the downcomer approaches the bottom of the cold-leg height. An azimuthal steam flow is directed from the intact cold legs to the broken cold leg within the downcomer, which may entrain water from the downcomer through the broken cold leg. Consequently, the downcomer liquid head, which is the driving force for core reflood, is reduced.

The distribution of the down-flowing ECC water in the downcomer has an influence on the water delivery to the LP which may be limited by steam upflow.

Multidimensional effects are ranked as being important during LBLOCA refill (OECD/NEA/CSNI, 1989; Wilson et al., 1990).

A reduction of the downcomer liquid head may reduce the driving force for core reflood.

During subcooled ECC water injection into hot water, the temperature distribution in the downcomer may be of interest for evaluating the PTS stress on the reactor vessel wall.

The PH is identified as S-22-GM2 (or No 42 in the list of 116 THPs) by Aksan et al. (2018a), and D'Auria et al. (2021b).

6.3.11.4 Global multidimensional fluid temperature, void, and flow distribution—steam generator secondary side

In the SG secondary side, there are several areas where multidimensional effects are of high importance and will influence the overall behavior of the primary to secondary side HT. The flow rate on the secondary side is determined by NC, and small changes to void and temperature distributions can have a substantial impact on the flow conditions and hence on the local HT. This has been demonstrated, for example, at feed and bleed operation of the auxiliary feedwater flow in the case of loss of main feedwater, where subcooled auxiliary feedwater is mixed with the vapor in the SG downcomer (D'Auria et al., 1993).

At the top of the downcomer subcooled feedwater is mixed with saturated liquid which is returning from the separators and dryers. Some vapor will be present in this separated liquid flow (carryunder) which will be condensed on contact with the feedwater. The feedwater is provided fairly uniformly around the top of the downcomer resulting in rather uniform downcomer flow conditions.

At the bottom of the downcomer, the flow enters the riser where the liquid is heated to saturation and boiling starts. Baffle plates will force the flow to pass the tube bundle in a transverse direction along with its main axial direction. This increases the HT coefficient on the tube's outer surface.

The two-phase flow enters the steam separators located at the top of the riser. The number of separators can vary from only a few to a vast number in parallel. They all rely on separation due to centrifugal forces. The two-phase mixture is set into rotation through swirl vanes at the separator inlets. The liquid will then be "spilled" over the separator's top edge and will fall back to the downcomer.

The vapor with some liquid droplets (carryover) then enters the dryers in the steam-generator dome. In the dryers, the remaining liquid is separated from the vapor and is feedback to the downcomer. The vapor passes through the SG outlet nozzle and enters the steam line.

Multidimensional effects prevail at almost every location in the flow path due to the system geometry, and the mixing of liquid and vapor. However, the most important parts are the riser and the steam separators.

In the riser, the baffle plates force the two-phase flow into a zigzag flow path. Behind the plates and tubes eddies will be created, which will strongly influence the two-phase flow pattern and the interfacial area distribution. A dispersed regime will likely be promoted even under conditions where other flow regimes normally would prevail. As mentioned above, HT on the outer surface of the tubes will be augmented over ordinary pipe flow because of the cross-flowing nature of the flow.

In the separators, three-dimensional effects will be introduced through the circulation of the flow. These effects will govern the actual separation of the two-phase mixture and will determine the carryunder and carryover fractions in the separated flow. The interfacial area and the slip are of essential importance for the separation behavior. The presence of a preheater in some SG designs produces additional multidimensional effects.

During intermediate-break LOCAs and SBLOCAs, it is essential to keep the heat sinks as intact as possible to mitigate the consequences of the transients. However, the detailed conditions of the SG secondary side are not of crucial importance in this respect, and neither are the two-phase multidimensional characteristics of the secondary flow in many cases. It has been found, though, that the conditions in the downcomer in some cases have influenced the overall SG behavior.

The PH is identified as S-23-GM3 (or No 43 in the list of 116 THPs) by Aksan et al. (2018a), and D'Auria et al. (2021b).

6.3.12 Heat transfer

The HT between wall structures and fluid is an essential heat transport mechanism in LWR for transporting the heat generated in the core to the turbines. In the first step, the heat is transferred in the core from fuel to fluid. The SG tubes transmit the heat between the high-pressure primary side and the low-pressure secondary side. The safety aspects of the HT are mostly related to the core region.

The fuel cladding temperature is usually near the saturation temperature of water. During an accidental increase in power or decrease in flow and pressure, a deterioration in the HT process is possible. The surface temperature increases to such a high level that the heated surface can no longer support continuous liquid contact. This PH is called a boiling crisis or dryout depending on the mechanism leading to the event. The key question is how efficiently the deteriorated HT mechanism can remove heat generated in the core after the reactor shutdown. The HT problems are both global in the case of core uncovery and local for high-power rods.

Due to the inadequate understanding of HT mechanisms, an empirical approach is often necessary. This approach requires the derivation of empirical HT correlations for each HT configuration, resulting in numerous correlations for each HT mode. Some correlations are based on the phenomenological basic formulations for explaining the parametric tendencies, but typically some key parameters are required for fitting against the experimental data.

At present, quite a large selection of HT correlations exists in the world literature and many correlations are derived from a large database. Most of the HT correlations are based on tube data since the simplest test geometry is a heated tube. Extrapolation of the tube correlations to other geometries (bundles, annuli) is common practice in reactor safety analysis. In principle, no new SETs are needed for the more accurate calculations of HT in expected incidents. The problem for model development is inadequate measurements of local parameters needed for developing models of HT mechanisms for two-fluid codes.

6.3.12.1 Natural or forced convection in the core, in the steam generator, and at structures

Either FCO or natural convection (NCO) can occur when single-phase liquid, vapor, or two-phase flow is present. During normal operation of a PWR the FCO to single-phase water is the main HT regime in the primary coolant system. In the core, subcooled boiling may also occur. During normal operation of BWR FCO, single-phase liquid in the core occurs only in the lower entrance region, while two-phase forced convective HT is the HT regime in the upper part of the core.

NC of the single-phase water may occur if the reactor is scrammed and the main coolant pumps are switched off due to different water densities in the hot part (core, UP, and hot part of SGs) and the cooler part of the primary system (cold side of SG U tubes and downcomer). In this situation forced or NCO HT (depending on the Reynolds number of the flow) can occur at any of the components involved.

Two-phase natural convective HT is important for the HT from the core to the SG secondary side in the case of small leak scenarios.

Two-phase convective HT is described in Sections 6.3.11.2–6.3.11.4.

Cooling of the core and the primary side fluid inventory via the SG has the highest priority in nuclear reactor safety. Apart from the core and SG HT an additional heat input may be relevant from hot structure material to the primary side water or steam during blowdown, for example, and the heat loss from the primary side to the environment through the vessel and piping structure material may affect the system behavior during long transients.

The PH is identified as S-26-HT1 (or No 47 in the list of 116 THPs) by Aksan et al. (2018a), and D'Auria et al. (2021b).

6.3.12.2 Subcooled and nucleate boiling heat transfer

The transition from single-phase liquid HT to nucleate boiling takes place once the wall temperature exceeds the temperature for the onset of nucleate boiling. According to one hypothesis, the condition for nucleation and thus the onset of subcooled boiling is the point where the convective heat flux and boiling heat flux are equal. A more refined treatment of nucleation can be derived by considering the temperature profile in the region adjacent to the wall.

Fully developed subcooled boiling does not exist immediately after the surface exceeds the saturation temperature. A region of partial boiling exists between the subsaturation zone and fully developed subcooled boiling.

The heat flux/surface temperature relationship in the fully developed region is independent of velocity and subcooling and may be correlated by the equations of the type obtained from pool boiling experiments, where the constant coefficients are mainly functions of the pressure, and the heat flux is a function of the wall superheating with an exponent $n = 2, \ldots, 4$.

Saturated nucleate boiling starts when the bulk liquid temperature equals saturation temperature. In saturated boiling the HT consists of two contributions, a nucleate pool boiling term and a convective term. The Chen correlation (Hsu and Graham, 1986) is an example of this kind of formulation. The FCO contribution is multiplied by a two-phase factor F and the nucleate boiling term by the suppression factor S when the local quality increases.

The nucleation term represents the normal nucleation caused by the nucleation sites on the wall. The convective term is an effect of the forced vaporization through the annular liquid layer on the wall.

The nucleate boiling region is an essential part of the HT curve. The onset of nucleate boiling has an effect on the void profile and thus on the nuclear power profile in the BWR core. In the PWR core, no bulk boiling is expected, but subcooled nucleate boiling (SNB) can occur in the top third of the core. The possible local voiding thus generated has the unwanted side effect of disturbing the neutron power profile.

Nucleate boiling is particularly relevant in special cases such as ATWS in BWR.

The PH is also identified as S-26-HT1 (or No 47 in the list of 116 THPs) by Aksan et al. (2018a), and D'Auria et al. (2021b), i.e., see Section 6.3.12.1.

6.3.12.3 Departure from nucleate boiling/dryout

A large number of terms have been used in the literature to denote the occurrence of CHF: burnout, dryout, boiling crisis, CHF, and DNB. The term DNB is usually reserved for the occurrence of CHF in the subcooled region caused by microlayer evaporation under a bubble. Typical mechanisms creating this kind of situation are the depressurization of the flow channel increasing the wall superheating, decreasing the liquid subcooling, and perhaps creating a bulk boiling in the liquid together with a strong nucleate boiling.

Dryout refers to the CHF occurrence caused by evaporation of the wall liquid film in the annular flow regime. When three fluid fields are considered in an annular flow pattern, vapor, liquid film, and droplets entrained from the liquid core, the hypothesis is that the boiling crisis in annular film occurs when the liquid film is depleted by evaporation and entrainment. A typical situation creating this kind of CHF pattern is a stationary boiling channel in stationary forced flow conditions, where an annular flow pattern exists.

The following observed asymptotic trends are valid for the CHF.

- local subcooling increases CHF,
- CHF approaches pool boiling CHF at low flows,
- CHF approaches zero for quality (void fraction) approaching unity,
- CHF reaches a maximum in the pressure (P) range of 3–4 MPa,
- CHF approaches zero when P approaches critical pressure,
- the existence of spacers in the bundle increases the CHF rate.

For the steady-state operation of the reactor, the avoidance of CHF is one of the main criteria for acceptance of the plant power level. Taking into account the peaking factors for the maximum power fuel bundle and for the maximum power fuel rod, no

CHF is allowed during steady-state conditions. Rods experiencing CHF may be damaged.

The criterion is also used in anticipated transients when the plant protection systems are operating normally.

In accidents including LOCA or in transients with an additional failure of plant protection systems, CHF at least in the hottest rods may be expected, and as a consequence, the rods may be damaged causing a release of radioactivity to the coolant. During LOCA the occurrence of CHF initiates the core heatup period.

The PH is also identified as S-26-HT1 (or No 47 in the list of 116 THPs) by Aksan et al. (2018a), and D'Auria et al. (2021b), i.e., see Section 6.3.12.1. Current predictive capabilities are discussed by D'Auria et al., (2021b).

6.3.12.4 Post–critical heat flux heat transfer

The post-CHF regime means in general the regime where the hot wall temperature or lack of coolant does not allow liquid to contact the surface. The HT models considered under this title are transition boiling, film boiling, HT to the droplet-dispersed mixture, and convection to single-phase steam. Above the dryout type of CHF, the coolant in the flow channel consists of a mixture of droplets and steam when the annular liquid film has evaporated. After the DNB type of CHF, the hot surface temperature prevents the turbulent mixture of coolant and steam from wetting the wall. During quasi-stationary conditions in the core after core uncovery only steam exists above the swell level. All these modes are considered post-CHF HT.

Transition boiling is a special HT mode because the heat flux generally declines with an increase in surface temperature. The HT area is bounded by the CHF and minimum heat flux. Transition boiling can be understood as a combination of unstable film boiling with low HT capability and nucleate boiling with high HT capability alternating on the heated surface. In high-quality regions, most transition boiling is due to convection to the droplets impinging on the superheated steam layer in the vicinity of the wall. The increase in mass flux increases the transition boiling rate, and liquid subcooling has a similar effect.

During film boiling, the heated surface is cooled by radiation, by FCO to vapor, and by the interaction of the liquid and the heated surface. The vapor may become highly superheated, and its temperature is controlled by HT from the wall and to the liquid phase in the chain wall-vapor-liquid. The liquid in the form of a dispersed spray of droplets is encountered at void fractions more than 80% (liquid deficient flow regime) and in the form of a continuous liquid core (inverted annular film boiling) usually encountered at void fractions below 30% (inverted annular flow regime). The transition regime between these two flow regimes takes place usually in the form of a churn turbulent liquid dispersed flow.

Of the film boiling regimes, the liquid dispersed regime with droplets (after dryout, above liquid swell level) is most commonly encountered and has been well studied. In a phenomenological model, it is assumed that HT takes place in two steps: from the heated surface to vapor and from the vapor to the droplets. The axial gradients of the droplet size are evaluated from the heat balance data and the droplet size is used for the interfacial drag and HT correlations. The wall temperature is found from the vapor temperature using a superheated steam HT correlation. In the model, a direct droplet-wall interaction can be modeled, as well. The droplet-dispersed HT regime exists during ECC above the swell level as a consequence of droplet entrainment. In BWR HT, the cooling effect of spray droplets is essential.

Phenomenological models have also been developed for the inverted annular flow regime. In the basic formulation, the balance equations are written for the vapor film between the surface and liquid core. The wall-steam friction is calculated for the laminar or turbulent steam layer and an average HT coefficient of the length is integrated from the beginning of the film to the start of unstable wave formation in the liquid-steam interface.

Phenomenological correlations are also developed to predict the degree of thermal nonequilibrium. The degree of the nonequilibrium is expressed by actual and equivalent equilibrium qualities.

For steady-state operation, the prediction of the post-CHF HT is not essential, because the existence of CHF is itself interpreted as a cause of the fuel damage. For BWR some attempts have been made to allow for local CHF and to demonstrate safety by using post-CHF HT models. After some fuel rod damage, as a consequence of CHF in the BWR bundle, the avoidance of CHF has been retained as an acceptance criterion. In transient analysis, the evaluation of the post-CHF HT is needed but also in this case the existence of CHF in the hot channel/hot rod is interpreted as a cause of the rod damage.

In accidents with a loss of reactor coolant inventory, the HT in the post-CHF mode controls the maximum temperature of the fuel. If the temperature exceeds 800°C, the heat production by cladding oxidation becomes a significant contributor to the further core heat-up rate. During a quasi-stationary core uncovery, the HT capability of steam depends on the boiling rate below the swell level and in this case, the maximum core temperatures depend on simple energy balance rules.

During ECC after an LBLOCA, the cooling by the droplet-steam mixture is essential both for bottom reflooding and UP spray injection. Quite soon after the reflooding water enters the core, the temperature increase in the upper part may be stopped due to the steam cooling caused by the droplets. In addition, as soon as spray droplets impinge on the hottest parts of the core, the cladding temperature starts decreasing. Thus the post-CHF HT is the main contributor to limiting maximum cladding temperatures in the core and preventing damage by oxidation.

The PH is also identified as S-26-HT1 (or No 47 in the list of 116 THPs) by Aksan et al. (2018a), and D'Auria et al. (2021b), i.e., see Section 6.3.12.1.

6.3.12.5 Radiation heat transfer

Thermal radiation will transfer energy both from surface to surface and from surface to the two-phase flow. The exchange of thermal radiation with the two-phase flow will mainly consist of absorption in the two-phase flow, mainly in the droplets. Due to the relatively low temperature the emission of the two-phase mixture is negligible. The radiation HT becomes important in addition to other HT modes when the structure temperature locally exceeds the saturation temperature by 200 K. These kinds of high temperatures are possible only in the core with pure steam, with a droplet-steam mixture, or with an inverted annular flow regime. Typically the net radiative heat flux streams from the highest temperature regions into colder parts, being partially absorbed in the fluid.

The radiation flux from the solid surface consists of two contributions: a radiated flux which is a function of the surface temperature to the fourth power, and a partial reflection of the incoming radiation.

Steam and liquid emit radiation in principle, but the emission is small compared to that of surfaces. Both liquid and steam absorb radiation. In practice the absorption by steam is negligible. The density of liquid water is sufficient to absorb all radiation in the space between fuel rods, for instance, if an inverted annular flow regime with a continuous liquid core fills the flow channel. For droplet dispersed mixture only a part of the radiation is absorbed and the rest streams to the next surface.

Before emergency cooling water enters the core during an LOCA during which the core has become uncovered, the radiation HT between structures is the most significant HT balancing the local temperature differences. The net heat flux is directed from the hottest rods to the cooler ones. In the BWR bundle the channel wall finally absorbs the heat streaming from the central parts of the fuel element. This heat flux chain between different rods and channel walls creates inside a fuel bundle a temperature profile, where the central rods are at a higher temperature than the peripheral ones. Due to its strong dependence on the fuel temperature the radiation may effectively prevent the temperature rise and prevent cladding oxidation.

The PH is identified as S-27-HT2 (or No 48 in the list of 116 THPs) by Aksan et al. (2018a), and D'Auria et al. (2021b).

6.3.12.6 Condensation in steam generator tubes and on structures

The condensation of steam on structures whose surfaces are below saturation temperature can be an efficient means of removing heat from a coolant system. On vertical and inclined surfaces the condensation of steam produces a thin film of liquid which is continuously removed by gravity. The rate of HT is dependent on the subcooling of the surface, and the thickness that the liquid film attains—HT is limited by conduction through the film. The HT coefficient is also affected by the steam flow rate to some degree, through its influence on the film thickness and state. The presence of small amounts of noncondensable gas in the vapor can result in a drastic reduction in condensation rates under stagnant conditions. This is because the flow of steam toward the condensing surface produces an increase in the concentration of the noncondensable gas at the interface, and the presence of the noncondensable lowers the saturation temperature of the steam.

In parallel channels, condensation can give rise to instabilities with different tubes going through a refilling and siphoning cycle at different times (D'Auria et al., 1990).

Condensation within SG tubes is an important mechanism for the removal of decay heat. Provided that the secondary side of the SG is cooled the primary pressure will closely follow the saturation pressure of the secondary side when the SG tubes contain steam. This enables reflux cooling to remove decay heat satisfactorily following an LOCA, and permits primary depressurization via secondary side cooldown when the primary circuit is voided above the core. Because of the large area available within the SG tubes, however, the exact magnitude of the condensation HT coefficient is not critical at decay heat levels. Only when the secondary side coolant level was very low or substantial quantities of noncondensable gas were blocking the SG tubes would the condensation rate be expected to limit the overall HT.

Condensation on external structures such as pipe walls or within the PRZ steam space can occur in transients where the pressure rises, with the effect of reducing the rate of pressure rise.

The PH is identified as S-28-HT3 (or No 49 in the list of 116 THPs) by Aksan et al. (2018a), and D'Auria et al. (2021b).

6.3.13 Quench front propagation/rewet

The reflooding and quenching phenomena are important for LWR. Indeed, it is usually postulated that the core is uncovered and overheats due to decay heat from the fission products and the energy stored before LOCA during the so-called blowdown phase of the LOCA. In PWR, the core is recovered, that is, the overheated rods are quenched and adequate HT is reestablished, by reflooding from below. ECC injection at the top of the core, or combined top and bottom injection are, however, also practiced in certain types of PWR. In BWR, the hot fuel rods may be quenched by a spray that forms a liquid film flowing down and cooling the rods as well as the walls of the "channels" containing the rod bundles and the water rods in the rod bundles.

Water is forced into the core by the ECCS via bottom flooding (PWR) or top spray (BWR) to stop the overheating of the fuel rods and reestablish cooling. Otherwise, cladding oxidation, zircaloy-water chemical reaction, or clad melting and the consequent release of radioactive products can occur.

Reflooding refers to a particular mode of postburnout cooling of a hot channel by refilling it with coolant. Quenching or rewetting of the hot surface occurs during reflooding. Quenching refers to the transition from a mode of HT characterized by total or almost total absence of liquid contacts with the wall to one where the wall is essentially wetted by the liquid. The HT coefficient increases dramatically following quenching.

As a result of the high temperatures attained by the clad before the emergency coolant arrives, water does not initially wet the hot clad surface. Rewetting or quenching of hot surfaces occurs when the coolant reestablishes contact with the dry surface. The surface temperature corresponding to the achievement of liquid-solid contact is the rewetting or quenching temperature. The temperature of the fuel pellets and fuel clad is reduced by heat conduction and convection to the coolant. As the coolant rises in a hot channel or the overheated nuclear fuel rod bundle, complex HT and two-phase flow phenomena take place and the succession of regimes moves gradually up the rod bundle channels. The hot surfaces along the channels experience in turn free- or forced-convection cooling by steam, dispersed-flow film boiling, inverted-annular film boiling, transition boiling, nucleate boiling, and finally single-phase convection to the liquid. Almost, all the HT modes listed under HT (Section 6.3.12) are encountered during the reflooding and quenching phase.

During reflood multidimensional flow patterns occur in the core and UP due to: Flow rates and flow regimes being such that gravitational forces are of the same order as inertial forces; nonuniformities in core power distribution and differences in resistance to flow through the intact and broken hot legs tend to promote multidimensional effects; as the flow passes through the UP it must flow around several structural elements and additionally the flow behavior of a collected pool in the UP may be highly three-dimensional.

Another important PH observed in integral PWR system LBLOCA experiments is core-wide cooling of the fuel cladding and quench during the blowdown phase. This PH is important regarding the heat up of the core because it removes a large part of the stored energy from the fuel during the early phases of the transient. The blowdown-phase quench was caused by the hydraulic response of the primary system during the transition from subcooled to saturated choked flow at the break and by the operational characteristics of the primary system coolant pumps. The rapid cooling of the cladding is primarily a result of low-quality, high upward core flow at a 7 MPa pressure range. The correlations currently used in the advanced codes cannot handle adequately such blowdown quench phenomena and no directly applicable separate effects—type experimental database exists.

QF propagation/rewet on fuel rods and, also channel walls and water rods (BWR) will be presented below, in some detail. Current predictive capabilities for this PH are discussed by D'Auria et al. (2021b).

6.3.13.1 Quench front propagation/rewet on fuel rods

QF progression has been an interesting physical PH studied intensively over the last 4 decades. After an LBLOCA has occurred the core is empty and adiabatic heat up is possible. Emergency coolant injection by bottom reflooding creates a rising quenching front, and upper head injection or core spray injection a falling film, and sputtering front. Many experiments in the 1970 and 1980s were done for understanding and visualization of the reflooding front movement. Visual demonstrations with an overheated, red-colored rod enclosed in a glass tube demonstrated clearly that quenching to the saturated temperature takes place at a distance of a few millimeters.

In the boiling curve, QF propagation means a movement from the film boiling HT mode through transition boiling to the nucleate boiling regime. In a quenching front, the axial temperature gradient of the cladding may be so large that a significant part of the heat in the unwetted part is removed by the axial conduction in the cladding. For a fast front movement, this gradient may be quite different between the inner and outer cladding surfaces. But other mechanisms may also support conduction-controlled quenching or be a main contributor to the cooldown close to the saturation temperature. These are transition boiling caused by the collapse of the liquid-vapor interface in the inverted annular film boiling regime and droplet impingement through the steam layer in the transverse direction toward the wall.

During the LBLOCA the first cladding heat up is caused by the stagnation of flow going through the core. Soon after this large mass flow rates may exist in the core first in the negative flow direction and later in the positive direction. This high mass flux rate may be sufficient for cooling the cladding along the whole core length without a specific quenching front (Aksan et al., 1989).

The conduction-controlled quenching mechanisms for a rising reflooding front and a falling film front are quite similar. In the wetted region, the heat is transferred in both cases by nucleate boiling and later by convection. In the unwetted part, the heat is removed by convection to vapor or to droplet-vapor mixture or by inverted annular flow film boiling during bottom reflooding. Precursory cooling effects support the QF movement.

The rewetting characteristics of the overheated core after the large LOCA was one of the most interesting research topics in the 1970 and 1980s and still has a significant influence on acceptance criteria in licensing and probabilistic safety assessment safety analyses. The main interest is related to the maximum temperature in the core, but this turnover temperature is determined by the liquid dispersed flow well before quenching. Depending on the amount of water available the cooldown takes place earlier or later.

The large temperature gradient in the cladding gives rise to mechanical stress on the cladding, and it may affect fuel damage and radioactivity leakages. The rapid temperature drop is also associated with strong steam generation and this may have an effect on system characteristics including.

- liquid entrainment rate,
- CCFL in the UTP,
- steam binding in the SG,
- multidimensional flow distribution in the core.

The PH is identified as S-49-QF1 (or No 88 in the list of 116 THPs) by Aksan et al. (2018a), and D'Auria et al. (2021b).

6.3.13.2 Quench front propagation/rewet on channel walls and water rods (boiling water reactor)

Following hypothetical complete emptying of the core in BWR-type reactors (e.g., as a consequence of an LBLOCA in one circulation line), both rod cladding and fuel box surfaces attain temperatures well above the saturation temperature of the fluid. (Usually, system pressure is around a few tenths of an MPa in these conditions.)

At a given time, the reflood process starts, caused usually both by the injection of cold liquid through the LPCI (low-pressure coolant injection) and the LPCS. The former system injects water into the downcomer or inside the core bypass region; in both cases, the liquid reaches the LP and enters the core from the bottom. In the latter system, the liquid is sprayed at the top of the core and creates a QF traveling from top to bottom.

Phenomena like CCFL, entrainment, PD, etc., are very important for predicting the transient evolution in these conditions.

The reflood PH inside each fuel bundle follows the same mechanisms described in Section 6.3.13.1. The fuel box walls play a special role that can be outlined as follows.

(a) Because there is essentially no power production and initial temperatures before reflood are slightly lower than cladding temperatures, reflood occurs earlier along the fuel box walls than adjacent to the fuel rods (the thickness of fuel boxes, which is far greater than the rod cladding thickness, does not appear to have a significant role in relation to the different rewet times between the rods and the fuel box walls).
(b) Because they cool earlier, the walls constitute a heat sink for the rods and make the reflood in the rods themselves easier; radiation HT plays a role in this regard.
(c) As an additional consequence of the first item, the liquid injected by the LPCI or sprayed by the LPCS can penetrate the LP preferentially along the fuel box walls and contribute to the reflooding of the core rods.

The local mechanism of reflood, including the two-dimensional nature of the PH, is the same as described in Section 6.3.13.1.

The rewetting of fuel box walls makes an important contribution to core reflood. In this sense, the present PH is as relevant as the reflood in the core rods. The same considerations as in the previous paragraph apply in this case.

The PH is identified as S-50-QF2 (or No 89 in the list of 116 THPs) by Aksan et al. (2018a), and D'Auria et al. (2021b).

6.3.14 Lower plenum flashing

When a pool containing saturated subcooled liquid is subjected to depressurization, the system pressure acting on the liquid itself may reach the saturation pressure corresponding to the initial liquid temperature. In this case, a more or less violent vaporization (flashing) may occur depending on the depressurization rate. (High-purity water bounded by a surface with negligible nucleation sites—a very flat surface—may behave differently; in particular, several dozen degrees of liquid superheating can be expected. This is not of interest in practical reactor situations.)

During flashing (nearly) homogeneous nucleation occurs in the bulk of the liquid, and the gas bubbles produced may entrain a large mass of liquid toward the upper parts of the pool (LP in this specific case).

The PH is significantly affected by the geometry of the system, with shallow pools giving rise to more violent flashing than tall ones containing the same volume of liquid.

The liquid entrained in the steam flowing toward the upper part of the vessel both in BWR and PWR systems, in most of the situations of practical interest, may give a significant contribution to core cooling. In the case of LBLOCA, the PH may contribute to the early core quench observed in the experiments performed in ITFs. The LP vaporization occurring late in the transient in the case of a large break may lead to the occurrence of flooding, thus preventing or limiting the flow of injected liquid into the downcomer. The last PH is more relevant for PWR systems. In the case of BWR, rapid boiling in the LP may prevent the fallback of liquid from the core making rod cooling more efficient.

The PH is identified as S-41-LPF (or No 67 in the list of 116 THPs) by Aksan et al. (2018a), and D'Auria et al. (2021b).

6.3.15 Guide tube flashing (boiling water reactor)

As indicated in the heading, the PH is typical of a BWR. From a phenomenological point of view, the substance of Section 6.3.14 is also applicable here. The particular geometry and the presence of relatively small holes (orifices) at the top of the guide tubes govern the development of the PH and make it distinct from LP flashing.

The liquid initially contained in the CRGTs is slightly more subcooled than the remaining LP liquid. For this reason, flashing during a given transient is delayed with respect to LP flashing. Level swell is also limited by the abovementioned orifices.

Given the above, this PH could be considered less important than the previous one as far as nuclear reactor safety is concerned. Nevertheless, it may affect core thermal-hydraulic behavior later in a transient (with respect to LP flooding).

The PH is identified as S-13-CRGT (or No 29 in the list of 116 THPs) by Aksan et al. (2018a), and D'Auria et al. (2021b).

6.3.16 One- and two-phase impeller pump behavior

In single-phase conditions the dimensionless pump characteristics are almost independent of the fluid (liquid, gas, steam). The small variations are mainly due to viscous properties connected to recirculation losses, shock losses on the impeller, and fluid friction (Comolet, 1976).

In two-phase flow the torque and head characteristics are affected by so-called degradation which is a function of void fraction, pressure, and flow rate. The main causes of the degradation seem related to phase separation in the impeller due to centrifugal forces, with liquid phase acceleration and relative deceleration of the gas or steam. The pre- and postrotations of fluid in the inlet and outlet pipes play a part in this issue. All these phenomena depend on the type (centrifugal, axial, mixed flow) and geometry of the pump.

For all LOCA with running pumps, the pump performance is important. For large LOCA the behavior of the intact loop pumps plays a role in determining the position of the stagnation points in the primary circuit.

For SBLOCA the pump behavior plays a role in steam-water mixing and core cooling.

The PH is identified as S-29-IMPU (or No 51 in the list of 116 THPs) by Aksan et al. (2018a), and D'Auria et al. (2021b).

6.3.17 One- and two-phase jet pump behavior

In normal operating conditions the jet pump operates as a single-phase ejector and is installed in the lower part of the downcomer (BWR-3, 4, 5, and 6). The driving flow provided by the external recirculation pumps is injected into the upper part of the ejector (converging section) and creates depressurization in the throat. The downcomer fluid is accelerated by the pressure difference created and mixes with the driving flow in the diverging part of the jet pump.

When recirculation pumps stop, as in most transients, the jet pumps behave exactly as ventures connecting the LP with the downcomer. In the case of main coolant pump cavitation, steam may flow out of the top nozzles completely preventing the correct operation of the device.

The component has a critical relevance in terms of nuclear reactor safety in the case of LOCA involving the break of a recirculation line. Owing to its axial elevation, the jet pump prevents mass depletion from the core region; so its mechanical integrity is of vital importance.

In the case of off-normal conditions, the jet pump simply introduces an extra PD in the flow path from LP to the downcomer and vice versa. Consequently, no new phenomena are expected in this case.

An exception may be the jet pump behavior during pump coast down. Possible unstable behavior may cause neutron flux instabilities or interact with the neutron flux. This last aspect may have some safety relevance.

The PH is identified as S-31-JPU (or No 56 in the list of 116 THPs) by Aksan et al. (2018a), and D'Auria et al. (2021b).

6.3.18 Separator behavior

Separators are installed in the vessel of a BWR and the secondary side of SGs equipped with U-tubes (PWR). The basic function is to separate the steam from the two-phase mixture coming either from the core (BWR) or U-tube risers (PWR), allowing (near) pure steam to flow to the turbine.

In all cases, the separation process is based on the larger inertia of the liquid phase with respect to the steam phase: the upward-flowing two-phase mixture is constrained to rotate. In this way, the steam can continue to flow upward while the liquid droplets impinge on the walls and are drained downward to the downcomer (of the main vessel or the SG in BWR and PWR cases).

Separation efficiency largely depends on the average velocity of the incoming mixture. At very low velocity the separator essentially behaves as a simple flow obstruction.

The component behavior has no direct relevance in terms of nuclear reactor safety for PWR and most cases for BWR. An exception, in the case of BWR, is that the PD introduced in the riser zone can facilitate the occurrence of core thermohydraulic-neutronic instabilities.

The PH is identified as S-51-SEP (or No 93 in the list of 116 THPs) by Aksan et al. (2018a), and D'Auria et al. (2021b).

6.3.19 Steam dryer behavior

The steam dryer allows the achievement of a vessel exit quality of near to one (0.999). The working principle is similar to the one adopted for the separator and exploits the different inertia of the light (steam) and heavy (liquid) phases. The two-phase mixture exiting from the separator (quality range between 0.98 and 0.99) flows horizontally in parallel flow paths created between adjacent undulated plates. The steam phase flows across, while the liquid droplets unavoidably impinge against the wall and are drained to the bottom of the dryer and toward the downcomer.

Considerations already detailed in Section 6.3.18 are also valid in this case. As in the case of the separator, the component behavior has no direct relevance in terms of nuclear reactor safety in PWR. The dryer does, however, contribute to the PD introduced in the riser zone of a BWR and therefore can facilitate the occurrence of core thermohydraulic-neutronic instabilities.

The PH is identified as S-55-SDR (or No 100 in the list of 116 THPs) by Aksan et al. (2018a), and D'Auria et al. (2021b).

6.3.20 Accumulator behavior

When, in an LOCA, the primary pressure in a PWR falls below the set point for accumulator injection, the driving gas within the accumulator can expand, delivering borated water (borate) coolant to the primary circuit. The rate of delivery of the flow depends on the resistance of the accumulator line (important in LBLOCA) and on the expansion characteristic of the driver gas (dominant effect for SBLOCA). The latter depends on the HT processes occurring between the gas and the accumulator structure and the liquid interface. The rate of accumulator delivery may also be influenced by feedback between the accumulator flow and the system pressure. Condensation on the injected liquid can lower the circuit pressure, increasing the flow and lowering the pressure further. On the other hand, accumulator injection can cause an increase in primary pressure (e.g., by reflooding the core and increasing heat input to the fluid), reducing the flow rate. The combination of local and system effects can give rise to instabilities in the accumulator flow rate.

When the liquid within the accumulator is exhausted the cover gas can flow into the primary circuit. The presence of a noncondensable gas in the primary can then modify the coolant distribution and the HT processes, particularly condensation. The emptying of the accumulator could be modified by the formation of a vortex at the outlet pipe.

Under the near adiabatic conditions occurring in the LBLOCA, the cover gas temperature can fall to subzero (°C) temperatures, leading to the possibility of frost forming in the accumulator lines.

The accumulators form a key element of the safety injection systems for dealing with LOCA. In an LBLOCA, they provide the principal means of limiting the peak clad temperature by refilling the vessel after blowdown and initiating reflood. In SBLOCA (particularly in beyond DBA scenarios where the HPIS is assumed to fail) the accumulators also play an important role in preventing or limiting core uncovery. The role of the accumulators in a range of accident scenarios is discussed in OECD/NEA/CSNI (1989).

The PH is identified as S-1-ACC (or No 1 in the list of 116 THPs) by Aksan et al. (2018a), and D'Auria et al. (2021b).

6.3.21 Loop seal filling and clearance

In most current PWR designs, the primary piping from the outlet of the SGs descends to a level roughly corresponding to halfway down the core before rising to enter the pump inlet from below. During an LOCA when the primary circuit is partially voided, liquid present in the U bend of this intermediate leg can form a barrier to the flow of steam around a loop. This liquid plug is said to form a loop seal. A pressure difference across the loop seal which is greater than the hydrostatic pressure arising from the height of the pump inlet side of the intermediate leg is necessary before steam can displace the loop seal.

The manner and extent of the displacement of the liquid from the loop seals depends on the steam supply characteristics. At reduced steam velocities and low liquid subcooling the expulsion of fluid will be incomplete. This aspect of loop seal behavior is closely connected with the PH of the CCFL. However, the pipe sizes of interest are very much larger than those studied in most SETs.

When the loop seals clear, their liquid contents are displaced into the reactor vessel, producing a rapid refill. The number of loop seals clearing, which depends for instance on the size of the break, affects the total amount of water added. How many loops need to clear is determined by the balance between the frictional PD around the cleared loops and the hydrostatic pressure difference corresponding to the depth of the loop seal.

It is possible for loop seals to reform after they have cleared. The backflow of ECCS from the cold legs through the pump is limited by the height of the impeller and the diffuser geometry but can occur if there is a general flow reversal, or if the level in the cold leg is sufficiently high. Condensate from the downside of the SG tubes also can contribute to the reforming of the loop seals.

The loop seals have a particularly significant effect on the behavior of a plant in a small cold-leg break LOCA. Until at least one of the loop seals has cleared steam produced in the core cannot reach the break, and the pressure in the primary system remains high, at a level determined by the secondary side pressure. Furthermore, as the liquid content of the primary side falls, the hydrostatic head of the liquid trapped in the pump inlet causes the liquid level in the core region to be depressed by an amount equal to the depth of the loop seal. This can lead to uncovering of the core. Loop seal reformation could impair long-term cooling after an LBLOCA by causing periodic depression of the mixture level in the core.

The PH is identified as S-39-LSC (or No 65 in the list of 116 THPs) by Aksan et al. (2018a), and D'Auria et al. (2021b).

6.3.22 Emergency core cooling bypass/downcomer penetration

Cold leg breaks for reactors with ECCS injecting in the cold leg only (at least, during the short-term initial phase of injection) are considered in this section.

The ECC bypass and downcomer penetration PH is a combination of three phenomena already described (Rohatgi et al., 1990).

- CCF or CCFL in the downcomer (see Section 6.3.10.5),
- liquid-vapor mixing with condensation in the downcomer (see Section 6.3.7.2),
- entrainment/deentrainment in the downcomer (see Section 6.3.6.3).

These three phenomena contribute to ECC bypass/downcomer penetration. This PH is listed separately because of its importance to ECC water injection performance. Complete ECC bypass occurs if the steam upflow rate is so high that it does not permit any liquid downflow through the downcomer. All the injected ECC water is carried out through the broken cold leg by the steam flow. This may be caused by high steam upflow in the downcomer (as CCFL), increased steam upflow due to pressure decrease at the condensation location in the cold leg and upper part of the downcomer, as well as high azimuthal steam velocities from the intact cold legs to the broken cold leg during LBLOCA blowdown (as entrainment). Downcomer penetration is partial or complete water downflow of the injected ECC water through the downcomer at lower steam upflow or lower azimuthal steam flow.

ECC water injected into the cold legs or directly into the downcomer has to flow downward through the downcomer to refill the LP and subsequently reflood the total core to prevent overheating. This water flow can be limited by upward and/or azimuthal steam flow. ECC bypass and penetration are ranked as very important during LBLOCA refill (OECD/NEA/CSNI, 1989; Wilson et al., 1990).

The PH is identified as S-14-ECCB (or No 31 in the list of 116 THPs) by Aksan et al. (2018a), and D'Auria et al. (2021b).

6.3.23 Parallel channel instabilities and other instabilities

The term "parallel channel instabilities" includes a broad range of instabilities that may occur in a system owing to the presence of a boiling boundary. Any boiling channel, even if operated in an open loop mode, may give rise to instabilities that can be detected as pressure, flow rate, temperature, void fraction, etc., oscillations (OECD/NEA/CSNI, 1990), see also Woo et al. (2007); Shi et al. (2015), Lopez de Bertodano and Fullmer (2015), and Wysocki et al. (2015).

There are several reasons for instabilities including static (e.g., Ledinegg) or dynamic (e.g., density wave) ones. The most studied instability mode is just called density-wave instability: a perturbation in a boiling channel inlet flow rate may result in a perturbation of the PD across the channel; this may reflect to the channel inlet, strengthening the initial flow rate variation (unstable system) or smoothing it (stable channel). A special case of "stable oscillations" may also occur.

The cases of interest essentially include a closed loop where the boiling channel is one of the parts of the system, and a single channel that constitutes the entire system, bounded by upper and LP with imposed conditions is the other; multiple parallel boiling channels are of considerable interest in the BWR technology. The core of a BWR can be considered a unique channel: in this case, parallel channels are constituted by the bypass and the downcomer. A single core channel can also be considered for instability purposes: in this case, the other channels are parallel channels that may interact to some extent with the other external channels (bypass and core).

This very short survey gives an idea of the number of thermal hydraulically unstable situations that may occur in a BWR. The complexity increases if one considers the neutron flux which may oscillate and interact with thermal-hydraulic instabilities (neutronic-thermal-hydraulic feedback).

So far, two general kinds of instabilities measured in BWR plants have attracted the attention of technologists.

(a) core wide oscillations: the neutron flux oscillates with the same frequency and amplitude in all parts of the core;
(b) regional oscillations: at least two parts of the core oscillate with different frequencies.

It seems easy to imagine that a very limited number of channels or even a single channel can oscillate at its characteristic frequency and amplitude in a core that may remain stable.

The PH is relevant in nuclear reactor safety. With reference to the BWR flow map (power versus core flow rate), a zone has been found, from operational experience, where the occurrence of instabilities is very probable. This is characterized by power and core flow rates around 40% and 30% of the nominal value, respectively. Operation in this zone may be necessary following a pump trip or during plant start-up transients.

The PH is identified as S-44-PCEI (or No 78 in the list of 116 THPs) by Aksan et al. (2018a), and D'Auria et al. (2021b).

6.3.23.1 Other instabilities

As mentioned above a wide variety of unstable situations may occur specifically when two-phase flows are involved in system thermal-hydraulic evolutions. Examples of these are the Rayleigh-Taylor instabilities, Pellegrini et al. (2015), and Kim and Kim (2015), the instabilities in downward flow, Ghione et al. (2018), and the Faraday waves affecting liquid levelss in large cross-sectional regions like downcomer of pressure vessel, Frepoli et al. (2015a). Those instabilities are relevant for nuclear thermal hydraulics, although they are not considered in the list of phenomena in this chapter (or the rest of the book).

6.3.24 Boron mixing and transport

Boron is a highly neutron-absorbing material and will consequently strongly affect the core power when inserted or removed from the core. Thus boron mixing and transport will not directly influence the thermal hydraulics (if one neglects possibly minor chemical effects) but will do so indirectly through the core power which is a thermal-hydraulic boundary condition. In PWR, boron is added to the coolant-moderator and its concentration will control the long-term variation of core reactivity. Because of the high liquid velocity in the RCS, the added boron will be mixed quite homogeneously with the coolant without steep gradients, and its transport through the RCS will closely follow the liquid transport.

However, there are low-flow conditions, for example, at idle reactor coolant pumps, when the pump power has temporarily been lost when inhomogeneous boron concentration can be attained in the loops with rather steep concentration gradients. In this case, the diffusion of boron within the liquid can be quite substantial and thus the boron transport will be different from the liquid flow.

The boron concentration will influence the PWR core power through reactivity feedback effects. The boron content impacts the moderator temperature reactivity coefficient and the boron concentration reactivity coefficient. For a fresh normal PWR core in equilibrium conditions (beginning of life) high boron concentration (1600–2000 ppm) will cause the former to be slightly positive at low moderator temperatures (300–400 K) and less positive at higher temperatures. At lower concentrations, it is always negative and more negative at higher temperatures. The boron concentration coefficient is always negative being more negative at low concentrations and temperatures.

Boron mixing and transport within the RCS is of crucial importance for core power control in LWR. The mixing and transport of boron influences through reactivity feedback the local core power during normal and off-normal operation of PWR and off-normal operation of BWR.

During normal operation of a PWR, the initial excess reactivity of a fresh core is compensated by adding boron to the coolant. This added boron is gradually decreased as the fuel burnup increases to keep core reactivity essentially constant throughout the core lifetime.

It is also a well-known fact that a PWR cannot be kept in a subcritical cold SHD condition immediately after refueling without the boron concentration in the coolant being above some value despite all the control rods being inserted. Thus the boron concentration and its mixing and transportation with the coolant is also of major concern at SHD conditions.

During an LBLOCA, the core fission power will almost immediately be terminated because of global core voiding and associated loss of moderation. However, during the reflood period the injected relatively cold ECCS water must have some level of boration to avoid possible local recriticalities of the core.

During an SBLOCA in a PWR with U-tube steam generators (UTSGs), it has been found that in the reflux condensation phase, there exists a mechanism that can accumulate boron-free condensate in the loop seal (Hyvarinen, 1992). Analyses of externally caused boron dilution transients suggest that the SBLOCA in this case could become a reactivity-initiated accident (RIA) initiator.

It has also been found that in some specific cases during SHD conditions in a PWR, there could be situations where pronounced heterogeneities in the boron concentration can develop and thus can potentially cause an RIA (Jacobson, 1992).

LBLOCA analyses do not address the boron crystallization concern during long-term cooling. This concern is specific to plants with short-term ECCS injection into cold legs only, then excluding plants with simultaneous cold legs/hot legs injection from the start of ECCS. In case of cold leg break with initial cold leg injection only, one can assume that steam only exits the core before reaching the break, then as boron concentration in steam is zero, the core boron concentration continuously increases, while the containment sump boron concentration decreases. Therefore, there is a risk of boron crystallization should the core concentration reach the corresponding precipitation limit. To eliminate such risk, the operator should switch over the long-term ECCS injection from cold legs to hot legs, in case of a cold leg break. The switchover time should avoid any boron crystallization in the core, and in the meantime, the sump concentration should not have been so depleted that a risk of core recriticality could occur due to too low boron concentration injected by the ECCS. This deals with cold leg breaks with initial safety injection into cold legs only. Previously mentioned problems should not occur in case of a hot leg break during a cold leg injection, with liquid from the core spilling over at the break. But after the ECCS switchover to hot legs, a similar problem would occur in this case of a hot leg break, not knowing accurately which hot leg ECCS part would reach the core and which part would be spilled at the break. Then, for hot leg breaks, it is necessary to keep part of ECCS injected into cold legs in the long term. Therefore, the long-term ECCS cooling includes either alternate cold leg/hot leg injection, or preferably to avoid numerous manual operations, simultaneous cold leg/hot leg injection. Accommodation of these risks, core boron crystallization and containment sump boron deconcentration, defines the switchover time to simultaneous hot legs/cold legs injection, as well as the initial boron concentration in the ECCS supply (initially contained in the Reactor Water Storage Tank).

The PH is identified as S-2-BO (or No 11 in the list of 116 THPs) by Aksan et al. (2018a), and D'Auria et al. (2021b).

6.3.25 Noncondensable gas effect (pressurized water reactor)

The presence of a noncondensable gas in the primary circuit plays a role in mechanical and thermal fluid behavior. Noncondensables are mainly nitrogen from the accumulator injection or hydrogen from clad oxidation.

In the case of nitrogen injection, the mechanical effects are related to gas expansion in the cold leg and the upper part of the downcomer. The momentum transfer between the gas, steam, and liquid increases the liquid velocity. This has an impact on the downcomer water penetration and water bypass to the break.

The thermal effect is related to the change in HT laws, particularly for the condensation process produced by noncondensables.

Both effects are associated with the end of accumulator injection. When nitrogen enters the cold leg, the mixing of gas with liquid stops oscillation due to plug formation. This is due to the compressibility of the gas—liquid mixture and reduced condensation rate.

Within the primary circuit, the gas migrates from its injection or generation point. There are several mechanisms for migration: the entrainment of gas by flowing steam, the gravity forces arising from density differences, and diffusion driven by the concentration gradient. In the SG, due to condensation in the tube inlet region and the flow of steam toward the SG, noncondensable gas stays at the top of the U-tubes and can limit NC.

This PH could have some influence on the general thermal-hydraulic behavior for LBLOCA but is of most relevance to accidents with NC phases.

The PH is identified as S-43-NCG (or No 76 in the list of 116 THPs) by Aksan et al. (2018a), and D'Auria et al. (2021b).

6.3.26 *Lower plenum entrainment*

In the case of an LBLOCA, the reverse flow in the core goes to the LP where saturated liquid is still present. Depending on steam flow and water level, liquid can be entrained to the downcomer, and then to the break. Oscillatory behavior can occur. This PH depends on the geometry and LP internal structures.

This PH has a direct impact on the nature of reverse flow in the downcomer during LBLOCA as well as on break flow and ECCS bypass.

The PH is identified as S-40-LPE (or No 66 in the list of 116 THPs) by Aksan et al. (2018a), and D'Auria et al. (2021b).

6.4 Thermal-hydraulic phenomena for integral facilities (pressurized water reactors, boiling water reactors, and Russian pressurized water-cooled and water-moderated energy reactors)

The SETF PH covers the overall spectrum of PH expected in accident scenarios (DBA envelope) of PWR and BWR; for instance, the ITF PH 'steam binding' considered by Aksan et al. (2018a), is discussed by SETF PH in sections 6.3.6.4, 6.3.6.5, 6.3.7.3, 6.3.11.1 and 6.3.13.1. Nevertheless, the consideration of PH occurring in ITF is essential for code validation; therefore, ITF THP identification and characterization have been done for PWR, BWR, and VVER within the OECD/NEA-CSNI framework. The results of this work have been described and introduced in different tables in Sections 6.4.1–6.4.3. Furthermore, the ITF phenomena part of the 116 PH list by Aksan et al. (2018a), and D'Auria et al. (2021b) are connected with the description hereafter.

6.4.1 *Thermal-hydraulic phenomena for integral facilities of pressurized water reactor*

The analysis and the classification of off-normal events that may happen in a real system are slightly different for PWR and BWR, although in both cases a simple classification based on rupture area has been preferred. For PWR the following classes of accidents were selected.

(1) Large breaks (rupture area greater than 25% of the maximum pipe area connected with the pressure vessel A_{max}); notably, the accident scenario following large breaks includes phenomena identified as I-3-BD, I-17-PFU, I-21-RE, and I-24-SBI (or, No 9, NO 83, No 91 and No 99 in the list of 116 THPs) by Aksan et al. (2018a), and D'Auria et al. (2021b); further characterization of steam binding (I-24-SBI) can be found in Chapters 8 and 11.

(2) Small and intermediate breaks with a UTSG (rupture area less than or equal to 25% A_{max}); core cooling occurs by NC or phenomena identified as I-10-NC1, I-11-NC2, and I-12-NC3 (or No 68, No 69, and No 70 in the list of 116 THPs) by Aksan et al. (2018a), and D'Auria et al. (2021b).

(3) Small and intermediate breaks in PWR with OTSG; related phenomena (including LBLOCA) are identified as I-4-NCBC, I-9-INC, I-14-NC5, I-20-RF, I-28-SH and S-58-OTSG (or No 10, No 54, No 72, No 90, No 106 and No 109 in the list of 116 THPs) by Aksan et al. (2018a), and D'Auria et al. (2021b).
(4) Transients, where upset conditions are created by single or multiple failures of one or more systems in the plant; one PH consequence of typical occurrences is identified as A-1-CHV-c (or No 4 in the list of 116 THPs) by Aksan et al. (2018a), and D'Auria et al. (2021b), see also section 6.5.11, below.
(5) Transients at shutdown conditions: selected related phenomena are identified as I-2-ASY-D, I-8-FO-c, and I-23-SIP (or No 3, No 40, and No 94 in the list of 116 THPs) by Aksan et al. (2018a), and D'Auria et al. (2021b).
(6) Accident management for a nondegraded core: selected related phenomena are identified as I-18-PRB, I-19-PRZ, I-22-RCM, and I-27-SULI (or No 86, No 87, No 92 and No 105 in the list of 116 THPs) by Aksan et al. (2018a), and D'Auria et al. (2021b).

LBLOCAs (item 1) are dominated by inertia effects and characterized by strong rather homogeneous steam/water flow distribution inside the primary loop, in particular during the depressurization period.

For SBLOCAs (item 2), gravity and stratification effects are important phenomena to be simulated. Intermediate break LOCA (item (2) as well) may contain features related to both LB and SB LOCA. The actual subdivision between items (1), (2), and (3) generally depends on the size of the plant, the position of the break, and the status of the pumps (on or off); the boundary selected (25% A_{max}) must be seen as approximate and somewhat arbitrary.

Transient with single or multiple failures (item 4) is mainly determined by the sequence of operation of valves, pumps, and engineered safety features.

Transients at shutdown conditions (item 5) are mainly related to the so-called mid-loop plant operations, which are characterized by the atmospheric pressure in the primary system (with openings or not) and the water level at the height of the loop connections to the reactor vessel.

Accident management for a nondegraded core (item 6) contains AM measures for a nondegraded core to prevent core degradation and damage.

In the term "phenomena" all the important thermal-hydraulic aspects expected to occur during the accident are included; "type of test" relates to the definition of the experiment simulating a characteristic accident scenario. Phenomena, test types, and test facilities are identified and selected essentially based on the personal experience of the experts assigned to perform the work.

The correlation between phenomena and test type in the tables is given three levels.

- Occurring, which means that the particular PH is occurring in that kind of test type (plus (+) sign in the matrix);
- Partially occurring: only some aspects of the PH are occurring (open circle (o) in the matrix);
- Not occurring (dash (−) in the matrix).

Widely used nomenclature has been adopted in the identification of phenomena and test types. The test types are identified and described in some detail in Appendix A of Annunziato et al. (1996), and the phenomena are defined also in some detail in Appendix B of the same report. For a better understanding of the tables, it is useful to describe one or two lines of each one of those tables.

Table 6.2 Cross-reference matrix for large breaks in pressurized water reactors.

Phenomena versus test type **+ occurring** **o partially occurring** **− not occurring**		**Test type**		
		Blowdown	**Refill**	**Reflood**
Phenomena	Break flow	+	+	+
	Phase separation (condition or transition)	o	+	+
	Mixing and condensation during injection	o	+	+
	Core wide void + flow distribution	o	+	+
	ECC bypass and penetration	o	+	o
	CCFL (UCSP)	o	+	+
	Steam binding (liquid carryover, etc.)	−	o	+
	Pool formation in upper plenum (UP)	−	+	+
	Core heat transfer incl. DNB, dryout, RNB	+	+	+
	Quench front propagation	o	o	+
	Entrainment (core, UP)	o	o	+
	Deentrainment (core, UP)	o	o	+
	One - and two-phase pump behavior	+	o	o
	Noncondensable gas effects	−	o	o

"The Large Break Matrix" is given in Table 6.2. As an example, consider the ECC bypass and penetration PH. It is considered of high importance for refill tests but of limited interest in blowdown and reflood tests. Another example is quench front propagation. This PH has been considered suitable for indirectly heated rods but not suitable for directly heated rods. Indirectly heated rods simulate better the heat capacity effect of the nuclear fuel rod, but they cannot take into account the effect of the gas gap of the nuclear fuel rod.

"The Small and Intermediate Breaks Matrices in PWR with UTSG and OTSG" are given respectively in Tables 6.3 and 6.4. It should be noted that, among the phenomena, the "structural heat and heat losses" PH [identified as I-26-SHH (or No 104 in the list of 116 THPs) by Aksan et al. (2018a), and D'Auria et al. (2021b)] has been considered to emphasize the notable distortions introduced by the heat release from structures in scaled-down facilities with respect to plant behavior; this is due to the larger values of structural masses and structure-to-fluid heat exchange areas relative to the volume-scaled values.

A relatively large number of test types, a total of seven, have been considered to emphasize that a number of different transients are possible under the general category of SBLOCAs.

Table 6.3 Cross-reference matrix for small and intermediate breaks in pressurized water reactors with U-tube steam generator.

Phenomenon versus test type + occurring o partially occurring − not occurring		Test type						
		Stationary test addressing energy transport on primary side	**Stationary test addressing energy transport on secondary side**	**Small leak overfeed by HPIS, secondary side necessary**	**Small leak without HPIS overfeeding, secondary side necessary**	**Intermediate leak, secondary side not necessary**	**Pressurizer leak**	**U-tube rupture**
Phenomena[b]	Natural circulation in one-phase flow, primary side	+	+	+	o	−	+	+
	Natural circulation in two-phase flow, primary side	+	−	o	+	+	o	−
	Reflux condenser mode and CCFL	+	−	−	+	+	−	−
	Asymmetric loop behavior	−	−	+	+	−	o	+
	Break flow	−	−	+	+	+	+	+
	Phase separation without mixture level formation	+	−	o	+	+	+	o
	Mixture level and entrainment in SG second side	−	+	+	+	+	+	+
	Mixture level and entrainment in the core	+	−	−	+	+	+	−
	Stratification in horizontal pipes	+	−	−	+	+	−	−

	Phase separation in T-junction and effect on break flow	−	−	−	+	+	−	−
	ECC-mixing and condensation	−	−	o	+	+	+	+
	Loop seal clearing	−	−	−	+	+	o	−
	Pool formation in UP/CCFL (UCSP)	+	−	−	o	+	+	−
	Core wide void and flow distribution	+	−	−	o	+	+	−
	Heat transfer in covered core	+	+	+	+	+	+	+
	Heat transfer in partly uncovered core	+	−	−	o	+	−	−
	Heat transfer in SG primary side	+	o	o	+	+	o	o
	Heat transfer in SG secondary side	o	+	+	+	+	+	+
	Pressurizer thermohydraulics	o	−	o	o	+	+	+
	Surge line hydraulics	o	−	−	o	+	+	o
	One- and two-phase pump behavior	−	−	−	o	+	−	−
	Structural heat and heat losses[a]	+	−	o	+	+	o	o
	Noncondensable gas effects	+	−	−	−	−	−	−
	Boron mixing and transport	+	−	+	+	+	+	+

[a]This phenomenon is related to scaled test facilities.
[b]For intermediate breaks, phenomena included in large-break reference matrix could also be important.

Table 6.4 Cross-reference matrix for small and intermediate breaks in pressurized water reactors (PWRs) with once-through steam generator (and matrix for PWRs with U-tube steam generator).

		Test type						
Phenomenon versus test type **+ occurring** **o partially occurring** **− not occurring**		**Stationary test addressing energy transport on primary side**	**Stationary test addressing energy transport on secondary side**	**Small leak overfeed by HPIS, secondary side necessary**	**Small leak without HPIS overfeeding, secondary side necessary**	**Intermediate leak, secondary side not necessary**	**Pressurizer leak**	**OTSG-tube rupture**
Phenomena[b]	Natural circulation in one-phase flow, primary side	+	+	+	o	−	+	+
	Natural circulation in two-phase flow, primary side	+	−	o	+	+	o	−
	Boiler condenser mode	+	−	−	+	+	−	−
	Asymmetric loop behavior	−	−	+	+	−	o	+
	Break flow	−	−	+	+	+	+	+
	Phase separation without mixture level formation	+	−	o	+	+	+	o
	Mixture level and entrainment in SG second side	−	+	+	+	+	+	+
	Mixture level and entrainment in the core	+	−	−	+	+	+	−
	Stratification in horizontal pipes	+	−	−	+	+	−	−
	Phase separation in T-junction and effect on break flow	−	−	−	+	+	−	−

	ECC-mixing and condensation	–	–	o	+	+	+	+
	Loop seal clearing	–	–	–	+	+	o	–
	Pool formation in UP/CCFL (UCSP)	+	–	–	o	+	+	–
	Core wide void and flow distribution	+	–	–	o	+	+	–
	Heat transfer in covered core	+	+	+	+	+	+	+
	Heat transfer in partly uncovered core	+	–	–	o	+	–	–
	Heat transfer in SG primary side	+	o	o	+	+	o	o
	Heat transfer in SG secondary side	o	+	+	+	+	+	+
	Pressurizer thermohydraulics	o	–	o	o	+	+	+
	Surge line hydraulics	o	–	–	o	+	+	o
	One- and two-phase pump behavior	–	–	–	o	+	–	–
	Structural heat and heat losses[a]	+	–	o	+	+	o	o
	Noncondensable gas effects	+	–	–	–	–	–	–
	Boron mixing and transport	+	–	+	+	+	+	+
	Intermittent two-phase natural circulation	–	–	–	+	+	o	o
	Natural circulation—core, vent valve, downcomer	+	o	+	+	+	+	+
	Refill of loops	–	–	+	+	o	+	–
	Superheating in secondary side	+	+	+	+	+	+	+

[a]This phenomenon is related to scaled test facilities.
[b]For intermediate breaks, phenomena included in large-break reference matrix could also be important.

Quench front propagation has been considered suitable for indirectly heated rods but not suitable for directly heated rods. As already mentioned above, indirectly heated rods simulate better the heat capacity effect of the nuclear fuel rod, but they cannot take into account the effect of the gas gap of the nuclear fuel rod [identified as I-21-RE (or No 91 in the list of 116 THPs) by Aksan et al. (2018a), and D'Auria et al. (2021b)].

"The matrix for small and intermediate breaks in PWR with once through steam generators (OTSG)" is given in Table 6.4. Twenty-seven phenomena and seven test types are included. The "structural heat and heat losses" PH is included by the same arguments as given above.

"The transient matrix" is given in Table 6.5. The following observations can be added to those in the preceding three sections.

- some of the phenomena discussed in the SB and IB matrix are relevant in the present case and they are not repeated in this table;
- thermal-hydraulic nuclear feedback is a PH that characterizes some transients: for this reason, the LOFT test facility (which was equipped with a nuclear core) was judged to be more suitable than the other experimental loops in operation;
- the experience gained from incidents occurring in power plants is potentially of great importance, as is that from the analysis of data from start-up, shutdown, and other maneuvers.

Table 6.6 is a cross-reference matrix dedicated to "transients at shutdown," that is, corresponding to accident scenarios with the plant in shutdown conditions (control rods inserted). A separate matrix has been prepared to emphasize the importance of these transients with respect to safety issues and the challenges for codes in low-pressure conditions.

Four test types have been considered for shutdown conditions (three of them relating to loss of RHR at mid-loop operation, and one to boron dilution at shutdown). The cold overpressure transients have not been included in the matrix because thermal hydraulics is quite simple (mostly single-phase liquid flow) and the phenomena associated are not expected to significantly challenge thermal-hydraulic codes.

Among the 24 phenomena mentioned in these tables for PWR, 20 have been taken from the SBLOCA and transient matrices and are not specific to transients at shutdown conditions. However, it has been considered that shutdown conditions required particular attention because of specific thermal-hydraulic conditions (atmospheric pressure) for which the validity of computer codes has to be demonstrated.

Table 6.7 is the cross-reference matrix for "Accident Management for a nondegraded core in PWR." With respect to the initial version of the report, this is a new matrix. However, the phenomena are those identified for SBLOCA and transients in PWR. This is because AM scenarios are initiated by the same events as SBLOCA and transients, but are terminated by appropriate operator actions according to AM procedures [e.g., this includes the PH identified as I-1-ASY-L (or No 2 in the list of 116 THPs) by Aksan et al. (2018a), and D'Auria et al. (2021b)].

The five test types selected in this matrix refer to the intrinsic nature of the AM procedure considered except one, which refers to failure in plant operation (SGTR with multiple failures).

Table 6.5 Cross-reference matrix for transients in pressurized water reactors.

Phenomenon versus test type + occurring o partially occurring − not occurring		Test type: ATWS	Loss of feedwater, non-ATWS	Loss of heat sink, non ATWS[c]	Station blackout	Steam line break	Feed line break	Reactivity disturbance	Over-cooling
Phenomena	Natural circulation in one-phase flow	+	+	+	+	+	+	o	o
	Natural circulation in two-phase flow	+	+	+	+	−	−	o	−
	Core thermohydraulics	+	+	+	+	o	o	+	o
	Thermohydraulics on primary side of SG	+	o	o	+	o	o	o	+
	Thermohydraulics on secondary side of SG	+	+	+	+	+	+	o	+
	Pressurizer thermohydraulics	+	+	+	+	o	o	o	+
	Surge line hydraulics (CCFL, choking)	+	+	+	+	o	o	o	o
	Valve leak flow[a]	+	+	+	+	+	+	+	+
	One- and two-phase pump behavior	+	+	+	+	o	o	o	+
	Thermohydraulic-nuclear feedback	+	−	−	−	−	−	+	−
	Structural heat and heat losses[b]	o	o	o	o	o	o	o	o
	Boron mixing and transport	−	−	−	−	o	−	−	o
	Separator behavior	o	−	−	−	+	−	−	−

[a]Valve flow behavior will be strongly design- and type-dependent, specific experimental data should be used, if available.
[b]This phenomenon is related to scaled test facilities.
[c]Isolation of one or more steam generators.

Table 6.6 Cross-reference matrix for transients under shutdown conditions in pressurized water reactors.

		Test type			
Phenomenon versus test type **+ occurring** **o partially occurring** **− not occurring**		**Loss of RHR with no opening**	**Loss of RHR with openings**	**Loss of RHR with dam in HL**	**Boron dilution at shutdown**
Phenomena	Pressurization due to boiling	+	+	+	−
	Reflux condenser mode and CCFL	+	+	o	−
	Asymmetric loop behavior	−	o	+	−
	Flow through openings (manways, vents)	−	+	+	−
	Mixture level formation in upper plenum and hot legs	+	+	+	−
	Mixture level and entrainment in the core	+	+	+	−
	SG siphon draining	−	−	+	−
	Asymmetry due to the presence of a dam	−	−	+	−
	Stratification in horizontal pipes	+	+	+	−
	Phase separation in T-junctions and effect on flow	−	+	+	−
	ECC mixing and condensation	+	+	+	−
	Loop seal clearing and filling	+	+	+	−
	Pool formation in UP/CCFL (UCSP)	−	−	−	−
	Core 3-D thermal-hydraulics	+	+	+	+
	Heat transfer in covered core	+	+	+	−
	Heat transfer in partially uncovered core	+	+	+	−
	Heat transfer in SG primary side	+	+	+	−
	Heat transfer in SG secondary side	+	+	+	−
	Pressurizer thermal-hydraulics[a]	−	x	x	−
	Surge line thermal-hydraulics[a]	−	x	x	−
	Structural heat and heat losses	−	−	−	−
	Noncondensible gas effects	+	+	+	−
	Boron mixing and transport	−	−	−	+
	Thermal-hydraulics-nuclear feedback	−	−	−	+

+, pressurizer manway open; −, pressurizer manway shut.
[a]Dependent on opening location.

Table 6.7 Cross-reference matrix for accident management of nondegraded core in pressurized water reactors.

		Test type				
Phenomenon versus test type **+ occurring** **o partially occurring** **− not occurring**		**High pressure, primary side feed and bleed**	**Low pressure, primary side feed and bleed**	**Secondary side, feed and bleed**	**RCP-restart in a highly, voided PCS**	**Primary to secondary break with multiple failures**
Phenomena	Natural circulation in one-phase flow, primary side	+	−	+	−	+
	Natural circulation in two-phase flow, primary side	+	+	+	−	+
	Reflux condenser mode and CCFL	−	−	+	−	+
	Asymmetric loop behavior	+	+	+	+	+
	Break flow	+	+	o	+	+
	Phase separation without mixture level formation	+	+	+	+	+
	Mixture level and entrainment in SG secondary side	−	−	+	−	+
	Mixture level and entrainment in the core	+	+	+	o	+
	Stratification in horizontal pipes	+	+	+	o	+
	Phase separation in T-junction and effect on break flow	+	+	o	−	+
	ECC-mixing and condensation	+	+	+	−	+
	Loop seal clearing	o	o	+	o	+
	Pool formation in UP/CCFL (UCSP)	+	+	+	−	+
	Core wide void and flow distribution	+	+	+	+	+

Continued

Table 6.7 Continued

	Test type				
Phenomenon versus test type **+ occurring** **o partially occurring** **− not occurring**	**High pressure, primary side feed and bleed**	**Low pressure, primary side feed and bleed**	**Secondary side, feed and bleed**	**RCP-restart in a highly, voided PCS**	**Primary to secondary break with multiple failures**
Heat transfer in covered core	O	o	+	−	+
Heat transfer in partly uncovered core	+	+	+	+	+
Heat transfer in SG primary side	−	−	+	o	+
Heat transfer in SG secondary side	−	−	+	o	+
Pressurizer thermohydraulics	+	+	o	o	+
Surge line hydraulics	+	+	o	o	+
One- and two-phase pump behavior	O	o	+	+	+
Structural heat and heat losses[a]	+	+	+	+	+
Noncondensable gas effects	o	+	+	+	+
Accumulator behavior	−	+	+	−	o
Boron mixing and transport	+	+	+	+	+
Thermohydraulic-nuclear feedback	−	−	−	+	−
Separator behavior	−	−	−	−	−

[a]This phenomenon is related to scaled test facilities.

6.4.2 Thermal-hydraulic phenomena for integral facilities of boiling water reactor

For BWR, almost all the plants now in operation are equipped with an Automatic Depressurization System (ADS), and therefore, even a SBLOCA is characterized by a fast depressurization following the actuation of this device. Thus, as a difference from the PWR case, only two tables have been considered: a LOCA table and a transient table. In the Transient table, start-up tests, frequent operational transients, and ATWS have been included. For both plant types, tables are composed of two elements, related to the following items: “Phenomena versus Test Types.”

The LOCA matrix is given in Table 6.8. The analysis of off-normal events in actual nuclear plants is potentially valuable, especially with reference to scaling and simulation problems. The “structural heat and heat losses” PH is included by the same arguments as given in Section 6.4.1.

A relatively large number of test types, six, has been considered to emphasize that a number of different transients are possible under the general category of blowdown.

The transient matrix is given in Table 6.9. The following observations can be made in addition to those in the preceding section on PWR.

- Almost all of the phenomena discussed in the LOCA matrix are also important for transients, but for brevity, they are not repeated in this table;
- Thermal-hydraulic nuclear feedback is the PH [identified as I-29-NTF2 (or No 107 in the list of 116 THPs) by Aksan et al. (2018a), and D’Auria et al. (2021b)] which characterizes some of the transients. Due to this reason, actual BWR plants have been judged much more suitable than the other experimental loops already in operation;
- The experience gained from off-normal events occurring in real plants is potentially of great importance, as is that from the analysis of data from start-up, shutdown, and other test or operational maneuvers.

Specific additional phenomena related to BWR are identified as I-5-BC, I-6-CLDO, I-7-COTH, S-30-IPU, I-16-NTF1, I-25-SLD and I-30-VCP (or No 18, No 19, No 28, No 55, No 77, No 101 and No 115 in the list of 116 THPs) by Aksan et al. (2018a), and D’Auria et al. (2021b).

6.4.3 Thermal-hydraulic phenomena of integral facilities of Russian pressurized water-cooled and water-moderated energy reactors

In addition, to set up validation matrices for WWER/VVER analyses, an international Working Group was formed on the initiative of the BMFT of the Federal Republic of Germany. A further evaluation of the WWER matrices was performed by a CSNI support group. Based on these CSNI matrices, the lists of phenomena have been reviewed and adapted to the characteristics of WWER-440 and WWER-1000 systems, respectively.

The topic was combined with the objective of cooperation to formulate an internationally agreed WWER-specific validation matrix as a supplement to the existing

Table 6.8 Cross-reference matrix for loss-of-coolant accidents in boiling water reactors.

Phenomena versus test type + occurring o partially occurring − not occurring		Test type: Large steam line break with fast depressurization	Large break below water level with fast depress	Small break without depress. Before ADS actuation	Intermediate break with slow depress.	Spray line break	Refill—reflood
Phenomena	Break flow	+	+	+	+	+	o
	Channel and bypass axial flow and void distribution	+	+	+	+	+	+
	Core-wide radial void distribution	o	o	+	+	+	+
	Parallel channel effects-instabilities	−	−	+	+	+	+
	ECC bypass	−	−	o	o	o	+
	CCFL at UCSP and channel inlet orifice	o	+	−	+	+	+
	Core heat transfer including DNB, dryout, RNB, surface to surface radiation	+	+	o	+	o	+
	Quench front propagation for both fuel rods and channel walls	−	−	−	−	−	+
	Entrainment and deentrainment in core and upper plenum	+	+	o	o	o	+
	Separator behavior incl. flooding, steam penetration and carryover	+	+	o	o	o	−
	Spray cooling	−	−	o	o	o	+
	Spray distribution	−	−	o	o	o	+
	Steam dryer—hydraulic behavior	+	−	o	o	−	−
	One- and two-phase pump recirculation behavior including jet pumps	o	o	+	+	+	o
	Phase separation and mixture level behavior	+	+	+	+	+	+
	Guide tube and lower plenum flashing	+	+	−	o	o	−
	Natural circulation—core and downcomer	−	−	+	o	o	+

	Natural circulation core bypass, hot and cold bundles	–	–	+	o	o	+
	Mixture level in core	–	–	+	o	o	+
	Mixture level in downcomer	+	+	+	+	+	+
	ECC mixing and condensation	–	–	+	o	+	+
	Pool formation in upper plenum	o	o	–	o	o	+
	Structural heat and heat losses[a]	o	o	o	+	+	+
	Phase separation in T-junction and effect on break flow	–	–	+	o	+	–

[a]This phenomenon is related to scaled test facilities.

Table 6.9 Cross-reference matrix for transients in boiling water reactors.

Phenomenon versus test type + occurring o partially occurring − not occurring		Test type: Stationary test measuring power flow map	Recirculation pump trip	Core stability	Loss of main heat sink	Feedwater flow or temperature disturbance (e.g., LOFW)	Loss of feedwater (LOFW) up to time of const. Pressure	Inadvertent increase in steam flow	ATWS	Station blackout (loss-of-offsite power, LOOSP)
Phenomena	Natural circulation in one- and two-phase flow	+	+	+	+	−	−	−	+	+
	Collapsed level behavior in downcomer	−	+	o	+	+	+	+	+	+
	Core thermal-hydraulics	o	+	+	+	o	o	o	+	+
	Valve leak flow	−	−	−	+	−	−	−	+	+
	Single-phase pump behavior[a]	o	+	o	+	o	o	+	+	+
	Parallel channel effects and instabilities (PCEIs)	−	+	+	o	−	−	−	+	+
	Nuclear thermal-hydraulic feedback including spatial effects	o	o	+	−	o	o	o	+	−
	Nuclear thermal-hydraulic instabilities	−	o	+	−	−	−	o	+	−
	Downcomer mixing	−	−	−	−	+	+	−	+	+
	Boron mixing and distribution	−	−	−	−	−	−	−	+	−
	Steam line dynamics	−	−	−	+	−	−	+	+	+

	Void collapse and temperature distribution during pressurization	–	–	–	+	–	–	–	+	+
	Critical power ratio	–	+	+	+	+	+	+	+	+
	Rewet after DNB at high pressure and high power including high core flow	–	+	–	+	–	–	o	+	o
	Structural heat and heat losses[b]	–	o	–	o	–	o	o	o	o

[a]Two-phase pump behavior is of interest for certain special ATWS and inadvertent increase in steam flow transients.
[b]This phenomenon is related to scaled test facilities.

OECD/NEA-CSNI code validation matrix for PWRs with UTSGs as presented in Annunziato et al. (1996).

Based on the CSNI cross-reference matrices, lists of phenomena have been reviewed and adapted to the characteristics of WWER-440 and WWER-1000 systems, respectively, and lists of test facilities suitable for code assessment have been assembled.

The above tasks were performed successfully by the Working Group under the leadership of GRS in close cooperation with IPSN during 1993–95, and the results were published by Liesch and Réocreux (1995).

The selection of tests from the large number of experiments proposed has to be continued to obtain those most suitable for code assessment with respect to a given PH or test type.

As a consequence, these activities will continue under the auspices of the OECD/NEA. Therefore, in June 1995 a new Support Group was installed to continue with the further evaluation of the matrices, concentrating on three tasks:

- Description of WWER-specific phenomena and safety relevance
- Optimization of the WWER-specific code validation matrices
- Development of criteria for the data bank storage of experimental data valid for the matrices

As a result of this work, the WWER validation matrix was completed and published in 2001 (OECD/NEA/CSNI, 2001). In the following three tables, as in the case of PWRs and BWRs, phenomena versus test types specific to VVER-440 and VVER-1000 will be presented for LBLOCAs, which include 16 phenomena and three test types (Table 6.10), small and intermediate leaks, which include 27 PH and eight test types (Table 6.11), and transients, which include 11 PHs and nine test types (Table 6.12). It should be noted that the majority of these phenomena for PWRs are also relevant to WWERs.

Specific additional phenomena related to WWERs are identified as S-32-LA, I-13-NC4, I-15-NC6, and S-57-HOSG (or No 57, No 71, No 73, and No 108 in the list of 116 THPs) by Aksan et al. (2018a), and D'Auria et al. (2021b).

6.5 Thermal-hydraulic phenomena related to advanced water-cooled reactors

This section will provide the relevant THPs for advanced light-water reactor (ALWR) designs including the reactor designs employing passive safety systems, in addition to the relevant THPs identified for the current generation of LWR in Sections 6.3 and 6.4.

IAEA-TECDOC-626 provides definitions for safety-related terms as applied to advanced reactors (IAEA, 1991). In that document, the concepts of passive and active safety systems are defined and discussed. The definition of a passive safety system is as follows: *Either a system which is composed entirely of passive components and*

Table 6.10 Cross-reference matrix for large breaks in Russian pressurized water-cooled and water-moderated energy reactors.

		Test type			Plant type	
Phenomena versus test type **+ occurring** **o partially occurring** **− not occurring** **Phenomena versus plant type** **+ fully specific to WWER** **o partially specific** **− not specific**		**Blowdown**	**Reflood**	**Refill**	**WWER-440/213**	**WWER-1000**
Phenomena	Break flow	+	+	+	−	−
	Phase separation (condition or transition)	o	+	o	+	o
	Mixing and condensation during injection	o	+	+	+	+
	Two-phase flow in SG primary and secondary side[a]	o	o	o	o	o
	Core wide void + flow distribution	o	+	+	+	+
	ECC bypass and penetration	o	o	+	+	+
	Upper plenum injection and penetration[a]	o	+	+	−	+
	CCFL (UCSP)	o	+	+	+	+
	Steam binding (liquid carryover, etc.)	−	+	o	o	o
	Pool formation in UP	+	+	−	+	+
	Core heat transfer incl. DNB, dryout, RNB	+	+	+	+	+
	Quench front propagation	o	+	o	+	+
	Entrainment (core, UP)[a]	o	+	o	+	o
	Deentrainment (core, UP)[a]	−	+	o	+	o
	Single- and two-phase pump behavior[a]	+	o	o	+	+
	Noncondensable gas effects[a]	−	+	+	o	o

[a]These phenomena are not covered in the OECD/NEA-CSNI Matrix for PWRs.

Table 6.11 Cross-reference matrix for small and intermediate leaks in Russian pressurized water-cooled and water-moderated energy reactors.

		Test type								Plant type	
Phenomenon versus test type **+ occurring** **o partially occurring** **– not occurring** **Phenomenon versus plant type** **+ fully specific to WWER** **o partially specific** **– not specific**		**Stationary test addressing energy transport on primary side**	**Stationary test addressing energy transport on secondary side**	**Small leak overfeed by HPIS, secondary side necessary**	**Small leak without HPIS, overfeeding, secondary side**	**Intermediate leak, secondary side not necessary**	**Pressurizer leak**	**SG-tube rupture**	**SG-header rupture**	**WWER-440/213**	**WWER-100**
Phenomena	Natural circulation in single-phase flow, primary side	+	+	+	o	–	+	+	+	–	–
	Natural circulation in two-phase flow, primary side	+	–	o	+	+	o	–	+	+	+
	Reflux condenser mode and CCFL	+	–	–	+	+	–	–	+	–	+
	Asymmetric loop behavior[a]	+	–	+	+	–	o	+	–	+	o
	Leak flow	–	–	+	+	+	+	+	+	–	–
	Phase separation without mixture level formation	+	–	o	+	+	+	o	+	+	o
	Mixture level and entrainment in SG (SS and PS)[a]	–	+	+	+	+	+	+	+	+	+
	Mixture level and entrainment in the core[a]	+	–	–	+	+	+	–	o	+	o

	Stratification in horizontal pipes	+	–	–	+	+	–	–	o	–	–
	Phase separation in T-junction and effect on leak flow	–	–	–	+	+	–	–	o	–	–
	ECC-mixing and condensation	–	–	o	+	+	+	+	+	+	+
	Loop seal clearance (CL)	–	–	–	+	+	o	–	–	–	–
	Pool formation in UP/CCFL (UCSP)	+	–	–	o	+	+	–	o	+	+
	Core wide void and flow distribution	+	–	–	o	+	+	–	o	+	o
	Heat transfer in covered core	+	+	+	+	+	+	+	+	o	–
	Heat transfer in partly uncovered core	+	–	–	o	+	–	–	o	o	–
	Heat transfer in SG primary side[a]	+	o	o	+	+	o	o	+	+	+
	Heat transfer in SG secondary side[a]	o	+	+	+	+	+	+	+	+	+
	Pressurizer thermohydraulics[a]	o	–	o	o	+	+	+	o	+	+
	Surge line hydraulics	o	–	–	o	+	+	o	o	o	o
	Single- and two-phase pump behavior[a]	–	–	–	o	+	–	–	o	+	+
	Structural heat and heat losses[b]	+	–	o	+	+	o	o	+	–	–
	Noncondensable gas effects	+	–	–	–	–	–	–	–	+	+

Continued

Table 6.11 Continued

	Test type								Plant type	
Phenomenon versus test type **+ occurring** **o partially occurring** **– not occurring** **Phenomenon versus plant type** **+ fully specific to WWER** **o partially specific** **– not specific**	**Stationary test addressing energy transport on primary side**	**Stationary test addressing energy transport on secondary side**	**Small leak overfeed by HPIS, secondary side necessary**	**Small leak without HPIS, overfeeding, secondary side**	**Intermediate leak, secondary side not necessary**	**Pressurizer leak**	**SG-tube rupture**	**SG-header rupture**	**WWER-440/213**	**WWER-100**
Natural circulation, core-gap-downcomer, dummy elements[a]	o	+	+	–	–	–	+	o	+	+
Loop seal behavior in HL[a]	+	–	–	+	+	+	–	o	+	–
Recirculation in the SG primary side[a]	+	–	–	+	o	o	–	–	+	+
Boron mixing and transport[a]	–	–	–	+	+	+	o	o	+	+
Water accumulation in SG tubes[a]	+	–	–	+	+	+	–	–	+	+

[a]These phenomena are not covered in the OECD/NEA-CSNI Matrix for PWRs.
[b]This phenomenon is related to scaled test facilities.

Table 6.12 Cross-reference matrix for transients in Russian pressurized water-cooled and water-moderated energy reactors.

		Test type									Plant type	
Phenomenon versus test type **+ occurring** **o partially occurring** **− not occurring** **Phenomenon versus plant type** **+ fully specific to WWER** **o partially specific** **− not specific**		**ATWS**	**Loss of feedwater, non-ATWS**	**Loss of heat sink, non-ATWS**	**Station blackout**	**Steam line break**	**Feed line break**	**Cooldown primary feed and bleed**	**Reactivity disturbance**	**Over-cooling**	**WWER-440/213**	**WWER-100**
Phenomena	Natural circulation in single-phase flow, primary side	+	+	+	+	+	+	+	o	o	−	−
	Natural circulation in two-phase flow, primary side	+	+	+	+	−	−	o	o	−	+	+
	Core thermohydraulics	+	+	+	+	o	o	o	+	o	+	o
	Thermohydraulics on primary side of SG[a]	+	o	o	+	o	o	+	o	+	+	+
	Thermohydraulics on primary side of SG[a]	+	+	+	+	+	+	+	o	+		
	Pressurizer thermohydraulics[a,b]	+	+	+	+	o	o	o	o	+	+	o
	Surge line hydraulics (CCFL, chocking)[b]	+	+	+	+	o	o	o	o	o	+	+
	Single- and two-phase pump behavior[a]	+	+	+	+	o	o	o	o	o	+	o
	Thermohydraulic-nuclear feedback	+	−	−	+	+	−	−	−	+	−	−
	Structural heat and heat losses[c]	o	o	o	o	o	o	o	o	o	−	−
	Boron mixing and transport	+	+	−	−	−	−	o	−	−	−	o

[a]These phenomena are not covered in the OECD/NEA-CSNI Matrix for PWRs.
[b]For phenomena requiring separate effects test, e.g., pressurizer behavior.
[c]This phenomenon is related to scaled test facilities.

structures or a system which uses active components in a very limited way to initiate subsequent passive operation. Four categories were established to distinguish the different degrees of passivity. THPs and related parameter ranges that characterize the performance of passive systems do not differ, in general, from phenomena that characterize the performance of systems equipped with active components. This is specifically true for transient conditions occurring during safety relevant scenarios. In other words, one can say that friction PD or HT coefficients are affected by local velocity and void fraction and not by the driving force that establishes those conditions, for example, gravity head or centrifugal pump. The same can be repeated for more complex phenomena like two-phase critical flow or countercurrent flow limiting.

Thus, a large number of THPs expected to occur in passive systems during transients and accidents are classified in OECD/NEA/CSNI documents as "separate effects" and "integral effects" as reported in Aksan et al. (1994) and Annunziato et al. (1996) hereafter. However, the specific layouts of passive systems and combinations of parameter ranges brought about the need to expand the original list of phenomena in the mentioned references. This was done in Aksan and D'Auria (1996), where mainly the passive systems proposed at the time of issuing of the report were considered.

As part of the IAEA's overall effort to foster international collaborations that strive to improve the economics and safety of future water-cooled NPP, an IAEA CRP was started in early 2004. This CRP, titled Natural Circulation Phenomena, Modeling and Reliability of Passive Safety Systems that Utilize Natural Circulation, focused on the use of passive safety systems to help meet the safety and economic goals of a new generation of advanced NPP. This CRP has been organized within the framework of the TWG-LWR and TWG-HWR and provided international cooperation on research work underway at the national level in several IAEA Member States.

The use of passive safety systems such as accumulators, condensation, evaporative heat exchangers, and gravity-driven safety injection systems, eliminates costs associated with the installation, maintenance, and operation of active safety systems that require multiple pumps with independent and redundant electric power supplies. As a result, passive safety systems are being considered for numerous reactor concepts (including in Generation III and III + concepts) and are expected to find applications in the Generation IV reactor concepts, as identified by the GIF Another motivation for the use of passive safety systems is the potential for enhanced safety through increased safety system reliability.

The CRP on NC and passive safety systems benefits from earlier IAEA activities that include developing databases on physical processes of significant importance to water-cooled reactor operations and safety, for example, in the area of THPs in AWCRs, recent IAEA activities have assimilated data internationally on HT coefficients and PD (IAEA, 2001); and have shared information on NC data and analytical methods (IAEA, 2001), and on experimental tests and qualification of analytical methods (IAEA, 2000a). To establish the progress of work in this CRP, an integrated research plan with a description of tasks addressing the objectives of the CRP was defined. These tasks are:

- Establish the state of the art in NC.
- Identify and describe reference systems.
- Identify and characterize phenomena that influence NC.
- Examine the application of data and codes to design and safety.
- Examine the reliability of passive systems that utilize NC.

The activity under the first task is aimed at summarizing the current understanding of NC system phenomena and the methods used experimentally to investigate and model such phenomena. In November 2005, the IAEA issued a technical document (IAEA, 2005), developed by the collaborative effort of the CRP participants and with major contributions from some selected experts in the CRP, aimed at documenting the present knowledge in six specific areas; advantages and challenges of NC systems in advanced designs, local transport phenomena and models, integral system phenomena and models, NC experiments, advanced computation methods, and reliability assessment methodology.

The activity for the third task is aimed at identifying and categorizing the NC phenomena of importance to advanced reactors and passive safety system operations and reliability. This task is the major link between the second and the fourth tasks. During the phase to perform the activities on the mentioned tasks, one of the unique features was to collect and include plant design descriptions with a strong emphasis on passive safety systems of the specific design. These descriptions of the passive safety systems were considered during the phenomena identification and characterization process by the group of experts. As a basis for PH identification in AWCRs, earlier works performed within the OECD/NEA framework during 1983–1997 were considered in detail. These are:

- Code validation matrix of thermal-hydraulic codes for LWR LOCAs and transients (OECD/NEA/CSNI, 1987)
- SOAR on thermohydraulic of ECC in LWR (Lewis et al., 1989)
- SET validation matrix for LWR (Aksan et al., 1994)
- Integral facility tests validation matrix for LWR (Annunziato et al., 1996)
- Status report on relevant thermal-hydraulic aspects of ARDs (Aksan and D'Auria, 1996)

Since Generation III and III + reactor designs contain technological features common to current-generation reactors, the phenomena identified during the work performed for the first to fourth items can be used as base knowledge. The fifth item provides important and relevant THPs for ARDs in addition to the relevant THPs identified for the current generation of LWRs. The list of relevant phenomena established in Aksan and D'Auria (1996) has been taken as the basis for AWCR work and has been modified according to the reactor types and passive safety systems considered. It should also be noted that in identifying the relevant THPs in the list provided in Table 6.13, expert judgment is the main contributor.

Table 6.13 Identification and characterization of phenomena for advanced water-cooled reactor designs including natural circulation and passive safety systems.

1	Behavior of large pools of liquid	Thermal stratification Natural/forced convection and circulation Steam condensation (e.g., chugging, etc.) Heat and mass transfer at the upper interface (e.g., vaporization) Liquid draining from small openings (steam and gas transport)
2	Tracking of noncondensable gases (essentially H_2, N_2, air)	Effect on mixture to wall heat transfer coefficient Mixing with liquid phase Mixing with steam phase Stratification in large volumes at very low velocities
3	Condensation on the containment structures	Coupling with conduction in larger structures
4	Behavior of containment emergency systems (PCCS, external air cooling, etc.)	Interaction with primary cooling loops
5	Thermo-fluid dynamics and pressure drops in various geometric configurations	3-D large flow paths, e.g., around open doors and stair wells, connection of big pipes with pools, etc. Gas–liquid phase separation at low Re numbers and in laminar flow Local pressure drops
6	Natural circulation	Interaction among parallel circulation loops inside and outside the vessel Influence of noncondensable gases Stability Reflux condensation
7	Steam liquid interaction	Direct condensation Pressure waves due to condensation
8	Gravity-driven cooling and accumulator behavior	Core cooling and core flooding
9	Liquid temperature stratification	Lower plenum of vessel Downcomer of vessel Horizontal/vertical piping
10	Behavior of density locks	Stability of the single interface (temperature and density distribution) Interaction between two density locks
11	Behavior of check valves	Opening/closure dynamics Partial/total failure
12	Critical and supercritical flow in discharge pipes and valves	Shock waves Supercritical flow in long pipes Behavior of multiple critical flow sections

Table 6.13 Continued

13	Behavior of heat exchangers and isolation condensers	Low-pressure phenomena
14	Stratification and mixing of boron	Interaction between chemical and thermohydraulic problems Time delay for the boron to become effective in the core
15	Core makeup tank behavior	Thermal stratification Natural circulation

The "expanded" OECD/NEA list of phenomena for passive systems was upgraded in the IAEA CRP on Natural Circulation Phenomena, Modeling and Reliability of Passive Safety Systems that Utilize Natural Circulation, considering recently proposed passive systems by the industry. The identification and characterization of additional (i.e., from the original separate effects and integral effects lists) phenomena for passive systems are presented in Table 6.13, which includes two main columns, other than the first column with numbering, which is consistent with the phenomena numbering in Aksan and D'Auria (1996):

a Column 1: phenomena number,
b Column 2: phenomena identification,
c Column 3: phenomena characterization based on individual phenomena descriptions.

The content of Table 6.13 is self-standing and directly understandable. In addition, a supportive description of individual phenomena is provided to clearly describe each PH. However, the following additional items should be noted:

- Specific geometry configurations or range of variations affecting thermal-hydraulic parameters justify the presence of phenomena at rows 2, 5, 6, 7, and 14 in both the present list and the list in Aksan et al. (1994). This is specifically true in the case of PH 6 (NC), which is expected to occur whenever a gravity environment exists.
- NC is also at the origin of core makeup tank (CMT) performance, the PH at row 15 in Table 6.13. However, the simultaneous presence of stratification in the tank, possible condensation with level formation inside the tank, the specific loop connection, and the values of boundary and initial conditions, suggest the consideration of a separate PH.

This section provides some details on important and relevant THPs for current-generation AWCRs (AP1000, ESBWR, EPR, VVER-1200, MASLWR, CAREM, SMART, etc.) in Table 6.13. Detailed descriptions of the 15 phenomena related to the listed AWCRs are also given.

6.5.1 Behavior in large pools of liquid

Large pools of water (e.g., up to several thousand cubic meters) at near-atmospheric pressure are incorporated into several advanced designs. These large pools provide a heat sink for heat removal from the reactor or the containment by NC, as well as a source of water for core cooling. Examples include the pressure suppression pool (wet well) of the ESBWR, the in-containment refueling water storage tank of the AP-1000, the pool of the emergency condenser of the SWR-1000, and the gravity-driven water pool of the AHWR.

Large pools may have a very wide spectrum of geometric configurations. HT in a limited zone in terms of volume (e.g., by condensing injected steam or by HT from an isolation condenser—IC) does not imply a homogeneous or nearly homogeneous temperature in the pool. Three-dimensional convection flows develop and affect the HT process, which results in temperature stratification.

Steam generated by HT or following injection may be released from the pool into the containment and influence the increase in containment pressure. Compared with homogeneous temperature distribution, the fluid at the top of the pool may reach the saturation temperature while the bulk fluid is subcooled. The evaporation from the top of the pool results in a pressure increase in the containment. Therefore, temperature stratification influences plant design. The three-dimensional nature of temperature stratification requires appropriate modeling.

The PH is identified as A-6-POO-c (or No 8 in the list of 116 THPs) by Aksan et al. (2018a), and D'Auria et al. (2021b).

6.5.2 Effects of noncondensable gases on condensation heat transfer

Condensation occurs when the temperature of vapor is reduced to below its saturation temperature. The presence of even a small amount of noncondensable gas (e.g., air, N_2, H_2, He, etc.) in the condensing vapor leads to a significant reduction in HT during condensation. The buildup of noncondensable gases near the condensate film inhibits the diffusion of vapor from the bulk mixture to the liquid film. The net effect is to reduce the effective driving force for heat and mass transfer. This PH is the concern of industrial applications and nuclear reactor systems.

In nuclear plants, the condensation of steam in the presence of noncondensable gas becomes an important PH during LOCAs when steam released from the coolant system mixes with the containment air. In addition, nitrogen gas in accumulators is a source of noncondensable gas, which can affect condensation HT inside the SG tubes of the NPP and may affect CMT performance. The effect of noncondensable gases on condensation HT is also relevant to certain decay heat removal systems in ARDs, such as the passive containment cooling system (PCCS).

The effect of noncondensable gases on the condensation of steam has been extensively studied for both natural and FCO flows. In each of them, geometries of interest (e.g., tubes, plates, annulus, etc.) and the flow orientation (horizontal, vertical) can be different for various applications. The condensation HT is affected by parameters such as mass fraction of noncondensable gas, system pressure, gas/vapor mixture Reynolds number, orientations of surface, interfacial shear, Prandtl number of condensate, etc. Multicomponent noncondensable gases can be present.

The PH is identified as A-9-HTCO-c (or No 50 in the list of 116 THPs) by Aksan et al. (2018a), and D'Auria et al. (2021b).

6.5.3 Condensation on containment structures

This PH involves heat and mass transfer from the containment atmosphere toward the surrounding structures. This PH would occur in existing reactors in the case of a coolant release into the containment. It also occurs in advanced designs where containment surfaces are cooled externally, usually by natural mechanisms. Good examples are the designs of the AP series by Westinghouse, where the steel containment is cooled externally by water flowing on its exterior surface from a reservoir above the containment, and by ascending air driven by buoyancy.

Steam condensation is largely affected by conditions that can be split into two groups depending on the relevance of the physical dimensions of the system. "Scale-independent factors" are variables like the fraction of noncondensables, pressure, gas composition, and so on, the effects of which could be well investigated through SETs. "Scale-dependent factors" are phenomena that require investigation with actual or scaled geometries (i.e., integral effects tests) since physical dimensions largely influence their quantitative effect. Examples of this kind are the NCO process at both sides of metallic structures and potential gas stratification.

Phenomena identification is considered in Section 6.8.

6.5.4 Behavior of containment emergency systems

Nuclear power reactor containments are equipped with safety systems that protect the containment integrity under various accident conditions. The focus of this PH is the NC cooling and HT in various containment passive cooling systems under accident conditions to remove the energy out of the containment by NC and condensation HT. Typical systems are the tube condensers such as the PCCS and external air cooling system or external liquid film cooling and internal condensation of steam in the containment by NC. The major purpose of these containment systems is to protect the containment under both DBA and severe accidents involving serious core damages and to prevent the significant release of radioactive materials into the atmosphere. These systems are required to remove the load on the containment from the LOCA and other accidents by removing the heat but containing the mass within the structure. Most of the load comes from the released steam from the primary coolant system due to the LOCA or venting of the pressure relief valves. The major part of the noncondensable gases consists of the original containment atmosphere such as air or nitrogen, however, with the core damage, hydrogen or fission gases can be also released into the containment atmosphere. The THPs of importance are tube surface condensation with noncondensable gases, NC of steam and noncondensable gases, degradation of condensation by the accumulation of noncondensable gases, and purging of noncondensable gases from condenser systems. The PCCS can be vertical or horizontal tube condensers in an external water pool, an exposed condenser tube system in the containment cooled by NC water through the tubes from the external pool or by external air circulation, and others.

Phenomena identification is considered in Section 6.8.

6.5.5 Thermofluid dynamics and pressure drops in various geometric configurations

PD is the difference in pressure between two points of interest in a fluid system. In general, PD can be caused by resistance to flow, changes in elevation, density, flow area, and flow direction. PD in NC systems plays a vital role in their steady state, transient, and stability performance.

It is customary to express the total PD in a flowing system as the sum of its components such as distributed pressure loss due to friction, local pressure losses due to sudden variations of shape, flow area, direction, etc., and pressure losses (the reversible ones) due to acceleration (induced by flow area variation or by density change in the fluid) and elevation (gravity effect). An important factor affecting the pressure loss is geometry. In a nuclear reactor, we have to deal with several basic geometric shapes (circular pipes, annuli, etc.) and several special devices like rod bundles, heat exchangers, valves, headers, plenums, pumps, large pools, etc. Other factors are concerned with the fluid status (single- or two-phase/one-component, two-component, or multicomponent), the flow nature (laminar or turbulent), the flow pattern (bubbly, slug, annular, etc.), the flow direction (vertical upflow, downflow, inclined flow, horizontal flow, CCF, etc.), flow type (separated and mixed), flow paths (one-dimensional or multidimensional, open or closed paths, distributor or collector), and the operating conditions (steady-state or transient).

An important focus of this PH is the geometric conditions that hinder the establishment of fully developed flow especially when the fluid in question is a mixture of steam, air, and water. This complex thermo-fluid dynamic PH warrants special attention. However, it is worth mentioning here that though in many systems like the primary system of an NPP, flow is mostly not fully developed, PD relationships used in these systems are invariably those obtained for developed flow. This practice is also experimentally proved to be more than adequate in most of the cases. However, in some specific cases like containment internal geometry, it is necessary to consider thermos-fluid dynamics in the developing region.

A final, very important, issue is concerned with the driving force depending on whether the flow is sustained by a density difference in the fluid (NC) or by a pump (FCO), or whether there will be feedback between the pressure loss and the extracted power or not. Normally the pressure loss inside a device depends on the nature of flow through the device and not on the nature of the driving head causing the flow. However, under some circumstances, because of local effects, the pressure loss may be influenced by the nature of the driving force.

The PH is identified as B-7-PD-c (or No 84 in the list of 116 THPs) by Aksan et al. (2018a), and D'Auria et al. (2021b).

6.5.6 Natural circulation

NC comprises the complex set of THPs that occur in a gravity environment when geometrically or materially distinct heat sinks and heat sources are connected by a fluid. No external sources of mechanical energy for fluid motion are involved when NC is established.

The above definition includes the situations of a heater immersed in a fluid, of an open flame in the air, of a chimney-driven fire, of an insurge of hot fluid into a pool of cold liquid, and of a heat source and sink (e.g., heater and cooler) consisting of separated mechanical components connected by piping and situated at different gravity elevations. NC also drives the occurrence of stratification in horizontal pipes.

Within the scope of this document, this PH involves the following system configurations.

(a) Heat source and sink of the primary loop constituted, respectively, by core and SG, or boiler, or primary side of the heat exchanger, with the core located at a lower elevation.
(b) Heat source and sink inside the pressure vessel, constituted, respectively, by core and (typically annular-like region of) vessel downcomer. "Steady-state" NC between core and downcomer occurs owing to continuous cooling of the downcomer fluid by a heat exchanger (boiler or SG) or by the continuous inlet of feedwater liquid at a temperature lower than the core outlet temperature.
(c) Cooling of the containment atmosphere by a closed loop.

In the current generation of nuclear plants, the NC core power removal capability is exploited for accident situations to demonstrate the inherent safety features of the plant (with the notable exception of the Dodewaard commercial BWR unit, shutdown in 1997). The NC is also occurring during various phases of the refueling.

In the future generation of nuclear plants, NC is planned to be used for ensuring the nominal operating conditions and for achieving safe cooling following accidents in a wider spectrum than foreseen for current generation reactors.

Related phenomena considered in the lists by Aksan et al. (2018a), and D'Auria et al. (2021b) are identified in Section 6.4.1, item (2).

6.5.7 Steam–liquid interactions

Large pools may have a very wide spectrum of geometric configurations. HT in one very limited zone in terms of volume (e.g., by condensing injected steam or by HT from a passive containment cooler) does not imply homogeneous or nearly homogeneous temperature in the pool. Many containment phenomena require a steam-liquid interface. Steam discharge into a suppression pool of BWR is a good example of this case; see also Section 6.3.8.3, PH S-12-CO4-c, or No 25 in the list of 116 THPs by Aksan et al. (2018a), and D'Auria et al. (2021b). After the breakup of the originally created bubbles in the suppression pool, the subsequent formation of bubble plumes takes place. Consequently, complete condensation occurs and this induces mixing in the pool, the process is determined by single- and two-phase NC. It is important to understand the breakup and plume-stirring process and mechanisms because the system pressure is ultimately controlled by the pressure in the vapor space above the water surface in the suppression chamber. This pressure is the sum of the partial pressures of steam and gas, the former controlled by the temperature at the pool surface. In turn, the pool surface temperature depends on the efficiency of steam condensation in the pool and the degree of mixing in the pool.

The following is a listing of steam-liquid interaction-related phenomena:

- Direct contact condensation of steam in pool water
- Bubble formation and breakup and the subsequent formation of bubble plumes
- Break-up and plume-stirring process and mechanisms inducing mixing in the pool

Examples of steam-liquid interactions are passive containment cooling that vents into the suppression pool of the ESBWR and the injection of steam–gas mixtures through a downcomer vent line into the suppression pool.

6.5.8 Gravity-driven cooling and accumulator behavior

Gravity-driven cooling provides ECC water by gravity draining, in events with a loss of coolant. This system requires a large volume of water above the core, plus additional depressurization capacity so that the primary coolant system can be depressurized to allow for gravity flow from the elevated suppression pool. Since there are no large reactor vessel pipes at or below the core elevation, this design ensures that the core will remain covered by water during all DBA. In general, the gravity-driven cooling concept is mainly based on the depressurization of the reactor pressure vessel (RPV) to a sufficiently low pressure to enable reflood of the core by gravity feed from an elevated pool. When the gravity-driven cooling operates, the gravity drain flow rate to the RPV depends on the piping geometry, the state of the fluid, and the pressure conditions in both the water pool and the RPV. Flow entering the RPV during the later stages of blowdown during a postulated LOCA must be sufficient to keep the nuclear core flooded. The system which provides gravity-driven cooling is a simple and economical safety system.

The following is a listing of the gravity-driven cooling-related phenomena:

- Depressurization of the RPV by discharging through depressurization valves into the drywell and increased pressure in the upper part of the containment
- Evaporation in the RPV due to depressurization
- Friction in the gravity-driven cooling system (GDCS) and injection lines, including the valves in these lines
- Large amounts of cold water immediately flood the lower parts of the RPV, causing:
  - collapse of voids
  - condensation of steam
  - suppression of boiling
  - increase in water level inside the RPV
- Condensing of the steam out of the RPV and drywell gas space until air accumulates on the primary sides of the PCCS, resulting in the termination of steam condensation

Examples of GDCS include the GDCS of ESBWR and the passive core flooding system (HA-2 hydraulic accumulators of the second stage) of WWER-1000/392 and passive core flooding systems (ECCS tank) of WWER-640/407.

The PH is identified as A-8-GDR (or No 45 in the list of 116 THPs) by Aksan et al. (2018a), and D'Auria et al. (2021b).

6.5.9 Liquid temperature stratification

Nuclear reactors that implement NC passive safety systems may produce large temperature gradients in their working fluid as a result of local cooling caused by ECC injection or local heating through steam condensation or heat exchanger HT. Thermal stratification arises because the low-flow condition typically encountered in an NC system greatly reduces the amount of fluid mixing that can occur. Examples of thermal stratification during ECC injection include the formation of cold plumes in the downcomer, and liquid thermal stratification in the LP, cold legs, and loop seals.

ECC injection into horizontal piping partially filled with steam also results in liquid temperature stratification. The cooler liquid condenses the steam forming a saturated layer of liquid water on top of the subcooled liquid layer. This saturated layer is at a higher temperature than the subcooled layer, resulting in a stratified temperature condition. The formation of the saturated layer may mitigate occurrences of condensation-induced water hammer events.

Liquid temperature stratification can also arise in passive safety systems such as the NC-driven CMT and the large liquid-filled tanks that serve as the heat sink for reactor core or containment passive cooling systems. Steam vented into the large safety tanks condenses in the cold liquid producing hot rising plumes that form thermal layers at the free surface of the tank. Thermal layers having different temperatures grow with time to create a large temperature gradient in the liquid.

The PH is identified as A-10-LTS-c (or No 58 in the list of 116 THPs) by Aksan et al. (2018a), and D'Auria et al. (2021b).

6.5.10 Behavior of density locks

The stability of the interface of the density locks, especially when two density locks are present, appears a critical aspect. Possible long-term variations of the static head in the pool (e.g., due to stratification or heating) may change the interface position in each density lock and consequently stability characteristics.

The PH is identified as A-4-DL-c (or No 6 in the list of 116 THPs) by Aksan et al. (2018a), and D'Auria et al. (2021b).

6.5.11 Behavior of check valves

Check valves connect the primary circuit with very large volumes through large pipes. Conditions in the piping with check valves may arise that cause rapid condensation on one side resulting in slam closure of the valve. The small diving forces may not be sufficient to reopen the valves again. The failure to open may prevent the coolant from flowing into the primary circuit; the failure to close may cause draining of the primary circuit; and in addition opening and closure cycles critical oscillations in the flow rates.

The PH is identified as A-1-CHV-c (or No 4 in the list of 116 THPs) by Aksan et al. (2018a), and D'Auria et al. (2021b).

6.5.12 Critical and supercritical flow in discharge pipes and valves

Under this item, shock waves, supercritical flow in long pipes, and behavior of multiple critical sections are included, see also Govier and Aziz (1976).

The PH is identified as A-7-CSC (or No 30 in the list of 116 THPs) by Aksan et al. (2018a), and D'Auria et al. (2021b).

6.5.13 Behavior of emergency heat exchangers and isolation condensers

The removal of decay heat from a nuclear core can be accomplished by passive means using either an emergency heat exchanger or an IC, depending on the system design. In some advanced PWRs, the emergency heat exchanger decay heat removal system consists of a closed loop that includes a shell and tube heat exchanger immersed in a large liquid pool that is elevated above the core. The relative elevation between the heat source and heat sink creates a buoyancy-driven NC flow that eliminates the need for a pump. Decay heat is removed from the core by convective HT from the fuel to the single-phase liquid in the reactor vessel. The heat stored in the liquid is carried by NC to the emergency heat exchanger. Heat is transferred from the fluid through the emergency heat exchanger tubes into the pool by three mechanisms: single-phase convective HT at the tube inside surface, heat conduction through the tube walls, and nucleate boiling at the tube outside surface. Some advanced PWRs use the SG as an intermediate emergency heat exchanger with a passively cooled, NC feedwater loop.

Some ABWRs use ICs as a means of removing core decay heat. The IC consists of a shell and tube heat exchanger immersed in a large liquid pool elevated above the core. In a BWR, core decay heat is removed by nucleate boiling. The steam generated by this process is condensed inside the IC tubes creating a low-pressure region inside the tubes which draws in additional steam. Thus the driving mechanism for the flow is steam condensation. Heat is transferred through the IC tubes into the pool by three mechanisms: single-phase steam condensation (phase change) at the tube inside surface, heat conduction through the tube walls, and convective HT at the tube outside surface. The condensate is returned as a single-phase liquid to the reactor vessel by gravity draining. The performance of the IC can be affected by the presence of noncondensable gases.

The following is a listing of the emergency heat exchanger-related local phenomena:

- emergency heat exchanger loop flow resistance
- buoyancy force
- single-phase convective HT

- shell-side nucleate boiling HT

The following is a listing of the IC-related local phenomena:

- IC loop flow resistance
- pressure steam condensation
- condensation HT in the presence of noncondensable gases
- shell-side convection HT
- condensate/steam CCFLs

One related PH is identified as A-5- PC-c (or No 7 in the list of 116 THPs) by Aksan et al. (2018a), and D'Auria et al. (2021b).

6.5.14 Stratification and mixing of boron

Boric acid is introduced into the reactor coolant to control long-term reactivity. Forced coolant circulation during normal operation ensures that the boric acid is homogeneously distributed in the RCS so that the boron concentration is practically uniform. A decrease in the boron concentration results in an increase in reactivity. Causes for decreasing boron concentrations are the injection of coolant with less boron content from interfacing systems (external dilution) or the separation of the borated reactor coolant into highly concentrated and diluted fractions (inherent dilution). Examples of external dilution are the injection of coolant of reduced boron concentration by the makeup system and the injection of low-boron pump sealing water into the primary system. Inherent dilution can occur after reflux condenser HT or backflow from the secondary system in the case of primary-to-secondary leakage accidents.

Operation in the reflux condenser mode over a lengthy period could occur in the event of an SBLOCA concurrent with limited operability of the ECCS. In such an event the condensate descending down the cold-leg SG tubing into the SG outlet plenum and from there into the pump seal could form slugs of low-boron water. On restoration of NC after refilling of the RCS such slugs would be transported toward the reactor core. However, on their way to the core, they would be mixed in the cold-leg piping, the RPV downcomer, and the LP and thus increase in boron concentration. Restarting a reactor coolant pump (RCP) after an SBLOCA or an SGTR is very unlikely to occur as such events can be identified based on measured data and starting an RCP is an action that would provide several individual actions and therefore some time. Assuming that an inadvertent demineralized water injection into one loop was to occur before the starting of the RCP in this loop despite the monitoring and measurement of the boron concentration of the water injected into the RCS, a slug of demineralized water moves toward the core inlet after pump start. Mixing the diluted slug with the ambient coolant of higher boron content provides the only mitigation mechanism before the slug enters the core.

The main mixing mechanism in the case of the low-boron water slug accelerated by the RCP start is turbulent mixing between the fluid flows having different velocities whereas in the case of reestablishing the NC after a reflux condensation phase, the main mixing mechanism is buoyancy-driven turbulent mixing. The density differences between the fluids are due to the temperature and the boron concentration differences.

The PH is identified as A-12- BO (or No 103 in the list of 116 THPs) by Aksan et al. (2018a), and D'Auria et al. (2021b).

6.5.15 Core makeup tank behavior

Several ARDs implement CMT to provide NC cooling to the core. CMT are elevated tanks connected to the reactor vessel and primary loop at the top and bottom of the tank. Special lines connect the bottom of the tank with the vessel and are termed direct vessel injection. In connection with this, an important interaction occurs between the CMT, the accumulator, and the IRWST also considering the actuation signal for automatic depressurization. The tanks are filled with cold borated water and can provide coolant injection at system pressure. The tanks are normally isolated from the reactor vessel by an isolation valve located at the bottom of the vessel. The fluid is always sensing full system pressure through the top connection line. In the event of an emergency, the bottom isolation valve is opened to complete the NC loop and permit cold borated water to flow to the core. The relative elevation between the core and the CMT and the density difference between the hot primary system water and the cold CMT water creates a buoyancy-driven NC flow that eliminates the need for a pump. Decay heat is removed from the core by convective HT from the fuel to the single-phase liquid in the reactor vessel. CMT behavior includes NC, liquid thermal stratification in the tank, and liquid flashing during plant depressurization.

The PH is identified as A-3-CMT-c (or No 5 in the list of 116 THPs) by Aksan et al. (2018a), and D'Auria et al. (2021b).

6.6 Thermal-hydraulic phenomena for supercritical water-cooled reactors

SCWR has been currently chosen by Generation IV International Forum as one of the reactor designs to be investigated. It is the only water-cooled NPP studied under the Generation IV framework. In Europe, investigations on the SCWR concept have been integrated into a joint research project, called a high-performance light-water reactor (HPLWR), which is an LWR operating at supercritical pressure (25 MPa). HPLWR is cofunded by the European Commission and has been worked out by a consortium of 10 partners from 8 European countries as their contribution to the Generation IV International Forum. The concept is based on an RPV similar to a PWR but designed for a higher pressure of around 25 MPa. Other plant characteristics of the HPLWR include also a coolant heat up from 280°C to more than 500°C (Squarer et al., 2003). At these conditions, water changes its phase continuously from liquid to steam without boiling, so that a boiling crisis in the core can be physically excluded. As in a BWR, the high-temperature steam is fed directly to the high-pressure turbine so that a closed primary cycle like that found in a PWR can be omitted. Steam separators or primary pumps, which are needed in BWR, are not required either.

High steam enthalpy increases the power density of the steam cycle by more than 40%. The envisaged net efficiency of 44% exceeds by far those of conventional LWR (Bittermann et al., 2004). It is expected that the HPLWR will be more economical than conventional LWR (Bittermann et al., 2003), it will have a higher efficiency and thus a better fuel utilization, which also produces less waste per kWh, but it still fulfills the high safety standards of third-generation nuclear plants, for example, EPR, SWR1000 (Aksan et al., 2003). The key advantages of this concept are a significant reduction in equipment, resulting in a smaller building volume and consequently lower plant erection costs compared with conventional PWR design, and lower fuel costs thanks to higher thermal efficiency while keeping the safety features at least as advanced as in third-generation LWRs.

During the safety analysis studies for the HPLWR, one of the initial phase activities was to identify the new additional THPs that needs to be considered for the thermal-hydraulic safety analysis. As a result of these investigations, a list of THPs for HPLWR conditions has been identified and selected in addition to the phenomena included in earlier studies mentioned in Sections 6.3 and 6.4. These are:

- thermal-physical properties near and at supercritical conditions
- critical flow during depressurization from supercritical conditions
- pressure loss characteristics of supercritical water (SCW)
- HT characteristics of SCW
- reflooding for tight lattice geometry at low pressure
- boil-off for tight lattice geometry at low pressure
- instability and NC in supercritical pressure systems

The above six identified THPs for HPLWR have also been confirmed for other SCWR designs (eight different concepts including SCWR using heavy water) during investigations and studies carried out (2007–12) by a group of scientists in an IAEA CRP published as *Heat Transfer Behaviour and Thermohydraulics Code Testing for Supercritical Water Cooled Reactors (SCWRs)* (IAEA, 2014a).

6.6.1 Thermal-physical properties near and at supercritical conditions

HT at supercritical pressure is mainly characterized by the thermal physical properties which vary strongly, especially near the pseudocritical line. Fig. 6.1 shows the specific heat as a function of pressure and temperature.

It is seen that at each pressure there is a local maximum of the specific heat capacity. In the subcritical pressure range the maximum specific heat is located on the saturation line. At the critical point ($P = 22.1$ MPa, $T = 374°C$) specific heat has its maximum value. In the supercritical pressure range, the line connecting the maximum values of the specific heat is called the pseudocritical line (PCL), as shown in Fig. 6.2.

It is seen that the pseudocritical temperature increases with increasing pressure. At a pressure of 25 MPa the pseudocritical temperature is 384°C. The specific heat at the

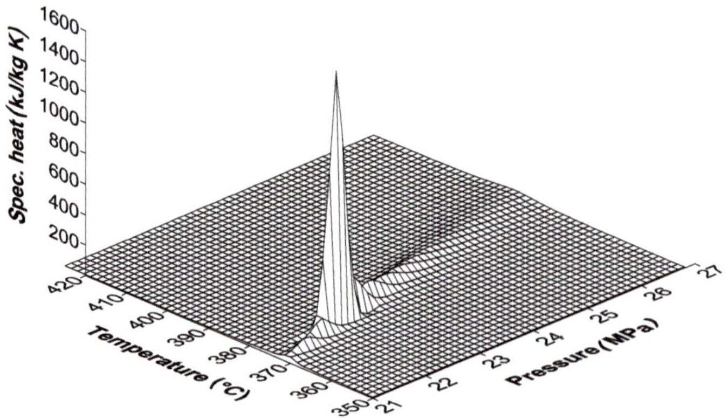

Figure 6.1 Specific heat of water (Wagner and Kruse, 1997).

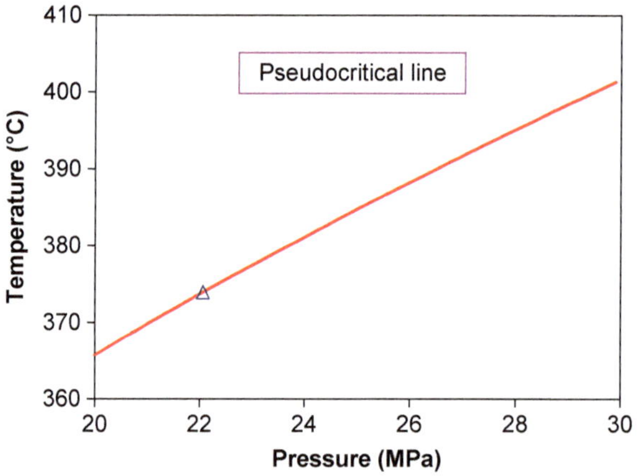

Figure 6.2 Pseudocritical line in a pressure–temperature diagram.

critical point is as high as 5600 kJ/kg K, which is more than 1000 times higher than that at room temperature. Near the pseudocritical line, the density decreases dramatically. There exists a large peak of the thermal expansion coefficient, which behaves very similarly to the specific heat. Thermal conductivity decreases with increasing temperature. It shows, however, a local maximum near the pseudocritical point. Beyond the pseudocritical temperature thermal conductivity decreases sharply. Similar behavior shows also dynamic viscosity. Due to the sharp increase in specific heat capacity, there exists a large peak of the Prandtl number at the pseudocritical point.

It has been found that near the critical point, thermal-physical properties obtained from different sources might deviate strongly from each other, especially as related to specific heat and thermal conductivity. This is due to the steep variation of the fluid properties, leading to a less accurate interpolation and a lower measurement accuracy. A comparison of two water-steam tables in use at EDF shows a deviation of up to 35% for thermal conductivity and up to 10% for heat capacity under a condition close to the critical point. If the physical properties show these types of deviations, the physical parameters, which describe certain PH, for example, the HT coefficient will also show some calculated differences. Consequently, it is important to use the same or at least similar steam tables for the application and use of different computer codes within a design or safety-related project when analysis work is performed.

Based on the experience gained and the critical review performed during the HPLWR project, it was recommended that the steam tables of IAPWS-97 and IAPWS-95, whichever is relevant were well accepted by international institutions and, therefore, should be taken as reference. Due to the large expenditure needed for the implementation of such a steam table into the existing codes, it is not required to use a unique steam table for each of the codes, but the deviation (compared with IAPWS-97 and IAPWS-95, respectively) of the steam tables used should be clearly indicated. The International Association for the Properties of Water and Steam (IAPWS) is an international, nonprofit association of national organizations concerned with the thermo-physical properties of water and steam, particularly those properties of high-temperature steam, water, and aqueous systems that are relevant to thermal power cycles and other industrial applications. The IAPWS released its formulation for general and scientific use in 1995. IAPWS-95 is a state-of-the-art, accurate representation and covers the entire range of light-water properties (Wagner and Kruse, 1997). At the limit boundaries, ideal gas behavior is satisfied and at the other "residual" behavior.

6.6.2 Critical flow during depressurization from supercritical conditions

Critical flow is defined as the maximum flow that can pass through an opening for a given thermodynamic condition. Under this definition, numerous configurations are possible: orifices or long tubes, single-phase flow or two-phase flow, rough or smooth tubes, sharp or round entrances, etc. All of these configurations affect the flow in their manner and thus must be characterized based on the specific application, in this case, an LOCA of an SCWR.

When a fluid discharges from a high-pressure and high-temperature system, a "choking" or critical condition occurs, and the flow rate no longer increases as the downstream pressure is further decreased. During a postulated LOCA of a PWR, the break flow will be subject to this condition (see Section 6.3.2 for details). An accurate amount of the critical flow rate is important for the evaluation of the reactor safety, because this flow rate controls the loss of coolant inventory and energy from the system, and thus has a significant effect on the accident consequence. A depressurization event or LOCA is a particular concern for SCW reactor safety because the coolant inventory is lower than that of a typical PWR for the same power output. This lower coolant inventory would result in a faster transient response of the reactor dominated by the discharge flow rate. Critical flow is coupled with drastic depressurization and vaporization, and the flow rate is heavily dominated by the vapor content or quality of the vapor, which is closely related to the onset of vaporization and the interfacial interaction between phases. This presents a major challenge for estimating the flow rate due to a lack of knowledge about those processes, especially under the conditions of interest for the SCWR. High-pressure depressurization leading to critical flow has been investigated extensively, in both experimental and theoretical ways. These studies have mostly been conducted with subcritical pressure or ideal gases well above the vapor dome, as presented in several comprehensive reviews. Selected examples of these reviews are given in Levy et al. (1982), Lee and Swinnerton (1983), and Saha (1978), but little work has been done near or above the thermodynamic critical point. To the author's knowledge, the available experimental data for SCW are those obtained by Lee and Swinnerton (1983) at a maximum temperature of 402°C and by Chen with a maximum temperature of 474°C (Chen et al., 2009). Based on the limited data of supercritical fluids, the critical flows at conditions above the pseudocritical point seem to be fairly stable and consistent with the subcritical HEM predictions. Further, in this regime, the flows tend to be lower than those in the two-phase region. Thus the major difficulty in the prediction of the depressurization flow rate remains in the region where two phases coexist, that is, within the two-phase portion of the vapor dome. In this region, the flow rate is strongly affected by the nozzle geometry and tends to be unstable. Various models for this region have been developed with different assumptions, for example, the HEM and Moody model (Moody, 1965), and the Henry-Fauske nonequilibrium model (Henry and Fauske, 1971) are currently used in subcritical pressure reactor safety design (Ardron and Furness, 1976). It appears that some of these models could be reasonably extended to above the thermodynamic pseudocritical point. The more stable and smaller flow observed in conditions above the pseudocritical point suggests that even though SCWR designs have a smaller coolant inventory, the safety implications of an LOCA and the subsequent depressurization from supercritical conditions may not be as severe as expected and the limiting accident scenario may be driven by the critical flow within the subcritical pressure region.

6.6.3 Pressure loss characteristics of supercritical water

Pressure losses in the heat transport system have a direct impact on the pump capacity and the power output, affecting the capital and operating costs. In addition, these losses affect the local flow conditions along the fuel assembly impacting the HT characteristics. Depending on the design, the pressure loss components consist of frictional loss, form losses (such as valves and elbows), head loss due to gravity, and loss due to acceleration. A significant portion of the overall pressure loss occurs in the core region (particularly over the bundle assembly). Therefore, it is important to evaluate the core pressure loss accurately (see also Sections 6.3.1.7 and 6.3.1.8).

6.6.4 Heat transfer characteristics of supercritical water

In supercritical fluids, the properties change rapidly with the temperature, especially in the pseudocritical region. Here, the local HT coefficient varies significantly, depending on flow conditions and the value of the heat flux. The effect of heat flux, together with the thermo-physical properties variations needs to be understood, based on experimental, numerical, and theoretical data available to be able to predict the onset of heat transfer deterioration (HTD).

As already indicated in Section 6.6.1, a large variation of thermal physical properties occurs near the pseudocritical line. This would lead to a strong variation in the HT coefficient. Just considering simply for single-phase steam flow near and at supercritical conditions, the Dittus-Boelter equation gives at the pseudocritical point ($T = 384°C$) an HT coefficient of about 40 kW/m^2K, more than twice that at low temperature (e.g., 300°C) and five times that at high temperature (e.g., 500°C). This shows clearly that due to the variation in thermal physical properties, the HT coefficient varies strongly near the pseudocritical line. The closer the pressure to the critical point is, the higher the peak of the HT coefficient.

As indicated and agreed in the open literature, the actual HT coefficient deviates from the Dittus-Boelter equation, especially near the pseudocritical line. At low heat fluxes, the HT coefficient is higher than the values predicted by Dittus-Boelter. This PH is called HT enhancement. At high heat fluxes the HT coefficient is lower than that computed by the Dittus-Boelter equation. At a low heat flux, the wall temperature behaves smoothly and increases with increasing bulk temperature. The difference between the wall and the bulk temperature remains small. At a high heat flux a similar behavior of the wall temperature is expected, except for the bulk temperature approaching the pseudocritical value. In this case a sharp increase in the wall temperature can occur. The wall temperature decreases again when the bulk temperature exceeds the pseudocritical value. This large increase in the wall temperature is referred to as *heat transfer deterioration*. In the literature there is still no unique definition for the onset of HTD, because the reduction in HT coefficient, or the increase in the wall temperature behaves rather smoothly, compared with the behavior of the boiling crisis, in which a much sharper increase in the wall temperature takes place.

6.6.5 Reflooding for tight lattice geometry at low pressure

The concept and the geometric arrangement of the HPLWR core, as an example of SCWR, differs drastically from conventional LWR concepts by the use of SCW to achieve higher core outlet temperatures. The fuel assembly design examples are based on tight lattice geometry and have different configurations (Hofmeister et al., 2005). Considering these types of fuel assembly arrangements, tight lattice geometry and special arrangements of fuel rods are considered a solution. In general, the quenching and rewetting phenomena process defined and described in Sections 6.3.13.1 and 6.3.13.2 for SET is valid also for tight lattice geometries. But it must be considered that specific reflooding modeling and experiments are needed for selected geometric consideration.

6.6.6 Boil-off for tight lattice geometry at low pressure

As already mentioned and considered in Section 6.6.6, CHF and post-CHF HT phenomena defined for SET in Sections 6.3.12.3–6.3.12.5 are valid also for tight-lattice geometries. But it must be considered that specific CHF and post-CHF HT models and experiments are needed for selected geometric consideration.

6.6.7 Instability and natural circulation in supercritical pressure systems

The design of SCW-cooled reactors poses new challenges in relation to the prediction of some basic phenomena, whose characteristics are relatively well known at subcritical pressure. In particular, stability and NC phenomena observed in single-phase conditions at subcritical pressures become more complex at supercritical pressures because of the strong variability of thermodynamic and thermophysical properties. Moreover, categories adopted for dealing with two-phase flow phenomena must be updated to deal with supercritical pressure flow regimes, having in principle a simpler structure than the corresponding two-phase flow ones, though they anyway involve uneven distributions of lighter and heavier fluids within the channel cross section. Flow instabilities and NC, being dealt with in this section, have been the subject of major experimental and modeling efforts at subcritical pressures. Some of these studies belong to both aspects since a quite populated intersection of the two categories is represented by the interesting domain of NC instabilities. In single-phase flow, in particular, apart from turbulence and acoustic effects, which are beyond the scope of this section, instabilities mainly occur in NC systems, being driven by a mechanism explained with effective elementary arguments by Welander in 1967 (Welander, 1967). In that work, it was suggested that pockets of fluid with perturbed temperature may emerge from the heat source and/or the sink, affecting the fluid residence time along the loop and being damped or amplified as a consequence of the subsequent passages through the heater and the cooler. This mechanism was proposed based on numerical computations and was confirmed later on by further experimental and computational studies, focusing on different interesting aspects affecting the PH (see, e.g., Creveling et al., 1975). A remarkable feature of the observed behavior is its chaotic character, mainly caused by the fact that the growth of flow perturbations frequently leads to flow reversal; as a consequence, the system is seen to continuously switch from clockwise to counterclockwise flow directions and vice versa, showing a high sensitivity to initial conditions. Two-phase flow instabilities are mainly connected with the behavior of boiling channels that are recognized to be susceptible to different instability mechanisms, having a "static" or a "dynamic" nature (see, e.g., Lahey and Moody, 1993). The discrimination among the two categories of instabilities is introduced considering whether their occurrence can be explained based on the static internal PD versus flow characteristic of the system. The Ledinegg "excursive" instability mechanism (Ledinegg, 1938) is representative of the category of static instabilities, while density wave, PD oscillations, and thermal oscillations are mechanisms that belong to the second category. Of these instabilities, density waves represent the most relevant instability mechanism for nuclear reactor applications since, in combination with neutronic feedback, it is responsible for observed space and time oscillations of BWR core power. These instabilities have been the subject of extensive studies, summarized in review papers and SOARs (March-Leuba and Rey, 1993). Basic experimental investigations (see, e.g., D'Auria et al., 1997a) allowed grasping the main features of involved phenomena, gaining the basic understanding necessary for analytical developments. An interesting body of information about reactor plant events is also available (see D'Auria et al., 1997a). Models for predicting the unstable behavior of boiling systems have been set up and qualified by applications ranging from the scale of experimental facilities to the full reactor scale (see, e.g., D'Auria et al., 1997a). Both time-domain (i.e., transient) and frequency-domain (i.e., linearized) models are available to estimate the stability thresholds of even very complex systems. NC phenomena have also been thoroughly studied at subcritical pressures, both in single- and two-phase flow conditions. The relevance of this PH for nuclear reactors is well known since NC involves several regimes of reactor operation. As previously mentioned, NC and stability phenomena have an extensive intersection in single- and two-phase flows. A limited list of relevant aspects relating to NC at subcritical pressures is reported hereafter:

- Different single- and two-phase NC modes appear during start-up and under normal operating conditions and postulated LWR accidents, as observed in experiments and code predictions (see, e.g., D'Auria and Frogheri, 2002).
- NC is relied upon for decay heat removal from the nuclear reactor core and/or the containment system; in particular, passive decay heat removal by NC represents one of the most important features in advanced (Gen III) and innovative (Gen IV) reactors.
- Different modes of flow instabilities are recognized in NC boiling systems, adding additional phenomena to be addressed to those already observed in boiling channels (see, e.g., D'Auria and Frogheri, 2002).
- The vast body of information from experimental and theoretical research on stability and NC performed at subcritical pressures constitutes a useful background also for SCW reactor applications, at least for two main aspects.
- The knowledge acquired on fundamental phenomena can help support the new developments required for supercritical pressure conditions, provided that analogies and differences between the two operating regions (subcritical and supercritical) are properly identified.
- Part of the operational and perturbed evolution of SCW reactors will occur at subcritical pressures, directly involving operating conditions for which previous knowledge is immediately applicable.

In particular, stability has been addressed as an issue of great importance for SCWR design, because heated channels with supercritical pressure fluids are assumed to be susceptible to similar instability phenomena as observed in boiling channels. In fact, the transition across the pseudocritical temperature can be considered a sort of "pseudoboiling" PH, leading to denser fluid at the channel inlet and a lighter one at the outlet. This is one of the reasons why the basic understanding and the numerical tools developed for two-phase flow instabilities are found immediately available for being converted for supercritical pressure instabilities. The scarcity of relevant experimental data on instabilities in supercritical pressure systems is presently a problem to be coped with to ascertain that this process of knowledge transfer from one research field to another is made without forgetting any important difference. On the other hand, NC at supercritical conditions has been the subject of a number of experiments performed in past and recent times. Experimental facilities working at supercritical pressures, in particular, sometimes made use of NC even when the primary focus was on different aspects (e.g., HT). However, in these cases, their data on NC may be insufficiently detailed to allow for meaningful analytical conclusions. Though relevant experiments and modeling tools are already available in this field, there is still the need for further understanding and specific assessment.

6.7 Phenomena in the CANDU and RBMK research reactors

The thermal-hydraulic system codes that were developed for PWR and BWR applications, i.e., validated based on the PH discussed in previous sections, were used for applications in any type of water-cooled reactors including CANDU and RBMK research reactors (RRs). A preliminary analysis completed by Aksan et al. (2018a), looking at thermal-hydraulic PH, demonstrated the applicability of validation processes for those codes to the non-LWR water-cooled reactors, although noting that differences in parameter ranges may require additional investigations. For instance, in the case of RR, low-pressure (operating) conditions are those expected for shutdown transients in LWR, section 6.4.1, item (5), and heat transfer in plate-type fuel elements is characterized by specific geometry (e.g., covered by different parameters ranges of existing PH).

Additional phenomena for validation were considered for CANDU and RBMK:

- The behavior of headers or fluid collectors, typically installed upstream and downstream main coolant pumps
- The stratification in horizontal channels where (horizontal) nuclear fuel rods are present
- The behavior of steam drums in the case of RBMK
- The behavior of the specific UTSG design in the case of CANDU (e.g., with feed-water injection at the bottom of U-tubes, see Chapter 23)

Headers (or fluid collectors, or fluid distributors) and steam drums phenomenology is covered by horizontal SG- and UTSG-related PH (first and third bullet items above); no new PH, in addition to specific parameter ranges, was identified in the case of CANDU UTSGs (fourth bullet item above). Therefore, only in the case of stratification in horizontal channels with active nuclear fuel a specific PH was added. This is identified as S-25-HOHT (or No 46 in the list of 116 THPs) by Aksan et al. (2018a), and D'Auria et al. (2021b).

6.8 Phenomena in core subchannels and containments

The thermal-hydraulic codes developed for system analysis, i.e., in relation to which a PH was identified for validation, are also applied to the core subchannel and containment analysis. The reader may find a description of thermal-hydraulic characteristics and transient performance, notably in the case of subchannels and containment in Chapters 18 and 19, respectively.

No specific PH needed for validation of system thermal-hydraulic codes was added in relation to core subchannels. Notably, PH S-21-GM1, section 6.3.11.2 deals with the three-dimensional core flow, i.e., a target and a challenge for subchannel modeling.

Otherwise, in addition to PH B-7-PD-c, already introduced in this chapter (sections 6.3.1.8 and 6.5.5), seven phenomena are considered to prove code capabilities (or to validate) for containment conditions also considering components and expected accident scenarios in selected (see Chapter 15) advanced reactors. These are identified as A-2-CE-c, S-62-CPC-c, A-11-NC, S-42-NCO-c, S-63-SPR-c, A-13-NCG and S-64-VEB-c (or No 26, 27, 74, 75, 95, 113 and 114 in the list of 116 THPs) by Aksan et al. (2018a), and D'Auria et al. (2021b). A short phenomena description is as follows:

- A-2-CE-c deals with passive cooling of containment by heat exchangers.
- S-62-CPC-c considers the containment role for long-term periods of accidents, including sump recirculation.
- A-11-NC deals with NC between the primary coolant system and containment (e.g., in case of accident scenarios for AP-1000), notably involving the in-containment water storage pool for fuel elements.
- S-42-NCO-c deals with NC (or three-dimensional convection) flows inside containment targeting hydrogen diffusion and stratification.
- S-63-SPR-c deals with containment spray cooling.
- A-13-NCG deals with the impact of air on condensation and consequently on three-dimensional convection flows.
- S-64-VEB-c deals with the impact of ventilation blowers on the thermal-hydraulic conditions of containment, as suggested by Fischer (2018).

6.9 Conclusions

The knowledge of THPs is the backbone and essential element for the development of the nuclear thermal hydraulics subject area. Consequently, THPs have been collected from OECD/NEA-CSNI and IAEA activities which covered design basis accidents or courses of events "before loss of core integrity" for LWR, AWCRs, and SCWR. The same phenomena may be important in different evolutions for single- and two-phase flows. This collection of phenomena can be used to prove the applicability and quality of best-estimate thermal-hydraulic system computer codes.

In this chapter, about 12 water-cooled reactor designs were considered. These are PWR-UTSG, PWR-OSTG, PWR-horizontal steam generator (VVER-440 and VVER-1000), BWR, AP1000, APR1400, EPR, ESBWR, ABWR, small modular reactors (e.g., SMART, MASLWR CAREM), and SCWR.

The phenomena are identified from experiments and through the expertise of the international experts who were involved in the various related activities. A complementary view is as follows: (1) phenomena must correspond to experiments, or experiments should exist that represent each PH (if not, new experiments should be planned and executed); (2) scaling should be considered in designing SETFs, ITFs, and related experiments; (3) databases from experiments should be sufficiently qualified (suitable qualification of experiments constitutes a challenge not discussed here and includes parameter ranges): (4) codes should be applied to compare experimental data, and suitable accuracy should constitute an achieved target (see Chapters 13 and 22); and (5) whenever possible, uncertainty in code predictions for reactor applications should consider accuracy or errors derived by comparing measured and calculated data.

The results of the work performed cover at least 40 years of activity performed at various times within OECD/NEA-CSNI and IAEA. A list of tables including THPs applies to the entire class of LWRs. All identified phenomena should be modeled according to state-of-the-art knowledge. System thermal hydraulics codes constitute the source of expertise associated with considered phenomena and the best and unique tool to calculate accident scenarios, including the course of events. In addition, these best estimate computer codes are the depository of the collected knowledge with all available elements, models of the phenomena, experiments for validation, etc.

Comprehensive experimental and code development research activities have been conducted, also very intensely at an international level, in the past three to 4 decades in relation to the understanding of THPs and for establishing related code predictive capabilities for existing nuclear power reactors. In the same context, research activities also addressed some of the phenomena for passive systems and SCWR. However, a systematic effort to evaluate the level of understanding of THPs for passive systems and supercritical water conditions with connected code capabilities appears to be limited and in general lacking.

Knowledge and understanding of phenomena are a prerequisite for performing meaningful accident analysis. The presented information can be used as part of the qualification process for system code calculations.

Exercises and questions

Exercises and questions are proposed hereafter with solutions (or hints to obtain solutions) given in Appendix 1 (Volume III of the book). Answers and solutions are expected to be provided by readers before reading Appendix 1. Please provide extended answers.

6.1 What is the motivation and the objective for the identification and characterization of phenomena in nuclear thermal hydraulics?

6.2 What is the difference between an SETF and an ITF? (in providing the answer, please consider and discuss the role of scaling).

6.3 What is the difference between a basic experiment and an SETF (or an ITF) experiment?

6.4 Please discuss the current status in relation to predictive capabilities by codes of the PH TPCF.

6.5 Please discuss the current status in relation to predictive capabilities by codes of the PH CHF.

6.6 Please discuss the current status in relation to predictive capabilities by codes of the PH CCFL (please include in the answer a discussion of the importance of scaling).

6.7 Referring to phenomena (also considering information in Chapter 17), please discuss the capability of system codes in predicting containment performance.

6.8 Referring to phenomena (also considering information in Chapter 18), please discuss the capability of system codes in predicting core subchannel conditions.

6.9 Please discuss the connection between accident scenarios and phenomena and parameters (see also Chapter 15).

6.10 Please discuss the relevance and the challenges for PIRT.

Acknowledgments

Special thanks to Paul Scherrer Institut, Villigen and Würenlingen, Switzerland (especially, Dr. G. Markoczy, Prof. Dr. G. Yadigaroglu, G. Varadi, and Dr. J. Dreier) for providing me the means and support to carry out the work in this writing. Further appreciations are to International Organizations, OECD/NEA/CSNI, Paris, France (especially, David Bessette, Heikki Holmstrom, Javier Reig), IAEA, Vienna, Austria (especially, John Cleveland, Sama Bilbao y Leon, Katsumi Yamada), and US NRC, Rockville, Maryland, USA for providing the necessary international working environment and support. The author is grateful to F. D'Auria (University of Pisa, Italy), H. Glaeser (GRS Garching, Germany), K. Wolfert (GRS Garching, Germany), C. Richards (AEA Technology, Winfrith, UK), J. Lillington (AEA Technology, Winfrith, UK), R. Pochard (IPSN: DPEI-CEA/FAR, Fontenay-Aux-Roses, France), and A. Sjoberg (Studsvik Eco & Safety AB, Nykoping, Sweden), A. Annunziato (JRC, Ispra, Italy), P. Marsili (ENEA, Rome, Italy), C. Renault (CEA, Cadarache, France), who are also contributors to separate effects (SET) and integral test facilities (ITF) Matrices work. The contribution of J-L. Gandrille in reviewing the Chapter 6 as in the first edition of the book is gratefully acknowledged.

Heat transfer in nuclear thermal hydraulics

7

Pavel L. Kirillov[1] *and Hisashi Ninokata*[2,3]
[1]IPPE, Obninsk, Russia; [2]Tokyo Institute of Technology, Tokyo, Japan; [3]Polytechnic University of Milan, Italy

Foreword

In recent years, the tendency has been to use numerical experiments in connection with rapid developments in computer technology. Admittedly, it is the right direction to pursue, but it does not relieve engineers from the obligation to gain a deep understanding of the physical processes, for "a correct statement of the problem—half the solution." Many try to solve the problems of computational methods without experimental work. The quality of direct numerical simulation (DNS) is almost equivalent to that of an experiment. However, DNS requires far more computing resources and speed than is available now or in the near future. Computational approaches besides DNS require physical models that must be constructed based on experimental information. Here, caution is needed. Einstein once said that *Mathematics are well and good but nature keeps dragging us around by the nose*. One can fairly accurately carry out all the calculations, and their results will differ from what happens in experiments, not by 5% or 10% but by 100%. But whereas the theoretician can be led by the nose alone, in the engineering world, we talk to hundreds and thousands of people to construct nuclear power plants! The expression by Pyotr L. Kapitsa, "Experiment is forever!" fits with thermal hydraulics perfectly.

Another trend of the modern approach to the design of advanced nuclear power—the transition from the calculation of average values for local, which is particularly important when designing intense nodes of the reactor core, steam generators, heat exchangers (Schlünder et al., 1983). This requires detailed knowledge of distributed parameters, detailed coolant flow and temperature distributions, the surface temperature distribution, and uncertainties to consider the account for possible deviations from the calculated actual values. Such an idea is constantly being implemented and put into practice guide authors.

In the preparation of the manuscript, data are based on not only known and proven works, reference books, but information available from the scientific and technological IAEA reports, international conferences, scientific information databases leading domestic and foreign organizations. While at the end of this chapter is a list of reference sources, useful reference books, and reports in connection to Part 1 are listed here.

Handbook on Thermal Hydraulics in Water-Cooled Nuclear Reactors
https://doi.org/10.1016/B978-0-323-85606-5.00010-6

General:

Birds, R.B., Stewart, W.E., Lightfoot, E.N., 2007. Transport Phenomena, second ed. John Wiley & Sons, Inc., New York.
Todreas, N.E., Kazimi, M.S., 2012. Nuclear Systems—I. CRC Press, Taylor & Francis.

Heat conduction:

Berman, B.R., 1976. Thermal Conduction in Solids. Oxford University Press, New York.
Grimvall, G., 1999. Thermophysical Properties of Materials. Enlarged and revised ed. Elsevier.
Chaikin, P.M., Lubensky, T.C., 1995. Principles of Condensed Matter Physics. Cambridge University Press.
Carslaw, H.S., Jaeger, J.C., 1959. Conduction of Heat in Solids, second ed. Clarendon Press, Oxford, UK.
Schneider, P.J., 1973. Conduction. In: Rohsenow, W.H., Hartnett, J.P. (Eds.), Handbook of Heat Transfer. McGraw-Hill Book Co.

Convective heat transfer:

Kaviany, M., 2001. Principles of Convective Heat Transfer, second ed. Springer.
Kays, W.M., Crawford, M.E., 1993. Convective Heat and Mass Transfer, third ed. McGraw-Hill Science.
Mills, A.F., 1999. Heat Transfer, second ed. Prentice-Hall, New Jersey.
Raithby, G.D., Hollands, K.G.T., 1985. Natural convection. In: Rohsenhow, W.M., Hartnett, J.P., Ganic, E.N. (Eds.), Handbook of Heat Transfer Fundamentals, second ed. McGraw-Hill.

Condensation heat transfer:

Collier, J.G., Thome, J.R., 1996. Convective Boiling and Condensation. Oxford University Press, New York.
Rohsenhow, W.M., Griffith, P., 1985. Condensation. In: Rohsenhow, W.M., Hartnett, J.P., Ganic, E.N. (Eds.), Handbook of Heat Transfer Fundamentals, second ed. McGraw-Hill.
Mills, A.F., 1999. Basic Heat & Mass Transfer, second ed. Prentice-Hall.
Kaviany, M., 2002. Principles of Heat Transfer. John Wiley & Sons, Inc., New York.

Part 1: Principles and modeling

Heat conduction, convection, and radiation are the major mechanisms we encounter in heat transfer phenomena in nuclear thermal hydraulics and are the subject of many books and articles in professional journals. This chapter is dedicated to the fundamental aspects of thermal hydraulics engineering in nuclear reactors based on well-established disciplines rather than recent cutting-edge technology developments in the area. Its content is mostly designed for professionals, as well as students performing coursework and pursuing degree programs. The remainder of Part 1 focuses on the heat conduction, condensation, radiation, and complex heat transfer problems occurring during the heat transfer process inside a nuclear fuel rod. We then delve into boiling heat transfer in Part 2.

7.1 Heat conduction

7.1.1 Basic principles and definitions

7.1.1.1 Temperature field and heat flux

The temperature field is described by a function T in space (x, y, z) and time τ and is written as (Berman, 1976; Carslaw and Jaeger, 1959; Schneider, 1973):

$$T = T(x, y, z, \tau),$$

or typically in a general coordinate system.

A set of points of the same temperature forms a curved *isothermal surface*, which we call an *isotherm*. We call the amount of heat transferred per unit time through an isothermal surface area A, *heat flow* Q. The heat flow per unit area is the *heat flux*,

$$q = dQ / dA \; \left(\mathrm{W/m^2}\right),$$

which is written by the Fourier's law, in proportion to the temperature gradient (Fig. 7.1)

$$\vec{q} = -\lambda \operatorname{grad} T = -\lambda \cdot \vec{n}_0 \cdot \frac{\partial T}{\partial n}, \tag{7.1}$$

where λ is the thermal conductivity (W/(m K)).

The Fourier's law assumes infinite speed of the heat propagation. In most cases of technical importance, this assumption is accepted. In the case of a high thermal conductivity, the heat flux is as follows:

$$q = -\lambda \frac{\partial T}{\partial n} + \frac{\lambda}{c\rho \cdot w_q^2} \frac{\partial q}{\partial \tau}, \tag{7.2}$$

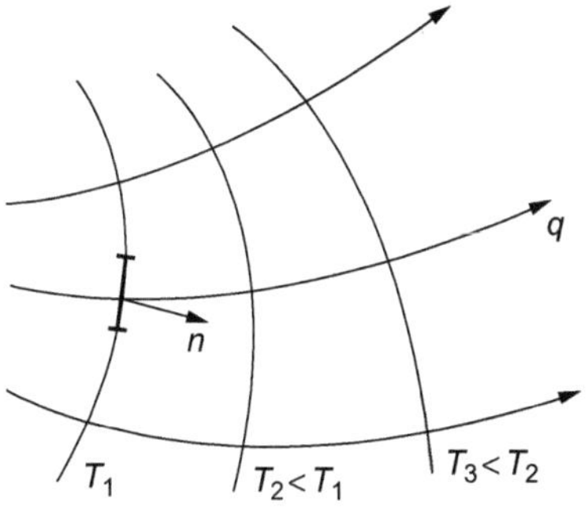

Figure 7.1 Temperature field and heat flux lines.

Table 7.1 Typical q''' in various power units.

Unit	Average power density q''' (MW/m^3)
Steam boilers	10
RBMK	2–4
BWR, BK	30–50
PWR, VVER	80–150
SFR, BN	500–600
BOR-60	1250
Rocket combustion chambers	20,000

where c and ρ are the heat capacity and density of the medium; w_q is the heat propagation speed; τ is the time; $\lambda/\left(\rho c w_q^2\right) = \tau_r$ is the relaxation time. $\lambda/(\rho c) = a$ (m^2/s) is called the thermal diffusivity. If $\tau_r << 1$, we arrive at Eq. (7.1).

In many textbooks and reference books (e.g., Todreas and Kazimi, 2012), it is a common agreement to express *heat flux* with a double dot "i.e., q" in W/(m^2 K). In the context that follows, however, we use q for the heat flux since there will be no mixed up with the expressions for the quantity per unit length or per unit volume. The heat generation per unit length (cylinder or tube) is the *linear heat rate* q' in W/m. The density of the heat generation in power plants and nuclear reactors is often expressed by q''' or q_V in W/m^3, the amount of heat generated per unit time (s) per unit of volume (m^3)—*volumetric heat generation rate* (see Table 7.1).

7.1.1.2 Heat conduction mechanisms

Heat transfer medium is always a set of discrete particles, and heat propagation is a reflection of the mechanisms of interaction of the particles. Ratio of the average kinetic energy E_k of a particle to the binding energy U between particles is the major criterion to determine the state of the substance, i.e., gas, liquid, and solid (see Fig. 7.2).

The energy transfer in gases is carried out by particle collisions and translational motion. From the molecular kinetics theory of molecular dynamics, the thermal conductivity is known to be written for the monatomic gas at normal temperatures and pressures as

$$\lambda = \frac{1}{3}\rho c_v l \overline{w},$$

where ρ is the density; c_v is the specific heat at constant volume; l the mean free path of the molecules; $\overline{w} = \sqrt{3RT/M}$ the average velocity of the molecules with R the universal gas constant and M the molecular weight. The mean free path of molecules is in inverse-proportion to the pressure ($l \propto 1/p$), and the density in proportion to the pressure ($\rho \propto p$). Hence $\rho \cdot l \sim$ const. and the gas thermal conductivity is weakly dependent on pressure. Thermal conductivity of gas is in the range 0.005–0.4 W/(m K).

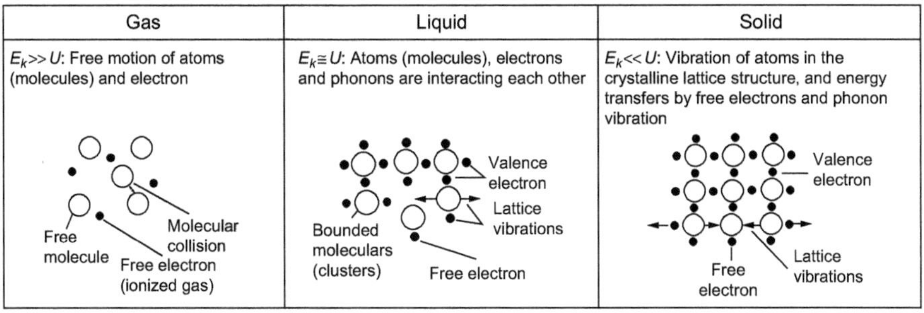

Figure 7.2 Heat conduction mechanisms.

It is noted that the highest thermal conductivity of pure gases is attained by hydrogen and helium, which are 0.15 and 0.125 W/(m K), respectively. The thermal conductivity of air is of the order of ~ 0.03 W/(m K). The thermal conductivity of gases increases with increasing temperature.

In liquids, the energy is transferred in the process of elastic collisions of oscillating particles. Exceptions are liquid metals in which the thermal conductivity close to that of solid metals in which heat is transferred not only by vibrations from one particle to another, but also with the free electrons (Kokorev and Farafonov, 1990). The thermal conductivity of the water is of the order of 0.6 W/(m K) at 20°C, and liquid sodium 75 W/(m K) at 300°C.

In solids, the energy transfer mechanism is associated with vibrations of the atoms that constitute the solid. The vibrations of atoms are independent of each other and can be transmitted at the speed of sound from one atom to another. The solid body can be regarded as a vessel containing a cloud of electrons and fictitious particles, i.e., phonons. The highest thermal conductivity of metal is for silver ~430 W/(m K) and copper ~ 400 W/(m K) (Fig. 7.3). Impurities reduce the thermal conductivity of pure metals.

7.1.1.3 Heat transfer and heat exchange

The term heat transfer refers to the process of transferring heat from the heated surface to the coolant (liquid or gas) or from the hot coolant to the cold surface (Fig. 7.4a).

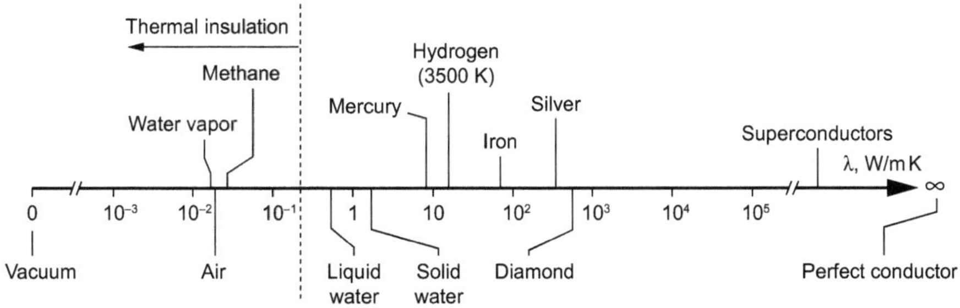

Figure 7.3 Typical values of thermal conductivity from thermal insulators to superconductors.

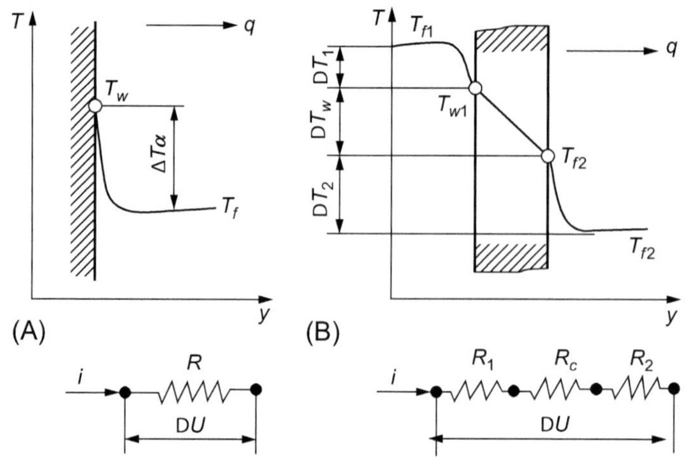

Figure 7.4 (A) Heat transfer; (B) heat exchange; and electrical analogy with Ohm's law: $\Delta T = qR$; $\Delta T = \Delta T_1 + \Delta T_2 + \Delta T_3$; $\Delta U = iR$; $\Delta U = i(R_1 + R_2 + R_3)$.

In engineering practice, finding the temperature gradient $(\partial T / \partial n)$ according to Eq. (7.1) in its various points is very difficult and this formula is replaced by Newton's law:

$$q = Q / A = \alpha \cdot \Delta T \left(\mathrm{W} / \mathrm{m}^2\right), \text{where } Q = \alpha \Delta T \cdot A \text{ in} | \mathrm{W}. \tag{7.3}$$

where $\Delta T = T_w - \overline{T}_f$ is the difference between the wall surface temperature and the average liquid temperature (bulk temperature); A is the heat transfer surface area; and α is the heat transfer coefficient in W/(m^2 K).

The term referring to heat exchange is the heat transfer from one (hot) heat transfer medium to another (cold) through a solid wall (Fig. 7.4b). The amount of heat transferred to the coolant from one another if they are separated by a solid wall, is from expression:

$$Q = k \cdot \Delta T \cdot A. \tag{7.4}$$

where $\Delta T = T_1 - T_2$; k is the overall heat transfer coefficient (W/(m^2 K)). The heat flux is

$$q = k \cdot \Delta T. \tag{7.5}$$

Eqs. (7.3) and (7.4) can be rewritten as: $q = \frac{T_w - \overline{T}_f}{1/\alpha}$; $q = \frac{\overline{T}_1 - \overline{T}_2}{1/k}$. It is an obvious analogy with Ohm's law in electrical engineering: $I = \frac{U_1 - U_2}{R}$, where R is the electrical resistance. By analogy, $1/a$ is called the thermal resistance of the heat transfer, and value $1/k$ is the overall heat transfer resistance.

7.1.2 Basic equations of thermal conduction: Conditions of uniqueness

7.1.2.1 Differential equation

The differential heat conduction equation is a mathematical expression of the law of conservation of energy. The most common expression of the time-dependent conduction equation for a fixed isotropic medium, which accounts for the dependence of λ, c_p, ρ on the temperature and the distribution of the volumetric heat sources q_V, is as follows:

$$c_p\rho\frac{\partial T}{\partial \tau}=\mathrm{div}(\lambda\cdot\mathrm{grad}\,T)+q_V. \tag{7.6}$$

For moving medium with the velocity components (v_x, v_y, v_z) in the Cartesian coordinate system, we have the total time derivative:

$$\frac{dT}{d\tau}\equiv\frac{\partial T}{\partial \tau}+v_x\frac{\partial T}{\partial x}+v_y\frac{\partial T}{\partial y}+v_z\frac{\partial T}{\partial z},$$

which is equal to the substantial time derivative $DT/D\tau$. At constant material properties of the substance, i.e., λ, c_p, and ρ are *constant*, Eq. (7.6) takes the form

$$\frac{\partial T}{\partial \tau}=a\nabla^2T+\frac{q_V}{\rho c_p}, \tag{7.7}$$

where $a=\lambda/c_p$ is the thermal diffusivity, the quantity which is proportional to the propagation velocity of the isothermal surface and $1/a$ characterizes the inertial properties of the body for the distribution of temperature field.

For stationary process, obviously,

$$\nabla^2T+\frac{q_V}{\lambda}=0. \tag{7.8}$$

In the absence of internal heat sources ($q_V=0$) at a steady temperature regime, the following Laplace equation governs the temperature field:

$$\nabla^2T=0. \tag{7.9}$$

Thus, the shape of the steady-state temperature distribution without heat sources at constant material properties does not depend on its properties.

The Laplace operator ∇^2 in different coordinate systems is as follows in Cartesian coordinates (x, y, z):

$$\nabla^2=\frac{\partial^2}{\partial x^2}+\frac{\partial^2}{\partial y^2}+\frac{\partial^2}{\partial z^2};$$

cylindrical coordinates (r, θ, z):

$$\nabla^2 = \frac{1}{r}\frac{\partial}{\partial r}\left(r\frac{\partial}{\partial r}\right) + \frac{1}{r^2}\frac{\partial^2}{\partial \theta^2} + \frac{\partial^2}{\partial z^2};$$

and spherical coordinates (r, θ, ϕ):

$$\nabla^2 = \frac{1}{r^2}\frac{\partial}{\partial r}\left(r^2\frac{\partial}{\partial r}\right) + \frac{1}{r^2 \sin\theta}\frac{\partial}{\partial \theta}\left(\sin\theta\frac{\partial}{\partial \theta}\right) + \frac{1}{r^2 \sin^2\theta}\frac{\partial^2}{\partial \phi^2}.$$

7.1.2.2 Conditions of uniqueness

Eqs. (7.6)–(7.8) are solved only if they are provided by appropriate initial and boundary conditions including geometry (shape, size of the body); physical characteristics (thermal properties, heat source distribution q_V (x, y, z, τ)); initial conditions (temperature distribution); boundary conditions (processes at the borders, in general, the conjugate conditions of temperature fields).

There are three kinds of boundary conditions given by

I. Temperature distribution at the borders;
II. Heat flux distribution (otherwise the temperature gradient distribution); and
III. Heat exchange conditions with the environment (i.e., the ambient temperature and heat transfer coefficient). This boundary condition can be represented as

$$-\lambda\frac{\partial T}{\partial n} = \alpha\left(T_w - \overline{T}_f\right). \tag{7.10}$$

Solutions of stationary heat conduction problems for simple geometry are straightforward and are found in basic textbooks.

7.1.3 Steady-state process

7.1.3.1 Bodies of simple form in one-dimension

This section displays analytical solutions obtained rather simply by integrating one-dimensional form of Eq. (7.8) with appropriate boundary conditions and initial conditions.

7.1.3.2 Plane wall

Given the wall thickness 2δ, internal volumetric heat source q_V, the thermal conductivity λ, and the same heat transfer coefficient α on the boundaries at $\pm\,\delta$:

$$\mp\lambda\left(\frac{dT}{dx}\right) = \alpha\left(T_w - \overline{T}_f\right),$$

where T_W is the wall surface temperature and $\overline{T}_f$ the fluid bulk temperature, the equation to solve is

$$\frac{dT^2}{dx^2}+\frac{q_V}{\lambda}=0,$$

and its solution is given by:

$$T(x)=\overline{T}_f+\frac{q_v\delta}{\alpha}+\frac{q_v\delta^2}{2\lambda}\left[1-\left(\frac{x}{\delta}\right)^2\right]. \tag{7.11}$$

The maximum temperature is at $x=0$:

$$T_{\max}=\overline{T}_f+\frac{q_V\delta}{\alpha}+\frac{q_V\delta^2}{2\lambda} \tag{7.12}$$

7.1.3.3 Cylindrical wall

For an annular body of the inner diameter r_1 and outer diameter r_2 with constant thermal conductivity, the equation to solve is

$$\frac{1}{r}\frac{\partial}{\partial r}\left(r\frac{\partial T}{\partial r}\right)+\frac{q_V}{\lambda}=0$$

Integrating the previous equation twice, we have

$$T(r)=-\frac{q_V r^2}{4\lambda}+C_1\, ln\, r+C_2. \tag{7.13}$$

Constants C_1 and C_2 are determined from the boundary conditions. For example, if the heat is removed through the outer surface of the cylinder by the coolant of bulk temperature T_2, the boundary conditions:

$$\text{At } r=r_1,\quad q''=0 \text{ and } \frac{dT}{dr}=0;\ \text{and}$$

$$\text{At } r=r_2,\quad -\lambda\left.\frac{dT}{dr}\right|_{r_2}=-\alpha(\mathrm{T}_{W2}-\mathrm{T}_2),$$

give the following result:

$$C_1=\frac{q_V r_1^2}{2\lambda};\ C_2=T_2+\frac{q_V r_2}{2\alpha}+\frac{q_V r_2^2}{4\lambda}-\frac{q_V}{2\alpha}\frac{r_1^2}{r_2}-\frac{q_V r_1^2}{2\lambda}\ln r_2. \tag{7.14}$$

As a consequence, we have

$$T_{W2} = T(r_2) = T_2 + \frac{q_V r_2}{2\alpha}\left[1 - \left(\frac{r_1}{r_2^2}\right)^2\right], \text{ and}$$

$$T(r) = T_{W2} + \frac{q_V r_2^2}{4\lambda}\left(1 - \frac{r^2}{r_2^2}\right) + \frac{q_V r_1^2}{4\lambda}\, ln\left(\frac{r}{r_2}\right)^2.$$

Since the linear heat rate $q' = q_V \pi\left[r_2^2 - r_1^2\right]$, the maximum temperature $T(r)$ is at $r = r_1$:

$$T_{\max} = T(r_1) = T_{W2} + \frac{q'}{4\pi\lambda}\left[1 - \frac{ln\,(r_2/r_1)^2}{(r_2/r_1)^2 - 1}\right]. \tag{7.15}$$

Note that the maximum temperature of solid cylinder is at the centerline given by $\int_{T_{W_2}}^{T_{CL}} \lambda dT = \frac{q'}{4\pi}$ and for the constant thermal conductivity, $T_{\max} = T_{CL} = T_{W2} + \frac{q'}{4\pi\lambda}$. Eq. (7.15) suggests that, for the same linear heat rate, the highest temperature of the annular pellet is lower than that of the solid fuel pellet by the factor $1 - \frac{ln\,(r_2/r_1)^2}{(r_2/r_1)^2 - 1}$.

7.1.3.4 Sphere

In case of spherical geometry, assuming no angular dependency of the thermal conditions, the differential equation is

$$\frac{1}{r^2}\frac{d}{dr}r^2\frac{dT}{dr} + \frac{q_V}{\lambda} = 0. \tag{7.16}$$

The solution is $T(r) = -\frac{q_V r^2}{6\lambda} + \frac{C_1}{r} + C_2$, and from the boundary conditions at $r = 0$, $dT/dr = 0$; and at $r = r_0$, $-\lambda dT/dr = \alpha(T_w - T_f)$, we have $C_1 = 0$; and $C_2 = T_f + \frac{q_V r_0}{3}\left(\frac{1}{\alpha} + \frac{r_0}{2\lambda}\right)$. Therefore, the maximum temperature is $T_f + \frac{q_V r_0}{3}\left(\frac{1}{\alpha} + \frac{r_0}{2\lambda}\right) = T_f + \frac{q_V d}{6}\left(\frac{1}{\alpha} + \frac{d}{4\lambda}\right)$ at $r = 0$. d is the diameter of the sphere and the temperature difference between the surface to the sphere center is obviously

$$\Delta T = T(r = 0) - T_w = \frac{q_V r_0^2}{6\lambda} = \frac{q_V d^2}{24\lambda}. \tag{7.17}$$

And the heat flux at the surface is: $q = \frac{q_V d}{6}$.

7.1.3.5 Critical diameter of thermal isolation

The heat flow through the pipe wall with thermal insulation can be written as $q_l = k_l\pi(T_{f1} - T_{f2})$ (W/m), where T_{f1} and T_{f2} are temperatures of the coolant on both side of the tube wall. With the subscripts i, 1, and 2 referring to the insulation, inner, and outer boundaries of the cylinder, respectively, d_i, d_1, and d_2 are the diameters and α_1 and α_2 the heat transfer coefficients. Then, with $d_i - d_2$ the thickness of the insulator wrapped around the cylinder, the overall linear heat transfer coefficient k_l can be calculated as follows:

$$\frac{1}{k_l} = \frac{1}{\alpha_1 d_1} + \frac{1}{2\lambda}\ln\frac{d_2}{d_1} + \frac{1}{2\lambda_i}\ln\frac{d_i}{d_2} + \frac{1}{\alpha_2 d_i}\ (\text{W}/(\text{m K})). \tag{7.18}$$

From Eq. (7.18) it is clear that an increase in the diameter of the thermal insulation (d_i) increases the thermal resistance of the layer $\left(\frac{1}{2\lambda_i}\ln\frac{d_i}{d_2}\right)$ and decreases the thermal resistance of the convection at the outer surface, i.e., to the environment $\left(\frac{1}{\alpha_2 d_i}\right)$.

Consequently, there exists a critical value of d_i that gives the smallest total thermal resistance ($1/k_l$) and therefore the highest heat transfer rate. A schematic view of this behavior is shown in Fig. 7.5. This diameter d_i is called the critical diameter $d_{i,cr}$ or the critical thickness of insulation ($d_{i,cr} - d_2$). It is obtained by solving $\partial(1/k_l)/\partial d_i = 0$:

$$\frac{\partial}{\partial d_i}\left(\frac{1}{2\lambda_i}\ln\frac{d_i}{d_2} + \frac{1}{\alpha_2 d_i}\right) = \frac{1}{2\lambda_i d_i} - \frac{1}{\alpha_2 d_i^2} = 0. \tag{7.19}$$

From Eq. (7.19), we have

$$d_{i,cr} = \frac{2\lambda_i}{\alpha_2}, \tag{7.20}$$

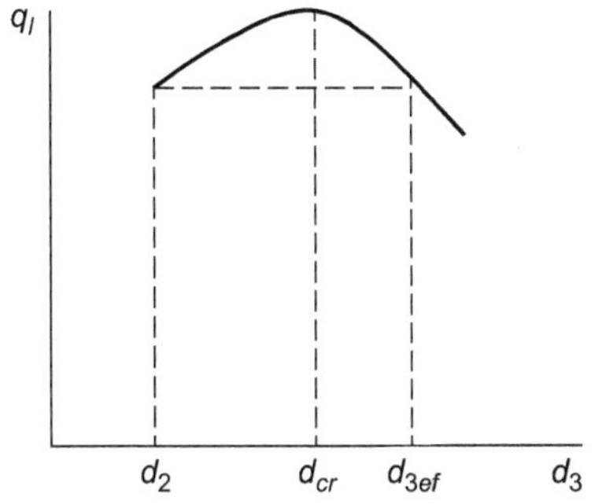

Figure 7.5 Dependence of thermal losses on a thickness of insulation imposed on a tube ($d_3 = d_i$).

that says the Biot number $Bi_{cr} = \frac{\alpha d_{i,cr}}{\lambda_i} = 2$. With $d_i < d_{i,cr}$, the thermal insulator loses its role, i.e., the heat transfer rate of insulated pipe will be more than that from the same pipe without insulation, while the thicker insulation than that of the critical diameter ($d_i > d_{i,cr}$) results in the less heat transfer rate than that from the same pipe without insulation but the increase in the outer insulation surface area will reduce the thermal resistance, i.e., increasing the convective heat transfer, which is preferable for the electrical insulation of wires (Ingersoll et al., 1954; Zvetkov and Grigor'ev, 2005).

7.1.3.6 Temperature dependence of thermal conductivity

The nuclear reactor fuel (UO_2) temperature under normal operating conditions may vary in the range 500−2000°C. In this range, the thermal conductivity of UO_2 changes about twice and is described by a nonlinear function of temperature (Fig. 7.6). Therefore, it is necessary to account for this temperature dependency $\lambda(T)$ in calculations (Kirillov and Bogoslovskaya, 2000; Kirillov, 2008). The approximate equation in one-dimension under stationary conditions to solve is

$$\frac{d}{dx}\left[\lambda(T)\frac{dT}{dx}\right] = -q_V. \tag{7.21}$$

The solution of this equation is simplified with the introduction of the Kirchhoff integral for the variable thermal conductivity, which has the dimension (W/m):

$$\Lambda(T) = \int_0^T \lambda(T)dT. \tag{7.22}$$

Note that $d\Lambda/dT = \lambda(T)$ and $\frac{d\Lambda}{dx} = \frac{d\Lambda}{dT}\frac{dT}{dx} = \lambda(T)\frac{dT}{dx}$. Substituting this expression in Eq. (7.22), we obtain

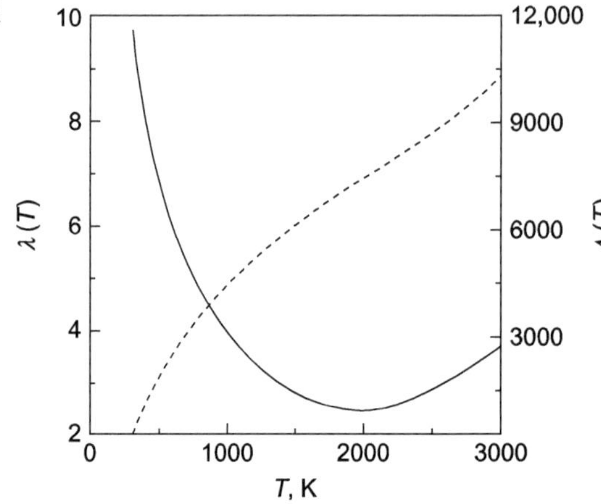

Figure 7.6 Dependency of λ_{UO_2} and Λ_{UO_2} (Eq. 7.22) on temperature.

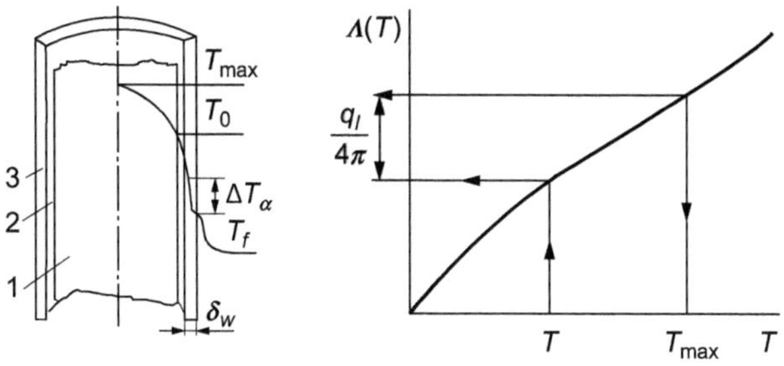

Figure 7.7 Scheme of calculation of maximum fuel pin temperature: 1-fuel, 2-gas gap, 3-cladding.

Table 7.2 Variations in integral thermal conductivity of fuel for various geometries.

	Geometry	$\Delta\Lambda$ (W/m)
1.	Flat plate fuel (q_V, W/m^3) with thickness 2δ. The surface temperatures on both sides are the same and the maximum temperature in the center of the plate.	$\frac{q_V\delta^2}{2}$
2.	Cylindrical fuel rod with linear heat rate (q', w/m).	$q'/4\pi$
3.	Spherical fuel of radius R with volumetric heat source (q_V, W/m^3).	$q_V R^2/6$

$$\frac{d^2\Lambda}{dx^2} = -q_V. \tag{7.23}$$

The fuel temperature calculation scheme is shown in Fig. 7.7.

Integrated thermal conductivity corresponding to the maximum temperature is given by the equation

$$\Lambda_0 = \Lambda_w + \Delta\Lambda \ (\mathrm{W/m}) \tag{7.24}$$

where Λ_w is the integrated thermal conductivity at the surface, $\Delta\Lambda$ is the value of which depends on the geometry and the volumetric heat source q_V (W/m^3) (see Table 7.2).

7.1.4 Transient heat conduction

7.1.4.1 Heating (cooling) of body

Transient heat conduction processes are divided into two groups (Ingersoll et al., 1954; Lykov, 1961, 1978).

(1) Body tends to become thermally equilibrium with the environment when it is heated (cooled). The heat exchange with the environment is carried out according to Newton's law (Figs. 7.8 and 7.9). A special case is the process $\alpha \to \infty$ when the surface temperature is ambient temperature.

(2) The body temperature undergoes regular periodic changes.

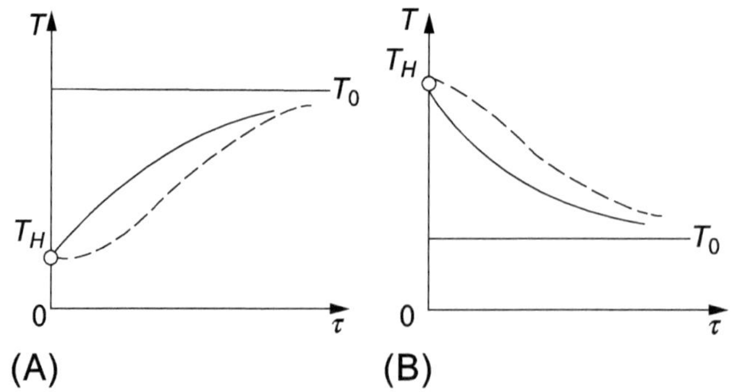

Figure 7.8 (A) Heating and (B) cooling of the body of arbitrary shape in medium with constant temperature T_0: surface temperature; - - - temperature inside the body.

Figure 7.9 Heating of semiinfinite body: T_0 is the initial temperature; T_W is the surface temperature; 1, 2 denote different time.

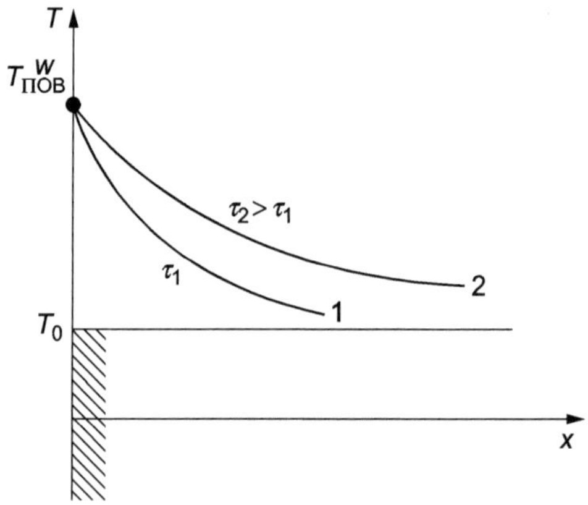

Nondimensional form of the differential equation that describes the time-dependent thermal conduction is as follows:

$$\frac{\partial \theta}{\partial \tau} = a\left(\frac{\partial^2 \theta}{\partial r^2} + \frac{n}{r}\frac{\partial \theta}{\partial r}\right), \tag{7.25}$$

where $\theta \equiv \frac{T - T_f}{T_0 - T_f}$; T_0 is the initial temperature of the body; T_f the ambient temperature; $n = 0,\ 1,\ 2$ designate an infinite plate, cylinder, and sphere, respectively. The dimensionless temperature field is determined by the Biot number ($Bi = \alpha l/\lambda$) where l is the characteristic dimensions of bodies, the Fourier number ($Fo = a\tau/l^2$), the body properties (a and λ), and the dimensionless coordinate ξ:

$$\theta = \sum_{n=1}^{\infty} A_n \cdot U(\mu_n \xi)\exp\left(-\mu_n^2 \cdot Fo\right), \tag{7.26}$$

where μ_n is the roots of the characteristic equation $\mu = \mu(Bi)$.

7.1.4.2 Heating and cooling of body at (Bi << 1)

A small value of Biot number $Bi = \alpha l/\lambda$ means any small l, or high thermal conductivity. In this case, the temperature inside the body is practically uniform:

$$\theta = \frac{T - T_f}{T_0 - T_f} = \exp[-(n+1)Bi \cdot Fo], \tag{7.27}$$

where T_0 is the initial temperature of the body; T_f the ambient temperature; $n = 0, 1, 2$ designate an infinite plate, cylinder, and sphere, respectively.

The time for cooling (and heating) bodies to the temperature θ is given by

$$\tau = K \ln \theta / Bi, \tag{7.28}$$

where $K = \delta^2/a$ for a plate; $K = R^2/2a$ for a cylinder, $K = R^2/3a$ for a sphere. If the plate, cylinder, and dimension, then a cylinder is heated (cooled) in two times, and the sphere three times faster than the plate, which corresponds to the ratio of surface area to volume of bodies (1:2:3).

7.1.4.3 Heating and cooling of body at (Bi >> 1)

$Bi = \alpha l/\lambda >> 1$ suggests that the body has a low thermal conductivity, and a sufficiently high heat transfer coefficient can be assumed to the environment. In this case the body surface temperature will be close to the ambient temperature. Correlations for the calculation of a dimensionless temperature in the middle of the bodies and the heating time up to a certain temperature θ_m are given in Table 7.3.

7.1.4.4 Heating and cooling of the finite body

On the surface of a semiinfinite body ($x = 0$), the temperature of surface T_s is assumed equal to the temperature T_0 at the time $\tau = 0$. Then the time-dependent temperature field in the body for $\tau > 0$ is described by the relation

Table 7.3 Correlations for dimensionless temperature in the middle of bodies with heating time at $Bi >> 1$.

Shape of the body	Dimensionless temperature at the middle of the body, θ	Time to heat up to the dimensionless temperature θ_m
Infinite plate	$\theta = 1.27e^{-2.47Fo}$	$\tau = \frac{\delta^2}{2.47a} \ln \theta_m$
Infinite cylinder	$\theta = 1.6e^{-5.79Fo}$	$\tau = \frac{R^2}{5.79a} \ln \theta_m$
Sphere	$\theta = 2.0e^{-9.87Fo}$	$\tau = \frac{R^2}{9.87a} \ln \theta_m$

$$\theta(x,\tau) = \frac{T(x,\tau) - T_s}{T_0 - T_s} = 1 - \operatorname{erf}\left(\frac{x}{2\sqrt{a\tau}}\right). \tag{7.29}$$

where $\operatorname{erf}\left(\frac{x}{2\sqrt{a\tau}}\right)$ is the error function.

The heat flux on the surface of the body varies as

$$q(0,\tau) = \frac{\lambda(T_0 - T_s)}{\sqrt{\pi a \tau}}. \tag{7.30}$$

The heat flux through the plane ($x > 0$), parallel to the surface of the semiinfinite body, is

$$q(x,\tau) = \frac{2\lambda(T_0 - T_s)}{\sqrt{\pi}} \exp\left(\frac{x^2}{4a\tau}\right). \tag{7.31}$$

The total amount of heat passing through the surface between the time τ_1 and τ_2, is equal to

$$Q = \int_{\tau_1}^{\tau_2} q(x,\tau)d\tau = \frac{2\lambda(T_0 - T_s)}{\sqrt{\pi a}}\left(\sqrt{\tau_2} - \sqrt{\tau_1}\right). \tag{7.32}$$

7.1.4.5 Surface temperature at two bodies contact

Assume two infinite bodies (1, 2) of an initial temperature $T_1 > T_2$, properties λ_1, λ_2, a_1, a_2, are in contact each other. The temperature at the contact location is assumed T_c ($T_1 > T_c > T_2$). Then the heat flow from body 1 to the contact position is, with $b \equiv \sqrt{\rho c_p \lambda}$,

$$q_1 = \frac{\lambda_1(T_1 - T_c)}{\sqrt{\pi a_1 \tau}} = \sqrt{b_1}\,\frac{(T_1 - T_c)}{\sqrt{\pi\tau}} \tag{7.33}$$

which equals the heat flow from the contact points to body 2:

$$q_2 = \frac{\lambda_2(T_c - T_2)}{\sqrt{\pi a_2 \tau}} = \sqrt{b_2}\,\frac{(T_c - T_2)}{\sqrt{\pi\tau}}.$$

Hence we obtain the temperature at the contact point as $T_c = \frac{\sqrt{b_1}T_1 + \sqrt{b_2}T_2}{\sqrt{b_1} + \sqrt{b_2}}$.

7.1.4.6 Regular thermal regime

If the dimensionless temperature profile does not change in time, we call this unsteady heat conduction process "*regular thermal regime*" according to Kondrat'ev (1954) and its generalization in Kudryashev and Zhemkov (1963) and Zvetkov and Grigor'ev

(2005). The temperature fields in the bodies of various shapes (sheet, cylinder, sphere, etc.), which are cooled in an environment at constant temperature with the constant heat transfer coefficient at the surface, can be represented by the series

$$\vartheta = [T_0 - T(x,y,z,\tau)] / (T_0 - T_f) = \sum_n A_n(Bi)\Phi(x,y,z,Bi)e^{-\mu^2 Fo},$$

where $Fo = a\tau/l^2$. T_0 and T_f are the initial body temperature and the ambient temperature. Cooling process of the body can be grouped into two stages. The first stage is characterized by the influence of the initial temperature distribution, when the rate of temperature change is different in different parts of the body. The second stage begins with the moment when the cooling rate does not depend on the initial conditions and determined only by the conditions of heat transfer at the interface, the physical properties of the body, its geometry and dimensions. In this case, the temperature field is described by

$$\frac{\theta}{\theta_0} = A\ \Phi(x,y,z)e^{-m\ \tau}, \tag{7.34}$$

where $\theta \equiv T(x,y,z,\tau) - T_f$; $\theta_0 \equiv T(x,y,z,0) - T_f$; and $m = \mu_1^2 a/l^2$. $\frac{1}{\theta}\frac{\partial\theta}{\partial\tau} = -m$ is the rate of cooling and indicates that the relative rate of temperature change does not depend on time or space, and is a constant. In Fig. 7.10 the time dependency of $\ln\theta = f(\tau)$ is shown at two points of the body (x_1, x_2). The rate of cooling is found by the slope of the line: $m = -\tan\phi = (\ln\theta_1 - \ln\theta_2)/(\tau_2 - \tau_1)$.

7.1.4.7 Periodic changes of temperature

Assume a boundary condition of the surface temperature of the semiinfinite body is prescribed as $t_w = t_m\cos(\omega\tau)$, where t_m is the amplitude; $\omega = 2\pi/T$ is the angular

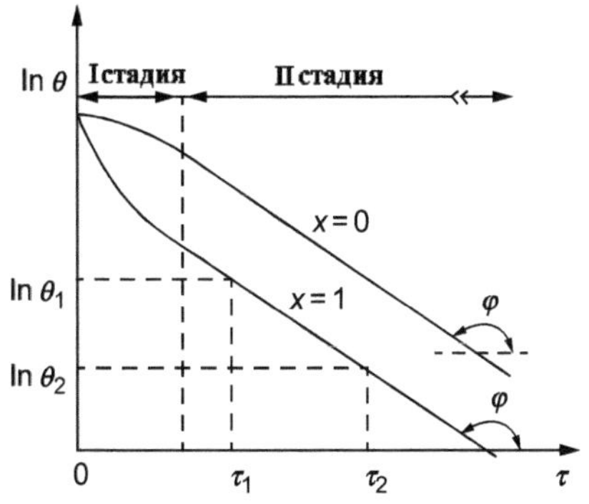

Figure 7.10 Schematics of dimensionless temperature versus time. Stage I: Initial unordered mode; stage II: Regular thermal regime.

frequency; and as $x \to \infty$ the temperature $t \to 0$. Note that t is temperature to distinguish from the period of oscillation T.

The solution of the time-dependent diffusion equation $\frac{\partial t}{\partial \tau} = a\frac{\partial^2 t}{\partial x^2}$ under the above boundary conditions is given by the temperature wave

$$t = t_m \exp\left(-\text{ч}\sqrt{\frac{\omega}{2a}}\right) \cdot \cos\left(\omega\tau - \text{ч}\sqrt{\frac{\omega}{2a}}\right). \tag{7.35}$$

The lower the temperature fluctuation frequency (longer period), the larger the oscillation amplitude at a given point of the body.

The heat flux on the surface will change over time according to Fick's law:

$$q = -\lambda\frac{\partial t}{\partial x}\bigg|_{x=0} = t_m b\sqrt{\omega}\cos\left(\omega\tau + \frac{\pi}{4}\right), \tag{7.36}$$

where $b = \frac{\lambda}{\sqrt{a}} = \sqrt{\rho c_p \lambda}$. The amount of heat diffusing into the environment in half a period $T/2$ through the unit surface area is found to be

$$Q = \int_0^{T/2} q d\tau = t_m b\sqrt{\frac{2T}{\pi}}. \tag{7.37}$$

The method of temperature waves is used to measure the thermal diffusivity. If we measure the delay of the wave $\Delta\tau$ at a depth x, then, knowing the period T, it is easy to determine a (thermal diffusivity).

7.1.5 Thermal conductivity of composite and dispersed media

Composites, laminates, fibrous, granular materials, dispersion media are frequently encountered in engineering. In the simplest case, e.g., a two-component system composed of a continuous phase matrix and inclusions with different thermal properties (Chudnovsky, 1954; Ingersoll et al., 1954; Kaviany, 2002; Lykov, 1967, 1978; Odelevsky, 1951).

Let us denote the volume fraction of one inclusion by φ and its thermal conductivity by λ_0. The continuous medium, such as gas, liquid, or solid in which a disperse phase is distributed, is called the dispersion medium. For instance, filled with solid particles, a liquid medium, or a gas is referred to as the dispersion medium. Heat transfer takes place in this dispersed medium by: (1) the thermal conduction inside an individual particle; (2) conduction through particle contact; (3) conduction in the continuous medium; (4) conduction through the contact of the medium with the particles; (5) convection heat transfer by gas and steam between the particles; and (6) the radiation from particle to particle.

Calculation of the individual processes and their additional combination is feasible only when one or two processes are crucial. Otherwise, such an approach requires

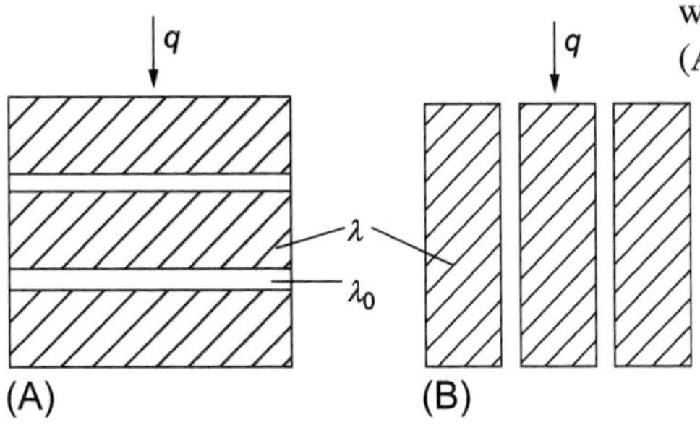

Figure 7.11 Two limiting arrangements of continuous phase matrix and inclusions with different thermophysical properties: (A) transverse; (B) longitudinal.

clarification of many difficult circumstances to be taken into account, knowledge on the microscopic structure and its characteristics of the particle material and the continuous medium, the processes of mutual influence on each other, etc.

Another approach is regarded as a quasihomogeneous dispersion medium with some effective features (mixture density, effective thermal conductivity, etc.), which is applicable to the heat transport equation. Then, the thermal characteristics (λ, a, ρ, c_p, etc.) are functions of the temperature, porosity, time coordinates (Yang and Nakayama, 2010; Dul'nev and Zarichniak, 1974; Dul'nev and Novikov, 1991; Fedotovsky, 1994; Kaviany, 2002). Two limiting cases are shown in Fig. 7.11.

If the ratio of the conductivity of the inclusions λ_0 to that of the matrix (continuous phase) λ is much smaller than unity, i.e., $\lambda_0/\lambda \ll 1$, the greatest effect is achieved in the insulation case of transverse heat flow (Fig. 7.11a). Heat conduction calculation can be carried out by the method of multilayer walls, for the known geometry, and using the effective thermal conductivity of the array (λ_{eff}) defined with the volume fraction of inclusions (φ) and the thermal conductivity (λ and λ_0):

$$\frac{1}{\lambda_{\text{eff}}} = \frac{1}{\lambda}(1 - \varphi) + \frac{1}{\lambda_0}\varphi. \tag{7.38}$$

In the case of $\lambda_0/\lambda << 1$, the least insulation effect is observed obviously at the longitudinal arrangement (Fig. 7.11b). In this case,

$$\lambda_{eff} = \lambda(1 - \varphi) + \lambda_0\varphi. \tag{7.39}$$

Bicomponent regular structure systems of spheres may be provided in the cubic, tetrahedral (hexagonal), or random packing arrangement (Fig. 7.12). The porosity of these structures, i.e., $1 - \varphi$ is 0.4764 for the cubic and 0.2595 in the hexagonal close-packed structure.

The effective thermal conductivity of the dispersion medium with the spherical inclusions is discussed in Meredith and Tobias (1961), Dul'nev and Zarichniak (1974), Fedotovsky (1994), Yamada and Ota (1980) and can be represented by the formula

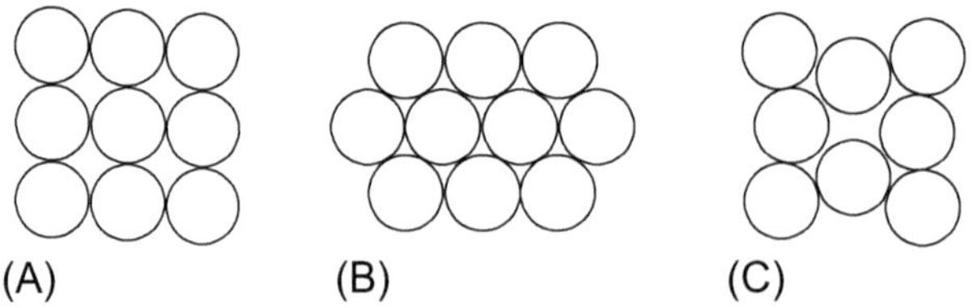

Figure 7.12 Regular packing of spheres: (A) cubic; (B) hexagonal; (C) random.

$$\frac{\lambda_{\text{eff}}}{\lambda}=\frac{(1-\varphi)+(\lambda_0/\lambda)(1+\varphi)}{(1+\varphi)+(\lambda_0/\lambda)(1-\varphi)}, \tag{7.40}$$

where λ and λ_0 are the thermal conductivity of the matrix and inclusions; φ is the volume fraction of inclusions. The formula is valid for $\varphi = 0\text{–}0.5$.

Similar results will give the formula of Maxwell-Aikin:

$$\frac{\lambda_{\text{eff}}}{\lambda}=\frac{2(1-\varphi)+(\lambda_0/\lambda)\ (1+2\varphi)}{(2+\varphi)+(\lambda_0/\lambda)\ (1-\varphi)}. \tag{7.41}$$

A rather new formula is found in a more recent publication by Kaviany (2002) for spherical inclusions in the matrix thermal conductivity λ:

$$\frac{\lambda_{\text{eff}}}{\lambda}=\frac{1}{1-(6\varphi/\pi)^{1/3}\cdot F_i(\gamma)}, \tag{7.42}$$

where $\gamma = \left(\frac{16}{9\pi\varphi^2}\right)^{1/3}\frac{\lambda/\lambda_0}{1-\lambda/\lambda_0}$ for $\varphi < 0.3$. Function $F(\gamma)$ is different for different relationships λ/λ_0. When $\lambda/\lambda_0 < 1$, e.g., with the ceramic, metal inclusion: $F_1(\gamma) = 1 - \frac{\gamma}{2(\gamma+1)^{1/2}}\, ln\, \frac{(\gamma+1)^{1/2}+1}{(\gamma+1)^{1/2}-1}$; and for $\lambda/\lambda_0 > 1$, e.g., examples with gases or low conductivity inclusion in highly conducting matrix:

$$F_2(\gamma)=1-\frac{\gamma}{(-\gamma-1)^{1/2}}\tan\frac{1}{(-\gamma-1)^{1/2}}.$$

For cubic form inclusions:

$$\frac{\lambda_{\text{eff}}}{\lambda}=\frac{1+\varphi^{2/3}(\lambda_0/\lambda-1)}{1+\varphi^{2/3}(1-\varphi^{1/3})(\lambda_0/\lambda-1)}. \tag{7.43}$$

If $\lambda_0/\lambda << 1$, $\lambda_{\text{eff}} = \lambda(1-\varphi^{2/3})/(1-\varphi^{2/3}+\varphi)$. When $\varphi = 0$, $\lambda_{\text{eff}} = \lambda$; when $\lambda_0 = \lambda$, $\lambda_{\text{eff}} = \lambda$.

7.2 Condensation

7.2.1 Physics of the processes

7.2.1.1 Condensation of pure steam

Condensation is a process of the transition from the vapor phase to the liquid phase under subcritical parameter conditions ($p < p_{cr}$, $T < T_{cr}$) (Chaikin and Lubensky, 1995). It takes place in the bulk or on the surfaces in contact with the steam. Condensation nuclei are solid or liquid particles, charged particles, or surface areas. The condensation of pure vapor in the volume, if there are no condensation nuclei available (homogeneous condensation), usually occurs when the vapor is supercooled relative to the saturation temperature $\Delta \mathrm{T} = \mathrm{T}_g - \mathrm{T}_S$. The degree of supersaturation is characterized by the ratio of the supercooled vapor pressure to the saturation pressure p/p_S. The formation of droplets is due to fluctuations in vapor density (Kutateladze et al., 1979). The droplet is an embryo of the liquid phase if its radius equals

$$R_{cr} = \frac{2\sigma T_S}{r\rho_f \Delta \mathrm{T}} \tag{7.44}$$

With increasing supercooling of the vapor (increase in ΔT or p/p_S), the critical radius of the droplets decreases, resulting in the increase in the likelihood of droplets formation. Droplets of $R < R_{cr}$ are unstable and evaporate quickly. If $R > R_{cr}$, the droplets are growing due to the influx of vapor condensation. The condensation begins typically on the surface with a slight supercooling of 0.01–0.05 K.

Interaction of the condensate with the surface is determined by the wettability, which is characterized by the contact angle θ, reckoned in the direction of the fluid. When $\theta < 90$ degrees (Fig. 7.13a) the liquid wets the surface, $\theta > 90$ degrees shows dropwise condensation (Fig. 7.13b).

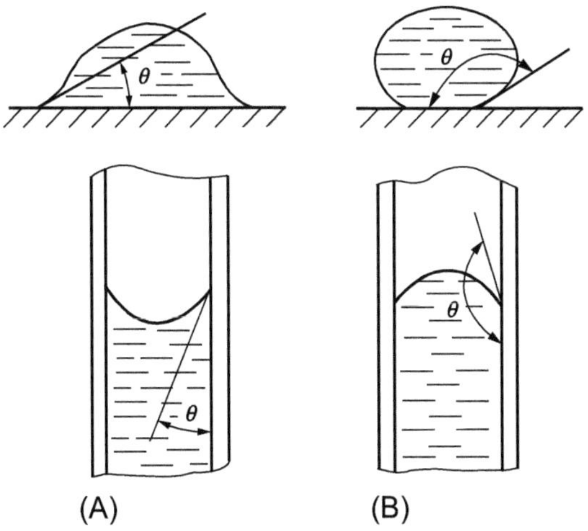

Figure 7.13 Contact angle of liquid wetting: (A) surface wetted by liquid; (B) surface not wetted.

There are two kinds of condensation, filmwise and dropwise. In either type of condensation, since the condensed phase has a much smaller specific volume, an intensive mass transfer is taking place from the bulk to the site of condensation. Because the vapor condensation releases the heat of the phase transition (latent heat of condensation), the condensation is always accompanied by the simultaneous transfer of not only the mass, but also the energy.

The vapor phase transition to the liquid phase could be represented by a series of individual processes for convenience of analysis, the impact of which can be different in different cases:

- Vapor supply to the interface surface by diffusion and convection;
- The actual phase transition process, which consists of collisions of vapor molecules on a solid surface or a liquid surface;
- Removals of heat released during condensation through the film or drops;
- Drainage to fluid accumulated on the surface.

The first phenomenon is characterized by the vapor velocity in the normal direction to the surface of condensation. The velocity can be determined from the heat flux q with mass flux $G_g = \rho_g w_g$ and $q = G_g \Delta h_{fg}$, therefore $w_g = \frac{q}{\rho_g \Delta h_{fg}}$. This relationship is suitable for low pressures at which $\rho_g/\rho_f \ll 1$. For high pressure $w_g \sim [q/(\rho_g \Delta h_{fg})](1 - \rho_g/\rho_f)$. For the vapor at a pressure of 1 atm $\approx$0.1 MPa and heat flux of 0.25 MW/m^2, we have $\rho_g = 0.587\ \text{kg/m}^3$, $\Delta h_{fg} = 2.164\ \text{MJ/kg}$ and $w_g = 0.2$ m/s.

7.2.1.2 Coefficient of condensation

Upon collision with the surface molecules (liquid or the solid wall), part of the vapor molecules will be captured, and the other part elastically bounced back. The fraction of molecules captured by the liquid film at the surface, relative to the total number of molecules colliding with the surface, is called condensation coefficient β.

The kinetic theory of gases allows us to calculate the mass flux (Hertz-Knudsen formula):

$$G \cong \frac{2\beta}{2-\beta}\left(\frac{p_g}{\sqrt{2\pi R T_g}} - \frac{p_w}{\sqrt{2\pi R T_w}}\right), \tag{7.45}$$

where R is the gas constant. The rest of the parameters are indicated in Fig. 7.14. The dependence of the coefficient of condensation β from steam pressure shown in Fig. 7.15a and b.

The vapor temperature T_g at the surface is higher than the condensate temperature T_w[1] due to the fact that the reflected molecules have an average energy higher than the entrapped vapor molecules and is lower than the bulk vapor temperature T_{g1} due to a

[1] Here we use the T_w as the temperature of the outer surface of the condensate film.

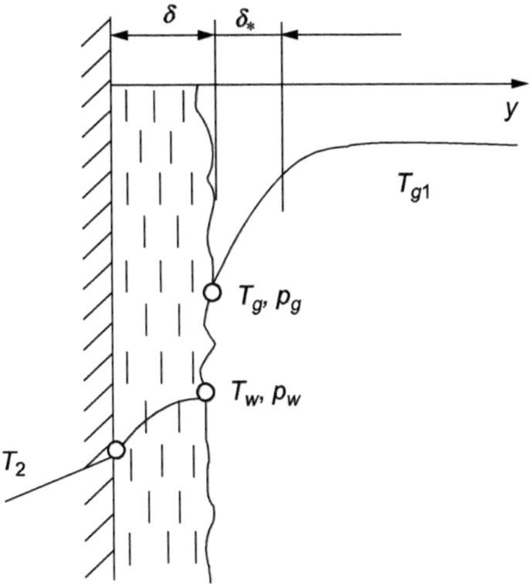

Figure 7.14 Typical temperature distribution in the film on the surface where the vapor condensates: $\delta^* \approx 10l$, l—the mean free path of molecules in the superheated steam; δ—condensate film thickness; δ^*—the boundary layer thickness near the surface of steam condensate film; T_g, p_g—temperature and pressure of vapor; T_w, p_w—temperature and pressure of the film of condensate on the surface. Note that T_w is not the wall temperature here.

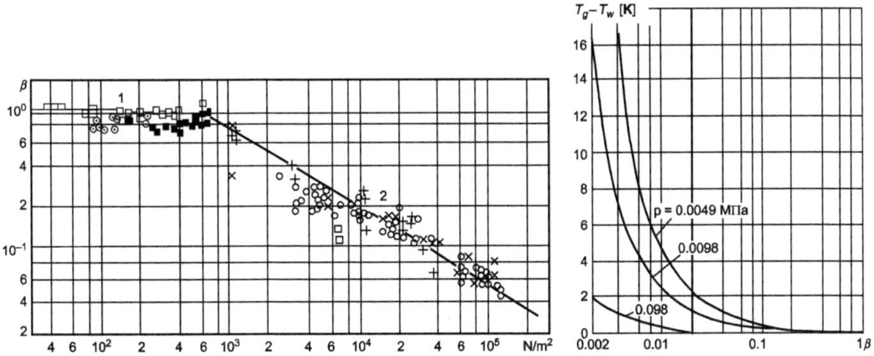

(A) (B)

Figure 7.15 (A) Dependency of the coefficient of condensation on vapor pressure, water, mercury, sodium, and potassium (Isachenko, 1977): 1—$\beta = 1$ for $p = 0.3$–700 Pa; 2—$\beta = 35p^{-0.56}$ for $p = 700$–10^5 Pa; ○, •, + and ■ are for K; □: Na; ⊙: Hg; ×: H_2O. (B) Influence of the coefficient of condensation and vapor pressure on the temperature jump ($T_g - T_W$): $q \sim 3 \times 10^4$ W/m^2; $p = 5 \times 10^{-3}$; 1×10^{-2}; 1×10^{-1} MPa (1, 2, 3).

heat removal by convective heat transfer from the vapor to the condensation film. Referring to the heat flux $q = G\Delta h_{fg}$, $T_g - T_w$ is related to the heat flux and the thermal resistance of the phase transition, i.e., $R_Ф = \frac{T_g - T_w}{q}$. Comparing this with Eq. (7.45), we have

$$R_Ф \cong \frac{2\beta}{2-\beta} \cdot \frac{\sqrt{2\pi R}}{\Delta h_{fg}} \cdot \frac{T_g - T_w}{\frac{p_g}{\sqrt{T_g}} - \frac{p_w}{\sqrt{T_w}}}. \tag{7.46}$$

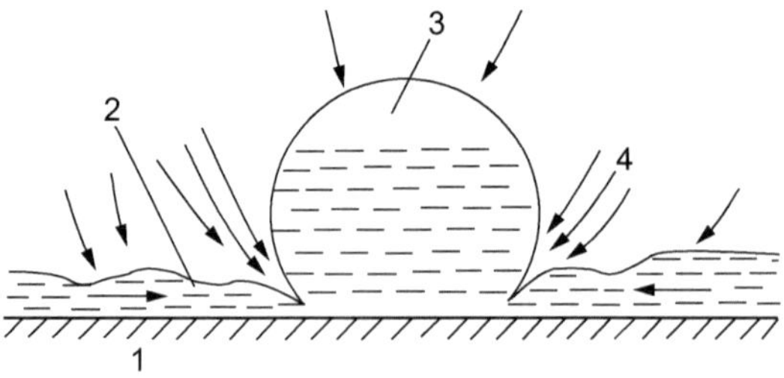

Figure 7.16 Dropwise condensation processes: 1—condensing surface; 2—microfilm; 3—a drop; 4—the main stream of the condensing vapor.

In Fig. 7.15b, the temperature drop ($T_g - T_w$) will be 2–20 K.

The latent heat released at the interface is transmitted to the wall. This process differs depending on the type of condensation. In the filmwise condensation, the thermal resistance of the condensate film R_ϕ depends on the nature of film flow. In laminar flows, R_ϕ is determined by the thermal conductivity of the fluid. In turbulent flows, R_ϕ plays an important role in heat transfer. The wavy nature of the film motion changes the velocity distribution in it, and its thickness, and therefore, affects the thermal resistance. These characteristics also influence vapor phase velocity. The rate of vapor condensation in flow channels varies along the length of the channel, which leads to various modes of vapor and condensate flows.

The thermal resistance of the heat exchange with the condensing droplet is smaller than that of the film condensation. This is due, firstly, to the fact that the droplets have a larger surface area of contact with the vapor, and secondly, to the movement of condensate microfilm formed on the surface which is contracted into the droplets as illustrated in Fig. 7.16. Therefore, the thermal resistance of the heat transfer at the dropwise condensation is affected by the proportion of the surface occupied by drops, the thermal resistance of the phase transition, and the thermal resistance of drops. With an increase in the supercooling, modes of surface heat transfer change.

These effects cannot be taken into account with ease, and the average heat transfer coefficient is used in engineering practices.

7.2.1.3 Condensation curve

The dependency of the condensation heat flux on the degree of supercooling of the surface is called the *condensation curve* (Fig. 7.17). The heat flux increases with increasing supercooling ($\Delta T = T_S - T_g$) and takes a maximum at a certain ΔT. With increasing supercooling ΔT, the entire surface is covered with a film of condensate. This phenomenon is observed on a vertical surface when the condensation rate exceeds the rate of droplet separation. With further increase in ΔT the surface temperature becomes below the freezing point, the ice layer formed on the surface and the vapor condenses on this layer. Increasing the steam condensation rate curve shifts to higher heat flux densities.

The system of differential equations describing the processes in the condensation include equations of energy, motion, continuity, mass transfer between the phases. If you take the temperature of the condensate T_w (in Fig. 7.14) equal to the saturation

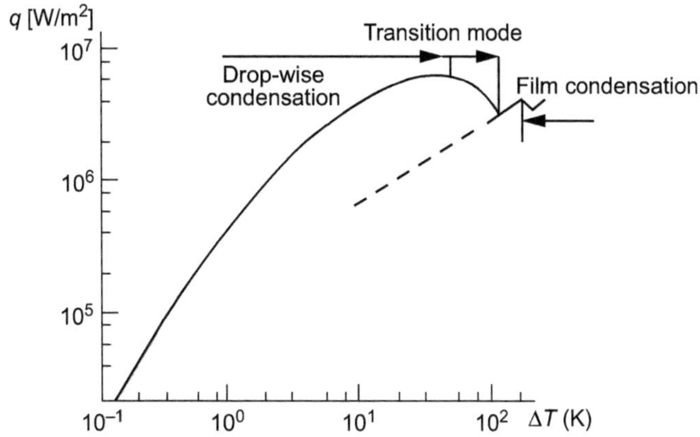

Figure 7.17 Condensation curve.

temperature of the surface, the heat flux through the condensing surface and through the cooling surface will vary due to the supercooling of condensate below T_S.

Bringing up all the equations to dimensionless form, you can select the criteria for determining the condensation process for each individual case.

7.2.2 Dropletwise condensation

When the condensate does not wet the heat transfer surface, the surface will be covered with drops but separated from the surface. Initially the vapor in contact with the surface forms an adsorption layer, and then a thin (a fraction of micrometer) film of liquid. This film, which is in a force field of molecular and rigid wall of vapor layer, is exposed to a disjoining pressure field. With a film breakup, droplets will be formed. The droplets grow, merge, fall down under the gravity force or break in moving vapor. Centers of condensation are micro cavity surface. The typical size of condensation nuclei is determined by the formula

$$R_{cr} = \frac{2\sigma T_S}{\Delta h_{fg} \rho_f \Delta \mathrm{T}}. \tag{7.47}$$

where $\Delta T = T_S - T_f$ is the difference between saturated vapor temperature and liquid temperature of the surface layer. With $\Delta T > \Delta T_{cr}$, condensation is possible. With $\Delta T < \Delta T_{cr}$ evaporation of the liquid layer takes place.

7.2.2.1 Interphase thermal resistance

The temperature difference (see Fig. 7.14) allows us to introduce a concept of interfacial thermal resistance R_{ϕ} and accordingly a conditional heat transfer coefficient α_{ϕ}. If we denote R_{λ} the thermal resistance of drops, the total thermal resistance of the heat exchange with the condensate drop will be

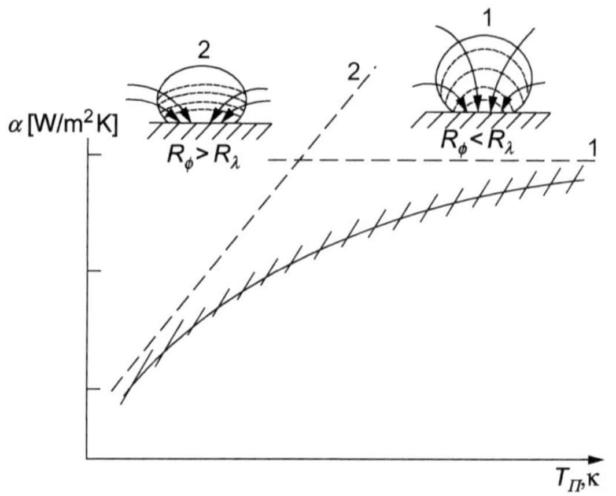

Figure 7.18 Dependency of the heat transfer coefficient of the dropwise condensation on the vapor temperature; showing isotherms and heat flow lines in the drop.

$$R = \frac{1}{\alpha} = \frac{1}{\Phi}\left(R_{\Phi} + R_{\lambda,\alpha_{\Phi} \to \infty} + \Delta R_{\lambda}\right). \tag{7.48}$$

where Φ is the fraction of the surface occupied by the drops; ΔR_{λ} is the increment of the thermal resistance of drops due to the increase in R_{Φ}.

Two extreme cases are illustrated in Fig. 7.18.

(1) The temperature and vapor pressure are sufficiently high—heat transfer is mainly determined by the thermal resistance of drops (Fig. 7.18, line 1) $R_{\Phi} < R_{\lambda}$ and

$$\alpha \cong \frac{\Phi}{R_{\lambda,\alpha_{\Phi} \to \infty}}, \tag{7.49}$$

where $R_{\lambda,\alpha_{\Phi} \to \infty} = \overline{d}/\left(0.834\lambda_f\right)$ is the thermal resistance of drops of average diameter $\overline{d}$;

(2) The temperature and vapor pressure are low enough—heat transfer is determined by the thermal resistance of the phase change (Fig. 7.18, line 2) (Rose, 1997, 2002; Borishansky et al., 1976; Ivanovsky et al., 1966; Ivanovsky et al., 1976):

$$\alpha \approx \alpha_{\Phi}\Phi. \tag{7.50}$$

For dropwise condensation to achieve high heat transfer coefficients, however, the use of the process is limited in practice because dropwise condensation generally occurs only at a low vapor pressure and on nonwettable surfaces (Zhao et al., 1990).

7.2.3 Filmwise condensation of a pure steam

7.2.3.1 Film flow regimes

Filmwise condensation heat transfer is associated mainly with the thermal resistance of the liquid film, which depends on its thermal conductivity, thickness, and velocity

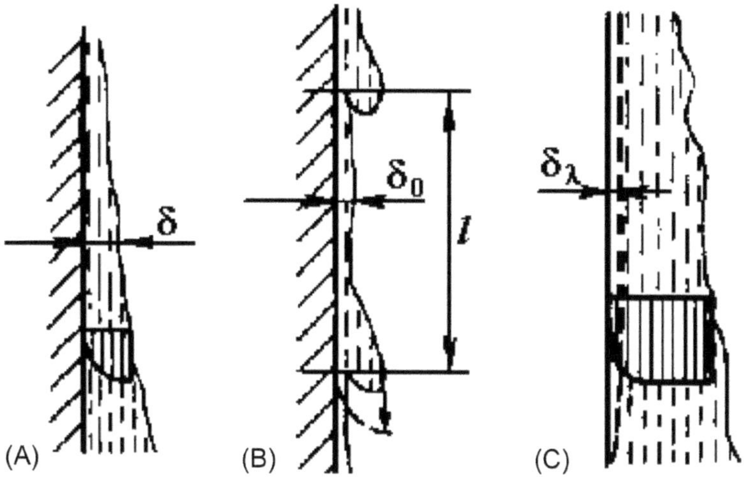

Figure 7.19 Film flow regimes: (A) Laminar (δ—film thickness); (B) laminar wavy (δ_0—residual thickness, l—wave length); (C) turbulent (δ_λ—laminar sublayer thickness).

distribution. Thickness and velocity distribution are determined by the mode of film flow on the wetted surface, which is determined by a balance of the gravity force, friction, inertia forces, and the surface tension. Assume a condensation on a vertical wall surface. Thickness of the condensate film and velocity are small on the upper part of the surface, where a laminar mode would prevail. As the flow rate in the film increases in the lower parts, there are wavy motions, and the turbulent regime prevails. Accordingly, there are three modes of film flow along the vertical or inclined surface (Fig. 7.19): Laminar ($Re < Re_b$) wavy ($Re_b < Re < Re_{cr}$) and turbulent (turbulent wavy) ($Re > Re_{cr}$). More detailed discussions on the film flow wavy motions are found in Spalding and Taborek (1983), Alexeenko et al. (1992), and Kutateladze and Nakoriakov (1982).

In the wave mode, two regions can be identified: a developing region of the wave mode ($Re_b < Re < Re_*$) and a developed region ($Re_* < Re < Re_{cr}$). All these mode boundaries are given by $Re_b = 7\left(\frac{Ar_*}{K^3 Pr^3}\right)^{1/8}$, $Re_* = 2.3 Ar_*^{1/5}$ and $Re_{cr} = 35 Ar_*^{1/5}$, where $Ar_* = \left(\frac{\sigma^3}{\nu^4 \Delta\rho g}\right)^{0.5}$, $Re = \frac{\overline{q}L}{\Delta h_{fg}\mu} = \frac{\rho w L}{\mu} \equiv \frac{V}{\nu}$, L is the length or diameter of the section [m]; V is the volumetric flow per unit length [m^2/s]; q is the heat flux [W/m^2]; and $K = \Delta h_{fg}/(c_{pf}\Delta T)$.

Heat transfer mechanism in a laminar mode depends on the conductivity and film thickness of the condensate. In the wave development region, the heat transfer is enhanced due to the convective mixing and increase in the interfacial surface area. Within a zone where there is a periodic wavy motion, the heat transfer is determined by the thermal resistance of the residual liquid film layer, in which the waves move.

7.2.3.2 Vertical surfaces: Laminar flow

Consider condensation of dry saturated steam fixed on a vertical surface, whose temperature is T_w. Theory of the heat transfer process is established by Nusselt.

The following assumptions are made:

- inertia forces are small compared with the forces of gravity and viscosity;
- heat through the film is transferred only by conduction, convective transfer, and thermal conductivity along the film are not considered;
- the temperature of the outer surface of the film is equal to T_s;
- physical parameters of the film are independent of temperature;
- friction at the interface of the liquid-vapor is absent, assuming the smooth film surface and small surface tension forces.

Heat transfer is carried out only by thermal conduction. Then, with the equations $q = \frac{\lambda}{\delta(x)}(T_S - T_w)$ and $q = \alpha(T_S - T_w)$, we have

$$\alpha(x) = \frac{\lambda}{\delta(x)}. \tag{7.51}$$

Thus, the problem of determining the heat transfer coefficient is reduced to finding the film thickness $\delta(x)$, which is obtained from the equations of motion and heat balance equation; x is the coordinate.

The film thickness is given by

$$\delta = \sqrt[4]{\frac{4\lambda_f \mu_f (T_S - T_w)x}{\Delta h_{fg} g \rho_f^2}}. \tag{7.52}$$

Hence, the local heat transfer coefficient is

$$\alpha_x = \sqrt[4]{\frac{\lambda_f^3 g \rho_f^2 \Delta h_{fg}}{4\mu_f (T_S - T_w)x}}. \tag{7.53}$$

The average heat transfer coefficient is

$$\overline{\alpha} = \frac{1}{h}\int_0^h \alpha_x dx = \frac{4}{3}\sqrt[4]{\frac{\lambda_f^3 g \rho_f^2 \Delta h_{fg}}{4\mu_f \Delta T \cdot h}} = 0.943\sqrt[4]{\frac{A}{\Delta T \cdot h}}, \tag{7.54}$$

where h is the height of plate, $A = \frac{\lambda_f^3 g \rho_f^2 \Delta h_{fg}}{\mu_f}$. From this equation, it is well understood that the average heat transfer coefficient decreases with an increase in the height h and the temperature difference ΔT.

Eq. (7.54) is expressed in terms of the Nusselt number:

$$\overline{Nu}=\frac{\overline{\alpha}h}{\lambda}=0.943\frac{h}{\lambda}\sqrt[4]{\frac{\lambda_f^3 g\rho_f^2\Delta h_{fg}}{\mu_f\Delta T\cdot h}}=0.943(Ar\cdot Pr\cdot K)^{1/4} \tag{7.55}$$

with $Ar=\left(gh^3/\nu^2\right)\left(1-\rho_g/\rho_f\right)$, $Pr=c_p\mu/\lambda$, $K=\Delta h_{fg}/c\Delta T$. Eq. (7.55) is valid when $\rho_f w_f \delta/\mu_f < 200$, $Pr > 0.5$ and $\rho_g \ll \rho_f$.

7.2.3.3 Laminar-wavy flow

In the laminar-wavy film flow, a large quantity of liquid mass is transported in the waves, and major thermal resistance is at the thin film of condensate with "residual" thickness δ_0 (see Fig. 7.19b). The residual thickness δ_0 virtually does not change with the condensate flow, i.e., a flow rate increase does not increase δ_0, but reduces the wavelength l. This wavy regime does not develop within the short period time of existence of this thin layer, which is of the order of the wave oscillation period. In this case, as for the laminar flow of the film $\alpha=\lambda/\delta_0$. The average heat transfer coefficient of the film flow of the laminar wave character is given by

$$\left(\frac{\alpha}{\lambda}\right)\left(\frac{\nu^2}{g}\right)^{1/3}=\frac{1}{\delta_0}\left(\frac{\nu^2}{g}\right)^{1/3}=0.527Ar_*^{-1/15} \tag{7.56}$$

where $Ar_*=\left(\frac{\sigma^3}{\nu^4\Delta\rho g}\right)^{0.5}$. The relation holds for $Re_* < Re < Re_{cr}=35Ar_*^{1/5}$.

In the wave development within the condensate film, the heat transfer coefficient is independent of the geometrical dimensions and is a function of the physical properties of the liquid film. The model will cover a range of $Re=200-10^3$.

The influence of the wavy film motion and a variation of the physical parameters in temperature on the heat transfer coefficient is accounted for by two correction factors to α_0:

$$\alpha=\alpha_0\varepsilon_\nu\varepsilon_T. \tag{7.57}$$

where α_0 is the heat transfer coefficient calculated by Eqs. (7.51)–(7.53), with physical parameters evaluated at T_s; $\varepsilon_\nu=Re^{0.053}$ is the correction due to the wave motion for $Re<10^2$ and may be given by another expression $\varepsilon_\nu=\left(\alpha h\Delta T/\Delta h_{fg}\mu_S\right)^{0.04}$; $\varepsilon_T=\left[(\lambda_w/\lambda_s)^3(\mu_s/\mu_w)\right]^{1/8}$ is the amendment due to a variation of the fluid properties in the film section. These correlations are valid for $0.5<\lambda_s/\lambda_w<2$ and $0.1<\mu_s/\mu_w<1$. Here, we assumed a wall suitable for transferring the heat released on the surface of the film due to condensation. The conclusion remains valid if, instead of the heat of condensation, we use a substitute of an effective

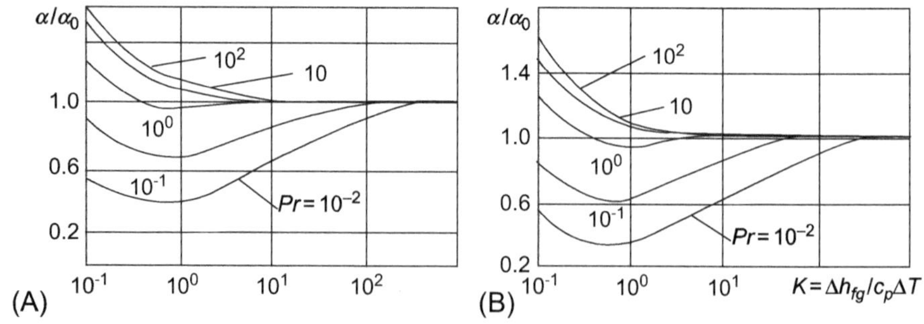

Figure 7.20 Influences of inertia forces of condensate film laminar flows on the convective heat transfer: (A) Flat vertical surface; (B) horizontal circular tube; α_0 is the heat transfer coefficient, from the Nusselt number calculated by Eqs. (7.51) and (7.53).

heat of condensation, which accounts for the liquid film supercooling and changes the temperature profile: $\Delta h_{fg}^{*} = \Delta h_{fg} + \frac{2}{3}c_{pf}(T_S - T_w)$.

Another prerequisite consideration for the solution has been the neglect of the inertia forces in the film and in the vapor. The influence of these forces may not be neglected, at least for two reasons: (1) steady-state vapor flow should show acceleration to the speed of the film interface boundaries; (2) the condensate flow would accelerate when it flows downward. When $K = \Delta h_{fg}/\left(c_{pf}\Delta T\right) > 5$ and $Pr = 1{-}100$, the inertial forces can be ignored since their effect on the heat transfer coefficient would be of only a few percent. Consideration of inertial forces is important when $K < 5$ for fluids of $Pr > 1$ and liquid metals of $Pr \ll 1$, i.e., in the critical region with large temperature differences, and in the case of condensation of metal vapor (Fig. 7.20).

7.2.3.4 Turbulent-wave flow

If greater than the critical Reynolds number ($Re_{cr} \approx 1600$), the film flow becomes turbulent. Increase in the hydraulic flow resistance of the film and increase in its thickness should cause a reduction of α. However, turbulence increases the effective thermal conductivity of the film flow (Isachenko, 1977; Labuntsov, 1960). When $Pr \geq 1$ the turbulence factor is more important and leads to higher heat transfer coefficient. For $Pr \ll 1$ (liquid metals), turbulence of the film does not play a big role in increasing the *effective thermal conductivity* ($\lambda_T < \lambda$) and the main factor is the increase in the film thickness that causes a reduction in heat transfer, even when compared with laminar flow (in Borishansky et al., 1976).

In the case of condensation on inclined surfaces, the gravitational acceleration g is modified to be its projection onto the axis x, i.e., $g \cos \varphi$. Here, φ is the angle between the vertical axis (gravity force direction) and x:

$$\alpha = \alpha_{\text{vertical}} \cdot (\cos \varphi)^{0.25}. \tag{7.58}$$

7.2.3.5 Horizontal tubes and rod bundles

Condensation on the horizontal circular pipes (φ-variable). Noting δ << d (= pipe diameter) and under the conditions of the laminar condensate film flow and with T_w = const, we have an approximation of the average heat transfer coefficient:

$$\overline{\alpha} = 0.728 \sqrt[4]{\frac{\lambda^3 \rho^2 g \Delta h_{fg}}{\mu d \Delta T}}. \tag{7.59}$$

or in the form of the Nu-number correlation when $\rho_g \ll \rho_f$:

$$\overline{Nu_N} = \overline{\alpha} d/\lambda = 0.728(Ar \cdot Pr \cdot K)^{1/4}. \tag{7.60}$$

Eqs. (7.59) and (7.60) are approximate. A more exact calculation can be made by incorporating the laminar and turbulent wave modes as suggested in Sukhatme (1990), Labuntsov (1960), and Tanasawa (1994). The wave mode to horizontal pipes develops when $Re > Re_b = \left(\frac{43}{\pi} Ar_*^{1/15}\right)^{1.06}$ and if $\frac{gd\rho}{\sigma} = We^{-0.5} > 4$. When q = const, 0.728 in Eqs. (7.58) and (7.59) is set equal to 0.693.

Average heat transfer coefficient of the stationary vapor flow in a bundle of horizontal pipes depends weakly on the relative distance between the tubes, dripping characteristics from the pipes and other parameters, and generalized dependencies mentioned previously (see Fig. 7.21).

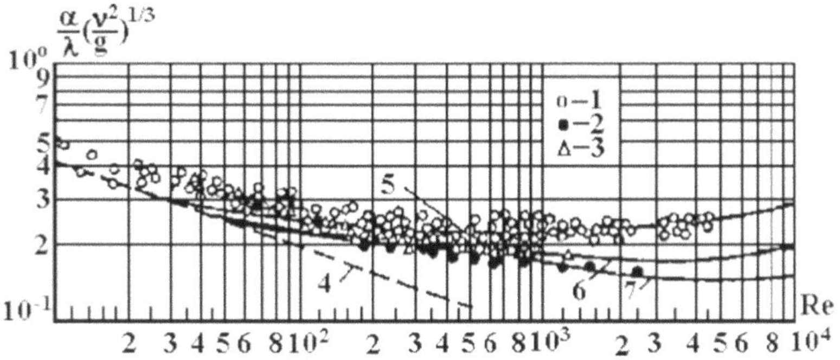

Fluid	Experiments	Calculation
H_2O vapor Pr = 3,5	1	5
H_2O vapor Pr = 1,7	2	6
Freon-21 Pr = 1,1	3	7
4–calculated by Nusselt		

Figure 7.21 Intensity of heat transfer by condensation.

The heat transfer rate on a horizontal tube surface is higher than on a vertical surface since the film thickness is less than the average. Condensation of the vapor occurs on the horizontal tube banks, condensate flows from the upper to the lower pipe, and the liquid layer is increased on the lower pipes and tends to decrease the heat exchange, but the liquid film is turbulent-induced by falling jets on the surfaces of lower rows of the tube bank. Therefore, the nature of heat transfer depends on the steam supply, its speed, and the layout of the beam.

7.2.3.6 Condensation of superheated and wet vapor

When the superheated (δT) vapor condensates, the condensation latent heat Δh_{fg} are transferred with the heat of superheating $c_{pf}\delta T$, then Δh_{fg} is replaced by $\left(\Delta h_{fg} + c_{pg}\delta T\right)$, so the heat transfer coefficient of the superheated vapor will be

$$\alpha = \alpha_S \sqrt[4]{1 + \left(c_{pg}\delta T\right)/\Delta h_{fg}}. \tag{7.61}$$

where α_S is the heat transfer coefficient of dry saturated vapor.

Condensation of the wet steam having a quality $x < 1$ releases an amount of the heat equal to $\Delta h_{fg}x$ per kg. Therefore, as an approximation, the following can be accepted:

$$\alpha \approx \alpha_S \sqrt[4]{x}. \tag{7.62}$$

In many cases of practical importance, with small overheating or at low humidity, where $(1 - x) < 0.1$, the heat transfer coefficient is $\alpha \approx \alpha_S$.

7.2.3.7 Condensation of metal vapor

Intensity of the heat transfer with metal vapor condensation depends little on the nature of the flow of the condensate film due to the high thermal conductivity of the metals. The decisive role in this case is played by the following two effects:

(1) Dominant thermal resistance, which is the phase transition (at relatively low temperatures and pressures, the vapor condensation rate $\beta < 1$);
(2) The presence of noncondensable gases in the vapor, whose effect is quite difficult to estimate.

When $p < 10^3$ Pa, $\beta = 1$. With increasing pressure condensation coefficient decreases according to $\beta = 20/\sqrt{p}$.

7.2.4 Film condensation of flowing steam

7.2.4.1 Vertical surface

Vapor flow has a dynamic effect on the phase interface, speeding up or slowing down the condensate film flow. If the steam flows downward (the direction coincides with the direction of gravity), the film speed increases, the thickness decreases, and the

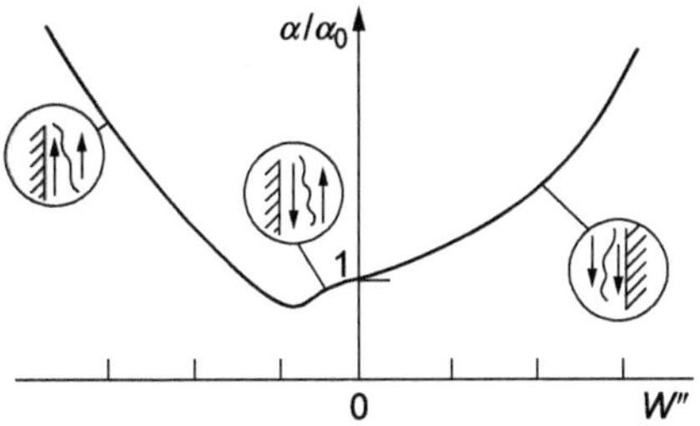

Figure 7.22 The change of heat transfer coefficient with vapor velocity w_g.

heat transfer rate increases. When the steam flows upward, hampering the condensate film, and increasing its thickness, there is a wave mode, and the heat transfer coefficient can be reduced. With a further increase in the vapor-phase velocity, the interphase friction force increases and drags the film upward, its thickness is reduced and the heat transfer coefficient (Fig. 7.22) increases again. At this high speed of the vapor, a break-up of the film is likely and the liquid droplets are entrained into the vapor flow, which leads to a reduction in the film thickness and a sharp increase in the heat transfer coefficient (Gogonin, 2006; Labuntsov, 1960a,b).

Steam-vapor flow acts on the condensate film flow so that it not only changes the film thickness but also leads to a perturbation of the film flow, and accelerates a transition from laminar to turbulent flow. Therefore, a molecular heat transfer is supplemented by turbulent transfer even at low *Re*.

7.2.4.2 Horizontal tubes and rod bundles (Breber et al., 1980; Labuntsov, 1957; Sukhatme, 1990)

Calculation of heat transfer in a horizontal tube bundle is complicated due to: (1) a decrease in steam velocity due to partial condensation as the steam moves in the tube bundle; (2) an increase in the condensate film thickness on the lower tube bank by a drainage from the upper to the lower tubes, which largely depends on the geometry of the tube array configuration. Usually tube bundle layouts of condensers are complex, making it difficult to put theoretical and even experimental results into practice. Therefore, it is often the case that condensation calculations in tube bundles are made based on the correlations and models developed in industry.

7.2.4.3 Condensation inside tube

Vapor condensation inside a tube is more complex phenomena than the stationary or moving steam condensation on the surface (Spalding and Taborek, 1983; Miropol'sky, 1962). The film of condensate is exposed to the steam flow in a pipe. The rate of vapor condensation decreases in length of the tube and the condensate flow rate increases. Assume steam quality $x > 1$ (saturated or superheated steam) in the inlet portion of the pipe, the whole pipe cross section is occupied by the steam. Downstream two-phase flow structure depends on the balance among the gravity and friction, vapor

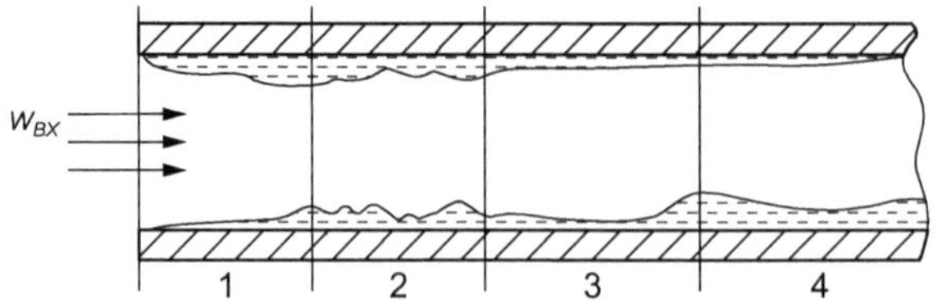

Figure 7.23 Scheme condensation and flow conditions inside the horizontal pipe (by Breber et al., 1980): 1—initial portion; 2—section of the turbulent film movement; 3—section with laminar film movement; 4—stream.

flow acting on the condensate film, and volumetric relationship of fluid and steam. Single-phase vapor flow at the tube inlet is usually in the turbulent flow regime. It can be replaced on the laminar flow. Although laminar, the condensate films may show turbulent motions near the inlet due to increased condensate flow rate and enters again into the laminar flow regime, where the steam flow velocity will drop and become low. Sketch of a horizontal pipe flow regimes of vapor and liquid is shown in Fig. 7.23.

In horizontal tubes a strong nonuniformity will be found in the condensate distribution over the cross section, leading to a considerable temperature difference of top and bottom of the tube.

Theoretical solutions would be obtained, typically applicable for a narrow range of parameters. Therefore, heat transfer calculations are carried out, knowing the steam velocity w_0 at the inlet or total mass flow rate, $w_0 = 4\rho w / \rho_f \pi d^2$ where ρ and w are the mixture density and velocity. The latter, although it is convenient for calculations, but is rather notional.

Condensing vapor behaviors in a vertical tube differ depending on the direction of flow of the vapor. In the case of countercurrent flow at low speeds (10–5 m/s), flooding may occur. At high speeds (15–30 m/s), the liquid portion breaks down from the film surface (Gogonin, 2006).

Under superheated vapor complete condensation in a pipe, where the vapor enters and exits as a supercooled liquid, it is possible to distinguish four characteristic sections along the tube as shown in Fig. 7.24:

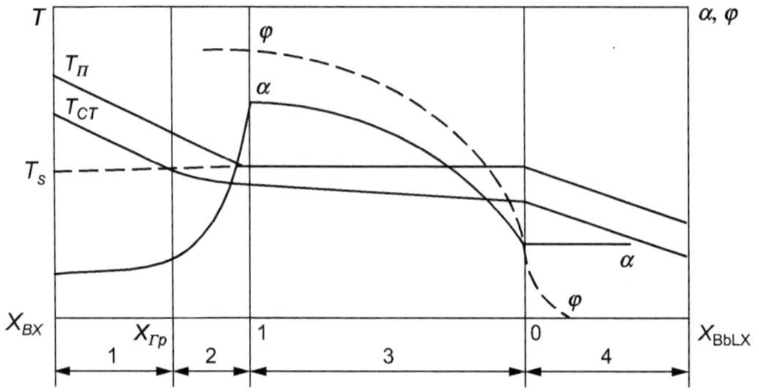

Figure 7.24 Modes and heat transfer characteristics of complete condensation of the superheated steam inside the tube (φ is the vapor void fraction).

(1) cooling of the superheated steam by condensation heat transfer $T_g > T_S$, $T_w \geq T_S$, $x_{\text{Вч}} \geq x > x_{\Gamma \text{p}}$. In this section, $T_w = T_S$, $x = x_{\Gamma \text{p}}$
(2) condensation of superheated vapor $T_g \geq T_S$, $T_w \leq T_S$, $x_{\Gamma \text{p}} \geq x \geq 1$
(3) Condensation of saturated vapor $T_g = T_S$, $T_w < T_S$, $1 \geq x \geq 0$
(4) condensate cooling $T_g < T_S$, $T_w < T_S$, $x < 0$

Here, $x = \left(i_{\text{CM}} - i_f\right) / \Delta h_{fg}$ is the relative enthalpy with i_{CM} the mixture enthalpy, and $x_{\Gamma p}$ is the relative enthalpy at which vapor condensation begins.

7.2.5 Condensation of vapor-gas mixture

7.2.5.1 Mechanisms of heat transfer: Three thermal resistances

A presence of noncondensable gases in the vapor-steam changes the heat transfer mechanism significantly: e.g., see details in the papers by Sawant et al. (2015), Dehbi (2015), and Zhang et al. (2018). Pressure of the gas-vapor mixture is approximately uniform in the cross section of the volume, but since the vapor partial pressure in a condensation surface is less than in the core flow, the gas concentration at the surface is higher than in the bulk (Fig. 7.25; Isachenko, 1977). As a result, there are two

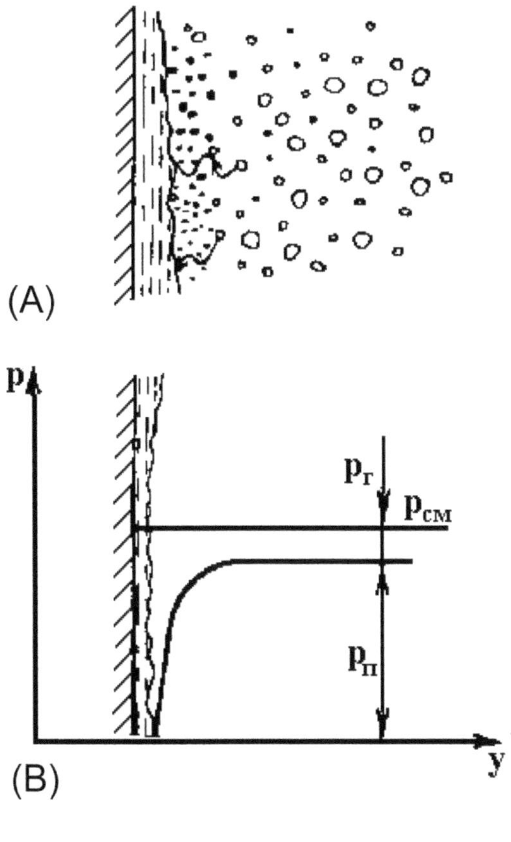

Figure 7.25 Condensation surface conditions: (A) Scheme of distribution of molecules at the condensation surface: •—gas molecules; ○—vapor molecules. (B) Distribution of partial pressures of steam and gas at the surface.

opposite diffusion processes: one diffusion of the gas mixture toward the film and the other diffusion of the gas away from the wall. The resulting mixture flow is zero but the vapor flow is not zero. The films of noncondensable gas are a barrier to incoming vapor to the film. The heat transfer coefficient in the presence of noncondensable gases decreases sharply and depends on the three thermal resistances—condensate film resistance R_{film}, interfacial thermal resistance R_{ϕ}, and the diffusion thermal resistance (resistance to steam supply to the condensation surface) R_{dif}:

$$1 / \alpha_{\text{CM}} = R_{film} + R_{\text{ф}} + R_{diff}. \tag{7.63}$$

This thermal resistance is a sum of the temperature differences ΔT_{film}, ΔT_{ϕ}, and ΔT_{diff}: i.e., $\Delta T = \Delta T_{film} + \Delta T_{\text{ф}} + \Delta T_{diff}$.

At moderate (not too low) pressure we can neglect the thermal resistance of the phase transition:

$$1 / \alpha = R_{film} + R_{diff}. \tag{7.64}$$

When calculating the thermal resistance of the diffusion account for the intensity of the supply of heat by convection $\alpha_C(T_S - T_w)$ and that by condensing vapor, where $G = \beta(c_S - c_w)$ is the vapor mass flux to the condensation surface; β is the mass transfer coefficient (m/s); c_S and c_w are the vapor concentration in kg/m^3. Hence, we come up with

$$\frac{1}{\alpha} = R_{film} + \frac{1}{\alpha_b + \Delta h_{fg}\beta \dfrac{c_S - c_W}{T_S - T_w}}. \tag{7.65}$$

where α is the conditional that accounts for the processes of heat and mass transfer in the gas-vapor mixture and the convective heat transfer through the film. Fig. 7.26 shows how the temperature field is distributed near the condensation surface with increasing number of noncondensable gases in the steam.

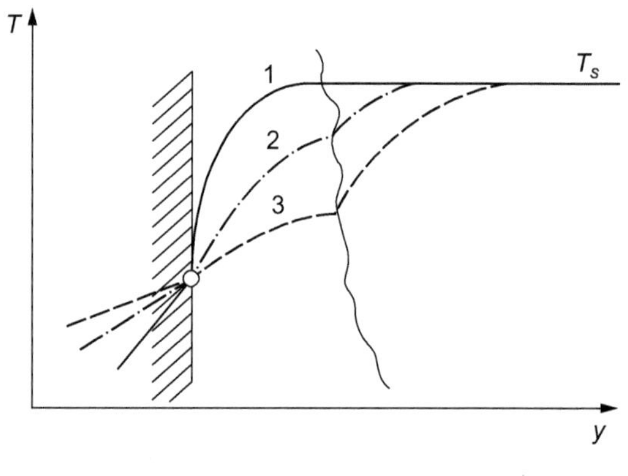

Figure 7.26 Changes in temperature field near the condensation surface with increasing amount of noncondensable gases (1 < 2 < 3).

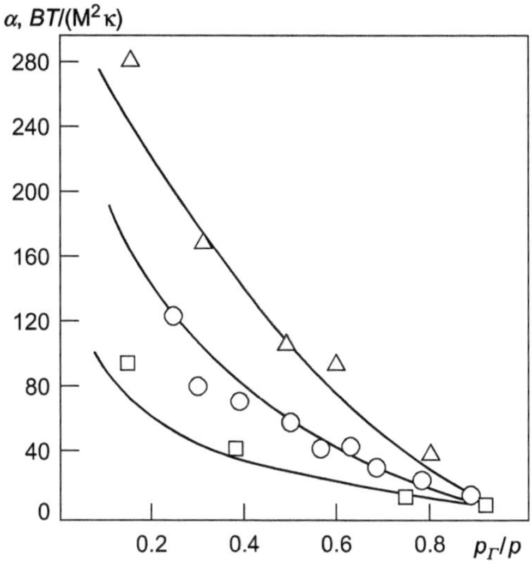

Figure 7.27 Effect of pressure and partial pressure of the gas in the condensation heat transfer coefficient of the steam from the steam-air mixture, □ $p = 0.1-0.3$ MPa; ○ $p=0.4-0.8$ MPa; △ $p=2.2-3.0$ MPa.

The intensity of condensation increases with increase in pressure. In Fig. 7.27 shown are the effects of the volume of gas content and pressure on the heat transfer coefficient.

7.2.6 Condensation at direct contact of the phases

Relevant to severe accidents evaluation of nuclear power reactors, it is important to consider the processes of condensation in direct contact of two phases, which play an important role in nuclear reactor safety analysis. These include:

(1) Condensation of large vapor bubbles in the fluid (vapor and water mixing under the ECCS operation modes);
(2) Condensation of vapor in the liquid (water and vapor mixing in the reactor vessel);
(3) Condensation of steam on the subcooled droplets (ECCS spray);
(4) Condensation on free surfaces.

A major difficulty in these problems is associated with defining the interface area per unit volume.

7.2.6.1 Condensation in subcooled liquid

Vapor condensation in the subcooled liquid flow is popular and observed, e.g., in the channels and piping of boiling water reactors (BWRs), steam generators, and mixers of vapor and liquid.

There are two stages in the collapse of vapor bubbles: at a large radius, the rate of change in the radius of a bubble is determined by the intensity of the interface heat transfer, while at a small radius, the main role is played by the inertia force. The first stage accounts for most of the lifetime of the vapor bubble, so it is more important.

A detailed discussion on the bubble radius change in the subcooled liquid is found in Florschuetz and Chao (1965).

The condensation of vapor bubbles flowing out of the nozzle into the subcooled turbulent flow, can be estimated as in the following. The rate of collapse of the vapor bubble is described by $\frac{dR}{d\tau} = \frac{q}{\rho_g \Delta h_{fg}} = \frac{\alpha \Delta T}{\rho_g \Delta h_{fg}}$, where the heat transfer coefficient can be found from the approximate relations for heat transfer solid sphere:

$$Nu = \frac{2\alpha R_0}{\lambda_f} = 0.37\left(\frac{2R_0\dot{R}}{\nu_f}\right)^{0.6}, \tag{7.66}$$

where $\dot{R} \equiv \frac{dR}{d\tau}$ is the surface velocity. This relationship gives satisfactory results for $\Delta T_{sub} = 15{-}100$ K, $p = 1{-}6$ MPa, $Re = 2000{-}3000$.

When the vapor jet injection is made into a large amount of subcooled liquid, condensation is characterized by an unstable boundary and changes in transient velocity profile. Evaluation of heat exchange between the vapor flow and liquid volume can be made by empirical relations $Nu = F(Re, Pr)$.

7.2.6.2 Condensation on liquid subcooled jet

Suppose a large volume is filled with vapor at a saturation temperature. A liquid jet at a temperature T_0 is lower than the saturation temperature ($T_0 < T_S$) and is fed through a hole of diameter d into this volume with a mass flow W_f [kg/s]. If the jet of a length $x = l$ condenses the amount of vapor G_k [kg/s], the heat balance equation can be written as: $G_k \Delta h_{fg} + G_k c_{pf}(T_S - T_l) = W_f c_{pf}(T_l - T_0)$; here T_0 and T_l are the average temperature of liquid-phase on the cross section at $x = 0$ and $x = l$. Condensation of the vapor will stop when the temperature of the liquid at the end of the jet reaches T_S. The quantity of heat transmitted by the vapor condensation to the jet, can be computed as $Q = G_k \Delta h_{fg} = \overline{\alpha}\overline{\Delta T}A$ where $\overline{\alpha}$ is the average heat transfer coefficient along the length; A is the area of the surface of the jet; $\overline{\Delta T} = \frac{T_l - T_0}{ln((T_S - T_0)/(T_S - T_l))}$ is the mean log temperature difference. Because $W_f = \rho_f w_0 \left(\pi d^2/4\right)$, the relationship between the coefficient of heat transfer and heating jets is of the form

$$\overline{\alpha} = \frac{1}{4}\rho_f c_{pf} w_0 \frac{d}{l}\, ln\, \frac{(T_S - T_0)}{(T_S - T_l)}\left(1 + \frac{1}{K}\frac{(T_S - T_l)}{(T_S - T_0)}\right)^{-1} \tag{7.67}$$

The heat transfer theory for the case of condensation of steam on the liquid jet (laminar and turbulent) has been developed only with considerable simplifications.

7.2.6.3 Condensation on dispersed liquid spray jet

Vapor condensation on the spray jet is considerably more intense than that in the continuous stream case due to a sharp increase in the surface area of the liquid-vapor contact. This process of droplet condensation can be represented with two stages.

In the first stage, the droplet surface is heated from its initial temperature T_0 to T_S. The rate of process is determined by the thermal resistance at the interface, which depends on the pressure, $(T_S - T_w)$ and the condensation coefficient β. For water $R_{\Phi} = 10^{-7}$ [(m^2 K)/W].

The rate of condensation in the second stage is determined by the heat removal process from the surface to inside the droplet, i.e., thermal resistance of drops. Influence of the convection on heat transfer inside the drop is low and, therefore, the resistance of the drop depends on the thermal conductivity of the liquid and the radius of the droplet. Hence, to obtain the heat transfer of drops, a classical solution approach is applied to the heat conduction problem for a solid sphere:

$$\alpha = \frac{1}{R_\lambda} = \frac{q_f}{T_S - T_K} = \frac{\pi^2 \lambda_f}{3R}. \tag{7.68}$$

Solving the heat transfer problem on the dispersed jet is regarded not yet completed due to a lack of a number of factors to be considered (spectrum of droplets, deformation, different types of coagulation or decay of the drops, etc.). For more information, see Berman et al. (1983) and Isachenko (1977).

7.2.7 Enhancement of heat transfer at condensation

When the thermal resistance of the condensate film (R_{film}) is large compared with other thermal resistances, for the enhancement of heat transfer it is necessary to reduce this resistance. One method of reducing this resistance is to make a transition to dropwise condensation. Other methods are based on an artificial reduction of the condensate film thickness by the employment of vertical pipes with ribbed or corrugated surfaces instead of smooth pipes. The ribs may be of very different forms—two-dimensional or three-dimensional (Fig. 7.28). The issue has been extensively discussed in open literature. In the review by Sukhatme (1990), for example, you can find the information to optimize the geometry and dimensions of the finned tubes. The presence of corrugations leads to the fact that the condensate film curved the surface.

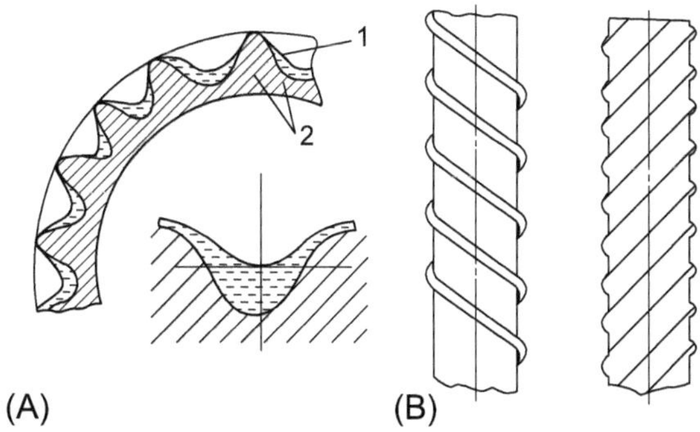

Figure 7.28 Ribbed and corrugated pipes for heat transfer enhancement: (A) Finned tube cross section: 1—condensate film; 2—tube wall; (B) types of protrusions on the tube surface.

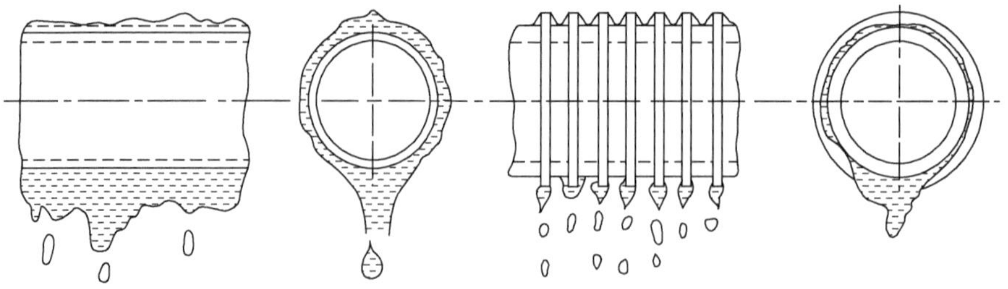

Figure 7.29 Dripping films with a horizontal smooth pipe and ribbed pipe.

Under the influence of the surface tension forces, the condensate moves from a crest to a trough, and the film thickness decreases for a greater portion of the condensate film surface, and therefore, the thermal resistance of heat transfer (Fig. 7.28a). Twisted ribs on the vertical tubes contribute to further enhancement of the heat transfer coefficient because the liquid film is formed on a short stretch, before drainage through the spiral protrusions (Fig. 7.28b).

For the vertical tubes, the heat transfer will increase until the area occupied by the ribs is around 10% of the total surface area. A further increase in the number of ribs has little effect. Enhancement of heat transfer with the ribs on vertical tube surfaces is effective as long as the ribs are not flooded with the condensate.

For the horizontal tubes, the presence of transverse or cross-ribs facilitates the withdrawal of the condensate film and improves the heat transfer coefficient. This effect is associated with a decrease in film thickness on the edge, while the main mass of the condensate quickly drains down along the troughs (Fig. 7.29). The effectiveness of cross-ribs sharply decreases with decreasing distance between the ribs.

One of the attractive methods that have been reported recently is coating tube surfaces. It provides a constant dropwise condensation on the surface of copper tubes (Zhao et al., 1990). This is achieved by coating the surface with the Cu-Cr-N 0.9 mm thick alloy layer. The coating is carried out by the ion implantation of Cr and N so that the crystal structure of the surface layer of copper is transformed into amorphous. The thermal resistance of the coating has become negligible and the heat transfer coefficient is 20–40 times higher than that for the film condensation.

7.3 Radiative heat transfer

7.3.1 Basic principles

Thermal radiation is a portion of the process of energy propagation by means of electromagnetic waves. When it is in thermodynamic equilibrium with a substance, we call it an equilibrium radiation. The radiation results from changes in electronic, vibrational, and rotational states of the atoms-molecules and the *emission* of the radiant energy takes place as a result of irregular deceleration of charged particles (electrons, ions) in the media. The reverse process is then what occurs when radiant energy impinges on a solid surface and causes its temperature rise (*absorption*).

From a corpuscular point of view, the radiant energy is emitted and absorbed in discrete forms of the electromagnetic waves, which could be described as if particles with mass zero and charge zero, i.e., photons are carrying energy with the speed of light c. The photon energy is $h\nu$, where $h = 6.626 \times 10^{-34}$ J s is the Planck constant, ν is the frequency of an electromagnetic wave. Note that the photon has momentum $h\nu/c$. The wavelength λ, characterizing an electromagnetic wave, is related to its frequency ν by the equation $c = \nu\lambda$. The speed of light $c = c_0$ (=299,792,458 m/s exactly) in vacuum, $c \approx c_0$ in gases, and in other media $c < c_0$. The ratio $n = c/c_0$ is refractive index. The frequency does not change when the electromagnetic wave passes through from one medium to another, but the wavelength because the light speed changes. Distribution of radiation energy per frequency (or wavelength) is radiation spectrum.

The spectrum of equilibrium radiation is independent of the nature of the substance and determined by Planck's law of radiation. Thermal radiation field in the electromagnetic spectrum covers a range of wavelengths $\sim 10^{-7}$–10^{-4} m (see Fig. 7.30), i.e., in the visible and infrared regions (Hackford, 1960). The visible part of the spectrum covers a wavelength of 0.40–0.76 μm. The infrared region of the spectrum consists of a near-infrared region (0.76–25 μm) and the far infrared region (25–1000 μm). A main portion of the thermal radiation energy falls on the wavelengths region of 0.76–15 μm, i.e., lies in the near infrared. Radiation in the visible region of the spectrum is significant only at very high temperatures.

Thermal radiation can be viewed as a surface phenomenon because thin surface layers (–0.001–1 mm) on the body are involved in the radiative heat transfer.

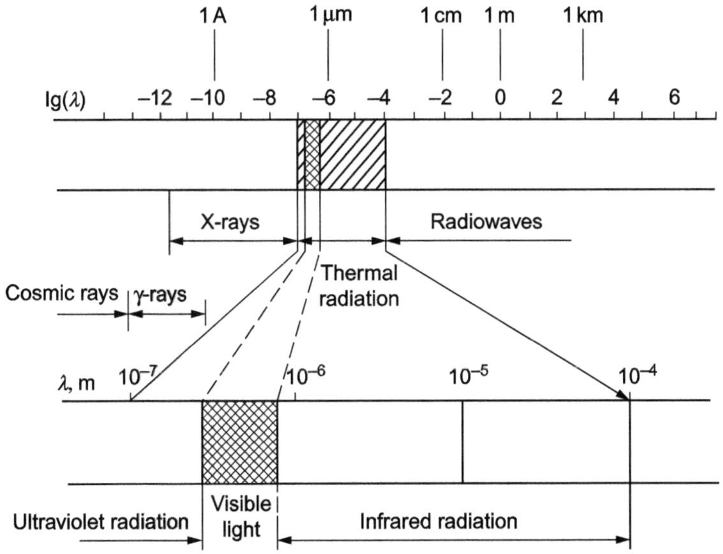

Figure 7.30 Portion of the thermal radiation in the electromagnetic spectrum.

Bodies can radiate energy of all wavelengths $\lambda = 0-\infty$ (a continuous spectrum) or selectively to a specific wavelength range (selective spectrum). With a change in temperature, not only the intensity but also structure of the radiation (spectrum) varies. There are monochromatic and full (integral) radiation modes. *Monochromatic* radiation corresponds to that of a narrow wavelength range $\lambda \sim (\lambda + d\lambda)$.

Integral radiation (full spectrum) is associated with the total radiant energy emitted from a body Q (W); in the entire wavelength range $0 < \lambda < \infty$, the integral flux density of the (hemispherical) radiation, i.e., the radiation flux emitted from the unit area per unit time we call *emittance* (or *emissive power*) $E = dQ/dA$ (W/m^2). Thus $Q = \int_A EdA$.

Emittance in an infinitesimal range of wavelengths divided by this interval is called the spectral density of radiation flux (spectral or monochromatic emittance): $J_\lambda = dE/d\lambda$ (W/m^3).

The amount of energy, emitted per unit time in the angular direction ψ the elementary area dA, per unit solid angle ω and per unit area of the projection onto a plane perpendicular to the direction of radiation, is called the brightness (intensity) of radiation:

$$I = \frac{d^2Q}{d\omega dA_n} = \frac{d}{d\omega}\left(\frac{1}{\cos\psi}\frac{dQ}{dA}\right) = \frac{dE_n}{d\omega}, \text{ with } dA_n \equiv dA\cos\psi \text{ and } E_n = \frac{dQ}{dA_n}. \tag{7.69}$$

$\frac{dI}{d\lambda} = I_\lambda$ is called the spectral brightness.

Thermal radiation energy impinging on the body can be absorbed, reflected, or can pass through the body:

$$Q_A + Q_R + Q_D = Q, \text{ thus } \frac{Q_A}{Q} + \frac{Q_R}{Q} + \frac{Q_D}{Q} = 1 \text{ or } A + R + D = 1, \tag{7.70}$$

where A, R, and D correspond to the absorbance, reflectance, and transmittance (transparency) of the material body. Here note that A is a fraction of absorption and should be distinguished from the area A in m^2. Particular cases of this equation leads us to concepts of ideal bodies, i.e.,

$A = 1$; $R = D = 0$—absolutely blackbody;
$D = 1$; $A = R = 0$—completely transparent;
$R = 1$; $A = D = 0$—absolutely mirror.

Dry air, mono- and di-atomic gases at temperatures below 3000 K can be regarded as transparent (diathermic).

Most of the solids are opaque, so that $D = 0$ and $A + R = 1$. In this case, the reflectivity and absorptivity of the body are interconnected. A blackbody does not exist, and usually $A < 1$ (gray body). In general, the absorbance depends on the wavelength. Such substances show selective absorption (see Fig. 7.31). Classification of different types of radiation is shown in Fig. 7.32. There, E is the *own* radiation; E_{inc} is the

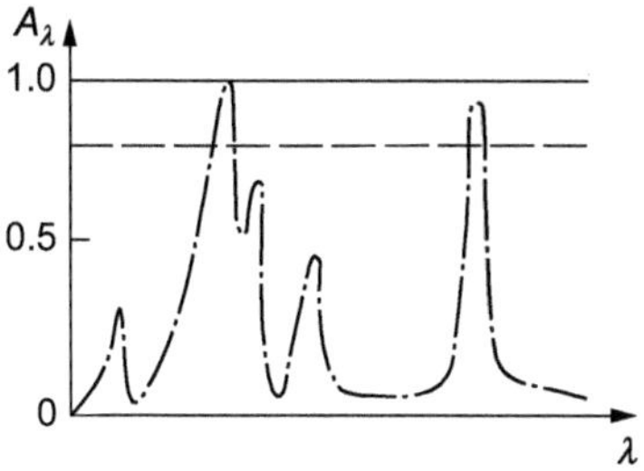

Figure 7.31 Absorption of different material bodies: — ideal blackbody; - - - gray body; -·-·-· selective absorption body.

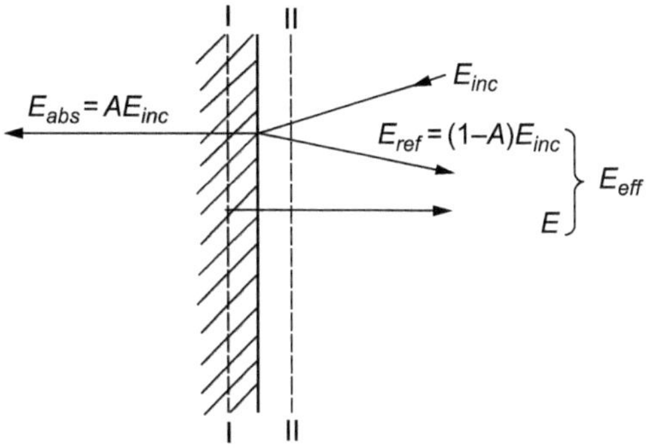

Figure 7.32 Classification of different types of radiation; I and II are the layers infinitesimally close to the surface.

incident of radiation on the body; $E_{abs} = AE_{inc}$ is the *absorption* of radiation; $E_{ref} = (1 - A)E_{inc}$ is the *reflected* radiation. A sum of the own and reflected radiations is called the *effective* radiation $E_{eff} = E + E_{ref} = E + (1 - A)E_{inc}$. It is noted that E_{abs}, E_{ref}, and E_{eff} are linear functions of E_{inc}. Properties (spectral content) of their own and the reflected radiations may be different.

Thermal radiation flux can be given from the heat balance. For the plane I—I (Fig. 7.32), we have

$$q_{res} = E - E_{abs} = E - A \cdot E_{inc}, \tag{7.71}$$

For the plane II—II in Fig. 7.32:

$$q_{res} = E_{eff} - E_{inc}, \tag{7.72}$$

Eliminating E_{inc} from these equations yields

$$E_{eff} = q_{res}\left(1 - \frac{1}{A}\right) + \frac{E}{A}$$

or

$$q_{res} = \frac{E}{1-A} - \frac{A}{1-A} E_{eff}. \tag{7.73}$$

More discussions on radiation heat transfer are provided in many monographs and textbooks, e.g., by Adrianov (1972), Blokh (1962), Cess (1964), Klimenko and Zorin (2001), Spalding and Taborek (1983), and Sparrow and Cess (1978).

7.3.2 Laws of thermal radiation

Thermal radiation laws apply to the equilibrium radiation of blackbody.[2] When all the bodies that are involved in the system take the same temperature, $q_{res} = 0$ and the radiation is called equilibrium (black). A heated blackbody emits photons, the spectrum of which depends only on the temperature. In the case of a gray body, radiation spectrum depends, besides the wavelength, on physical properties and surface conditions of the body, and can only be determined experimentally.

7.3.2.1 Planck's law (Fig. 7.33)

Spectral distribution of the radiation flux of a blackbody is expressed by a function of wavelength and temperature:

$$J_{0\,\lambda} = \frac{C_1}{\lambda^5 \left(e^{C_2/\lambda T} - 1\right)}, \tag{7.74}$$

where C_1—first radiation constant; C_2—second radiation constant, λ—wave length (m); T—absolute temperature (K). All the fundamental constants are shown in Table 7.4.

Another expression for this law uses $\lambda = nc_0/\nu$:

$$J_{0\nu} = \frac{2\pi h^2 n^2 \nu^3}{c_0^2 \left(e^{h\nu/kT} - 1\right)}, \tag{7.75}$$

where h is the Planck's constant; k is the Boltzmann constant; c_0 is the speed of light in vacuum and $n = c/c_0$ is the refractive index of the medium.

With the Stefan–Boltzmann constant σ, the law in the form of

$$\frac{J_{0\,\lambda}}{\sigma n^3 T^5} = \frac{C_1/\sigma}{(n\lambda T)^5 \left(e^{C_2/n\lambda T} - 1\right)}, \tag{7.76}$$

leads to a conclusion that $J_{0\,\lambda}/(\sigma n^3 T^5)$ is a function of only one variable $n\lambda T$.

[2] Values characterizing the blackbody radiation will be designated by the subscript 0.

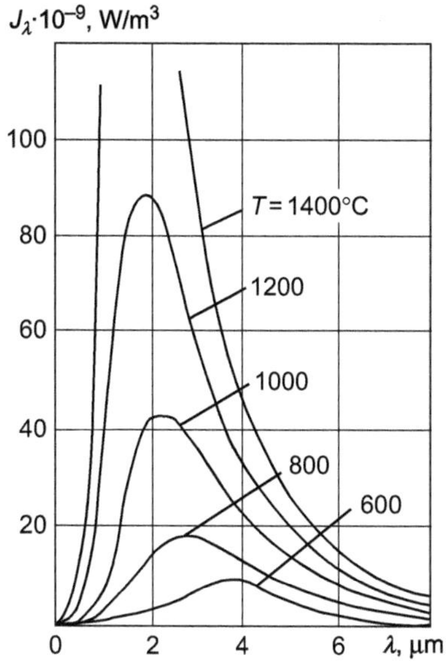

Figure 7.33 Graphical representation of the Planck's law.

Table 7.4 Fundamental constants in the law of radiation.

Planck's constant	h	6.6260×10^{-34}	J/s
Boltzmann constant	k	1.38066×10^{-23}	J/K
Speed of light in vacuum	c_0	2.99792×10^{8}	m/s
Stefan–Boltzmann constant	$\sigma_0 = \frac{2k^4\pi^5}{15h^3c_0^2}$	5.6705×10^{-8}	W/(m² K⁴)
First constant of radiation	$C_1 = 2\pi h c_0^2$	3.7418×10^{-10}	W m²
Second constant of radiation	$C_2 = \frac{hc_0}{k}$	1.4388×10^{-2}	m K

Two limiting cases of the Planck's law are in the following.

(1) with $\lambda T \gg C_2$ (or $h\nu \ll kT$), i.e., at high temperatures or long wavelengths $\lambda T > 1.4\times 10^{-2}$ (m K),

$$J_{0\lambda} = \frac{C_1}{C_2}\cdot\frac{T}{\lambda^4} = C_3\cdot\frac{T}{\lambda^4},$$

where $C_3 = 2.601\ 10^{-8}$ W m/K. This expression is known as the *Rayleigh-Jeans law*.

(2) with $\lambda T \ll C_2$ (or $h\nu \gg kT$), i.e., at low temperatures or short wavelengths $\lambda T < 1.4\times 10^{-2}$ (m K), the Planck's law becomes the *Wien's law*:

$$J_{0\lambda} = \frac{C_1}{\lambda^5}e^{-C_2/\lambda T}. \quad (7.77)$$

The position of maximum of the radiation intensity is obtained from Eq. (7.74), equating $dJ_\lambda/d\lambda$ zero. As a result,

$$(\lambda T)_{\max} = 2.8978 \times 10^{-3}\ (\text{mK}) \text{ or } h\nu_{\max}/kT = 2.8214. \tag{7.78}$$

These expressions are called the *Wien's displacement law*. For example, maximum solar emission is observed at $\lambda_{\max} = 0.475 \times 10^{-6}$ m, which suggests that the sun's surface temperature is about 6100 K.

Maximum radiation intensity corresponding to the wavelength is determined from the relation

$$J_{0\lambda}^{\max} = C_4 T^5, \tag{7.79}$$

with $C_4 = 1.33 \times 10^{-5}$ W/(m^3 K^5).

Stefan–Boltzmann law: A total amount of energy radiated by a blackbody per unit time per unit area in the entire range of wavelengths (frequencies), i.e., the total blackbody radiation flux is obtained by an integral of spectral distribution of Planck's law, Eq. (7.74):

$$E_0(T) = \int_0^\infty J_\lambda d\lambda = \int_0^\infty \frac{C_1}{\lambda^5 \left(e^{C_2/\lambda T} - 1\right)} d\lambda = E_0 = \sigma_0 T^4, \tag{7.80}$$

where $\sigma_0 = \frac{2\pi^5 k^4}{15 c_0^2 h^3} = \frac{C_1}{15}\left(\frac{\pi}{C_2}\right)^4 = 5.67 \times 10^{-8}$ W/(m^2 K^4) is the Stefan–Boltzmann constant (radiation constant of absolute blackbody).

For gray bodies, own radiation is proportional to the fourth power of the temperature. Then the Stefan–Boltzmann law takes the following form:

$$E = \varepsilon E_0 = \varepsilon \sigma_0 T^4. \tag{7.81}$$

where $\varepsilon = E/E_0$ is the thermal emissivity (integral emissivity or emissivity).

Kirchhoff's law establishes a relationship between emittance and absorbance of any material (gray and black) bodies in thermodynamic equilibrium. The ratio of emittance (E) of the gray body to its dimensionless coefficient of absorption A given by Eq. (7.70) is constant for a given temperature and equal to the emittance of the blackbody:

$$E/A = E_0 = f(T). \tag{7.82}$$

Because $E/E_0 = \varepsilon$, the essence of Kirchhoff's law reduces to the statement that, in thermodynamic equilibrium, absorbance and the thermal emissivity of the surface are numerically equal, i.e.,

$$A = \varepsilon. \tag{7.83}$$

For majority of the surfaces encountered in engineering processes, the temperature dependence of ε(T) is weak in general. Detailed characterization of the radiating surface is a spectral thermal emissivity:

$$\varepsilon_\lambda = \frac{J_\lambda(T)}{J_{0\,\lambda}(T)} \text{ or } \varepsilon_\nu = \frac{J_\nu(T)}{J_{0\,\nu}(T)} \tag{7.84}$$

which is a ratio of the spectral radiation flux of the material to that of the blackbody.

The emissivity of some materials of nuclear engineering is shown in the following:

UO_2	$\varepsilon_\lambda = 1 - 0.16 \exp[-3.79 \times 10^{-4}(T - 3120) - 3.27 \times 10^{-7}(T - 3120)^2]$ for $\lambda = 0.65$ μm; T in K
PuO_2	$\varepsilon = 0.548 + 1.65 \times 10^{-4}\,T$, where ε is the hemispherical emissivity; T in K
Alloy E-110	$\varepsilon = 0.167–0.184$ for $T = 80–200$°C
Alloy E-110	$\varepsilon = 0.167–0.184$ for $T = 80–200$°C
Alloy E-635	$\varepsilon_\lambda = 1.02 + 3 \times 10^{-4}\,T$ for $\lambda = 0.65$ μm; T in K
ZrO_2	$\varepsilon_\lambda = 0.48–0.50$

Corium (core melt) $\varepsilon = 0.85$ for $T_{m.p.} = 2850$ K, where ε is the integrated emissivity (Harding et al., 1989).

Lambert's cosine *law* determines the distribution of radiant intensity of a blackbody in directions. The radiation flux per unit area in a certain direction (E_ψ) is proportional to the radiation flux in a direction normal to the surface (E_n) and the cosine of the angle (ψ) between two directions:

$$E_\psi = E_n \cos \psi. \tag{7.85}$$

When the radiation obeys Lambert's law, the brightness does not depend on the direction. Fig. 7.34 shows a qualitative distribution of the relative emittance of different bodies in all directions.

A body, that obeys Lambert's law, is called the diffuse emitter, while the radiation of actual solid materials obeys Lambert's law only partially. Metals exhibit a maximum emission intensity at $\psi = 40–80$ degrees, i.e., observed from the surface

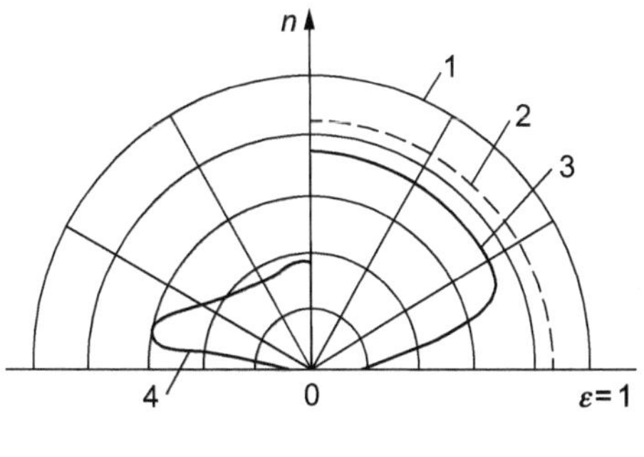

Figure 7.34 The distribution of the relative emittance of different bodies depending on the direction: 1—blackbody; 2—gray body; 3—oxidized metals, dielectrics, wood, paper; 4—polished metals.

at a considerable angle. Dielectrics give the greatest intensity of radiation in the normal direction in the range $\psi = 0-60$ degrees and low value in the range of high angles ψ.

In radiation characteristics of actual bodies, a conventional temperature is introduced along with the degree of blackness (thermal emissivity) as shown in the following.

Radiation temperature (T_{rad}) is a conventional temperature of a blackbody for the same radiation flux as that from the body under consideration: $E_0(T_{rad}) = E(T)$. From the Stefan–Boltzmann law $\sigma_0 T_{rad}^4 = \varepsilon\, \sigma_0 T^4$. Therefore,

$$T_{rad} = T\sqrt[4]{\varepsilon}. \tag{7.86}$$

Color temperature (T_c) is a conventional blackbody temperature at which the ratio of the spectral intensities at two wavelengths (colors) is the same as that of the considered body $\frac{J_{0\lambda_1}}{J_{0\lambda_2}} = \frac{J_{\lambda_1}}{J_{\lambda_2}}$. Using the Wien's law, we obtain

$$\frac{1}{T_c} = \frac{1}{T} - \frac{\ln\left(\varepsilon_{\lambda_1}/\varepsilon_{\lambda_2}\right)}{C_2\left(\frac{1}{\lambda_1} - \frac{1}{\lambda_2}\right)}. \tag{7.87}$$

Bright temperature (T_b) is determined by equating the brightness of blackbody and the body under consideration $I_{0\lambda}(T_b) = I_\lambda(T)$, i.e.,

$$\frac{1}{T_b} = \frac{1}{T} - \frac{\lambda \ln \varepsilon_\lambda}{C_2}. \tag{7.88}$$

Relations linking T_{rad}, T_c, and T_b with T are the basis of the methods to measure high temperatures by optical methods (pyrometers—radiant color luminance).

7.3.3 Radiation characteristics

In describing the radiative properties of actual bodies, the use is made of relative dimensionless quantities and properties in linking real and ideal bodies.

Relative emissivity ε (thermal emissivity) is a ratio of the real radiation surface energy of to that of an ideal blackbody at a same temperature and other conditions being equal.

Absorptivity A is a ratio of the energy absorbed by the surface to the energy absorbed by the blackbody irradiated by a same incident radiation flux, and other conditions being equal.

Reflectivity R is a ratio of the energy reflected by the surface to the energy reflected by ideally reflecting (mirror) surface irradiated by the same incident radiation flux and other conditions being equal.

Transmissivity D (transparency) of the material is a ratio of the energy transmitted through the material to the incident radiation energy.

In practice, it is more convenient to work with the averaged radiation properties over all directions. Thus, considering the energy emitted by the element of the body surface into all directions over that element suggests a hemispherical emittance or the degree of blackness. Depending on whether it is over an entire range or a narrow range of wavelengths, we have the total (integral) or monochromatic (spectral) hemispherical emissivity. Similarly, corresponding radiation properties are defined for reflectivity and absorptivity of the surface.

Radiation characteristics of real bodies depend on the wavelength of the radiation, surface temperature, and the direction of the emitted or incident radiation. The surface, which has the same emissivity and absorbance at all wavelengths refers to gray bodies: $\varepsilon_\lambda = \text{const}$ and $A_\lambda = \text{const}$. The gray body is an idealized concept just because in real bodies, these radiation properties are dependent on the wavelength.

The integral emissivity of metals is

$$\varepsilon = 0.576\sqrt{\rho_e T} - 0.124\,\rho_e T. \tag{7.89}$$

where ρ_e is the electrical resistivity (Ω cm); and T is in K.

Spectral emission of pure metals in a short wavelength region of the spectrum ($\lambda < 0.5$ μm) is more than half of the blackbody radiation at the same temperature. When $\lambda > 1$ μm, the spectral emissivity of most metals $\approx 0.05-0.1$. For example, to improve the thermal insulation properties, the walls of cylinders, flasks, and Dewar vessels are coated with silver ($\varepsilon = 0.02-0.03$) or aluminum ($\varepsilon = 0.04-0.06$).

Water and ice have a higher emissivity for the longwave thermal radiation. Mono- and di-atomic gases (He, H_2, N_2, etc.) are practically transparent to thermal radiation, i.e., neither absorb nor emit radiation. Triatomic gases (H_2O, CO_2, SO_2, NO_2, etc.) have high emissivity and absorptivity. They absorb and emit at specific intervals (bands) of wavelengths and transparent at the remaining of the spectrum.

Emission from gas deviates from the fourth power law. For example: $E_{CO_2} = 3.5 \times 10^{-7}(pl)^{0.33}T^{3.5}$; $E_{H_2O} = 3.5 \times 10^{-6}p^{0.8}l^{0.6}T^3$ where p is the partial pressure and l is the path length along the propagation. However, in engineering applications, we use the fourth power law, for the sake of convenience, and the deviation is taken care of by the relative emissivity of gas:

$$E = \varepsilon(T, pl)\sigma_0 T^4. \tag{7.90}$$

$\varepsilon(T, pl)$ are given in the form of graphs for each gas. Issues related to the radiation heat transfer in high-temperature gases are discussed in detail in Golovnyov et al. (1984).

Reflectivity of the body ($R = 1 - \varepsilon$) depends on directions of the incident and reflected radiation (Kitzel', 1973). Extreme cases of reflections from surfaces are diffuse and specular reflection. A diffuse surface reflects the incident beam so that the intensity of reflected radiation is the same for all the reflection angles (Fig. 7.35). A specular surface reflects the beam at the same angle just in the same way as a mirror does. The surface is a mirror when its roughness is much smaller than the wavelength of the incident beam. Thus, the same surface can become diffuse or mirror, depending on the incident radiation wavelength.

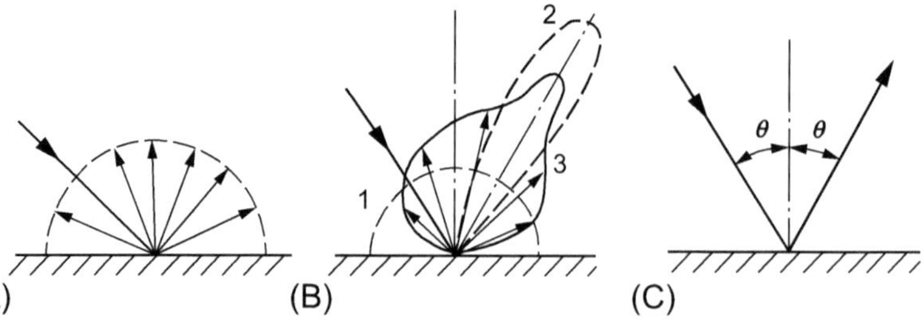

Figure 7.35 Types of reflective surfaces: (A) diffuse; (B) actual; (C) specular. 1—Reflectance for short wavelength; 2—reflectance for long wavelength; 3—integrated reflectance (over all wavelength).

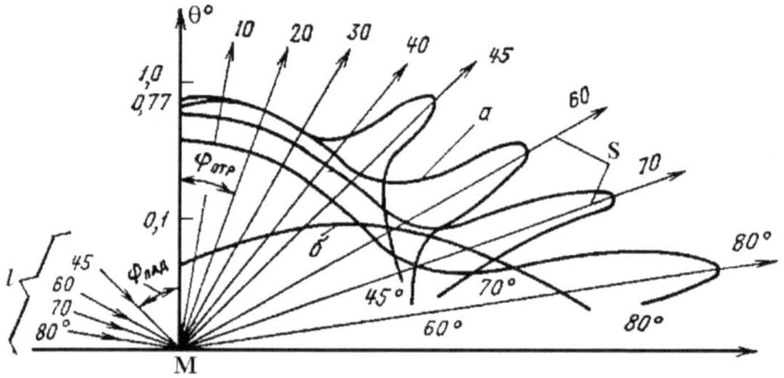

Figure 7.36 Dependence of the brightness coefficient of the coatings on the angles between incident and reflection beams: a—white shiny surface; b—black matte finish.

Fig. 7.36 shows the dependence of the reflection coefficient of the paint at different angles of incident and reflected radiation.

White (color) materials can have a high reflection coefficient for short-wave radiation and are "black" for the longwave. Due to the selectivity of reflectance bodies, the latter depends essentially on the temperature of the incident radiation source.

Transparency (transmissivity) of the bodies for thermal radiation may be visible even for relatively thick layers. It is so-called greenhouse effect due to high-transparency glass for shortwave solar radiation and extremely low opacity to the long-wave infrared radiation. Bodies that are completely opaque in the visible spectrum at normal temperatures can be transparent to the thermal radiation at high temperatures (e.g., porcelain, BeO, etc.).

7.3.4 Radiation heat transfer between bodies

First let us consider the heat exchanges between two parallel plates of infinite length (1) and (2), which are separated by infinite transparent medium. Calculation of radiation heat transfer is carried out according to the formula

$$q_{1,2} = \sigma_0 \varepsilon_{red} \left(T_1^4 - T_2^4 \right). \tag{7.91}$$

where $\varepsilon_{red} = \left(\frac{1}{\varepsilon_1} + \frac{1}{\varepsilon_2} - 1 \right)^{-1}$ is the reduced emissivity of the system; ε_1. ε_2 are the surface emissivity; T_1 and T_2 are the surface temperatures (K).

The total amount of heat transferred from surface 1 to surface 2 is

$$Q = q_{1,2} A = \sigma_0 \varepsilon_{red} \left(T_1^4 - T_2^4 \right) A. \tag{7.92}$$

where A is the surface area (m^2).

7.3.4.1 Arbitrary arrangement of two surfaces

The geometry of the system is taken into account by means of integral angular radiation coefficients (irradiance coefficients). The angular coefficient determines the amount of diffusely distributed radiation energy, which is transferred from one surface to another. The physical meaning of the angular coefficient can be understood by the following example. The radiation flux leaving the surface 1, is evenly distributed over the surface of hemisphere. In this space over surface 1, surface 2 is located with a certain area which makes a solid angle φ_{12}. Then, this angle determines the portion of the radiant energy that falls from surface 1 to surface 2: $\varphi_{12} = \frac{Q_{1 \to 2}}{Q_1}$ and $\varphi_{21} = \frac{Q_{2 \to 1}}{Q_2}$. The first subscript refers to the emitting surface and the second to the irradiated surface. The angular coefficients φ possess the reciprocity with the exposed area A:

$$\varphi_{12} \cdot A_1 = \varphi_{21} \cdot A_2. \tag{7.93}$$

The angular coefficient is dimensionless ($\varphi < 1$), reflecting the purely geometrical features of the placement of the two bodies in space. There are graphic-analytical and experimental methods for determining the angular coefficients.

When the surface area receiving the heat is small, corresponding angular coefficient is also small. In a closed system, all the radiation energy originating from any surfaces is distributed over the rest of the surfaces of the system so that $\sum_{j=1}^{n} \varphi_{ij} = 1$, where n is the number of surfaces, forming a closed system.

The heat flow between two gray bodies with temperatures T_1 and T_2 according to the Stefan–Boltzmann law $Q_{12} = \sigma_0 \left[T_1^4 A_1 \varphi_{12} - T_2^4 A_2 \varphi_{21} \right]$ or using the property of reciprocity,

$$Q_{12} = \sigma_0 \left(T_1^4 - T_2^4 \right) A_1 \, \varphi_{12} = \sigma_0 \left(T_1^4 - T_2^4 \right) A_2 \varphi_{21}. \tag{7.94}$$

Full radiant flux for a system consisting of a gray body is given by

$$Q_{12} = \sigma_0 \varepsilon_{red} \left(T_1^4 - T_2^4 \right) A_1 \varphi_{12}. \tag{7.95}$$

where $\varepsilon_{red} = \dfrac{1}{1+\varphi_{12}\left(\dfrac{1}{\varepsilon_1}-1\right)+\varphi_{21}\left(\dfrac{1}{\varepsilon_2}-1\right)}$ is called a reduced degree of blackness of a pair of bodies and is an optical and geometrical parameter. The value of $A_1 \times \varphi_{12} = H_{12}$ is called the mutual radiation surface of bodies. Q_{12} are in (W).

7.3.4.2 Heat transfer between body and shell

Assume one body is in a cavity of another body and their sizes are given. With the surface areas (A_1, A_2) and temperature (T_1, T_2) being given and noting the absorption coefficients can be set equal to the emissivity (ε_1 and ε_2), then the calculation of an amount of heat transmitted from one surface (1) to the other surface (2) will be

$$Q_{12} = \frac{\sigma_0\left(T_1^4 - T_2^4\right)A_1}{\dfrac{1}{\varepsilon_1} + \left(\dfrac{1}{\varepsilon_2} - 1\right)\dfrac{A_1}{A_2}}. \tag{7.96}$$

If the surface area of the body in the cavity is much less than the surface of the shell ($A_1 \ll A_2$) or $\varepsilon_2 = 1$, Eq. (7.96) takes the form

$$Q_{12} = \sigma_0\, \varepsilon_1 \left(T_1^4 - T_2^4\right)A_1. \tag{7.97}$$

If $A_1 \cong A_2$ we will arrive at the parallel plate formula, Eq. (7.92).

Many other specific problems and solutions are found in Siegel and Howell (1981).

7.3.5 Heat transfer in absorbing and emitting environments

Electromagnetic wave, as it passes through a medium, interacts with the particles of the medium and undergoes scattering and absorption processes. Interactions of the wave occur with the molecules or atoms with different structure, solid particles, or liquid droplets in gases. The medium may have its own emission sources; In addition, the radiation also causes some brightness of scattering medium. The radiation energy being absorbed is turned into heat, changes the thermal state of the medium and will be reemitted.

The overall change in spectral density of the radiation flux over a spatial length dx due to the incident radiation J_λ and its own intrinsic $I_{\lambda s}$ is expressed by the equation

$$dJ_\lambda = -\beta_\lambda J_\lambda dx + \varepsilon_\lambda I_{0\lambda s} dx. \tag{7.98}$$

where $\beta_\lambda = \beta_\lambda^A + \beta_\lambda^S$ is the spectral attenuation coefficient of radiation due to absorption (β_λ^A) and due to scattering (β_λ^S); ε_λ is the spectral emissivity of thermal radiation of the medium.

If its own intrinsic radiation of the medium is much less than the attenuation of the incident radiation, i.e., $\varepsilon_\lambda I_{\lambda s} \ll \beta_\lambda J_\lambda$, the transport equation becomes: $\frac{dJ_\lambda}{J_\lambda} = -\beta_\lambda J_\lambda dx$.

From this we have $\frac{J_\lambda^S}{J_\lambda^A} = \exp\left(-\int_0^s \beta_\lambda dx\right)$; $\frac{J_\lambda^A - J_\lambda^S}{J_\lambda^a} = 1 - exp\left(-\int_0^s \beta_\lambda dx\right) = A_{\lambda s}$, where $A_{\lambda s}$ signifies the absorption (and scattering) ability of the layer of thickness s for the beam of wavelength λ. The ratio generally depends on the physical properties of the medium (absorption and scattering, the concentration of the absorbing substance, wavelength, temperature, thickness, etc.).

Effect of concentration of absorbing material in the medium on the optical properties is established in the first approximation by the Bouguer-Beer law, in which the relative attenuation of the beam of rays is proportional to the volume concentration of the absorbing substance μ.

Absorption ability or absorptivity of such a medium can be determined by the formula

$$A_\lambda = 1 - \exp(-\beta_\lambda\, \mu\, s). \tag{7.99}$$

This law holds only approximately for monochromatic radiation of low concentrations. Deviations from the law are caused by the interactions among the particles or molecules of absorbing materials. For nonmonochromatic radiation and for high concentrations of absorbing substances, the absorption coefficient depends on the concentration and length of the beam path.

The value given by

$$L_\lambda = \int_0^s \beta_\lambda dx \tag{7.100}$$

is called the *optical thickness* of the medium. If the absorption coefficient does not vary in distance, then $L_\lambda = \beta_\lambda s = \frac{s}{1/\beta_\lambda}$. Here s is the characteristic length, e.g., the layer thickness, and $1/\beta_\lambda$ can be regarded as the mean free path of photons. If β_λ is large (shorter mean-free-path), penetration depth is small, while for small β_λ (longer mean-free-path) the penetration depth is deep. In this way, the optical thickness is the ratio of the characteristic dimensions of the length of the radiation penetration. When the optical thickness $L_\lambda \gg 1$, the medium is said optically thick (the characteristic size of the system is significantly larger than the mean photon path length). If $L_\lambda \ll 1$ the medium is called optically thin and if $L_\lambda \to 0$ the medium is not involved in heat exchange (transparent or diathermic medium).

Engineering calculation methods of radiative heat transfer often use the *effective thickness* of the emitting layer (l_{eff}). For a medium with a volume V and a small optical thickness ($L_\lambda \ll 1$),

$$l_{eff} = \frac{4V}{A}. \tag{7.101}$$

where A is the surface area.

In many practical situations we encounter, we find Eq. (7.101) often overestimates the optical thickness of layers l. Therefore, we decided to use formula

$$l_{eff} = \frac{3.6V}{A}, \tag{7.102}$$

which reflects a real balance between the effective values l_{eff} and the equivalent thickness layer l_0 (Blokh, 1991).

The effect of optical layer thickness L on the effective thickness l is accounted for as follows:

$$l_{eff} = \frac{3.6V}{A} \frac{Le^{-L}}{L + e^{-L} + Le^{-L} - 1}. \tag{7.103}$$

For $L = 0$, the effective thickness of the layer l_{eff} becomes l_0 and defined by Eq. (7.102).

7.4 Complex heat transfer

7.4.1 The complex heat transfer and basic principles of calculation

Two or more heat transfer mechanisms often act simultaneously, for example: thermal conduction and radiation (λ-r), convection and radiation (c-r), convection-conduction (c-λ), or even all three together. The usual method of complex heat transfer calculations is a summation of the heat fluxes of each heat exchange mechanism:

$$\begin{aligned} q &= q_\lambda + q_c \\ q &= q_c + q_r \\ q &= q_\lambda + q_c + q_r \end{aligned} \tag{7.104}$$

The approach is valid only if each mechanism shown on the right-hand-side of the equation does not affect the other(s), i.e., if there is no interactions among them. This usually occurs when a magnitude of one component of the heat flux is significantly different from the other(s) (Özisik, 1973; Kokorev and Kharitonov, 1986)

There are two types of interaction effects of concern:

1. The effect of radiation in an absorbing medium appears in a similar form to heat sources or sinks in heat conduction or convection. This problem becomes nonlinear;
2. The interaction of radiation can change the boundary conditions.

A special case of complex heat transfer is the simultaneous transport of heat by radiation (q_r) and directional flow of electron flows direction that are emitted from a heated surface (cathode) to the other, colder (anode) surface (q_e):

$$q = q_r + q_e \tag{7.105}$$

7.4.2 Radiation heat transfer between flow of emitting gas and wall

7.4.2.1 Radiation and conduction

We now consider here a stationary system involving the absorption and conduction by the gas medium. The problem is characterized by the following parameters: *optical thickness* of the medium (L, see Section 7.3.5); emissivity of the surfaces (ε_1 and ε_2); relative temperature of the cold surface to the hot surface ($\theta = T_2/T_1$) and the parameter $N = \frac{\lambda a}{4\sigma_0 T_1^3}$, where λ is the thermal conductivity of the gas; a is the gas absorption coefficient; and σ_0 is Stefan–Boltzmann constant. The parameter N characterizes the interaction intensity of heat transfers by conduction and radiation: if $N \to \infty$ (T_1 at lower temperature), heat is transferred only by thermal conductivity; if $N \to 0$ (at higher temperature T_1), heat is transferred only by radiation.

Here are two extreme cases:

- Optically thin gas layer ($L \to 0$). The radiation heat is not absorbed by the gas but practically by every surface element of the medium that interacts with the radiation coming directly from the boundary surfaces. The total heat flux will be given, with $q_r = \alpha_r(T_1 - T_2)$ and $q_\lambda = \alpha_\lambda(T_1 - T_2)$, by $q = q_r + q_\lambda = (\alpha_r + \alpha_\lambda)(T_1 - T_2)$, which is also written as

$$q = \varepsilon_{red}\sigma_0\left(T_1^4 - T_2^4\right) + \frac{\lambda}{\delta}(T_1 - T_2). \tag{7.106}$$

The first term on the right-hand-side of Eq. (7.106) is Eq. (7.91) with the reduced emissivity defined in Section 7.3.4.

- Optically thick gas layer ($L \to \infty$). The radiation heat is readily absorbed by the gas. Radiation properties of the surfaces affect only the layers of gas within a distance much smaller than the size of the system. In this case, the radiation heat transfer approaching the diffusion process, the analogy with the thermal conductivity of the gas can be made. The heat flux is expressed by

$$q = \frac{4}{3}\frac{\sigma_0}{a\lambda}\left(T_1^4 - T_2^4\right) + \frac{\lambda}{\delta}(T_1 - T_2). \tag{7.107}$$

In general, the radiation beam penetrates the surface of the medium volume and is partially absorbed. This affects the temperature distribution within the volume and thus changes the thermal conductivity under the heat transfer conditions. This is the most complicated case, and the total heat flux q cannot be obtained readily by adding q_λ and q_r because of their strong interconnected behaviors.

The temperature level of the nuclear fuel depends on the thermal conductance λ/δ at a radial clearance δ between the fuel pellet and cladding as it determines the temperature drop. The value λ/δ depends on the design of fuel element (cladding diameter, gap, etc.), the properties of the cladding and fuel materials, surface conditions (roughness, coatings, presence of the oxide films), properties of the fission gas and fill-in gas,

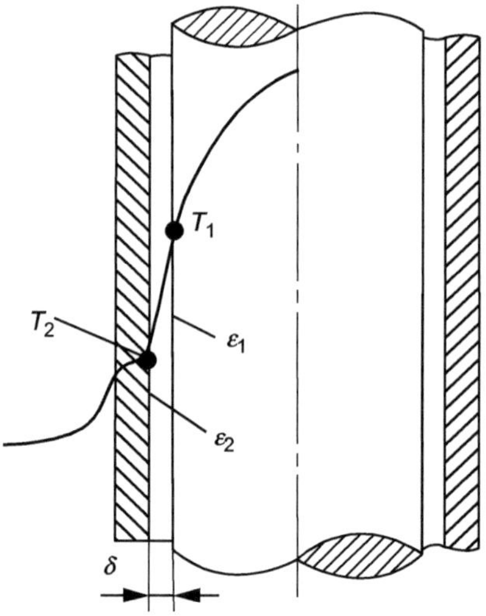

Figure 7.37 Calculation of complex heat transfer in the gaseous fuel rod gap.

its pressure, and the metallurgical and many other processes occurring in the fuel during irradiation (Fig. 7.37).

The radiation heat flux is estimated by $q_r = \alpha_r(T_1 - T_2)$, where α_r is the gap thermal conductance given by

$$\alpha_r = \varepsilon_{red} 4\sigma_0 \overline{T}^3, \tag{7.108}$$

where $\varepsilon_{red} = \left(\frac{1}{\varepsilon_1} + \frac{1}{\varepsilon_2} - 1\right)^{-1}$ and $\overline{T} = \frac{T_1 + T_2}{2}$ is the average temperature in the gap region.

For a temperature of the inner surface of cladding 350–700°C and the temperature difference in the gap 600–700°C, you may find $\alpha_r \approx 50$–150 W/(m^2 K), i.e., *the heat transfer by radiation under the operating conditions is in most cases less than 10% of the total heat flux.* In evaluating the gap conductance, it is important to take into account the changes in many parameters over the nuclear reactor operation period, including the gap size, thermal conductivity of the gas in the gap, and the mass transfer of fuel due to evaporation/condensation.

7.4.2.2 Radiation and convection

Consider we have both radiation and convection in practical engineering problems. In most cases, the two heat fluxes are assumed to be independent and dealt with separately: $q = q_r + q_c$. This assumption is accurate enough if one of the two heat fluxes is significantly smaller than the other.

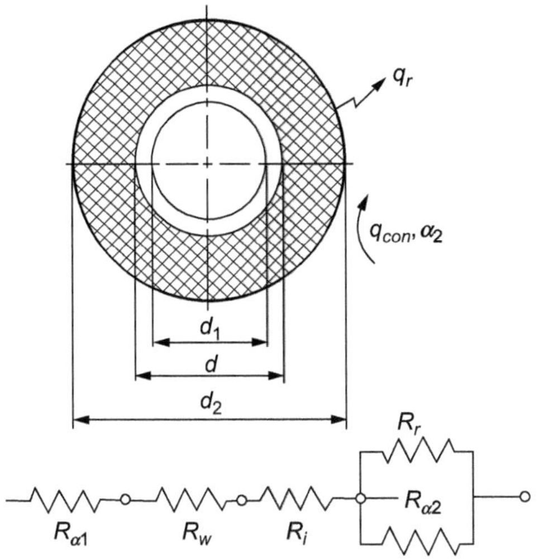

Figure 7.38 Calculation of heat transfer from the coolant through a conduit with insulation: its equivalent electric circuit diagram.

Sometimes there are cases where three mechanisms of heat transfer (heat conduction, convection, and radiation) are to be considered. An example is the heat transfer through an insulated circular pipe with an outer diameter of the insulation d_2, heat transfer coefficient α_c by convection and α_r by radiation to the environment (Fig. 7.38). Temperatures on the insulation surfaces are T_{1i} and T_{2i}. The heat flow from the unit length q' (W/m) will be equal to

$$q' = \frac{\pi(T_{1i} - T_{2i})}{\frac{1}{\alpha_1 d_1} + \frac{1}{2\lambda_w} \ln \frac{d}{d_1} + \frac{1}{2\lambda_i} \ln \frac{d_2}{d} + \frac{1}{\alpha\, d_2}}. \tag{7.109}$$

The thermal resistance of heat transfer inside the pipe is $1/(\alpha_1 d_1)$; the thermal resistance of the pipe wall $\frac{1}{2\lambda_w} \ln(d/d_1)$ can be neglected for simplicity. The thermal resistance of the heat transfer at the outer insulation surface is due to the radiation from the insulation surface and the convection:

$$\frac{1}{\alpha} = \frac{1}{\alpha_r(T_{2i})} + \frac{1}{\alpha_c}; \tag{7.110}$$

Therefore, three parameters in the two Eqs. (7.109) and (7.110), i.e., q' as the linear heat rate of the insulated pipe, α_r as the radiation heat transfer coefficient with the surface insulation, and T_{2i} as the surface temperature of the insulation, depend on each other. When the ambient temperature is provided, the problem is solved by successive approximation.

7.4.3 Contact heat transfer

7.4.3.1 Contact resistance

Nuclear fuel elements usually take the form of fuel pellets inside the cladding, where with a gap between them the mechanical contacts are assumed. The thermal conductance of such contact is of the order of 10^3–10^4 W/(m^2 K). When the heat flux is ~10^6 W/m^2 and a temperature difference across the contact could be 100 degrees or more. However, as noted in Section 7.4.2, radiation heat transfer hardly affects this temperature difference. For more information, see Holm (1967), Shvetz and Dyban (1964), Shlykov (1965), Shlykov et al. (1977), and Kharitonov et al. (1973).

Thermal contact resistance plays an important role in problems of semiconductor thermoelectric generators, thermionic converter of nuclear energy into electrical energy, in devices operating in the vacuum of space. For example, see Clausing and Chao (1965) and Fried and Costello (1962).

Conventional nuclear reactor fuel materials are metal alloys, oxides, carbides, nitrides, uranium, and plutonium metals (Boeschofen, 1955; Miller, 1962; Avduevsky and Koshkin, 1992; Özisik, 1973; Popov, 1977; Spiridonov et al., 1974; Rapier et al., 1963; Waltar and Reynolds, 1981). The thermal contact resistance between the metallic cladding and the ceramic fuel (UO_2, UC, UN, etc.) is particularly a difficult issue (Miller, 1962) for the following reasons: first, the ceramic fuel does not come into close contact (i.e., plane mode) with the cladding inner surface; second, because of the ceramics' poor thermal conductivity and high hardness, the conductivity of the actual contact points may be small.

7.4.3.2 Rough surfaces

The actual contact area between two solid surfaces is essential information to evaluate contact resistance (Dyomkin, 1962; Dul'nev, 1963; D'iachenko et al., 1963; Kaganer and Zukova, 1966; Miller, 1962; Bahrami et al., 2006). The contact resistance is a relative concept because the actual surface roughness and waviness of the contact surfaces between solids are discrete. Some of the heat goes through the contact spots of solids and others through the gaseous gaps. Since the thermal conductivity of the contact spots is higher than that of the gaseous medium in the gap, the heat flow streamlines are highly congested toward these local spots with the thermal contact resistance $R = \Delta T/q$, where ΔT is the temperature difference at the contact position.

For many surfaces encountered in engineering practice, a population density of the contact points, therefore, a degree of concentration of lines of the heat flow, has a definite influence on the thermal contact resistance (see Fig. 7.39a; Latyev et al., 1974).

The inverse of the thermal contact resistance $1/R$ is the thermal conductance ($=\alpha$) of the contact. The total contact resistance is made up of two resistances in parallel—one of the gaseous medium in the gap (R_g) and the other of the mechanical contact spots (R_m):

$$\frac{1}{R}=\frac{1}{R_g}+\frac{1}{R_m}, \tag{7.111}$$

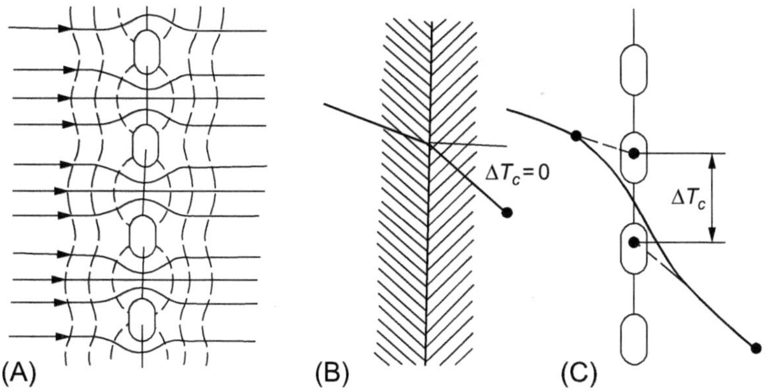

Figure 7.39 Heat flow line (—) and isotherms (- - -) at the contact between two plates (A); temperature distribution in an ideal location (B); and real (C) at the contact.

where R_g is determined by the gap width and conductivity λ_g of the medium filling the gap; R_m is the actual contact resistance based on the contraction of the heat flow streamlines. When the gap is 20 μm wide and filled with air, $R_g = \frac{\delta}{\lambda} = \frac{20 \times 10^{-6}}{3.2 \times 10^{-2}} = 6.2 \times 10^{-4}$ (m^2 K/W). It is noted that this resistance is equivalent to the resistance of ST-3 steel plate of thickness 33 or 150 mm thick aluminum.

Eq. (7.111) is rewritten in terms of the thermal conductance (α_g and α_m) and the fraction of actual contact area S to the total surface area $\eta = S/A$:

$$\alpha = \alpha_g(1 - \eta) + \alpha_m \eta, \tag{7.112}$$

Variables η and S depend on the characteristics of roughness and waviness of the surfaces, their mechanical properties, compression force and other parameters.

In calculations, it is usually assumed that the actual contact area S formed by a plurality of circular spots of radius a, in which the heat flow streamlines are "pulled together" (Fig. 7.39a). The corresponding additional thermal resistance associated with the curvature of the heat flow streamlines is taken into account by introducing a correction factor ψ_t.

If the mean free path of the gas molecules in the gap is commensurable with size of the gap, the gas may not be regarded as a continuous medium. Then, the discreteness of gas properties leads the temperature, instead of a linear profile, to have a distribution in the planar gap region, resulting in a temperature drop (jump) as shown in Fig. 7.39c. To characterize this distortion, we introduce a concept *length* of "a temperature jump" ($\Delta T_1 + \Delta T_2$), which is equivalent to an increase in the thickness of gas layer ($l_1 + l_2$). The length of a temperature jump would be $l \approx \ell/\xi$, where ℓ is a mean free path length of the molecules, ξ is the accommodation coefficient (see below).

In view of the above effect, the thermal resistance at the gas gap can be written as

$$R_g = \frac{1}{\alpha_g} = \frac{\delta_e}{\lambda} + \frac{l_1 + l_2}{\lambda} = \frac{\delta_e}{\lambda} + \frac{1}{\alpha_*}, \tag{7.113}$$

where $\alpha_* = \xi_e \frac{\gamma+1}{\gamma-1} \cdot \frac{p}{2} \sqrt{\frac{k}{2\pi MT}}$; $\gamma = \frac{c_p}{c_v}$; p is the pressure (Pa); k is Boltzmann constant ($=1.38 \times 10^{-23}$ J/K); T is the temperature (K); M is the molecular mass of the gas; ξ_e is the equivalent accommodation coefficient associated with the accommodation coefficients on surfaces 1 and 2 by the relation $1/\xi_e = 1/\xi_1 + 1/\xi_2 - 1$.

7.4.3.3 Accommodation factor

Accommodation coefficient characterizes the energy transfer efficiency of the gas molecules to the surface and depends on the nature of the gas and the surface material. The higher the molecular weight of the gas, the higher the accommodation coefficient:

$$\xi = \frac{\mu}{2(1+\mu)}, \tag{7.114}$$

where $\mu \equiv M_g / M_W$ is the ratio of atomic (molecular) mass of the gases and the surface material.

The amount of heat conduction through the gas layer between the fuel pellet and cladding in the nuclear fuel rods are not strongly dependent on the composition of the gas (Fenech and Rohsenow, 1963). Thickness of the gas layer, in which the temperature jump is observed at the boundary of gas and solid body, is given by $\delta_g \approx \ell / \xi$, where ℓ is the mean free path, ξ is the accommodation coefficient, the fraction of the molecular energy that is exchanged with the wall per one collision.

Thermal conductivity at the gap, then, may be given by

$$\alpha \sim \frac{\lambda}{\delta_g} = \frac{\lambda \xi}{l}. \tag{7.115}$$

From $\lambda \sim 1/\sqrt{M_g}$, $\xi \sim \sqrt{M_g}$, where M_g is the atomic mass of the gas in the gap, we can see that α is weakly dependent on the type of gas. The heat transfers through the actual contact point and even due to radiation reduce the effect of this dependence. However, as the burnup increases and hence the pressure in the fuel element, this relationship given by Eq. (7.115) may be slightly modified as indicated in the next table:

Gas	He	Ne	Ar	Kr	Xe
$\lambda\xi$ (MW/m K)	4.6	3.2	4.0	3.1	4.1
α (kW/m^2 K)	1.5	2.0	3.4	2.9	4.0

The thermal resistance R (in Eq. 7.111) decreases with increasing contact pressure of the contact surfaces. While at low contact pressures and temperatures, almost 80% of the heat is passing through a gas layer, at high temperatures and high contact pressures, more than 90% of the heat goes through the contact points.

Taking into account the factors mentioned just above, the relationship of the contact conductivity Eq. (7.112) can be written as

$$\alpha = \frac{\lambda}{\delta_e}(1-\eta) + \frac{1}{\pi a^2 R \psi_t \psi_k}\eta, \tag{7.116}$$

where λ is the thermal conductivity of the medium in the gap; R is the thermal resistance of an ideal contact between two infinite surfaces; $\delta_e = \frac{A}{\int_A \frac{dA}{\delta(A) + 2l_m}}$ is an equivalent thickness of the gap based on the lengths of the temperature jump; $\int_A \frac{dA}{\delta(A)+2l_m} \approx \left(\overline{h_1} + \overline{h_2}\right) + 2l_m$; $\overline{h_1}$ and $\overline{h_2}$ are the average height of roughness; l_m—the length of temperature jump; ψ_k—correction factor taking into account tapering the roughness; a—the average radius of the contact points.

Data evaluation and modeling of the equivalent gap are controversial, due to the influences of the distribution function of height and shape of roughness. The equivalent gap may be defined by the profilograph surface.

The complexity of calculating the contact heat transfer with Eq. (7.116) is associated with many difficulties in determining the quantities involved in the relationship (δ_e, η, a, ψ_t, ψ_k).

There are two popular theories (models) for contact resistance. One theory suggests an *elastic* deformation of roughness elements on the contact surfaces. Other is their *plastic* deformation. For further reading, see McWald and Marschall (1993), Rapier et al. (1963), Sanderson (1961), Yovanovich (1986, 1998), Yovanovich and Marotta (2003), Skipper and Wootton (1958), and Skipper (1958).

The final word, as always, experimental results play crucial roles in this area.

Part 2: Boiling heat transfer

In this second half of this chapter, we concentrate on boiling heat transfer phenomena first focusing on nucleate and film boiling in a nonflowing system, i.e., in a pool of a large volume, and then on convective boiling in channels, focusing on the fundamental aspects of heterogeneous nucleation and heat transfers, i.e., boiling two-phase flow heat transfer from the heated surface to fluids. It is designed again for professionals as well as for students, being aimed at helping readers to better understand the phenomena and at providing the basis on which physical modeling of the phenomena is to be made for the system safety analysis and subchannel analysis of nuclear reactors.

7.5 Boiling process and heat transfer

7.5.1 Advantages of the boiling process

Boiling is a process of evaporation of the bulk liquid. The main advantage of using a boiling process in nuclear engineering is due to a fact that quite high heat fluxes are achieved without increasing significantly the surface temperature. Mechanisms of heat removal from the heated surface is associated with conduction, liquid- and vapor/gas-phase convection, phase change of liquid into vapor with the latent heat of vaporization, and radiation. The liquid-phase motion near the surface is largely determined by the dynamics of growth of vapor bubbles. The bubble dynamics is associated with the temperature field in the region surrounding the vapor bubble; and the temperature field is a result of interaction with the bubble dynamics itself. The number of nucleation sites is determined by the conditions at the surface (surface roughness, wettability, etc.). A large number of governing parameters (geometry, properties of the liquid, vapor, surface properties, and the boundary conditions of each phase) complicate the theoretical analysis of the boiling process. The intensity of the boiling heat transfer will be dependent on many factors difficult to control (surface roughness, the amount of dissolved gases in the liquid, wettability, etc.), which cannot be fully taken into account in theory and experiment. Insights can be found in the papers by Ose and Kunugi (2015), Su et al. (2018), and Zhang J. (Jinming) et al. (2022).

There are at least three reasons for the heat transfer enhancement in the boiling. First, the strong turbulence agitation to the boundary layer at the expense of growth and separation of vapor bubbles. Second, the heat transfer by evaporation of the liquid microlayer at the bottom of the vapor bubble. Third, the transfer of the latent heat of vaporization into the growing vapor bubble. The absence of a closed mathematical description of the process of boiling heat transfer and an abundance of factors influencing the process make choices of generalized variables rather arbitrary. Correlations obtained as a result of processing and synthesis of experimental data are inevitably semiempirical, and able to describe only a certain average level of the heat transfer.

> *... The fact of the successful approximation .... Experimental data of some empirical formula does not give grounds for approval of the true extent of influences of various properties on the heat transfer during boiling*
>
> *Labuntsov (1972).*

7.5.2 Onset condition of boiling on a heated surface

Formation of the vapor phase in a volume of liquid is due to density fluctuations caused by the random motion of molecules.

For a spherical bubble to stay in a volume of liquid and not to collapse under the forces of surface tension, the pressure therein must be greater than the pressure of the surrounding fluid, and its size must be larger than a certain critical radius R_{cr}. This value is calculated from the Laplace's pressure $\Delta p_1 = 2\sigma/R$, and Thomson modification, taking into account the change in the vapor pressure of the concave curved surface of radius R compared with a flat surface: $\Delta p_2 = \frac{2\sigma}{R} \cdot \frac{\rho_g}{\rho_f - \rho_g}$. A sum of pressure differences $\Delta p = \Delta p_1 + \Delta p_2$ must comply with the Clausius-Clapeyron equation: $\frac{\Delta p}{\Delta T} = \frac{\Delta h_{fg}}{T(v_g - v_f)}$. Hence the critical (minimum) radius of the nucleus R_{cr} of the vapor can be obtained from

$$R_{cr} = \frac{2\sigma T(v_g - v_f)}{\Delta h_{fg} \Delta T} \frac{\rho_f}{\rho_f - \rho_g}. \tag{7.117}$$

For $\rho_g \ll \rho_f$ or $(v_f \ll v_g)$, we have

$$R_{cr} \approx \frac{2\sigma T_S}{\Delta h_{fg} \rho_g \Delta T} = \frac{2\sigma \Re T_S^2}{\Delta h_{fg} p \Delta T}, \text{ with } p = \rho_g \Re T_S, \Re: \text{gas|constant}. \tag{7.118}$$

Required wall superheating can be obtained for the formation of a vapor bubble of radius R by

$$T_w - T_S = \Delta T \approx \frac{2\sigma T_S}{\Delta h_{fg} \rho_g R} = \frac{2\sigma \Re T_S^2}{\Delta h_{fg}\, pR}. \tag{7.119}$$

This equation can be used for the subcooled liquid to its saturation temperature T_S (Hahne et al., 1990).

Boiling begins on the heated surface where the surface is at an excess temperature and the liquid at saturation. The birth of a vapor bubble on a solid surface takes place at the nucleation centers where the surface energy for the vapor formation is lowest. Usually this recess is portions with weak molecular adhesion between liquid and solid. For the vapor bubble formation to occur on the surface, also a certain degree of superheating of the liquid above the saturation temperature is required. It depends on the purity of the liquid and the presence of vaporization centers, i.e., surface conditions. The centers of vaporization—cavities on the surface, are filled with gas or vapor. Under ideal conditions, superheat in the bubble of radius R is described by classical theory:

$$\Delta T_S = T_g - T_S = \frac{pT_S(v_g - v_f)}{\Delta h_{fg}} \ln\left(1 + \frac{2\sigma}{pR}\right).$$

Typically, for $\frac{2\sigma}{pR} << 1$, $\ln\left(1 + \frac{2\sigma}{pR}\right) \cong \frac{2\sigma}{pR}$ with sufficient precision. R may be regarded as the maximum radius of the cavity—the center of vaporization would be 0.2–0.4 μm. Superheating ΔT_w and the surface heat flux corresponding to the beginning boiling (q_{bb}[3]) in general can be calculated by the relationships

[3] bb: beginning of boiling.

$$\Delta T_w = T_w - T_S = \frac{pT_S(v_g - v_f)}{\Delta h_{fg} Re^{0.2} Pr^{0.5} (p/p_{cr})};$$

and

$$q_{bb} = \alpha(T_w - T_f) = \alpha[(T_S - T_f) + \Delta T_w],$$

where the heat transfer coefficient α can be obtained from single-phase convective heat transfer correlations.

Size of the vapor embryos depends on the surface roughness and wettability conditions. The number and size of nucleation sites define the boiling initiation. If the nucleation sites are many and large, the boiling initiation requires small superheating. Good wettability liquid fills the surface cavity and the excess temperature rise of the surface temperature over saturation, i.e., ($T_w - T_S$) must be higher than in the case of poor wettability fluid. Good wettability fluid leads to hysteresis phenomena during boiling—a significant increase in the surface temperature before boiling and its subsequent decline from the start of boiling.

For good wettability liquids which begin to boil as the steam embryo of the critical radius:

$$R_{cr} = \left[\frac{2\sigma\lambda_f T_S(v_g - v_f)}{q\Delta h_{fg}}\right]^{0.5},$$

A recommended correlation for the amount of wall surface superheat is

$$T_w - T_S = \frac{2\sigma}{R_m} \frac{T_S(v_g - v_f)}{\Delta h_{fg}},$$

where R_m is the radius of the largest cavity ($R_m < R_{cr}$).

The heat flux corresponding to the beginning boiling (q_{bb}):

$$q_{bb} = \alpha\left[(T_S - T_f) + \left(\frac{2\sigma}{R_m}\right) \frac{T_S(v_g - v_f)}{\Delta h_{fg}}\right].$$

Magnitude of $2\sigma/R_m$, for example for water at 0.1 MPa ($\sigma \sim 60$ mN/m) and $R_m \approx 0.4$ μm, is 1.5×10^3 Pa. With good wettability of surfaces and without impurity (e.g., liquid metals, liquids with low surface tension, and the like), the high superheating is usually realized. Under these conditions, growth rate of the vapor volume is limited by (Gavrilov, 1999):

$$R = \sqrt{\frac{6Mc_f}{k} \frac{\rho_f}{\rho_g} a\tau}, \tag{7.120}$$

where M is the molecular weight; $k = 1.38 \times 10^{-23}$ J/K is Boltzmann constant; a is the thermal diffusivity in m^2/s; c_f is the specific heat capacity of the liquid in J/(kg K) and the time τ is in s. Eq. (7.120) implies that the growth rate is independent of the magnitude of superheating.

In general, the interface velocity is determined by solving Rayleigh's equation, transient energy and mass conservation equations simultaneously, considering the energy-mass balance conditions at the interface.

7.5.3 Types of boiling

Depending on the heat flux and other factors of the wall surface, vapor bubbles or a continuous layer of vapor is formed, which is accordingly called bubbly or film boiling. In the nucleate boiling, the liquid is in direct contact with the nucleation site on the surface. In the film boiling, the vapor layer separates the liquid from much of the heated wall surface. The vapor layer covering the heated wall surface transfers the heat by conduction and reduces the heat transfer rate by many times, so a transition from the nucleate boiling to the film boiling takes place at a constant heat flux and may be accompanied by a sharp increase in the surface temperature of the heating wall (heat transfer crisis or burnout, or simply "crisis").

The various types of boiling are classified as follows:

- By the nature of the boiling—nucleate boiling and film boiling;
- The type of convection—the boiling under natural convection (in a large volume), and forced convection;
- Magnitude of the average fluid temperature ($\overline{T}_f$) relative to the saturation temperature (T_S)—subcooled boiling if $\overline{T}_f < T_S$; liquid temperature close to the saturation temperature ($\overline{T}_f \approx T_S$)
- On the surfaces—horizontal, vertical, inclined;
- The nature of the boiling—developed, developing, or undeveloped and unstable.

Nucleate boiling is developed for a large number of nucleation sites and undeveloped for a small number of nucleation sites. In the latter case, a large proportion of the heat is removed by the single-phase liquid flow convection. The boiling process becomes unstable, if the nucleate boiling regime is replaced by a liquid-phase random natural convection, and then back to nucleate boiling again.

7.5.4 Boiling curve

Dependence of the heat flux on the temperature difference ($T_w - T_S$) is the boiling curve (Fig. 7.40), originally proposed by Nukiyama (1934). With increasing the temperature difference, the heat flux reaches a maximum and then decreases and increases again (curve ABB′C′CDE). On this curve, you can highlight areas of different modes of heat transfer and the characteristic points (see Fig. 7.40).

For the region of free convection (ABK), the heat transfer coefficient $\alpha \sim \Delta T^m$, typically with $m = 1/4$ for the laminar region and $m = 1/3$ for turbulent flow regime. Consequently, $q = \alpha \cdot \Delta T \sim \Delta T^{5/4 \sim 4/3}$. At point B, boiling begins on the surface. If

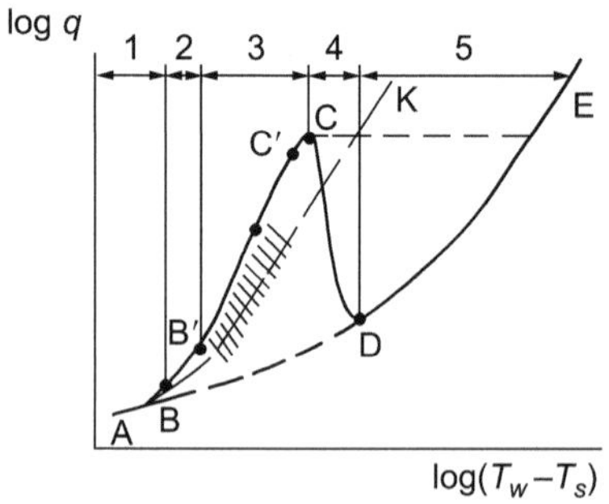

Figure 7.40 Boiling curve for a large volume (Nukiyama curve). *Different modes of heat transfer*: 1—gravitational convection; 2—undeveloped boiling; 3—developed boiling; 4—boiling transition (from bubble to film); 5—film boiling; \\\\\—region stable boiling. *Typical modes*: AB—natural (free) convection; B—beginning of nucleate boiling; B′—beginning of the developed nucleate boiling; C—point of crisis; D—Leidenfrost point; C′—point of deviation from the mode of nucleation.

the temperature of the liquid in a volume is lower than T_S, the bubbles within the liquid condense. If the fluid temperature is equal to T_S, the bubbles will rise to the liquid surface without condensing.

In region 2 (BB′) with increasing heat flux or temperature difference increases the number of nucleation sites and increases the heat transfer rate, although still a considerable amount of heat removed by convection. With further increase in q or the number of nucleation sites continuously increases, an increasing portion of the heat is removed by evaporation.

In the region (B′C′), $q \sim \Delta T^{3 \sim 4}$; Point C′ corresponds to the deviation from the nucleate boiling regime; and the point C is reached the maximum heat flux, called critical heat flux (CHF) because a further (even slight) increase in the q results in a dramatic increase in the wall surface temperature according to the line (C−E). If the surface temperature can be controlled, it will lead to an increase in vapor blanketing the surface and the heat flux decreases (CD). For the vapor flowing along the surface, it requires a significantly higher wall surface temperature (DE) to achieve the same heat flux of the nucleate boiling.

In the region of sustainable film boiling, the heat is transferred from the heated wall through the vapor by the evaporation of a thin liquid layer, steam convection, and radiation. The contribution of the radiation increases with ΔT. In the region (DE) $q \sim \Delta T^3$.

Influences of various factors on the boiling curve for pool boiling is illustrated qualitatively by a series of charts as illustrated in Fig. 7.41 (Chai and Shoji, 2001; Shoji, 1998; Soloviev, 1998).

The boiling curve shifts to the right with the increase in the system pressure and the heat transfer coefficient increases with the pressure in the nucleate and film boiling regime (A). At very low pressures (R_{cr} is large), there is a delay to start boiling in the process of natural convection that requires more fluid to superheated ($T_w - T_S$).

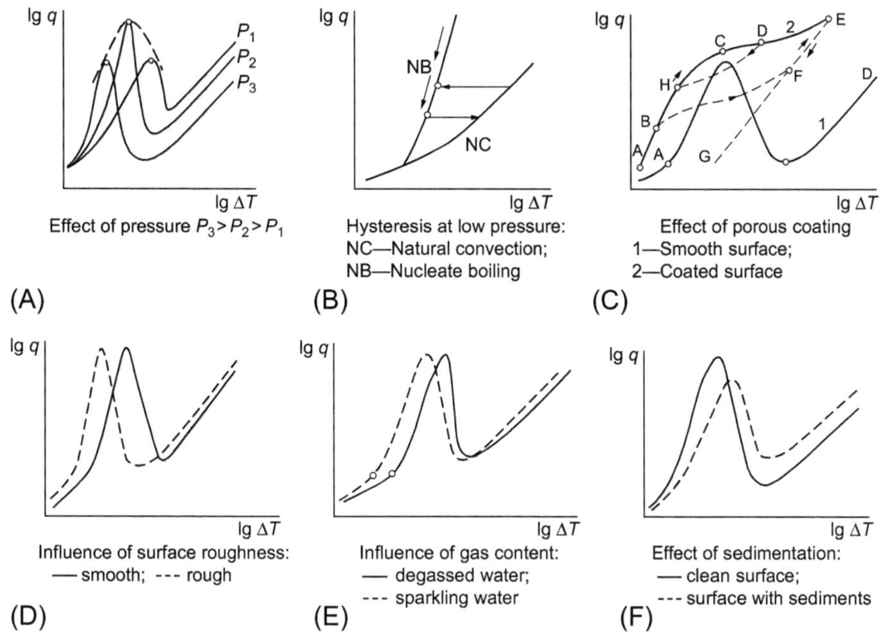

Figure 7.41 Influence of various factors on the process of the boiling (boiling curve) in bulk fluids.

After the boiling initiation, this superheating will be reduced. With a reduction in the heat flux, a transition can occur from nucleate boiling (NB) again to the natural convection (NC)—hysteresis phenomenon (Fig. 7.41b). At sufficiently high heat flux with increasing surface superheating, a possible transition could take place from nucleate to film boiling, bypassing the transition boiling regime. On the other hand, with a decrease in the film boiling surface (or reduction of the heat flux), the system can switch to the nucleate boiling regime, also bypassing the transitional regime—hysteresis II.

Issues related to the transition boiling regime have been studied in the last decades intensively (Gavrilov, 1999; Haddad and Cheung, 1998; Klimenko and Snytin, 1988), although a satisfactory understanding of the process has not yet been reached. Fig. 7.41c schematically shows a difference of the boiling curve on a smooth surface and a porous surface coating. On the porous surface, boiling occurs at lower heat flux density (A) than on the smooth surface (A′). If the coating has high permeability and high thermal conductivity, then the boiling curve would be ABCD. If the vapor encounters resistances in coming out of the porous structure, the process while reducing the q is on the line DH or EFG. One of the reasons for the observed hysteresis is due to the difficulty for the liquid to penetrate from the small pores into the large space. In this case, the heat transfer coefficients at the EFG line are determined from the thermal resistance of the porous coating δ/λ. This mode of nucleate boiling is considered to have the vapor film inside the porous layer. With low permeability porous coating, a transition to the film boiling is observed with the vapor available on the BF line at $q < q_{cr}$. The position of the curve (2) depends on the coating thickness, permeability, and thermal conductivity. Data in some areas of the curve (2) for coatings can be found in industrial practices. An overview and the method of heat transfer calculation at boiling on the porous surfaces is found in Soloviev (1998).

The presence of surface roughness increases the heat transfer rate (Fig. 7.41d). The presence of gas in the liquid enhances boiling at low heat fluxes and the heat transfer during the nucleate boiling (Fig. 7.41e). Effects of deposits (e.g., corrosion products) typically come from the low thermal conductivity of the deposits, resulting in an increase ΔT and the boiling curve to shift to the right (Fig. 7.41f).

Bubbly boiling structure on a horizontal surface will change as the heat flux increases from single large steam bubbles to vapor mushroom formations (Gertner, 1965) (see Fig. 7.42).

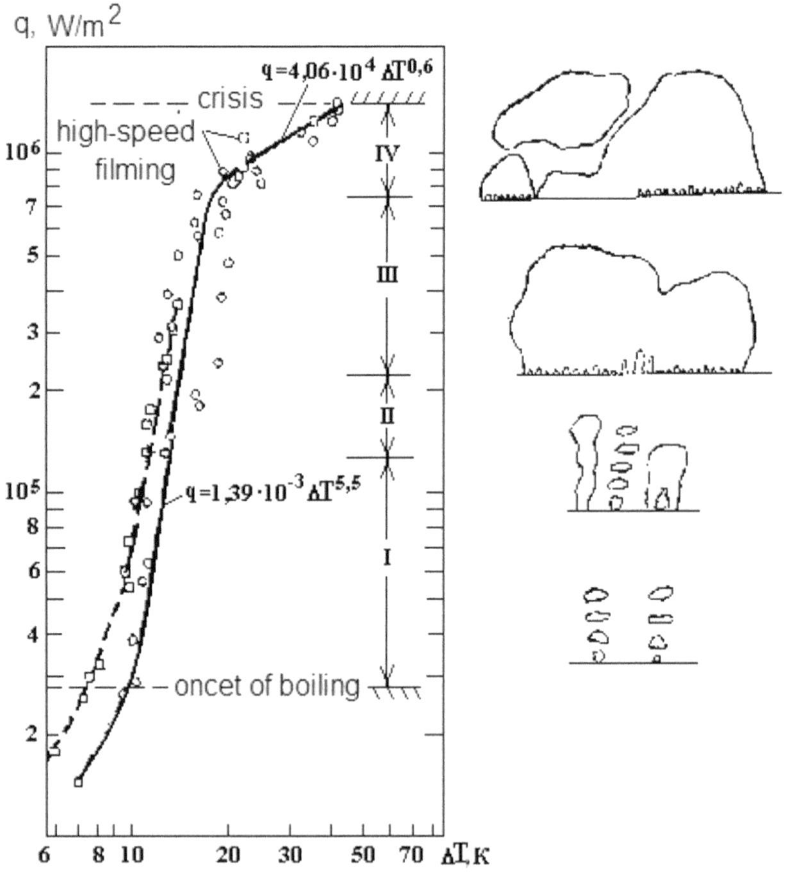

Figure 7.42 Boiling curve and vapor bubble structure depending on heat flux: I—individual bubbles region; II—the first transition region; III—mushroom region; IV—the second transition region.

7.5.4.1 Minimum film boiling temperature

On the boiling curve illustrated in Fig. 7.43, point D, corresponding to the minimum heat flux q_{min} and temperature T_{min}, it is called the Leidenfrost point or minimum film boiling (MFB) temperature T_{MFB}. Various names have been given to the T_{MFB} including rewetting temperature, quench temperature, Leidenfrost temperature, film boiling collapse temperature and others. Heat transfer characteristics in the vicinity of the T_{MFB} point are important in assessing the cooling process of the heated surfaces (hardening treatment by quenching) and in evaluating the ECCS for loss-of-coolant accidents (LOCAs). This point separates the transition boiling and film boiling modes: in the region $T > T_{MFB}$ the liquid ceases to wet the heated surface. A conventional technique to define T_{MFB} is associated with the measurement of the lifetime (evaporation) τ of drops on the hot surface heated up to T_w. Fig. 7.44 shows a comparison of the curve $\tau = f(T_w)$ as if it were inverted boiling curve.

It is well known that the mode of liquid contact with the heated surface is dependent not only on the temperature, but also on the liquid-wall interface conditions (surface smoothness, surface roughness, surface geometry, subcooling, surface tension, boundary conditions, etc.). It is noted that, in Fig. 7.43, the deterioration of the wettability will shift the beginning of film boiling to T_E ($<T_{min}$) while improved wettability to T_A ($>T_{min}$), where EDA is the transition region to the film boiling regime and the region AF is the mode of pure film boiling.

Trying to take into account the thermal properties of the heated surface and the wettability conditions of the liquid-solid system, Baumeister and Simon (1973) arrived at the following semiempirical expression:

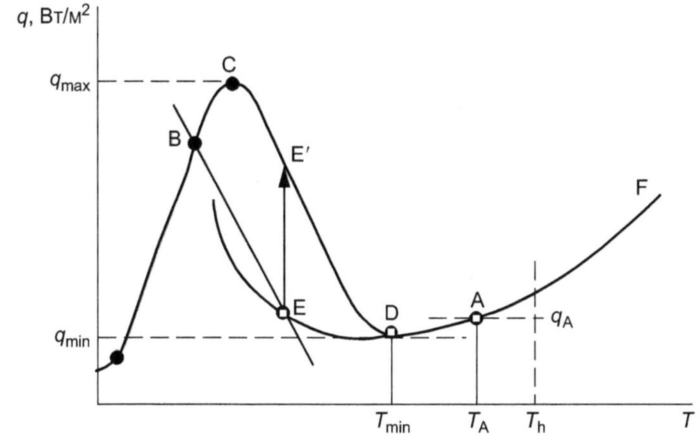

Figure 7.43 Boiling curve: AF—film boiling; CE′—transition boiling from nucleate to film boiling; EDA—transition from film boiling; EE′—heat flux jump with improved wetting; T_A, q_A—temperature and heat flux in the presence of liquid contact; C initiation of transition boiling.

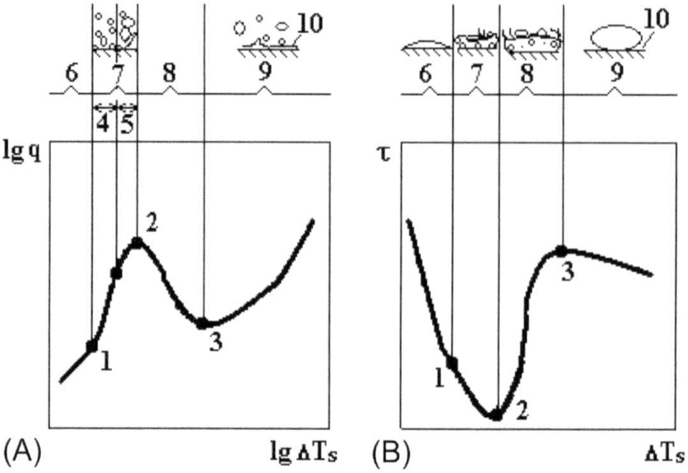

Figure 7.44 (A) Boiling curve at saturation temperature and (B) liquid droplets evaporation time on a heated surface: 1—nucleate boiling initiation; 2—critical heat flux; 3—minimum heat flux (Leidenfrost point); 4—separated bubbles; 5—jets and steam column mode; 6—single-phase convection mode; 7—bubble-film coexisting boiling regime; 8—transition boiling regime; 9—film boiling regime; 10—vapor layer; τ—the evaporation of droplets.

$$T_{\min} = T_f + \left\{0.844T_{cr}\left[1 - \exp\left(-0.016\left[(\rho_s/M_s)^{1.33}/\sigma\right]^{0.5}\right)\right] - T_f\right\}$$
$$\left[\exp\left(3.066\times10^6\beta\right)\, erfc\left(1758\sqrt{\beta}\right)\right]^{-1},$$

where the subscript s refers to the surface material; $\beta = 1/(\rho_s c_{ps}\lambda_s)$; σ is the surface tension; M_s is the atomic weight of the surface materials; λ_s is the thermal conductivity; ρ_s is the density; c_{ps} is the specific heat of the material. It is logical to assume that the influence of ΔT_{sub} (subcooling) on $T_{\min}$ will be similar to the effect on the CHF q_{cr}, i.e., $q_{cr} \approx q_0 + \Delta T_{\text{sub}}$. Experiments were carried out with the hot spheres (1−100 mm in diameter) that are immersed in various fluids (water, Freon, sodium, etc.) by many researchers. For water, obtained empirical relationships are summarized by the formula

$$T_{\min}(^\circ\text{C}) = A + B\Delta T_{\text{sub}},$$

where $A = 200{-}220^\circ$C; $B = 6.3{-}8$. Quenching cooling mode at high ΔT_{sub} will lead to a sudden achievement of higher wettability resulting in a jump in the heat flux. It is commonly accepted that during the initial stages of the drop-surface contact, the interface temperature between the liquid and solid is dictated by the thermal properties of the liquid and solid as well as by their initial temperatures. This interface temperature T_i is given by the solution to the one-dimensional energy equation with semi-infinite body boundary conditions (Eckert and Drake, 1972):

$$T_i = \left[(\lambda \rho c_p)_s^{0.5} T_0 + (\lambda \rho c_p)_f^{0.5} T_f\right] / \left[(\lambda \rho c_p)_s^{0.5} + (\lambda \rho c_p)_f^{0.5}\right].$$

The limiting temperature at which the initial contact can stay nonequilibrium or metastable, i.e., superheating of a liquid above its boiling point, between the liquid and the drop surface, is the *homogeneous nucleation temperature* T_{hn}. A detailed theory of this process was developed and reported by Skripova (1972). An empirical relationship for T_{hn} is established by Ramilison and Lienhard (1988):

$$T_{hn} = T_{cr}\left[0.923 + 0.077(T_s/T_{cr})^9\right].$$

It is necessary to distinguish between T_{hn} ($\equiv T_{\lim}$) and $T_{\min}$ ($=T_{\mathrm{MFB}}$). $T_{\lim}$ ($\equiv T_{hn}$) is determined purely by the thermodynamic conditions. $T_{\min}$ is affected by not only thermodynamic factors, but also by many other factors such as actual fluid periodical contacts with the wall, insulation of the surface, etc. Experimental data indicate that the pressure is an important parameter for thermophysical properties of the fluid, surface tension, contact angle, and surface material which can play an important role in determining $T_{\min}$.

In accordance with Klimenko and Snytin (1988):

$$\frac{T_{\min} - T_s}{T_{\lim} - T_s} = \frac{\Delta T_{\min}}{\Delta T_{\lim}} = \left[8.2 + 4\frac{(\rho c\lambda)_f}{(\rho c\lambda)_W} + 60\left(\frac{(\rho c\lambda)_f^{1/2}}{(\rho c\lambda)_W}\right)\right]\cdot f\left(\frac{\rho_f}{\rho_g}\right),$$

where $f(\rho_f/\rho_g) = \begin{cases} (\rho_f/\rho_g)^{-0.52} & \text{for } (\rho_f/\rho_g) < 360; \\ 0.047 & \text{for } (\rho_f/\rho_g) > 360. \end{cases}$

This relationship is obtained in the range of $p/p_{cr} = 9 \times 10^{-4}$–$0.95$; $(\rho c\lambda)_f/(\rho c\lambda)_W = 6.7 \times 10^{-5}$–$4.9$; and $\rho_f/\rho_g = 1.9$–7550.

A detailed overview of the temperature $T_{\min}$ is available in Bernardin and Mudawar (1999), which is excerpted in Table 7.5. It shows the diversity of mechanisms related to the Leidenfrost phenomena.

Two theories have been proposed for the analytical prediction of the MFB temperature. One theory says that the minimum temperature is a thermodynamic property of the fluid (i.e., maximum liquid temperature) and thus is primarily a function of pressure. The other theory suggests that rewetting commences due to hydrodynamic instabilities which depend on the velocities, densities, and viscosities of both phases as well as the surface tension at the liquid-vapor interface (IAEA, 2001).

During fast transitions, where insufficient time is available to fully develop the hydrodynamic forces, rewetting is expected to be thermodynamically controlled, while for low flows and low pressures, where sufficient time is available and the volumetric expansion of the fluid near the wall is large, rewetting is more likely to be hydrodynamically controlled. Once rewetting has occurred locally, the rewetting front can then propagate at a rate which is primarily controlled by axial conduction. These theories can be modified to include the thermal properties of the surface (IAEA, 2001).

Table 7.5 Possible mechanisms of Leidenfrost phenomenon (Bernardin and Mudawar, 1999).

Model	Pictorial description	Relevant correlations
Hydrodynamic instability	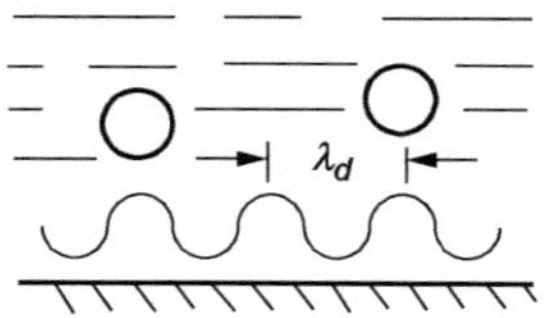	Most dangerous wavelength: $\lambda_d = 2\pi\sqrt{\frac{3\sigma_f}{g(\rho_f - \rho_g)}}$
Metastable liquid mechanical stability	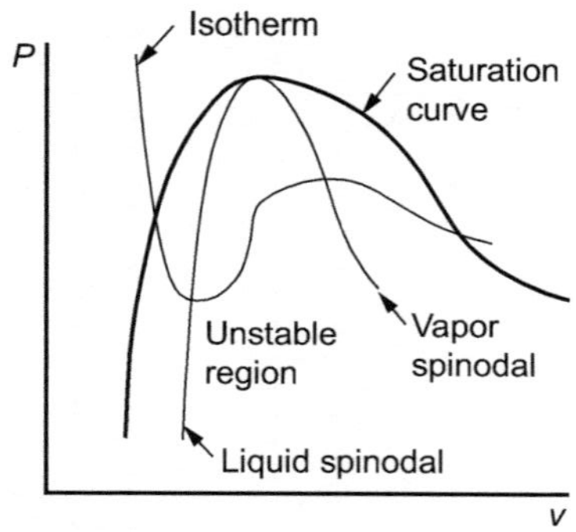	Mechanistic stability condition $\left(\frac{\partial p}{\partial v}\right)_T = 0$ Spinoidal or liquid superheat limit: (using van der Waals equations) p/p_{cr} << 1 $T_{hn} = 0.844\ T_{cr}$

Metastable liquid-kinetic stability	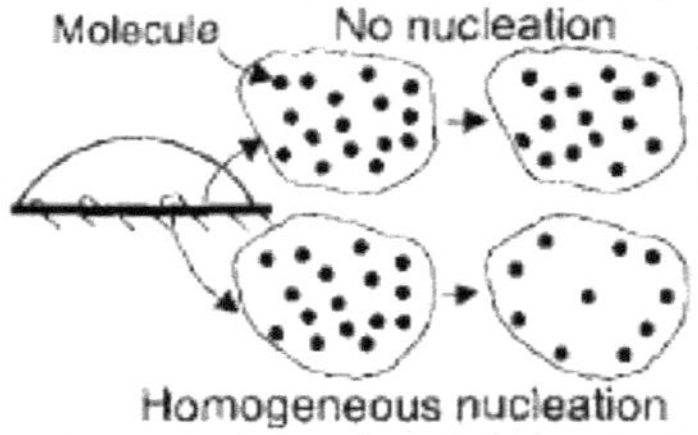	Homogeneous nucleation limit $J = N_f \left(\frac{3\sigma}{\pi m}\right)^{0.5} \exp\left\{\frac{-16\pi\sigma^3}{3k_b T_f \left[\eta \rho_{sat} T_f - p_f\right]^2}\right\};$ $\eta = \exp\left\{\frac{v_f\left[\rho_f - \rho_{sat}(T_f)\right]}{RT_f}\right\}$
Thermomechanical effect	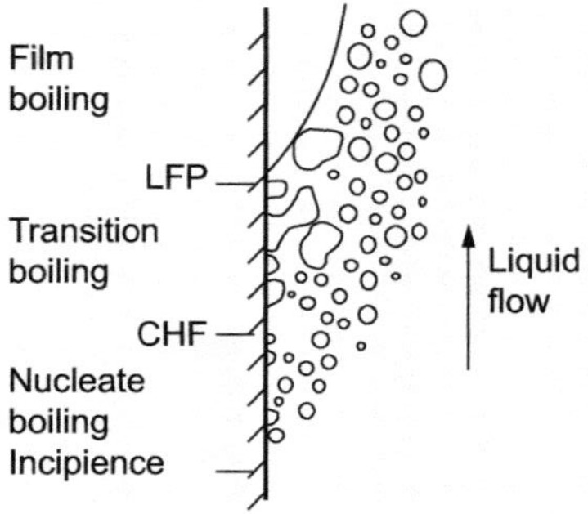	Energy balance for section *D* $h_g(T_g) - h_f(T_{leid}) = 0.5\left[v_g(T_g) - v_f(T_{leid})\right] \times \left[p_{sat}(T_{leid}) - p_{sat}(T_g)\right]$
Wettability-contact angle	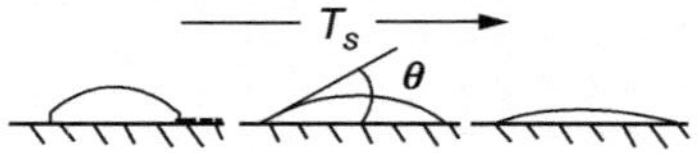	Contact angle temperature dependence $\cos\theta = 1 + C(T_c - T)^m$
Wettability-surface adsorption	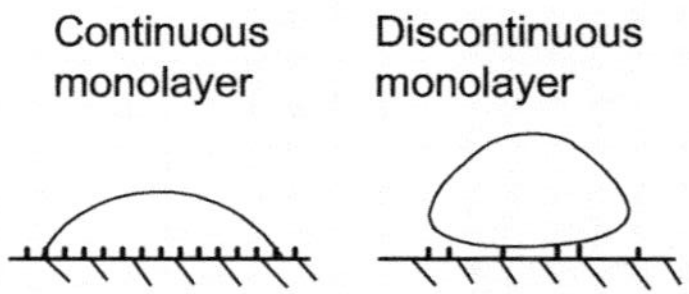	Monolayer molecular surface coverage temperature dependence $\frac{\Gamma}{\Gamma_0} = \frac{\exp\left(\frac{Q_0}{RT_i}\right)}{\left[\frac{(2\pi MRT)^{0.5}\Gamma_0}{Na\ Pr_0}\right] + \exp\left(\frac{Q_0}{RT_i}\right)}$

Table 7.6 Leidenfrost temperature T_{LFP} for saturated fluids that are in contact with the smooth clean surfaces at atmospheric pressure (Soloviev, 1998).

Liquid	Test conditions	T_{LFP} (K)
Sodium	Tantalum sphere in bulk	1600
Potassium	Tantalum plate and drop	1588
	Stainless steel plates in a large volume	874
Mercury	Tantalum plate and drop	841
	Copper plate and drop	964
Water	Aluminum and drop	426
	Platinum wire in a bulk water	423
Nitrogen	Sphere of the copper in a bulk	96
	Sphere of Teflon in a bulk	200

7.5.4.2 Leidenfrost temperature

The evaporation of superheated droplets on the surface depends on the temperature. Thus, a drop of the radius of 2 mm on the brass plate evaporates over time:

at ~ 100°C—35 s; at ~ 170°C—a few seconds (the first critical temperature);
at ~ 240°C—130 s (the second critical temperature—the temperature of the Leidenfrost);
at ~ 500°C—60 s (spheroidal state).

In the last case, the drop is separated from the heated surface and floating on the superheated vapor layer; for estimates the following can be used:

$$T_{\min} = (27/32)T_{cr},$$

where T_{cr}—thermodynamic critical temperature in K (see Table 7.6).

7.6 Nucleation pool boiling

7.6.1 The mechanisms of the nucleation process

Complete theory of boiling has never been established that accounts for all the factors affecting the magnitude of the heat transfer. They include primarily physical properties of the fluid, complex surface conditions (surface roughness, its size distribution, surface deposits, wettability, surface material properties, etc.). The system of differential equations and boundary conditions, describing the boiling two-phase flows, include the following:

- Momentum conservation equation for the liquid and vapor phases;
- Mass continuity equation for the liquid and vapor phases;
- Energy conservation equation for the liquid and vapor phases;

- and appropriate initial and boundary conditions including those associated with the diffusion processes:
- Thermal and mechanical interactions at the phase interface;
- Continuity of the energy flow (heat flux) across the phase interfaces; $q_{ip} = -\lambda(\partial T/\partial n)_{ip} = w_g \rho_g \Delta h_{fg}$ where w_g is the local vapor-phase velocity on the interface and is assumed to be the speed of evaporation;
- Equal temperatures at the phase interface.

Mathematical description of physical models of the boiling process can only be formal because of the uncertainties of many factors involved in closure equations, and the lack of accurate boundary conditions, in particular at the interfaces between the vapor and liquid phases. Therefore, these relations serve only to establish a kind of generalized variables, which make processing of experimental data possible. A capillary constant is often used either as a reference length $l = \sqrt{\sigma/g(\rho_f - \rho_g)}$, a quantity related to the size of the critical (minimum) radius of the nucleus of vapor $R_{cr} = \frac{\sigma T_S}{\Delta h_{fg}\rho_g(T_w - T_S)} = \frac{2\sigma T(v_g - v_f)}{\Delta h_{fg}\Delta T} \cdot \frac{\rho_f}{\rho_f - \rho_g}$ (Eq. 7.117) (Lykov, 1954). The characteristic velocity is the speed of evaporation $w_g = q/(\Delta h_{fg}\rho_g)$ under the thermal equilibrium condition, which determines the vapor volume withdrawn from the unit interphase surface area per unit volume and per unit time (m/s).

The high heat flux density at boiling is mainly associated with the following two processes: (1) agitation of liquid steam bubbles near the surface of the heat exchange (q_{conv}); and (2) intensive evaporation of the micro liquid layer at the bottom of the vapor bubble (q_{eva}). Thus, $q = q_{conv} + q_{eva}$.

Formation of one vapor bubble of diameter d_0, i.e., the volume $V = \pi d_0^3/6$ consumes the heat $Q_1 = \Delta h_{fg}\rho_g V = \Delta h_{fg}\rho_g \frac{\pi d_0^3}{6}$. Let the number density of the existing vaporization centers per unit area be N (m^{-2}), the total heat transfer area A, the separation frequency f_i, the volume of each bubble V_i, then the amount of heat removed by evaporation of the liquid will be given by

$$q_{eva} = \frac{\sum_1^N Q_i f_i}{A} = \Delta h_{fg}\rho_g \frac{\sum_1^N f_i V_i}{A} = \Delta h_{fg}\rho_g N f \overline{V} = \Delta h_{fg}\rho_g N \frac{\pi}{6} f \overline{d_0}^3.$$

As the heat flux or temperature difference (ΔT) increases, the number of vaporization centers grows ($N \sim q$; $N \sim \Delta T^3$). The average frequency f and the detachment diameter d_0 are virtually independent of q. Therefore, the average production of vaporization centers under the developed boiling is approximately constant. In the developed boiling, the fraction of the heat transported by the evaporation (q_{eva}/q) increases as q increases and approaches to 100%, i.e., $q \approx q_{cr}$.

7.6.2 Influence of various factors on heat transfer in nucleate boiling

The effect of pressure is shown in Fig. 7.45 (Borishanskii, 1961; Borishanskii et al., 1964). With increasing pressure, the heat transfer coefficient increases, slowly at first

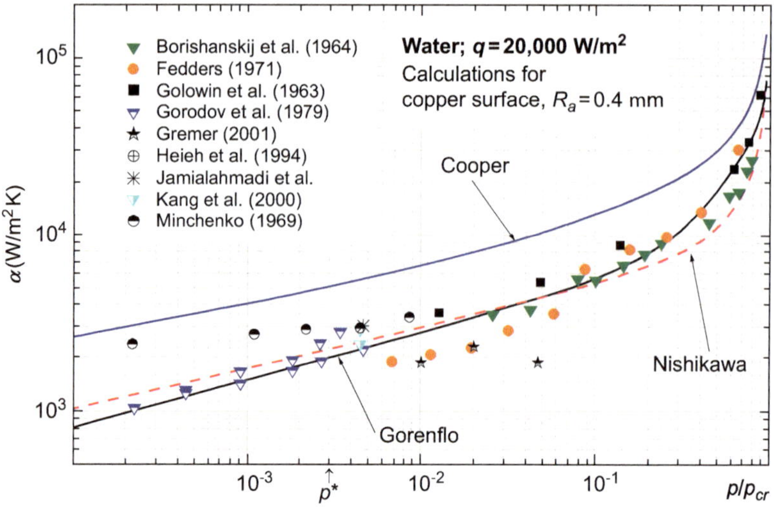

Figure 7.45 The effect of pressure on the heat transfer coefficient of nucleate boiling (Borishanskii, 1961; Borishanskii et al., 1964): p_{cr}—critical pressure, p^*—conditionally selected pressure, $p^* = 0.03p_{cr}$.
Adapted from Feldmann, H., Luke, A., 2008. Nucleate boiling in water for different pressures. In: International Refrigeration and Air Conditioning Conference. Paper 982.

(up to $\pi = p/p_{cr} \approx 0.2$), then (for $\pi > 0.2$) the influence of pressure increases sharply.

The reason for this pressure dependency is explained from the nature of changes in physical properties of the fluid, i.e., first of all, Δh_{fg}, σ, ρ_f, and ρ_g. Growth of the vapor bubbles at high pressure is explained by the heat transferred from liquid to the vapor bubble through the interface surface in the base layer (Labuntsov, 1960). An increase in the pressure reduces the surface tension and the heat of evaporation, facilitating nucleation and growth of the vapor bubbles, reduction in R_{cr}, and the growing number of vaporization centers. At low pressures ($\pi = p/p_{cr} < 5 \times 10^{-4}$, $p < 0.01$ MPa), boiling initiation becomes difficult, expanding the convective heat transfer region, decreasing the bubble detachment frequency, and increasing R_{cr}.

The influence of surface roughness on heat transfer is evaluated based on information about active centers of vaporization (cavities), including number density and both the value of protrusions and depressions and their size distribution, and even the shape of cavities. In general, with an increase in surface roughness for the same heat flux, we have less wall superheat, i.e., an increase in heat transfer. However, this observation is not universal. Depending on the size and shape distributions of the cavities (caverns), the boiling curve varies. For example, Fig. 7.46b shows boiling curves for cavities within the normal distribution of sizes, of equal size, and for a specially selected size distribution of cavities. Thus you can control the shape of the boiling curve (i.e., the flux) with new technology for producing special surfaces with cavities and hollows (Fig. 7.46).

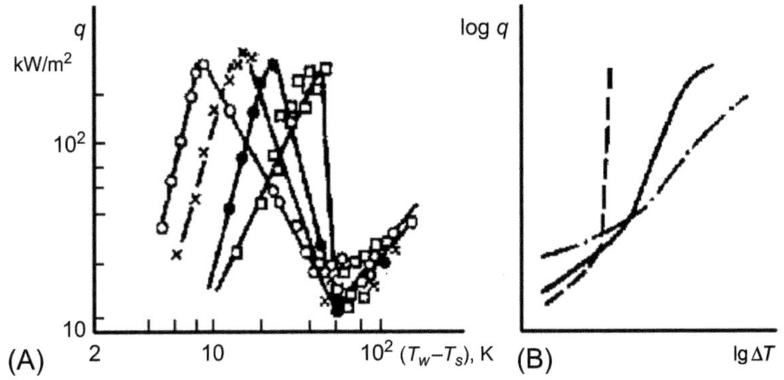

Figure 7.46 Influence of surface roughness on the boiling curve: (A) □ → • → × → ○—increase roughness; (B) for the normal distribution of the cavity sizes; - - - all cavities of the same size; -·-·- for cavities of specially selected size distribution.

Impurities in the water cause deposits on the heating surface, changing the surface roughness. On the one hand, this leads to the formation of new cavities and consequently to increases in the heat transfer coefficient. On the other hand, the additional thermal resistance of these deposits reduces the heat transfer coefficient. In practice, however, due to the range of operating conditions, it is difficult to estimate the thickness and thermal conductivity of the deposits. Thus, we rely on the thermal resistance of the deposits provided from industry experience.

The formulas that follow are suitable for smooth surfaces with no deposits. During boiling, surface deposits should be considered a sediment layer with thermal resistance. In this case, $1/\alpha = 1/\alpha_p + R$, where α_p is the heat transfer coefficient obtained from the equations that follow, and R is the thermal resistance of the deposits.

Material thermal properties and the thickness of the heat-transfer wall may affect the effective heat-transfer coefficient. The reason for this lies in the unsteady heat conduction that develops in the vicinity of the cavity center of vaporization. If the wall material has high thermal conductivity and high heat capacity c_p, it will facilitate the influx of heat of vaporization to the cavity centers and provide a high heat transfer rate. It should be noted that when the deposit layer is thick ($\delta > 1$ mm), which is not unusual in normal industrial operations, the boiling heat transfer rate does not practically depend on the material and its conditions but is greatly influenced by the surface roughness and the presence of low conduction deposits on it. It should also be noted that the heat transfer coefficient of the nucleate boiling in a large volume is essentially independent of the orientation of the heat-transfer surface if the detachment of steam bubbles is ensured on the surface at any angles.

Modeling of boiling heat transfer of pure liquid and binary mixtures generally includes the following processes (Balakrishnan, 1998):

1. Microconvection in the boundary layer due to an increase in bubbles at the heating surface;
2. The flow of fluid as a result of the growth and detachment of the bubble, the flow of the cold liquid toward the heating surface;

3. Transient heat conduction in a cooling fluid as it becomes warm;
4. Vaporization of the microlayer during bubble growth;
5. Agitation of the liquid layer due to bubble growth, bubble detachment, and "cold" flow to the surface;
6. Turbulence generated by natural convection in the region adjacent to the center of evaporation, where there is no evaporation.

Heat transfer in binary mixtures is usually more intense than in pure liquids for easier evaporation of the volatile component.

Thus, the pool boiling process in a large volume is associated with the active effects of vapor bubbles and the rate of heat transfer, which can be considered the sum of four components:

(1) Microconvection occurring near the surface during the growth and detachment of vapor bubbles;
(2) Evaporation of the liquid layer underneath the vapor bubbles, taking into account the transfer of latent heat of vaporization to inside the vapor bubble;
(3) Natural convection in the large volume;
(4) Thermocapillary flow (Marangoni flow).

The last component is connected to longitudinal movement along the surface caused by a temperature gradient and surface tension at the liquid-vapor interface. It does not play a big role in boiling at the saturation temperature but essentially at high subcooling. In most simple models of boiling heat transfer, heat flux appears to be a sum of only the two heat fluxes associated with convection (q_{conv}) and evaporation (q_{eva}): $q = q_{conv} + q_{eva}$.

7.6.3 Correlation estimates

Heat transfer in the developed boiling region is calculated on the clean surface in a large volume of water by the following heat transfer coefficient:

$$\alpha = 4.34\, q^{0.7} \left(|p^{0.14} + 1.35 \times 10^{-2} p^2 \right), \tag{7.121}$$

where q is expressed in W/m^2; p in MPa; and α in W/(m^2 K). The range of applicability of Eq. (7.121) will be: $p = 0.1-20$ MPa; $q < 0.4$ MW/m^2. The equation is applicable at temperatures from the subcooled to saturation range. The temperature difference in this case is defined as the difference between the surface temperature and the saturation temperature. The dependence of the heat transfer coefficient on the saturation temperature of water is expressed empirically by the following equation:

$$\alpha = 10.45 q^{0.7} [3.3 - 0.0113 |(T_S - 373)]^{-1}, \tag{7.122}$$

where T_S is in K. The same applicability range for Eq. (7.121) is applied. Assuming that the same boiling heat transfer process can be determined by the internal flow mechanisms, Labuntsov (1960) proposed the correlation

$$Nu = C\, Re^n Pr^{1/3}, \tag{7.123}$$

where $Nu = \frac{\alpha l}{\lambda_f}$; $Re = \frac{wl}{\nu_f} = \frac{ql}{\nu_f \Delta h_{fg} \rho_g}$ since $w = \frac{q}{\Delta h_{fg} \rho_g}$; Pr is the Prandtl number of the liquid; and $l \equiv \frac{c_{pf} \rho_f \sigma T_S}{\left(\Delta h_{fg} \rho_g\right)^2}$.

To calculate the heat transfer during nucleate boiling in a large volume, the following fairly simple equation could be used (Minchenko, 1960):

$$Nu = 0.55\left(PeK_p\right)^{0.7} Re^n Pr^{1/3}, \tag{7.124}$$

where $Nu = \frac{\alpha l}{\lambda_f}$; $Pe = Re\, Pr = \frac{ql}{\nu_f \Delta h_{fg} \rho_g} \frac{\nu_f}{\alpha} = \frac{ql}{\Delta h_{fg} \rho_g \alpha}$; $K_p = \frac{p}{\sqrt{\sigma g\left(\rho_f - \rho_g\right)}}$; and $l = \sqrt{\frac{\sigma}{g\left(\rho_f - \rho_g\right)}}$.

A more widely used correlation is by Rohsenow (1952) and based on the idea that the heat transfer coefficient of boiling mainly depends on the degree of turbulence in the boundary layer and the increase in uprising steam bubbles (i.e., the main component of the heat flux is contributed by convection). Then,

$$Nu = C_{sf}\, Re^m Pr^n,$$

where C_{sf} is the coefficient that depends on the thermal properties of the fluid at a temperature of saturation and surface conditions. Another form of the equation is

$$\frac{Re\, Pr}{Nu} = \frac{c_p \Delta T}{\Delta h_{fg}} = C_{sf}\left[\frac{q}{\Delta h_{fg} \mu} \sqrt{\frac{\sigma}{g\left(\rho_f - \rho_g\right)}}\right]^{0.33} \left(\frac{c_p \mu}{\lambda}\right)^n,$$

where $\Delta T = T_w - T_S$. For water at $T_S = 100–350°C$, $q = 3–3200\ \mathrm{kW/m^2}$; $n = 1$; $C_{sf} = 0.013–0.015$. For low pressure $C_{sf} = 0.007–0.008$, as shown in Pioro (1999); for hydrocarbons, freon, C_{sf}, is widely spread in 0.0014–0.0018.

7.6.4 On the hydrodynamic theory of similarity of boiling heat transfer

The thermodynamic similarity principle provides a formula for the heat transfer coefficient of boiling up to a constant coefficient, which is determined experimentally and is a constant for thermodynamically similar liquids. Similitude substances based on this principle were first suggested and applied to the problems of heat transfer and the crisis at boiling by Novikov (1961) and developed further by Borishanskii et al. (1964, 1973). Since the boiling heat transfer coefficient is a function of the developed heat flux and physical properties of liquid and vapor, its first approximation is represented as a product of the powers of these values:

$$\alpha \sim q^n r^{n_1} \sigma^{n_2} \lambda^{n_3} \mu^{n_4} \ldots \tag{7.125}$$

On the saturation line, any physical parameter can be determined from the form given by

$$\chi_i = \left(\frac{M}{g}\right)^{m_{1i}} T_{cr}^{m_{2i}} p_{cr}^{m_{3i}} R^{m_{4i}} f_i\left(\frac{p}{p_{cr}}\right), \tag{7.126}$$

Therefore, formula (7.125) takes the form

$$\alpha = q^n \left(\frac{M}{g}\right)^{m_1} T_{cr}^{m_2} p_{cr}^{m_3} R^{m_4} F_1\left(\frac{p}{p_{cr}}\right). \tag{7.127}$$

Several experiments have shown and established that $\alpha \sim q^{2/3}$. Therefore for the R.H.S. of Eq. (7.127) to have the dimension of the heat transfer coefficient, $m_1 = -1/6$; $m_2 = -5/6$; $m_3 = 1/3$; $m_4 = 1/6$. Thus, the final form is

$$\alpha = B\left(\frac{gR}{M}\right)^{1/6} p_{cr}^{1/3} T_{cr}^{-5/6} q^{2/3} F_2\left(\frac{p}{p_{cr}}\right). \tag{7.128}$$

The magnitude B and the function F_2 are determined from experiments. Estimates of the heat transfer coefficients for the boiling, for which there are no direct measurements, can be performed using the following equation:

$$\alpha = 882\, p_{\text{K}}^{1/3} T_{\text{K}}^{-5/6} M^{-1/6} \pi^{0.1} \left(1 + 4.65\pi^{1.13}\right) q^{2/3}, \tag{7.129}$$

where $\pi = p/p_{cr}$.

7.6.5 Boiling on a coated porous surface

Fig. 7.47 shows that the use of a porous coating of high permeability and a high thermal conductivity layer allows for significantly increasing the heat transfer coefficient of nucleate boiling and CHF, where boiling initiates on the porous surface at a lower heat flux (A) than on the smooth surface (A′). In the figure, boiling curve 2 is for the coated surface, and line 1 is for a smooth surface. The position of curve 2 has a large uncertainty and may shift by dozens or hundreds of degrees (temperature) depending on the characteristics of the porous structure. At complicated exhalation of the vapor from the capillary-porous structure and with a decrease in the heat flux, the boiling curve can be represented by BFE (line 3). The heat transfer coefficient on line 3 is determined by the thermal resistance of the porous coating, i.e., $\alpha \approx \lambda_s/\delta$, where λ_s is the thermal conductivity, and δ is the thickness of the porous layer.

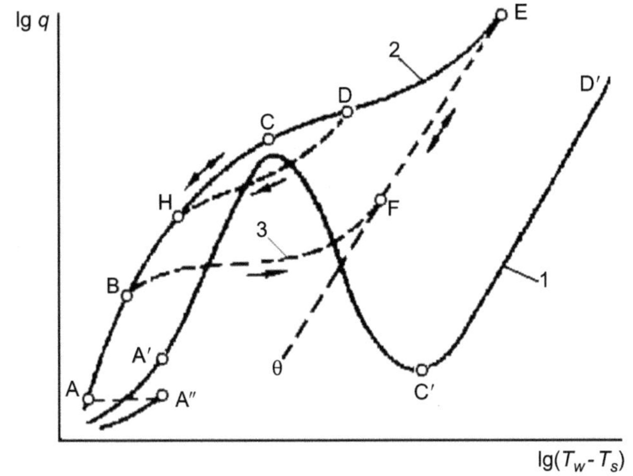

Figure 7.47 Generalized boiling curves by Borishanskii et al. (1973): 1—to a smooth surface; 2—on the surface of a capillary-porous coating.
Data from Kovalev, S.A., Solovev, S.L., 1989. Evaporation and Condensation in Heat Pipes, Nauka, Moscow (in Russian); Kovalev, S.A., Solovev, S.L., Ovodkov, O.A., 1985. Liquid boiling on porous surfaces. In: Proc. of School – Seminar: Heat and Mass Transfer Processes at Two-Phase Changes and at Two- Phase Flows, pp. 26–38 (in Russian); Kovalev, S.A., Solovev, S.L., Ovodkov, O.A., 1987. Liquid boiling on porous surfaces. Heat Transfer Soviet Research 19 (1), 109–120.

The physical model of heat transfer at boiling on the capillary-porous surface is shown in Fig. 7.48.

Modeling is based on a phasic description of motion and heat transfer in a porous layer. We assume a stable stationary regime of evaporation inside the capillary-porous structure, that vapor and liquid flows move along separate channels, that heat is transferred by heat conduction through the skeleton of the porous coating to the liquid meniscus where the main evaporation takes place, that the liquid completely wets the porous material of the coating, and that the contact angle θ equals zero. Results from numerical experiments are presented by Soloviev (1998) for different characteristics of the capillary-porous structure. Capillary-porous surface boiling under forced convection cooling conditions is discussed in Section 7.8.5.

7.6.6 Bulk boiling in tubes

The boiling process is often used in the design of heat exchangers and steam generators in vertical or horizontal pipes. Typical examples are pressurized water reactor (PWR) steam generators and heat exchangers for passive heat removal systems. In many cases, the temperature difference $\Delta T = T_w - T_S$ is small and does not exceed 20–30 K.

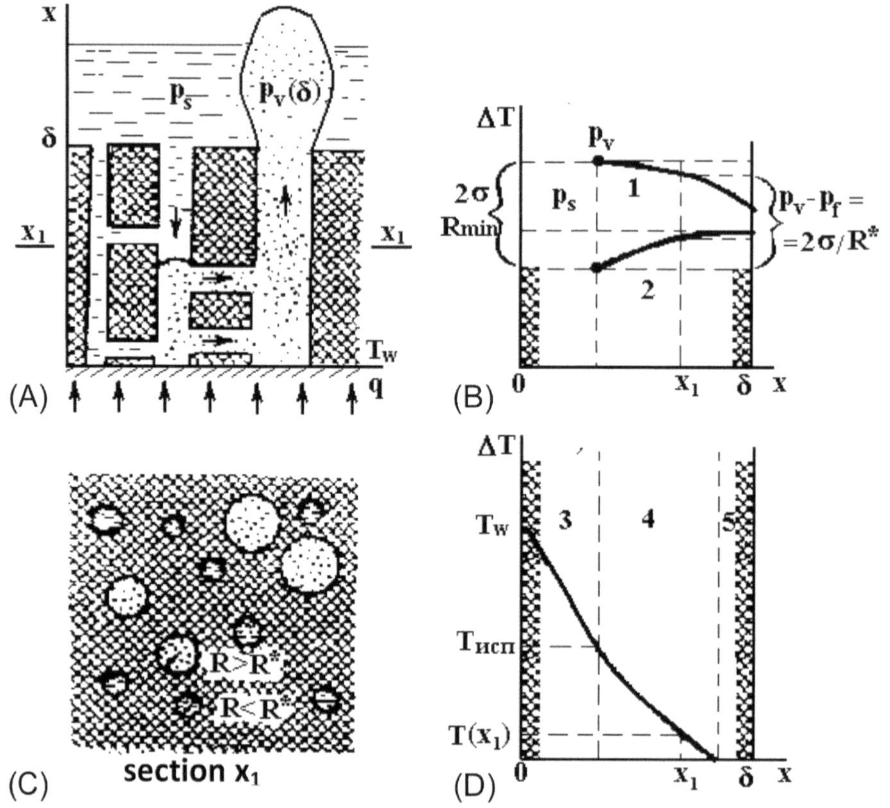

Figure 7.48 Heat transfer model with liquid boiling on the capillary-porous surface (from Soloviev, 1998). (A) phasic motion in the porous structure; (B) pressure changes in the thickness of the porous layer $0 < x < \delta$; (C) in the—horizontal section x_1—x_1 porous structure; (D) temperature variation of the skeleton of the porous coating thickness $0 < x < \delta$. 1—vapor; 2—liquid; 3—dried layer; 4—saturation.

7.6.6.1 Low-pressure condition (0.1–0.3 MPa)

In this pressure range, there are two major heat exchange mechanisms. The first is heat transfer increases due to liquid-phase turbulence and vapor bubble generation (for $q < 50$ kW/m^2); the second is heat exchange deterioration due to the formation of a large vapor bubble conglomerate covering the heated wall surface with $q > 50$ kW/m^2.

Based on a review of publications until 1997 by Parlatan and Rohatgi (1997) for all models of vertical tube heat exchangers, simple correlations are recommended for the pool boiling heat transfer coefficients so that they are easy to use in codes for nuclear system safety analysis and subchannel analysis:

$$q = A\Delta T_S^b, \tag{7.130}$$

Table 7.7 Coefficients in Eq. (7.130).

Mode	ΔT_S (K)	A (W/m^2 K^b)	b
Free convection	$\Delta T_S \leq 3$	365	1.333
Transitional regime from single-phase convection to nucleate boiling	$3 \geq \Delta T_S \leq 6$	775	1.52
Nucleate boiling	$6 \geq \Delta T_S \leq 22$	52.4	3.058
	$22 \geq \Delta T_S \leq 30$	7800	1.44
Transition boiling	$30 \geq \Delta T_S \leq 70$	5.58×10^{12}	−4.555
Film boiling	$70 \geq \Delta T_S \leq 400$	857	0.75

Table 7.8 Coefficients in Eq. (7.131).

Pipe arrangement	C	m	n	k
Horizontal	0.015	4.508	0.084	−1.318
Vertical	0.024	3.862	0.672	−1.658

where q is in W/m^2; $\Delta T = T_w - T_S$ in K. Modes of boiling and wall superheating with respect to the saturation temperature ΔT, A, and b are provided in Table 7.7.

Nucleate boiling heat transfer under atmospheric pressure conditions (0.1 MPa) attracts attention in passive heat removal systems design. A study by Chun and Kang (1998) shows a heat transfer coefficient for vertical and horizontal tubes that depends on surface roughness Δ:

$$\alpha = C \cdot \Delta T_S^m \cdot \Delta^n \cdot d^k, \tag{7.131}$$

where α is in kW/m^2 K; ΔT_S in K; Δ in nm (10^{-9} m); and d is the tube diameter in mm (Table 7.8).

Eq. (7.131) is applicable for the range of $d = 9.7–25.4$ mm; roughness $\Delta = 15.1–60.9$ nm; p at 0.1 MPa.

Surface roughness is substantially more effective for vertical pipes than for horizontal pipes, although in both cases an increase in roughness causes an increase in heat transfer coefficients. Instead of Eq. (7.131), you could use a less precise but simpler empirical relationship:

$$\alpha = 0.266 q^{0.812} \pm 25\%, \tag{7.132}$$

where α is in kW/m^2 K and q in kW/m^2.

7.6.6.2 High-pressure region (p > 0.3 MPa)

Under high-pressure conditions, heat transfer coefficients are quite high (2.5×10^4 to 2.5×10^5 W/m^2 K). Many authors have recommended several formulas for pool boiling. For instance, see IAEA TECDOC 1203 (2001). The formula by Rassokhin (1980) is perhaps one of the most preferred and provides some leeway in calculation:

$$\alpha = 5.51 q^{2/3} p^{0.25} \text{ for } |0.1 < p < 8 \text{ MPa}; \tag{7.133}$$

$$\alpha = 0.58 q^{2/3} p^{1.33} \text{ for } |8 < p < 20 \text{ MPa}; \tag{7.134}$$

where q is in W/m^2 and p in MPa.

7.6.6.3 Effect of oxide films and deposits on the outer surface of the pipe

Oxide films and deposits formed on tube surfaces reduce the heat transfer coefficient. Determination of the composition and structure of deposits, taking into account the impact of these factors, is a complicated task. Oxide films are usually composed of hematite (Fe_2O_3) or magnetite (Fe_3O_4). The formation of oxide films on the surface of the pipe is subject to the law (Glebov et al., 1983; Man'kina, 1977)

$$M = A\, exp(-B / T)\tau^c, \tag{7.135}$$

where M is in g/m^2, T in K, and τ in h. A, B, and C in Eq. (7.135) are found in Table 7.9.

The thermal conductivity of hematite (Fe_2O_3) and magnetite (Fe_3O_4) can be described by the following:

$$\lambda(Fe_2O_3) = 13.215 - 0.375 \cdot 10^{-1} T + 0.379 \cdot 10^{-4} T^2; \tag{7.136}$$

$$\lambda(Fe_3O_4) = 4.133 - 0.852 \times 10^{-2} T + 0.757 \cdot 10^{-5} T^2; \tag{7.137}$$

where λ is in W/(m K) and T in °C. Deposits usually consist of two layers: (1) firmly adjoining the surface; and (2) outer and more porous. They differ in their composition and porosity.

Table 7.9 Coefficients for Eq. (7.135).

pH	Type of steel	*A*	*B*	*C*
7.8	High-chromium[a]	3.91×10^6	8590	0.25
	Austenite[b]	1600	5720	0.38
9–10.5	High-chromium[a]	6.57×10^5	7830	0.26
	Austenite[b]	152	6680	0.41

[a]C—0.12; Cr—1; Mo; V.
[b]C—0.12; Cr—18; Ni—12; Ti.

Layer	(1)	(2)
Density (g/cm^3)	4.4–4.7	1–4
Chemical composition at pH 9	Fe_3O_4	Fe_3O_4
Porosity (%)	10	20–75
Chemical composition at pH 9	Fe_2O_3	Fe_3O_4
Porosity (%)	4	50

More details about the influence of deposits on the heat transfer coefficient for steam generators are found in Trunov et al. (2001) for the PGV-1000 of a VVER.

7.7 Film pool boiling

7.7.1 Stationary film boiling

In the film boiling mode, the liquid is separated by vapor film from the heating surface. The intensity of heat transfer is determined by convection, while in a high-temperature environment, by $\Delta T = T_w - T_S$ and radiation heat transfer through the vapor film. The mechanisms of vapor detachment out of the cavity on the wall surface are related to the thickness and the vapor film flow mode (laminar wave, turbulent, mixed). Therefore, the configuration and orientation of the heating surface and the physical properties of liquid and vapor have a certain importance in the gravity field. Theoretical analysis of film boiling is simplified compared with the analysis of nucleate boiling because the process takes place only at the interface of the two phases and is in quite an ordered pattern. However, analysis of the instability of the vapor film boundaries and the presence of irregular waves is suitable only for qualitative assessment, which helps us understand the physical picture of the phenomenon but cannot provide quantitative results.

In this boiling regime, total heat flux can be represented by the sum of convection (q_c) and radiation (radiant) (q_r) heat transfer: $q = q_c + q_r$. If the heat transfer coefficient refers to the temperature difference ($T_w - T_S$), then the total heat transfer coefficient of film boiling is determined to a first approximation by the sum of the convection (α_c) and radiation (α_r) components:

$$\alpha = \alpha_c + \alpha_r, \tag{7.138}$$

where $\alpha_r = \varepsilon_l \cdot 5.66 \cdot 10^{-8}(T_w^4 - T_S^4)/(T_w - T_S)$, W/m^{2}K; $\varepsilon_l = 1/(1/\varepsilon_w + 1/\varepsilon_S - 1)$ is the reduced emissivity; $\varepsilon_w \approx 0.75$ is the surface emissivity, and $\varepsilon_S = 1$ is the emissivity of the water surface.

The convective heat transfer coefficient α_c of the film boiling of saturated water and other liquids is from the empirical relation by Klimenko (1984):

$$Nu = CAr^m Pr^n f(K) \pm 25\%, \tag{7.139}$$

where $Nu = \alpha\delta/\lambda_g$; $Ar = g\delta^3(\rho_f/\rho_g - 1)/(\nu_g)^2$; $Pr = \nu_g/a_g$; $K = \Delta h_{fg}/c_{pg}\Delta T$; and $b = \sqrt{\sigma/g(\rho_f - \rho_g)}$.

σ, Δh_{fg}, and ρ_f are evaluated at the saturation temperature T_S, other material properties are determined, and an average temperature of the vapor is given by $\overline{T} = T_S + (T_w + T_S)/2$. C, m, n, and the form of the function f (K) in the equation are given in Table 7.10.

Eq. (7.139) gives an average value of the heat transfer coefficient across the flat wall surface. On the curved surfaces, local values of α may differ by 20%–30%.

7.7.2 Rapid cooling of very hot bodies (hardening modes)

When a heated solid metal body is immersed in cold water, there is a series of rapid-mode cooling called quenching that changes the characteristics of metals: hardening steels to acquire special steel structure, hardness, and other characteristics; production of aluminum; hardening aluminum alloys, and so on. Knowledge of these processes is necessary for the evaluation of residual stress and fatigue.

For nuclear reactor safety, investigation of similar phenomena is important for the analysis and calculation of reactor core cooling data in the event of accidents.

In the metallurgical process, errors with heat transfer calculations may be corrected quite easily by conducting repeated "integrated" experiments and applying appropriate corrections. However, in the nuclear power reactor, it is difficult to do so or in many cases impossible. Therefore, the importance of the calculation of such events cannot be overestimated for the nuclear power industry. Rapid cooling of hot objects in the water involves several modes of heat transfer that replace one another, i.e., modes of film boiling, nucleate boiling, convection (forced and natural), and thermal conduction. As shown in Section 7.1.4, a boiling curve has three characteristic points: (1) the MFB temperature or Leidenfrost temperature T_L (T_{MFB} or T_{LFP}); (2) the surface temperature corresponding to the CHF (q_{cr}) T_c; and (3) the onset boiling point T_b for water under normal conditions (0.1 MPa, 20°C). Consider again at this point the boiling curve (Fig. 7.49) in a large volume.

A hot rod submerged in cold water will not be wetted by water due to blanketing by the low conductivity vapor around the rod surface. Heat is transferred to the water from the hot surface by radiation (Q_R) and vapor-phase thermal conduction (Q_λ). Heat flow (Q_λ) through a continuous vapor film is determined by the temperature difference ($T_w - T_f$). The contribution of radiation decreases rapidly as $Q_R \sim T_w^4$:

1. If the water is heated at the contact with the hot body to T_S ($T_f \approx$ 10–20°C), the vapor bubbles generated at the interface with the liquid surface quickly condense, providing a stable thin film of vapor (no bubbles!) (Fig. 7.50a).
2. For water temperature at about $T_f \approx$ 35–40°C; at the bottom of the rod, there is the film boiling; near the liquid level, an intensive transitional or nucleate boiling, driving to the rapid formation of large quantities of steam (explosive boiling) and then a sharp cooling of the rod (Fig. 7.50b).

Table 7.10 Data necessary for Eq. (7.139).

The shape and orientation of the surface	Size δ	Range of applicability		C	m	n	$f(K)$
Sphere	Diameter D	$D/b \geq 6$	$3 \times 10^4 < Ar_D < 2 \times 10^7$	0.70	1/4	1/3	1.0 for $K \leq 1.4$; 0.92 for $K^{1/4} > 1.4$
			$Ar_D > 2 \times 10^7$	0.175	1/3	1/3	1.0 for $K \leq 1.6$; 0.85 for $K^{1/3} > 1.6$
Horizontal cylinder	Diameter D	$D/b \geq 2$	$Ar_D < 2 \times 10^7$	0.64	1/4	1/3	1.0 for $K \leq 1.2$; 0.96 for $K^{1/4} > 1.2$
			$Ar_D > 2 \times 10^7$	0.16	1/3	1/3	1.0 for $K \leq 1.4$; 0.89 for $K^{1/3} > 1.4$
Horizontal flat surface	$l = 2\pi b$	$D^*/b \geq 31$	$Ar_D < 1 \times 10^8$	0.19	1/3	1/3	1.0 for $K \leq 1.4$; 0.89 for $K^{1/3} > 1.4$
			$Ar_D > 1 \times 10^8$	0.0086	1/2	1/3	1.0 for $K \leq 2.0$; 0.71 for $K^{1/2} > 2.0$
Vertical surface	Length l	$Ar_L > 2 \times 10^7$ For cylinders $D/b \geq 1.8$		0.20	1/3	1/3	1.0

D—the characteristic size of the flat surface (disc diameter, the width of the tape, etc.).

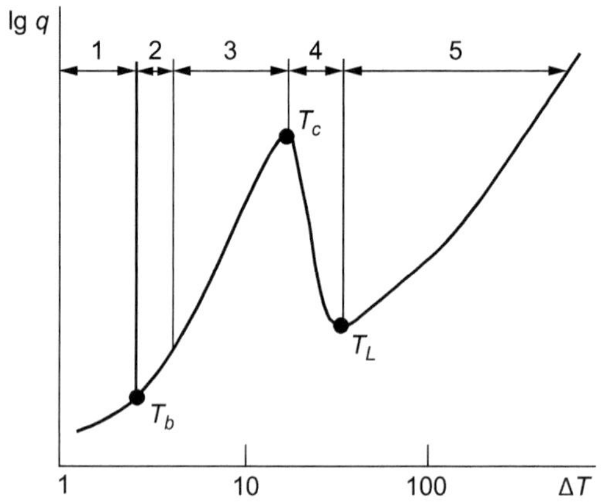

Figure 7.49 Boiling curve in a large volume of water (0.1 MPa). Regions of various modes of heat transfer: 1—free convection; 2—undeveloped boiling; 3—developed boiling; 4—transition boiling (from bubbly to film); 5—film boiling; T_b—initial boiling point; T_c—wall surface temperature corresponding to the critical heat flux q_{cr}; T_L—Leidenfrost temperature (minimum film boiling temperature T_{MFB}).

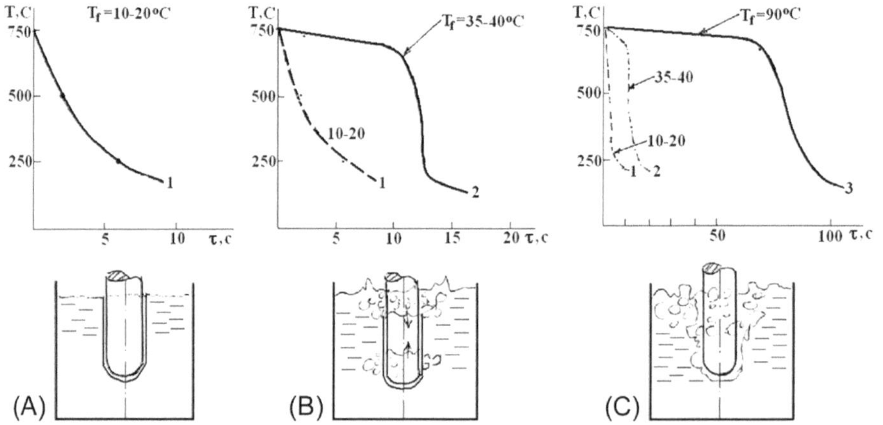

Figure 7.50 Cooling mode of a hot rod (0.1 MPa). (A) Thin film of vapor. Quiet and rapid cooling. (B) Wetting front movement; explosive boiling. (C) Film mode with blanketing and bubbly mode. Slow cooling.

3. For water temperatures close to T_S ($T_f \approx$ 80–90°C), the film boiling regime is realized and covered with thick vapor film. As the cooling of the rod proceeds to transition, the bubbly boiling regime prevails (Fig. 7.50c).
4. The transition boiling regime is realized for surface temperatures from T_S to T_c (temperature corresponding to the CHF).
5. For $T < T_c$, the bubbly mode becomes dominant with intensive heat removal and is then replaced by the natural or forced convection mode of cooling.

Below is a flow calculation of hot body cooling immersed in water. This referenced scheme is in accordance with the modified code RELAP5/MOD2 and CFX4.4 (Yeoh and Tu, 2002) in the quasi-approximation.

The total amount of heat (Q) is assumed to consist of four components:

Q_R—heat dissipation by radiation from the body to the water;
Q_λ—heat removal by conduction to the vapor;
Q_e—heat withdrawn through liquid evaporation; and
Q_c—heat allocated to turbulent convection.

$$Q = Q_R + Q_\lambda + Q_e + Q_c. \tag{7.140}$$

The individual components are calculated as follows.

(1) The amount of heat transferred by radiation:

$$Q_R = \varepsilon_{1,2}\sigma_0\left(T_w^4 - T_f^4\right)F, \tag{7.141}$$

where T_w and T_f are temperatures of the body and fluid; $\sigma_0 = 5.668 \times 10^{-8}$ W/(m^2 K^4); and $\varepsilon_{1,2}$ is the emissivity of the surface of the body system (ε_1) and of the water (ε_2). F—body surface area (m^2).

(2) The amount of heat transferred by conduction through the vapor layer to the water:

$$Q_\lambda = \left[\frac{2}{\sqrt{\pi}}\sqrt{\lambda_f \rho_f c_{pf}}\,\sqrt{f}\right](T_w - T_f)F_1, \tag{7.142}$$

where f is the frequency of the bubble formation (Kocamustafaogullari and Ishii, 1995) given by

$$f = \frac{1.18}{d}\left[\frac{\sigma g \Delta\rho}{(\rho_f)^2}\right]^{0.25}; \tag{7.143}$$

where d is the detachment diameter of the vapor bubble,

$$d = \frac{2.42 \times 10^{-5} p^{0.709} a}{\sqrt{b\Phi}}; \tag{7.144}$$

where p—pressure; $b = \frac{T_S - T_e}{2(1 - \rho_g/\rho_f)}$ is the coefficient taking into account the effect of pressure and subcooling; $\Phi = \begin{cases} (w_e/0.61)^{0.47} & \text{for } w_e \geq 0.61 \text{ m/s} \\ 1.0 & \text{for } w_e < 0.61 \text{ m/s} \end{cases}$ is the coefficient reflecting the effects of the detachment speed and diameter d; w_e—velocity of the water; $F_1 = n(\pi d^4/4)$ is the fraction of the surface projected by the vapor bubbles; n—number of vapor bubbles on the unit surface: $n = \frac{1}{d^2}\left[\frac{2\sigma T_S}{(T_W - T_S)\rho_g \Delta h_{fg}}\right]^{-4.4} f(\rho^*)$,

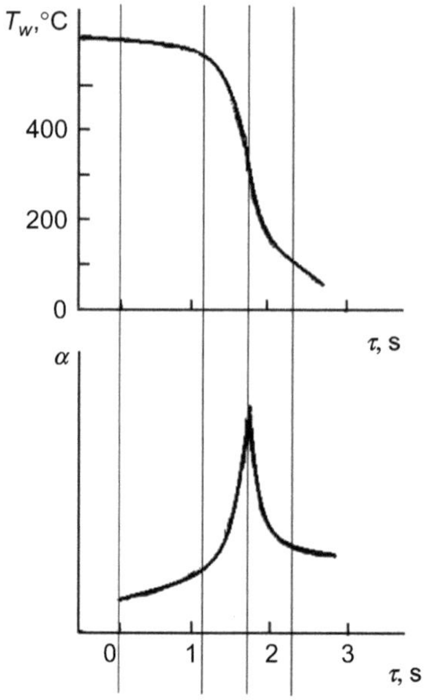

Figure 7.51 Dependence of $T_w(\tau)$ and $\alpha(\tau)$: 1—film boiling mode; 2—bubbly mode; 3—convective cooling mode.

where $\rho^* = (\rho_f - \rho_g)/\rho_g = (\rho_f/\rho_g) - 1$; Δh_{fg} is the heat of vaporization; $f(\rho^*) = 2.157 \times 10^{-7}(\rho^*)^{-3.2}(1 + 0.0049\,\rho^*)^{4.13}$; and $a = \frac{[q - \alpha_f(T_w - T_e)]^{1/2}\lambda}{2c^{1/3}\Delta h_{fg}\sqrt{\pi\lambda_f/(\rho_f c_{pf})}\,\rho_g}\sqrt{\frac{\lambda_W \rho_W c_{pW}}{\lambda_f \rho_f c_{pf}}}$; $c = \frac{\Delta h_{fg}\mu_f\left[c_p/0.013\ \Delta h_{fg}\cdot Pr^{1.7}\right]^3}{(\sigma/\Delta\rho\ g)^{0.}}$.

(3) The amount of heat used in evaporation:

$$Q_e = nf\left(\frac{\pi}{6}d^3\right)\Delta h_{fg}\rho_g. \tag{7.145}$$

(4) The amount of heat carried by turbulent flow natural convection:

$$Q_c = St\cdot\rho_f c_{pf} w_e (T_w - T_f)(1 - F_1). \tag{7.146}$$

From experiments, $T_w = f(\tau)$—the transient surface temperature measurement enables calculation of the heat transfer coefficient $\alpha = f(\tau)$ due to $\alpha \sim \frac{\partial(T_w - T_f)}{\partial\tau}$. Qualitatively, a variation of the heat transfer coefficient is shown in Fig. 7.51 (Gilles et al., 2002).

7.8 Convective boiling heat transfer in channels

7.8.1 Boiling modes in flow channels

Various flow regimes and modes of heat transfer in a vertical heated tube are shown in Figs. 7.52 and 7.53. Fig. 7.52 shows the flow conditions and heat transfer modes in a

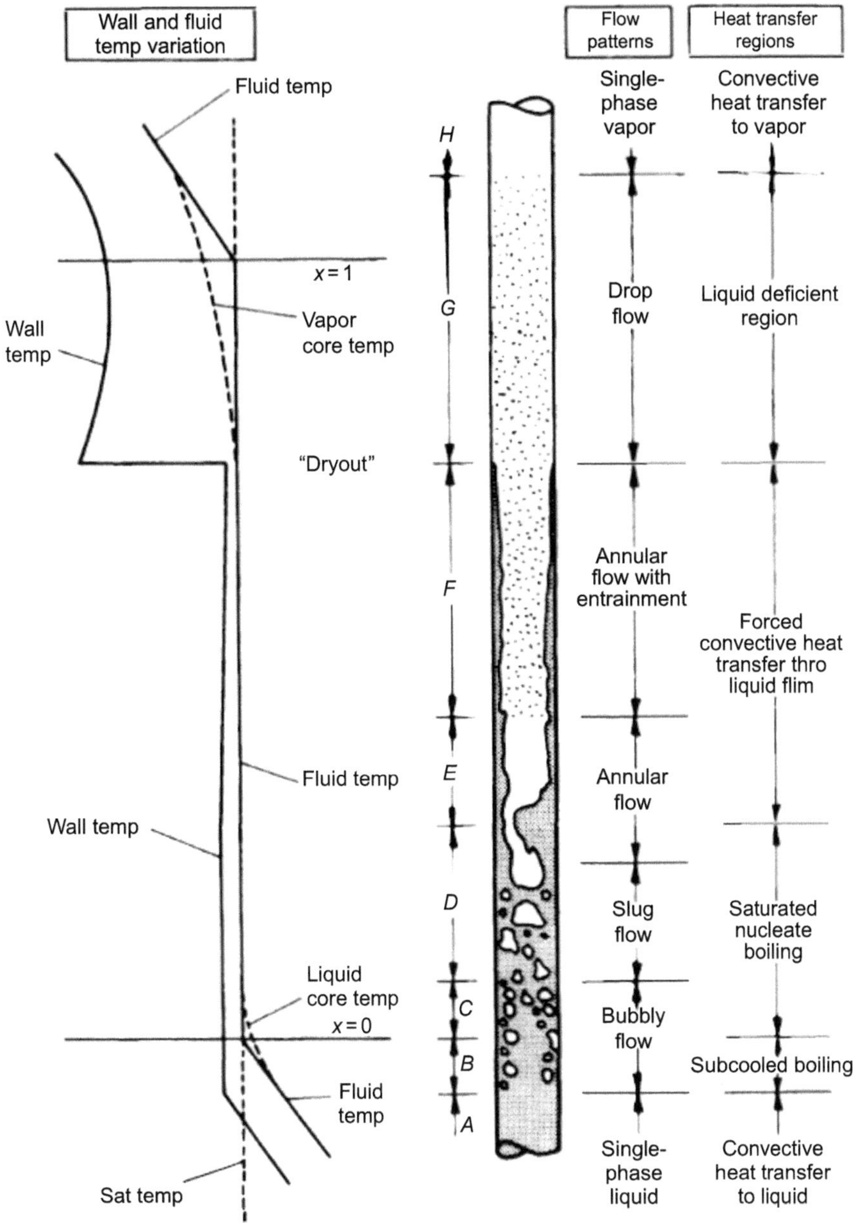

Figure 7.52 Flow patterns and modes of heat exchange ($q = \text{const}$) (Tong and Tang, 1997). Regions of heat transfer in convective boiling in a heated channel with constant heat flux; the regions A–H are further described in Fig. 7.54.
From Ahmada, M. et al., 2013. Phenomenological modeling of critical heat flux: the GRAMP code and its validation. Nuclear Engineering and Design 254, 280–290.

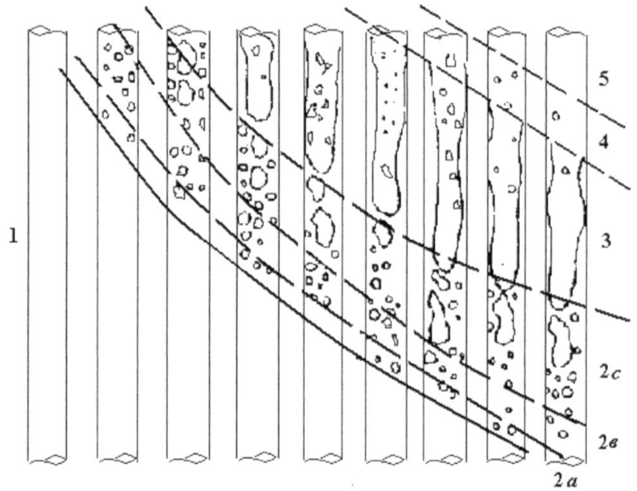

Figure 7.53 Different flow patterns of two-phase flow in a long pipe (constant: pressure, mass flow rate, inlet enthalpy) by Hewitt and Hall Taylor (1972): 1—single-phase liquid flow; 2—bubbly, slug; 3—annular-dispersed; 4—mist; 5—single-phase vapor flow.

uniformly heated tube ($q = \text{const}$). The figure has been cited in many reports and books on two-phase flow and heat transfer (e.g., Wallis, 1972; Mandhane et al., 1974; Butterworth and Hewitt, 1977) and was originally adopted from Tong (1965). Fig. 7.52 is from the second edition of Tong and Tang (1997) and is self-explanatory, as it contains all the necessary information.

Fig. 7.53 shows changes in the flow patterns of two-phase flow in a long pipe with increasing heat flux from left to right, with the fluid inlet temperature, mass flow rate, and system pressure held constant. In the single-phase flow region (region 1), while the liquid on the wall surface may be heated to the saturation temperature, the wall temperature is still below the value required for vaporization in the presence of convective heat transfer. Just downstream of region 1, the fluid at the wall becomes superheated such that evaporation occurs in the microcavity centers, although bulk flow is still subcooled, i.e., below the saturation temperature. This is an area of nucleate boiling of the subcooled liquid (2a). The difference between the saturation temperature and the mass-average temperature of the liquid is called subcooling $T_S - \overline{T}_f = \Delta T_{\text{sub}}$. The upper boundary of region 2a in Fig. 7.53 corresponds to the vapor quality $x = 0$. This portion of the liquid is heated to the saturation temperature so that the average liquid-vapor flow enthalpy is equal to that of saturated liquid ($h_{f,sat}$). In the nucleate boiling of the saturated liquid region (2b) as well as in the subcooled boiling region, equilibrium steam quality (or relative enthalpy) at a distance z is determined by the heat balance $x(z) = \left[h(z) - h_{f,sat}\right]/\Delta h_{fg}$, where $h(z)$ is mixture enthalpy and Δh_{fg} is latent heat. With this definition, it may have a negative or positive value.

With increasing steam quality, plug or slug flow regime prevails (region 2c). The upper boundary of this region is in transition to the dispersed-annular flow regime (region 3) where there is a change in the mechanism of heat transfer. There, the process of boiling is mostly replaced by evaporation. While boiling can be observed within a thick liquid film, bubble generation in a thin liquid film is usually suppressed, and the process of heat transfer cannot be referred to as boiling. The heat is transferred by convection from the wall to the interface between the liquid and the vapor core where evaporation takes place. The upper boundary of region 3 in Fig. 7.53 corresponds to the complete evaporation of the liquid film and there dominates a dispersed flow mode with gradual evaporation of droplets in the superheated steam (region 4), which shifts into the single-phase vapor flow convection mode—region 5. At the location of film dryout, the heat transfer coefficient drops sharply, whereas with high heat flux, the wall surface temperature may reach unacceptable values (burnout, crisis). This is the CHF (q_{cr}). The line between areas 3 and 4 corresponds to CHF associated with the dryout of the liquid film.

When $q > q_{cr}$, the mechanism of heat transfer varies depending on the boiling regime, i.e., a departure from nucleate boiling (DNB), the formation of a conglomerate of bubbles of steam shielding the surface of the bulk liquid from the heated wall or the boiling transition, or liquid film dryout. Film boiling under the forced convection mode is similar to that observed in pool boiling. The heated surface is covered with a vapor film of low heat conduction through which the heat must be transferred.

The dependency of two-phase forced convection heat transfer regimes on quality (vertical axis) and heat flux (horizontal axis) is shown in Fig. 7.54 (taken from Collier and Thome, 1980).

It should be borne in mind that in Fig. 7.54, the equilibrium quality x only qualitatively reflects the border areas. Therefore, when $x > 1$ (area 6), a thermodynamically nonequilibrium vapor flow is possible in the presence of liquid droplets. Then the boundary between regions 9 and 6 moves to the right.

Since the void fraction (φ) in the flow determines the flow and heat transfer modes, it is advisable to consider the actual steam quality that accounts for the nature of the change in φ along the channel (for example, $q = \text{const}$) instead of the equilibrium steam quality x_e, as depicted in Fig. 7.55.

First, vapor bubbles appear at location "ib." Moving through, or migrating away from the thin boundary layer near the wall surface, the vapor bubbles condense in a bulk of highly subcooled liquid. Here, the void fraction is low, because the growth of the bubble layer is limited by the high level of subcooling. As a result, the subcooling decreases in the flow core and gradually attains the conditions under which the rate of vapor generation exceeds that of vapor condensation in the stream. From this point the void fraction begins to increase rapidly, which is considered the beginning of the vapor generation—nvg (net vapor generation).

The equilibrium quality can be zero ($x_e = 0$) in the nucleate boiling (beginning) region even with significantly high values of the void fraction (φ ›› 0). Thus, this mode is under thermodynamically nonequilibrium conditions, i.e., where the flow contains a significant amount of vapor (by volume) in the subcooled liquid as a whole. A simple calculation assumes that the undeveloped surface boiling starts in the section in which the surface temperature is saturation temperature T_S. Then a magnitude of the relative

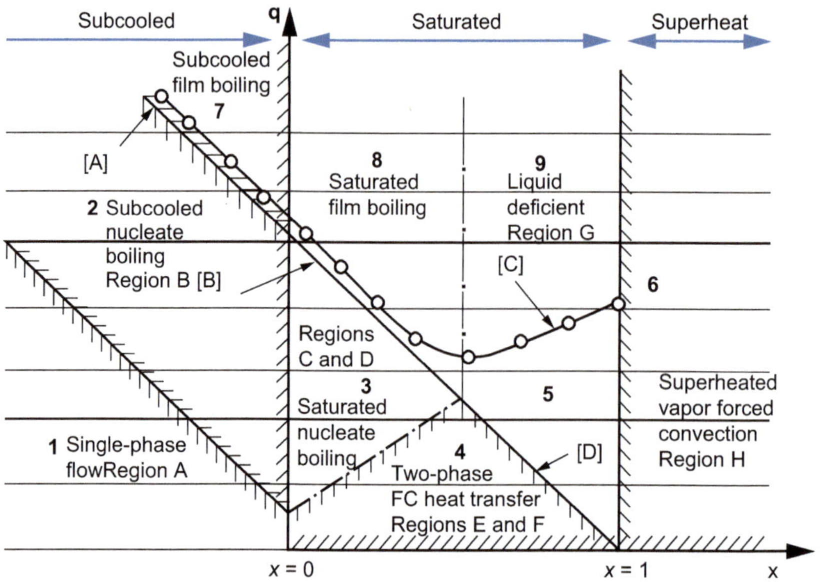

Figure 7.54 Heat transfer regions in forced convection; regions A—H in Fig. 7.52. 1—Single-phase fluid flow; 2—subcooled nucleate boiling ($T_f < T_s$); 3—saturated nucleate boiling $T_f = T_s$; 4—two-phase forced convection heat transfer; 5—liquid deficient region (degraded heat transfer); 6—superheated vapor forced convection; 7—subcooled film boiling ($T_f < T_s$); 8—saturated film boiling $T_f = T_s$; 9—liquid deficient and heat transfer crisis; [A]—subcooled CHF by DNB; [B]—saturated CHF by DNB ($T_f = T_s$); [C]—typical burnout locus; [D]—CHF by film dryout.

enthalpy $x_{ib} = \Delta h / \Delta h_{fg}$ (i.e., quality) can be evaluated by the convective heat transfer coefficients α_c:

- For longitudinal flow inside and outside a tube or pipe, $x_{ib} = \Delta h / \Delta h_{fg} = -qc_p / \{\alpha_c \Delta h_{fg}\}$ since $\Delta h = -q\{c_p / \alpha_c\}$, where q is the heat flux, c_p is the heat capacity, α_c is calculated on the basis of the formulas for the single-phase convective heat transfer, Δh_{fg} the latent heat; and
- For transverse flow across tube bundles, $x_{ib} = -6.17Re^{0.18} qc_p / \{\alpha_c \Delta h_{fg}\}$, where Reynolds number (Re) is calculated from the velocity at the narrow gap; α_c is the convective heat transfer coefficient for cross-flows in the bundle.

The recently proposed method of calculating heat transfer in boiling channels consists of the following three modes (Kandlikar, 1990, 1991, 1998):

- Partial (or undeveloped) boiling at qualities $x_{ib} < x < x_{dnb}$ (region 2 in Fig. 7.55), where x_{ib} corresponds to the quality at the beginning of nucleate boiling and x_{dnb} to the developed nucleate boiling;
- Fully developed boiling at $x > x_{dnb}$ (region 3 in Fig. 7.55);
- The mode of a high steam quality for $x > x_{nvg}$, starting from the point of beginning net vapor generation (x_{nvg}), where

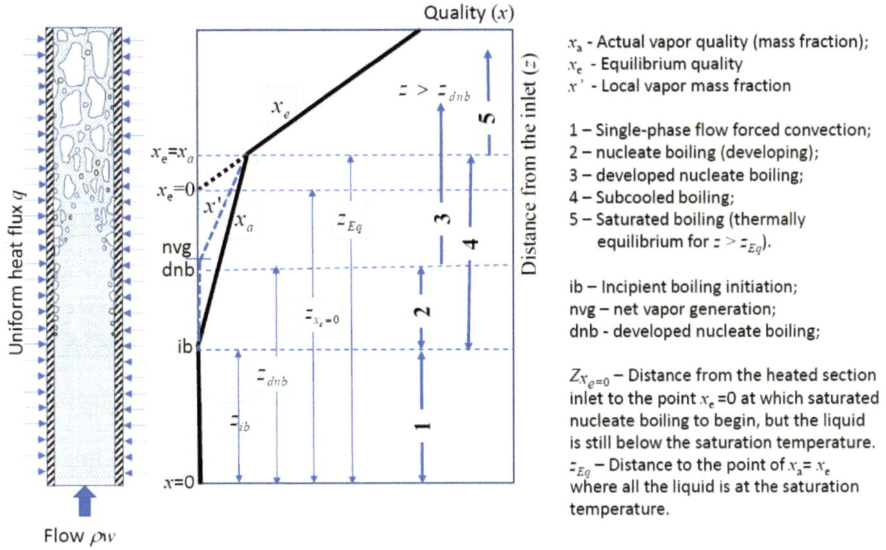

Figure 7.55 Vapor quality, flow conditions and heat transfer. x_a—actual vapor quality; x_e—equilibrium quality; 1—single-phase flow forced convection; 2—nucleate boiling (developing); 3—developed nucleate boiling; 4—subcooled boiling; 5—saturated boiling (thermally equilibrium for $z > z_{EQ}$). ib—incipient boiling initiation; nvg—net vapor generation; dnb—developed nucleate boiling; $z_{x_e=0}$—distance from the heated section inlet to the point $x_e = 0$ at which saturated nucleate boiling to begin, but the liquid is still below the saturation temperature. z_{Eq}—distance to the point of $x_a = x_e$ where all the liquid is at the saturation temperature.
Modified based on Collier, J.G., 1981. Convective Boiling and Condensation. McGraw-Hill, New York.

$$x_{nvg} = -0.0022\left(qd/\Delta h_{fg}\rho_f\lambda_f\right)Bo\ Re\ Pr_f \quad \textit{for } Re\ Pr_f < 7\times 10^4;$$

$$x_{nvg} = -154Bo \quad \textit{for } RePr_f > 7\times 10^4.$$

where $Bo = q/\left(\Delta h_{fg}w\right)$; d is the pipe diameter; and Re and Pr_f are for the liquid phase. Points "dnb (developed nucleate boiling)" and "nvg (net vapor generation)" are sufficiently close to each other. While these techniques are not yet widely used, logic and additional studies may allow us to abandon the simplified representations.

7.8.2 Onset of subcooled nucleate boiling

Flow regimes and temperature changes along the inner surface of a circular pipe for a uniform heat flux distribution are shown in Fig. 7.52. The mean liquid-phase temperature varies linearly along the heated length in the convective region away from the inlet according to the heat balance. In addition, the wall temperature along the inner surface of the pipe varies linearly: $T_w(z) = \overline{T}_f + q/\alpha_c$, where α_c is the single-phase flow heat transfer coefficient. At a certain wall superheating of $T_w(z) - T_S = \Delta T_{ob}$, water begins to boil on the surface. The subscript *ob* designates the onset of nucleate boiling. The magnitude of the wall superheating could be found by the following empirical relation in Petukhov and Shilov (1987):

$$T_w - T_S = 22.65q^{0.5}\exp(-p/8.7), \tag{7.147}$$

where q is in MW/m^2 and p in MPa. The range of application of formula (7.147) is $d = 3.6-5.7$ mm; $p = 0.7-17$ MPa; $\overline{T}_f = 115-340°C$; $G = 11-10^4$ kg/(m^2 s); and $q < 12$ MW/m^2.

In the case of flows in narrow annular channels, the onset of surface boiling occurs with wall-superheat, determined by the formula

$$T_w - T_S = 20.7qp^{0.43}d_e^{0.23}/w_0. \tag{7.148}$$

where d_e is the equivalent hydraulic diameter (mm); w_0 is the flow velocity (m/s); q is the heat flux (MW/m^2); and p is the pressure (MPa). The onset of surface boiling is related to the flow enthalpy at the onset of boiling h_{ob}, which is determined from the equation (Tarasova and Orlov, 1964)

$$h_f - h_{ob} = 0.3q^{1.1}d_e^{0.2}G^{-0.9}\left(\rho_g/\rho_f\right)^{0.3} \pm 15\%, \tag{7.149}$$

where h is enthalpy (kJ/kg); q is heat flux (MW/m^2); d_e is the equivalent hydraulic diameter of the channel (m); and G is the mass flow rate (kg/m^2/s). The formula is applicable in the range $p = 4.9-19.6$ MPa; $G = 1300-11{,}000$ kg/(m^2 s); $q = 0.43-1.3$ MW/m^2; and $d_e = 2.9-6.3$ mm.

The enthalpy at the beginning of surface boiling can be calculated through vapor quality ($x_{ob} < 0$):

$$h_f - h_{ob} = -\Delta h_{fg}x_{ob},$$

where $x_{ob} = -0.49Re_1^{0.7}Re_2^{-0.3}\pi^{0.15}$; $Re_1 = qd_\sigma/\Delta h_{fg}\mu_f$; $Re_2 = Gd_\sigma/\mu_f$ with $d_\sigma \equiv \sqrt{\sigma/g(\rho_f - \rho_g)}$; and $\pi = p/p_{crt}$, where p_{crt} is the critical pressure. In general, subcooling at boiling initiation can be found approximately by the equation $\Delta T_S = T_S - \overline{T}_f = q(1/\alpha_c - 1/\alpha_b)$, where α_c and α_b are the heat transfer coefficients in single-phase flow convection and developed boiling, respectively. In the presence of deposits on the surface having thermal resistance R (m^2 K/W), the value of subcooling at the start of boiling will equal $\Delta T_S = q(1/\alpha_c - 1/\alpha_b + R)$.

7.8.3 Subcooled boiling, developing region

Heat transfer in the channels under the conditions of undeveloped subcooled boiling of water, corresponding to region 2 in Fig. 7.55, is calculated by the formula

$$\frac{\alpha}{\alpha_0} = \left\{1 + \left[\alpha_0\left(\frac{1}{\alpha_c} - \frac{T_S - \overline{T}_f}{q}\right)\right]^{-3/2}\right\}^{2/3}$$

where α_0 is calculated by the formulae given in Section 7.6 for pool boiling in a large volume; and α_c is the single-phase flow convective heat transfer coefficient.

7.8.3.1 Features of subcooled boiling at low pressures (0.1–1.0 MPa) (Wadekar, 1990)

Subcooled boiling at low pressures is characterized by large specific vapor volumes and, hence, high velocities. In this situation, the saturation temperature decreases toward the downstream region of the channel according to the decrease in pressure; and in the stabilized convective heat transfer section (with $q = \text{const}$), it increases according to the increase in the average bulk liquid temperature.

Fig. 7.56a shows the axial temperature profile for the onset of boiling of flows with the subcooled bulk liquid. In this figure, after the surface temperature reaches the desired superheat (point A), the subcooled liquid water starts boiling and the surface temperature decreases rapidly because of the high heat transfer coefficients. With an

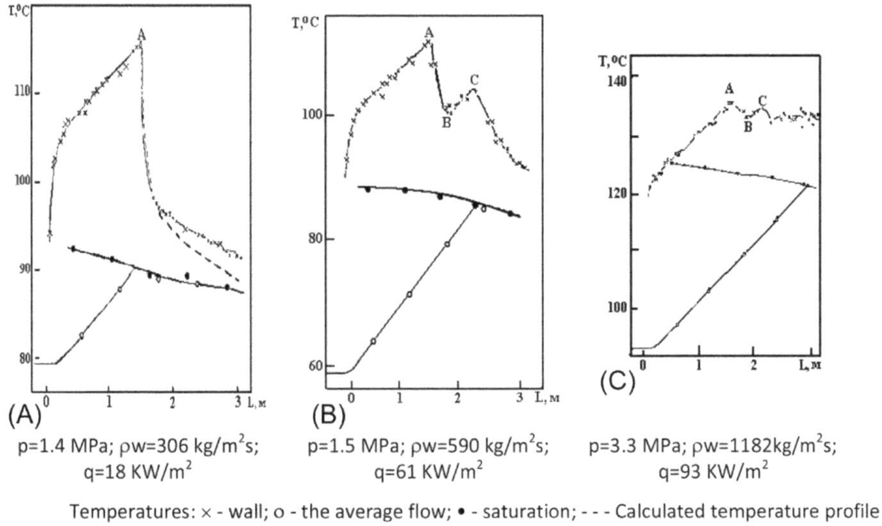

Figure 7.56 Boiling onset with different wall temperature peaks (Wadekar, 1993). Temperatures: ×—wall; o—the average flow; •—saturation; - - - calculated temperature profile.

increase in the mass flux (Fig. 7.56b), the boiling is suppressed and the surface temperature decreases (AB). After the point B the surface temperature T_w is observed to rise (up to Point C) according to the increase in the bulk fluid temperature $\overline{T}_f$. This indicates that while the true vapor quality does not change, the 2 T_W peaks appear. Fig. 7.56c shows a wavy surface temperature profile, which should be corresponding to T_S changes along the length of the channel.

Boiling flow heat transfer in a vertical pipe with subcooling has other features (Dougherty et al., 1990). If there is no boiling, the heat transfer coefficient α_i at the pipe inlet can be found from the correlation

$$Nu = \frac{\alpha_i d}{\lambda_f} = 0.166 Re^{0.66} Pr^{0.33} \left(\mu_f / \mu_w\right)^{0.14}.$$

The heat transfer coefficient for the flow of subcooled ($x = -0.3$–0) fluid is given by

$$\alpha / \alpha_i = 4.13 + 53.89x + 381.38x^2 + 1257x^3 + 1572.7x^4.$$

From this it follows that when $x = 0$, $\alpha = 4.13\alpha_i$; at $x = -0.1$, $\alpha = 1.44\alpha_i$; at $x = -0.2$, $\alpha = 1.04\alpha_i$.

7.8.4 Developed boiling in channels in subcooled to saturated region

The heat transfer coefficient of the forced flow boiling is influenced mainly by two processes—convection and boiling. It is well understood that the former process can suppress the latter. There, a magnitude of the heat transfer coefficient can be similar to that of the convection heat transfer coefficient α_c at low heat flux and high velocity, and the pool boiling heat transfer coefficient α_0 at high heat flux and moderate velocity.

The method proposed by Kutateladze (1979) allows you to take into account the influences of the flow velocity and heat flux without relying on complicated criteria. First, heat transfer rates α_0 at boiling and α_c for turbulent flow convection of the liquid at T_S are compared. That is, depending on the ratio α_0/α_c the following equations are to be used:

$$\alpha_0 / \alpha_c < 0, \alpha = \alpha_c; \tag{7.150}$$

$$0.5 < \alpha_0 / \alpha_c < 3, \alpha / \alpha_c = \sqrt{1 + (C\alpha/\alpha_c)^2}; \tag{7.151}$$

$$\alpha_0 / \alpha_c > 3, \alpha = \alpha_0 \text{ or } \alpha = 0.9\alpha_0. \tag{7.152}$$

where α is the heat transfer coefficient for the considered mode, α_c is the convective heat transfer coefficient calculated according to the formulae. In Eq. (7.151), C is a function of p and, for $p < 15$ MPa, is 0.8–0.9. According to Kutateladze, Eq. (7.152) is applicable to the nucleate boiling of water at T_S and to the subcooled boiling. Eqs. (7.150)–(7.152) are applicable for $p = 2–20$ MPa and are accurate within ±35%. A similar approach is found in Wadekar (1990).

Instead of formula (7.151), Yagov et al. (1998) recommended a formula that gives the best agreement with the experiment for the case of boiling in a large volume of water:

$$\alpha / \alpha_c = \sqrt[3]{(\alpha_0/\alpha_c)^3 + (1 + \Delta T_S/\Delta T)^3}; \tag{7.153}$$

$$\Delta T_S = T_S - \overline{T}_f;\ \Delta T = T_w - \overline{T}_f.$$

When a two-phase mixture flows in a pipe or annular channel, heat transfer coefficients are calculated according to the formula obtained in CKTI[4] by Borishanskii et al. (1973) and Handbook (1986):

$$\alpha = \alpha_1\sqrt{1 + 7 \times 10^{-9}\left(\frac{\Delta h_{fg}\rho_f J_{CM}}{q}\right)^{3/2}\left(\frac{0.7\alpha_0}{\alpha_1}\right)}, \tag{7.154}$$

where $\alpha_1 = \sqrt{\alpha_c^2 + (0.7\alpha_0)^2}$, α_c is the heat transfer coefficient of single-phase flow at velocity w_0 in a pipe or channel; α_0 is the heat transfer coefficient of pool boiling; $J_{CM} = J_{f0} + J_{g0}$ is the mixture volumetric flux where $J_{f0} = G(1 - x)/\rho_f$ and $J_{g0} = Gx/\rho_g$; x is the mass quality. The range of application for water is $p = 0.2–17$ MPa; $J_{CM} = 1–300$ m/s; and $q = 0.08–6$ MW/m^2.

The influence of mixture velocity becomes notable when the following relation is true: $(\Delta h_{fg}\rho_f J_{CM}/q)\cdot(0.7\alpha_0/\alpha_1)^{4/3} > 5 \times 10^4$. In this case, the heat transfer coefficient is in proportion to the mass flux to the 0.8th power, i.e., $\alpha \sim W^{0.8}$. At lower values of this combination, $\alpha \sim q^{0.7}$ and does not depend on J_{CM}, and then $\alpha = \alpha_1$ (see Fig. 7.57).

It would be sufficient to use a simple formula to evaluate boiling heat transfer for different fluids, as proposed by Borishanskii et al. (1973):

$$\alpha / \alpha_c = F; \tag{7.155}$$

$$F = 3.2Y + 1;\ Y = [x/(1 - x)]^{0.75}(p_c/p)^{0.375}M^{0.125},$$

where M is molecular mass, the limit of applicability is $Y = 1–20$, the accuracy is, for $F < 1$, ±35%; for $F = 0$, ±20%; for $F = 40$, ±5%.

[4] Former CKTI, Central Boiler and Turbine Institution in Russia (http://english.ckti.ru/about.html), Currently, I.I. Polzunov Scientific and Development Association on Research and Design of Power Equipment.

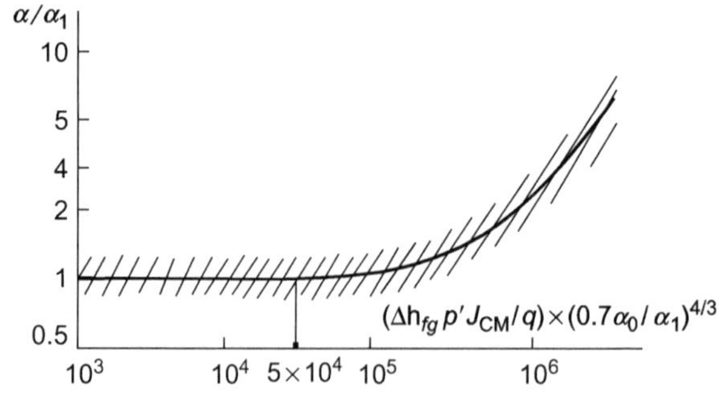

Figure 7.57 Heat transfer in two-phase flow in channels.

7.8.5 Boiling on porous structures under forced convection conditions

7.8.5.1 Subcooled boiling on capillary-porous surfaces

The concept of the process of subcooled boiling subcooled for this case is presented in Soloviev (1998).

Depending on the magnitude of the heat flux and degree of subcooling, there are three modes of heat transfer: (1) microtube mode, (2) bubble roughness mode, and (3) bubbly surface roughness mode (see Fig. 7.58). The calculation of the heat transfer coefficient for mode 2 can be performed by the method proposed by Avdeev (1982, 1985):

$$\alpha = \rho_f w c_p \{ |[5.67 \ln(D/d) + 4.92]|[5.67 \ln(D/d) - 3.58]| \}^{-1} \pm 30\%, \quad (7.156)$$

where D is the channel diameter, and d is the bubble diameter, defined by the formula

$$d = 1.24 d_s \varepsilon^{-1.18} \sqrt{q_l / q_c},$$

where d_s is the pore diameter; ε is the porosity; and q_l and q_c are heat flux associated with evaporation and convection, given by

$$q_l = \int_0^\delta \alpha_g [T_W - T_g(x)] dx; \; q_c = \alpha_c (T_S - T_f)$$

Heat transfer for mode 3 can be determined by the following (Avdeev, 1982, 1985):

$$\alpha = \rho_f w c_p \{ |[5.67 \ln(D/d) + 4.92]|[5.67 \ln(D/d) + 7.08]| \}^{-1} \pm 30\%, \quad (7.157)$$

where d is the detachable bubble diameter (see above).

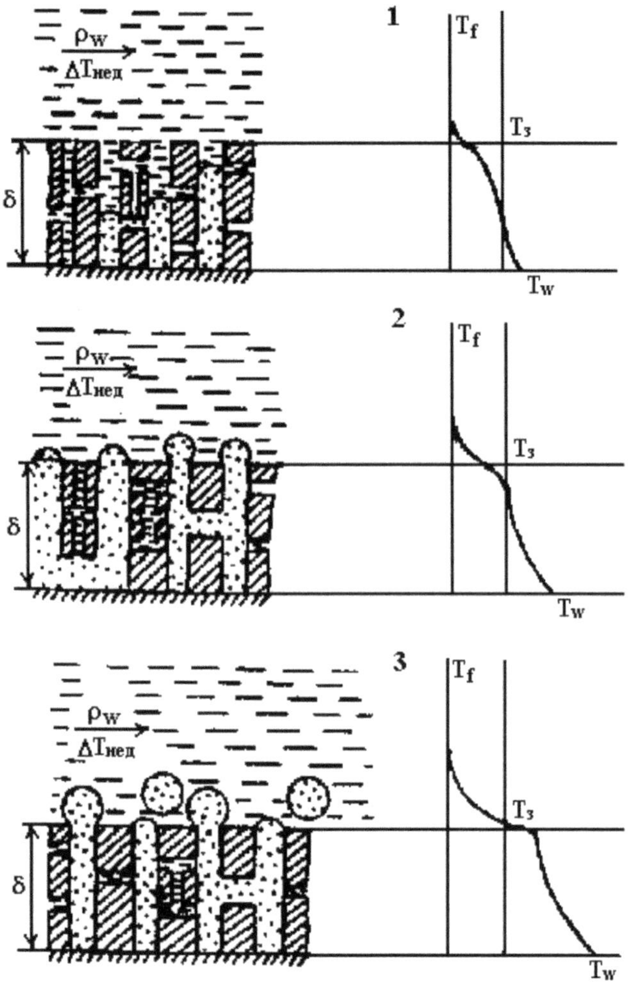

Figure 7.58 Subcooled liquid boiling model at the capillary-porous surfaces. Mode 1—microtube mode, in which heat is transferred from the surface through pores that act as heat pipes (evaporation occurs at the base and condensation at the top); Mode 2—bubble roughness mode, in which at the outer boundary of the porous layer formed stationary steam bubbles on top of which there is a vapor condensation; Mode 3—bubbly surface roughness mode, in which at the outer boundary of the porous layer, the stationary bubbles are formed and condense on the top.

7.8.6 Heat transfer in film boiling and superheated vapor flow in channels

Film boiling heat transfer was extensively investigated during the last half-century. Excellent reviews are found in textbooks by Tong (1965), Tong and Tang (1997), Collier and Thome (1980), Delhaye et al. (1981), Butterworth and Hewitt (1977), a handbook by Hetsroni (1982), and articles by Ganic and Rohsenov (1977), Mayinger and Langer (1978), Tong (1978), Sergeev (1978, 1987, 2007), Groeneveld and Snoek (1986), Groeneveld (1992), Yadigaroglu and Andreani (1989), Sakurai (1990), Andreani and Yadigaroglu (1989, 1994), and in the proceedings of the first International Symposium on Fundamental Aspects of Post-CHF Heat Transfer (1984). A good summary of the literature survey was published in 2001 in the IAEA's Coordinated Research Project on Thermohydraulic Relationships for Advanced Water Cooled Reactors (IAEA, 2001).

In this section, we discuss some two-phase flow heat transfer regions depicted previously in Fig. 7.54 by region 5—liquid deficient region (degraded heat transfer), region 6—superheated vapor forced convection, region 7—subcooled film boiling ($T_f < T_S$), region 8—saturated film boiling $T_f = T_S$, and region 9—liquid deficient and heat transfer crisis; in Fig. 7.53 by region 3—dispersed-annular, region 4—mist and region 5—single-phase superheated vapor; and in Fig. 7.52 by region G (liquid deficient) and region H (superheated vapor).

When boiling two-phase flows in a heated channel, a drastic deterioration of heat transfer (heat transfer crisis) may occur, which appears as an increase in temperature on the heat-transfer surface or a reduction in the heat flux q. The heat transfer mode between the onset location of the crisis and its downstream region where the flow becomes single-phase superheated vapor flow is called film boiling. In this regime, at a constant heat flux, an increase in surface temperatures is observed and offset the crisis in the region of low steam quality with increasing q (see Fig. 7.59).

The film boiling is characterized by an aggravated mode and temperature fluctuations due to the variable surface contact modes of liquid and vapor. This regime is implemented in the steam generator but is to be avoided in the case of nuclear fuel rod bundle cooling. Nevertheless, some experiments show that a short-term operation of fuel rods in such a regime under certain parameters is possible if the excess temperature of the fuel cladding surface and the amplitude of the oscillations do not exceed

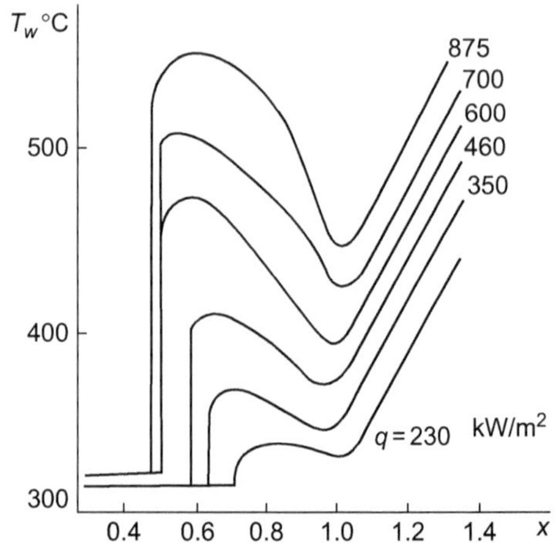

Figure 7.59 Wall temperature dependence on the vapor quality and heat flux at $\rho w = 1400$ kg/(m^2 s); $P = 9.8$ MPa.

permissible limits. From this point of view, perhaps a more accurate calculation of heat transfer is necessary. The temperature jump at a crisis decreases with increasing mass flow rate and pressure and increases with increasing heat flux. The two-phase flow structure in this region varies with the nature of the crisis, i.e., DNB ([A] in Fig. 7.54) or film dryout ([D] in Fig. 7.54).

If a crisis occurs in the subcooled two-phase flow or a two-phase flow of low steam quality, i.e., due to DNB, the realized two-phase flow mode is called "*inverted annular flow*" where the superheated vapor flow flowing along the wall separates the subcooled layer from the wall. The inverse-annular flow regime is preceded by the transition boiling section in the upstream region which is bounded by the DNB point (CHF) and the quench front or the minimum heat flux of the film boiling on the boiling curve.

As described in Section 7.5.4 on the boiling curve in the pool boiling, the location of the MFB temperature (T_{MFB}) separates the high-temperature region of the inefficient film boiling or vapor cooling, from the lower-temperature region of the more efficient transition boiling region. Thus it provides a limit to the application of transition boiling and film boiling correlations. T_{MFB} also represents a temperature boundary beyond which surface properties and surface conditions generally do not affect the heat transfer. Wettability or the contact angle, although important in nucleate and transition boiling, is not applicable in the film boiling regime, and conduction along the surface becomes less important when nucleate and film boiling no longer occur one by one.

It is noted that, while the CHF has been extensively studied and can be predicted fairly well by a variety of correlations, minimum heat flux under forced convective conditions has undergone less investigation. There is no consensus on the effects of various system parameters on T_{MFB} under these conditions, but it should be affected by flow, pressure, fluid properties, and heated surface conditions in the same way as in the pool boiling described in Section 7.5.4.

In the "*dispersed-annular flow*" regime, the crisis is usually associated with liquid film dryout on the heated surface. In this case, the heated wall is cooled by the vapor flow, which is superheated while flowing along the channel. In this vapor flow region, liquid droplets exist and gradually evaporate exchanging heat with the superheated vapor.

In general, dispersed two-phase flow in this zone is in thermal nonequilibrium, and the wall surface temperature is determined by several processes: convective heat transfer by vapor flow, radiation, and evaporation of the droplets. Actual flow parameters depend on the evaporation of droplets and the heat-transfer area in vapor flow, whose temperatures may exceed the saturation temperature. In simple calculations, thermal equilibrium between liquid and vapor phases is often assumed. The boundary between this region and a superheated vapor flow region of the channel is considered where the equilibrium steam quality $x = 1$.

Depending on the operating conditions and channel geometry, flowing vapor in the post-CHF zone may be superheated in the inverse-annular regime where the liquid is separated from the heated surface by the vapor film (downstream of the DNB point and the transition boiling region) or in the dispersed-annular flow regime (downstream of the film dryout point) where the liquid droplets are distributed in the vapor core flow.

The structure of two-phase flow in the postminimum heat flux region is stable and most of the liquid phase decays into droplets toward the downstream region and the flow becomes dispersed which is a dominant flow regime in the heat transfer crisis. The higher surface temperature in this zone arises because superheated vapor is primarily in contact with the wall.

7.8.7 Inverse-annular flow boiling and dispersed-annular flow film boiling

There are few reliable guidelines for the calculation of the inverse-annular mode heat transfers. A large number of semiempirical models have been developed, most of which are quite complicated. It is difficult to check many of the parameters used in these models and match them with the final results from individual experimental data, and their validity of generalization has not yet been demonstrated. Furthermore, experimental data do not always provide sufficient accuracy and are adopted with the backgrounds of models often facing difficult justification.

Heat transfer in the film boiling of the subcooled liquid region is made up of three mechanisms:

(1) Convective heat transfer from the heated surface by vapor flow convection in the vapor film along the surface;
(2) Conduction and laminar convection heat transfer from the vapor film to the vapor-liquid interface; and
(3) Thermal radiation.

Heat transfer in the dispersed-annular flow mode, which appears most often in the superheated zone of high steam quality ($x > 0.2$), generally comprises the following components:

(1) Convective heat transfer from the wall to the bulk vapor flow;
(2) Conductive heat transfer from the wall to liquid droplets in contact with the thermal boundary layer (dry contact) and that impact the wetted (or not wetted) wall;
(3) Radiation heat transfer from the wall to both bulk vapor and liquid droplets;
(4) Convective heat transfer by bulk vapor flow to suspended droplets in the vapor core.

The majority of heat is transported by convection. The presence of superheated vapor and liquid droplets at saturation temperature T_S results in the actual vapor flow quality x_a being lower than the equilibrium vapor quality x. Let us define a degree of thermodynamic nonequilibrium by the ratio of these two values $x_a/x = n$ (≤ 1 in the dispersed-annular regime). Then the mean temperature and enthalpy of the superheated vapor in this case are expressed by the relations

$$\overline{T}_v = T_S + \frac{\Delta h_{fg}}{c_p}\frac{(1-n)}{n} = T_S + \frac{\Delta h_{fg}}{c_p}\cdot\frac{x - x_a}{x_a};$$

$$\overline{h}_v = h_g + \Delta h_{fg}\frac{(1-n)}{n} = h_g + \Delta h_{fg}\frac{x - x_a}{x_a}.$$

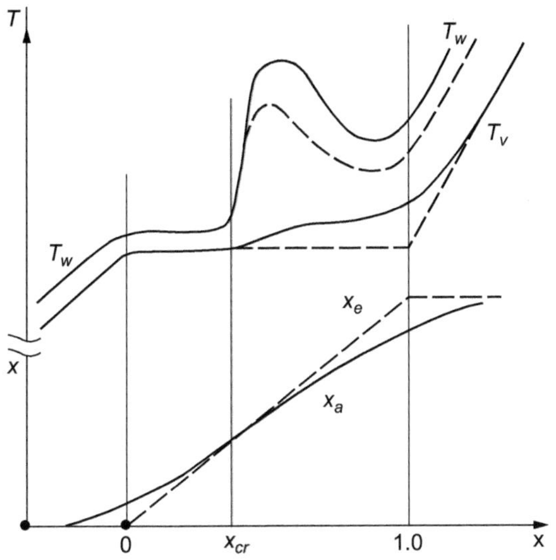

Figure 7.60 Temperature behaviors to the vapor quality and the actual vapor quality in the crisis zone (q = const and $x > x_{cr}$): — thermally nonequilibrium flow; x_a is the actual quality; - - - thermally equilibrium flow; x_e is the equilibrium quality. T_v and T_w are the vapor phase and wall temperatures.

When the mass flux of the mixture is higher than 1000 kg/(m^2 s), the thermodynamic nonequilibrium will be small and $n \approx 1$. The nature of the temperature distribution and the actual vapor quality in the crisis zone is shown in Fig. 7.60. The crisis zone could be defined as the region where the actual vapor quality x is higher than the critical steam quality x_{cr} at which DNB would occur. The heat transfer coefficient is usually referred to as the temperature difference ($T_w - T_S$).

7.8.7.1 Heat transfer in a dispersed droplets-vapor mixture flow

When the dispersed-annular flow structure changes to dispersed droplet-vapor mixture flow in channels, heat transfer can be calculated by accounting for the thermal nonequilibrium of the flow (Sergeev et al., 1985; Sergeev, 2002). There, the two-phase flow in the crisis region is treated by a homogeneous mixture of superheated vapor and liquid droplets. Heat is considered transferred by convection from the heating surface to vapor and from vapor to droplets (droplets evaporating on the wall surface, with radiation heat transfer neglected). The heat transfer coefficient from the wall to the vapor and the critical steam quality (x_{cr}) are assumed to be known.

The method can be applied for flows in pipes, annular channels, and nuclear fuel rod bundles with spacer grids, and their hydraulic diameter would be 4–20 mm, and $p = 1.1-19.6$ MPa, $\rho w = 70-2000$ kg/(m^2 s), $q \leq 1$ MW/m^2, $T_w - T_S \leq 500$°C, and $x > x_{cr} \geq 0.1$.

Variations in actual vapor quality x_a in *the post-CHF zone* in the channel can be determined by the following differential equation:

$$\frac{dx_a}{dx_e} = Cm\frac{\lambda_g}{\sigma\rho_f}\frac{P_W(\rho w)^2}{\sum_j(qP_h)_j}x_a(1-x_a)\left(\frac{x_e-x_a}{x_a}\right)^n, \tag{7.158}$$

where x_e is the thermodynamically equilibrium mass quality; C is an empirical constant ($C = 1.5$), P_W and P_h are the wetting and heating perimeters, respectively, in m; q is the local heat flux on the jth heat transfer surface (W/m^2); ρw is the mass flux (kg/m^2 s); and parameters m and n depend on pressure.

For $p < 6$ MPa:	$m = 1149.1 - 28.2534p + 0.995p^2 - 1.8421 \times 10^{-2}p^3 + 1213 \times 10^{-4}p^4$, °C;
	$n = 1.0003 + 4.3389 \times 10^{-3}p - 2.5466 \times 10^{-5}p^2$;
For $p \geq 6$ MPa:	$m = 864.664 - 4.0128p$, °C;
	$n = 1.0861 + 8.6654 \times 10^{-4}p + 1.0772 \times 10^{-5}p^2$,

The pressure p is in bars.

For flows in pipes and rod bundles, $C = 1.5$. For annular channels with one-sided heating, $C = 3$. For annular channels with two-sided heating, heat flux to the vapor at the second surface is assumed to be zero, and the constant $C = 3$. In the downstream of the crisis section, C is assumed to equal 1.5 (Sergeev et al., 1985). The "actual" vapor quality distribution x_a is obtained by integration of Eq. (7.158) for the heat-transfer crisis section, which depends on x_{cr} and $\overline{T}_v = T_S$.

For $n = 1$ (highly superheated vapor flows, $x_a = x_e$), Eq. (7.158) has an analytical solution:

$$x_a = 1 - \frac{exp\left(-\xi^2/2\right)}{B + \Phi(\xi)\sqrt{\pi A/2}}, \tag{7.159}$$

where $= Cm\lambda_g \frac{1}{\sigma\rho_f} \frac{(\rho w)^2 P}{\sum_i q_i(z) P_{Ti}}$, B is the constant determined from the condition $x = x_{cr}$, $\xi = \sqrt{A}(x - 1)$, and $\Phi(\xi)$ is the probability integral. Eq. (7.159) could be used for testing the numerical integration scheme of Eq. (7.158).

7.8.7.2 Procedure for calculating heat transfer in the crisis zone

- By solving the differential Eq. (7.158) or by formula (7.159), the actual vapor quality x_a is obtained at a given vapor equilibrium quality x_e.
- Then, with x_e and x_a, the enthalpy of superheated vapor is given by $\overline{h}_v = h_g + \Delta h_{fg}\left[\frac{x_e - x_a}{x_a}\right]$ in a specified section of the channel.
- With the given enthalpy of superheated vapor and the saturation enthalpy from the steam table, $\overline{T}_v = T_S + m\left[\frac{x_e}{x_a - 1}\right]^n$ determines the superheated steam temperature.
- The wall temperature on the ith heat-transfer surface is calculated by: $T_W = \overline{T}_v + \frac{q_i}{\alpha}$, where α is the convective heat transfer coefficient to the superheated vapor; $\overline{T}_v$ is the temperature of the superheated vapor (K); and T_W is the wall temperature (K).

Thermal properties refer to those at the superheated vapor temperature $\overline{T}_v$.

7.8.7.3 Annular channels

The heat transfer coefficient for annular flow channels is obtained from the following equations (Handbook PTM.24.031.05-72, 1974):

- when the inner wall of the annular channel is heated:

$$Nu_1 = 0.86Nu(d_1/d_2)^{-0.16};$$

- when the outer wall of the annular channel is heated:

$$Nu_2 = Nu\left[1 - 0.14(d_1/d_2)^{0.6}\right];$$

- when both inner and outer walls are heated:
  For the inner wall:

$$Nu_1^* = Nu_1 / [1 + Nu_1\theta_1(q_2 / q_1)];$$

For the outer wall:

$$Nu_2^* = Nu_2 / [1 + Nu_2\theta_2(q_1 / q_2)],$$

where $Nu = 0.023Re^{0,8}Pr^{0,4}\left[\overline{T}_v/T_w\right]^{0.5}$; $\theta_1 = 32[0.16(d_1/d_2)]Re^{-0.88}$; $\theta_2 = \theta_1 \cdot d_1/d_2$; $d_h = d_2 - d_1$; $Nu = \frac{\alpha d_h}{\lambda}$; $Re = \frac{(\rho\, w)xd_h}{\mu}$; $Pr = \frac{\nu}{\alpha}$; $Nu_1 = \frac{\alpha_1 d_h}{\lambda}$; and $Nu_2 = \frac{\alpha_2 d_h}{\lambda}$. Subscripts 1 and 2 of the thermal properties at $\overline{T}_v$ refer to the inner and outer walls of the annular channel, respectively.

7.8.7.4 The rod bundles

The heat transfer coefficient can be represented by (Subbotin et al., 1975)

$$Nu = C\, Re^{0.8}Pr^{0.4}, \tag{7.160}$$

where $C = 0.0165 + 0.02\left[1 - \frac{0.91}{(P/D)^2}\right]\left(\frac{P}{D}\right)^{0.15}$, and *P/D* is the rod pitch-to-diameter ratio.

7.8.7.5 Heat transfer at the entrance region of a crisis zone

Heat transfer to the superheated vapor flow in the region from the entrance to downstream of the crisis zone, where the thermal boundary layer is formed an approximate length of 15 hydraulic diameters, is calculated by

$$Nu = 1.38\left(\frac{d_h}{\Delta z}\right)^{0.12} Nu_0, \tag{7.161}$$

where Nu_0 is the Nusselt number for steady-state heat transfer, and Δz is the distance from the cross section of the heat-transfer crisis (Isachenko et al., 1981).

7.8.7.6 Influences of spacer elements

In this section, integration of Eq. (7.158) is carried out with a new boundary condition given by Eq. (7.162) that accounts for an increase in vapor quality due to additional evaporation resulting from interactions of flow with the spacer elements (Sergeev et al., 1990):

$$\Delta x = (1 - x)k(\Delta A / A), \tag{7.162}$$

where ($\Delta A/A$) is the extent of the flow channel area blockage by the spacer, and k is a multiplication factor that depends on the geometry of the spacer and its mode of settings ($0 < k < 1$). For simple spacer elements without swirl-enhancing tabs, the factor k is given by

$$k = (1 - \rho w / 3000) \pm 0.15, \text{for } \rho w \leq 1500 \text{ kg} / \left(\text{m}^2 \text{ s}\right).$$

If the calculated vapor quality in the spacer element section by Eq. (7.162) exceeds the value of relative enthalpy (quality) at the same location, the difference $(x_a - x)$ is considered due to the evaporation of the liquid film, which is formed as a result of flow interactions with the spacer and covers the heated wall surface on the segment length:

$$\Delta z = G \cdot \Delta h_{fg} \cdot (x_a - x) / \sum_i (q P_{\text{T}i}), \tag{7.163}$$

In this spacer region, heat transfer is calculated using the formulas for boiling, after which the procedure described above with the solution of Eq. (7.158) is repeated, starting from the section located at a distance Δz from the spacer, where the flow is considered to be at thermal equilibrium.

The following accounts for single-phase flow heat transfer enhancement by spacers, as suggested by Yao et al. (1982):

$$Nu / Nu_0 = 1 + 5.55 \cdot (\Delta A/A)^2 \cdot e^{-0.13z/d_h}, \tag{7.164}$$

where Nu_0 is the Nusselt number in the absence of the spacer; Nu is the Nusselt number in the presence of the spacer; z is the distance from the spacer; and d_h is the equivalent hydraulic diameter of the flow channel.

7.8.7.7 Consideration of radiation

The total heat transfer by vapor flow convection and radiation to the droplets is expressed by Sergeev et al. (1990):

$$q_i = \alpha\left(T_w - \overline{T}_v\right) + \sigma\varepsilon\left(T_{wi}^4 - T_S^4\right), \tag{7.165}$$

where σ is the Boltzmann constant, and ε is the reduced emissivity of the blackbody-droplet system.

The wall temperature is obtained following the above procedure of solving Eqs. (7.158) and (7.165). For a given wall temperature, the reduced heat transfer coefficient is calculated by $\alpha_i = q_i/(T_{wi} - T_S)$.

The units of various parameters used in the formulas are the following: A—W/(m^2 K); T—in °C; q—W/m^2; p—MPa; h—J/kg; ρw—kg/(m^2 s); Δh_{fg}—J/kg; μ—N s/m^2; $d_h = (d_2 - d_1)$—m; d_1, d_2—m; λ—W/(m K); σ—N/m; and ρ—kg/m^3.

The range of applicability of the method is as follows:

$p = 1.1\text{–}19.6$ MPa; $T_w - T_f \leq 500°\text{C}$; $\rho w = 70\text{–}2000$ kg/(m^2 s); $q \leq 1$ MW/m^2; $x_b > x_{cr} \geq 0.1$; and $d_h = 4\text{–}20$ mm.

The maximum accuracy in calculating α is $\sigma \pm 25\%$. The mean accuracy for circular tubes with uniform and cosine shape heat fluxes is $\sigma \pm 10\%$.

7.8.7.8 Simplified calculation method

Heat transfer is calculated in circular tubes and annuli by the following formula (Miropol'sky, 1975):

$$Nu_g = 0.023\left\{Re_g Pr_{gW}\left[x + \frac{\rho_g}{\rho_f}(1 - x)\right]\right\}^{0.8} \cdot Y,$$

where $Y = 1 - 0.1\left(\rho_f/\rho_g - 1\right)^{0.4}(1 - x)^{0.4}$; $Nu_g = \alpha d/\lambda_g$, $Re_g = Gd/\mu_g$. The range of application of the formula is $p = 4\text{–}20$ MPa; $\rho w = 1000\text{–}2000$ kg/(m^2 s). For a wider range of parameters, you may use the formula by Groeneveld (1973a):

$$Nu_g = 0.00327\left\{Re_g\left[x + \frac{\rho_g}{\rho_f}(1 - x)\right]\right\}^{0.9}\left(Pr_{gW}\right)^{1.32} Y^{-1.5}, \tag{7.166}$$

The range of application of the formula is given in Table 7.11.

At lower mass flux than 1000 kg/(m^2 s), superheated vapor flow is characterized by significant thermodynamic nonequilibrium in the presence of liquid droplets in the vapor flow stream. In this case, the heat transfer coefficient refers to the temperature difference $(T_w - \overline{T}_v)$ instead of $(T_w - T_S)$ as commonly practiced for thermodynamic

Table 7.11 The range of application of formula (7.166) for pipes and annular channels.

Parameters	Circular tube	Annular channel	Parameters	Circular tube	Annular channel
d (mm)	2.5–25	1.5–6.3	q (W/cm^2)	12–210	45–225
p (MPa)	6.8–22	3.4–10	Y	0.706–0.975	0.610–0.953
ρw (kg/(m^2 s))	400–5300	800–4100	Pr_{gw}	0.88–2.21	0.91–1.22
x	0.1–0.9	0.1–0.9	Nu_g	95–1700	160–640

equilibrium flow. As a result, $T_w = \overline{T}_v + q/\alpha$. The superheat is calculated from a steam table, knowing $\overline{h}_v$:

$$\overline{h}_v = h_g + \Delta h_{fg}(x - x_a)/x_a, \tag{7.167}$$

where x is the equilibrium vapor quality; x_a is the true vapor mass quality; Δh_{fg} is the heat of vaporization; and h_g is the saturation enthalpy of vapor.

7.8.7.9 Heat transfer in the superheated zone of the helical coil pipes

The helical coil heat exchanger has many features that differ from those of straight tube heat exchangers. For instance, the flow in the helical coil tube shows the influence of centrifugal force, and the heat transfer crisis occurs not uniformly along the perimeter of the tube. There exists a large variation in the heat transfer coefficient around the perimeter and, as a result, large temperature gradients. Nonequilibrium effects at a crisis zone are less than those of the straight tubes/pipes of identical parameters. With d the tube inner diameter and D the coil diameter, the calculation formula takes the form (in Handbook, 1983) for $d/D \geq 0.015$:

$$Nu = \alpha d/\lambda_g = C\, Re_m^{0.8} Pr_w^{0.8} z \pm 25\%,$$

where $C = 0.017$ for helical coil tubes; $C = 0.017(1 + 3.18d/D)$ for pipes bent at 90 degrees; $z = 1 + y(1 - x)$, $y = 0.5(\rho_f/\rho_g - 1)^{0.8}$ for $\rho_f/\rho_g \leq 480$; and $y = 70$ for $480 < \rho_f/\rho_g < 1300$. For $d/D < 0.015$,

$$Nu = 0.023 Re_m^{0.8} Pr_w^{0.8}\left[1 - 0.1\left(\frac{\rho_f}{\rho_g} - 1\right)^{0.4}(1 - x)^{0.4}\right].$$

The formula is based on experimental data obtained in the steam-water mixture in the range of $Re_m = 5.4 \times 10^3$ to 4×10^5; $d/D = 0.007–0.14$; $\rho w = 100–2000$ kg/(m^2 s); $p = .1–21.5$ MPa; $q = 0.1–1.1$ MW/m^2; and $x = 0–1.0$.

Another formula for calculating heat transfer is given by

$$Nu = Nu_0\left[1 + 0.1\left(\frac{\rho_f}{\rho_g} - 1\right)^{0.4}(1 - x)\right]\left[x + \left(\frac{\rho_f}{\rho_g}\right)(1 - x)\right]^{0.8}.$$

where Nu_0 is the Nusselt number for single-phase flow in coiled pipes.

7.8.7.10 Heat transfer of superheated steam

The magnitude of the heat transfer rate to superheated vapor as well as to gases is significantly lower than that to liquids. Therefore, the temperature difference $T_w - T_f$ becomes significant, which solicits the need to consider heat transfer coefficients applicable to a wide spectrum of the physical properties of the flow section.

Calculating the average heat transfer rate (for $l/d \geq 50$) in the vapor flow away from the critical region can be performed according to the formulae, which account for variations in the physical properties of simplices ρ_w/ρ_f or $\overline{T}_v/T_w$:

For $q \leq 0.5$ MW/m^2 (Eq. 7.168), Miropol'sky (1975), Eq. (7.169), Lel'chuk and Elfimov (1968),

$$\overline{Nu} = 0.028\overline{Re}^{0.8}\ \overline{Pr}^{0.4}(\rho_W/\rho_v)^{1.15}; \tag{7.168}$$

$$\overline{Nu} = 0.023\overline{Re}^{0.8}\ \overline{Pr}^{0.4}\left(\overline{T}_v/T_W\right)^{0.5} \pm 10\%, \tag{7.169}$$

where all the properties are selected based on the average temperature of the superheated vapor $\overline{T}_v$, $\overline{Re} = \frac{(\rho w)xd}{\mu}$, $\overline{Pr} = \frac{\mu c_p}{\lambda}$. These equations are applicable for $\rho w = 300-1000$ kg/(m^2 s) and $p = 4-20$ MPa. For $q > 0.5$ MW/m^2, heat transfer to the vapor is obtained based on the arithmetic mean of the results from Eqs. (7.168) and (7.169).

The local heat transfer coefficients for pipes can be found in Petukhov and Kurganov (1974) and Petukhov (1987):

$$\frac{Nu}{Nu_0} = \exp\left\{-K\left[0.013f\left(x/_d\right) + 1.18K\frac{1.25\times10^{-4}(x/d)^2}{1 + 10^{-4}(x/d)^{2}}\right]\right\}, \tag{7.170}$$

where x is the distance from the entrance of the test section; and $K = \frac{qd}{\lambda_f \overline{T} Nu_0} = \frac{T_w - \overline{T}_f}{\overline{T}_f}$ is the coefficient characterizing the temperature variation in the flow when the heat flux is changed:

$$Nu_0 = Nu_\infty \varepsilon_l, \text{ where } \varepsilon_l = 1 + \frac{0.48\left[1 + 3600/\left(Re\sqrt{x/d}\right)\right]}{(x/d)^{0.25}} e^{-0.17x/d}; \quad (7.171)$$

$$Nu_\infty = \frac{(\xi/8)Re\,Pr}{1 + 900/Re + 12.7\sqrt{\xi/8}\left(Pr^{2/3} - 1\right)}; \text{ and} \quad (7.172)$$

$$\xi = [1.82 \ln(Re/8)]^{-2}. \quad (7.173)$$

For $Pr = 0.65-1$, good results are obtained by a simple formula:

$$Nu_\infty = 0.0225 \mathrm{Re}_f^{0.8} Pr_f^{0.6} \pm 2\% \quad (7.174)$$

Eq. (7.170) can be rewritten as

$$\frac{T_w}{\overline{T}_f} = 1 + K \exp\left\{K\left[0.013 f\left(x/_d\right) + 1.18\,K \frac{1.25 \times 10^{-4}(x/d)^2}{1 + 10^{-4}(x/d)^2}\right]\right\}, \quad (7.175)$$

which allows immediate calculation of the local wall temperature for q, d, G, and T_f.

Eq. (7.175) can be simplified for $x/d > 50$, in which $f(x/d) = 1$:

$$\frac{T_w}{\overline{T}_f} = 1 + K \exp\{K[0.013 + 1.4\,K]\} \quad (7.176)$$

7.9 Boiling at surface deposition

7.9.1 Key processes and mechanisms

Deposits on the surfaces of fuel elements and steam generators are due to impurities, which enter the water mainly

(1) with suction of cooling water in the condensers;
(2) with compensation water to recover losses due to leakages of water and steam and blow-down of the steam generator;
(3) with impurities introduced into the water loop for adjusting changes in the water level;
(4) as the corrosion products of equipment components;
(5) in the form of fission products and particles of nuclear fuel (with infringement of hermetic fuel cladding tightness).

Deposits formed on the steam-generator tube surfaces are called scales. They differ in the chemical-phase compositions and structures. In the energy sector, the following types of scale (as the dominant component) are representative: (1) calcium and magnesium; (2) iron oxide; (3) iron phosphate; (4) ferromagnetic and aluminosilicate; and (5) copper. Deposits caused by the precipitation of impurities in the BWR fuel bundle are discussed by Le Corre et al. (2022); clogging in the secondary circuit of PWR is also discussed by Muller et al. (2015) (see also Chapter 8).

There are two mechanisms of deposit formation on the surface during boiling: first—connected with the processes of crystallization of impurities that fall from a supersaturated solution of water; and second—sticking to the surfaces of colloid or particulate solids (or sludge crud[5]). A result of the first mechanism is the formation of a dense layer of the solid phase and a little of the porous surface. The second provides the formation of a sufficiently loose sticky layer between itself and the particle surface. These particles may come off into the fluid flow and fall back again onto surfaces, depending on the hydraulic and temperature conditions in a different part of the circuit. In nuclear power plant (NPP) systems, formation, entrainment, and migration of particles, which are sparingly soluble corrosion products, along coolant circulation systems determine, to a large extent, the radiation dose distribution at the NPP (e.g., Voronov et al., 2006; British Nuclear Energy Society, 2000).

Deposits of corrosion products in nuclear power circuits are poorly soluble in many cases, even in concentrated acids and alkalis. One reason for this is the deep hydration precipitation at high temperatures and pressures. Another reason is the presence of metal silicate sediments. Silicic acid enters the coolant circuit from the feed water or structural materials (for 1H18N10 type steels, Si content may be up to 1%).

During the circuit loop operation, usually 20%–80% of the metal is corroded in the circulating water. The intensity of the deposits and their physicochemical properties are dependent on the concentration of impurities (e.g., iron oxide) and the dispersion spectrum of particle sizes. Most of the particles (65%–80%) have a size of 0.9–1.3 μm. The precipitation of iron oxides cannot be described simply by a single system since processes exist with different natures at the same time. For example, the results of industrial research show that the formation time of iron oxides in the boiler tubes is subject to

$$M = A \exp\left(-\frac{B}{T}\right)\tau^{c}, \tag{7.177}$$

where M is in g/m^2; T is in K; and τ is in hours. The constants A, B, and C in this formula depend on the steel grade and the water pH (Table 7.12).

In modern power plants, which are constructed with different grades of steel, strict water quality control is required. There, the basic substances dissolved in water are primarily iron compounds in the form of magnetite (Fe_3O_4) and hematite (Fe_2O_3). Since the solubility of magnetite is higher in water than in vapor, magnetite aerosol

[5] Crud—chalk river unidentified deposit.

Table 7.12 Constants *A*, *B*, and *C* in Eq. (7.177).

pH	Steel grade	*A*	*B*	*C*
pH 7–8	12X1M[Fcy]	3.91×10^6	8590	0.25
	12X18H12T	1600	5720	0.38
pH 9–10.5	12X1M[Fcy]	6.57×10^5	7830	0.26
	12X18H12T	152	6680	0.41

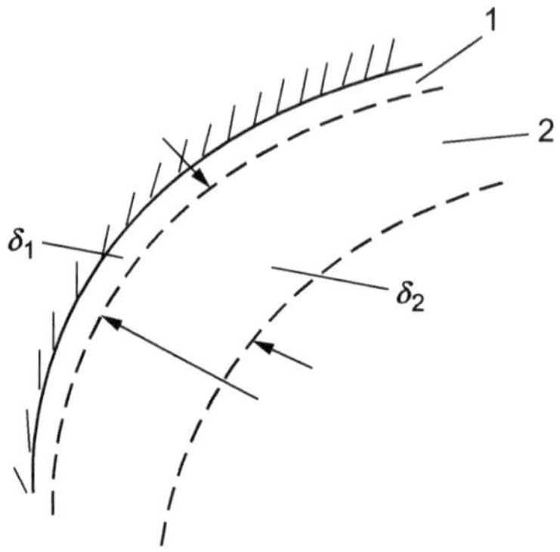

Figure 7.61 Two layers of the deposits on the surface (internal—δ_1 and external—δ_2).

Table 7.13 Characteristics of Layers 1 and 2.

pH	Characteristics	1	2
pH 9	Density (g/cm^3)	4.4–4.7	1–4
	Chemical composition	Fe_3O_4	Fe_3O_4
	Porosity (%)	10	20–75
pH 7	Chemical composition	Fe_2O_3	Fe_3O_4
	Porosity (%)	4	50

formation will be prompted out of its solution in a vapor environment. Individual particles (aerosols) of magnetite are in the shape of a sphere with a diameter of 0.1–0.5 μm, which easily stick together to form deposits on the porous surface.

Deposits form two layers (Fig. 7.61) on the surface. The inner layer has a dense structure. It is formed by oxidation of the steel surface, and its thickness varies parabolically with time. The outer, thicker layer has a loose structure and is formed by deposits at the boiling of flowing water. Some layer characteristics are shown in Table 7.13.

Hereafter we focus on the influences of this surface deposition on heat transfer.

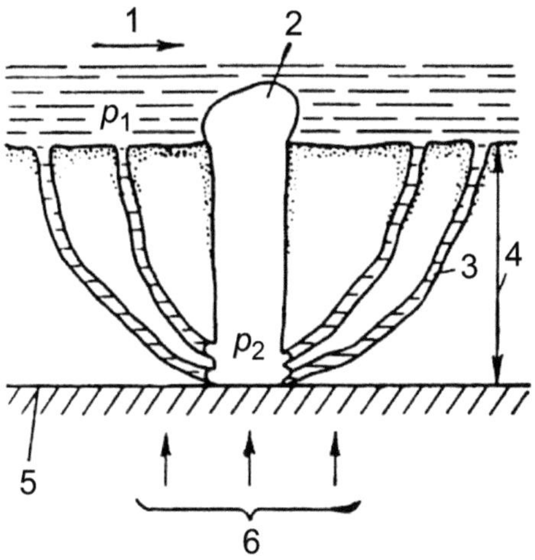

Figure 7.62 Macbeth Model—flow circulation in the deposits with high rate of evaporation of the readily soluble compounds in vapor sediments: 1—the flow of water under pressure p_1; 2—exit of steam "steam trunks or steam pipe" with pressure p_2 as a result of the consistent formation and the separation of the vapor bubbles; 3—capillary channels (wicks), sucking liquid to the bottom of the "steam pipe"; 4—the thickness of the sludge sediments; 5—the surface of the metal; 6—heat flux direction.

A characteristic feature of a typical layer of deposited materials is the presence of large porous openings ("steam trunks" about 5 μm); the density of such holes reaches 5000 pores/mm^2. This structure provides a so-called "wicks boiling" (see Fig. 7.62).

Dissolved substances in water (compounds of Cr, Ni, Mn, Zr, Cu, Al, and others) form thick solid structures that are stronger magnetite but can be peeled off from the surface at cyclic temperature changes. The porosity of the magnetite scale is ~65%–70%, and "steam channels" occupy ~10%. When pores are filled with other compounds, the process of matchlock boiling abates. Notably, the majority of heat transfer in this case is carried out by thermal conduction through the layer, which causes an increase in the heated surface temperature (while maintaining the values of heat flux).

7.9.2 Composition and thermal conductivity of the sediments

Water may be present in ferrous iron compounds comprising FeO, compounds containing di- and trivalent iron forming Fe_3O_4, and ferric compounds turning into Fe_2O_3. The main parameters that determine the shape of iron oxides are the pH, temperatures, and redox potential of the system.

At a temperature ≤400°C, the thermodynamically stable oxide is magnesia. Under conditions of high oxidation potential (high oxygen concentration) and higher temperature, the stable oxides may be hematite or its hydrated forms.

The theoretical density of the magnetite Fe_3O_4 is 5180 g/cm^3, and the density of the outer (loose) layer is 1.25–4.1 g/cm^3, corresponding to a porosity of 75%–20%.

The thermal conductivity of porous sediments can be approximated by the following at the location perpendicular to the direction of the heat flux $\frac{1}{\lambda} = \frac{1-\varphi}{\lambda_1} + \frac{\varphi}{\lambda_2}$, where λ_1 and λ_2 are the thermal conductivity of the sediments in the pores, and

φ is the porosity at the location parallel to the direction of the heat flux (usual direction):

$$\lambda = \frac{\lambda_1 \nu \left[1 - \phi(1-\nu)\right]}{\nu A + (1-A)[1-\phi(1-\nu)][\nu + \phi(1-\nu)]}, \tag{7.178}$$

where $\nu = \lambda_1/\lambda_2$, and A is the empirical coefficient and is 0.94.

A more sophisticated method is described in Chudnovskaya et al. (1978), where the thermal conductivity of the compounds hematite (Fe_2O_3) and magnetite (Fe_3O_4) is given by

$$\lambda = 13.215 - 0.375 \times 10^{-1} T + 0.379 \times 10^{-4} T^2 \quad \text{for } Fe_2O_3 \tag{7.179}$$

$$\lambda = 4.133 - 0.852 \times 10^{-2} T + 0.757 \times 10^{-5} T^2 \quad \text{for } Fe_3O_4 \tag{7.180}$$

where λ is in W/(m K), and T is in °C.

7.9.3 Effects of deposits on flow resistance and heat transfer coefficient

Sediments on the surface of the flow circuit lead to (1) a reduction in the cross-sectional flow area of the channels, which becomes particularly notable for channels with a small hydraulic diameter; (2) an increase in surface roughness. Both effects are not difficult to consider in the calculation of hydraulic resistance if the thickness and roughness of the deposits are known, although such parameters are difficult to predict in advance due to the large number of factors associated with them. Therefore, it is usual to rely on special experiments, empiricism, or industrial experience.

If the heat flux (q) and heat transfer coefficient for a "clean" environment α_0 are specified, we can calculate the temperature of the outer surface of the heat-wall:

$$T_w = \overline{T}_f + q/\alpha_0,$$

where $\overline{T}_f$ is the average temperature of the bulk coolant.

For surface conditions with deposits, we have a similar relationship: $T_w = \overline{T}_f + qR$, where $R = \frac{1}{K} = \frac{1}{\alpha} + \frac{\delta}{\lambda}$ is the total thermal resistance at the surface and consists of two parts (heat transfer resistance $1/\alpha$ and thermal resistance of sediments δ/λ). There, δ and λ are the thickness and thermal conductivity of the scale, respectively; and α is the coefficient of heat transfer from the surface of the scale to the flowing liquid.

If we keep in mind the roughness of surface sediments and the ability for "wick boiling" the layer (see above), we should assume $\alpha > \alpha_0$. Considering that $\alpha = \alpha_0$, we then obtain higher values of T_w that can go into conservative calculations. The calculation T_w is possible, but its results should be treated with caution, i.e., it is possible to predict the value of δ based on experimental data, as well as the values of λ based on analysis of the composition and characteristics of the sediments. This approach can only serve to estimate, and a similar results must be obtained based on an experiment or industrial experience.

7.10 Unstable boiling modes

7.10.1 Types of instabilities

1. *Unstable pool boiling* occurs in the process of surface cooling in a large volume of fluid, which may be followed by a transition from the liquid-phase convection to the nucleate boiling regime and again to the convection. The nucleate boiling can be replaced by the film boiling and return to the nucleate boiling regime. All of these transitions at a constant heat flux lead to unstable regimes—temperature fluctuations/oscillations at the cooled surface—that adversely affect structural integrity.
2. *Instability of boiling in a single channel* has a different (thermohydraulic or thermal fluid dynamic) nature. If boiling occurs in a separate part of the channel, its pressure drop changes. As a result, it may cause a change in the flow rate. Under certain circumstances, fluctuations in flow rate and pressure can be periodic (pulsating mode).

   To avoid unstable regimes when operating the steam-generating channel, its hydraulic characteristics should be known. The latter reflects the dependence of the total channel pressure drop on the flow rate (or mass flux) at the steady-state regime: $\Delta p = f(G) = f(\rho w)$.

   For stable operation of the channel, this characteristic must be unambiguous (Fig. 7.63). Characteristics that show the same pressure drop value for different flow rates in a certain range of Δp are considered *unstable* (Habensky and Gerliga, 1994). Furthermore, it is necessary to know the enthalpy distribution along a channel length. Pulsations of the flow rate occur when the ratio $\Delta h_{ec}/\Delta h$ is below a certain value (here Δh_{ec} is the enthalpy growth within the economizer section; Δh is the total increase in enthalpy).
3. *Boiling instabilities in parallel channels.* In the case of flow boiling, separate parallel tubes have different pressure drop characteristics and are under different heating conditions. Assume these

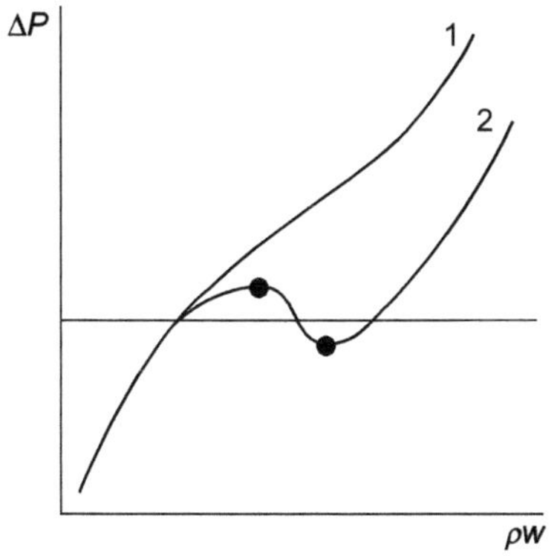

Figure 7.63 Typical hydraulic characteristics. 1—single-valued, stable; 2—multiple-valued, unstable: •—points of extremum.

parallel flow channels are connected at an inlet plenum and an outlet plenum, i.e., share the inlet and outlet in common. In this parallel channel system, flow and temperature redistributions take place, showing a different enthalpy increase in different channels. It is characterized by the ratio of the enthalpy increase in the ith single tube (Δh_i) to the mean value of enthalpy for the whole tube bundle ($\Delta \overline{h}$), i.e., $K_i = \Delta h_i / \Delta \overline{h}$ and a hydraulic nonuniformity factor $\eta_i = \zeta_i / \zeta_B$ a ratio of the hydraulic resistance (drag) coefficient of the ith channel $\zeta_i = \frac{\Delta p_i \cdot 2\rho_i}{(\rho_i w_i)^2}$ to that of the whole bundle $\zeta_B = \frac{\Delta \overline{p} \cdot 2\overline{\rho}}{(\overline{\rho w})^2}$; ρ_i is the coolant density in a tube; and $\overline{\rho}$ is the mean density in the bundle. Large values of these ratios in a bundle of parallel tubes result in unstable modes.

7.10.2 Unstable pool boiling

At low reduced pressures $p/p_{cr} < 2 \times 10^{-3}$ and good wettability of surface cavities (centers of vapor generation) significant liquid superheating occurs with regard to the saturation temperature. At low pressures, the critical radius of the vapor nucleus reaches its high value ($R_{cr} \sim 1/\rho_g$), boiling appears to be complicated, and natural convection is delayed. This phenomenon is also due to the decrease in the number of vapor generation centers (Avsentuk and Kutateladze, 1977). In some cases, a transition from natural convection to film boiling is possible while bypassing nucleate boiling (Fig. 7.64).

The S-shaped form of the boiling curve when q is the independent variable (for example, electric heating) determines the number of vapor generation centers per area unit (N/A) and surface superheating relative to saturation temperature (ΔT), with all other factors being equal:

$$q \sim (N/A)^n (\Delta T)^m.$$

The values of (N/A) and (ΔT) are related. So an increase in ($q > q_A$) also causes an increase in (N/A), which causes ΔT to decrease (curves 2 and 3 in Fig. 7.64). Usually, $n \sim 0.5$–0.7 and $m \sim 3$.

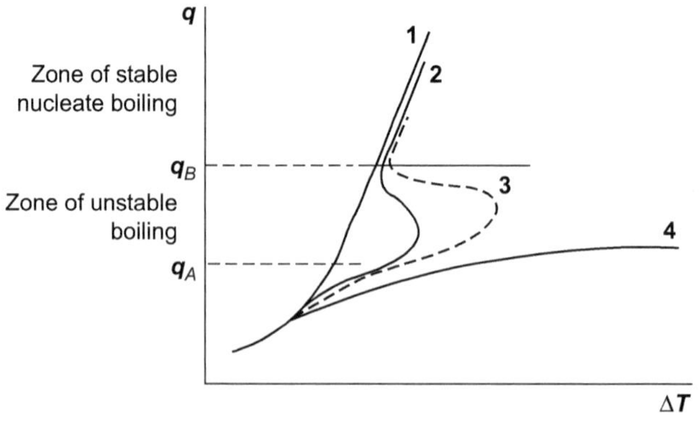

Figure 7.64 Boiling curves at different pressures ($p_1 > p_2 > p_3$). 1—High pressures; 2, 3—low pressures; 4—transition from natural convection to the film boiling.

The mechanism of film boiling bypassing nucleate boiling has a cavitational character—i.e., an explosive emergence of vapor bubbles is observed on the surface, as well as the formation of a vapor film (curve 4).

7.10.3 Instability of channel boiling

In a single channel, boiling instability is primarily associated with hydrodynamic instability, especially in circuits with natural circulation. Despite the constant power applied to the channel, oscillations of the flow rate can be observed after the onset of boiling. There are several explanations for this phenomenon. Let us suppose that as a result of some disturbance, the flow rate through the channel will increase slightly. This causes all factors to equal the decrease in vapor generation, which will reduce the value of the lifting force and the flow rate. However, the flow rate decrease will cause an increase in quality and lifting forces; therefore, it should result in a further increase in the flow rate. The sequence of each process takes place with some delay and causes flow rate pulsation. If the ratio of two successive amplitudes $D = A_1/A_2 < 1$, a damping of oscillations has occurred, and at $D > 1$, a pulsation gain will be observed.

All these processes have, to a large extent, nonlinear relationships since hydraulic resistance is a nonlinear function of water and vapor velocities and the nonlinear function of the void fraction, which in turn depends on the applied power and flow rate.

The hydrodynamic characteristic of a single steam-generating pipe with a certain diameter and length is described by function $\Delta p = f(\rho w)$. Here Δp is the full hydraulic resistance of the channel, which consists of two components: Δp_1 is the resistance of the economizer section; Δp_2 is the resistance of the evaporator section; and $\Delta p = \Delta p_1 + \Delta p_2$:

$$\Delta p_1 = \xi \frac{l_1}{d} \frac{(\rho w)^2}{2\rho_f}; \ \Delta p_2 = \xi \frac{l - l_1}{d} \frac{(\rho w)^2}{2\rho_f} \left[1 + \overline{x}\left(\rho_f / \rho_g - 1\right)\right]\overline{\psi}. \tag{7.181}$$

The thermal balance equation for the economizer and evaporation sections is as follows:

$$\pi d l_1 \overline{q} = \frac{\pi d^2}{4} \rho w \left(h_{f,sat} - h_{in}\right);$$

$$\pi d (l - l_1)\, \overline{q} = \frac{\pi d^2}{4} \rho w \left(h_{out} - h_{f,sat}\right);$$

where $l_1 = \frac{\Delta h \cdot d\rho w}{4\overline{q}}$; $\Delta h_{in} = h_{f,sat} - h_{in}$; $x_{out} = \frac{4\overline{q}\, l}{\Delta h_{fg} d\rho w} - \frac{\Delta h_{in}}{\Delta h_{fg}}$; and $\overline{x} = \frac{x_{out}}{2}$.

After substitution of all the relations in Eq. (7.181), by setting $\overline{\psi} = 1$, we have

$$\Delta p = A(\rho w)^3 + B(\rho w)^2 + C(\rho w). \tag{7.182}$$

where $A = \frac{\xi \, \Delta h_{in}^2}{16 \, \Delta h_{fg} \, \rho_f \overline{q}} \left(\frac{\rho_f}{\rho_g} - 1 \right)$; $B = \frac{\xi l}{2\rho_f d} \left[1 - \frac{\Delta h_{in}^2}{\Delta h_{fg}} \left(\frac{\rho_f}{\rho_g} - 1 \right) \right]$; and $C = \frac{\xi \, l^2 \, \overline{q}}{\rho' d^2 \Delta h_{fg}} \left(\frac{\rho_f}{\rho_g} - 1 \right)$.

In general, the solution of this equation provides one or three real roots. The extremums of the curve in Fig. 7.64 are determined by differentiation of Eq. (7.182):

$$\frac{d(\Delta p)}{d(\rho w)} = 3A(\rho w)^2 + 2B(\rho w) + C = 0, \tag{7.183}$$

Eq. (7.183) yields $(\rho w)_{in} = \frac{-B \pm \sqrt{B^2 - 3AC}}{3A}$. If Eq. (7.183) has no real roots, then it means that the hydrodynamic characteristic is stable. The stability condition $B^2 - 3AC < 0$ leads us to an approximate relationship:

$$\Delta h_{in} \leq \frac{7.46 \, \Delta h_{fg}}{(\rho_f / \rho_g) - 1}. \tag{7.184}$$

Eq. (7.184) suggests that the region of Δh_{in} for stability increases as pressure increases because the denominator decreases faster than the numerator. Therefore, with a decrease in pressure, the probability of unstable regimes increases.

7.11 Augmentation of heat transfer of surface with boiling water

7.11.1 Method of augmentation

Nuclear technology is in a constant search for new methods of heat transfer enhancement applied to water-cooled reactors. Along with traditional methods, newly developed methods suggesting improved fluid mixing in the channel may seem exotic. Some provide possibilities in principle to increase CHF by twice or more, which will increase the specific power of the reactor, or to provide a larger margin to thermal limits of the heat transfer crisis than at present, that is, to increase the reliability of safety system installation (Perepelitsa and Sapankevich, 1976; Kalinin et al., 1986).

In general, the heat transfer coefficient during boiling is sufficiently high, of the order of 10^4–10^5 W/(m^2 K), and there seems no sense in managing a further increase. However, in the case of transition from a nucleate boiling to a film boiling mode (boiling crisis, CHF), the heat transfer coefficient can be reduced by one to two orders of magnitude. To prevent burnout while maintaining constant heat flux, it is necessary to apply the device and methods of heat transfer augmentation. In this regard, promising improvements may be achieved by using means that provide the ability to increase either permissible reactor power or margins before the heat transfer crisis.

Classification of heat transfer enhancement methods is rather conditional. Methods vary according to the principles of physical phenomena that determine process augmentation (Thome, 1990; Holpanov and Shkadov, 1990; Kaviany, 1999):

1. Provide better flow mixing by installation of special geometry devices such as twisted tubes, twisted ribs on the surfaces of fuel rods, fuel rod spacer grids, etc.;
2. Change surface conditions (roughness, porous coatings, and others);
3. Improve surface wettability;
4. Use water additives in very low concentrations (nanofluids) that do not change water properties but affect transport processes in the boundary layer.

7.11.2 Augmentation of heat transfer by agitation of the swirl flow

Additional two-phase flow turbulence can be generated through transverse or helical corrugations, ribs, inserts of various braided tapes, screws, etc., or by knurling the pipes (see Fig. 7.65; Dzubenko et al., 1988). Also see Zysin and Dorfman (1986) and Kalinin et al. (1990).

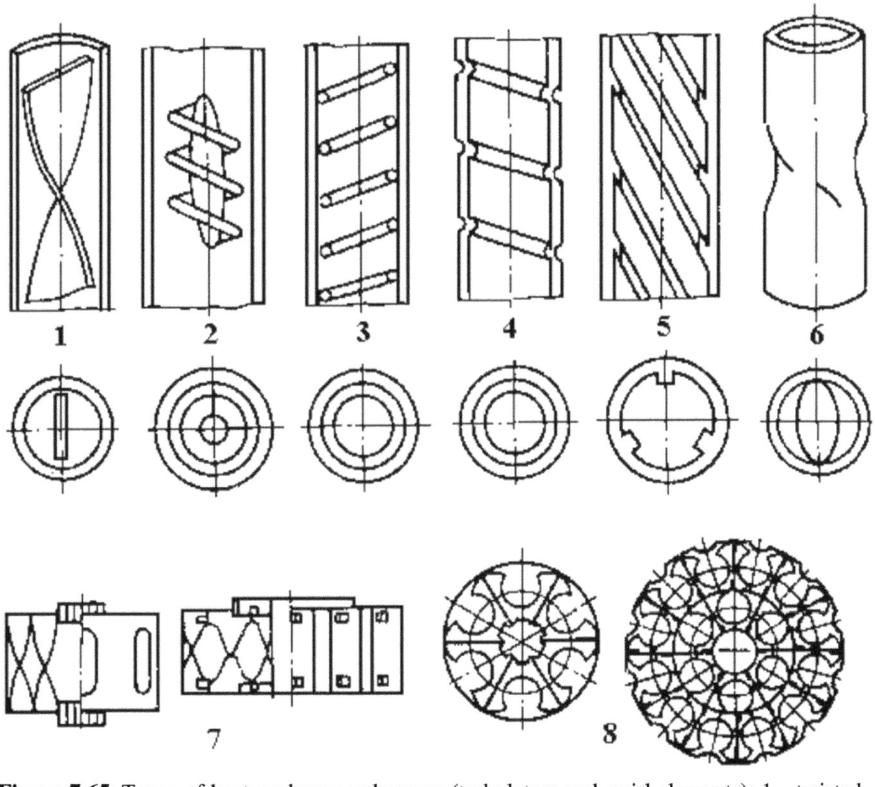

Figure 7.65 Types of heat exchange enhancers (turbulators and swirl elements): 1—twisted ribbons; 2—swirl insert (propellers, screws); 3—helical wire; 4—corrugations (ridge) on the surface; 5—the internal edges; 6—elliptic coiled tubing; 7, 8—lattice spacer with swirlers.

In a bubbly flow regime of low quality, devices such as baffles do not provide any advantages. In the dispersed-annular flow regime, especially at high mass flux, CHF will be increased by enhancing the turbulence intensity. In the case of an incorrectly chosen design parameter value, e.g., resulting in a narrower flow area and leading to an increase in velocity at the heated wall, installing vortex generators may reduce the critical power limit and steam quality due to the liquid film thinning effect.

In the rotating flow field by centrifugal force, the distribution of liquid droplets will be shifted toward the outer wall, which means a higher concentration of the liquid phase around it and an increase in CHF. It should be noted that the increase in CHF depends on the fluid motion, i.e., mainly on the magnitude of centrifugal acceleration (w^2/R) and not on subcooling. As a first approximation, $\alpha \sim \left(w^2/R\right)^{0.25-0.3}$. The more intensive the twisted flow (i.e., with the smaller pitch arrangement of the swirl element) and the lower the steam quality in the dispersed-annular flow regime, the more stable the liquid film surface becomes. A twisted flow is realized by installing swirl elements (belt, propeller insert, helical wire, helical ribs, twisted elliptic pipe, etc.). These elements generate much the same twisted flow field, yielding the same order of increase in the heat transfer coefficient as a first approximation.

Heat transfer augmentation in the film flow regime in channels has been discussed by many authors (Kalinin et al., 1990; Kamenchshikov et al., 1984; Shchukin and Halatov, 1982; Shchukin, 1980; Mitrophanova, 2002; Webb, 1993). In brief, it can be carried out in four ways: (1) agitation of the wall vapor layer; (2) forced wall irrigation (water supply to the liquid film on the wall); (3) providing good wettability of the wall at high temperature differences (transition to nucleate boiling); and (4) decreasing the thermal nonequilibrium stream by dispersing the liquid phase.

With the help of ring orifices or knurling on the surface, the forced irrigation of fluid film on the wall will be effective in reducing the thermal nonequilibrium of the flow. Then the heat flux from the wall is divided into two, with one to the vapor phase (q_g) and the other to the droplets depositing onto the wall (q_f):

$$q = q_g(1-\omega) + q_f\omega = \alpha_g\left(T_w - T_g\right)(1-\omega) + \alpha_f(T_w - T_S)\omega, \tag{7.185}$$

where ω is the proportion of the wall surface droplets irrigated; α_g and α_f are the heat transfer coefficients for the vapor phase and the liquid droplets; and T_g is the average temperature of the superheated vapor.

If $\alpha_f > \alpha_g$ (the heat flux), the liquid layer becomes significant such that $T_w \rightarrow T_S$, and there is nucleate boiling but no superheating of the vapor phase.

Heat transfer augmentation in the boiling tubes could be achieved by periodically arranged projection-type (protrusions) turbulence promoters such as carving spirals but at the cost of an increase in the flow resistance. Fig. 7.66 shows, for example, a scheme with annular projections in a tube:

$$\alpha = F(d/D, l/D, h, \alpha_0, \alpha_1, \alpha_2), \tag{7.186}$$

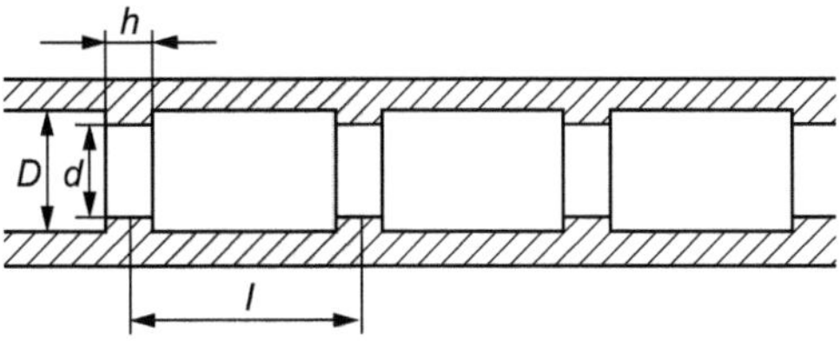

Figure 7.66 Geometry of the tube with annular projections.

where α_0 is the heat transfer coefficient at the boiling surface of a smooth pipe diameter D; and α_1 and α_2 are the heat transfer coefficients of the pure convection and the pool boiling in a large volume. α_0 can be calculated by the following equation (D.A. Labuntsov):

$$\frac{\alpha_0}{\alpha_1} = \frac{4\alpha_1 + \alpha_2}{5\alpha_1 - \alpha_2}. \tag{7.187}$$

The following equations are suggested by Kalinin et al. (1990) and are recommended to find α_1 and α_2:

$$Nu_1 = 0.023 Re_f^{0.8} Pr_f^{0.4} \left(\frac{Pr_f}{Pr_W} \right)^{0.06}. \tag{7.188}$$

where the fluid temperature T_f is defined by $T_f = \dfrac{T_{in} + T_{out}}{2}$

$$Nu_2 = 0.125 Re_*^{0.65} Pr^{0.33}, \tag{7.189}$$

where $u_2 = \alpha_2\, l/\lambda$; $Re_* = q\, l/\Delta h_{fg}\, \rho_g \nu$; $l = c_p \rho_f \sigma T_S / \left(\Delta h_{fg} \rho_g\right)^2$; and ν, λ, c_p, and Pr are all evaluated at temperature T_S.

Eq. (7.187) is proposed when $0.5 \le \alpha_2/\alpha_1 \le 2$. If $\alpha_2/\alpha_1 < 0.5$, convection heat transfer is determined by $\alpha_0 = \alpha_1$. When $\alpha_2/\alpha_1 > 2$, heat transfer is determined by boiling and $\alpha_0 = \alpha_2$.

Shown in Fig. 7.66 are the annular projections (orifice) that increase the heat transfer coefficient by 30%–40%, while the stronger, the smaller the ratio d/D, l/D (α_2/α_1), i.e., the greater the contribution to the overall forced convection heat transfer:

$$\frac{\alpha}{\alpha_0} = 1 + 1.35\left(1 - 0.371\frac{l}{D}\right)\left(1 - \frac{d}{D}\right)^{0.418}\left(1 - 0.104\frac{\alpha_2}{\alpha_1}\right). \tag{7.190}$$

Eq. (7.190) is valid for $(l/D) = 0.25–0.6$; $(d/D) = 0.88–0.94$; $\alpha_2/\alpha_1 = 0.5–2.4$; $Re = (2–6) \times 10^4$; and $q = (1.8–8) \times 10^4$ W/m^2.

7.11.3 Special surface: Porous structures

With a specially selected distribution of cavity sizes on the surface, you can change the boiling curve (i.e., the heat exchanges) as shown earlier. Special surface treatment technology has been developed, some forms of which are seen in Fig. 7.67. Porous coatings typically have an irregular structure and porosity.

Surface heat transfer by boiling water is determined by interactions between the vapor flow formed by evaporation and fluid flow that flows toward the surface. Displacement of the liquid by vapor determines the first limiting condition—the hydrodynamic limit.

When the liquid flows toward the surface, a second hydrodynamic limit occurs that is defined by the hydraulic resistance of viscous flow in a porous media structure.

The influence of coatings on the heat transfer coefficient is explained by a combination of factors, i.e., movements in capillary-porous structures, increases in the number of vaporization centers, and improvements in the evacuation of the conglomerates of vapor through the pores of the porous layer (Thome, 1990; Soloviev, 1998). All these effects depend on the thermophysical properties of the fluid and the surface geometry and structure of the coatings (film thickness, pore size distribution, etc.).

In general, the liquid evaporation mechanism limits heat flux, with the limitation depending on (1) the conditions of liquid supply to the surface and (2) removals of vapor from the surface and thus depends on vapor-liquid interaction and hydraulic resistance in the case of countercurrent flows.

When thick porous coatings have additional hydraulic resistance to fluid flows, the vapor phase may reduce the possibility of heat transfer enhancement and may even lead to the opposite effect. Thin porous coatings show the possibility of enhancement by supplying liquid to the surface by capillary forces and reducing interactions between the vapor and liquid streams. A porous coating specially created to supply liquid arteries is even more effective. For example, a specially designed porous coating of variable thickness accommodates capillary structures with arteries, such as heat pipes, thus increasing the heat transfer rate at boiling in a large pool by three times or more compared with a flat surface.

7.11.4 The effect of improving the wettability of the surface (superhydrophilicity)

There are proposals to use superhydrophilicity to enhance heat exchange at boiling and condensation. Superhydrophilicity promotes liquid spreading over the heated surface

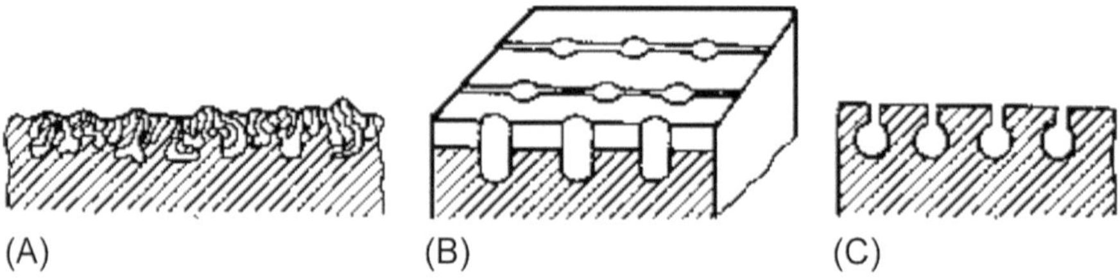

Figure 7.67 Special (structured) heat transfer surface: (A) high flux; (B) thermoexcel; (C) Gewa.

even at high temperatures. Photocatalysis is a method to decompose various organic substances using powerful ultraviolet irradiation. Titanium dioxide (TiO_2) is a substance with photocatalytic ability and unique characteristics. Another characteristic associated with TiO_2 published recently is superhydrophilicity. It was found that (1) CHF at the surface of a heated metal sample coated with TiO_2 is two times higher than that for uncoated surfaces; (2) the onset temperature of wetting (Leidenfrost temperature) increases with decreases in the contact angle.

The creation of very thin continuously stable falling films at extremely low costs has become possible; it was not previously possible with conventional surfaces. Measurement of the rate of cooling of the heated metal sample in boiling water showed that the sample coated with TiO_2 cooled much faster than the uncoated sample, as the first film boiling regime was destroyed and followed by a transition to nucleate boiling heat transfer with a significantly higher heat transfer rate.

The influence of wettability on the minimum heat flux point was observed and reported by many investigators. Leidenfrost temperature increases with a decrease in contact angle and increases in sets with improved surface wettability.

7.11.4.1 The aqueous slurry: Augmentation using additives to water

Suspensions of dispersed particles of sizes 10–100 nm in water or other liquids are referred to as nanofluids. Studies of boiling water with small additions of ($\leq$0.001%) nanoparticles showed that the rate of heat transfer and CHF could be increased by two times. Although these experiments are not directly related to nuclear reactor applications, the possibility of this kind of heat transfer augmentation cannot be excluded (Table 7.14).

It is estimated that applying this approach to a Westinghouse-type conventional fuel assembly (17 × 17) may allow the reactor core outlet temperature to increase by 7°C (20%), and in combination with increasing the mass flow by 18%, the power could be increased by ~ 40% (1.18 × 1.2–1.4).

Table 7.14 Typical thermal-hydraulic characteristics of pressurized water reactors (PWRs) and fuel assemblies.

Parameters	PWR (conventional)	PWR (power increased)
Thermal power (MW)	3411	4775
Mass flow (kg/s)	17,700	21,240
Lattice	17 × 17	21 × 21
Average volumetric power density (kW/L)	104.5	146.3
Linear power (kW/m)	46.6	42.8
Maximum fuel temperature (°C)	1870	1830
Coolant velocity (m/s)	5.4	6.4
Pump power (MW)	2.5	5.0

While nanofluid technology has not been fully mastered even in prototype installations, constant research and development efforts are in motion for new methods of heat transfer augmentation to apply to water-cooled nuclear reactors. Along with traditional methods, including improved fluid mixing in and between subchannels, the newly developed methods may yet seem exotic.

7.12 Critical heat flux and reflood

CHF, DNB, and related parameters (see Chapter 3) constitute key phenomena for accident performance and analyses in WCNR, as discussed in Chapters 6 and 15. In this chapter, Sections 7.5.4, 7.8.1, and 7.8.7 deal with fundamental aspects of CHF, and Sections 7.5.4, 7.7.2, and 7.8.7 connect with fundamental aspects of reflood. Furthermore, Section 3.3.2.4 (Chapter 3) deals with the phenomenological distinction between CHF and DNB.

Hereafter, attention goes to the discussion of aspects closely connected with the nominal operation (i.e., maximum removable thermal power from nuclear fuel, see Chapter 23) and transient performance of the cores of nuclear reactors in the case of CHF or DNB, and accident/transient performance (only) in the case of reflood.

The ballooning, burst, fuel relocation, and reflood of ballooned fuel imply consideration of nuclear fuel behavior. including structural mechanics performance, and do not fall within the boundary of interest for the present book, although Chapter 24 includes a description of current trends and perspectives in those areas.

7.12.1 Critical heat flux and departure from nucleate boiling

Textbooks, journals, and conferences focus on CHF and DNB. Starting in 1996 with a Special Issue of J NED, e.g., Groeneveld (1996), [*who mention that by 1990 over 700 CHF correlations were available … and this is a cause for concern …*], the publication of thousands of papers and documents followed, making impossible, today (i.e., 2022), a comprehensive summary of all contributions in the area with reasonable work force. Excellent reviews may be found in textbooks by Collier and Thome (1980), Tong (1965), Tong and Weisman (1970), Hewitt and Hall-Taylor (1970), and Hetsroni (1982), and review articles by Bergles (1977), Tong (1972), Groeneveld and Snoek (1986), Weisman (1992), and Katto (1994). Insights into the topic can be found in works by Wieckhorst et al. (2015), Herer (2015), Liu and Shan (2015), Chu et al. (2015) Kossolapov et al. (2022), and Bucci (2022).

Relevant information hereafter is adapted from a document issued following an IAEA Coordinated Research Project (CRP), Aksan et al. (2001), (or IAEA Tecdoc 1203, IAEA, 2001), see also Groeneveld et al. (1999), where the authors of the present book provided contributions. The objective is to describe CHF methods suitable for the prediction of critical power under both normal and accident conditions in systems codes (see Chapter 11).

The two main applications for CHF predictions are as follows:

(i) To set the operating power with a comfortable margin to avoid CHF occurrence, as shown in Fig. 7.68. The margin to avoid CHF can be expressed in terms of

- Minimum critical heat flux ratio (MCHFR, ratio of CHF to local heat flux for the same pressure, mass flux, and quality),

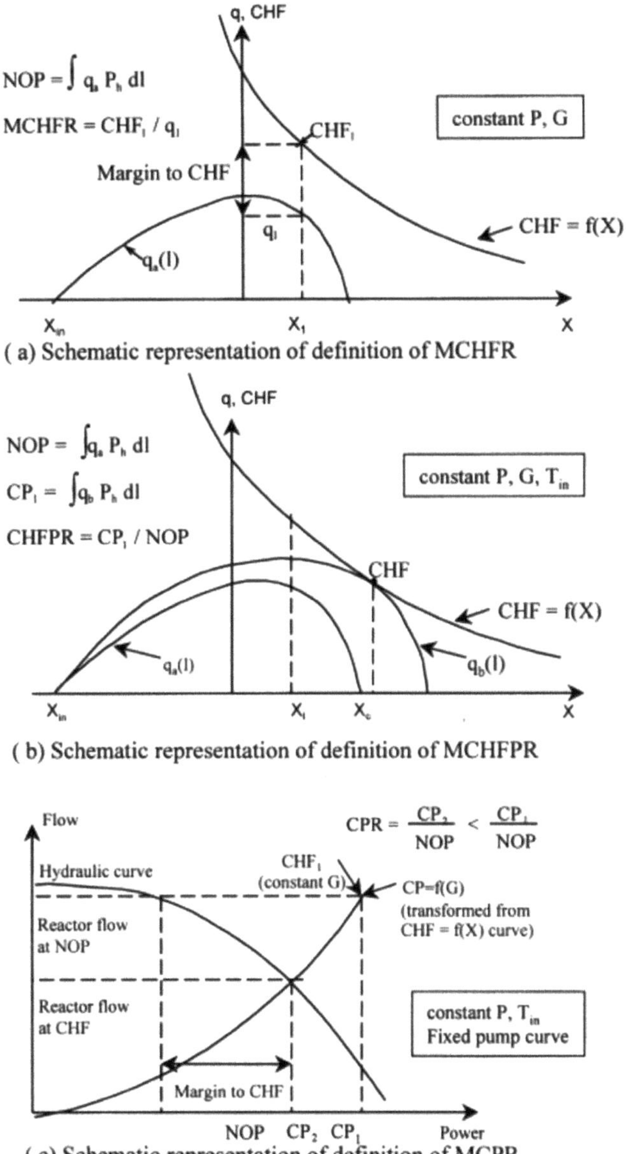

(a) Schematic representation of definition of MCHFR

(b) Schematic representation of definition of MCHFPR

(c) Schematic representation of definition of MCPR

Figure 7.68 Definition of margins to CHF as defined by MCHFR, MCHFPR and MCPR.

- Minimum critical heat flux power ratio (MCHFPR, the ratio of power at initial CHF occurrence to the operating power for the same pressure mass flux and inlet temperature),
- Minimum critical power ratio (MCPR, the ratio of reactor or fuel channel power at initial dryout occurrence to normal operating power for the same system, pressure, and inlet temperature).

(ii) To evaluate the thermal-hydraulic and neutron physics response to CHF occurrence in a reactor core. This requires knowledge of how CHF spreads in the reactor core, which in turn requires a best-estimate prediction of the average CHF for a section of the core or prediction of variation in the fuel surface area in dryout as a function of power (as discussed e.g., in Chapters 6 and 15).

In the following subsections, we discuss the following:

a mechanisms,
b experimental databases,
c steady-state prediction methods for both tubes and bundle geometries, ranging from correlations to subchannel codes, analytical models, and lookup tables,
d transient prediction methods.

7.12.1.1 Critical heat flux mechanisms

In forced convective boiling, a boiling crisis (or burnout, dryout, or DNB) occurs when heat flux increases to such a high level that the heated surface can no longer support continuous liquid contact. We identify this as the CHF. It is characterized either by a sudden rise in surface temperature caused by blanketing of the heated surface by a stable vapor layer or by small surface temperature spikes corresponding to the appearance and disappearance of dry patches. CHF normally limits the amount of power transferred, both in nuclear fuel bundles and in conventional boilers. Failure of the heated surface may occur after exceeding CHF. This is especially true under high-pressure, highly subcooled CHF conditions. With high flows and positive dryout qualities, postdryout heat transfer is reasonably effective in keeping the heated surface temperatures at moderate levels, i.e., without challenging the structural integrity of the heater. In flow boiling, CHF mechanisms depend on flow regimes and phase distributions that are in turn controlled by pressure, mass flux, and quality. For nuclear reactor conditions of interest, flow quality generally has the strongest effect on CHF; CHF decreases rapidly with increases in quality.

Departure from nucleate boiling

DNB can be a nucleation-induced mechanism for a boiling crisis. This is encountered at high subcooling (negative flow qualities) where heat is transferred efficiently by nucleate boiling. Here the bubbles grow and collapse at the wall; convection and conduction heat transfer takes place in the tiny space around the bubbles. Under those conditions, Collier and Thome (1980) and Tong (1972) suggest that the CHF occurrence is due to the spreading of a dry patch following microlayer evaporation under a bubble and coalescence of adjacent bubbles (although no definite proof of this is yet available). The occurrence of CHF here only depends on local surface heat flux and flow conditions, and the upstream heat flux distribution does not affect the phenomenon.

Bubble clouding can be another mechanism for DNB. In subcooled and saturated nucleate boiling (approximate quality range of −5% to +5%), the number of bubbles generated depends on the heat flux and the bulk temperature. Tong (1965) and Weisman and Pei (1983) suggest that bubble population density near the heated surface increases with increasing heat flux, and a so-called bubble boundary layer often forms a short distance from the surface. If this layer is sufficiently thick, it can impede the flow of coolant to the heated surface. This in turn leads to a further increase in bubble population until the wall becomes so hot that a vapor patch forms over the heated surface.

In either condition depicted above, the surface temperature excursion following DNB is typically rapid (fast dryout) at low pressure and results in failure of the heated surface.

Critical heat flux by Helmholtz instability

In saturated pool boiling, the maximum vapor removal rate controls CHF. Zuber's theory of CHF [as reported by Hsu and Graham (1986)] assumes a heated surface covered by a rising vapor column with countercurrent liquid jets flowing downward to compensate for the removal of liquid by evaporation. Ultimately, at very high heat flux levels (vapor removal rates), the relative velocity between liquid and vapor will be so high that an unstable flow situation occurs, resulting in a CHF condition. Kutateladze (1952) proposed this explanation based on his hydrodynamic theory of instability. A similar situation occurs at very low flow rates or at flow stagnation conditions.

Critical heat flux following annular film dryout

In the annular dispersed flow regime (high void fraction and mass flow), the liquid will be in the form of a liquid film covering the walls, and the entrained droplets move at a higher velocity in the center of the flow channel. Continuous thinning of the liquid film takes place due to the combined effect of entrainment and evaporation. Near the dryout location, the liquid film becomes very thin, and due to the lack of roll waves (which normally occur at higher liquid film flow rates), entrainment is suppressed. If the net droplet deposition rate does not balance the evaporation rate, the liquid film must break down. The temperature rise accompanying this film breakdown is usually moderate (stable dryout). One of the following mechanisms may promote the liquid film breakdown:

(i) Thermocapillary effect: If a significant amount of heat transport occurs by conduction through the liquid film and the interface is wavy, the temperature of the liquid–vapor interface will have a maximum in the valley of the wave, and large surface tension gradients will be present. The surface tension gradients tend to draw liquid to areas of high surface tension. Under the influence of the thermocapillary effect, the liquid film will eventually break down in the valley of the wave. This mechanism is important for low flows and high qualities.

(ii) Nucleation-induced film breakdown: Hewitt et al. (1963) noted that nucleation and surface evaporation could occur simultaneously in the annular flow regime. If the liquid film thickness is close to the maximum bubble size, the bubble may rupture the liquid–vapor interface, and a momentary dry patch could occur. At high heat flux levels, a high dry-patch temperature may prevent local (spot) rewet by the liquid film. This mechanism will only occur for local heat flux spikes or nonuniform axial heat flux distribution.

Unstable or periodic dryout

The CHF may be considerably reduced by the hydrodynamic characteristics of the experimental equipment. Flow oscillations occur in parallel channels, channels experiencing slug flow, or systems having a compressible volume near the inlet. During oscillation, the velocity at the wall periodically slows down, thus permitting the boundary layer to superheat; this may lead to the premature formation of a dry patch. Oscillations in the surface temperature occur together with unstable dryout.

Slow dryout

During a slow dryout, the heated surface does not experience the usual steep temperature excursions; instead, a gradual increase in surface temperature with power occurs. A slow dryout usually occurs in flow regimes where the phases are distributed homogeneously, such as froth flow or highly dispersed annular flow at high mass velocities (>2.7 Mg m^{-2} s^{-1}) and void fractions $>80\%$. At these conditions, liquid-wall interaction is significant, thus limiting the temperature rise following dryout. Calculations based on cooling by vapor flow only indicate that post-CHF temperatures are below the MFB (Leidenfrost) temperature; hence, depositing droplets may wet the surface, thus increasing the heat transfer coefficient.

7.12.1.2 Critical heat flux database

Since CHF usually limits power output in water-cooled reactors, accurate CHF values are required. CHF measurements are available extensively in simple geometries such as directly heated tubes. Such measurements have helped us understand CHF mechanisms. However, to obtain accurate CHF values at reactor conditions of interest, experiments in test sections closely simulating the reactor fuel bundles are required. Such experiments are very expensive; e.g., CHF tests in Canada alone cost more than 30 million dollars over the 1980–2000 period.

To reduce the expense and complexity of CHF testing of full-scale fuel bundles with high-pressure steam-water, low-latent-heat modeling fluids have been used. Many laboratories have adopted Freon as a modeling fluid for simulating the CHF of water. Reliable CHF measurements result at considerably lower pressures (e.g., 1.56 MPa in Freon-12 compared with 10 MPa in water), temperatures (e.g., 50°C in Freon-12 compared with 300°C in water), and powers (e.g., 685 kW in Freon-12 compared with 10 MW in water), resulting in cost savings of around 80% compared with equivalent experiments in water.

Aksan et al. (2001) distinguish tube and bundle databases. The tube database is the most complete and covers a much wider range of flow conditions than any other geometry. Otherwise, bundle geometries can be broken down into subchannels (see Chapter 18) that behave (better, that researchers assume behave) as tubes with correction factors applied to account for specific effects.

Fig. 7.69 provides an excerpt from Table 3.1 of Aksan et al. (2001), which summarizes data collected jointly by AECL and IPPE and is at the basis of the development of CHF prediction methods, including the CHF lookup table, Groeneveld et al. (1996).

Diameter (mm)	Length (m)	Pressure (MPa)	Mass Flux ($Mg.m^{-2}.s^{-1}$)	Dryout Quality (-)	Inlet Subcooling ($kJ.kg^{-1}$)	Critical Heat Flux ($MW.m^{-2}$)	No. of Data
10.0	1.000–4.966	9.80–19.6	0.216–7.566	−0.866–0.944	57–1398	0.134–4.949	1108
3.94–20.1	0.400–3.750	0.22–8.97	0.100–3.183	−0.069–1.054	−50–1640	0.278–7.477	2664
3.93–37.5	0.216–3.750	1.13–9.91	0.160–5.586	−0.005–0.993	−16–2711	0.503–6.620	1343
2.40–36.0	0.500–1.880	3.05–7.10	0.093–2.725	0.207–0.903	371–1065	1.026–5.130	116
10.0	1.000–4.966	3.00–20.0	0.156–8.111	−0.866–1.061	26–1414	0.135–5.476	1496

Figure 7.69 Excerpt from Table 3.1 by Aksan et al. (2001), reporting ranges of conditions covered by various sets in the AECL databank.

From the same document, Fig. 7.70 shows that the conditions covered, although extensive, do leave several gaps in the data.[6]

The parameters controlling CHF in tubes (for steady-state conditions and uniform heat flux distributions) are the following:

(i) Primary: thermodynamic quality, mass velocity, pressure, and diameter;
(ii) Secondary: heated length, surface roughness, conductivity, and tube wall thickness.

Secondary parameters usually have an insignificant effect on CHF for conditions of interest.

A large number of CHF experiments in bundles range from three-rod bundles to full-scale simulations of actual fuel bundles. The following parameters are important in controlling CHF in fuel bundles:

(a) Flow parameters (pressure, mass flow, and quality). These include average cross-sectional flow conditions and the distribution of flow parameters (i.e., distribution of enthalpy and flow across a bundle as evaluated by subchannel codes or other empirical means).
(b) Bundle geometric parameters (number of rods, rod spacing, unheated flow boundary, and heated length);
(c) Spacer grids and CHF enhancement devices (grids, appendages, and mixing vanes) and their axial spacing;
(d) Heat flux distribution (axial and radial heat flux distributions and flux tilt across elements).

Because of the proprietary nature of bundle CHF data, reviews are usually restricted. Hughes et al. (1974) provide a compilation of bundle CHF data sources, and Macek (1998) contributed the ranges of conditions covered by specific bundle data for VVER, see Fig. 7.71. However, as most bundle experiments still use fixed thermocouples, the reliability of experimental CHF data as representing the initial

[6] The nonproprietary part of the CHF databank, containing over 30,000 CHF data, obtained in directly heated tubes, has been deposited in the International Nuclear Safety Center (INSC) Database at Argonne National Laboratory (Aksan et al., 2001, provide details for the INSC foundation and objectives and structure).

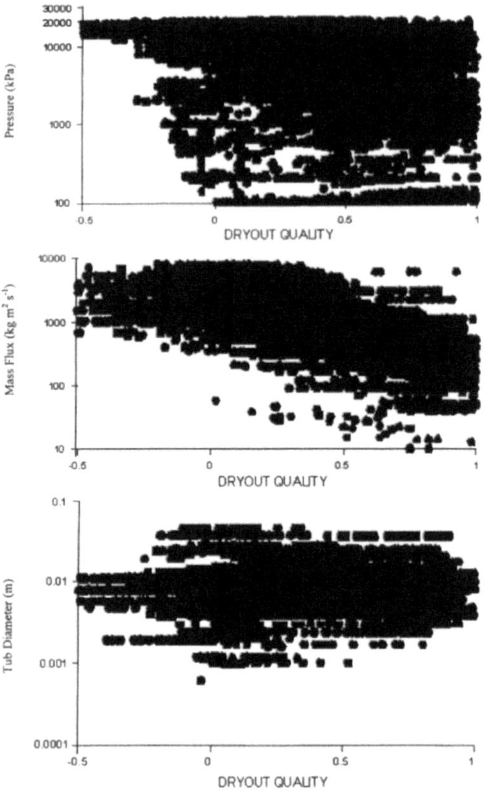

Figure 7.70 Detail from Figure 3.2a by Aksan et al. (2001), reporting the ranges of test conditions for the combined AECL-IPPE tube-CHF data bank.

occurrence of CHF may be too optimistic (i.e., overpredict CHF). Schenk and Groeneveld (1990) demonstrated that large differences (up to 20%) in bundle CHF could occur around the circumference of the most critical rod at the axial location corresponding to the initial CHF.

7.12.1.3 Critical heat flux prediction methods

The complexity of predicting CHF in a nuclear fuel bundle may be best understood by first considering the prediction of CHF of the simplest experimental setup—a uniformly heated tube cooled internally by a fluid flowing at a steady rate vertically upward. Here CHF is a function of the following independent variables:

$$CHF = f(L_H, D_e, G, H_{in}, P, E) \tag{7.191}$$

where, $L_H, D_e, G, H_{in}, P, E$ are heated length, equivalent diameter, mass flow rate, channel inlet enthalpy, pressure, and a parameter to account for the effect of the heated surface, i.e., surface roughness, thermal conductivity, and wall thickness. Yang et al. (2019) provide an overall view of CHF complexity, including subchannel code applications.

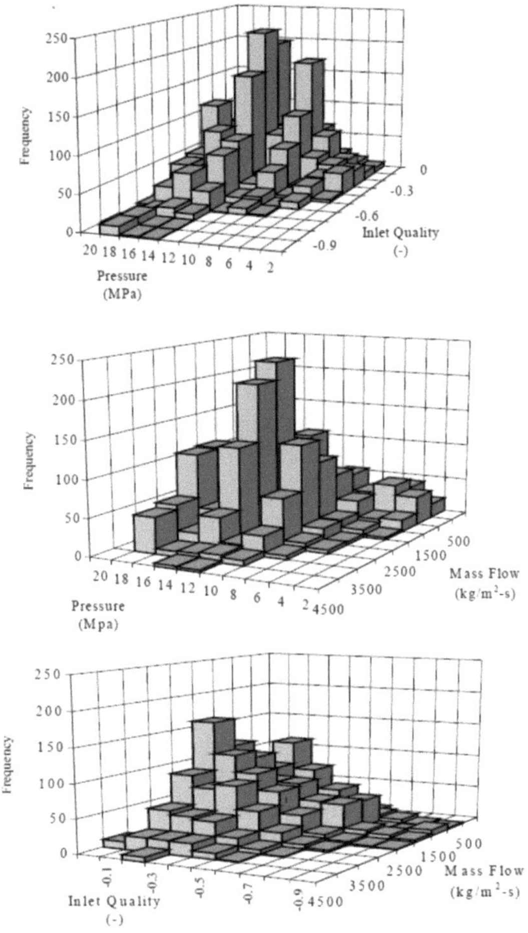

Figure 7.71 Detail from Figure 3.2b by Aksan et al. (2001), reporting the ranges of test conditions for the VVER-CHF data bank.

Analytical or mechanistic models

Analytical or mechanistic CHF models look at the physical mechanisms and satisfy the conservation equations. They generally require a two-fluid model approach but occasionally must use a three-field approach (e.g., dispersed annular flow). Because of a limited understanding of the mechanisms involved and a lack of measurements of interfacial parameters, the models are less accurate than empirical correlations over the range of their database; see the review by Weisman (1992). We summarize three key approaches for mechanistic modeling; see Liang and Mudawar (2018) for phenomenological details in the case of pool boiling.

The annular film dryout model considers the mass balance on the liquid film in an annular flow regime and postulates that CHF corresponds to depletion of the liquid film. Equations for droplet entrainment and deposition are typically necessary. The model provides a reasonable prediction of CHF for the annular flow at medium to high pressures and flows and void fractions exceeding 50%, Hewitt and Hall-Taylor (1970).

The bubbly layer model postulates that CHF occurrence in the lower quality regime first occurs when the bubble layer covering the heated surface, becomes so thick and saturated with bubbles that liquid mixing between the heated surface and the cooler core liquid becomes insufficient. Weisman and Pei (1983) proposed the model; Ying and Weisman (1986) appear to predict CHF with reasonable accuracy under high pressure, high flow, and low quality conditions.

The Helmholtz instability model considers the boiling crisis in pool boiling. The boiling crisis occurs when the flow of vapor leaving the heated surface is so large that it prevents a sufficient amount of liquid from reaching the surface to maintain the heated surface in the wet condition. The phenomenon that limits the inflow of liquid is Helmholtz instability, which occurs when a countercurrent flow of vapor and liquid becomes unstable. Zuber (1959) and Kutateladze (1952) have derived equations for CHF based on the Helmholtz instability theory; their predictions agree with CHF values measured in pool boiling systems. For very low flows, a modified version of this model as expressed by the Zuber-Griffith CHF correlation:

$$CHF = CHF_{PB}(1 - \alpha) \tag{7.192}$$

which appears reasonable for upflows and downflows less than 0.1 Mg m^{-2} s^{-1} and $\alpha < 08$. However, for $\alpha > 0.8$, this correlation significantly underpredicts CHF. Under these conditions, the $(1 - \alpha)$ correction of Eq. (7.192) is not recommended.

Empirical critical heat flux prediction methods

We distinguish empirical CHF prediction methods into those based on inlet conditions and those based on local cross-sectional average (CSA) conditions.

The inlet-conditions-type prediction methods are all in the form of empirical correlations based on CSA inlet conditions (P, G, H_{in}) and usually assume the "overall power" hypothesis. This hypothesis states that for a given geometry and inlet conditions, the critical power "no dryout" (NDO, a power corresponding to the first occurrence of CHF for that geometry) is independent of axial or radial heat flux distribution, or

$$Q_{NDO} = f_{CSA}(P_{in}, G_{in}, H_{in}, L_H) \tag{7.193}$$

This permits the use of CHF correlations derived from uniformly heated bundle data for the prediction of dryout power in nonuniformly heated bundles of identical geometry (i.e., identical cross section and heated length). This technique gives a reasonable estimate of dryout power in the annular flow regime for symmetric flux profiles and form factors ($=q_{max}/q_{avg}$) close to unity. However, we do not recommend them for form factors significantly different from unity. This approach is reasonable to predict the critical power of fuel channels with a fixed cross section, heated length, axial flux distribution (AFD), and radial flux distribution (RFD), irrespective of the form factor. If the experimental AFD and RFD represent the worst flux shapes from a CHF point of view, the empirical correlations provide lower-bound predictions.

The inlet-conditions method cannot be used for predicting the location and magnitude of CHF except when CHF initially occurs at the downstream end.

The local-conditions-type prediction methods follow the local-conditions hypothesis, which states that the local CHF is dependent only on local conditions and not on upstream history. In principle, the local-conditions hypothesis is sound if it depends on true local conditions (which must include the radial distribution of void, liquid and vapor velocity, liquid temperature, and turbulent velocity fluctuation near the wall). Hence, ideally

$$CHF = f_{CSA}(P_{in}, G_{in}, X_{DO}) \tag{7.194}$$

In practice, only the local cross section average pressure, flow, and quality are known; the usual assumption is

$$CHF = f_{CSA}\left[\alpha(z), H_l(z), H_g(z), P\right] \tag{7.195}$$

This form is more convenient than Eq. (7.191) since it depends on fewer parameters and permits the prediction of the location of CHF. One complication with this method is its lack of capability to account for the effect of AFD.

Two methods attempting to account for the effect of a nonuniform AFD on CHF are the boiling-length-average (BLA) approach and the F-factor approach; see Tong (1965, 1972) and Kirillov and Yushenko (1996). The F-factor approach tends to modify CHF correlations designed for uniform heating, while the BLA approach modifies the heat flux distribution. Lahey and Moody (1993) have shown that the two techniques are similar, yield similar answers, and are reasonably successful in predicting CHF under various nonuniform AFD conditions.

To overcome the main disadvantage of local-conditions-type prediction methods, i.e., their limited range of application, the CHF table lookup method was proposed. The initial attempt to construct a standard table of CHF values for a given geometry was made by Doroshchuk et al. (1975) using a limited database of 5000 data. The CHF table approach, which is basically a normalized databank, has been continued at CENG (Grenoble), the University of Ottawa, IPPE, and Chalk River using a much more extensive database (30,000 data). The International CHF table lookup method, Groeneveld et al. (1996), provides CHF values for water-cooled tubes at discrete values of pressure (P), mass flux (G), and quality (X) covering ranges of 0.1–20 MPa, 0–7500 kg m−2 s−1, and −50% to 100%. Here, zero flow refers to pool-boiling conditions, and negative vapor qualities refer to subcooled conditions. Linear interpolation between table values is used to determine CHF. Extrapolation is usually not necessary as the table covers a wide range of conditions. Compared with other available prediction methods, the tabular approach has the following advantages (which require continuous confirmation):

(i) greater accuracy,
(ii) wider range of applications,
(iii) correct asymptotic trend,

(iv) less computing time,
(v) can be easily updated if additional data become available.

Although tabular techniques initially made reference to tubular geometries and found application in subchannel codes, their greatest potential for application is in predicting the consequences of postulated LOCAs. To apply the tables to transient heat transfer in bundles, the use of adjustment factors is necessary to correct for geometry, flux shape, and possible transient effects. The lookup table approach has certain disadvantages:

(i) It is a purely empirical prediction method and hence does not reflect any of the physics.
(ii) It could introduce erroneous trends if the underlying database is subject to experimental errors.

Application to bundle geometries

The prediction of critical power in untested fuel bundle geometries remains unreliable. Effects of flux distribution, grid spacers and bundle array dimensions may need further investigations.

Empirical CHF prediction methods use cross-sectional average conditions to predict CHF or critical power and are designed for tubes or bundles. For bundles for which experimental data can be obtained (using an electrically heated fuel bundle simulator with a fixed axial and a radial flux distribution), a variant of the following methodology is frequently employed:

a To obtain sufficient data for deriving an empirical CHF correlation for conditions of primary interest;
b To extrapolate the empirical correlations (which are usually based on a given axial and radial flux distribution) to other flux distributions of interest using the change in CHF as predicted by subchannel codes or empirical methods to account for changes in the upstream flux shape;
c To similarly extrapolate to other conditions not tested in the full-scale simulation tests using trends observed in simpler geometries, or as predicted by subchannel codes.

The subchannel approach methods are different from the empirical approach, as they predict the axial variation in flow and enthalpy for each subchannel. It is particularly useful for bundles for which no direct experimental data are available. One may pursue the following methodology:

a To employ subchannel codes to predict the flow and enthalpy predictions across the bundle;
b To employ subchannel CHF models (basically modified tube CHF prediction methods) for predicting the initial CHF occurrence anywhere in the bundle.

Two definitions of subchannels are currently in use (see Chapter 18 for details).

The conventional approach defines subchannel boundaries by lines between rod centers and is used in subchannel codes such as ASSERT, Carver et al. (1993), and COBRA, Rowe (1971). The rod-centered approach defines subchannel boundaries by lines of zero stress between rods and is used primarily to predict CHF in the annular flow regime, e.g., using the annular flow model by Hewitt and Hall-Taylor (1970), annular flow model, or an equivalent CHF correlation; see Weisman and Bowring (1975).

The enthalpy imbalance approach constitutes the alternative subchannel approach, as discussed in the following:

- McPherson (1971) applied it to various bundle geometries contained in pressure tubes.
- Bobkov et al. (1995, 1997) applied it to eccentric annuli and bundle subchannels.
- Leung (1997) applied it to 37 element bundle CHF predictions.

The approach of Kirillov et al. (1996) considers differences in enthalpy growth rates among bundle subchannels and based on this defines a quality imbalance, ΔX, for that bundle. This quality imbalance (a variation of this is the enthalpy imbalance number specified by McPherson (1971)), represents the difference in qualities between the cross-sectional average bundle quality and the maximum bundle subchannel quality for a given cross section. The difficulty is in predicting the ΔX value; no general expression for the enthalpy imbalance is yet available.

The lookup table

The tube CHF lookup, Groeneveld et al. (1996), upgraded by Groeneveld et al. (2007), see also Groeneveld (2019), must be converted into a prediction method for bundle geometries. Two approaches are mentioned for doing this: the subchannel-based approach and the cross-sectional average bundle approach.

A subchannel code (Chapter 18) is necessary to predict CHF for bundle geometry. The following correction factor methodology applies to evaluate the bundle or subchannel CHF:

$$CHF_{BUNDLE} = CHF_{TABLE}\, K_1 K_2 K_3 K_4 K_5 K_6 K_7 K_8 K_9 \tag{7.196}$$

where nine complex factors are necessary. The factor definition is as follows; see also Choi (2017):

K_1 = Subchannel or tube-diameter cross-section geometry factor. This includes the observed diameter effect on CHF. This effect is slightly quality-dependent.

K_2 = Bundle geometry factor. This depends on the bundle pitch and is a weak function of P, G, and X.

K_3 = Grid spacer factors. It depends upon the pressure loss coefficient and the distance from the grid.

K_4 = Heated length factor. The consideration of homogeneous void fraction allows a reasonable prediction of the diminishing length effect at subcooled conditions.

K_5 = Axial flux distribution factor. The F-factor method proposed by Tong (1972) is applicable within narrow ranges of conditions.

K_6 = Radial or circumferential flux distribution factor. It accounts for the effect of flux tilts across the bundles; see, e.g., Yin et al. (1991).

K_7 = Flow orientation factor (or horizontal flow factor). Wong et al. (1990) proposed a formulation based on a balance of turbulent and gravitational forces.

K_8 = Vertical flow factor. This includes pool boiling and downward flow.

K_9 = Pressure out-of-range factors proposed by Choi (2017).

In the case of research reactors and rectangular channels, Song et al. (2020) proposed a specific methodology for deriving the factors.

7.12.1.4 Critical heat flux and transient conditions

Accuracy of CHF prediction methods is essential to determine margins to CHF, Fig. 7.68, and allowable (linear) power of bundles and core in nominal operating conditions of nuclear reactors (Chapter 23). Development of subchannel codes (Chapter 18), as already mentioned, aims at supporting the design including the interpretation of specific bundle CHF experiments.

CHF in transient conditions occurs following a variety of design basis accidents (Chapter 15): system thermal-hydraulic codes (Chapter 11) are essential to investigate those conditions, although subchannel codes may support related investigations. A comprehensive picture of CHF occurrences in DBA conditions is well beyond the scope here, although readers may get relevant information in Chapter 6 and 15. To show the complexity of CHF evaluation, we provide a few notes below distinguishing small-break LOCAs (SBLOCAs) and power excursions.

Small-break loss-of-coolant accidents and critical heat flux

Experimental SBLOCA counterpart test (CT) scenarios are available from various ITFs; these are suitable for validation and scaling confirmation; see e.g., Chapter 21, Section 13.2.2.6 (Fig. 13.3) of the book and D'Auria et al. (1992). Here we consider LOBI and BETHSY CT data. Although CHF is not the parameter shown in Figs. 7.72 and 7.73, CHF occurrences are at the origin of the diagrams derived from experiments. Neglecting similarities and differences between LOBI and BETHSY diagrams (eventually, related investigations have included a scaling study), one may note the following:

- Temperatures greater than about 570 K indicate the occurrence of a CHF condition, typically a few dozen seconds before the picture is snapped.
- The prediction of (reported-measured) rod surface temperatures needs a post-CHF heat transfer model: a CHF occurrence typically triggers the post-CHF phenomenology.[7]
- The cosine axial power profile characterizes both (i.e., LOBI and BETHSY) electrically heated fuel bundles. However, in the former case (Fig. 7.72), the majority of CHF conditions occur at the top of the fuel bundle, while in the latter case (Fig. 7.73), a higher number of CHF conditions occur within the high-power region.
- Axial and radial disuniformities characterize the CHF condition in both cases: one may observe no dryout at the bottom, i.e., at any axial location, and at the same time, rods in post-CHF and nucleate boiling conditions, i.e., at the middle and the top of fuel rod simulators.

Three-dimensional flow behavior takes place in the core of "thin and long" experimental facilities sometimes referred to as one-dimensional ITFs. We call the phenomenon "Global multi-D fluid temperature, void and flow distribution-core" in Chapter 6 and in Table 15.2 (Chapter 15).

[7] This is the connection between the current top diagrams in Figs. 7.72 and 7.73 and CHF. Time of occurrence of CHF in horizontal axes, would show a more direct connection with the topic of the present section.

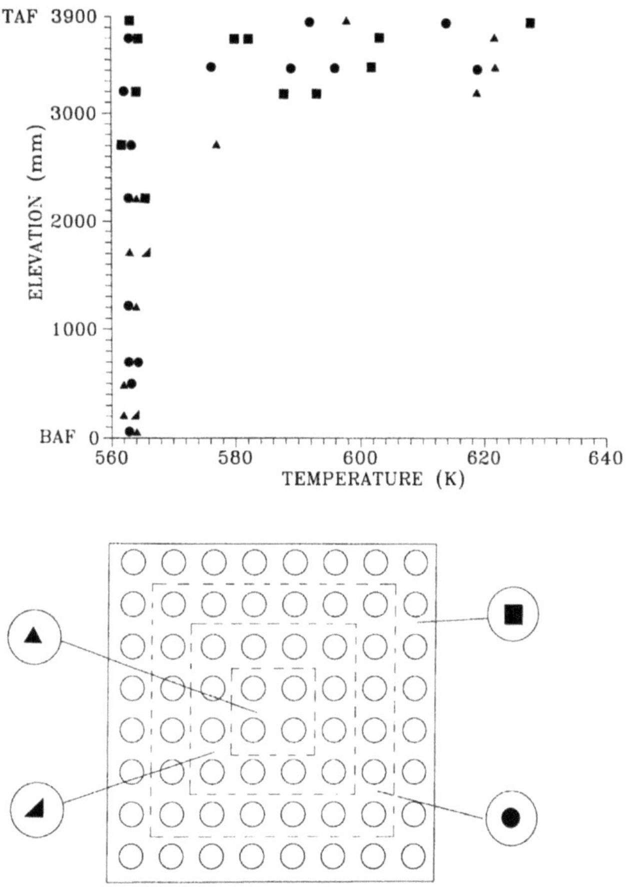

Figure 7.72 Snapshot of measured rod surface temperatures at the time of the first PCT during the LOBI small-break loss-of-coolant accident counterpart test, Aksan et al. (2001).

System thermal-hydraulic codes with the capability of calculating overall system behavior are necessary for detailed simulation of the local phenomena in the core, i.e., those that derive from measurements during SBLOCA CTs in LOBI and BETHSY. Subchannel code capabilities under transient conditions are necessary to distinguish individual subchannel behaviors. At the time of issuing the present book, this capability is not yet established, although attempts to use system codes as subchannel codes are available (e.g., Petruzzi et al., 2004), and comparisons of predictions of experiments by system codes and subchannel codes are also available (e.g., Cherubini et al., 2009, 2010).

Critical heat flux following power excursion

Fission power excursions may occur in case of accidents in water-cooled nuclear reactors, e.g., Pecchia et al. (2015), and Hernandez et al. (2020), including LOCAs in reactors having positive void reactivity, e.g., D'Auria et al. (2021a). In addition, in

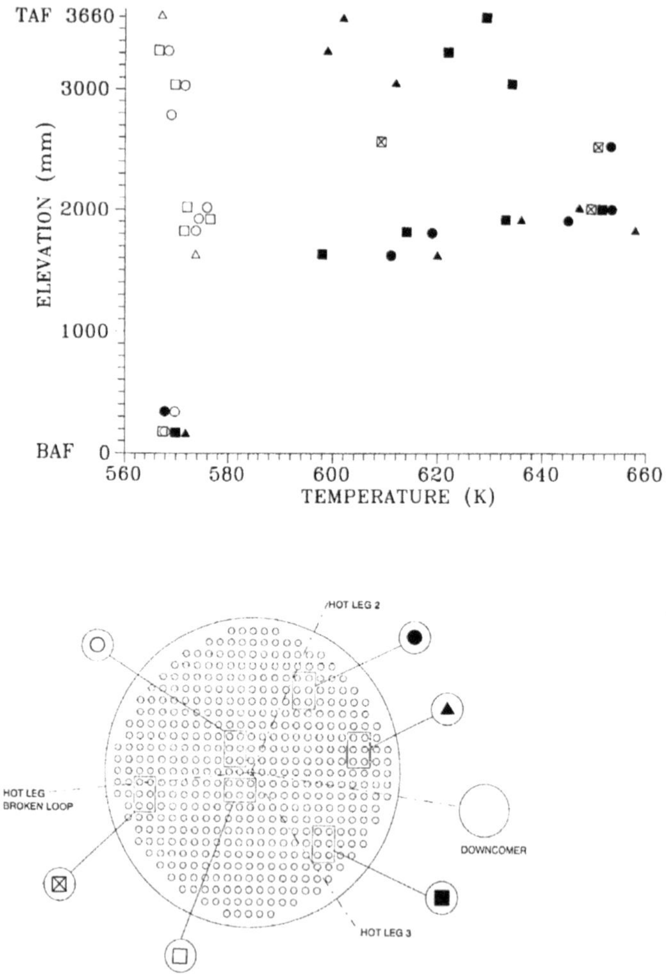

Figure 7.73 Snapshot of measured rod surface temperatures at the time of the first PCT during the BETHSY small-break loss-of-coolant accident counterpart test, Aksan et al. (2001).

the case of control rod ejection, Alzaben et al. (2019), a local increase in neutron flux causes subcooled void production: the heat transfer coefficient (HTC) for subcooled nucleate boiling under transient conditions may reveal the higher-than-HTC values found for nearly steady-state conditions. Thereafter, void production affects local moderation (not only due to a change in moderator density but also due to displacement of boron if present) and the buildup of pressure in the reactor pressure vessel in a period of the order of 10^{-1} s.[8] Then, the correct evaluation of HTC constitutes a safety issue.

[8] Smoothing of the pressure pulse typically occurs in the pressurizer of PWR with a properly designed surge-line.

CHF is not necessarily involved in the scenario consequent to control rod ejection; however, CHF occurrence and HTC at CHF occurrence may be part of the concerned safety issue. He et al. (2021), in their extensive review (of transient CHF), present data from 1968 showing transient CHF flux increases (heated wire diameter in the range 0.01−0.1 mm) up to three times related to steady-state CHF with a power ramp period in the range of 1−10 ms. A longer power ramp period negatively affects CHF value, e.g., results qualitatively consistent with the findings by Walunj and Sathyabhama (2018) in the case of heated wires in pool boiling. The same authors, He et al. (2021), did not report significant differences in transient CHF related to steady-state CHF in the case of rod bundles, although an inequality $(CHF_{transient}/CHF_{steady\ state}) > 1$ prevails in situations of interest to nuclear reactor safety.

7.12.2 Reflood

Let us introduce the terminology first: reflood, rewet, quench front propagation, and return to nucleate boiling (RNB) indicate the same phenomena, although differences may apply. Reflood deals with a phenomenological window (Chapter 15) expected at a system level in WCNR following large-break LOCAs (LBLOCAs). Rewet indicates the phenomena taking place in the core during reflood, which imply the jump of surface temperature from film boiling (steam on the surface of fuel rods) to nucleate boiling (liquid on the surface) conditions. Quench front refers to the progression of the liquid, either bottom-up or top-down, wetting the surface. RNB shall be takes as the opposite of DNB. The occurrence of the MFB temperature is a condition for reflood, rewet, quench front propagation, or RNB. Furthermore, we may consider rewet, quench front propagation, and RNB as aspects of reflood.

As in the case of CHF, textbooks, journals, and conferences focus on reflood, and it is not the objective here to duplicate the related information. Rather, the objective hereafter is to provide a picture for reflood (or better, rewet) connecting the heat transfer phenomenology (present chapter) with reflood discussed in Chapters 6 and 15.

7.12.2.1 Phenomenology and predictive capabilities

Following an LOCA, typically an LBLOCA, the primary system, notably a PWR or BWR, depressurizes and the mass inventory of coolant achieves a minimum value around 10%−20% of initial value, i.e., toward the end of the Blowdown phenomenological window. Refill of the vessel starts following intervention of early ECCS and, when the liquid level crosses the bottom fuel plate, or when hot leg injection (case of some PWR) or core spray (case of BWR) starts, the reflood phenomenological window occurs. The reflood implies the local rewet of fuel rod surfaces and the QF propagation. We distinguish bottom-up and top-down reflood, depending upon upward or downward QF propagation. Fig. 7.74, or first step of reflood phenomenology provides a view of the reflood at the vessel and the fuel rods levels.

Bottom up reflood is gravity driven: primarily, ECC pumps establish the level in down-comer of PWR (otherwise, in cases of BWR the top of jet pumps fixes the down-comer level, Fig. 23.3 in Section 23.3.1) and natural circulation flow allows

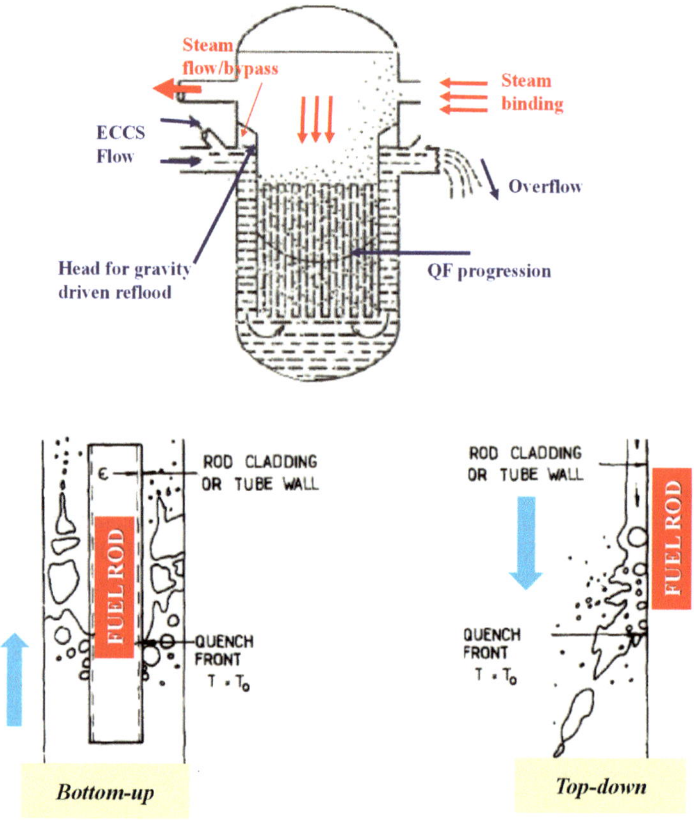

Figure 7.74 Step 1 for reflood phenomenology. Top: picture for the reflood scenario in vessel. Bottom: picture for fuel rod rewet; upward QF propagation (or bottom-up reflood), left and downward QF propagation (or top-down reflood), right.

the QF propagation. Steam binding (e.g., Sections 6.3.6.4 and 6.4 in Chapter 6) puts an obstacle to the advancement of the QF.

Unavoidably in a gravity environment, following downward QF propagation, steam production occurs due to heat transfer from nuclear fuel rods. Steam (unavoidably) flows upward and flooding and CCFL (e.g., Section 6.3.10) put an obstacle to the advancement of the QF.

Downward and upward QF propagations may occur at the same time, with some details discussed in Section 7.12.2.3.

Fig. 7.75 derived for bottom-up reflood represents the second step of reflood phenomenology. This is qualitatively valid for the top-down reflood, namely the left sketch in the figure.

The intensive heat transfer at the QF location together with the large difference in rod surface (and interior) temperatures between locations upstream and downstream

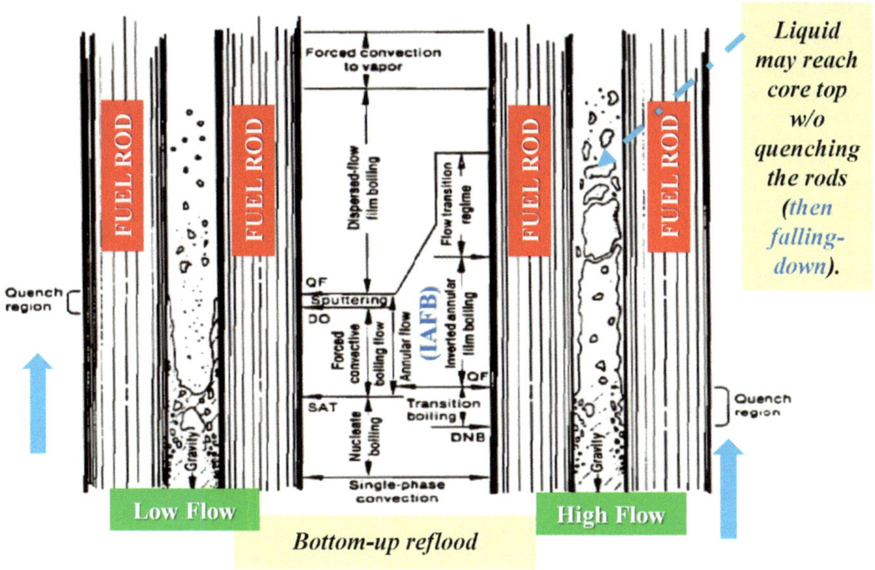

Figure 7.75 Step 2 for reflood phenomenology: connection between heat transfer and flow regimes. Left: low flow condition. Right: high flow condition.

the QF, makes not applicable the convection HTC (empirical) equation or definition, as provided in Section 3.3.2.3 (Chapter 3). Namely, an important portion of total thermal power transport occurs axially from downstream to upstream the QF, making invalid the assumption of heat transfer occurring orthogonal to the heated wall. The concerned axial temperature difference develops over a distance in the order of millimeters across the QF location and may have a value up to some 100 K.

The deep consideration by the reader of Fig. 7.75 replaces any discussion here. One may note (connecting with definitions in other parts of the book):

(a) droplet formation at the sputtering region: these induce the precursory cooling, typically after the occurrence of peak cladding temperature (PCT) and steam binding;
(b) arrival of droplets at the top of the core induces a possible level formation above the upper tie plate (UTP) and the consequent liquid fall down (or top-down reflood);
(c) the peculiar (specific for the reflood/rewet phenomenon) occurrence of the inverted annular film boiling (IAFB) heat transfer regime.

The MFB temperature (T_{MFB}) has different definitions in the literature, e.g., Section 7.5.4.1, connected or less with the Leidenfrost temperature. Here, T_{MFB} is the temperature at which the rewet occurs or the jump from film boiling (limited convection heat transfer owing to steam on the heating surface) to nucleate boiling heat transfer (efficient convection heat transfer with liquid wetting the heating surface). The RNB occurrence is an associated consequence, assuming an earlier DNB occurrence (see above). T_{MFB} is the rod surface temperature (nuclear fuel) at the QF location.

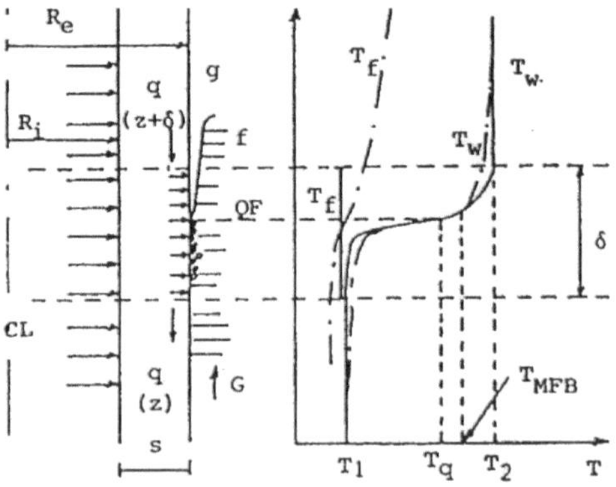

Figure 7.76 Step 3 for reflood phenomenology: qualitative sketch and relevant physical quantities for (bottom-up) reflood/rewet characterization.

The notes above allow the step 3 of reflood phenomenology, Fig. 7.76: the δ length is a few millimeters and the jump of Tw across the QF can be in the order of some 100 K, as already stated; T_{MFB} is in-between the values of rod surface temperature (Tw) downstream and upstream the QF.

7.12.2.2 Historical remarks and modeling

Pioneering publications in the area, e.g., Duffey and Porthouse (1973), see also Piggot and Duffey (1975), and Pearson et al. (1977), identify Semeria and Martinet (1965)[9] as originators for the reflood modeling idea, although others, e.g., Lahey and Moody (1993), report Yamanouchi (1968) as the originator of the same idea.

The Fourier time dependent conduction heat transfer for an isotropic medium, Eq. (7.6), and the sketch for the integration domain for the Fourier equation in Fig. 7.77 (derived from the phenomenological picture in Fig. 7.76), Cioni et al. (1986), lead to the still "… *unchanged*[4] …" reflood modeling idea.

Making reference to Fig. 7.77, Eq. (7.6), written in the two dimensions, "*z*" and "*y*" assuming a flat slab for the clad wall (this is justified by $s/R \ll 1$, where "*s*" and "*R*" are the clad thickness and the pin radius, respectively), and $q_V = 0$ (or $q''' = 0$, i.e., no heat generation in the solid), becomes (see also D'Auria et al., 2021b)

$$\frac{\partial^2 T}{\partial y^2} + \frac{\partial^2 T}{\partial z^2} = \frac{\rho c_p}{\lambda}\frac{\partial T}{\partial t} \tag{7.197}$$

Introducing (the "idea")

$$\frac{\partial T}{\partial t} = U\frac{\partial T}{\partial z} \tag{7.198}$$

[9] Romney Duffey wrote "… *there has been a zillion papers since (especially recently from the steel industry, plus India and China) but the controlling [modeled] physics remains unchanged,*" Duffey (2021).

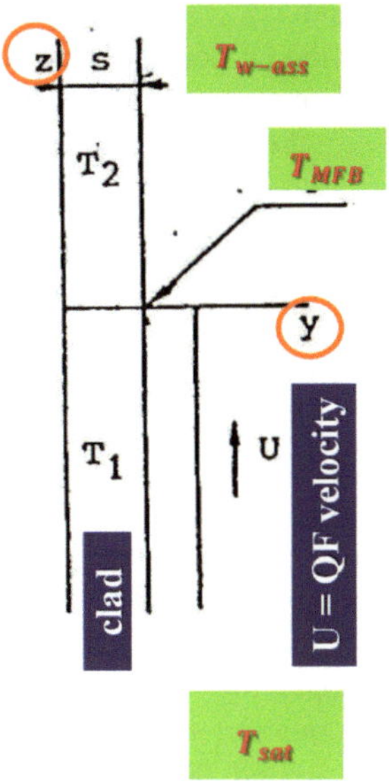

Figure 7.77 Sketch for the reflood modeling idea: the integration domain in the clad and in the fluid region.

where U is the (unknown) QF velocity, and Eq. (7.197) becomes

$$\frac{\partial^2 T}{\partial y^2} + \frac{\partial^2 T}{\partial z^2} = \frac{\rho c_p}{\lambda} U \frac{\partial T}{\partial z} \tag{7.199}$$

Time ("t") disappears in Eq. (7.199). Let us introduce the Fourier law ($q = q'' = -\lambda \frac{\partial T}{\partial z}$) and associated assumptions:

$$\frac{\partial^2 T}{\partial y^2} \approx \frac{\partial T}{\partial y} \frac{1}{s} \approx -\frac{1}{\lambda s} h(z)(T_w - T_{sat}) \tag{7.200}$$

The QF moves at a velocity U (different ant any z location), and the picture sketch in Fig. 7.77 reproduces at each axial location. Considering the assumptions in Eq. (7.200), Eq. (7.199) becomes an ordinary differential equation of the second order:

$$\frac{d^2 T}{dz^2} + \frac{\rho c_p}{\lambda} U \frac{\partial T}{\partial z} - \frac{1}{\lambda s} h(z)(T_w - T_{sat}) = 0 \tag{7.201}$$

Eq. (7.201) is solved under the green boundary conditions underlined in Fig. 7.77, i.e.,

$$T\ (z \gg 0) = T_{w-ass}, \text{and } T\ (z \ll 0) = T_{sat} \tag{7.202}$$

where $z = 0$ is the QF location and T_{w-ass} and T_{sat} are the rod surface temperatures (assumed known, independent of the reflood/rewet model) downstream and upstream of the QF. T_{MFB} is also assumed to be known independent of the reflood/rewet model.

By defining $h(z) = HTC_{\text{reflood}}$, the solution of Eq. (7.201) is

$$HTC_{\text{reflood}} = \left(U^2\rho^2c_p^2\ \frac{s}{2\lambda}\right)\left\{\left[\frac{2(T_{w-ass}\ -\ T_{MFB})}{(T_{MFB}\ -\ T_{sat})}+1\right]-1\right\} \tag{7.203}$$

Namely, assuming to know the wall temperature upstream and downstream the QF and the MFB temperature (from an independent correlation) from the measurement of U one may calculate HTC_{reflood}, i.e., $HTC_{\text{reflood}} = F\ (U,\ T_{MFB})$, primarily.

Reflood reviews are due to Yadigaroglu et al. (1993), Hochreiter et al. (2000), and Zeng et al. (2010), who also discuss the ballooning condition. In their reflood review, Yang et al. (2021) explain current frontiers in reflood modeling, e.g., droplet entrainment, deposition and break-up, in addition to the need for liquid-vapor interface measurement; they conclude "*Research on the reflood of a reactor core remains one of the top priorities for nuclear reactor thermal—hydraulic and safety analysis.*"

Current developments "rotate around" the modeling algorithm above outlined or discuss experiments and code applications.

Advancements in reflood modeling discussed by Yoo et al. (2021) focus on a new droplet deposition model including droplet-wall heat transfer, enhancement factor, and droplet contact heat transfer efficiency correlations. Jin et al. (2020) propose a specific correlation for spacer grid pressure drop based on reflood experiments in RBHT test facility.

Lutsanych et al. (2017) and Clifford and Ferroukhi (2021) discuss the results of applications of three-dimensional system codes CATHARE and TRACE, respectively, to reflood experiments.

Experimental data are discussed by Stuckert et al. (2011), and Chikhi and Fleurot (2012), who focus on the interaction of reflood with H2 production.

7.12.2.3 Issues and challenges

Today relevance and interest toward reflood comes from observing Fig. 13.17 (Section 13.4.3.2). Results from analysis of an LBLOCA in a PWR from several participants (i.e., a dozen+) are compared within the BEMUSE project framework of OECD/NEA/CSNI that was completed in 2012 (see Chapter 13 for details). Uncertainty investigations allowed demonstrating that the results of any code calculation bound the best estimate results of any remaining calculation. This includes the pressure, the break flow, and the PCT following CHF occurrence during the blow-down phenomenological window. The same conclusion (i.e., uncertainty bounding ...) does not apply to the reflood period for which a wide spread of calculation results was observed. This, on the one hand, prevented the issue of a positive message about the robustness of uncertainty evaluations, on the other hand, it showed inadequacies in reflood modeling and triggered new research in the area,[10] which are still in progress, OECD/NEA/CSNI (2022).

Further reflood issue-challenges are summarized below. Two of those derive from Figs. 7.78 and 7.79.

Data measured in Halden nuclear reactor in prototypical nuclear fuel rods, Fig. 7.78, Svanholm et al. (1995), reedited by D'Auria et al. (2021b), show several quench temperatures, $T_{QF} = T_{MFB}$, corresponding to one given PCT, T_P in the diagram, where T_{QF} triggers the (local) rewet. This makes ambiguous the applicability of Eq. (7.203) when experimental data are not available.

Fig. 7.79 shows a reflood-rewet scenario consistent with Fig. 7.78: experimental data and code calculation results are reported for the progression (bottom-up and top-down) of the QF. The applicability of Eq. (7.203) is questionable in the concerned situation, too. "Same-time" reflood, or homogeneous reflood, is a possible occurrence detected in various experiments, typically at the core top and toward the end (and completion) of the reflood, D'Auria et al. (2021b). Quench front propagation, i.e., the solid black line, showing results of top-down QF progression from code prediction in Fig. 7.79, does not match the almost instantaneous reflood occurring at about 220 s in the experiment (red crosses inside the green ellipse in the sketch). In different terms, the jump of rod surface temperatures from film boiling to nucleate boiling appears to

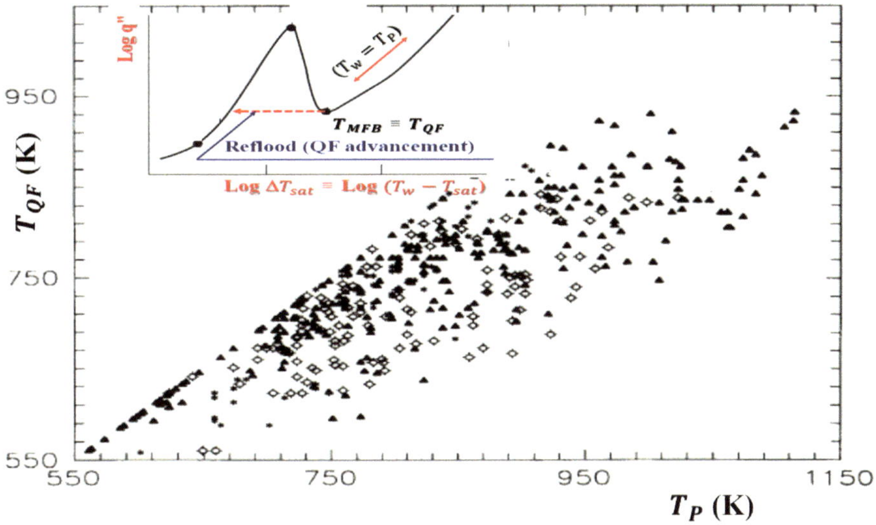

Figure 7.78 Reflood issue-challenge 1, from Halden experiments, Svanholm et al. (1995): at a given peak clad temperature, T_P, various temperatures $T_{MFB} = T_{QF}$, trigger the "local" rewet.

[10] Spread of results in reflood predictions (Fig. 13.17) may have different origins like insufficient nodalization qualification (and user effect) and modeling of pressure drops (e.g., at geometric discontinuites) and CCFL. However, triggering of new research on reflood resulted in an outcome from the BEMUSE project.

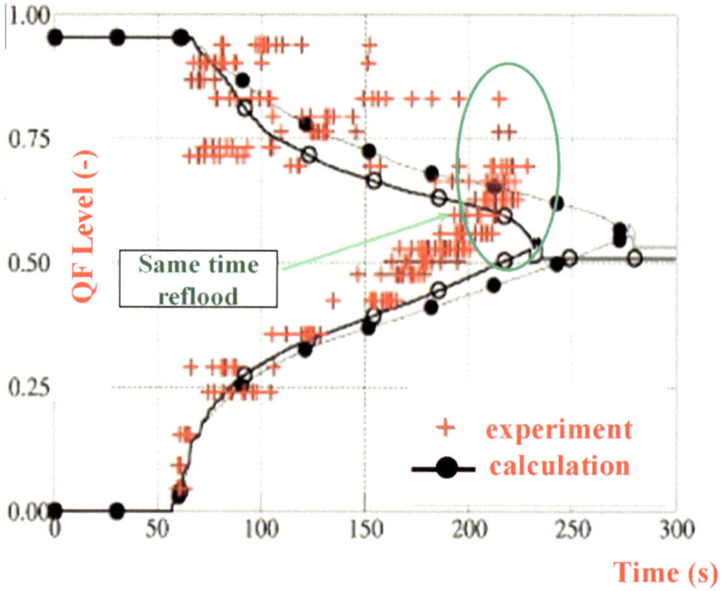

Figure 7.79 Reflood issue-challenge 2, generic comparison between measured and calculated QF propagation velocity. "Same time" rewet-reflood occurs in experiment, not predictable by code modeling, D'Auria et al. (2021b).

occur in experiments following the sudden void collapse, rather than following the QF propagation as predicted by the code model.

Similar to "same-time" reflood or homogeneous reflood is the phenomenon identified by Yang et al. (2019), and Liu et al. (2019); in the case of CHF experiments in uniformly heated rod bundle, they observe the exit reflood quenching phenomenon.

Kim et al. (2018) and Bang et al. (2022), discuss the interaction between reflood, ballooning, and fuel relocation. The authors of the last paper, when modeling reflood, introduce correlations for dispersed flow film boiling and inverted slug film boiling, in addition to the IAFB regime, Fig. 7.75. The nuclear fuel behavior during LOCA including reflood and high burn-up conditions is considered by Capps et al. (2021).

A summary of proposals for improvements in reflood modeling can be found in Section 24.2.4.2 (Chapter 24).

Exercises and questions

Exercises and questions are proposed hereafter with solutions given in Appendix 1 (Volume III). The readers should answer the questions or provide solutions to exercises before reading Appendix 1. Please provide extended answers.

7.1 What is the meaning of "conjugate heat transfer"?

7.2 Please explain the difference between Fourier equation and Fourier law.

7.3 An approximate solution of the Fourier equations allows the (demonstration of a linear) connection between linear power (w/m, derived from neutron flux and uranium fissions), fluid temperature, and uranium pellet centerline maximum temperature. Please derive such a solution (literature search) and discuss the key information and implications.

7.4 Discuss why (typically, under similar system conditions, pressure, velocity, etc.) the nucleate boiling heat transfer coefficient is (much) larger than the forced convection heat transfer coefficient.

7.5 Discuss the ideal existence of a multi-D heat transfer coefficient surface (or hypersurface). Discuss the advantages and the difficulties of deriving such a surface.

7.6 Discuss why rewet (local) process (part of reflood at a system level) modeling may not make use of definitions and derivations of heat transfer coefficient in heat transfer regimes like nucleate boiling or forced convection. Also, consider the discussion of reflood in Chapter 11.

7.7 Make a possible distinction between natural convection, heat transfer regime, (distinguishing pool boiling and flow boiling), forced convection (primarily, flow boiling heat transfer regime), and natural circulation (that involves the need for a heat sink and a heat source).

7.8 Identify the interference between radiation heat transfer, nucleate boiling heat transfer (weak interaction), and film boiling heat transfer (possible strong interaction). Discuss possible means to separate the impact of radiation heat transfer in an experiment where film boiling occurs and film boiling heat transfer is the topic of measurement.

7.9 Discuss why (early) critical heat transfer correlations (e.g., Rohsenow) may calculate CHF values much greater than the same value predicted by correlations/approaches like the lookup table under similar conditions (namely pool boiling).

7.10 Discuss the relevance of surface conditions (and of chemical/physical interactions between fluid and wall) upon heat transfer. Please provide a quantitative example (e.g., heat transfer coefficient on a clean surface and a surface with oxide and crud).

7.11 Please describe what CRUD is and how much CRUD and oxide layer may affect the heat transfer coefficient in a nuclear fuel rod's typical working condition. This affects the centerline pellet temperature in steady-state conditions. Please connect the answer to Question 7.3.

7.12 Considering the answers to questions 7.3 and 7.11, please inform about the impact of heat transfer coefficient, in nominal operation condition, upon the PCT should an LBLOCA (opening time of the rupture, unphysically taken as few milliseconds) occur in a current PWR. Please consider conditions that are established a few seconds after break opening (typically first PCT or blowdown PCT).

Acknowledgments

Sections 7.1−7.4 of the current chapter are based on Chapters 5, 6, 7, 11, 12, and 13 of the book by Kirillov et al. (Eds.), (2010). One of the authors of the present chapter, Prof. Kirillov, acknowledges the contributions of other editors (Bobkov, Zhukov and Yur'ev), in particular of the co-lead editor, Dr. Bobkov. Prof. Kirillov acknowledge the help of Dr. G. Bogoslovskaya in translating Russian text into English: without her help, the present chapter was not possible. Also noting that the original book (currently in three volumes, see general list of references) was contributed by many colleagues in the thermal hydraulics community in Russia, Prof. Kirillov and Prof. Ninokata express sincere gratitude to the staff of thermal and other departments

IPPE, Obninsk Institute of Atomic Energy—National Nuclear University (MEPhI) whose research results are reflected.

F. D'Auria (one of the lead authors of the document IAEA, 2001; see also Aksan et al., 2001) adapted Section 7.12.1 from the outcome of an IAEA CRP. The chairperson of the CRP, Dr. Aksan, the lead authors (Groeneveld, D., Kirillov, P., Saha, D., Badulescu, A. and Cleveland, J.) and the contributors (Akimoto, H., Baek, W.P. Bobkov, V.P., Cevolani, S., Chang, S.H., Chen, Y.Z. Cheng, X., Chung, M.K., Ivashkevitch, A.A., Leung, K.H., Macek, J., Pilkhwal, D.S., Roglans-Ribas, J., Smogalev, I.P., Tanrikut, A., Venkat Raj, Vijayan, P.K., Vinogradov, V.N. and Yesin. O.) are gratefully acknowledged.

Pressure drops in nuclear thermal hydraulics—Principles, experiments, and modeling

8

Klaus Umminger[1] *and Francesco D'Auria*[2]
[1]Framatome, Erlangen, Germany; [2]University of Pisa, Pisa, Italy

Foreword

Pressure drops and heat transfer are primary concerns in system thermal hydraulics (SYS TH). This reflects the importance of related modeling and the results from system codes that are applied to predict nuclear power plant (NPP) accident scenarios. Although pressure drops and heat transfer constitute key modeling aspects of SYS TH codes, as discussed in Chapters 5 and 9, some chapters give special devotion to these topics, namely Chapter 7 and the current chapter, which provide detailed information about heat transfer and pressure drops, respectively.

Pressure drop models have a long history dating from the beginning of the previous century, as discussed in Chapter 2 and outlined in this chapter. Furthermore, pressure drops directly affect the phenomena and accident scenarios discussed in Chapters 6 and 15. The discussions in this chapter are complementary and linked to the information within those chapters. Code validation is discussed in Chapter 13 and includes pressure drop models directly.

8.1 Introduction

The importance of pressure drops in nuclear reactor thermal hydraulics (NPP design and safety aspects) constitutes the motivation for this chapter. The objective is to present the current status and the vision for the future in this area of thermal hydraulics science.

Aside from the formulations of the laws of conservation of mass, energy, and momentum, the equations for pressure loss and heat transfer are the most important formulations in the design of thermal-hydraulic systems.

To solve the basic conservation equations in SYS TH codes, additional constitutive equations are required simply because the number of unknowns is greater than the number of equations.

For NPPs, knowledge of the pressure loss in and between individual systems and components is decisively important during both "normal" power operations and especially the analysis and assessment of accident situations.

Handbook on Thermal Hydraulics in Water-Cooled Nuclear Reactors
https://doi.org/10.1016/B978-0-323-85606-5.00003-9

During the power mode, power generated in the core must be safely removed by the coolant mass flow rate. Reactor coolant pumps (RCPs) are designed to provide the required coolant mass flow rate in a steady-state power mode. Here the pressure buildup across the pumps equals the total pressure loss across the cooling system. This means that for the design of a reactor plant, knowledge of the pressure losses across the individual components of the plant is required, as summarized in the following example.

Many of the important phenomena expected to occur in accident situations are highly influenced by prevailing pressure drops, e.g.:

- Break flow following loss-of-coolant accidents (LOCAs).
- Emergency core cooling (ECC) injection occurring from passive injection systems (e.g., accumulators)
- ECC bypass inside the reactor pressure vessel (RPV) occurring primarily during the early phase of blowdown (BD) in the case of LOCAs
- Loop seal formation and clearing (LSC) affecting overall system performance

The various components of the pressure drop (static head, acceleration, friction, and local pressure loss) have a large impact on every natural circulation (NC) situation that may occur during normal operations or accident conditions with respect to stability or the existence/establishment or stagnation/interruption of NC within the entire system or in individual components/subsystems.

The reproduction of typical reactor pressure losses is an important scaling aspect for integral test facilities (ITFs). Both overall system pressure drops and adequate simulations of pressure drops in individual components have tremendous importance (especially under NC conditions).

Pressure drop tests for specific components, such as spacer grids and fuel bundles, are typically conducted in separate effects test facilities (SETFs). So-called basic experiments are performed to systematically investigate the pressure drop as a function of the geometry of the flow channel and the flow structure.

Examples of pressure drop–related topics not considered in this chapter are:

- Three-dimensional pressure drop tensors, e.g., Vivaldi (2022)
- Pressure drops in spacer grids, e.g., Park and Wang (2018) (further details in Chapter 18)
- Determination of pressure loss coefficients in complex geometries, e.g., Sheng and Seidl (2015)
- Clogging impacts on pressure drops, e.g., Muller et al. (2015)

Finally, the close coupling of pressure drops with heat transfer (Chapter 7) and most phenomena relevant to nuclear thermal hydraulics (Chapter 6 and Aksan et al., 2018a) must be mentioned.

8.2 Pressure drop components

The starting point for pressure drops equations is the theorem of the conservation of momentum. Accordingly, the total static pressure drop—the drop to which a fluid is exposed when it flows on a flow path under steady-state conditions—comprises the following components:

$$\Delta p = \Delta p_e + \Delta p_a + \Delta p_f + \Delta p_l \tag{8.1}$$

where the reversible components are Δp_e, the pressure drop due to static head caused by different heights, and Δp_a, the acceleration pressure drop caused by a change in the cross-sectional area or density of the fluid (which may be partly reversible or reversible in principle); and the irreversible components are Δp_f, the friction pressure drop caused by shear stress on the walls, and Δp_ℓ, the local pressure drop caused by changes in flow geometry and direction.

The total pressure drop therefore comprises the reversible components of the height (head) and (partly of) acceleration and a permanent pressure drop (referred to as pressure loss in the following), which in turn is made up of friction pressure loss and local form (shape) loss. As fluid conditions change over time, the acceleration pressure drop also includes a transient term. This term, although part of the equations as the basis of SYS TH codes (see Chapters 5 and 11), has a negligible impact on predictions of system performance for most situations of interest.

The permanent pressure drop caused by the dissipation of kinetic or mechanical energy arises both from friction on walls and structures and especially from two-phase water/steam mixtures due to irreversible momentum exchange between the phases.

Eq. (8.1) derived from the theorem of the conservation of momentum, also expresses the conservation of mechanical energy along a flow path including the reduction in energy due to dissipation.

Pressure drop modeling, experimental data, and results from correlations are unavoidably affected by scaling, e.g., mainly the size of the pipe in the present case, as well as other phenomena, see Chapter 21; Kong et al. (2018) provide pressure drop data and discuss the data measured for two pipes of different diameters while considering air–water flows.

8.3 Overview of pressure drop models and related phenomena

Pressure drops are typically the result of balance equation (Chapter 5) and constitutive equation (Chapter 9) parts of system codes (Chapter 11). The model bases of those balance and constitutive equations and system codes are discussed hereafter.

8.3.1 Friction pressure drop

The reader may consider two introductory issues: (1) the existence of two formulations for friction pressure drops, as discussed in Section 8.5.1.1, and (2) the difficulty of experiments to measure friction pressure drop alone (i.e., without the acceleration pressure drop or gravity components). This might provide advance justification for errors that nowadays characterize the predictability of this quantity.

8.3.1.1 Single-phase flow

Modeling friction pressure drops in single-phase flows follows an empirical approach widely validated by experiments and theoretical data; see, e.g., Nikuradse (1932). The irreversible pressure drop caused by sheer stress at the wall, sometimes called distributed pressure drop because it occurs all along the lengths of the flow channel, is determined by:

$$\Delta p = f\frac{L}{D_h}\cdot\frac{W^2}{2\rho A^2} \tag{8.2}$$

where f is the friction factor and D_h is the hydraulic diameter (equal to four times the flow area/wetted perimeter). Pressure drops across tubes, rectangular channels, and bare rod bundles (without spacers) are typical examples of this pressure drop component.

The friction factor depends on the Reynolds (Re) number within the flow channel and on the roughness of the wall under turbulent flow conditions, e.g., Tikadar et al. (2018). Commonly used correlations, some of which are discussed in the following, apply to fully developed steady-state flow.

Adiabatic single-phase flow

For fully developed laminar flow, the friction factor can be expressed as follows:

$$f = \frac{64}{Re} \tag{8.3}$$

This is valid for Re numbers <2300 (roughly).

For turbulent flow in smooth pipes, many friction factor correlations are available in the literature. A few that are commonly used are mentioned in the following.

According to Blasius (1913), the following simple formula applies for the range $3000 < Re$ number $< 10^5$

$$f = 0.3164\cdot Re^{-0.25} \tag{8.4}$$

The Hermann equation is appropriate for the higher range from $Re \tilde{} 2\times10^4$ to 2×10^6

$$f = 0.0054 + 0.3964\cdot Re^{-0.3} \tag{8.5}$$

Colebrook and White (1937) (see also Colebrook, 1939) proposed an implicit equation valid for smooth and rough pipes (considering the pipe roughness, e) over the entire range of Re numbers above 3000:

$$\frac{1}{\sqrt{f}} = 0.86\ ln\left(\frac{e/D}{3.7} + \frac{2.5L}{Re\sqrt{f}}\right) \tag{8.6}$$

The explicit equation provided by Filonenko (1948):

$$\frac{1}{\sqrt{f}} = 1.82 \cdot log() - 1.64 \tag{8.7}$$

is a good approximation of the Colebrook–White equation for smooth tubes in the range of $4 \times 10^3 < Re < 10^{12}$.

Correlations for circular pipes are normally also used to calculate single-phase pressure drops in rectangular channels, in annuli, without the existence of experimental data, and in bare rod bundles applying the hydraulic diameter concept. One commonly used correlation for bare rod bundles is proposed by Kays (1975):

$$f = f_{cir} \cdot K_1 \tag{8.8}$$

where K_1 is a function of p/D (pitch to diameter) based on work by Diessler and Taylor (1956), and f_{cir} can be calculated using correlations for circular pipes.

8.3.1.2 Two-phase flow

While the pressure drop in a single-phase flow is characterized by the level of turbulence, the two-phase pressure drop is connected to momentum exchange between the two phases and is characterized by the two-phase flow pattern. Several methods for predicting two-phase pressure drop can be found in the literature. The two-phase multiplier approach is the basis for most cited methods and is the accepted engineering model to account for the effect of a two-phase mixture in a flow channel. The fundamental idea behind this is to calculate the pressure drop of one phase (gas or liquid), Δp_{L}, first. To determine the two-phase pressure drop, Δp_{TP}, the single-phase pressure drop is multiplied by a two-phase multiplier $\Phi^2_{L,G}$ to consider the influence of the second phase. Four basis methods to define the two-phase multiplier are commonly used:

- Two methods that assume the liquid or gas phase flowing alone in the flow channel
- Two methods that assume the entire mixture flowing as liquid or gas only

Applying these basis definitions for the two-phase multipliers, the two-phase friction pressure gradients can be expressed as follows:

$$\begin{aligned}
\left\{\frac{\Delta p}{\Delta z}\right\}_{TPF} &= \left\{\frac{\Delta p}{\Delta z}\right\}\Phi^2_L \text{ for the liquid alone case} \\
\left\{\frac{\Delta p}{\Delta z}\right\}_{TPF} &= \left\{\frac{\Delta p}{\Delta z}\right\}\Phi^2_G \text{ for the gas alone case} \\
\left\{\frac{\Delta p}{\Delta z}\right\}_{TPF} &= \left\{\frac{\Delta p}{\Delta z}\right\}\Phi^2_{LO} \text{ for the entire mixture flowing as liquid only and} \\
\left\{\frac{\Delta p}{\Delta z}\right\}_{TPF} &= \left\{\frac{\Delta p}{\Delta z}\right\}\Phi^2_{GO} \text{ for the entire mixture flowing as gas only}
\end{aligned} \tag{8.9}$$

The Φ_L^2, Φ_G^2, Φ_{LO}^2, and Φ_{GO}^2 terms constitute the two-phase pressure drop multiplier to be determined. The numerous approaches found in the literature on calculating two-phase pressure drops can be categorized into two general groups:

- Modifications of the single-phase friction factor based on the homogeneous model
- Empirical calculations that apply the two-phase multiplier concept based on a separated flow model

Some direct empirical models also exist (e.g., Lombardi and Pedrocchi, 1972). According to this approach, the two-phase pressure drop is directly determined as a function of mass flux, mixture density, the geometry of the flow channel, etc. without reference to a single-phase pressure drop. Furthermore, flow pattern–specific models have been developed to provide flow pattern–specific correlations.

Finally, it must be mentioned that modern computer codes based on two-fluid models require additional correlations for the partitioning of the wall friction between the two phases (wall-to-phase for each phase) and interfacial friction models. Some widely used and often cited correlations are summarized in IAEA (2001), and a few typical examples are outlined in the following.

Homogeneous model

Applying the homogeneous model, the two-phase flow is treated as a pseudofluid characterized by suitable averaged properties of the liquid and gas phases. The first approaches to determine the frictional pressure drops of two-phase flow were based on modifications of the friction factor for single-phase flow. For this approach, the two-phase pressure drop caused by shear stress at the wall can be expressed as:

$$\Delta p_{TP} = f_{TP}\frac{L}{D_h}\cdot\frac{W^2}{2\rho A^2} \tag{8.10}$$

where f_{TP} is the friction factor for two-phase flow and ρ is the density of the mixture. The friction factor is calculated by applying one of the equations for single-phase flow, Eqs. (8.3)–(8.7) and using the mixture viscosity to determine the Re number. Several two-phase flow viscosity models are available; the most common ones have been introduced by McAdams et al. (1942):

$$\frac{1}{\mu} = \frac{x}{\mu_G} + \frac{(1-x)}{\mu_L} \tag{8.11}$$

and Cicchitti et al. (1960):

$$\mu = x\mu_G + (1-x)\mu_L \tag{8.12}$$

Additional examples of two-phase viscosity models for use in the homogeneous approach are given in IAEA (2001). Many of these models do not lead to significantly different results. Homogeneous models are expected to provide good results for high mass flux flows. All homogeneous models can also be formulated in terms of two-phase friction multipliers.

Two-phase multiplier for separated flows

The frictional pressure drop in two-phase flow is typically predicted using separated flow models. In the separated flow model, the phases are considered to flow separately in the flow channel, with each having a given velocity and capturing a certain fraction of the channel's cross-sectional area.

Two-phase pressure drops have been the subject of numerous investigations over the last 6 decades. Most developed methods are based on empirical calculations that apply the two-phase multiplier approach. The methods by Martinelli and Nelson (1948) and Lockhart and Martinelli (1949) are considered the origin of the two-phase friction multiplier approach. Lockhart and Martinelli (1949) defined a parameter X, known as the Martinelli parameter:

$$X^2 = \frac{\Phi_G^2}{\Phi_L^2} = \frac{(dp/dz)_L}{(dp/dz)_G}$$

Based on the idea of Lockhart–Martinelli, other researchers have developed further correlations to account for specific geometries or flow conditions, e.g., see Fitzsimmons (1964) and Friedel (1980). The Martinelli–Lockhart–Nelson approach does not consider interactions between the phases or the influence of mass flux on the two-phase multiplier.

These factors were addressed in correlations developed by Baroczy (1966) and Friedel (1979) that consider the effect of mass flux on the two-phase multiplier and the method introduced by Chisholm (1973), see also Chisholm (1968) and Chisholm and Sutherland (1969), that accounts for interactions between the phases.

Two-phase multiplier for separated flows—bundle

A derivation is summarized below that confirms the empiricism level still necessary nowadays to determine two-phase pressure drops in situations of interest in nuclear thermal hydraulics. Gui et al. (2022), proposed correlations for the single-phase friction coefficient and two-phase multiplier in the case of the fuel bundle of a typical steam generator (SG) in a pressurized water reactor (PWR).

The (single-phase) friction factor is given by (the reader may find symbols that are not defined below in the book's general list of symbols):

$$f = 1.53\, Re^{-0.34},$$

applicable for Re numbers in the range of 17,000–65,000. The two-phase multiplier is given by

$$\Phi_{LO}^2 = 1 + \left(C\Gamma^2 - 1\right)\left\{N_{conf}[x(1-x)]^{0.875} + x^{1.75}\right\},$$

where Γ^2 and N_{conf} derive from an earlier work by Tran et al. (2000). Namely, $\Gamma^2 = \frac{(dp/dz)_{GO}}{(dp/dz)_{LO}}$ and $N_{conf} = \frac{\left[\frac{\sigma}{g(\rho_L - \rho_G)}\right]^{0.5}}{D}$ (as previously introduced by Cornwell and Kew, 1993). Notably, N_{conf} accounts for bubble confinement within a cross section orthogonal to the flow direction. Furthermore, C is proposed (by Gui et al., 2022) as $C = 0.49 Fr_{LO}^{-0.63}\left(\frac{p}{p_{crit}}\right)^{-0.81}\left(\frac{1-x}{x}\right)^{0.11}$, where $Fr_{LO} = \frac{G^2}{gD\rho_L^2}$. The correlations have been validated in the following key parameter ranges: p = 5−9 MPa, G = 100−350 kg/m^{2}s, and x = 0−0.41.

The authors compared the results from their correlation with experimental data and results from other correlations available in the literature. Based on a database of experiments (several dozen points) independent of those used to develop the correlation, errors not smaller than 30% characterize the new correlation (much larger errors are found for previous correlations). This justifies its further attention in the concerned research area (see Chapters 2 and 24).

8.3.1.3 Concluding remarks

A not comprehensive (as mentioned) overview of friction pressure drops has been given in this section. Additional snapshot-like information that may be useful for the reader for connecting friction pressure drop with various topics in the book or applications of nuclear thermal hydraulics are listed in the following:

- All textbooks in nuclear thermal hydraulics (see references in Chapters 1 and 2) provide key information about pressure drops. A pioneering, still valid, and applicable document mostly focusing on local pressure drops is by Idel'Chik (1960) (see also Idel'chik, 1979), and a more recent effort to summarize current technological understanding in the area of water-cooled nuclear reactors has been performed within the framework of the International Atomic Energy Agency (IAEA) activities (IAEA, 2001).
- The principle approach to the calculation of friction pressure drops is based on the definition of "τ_w" as mentioned in Chapters 5 and 11 (see also Chapter 9). The approach suffers from the difficulty in determining the derivative of fluid velocity in the direction perpendicular to the wall where the friction is occurring. The difficulty occurs for Newtonian fluids and may be revealed as higher in the case of non-Newtonian fluids. Computational fluid dynamics (CFD) codes or approaches (see Chapter 12) have not yet eliminated the problem. The engineering approach widely used is based upon the definition of "*f*" (friction coefficient) and the computation of hydraulic diameter and Re number (also a function of the hydraulic diameter).
- The concept of equivalent diameter, and the situations of "fully developed flow" and "laminar or turbulent velocity profile" in the cross sections, largely affect the computation of friction pressure drops. The related evaluations are not immune to errors and approximations (Del Frate et al., 2016) (see also the Ph.D. thesis of Del Frate, 2016). Furthermore, the condition of fully developed flow is needed to establish any meaningful/qualified correlation (e.g., for the friction coefficient); however, in no region of a water-cooled reactor—either a PWR or a boiling water reactor (BWR)—except for the primary side of vertical parts of tubes in SGs, do fully developed conditions occur. So errors are expected in any calculation of practical interest: this is an issue that generates the need for uncertainty quantification (Chapter 13).

- Experimental data are needed to circumvent the limitations and deficiencies outlined in the previous bullet items. However experimental data are not immune to problems: measurement errors, reproducibility of test conditions, measured data, and lack or insufficient knowledge of boundary conditions are reasons that give rise to the scatter of data; this requires an interpretation by the analysts. Powerful statistics-based methods are available that do not reduce to zero the arbitrariness, e.g., when formulating a correlation (for instance, the exclusion of singular points that do not fall into a selected category may not be based on fundamental principles of physics).
- The validity range for a correlation is at the origin of an additional problem for any empirical model based on the analysis of experimental data. A strict interpretation of the information from a correlation would imply the use of the assigned correlation only within the ranges of parameters that are used to develop or qualify the correlations. However, if one considers
  - combinations of the ranges of parameters (typically, many more than three parameters appear explicitly or implicitly, including boundary and initial condition values) used in development or available to demonstrate the qualification level of the correlation, and
  - expected ranges of interest in water-cooled reactors, namely high pressure and large geometry,

the only applicable conclusion is that the concerned correlation is no applicable for predictions in water-cooled reactor conditions. This constitutes the subject for scaling analysis and contributes to the uncertainty expected for any calculation (see Chapter 13).

8.3.2 Local pressure drop

When local obstacles connected with flow areas or flow direction changes occur in the flow channel, additional irreversible pressure losses known as local or form losses must be considered.

Pressure drops across area changes (abrupt or continuous), bends, valves, elbows, and tees are typical examples, e.g., Kim and Kong (2020). In NPPs, the geometries of main interest for local pressure drops are the spacer grids, tie plates, and transition zones between individual components related to changes in the flow area (e.g., SG U-tube inlet/exit or various bypasses in the PWR RPV) or direction [e.g., hot leg to SG inlet chamber, loop seal (LS) bends].

The local pressure drop is generally given by:

$$\Delta p = K\frac{W^2}{2\rho A^2} \tag{8.13}$$

where K is the local loss coefficient, which is different for single-phase and two-phase flows depending on the geometric details of the obstacle.

8.3.2.1 Area changes

Single-phase flow

A common flow obstacle is a sudden cross-sectional area change. Fig. 8.1 shows the flow structure for a single-phase flow through a sudden expansion (upper diagram).

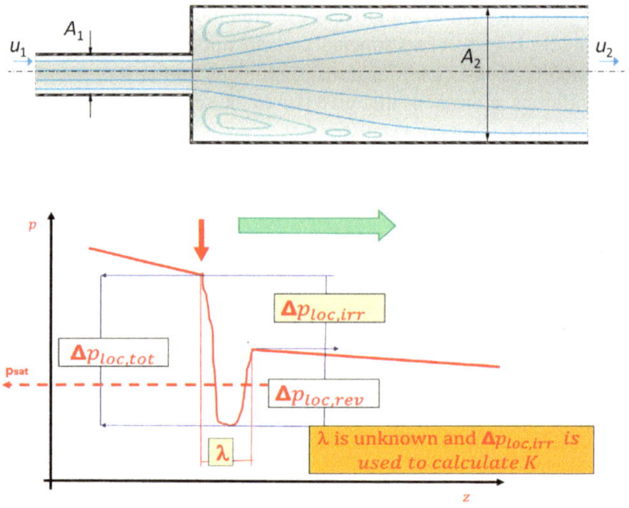

Figure 8.1 Flow through sudden expansion and qualitative pressure line along flow direction.

The lower diagram shows the repartition of pressure drop components at a typical geometric discontinuity. Total, reversible, and irreversible contributions are defined in the diagram. One may note that only the irreversible contribution is necessary and is currently adopted for computation. Thus, the total and reversible components do not appear in modeling. In most experiments, the total pressure drop and the "λ" distance are not measured; the difficulty can be envisaged of measuring those quantities for a suitable range of Re numbers, e.g., (for a nuclear reactor vessel, as mentioned in the previous paragraph) in connections between the cold leg and downcomer (DC), the DC and lower plenum, and the lower plenum and core inlet plate. Furthermore (the situation depicted in the bottom sketch of Fig. 8.1), total pressure drops may pull the local pressure below the saturation value; this causes vaporization that cannot be predicted by any model because the models solely consider the irreversible contribution.

Turbulent recirculation flow zones with mechanical energy losses form behind the opening of the cross-sectional area. After a transition length (8−10 times the larger tube diameter) behind the enlargement, the flow again becomes fully developed (VDI Heat Atlas, 2010). Based on a balance of mass, momentum, and kinetic energy and applying the Borda−Carnot assumption, the irreversible local pressure loss can be expressed as follows:

$$\Delta p = \left(1 - \frac{A_1}{A_2}\right)^2 \frac{\rho u_1^2}{2} \tag{8.14}$$

and according to Eq. (8.13), the local loss coefficient is as follows:

$$K = \left(1 - \frac{A_1}{A_2}\right)^2 \tag{8.15}$$

Eq. (8.15) is valid for turbulent flow. Loss coefficients for the pressure drop in laminar flow have been presented by Kays (1950).

The flow process at a sudden contraction can be treated similarly with the exception that an additional effect must be considered. Behind the point of area reduction, the flow continues to contract and forms a so-called vena contracta at location C in Fig. 8.2.

The local pressure loss and the loss coefficient can be expressed as follows:

$$\Delta p = \left(1 - \frac{A_2}{A_c}\right)^2 \frac{\rho u_2^2}{2} \tag{8.16}$$

$$K = \left(\frac{A_2}{A_c} - 1\right)^2 \tag{8.17}$$

The smallest cross-sectional flow area A_c is not exactly known and is an empirical function of A_2/A_1.

Two-phase flow

The presence of two fluids, notably liquid and steam, increases the complexity of the pressure drop phenomenon. The "*K*" factor (as defined by Eq. 8.13) in two-phase flow is primarily a function of the geometric characteristics of the system (this is at the origin of the definition of the form loss factor for "*K*") and the void fraction, other than the fluid velocity or Re number. Related correlations necessarily have an empirical nature (see, e.g., Van Rooyen and Thome, 2013). Recently, Wang et al. (2020) investigated the branch pressure drop in the case of bubbly flow. They proposed a modeling approach based on a mechanistic formulation for Φ_{LO}^2 and found "... frictional DP between the measured and the predicted results are located within approximately 40%."

The following should be considered before application:

(1) Range of validity, part 1: the range of void fraction considered in the experiments is important together with the number of boundary values used for void fraction: in other terms, extrapolation of "*K*" between two void fraction measurements is not appropriate when the distance between two points is greater than 0.05.

(2) Range of validity, part 2: the way to measure void fraction and the experimental error in the measurement should be documented.

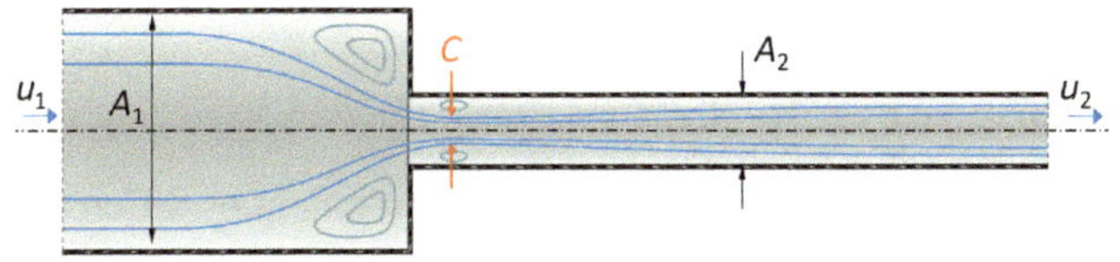

Figure 8.2 Flow through sudden contraction.

(3) Flow reversal condition in the same geometric configuration may give rise to a dramatically different dependence of "K" on void fraction and Re number than measured for direct flow.
(4) Use of the correlation shall be connected with a suitable model to predict void fraction (i.e., consistent with the model used to generate the correlation).

8.3.2.2 Concluding remarks

Pressure drops at geometric discontinuities either in single- or two-phase conditions are currently modeled by the application of empirically based coefficients. Assumptions are needed to determine those coefficients from experiments and, usually, no dependence is available from Re numbers and void fractions: the impact of assumptions and the lack of specification of dependencies on results of prediction is unknown. This may largely contribute to the uncertainty of SYS TH code predictions, as discussed in Chapter 1.

8.3.3 Acceleration pressure drop

The formulation of the momentum equation or force balance for a control volume, in the case of two-phase separated flows (or a UVUT model), brings to the following time- (or temporal) and space- (or spatial) acceleration terms (Lahey and Moody, 1993):

Time acceleration pressure drop:

$$\frac{\vartheta p}{\vartheta t}=\frac{\vartheta}{\vartheta t}\left[\rho_f(1-\alpha)w_f+\rho_g\alpha w_g\right]$$

Space acceleration pressure drop:

$$\frac{\vartheta p}{\vartheta z}=\frac{\vartheta}{\vartheta z}\left[\rho_f(1-\alpha)w_f^2+\rho_g\alpha w_g^2\right]$$

Time and space acceleration pressure drops constitute the result of the integration of mass, momentum, and energy equations along the fluid path in the concerned control volume (Δz length). The averaging signs of the original formulation are omitted.

Furthermore, see also IAEA (2001).

- Space-acceleration pressure drops may occur in steady-state flow conditions (also identified as space-acceleration or spatial-acceleration).
- Space-acceleration pressure drops are zero in the case of a steady flow of incompressible fluid.

8.3.4 Elevation pressure drop

Elevation pressure drops are caused in a gravity environment by different elevations of two points considered for evaluation (i.e., the established 'ρgh' expression). Under two-phase flow conditions, elevation pressure drops (like acceleration pressure drops) largely depend on void fraction: a suitable void fraction model is needed for the evaluation (e.g., qualified SYS TH code, Chapter 11).

8.4 Relations among pressure drops, phenomena, and accident scenarios

The thermal-hydraulic phenomena (e.g., pressure drops) under the accident conditions expected in water-cooled nuclear reactors, as mentioned, are discussed in Chapter 6. Furthermore, the cross-link between phenomena, including pressure drops and accident scenarios, can be found in Chapter 15. Insights related to selected topics, notably the large-break loss-of-coolant accident (LBLOCA) scenario, are discussed in this section.

8.4.1 Pressure drops in large-break loss-of-coolant accidents

The distribution of Δp and flow resistances in the reactor coolant system (RCS) plays an important role in all phases of a hypothetical LBLOCA:

- Blowdown (BD),
- Refill (RF),
- Reflood (RE).

The chronology of events is described in detail in Chapter 15.

The following sections focus especially on the relation between pressure drops and the occurring flow phenomena and their impact on heat transfer. As described in Chapter 15 the double-ended guillotine break (DEGB) in the cold leg between RCP and RPV is considered limited or worst case concerning peak cladding temperature (PCT) performance and is used as a reference scenario for the explanations that follow. Most of the descriptions refer to PWR, with a preference for cold leg ECC injection.

8.4.1.1 Stagnation point occurrences in blowdown and refill phases

With the occurrence of the break, BD starts with pressure wave propagation/jet impingement causing load on RPV internals during the first 100 milliseconds (this will not be described in detail here). At first subcooled critical flow occurs at both ends of the break. High mass flow rates at the break result in a sharp primary pressure decrease and saturation conditions in the hottest regions of the RCS (upper part of the core, upper plenum (UP), hot legs) are reached within the first seconds.

Due to flashing, the depressurization rate decreases; however, the "saturation front" further expands to constitute the entire primary system within the next few seconds. The flow phenomena (flow direction) in the individual components are mainly influenced by the flashing processes, by the prevailing flow resistances and the critical flow via both ends of the break, and to some extent by the performance of the RCP.

The loss of subcooling at the break and the reduced pressure result in a remarkable reduction in break flow rate and therefore in a reduced depressurization rate. In any case, during the (very) first period after the occurrence of the break, flow stagnation in the core region and reverse flow at the core inlet has to be expected (Fig. 8.3). The RPV or the core region provides a rather large volume at high temperatures and therefore a significant source/reservoir for pronounced flashing processes.

On the other hand, a simplified pressure balance analysis can demonstrate that the core is a favorite candidate for the stagnation point. Starting from the stagnation point, there are two or more (depending on the number of intact loops) flow paths to the RPV side of the break. One leads directly via the lower plenum, the DC, and the cold leg of the broken loop to the break. In parallel, flow paths exist via the UP of the SG and the RCP of the intact loops and via the DC to the break. The total break mass flow rate at the RPV side (critical flow conditions) is composed of both the flow via the lower plenum and the flow from the cold legs of the intact loops. The flow distribution and thus the location of the stagnation point between these individual flow paths are self-regulating adjusted in such a way that the corresponding pressure drops between the stagnation point and the break location are identical. For an overall pressure balance analysis also the second main flow path via the UP, the SG and RCP of the broken loop to the pump side of the break must be considered.

The break flow at both sides of the break is characterized by critical flow in the very first phase under subcooled and then under two-phase flow conditions. The mass flow rate via the RPV side is significantly higher compared with the pump side flow rate, as the flow resistance via the SG and especially the RCP of the broken loop is generally higher as the flow resistance via the other flow paths to the pump side of the break.

With the flow stagnation and reversal occurring within 1 s after the break, critical heat flux (CHF) conditions are reached in the core causing the fuel rods to experience DNB. The reduced heat transfer (film and transition boiling, see Chapter 7) leads to rise in cladding temperatures and to the occurrence of the first PCT (BD PCT) even

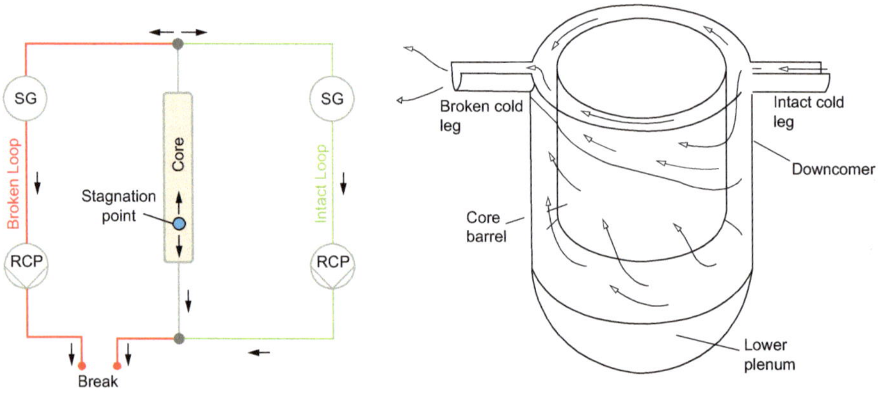

Figure 8.3 Stagnation point in the reactor pressure vessel during blowdown and refill phases. *RCP*, reactor coolant pump; *SG*, steam generator.

though the reactor is rapidly shutdown either by reactor trip (low pressure set point is reached) or due to core coolant voiding (decay power).

If the RCP in the intact loops continue to supply water to the RPV, a short period of positive (upward) flow at the core inlet may occur. This is going to happen when the total mass flow rate from the intact loops exceeds the break flow rate on the RPV side. The RCPs are either tripped directly with the occurrence of the break in the case of LOOP or shortly afterward on indication of certain RCP trip signals (e.g., low system pressure, low Δp over the pump). The amount of water supplied to the RPV from the intact loops depends on the T/H conditions at the RCPs and the operation/performance of the pump (timing of RCP trip, coast-down behavior, and inertia of pump masses) and the number of intact loops (tendency for positive flow more pronounced in four-loop plants).

The positive flow into the core may improve the heat transfer in the core terminating the further cladding temperature increase and causing partial core rewet during this period (bottom-up cooling around 10 s after break occurrence). With the occurrence of two-phase conditions at the RCP their effectiveness is reduced, the positive core flow is terminated, and once again flow reverses.

In the following time some additional processes, such as flashing and delivery of water from the pressurizer (PRZ), upper head, or UP to the core may also contribute to the core cooling during the BD phase (top-down cooling). At the end of BD, the RCS is completely empty except for some residual water in the lower plenum of the RPV. According to the flow resistance of the individual flow paths, stagnation points are expected to occur in the region of the hot legs. From these points, steam is flowing in negative direction via the RPV to the RPV side of the break. In the intact loops steam is also flowing from these points via the SG, RCP, and the DC (positive direction) to the RPV side of the break. In the broken loop the steam directly flows from the hot leg via the broken loop to the pump side of the break. The negative steam flow through the core ends as the level in the UP reaches the upper end of the core barrel during the RF phase.

8.4.1.2 Pressure drops and reflood phase

As described in more detail in Chapter 15, the RE phase begins when the water level reaches the lower end of the core. During the RE phase the core is filled with coolant and the core cooling is reestablished (quenching) throughout the entire core region. Depending on the boundary conditions, the RE phase may last between <100 s and several minutes. There is a strong coupling between the reflooding process and the encountering heat transfer modes and thereby with the quench front propagation and rewetting processes in the core.

During the early RE phase, coolant from the accumulators (rather high mass flow rates) enters the core initiating or intensifying the quenching process at the bottom of the core. Heat release from the hot fuel rods to the surrounding water leads to intensive steam production. The creation of steam results in an increase in the pressure drop along the steam flow paths from the core to the break and increased pressure in the core region compared with the pressure at the break or in the DC. This local pressure buildup above the water or water/steam mixture level in the core has a delaying effect on the further flooding of the core and accelerates the filling of the DC leading to a remarkable difference in water levels between the core region and the DC region. Fig. 8.4 shows corresponding results from a PKL experiment. The level in the DC is of course the driving force for the flooding process in the core. However, the flooding rate and thereby the propagation of the quench front also depends on the evolution of the absolute pressures above the water levels in the core and in the DC, i.e., the pressure difference between these locations.

The pressure in the core is on the one hand a function of the amount of generated steam, which again depends on the decay power, power distribution, and quench front propagation itself, and on the other hand, is a function of flow resistance from the core to the break location or the condensation location in the cold legs.

Steam flowing from the core is divided within the UP according to the pressure losses of individual loops in the direction of the two ends of the break. Here in particular, pressure losses through RCPs are important. On its way to the break location, the steam is heated up when passing SGs that are still under hot conditions. In the broken loop, it can be assumed that the steam flows at the pump side of the break under superheated conditions into the containment. In the intact loops, steam condenses at ECC injection points in the cold legs, depending on the amount of injected ECC, partially or completely. The thus-heated ECC water flows toward the DC and from there partly toward the lower plenum and contributes to flooding the core. A partial flow also

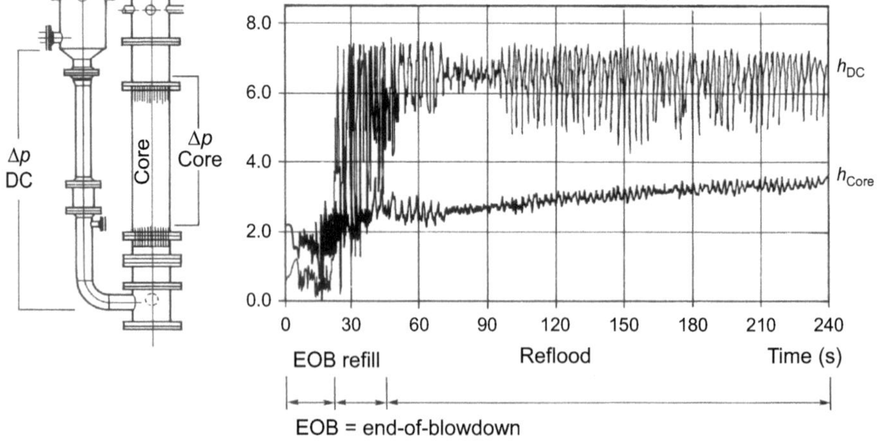

Figure 8.4 PKL II facility: RPV level evolution during reflood. *DC*, downcomer; *UP*, upper plenum.

passes through the DC to the RPV side of the break and is discharged there. The pressure at the top of the DC is finally determined by the pressure loss in the primary side loops and by condensation processes in the region of the ECC injection. In addition to design-related pressure losses due to friction and form losses and local and temporal acceleration due to changes in cross-sectional areas or by evaporation and condensation processes, also certain thermal-hydraulic processes/phenomena can cause an increase in the pressure drops in the primary circuit, e.g.:

- water entrainment and steam binding (S.B.) (see Section 8.4.1.3),
- countercurrent flow limitation (CCFL) phenomena (see Section 8.4.2.1),
- deposition of boron acid in the SG U-tubes (see Section 8.4.1.4),
- backward flow of ECC water via the RCP and accumulation in the LS (see Section 8.4.1.4).

The last two mentioned phenomena are more expected in the long-term cooling phase.

Generally, pressure losses in the primary circuit have a major impact on the flooding rate and on the progression of the quench front and therefore on the PCT and the timing of the turnaround of the cladding temperatures and the complete quenching. Depending on the boundary conditions differential pressures of more than 100 mbar in configurations with cold leg breaks and cold leg injections may occur.

Fig. 8.4 shows exemplarily typical results of an experiment from the PKL II B test series with a double-ended break in the cold leg and only cold-side ECC supply. The experiments of the PKL II B series include starting from a primary pressure of 40 bar, the end of the BD phase, and refilling and reflooding the core. Shown are the measured levels in the RPV DC and the core area and the measured differential pressure between the UP and the top of the DC. It can be seen that with the beginning of the flooding process, the levels in the DC and the core significantly diverge. With decreasing decay power during the flooding process also the pressure difference decreases and the levels equal to itself. That is, with already filled DC water, inventory in the core steadily increases. The flooding process itself and the potential variations/oscillations in level are subjected to a self-regulating mechanism. The rapid progress of the quench front causes intensive steam formation associated with an increase in pressure drop, and thus a slowdown of the flooding process. This in turn means a reduced steam formation, resulting in a renewed acceleration of the flooding process.

When interpreting the test results, it should be noted that the illustrated collapsed levels were directly derived from differential pressure measurements along the core or the DC neglecting other pressure loss components (friction, form losses, acceleration). It is assumed that the hydrostatic component is predominant in the measured differential pressure and the so-derived collapsed levels represent the water inventory in the respective components with a good approximation. It must however be remembered in particular for the interpretation of the temporal variations that dynamic effects are included in the displayed measurement signals that partly can even lead to a phase shift between the measured differential pressures and the actual inventory fluctuations.

The course of the mixture level in the core region can be well reproduced through cascaded differential pressure measurements along the core (Fig. 8.5), at least by trend (at certain times). The passing of the mixture level at the respective lower pressure

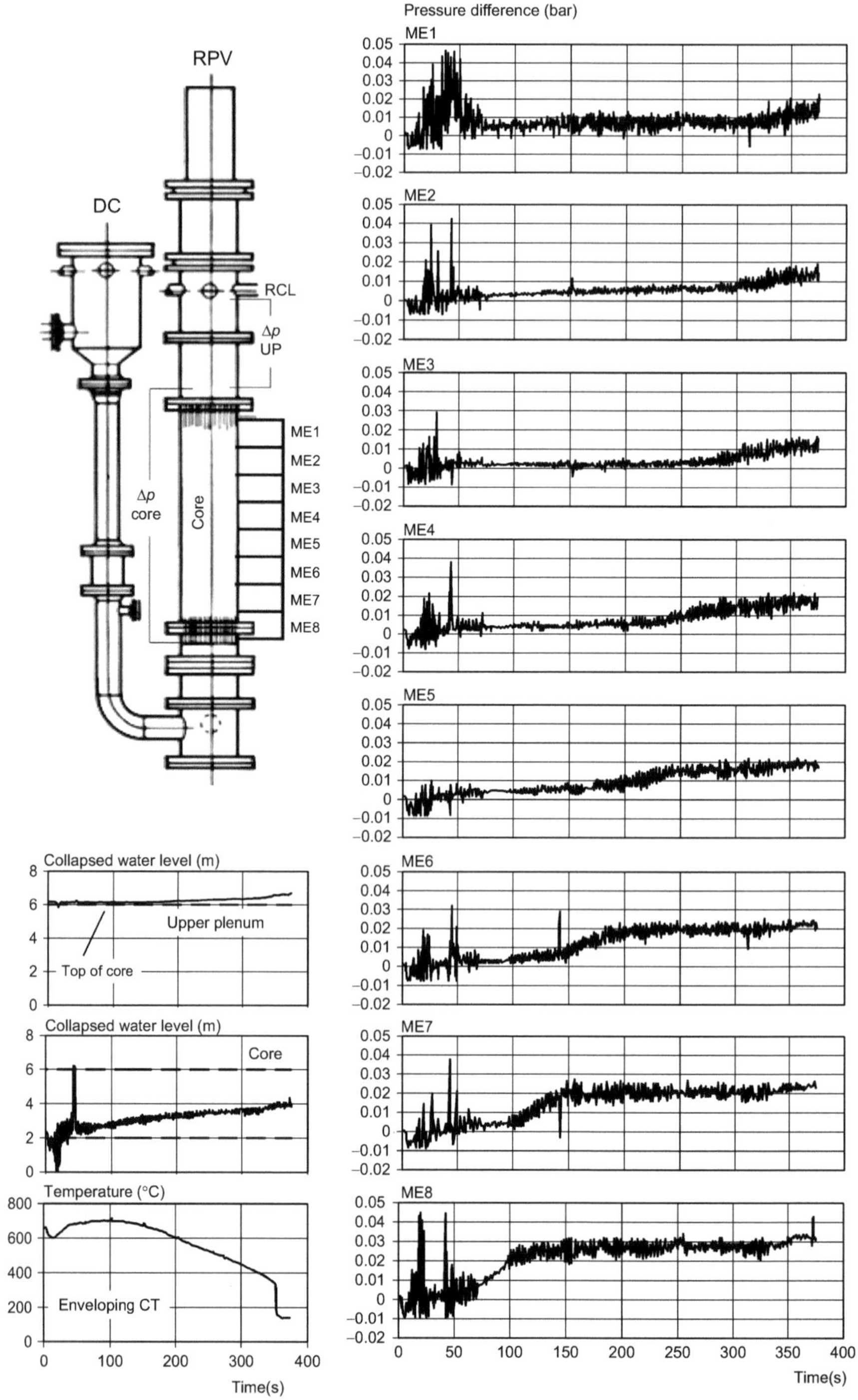

Figure 8.5 PKL II facility: Δ*p* along the core during reflood. *CT*, cladding temperature; *DC*, downcomer; *RCL*, reactor coolant loop; *RPV*, reactor pressure vessel; *UP*, upper plenum.

tapping point is indicated by a significant increase in the measured differential pressure in the respective measuring section. The progress of the quench front is determined by measuring the surface temperature detected at the fuel rod simulators (lowering the temperature to near saturation). The relation between the fictitious collapsed level (water inventory), the mixture level, and the quench front depends on the system pressure, stored energy in the rods, the decay power, and the flow pattern developing in the core above the quench front (Chapter 7).

The pressure drops at spacer grid locations during RE are investigated by Jin et al. (2020), mainly connected with the dispersed flow film boiling regime. Those authors provided detailed measurements of pressure drops in cells below, above, and across the spacer grid, together with specific correlations. The comparison between calculated and measured data (a few dozen data points) showed errors in predictions not bounded by $\pm 25\%$.

In summary, the pressure losses in the primary circuit lines have a major impact on the flooding process in the core. As part of the PKL I test series, this was confirmed quantitatively by varying the flow resistance of the RCPs (RCP simulators were used in PKL). Higher pump resistances lead to higher PCT and longer quench times. This effect is especially relevant when configurations with cold-side break and cold-side ECC injection are considered. In other constellations, in terms of the breaking and ECC injection locations, this influence is less important.

8.4.1.3 Steam binding

During the RE phase of LBLOCA large vapor quantities are generated in the core when the quench front is propagating along the core. Droplets may be entrained out of the core by high steam velocities and evaporate at the hot structures in the UP, the hot legs, and especially in SG U-tubes because SGs are normally at higher temperature levels than the primary side during this phase.

The evaporation of liquid carryover increases the pressure above the core, which again opposes the driving pressure for the displacement of water from the DC to the core, reduces the core flooding rate and thus adversely affects core cooling. This effect is called "steam binding." While more commonly "S.B." is defined as the impact of core steam generation and water carryover on the general increase in UP pressure, in the literature, it mainly refers to the contribution of the pressure increase generated by the evaporation of entrained water in SG U-tubes. In the following explanations, "S.B." will be used in the latter sense. The effect of S.B. can be directly expressed by the increase in differential pressure over the loops and over the SGs, respectively.

As S.B. is directly connected with the entrainment of water, the velocity of the steam and thus the system pressure is of great importance to the S.B. effect. Higher containment pressure and thus higher system pressure tends to reduce S.B. in the SG due to higher steam density. On the other side high pressure on the SG secondary side tends to increase S.B. due to the more effective evaporation process. In this connection, as mentioned, the evaporation leads to an increase in steam velocity and therefore an increase in pressure drop, and the rising temperature (lower density, higher velocity) of the steam results in an increase in pressure drops over the SGs.

The flow at the SG outlet mainly consists of superheated steam whereas the entrainment of a certain amount of water toward the SG outlet even under superheated steam conditions cannot be excluded as experiments in the FLECHT-SEASET test facility have indicated. This amount of water does not contribute to S.B. in the SGs just as well as the water deentrained on its way from the core toward the SG in the UP, the hot legs, and the SG inlet chambers.

A number of experiments in SETFs (e.g., FLECHT-SEASET) and ITFs (e.g., BETHSY and PKL II) related to water carryover and S.B. have been performed, some of them several decades ago. Separate effects tests (SETs), especially on water entrainment and deentrainment in the UP, hot legs, and SG inlet chamber, have also been performed under real plant dimensions in the UPTF facility.

Results from FLECHT-SEASET demonstrate the effect of S.B. on the pressure drop over the SG observed in an SET. Code assessment calculations against FLECHT-SEASET experiments demonstrated that the S.B. effect is overpredicted.

S.B. and the impact on core cooling was also an aspect of the already cited integral experiment performed within the PKL II B program. In this test, the entrainment of water from the core was only observed in the first phase of the RE phase, when high steam velocities occurred due to intense evaporation processes at the bottom of the core. In the predominant part of the RE phase, almost no water entrainment (and evaporation in the SGs) took place.

The increased pressure drop in the SG is mainly due to the heat-up of steam and the consequential increase in steam velocity (lower density). However, the contribution of the pressure drop over the SG to the entire pressure drop over the loop is relatively small (Fig. 8.6, top). This rather limited effect of the S.B. and entrainment of water may be attributed to the relatively high system or containment pressure of 4 bar postulated for the assumed scenario. An assessment of RELAP5 using this PKL experiment showed the tendency of the code to overpredict the S.B. phenomena (overprediction of pressure drop across the SG) by overpredicting the liquid entrainment from the core, finally resulting in higher core cladding temperatures in the calculation.

In general, it seems that the existing experimental database is not sufficient to completely cover the entire range of relevant parameters including all relevant phenomena associated with S.B. (water entrainment, deentrainment, heat transfer in SG U-tubes with hot walls and entrained water, possible entrainment of water into the SG outlet chamber, resulting pressure drop across the SG). Additional integral tests or SETs, especially in the lower pressure range with adequate instrumentation, are therefore desirable.

8.4.1.4 Pressure drop during long-term cooling

Continuous boiling in the RPV is expected during the long-term phase of a large cold leg break LOCA in PWR plants with sole cold leg ECC injection. In this state, the reactor core is rewetted and covered by a two-phase mixture of water and steam, but due to the mixing of cold ECC water with steam in the upper DC, the coolant fed to the core in compensation for the evaporation is almost at saturation temperature. The limitation of the ECC flow from the cold side via DC to the core is a consequence

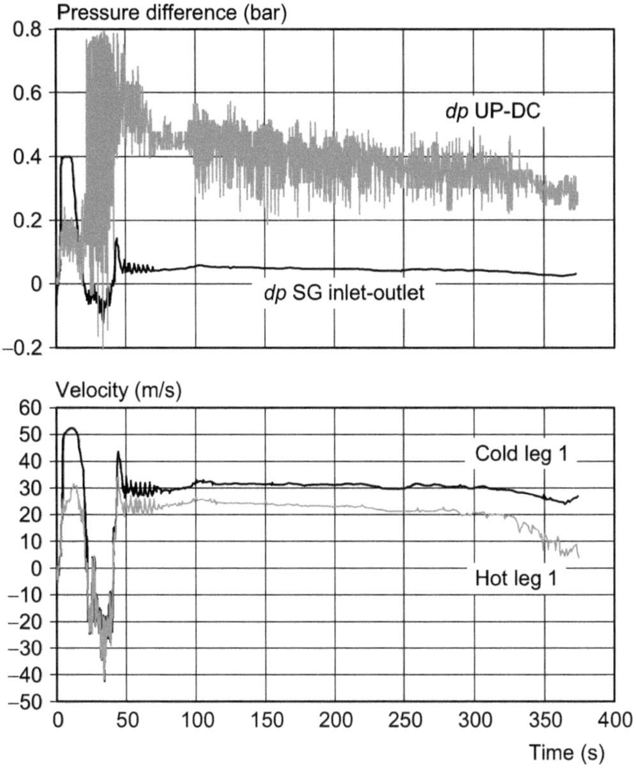

Figure 8.6 PKL II facility: Δp across steam generator (*SG*, steam binding) and loop velocities during reflood. *DC*, downcomer; *UP*, upper plenum.

of the pressure balance between the hot side (core, UP) and the cold side (break location); hot side UP pressure acts against the flow and thereby avoids subcooling of the core volumes. In doing so, the ECC flow theoretically only compensates for the evaporation rate in the core, regardless of the cold-side ECC flow rates as any excess ECC is lost via the break.

Thus, the boiling process in the core continues in the long term. The continuous replacement of almost boron-free steam by borated ECC water leads to a continuous increase in boron concentration in the core and the temperature-dependent limit of boron solubility in water may be reached. Locally initiated crystallization then effectuates the formation of solid boron particles that may compromise core cooling by impeding the heat transport from the rods through plate-out of boric acid at the rod cladding surfaces. An effective measure to avoid or reverse a present or developing boron enrichment process in the core is the switch-over to hot-leg ECC injection. A hot-leg ECC flow high enough may create a net flow that displaces the high boron coolant inventory from the RPV riser side via DC toward the break ("flushing" of the core).

The maximum switching time (from cold to hot leg injection) depends on the speed of the boron enrichment process and thus on the size of the mixing volume, i.e., the amount of liquid water in the RPV available for mixing with the continuously accumulating boric acid. The available mixing volume, i.e., the mixture level on the hot side again is mainly influenced by the flow resistance along the loops and its possible increase during the transient. In a cold-leg break scenario, a high flow resistance between UP (close to the steam source) and cold side (break, heat sink) results in a high differential pressure between UP and break location. The hot-side swell level accounts for this pressure differential: The higher the flow resistance, the higher the pressure differential between UP and cold leg, and the lower the mixture level above the lower plenum and the mixing volume available.

Of course, the parameters for UP pressure are mostly given by the PWR design concept:

- bypass flow from the upper head/UP to the DC
- pressure losses along loops determined by SG U-tubing and RCPs at a standstill

In addition, certain effects that possibly occur during the transient may contribute to the increase in the overall pressure differential between hot and cold sides. Refilling of LSs by ECC backflow via the RCPs resulting in partial or complete blockage of one or multiple crossover legs by water or two-phase mixture increases the flow resistance (in contrast to LSC, which reduces flow resistance; see Section 8.4.2.3). S.B. (see Section 8.4.1.3), as well as the plate-out of boric acid in SG U-tubes resulting from the evaporation of coolant entrained into the U-tubes (the secondary side is at a higher temperature), may also slowly increase the flow resistance across the SG and contribute to the overall pressure differential. In this way, high steam loads (comparably high core power in the early long-time-cooling phase) in conjunction with built-up flow resistances along the loops reduce the swell level within the core (swell level depression) and thereby significantly reduce the mixing volumes within the RPV (Fig. 8.7).

Systematic experimental studies on the size of the mixing volume in the RPV as a function of the decay power and the flow resistance in the crossover legs (geometrically or by refilling with ECC water) were performed within previous PKL test programs. Additional concluding investigations are still ongoing. As an example, Fig. 8.8 shows schematically the impact of the refilling of one LS on the pressure difference between the UP and the upper DC and the following decrease in collapsed level in the RPV as observed in the PKL experiments.

8.4.1.5 Pressure drops at spacer grids and critical heat flux

For some time (decades), the effect of pressure drops across spacer grids on CHF has been known. Spacer grids, besides ensuring tightness for the fuel bundle and reducing the vibration of fuel pins, enhance local turbulence, which increases the CHF value downstream of the location of the grid. This allows higher values for linear heat flux and consequently for core power. What is stated above applies to the nominal operating conditions of the cores of water-cooled reactors. Cooling enhancement is also expected under transient conditions; this is the motivation for keeping this section in a heading section dealing with LBLOCA.

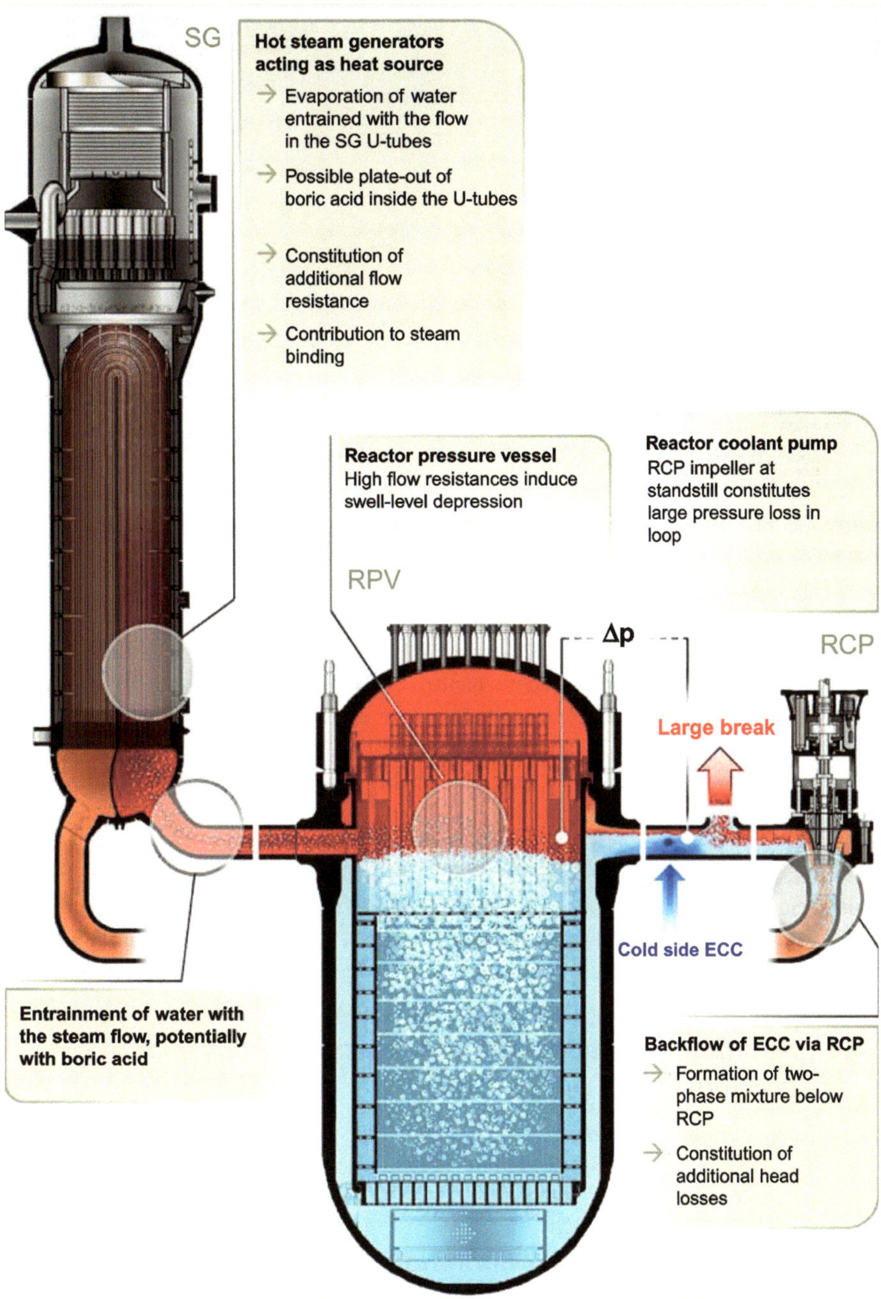

Figure 8.7 Increase in Δp along loops by boron deposition and loop seal refilling. *ECC*, emergency core cooling.

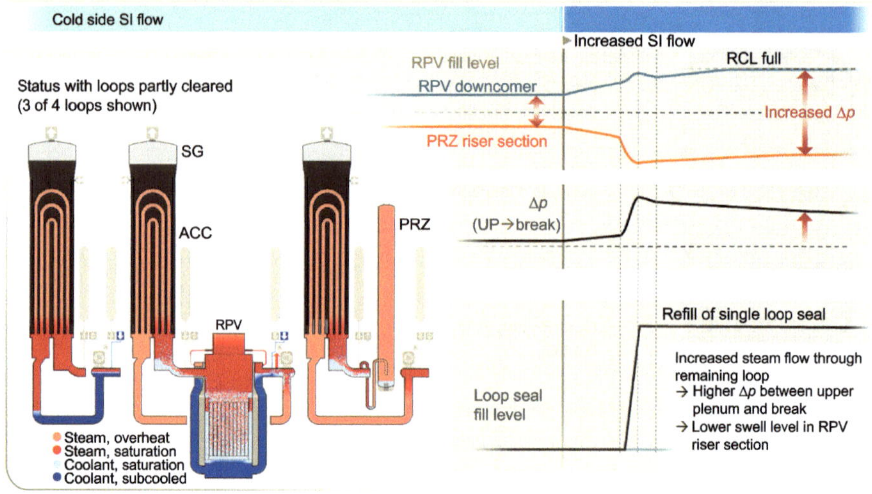

Figure 8.8 PKL III facility increase in Δp across loop due to loop seal refilling. *ACC*, accumulator; *PRZ*, pressurizer; *RCL*, reactor coolant loop; *RPV*, reactor pressure vessel; *SI*, safety injection; *UP*, upper plenum.

Pressure drops across spacer grids and CHF enhancement by spacer grids are the topics of many investigations; recent papers by Batta and Class (2017) (related to spacer grids in metal-cooled reactors), Schikorr et al. (2010), Agbodemegbe et al. (2015), Ikeda (2014), and Jayanti and Reddy (2013) (the last three papers focus on CHF) are included in the list of references for this chapter, which readers can consult for further details.

All the concerned investigations (with the notable exception among the papers listed above of the work by Schikorr et al. (2010), where the Rehme correlation is at the center of attention) focus on CFD modeling and applications. Therefore, the (obvious) conclusion here is that CFD is the proper technological area within nuclear thermal hydraulics for pressure drops and CHF in spacer grids (see, e.g., Chapter 12).

8.4.2 Additional relations between pressure drops and phenomena

As expected, pressure drops have a role in most of the phenomena identified in Chapter 6, see also Aksan et al. (2018a), and in the accident scenarios discussed in Chapter 15. A few examples are provided in Sections 8.4.2.1–8.4.2.5.

8.4.2.1 Pressure drops and flooding at countercurrent flow limitation location

Countercurrent flow (CCF) with upward-oriented steam flow and gravity-driven water downflow in vertical flow sections is expected in many accident sequences. CCF may also appear in inclined or even horizontal flow sections if a connection exists with inclined or vertical sections experiencing CCF. At high steam flow rates or steam velocities, water flow in the opposite direction may be impeded or even terminated/diverted in the same direction as the steam flow (cocurrent flow) due to momentum exchange between the phases. This phenomenon, called CCFL, may lead to the accumulation of water, i.e., flooding or water holdup at the CCFL location. CCFL is considered a key thermal-hydraulic phenomenon in reactor safety analyses because it can prevent or at least deteriorate the backflow of water to the reactor core. CCFL is generally associated with an increased pressure drop at the CCFL location. It therefore leads to a displacement of water from the core to the CCFL location and possibly causes a certain amount of suppression of the core level.

CCFL is likely in complex geometries characterized by flow restrictions, bends, or significant reductions in flow areas, such as the upper core tie plate, hot leg bend, the inlet of the SG header, or the inlet of the SG tubes. Other possible CCFL positions are the RPV DC, the PRZ surge line, and the ascending part of the crossover leg. Specific local modeling is necessary with system codes to obtain reliable predictions. Flooding correlations are established from experimental data, and the parameters of these correlations are part of the code input deck.

Extensive experimental investigations in SETFs and ITFs have been performed in recent decades to address the CCFL issue and to provide an experimental database for the development and validation of the corresponding models. Comprehensive experiments in full-scale PWR geometry on CCFL in the hot legs during reflux condenser mode in small-break LOCA (SBLOCA) conditions have been performed in UPTF. Additional investigations in UPTF addressed CCFL at the upper core tie plate and in the RPV DC during LBLOCA. The UPTF test results and the comparison with results from smaller test facilities demonstrated the dependency of CCFL occurrence on the geometric dimensions and the importance of scaling for the investigated cases. Pronounced heterogeneous 3D flow distributions with water downflow (breakthrough) and steam upflow paths have been observed and, for example, nonuniform distribution of vertical differential pressure across the tie plate has been detected (Mayinger et al., 1993).

CCFL in SG U-tubes has been mainly addressed by systematic investigations in ITFs (e.g., LSTF, BETHSY, and PKL). As mostly PWR-prototypical SG tubing with original dimensions of the individual U-tubes is used in the test facilities, the experimental results are largely representative of the PWR.

CCFL under reflux condenser conditions has been extensively investigated for instance in the PKL III test facility. Within parametric studies, the influences of the specific SG load (heat to be removed by the SG), primary pressure on the occurrence of CCFL, and the amount of water accumulated in U-tubes due to CCFL were systematically analyzed under quasi-steady state and transient conditions (secondary side cool down). As an example, Fig. 8.9 shows typical results from an experiment on CCFL under reflux condenser conditions performed with a stepwise increase in the SG load from 4% to 20% of the nominal SG power at a constant primary pressure of 40 bar (realized by controlling the secondary side pressure). The measured Δp, between the in—or outlet chamber and the outlet chamber and the top of the U-tubes for three tubes with different heights as well as the measured Δp between the SG inlet and

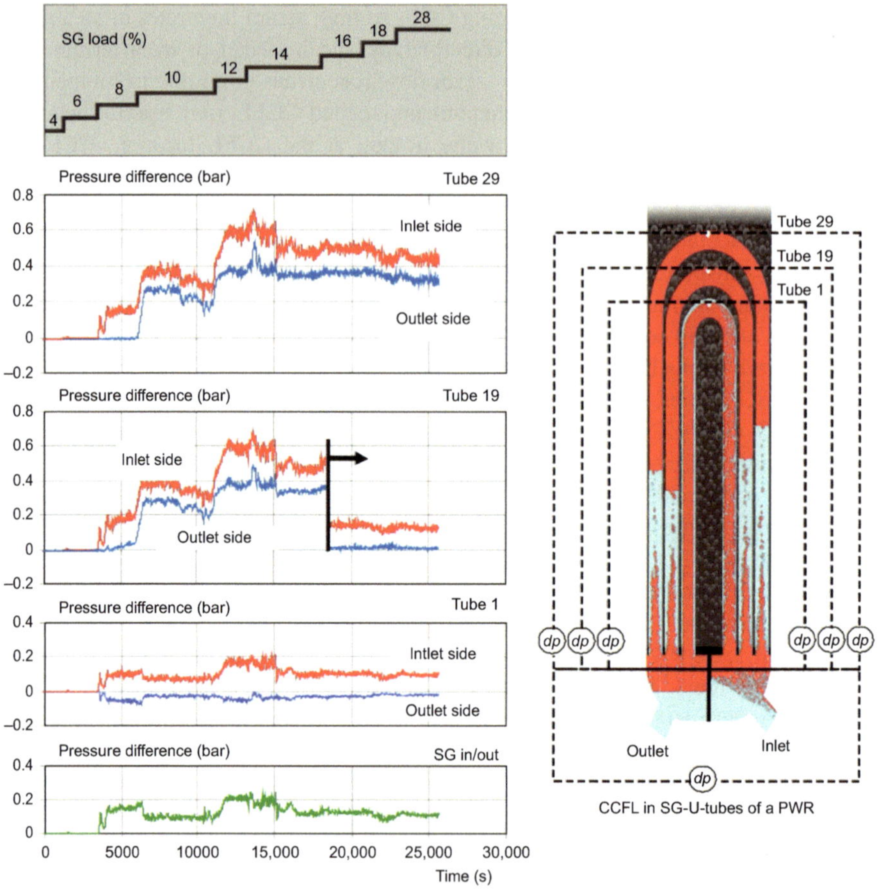

Figure 8.9 PKL III facility—Countercurrent flow limitation (*CCFL*) in steam generator (*SG*) U-tubes during reflux condenser mode of operation. *PWR*, pressurized water reactor.

outlet chamber are shown in Fig. 8.9. In the PKL test facility 3 out of 30 U-tubes are equipped with Δp measurements on the inlet and outlet sides of each of the four SGs.

The test started in "pure" reflux condenser condition, i.e., the mixture level in the RPV was between the core exit and the lower edge of the RCL and there was no water accumulation, neither in the hot legs nor in the SG inlet chamber or the U-tubes. Under these conditions, the measured Δp on the inlet side (CCF) and outlet side (concurrent flow) is close to zero. The onset of CCFL in the U-tubes is indicated by a significant increase in Δp on the SG inlet side and by an increase in Δp between the SG inlet and outlet chamber at an SG load of 6%. It has to be mentioned that in the SG inlet chamber liquid hold-up occurred at the same power level, however shortly after the occurrence of liquid holdup in the U-tubes. Consequently, liquid holdup in the U-tubes was the result of CCFL and not simply a continuous rise of the steam/water mixture level from the inlet chamber into the U-tubes.

Fig. 8.9 also reveals a rather heterogeneous flow behavior among the individual U-tubes. While the longer tubes experience water holdup, condensate is carried over to the outlet side in the short U-tube creating a pressure drop (due to high steam/water velocities) from the top of the U-tubes to the SG outlet chamber (cocurrent water/steam flow) indicated by the negative value for the pressure difference in the outlet side of the short tube. The steam is not completely condensed in the shorter tubes, and the remaining steam flows via the SG outlet chamber into the outlet sides of other tubes experiencing CCFL in their inlet sides. As a result, CCFL may also occur in the outlet sides of longer tubes if this steam flow rate is large enough. The entirety of pressure drops on the inlet side (carryover of condensate) and the outlet side (cocurrent flow of steam and condensate) correspond to differences in pressure drops between the inlet and outlet sides of longer tubes that are mainly caused by the hydrostatic head of accumulated water, whereas friction pressure drop, local pressure drop, and possible dynamic effects are also included in the measured Δp.

8.4.2.2 Pressure drops and two-phase critical flow

When a fluid passes from a high-pressure region with a given thermal-hydraulic state (pressure, temperature, void) via a restriction (e.g., pipe, orifice, valve) to a lower-pressure environment, the mass flow is determined by the pressure difference between the two regions as far as a certain limit for the pressure difference is not exceeded. Under these conditions a pressure reduction in the downstream region results in an increase in mass flow. However, if this pressure falls below a certain value (i.e., Δp exceeds a certain value), no further increase in mass flow takes place (Fig. 8.10).

The resulting flow (called critical or choked flow) is then independent of the conditions in the downstream region and is for a given geometric configuration, only a result

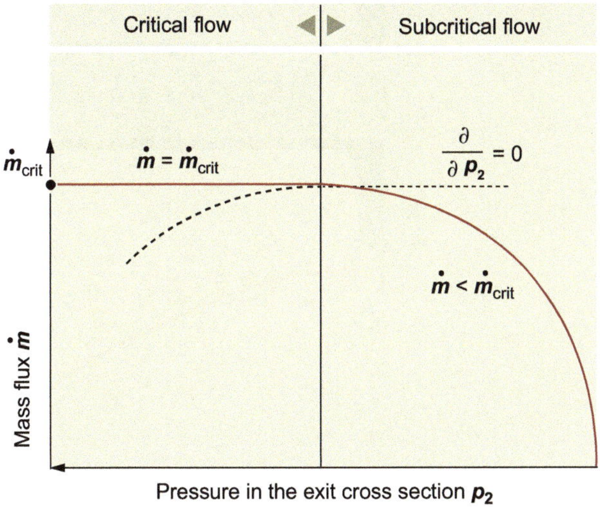

Figure 8.10 Relation between mass flow and pressure in the exit cross section.

of the conditions (e.g., pressure, temperature, void, quality, flow region) in front of the restriction. Fig. 8.11 shows schematically the pressure profile along the flow path from the inlet to the outlet of a flow duct (restriction). The pressure reduction with increasing gradient is connected to an acceleration of the two-phase mixture. The reason for critical flow is the strong expansion of the gaseous phase, which can be explained based on energy conservation (transposition from potential to kinetic energy). For subcritical flow, the pressure in or just upstream of the exit cross section corresponds to the pressure in the downstream region p_2. In the case of critical flow, a pressure jump is observed in the exit cross section. The minimum achievable pressure p_1 is called critical pressure p_c and cannot be decreased further even if the pressure p_2 is reduced.

As outlined in Chapter 6, knowledge of two-phase mass flow under critical flow conditions is a very important issue in reactor safety analysis concerning a broad range of accidents/events and is relevant for different components/structures in reactor systems, such as the following:

- Loss of inventory through breaks/leaks ranging from small cracks to DEGBs
- Restrictions inside a system (e.g., surge line or locked RCP after DEGB)
- Critical flow through valves, e.g., during
  - Cycling of safety and relief valves to control primary and secondary side pressures, e.g., in SBO, LOFW, ATWS, and SBLOCA transients
  - Depressurization of primary and/or secondary side by fully opening of corresponding valves (as AM measures)

Numerous models for critical two-phase flow have been developed for use in thermal-hydraulic system codes. The detailed description of these models, which differ

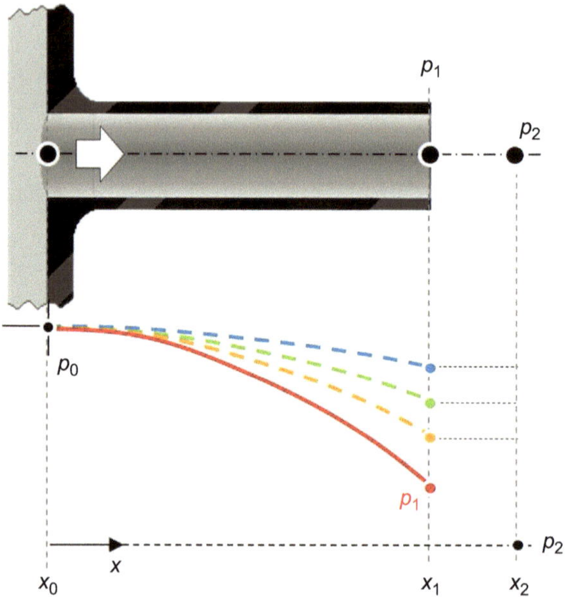

Figure 8.11 Pressure along the flow path as function of the exit pressure.

from others in their complexity or level of assumed disequilibrium (thermal and mechanical) can be found in the open literature and is not subject to this chapter. Such critical flow models have been assessed against an extensive set of data from critical flow experiments mainly conducted in SET. Several studies based on existing experimental results have been performed in recent years (e.g., Sokolowski et al., 2012; Bartosiewicz et al., 2011), to validate the models in the currently used codes, to investigate the range of applicability of the individual models and to identify options for further improvements in the calculation of the critical flow. None of the existing models can cover the entire parameter range expected to occur under accident situations with respect to the geometric configuration and the thermal-hydraulic conditions in front of the critical flow area.

A large number of SETs on critical flow, with the first ones performed during the 1970s, can be found in the literature. The OECD/NEA has selected some experiments as reference tests for critical flow during LOCA transients (see Chapter 6). One of the most prominent test campaigns used as a source of experimental data for validation of two-phase critical flow models was performed within an international project in the Swedish Marviken Power Station facility under nearly real plant conditions (Ericson and Hall, 1978) in the years 1978–79.

Typical results of another experimental campaign on two-phase critical BD from a pressure vessel following a pipe break, performed at the University of Pisa, are shown in Fig. 8.12. The illustrated curves represent the time trends of the pressures measured in the pressure vessel and the exit pipe at two different positions. A method for analyzing the most important phenomena in the BD transient was applied by D'Auria and Vigni (1983) to these experimental signals to obtain values for flow rate as well as spatial and temporal trends of quality and velocity in the exit pipe.

With the spatial profiles of the pressures (input) Fig. 8.13 shows the calculated velocity and quality profiles along the exit pipe at a certain point in time ($t = 6$ s). The

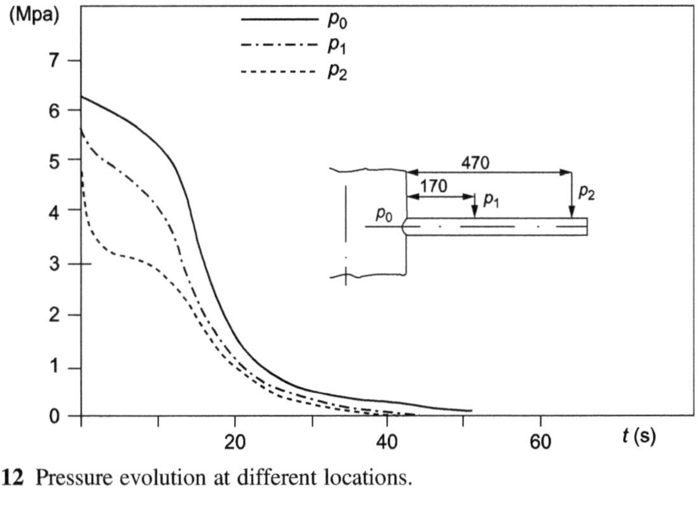

Figure 8.12 Pressure evolution at different locations.

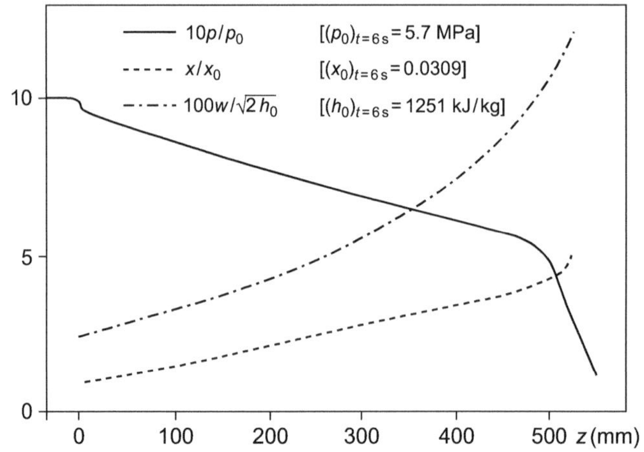

Figure 8.13 Pressure, quality, and velocity versus distance from the nozzle inlet.

pressure decrease in the pipe causes, in this case, evaporation of water entering the nozzle and then a continuous increase in velocity. In the applied theoretical method, axially symmetric flow under thermodynamic and mechanic equilibrium conditions was assumed. The pressure drop due to local discontinuity and wall friction was considered by corresponding input data (pressure loss coefficient at pipe inlet, distributed friction factor between pipe wall and fluid, and two-phase flow multiplier).

In general, most of the calculation methods for critical mass flow assume in a first approximation isentropic changes of the conditions from the inlet to the exit of the flow duct. Under these conditions (typical for an ideal nozzle) the pressure drop along the flow duct only consists of acceleration effects (wall friction is neglected). Depending on the geometry and the operating conditions, the friction pressure drop may not be neglected. In constant cross section pipes with choking occurring at the pipe exit, the critical flow may be significantly influenced by the entrance losses and the wall friction in the pipes. The length of the pipe or the L/D ratio is one decisive factor in quantifying the impact of friction pressure drop on the critical flow rate.

In the case of narrow wall penetrating cracks in pressurized components, the hydraulic resistance in the flow path is very important for the determination of the leakage rate. As an example, Fig. 8.14 shows the pressure distribution through a crack, calculated with a finite volume program based on geometric characteristics of the leakage path (Keim et al., 1999).

The analysis of leak flow rates of through-wall cracks is also relevant in connection with the qualification of leakage detection and surveillance systems used in NPP and with respect to break preclusion concepts including leak-before-break.

8.4.2.3 Loop seal formation and sealing

Depending on the break size and the availability of the ECC system, a loss of coolant with a small or intermediate break may lead to a significant reduction in primary

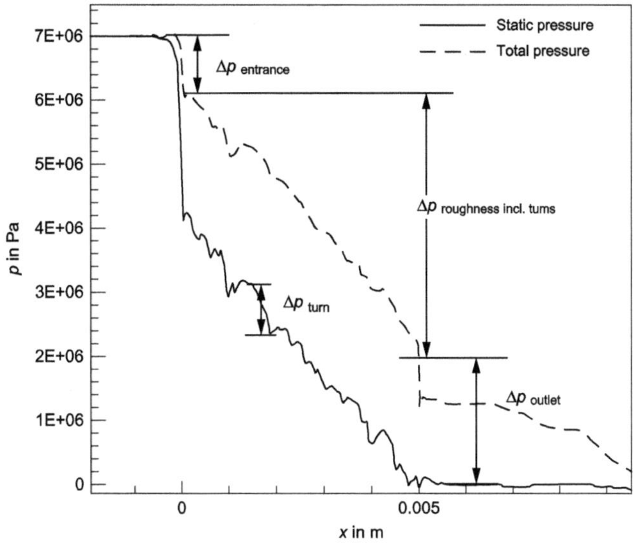

Figure 8.14 Pressure within the crack.

system inventory with interruption of NC and under certain boundary conditions even to core uncovery and cladding temperature increase. The occurrence, extent, and termination of core uncovery are closely linked with the occurrence of LSC. In contrast to an LBLOCA where the LSs are already completely emptied during the BD phase, they remain initially at least partly filled with water in SBLOCA or intermediate-break LOCA (IBLOCA) transients.

In the case of cold leg break the core level depends on the pressure drop of the steam flowing from the core to the break. As long as the LSs are filled with water the flow path via the SG and the crossover legs are blocked and the only available flow paths to the break are via the upper head—DC bypass (and via the nozzle bypass) leading to a relatively high pressure difference between the UP and the cold leg (break location).

In the case of SBLOCA a certain part of the decay power is transferred to the secondary side under so-called reflux condenser conditions so that the primary pressure is mainly determined by the secondary pressure. In the case of an IBLOCA, the primary pressure typically drops below the secondary pressure due to the larger break size and there is no heat transfer from the primary to the secondary side for a major part of the transient. Continued loss of inventory (assuming no or small ECC injection rates compared with the break flow rate) results in a continuous level reduction in the core.

In parallel, a continuous displacement of water in LSs from the SG outlet side to the upward part, i.e., toward the break in the cold leg, occurs. Assuming that the pressure drop from the core outlet via the SG to the SG outlet is small, the pressure difference between both sides of the LSs and between the core and DC are similar so that the level reduction on the SG side of the LS proceeds synchronously with the level in the UP/ core region (manometer balance). As soon as the water level in the downward vertical part of the LS reaches the upper edge of the horizontal section, steam can flow through the corresponding loop to the break and can also push water in the upward vertical part of the LS toward the cold leg. An additional steam flow path from the core to the break is generated in this way. This effect is called "LS clearing."

As a consequence, the differential pressure between the UP and the DC decreases and the core and DC water levels equalize. A potentially preceding core uncover may be terminated in this way. This effect can be supported by additional displacement of water from the LSs to the DC and the core. The described phenomena could also be observed in a PKL test simulating an SBLOCA transient as shown in Fig. 8.15. Core uncover occurred before LSC (situation 2). Due to the relative high-pressure difference between the UP and the upper DC, the core level was "suppressed." With LSC and subsequent pressure equalization between UP and DC, water from the DC and the LS was displaced into the core, and the core cooling was reestablished.

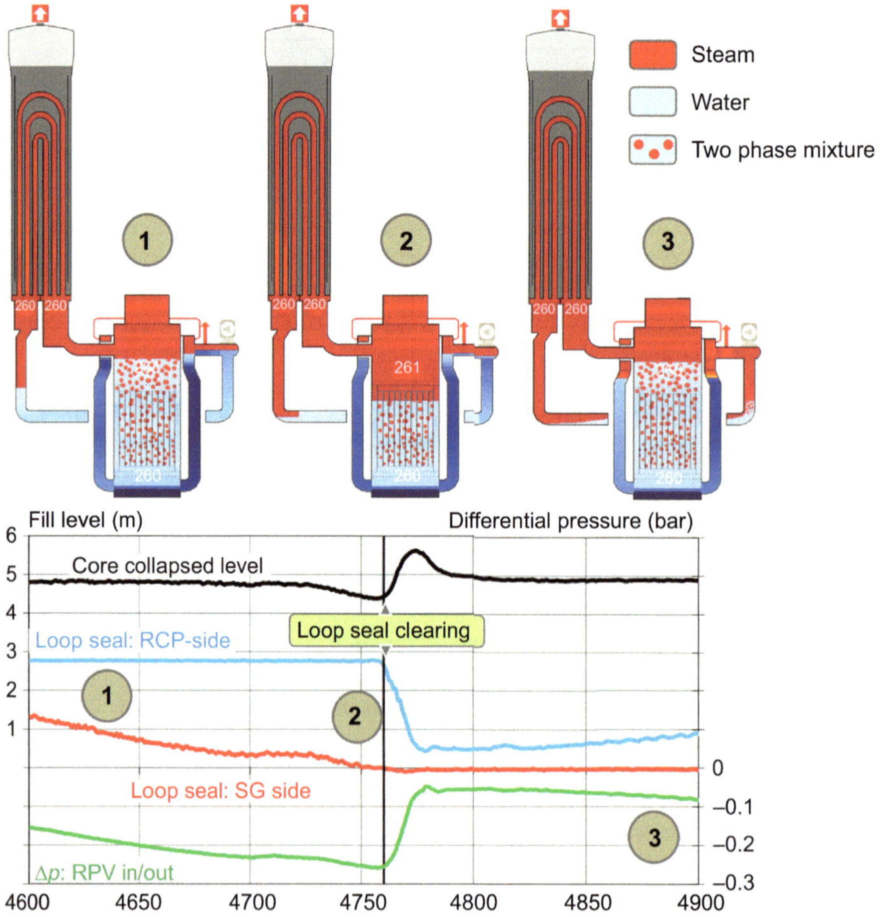

Figure 8.15 PKL facility test of loop seal clearing. *RCP*, reactor coolant pump; *RPV*, reactor pressure vessel; *SG*, steam generator.

Insofar as LSC is related to core uncover/recovery in SBLOCA and IBLOCA transients. The occurrence of LSC related to the occurrence of core uncover is mainly determined by the difference in elevation between the top of the core and the horizontal part of the LS. The timing, the dynamic of LSC as well as the number of loops involved in LSC depend on additional parameters such as break size, and flow resistance via UP—DC bypass and by the operation of the SG secondary side (e.g., cool-down rate). LSC may also be influenced or even initialed by the injection of ECC water into the cold legs causing condensation at the injection location.

In addition, SETs on LSC were performed in the 1:1 scaled UPTF in a single loop configuration to investigate the residual water level, the flow pattern, and the pressure drop across the LS during and after its clearance (Liebert and Emmerling, 1998). As an example, Fig. 8.16 shows the pressure drop across the LS for varying steam flow without droplet entrainment for two pressure levels. The amplitudes of differential pressure for three different steam flow rates are also illustrated.

8.4.2.4 Flow reversal in steam generator U-tubes

NC plays an essential role in decay heat removal for many operational transients and accidents in PWRs characterized by loss of forced circulation in the primary system. Loss of forced circulation occurs under loss of off-site power conditions (RCPs are no longer available) or due to automatic or operator-initiated shutdown of the RCP, e.g.,

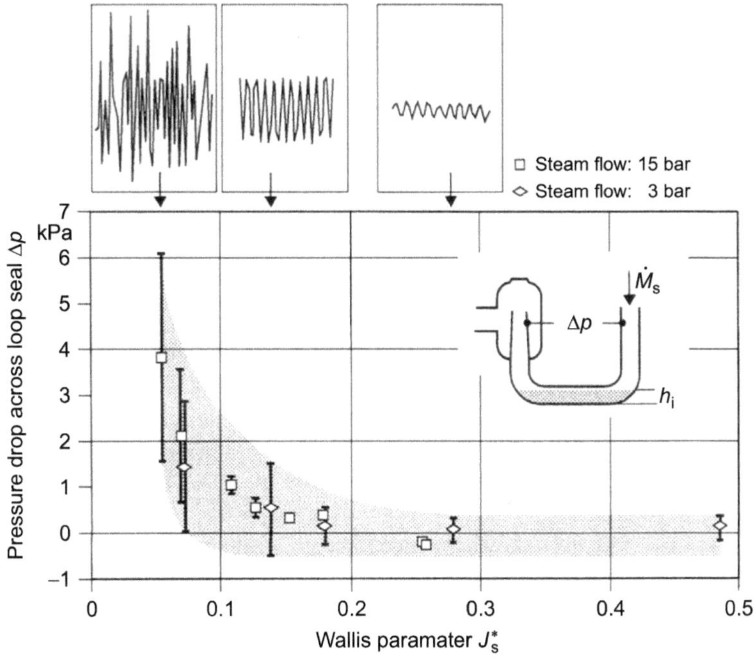

Figure 8.16 UPTF facility test of loop seal clearing (Liebert and Emmerling, 1998).

in the case of LOCA. Under NC conditions, decay power is transported from the core (heat source) to the SGs (heat sink) at flow rates that are significantly lower than the flow rates in normal (forced flow) operation (in the order of some percent, e.g., the core flow rate under NC conditions: about 600 kg/s, for forced flow: about 20,000 kg/s).

The driving forces for NC are the temperature/density differences in the RPV between the DC and the core region and in the SGs between the inlet and the outlet side. Under steady-state conditions, the NC flow mass flow rates in the individual loops are adjusted in a self-regulating way so that the driving forces are equalized by the irreversible pressure drops due to friction and local flow resistances. The performance of the SGs in general, and the individual U-tubes in particular, is of great importance for the overall system behavior under NC. The flow behavior of the individual U-tubes is a rather complex mechanism, mainly because of the following two aspects.

- The relatively low driving forces between the inlet and outlet plenum of the SG, typical for NC conditions.
- The existence of parallel flow channels between the inlet and outlet plenum of the SG comprising several thousand U-tubes of various lengths.

Extensive investigations on the flow behavior of SGs performed in ITFs over the last decades revealed heterogeneous flow distribution among the U-tubes for many NC situations. In these tests, flow reversal was observed in some U-tubes leading to a flow from the SG outlet to the SG inlet plenum in such tubes while at the same time, other U-tubes still experienced forward flow from the SG inlet to the SG outlet plenum. This heterogeneous behavior was observed under single and two-phase flow conditions and in the transition phases from two-phase flow to reflux condensation and from reflux condenser conditions to the restart of two-phase NC. Fig. 8.17 shows a typical situation with heterogeneous flow behavior for single-phase NC. In the illustration, the short and medium tubes represent U-tubes with forward flow and heat transfer from the primary to the secondary side mainly on the lower part of the inlet side. The long U-tube represents U-tubes with reverse flow. In such tubes, the temperatures over the entire length correspond to the secondary side and there is almost no heat exchange between the primary and secondary sides.

In the experiments, using temperatures measured inside different U-tubes along the tubes' lengths is an appropriate method to detect the flow direction in these tubes. If a tube is subject to continuous forward flow, it displays a temperature profile that decreases from the primary side temperature at the tube inlet to the secondary temperature toward the outlet. In contrast, an equalization of the primary and secondary temperatures along the active length (no temperature profile) is an indication of flow stagnation or flow reversal.

As an example, Fig. 8.18 shows the time trends of primary side temperatures on the inlet sides of two different U-tubes of the same SG observed in a PKL experiment during a phase with steady-state single-phase NC conditions. While the evolution of the temperatures in tube two reflects continuous stable forward flow, the drop in the primary side temperatures to the secondary temperature is a clear indication of the

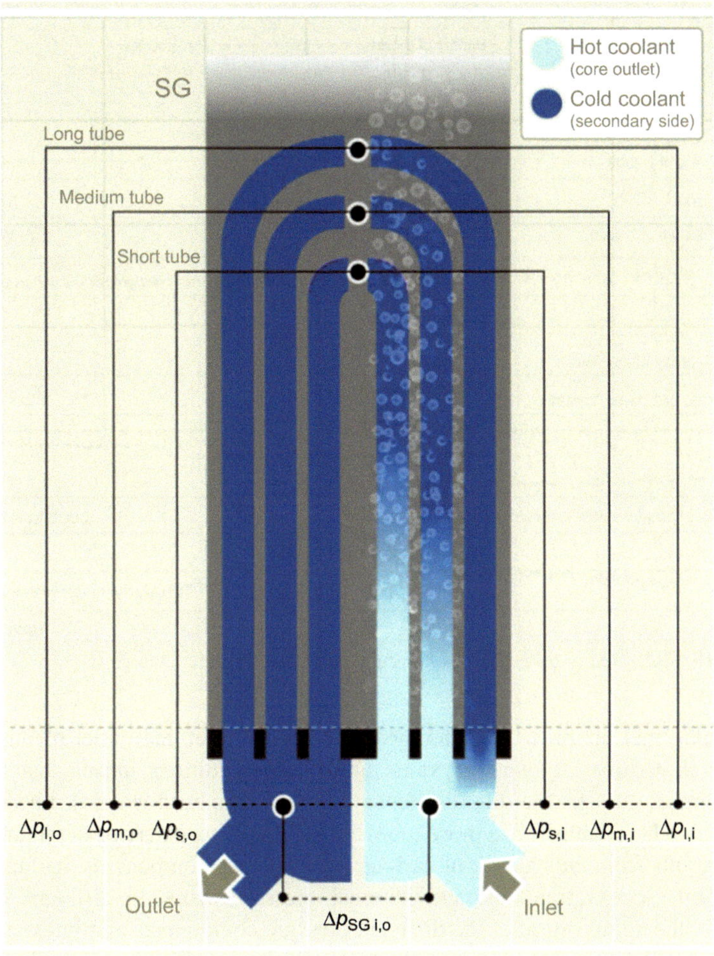

Figure 8.17 Heterogeneous behavior in steam generator (*SG*) U-tubes (forward and reverse flow).

interruption of forward flow in tube 1. The change in flow distribution within the SG is reproduced by a change in differential pressure between the SG inlet and outlet plenum and a slight reduction in the loop mass flow rate.

In NC conditions with heat transfer from the primary to the secondary side the pressure difference between the inlet and the outlet plenum is normally negative (the outlet pressure is higher than the inlet pressure). This results from the difference in gravitational heads between the downflow side (cold) and the upflow side (hot) generally exceeds the frictional and local pressure drops along the tube at low flow rates typical for NC. At higher flow rates the frictional pressure drop exceeds the gravitational component and the resulting pressure difference over the SG reaches a positive value.

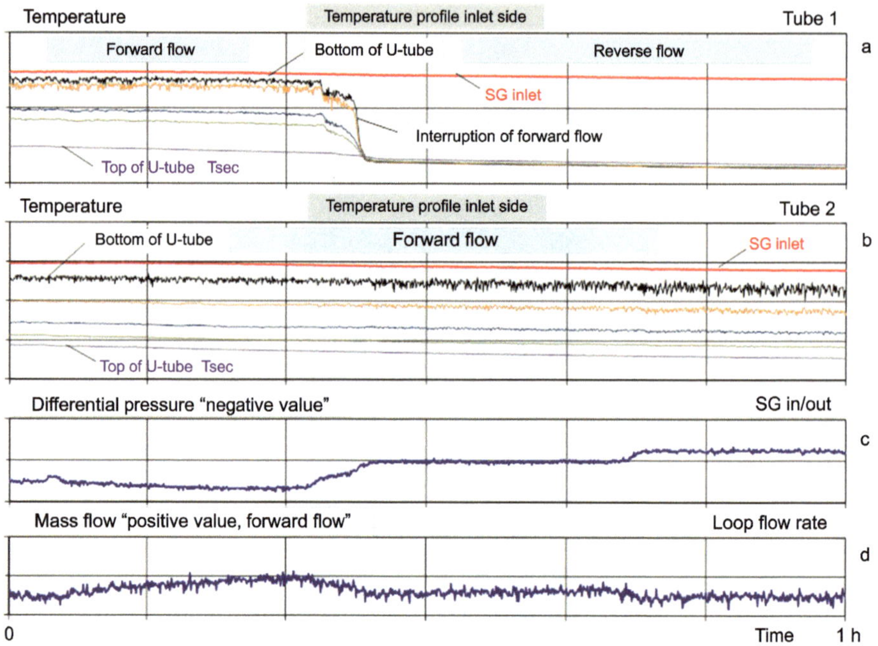

Figure 8.18 PKL facility test of natural circulation. *SG*, steam generator.

Assuming that the pressure difference between the inlet and outlet plenum is the same for all U-tubes, a negative value under NC conditions means that in tubes with no temperature profile (the difference in hydrostatic head is zero), flow reversal is expected to be established in those tubes. In general, flow behavior in the individual U-tubes results from the balance of various pressure drop components, including frictional, local, acceleration, and gravitational pressure drops in different U-tubes. Because of the composition of the different pressure components and interactions between the parallel connected U-tubes, the "normal" forward flow in some tubes may be disturbed, e.g., by a change in boundary conditions. In the abovementioned PKL test, a slight change in secondary pressure was the reason for the termination of the forward flow in at least 1 (of 28) U-tube.

In addition, mathematical models and solutions using the characteristic flow curves (differential pressure over mass flow rates) for forward and reverse flows in single U-tubes based on momentum and energy balance equations have been developed by several researchers (see also Chapter 5 and 11) in an attempt to interpret/confirm the observed heterogeneous flow behavior in U-tubes, demonstrate the required conditions for reverse flow, and identify preferred U-tube positions for occurrences of reverse flow. More details on the applied methods can be found in the corresponding literature.

It must be mentioned that despite the possible occurrence of flow stagnation and reversal in some U-tubes under NC conditions, the overall net forward flow will always remain stable as long as the corresponding SG is available as a heat sink. The impact on overall heat transfer is also rather small, as the loss of heat transfer in the tubes with a reverse flow is compensated by increasing heat transfer in tubes with continuous forward flow (SGs are designed for much higher loads). However, the total mass flow rate through the SG and therefore through the core may decrease because of reverse flow in some U-tubes. This again leads to a higher core outlet temperature. To address this aspect of code calculations, the application of multitube modeling is important. It must be noted that flow behavior in U-tubes could be even more unstable and more heterogeneous under two-phase flow conditions. Heterogeneous water/steam distribution at the tube sheet with a preferred steam supply to some tubes may promote heterogeneous flow behavior in individual U-tubes.

8.4.2.5 Stability in boiling channels

Focus is given in this section to the qualitative role and the importance of pressure drops in the characterization of stability of fluid systems (single- and two-phase).

Currently, more than 80 BWRs operating worldwide provide an important contribution to the overall electricity supply from nuclear sources. Water boiled in the core produces steam that is directly used to rotate turbine(s), thus constituting the simplest industrial system built so far that is suitable for transforming fission power into electrical power. However, concerns about fundamental aspects of interactions between SYS TH and neutron kinetics have so far obstructed or delayed (or made less effective) the exploitation of this type of reactor concept. In the past, the BWR stability issue (D'Auria et al., 1997a,b) caused many concerns and required specific attention from technologists and researchers.

The classification of instability mechanisms provides an idea of the topic.

Different classifications can be found in the literature. These are based on (a) fundamental mechanisms, (b) patterns of core oscillations, and (c) system interactions:

(a) Classification based on fundamental thermal-hydraulic mechanisms

The static instabilities are
- flow excursion (Ledinegg type)
- boiling crisis
- relaxation types, including flow pattern transition, geysering, and chugging

The dynamic instabilities are
- density wave oscillations
- pressure drop oscillations
- acoustic oscillations
- thermal oscillations

(b) Classification based on geometric features of the oscillations

In this case the relationships between the phases and amplitudes of waves at different axial and azimuthal positions in the core of a reactor or reactor simulator are considered. Coupled neutron physics/thermal-hydraulic and pure thermal-hydraulic oscillations can be distinguished using the following classification:
- Core-wide or in-phase oscillations
- Regional or out-of-phase oscillations
- Single-channel or local oscillations
- Punctual or not propagating oscillations (their existence can only be envisaged from the performed activity and considered documentation)

Most phenomena above can be explained by referring to the concept of "harmonic modes" of reactor diffusion equations.

(c) Classification based on system design

This classification is valid primarily for coupled nuclear–thermal-hydraulic systems (BWR plants) but may also occur in pure thermal-hydraulic systems. The overall system performance and the origin of oscillations are concerned. The following classification can be introduced:

- Oscillations involving core-bypass
- Oscillations involving vessel DC
- Oscillations involving the entire primary loop (also referred to as loop-type oscillations)
- Oscillations induced by balance of plant, including control systems

The analysis of instability phenomena may constitute a multidisciplinary subject, involving different technological areas like transient thermal hydraulics, neutronics, fuel behavior and fuel cycle management, instrumentation, plant control and monitoring, and detailed knowledge of plant features and operating conditions.

A wide-ranging body of literature exists in this area, see, e.g., Ishii and Zuber (1970), Yadigaroglu and Chan (1979), D'Auria et al., 1997a,b, Vijayan (2002), and D'Auria et al. (2016). Hereafter there is no intention to duplicate the information in reports and papers devoted to the stability either in BWR and/or in boiling/condensing systems. However, it should be mentioned that the measurement, the modeling, and the prediction of pressure drops have a key role in this area. Sample reference data for pressure drops during stability events are given in Figs. 8.19 and 8.20: these constitute typical diagrams adopted to characterize the stability performance of a system. Change of sign or passing from positive to negative value for pressure drops (Figs. 8.19 and 8.20) implies the occurrence of flow reversal. Amplitudes and frequency of pressure drop oscillation (Fig. 8.19) correspond to amplitude and frequency for mass flows.

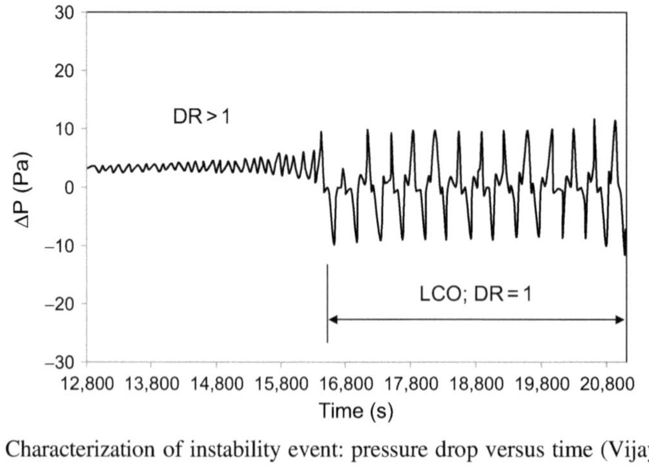

Figure 8.19 Characterization of instability event: pressure drop versus time (Vijayan, 2002). *DR*, decay ratio; *LCO*, limiting condition for operation.

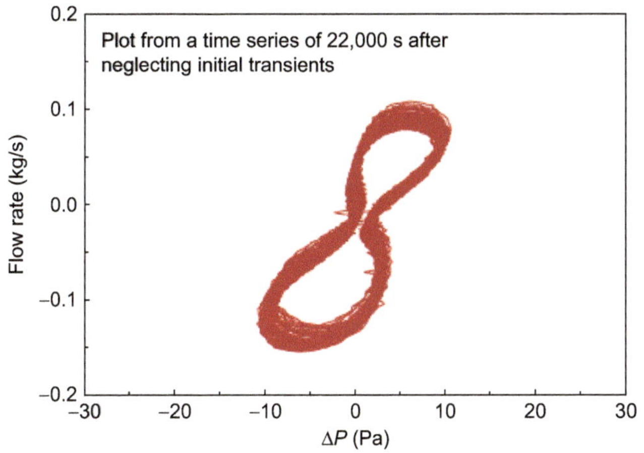

Figure 8.20 Characterization of an instability event: pressure drop versus mass flow rate (Vijayan and Nayak, 2012).

8.5 Modeling pressure drops and applications in thermal-hydraulic system codes

System codes (Chapter 11) can be considered the basket of expertise gained from experiments and modeling in nuclear thermal hydraulics. Therefore, information in Sections 8.2−8.4 is relevant hereafter. The importance of system codes in applications is suggested in the introduction of the notes below.

8.5.1 System thermal-hydraulic code modeling

SYS TH codes constitute the tip of the pyramid of knowledge in nuclear thermal hydraulics as discussed in various chapters of the book. Insight into code modeling can be found in Chapters 5, 10, and 11), including modeling for pressure drops (i.e., the momentum equations). A few comments about modeling features are provided hereafter rather than insights into the features (notably including numerical aspects) of the pressure drop models.

8.5.1.1 Wall friction

Before introducing more details, let us consider the difference between the empirical formulation for Δp_{fr} (e.g., Eq. 8.2) and the "first-principle" based formulation where τ_w is taken at the origins of friction pressure drops. The latter formulation derives from a force balance for the control volume of fluid, adopting the mechanics of the continuum approach: the (so far insurmountable) problem is the availability of a suitable value for the derivative $\frac{\vartheta w}{\vartheta y}$ (w is the fluid velocity and "y" is the space coordinate orthogonal to the fluid direction) at the fluid-wall interface, e.g., Alves Portela and Sandham (2020) give an idea for its implementation to solve practical problems.

The former formulation (or Eq. 8.2) is empirical and is still adopted in codes within nuclear thermal hydraulics, 90 years after its introduction.

With reference to the RELAP5 code, the wall friction model consists of two main parts:

- Computing the overall two-phase friction pressure drop,
- Apportioning the total wall friction into liquid and vapor components.

The wall friction model is based on a two-phase multiplier approach calculated from the HTFS-modified Baroczy correlation, which is independent of the flow regime. The individual phasic wall friction components are calculated by apportioning the two-phase friction between the phases using a technique derived by Chisholm from the Lockhart-Martinelli model. The calculated phasic factors depend on flow regimes. The comments are as follows:

- Notwithstanding highly (or relatively) sophisticated mass, momentum, and energy balance equations, i.e., the so-called unequal velocities, unequal temperatures, or six partial-derivative sets of equations, single-phase and two-phase pressure drops are based on empirical correlations (basically those discussed in Section 8.3) developed during the 1960s.
- The process "of apportioning" mentioned above implies that one overall data (typically the total friction pressure drop available from experiments) is split into two or more components, according to the structure of the balance equations. In the majority of the conditions of interest there is no possibility of estimating the correctness of, i.e., to validate, the apportioning process.
- Pressure drops depend highly on flow regimes: flow regimes depend highly on geometry and the condition of the fully developed flow. In the case of large diameter pipes (up to around 1 m) like those constituting the hot legs and the cold legs of RCS in PWR, experiments in two-phase conditions have not been performed. Necessarily simplified flow regimes are considered in modeling that are derived from small-scale experiments (scaling issue occurs as discussed in Chapter 13).

From the viewpoint of the code user (or analyst), one may add the following:

- *Roughness*: the related value must be supplied by the analyst and may not constitute available data for the system to be simulated by the code; furthermore, roughness may vary during the lifetime of the system, e.g., due to corrosion, erosion oxide formation, and crud deposit.
- *Equivalent diameter*: the related value must be supplied by the user; an engineering formulation is available that is directly connected with simple geometry without possible three-dimensional effects; the value of the equivalent diameter is therefore used to adapt the results of code prediction, notably pressure drops, to measured data if available; the process is prone to some arbitrariness from the code user side; the fixed value for equivalent diameters also affects the calculation of the heat transfer coefficient in the concerned region.

One may summarize that modeling for distributed friction pressure drops is part of the code and user effect is restricted to the (sometimes unknown) values of roughness and equivalent diameter.

8.5.1.2 Form losses

Losses due to abrupt area change are calculated using form loss models and originate from the geometric discontinuities discussed earlier. Notably, connections between pipes (relatively small diameter) and plena (relatively large diameters), changes in the direction of the flow in plena, and the entrance-exit of a plate (e.g., the core support plate) are characterized by form loss coefficients that constitute inputs for the code.

The dependence of the *K* factor on the *Re* number (or fluid velocity) is typically allowed in correlations to be supplied by the code user; however, no dependence for the *K* factor is envisaged in available models on void fraction. The comments are as follows:

- The first two bullet items in the previous paragraph, i.e., current sophistication level of modeling and apportioning procedure, also apply here.
- *K* factor values (very) strongly depend on details, e.g., rounded or sharp edge, which are not available in all locations of an RCS in water-cooled nuclear reactors.
- At every discontinuity, whatever its geometric complexity, different flow situations may occur, namely in two-phase conditions: cocurrent flow of steam and liquid (with steam and liquid flow configurations affected by flow pattern determined by upstream conditions), CCF of steam and liquid and flow reversal (i.e., liquid and steam flowing together in a direction opposite to the direction that characterizes the nominal operating conditions of the RCS).
- With the partial exception of the core region in all water-cooled reactors and the SG tube region in PWR, local pressure drops constitute the largest part of total pressure drops in RCS at any time during a transient. (*Note*: a partial exception is introduced in the core region because of the importance of the bottom, top, and spacer grids as far as total pressure drops in the core are concerned).

From the viewpoint of the code user (or analyst), one may add the following:

- The code user is responsible for supplying suitable data for *K* factors as calculation input parameters: unavailability of data (typical situation) should correspond to the execution of proper sensitivity and uncertainty analyses (this is not always possible).
- The user must consider the possibility of large differences between direct and reverse *K* factor values.
- CFD codes may provide valid support to analysts for systematic estimation of *K* factors at any geometric discontinuity (Chapter 12, e.g., Section 12.4.1; see also section "Form losses predicted by computational fluid dynamic" immediately below), although those results are not immune to errors.

So one may expect highly approximate calculation results: the degree of approximation changes with void fraction. An even larger impact of *K* factors on calculation results may be expected if one considers the combination of fluid velocities and void fractions.

One may summarize that modeling for local pressure drops is not part of the code and user effect is decisive for attaining reasonable results from code applications.

Form losses predicted by computational fluid dynamics

A state-of-the-art application of CFD codes to predict pressure drops in (research) nuclear reactor-relevant geometry is discussed by Lopes et al. (2021). Those authors applied k-ε and k-ω models to geometric discontinuities in concerned reactors having available experimental data. Differences in calculated pressure drops derived from k-ε and k-ω models are up to 10% (of an average value from those two models).

Differences related to experimental data (in this case friction pressure drops are considered together with pressure drops at geometric discontinuities) are typically around 10% of the experimental values with maximum values larger than 30%.

8.5.2 Pressure drops and steady-state qualification of computational model

The establishment by the code user of correct steady-state conditions for pressure (based on pressure drops) along the loop is a mandatory prerequisite for the prediction of any transient scenario and NPP performance. Acceptable errors are proposed by Bonuccelli et al. (1993). An example of the current situation is discussed in the following, based on a code application activity performed within an international framework in 2006 (Petruzzi et al., 2006).

The calculated normalized (with respect to the hot leg inlet position) pressure distribution curve versus loop length related to the LOFT nuclear facility is shown in Fig. 8.21. The activity was performed by a dozen groups of analysts using different SYS TH codes; it constitutes the second phase (of six) of the OECD/NEA/CSNI BEMUSE project, see also Glaeser et al. (2011). The following considerations apply:

- Almost all calculated values fall within the experimental uncertainty bands in correspondence with the hot leg outlet position and the LS.
- In correspondence with cold leg inlet and outlet positions, the major portion of the calculations falls outside the uncertainty bands.
- The positions characterized by large discrepancies among the calculated results of the participants are (1) SG U-tube top elevations, (2) the pump inlet and outlet, and (3) the lower plenum. No experimental value was available for those locations.

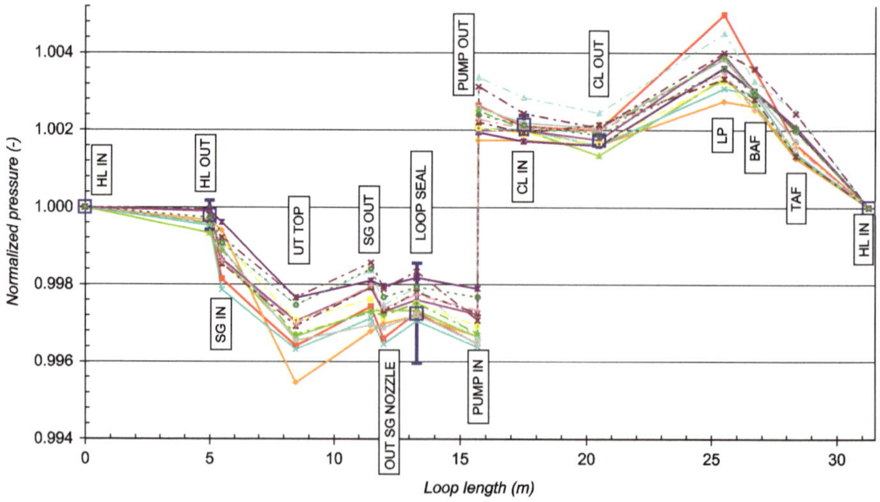

Figure 8.21 Comparison of normalized pressure curves along the loop, BEMUSE phase II. *BAF*, bottom of active fuel; *CL*, cold leg; *HL*, hot leg; *LP*, lower plenum; *SG*, steam generator; *TAF*, top of active fuel; *UT*, U-tubes (of steam generator).

Not surprisingly, large differences were observed in comparisons of various calculated data (as well as comparisons between experimental data and code calculations) for the concerned LBLOCA. Full documentation is available in Petruzzi et al. (2006) (see also de De Crécy et al., 2008).

8.5.3 Pressure drop predictions/measurements during transients

In nuclear thermal hydraulics, pressure drops during a transient do not necessarily imply transient pressure drops or time-acceleration pressure drops. Namely, quasi-steady conditions and related assumptions can be used to predict and characterize pressure drops for most of the NPP design basis accident (DBA) scenarios discussed in Chapter 15. Fast-changing conditions that require considering the time acceleration pressure drop include, for instance, the initial tenths of seconds of an LOCA, or in the case of a control rod ejection (CRE) event, a portion of a reactivity-initiated accident.

A sample list of items connecting pressure drops and pressure drop predictions with transient conditions in NPP is provided below not in an order of importance. In this case, the word transient refers to accident scenarios in water-cooled nuclear reactors within the DBA category and simulated accident scenarios in ITF.

- LS. The LS is the U-shaped part of the cold leg in PWR upstream of the main coolant pumps (MCPs). The elevation of the horizontal axis of the LS is lower than the elevation of the top of active fuel (TAF) for the core. This induces the possibility of the formation of a plug of liquid in the LS during accident scenarios. The plug formation constitutes a seal for the coolant flowing in the loop, hence the name LS. The depth of the LS vertical part is close to 1 m. LSC may induce level recovery in the core and in some cases quenching of a previous dryout condition. Measurement of pressure drop across the vertical part provides information about the LS status and possible clearing occurrence.
- Bypass. Several bypass flow paths inside the vessel of a PWR play a role primarily in the case of SBLOCA. Pressure drop measurements indicate the direction of the flow and may contribute to determining the bypass flow rate.
- Flow reversal. Direct flow at any location in the primary system of either a PWR or a BWR is connected with nominal operating conditions. However, flow reversal may occur almost anywhere in the circuit during transient conditions. The predictability of flow reversal (i.e., very important to calculate system performance in the case of transients) is connected with the knowledge of related loss coefficient at abrupt discontinuities. Those coefficients may be different from "direct" coefficients at the same location and their value may reveal (much) more uncertain than the value characterizing the direct flow.
- Stability of single-phase and two-phase or boiling condensing systems. In addition to BWR stability mentioned in Section 8.4.2.5, flow oscillations, thus stability issues, are expected in one- and two-phase NC conditions in PWR (see also Chapter 15). Namely, it is well established that selected groups of U-tubes in SGs may experience flow oscillations including flow reversal, thus reducing the overall transit flow area for the coolant in the core. Pressure drops are the key indicator of this phenomenon.
- Emptying and filling and DC level (SG and RPV in PWR and RPV in BWR). Change of level, namely in the PRZ of a PWR, implies changes in the dynamic response characteristics

of the overall system. Empty or overfilled PRZ signifies a loss of control capacity for the pressure in the system. The knowledge of the narrow range and wide range of levels in SGs and RPVs of BWRs allows the control of the core power. A pressure drop signal is used for detecting the level.

- Advancement of quench front. Specifically in the case of an LOCA in either a PWR or a BWR, core uncovery is expected and must be followed by a proper recovery that is mainly governed by a level rise in the DC of the vessel. Then quench front advancement is detectable by pressure drop measurement.
- MCPs. Values for the MCP head, the head during MCP rundown (coast-down), and the head under locked or free rotor conditions constitute essential parameters affecting coolant flow in the core. Related pressure drop measurements are essential for determining system responses in the case of transients.
- Sump recirculation and core pressure drops. Long-term cooling implies a containment sump (PWR) or pressure suppression pool recirculation (BWR). In both cases "small" debris accumulated at the bottom of the containment may pass through filters and recirculation pumps and produce partial core blockage. Ballooning of fuel rods may also cause partial core blockage. Cross-correlated pressure drop signals from different regions of the vessel may allow the detection of core blockage. Related techniques may need to be developed.
- Measurement issue. At least three issues must be carefully addressed for a qualified transient pressure drop measurement:
  - The absolute value for the range of variation expected for pressure drop measurement may be several orders of magnitude lower than the absolute pressure acting in the system at the location of pressure drop transducer terminals (i.e., the port connecting the measurement pipeline and the concerned "high" pressure system).
  - The distance between the location of pressure drop transducers and the terminals (or taps) is typically of the order of several dozen meters. A small (typically 5−10 mm internal diameter) measurement pipeline (hereafter called pipeline) connects the terminals with the pressure transducers. Usually bends and U-configurations cannot be avoided. Filling of the pipeline by incompressible fluid must be ensured by proper actions before the start of the transient (e.g., an experiment in ITF).
  - Very small leakages in the pipelines, gas liberated from liquid inside the pipelines when the absolute value of the system pressure decreases, initially undetected gas bubbles, regions of high fluid temperature in the pipe such that liquid flashing may occur when the absolute value of the system pressure decreases, may produce unqualified pressure drop signals with very few possibilities to detect the failure or to recover the right signal.

The provided overview confirms the role and stresses the importance of pressure drops in nominal and transient operation of water-cooled reactors: pressure drops are not measured in all cases in NPP; however, predictive capabilities for pressure drops must be ensured in all cases and, in all cases, pressure drops are measured in ITF.

8.6 Measurement techniques

The measurement of pressure drops constitutes the focus of this chapter. Here one may state the following:

(a) Measurable quantities exist that are derived from the pressure drop type of measurement (e.g., liquid levels).

(b) Quantities can be measured that are directly connected with pressure drops (e.g., fluid velocities and mass flow rates).

The techniques considered hereafter deal with quantities in categories (a) and (b) in addition to direct measurements of pressure drops. Furthermore, measurements in experimental loops (or test facilities) are distinguished from the measurements adopted in NPPs. Namely, the key measured quantities associated with pressure drops are the following:

- In test facilities: local pressure drop; collapsed/swell level; void fraction; density; mass flow
- In NPP systems: RPV level/loop level; Δp over MCP; mass flow

Hereafter, some comments are provided on the current state of the art in performing pressure drop measurement rather than discussing (important) details of experimental techniques including techniques for interpreting raw data from transducers. When considering the issues (or measurement drawbacks) in the forthcoming sections, measurement complexity should be considered (high temperature, high pressure, and in the case of NPPs, the presence of radiation fields).

8.6.1 Direct measurement of pressure drops

Reference is made to direct measurements for pressure drops in ITF and in NPP. The typical configuration of interest below is constituted by two pressure taps connected by two hydraulic lines to two terminals of a pressure transducer. Pressure transmission lines have a diameter of 6–8 mm and are filled by incompressible fluid, typically liquid water. It must be ensured that liquid does not vaporize at contact with the high-temperature fluid and that (reasonably) noncondensable gases are not present to form bubbles or not even dissolved in the fluid. Dissolved gas in the fluid (as well as steam produced at the interface between measuring liquid and hot process fluid, and gas produced during a depressurization process) may heavily affect the measured values.

8.6.1.1 Distributed (friction and acceleration)

The presence of gas or vapor in the transmission line is one reason of concern for the quality of measurement: this is also connected with the length of pressure lines (up to several dozen meters) and to the many elbows and curves necessarily part of the construction of those lines. The main additional issues that may cause measurement errors for distributed friction pressure drops (acceleration pressure drop issues can be inferred as a consequence) are.

- Static pressure target. The goal of the measurement is the static pressure in the majority of cases; problems may arise because of the possible influence of dynamic pressure, e.g., caused by imperfectly aligned pressure taps.
- Elevation pressure drops. It is necessary to separate the contribution of elevation pressure drops: this is not straightforward (see also below) when two-phase flows are involved.
- Compressible fluids including two-phase. It is necessary to separate the contribution of (spatial) acceleration pressure drop: this is not straightforward (see also below) when two-phase flows are involved.

- Two-phase conditions. An interpretation model is needed to estimate the average void fraction between the two taps. The model is necessarily imperfect, and its quality depends on the flow regime experienced by the process fluid.
- Long piping for transmitting pressure drop signals may generate errors in measurement in very fast transients in addition to errors possibly originated by the presence of gas or steam.

8.6.1.2 Locally or at geometric discontinuities

The issues described for distributed friction pressure drops also apply to the measurement of local pressure drops at geometric discontinuities, e.g., signals used to characterize the *K* factors. In addition, the following issues are specific to local pressure drop measurements.

- Irreversible portion 1. Any local pressure drop implies the occurrence of a total pressure drop that is formed by a reversible and an irreversible portion. Only the irreversible portion is of technological interest and contributes to the *K* factor. Typically, the total pressure drop occurs immediately downstream of the geometric discontinuity: the reversible portion brings a recovery in pressure at some distance downstream. The value of the "distance downstream" causes the issue: the distance is not known and depends on the geometric configuration and the flow conditions.
- Irreversible portion 2. The measurement of the "distance downstream" is troublesome and is possible only in selected reference situations. Hence unavoidable errors must be expected from measurements.
- Special geometry. Local pressure drops also occur inside the plena, e.g., lower plenum or UP in the case of PWR RPV. In those situations, it is impractical to install transducers along the actual flow path of the fluid. This gives origin to a hidden error when attempting to simulate fluid flow behavior in those conditions.

8.6.2 Quantities connected with pressure drops

Pressure is a scalable physical variable, which describes the force on a unit of area.

Together with the measurement of temperature, the measurement of pressure is one of the first and most used measurements in science. Pressure measurement, which is one of the so-called "conventional measurements," continues to play a central role in the detection and recording of flow processes in thermohydraulic systems. In nuclear engineering, this applies not only to the detection and recording processes in NPPs but also to the test facilities used for analyzing and modeling important thermohydraulic processes.

In practice, the pressure at certain measuring points in the system to be investigated is forwarded to a sensor via pressure tapping points and so-called pressure sensing lines. The sensor responds to the pressure exerted on it or pressure difference with a displacement or deformation (e.g., a diaphragm). The deformation is converted via appropriate sensors (capacitive, inductive, etc.) into electrical signals proportional to the differential pressure.

Due to this measurement principle, pressure measurements generally give a pressure difference between two points or the difference in pressure between a measurement point and a reference pressure.

If the reference pressure is zero (i.e., complete vacuum) the measured value corresponds to the so-called absolute pressure, if the reference pressure equals the atmospheric pressure, the measured pressure difference corresponds to the so-called overpressure (aka positive or gauge pressure) at the measuring point concerned. However, if both measuring points are located within the system, the measurement registers the pressure difference between these two points, i.e., the pressure drop along the flow path that is mainly of interest later.

As described in Section 8.2, the total measured pressure drop is made up of several components including gravity, acceleration, and pressure losses.

$$\Delta p = \Delta p_e + \Delta p_a + \Delta p_f + \Delta p_l \tag{8.18}$$

where Δp_e is the pressure drop due to static head. Δp_a is the acceleration pressure drop due to a change in flow velocity along the flow path as a result of a change in the cross-sectional area or density. In non-steady-state processes, another term is added to this convective acceleration, which expresses the local acceleration over time.

Under certain conditions or if certain influencing factors are ignored, the differential pressure measurement can be used both to determine the permanent pressure loss and to derive other important physical variables, such as

- level (swell level/collapsed level)
- mean fluid density
- mean void fraction

In steady-state flows of incompressible fluids in flow paths with a constant cross section, the acceleration term is omitted completely. For horizontal flows, the head component of Eq. (8.18) is also omitted and the measured differential pressure in fully developed flow is the remaining pressure loss in the flow path ($\Delta p_f + \Delta p_e$).

8.6.2.1 Filling level

Using the static pressure of water columns to take differential pressure measurements of the water filling level or water reserves is a common method used both in NPPs and associated testing systems. If tanks are filled with water and steam (idle system) and the two phases exist in separate forms, the differential pressure measurement is a simple and very precise method for determining the filling level; this method requires knowledge of the densities of the two phases (e.g., by means of additional absolute pressure and temperature measurements). The water filling level can be determined via the following designation.

Wherever there are high-pressure levels, the static pressure of the steam can no longer be neglected and should be taken into account when determining the water filling level.

8.6.2.2 Collapsed level

It is not possible to directly determine the swell level (separation between two-phase mixes with a predominant steam proportion) by measuring the differential pressure alone; additional measuring information is required. However, conducting several differential pressure measurements in succession can pinpoint the position of the swell level to at least a segment of the corresponding differential pressure measurement. In addition, if the mixture level changes (increase/decrease), it is possible to at least trace the trend of the temporal progression of the mixture level. For instance, usually information can be derived from the Δp measuring signal curve about the passing of the swell level at a pressure removal point and therefore the transition to a predominant steam proportion within the associated Δp measurement (when the swell level decreases).

If a measuring range section is filled with a two-phase mixture and if the water and steam density is known, it is possible to derive from the measured differential pressure a constant assumed mean density of the mixture or the mean *void fraction*.

Strictly speaking, these considerations apply only to states or processes in which pressure drop due to dynamic effects (e.g., wall friction, two-phase friction) can be neglected.

This usually also applies to flow conditions in which a specific steam quantity flows from bottom to top through standing water in tanks or vertical pipes.

The proportion of the pressure loss (due to friction) of the measured differential pressure can be neglected, especially in small steam mass flows and higher pressure levels compared with the height (head) component. Generally, the height (head) component can be determined by also measuring the volumetric steam proportion using special two-phase measuring equipment (see below, *will be included*). The pressure loss proportion is the difference between the measured differential pressure and the height (head) component determined by measuring the steam content.

8.6.2.3 Practical applications for differential pressure measurement in nuclear power plants

Absolute pressure and differential pressure measurements are, in combination with temperature measurements, the main measuring methods used in NPPs and are used both as operating instruments and for incident instrumentation purposes. For instance, differential pressure measurements taken via MCPs in a BWR not only supply information about the number of operational pumps but also are used to determine the loop flow rate using the pump characteristic curve, as well as helping to protect the aggregates. During commissioning tests, pressure loss measurements via other plant parts are taken in some cases to obtain an overall picture of the total pressure loss and establish the status of individual plant components; they are also used to confirm computed values on which the plant was designed.

8.7 Envisaged developments in pressure drops

The phenomena associated with the loss of pressure along with those associated with the heat exchange gave rise to major technological problems since the advent of thermal machines in the 18th century (the identification and the characterization of pressure drops and heat transfer mechanisms came much earlier during the human civilization history), as already discussed; see also D'Auria (2012) and Chapter 24. Furthermore, within the areas of pressure drops and heat transfer, a tight connection exists between theoretical modeling and the actual configuration of the systems.

Insights addressing the above statements can be found in this chapter and other parts of the book. Furthermore, Reynolds, Moody, Colebrook, Darcy, Idel'chik, Martinelli, Nelson, etc. (see also Chapter 1), are eminent scientists whose name is associated with the calculation of pressure drops.

The target for this section is to identify directions for possible improvements in the knowledge in the area of pressure drops.

Looking at the future may benefit from a view of the past. A few sentences from the well-known, more than 500 pages Idel'chik Handbook (Idel'chik, 1960) are reported:

> *There is almost no branch of engineering which is not somehow concerned with the necessity of moving fluids through pipes, conduits, or machinery. The degree of complexity of a hydraulic or gas-air line can be quite varied. In some cases these are largescale systems of pipes, gas mains, water conduits, steam pipes, air ducts, etc., while in other cases these are pipelines of relatively small length, but having a large number of fittings and branches, various obstructions such as throttles and adjusting devices, grids, protruding parts, etc. … Until recently only restricted data were available on the subject, and these were scattered among various textbooks … and in scientific papers. In many cases these data are contradictory or dated, and deal with only a limited number of local resistances. Furthermore, the coefficients of local resistances generally were given only for special geometric and physical characteristics. In order to fill this gap, in 1954 the author published a book … We present here in the same spirit this special handbook on the local fluid and friction resistances.*
>
> *The writing of this handbook presented considerable difficulties, mainly due to the range of local resistances, their geometric boundaries, and the states of flow in them, which are much narrower than required by practice. Furthermore, much of the data obtained is insufficiently accurate and reliable; this is particularly true of the coefficients of local resistances. Therefore, it would have been better to delay the publication of this handbook until all coefficients of local resistances could have been checked and refined experimentally by some standard method, based on the contemporary level of metrology. Unfortunately, it seems unlikely that such a series of experiments would be completed in the near future.*

Those words are perfectly true today, after more than half a century. Namely:

- More experiments have been done, but they number far fewer than needed to satisfy the geometric and operating conditions of water-cooled reactors, specifically where two-phase conditions are concerned; errors, which are not always documented, are the result of experiments; measurement errors are expected to be larger under two-phase conditions;
- CFD codes have been made available that, at least under single-phase conditions, could be used to determine "all coefficients of local resistances," but, errors in computed results (a bit less than $|10\%|$ of the actual value) may reveal higher than precision needs; and

- Precision targets from the side of NPP designers or safety analysts have not been fixed: these could be a function of the *Re* number and void fraction and could be less than |1%| of the actual value, i.e., 10 times below established computational capabilities, and possibly much below attainable accuracy from experiments.

As defined above is the context in which an attempt is made to identify some lines of development in the evaluation of pressure losses.

8.7.1 Extension of experimental database

A nonexhaustive list of keywords associated with pressure drops includes:

- Laminar or turbulent
- Fully developed flow or nondeveloped flow
- Transient or steady-state and acceleration (in space or time)
- Compressible or incompressible fluid
- Single- or two-phase and (in the case of two phases) two-field or multiple-field
- Distributed (friction) or local (i.e., at geometric discontinuity)
- Wall roughness
- Sharp or smooth edge (in the case of geometric discontinuity)
- Viscosity of fluid, Newtonian fluids, and non-Newtonian fluids
- Velocity profile (single-phase flow) or fluid velocities and void fraction profile (two-phase flow)

The items above can be used in designing new experiments and planning developments in the area to (1) identify the affecting variables-parameters and their related ranges and (2) build a multidimensional table or several tables covering the identified topics and the ranges of relevant parameters. It is also clear that the characterization of items or conditions like "nondeveloped flow," "viscosity," and "level of smoothness" may require an infinite variety of situations. In this case, an attempt should be made to group the envisaged situations into a manageable number.

It is beyond the objective of the present book to pursue either target (1) or target (2) just mentioned. To this aim, groups of experts, from nuclear regulators, designers, and scientific institutions involved with thermal-hydraulic programs, should be convened. Here, the importance of such activities is recognized, and further guidelines are provided in Section 8.6.2 related to specific topics.

8.7.2 Developments in two-phase flow modeling

It is clear from the picture given in the entire chapter and from the snapshot views of previous sections that two-phase flows deserve or are expected to deserve the largest attention as far as possible developments in the area of pressure drops are concerned.

Experimental and modeling research in the 1960s, even before the advent of computers, demonstrated the existence of a situation as depicted in the sketch of Fig. 8.22. Under the condition of equal mass flow rate (i.e., kg/s) the friction pressure drops for liquid get a value much lower than in the case of steam (or gas) pressure drops. The two-phase pressure drop curve is assumed to follow a trend bounded by the two curves

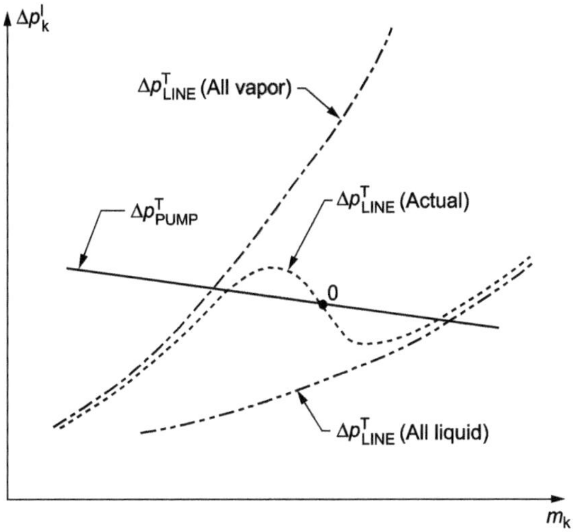

Figure 8.22 Friction pressure drop in horizontal channel as a function of mass flow rate for the single-phase (liquid and vapor) and two-phase conditions.

relating to single-phase conditions. Furthermore, the two-phase curve does not show a monotonous trend, and the well-known Ledinegg instability is expected.

8.7.2.1 Expected desirable developments

Significant examples of expected desirable developments, without having the target of a systematic and comprehensive line of attack to the overall issue and assuming proper support from the execution of experiments, are:

(A) Related to distributed pressure drops. Substitution of the two-phase multiplier approach with an improved mechanistic approach where wall drag, interfacial drag, and flow regimes play the key role. Fluid velocities, void fraction, absolute pressure, and diameter size (in addition to roughness and equivalent diameter) should be considered with suitable ranges of variations. The new approach should directly account for nondeveloped flow conditions.

(B) Related to local pressure drops. An attempt should be made first to distinguish among groups of situations: e.g., abrupt area changes in a pipe-like configuration, bends, branches, connections of pipes with reservoirs, changes in flow-path directions inside plena, component related, special situations like the interface between the jet pump and DC region in a BWR. Then, for each group, key geometry-related parameters should be identified. Then the fluid-dynamic parameters should be identified with suitable ranges of variations. Finally, correlations, as far as possible of mechanistic type (i.e., considering the balance equation) should be developed dealing with functional relationships between local loss coefficients and the parameters belonging to the different groups of situations.

8.7.3 Modeling improvements in thermal-hydraulic system codes

SYS TH codes, as discussed in Chapters 5, 9−11, 13, and 15 (balance equations, constitutive equations, special models, structure, validation, and application, respectively), constitute complex computational tools that can be imagined as repositories for the knowledge in different areas of thermal hydraulics. This also applies to the modeling of pressure drops.

The key issues related to pressure drop prediction within the validation process of SYS TH codes, i.e., comparing measured and calculated values, appear to be:

- Lack of or insufficient information about the experimental error (e.g., experimental values are also needed to fix suitable steady-state conditions when transient calculations for closed loop are performed)
- Occurrence of non-fully developed flow conditions (within the code, these are dealt with by correlations developed or qualified for fully developed flow conditions)
- Lack of or insufficient information about flow regimes or flow regime changes
- Need for the code user to supply information related to local pressure drop coefficients

The first two issues apply to single- and two-phase conditions and the third issue is valid for two-phase conditions.

8.7.3.1 Envisaged developments

Traceable improvements for SYS TH codes imply resolving the listed issues by considering developments in the areas of experiments and two-phase flow modeling (Sections 8.6.1 and 8.6.2). Solution of the issue at the last bullet item (reduction of user effect), may bring to the need of creating a database for local pressure drop coefficients for NPP geometric conditions. In this connection, flow reversal conditions (whose consideration is essential to determining system performance in the case of transients like LOCAs), components (MCPs, valves, separator dryers, accumulator lines, etc.), and local pressure drops should receive proper attention.

8.7.4 Trends for computational fluid dynamics applications

CFD development is driven by a context much wider than nuclear thermal hydraulics as also discussed in Chapter 12. CFD codes are widely used in various areas of nuclear thermal hydraulics including design and safety aspects. CFD codes are also coupled with SYS TH codes and are used to support the results of related calculations in selected fields. A snapshot picture connecting CFD codes and pressure drops in nuclear thermal hydraulics can be given as follows.

(A) Validation studies (for CFD codes) focus on three-dimensional phenomena like the mixing of different fluid streams and characterization of local conditions, namely velocities densities or even void fraction when two-phase flows are involved; in very few cases, a check is made that pressure drops along the fluid paths are close enough to experimental data: in the majority of case are not even measured properly or available from experiments. Then, CFD computational tools can be found qualified in computing three-dimensional phenomena in the case of some experiments without proving consistency with experimental data of pressure drops in the same experiments.

(B) CFD computational tools have the intrinsic and unique capability to predict pressure drops at geometric discontinuities: therefore, it is also possible to predict the relationship between local pressure drop coefficients in various situations and Re number. As already discussed, this relationship constitutes an important input parameter for SYS TH codes.

Then, in some cases mesh convergence of CFD code results can be demonstrated as part of the adoption of best practice guidelines.

8.7.4.1 *Envisaged developments*

Envisaged developments for CFD codes in the area of pressure drops can be derived, as follows.

- First: precision targets consistent with current numerical capabilities and possibly with needs in nuclear thermal hydraulics should be established, e.g., related to pressure drops.
- Second: the capability to predict pressure drops in experiments dealing with any phenomenon, whatever one-dimensional or three-dimensional, item (A), mentioned previously, should be demonstrated. This must become a necessary part of the CFD validation process.
- Third: systematic application of CFD codes to the prediction of local loss coefficients for the whole variety of situations of interest in nuclear thermal hydraulics should be promoted. Prediction errors should be documented and, possibly, found to be lower than acceptable errors, item (B), mentioned previously. Mesh convergence tests should be documented together with guidelines for controlling or further reducing the errors.

8.8 Conclusions

The motivation for this chapter is the importance of pressure drops in nuclear thermal hydraulics and in reactor design and safety technology. The content of the chapter depicts the current status and the importance and the role of the prediction of pressure drops (or losses).

Furthermore:

(a) Pressure drops are considered in all textbooks on thermal hydraulics; as much as possible, complementary information has been selected and is provided in this chapter.
(b) In the area of two-phase flow it is noted again that current modeling capabilities are based on the two-phase multiplier approach, i.e., an averaged approach proposed and developed during the 1960s or in the previous century, i.e., at a time when computers did not exist and partial differential equations of two fluid models could not be handled.
(c) Recommendations for the future and foreseen developments constitute the objective of Section 8.7 in this chapter and are not repeated here.

Finally, in addition to other investigations dealing with errors in predicting the pressure drops considered in the above sections, we cite Yang et al. (2016), who discuss errors or differences between CFD code predictions and experimental data relevant to nuclear technology (fuel bundle). Whatever CFD model is adopted, errors inconsistent (i.e., too high, basically larger than 10% of the measured value) with the needs of nuclear thermal hydraulics can be found.

In relation to both the "CFD" and the "two-phase system thermal hydraulics" areas, the achievement of suitable or satisfactory (to the code user) predictions, e.g., in calculating mixing or transient scenarios, respectively, without the demonstration of adequate calculation of pressure drops is not acceptable.

Exercises and questions

Exercises and questions are proposed hereafter with solutions (or hints to get solutions) given in Appendix 1 (Volume III of the book). Answers and solutions are expected to be provided by readers before reading Appendix 1. Please provide extended answers.

8.1 Discuss the interconnection between pressure drops and momentum or force balance equations.

8.2 Discuss the motivation, difference, and implication of friction pressure drop term based on "f" and "τ_w."

8.3 Why is Φ_{LO}^2 defined as a quadratic term?

8.4 Considering the equation for acceleration pressure drop and making reference to a horizontal pipeline under steady-state conditions where gas is flowing in the presence of friction pressure drop and heat transfer to the walls, declare whether inlet velocity is larger than outlet velocity (and what assumptions are needed to answer this question).

8.5 What pressure drop is needed to create LS clearing under the typical conditions of an SBLOCA in a PWR?

8.6 What is the typical singular pressure drop caused by a spacer grid? (Please consider the information in this chapter and the literature on steady-state flow in reactor conditions.)

8.7 Discuss the occurrence of possible cavitation conditions and consequent steam generation at geometric discontinuities.

8.8 Two-phase pressure drops are typically measured in experiments in steady state and fully developed flow conditions: discuss the related implications in calculating pressure drops in reactor conditions during an accident.

8.9 In what way is a single pressure drop measurement (or measured data) transferred to a two-fluid UVUT model? What assumptions and procedures are introduced?

8.10 What is the typical error in calculating pressure drops (notably, two-phase flow, friction, and located at a geometric discontinuity)?

8.11 Why are pressure drops at geometric discontinuities not a part of the balance equations implemented into SYS TH codes? What are the challenges in these connections for code users?

8.12 What is the (superior) role of CFD codes related to SYS TH codes relative to pressure drops?

Acknowledgment

J-L. Gandrille revised the original version (Edition 1) of this chapter; he is gratefully acknowledged.

Constitutive equations

9

Jae Jun Jeong
Pusan National University, Pusan, Korea

Foreword

The concept of constitutive equations is strictly linked to the modeling of two-phase nuclear thermal hydraulics. Balance equations derived from basic principles, as discussed in Chapter 5, cannot be solved without closure or constitutive equations. Here one may distinguish, within the broad category of closure equations (i.e., all the equations needed to solve balance equations), constitutive equations as equations derived from experiments. In this case, constitutive equations have an empirical or a semiempirical nature. This chapter discusses constitutive equations with specific reference to the models adopted in existing system thermal-hydraulic codes.

Understanding this chapter is essential for readers to develop ideas about the capabilities and limitations of the existing thermal-hydraulic system codes discussed in Chapter 11. Constitutive equations and related modeling are associated with system geometry and are part of the scaling issue in system thermal hydraulics (also discussed in Chapter 11).

Furthermore, constitutive equations represent (1) one motivation and origin of uncertainty, as discussed in Chapter 13, and (2) an area that requires continuous future development, including the planning of new experiments.

9.1 Introduction

Various mathematical models have been developed for two-phase flow: homogeneous equilibrium, drift-flux, two-fluid—known as the unequal-velocity unequal-temperature (UVUT) or six-equation model—two-fluid three-field (Frepoli et al., 2003), and two-fluid four-field (Lahey and Drew, 2001). Each model was derived from the first principles of thermodynamics and mechanics under specific assumptions for its simplification. Among these models, the one-dimensional (1D) two-fluid six-equation model has been adopted in most state-of-the-art system TH codes, such as RELAP5/MOD3 (USNRC, 1998), RELAP5-3D (Idaho National Laboratory, 2012), TRAC-PF1 (Liles et al., 1988; Spore et al., 1993), TRACE V5 (USNRC, 2013), CATHARE 2 (Bestion, 1990), and MARS (Jeong et al., 1999).

In the two-fluid six-equation model, the mass, energy, and momentum balance equations for liquid and vapor/gas mixture phases are established separately. The six equations are represented in differential form, with time and one space dimension

Handbook on Thermal Hydraulics in Water-Cooled Nuclear Reactors
https://doi.org/10.1016/B978-0-323-85606-5.00013-1

as independent variables, in terms of time and volume-averaged dependent variables. Each phase in the two-fluid model has two interfaces (see Fig. 9.1a):

- A variable interface between the two phases where heat, mass, and momentum transfer occur
- An interface with the fixed wall where heat and momentum transfer occur

The mathematical models used to represent interface transfer phenomena in terms of time- and volume-averaged dependent variables are called **constitutive equations**. The term "constitutive equations" is associated with a specific set of governing equations. For the two-fluid model, they link the six governing equations, leading to "mathematical closure" together with **state equations** (Meyer et al., 1967; Haar et al., 1984; Watanabe and Dooley, 1996; Harvey et al., 2000; Lemmon et al., 2010).

In this chapter, the constitutive equations of two-fluid models, selected from RELAP5/MOD3 (hereafter, RELAP5), RELAP5-3D, TRACE V5, or CATHARE 2, are briefly presented and discussed. For readers wishing to compare these equations with those of other system codes, including RELAP-7, WCOBRA/TRAC, TRAC-M, and ATHLET, the review papers by Roth and Aydogan (2014a,b) are recommended. They provide a systematic comparison of various system TH codes and discuss their governing equations, closure relations, numerical solutions, code validation, and limitations.

Some system TH codes, such as WCOBRA/TRAC (Thurgood, 1983), SPACE (Ha et al., 2011), and CATHARE 3 (Vallette et al., 2011; Bassenghi, 2013), use the two-fluid three-field model for two-phase flows, which has three additional interfaces, as shown in Fig. 9.1b, because of the separate modeling of entrained liquid (i.e., droplets). Thus, additional constitutive equations are necessary for droplet entrainment/deposition and the mass, energy, and momentum exchanges at the droplet-to-vapor interface. These are discussed in Section 9.8.

9.2 Two-fluid equations and constitutive equations

The basic field equations for a 1D two-fluid model consist of two-phasic continuity, momentum, and energy equations. The following equations, which may differ for

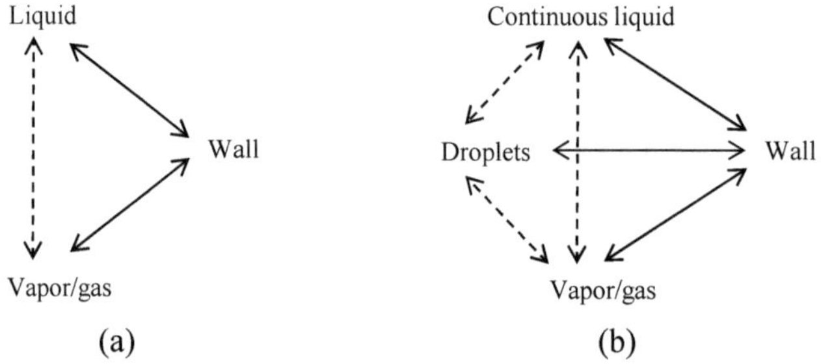

Figure 9.1 Interfaces in two-fluid and two-fluid three-field models. (a) Two-fluid model. (b) Two-fluid three-field model.

various system TH codes, were selected to illustrate both the basic form and the constitutive terms of a two-fluid model.

- Continuity equation of the k-phase:

$$\frac{\partial}{\partial t}(\alpha_k\rho_k)+\frac{\partial}{\partial x}(\alpha_k\rho_k V_k)=\Gamma_k, \tag{9.1}$$

where $k = f$ or g (for liquid or vapor, respectively).

- Momentum equation of the k-phase:

$$\begin{aligned}\frac{\partial[\alpha_k\rho_k V_k]}{\partial t}+\frac{\partial}{\partial x}(\alpha_k\rho_k V_k V_k)= & -\alpha_k\nabla P+F_{ik}+F_{vm,k}\\ & -F_{wk}+\alpha\rho_k g+\Gamma_k V_{ik}+p_i\frac{\partial}{\partial x}\alpha_k\end{aligned} \tag{9.2}$$

- Energy equation of the k-phase:

$$\frac{\partial}{\partial t}(\alpha_k\rho_k u_k)+\frac{\partial}{\partial x}(\alpha_k\rho_k V_k u_k)=-P\frac{\partial\alpha_k}{\partial t}-P\frac{\partial}{\partial x}(\alpha_k V_k)+Q_{wk}+Q_{ik}+\Gamma_k h_{ik}^* \tag{9.3}$$

On the right-hand side of Eqs. (9.1)—(9.3), we must identify 14 transfer terms that at a minimum require constitutive relations (D'Auria et al., 2012). These include

- Wall transfers
  - Liquid—wall friction (F_{wf}),
  - Gas—wall friction (F_{wg}),
  - Wall-to-liquid heat transfer (Q_{wf}),
  - Wall-to-gas heat transfer (Q_{wg}).
- Interfacial transfers
  - Interfacial mass transfer (Γ_k, vaporization or condensation; $\Gamma_f = -\Gamma_g$),
  - Liquid-to-interface heat flux (Q_{if}),
  - Gas-to-interface heat flux (Q_{ig}),
  - Energy transfer to liquid due to interfacial mass transfer $\left(\Gamma_f h_{if}^*\right)$,
  - Energy transfer to gas due to interfacial mass transfer $\left(\Gamma_g h_{ig}^*\right)$,
  - Interfacial momentum transfers:
    - · Interfacial friction (F_{ik}),
    - · Virtual mass force $\left(F_{vm,k}\right)$,
    - · Differential terms of momentum equations related to pressure differences between phases and interface ($p_i\frac{\partial}{\partial x}\alpha_k$),
    - · Momentum transfer to liquid due to interfacial mass transfer $\left(\Gamma_f V_{if}\right)$,
    - · Momentum transfer to gas due to interfacial mass transfer $\left(\Gamma_g V_{ig}\right)$.

When noncondensable gases (NCGs) are present, they are assumed to move with the same velocity and have the same temperature as the vapor phase. By assuming that all properties of the gas phase (subscript g) are mixture properties of the steam/noncondensable mixture, the continuity equation of each phase, Eq. (9.1), is unchanged. However, it is necessary to add a mass conservation equation for the total NCG component, given by

$$\frac{\partial}{\partial t}\left(\alpha_g \rho_g X_n\right) + \frac{\partial}{\partial x}\left(\alpha_k \rho_k X_n V_k\right) = 0, \tag{9.4}$$

where X_n is the total NCG mass fraction in the gas phase. It is also necessary to modify the energy equations to include the sensible interface heating term Q_{gf}, which is the heat transfer at the NCG–liquid interface. This term is required because the interfacial terms use saturation temperature based on the bulk steam partial pressure rather than the saturation temperature based on the local interface steam partial pressure, which is discussed in Section 9.4. The resulting energy equations are

$$\begin{aligned}\frac{\partial}{\partial t}\left(\alpha_g \rho_g u_g\right) + \frac{\partial}{\partial x}\left(\alpha_g \rho_g V_g u_g\right) = & - P\frac{\partial \alpha_g}{\partial t} - P\frac{\partial}{\partial x}\left(\alpha_g V_g\right) + Q_{wg} \\ & - Q_{gf} + Q_{ig} + \Gamma_g h_{ig}^*\end{aligned} \tag{9.5}$$

$$\begin{aligned}\frac{\partial}{\partial t}\left(\alpha_f \rho_f u_f\right) + \frac{\partial}{\partial x}\left(\alpha_f \rho_f V_f u_f\right) = & - P\frac{\partial \alpha_f}{\partial t} - P\frac{\partial}{\partial x}\left(\alpha_f V_f\right) + Q_{wf} \\ & + Q_{gf} + Q_{if} + \Gamma_f h_{if}^*\end{aligned} \tag{9.6}$$

The two-fluid equations listed above are formulated in one dimension in the direction of the flow. This means that cross section–averaged flow variables are used in the governing equations. Thus, the constitutive equations for thermal-hydraulic phenomena at the wall and the two-fluid interface should be developed in terms of cross section–averaged properties.

All constitutive equations in a system TH code should be developed for application within the wide range of flow conditions (USNRC, 1987) that may occur during design basis accidents in light-water reactors:

- Pressures from 0.1 to ~20 MPa,
- Temperatures up to 1500 K,
- Various flow passage geometries, including pipe, rod bundle, and annulus with equivalent diameters from 0.01 to ~1.3 m,
- Subcooled liquid (SCL), two-phase mixture, and superheated vapor flows,
- Various flow regimes in horizontal, vertical, and inclined ducts,
- Pre- and post-CHF wall heat transfer.

In general, most constitutive equations have been developed using experimental data from a limited range of flow conditions and parameters in reduced-scale facilities, leading to issues of scaling and uncertainty. Therefore, constitutive equations are one of the key modules in system TH code for determining its overall accuracy (Ahn et al., 2015). For example, about 50% of the uncertainties about the peak cladding temperature for a loss-of-coolant accident (LOCA) in a pressurized water reactor (PWR) come from interfacial and wall heat transfer models (Choi et al., 2012).

Among the transfer terms above, the virtual mass force (or added mass force) and differential terms of momentum equations related to pressure differences between phases and interfaces are discussed first.

In the CATHARE 2, TRACE, and RELAP5 codes, the virtual mass terms are implemented as follows:

$$F_{vm,k} = \begin{cases} (-1)^k C_{vm}\alpha_f\alpha_g\rho_m\left(\dfrac{dV_g}{dt} - \dfrac{dV_f}{dt}\right) \text{ in CATHARE 2,} \\ (-1)^k C_{vm}\alpha_f\alpha_g\rho_m\left(\dfrac{\partial V_g}{\partial t} + V_f\dfrac{\partial V_g}{\partial z} - \dfrac{\partial V_f}{\partial t} - V_g\dfrac{\partial V_f}{\partial z}\right) \text{ in TRACE,} \\ (-1)^k C_{vm}\alpha_f\alpha_g\rho_m\left(\dfrac{\partial V_g}{\partial t} - \dfrac{\partial V_f}{\partial t}\right) \text{ in RELAP5,} \end{cases} \tag{9.7}$$

where $(-1)^k = \begin{cases} -1\, for\ k = g \\ 1 \text{ for k} = 1 \end{cases}$, and C_{vm}: virtual mass, which depends on the flow regime:

$$\rho_m = \alpha_g\rho_g + \alpha_f\rho_f,$$

where *d/dt*: the material derivative.

In RELAP5, the spatial derivative terms of virtual mass force were ignored (Anderson et al., 1977; USNRC, 1998a,b). This is because inaccuracies in approximating the spatial derivative portion of the term for the relatively coarse nodalizations used in system representations can lead to nonphysical characteristics in the numerical solution. Meanwhile, in TRACE (USNRC, 2013), the virtual mass terms are ignored in the momentum equations because they show no evidence of significance in reactor safety problems and create significantly increased complexity for the solution procedure. The TRACE code developers will reconsider virtual mass terms if higher-order numerical methods are introduced into a future version of TRACE and a well-posed set of partial differential equations (PDEs) becomes important. However, for the critical flow model, the effects of virtual mass terms are considered separately. In CATHARE 2, the virtual mass term was well preserved for the well-posedness of two-fluid governing equations (Bestion, 1990).

These differences among the three codes indicate that virtual mass is not a primary concern from a physical point of view. Virtual mass primarily affects the mixture sound speed and temporal acceleration of phasic slip. The simplified form might be adequate in the system of governing equations, or it may be disregarded.

Most system TH codes, such as RELAP5, TRACE, and CATHARE 2, assume the two phases have equal pressures in the mass and energy equations. This assumption is relaxed in the momentum equations in the case of horizontally stratified flow in RELAP5 and TRACE. However, CATHARE 2 always considers the effect of unequal pressures in the momentum equations by adding (or subtracting) $p_i A \partial \alpha_g / \partial z$ into the interfacial stress terms, where

$$p_i = \begin{cases} \alpha_g \alpha_f (\rho_f - \rho_g) g D & \text{for the stratified flow,} \\ \dfrac{\alpha_g \alpha_f \rho_f \rho_g}{\alpha_g \rho_f + \alpha_f \rho_g} (V_g - V_f)^2 & \text{for the non} - \text{stratified flows.} \end{cases} \tag{9.8}$$

In nonstratified flows, the expression for p_i was simply chosen to provide hyperbolicity in the system of equations. It is known that the term $p_i A \partial \alpha_g / \partial z$ has less physical significance but enhances stability in the CATHARE numerical solution method.

The remaining 13 terms of the 15 listed above can then be reduced to the following four groups: (1) interfacial heat and mass transfer, (2) interfacial drag, (3) wall heat transfer, and (4) wall friction. The last two were already discussed in Chapters 7 and 8, respectively, in terms of physical phenomena and models. In the remainder of this chapter, they are discussed again as constitutive equations.

Two-phase flow may exist under various flow regimes characterized by specific phase repartitions. Every flow regime has an internal structure and transfer mechanisms. This makes it necessary to use a flow regime map (or flow pattern map) and develop flow-regime-dependent constitutive relations for interface and wall transfers. Thus, the flow regime map is discussed first in Section 9.3, followed by a discussion of the interfacial heat and mass transfer, interfacial drag, wall heat transfer, and wall friction.

9.3 Flow regime map

The extensive studies of gas–liquid two-phase flow have their origins in the petrochemical and chemical process industries. Those studies were then further investigated for specific uses in the nuclear power industry. A complicated feature of two-phase flow is its liquid–gas interface structure. A particular type of geometric distribution of liquid and gas phases in a two-phase flow is called a flow pattern or flow regime (Collier and Thome, 1994). It has an important effect on the prediction of void fraction, wall and interfacial heat transfer, interfacial drag, and pressure drop; for specific details, see, e.g., Oh et al. (2015), Rossi et al. (2015), Saxena and Prasser (2018).

In most system TH codes, the flow regimes are divided into pre- and post-CHF flow depending on the wall temperature. The post-CHF flow regime is used when the temperature of a wall surface is above the minimum stable film-boiling temperature and, thus, the liquid phase cannot contact the hot surface. Otherwise, the pre-CHF flow regime is used.

Pre-CHF flows are typically subcategorized as vertical, horizontal, and inclined flows. Fig. 9.2 shows a schematic representation of vertical flow regimes in pipes (Collier and Thome, 1994). In bubbly flow, small bubbles are dispersed as discrete bubbles in a continuous liquid phase. With an increasing void fraction, small bubbles coalesce into larger slugs, whose diameters approximate the pipe diameter, resulting in slug flow. The nose of the slug bubble has a spherical cap. The length of the main gas bubble can vary considerably. Also of note is that the flow structure becomes unstable with increasing flow velocity. Although the net flow remains upward, flow components travel in both directions, with instability coming from the balance of gravity and shear forces acting in opposing directions on the thin liquid film of Taylor bubbles. Slug bubbles are more likely in smaller pipes, while Taylor cap bubbles are more common in larger ones. Churn flow may be formed by the breakdown of large bubbles in slug flow. The annular mist regime follows increased flow rates. The interfacial shear of high-velocity gas on the liquid film dominates gravity in the annular flow regime, e.g., Pan et al. (2015). When heat transfer from the wall increases further, liquid film disappears or dries out, finally becoming a mist flow.

Fig. 9.3 shows horizontal flow regimes. Their characteristics vary because of phase asymmetry resulting from the effects of gravity (Collier and Thome, 1994). The bubble flow pattern is like that of vertical flow; at low liquid velocities, gas bubbles tend to travel within the upper half of the pipe, whereas at high liquid velocities, bubbles are dispersed. With increasing vapor flow, plug flow forms. This is similar to slug flow in a vertical pipe. Stratified or stratified wavy flow only occurs at low liquid and gas velocities where the two phases flow separately with a relatively smooth interface. As the gas velocity is increased, the interface becomes disturbed by waves traveling in the direction of the flow. A further increase in gas velocity causes waves at the

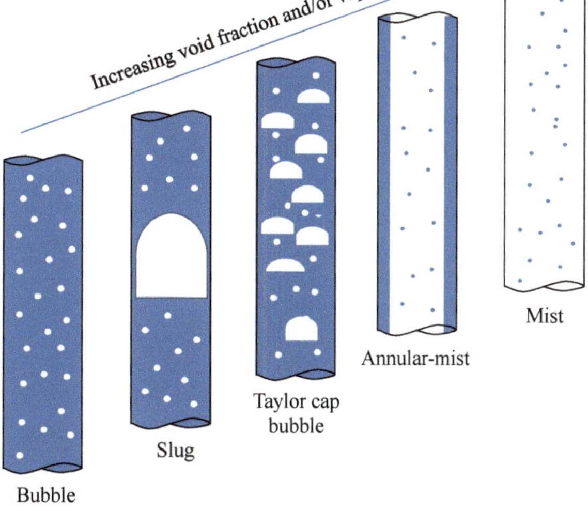

Figure 9.2 Schematic of vertical flow regimes.

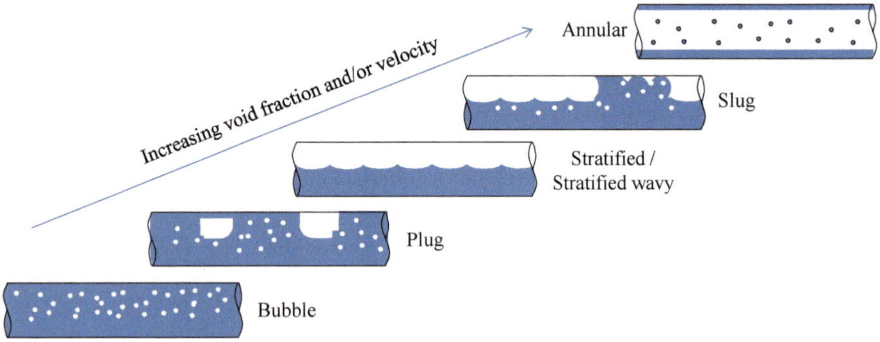

Figure 9.3 Schematic of horizontal flow regimes.

interface to bridge the channel and form a frothy slug that is propagated with a high velocity. A still-higher gas velocity results in the formation of a gas core with a liquid film around the periphery of the pipe. This annular regime is usually accompanied by entrained droplets due to the high gas velocity.

In post-CHF flows, the liquid phase cannot contact the hot surface. This leads to inverted flow regimes, with the gas phase in contact with the wall, as illustrated schematically in Fig. 9.4. In inverted annular flows, there is a core of liquid in the center of the pipe surrounded by an annulus of vapor. In inverted slug flow, large slugs of liquid

Figure 9.4 Schematic of inverted flow regimes under post−critical heat flux (*CHF*) conditions (USNRC, 2013).

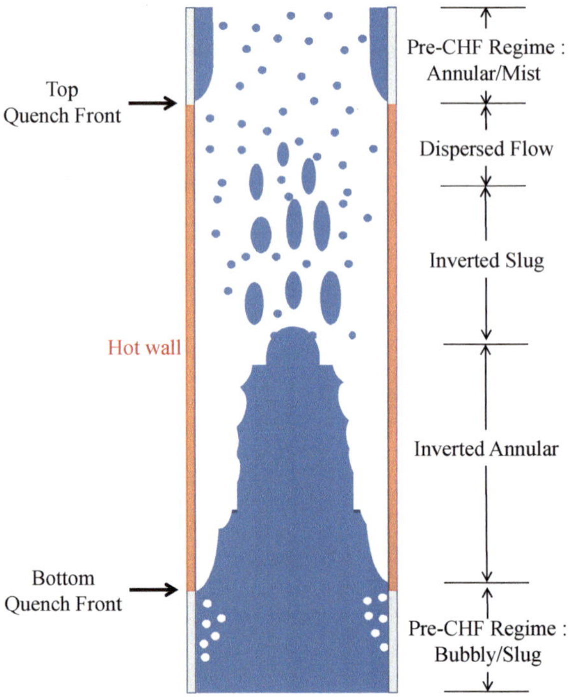

surrounded by vapor are present. In dispersed flows, droplets are present in a vapor flow with no liquid in contact with the wall. These flow regimes may exist in the reflood phase of an LOCA in a PWR.

The flow regime map and transition criteria have been developed to identify the flow regimes for local flow conditions. Some old flow regime maps that are still in use are summarized in Table 9.1. The complexities and deficiencies in understanding various flow regimes can be seen indirectly in the various flow variables used for flow regime identification. This complexity results from difficulties in observation and measurement: flow regimes are very sensitive to flow geometry, and the transition mechanisms are not clearly known.

Traditional two-phase flow-regime criteria based on superficial gas and liquid velocities may be unsuitable for analyzing rapid transient or entrance flows using the two-fluid model. Mishima and Ishii (1984) assumed that direct geometric parameters, such as the void fraction, were conceptually simpler, and therefore, parameters more reliable than those traditionally used could be employed in flow regime criteria. They developed new flow-regime criteria for upward gas–liquid flow in vertical tubes considering the mechanisms of flow regime transitions. These became the basis for the RELAP5 and TRAC-PF1 flow regime maps.

In the remainder of this section, the vertical and horizontal flow regime maps of RELAP5 are briefly introduced first, and those of other system TH codes are also discussed. The RELAP5 flow regime maps are similar to those of the RELAP5-3D, TRACE, and SPACE codes. It should be noted that in RELAP5, volume and junction flow regime maps are provided separately. The volume flow regime map is based on

Table 9.1 Old flow regime maps.

Authors	Geometry	Flow regimes	Flow variables
Baker (1954)	Horizontal pipe	Bubble, forth, plug, slug, stratified, wavy, annular	Superficial mass Velocities of the vapor and liquid phases
Taitel and Dukler (1976)	Horizontal pipe	Bubble, bubble-slug, slug, churn, wispy-annular, annular	Four dimensionless numbers: T, F, K, X
Fair (1960)	Vertical tube	Bubble, slug, annular, mist	Mass flux and a dimensionless parameter, $\left(\frac{x}{1-1}\right)^{0.9}\left(\frac{\rho_l}{\rho_g}\right)^{0.5}\left(\frac{\mu_l}{\mu_g}\right)^{0.5}$
Hewitt and Roberts (1969)	Vertical tube (Small diameter 1–3 cm)	Bubble, plug/slug, annular, wavy, stratified	Superficial momentum fluxes of the liquid and vapor phases
Taitel, Bornea, and Dukler (1980)	Vertical tube (5 cm diameter)	Bubble, slug, churn, finely dispersed bubble	Superficial velocities of the vapor and liquid phases

volume quantities and used for interfacial heat and mass transfer, wall friction, and wall heat transfer. The junction flow regime map is based on both junction and volume quantities. It is used for interphase drag and shear, as well as the coefficient of virtual mass. Conceptually, the volume and junction flow regime maps are the same.

9.3.1 Vertical flow regime map

Fig. 9.5 shows the schematic of the RELAP5 vertical flow regime map (USNRC, 1998). This map, for both upflow and downflow, is used with volumes whose elevation angle ϕ is such that $45 < |\phi| < 90$ degrees. It should be noted that in RELAP5-3D (Idaho National Laboratory, 2012), nearly the same vertical flow regime map is used for $60 \leq |\phi| < 90$ degrees anand horizontal flow regime map for $0 < |\phi| < 30$ degrees. An interpolation region between vertical and horizontal flow regimes is used for volumes whose absolute value of vertical inclination angle is between 30 and 60 degrees. Kim et al. (2007, 2010) showed that the flow regime is very sensitive to flow geometry and angle. Clearly, the approaches for the two codes may involve great uncertainties.

The vertical flow regime map in Fig. 9.5 is modeled as nine regimes; four for pre-CHF heat transfer, four for post-CHF heat transfer, and one for vertical stratification. For pre-CHF heat transfer, the regimes modeled are the bubbly, slug, annular-mist, and mist-pre-CHF regimes. The mist-pre-CHF regime was added for smoothness. For post-CHF heat transfer, the bubbly, slug, and annular-mist regimes are transformed into the inverted annular, inverted slug, and mist regimes, respectively, as suggested

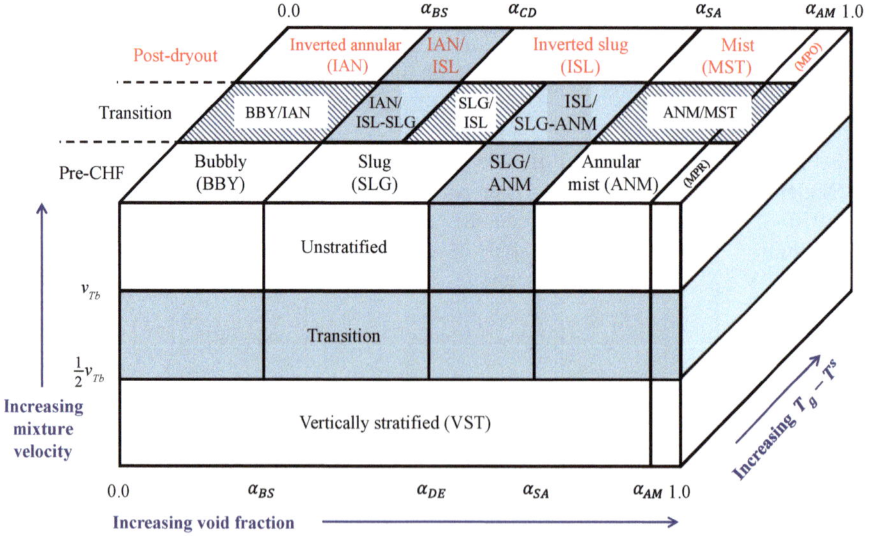

Figure 9.5 Schematic of RELAP5 vertical flow regime map (USNRC, 1998). *CHF*, critical heat flux.

by Ishii and Jarlais (1982). The mist post-CHF regime was added for symmetry with the mist pre-CHF regime.

The schematic in Fig. 9.5 is three-dimensional (3D) to illustrate flow regime transitions as functions of void fraction (α_g), average mixture velocity (v_m), and boiling regime (pre-CHF, transition, and postdryout), where

$$v_m = \frac{G_m}{\rho_m}, \tag{9.9}$$

$$G_m = \alpha_g \rho_g |V_g| + \alpha_f \rho_f |V_f|. \tag{9.10}$$

When the mixture velocity is sufficiently low (~50%) compared with the Taylor bubble velocity (v_{Tb}), the flow is assumed to be vertically stratified. v_{Tb} is given by the correlation

$$v_{Tb} = 0.35 \left[\frac{gD(\rho_f - \rho_g)}{\rho_f} \right]^{1/2}. \tag{9.11}$$

In pre-CHF regimes, the void fraction of the bubbly-to-slug transition (α_{BS}) is given by

$$\alpha_{BS} = \begin{cases} \alpha_L & \text{for } G_m \leq 2,000\ kg/m^2 s, \\ \alpha_L + 0.001(0.5 - \alpha_L)(G_m - 2,000) & \text{for } 2,000 \leq G_m \leq 3,000\ kg/m^2 s, \\ 0.5 & \text{for } G_m \geq 0, \end{cases} \tag{9.12}$$

where $\alpha_L = 0.25 \min\left[1.0, \left(\frac{D^*}{22.22}\right)^8\right]$, and

$$D^* = D\left[\frac{g(\rho_f - \rho_g)}{\sigma}\right]^{1/2},$$

where D: pipe diameter.

The slug-to-annular transition void fraction (α_{SA}) is given by

$$\alpha_{SA} = 0.25 \min\left(\alpha^f_{crit}, \alpha^e_{crit}\right) \tag{9.13}$$

$$\text{where } \alpha^f_{crit} = \begin{cases} \frac{1}{V_g}\left[\frac{gD(\rho_f - \rho_g)}{\rho_g}\right]^{1/2} & \text{for upflow,} \\ 0.75 & \text{for downflow and countercurrent flow,} \end{cases}$$

$$\alpha_{crit}^{e} = \frac{3.2}{V_g}\left[\frac{g\sigma(\rho_f - \rho_g)}{\rho_g{}^2}\right]^{1/4}.$$

The transition region between the slug flow and annular mist flow regimes is defined by α_{DE} and α_{SA}, where

$$\alpha_{DE} = \max(\alpha_{BS}, \alpha_{SA} - 0.05). \tag{9.14}$$

The minimum void fraction for annular mist flow, α_{SA}, is constrained to lie between 0.5 and 0.9. For the transition between annular mist and pre-CHF mist, the value of α_{AM} is

$$\alpha_{AM} = 0.9999. \tag{9.15}$$

If the hydrodynamic volume has heat flux from a surrounding wall to the vapor or the reflood model is turned on for that wall and the vapor temperature is more than 1 K superheated, then the flow regime is a post-CHF regime, in which case the previous α_{BS} and α_{SA} are still effective. The transition region between the inverted slug regime and inverted annular flow regime is defined by α_{BS} and α_{CD}, where

$$\alpha_{CD} = \alpha_{BS} + 0.2. \tag{9.16}$$

9.3.2 Horizontal flow regime map

The horizontal flow regime map is for volumes whose elevation angle ϕ is such that $0 < |\phi| < 45$ degrees. This map is like the vertical flow regime map except that the post-CHF regimes are not included, and a horizontally stratified regime replaces the vertically stratified regime. The horizontal flow regime map therefore consists of horizontally stratified, bubbly, slug, annular-mist, and mist pre-CHF regimes. A schematic for the horizontal flow regime map is shown in Fig. 9.6 (USNRC, 1998). The transition criteria for the bubbly to slug and the slug to annular mist regimes are similar to those for the vertical map.

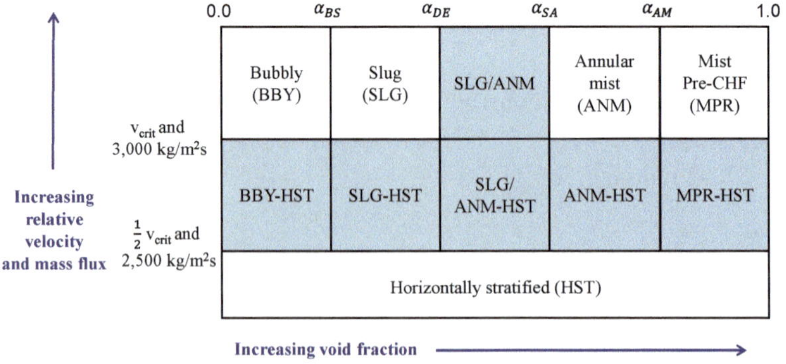

Figure 9.6 Schematic of RELAP5 horizontal flow regime map (USNRC, 1998).

The bubbly–slug transition criterion is

$$\alpha_{BS} = \begin{cases} 0.25 & \text{for } G_m \leq 2,000 \, kg/m^2 s, \\ 0.25 + 0.00025\,(G_m - 2,000) & \text{for } 2,000 < G_m < 3,000 \, kg/m^2 s, \\ 0.5 & \text{for } G_m \geq 3,000 \, kg/m^2 s. \end{cases} \tag{9.17}$$

The minimum void fraction for annular mist flow, α_{SA}, is the constant

$$\alpha_{SA} = 0.8, \tag{9.18}$$

and the transition region between the slug flow and annular mist flow regimes is defined by α_{DE} and α_{SA}, where

$$\alpha_{DE} = 0.75. \tag{9.19}$$

The annular mist to pre-CHF mist transition criterion is

$$\alpha_{AM} = 0.9999. \tag{9.20}$$

The flow is horizontally stratified when the difference in phasic velocity satisfies the condition

$$|V_g - V_f| < V_{crit}$$

and mass flux satisfies the condition

$$G_m < 3,000 \text{ kg/m}^2\text{s},$$

where $V_{crit} = \frac{1}{2}\left[\frac{(\rho_f - \rho_g) g \alpha_g A}{\rho_g \sin\theta}\right]^{1/2} (1 - \cos\theta)$, with angle θ as defined in Fig. 9.7. If the above two conditions are not met, the flow field undergoes a transition to a bubbly, slug, annular mist, or pre-CHF mist flow regime.

In addition to the vertical and horizontal flow regime maps, RELAP5 provides a high mixing flow regime map, which is used in pumps, and an ECC mixer volume flow regime map.

9.3.3 CATHARE 2 and TRACE flow regime maps

In the CATHARE 2 code, only two flow regime transition criteria are explicitly written, and they are used in several closure terms (Bestion, 1990; Serre et al., 2011; Bassenghi, 2013). The first transition is between stratified and no stratified flow, which is associated with the parameter R in Fig. 9.8. This transition depends on two criteria: the

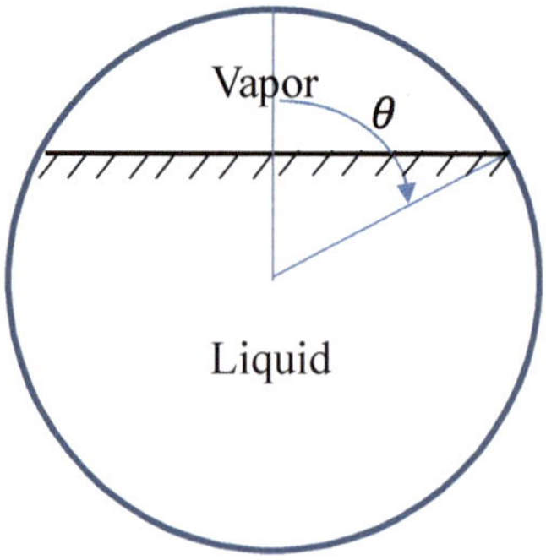

Figure 9.7 Horizontally stratified flow.

first is based on the Kelvin–Helmholtz instability threshold, and the second depends on the relative effects of bubble sedimentation and bubble turbulent mixing. The second transition occurs between annular and droplet flows and is associated with the parameter *E* in Fig. 9.8. The parameters *R* and *E* are used in interpolation to establish smooth flow-regime transitions. The post-CHF flow regimes in CATHARE 2 are inverted annular and dispersed flow (liquid slugs and/or droplets).

The TRACE code considers three distinct classes of flow regimes (USNRC, 2013):

(i) The pre-CHF flow regime, which comprises the bubbly/slug and annular/mist regimes,
(ii) The horizontal stratified flow regime, which is available for horizontal or inclined 1D components,
(iii) The post-CHF flow regime, which encompasses inverted flow regimes that occur when the wall is too hot for liquid–wall contact; the TRACE pre-CHF flow regime map, which is used for interfacial heat transfer and is a TRAC-PF1/MOD2 (Spore et al., 1993); as shown in Fig. 9.9, this flow regime is based on void fraction and mass flux, with the mass flux range limited for the Taylor cap or slug bubble flow regime; its map is like that of RELAP5 except with no vertically stratified flow regime.

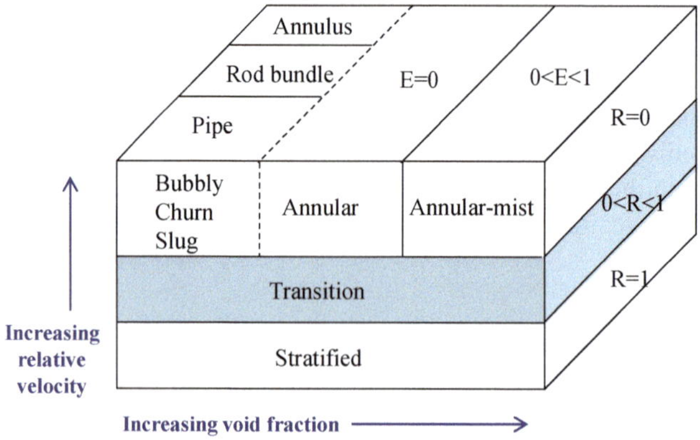

Figure 9.8 Schematic of CATHARE 2 pre–critical heat flux flow regime map (Bassenghi, 2013).

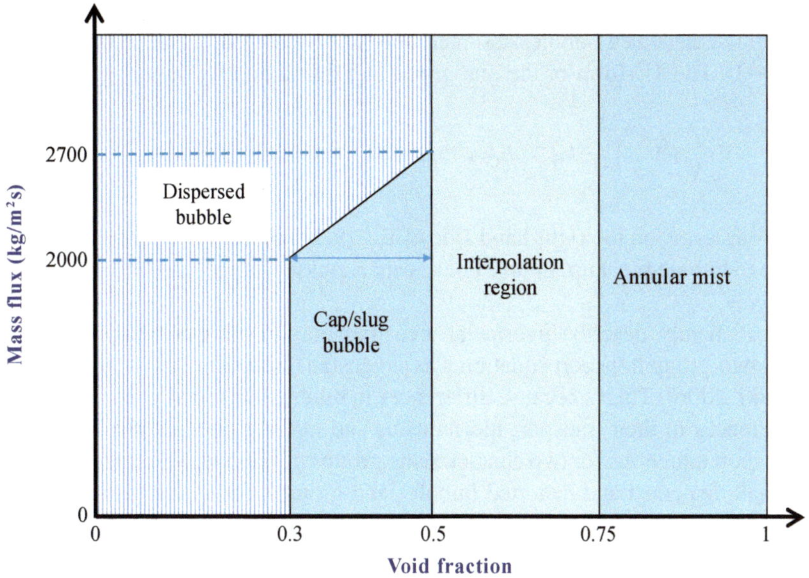

Figure 9.9 Schematic of TRACE pre–critical heat flux flow regime map (USNRC, 2013).

9.3.4 Limitations of the flow regime map approach and recent improvements

The flow regime map is quite important because it determines the flow regime that applies for a given local flow condition and thus determines the interfacial transfer model and interfacial area per unit volume (or interfacial area concentration—IAC).

The shortcomings caused by traditional two-phase flow regimes and regime transition criteria were pointed out by Ishii (1975) and coworkers (Ishii and Mishima, 1980; Hibiki and Ishii, 2002; Ishii, Wu, Assad, and Uhle, 1998; Uhle, Wu, Ishii, 1998) and can be summarized as follows:

- The flow regime transition criteria are algebraic correlations typically developed from steady-state, fully developed flows. Hence, the effects of the entrance and developing flow cannot be correctly accounted for, nor can gradual transitions between regimes.
- Transition criteria and flow-regime-dependent IAC correlations are valid for limited parameter ranges under certain operational conditions and geometries. Most are obtained from simple experiments and phenomenological models. Often the scale effects of geometric and fluid properties are not adequately considered.
- The method based on flow regime transition criteria is a two-step method that requires flow configuration transition criteria and IAC correlations for each flow configuration. The compound errors from transition criteria and IAC correlations can be quite significant.

To solve such problems, the interfacial area transport equation (IATE) was introduced (Ishii, 1975; Ishii and Kojasoy, 1993; Kocamustafaogullari and Ishii, 1995). The IATE can be obtained by considering a fluid particle number density transport equation analogous to Boltzmann's transport equation (Wu, Kim, Ishii, and Beus,

1998; Morel, Goreaud, and Delhaye, 1999). A one-group formulation was suggested first, with the dispersed bubbles assumed to be spherical (Ishii et al., 2000; Hibiki and Ishii, 2002). The 1D form of the one-group IATE in a bubbly flow is given by

$$\frac{\partial a_i}{\partial t} + \nabla \cdot \left(a_i \vec{V}_g \right) = \Phi_B - \Phi_C + \Phi_P + \Phi_E. \tag{9.21}$$

The four terms on the right-hand side of Eq. (9.21) represent the rates of change in IAC due to bubble breakup, bubble coalescence, phase change, and bubble expansion, respectively.

To realistically describe interfacial area transport in various two-phase flow regimes, a two-group transport equation was suggested (Ishii, Kim and Kelly, 2005; Ishii and Hibiki, 2006). This is because differences in bubble sizes or shapes cause substantial differences in their transport mechanisms and interaction phenomena. Therefore, two transport equations for two characteristic groups of bubbles were suggested: group 1 for small dispersed and distorted bubbles and group 2 for cap/slug/churn-turbulent bubbles. The maximum bubble diameter of the group-1 bubble is given by

$$D_{d,\max} = 4\sqrt{\sigma/(g\Delta\rho)}. \tag{9.22}$$

Two-group IATEs have more complicated source and sink terms, including additional intergroup transfer terms.

In the development of two-group IATEs, major difficulties arise from the existence of multiple types of bubbles and their complicated interactions (Ishii, Kim and Kelly, 2005). Interactions between bubbles in the same group and between bubbles from the two different groups should be modeled along with intergroup transfer at the group boundary due to expansion and compression. An important aspect of the two-group transport formulation is that the particle distribution function must be properly averaged due to the existence of several types of bubbles in the two-group transport equation. The source and sink terms of the interfacial area are still being developed and strongly depend on flow conditions and geometries. So far, most experiments for interfacial area research have been performed in round tubes (Revankar and Ishii, 1992; Grosstete, 1995; Hibiki et al., 1998, 2001; Hibiki and Ishii, 2000, 2002; Fu, 2001; Smith, 2002; Ishii and Kim, 2004; Ishii et al., 2005; Yao and Morel, 2004; Smith et al., 2012), with a few in annuli (Hibiki et al., 2003; Jeong et al., 2008; Ozar et al., 2013a,b; Brooks et al., 2014) and rod bundles (Yang et al., 2013). Experiments on sudden area changes, elbows, bends, and large volumes are still scarce.

Considering the wide range of flow conditions that the constitutive equations of a system TH code should cover, it is somewhat premature to use the interfacial area transport model in a system TH code because of the lack of sufficient data and validated models. Nevertheless, RETRAN-3D (CS&A, 2001) attempted to adopt the dynamic flow regime based on an IATE different from Eq. (9.21) (Stuhmiller, 1986). However, there has not been sufficient validation work to recommend the model for general use. The TRACE development road map also includes the IATE. In

CATHARE 3, the IATE is under development (Emonot et al., 2009) and is to be used when the results are validated and reliable. It seems that completely replacing the flow regime map with an IATE will require substantial work and take a very long time.

A two-step approach for the future development of IATE is desirable. First, we must focus on the in-depth development of the one-group IATE, which is limited to a dispersed bubbly flow but provides enhanced accuracy. Thereafter, we may proceed with the second step, the development of two-group IATEs, including the source and sink terms, in parallel with interfacial area transport by droplets (Thurgood et al., 1983), that cover all flow regimes. In this step, the three-field model is naturally used at a minimum. Additional efforts are needed to fit two-group IATEs into two-fluid or three-field models (Lee et al., 2013) because only one vapor velocity is defined in the two-fluid model but the vapor velocities of two-group bubbles are required by two-group IATEs. Sun et al. (2003) and Wang et al. (2011) have suggested models to implement the IATE in two-fluid models. These combined efforts should be able to replace the conventional flow regime map approach with IATEs.

9.4 Interfacial heat and mass transfer

Interfacial heat and mass transfer may occur in the bulk and/or in the thermal boundary area near the wall (e.g., see Fig. 9.10). The former results from energy exchange between the two phases, determining the level of thermal nonequilibrium, while the latter is due to heat transfer at the wall. Even when the bulk liquid temperature is lower than the local saturation temperature, the liquid in the thermal boundary layer near the wall can be superheated, which may lead to subcooled boiling. The same is true for the filmwise condensation at a subcooled wall. A special model is needed for this and is discussed in Section 9.4.

This section primarily addresses interfacial heat and mass transfer in the bulk fluid. This two-phase flow phenomenon involves great uncertainties because the fundamental mechanisms are not yet well understood, and direct measurement is nearly impossible.

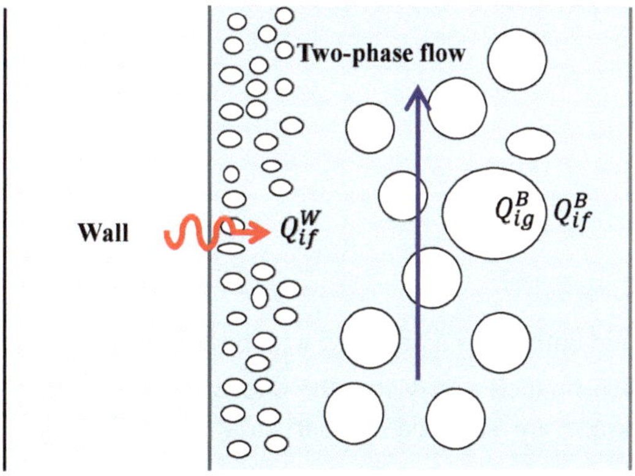

Figure 9.10 Interface heat transfer in the bulk and near the wall for subcooled boiling (USNRC, 1998).

9.4.1 Bulk interfacial heat and mass transfer

In the two-fluid six-equation model in RELAP5 (USNRC, 1998), the total volumetric vapor generation rate or interfacial mass transfer rate is given by

$$\Gamma_g = \Gamma_{ig} + \Gamma_w \tag{9.23}$$

where $\Gamma_g = -\Gamma_f$, Γ_{ig} is the vapor generation rate in the bulk fluid, and Γ_w is the rate in the thermal boundary area near the wall. Similarly, the vapor-side interfacial heat transfer rates are given by

$$Q_{ig} = Q_{ig}^B + Q_{ig}^W. \tag{9.24}$$

Liquid–vapor-side interfacial heat transfer rates are given by

$$Q_{if} = Q_{if}^B + Q_{if}^W. \tag{9.25}$$

Bulk interface heat transfer occurs at the steam–liquid interface in the bulk. This represents a thermal energy exchange between the fluid interface at the saturation temperature T^s and the bulk fluid state. For gas, bulk interface heat transfer is given by

$$Q_{ig}^B = H_{ig}(T^s - T_g) \tag{9.26}$$

where H_{ig} is the gas interface heat transfer coefficient per unit volume, and T_g is the gas temperature. For liquid, the bulk interface heat transfer is given by

$$Q_{if}^B = H_{if}(T^s - T_f) \tag{9.27}$$

where H_{if} is the liquid interface heat transfer coefficient per unit volume, and T_f is the liquid temperature.

In general, it is assumed for simplicity that the bulk interfacial heat transfer terms and the near-wall energy exchange terms each sum independently to zero. Then,

$$Q_{ig}^B + Q_{if}^B + \Gamma_{ig}\left(h_g^* - h_f^*\right) = 0 \tag{9.28}$$

and

$$Q_{ig}^W + Q_{if}^W + \Gamma_w\left(h_g^{'} - h_f^{'}\right) = 0 \tag{9.29}$$

where h_g^* and h_f^* are the vapor and liquid enthalpies associated with bulk interface mass transfer, respectively. They are defined in such a way that the interface energy jump conditions at the liquid–vapor interface are satisfied. In particular, h_g^* and h_f^* are chosen to be h_g^s and h_f, respectively, for vaporization, and h_g and h_f^s, respectively, for condensation, where the superscript s indicates saturation. The same is true for the phasic enthalpies $h_g^{'}$ and $h_f^{'}$ associated with wall (thermal boundary layer) interface mass transfer.

It is assumed that vapor appears (or, is generated) at saturation, and it follows that $Q_{ig}^W = 0$ for boiling processes in the boundary layer near the wall (i.e., subcooled boiling). Thus, Eq. (9.29) can be used to solve for the interface vaporization rate in the boundary layer near the walls (see Section 9.4):

$$\Gamma_w = \frac{-Q_{if}^W}{h_g^{'} - h_f^{'}}. \tag{9.30}$$

Similarly, since it is assumed that liquid appears at saturation, it follows that $Q_{if}^W = 0$ for condensation processes in the boundary layer near the wall. Eq. (9.29) can thus be used to solve for the interface condensation rate in the boundary layer near the walls:

$$\Gamma_w = \frac{-Q_{ig}^W}{h_g^{'} - h_f^{'}}. \tag{9.31}$$

Using the above equations, the total interfacial mass transfer rate is represented by

$$\Gamma_g = -\frac{H_{ig}(T^s - T_g) + H_{if}(T^s - T_f)}{h_g^* - h_f^*} + \Gamma_w. \tag{9.32}$$

Eq. (9.32) shows that the interfacial mass transfer rate is directly related to the interfacial heat transfer rate.

When NCG is present, the energy equations are modified to include the sensible interface heating term Q_{gf}. This term is given by

$$Q_{gf} = \frac{P - P_s}{P} H_{gf}(T_g - T_f) = \frac{P_n}{P} H_{gf}(T_g - T_f) \tag{9.33}$$

where H_{gf} is the sensible heat transfer coefficient per unit volume. This makes use of Dalton's law ($P = P_s + P_n$), where P_s and P_n are the steam and NCG partial pressures, respectively. The multiplier in Eq. (9.33) is an ad hoc function used to turn this term off when no NCG is present in the control volume.

The interfacial heat transfer and mass transfer terms are also modified when NCG is present. Gas bulk interface heat transfer, Eq. (9.26), is modified as follows:

$$Q_{ig}^B = \frac{P_s}{P} H_{ig}\left[T^s(P_s) - T_g\right] \tag{9.34}$$

where $T^s(P_s)$ is the saturation temperature corresponding to the partial pressure of steam. All the saturation properties in Eq. (9.26) through Eq. (9.32) should be replaced with those based on the partial pressure of steam instead of the total pressure. The total interfacial mass transfer rate, Eq. (9.32), is then expressed as

$$\Gamma_g = -\frac{\frac{P_s}{P} H_{ig}\left[T^s(P_s) - T_g\right] + H_{if}\left[T^s(P_s) - T_f\right]}{h_g^* - h_f^*} + \Gamma_w \tag{9.35}$$

9.4.2 Interfacial area concentration

Note that the interfacial heat transfer coefficient (IHTC) per unit volume (H_{ik}) in Eq. (9.35) is generally given by

$$H_{ik} = h_{ik} a_i \tag{9.36}$$

where h_{ik} is the IHTC, and a_i is the IAC. The former requires special models: empirical, semiempirical, or ad hoc. The IAC is mechanistically modeled, which depends on the flow regime. For combined flow regimes, such as slug flow or annular-mist flow, the effects of the two flow regimes are considered separately. For example, in the case of annular-mist flow, H_{ik} is given by

$$H_{ik} = h_{ik}^{film} a_i^{film} + h_{ik}^{drop} a_i^{drop}. \tag{9.37}$$

where the terms on the right-hand side indicate the contributions of liquid film and droplets, respectively.

The IAC models are similar in various system TH codes. For bubbly and mist flow regimes, the model is rather clear. According to Wallis (1969) and Shapiro (1957), the dispersed bubbles or droplets can be assumed to be spherical particles with a size distribution of the Nukiyama—Tanasawa form (USNRC, 1998). Using the Nukiyama—Tanasawa distribution function, the IAC can be obtained as

$$a_i = \frac{3.6\overline{\alpha}}{d_o} \tag{9.38}$$

where $\overline{\alpha} = \alpha_g$ for bubbles and $\overline{\alpha} = \alpha_f$ for droplets. The average diameter d_o is obtained by assuming that $d_o = d_{max}/2$. The maximum diameter, d_{max}, is related to the critical Weber number, We, by

$$We = \frac{d_{max}\rho_c\left(v_g - v_f\right)^2}{\sigma} \tag{9.39}$$

where σ is the surface tension, and ρ_c is the continuous-phase density. The values for We are presently taken as

$$We = \begin{cases} 10 & \text{for bubbly flow,} \\ 3 & \text{for mist} - \text{pre} - \text{CHF flow,} \\ 12 & \text{for mist and mist} - \text{post} - \text{CHF flow.} \end{cases} \tag{9.40}$$

Similarly, IACs can be modeled for other regimes. For the slug flow regime, Taylor bubbles and small bubbles are modeled separately. For the annular-mist flow regime, liquid film and droplets are modeled separately, and the IAC by droplets is obtained using Eqs. (9.38) and (9.39). For inverted flow regimes, the effects of liquid core, liquid slug, and droplets are considered separately. Table 9.2 summarizes the IAC model in the RELAP5 code (USNRC, 1998).

9.4.3 Interfacial heat transfer coefficient

IHTCs are rather complicated because of fundamental difficulties in experimental observation and modeling. In the RELAP5 code, interfacial heat transfer for each flow regime is split into four cases based on the fluid temperature: superheated liquid (SHL), SCL, superheated gas (SHG), and subcooled gas (SCG). Correlations are provided for the four cases in each regime, and some correlations are reused.

In the bubbly flow regime, for a condition of SHL, the interfacial heat transfer is the larger of the model for bubble growth developed by Plesset and Zwick (1954) or the model for convective heat transfer for a spherical bubble (as modified by Lee and Ryley, 1968).

For the bubbly flow regime with a condition of SHG, the assumed IHTC is high enough that the vapor temperature will relax toward the equilibrium (saturation) condition; h_{ig} is set to 10^4 W/m^{2}K. The same model is used for the bubbly flow regime with a condition of SCG.

Analogously, in the mist flow regime for the condition of SHG, a convective heat transfer model for a spherical droplet (Lee and Ryley, 1968) is used for the IHTC. For mist flow with SHL, it is assumed the IHTC is high enough so that the liquid temperature will relax toward the equilibrium (saturation) condition.

In the bubbly flow regime for the SCL condition, interfacial heat transfer is calculated by the modified Unal bubble collapse model (Unal, 1976; Riemke, 1993) and the Lahey model (1978).

In the slug flow regime, interfacial heat transfers at Taylor bubbles and small bubbles are modeled separately. The large IHTC is used for Taylor bubbles to bring the vapor phase to saturation quickly. In the annular-mist flow regime, the interfacial heat transfers at the annular film and droplets are separately modeled. For the SCL condition, the interfacial mass transfer is calculated by the modified Brown droplet model (Brown, 1952) for drops and the modified Theofanous interfacial condensation model (Theofanous, 1979) for film.

Table 9.3 summarizes IHTC correlations in the RELAP5 code (USNRC, 1998). As shown in Table 9.3, some of the IHTC models are on an ad hoc basis, and these need to be assessed against experimental data. Very large IHTC puts the liquid and vapor phases in thermal equilibrium. This can lead to making the system TH code numerically more stable, but with less physical results.

Table 9.2 Interfacial area concentration model in RELAP5 code.

Flow regime	Interface	Interfacial area concentration (a_i)
Bubbly	Small bubble, a_i	$a_i = \frac{3.6\overline{\alpha}}{d_0}, \overline{\alpha} = \alpha_g$ for bubbles and $\overline{\alpha} = \alpha_f$ for droplets where d_0 = average bubble diameter $= (1/2)d_{max}$, $d_{max} = \frac{We\sigma}{\rho_c(V_g - V_f)^2}, We = 10.0.$
Slug	Taylor bubble, $a_{i,Tb}$	$a_{i,Tb} = \frac{\alpha_b}{L} + \left(\frac{3.6\alpha_{gs}}{d_0}\right)(1 - \alpha_b)$ where $\alpha_b = \frac{\alpha_g - \alpha_{gs}}{1 - \alpha_{gs}}$, L : cell length,
	Small bubble, $a_{i,bub}$	$\alpha_{gs} = \alpha_{BS}\, exp\left[-8\left(\frac{\alpha_g - \alpha_{BS}}{\alpha_{SA} - \alpha_{BS}}\right)\right].$ $a_{i,bub} = \frac{3.6\overline{\alpha}}{d_0}$
Annular-mist	Film and droplets	$a_i = \left(\frac{4C_{ann}}{D}\right)(1 - \alpha_{ff})^{1/2} + \left(\frac{3.6\alpha_{fd}}{d_0}\right)(1 - \alpha_{ff})$ where C_{ann} : roughness parameter $\alpha_{fd} = \frac{\alpha_f - \alpha_{ff}}{1 - \alpha_{ff}}$, $\alpha_{ff} = \begin{cases} \alpha_f C_f\, exp\left[-7.5 \times 10^{-5}\left(\frac{\alpha_g v_g}{u_c}\right)^6\right] & \text{for vertical flow,} \\ \alpha_f C_f\, exp\left[-4.0 \times 10^{-5}\left(\frac{\lvert V_g - V_f\rvert}{v_{gL}}\right)^6\right] & \text{for horizontal flow,} \end{cases}$ $C_f = 1.0 - 10^{-4}\left(\alpha_f \rho_f \lvert v_f\rvert \frac{D}{\mu_f}\right)^{0.25}$, u_c : the critical entrainment superficial velocity, v_{gL} : *the horizontal stratification critical velocity*

Horizontal stratified	Horizontal surface, a_i	$a_i = 4C_{st}\frac{\sin\theta}{\pi D}$ where C_{st} : roughness parameter
Inverted annular	Small bubble, $a_{i,bub}$	$a_{i,bub} = \frac{3.6\alpha_{gb}}{d_0}(1 - \alpha_B)$ where $\alpha_{gb} = \frac{\alpha_g - \alpha_B}{1 - \alpha_B}$, α_{gb} : the vapor/gas void fraction in the liquid core, d_0 : the bubble diameter, $We = 10.0$, α_B : the fraction of the total area occupied by the vapor/gas annulus.
	Liquid core, $a_{i,ann}$	$a_{i,ann} = \frac{4}{D}(1 - \alpha_B)^{1/2}$
Mist	Droplets, a_i	$a_i = \frac{3.6\overline{\alpha}}{d_0}$ where $We = 3.0$ and $\mu_m = \frac{\mu_g}{(\alpha_g)^{2.5}}$ for mist pre-CHF, $We = 12.0$ and $\mu_m = \mu_g$ for mist and mist post-CHF

Table 9.3 Interfacial heat transfer coefficient (HTC) correlations in RELAP5 code.

Flow regime	Inter-face	Superheated liquid (SHL)	Subcooled liquid (SCL)	Superheated vapor/gas (SHG)	Subcooled vapor/gas (SCG)
Bubbly	Small bubble	Max (Plesset and Zwick, 1954), modified Lee-Ryley (1968)	Modified Unal (1976), Lahey (1978)	$h_{ig} = 10^4 W/m^2K$ a_i: Bubbly SHL	Bubbly SHG
Slug	Taylor bubble	Large HTC	Special model	Modified Lee-Ryley (1968)	h_{ig}: Bubbly SHG
	Small bubble	Bubbly SHL	Bubbly SCL	h_{ig}: Bubbly SHG $a_{i,bub}$: *slug SHL*	Slug SHG
Annular mist	Film	Large HTC	Modified Theofanous (1979)	Dittus-Boelter (1930)	h_{ig}: Bubbly SHG $a_{i,ann}$: Annular mist SHL
	Droplets	Quadratic function	Modified Brown (1952)	Lee-Ryley (1968)	h_{ig}: Bubbly SHG $a_{i,drp}$: Annular mist SHG
Inverted annular	Small bubble	Bubbly SHL	Bubbly SCL	h_{ig}: Bubbly SHG $a_{i,bub}$: Inverted annular SHL	Inverted annular
	Liquid core	Large HTC	Dittus-Boelter (1930)	$h_{ig} = \frac{k_g}{D}$ $a_{i,ann}$: Inverted annular SHL	SHG
Inverted slug	Liquid slug	Large Nu	Brown (1952)	$h_{ig} = \frac{k_g}{D}$ $a_{i,ann}$: Inverted slug SHL Modified Lee-Ryley	Inverted slug SHG
Dispersed	Droplets	Large Nu	Brown (1952)	Modified Lee-Ryley	h_{ig}: Bubbly SHG a_{gf}: Dispersed SHG
Horizontally stratified		Large Nu	Dittus-Boelter (1930)	Dittus−Boelter and large Nu	h_{ig}: Bubbly SHG a_{gf}: Horizontally stratified SHG
Vertically stratified		$H_{if} = H_{if,REG}$ (unclear)	McAdams	McAdams	$H_{if} = H_{if,REG}$ (Unclear)

9.4.4 Interfacial heat transfer in the TRACE code

The formulation of interfacial heat transfer in the TRACE code (USNRC, 2013) is basically the same as that in RELAP5. However, for the special case where liquid is superheated with respect to saturation at the total pressure, rapid nucleation will occur at any surface, and the resulting IHTC will increase dramatically as the amount of SHL increases. Under these conditions, Eq. (9.27) is modified to add a flashing term, which is

$$Q_{if}^{B} = h_{if} a_i \left(T^s(P_v) - T_f\right) + h_{flash} a_i \left(T^s - T_f\right). \tag{9.41}$$

When NCGs are present, several modifications are made to the IHTC models to consider the additional heat transfer resistance associated with the diffusion of the NCG.

In the TRACE code, the IHTCs (h_{ig}, h_{if}, and h_{flash}) and IAC (a_i) depend on the flow regime. Each flow regime has its correlations for a_i, h_{ig}, h_{if}, and h_{flash}. Unlike the RELAP5 code that uses different IHTCs according to fluid temperature, one set of correlations is used for each flow regime and, if needed, the effect of fluid temperature is considered. Tables 9.4 and 9.5 summarize the IAC and IHTC correlations of the TRACE code.

9.5 Interfacial drag

The equations of motion in the two-fluid six-equation model are coupled by two interfacial terms: one resulting from the interfacial drag force between the phases (F_{ik}) and the other from the momentum transfer associated with mass transfer ($V_{ik}\Gamma_{ik}$). The interfacial drag determines the slip between liquid and vapor phases, phase separation, two-phase mixture level, etc. This term is very important in a slow two-phase flow transient, such as a small-break LOCA.

The interfacial drag between the phases is generally represented as

$$F_{ik} = (-1)^k C_i \left|V_g - V_f\right| \left(V_g - V_f\right), \tag{9.42}$$

where C_i is the interfacial drag coefficient. Two models are widely used for the phasic interfacial drag computation in system TH codes: the drift flux method and the drag coefficient method.

9.5.1 Drift flux method

In RELAP5, TRACE, and CATHARE 2, the drift flux approach is used in the bubbly and slug flow regimes for vertical flow. This is because lots of reliable drift flux data are available in vertical bubbly and slug flows.

The drift flux model specifies the distribution coefficient (C_o) and the vapor drift velocity (V_{gj}). By using appropriate assumptions, these two quantities are converted

Table 9.4 Interfacial area concentration model in TRACE code.

Flow regime	Interfaces	Interfacial area per unit volume
Bubbly	Dispersed bubble, $a_{i,DB}$	$a_{i,DB} = \frac{6\alpha}{d_{DB}}, \alpha_g < \alpha_{DB}$ where $d_{DB} = 2\sqrt{\frac{\sigma}{g\Delta\rho}}$, $\alpha_{DB} = \begin{cases} 0.3 & \text{for } G \leq 2000\left(\frac{kg}{m^2 s}\right), \\ 0.3 + 0.2\left(\frac{G - 2000}{2700 - 2000}\right) & \text{for } 2000 < G < 2700\left(kg/m^2 s\right), \\ 0.5 & \text{for } G \geq 2700\left(\frac{kg}{m^2 s}\right) \end{cases}$
Cap bubble/ Slug	Large bubble, $a_{i,LB}$	$a_{i,LB} = \frac{C^*}{D^*}\left(\frac{\alpha - \alpha_{DB}}{1 - \alpha_{DB}}\right), \alpha_{DB} < \alpha_g < 0.5$ where $C^* = \begin{cases} 4.5 \ ; D_h < D_{h,crit} \\ 16 \ ; D_h \geq D_{h,crit} \end{cases}$ $D^* = \begin{cases} D_h \ ; D_h < D_{h,crit} \\ D_{h,crit} \ ; D_h \geq D_{h,crit} \end{cases}$ $D_{h,crit} \approx 50 \cdot La = 50 \cdot \sqrt{\frac{\sigma}{g\Delta\rho}}$
	Dispersed bubble, $a_{i,DB}$	$a_{i,DB} = \frac{6 - \alpha_{DB}}{d_{DB}}\left(\frac{1 - \alpha_g}{1 - \alpha_{DB}}\right), \alpha_{DB} < \alpha_g < 0.5$
Annular-mist	Film, a_i	$a_i = \frac{4}{D_h}\sqrt{\alpha_g f_{wet}}$ where f_{wet}: the fraction of the surface wetted by the film.
	Droplets, $a_{i,drop}$	$a_{i,drop} = \frac{6\alpha_g\alpha_d}{(1 - \alpha_d)d_d}$ where α_d: fraction of annular core region occupied by droplets.
Stratified		$a_i = \frac{S_i}{A}$, S_i: the width of the stratified two-phase interface 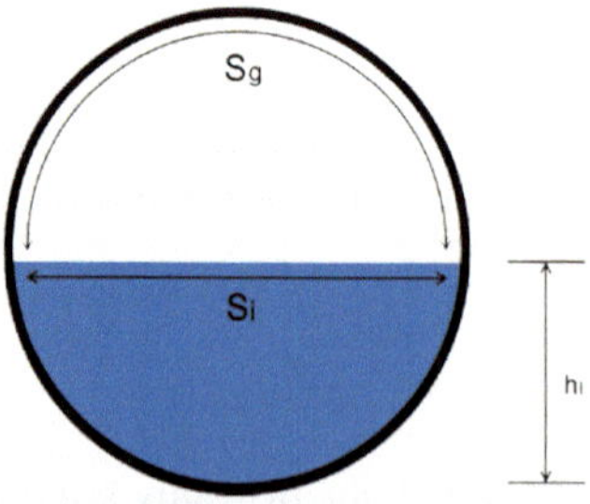

Table 9.5 Interfacial heat transfer coefficient (HTC) correlations in TRACE code.

Flow regime	Interface	Liquid side	Vapor/gas side
Bubbly	Dispersed bubble	Ranz and Marshall (1952)	Constant HTC $h_{vi} = 1000(W/m^2K)$
Cap bubble/ Slug	Large bubble	Ranz and Marshall (1952)	Constant HTC
	Dispersed bubble		$h_{vi} = 1000(W/m^2K)$
Annular-mist		$h_{li,film} = \left[(h_{li,lam})^2 + (h_{li,turb})^2\right]^{1/2}$	
	Film	Laminar–Kuhn (1994)	Max [Dittus–Boelter, 4.0]
		Turbulent–Gnielinski (1976)	
	Droplets	Kronig and Brink (1950)	Ryskin (1987)
Stratified		$h_{li} = Min\left[25000, \frac{k_l}{h_l} Nu_{li}\right]$ h_l: The height of liquid layer Nu_{li}: Nusselt nu. Used for liquid film in annular flow	Unclear

into a constitutive relation for the interfacial frictional force per unit volume (USNRC, 1988; USNRC, 2013). In RELAP5, the drift flux method leads to the following interfacial drag:

$$F_i = (-1)^k C_i \left| C_1 V_g - C_0 V_f \right| \left(C_1 V_g - C_0 V_f \right), \tag{9.43}$$

where $C_i = \frac{\alpha_g \alpha_f^3 (\rho_f - \rho_g) g}{V_{gj}^2}$, and

$$C_1 = \frac{1 - \alpha_g C_0}{1 - \alpha_g}.$$

The drift flux correlations (C_o and V_{gj}) from the literature based on Putney (1988a,b,c, 1989, 1991) and Chexal and Lellouche (1986) are used, which had been developed for rod bundles, small pipes (D < 0.018 m), intermediate pipes (0.018 m < D < 0.08 m), and large pipes (0.08 m < D) for a wide range of flow rates.

A similar approach is used in the TRACE code for bubbly/slug flow regimes in pipes and rod bundles. Meanwhile, Eq. (9.42) is used instead of Eq. (9.43) and the friction coefficient C_i is given by

$$C_i = \frac{\alpha_g \alpha_f^3 (\rho_f - \rho_g) g}{V_{gj}^2} P_S, \tag{9.44}$$

where $P_S = \frac{\left(\frac{1 - C_0 \alpha_g}{1 - \alpha_g} V_g - C_0 V_f \right)^2}{V_{gj}^2}$. The profile slip factor, P_S, can be considerably less than unity. The term comes from the fact that the velocities used in the field equations in TRACE are void-weighted area-averaged phase velocities (Brooks et al., 2012). The drift flux parameters C_0 and V_{gj} are prepared for dispersed bubbly/slug flow and Taylor cap bubble flow in pipes (Zuber and Findlay, 1965; Wallis, 1969; Ishii, 1977; Kataoka and Ishii, 1987) and for bubbly/slug flow in rod bundles (Bestion, 1985; Salay, 2001; Coddington and Macian, 2002). It is known that the Bestion model, $V_{gj} = \sqrt{g \Delta \rho D_h / \rho_g}$ and $C_o = 1.2$, did a reasonably good job for low-flow low-pressure conditions in rod bundles. Based on a study by Salay (2001), the Bestion model was extended to high-pressure boil-off conditions in TRACE. However, for the BWR conditions, the value of the distribution parameter C_o was set to unity, as recommended by Coddington and Macian (2002), to correct the underprediction of the void fraction.

CATHARE 2 also uses a drift-flux model to compute the interfacial drag on the bubbles in bubbly flow, where configuration-dependent parameters are additionally considered for flows in rod bundles, pipes, and annuli (Bestion, 1985, 1990).

9.5.2 Drag coefficient method

The drag coefficient method, using Eq. (9.42), is used in all flow regimes except for the flow regimes that use the drift flux method. Correlations for the drag coefficients and the IAC (see Tables 9.2 and 9.4) are needed in this method. For illustration, a description of the RELAP5 drag coefficient method follows below.

The constitutive relation for the frictional force on a body moving relative to a fluid is given by

$$F = \frac{1}{2}\rho V^2 C_D A \tag{9.45}$$

where

F = drag force,
ρ = fluid density,
V = velocity of the body relative to the fluid,
C_D = drag coefficient, and
A = projected area of the body.

Expressing the frictional force for a group of bodies moving relative to a fluid (e.g., bubbles moving through liquid or droplets moving through vapor) in terms of the frictional force for each body leads to the following constitutive relation for the frictional force per unit volume.

$$\begin{aligned} F_i &= \frac{1}{8}\rho_c C_D S_F a_i \left|V_g - V_f\right|\left(V_g - V_f\right) \\ &= C_i\left|V_g - V_f\right|\left(V_g - V_f\right) \end{aligned} \tag{9.46}$$

where

$C_i = \frac{1}{8}\rho_c C_D S_F a_i$,
ρ_c = density of continuous phase, and
S_F = shape factor.

The additional factor of 1/4 in Eq. (9.46) comes from the conversion of the projected area of spherical particles (πr^2) into the interfacial area ($4\pi r^2$) and the shape factor is included to account for nonspherical particles. As shown in Eq. (9.46), the drag is proportional to the IAC. The IAC models in Table 9.2 are used to obtain the IAC for a specific flow regime. This again shows that the flow regime map is very important for two-phase flow computations.

The drag coefficient to be used in nonvertical bubbly flow and all droplet flow situations is given by Ishii and Chawla (1979) for the viscous regime, as

$$C_D = \frac{24\left(1.0 + 0.1 Re_p{}^{0.75}\right)}{Re_p} \tag{9.47}$$

where the particle Reynolds number, Re_p, is defined as

$$Re_p = \frac{|V_g - V_f| d_0 \rho_c}{\mu_m}. \tag{9.48}$$

The density, ρ_c, is given by ρ_f for bubbles and ρ_g for drops. The mixture viscosity, μ_m, is for the continuous phase. It is given by $\mu_m = \mu_f / \alpha_f$ for bubbles and $\mu_m = \frac{\mu_g}{\alpha_g^{2.5}}$ for mist pre-CHF, mist, and mist post-CHF droplets. As shown in Eqs. (9.46) and (9.48), the average bubble (or droplet) diameter is very important for determining interfacial drag.

For nonvertical slug flow, slug flow for nonvertical geometry is modeled as a series of Taylor bubbles separated by liquid slugs containing small bubbles. The effects of Taylor bubbles and the small bubbles are separately modeled (See also Table 9.2 for the IAC). The drag coefficient for the Taylor bubbles in nonvertical slug flow is given by Ishii and Chawla (1979):

$$C_D = 10.9\left(\frac{D'}{D}\right)(1 - \alpha_b)^3 \tag{9.49}$$

where D' is the Taylor bubble diameter, and α_b is the void fraction of a single Taylor bubble in the total mixture. The drag coefficient for small bubbles in nonvertical slug flow is again given by Eq. (9.47).

Similarly, annular-mist flow is characterized by a liquid film along the wall and a vapor core containing entrained liquid droplets. The IAC model is presented in Table 9.2. The interfacial friction factor of the liquid film is described by a standard correlation in the laminar region and is based on Wallis's correlation in the turbulent region. The values of C_D are

$$C_D = \begin{cases} \dfrac{64}{Re_g} & \text{for } Re_g \le 500, \\ \left(\dfrac{1,500\text{-}Re_g}{1,000}\right)\dfrac{64}{Re_g} + \left(\dfrac{Re_g\text{-}500}{1,000}\right)0.02\left\{1 + 150\left[1\text{-}(1\text{-}\alpha_{ff})^{1/2}\right]\right\} & \text{for } 500 < Re_g < 1,500, \\ 0.02\left\{1 + 150\left[1 - (1 - \alpha_{ff})^{1/2}\right]\right\} & \text{for } Re_g \ge 1,500, \end{cases} \tag{9.50}$$

where $Re_g = \frac{\rho_g |V_g\text{-}V_f| D_g}{\mu_g}$, and $D_g = \alpha_g^{1/2} D$ is the equivalent wetted diameter.

The drag coefficient for droplets is again given by Ishii and Chawla in Eq. (9.47).

9.6 Wall heat transfer

The constitutive equations to represent wall heat transfer are of great significance in system TH codes. They link the governing equations for two-phase flow and the heat conduction equations for fuel rods, heat exchangers, and structures. The constitutive equations for radiation heat transfer are not discussed here, which becomes increasingly important in film boiling, particularly when the surface temperature is greater than ~ 1000 K.

Wall heat transfer models play a crucial role in the system TH codes, requiring continuous improvement. It is well known that the model uncertainties of post-CHF heat transfer and condensation under the presence of NCGs are still great, thus leading to the continuous development of new models (Chatzikyriakou et al., 2010; Chikhi and Fichot, 2010; Wua et al., 2012; Huang et al., 2015). The implementation of a new model into a system TH code, however, requires a very careful approach. The wall heat transfer package in a code consists of many specific models. For example, RELAP5-3D uses 31 heat transfer models. A change in one of the models, which was proved to be advanced based on the assessment against separate effects tests, may not yield expected results in a transient calculation for an integral effects test or a nuclear power plant because it may change the transient sequence, may not be used as intended, or may affect the wall heat transfer at a transition region by interpolation. Thus, a systematic assessment using integral effects tests is needed also. This causes a significant delay between a new model development and its implementation in a system TH code.

Wall heat transfer is dependent on the flow channel geometry, flow regime, relative wall temperature, etc., which results in very complicated logic for selecting appropriate heat transfer models. Roth and Aydogan (2014b) outlined the wall heat transfer models of various system TH codes. RELAP5-3D implements specific models for pipes, parallel plates, rod bundles, swirl tubes, horizontal plates, and annuli. TRACE has specific correlations for four geometries; rods, plates, cylinders, and tube bundles. RELAP5-3D uses 31 heat transfer models to represent 12 heat transfer modes. On the other hand, TRACE has 20 heat transfer models for 11 heat transfer modes. CATHARE 2 takes a relatively simple approach; eight heat transfer models for six heat transfer modes, considering three geometries of rods, plates, and cylinders. Considering the complexity of the wall heat transfer models, RELAP5 was selected here to illustrate how they are implemented in a system TH code.

In RELAP5, the total wall heat flux (q''_W) is the heat flux to the vapor plus the heat flux to the liquid:

$$q''_W = h_{wg}\left(T_W - T_{ref,g}\right) + h_{wf}\left(T_W - T_{ref,f}\right) \tag{9.51}$$

where

h_{wg} = heat transfer coefficient to gas,
h_{wf} = heat transfer coefficient to liquid,
T_w = wall temperature,
$T_{ref,g}$ = gas reference temperature, and
$T_{ref,f}$ = liquid reference temperature.

The reference temperatures can be the local gas or liquid temperature or the saturation temperature, depending on the heat transfer coefficient correlation being used. The RELAP5 wall heat transfer correlations are based mainly on internal flow in pipes.

Additional geometries considered are vertical parallel plates, vertical and horizontal tube bundles, and horizontal flat plates. RELAP5 has a reflood heat transfer model developed by the Paul Scherrer Institute in Switzerland (Analytis, 1996). For a reflood heat structure where heat conduction to both axial and radial directions is allowed, the wall heat transfer logic used is the same as for other surfaces.

In the TRACE code, the total wall heat flux (q''_W) is represented by

$$q''_W = h_{wf}(T_W - T_f) + h_{wsat}(T_W - T_{sat}) + h_{wg}(T_W - T_g). \tag{9.52}$$

The second term on the right-hand side is used to represent the subcooled boiling, where the Lahey model is adopted. In Eq. (9.51), the wall heat transfer to liquid contains the subcooled boiling effect. Thus, Eqs. (9.51) and (9.52) are the same. For the default CHF model in TRACE, the 1995 AECL-IPPE CHF lookup table (Groeneveld et al., 1996) was selected. It is based on an extensive database of CHF values obtained in tubes with a vertical upflow of a steam–water mixture and provides the value of the critical heat flux as a function of the local conditions. This method was selected because of its reasonably good accuracy and wide range of applicability.

A boiling curve is used in RELAP5 to select the wall heat transfer correlations when the wall surface temperature is above the saturation temperature (superheated relative to the saturation temperature based on total pressure). When a hydraulic volume is voided and the adjacent surface temperature is subcooled, steam condensation on the surface is predicted. If NCGs are present, the phenomena are more complex, because while boiling is a function of the wall superheat based on the total pressure, condensation is based on the partial pressure of steam. When the wall temperature is less than the saturation temperature based on total pressure but greater than the saturation temperature based on steam partial pressure, a convection condition exists. Fig. 9.11 illustrates these three regions.

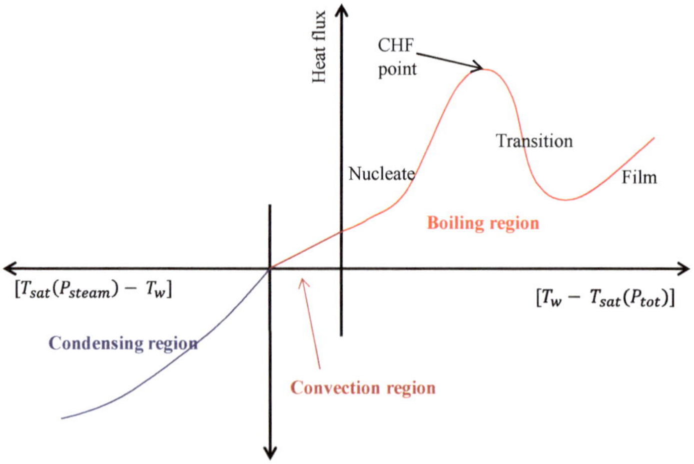

Figure 9.11 RELAP5 wall heat transfer curves. *CHF*, critical heat flux.

9.6.1 Convection region

In the convection region, the heat transfer coefficient to liquid is the maximum among those for forced convection, laminar convection, and natural convection. The correlations are by Dittus-Boelter (1930), Kays (1955), and Churchill-Chu (1975), respectively. For parallel plates, the Petukhov correlation (1970) is used in place of the Dittus−Boelter correlation, and the Elenbaas correlation (Incropera et al., 1990) is used instead of the Churchill−Chu correlation.

For vertical bundles, the turbulent convection coefficient is multiplied by the tube pitch-to-diameter ratio as suggested by Inayatov (1975). Horizontal bundles in RELAP5 differ from pipes in their nucleate boiling, critical heat flux, and natural convection prediction methods. Nucleate boiling predictions follow the Polley−Ralston−Grant method (1981), and the prediction of critical heat flux uses the Folkin−Goldberg method (1980). Horizontal bundles use the Churchill-Chu (1975a) horizontal cylinder correlation.

9.6.2 Boiling region

The boiling curve uses the Chen (1966) boiling correlation up to the critical heat flux point. A table lookup method developed by Groeneveld, Cheng, and Doan (1986) is used for the prediction of the critical heat flux. When the wall superheat exceeds the critical value, the heat flux for both the transition boiling and the film boiling regimes are calculated and the maximum value is used. This eliminates the need to predict a minimum film boiling temperature. The Chen−Sundaram−Ozkaynak correlation (1977) is used for transition boiling and a modified Bromley correlation (1950) is used for film boiling.

The post-CHF heat transfer regimes mainly determine the peak cladding temperature (PCT) at the reflood phase of an LOCA in a PWR (Choi et al., 2012). Besides their droplet behaviors at that time, they still have great uncertainty.

When the wall temperature exceeds that for the onset of nucleate boiling but with subcooled bulk liquid, subcooled nucleate boiling occurs. In this situation, wall heat can be transferred to the liquid phase only in the two-fluid formulation, but a portion of the wall heat flux is spent for vapor generation in the boundary layer near the wall. To obtain the fraction of the heat flux that causes vapor generation near a superheated wall, the Lahey method (1978) is used in RELAP5. The resulting vapor generation rate is given by

$$\Gamma_w = \frac{q'' A_w}{\left(h_g' - h_f'\right)} \mathrm{Mul}, \tag{9.53}$$

where q'' = total wall heat flux, A_w = wall surface area, V = cell volume, and

$$\mathrm{Mul} = \frac{h_f - h_{cr}}{\left(h_f^s - h_{cr}\right)(1+\varepsilon)},$$

where h_{cr} = critical enthalpy for net voids calculated using the Saha–Zuber correlation (1974), and $\varepsilon = \frac{\rho_f\left[h_f^s - \min\left(h_f, h_f^s\right)\right]}{\rho_g h_{fg}}$, a pumping factor.

It was shown by Hari and Hassan (2002) that Lahey's formulation underpredicts void fraction under low-pressure conditions. They recommended Bowring's and Zeitoun and Shoukri's pumping factor formulations (Collier and Thome, 1994) for low-pressure conditions.

9.6.3 Condensing region

Two methods are available for calculating condensation heat transfer coefficients for all geometries in the RELAP5/MOD3 code. The default method uses the maximum of Nusselt (1916) and Shah (1979). When NCGs are present, the default method uses the Colburn-Hougen (1934) iteration method to solve for the interface temperature between the steam and liquid and uses that value in the heat flux calculation. The alternate method uses Nusselt's laminar film coefficient modified for gas turbulence and NCGs, using the multipliers developed by Vierow and Schrock (1992).

It is well known that the presence of even a small quantity of NCG in the condensing vapor profoundly influences the overall heat transfer resistance during condensation (Collier and Thome, 1994). The NCG is carried with the vapor toward the interface, where it accumulates so that the partial pressure of the vapor at the interface decreases, thereby decreasing the driving potential for heat transfer across the liquid film at the condensing wall. The process limiting the heat transfer rate then becomes the mass transfer rate for the vapor diffusing through the gas–vapor mixture to the interface (See Fig. 9.12).

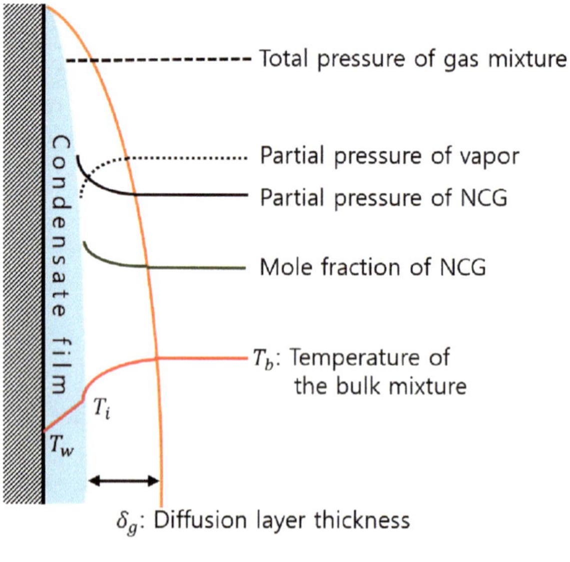

Figure 9.12 Schematic illustration of diffusion layer. *NCG*, noncondensable gas.

Recently, to improve the predictive capability of system thermal-hydraulic codes for the passive containment cooling system of advanced light-water reactors, new models for condensation in the presence of NCGs have been developed. They can be divided into two categories: (1) an extension of the heat and mass transfer analogy method and the diffusion method proposed by Colburn and Hougen (Herranz et al., 1998; Liao and Vierow, 2007; Lee et al., 2020; Benteboula and Dabbene, 2020), and (2) improved semiempirical correlations (Dehbi, 1991; Lee et al., 2022).

9.7 Wall friction

Frictional pressure losses in a system TH code are usually divided into wall friction and form loss (or called form drag). The former models the fluid—wall shear using a friction factor approach. The latter models geometry-specific pressure losses at a sudden area change (expansion and contraction), orifice, elbow, bend, or complicated flow passage geometries, using an additive loss coefficient, of which form loss factor is usually input by the user.

Various wall friction models were introduced in Chapter 8. This chapter emphasizes the method for implementing wall friction models into system TH code. In the two-fluid formulation, the phasic pressure gradient due to wall friction can be given by

$$\left(\frac{dP}{dx}\right)_{fr,f} = -C_{wf}|V_f|V_f, \tag{9.54}$$

and

$$\left(\frac{dP}{dx}\right)_{fr,g} = -C_{wg}|V_g|V_g \tag{9.55}$$

where C_{wf} and C_{wg} are the wall friction coefficients for wall—liquid and wall—gas shear, respectively.

9.7.1 Single-phase flow

In the case of a single-phase flow, the wall friction coefficients of the k-phase are defined by

$$C_{wk} = \lambda_D \frac{\rho_k}{2D} \tag{9.56}$$

where the subscript k indicates the fluid phase (f or g) and λ_D is the friction factor. To obtain the friction factor, RELAP5 uses correlations for laminar and turbulent flows with interpolation in the transition regime. It is calculated as

$$\lambda_D = \begin{cases} \dfrac{64}{Re\Phi_s} & \text{for } 0 \le \text{Re} \le 2200, \\ \left(3.75 - \dfrac{8250}{Re}\right)\left(\lambda_{T,3000} - \lambda_{L,2200}\right) + \lambda_{L,2200} & \text{for } 2200 < \text{Re} < 3000, \\ \left\{-2\, log_{10}\left\{\dfrac{\epsilon}{3.7D} + \dfrac{2.51}{Re}\left[1.14 - 2\, log_{10}\left(\dfrac{\epsilon}{D} - \dfrac{21.25}{Re^{0.9}}\right)\right]\right\}\right\}^{-2} & \text{for Re} \ge 3000, \end{cases} \tag{9.57}$$

where Re is the Reynolds number, and

Φ_S: a user-input shape factor for noncircular flow channels,
$\lambda_{L,2200}$: the laminar factor at a Reynolds number of 2200,
$\lambda_{T,3000}$: the turbulent friction factor at a Reynolds number of 3000,
ε: the surface roughness.

The turbulent friction factor is given by the Zigrang-Sylvester (1985) approximation to the Colebrook–White correlation (Colebrook, 1939). It has the advantage of being an explicit relation for the friction factor, while the Colebrook–White correlation is a transcendental function requiring iteration to determine the friction factor. The friction model applies to unheated surfaces. The isothermal friction factor can be corrected for variations in fluid viscosity near a heated surface using the relationship used in the VIPRE code (Stewart et al., 1985).

TRACE uses the Churchill formula (1977) for the friction factor because it applies to all three flow regimes—laminar, transition, and turbulent:

$$\lambda_D = 8\left[\left(\frac{8}{Re}\right)^{12} + \frac{1}{(a+b)^{3/2}}\right]^{1/12} \tag{9.58}$$

where $a = \left\{2.457 \cdot \ln\left[\dfrac{1}{\left(\dfrac{7}{Re}\right)^{0.9} + 0.27\left(\dfrac{\epsilon}{D_h}\right)}\right]\right\}^{16}$,

$$b = \left(\frac{3.753 \times 10^4}{Re}\right)^{16},$$

$$\text{Re}_k = \max\left(10, \frac{\rho_k |V_k| D_h}{\mu_k}\right).$$

Its predictions for the laminar regime are in agreement with Eq. (9.57). The predictions for transition flow are subject to some uncertainty. However, its accuracy for turbulent flow (Ebadian et al., 1998) shows a maximum deviation from the Colebrook–White equation of about 3.2% for $4000 < \text{Re} < 10^8$ and $10^{-8} < \epsilon/D < 0.05$.

9.7.2 Two-phase flow

For a two-phase flow, the wall friction model based on a two-phase multiplier approach is used in RELAP5. For normal pre-CHF flow regimes, the Lockhart–Martinelli model computes the overall friction pressure drop in terms of the liquid-alone wall friction pressure drop:

$$\left(\frac{dP}{dx}\right)_{fr,2\phi} = \phi_f^2 \left(\frac{dP}{dx}\right)_{fr,f} \tag{9.59}$$

or the vapor-alone wall friction pressure drop:

$$\left(\frac{dP}{dx}\right)_{fr,2\phi} = \phi_g^2 \left(\frac{dP}{dx}\right)_{fr,g} \tag{9.60}$$

where ϕ_f and ϕ_g are the liquid-alone and vapor-alone two-phase Darcy–Weisbach friction multipliers, respectively. The phasic wall friction pressure gradients are expressed as

$$\left(\frac{dP}{dx}\right)_{fr,f} = \lambda_f^{'}\left(Re_f^{'}\right)\frac{\dot{m}_f{}^2}{2D\rho_f A^2} \tag{9.61}$$

for the liquid alone, and

$$\left(\frac{dP}{dx}\right)_{fr,g} = \lambda_g^{'}\left(Re_g^{'}\right)\frac{\dot{m}_g{}^2}{2D\rho_g A^2} \tag{9.62}$$

for the vapor alone, where $\lambda_k^{'}$ indicates the liquid and vapor-alone Darcy–Weisbach friction factors calculated at the respective Reynolds numbers,

$$\mathrm{Re}_f^{'} = \frac{\dot{m}_f D}{\mu_f A} \tag{9.63}$$

and

$$\mathrm{Re}_g^{'} = \frac{\dot{m}_g D}{\mu_g A}. \tag{9.64}$$

The liquid and vapor mass flow rates, respectively, are defined as

$$\dot{m}_f = \alpha_f \rho_f V_f \mathrm{A} \tag{9.65}$$

and

$$\dot{m}_g = \alpha_g \rho_g V_g \mathrm{A}. \tag{9.66}$$

The two-phase multiplier ϕ_f and ϕ_g is calculated from the Heat Transfer and Fluid Flow Service–modified Baroczy correlation (Chaxton, 1972) in RELAP5. The individual phasic wall friction components are calculated by apportioning the two-phase friction between the phases using a technique derived by Chisholm (1967) from the Lockhart-Martinelli (1949) model. The partitioning model is based on the assumption that the frictional pressure drop may be calculated using a quasi-steady form of the momentum equation and the effect of individual flow regime is separately modeled. For example, in a bubbly flow at pre-CHF flow regime, the void fraction at the wall (α_{wg}) is approximated as the bulk vapor volume fraction (α_g). The wall friction coefficient for wall–gas shear (C_{wg}) is a function of α_{wg}. This indicates that the wall friction is partitioned to both liquid and vapor phases in a bubbly flow.

In TRACE, a similar approach is used for the calculation of wall friction. However, for the normal pre-CHF flow regimes (bubbly/slug and annular/mist), all wall drag is applied to the liquid phase alone. This is different from that in RELAP5. Thus,

$$\left(\frac{dP}{dx}\right)_{fr,f} = -C_{wf}\left|V_f\right|V_f, \tag{9.67}$$

or in terms of the two-phase multiplier for the liquid phase considered to flow alone in the channel,

$$\left(\frac{dP}{dx}\right)_{fr,f} = -\Phi_{\mathrm{f}}{}^2\left[\frac{f_{2\Phi,\mathrm{f}}\left|G_f\right|G_f}{2D_h\rho_f}\right] \tag{9.68}$$

where the friction factor, $f_{2\Phi,l}$, is calculated with

$$\mathrm{Re}_{2\Phi,\mathrm{f}} = \frac{\left|G_f\right|D_h}{\mu_f} = \frac{(1-\alpha)\rho_f\left|V_f\right|D_h}{\mu_f}. \tag{9.69}$$

The wall drag coefficient is then

$$C_{wf} = \Phi_f^2\left[\frac{f_{2\Phi,\mathrm{f}}(1-\alpha)^2\rho_f}{2D_h}\right]. \tag{9.70}$$

To complete the specification of wall drag for pre-CHF regimes, it remains to specify the two-phase multiplier, Φ_f^2, and the friction factor, $f_{2\Phi,f}$. In TRACE, the two-phase multiplier Φ_f^2 is a function of void fraction only, where the function depends on the flow regime. The friction factor is also dependent on the flow regime.

For two-phase flows with nucleate boiling, the pressure drop data of Ferrell and Bylund (1966) show that the two-phase multiplier when boiling is present is significantly higher and contains a pronounced mass flux effect. Thus, an additional multiplier is devised to account for the effect.

In horizontal stratified flow, both the liquid and gas phases are in contact with the pipe wall, so the wall drag coefficient is specified for both phases. For the "inverted" post-CHF flow regimes, the wall surface temperature is beyond the minimum stable film boiling temperature, so only the gas phase is in contact with the wall. Consequently, wall–fluid shear only occurs for the gas phase and C_{wf} should be zero.

For the pre-CHF flow regimes, the two-phase frictional pressure drop models of the RELAP5/MOD3 and TRACE codes are similar to each other, but the partition methods are different. In RELAP5/MOD3, the wall friction is partitioned to each phase. Meanwhile, in TRACE and CATHARE 2, the wall friction is applied to the liquid phase alone for all pre-CHF flow regimes. From a physical point of view, the latter approach seems reasonable because the vapor phase does not directly touch the wall. But this may lead to the prediction of unphysical results. For example, in the case of steady-state bubble flow in a horizontal pipe, simple algebraic operations of the phasic momentum equations lead to the result that vapor-phase velocity is greater than liquid-phase velocity if $C_{wg} = 0$ (Kim et al., 2014). The RELAP5/MOD3 approach, where C_{wg} is a nonzero coefficient, yields more reasonable results in this case. In this regard, Hibiki and Jeong (2022) theoretically derived a 1D two-fluid momentum equation for bubbly, slug, churn-turbulent, and mist flow regimes from a general two-fluid model (Ishii and Hibiki, 2011). The resulting equations for liquid and vapor phases are

$$\frac{\partial}{\partial t}\left(\alpha_g \rho_g u_g\right) + \frac{\partial}{\partial z}\left(\alpha_g \rho_g u_g^2\right) = -\alpha_g \frac{\partial P}{\partial z} - \alpha_g \rho_g g_z + M_{ig} - \alpha_g F_w, \tag{9.71}$$

and

$$\frac{\partial}{\partial t}\left(\alpha_f \rho_f u_f\right) + \frac{\partial}{\partial z}\left(\alpha_f \rho_f u_f^2\right) = -\alpha_f \frac{\partial P}{\partial z} - \alpha_f \rho_f g_z + M_{if} - \alpha_f F_w, \tag{9.72}$$

where F_w is the wall friction force and $\boldsymbol{M}_{ik}$ is the interfacial drag force. According to Eqs. (9.71) and (9.72), wall friction is proportionally partitioned to each phase based on the phasic volume fraction. This is consistent with the partition method in the SPACE code (Kim et al., 2014) and leads to a realistic relative velocity.

The two-phase wall friction models for bends, sudden area changes, and helically coiled tubes need further improvement (Hsu et al., 2015; Colombo et al., 2015), which requires experimental data and correlations for an extended range of validity.

9.8 Two-fluid three-field equations and additional constitutive equations

As mentioned in Section 9.1, the two-fluid three-field model is used in some system TH codes, such as WCOBRA/TRAC, SPACE, and CATHARE 3. The three fields represent a continuous liquid, an entrained liquid (i.e., droplets), and a vapor field

(Jeong et al., 1999; Frepoli et al., 2003; Jayanti et al., 2004). Basically, the two-fluid three-field model is very similar to the two-fluid model except that the entrained or dispersed liquid (i.e., droplets) is modeled separately. Thus, additional constitutive equations are needed for droplet generation rate, i.e., entrainment and deposition, and the mass, energy, and momentum exchanges at the droplet-to-vapor interface. It should be noted that the latter can be covered by those for the two-fluid model, and the former is unique to the two-fluid three-field model.

9.8.1 Two-fluid three-field equations

In the two-fluid three-field model, the mass, energy, and momentum equations for the three fields are separately established, and they are linked by the interfacial mass, energy, and momentum transfer models.

The continuity equation for the k-field is

$$\frac{\partial}{\partial t}(\alpha_k \rho_k) + \nabla \cdot \left(\alpha_k \rho_k \underline{u}_k\right) = \Omega_k, \tag{9.73}$$

where $k = l$, v, or d, and

$$\Omega_v = \Gamma_v, \tag{9.74}$$

$$\Omega_l = -(1-\eta)\Gamma_v - S_E + S_{DE}, \tag{9.75}$$

$$\Omega_d = -\eta\Gamma_v + S_E - S_{DE}, \tag{9.76}$$

$$\eta = \alpha_d / (\alpha_l + \alpha_d).$$

where $\Gamma_v, S_E,$ and S_{DE} are the respective volumetric vapor generation, droplet entrainment, and deposition rates. The momentum equation for the k-field is

$$\frac{\partial}{\partial t}\left(\alpha_k \rho_k \underline{u}_k\right) + \nabla \cdot \left(\alpha_k \rho_k \underline{u}_k \underline{u}_k\right) = -\alpha_k \nabla P + \nabla \cdot \left[\alpha_k\left(\tau_k + \tau_k^T\right)\right] + \alpha_k \rho_k \underline{g} + \underline{M}_{ik}, \tag{9.77}$$

where $\underline{M}_{ik}$ is the interfacial momentum transfer term. The energy equation for the k-field is

$$\frac{\partial}{\partial t}(\alpha_k \rho_k e_k) + \nabla \cdot \left(\alpha_k \rho_k e_k \underline{u}_k\right) = E_k^D - P\frac{\partial}{\partial t}\alpha_k - P\nabla \cdot \left(\alpha_k \underline{u}_k\right) + Q_{ik}, \tag{9.78}$$

where E_k^D includes the k-phase conduction, turbulent energy source, and viscous dissipation represented in terms of diffusion.

Eqs. (9.73), (9.77) and (9.78) involve terms requiring models and correlations for mathematical closure. The constitutive equations for the two-fluid model in Section 9.3 through 9.7 can be used except for droplet entrainment and deposition rates.

In a two-fluid model, the continuous liquid and droplet fields are modeled as a liquid field, and thus, the two fields in annular-mist or inverted annular-mist flows are indistinguishable. To resolve this problem, an entrainment fraction model is typically used with two-fluid codes, such as RELAP5/MOD3 (USNRC, 1998a,b), to provide the fraction of liquid flow entrained as droplets.

9.8.2 Droplet entrainment rate and deposition phenomena

Liquid droplets can be generated in annular-mist, inverted annular-mist, and horizontal stratified flows in the primary coolant system of a PWR. In addition, droplets can be produced at the quench front in the reflood phase of a large-break LOCA in a PWR.

Droplet entrainment occurs in various mechanisms. Fig. 9.13 illustrates some entrainment mechanisms (Ishii and Grolmes, 1975). First, droplet entrainment occurs when part of a roll wave is sheared by gas turbulent flow when a large-amplitude roll wave is formed on the liquid film. The second type of entrainment is caused by the liquid film being undercut by turbulent gas flow and is mainly observed when the liquid film flow rate is low. The third type is caused by a bubble burst. When the bubble comes out of the liquid film, a thin liquid film forms around the bubble. It is disintegrated by gas turbulent flow, and the thin liquid film breaks into droplets. The last type of entrainment is caused by the impingement of the droplets or liquid film wave to the liquid film interface. The key variables of droplet entrainment are different for each mechanism.

Droplet entrainment models for an annular-mist flow in a vertical pipe are introduced in Section 9.8.3, and those for a horizontal stratified flow (Ardron et al.,

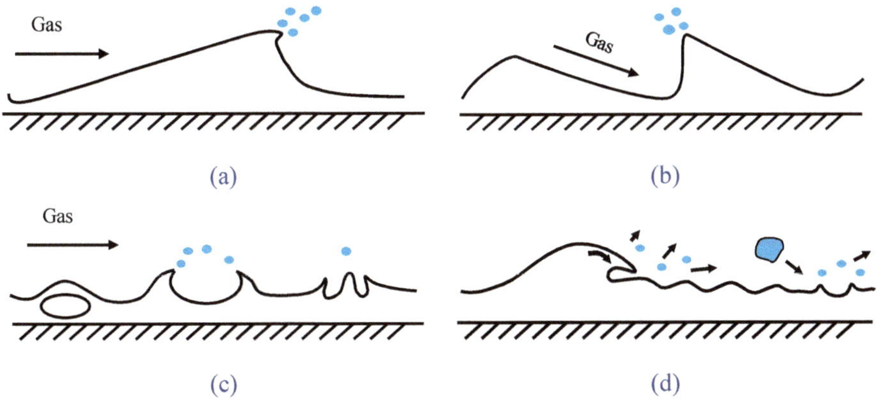

Figure 9.13 Mechanisms of droplet entrainment (Ishii and Grolmes, 1975). (a) Roll wave (b) wave undercutting (c) bubble burst (d) liquid impingement.

1990; Stevanovic et al., 1995; Höhne et al., 2015; Bae et al., 2021) are skipped here. Droplet entrainment models for the quench front are discussed in Section 9.8.4.

Many experimental studies (Jagota et al., 1973; Andreussi et al., 1983; Leman et al., 1985; Govan et al., 1989; Jepson et al., 1989; Fore and Dukler, 1995) have shown that droplet deposition has a similar relationship with diffusion. Based on that observation, many droplet deposition models (McCoy and Hanratty, 1977; Kataoka and Ishii, 1982; Hewitt and Govan, 1990; Okawa et al., 2003) have been developed, as discussed in Section 9.8.5.

9.8.3 Droplet entrainment model for vertical annular-mist flow

Although droplet entrainment has many possible mechanisms, the roll wave mechanism is the one primarily used in model development. Droplet entrainment caused by a roll wave is heavily influenced by interfacial frictional force and surface tension (Ishii and Grolmes, 1975). *The remainder of this section* introduces the droplet entrainment rate model based on the roll wave mechanism.

Hewitt and Govan (1990) developed the droplet entrainment rate model by adopting gas mass flux, liquid film mass flux, and surface tension as the main variables:

$$\dot{m}_E = 5.75 \times 10^{-5} G_g \left[\left(G_{lf} - G_{lfc} \right)^2 \frac{D_h \rho_l}{\sigma \rho_g^2} \right]^{0.316}, \tag{9.79}$$

where

$$G_g = \alpha_g \rho_g u_g, \tag{9.80}$$

$$G_{lf} = \alpha_{lf} \rho_l u_{lf}, \tag{9.81}$$

$$G_{lfc} = \frac{\mu_l}{D_h} exp(5.8504 + 0.4249) \frac{\mu_g}{\mu_l} \sqrt{\frac{\rho_l}{\rho_g}}. \tag{9.82}$$

Eq. (9.82) is the onset of entrainment as proposed by Owen and Hewitt (1987). It represents the minimum value of liquid film mass flux at which droplet entrainment can occur. The coefficient and exponent in Eq. (9.79) were obtained by fitting with experimental data.

Similarly, Okawa et al. (2003) assumed that interfacial frictional force causes droplet entrainment by deforming the roll wave, while surface tension suppresses droplet entrainment by maintaining the shape of the roll wave. Based on this assumption, Okawa et al. proposed a nondimensional variable proportional to droplet entrainment:

$$\pi_E = \frac{f_i \rho_g j_g^2 \delta}{\sigma}. \tag{9.83}$$

In addition, the density ratio (ρ_l/ρ_g) was adopted to apply the model to a wide range of pressures. Finally, they suggested the following droplet entrainment rate model:

$$\dot{m}_E = 4.79 \times 10^{-4} \rho_l \frac{f_i \rho_g j_g^2 \delta}{\sigma} \left(\frac{\rho_l}{\rho_g} \right)^{0.111} \quad \text{for } \mathrm{Re}_{lf} \geq 320, \tag{9.84}$$

where

$$\delta = \frac{1}{4} \sqrt{\frac{f_w \rho_l}{f_i \rho_g}} \frac{j_l}{j_g} D_h, \tag{9.85}$$

$$f_i = 0.005 \left(1 + 300 \frac{\delta}{D_h} \right), \tag{9.86}$$

$$f_w = \max \left(\frac{16}{\mathrm{Re}_{lf}}, 0.005 \right), \tag{9.87}$$

$$\mathrm{Re}_{\mathrm{lf}} = \frac{G_{lf} D_h}{\mu_l} = \frac{\rho_l j_{lf} D_h}{\mu_l}. \tag{9.88}$$

The entrainment models in Eqs. (9.79) and (9.84) were developed by assuming an equilibrium state. Meanwhile, Kataoka and Ishii (1982) proposed an entrainment model by dividing the annular-mist flow into the entrance region and the fully developed region:

$$\begin{aligned} \frac{\dot{m}_E D_h}{\mu_l} = {} & 1200 \mathrm{Re}_l^{-0.5} \mathrm{Re}_{lf\infty}^{-0.25} We^{-1.5} \left(\mathrm{Re}_{lf} - \mathrm{Re}_{lf\infty} \right)^2 + 6.6 \\ & \times 10^{-7} \mathrm{Re}_l^{0.74} \mathrm{Re}_{lf}^{0.185} We^{0.925} \left(\frac{\mu_g}{\mu_l} \right)^{0.26} \end{aligned} \tag{9.89}$$

where

$$\mathrm{Re}_l = \rho_l j_l D_h / \mu_l, \tag{9.90}$$

$$\mathrm{Re}_{\mathrm{lf}} = \mathrm{Re}_l (1 - E), \tag{9.91}$$

$$\mathrm{E} = \rho_l \alpha_d u_d \,/\, \left(\rho_l \alpha_{lf} u_{lf} + \rho_l \alpha_d u_d \right) \tag{9.92}$$

$$\mathrm{Re}_{\mathrm{lf}\infty} = Re_l (1 - E_\infty), \tag{9.93}$$

$$E_\infty = \tanh \left(7.25 \times 10^{-7} We^{1.25} \mathrm{Re}_l^{0.25} \right), \tag{9.94}$$

$$We = \frac{\rho_g j_g^2 D_h}{\sigma}\left(\frac{\Delta\rho}{\rho_g}\right)^{1/3}. \tag{9.95}$$

The first and second terms on the right-hand side of Eq. (9.89) represent the entrainment rate in the entrance region and the equilibrium state, respectively.

Lopez et al. (1997) improved the droplet entrainment model of Kataoka and Ishii (1982) by adding the density ratio ($\Delta\rho/\rho_g$) and the onset of entrainment (Re_{lfc}):

$$\dot{m}_E = k_E\frac{\mu_l}{D_h}\left[We_g\sqrt{\Delta\rho/\rho_g}\left(\mathrm{Re}_{lf} - \mathrm{Re}_{lfc}\right)\right]^{0.925}\left(\frac{\mu_g}{\mu_l}\right)^{0.26}, \tag{9.96}$$

where $\mathrm{We}_g = \rho_g j_g^2 D_h/\sigma$ and $\mathrm{Re}_{lfc} = 80$. Lopez et al. proposed $k_E = 4.47 \times 10^{-7}$ using vertical air–water flow experimental data.

Han et al. (2019) assessed the droplet entrainment models listed above and proposed a new one. The key variables of the droplet entrainment rate were deduced from analyses of experimental data and existing models. The entrainment rate is an increasing function of both liquid film mass flux and gas mass flux; for a given gas mass flux, a critical value exists for liquid film mass flux. Surface tension is also known to suppress droplet entrainment associated with roll waves. To consider the effects of surface tension and the liquid film inertia force, the new model includes the Weber number of the liquid film. In addition, the density ratio ρ_l/ρ_g must be included to cover the effect of pressure. Based on these changes, a new model was suggested as follows:

$$m_E = 3.987 \times 10^{-4}\left(G_{lf} - G_{lfc}\right)^{0.338} G_g^{0.662} We_{lf}^{0.155}\left(\frac{\rho_l}{\rho_g}\right)^{0.24}, \tag{9.97}$$

where $G_{lfc} = 8\left(\mu_l^2\rho_l\sigma/D_h^3\right)$. This model was assessed against vertical annular-mist flow experiments, for which the Okawa model (2003) was adopted for the deposition rate. The calculated entrained liquid fractions were compared with 455 air–water and steam–water experimental data points. The new model showed greater accuracy and precision than the four existing models.

9.8.4 Droplet entrainment model for quench front

In the reflood process, the postdryout flow regime is formed above the quench front. Unlike a normal boiling condition where a liquid phase covers the heating surface, the postdryout regime consists of a gas phase near the heating surface and a liquid phase with droplets at the center of the flow channel. As the fuel rods are quenched by the coolant, a large amount of steam is generated, and as the steam drags a portion of the continuous liquid phase, droplets are drastically generated. The total volume of droplets generated in the core during the reflood phase is about 10%–40% of the injected coolant (Damerell et al., 1993).

Yoo et al. (2019) quantitatively analyzed the reflood heat transfer of the FLECHT SEASET tests. They carried out sensitivity calculations with respect to the models for film boiling heat transfer, interfacial drag, and interfacial heat transfer as well as the droplet models. The results showed (1) the film boiling heat transfer plays the most dominant role in the peak cladding temperature and (2) the droplet entrainment at the quench front has the second largest impact on the peak cladding temperature and the greatest effect on the quenching time. Thus, the entrainment model is quite important for analyzing reflood heat transfer in a PWR.

Droplet entrainment in postdryout regimes, such as an inverted annular and an inverted slug formed downstream of the quench front, has been extensively studied. The model in the COBRA-TF code was developed based on the entrainment by vapor bubbling through a liquid pool in the early 1980s (Paik et al., 1985). In addition, the Yonomoto model (1996) was modified from the COBRA-TF model. The model in the CATHARE3 was empirically developed using the analysis result of the reflood experiments (Valette et al., 2011). Holowach et al. (2003) suggested a model with enhanced experimental and theoretical basis using the instability analysis of a liquid jet. These models use postulated droplet generation mechanisms near the quench front, such as vapor bubbling or liquid jet breakup, however, they have not properly considered the actual phenomena of the droplet formation in the postdryout regime.

Table 9.6 shows the four droplet entrainment correlations for the postdryout regimes formed downstream of the quench front. The droplet entrainment model of the COBRA-TF code (Thurgood et al., 1983; Salko and Avramova, 2015) has been assessed in the studies of Paik et al. (1985) and Ha et al. (1998) using the FLECHT SEASET reflood tests. In the study of Ha et al. (1998), the Yonomoto model with a partial modification in the COBRA-TF model was recommended based on the results of the FLECHT SEASET analysis. In the study of Valette et al. (2011), the reflood heat transfer experiments were simulated using the CATHARE3 code, in which a droplet entrainment model based on relative velocity was applied. These three models include several engineering factors as listed in Table 9.6, but the rationale for using these factors is not clear. Holowach et al. (2003) developed a model to predict the number of droplets generated at the quench front using vertical tube reflood experiment data and the Kelvin–Helmholtz instability analysis. They estimated the droplet entrainment rate in the experiments by applying the assumption that droplets exist after the quench front only. However, this assumption does not appropriately consider the actual phenomenon because the continuous liquid phases, such as a liquid slug or a liquid jet, exist simultaneously with the droplets. In addition, since the correlation is defined using the data of the tube experiments, it is expected that there might be some limitations in applying it to the rod bundle experiments. All four models hardly reflect the droplet generation mechanisms in the postdryout regime.

In this regard, Yoo et al. (2020, 2021) developed a new droplet entrainment model for reflood heat transfer. For the model development, they used the results of several visualization experiments observing the hydrodynamic behavior of the postdryout regime; the postdryout regime was divided into an inverted annular regime, an agitated regime, and a dispersed droplet/ligament regime. In the inverted annular regime, droplets detaching from the core liquid jet were observed. In the agitated regime, droplets

Table 9.6 Droplet entrainment models for postdryout regime.

Model	Correlations
COBRA-TF	$m_E = \min[1.0, 5.0\alpha_l] \times \max[0.0, m_{E1} - \max(0.0, m_{d,i})]$, $m_{E1} = 1.5 \min\left[2.5, \left(\frac{u_g}{u_{crit}}\right)^2\right] m_g$, $u_{crit} = \left(\frac{4We_d}{3C_D}\right)^{0.25}\left(\frac{\sigma g \Delta\rho}{\rho_g^2}\right)^{0.25}$, $We_d = 2.0, C_D = 0.45$.
Yonomoto	m_E and m_{E1}: same as the COBRA-TF model, $u_{crit} = \left(\frac{3.57}{C_D \alpha_g}\right)\left(\frac{\sigma g \Delta\rho}{\rho_g^2}\right)^{0.25} N_{\mu g}^{1/6}$, $N_{\mu g} = \frac{\mu_g}{\left(\rho_g \sigma \sqrt{\frac{\sigma}{g\Delta\rho}}\right)^{0.5}}$, $C_D = \begin{cases} \frac{24}{\text{Re}_d}\left(1 + 0.15\text{Re}_d^{0.687}\right) & \text{for Re}_d < 1000 \\ 0.44 & \text{for Re}_d \geq 1000 \end{cases}$.
CATHARE3 (Valette et al., 2011)	$m_E = 4 \times 10^{-6} Ai_l \rho_l (u_g - u_l) We^{0.625}$, $We = \frac{\rho_g (u_g - u_l)^2 D_h}{\sigma}$.
Holowach	$m_E = 1.46 \times 10^{-8} \frac{\lambda_{crit} \rho_l P_h \left(u_{g,crit}^2 + u_{g,bqf}^2\right)^{0.5}}{A_f} \text{Re}_{g,gen}^{1.83}$.

were generated from the breakup of the core liquid jet and the liquid sheet. Based on these observations, they proposed a new droplet entrainment model for the core liquid jet and the liquid sheet and assessed it using 7 FEBA tests and 14 FLECHT SEASET tests. The model yielded the smallest mean absolute errors for the peak cladding temperature and quenching time compared with those of the four existing models in Table 9.6. They further developed the model to improve the prediction of the droplet–wall heat transfer downstream of a flow blockage in the reflood phase (Yoo et al., 2021).

9.8.5 Droplet deposition model

The droplet deposition has been known to have similar characteristics with a diffusion process through many experimental studies. So, the droplet deposition models have been developed as the product of the product of the deposition coefficient (k_d) and the droplet concentration (C):

$$\dot{m}_D = k_d C, \tag{9.98}$$

where $C = \rho_d \frac{j_d}{j_d + j_g}$.

The droplet deposition coefficient tends to be inversely proportional to droplet concentration (Hewitt and Govan, 1990; Okawa et al., 2002, 2003) and droplet diameter (McCoy and Hanratty, 1977). To consider these, many researchers have proposed the deposition coefficient as an empirical correlation between droplet diameter and droplet concentration. Some important deposition models are presented in Table 9.7.

9.9 Validation of constitutive equations

The V&V process of system TH codes is discussed in detail in Chapter 13. In this Section, the validation of constitutive equations is briefly presented.

A system TH code usually adopts several mechanistic models and constitutive equations to represent various two-phase flow phenomena that may be expected to occur during design basis accidents or anticipated transients in an LWR. For example, RELAP5/MOD3.3 adopts more than 200 correlations to represent two-phase flow and heat transfer processes (USNRC, 1998b). These correlations have crucial effects on the applicability and accuracy of the code, and thus, all correlations should be validated appropriately.

As concerns the validation of the correlations, it should be noted that the calculation of any two-phase flow process usually needs several relevant correlations. For example, to predict the subcooled boiling in a vertical pipe, three correlations are involved at least; wall heat partition, interfacial drag, and interfacial heat transfer.

Table 9.7 Droplet deposition correlations.

Authors	Deposition correlation
McCoy and Hanratty (1977)	$\frac{k_d}{u^*} = \begin{cases} 3.25 \times 10^{-4}(\tau^+)^2 & \text{for } \tau^+ \leq 22.87, \\ 0.17 & \text{for } 22.87 \leq \tau^+ \leq 8340, \\ 20.7(\tau^+)^{-0.5} & \text{for } \tau^+ > 8340. \end{cases}$ where $\tau^+ = \frac{D_d^2 u^{*2} \rho_g \rho_d}{18\mu_g^2}$, $u^* = j_g\sqrt{f/2}$, and $f = 0.0791/\mathrm{Re}_g^{0.25}$.
Kataoka and Ishii (1982)	$k_d = 0.022\rho_g j_g Re_g^{-0.25}\left(\frac{C}{\rho_g}\right)^{-0.26}$, where $We_g = \rho_g j_g^2 D_h/\sigma$ and $\mathrm{Re}_g = \rho_g j_g D_h/\mu_g$.
Hewitt and Govan (1990)	$k_d = k\sqrt{\frac{\rho_g D_h}{\sigma}}$, where $k = \begin{cases} 0.18 & \text{for } C/\rho_g \leq 0.3, \\ 0.083(C/\rho_g)^{-0.65} & \text{for } C/\rho_g > 0.3. \end{cases}$
Okawa et al. (2003)	$k_d = 0.0632\sqrt{\frac{\rho_g D_h}{\sigma}}\left(\frac{C}{\rho_g}\right)^{-0.5}$

This indicates the validation or assessment of a correlation cannot be completely separate but rather synthetic. In addition, we need to be aware that

- **(i)** Each correlation was developed to represent a single phenomenon based on a limited experimental data set. This leads to the *limited range* of its applicability.
- **(ii)** Code developers adopt one or several correlations (and combine them) to represent a phenomenon at *all possible flow conditions* that may be expected to occur during a transient. Interpolation of two or more correlations and, although not desirable, extrapolated use of a correlation might be sometimes inevitable.
- **(iii)** Code users use codes for their own purposes. When they calculate a specific transient or accident, the physical phenomena of great interest may change as time goes on (that is, as the transient advances). This means *the importance of a correlation changes according to the transient advancement.*

These features imply that constitutive equations should be validated in a synthetic approach. In this regard, the validation of constitutive equations is carried out in the framework of the code validation using separate effects tests (SET), where key two-phase flow and heat transfer phenomena are identified first, and thereafter, the code (or the correlations in practice) is systematically assessed against a set of relevant experimental data.

An internationally agreed-upon SET validation matrix for system TH codes has been established by the OECD/NEA Committee on Safety of Nuclear Installations (OECD/NEA, 1993; Aksan, 2008). The construction of such a matrix was an attempt to systematically collect the best sets of openly available test data for code validation, assessment, and improvement, as well as for quantitative code assessment. The experts identified 67 phenomena relevant to two-phase flow in LOCAs and transients in LWRs and set up a catalog of information sheets on experimental facilities as a basis for the selection of facilities and specific tests. The 67 phenomena involve 9 basic phenomena: evaporation due to depressurization, evaporation due to heat input, condensation due to pressurization, condensation due to heat removal, interfacial friction in vertical flow, interfacial friction in horizontal flow, wall-to-fluid friction, pressure drops at geometric discontinuities, and pressure wave propagation. All are related to the constitutive equations introduced in Section 9.1 through 9.6. The other 58 phenomena are related to component-specific two-phase flow phenomena. Finally, they established the SET cross-reference matrix for the developmental assessment of system TH codes, consisting of 2094 tests from 187 test facilities. The SET validation matrix for a recent CATHARE-2 version has $\sim$1000 tests from $\sim$50 test facilities.

Again, it should be noted that validating constitutive equations requires a large amount of work and should thus be carried out systematically.

9.10 Summary remarks

This chapter describes constitutive equations suitable for solving a set of UVUT balance equations and developing system thermal hydraulics. The constitutive equations are relevant for fluid-to-wall and fluid-to-gas interactions. Fluid-to-wall-related equations are also covered in Chapters 7 and 8.

The empirical (or semiempirical) nature of the equations results from the present overview, as well as the widespread effort needed for their development in terms of time (several decades) and the geographical origins of authors (from everywhere in the world where nuclear technology is exploited).

Selected issues connected with the development and application of constitutive equations and of interest for future investigations are summarized as follows:

(a) The approach of a two-phase multiplier for the prediction of distributed friction pressure drops in two-phase flow may benefit from deep revision where fundamental terms like τ_i and τ_w are used to build a mechanistic model with a reduced (i.e., compared with current models) impact/role for empirical constants.

(b) A unified multidimensional convection heat transfer surface should be constructed where, in addition to the established heat transfer regimes, phenomena or aspects (not an exhaustive list) like reflood, radiation heat transfer, and transient boiling results of thermal power pulse are integrated.

(c) The development of terms including first- and second-order (and eventually higher) derivatives of unknowns should be promoted, eventually showing their importance (or their low impact) for solving UVUT equations, including the hyperbolicity issue.

(d) An attempt should be made to connect flow regimes to the concept of virtual mass.

(e) The pressure drop at geometric discontinuities (or abrupt area change) should also be considered a part of constitutive equation packages.

(f) A special effort should be made to integrate constitutive equation modeling and water properties modeling and formulation, with specific attention given to derivatives that may affect the numerical solution of algebraic UVUT models; the sophistication in computing water properties should be connected with the structure and validation of constitutive equations.

(g) It should be noted that the constitutive equations for the 3D module of system codes are not separately prepared in 3D analysis. Most are not truly 3D but rather are 3D extensions of 1D models. Thus, it should be noted that they work appropriately only when used at a system scale.

Exercises and questions

Exercises and questions are proposed hereafter with solutions given in Appendix 1 (Volume III). Answers and solutions are expected to be provided by readers before reading Appendix 1. Please provide extended answers.

9.1 Provide your own definition of constitutive equations and closure equations (or closure laws).

9.2 Discuss the meaning of "unknowns" in constitutive equations (associated with the unknowns of PDEs where constitutive equations are embedded).

9.3 Discuss the connection between the Jacobian and the formulation of constitutive equations when the implicit solution method is pursued to solve the PDE (including constitutive equations).

9.4 Distinguish (and discuss) between drift flux and slip ratio. Shortly describe the origin of drift flux.

9.5 What is the connection between constitutive equations and flow regimes (or flow patterns) for two-phase water flows?

9.6 Report typical values of interfacial areas (or better ranges of areas) per unit volume (after fixing a value for the diameter in a cylindrical pipe) for bubbly flow, slug flow, churn flow annular flow, and dispersed flow at a fixed pressure (or a pressure range you may select), distinguishing (if applicable) horizontal and vertical flows. Please also add the case of stratified flow in horizontal cylindrical channels.

9.7 Please discuss the "repartition" problem for constitutive equations (i.e., application in UVUT models). Please also discuss the complexity of "repartition" in the case of 3D modeling.

9.8 What is the connection (if any) between constitutive equations (and closure laws) and the hyperbolicity of the solution of PDEs?

9.9 Is it reasonable to identify the k-ε models [adopted in computational fluid dynamics codes] as constitutive equations? Please discuss why and identify the differences between the k-ε models and constitutive equations discussed in this chapter.

9.10 Discuss the meaning of scaling (and scaling capability) in closure laws.

List of symbols and abbreviations

Part I—Symbols

Symbols latin

$\boldsymbol{A}$ surface (m^2)
$\boldsymbol{A}$ cross-sectional area of a duct (m^2)
$\boldsymbol{a_i}$ interfacial area concentration (m^{-1})
$\boldsymbol{C}$ wall contour in a duct cross section
$\boldsymbol{C_s}$ pressure wave (sound) celerity (m/s)
$\boldsymbol{Cp}$ heat capacity at constant pressure (J/kg/K)
$\boldsymbol{C_d}$ drag coefficient
$\boldsymbol{C_f}$ friction coefficient
$\boldsymbol{C_i}$ interface contour in a duct cross section
$\boldsymbol{C_k}$ wall contour in a duct cross section in contact with phase k
$\boldsymbol{C_a}$ void wave celerity (m/s)
$\boldsymbol{C\mu}$ turbulent viscosity coefficient
$\boldsymbol{D}$ diameter (m)
$\boldsymbol{D}$ mass diffusivity (m^2/s)
$\boldsymbol{D_h}$ hydraulic diameter (m)
$\overrightarrow{F}$ specific external forces (usually limited to gravity: $\overrightarrow{F} = \rho \overrightarrow{g}$) ($N/m^3$)
$\boldsymbol{F_{ik}}$ interfacial momentum transfer to phase k per unit volume (N/m^3)
$\overrightarrow{g}$ gravity acceleration vector (m/s^2)
$\boldsymbol{g_i}$ gravity acceleration vector components (m/s^2)
$\boldsymbol{g_z}$ gravity acceleration projected along $\overrightarrow{O}_z$ axis (m/s^2)
$\boldsymbol{G}$ mass flux (kg/m^2/s)
$\boldsymbol{G_k}$ mass flux of phase k (kg/m^2/s)
$\boldsymbol{h}$ specific enthalpy (J/kg)
$\boldsymbol{h_k}$ specific enthalpy of phase k (J/kg)
$\boldsymbol{h_{ksat}}$ specific saturation enthalpy of phase k (J/kg)
$\boldsymbol{h_{lv}}$ latent heat (J/kg)
$\boldsymbol{H_k}$ time- and/or space-averaged specific enthalpy of phase k (J/kg)
$\boldsymbol{j_k}$ local superficial velocity of phase k (m/s)
$\boldsymbol{J}$ diffusive flux
$\boldsymbol{J_k^t}$ turbulent flux of ψ_k
$\boldsymbol{J_k}$ area-averaged superficial velocity of phase k (m/s)
$\boldsymbol{k}$ turbulent kinetic energy (m^2/s^2)
$\dot{\boldsymbol{M}}$ mass flowrate (kg/s)
$\dot{\boldsymbol{m}}$ mass flux (kg/m^2/s)
$\dot{\boldsymbol{M}}_k$ mass flowrate of phase k (kg/s)

$\boldsymbol{M_{ik}}$ interfacial force per unit volume (N/m^3)
$\boldsymbol{M_{ik}^{D}}$ drag force per unit volume (N/m^3)
$\boldsymbol{M_{ik}^{L}}$ lift force per unit volume (N/m^3)
$\boldsymbol{M_{ik}^{AM}}$ added mass force per unit volume (N/m^3)
$\boldsymbol{M_{ik}^{TD}}$ turbulent dispersion force per unit volume (N/m^3)
$\vec{n}_k$ unit vector normal to interface-oriented outside phase k
Nu Nusselt number
$\boldsymbol{p}$ pressure (Pa)
$\boldsymbol{P}$ space-time averaged mixture pressure (Pa)
Pr Prandtl number
$\vec{q}$ heat flux vector (W/m^2)
q_j conduction heat flux vector components (W/m^2)
q_{ext} volumetric external energy source, e.g., due to radiation (W/m^3)
$\vec{q}_k$ heat flux vector in phase k (W/m^3)
q_{kj} conduction heat flux vector components in phase k (W/m^2)
q_{ik} heat flux per unit volume from interface to phase k (W/m^3)
q_{idl} heat flux per unit volume from interface to droplet (W/m^3)
q_{ivdl} heat flux per unit volume from vapor to interlace of droplet (W/m^3)
q_{icl} heat flux per unit volume from interface to continuous liquid (W/m^3)
q_{ivcl} heat flux per unit volume from vapor to interface of continuous liquid (W/m^3)
q_{kj}^{t} turbulent heat diffusion vector components in phase k (W/m^2)
q_w wall heat flux (W/m^2)
q_{wk} wall heat flux to phase k (W/m^2)
$\boldsymbol{Q}$ volume flowrate (m^3/s)
$\boldsymbol{Q_k}$ volume flowrate of phase k (m^3/s)
$\boldsymbol{Q_{ik}}$ interface-to-phase k interfacial energy transfer rate per unit volume (W/m^3)
Re Reynolds number
$\boldsymbol{R_k}$ surface fraction of phase k (in 1-D equations)
$\boldsymbol{R_{ij}}$ Reynolds stress tensor
$\boldsymbol{s}$ specific entropy (J/K/kg)
$\boldsymbol{t}$ time (s)
$\boldsymbol{T}$ temperature (K)
$\boldsymbol{T}$ stress tensor (KPa)
$\boldsymbol{u}$ specific internal energy (J/kg)
$\vec{v}$ velocity vector (m/s)
v_j velocity vector components (m/s)
$\vec{v}_k$ velocity vector in phase k (m/s)
$v_{k,j}$ velocity vector components for phase k (m/s)
$\vec{v}_{ni}$ component of velocity of interface normal to the interface
$\boldsymbol{V}$ volume (m^3)
$\boldsymbol{V}$ time- and/or space-averaged velocity (m/s)
$\boldsymbol{V_k}$ time- and/or space-averaged velocity of phase k (m/s)
$\boldsymbol{V_{idl}}$ interfacial axial velocity at continuous liquid interface (m/s)
$\boldsymbol{V_{icl}}$ interfacial axial velocity at continuous liquid interface (m/s)
$\boldsymbol{Y_k}$ mass concentration of species k

Symbols greek

α_k space-time fraction of phase k
β added mass coefficient
γ slip ratio
δ_{32} Sauter mean diameter
ε_k local time fraction of phase k
ε dissipation rate (m^2/s^3 = W/kg)
λ heat conductivity ($W/m^2/K$)
μ dynamic viscosity (kg/s/m)
ν kinematic viscosity (m^2/s)
Γ_{ik} interface-to-phase k interfacial mass transfer rate per unit volume ($kg/m^3/s$)
Γ_{vcl} interfacial mass transfer rate per unit volume from continuous liquid to vapor ($kg/m^3/s$)
Γ_{vdl} interfacial mass transfer rate per unit volume from dispersed liquid to vapor ($kg/m^3/s$)
Γ_d mass transfer rate per unit volume by droplet deposition ($kg/m^3/s$)
Γ_e mass transfer rate per unit volume by droplet entrainment ($kg/m^3/s$)
σ surface tension (N/m)
ρ density (kg/m^3)
ρ_k density of phase k (kg/m^3)
ρ_m mixture density (kg/m^3)
σ surface tension (N/m)
τ_{ij} viscous stress tensor components (Pa)
τ_{wk} wall friction force applied to phase k (Pa)
ϕ porosity
ϕ volumetric source of ψ
X_f fluid characteristic (or indicator) function
ψ specific quantity
X_k phase characteristic (or indicator) function

Other scripts

T viscous stress tensor components (Pa)

Subscripts and superscripts

AM relative to the added mass force
c relative to continuous field
cl relative to continuous liquid field (in a 3-field 2-phase flow of a fluid with its vapor)
dl relative to dispersed liquid field (in a 3-field 2-phase flow of a fluid with its vapor)
d relative to dispersed field
cl relative to continuous liquid
dl relative to dispersed liquid
D relative to drag force
f relative to liquid phase in a gas-liquid two-phase flow
g relative to gas phase in a gas-liquid two-phase flow
k relative to phase *k* or field *k*
i relative to interface

i ith component
j jth component
ki relative to phase k or field k at interface
k relative to wall
l relative to liquid phase (in a two-phase flow of a fluid with its vapor)
L relative to lift force
m relative to a mixture
sat relative to a saturation
S relative to a surface
t relative to turbulence
TD relative to turbulent dispersion force
v relative to vapor phase (in a two-phase flow of a fluid with its vapor)
x, y, z coordinates
z curvilinear abscissa

Double averaged quantities in single-phase flow

$\varrho \triangleq \widehat{\overline{\rho}} = \langle \overline{\rho} \rangle^S$ space-time averaged density

$\mathrm{P} \triangleq \widehat{\overline{p}} = \langle \overline{p} \rangle^S$ space-time averaged pressure

$\mathrm{V} \triangleq \check{\tilde{v_z}} = \frac{\langle \overline{\rho v_z} \rangle^S}{\langle \overline{\rho} \rangle^S} = \frac{\langle \overline{\rho} \tilde{v}_z \rangle^S}{\langle \overline{\rho} \rangle^S}$ mass-weighted space-time velocity

$\mathrm{H} \triangleq \overline{h} = \frac{\langle \overline{\rho h} \rangle^S}{\langle \overline{\rho} \rangle^S} = \frac{\langle \overline{\rho} \tilde{h} \rangle^S}{\langle \overline{\rho} \rangle^S}$ mass-weighted space-time enthalpy

Double averaged quantities in single-phase flow

$\varrho_k \triangleq \widehat{\overline{\rho_k}}^k = \frac{\langle \overline{X_k \rho_k} \rangle^S}{\alpha_k}$ space-time phase average density

$\varrho_m \triangleq \alpha_v \varrho_v + \alpha_l \varrho_l$ space-time mixture average density

$P_k \triangleq \widehat{\overline{p_k}}^k \triangleq \frac{\left[\langle \overline{X_k p_k} \rangle^S \right]}{\alpha_k}$ space-time phase average pressure

$P \triangleq \alpha_v P_v + \alpha_l P_l$ space-time mixture average pressure

$V_k \triangleq \overline{v_{zk}}^k \triangleq \frac{\langle \overline{X_k \rho_k v_{zk}} \rangle^S}{\langle \overline{X_k \rho_k} \rangle^S} = \frac{\langle \overline{X_k \rho_k v_{zk}} \rangle^S}{\alpha_k \hat{\overline{\rho k}}^k}$ mass-weighted space-time phase average velocity

$j_k \triangleq \overline{X_k v_{zk}} = \varepsilon_k \overline{v}_{zk}^k$ local lime phase averaged superficial velocity

$j_k \triangleq j_v + j_l$ local time mixture-averaged superficial velocity

$J_k \triangleq \langle j_k \rangle^S = \langle \varepsilon_k \overline{v}_{zk}^k \rangle^S$ space-time phase average superficial velocity

$J_k \triangleq J_v + J_l$ space-time mixture average superficial velocity

$v_k \triangleq \frac{J_k}{\alpha_k} = \frac{\langle \overline{X_k v_{zk}} \rangle^S}{\langle \overline{X_k} \rangle^S}$ space-time phase average velocity

$G_k \triangleq \alpha_k \varrho_k V_k = \langle \overline{X_k \rho_k h_k} \rangle^S$ phase mass flux

$G \triangleq G_v + G_l$ mixture mass flux

$H_k \triangleq \overline{\check{h}_k}^k = \frac{\langle \overline{X_k \rho_k h_k} \rangle^S}{\langle \overline{X_k \rho_k} \rangle^S} = \frac{\overline{\hat{\rho}_k h_k}^k}{\hat{\overline{\rho}}_k^k}$ mass-weighted space-time phase average enthalpy

$H_m \triangleq \alpha_v \frac{\varrho_v H_v + \alpha_l \varrho_l H_l}{\varrho_m}$ mass-weighted space-time mixture average enthalpy

$\widehat{H}_k \triangleq \frac{\langle \overline{X_k \rho_k v_{zk} h_k} \rangle^S}{\langle \overline{X_k \rho_k v_{zk}} \rangle^S} = \frac{\alpha_k \langle \overline{\rho_k v_{zk} h_k} \rangle^S}{G_k}$ mass flux-weighted space-time phase average enthalpy

$\widehat{H}_m \triangleq \frac{G_v \widehat{H}_v + G_l \widehat{H}_l}{G}$ mass flux-weighted space-time mixture average enthalpy

$x \triangleq \frac{\alpha_v \varrho_v}{\varrho_m}$ static quality

$X \triangleq \frac{G_v}{G}$ dynamic quality

$X_{th} = \frac{\widehat{H}_m - h_{lsat}}{h_{vl}}$ thermodynamic quality (or equilibrium quality)

Non-dimensional quantities

Bo	Bond number (Eötwos)	$Bo = \frac{g\Delta\rho L^2}{\sigma}$
Fr	Froude number	$Fr = \frac{v}{\sqrt{gD}}$
J_k*	non-dimensional superficial velocity of phase k (m/s^2)	$J_k* = \frac{\sqrt{\rho k} J_k}{\sqrt{g\Delta\rho D}}$
K_k	Kutateladze number for phase k	$K_k = \frac{\sqrt{\rho k} J_k}{(g\Delta\rho D)^{1/4}}$
Re	Reynolds number	$Re = \frac{vD}{\nu}$
We	Weber number	$We = \frac{\rho \Delta V^2 L}{\sigma}$

Operators

∇ gradient operator

$\nabla.$ divergence operator

$\nabla^2 = \Delta$ Laplacian operator

$\vec{A}\ \vec{B}$ dyadic product of vectors (=tensor)

$\vec{A} \cdot \vec{B}$ scalar product of vectors (=scalar)

A*:*B double-dot product of 2 tensors (=scalar)

$^t\boldsymbol{T}$ transpose tensor

Time averaging operators

$\bar{f} \triangleq \frac{1}{T} \int_T f\, dt$ time averaging

$\tilde{f} \triangleq \overline{\rho f} / \bar{\rho}$ mass-weighted lime averaging

$\bar{f}^k \triangleq \overline{X_k f_k} / \overline{X_k}$ phase time averaging

$\tilde{f}_k^k \triangleq \frac{\overline{X_k \rho_k f_k}}{\overline{X_k \rho_k}} = \frac{\varepsilon_k \overline{\rho_k f_k}}{\varepsilon_k \overline{\rho_k}^k} = \frac{\overline{\rho_k f_k}^k}{\overline{\rho_k}^k}$ phase mass-weighted time averaging

Space averaging operators

$\langle f \rangle^L \triangleq \frac{1}{L} \int_L f\, dl$ space averaging over one dimension

$\langle f \rangle^S \triangleq \frac{1}{A} \int_A f\, ds$ area averaging

$\langle f\rangle^{V} \triangleq \frac{1}{V}\int_{V} f\, dv$ volume averaging

$\langle f(x,y,z,t)\rangle^{\delta v} \triangleq \iiint_{R^3} g\left(x' - x, y' - y, z' - z, t\right)\cdot f\left(x', y', z' t\right)dx' dy' dz'$ volume filtering

$\check{f} \triangleq \frac{\langle \rho f\rangle^{S}}{\langle \rho\rangle^{S}}$ mass-weighted area averaging

$\widehat{f}_{k}^{k} \triangleq \frac{\langle X_k f_k\rangle^{S}}{R_k}$ phase area averaging

$\check{f}_{k}^{k} \triangleq \frac{\langle X_k \rho_k f_k\rangle^{S}}{\langle \rho_k\rangle^{S}}$ phase mass-weighted area averaging

Space-Time averaging operators

$\widehat{\overline{f}} \triangleq \langle \overline{f}\rangle^{S}$ space-time averaging (time-area averaging)

$\overline{f} \triangleq \frac{\langle \overline{\rho f}\rangle^{S}}{\langle \overline{\rho}\rangle^{S}}$ mass-weighted space-time averaging

$\widehat{\overline{f}}_{k}^{k} \triangleq \frac{\langle \overline{X_k f_k}\rangle^{S}}{\alpha_k}$ phase space-time averaging

$\overline{f}_{k}^{k} \triangleq \frac{\langle \overline{X_k \rho_k f_k}\rangle^{S}}{\langle \overline{X_k \rho_k}\rangle^{S}} = \frac{\overline{\widehat{\rho_k f_k}}^{k}}{\overline{\widehat{\rho}}_{k}^{k}}$ mass-weighted phase space-time averaging

Symbols

a thermal diffusivity, m^2/s; the sound velocity, m/s; eccentricity, m; gas absorption coefficient
A cross-section area of the channel, m^2; absorbance
c specific heat, J/(kg K)
c_p speed of light, m/s
D diameter (shell), m; fuel rod diameter, m; transmittance
d diameter (pipe), m
f frequency, s^{-1}
g gravitational acceleration, m^2/s
G mass flux (=ρw, kg/m^2/s) or mass flow (=ρwA, kg/s)
h height (ribs, channel, etc.), m; enthalpy, kJ/(kg)
j volumetric fiux, m/s
L, l length, m
m volume porosity; added mass, kg
M mass, kg
n number of tubes, ribs, pumps, ..., etc.
q' linear heat rate, W/m
q* or *q'' heat flux, W/m^2
q_v **or** ***q'''*** volumetric power density, W/m^3
Q volumetric flow, m^3/s; total energy, kJ
p pressure, Pa
P wetted perimeter, m; fuel rod pitch, m
r, R radius, m
R thermal resistance, (m^2K)/W; reflectance; gas constant
s spacing, m (of rod or pipe arrangement); slip ratio; cross sectional area, m^2
T temperature, K
v specific volume, m^3/kg

V volumetric flow, m^3/s
w velocity, m/s
x vapor quality (static; flowing); also relative enthalpy
Z axial coordinate

Greek

α Heat transfer coefficient, $W/(m^2K)$; kinetic energy coefficient
β thermal expansion coefficient (1/K); angle; condensation coefficient
δ thickness; height roughness, m
Δ equivalent roughness, m
Δh_{fg} latent heat (heat of evaporation), J/kg
ζ local resistance coefficient
ε emissivity
η efficiency, coefficient of performance
Θ ratio (R_1/R_2; d_1/d_2 T_w/T_0)
λ thermal conductivity, W/(m K); wave length, m
μ viscosity, Pa s
ν kinematic viscosity, m^2/s; frequency of light, 1/s
ρ density, kg/m^3
σ surface tension, N/m^2
σ_0 Stefan-Boltzmann constant, $W/(m^2K^4)$
τ shear stress, N/m^2; time, s
φ vapor volumetric fraction (void fraction); angle
ψ coefficient of two-phaseflow inhomogeneity
ξ accommodation constant (defined by Eq. (7.114))
ω cross-sectional area, m^2; angular frequency (2n/), 1/s

Subscripts, superscripts and indices

0 black body
c convection
g gas (vapor-phase)
ib, ob onset boiling, boiling beginning
max maximum
min minimum
r, R radiation
S, sat saturation or saturated
sub subcooled
ν vapor
λ conduction
$+$ nondimensional *av* quantity average
f fluid (liquid-phase)
hn homogeneous
in nucleation entrance
MFB minimum film boiling

out exit
red reduced
s surface
T turbulent
W wall; condensate film surface
– averaging sign (over-bar)

Symbols

$j_x = \frac{\dot{M}_x}{\rho_x A_{DC}} \frac{\rho_x^{1/2}}{[gW(\rho_w - \rho_s)]^{1/2}}$ Wallis number = dimensionless superficial velocity based on down-comer annulus circumference
$K_x = \frac{\dot{M}_x}{\rho_x A_{DC}} \frac{\rho_x^{1/2}}{[g\sigma(\rho_w - \rho_s)]^{1/4}}$ Kutateladze number = dimensionless superficial velocity based on surface tension
$\mathbf{A_{DC}}$ flow cross section of down-comer
a × 100 (%) probability (quantile) of output quantity or of the F-distribution
b × 100 (%) confidence level that the maximum code result will not be exceeded with the probability a × 100 (%) of the corresponding output distribution
Bo Bond number
$\mathbf{d_{av}}$ downcomer average diameter
$\mathbf{d_{outer}}$ downcomer outer diameter
Fr Froude number
g gravity
$\mathbf{k_1 = 2m+2}$ degree of freedom of the F-distribution
$\mathbf{k_2 = 2n-2m}$ degree of freedom of the F-distribution
L characteristic geometrical length
M mass flow rate
m fraction of fuel rods not meeting the acceptance criterion
n number of code calculation runs
N number of total fuel rods, or fuel rods with the highest power or stored heat in the core
p probability for no exceedance of the acceptance criterion
P = m/n fraction of exceedance of the acceptance criterion
$\mathbf{P_{u,95}}$ upper 95% confidence limit for fraction of exceedance P
$\mathbf{P_{l\ 95}}$ lower 95% confidence limit for fraction of exceedance P
W average down-comer annulus circumference
Y(1…n) ordered output parameters of n calculation runs
$\Theta_{\mathbf{ECC\text{-}BCL}}$ angle between ECC injection leg and broken loop
ν kinematic viscosity
ρ density
σ surface tension

Subscripts

g vapour
l liquid
s steam
w water
x either steam or water

Symbols

a, b, c Steinhart-Hart parameters: $a = 1.40 \times 10^{-3}$, $b = 2.37 \times 10^{-4}$, $c = 9.90 \times 10^{-8}$
A cross-sectional area
$\boldsymbol{A_P}$ frontal projected area
AC transient or alternating current
B magnetic field vector
c specific heat, sound speed
$\boldsymbol{c_D}$ friction coefficient
C coefficient, capacitance
d, D diameter
Dk set of internal points of k-phase domain (k = G [gas] or L[liquid]) identified by position
DC direct-current
E potential
E electric field vector
Δf receiver frequency-shift
F frequency
F force vector
g gravitational acceleration
G electrical conductance (G = 1/R)
h heat transfer coefficient, height
I electric current
j electric current
k (k_I, k_R, k_T) are incident, reflected and transmitted (refracted) electromagnetic wave vectors
k thermal conductivity, first-order temperature coefficient of resistance
K rigidity constant, factor which takes into account the sensor volume and geometry
L capillary tube length, wire length, latency length ($L = u\Delta t_1$)
m, $\boldsymbol{m_P}$ mass
$\dot{m}$ mass flow rate
Δm mass balance
$\mathbf{M_A}$ atomic mass
M mass flow rate vector
n number of nuclei per unit volume
$\boldsymbol{n_1}$, $\boldsymbol{n_2}$ refractive indexes of the medium 1 and 2 (in vacuum $n = 1$)
N number of entering particles per unit time, N = Φ A
$\mathbf{N_{AV}}$ Avogadro's number
Nu Nusselt number, Nu $= \dot{q}D/[k(T_w - T_\infty)]$
p pressure
$\mathbf{P_0}$ reference pressure
$\mathbf{P_a}$ applied pressure
Δp differential pressure
$\mathbf{P_k}$ phase density function (k = G [gas] or L[liquid])
q electric charge
$\dot{q}$ heat flux
Q applied heat
r resistivity
R rotating pipe-arm length, bend radius, resistance
ΔR change in resistance

R_0 maximum diameter of the float
$R(z_0)$ conical tube radius
Re Reynolds number, Re = uD/ν
t time
Δt period
S, S1, S2 threshold levels in Fig. 17.5
S(T), S± temperature-dependent material property (Seebeck coefficient)
T observation time, temperature
T_w wall temperature
T_∞ ambient temperature
T_{meter} temperature obtained by meter
T_{ref} reference junctions' temperature
T_{sense} target temperature by sensing
ΔT temperature difference
∇T gradient of temperature
u, u velocity, velocity vector
V volume, voltage
$\dot{V}$ volumetric flow rate
∇V gradient of voltage
w wall thickness
x position vector
X imaginary part the reactance
z elevation of the float
Z electrical impedance

Greek

α temperature coefficient of resistance
α, α_n, α^k void fraction, n-dimensional void fraction, k-dimensional double-averaged void fraction
α_T temperature coefficient of resistance
β wire resistivity variation
ε emissivity
θ angle
λ interaction length
μ dynamic viscosity
ν kinematic viscosity
ξ_G, ξ_L dielectric constants of gas (G) and liquid (L)
ρ density
σ cross-section of the interaction, Stefan-Boltzmann constant
ϕ phase
Φ particle flux
ω electromagnetic wave frequency
ω vorticity vector
Ω angular velocity

Part II—Abbreviations/Acronyms

0D or 0-D zero-dimensional
1D or 1-D one-dimensional
2D or 2-D two-dimensional
3D or 3-D three-dimensional
AA accident analysis, or average accuracy
ABB Asea Brown Boveri
ABWR advanced boiling water reactor (BWR)
AC alternate current
ACC accumulator
ADI alternating direction implicit (numerical solution method)
ADIFOR Automatic differentiation of Fortran
ADS automatic depressurization system
AEA / AEAT Atomic Energy Agency / Technology (UK)
AEC Atomic Energy Commission (see USNRC)
AECL Atomic Energy of Canada Limited
AEKI Atomic Energy Agency (Hungary)
AERB Atomic Energy Research Body in India
AES VVER-1000 type advanced reactor
AESJ Atomic Energy Society of Japan
AFW auxiliary feed-water
AI artificial intelligence
AIAA American Institute of Aeronautics and Astronautics
AIRS Advanced Incident-Reporting System
ALARA as low as reasonably achievable
ALE Arbitrary Eulerian-Lagrangian
ALWR advanced light water reactor
AM accident management
AM-1 Atom Mimy (peaceful atom)—First Nuclear Reactor in Soviet Union
AMD computer identification
AMG Accident Management Guidelines
AMP accident management procedures
ANE Annals of Nuclear Energy (Journal)
ANL Argonne National Laboratory
ANS American Nuclear Society
ANSI American National Standards Institute
ANSYS see CFX
AOA axial offset anomaly
AOC agent of change (in scaling)
AOO anticipated operational occurrence
AP-1000 advanced PWR, 1000 MWe (see also AP1000) by Westinghouse
AP-600 advanced PWR, 600 MWe (see also AP600) by Westinghouse
APEX name of ITF in the United States
APEX-CE Advanced Plant Experiment – Combustion Engineering
APR1400 advanced pressurized water reactor (KEPCO design)
APRM Average Power Range Monitor
APROS SYS TH code developed in Finland

APS American Physical Society
ARD advanced reactor designs
ARN Autoridad Regulatoria Nuclear (in Argentina)
AS accident scenario
ASCLIM Consortium in EU dealing with CFD
ASFRE computer code (analysis of sodium flow in reactor elements) developed in Japan
ASME American Society of Mechanical Engineering
ASTAR Consortium in EU dealing with CFD
ASTM American Society for Testing and Materials
ATF accident tolerant fuel
ATHLET SYS TH code developed in Germany
ATI axial turbulence intensity
ATLAS name of ITF, PWR simulator in Korea
ATWS anticipated transient without scram
AWCR advanced water-cooled reactor
B boron, or bubbling
B & W Babcock and Wilcox
BAF bottom of active fuel
BC boundary condition, or bubble condenser (containment type) facility in Russia
BCP bottom core plate
BD blow-down (phase of a LOCA), or Russian Test Facility
BDBA beyond DBA
BDEE beyond design basis external event
BD-SAT saturated blow-down
BD-SBC sub-cooled blow-down
BE best estimate—also identification acronym for a series of conferences
BEAU best estimate Analysis and Uncertainty
BEB big encapsulating bubbles
BEEB big encapsulating elongated bubbles
BEM boundary element method
BEMUSE CSNI Project, Best Estimate Methods – Uncertainty and Sensitivity Evaluation
BEPU best estimate plus uncertainty
BETHSY name of ITF in France, PWR simulator
BF bubble flow
BFBT BWR full-size fine mesh bundle test
BFBT name of SETF in Japan
BFC Battelle-Frankfurt Containment experiment
BFECC back-and-forth error compensation and correction
BIC boundary and initial conditions
BI-CG see CG and Section 20.3.5.6
BI-CGSTAB BI-CG stabilized
BL broken loop
BMFT Federal Ministry for Research and Technology of the Federal Republic of Germany
BNL Brookhaven National Laboratory
BOC beginning of cycle and bottom of core (elevation of FA hydraulic entrance)
BOL beginning of life
BoP (or BOP) balance of plant
BOT Break Opening Time
BPG Best Practice Guidelines
BR broken (loop), see also BL

BSL baseline (also Standard "Menter baseline two-equation model", Menter, 1994)
BT Boiling Transition
BU bottom up
BWR boiling water reactor
BWRS BWR stability
C chugging, or cylindrical (Containment Shape)
CA compensating action
CANDU reactor designed by AECL in Canada, Canadian deuterium uranium reactor
CASL Research Project financed by DOE in the US
CATHARE SYS TH code developed in France
CBA PIE used for structure mechanics assessment
CCF counter current flow
CCFL countercurrent flow limitation
CCTF name of large-scale SETF in Japan
CCVM Computer (or CSNI) Code Validation Matrix
CCW Component Cooling Water
CD core damage
CD-Adapco see STAR-CCM+
CDF core damage frequency
CE Combustion Engineering
CEA Commissariat a L'Energie Atomique (France)
CERTA consolidation of integral system test experimental databases for reactor thermal-hydraulic safety analysis, EC project
CET Core Exit Thermocouples, or Combined Effect Test
CFD computational fluid dynamics
CFD4NRS Working Group for CFD (and set of activities / conferences) at OECD/NEA/CSNI
CFDS see FLOW 3D
CFL Courant Friedrichs Lewy
CFR Code of Federal Regulation
CFX CFD code
CG conjugate gradient (numerical algorithm)
CGNE conjugate gradient on the normal equations to minimize the error
CGNR conjugate gradient on the normal equations to minimize the residual
CGS CG squared
CHF critical heat flux
CHT conjugate heat transfer
CIAU code with capability of Internal Assessment of Uncertainty
CILC crud-induced localized corrosion
CIPS crud-induced power shift
CIWH condensation-induced water hammer
CJ conical jet
CL cold leg
CLL collapsed liquid level
CMFD computational multi-fluid dynamics
CMMI capability maturity model integration
CMT core makeup tank
CNRA NEA Committee on Nuclear Regulatory Activities
CO containment, or condensation oscillations
COBRA subchannel computer code developed in USA
CON Concrete (Containment Pressure boundary type)

CONT containment
CPH challenging phenomena
CPU central process unit
CR control rod
CRE control rod ejection
CRGT control rod guide tube
CRP Coordinated Research Project (inside IAEA)
CRUD Chalk River unidentified deposit
CS critical section
CSA (set of) PIE used for thermal hydraulics safety assessment
CSAU code scaling applicability and uncertainty
CSNI Committee on the Safety of Nuclear Installations
CST condensate storage tank
CSTF Containment Systems Test Facility
CT computerized tomography, or Counterpart Test
CT also Clad Temperature
CTF COBRA-TF
CV control volume
CVCS chemical and volume control system
CV & J control volume and (or plus) junction
CVTR Carolinas Virginia Tube Reactor Containment
D diameter, or droplet
DA discrete approximation, difference approximation
DBA design basis accident
DC direct current, or down-comer, or dual cycle (-BWR)
DCC direct contact condensation
DDES delayed DES
DDET Deterministic Dynamic Event Tree
DEC design extended conditions (-A or -B)
DEGB double-ended guillotine break
DEHL double-ended hot leg (break)
DES detached eddy simulation
DF design factor
DFFB dispersed flow film boiling
DG diesel generator
D-HTC droplet HTC
DiD defense in depth
DJ divergent jet
DM design margins
DMM dynamic memory management
DNB departure from nucleate boiling
DNBR DNB ratio
DNS direct numerical simulation
DOE Department of Energy (see US DOE)
DP differential pressure
DPA distributed parameter approach
DPV depressurization valve
DPw Decay Power
DR decay ratio
DRM distributed resistance model

DSA deterministic safety assessment (or analysis)
DSS dynamical system scaling
DTM distributed thermal model
DV designer and vendor (of nuclear power plants)
DVI Direct Vessel Injection (line)
DWI density wave instability
DWO density wave oscillation
E Elliptical (Containment Dome Shape)
EARSM explicit algebraic RSM
EASM explicit algebraic stress model
EB elliptic blending (related to RSM)
EBR-1 experimental breeder reactor—First Nuclear Reactor in the United States
EC European Commission
ECC emergency core cooling
ECC(S) emergency core cooling (system)
ECCI ECC injection
ECCS emergency core cooling system
ECDB external chugging with detached bubble
ECEB external chugging with encapsulating bubble
ECLN Equal CL number
ECORA consortium in EU dealing with CFD FLOWMIX-R
ECT electrical capacitance tomography
EDB Experimental Data Base
EDF (or EdF) Electricite de France, nuclear utility in France
EFW emergency feed-water
EI electrical impedance
EIT electrical impedance tomography
EJ ellipsoidal jet
ELES embedded LES
ELN Equal Loop Number
EM electromagnetic, or evaluation model
EMDAP Evaluation Model Development and Assessment Process
ENRL Equal Number Recirculation Loop
ENS European Nuclear Society
ENUSA Empresa Nacional del Uranio, SA (Spain)
ENV environment
EOB ellipsoidal oscillatory bubble
EOC end of cycle
EOHL end of heated length
EOP emergency operating procedure
EoS equation of state
EP electrical probe
EPR European Pressurized Water Reactor (AREVA / Framatome design)
EPRI Electrical Power Research Institute (in USA)
EPS emergency power system
ERCOFTAC Network of CFD specialists (see corresponding website)
ERDA Energy Research and Development Administration, see USDOE
ERI electrical resistivity imaging (see also ERT)
ERT electrical resistivity tomography, or Emergency Rescue Team
ERV electromagnetic relief valve

ERVCS External Reactor Vessel Cooling System
ESBWR economic simplified boiling water reactor designed by General Electric
ESF engineered safety features
ESP electric submersible pump
ET event tree
ETFS Experimental Thermal and Fluid Science (Journal)
ETN nuclear utility in Brazil
EU European Union
EUBORA consortium in EU dealing with CFD
EUR European Utilities Requirements
EUROFASTNET EC network for planning TH project
EVET equal velocities equal temperatures
EVETUG EVET and unequal Gibbs free enthalpy
EVM eddy viscosity model
EVUTUP equal velocities, unequal temperatures, unequal pressures
EVVD equal-volume-exchange turbulent mixing and void drift (model)
F Force, or Flat (Containment Dome Shape)
FA fuel assembly, or fine allocation
FAD fuel assembly distortion
FC fuel channel, or forced convection
FCB fuel channel blockage
FCM Fractional Change Metric
FCO forced convection
FEP fluorinated ethylene propylene
FESTA Facility for Experimental Simulation of Transients and Accidents in Korea (namely SMART-Integral Test Loop)
FFRD fuel fragmentation, relocation and dispersal
FFTBM Fast Fourier Transform Based Method
FH Full Height
FINAS Fuel Incident Notification and Analysis System
FIST name of ITF in the United States, BWR simulator (Full Integral Simulation Test)
FIV flow induced vibration
FIX-II name of ITF in Sweden, BWR simulator
FLOW 3D CFD code
FLOWMIX-R consortium in EU dealing with CFD
FLOWNET network of CFD specialists (see corresponding website)
FLUENT CFD code
FNPP floating NPP
FoM or FOM figure of merit
FONESYS network of SYS TH code developers (see the web)
FP fission products or fission power
FPr full pressure
FPw Full Power
FRC Fractional Rate of Change
FSA Fractional Scaling Analysis
FSAR Final Safety Analysis Report
FSI fluid structure interaction
FSM flow simulation methodology
FSTF Full-Scale test facility
FT fault tree

FTB fine transfer block
FTHEL test facility in Korea connected with SMART reactor design
FTUS upwind space finite difference (numerical method)
FW feed-water
FWIV feed-water isolation valve
FWLB feed water line break
GBY bypass flow-rate
GC core flow-rate
GDCS gravity-driven cooling system
GDF general design framework
G-DRM grid DRM
G-DTM grid DTM
GE General Electric
GERDA Geradrohr Dampferzeuger Anlage test facility in Germany
GF gravity feed
GFA fuel assembly flow-rate
GFR gas fast reactor
GGDH general gradient diffusion hypothesis
GIRAFFE name of ITF in Japan (Gravity-Driven Integral Full Height Test for Passive Heat Removal)
GIS grid-induced separation
GIST GDCS (Gravity Driven Cooling System) Integral System Test
GL liquid flow-rate (from SEP)
GMRES (numerical) algorithm defined in Section 20.3.5.5
GRNSPG Nuclear Research Group in San Piero a Grado (of University of Pisa)
GRS Gesellschaft für Anlagen- und Reaktorsicherheit (Germany)
GS steam flow-rate (from SEP)
GT-MHR gas-turbine modular high temperature reactor
GTRF grid to rod fretting
GUI graphical user interface
GUM guide to the expression of uncertainty
H hydrogen, or Hemispherical (Containment Dome Shape)
HA hydro accumulators
H2TS Hierarchical two-Tiered Scaling
HBO hemispherical bubble oscillation
HDR HeiBdampfreaktor, name of large-scale SETF in Germany
HEBR header break
HEM homogeneous equilibrium model
HEU highly enriched uranium
HF heat flux
H-F Henry-Fauske (TPCF model)
HFP hot full power
HL hot leg
H&M heat and mass
HOSG horizontal steam generator
HPC high performance computing
HPCI high-pressure coolant injection
HPCS high-pressure core spray
HPI high-pressure injection

HPIS high pressure injection system
HPLWR high performance light water reactor
HPSI high pressure safety injection
HRA human reliability analysis
HT heat transfer
HTA heat transfer area
HTC heat transfer coefficient
HTD heat transfer deterioration
HTGR high-temperature gas reactor
HTIR high temperature irradiation resistant (thermocouple)
HTS heat transfer surface
HWA hot wire anemometer
HWR heavy water reactor
HX heat exchanger
HZP hot zero power
I & C instrumentation and control
I imposed (related to TSE), or initial, or Integral design
I&BC Initial and Boundary Conditions
I/O Input/Output
IA interfacial area (concentration)
IAC Interim Acceptance Criteria, or interfacial area concentration
IADF inverted annular droplet field
IAEA International Atomic Energy Agency
IAFB inverted annular film boiling
IAT Interfacial area transport
IB-LOCA intermediate break LOCA (or IBLOCA)
IBM immersed boundary method
IC initial condition, or isolation condenser, or internal chugging
ICM interfacial capturing method (or model)
ICS Isolation Condenser System
IDDES improved DDES
IDPSA Integrated Deterministic and Probabilistic Safety Assessment
IEC International Electrotechnical Commission
IEEE Institute of Electrical and Electronics Engineers
IET Integral Effect Test (facility), see also ITF and IETF
IETF IET Facility
IHE intermediate heat exchanger
IJMF International Journal on Multiphase Flow
IL intermediate leg
ILOCA interfacing LOCA
INES International Nuclear and Radiological Event Scale
INKA Integral Test Facility Karlstein, in Germany
IOC interface oscillation condensation
IPA integral parameter
IPSN Institut de Protection et de Sûreté Nucléaire (France)
IRIDM Integrated Risk Informed Decision Making
IRS International Reporting System for Operating Experience
IRSN Institute for Radiological Protection and Nuclear Safety in France
IRSRR Incident-Reporting System for Research Reactors

IRT interface reconstruction technique
IRWST In-containment Refueling Water Storage Tank
ISB name of ITF in Russia, VVER-1000 simulator
ISB-VVER Russian Test Facility (see also ISB)
ISFB inverse slug film boiling
ISO International Organization for Standardization
ISP International Standard Problem
IST Integrated Systems Test (IST) Facility
IT Integral Test
ITDB Incident and Trafficking Database
ITF Integral Test Facility, see also IET and IETF
ITM interfacial (or interface) tracking model (or method)
ITS interface tracking simulation, or international temperature scale
IUQ input uncertainty quantification
J jetting
JAEA Japan Atomic Energy Agency
JAERI Japan Atomic Energy Research Institute
JHT Journal of Heat Transfer (of ASME)
J-HTC Jet HTC
JNES Japan Nuclear Energy Safety Organization
JP jet pump
JSME Japan Association of Mechanical Engineering
KAERI Korea Atomic Energy Research Institute
KEPRI Korean EPRI
KfK Research institute at Karlsruhe in Germany
K-H Kelvin Helmholtz
KINS Korean Institute for Nuclear Safety
k-kL two-equations RANS model (L is the integral scale length of turbulence), Rotta, (1972)
KMS Russian Test Facility
KNS Korean Nuclear Society
KORSAR SYS TH code developed by Russia
KPD coefficient for pressure drop at geometric discontinuity
KREM KEPRI realistic evaluation model
KWU KraftWerk Union (NPP designer Company)
LA level 'A' (code-user qualification)
La-C Large chugging / intermittent growth
LANL Los Alamos National Laboratory
LB Large Break or Level 'B' (code-user qualification)
LBLOCA large break LOCA
LBM lattice Boltzmann method
LC Level 'C' (code-user qualification)
LCC/SW loss of component cooling service water
LCSP lower core support plate
LDA laser Doppler anemometer
LDV laser Doppler velocimetry
LEIS large-eddy and interface simulation
LES large eddy simulation
LEU low enriched uranium
LEVM linear eddy viscosity model

LHGR linear heat generation rate
LHS Latin hypercube sampling, or left-hand side
LIME lightweight integrating multi-physics environment
LLM logarithmic law mismatch
LMFBR liquid metal fast breeder reactor
LMFR liquid metal fast reactor
LOBI name of ITF of EC installed in Italy, PWR simulator (Loop for Blow-down Investigation or Loop for Off-normal Behavior Investigations
LOCA loss of coolant accident
LOFA loss of flow accident (also MCP-trip)
LOFT name of ITF in United States, PWR simulator (Loss-of-fluid-test)
LOFW loss of feed-water
LOOSP loss of on-site and off-site power (or SBO plus DG failure)
LORHR loss of residual heat removal
LP lower plenum
LPA lumped parameter approach
LPCI low-pressure coolant injection
LPCS low-pressure core spray
LPI low-pressure injection
LPIS low-pressure injection system
LPRM local power range monitor
LRR Launder-Reece-Rodi (related to RSM)
LS loop seal
LSC loop seal clearing (phenomenon)
LSTF name of ITF in Japan, PWR simulator
LUT look-up tables (for CHF), or Lappeenranta University of Technology
LvK von-Karman length scale
LWR light water-cooled reactor
M Multi-compartment (Containment)
MANIVEL test facility (SETF type) used for CFD benchmark
MARNET network of CFD specialists (see corresponding website)
MARS SYS TH code developed in Korea
MARVIKEN name of large-scale SETF in Sweden
MASLWR Multi-Application Small Light Water Reactor, test facility in USA
MBLOCA medium break LOCA, see also IBLOCA
MCNP computer code (radiation transport) developed in USA
MCP main coolant pump
MCPR minimum critical power ratio
MCR main control room
MD-HTCS multi-dimensional heat transfer coefficient surface
MFB minimum film boiling
MHI Mitsubishi Heavy Industry
MI mass inventory
MILES monotone integrated LES
MIST name of ITF in the United States, PWR simulator, Multi-loop Integral System Test Facility
MISTRA Mitigation and Stratification facility in France
MIT Massachusetts Institute of Technology
MLHGR Maximum Linear Heat Generation Rate
MME mass, momentum and (thermal) energy

MMS method of manufactured solution
MOC middle of cycle, or method of characteristics
MOX mixed oxide (uranium and plutonium) nuclear fuel
MPI message passing interface
MPTR multiple pressure tube rupture
MRCP Main Reactor Coolant Pump
MSD modeled stress depletion
MSIV main steam isolation valve
MSLB main steam line break
MSM momentum source model
MSSV main steam safety valves
MVG mixing vane grid
NASA nuclear utility in Argentina
NB (saturated) nucleate boiling (see also SANB)
NC natural circulation, or no condensation, or nuclear core
nc non-condensable
NCFR natural circulation flow regime
NCG Non Condensable Gas
NCO natural convection (HT regime)
NDP non dimensional parameter
NEA Nuclear Energy Agency of OECD
NEB non encapsulating bubbles
NED Nuclear Engineering and Design (Journal)
NEPTUNE CFD code developed in France
NET Nuclear Energy and Technology (Journal)
NEWS Nuclear Events Web-based Systems
NIKIET NPP Design Institution in Russia
NISA Nuclear and Industrial Safety Agency
NIST NuScale integral system test (facility in USA)
NK neutron kinetics
NLEVM nonlinear eddy viscosity model
NMED Nuclear Material Events Database
NNC non nuclear core
NP Nominal power or neutron physics
NPARC network of CFD specialists (see corresponding website)
NPD nuclear power demonstration
NPP nuclear power plant
NPSH net positive suction head
NQA nuclear (grade) quality assurance
NRC Nuclear Regulatory Commission, see USNRC
NRI Nuclear Research Institute (Czech Republic)
NRS nuclear reactor safety
NRSR nuclear reactor safety and risk
NS, or N-S Navier-Stokes
NSC Nuclear Science Committee of NEA
NT Nuclear Technology (Journal of ANS)
NTC negative temperature coefficient
NTF nuclear thermal fluid
NTH nuclear thermal hydraulics
NURESAFE EC project follow-up of NURISP

NURESIM EC planning project based on EUROFASTNET
NURETH Nuclear Reactor Thermal-hydraulics (series of ANS conferences)
NURISP EC project follow-up of NURESIM
NVFP nonvolatile fission products
NVG net vapor generation
OB oscillatory bubble
OCJ oscillatory cone jet
ODE ordinary differential equations
OE onset of entrainment
OECD Organization for Economic Cooperation and Development
OI oscillation interface
OJ oscillatory jet
OMEGA test facility (SETF type) also used for CFD benchmark
ONB onset of nucleate boiling
OOP object oriented programming
ORNL Oak Ridge National Lab
OSU Oregon State University
OSV onset of significant void
OTIS Once-Through Integral System
OTSG once through steam generator
OUR origin of unreliability
P pressure
PACTEL name of ITF in Finland (originally, VVER-440 simulator), Parallel Channel Test Loop
PAHR post-accident heat removal
PANDA name of large-scale SETF in Switzerland (Passive Nachwärmeabfuhr- und Druckabbau-Testanlage
PCCS passive containment cooling system
PCEI parallel channel effects and instabilities
PCG preconditioned conjugate gradient
PCI pellet clad interaction
PCIS primary containment isolation system
PCMI pellet clad mechanical interaction
PCMWT passive core make-up tank
PCT peak clad temperature
PCV primary containment vessel
PD pressure drop
PDA phase Doppler anemometer, particle dynamics analysis
PDDM phase Doppler Difference Method
PDE partial differential equation
PDF Probability Density Function
PDGD pressure drop at geometric discontinuity
PDI phase Doppler Interferometry
PDPA phase Doppler Particle Analysis
PET positron emission tomography
PH phenomenon, or maximum impact overpressure
Ph.W phenomenological windows, or PHW
PHWR pressurized heavy water reactor (Atucha II) designed by KWU (G)
PIE postulated initiating event
PIPER-ONE name of ITF in Italy, BWR simulator

PIRT phenomena identification and ranking table
PIUS process intrinsic ultimate safety (reactor design concept)
PIV particle image velocimetry
PKL name of ITF in Germany, PWR simulator, (Primärkreislauf)
PLIF planar laser-induced fluorescence
PM probability of mission (for TPS)
PM-5 Russian Test Facility
PMF porous media formulation
PMK name of ITF in Hungary, VVER-440 simulator (also PMK-2)
Pn non condensable gas pressure
PNE Progress in Nuclear Energy (Journal)
PORV pilot-operated relief valve
PRA probabilistic risk assessment
PREMIUM project in area of uncertainty launched by CSNI
PRHR passive residual heat removal
PRISE primary to secondary leakage (in VVER)
PRZ pressurizer
PS primary side (of RCS) or primary system
PSA probabilistic safety assessment (or analysis)
PSB name of ITF in Russia, simulator of VVER-1000
PSBT name of SETF in Japan, PWR Subchannel and Bundle Tests
PSB-VVER Russian Test Facility, see also PSB
PSD power spectral density
PSI Paul Scherrer Institute (Switzerland)
PSP pressure suppression pool
PT pressure tube
PTC positive temperature coefficient
PTS pressurized thermal shock
PTV particle tracking velocimetry
PUMA name of large-scale SETF in the United States (addressing SBWR simulation), Purdue University Multi-dimensional Integral Test Assembly
PV pressure vessel
PW pressure wave
PWR pressurized water reactor
PWR-O PWR equipped with OTSG
PWR-V PWR equipped with HOSG (VVER type)
QA quality assurance, or qualitative analysis
QAM Quantity Accuracy Matrix
QCV quick-closing valve
QF quench front
QFP QF propagation
Q-factor quality factor of a resonant system
QNET-CFD network of CFD specialists
Q-SM quantity for (characterizing) safety margins
R & D research and development
R resulting (related to TSE), or condensation potential
RA relevant aspects, or regulatory authority
RANS Reynolds average Navier-Stokes
RB reactor building or radius of a bubble (also R_B)

RBMK Graphite-moderated, Boiling Water-Cooled, Channel-Type Reactor, NIKIET designer in Russia
RBMT rod bundle mixing tests
RC reactor cavity
RCA PIE used for assessment of radiological releases
RCIC Reactor Core Isolation Cooling
RCL reactor coolant loop (for ITF)
RCP reactor coolant pump
RCS reactor coolant system
RD-14M name of ITF in Canada, CANDU simulator
RE reflood (phase of a LOCA)
RELAP SYS TH code developed in the United States
REV representative elementary volume
REWET SETF in Finland
RF refill (phase of a LOCA)
RG Regulatory Guide (of USNRC)
RH Reduced Height
RHR residual heat removal
RHS right-hand side
RIA reactivity initiated accident
RIDM risk informed decision making
RIP internal recirculation pump
RL Recirculation Loop
RL recirculation loop
RLN Reduced Loop Number
RMS root mean square
RNG renormalization group (method to renormalize the NS equations)
RNJP Reduced Number of Jet Pump
ROSA Rig of Safety Assessment (in Japan)
ROSA-III name of ITF in Japan, BWR simulator
ROSA-LSTF ROSA- Large Scale Test Facility
RP recirculation pump
RPr reduced pressure
RPV reactor pressure vessel
RR research reactors
RSA relative scaling analysis
RSICC Radiation Safety Information Computational Center
RSM Reynolds stress model
RST rod surface temperature
RTA relevant thermal hydraulic aspect
RTD resistance temperature detector (or resistance thermometer)
S Spherical (Containment Shape)
SA Spalart-Allmaras (model), or severe accident
SAMG Severe Accident Management Guidelines
SAMP severe accident management procedure
SANB saturated nucleate boiling
SAR safety analysis report
SAS scale-adaptive simulation
SB Small Break; Russian Test Facility

S.B. steam binding
SBLOCA small break LOCA
SBO station blackout
SBWR simplified boiling water reactor
SC suppression chamber (or PSP), or stable condensation
SCA subchannel (code) analysis
SCB sub-cooled nucleate boiling (see also SNB)
SCOP test facility in Korea connected with SMART reactor design
SCRAM safety control rods activation mechanism
SCTF name of large-scale SETF in Japan
SCUP UPC scaling methodology (Kv-scaled type)
SCV steam cavity
SCWR supercritical water-cooled reactor
SD steam dome or steam drum
SDRU steam drum
SE steam escape
SEMISCALE name of ITF in the United States, PWR simulator
SEP separator(s)
SET separate effect test
SETF separate effect test facility
SETS stability enhanced two step (numerical method)
SF scaling factors
SFP spent fuel pool
SFR Sodium Fast Reactor
SG steam generator
SGDH simple gradient diffusion hypothesis
SGS sub-grid scale
SGTR steam generator tube rupture
SGTS Standby Gas Treatment System
SHD shutdown (accident conditions), or SH-D
SI safety injection
SILENCE Network of SYS TH experimentalists (see the web)
SIS safety injection system
SIT safety injection tank (see also accumulator)
SJ stable jet, or subsonic jet
SLB steam line break
SM safety margins
SM2A SMAP follow-up project
SMAP OECD/NEA/CSNI Safety Margin Project
SMART advanced reactor design in Korea, System-Integrated Modular Advanced ReacTor
SMC second moment closure
Sm-C small chugging
SMR small and modular (or medium) reactor
SNB sub-cooled nucleate boiling
SNUF Seoul National University Facility
SOAR State of the Art Report
SP standard problem (CFD exercise), or single phase
SPACE SYS TH code developed in Korea
SPD symmetric positive definite (matrix)
SPES name of ITF in Italy, PWR simulator (Simulatore PWR per Esperienze di Sicurezza)

SPL single phase liquid convection
SPV single phase vapor convection
SQE software quality engineering
S/R steam relief
SRI Stanford Research Institute
SRS Safety Report Series (of IAEA), or scale resolving simulation
SRV steam relief valve
SS secondary side (or secondary system)
SSC system, structures and components
SSG simple support grids or Speziale-Sarkar-Gatski (related to RSM), or Specific Safety Guide (of IAEA)
SSI soil structure interaction
S-SOAR Scaling SOAR
SST shear stress transport
ST similar test, or Steel (Containment Pressure Boundary Type)
STAR-CCM+ CFD code
STLC Steel lined-concrete (Containment Pressure Boundary Type)
STNI Science and Technology of Nuclear Installations (Journal)
STRESA Storage of Thermal REactor Safety Analysis Data
S&U sensitivity and uncertainty
SVP single valued parameter
SW Service Water
S-W Steam – Liquid Water
SWAT test facility in Korea connected with SMART reactor design (SMART ECC Water Asymmetric Two-phase Choking Test Facility
SWINTH Specialists Workshop on Advanced Instrumentation and Measurement
SYS System
SYS TH system thermal hydraulic (code)
SZL Shih, Zhu and Lumley (CFD model)
T temperature, or transition
TA total accuracy
TAF top of active fuel
TAMU Texas A&M University
TAV Time Accuracy Vector
TBL turbulent boundary layer, or Two Bundle (test) Loop
TC transition chugging, or thermocouple
T/C thermocouple
TCO transient chugging oscillation
TD top down (reflood)
TDC thermal diffusion coefficient
TE truncation error
TEC Toledo Edison Corporation
TECC thermal-hydraulics of ECCS
TEPCO Tokyo Electrical Power Company
TF test facility
TFQMR transpose free quasi minimal residual
TG turbo-generator
TH or T/H thermal-hydraulics
ThAI Thermal-hydraulics, Hydrogen, Aerosols, Iodine, test facility in Germany
THERFAT consortium in EU dealing with CFD

THTF name of large-scale ITF in the United States
TIA transport of interfacial area
TLTA Two-loop Test Apparatus in USA
TM Technical Meeting (of IAEA), or target mission
TMI-2 Three Mile Island (unit 2)
TNP Time Not Preserved (in scaling)
TNT Kok's k-ω model for turbulent/non-turbulent interfaces (Hellsten, 2004)
TOC top of core (top of core upper plate, above TAF)
TOI VVER-1000 type of advanced reactor
TP two-phase, or time preserved (in scaling)
TPCF two-phase critical flow
TPPD two-phase pressure drop
TPS thermal hydraulic (type of) passive system
TPTF SETF
TRACE SYS TH code developed in the United States
TRAN transition boiling
TransAT CFD code developed in Switzerland
TSE time sequence of events
TT turbine trip
TV tracking velocimetry
UCSP upper core support plate (see also UTP)
UH upper head
UJ unstable jet
UK United Kingdom
UM name of ITF in the United States
UMAE uncertainty methodology based on accuracy extrapolation
UMCP University of Maryland College Park (UMCP) 2×4 loop
UMS project in area of uncertainty launched and completed by CSNI
UNIPI University of Pisa (Italy)
UP upper plenum
UPC University of Barcelona (Spain)
UPTF name of large scale-SETF in Germany, PWR simulator (Upper Plenum Test Facility)
UQ uncertainty quantification
URANS unsteady RANS, also (U)RANS
US DOE US Department of Energy
US NRC United States Nuclear Regulatory Commission, also USNRC
US United States
USA United States of America
USAEC United States Atomic Energy Commission
USSR Union of Socialist Soviet Republics
UT U-tubes (of steam generator)
UTP upper tie plate (see also UCSP)
UTSG U-tubes steam generator
UVUT unequal velocity unequal temperature
UVUTUP UVUT plus unequal pressures
V & V verification and validation
V & V & C V & V and consistency
VHTR very high temperature reactor
VISTA-ITL Experimental Verification by Integral Simulation of Transients and Accidents – Integral Test Loop in Korea

VLES very large eddy simulation
VOF volume of fluid
VR virtual reactor (e.g., by CASL)
VS liquid slug velocity
VTT Technical Research Centre of Finland
VV vent valves
VVER (WWER) water-cooled water-moderated energy reactors designed by Gidropress in Russia (Vodo-Vodyanoi Energetichesky Reactor (see also WWER)
WALE wall adapting local eddy
WB IRS Web-Based Incident-Reporting System
WCNR water-cooled nuclear reactor
WDVB wet-well-to-drywell vacuum breakers
WENRA Western Association of Nuclear Regulatory Authorities
WG working group
W-Gama working group inside CSNI
WMLES wall-modeled LES
WMS wire mesh sensor
WNA World Nuclear Association
WWER see VVER, Water-Water-Energy-Reactor (see also VVER)
XS cross section
ZDES zonal DES

Questions and answers from Chapters 1 to 9

Francesco D'Auria[1], *Jae Jun Jeong*[2], *Tomoaki Kunugi*[3], *Jordi Freixa*[4], *Horst Glaeser*[5], *Francesc Reventos*[4], *Upendra Singh Rohatgi*[6]
[1] University of Pisa, Pisa, Italy, [2] Pusan National University, Pusan, Korea, [3] University of Kyoto, Kyoto, Japan, [4] Polytechnic University of Catalonia, Barcelona, Spain, [5] GRS, Garching, Germany, [6] Brookhaven National Laboratory, New York, NY, United States

List of contents

Introductory remarks

The exercises and questions that follow are subdivided into two parts: Part 1 (questions and texts of exercises) and Part 2 (answers). One may note that

- Part 1 reports the same texts given at the bottom of each chapter (from 1 to 9) of the book.
- Texts in Part 1 deal with nuclear thermal hydraulics: we did our best to allow the reader to find solutions or answers from reading the book and, surely, the book provides the basis to arrive at solutions.
- The minimum number of 10 exercises and questions applies to each chapter (reader may find a larger number of exercises for selected chapters).
- In some cases (we tried to minimize this), the connection between each exercise and the related chapter is weak. Still the exercise is relevant for the book content, and it is complementary to the concerned chapter.
- Multiple answers are possible for a number of exercises and not all of them are reported in Part 2. In answering questions, in case the question is not clear enough, the reader should reformulate the question (eventually reducing the scope of the original question), rewrite the reformulated question and answer the reformulated question.
- Some questions and related answers may deal with the same or a similar topic. In other words, repetitions of questions are not avoided (but only minimized). The reader may look at the same question in different contexts.
- An attempt is made in answering the questions, not in all cases, to use wording different from traditional and established glossary in nuclear thermal hydraulics. This should not confuse the reader; rather this should provide a broader picture of the concerned topic, which complements the discussion in the book.

- In some cases, partly connected with item above, an answer may raise more questions than the initial question. We hope that this brings the reader to investigate deeper into the concerned topic.
- The numbering of figures and tables in the present appendix is X.t-Y, where X is the chapter (i.e., 1−24), t is the question number (there is no distinction between question and answer) and Y can be A, B, C, … (until needed).
- In a few cases, references (from general list of references of the book) are called: we adopt a different citation style than in other parts of the book (e.g., no blue characters).

Any question and any answer may open to new questions including clarifications. Any answer may reveal imperfect (hopefully, not wrong). The reader is invited to be in touch with editors and authors in case he/she has doubts on answers or has a new question.

Part 1—Questions (Chapters 1−9)

Chapter 1

1.1 Provide a distinction between thermal hydraulics and thermodynamics.
1.2 Provide a distinction between thermal hydraulics and fluid dynamics.
1.3 Identify the key features of nuclear thermal hydraulics based on your knowledge and the reading of Chapter 1 of the book.
1.4 Why validation has a key role in nuclear thermal hydraulics?
1.5 Who (which institution) is fixing the DBA/DBC envelope?
1.6 What are the challenging parameters for modeling convective heat transfer?
1.7 What are the challenging parameters for modeling pressure drops?
1.8 What are the key differences between a SYS TH code and a CFD code?
1.9 What are the main similarities between a SYS TH code and a CFD code?
1.10 Discuss the role of phenomena in nuclear thermal hydraulics.
1.11 Why turbulence and averaging constitute current challenges in nuclear thermal hydraulics?
1.12 Can you distinguish the words Lagrangian-approach and Eulerian-approach in nuclear thermal hydraulics?
1.13 Summarize, possibly based on a web search, the role of Euler in the progressing of fluid models.
1.14 Provide examples of the use of nuclear thermal hydraulics in Probabilistic Safety Assessment (PSA).
1.15 Discuss the connection between nuclear thermal hydraulics and neutron physics (e.g., not only moderation).

Chapter 2

2.1 Discuss the importance USNRC Interim Acceptance Criteria for the ECCS design (issued in 1971 at a time when NRC was AEC) for nuclear thermal hydraulics.
2.2 In which international context the process code Validation for system thermal hydraulics was established?

2.3 Please provide suitable definitions for "scaling," "scaling issue," and "addressing the scaling issue."
2.4 Can you outline the history of scaling in nuclear thermal hydraulics?
2.5 Is V&V part of scaling or scaling part of V&V?
2.6 Why Uncertainty and Uncertainty Evaluation are relevant (essential) for the applications of nuclear thermal hydraulics to the WCNR technology?
2.7 What is the connection between Probabilistic Safety Assessment and nuclear thermal hydraulics?
2.8 What are the key elements for the Best Estimate Plus Uncertainty (BEPU) approach and how those elements are connected with nuclear thermal hydraulics?
2.9 Who are the stakeholders for nuclear thermal hydraulics?
2.10 Explain the cross connection between nuclear thermal hydraulics and design of passive systems.

Chapter 3

3.1 Consider a vertical pipe with an internal diameter of D = 0.6 m and bubbly flow conditions. If the linear interfacial drag area is 89.6 m and the mean bubble diameter is 1:5 mm, what is the void fraction?
3.2 Consider a vertical pipe with an internal diameter of D = 0.5 m where the flow conditions are inverted annular flow, and the void fraction is 0.06. What is the linear interfacial heat transfer area in a cut section?
3.3 Consider a vertical pipe with an internal diameter of D = 0.7 m where the flow conditions are annular flow with a film thickness δ = 0.005 m. Calculate the void fraction
3.4 A flow has a void fraction of 0.3. Use the Fauske void correlation model to obtain the quality if the density ratio, is $\rho_l/\rho_g = 900$.
3.5 A flow has a void fraction of 0.3. Use the HEM void correlation model to obtain the quality if the density ratio is $\rho_l/\rho_g = 900$.
3.6 Calculate the convection heat transfer coefficient in the core during normal operation (liquid forced convection). Consider the following information:

- the thermal conductivity of saturated water at 300°C is: $k_{H2O} = 0.55$ W/mK
- the dynamic viscosity of saturated water at 300°C is 8.6×10^{-5} N s/m^2
- the fluid density is 716 kg/m^3
- the specific heat is 5.7 kJ/kg K
- the Hydraulic diameter is 0.013 m
- the core flow velocity is constant and equal to 5 m/s

3.7 For each of the following processes, what is the (nondimensional) number that better represents/explain the phenomenon:

- ➢ CCFL
- ➢ Stratification
- ➢ Convection heat transfer
- ➢ Laminar flow
- ➢ Inverted annular film boiling

3.8 For each of the following processes, what is the (nondimensional) number that better represents/explain the phenomenon:

✔ Chocking flow

- Off-take model
- End of natural circulation
- Turbulence
- Annular film heat transfer

3.9 Consider the geometry of the FEBA experiment shown in Fig. 3.A where the values are expressed in millimeters.

(a) Calculate the hydraulic diameter for the flow path

(b) Calculate approximately the hydraulic diameter of a PWR reactor having 157 fuel assemblies with a distribution of fuel rods similar to the one in Fig. 3.9a

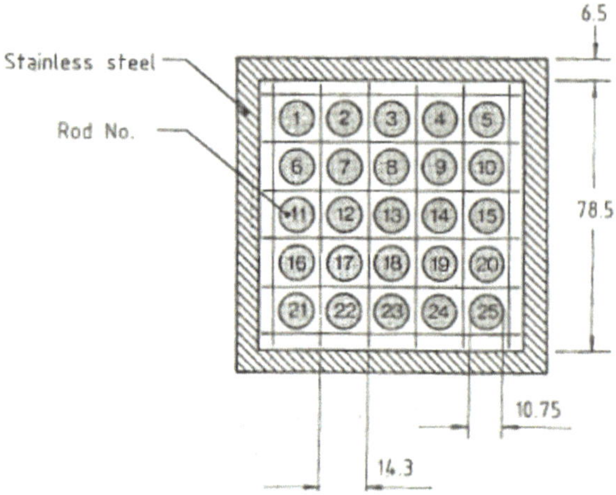

Figure 3.9a Cross section view of the FEBA fuel assembly experimental set-up.

3.10 Consider the geometry of the PSBT experiment shown in Fig. 3.10a where the values are expressed in millimeters.

(a) Calculate the wetted hydraulic diameter for the flow path

(b) Calculate the heated hydraulic diameter for the flow path

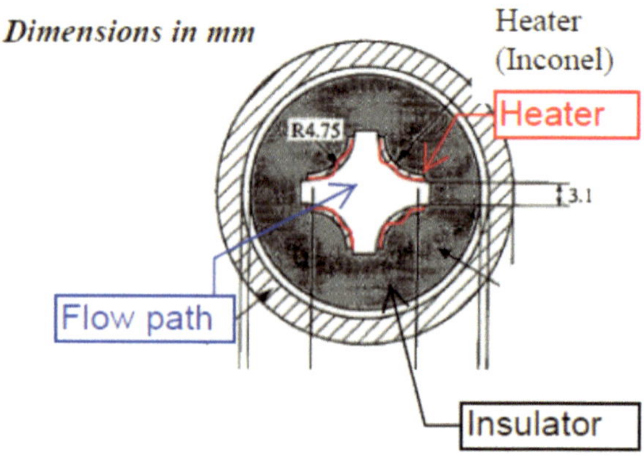

Figure 3.10a Cross section view of the PSBT subchannel experimental set-up.

3.11 Consider the geometry of the PSBT experiment shown in Fig. 3.11a where the values are expressed in millimeters.

(a) Calculate the wetted hydraulic diameter for the flow path

(b) Calculate the heated hydraulic diameter for the flow path

(c) This experiment represents a particular subchannel in a fuel assembly, draw a portion of the fuel assembly where this particular sub channel appears and identify it.

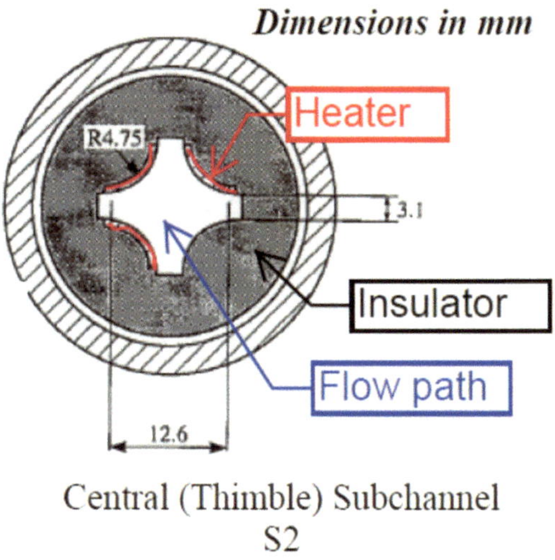

Figure 3.11a Cross section view of the PSBT subchannel experimental set-up (note the differences related to Fig. 3.10a).

3.12 Consider the geometry of the PSBT experiment shown in Fig. 3.12a where the values are expressed in millimeters.

(a) Calculate the wetted hydraulic diameter for the flow path

(b) Calculate the heated hydraulic diameter for the flow path

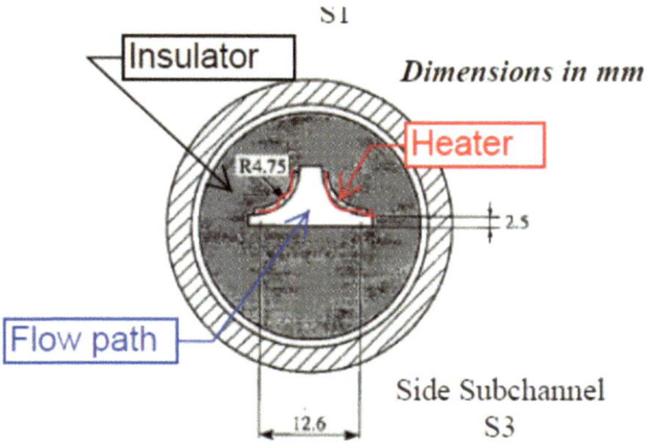

Figure 3.12a Cross section view of the PSBT subchannel experimental set-up (note the differences related to Figs. 3.10a and 3.11a).

3.13 Consider the geometry of the PSBT experiment shown in Fig. 3.13a where the values are expressed in millimeters.

(a) Calculate the wetted hydraulic diameter for the flow path

(b) Calculate the heated hydraulic diameter for the flow path

Figure 3.13a Cross section view of the PSBT subchannel experimental set-up (note the differences related to Figs. 3.10a, 3.11a and 3.12a).

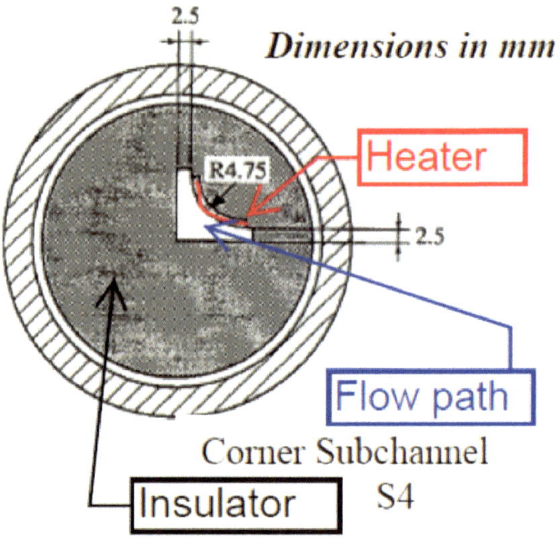

3.14 Fig. 3.14a displays a U-tubes steam generator section (related to a test facility simulating a PWR) at mid elevation.

Figure 3.14a Cross section view of a U-tubes SG (related to an experimental facility).

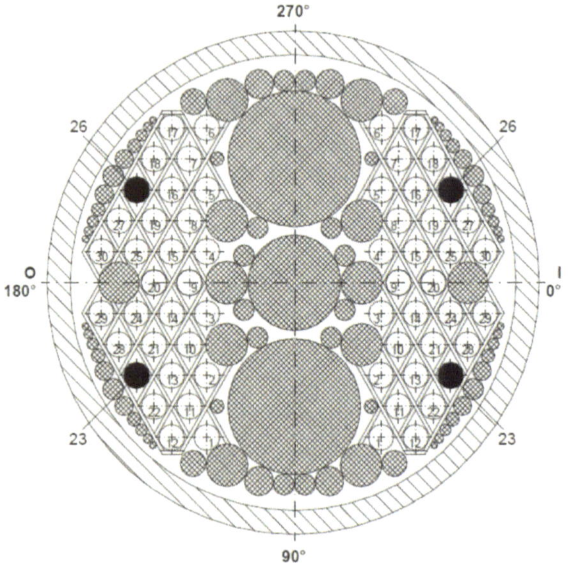

The numbered small circles represent the U-tubes, there are a total of 30 with an inner diameter of 19 mm. The crossed circles represent fillers (metallic structures). The inner diameter of the steam generator secondary side section is 0.38 m.

(a) Calculate the hydraulic diameter of the flow path through the primary side of the U-tube bundle.

(b) Estimate the hydraulic diameter of the secondary side flow path. Explain your hypotheses.

Chapter 4

4.1 Please, explain a PIRT and its connection with this book (e.g., Chapter 6). Discuss separately the words "phenomena," "phenomena identification," and "phenomena ranking."

4.2 Discuss challenges of a PIRT process, possibly including examples.

4.3 What are the elements for software adequacy? Where are these elements discussed in the book?

4.4 What is EMDAP? Identify connections between EMDAP and topics in different chapters of the book.

4.5 Discuss the connection between thermal hydraulics and nuclear reactor safety.

4.6 What are the bases to arrive at the "thermal hydraulic analysis envelope"?

4.7 Define with your own words: CFR, RG, SRP and FSAR. Please discuss the connection among the key objectives, which are embedded inside each of the acronyms.

4.8 What are the connections between the capabilities and the key features of system thermal hydraulic codes and the content of the chapter 4 of the book?

4.9 Discuss key differences and similarities between IAC (Interim Acceptance Criteria) for the design of ECCS (1971), the RG 1.70 (1978) and the RG 1.203 (2005). *Please note that the concerned documents are not homogeneous (in relation to the content and objectives).*

4.10 What is the role of experiments within the context of chapter 4?

Chapter 5

5.1 Please provide a distinction between HEM, EVUT, UVET and UVUT equations for system thermal hydraulics codes.

5.2 Please discuss the role of drift flux model into UVET and UVUT models.

5.3 Please discuss the features and the challenges of the UVUTUP model.

5.4 Please discuss the features and the challenges of the interfacial area transport (IAT) model.

5.5 Please discuss the challenges of developing three or multi-field models in nuclear thermal hydraulics. Please refer to the droplet field model, primarily.

5.6 What are the difficulties in developing and in qualifying 3D models for two-phase flow? Please refer to 3D equation models suitable for modeling selected components of reactors. For instance, explain why a 3D model exists for core and not necessarily for the hot leg; consider the case of steam generator 3D modeling (primary and secondary side).

5.7 Discuss the issue of heat flux partitioning for: (a) UVUT 1D model; (b) UVUT 3D model; (c) three field 1D model.

5.8 Discuss why balance or conservation equations currently suitable for modeling single and two phase flow do not include directly (another equation dealing with) the conservation of angular momentum?

5.9 Discuss the connection between balance equations (refer to a specific set of equations you are more familiar) and entropy increase, should a nonreversible situation (e.g., Blowdown) be the objective of the solution.

5.10 Explain why geometric discontinuities are not solved within balance equations (e.g., momentum equation may lack the term of pressure drop at geometric discontinuity). Please distinguish the case of abrupt discontinuities (e.g., an orifice in a pipe or the core inlet plate) and smooth discontinuities (e.g., a bend in the hot leg or the connection between Downcomer and lower plenum in the reactor pressure vessel of a BWR or a PWR). In the case of BWR, you have jet pumps: if possible consider in the answer.

5.11 Explain why 1D equations have the capability to simulate (reproduce experimental data) with sufficient accuracy in some cases (if any), 3D situations that occur in the core and imply the consideration of transversal flows (e.g., mixing in a fuel bundle).

Chapter 6

6.1 What is the motivation and the objective for the identification and characterization of phenomena in nuclear thermal hydraulics?

6.2 What is the difference between a SETF and an ITF? (in providing the answer, please consider and discuss the role of scaling).

6.3 What is the difference between a basic experiment and a SETF (or a ITF) experiment?

6.4 Please discuss the current status in relation to predictive capabilities by codes of the phenomenon TPCF.

6.5 Please discuss the current status in relation to predictive capabilities by codes of the phenomenon CHF.

6.6 Please discuss the current status in relation to predictive capabilities by codes of the phenomenon CCFL (please include in the answer a discussion of the importance of scaling).

6.7 Making reference to phenomena (also considering information in Chapter 17), please discuss the capability of system codes in predicting containment performance.

6.8 Making reference to phenomena (also considering information in Chapter 18), please discuss the capability of system codes in predicting core subchannel conditions.

6.9 Please discuss the connection between accident scenarios and phenomena and parameters (see also Chapter 15).

6.10 Please discuss the relevance and the challenges for PIRT.

Chapter 7

7.1 What is the meaning of "conjugate heat transfer"?

7.2 Please explain the difference between Fourier Equation and Fourier Law.

7.3 An approximate solution of the Fourier equations allows the (demonstration of linear) connection between linear power (w/m, derived from neutron flux and uranium fissions), fluid temperature and uranium pellet centerline maximum temperature. Please derive such solution (literature search) and discuss the key information and implications.

7.4 Discuss why (typically, under similar system conditions, pressure, velocity, etc.) nucleate boiling heat transfer coefficient is (much) larger than forced convection heat transfer coefficient.

7.5 Discuss the ideal existence of a multi-D heat transfer coefficient surface (or hyper-surface). Discuss the advantages and the difficulties of deriving such surface.

7.6 Discuss why rewet (local) process (part of reflood at a system level) modeling may not make use of definitions and derivations of heat transfer coefficient in heat transfer regimes like nucleate boiling or forced convection. Consider also the discussion of reflood in Chapter 11.

7.7 Make a possible distinction between natural convection, heat transfer regime (distinguishing pool boiling and flow boiling), forced convection (primarily, flow boiling heat transfer regime) and natural circulation (that involved the need of a heat sink and a heat source).

7.8 Identify the interference between radiation heat transfer, nucleate boiling heat transfer (weak interaction) and film boiling heat transfer (possible strong interaction). Discuss possible means to separate the impact of radiation heat transfer in an experiment where film boiling occurs and film boiling heat transfer constitutes the topic of the measurement.

7.9 Discuss why (early) Critical Heat Transfer correlations (e.g., Rohsenow, etc.) may calculate critical heat flux values much greater than the same value predicted by correlations/approaches like the Look-up Table in similar conditions (namely pool boiling).

7.10 Discuss the relevance of surface conditions (and of chemical/physical interactions between fluid and wall) upon heat transfer. Please provide quantitative example (e.g., heat transfer coefficient in a clean surface and in surface with oxide and crud).

7.11 Please describe what is CRUD and how much CRUD and oxide layer may affect heat transfer coefficient in a nuclear fuel rod typical working condition. This affects the centerline pellet temperature in steady state conditions. Please connected with the answer to question 7.3.

7.12 Considering the answers to questions 7.3 and 7.11, please inform about the impact of heat transfer coefficient, in nominal operation condition, upon the Peak Cladding Temperature—PCT, should a Large Break LOCA (opening time of the rupture, un-physically taken as few milliseconds) occur in a current PWR. Please consider conditions that establish a few seconds after break opening (typically first PCT or Blowdown PCT).

Chapter 8

8.1 Discuss the interconnection between pressure drops and momentum or force balance equation.

8.2 Discuss the motivation, the difference and the implication of friction pressure drops term based on "f" (friction of Moody diagram) and on "τ_w."

8.3 Why Φ_{LO}^2 is defined as a quadratic term in calculating two phase friction pressure drops?

8.4 Considering the equation for acceleration pressure drop and making reference to a horizontal pipeline in steady state conditions where a gas is flowing in the presence of friction pressure drop and heat transfer to the walls, declare whether inlet velocity is larger than outlet velocity (and what assumptions are needed to answer this question).

8.5 What is the pressure drop needed to create loop seal clearing in the typical conditions of a SBLOCA in PWR?

8.6 What is the typical singular pressure drop caused by a spacer grid? (please consider information in this chapter and in the literature for steady state flow in reactor conditions).

8.7 Discuss the occurrence of possible cavitation conditions and consequent steam generation at geometric discontinuities.

8.8 Two-phase pressure drop are typically measured in experiments in steady state and fully developed flow conditions: discuss about related implications in calculating pressure drops in reactor conditions during an accident.

8.9 In which way a single pressure drop measurement (or measured data) is transferred to two-fluid UVUT model? What assumptions and what procedure are introduced?

8.10 What is the typical error in calculating pressure drops (noticeably, two-phase flow, friction and at geometric discontinuity)?

8.11 Why pressure drop at geometric discontinuities are not part of balance equations implemented into SYS TH codes? What are the challenges for code-users in this connections?

8.12 What is the (superior) role of CFD codes related to SYS TH codes in relation to pressure drop?

Chapter 9

9.1 Provide your own definition of constitutive equations and closure equations (or closure laws).

9.2 Discuss the meaning of "unknowns" in constitutive equations (connect with the unknowns of PDE where the constitutive equations are embedded).

9.3 Discuss the connection between Jacobian and formulation of constitutive equations when the implicit solution method is pursued to solve the PDE (which include the constitutive equations).

9.4 Distinguish (and discuss) between drift flux and slip ratio. Shortly describe the origin of drift flux.

9.5 What is the connection between constitutive equations and flow regimes (or flow patterns) for two-phase water flows?

9.6 Report typical values of interfacial areas (or better ranges of areas) per unit volume (after fixing a value for diameter in cylindrical pipe) for bubbly flow, slug flow, churn flow annular flow and dispersed flow at a fixed pressure (or pressure range (you may select) distinguishing (if applicable) horizontal and vertical flows. Please also add the case of stratified flow in horizontal cylindrical channels.

9.7 Please discuss the "repartition" problem for constitutive equations (i.e., application in UVUT models). Please discuss also complexity of "repartition" in case of 3D modeling.

9.8 What is the connection (if any) between the constitutive equations (and closure laws) and hyperbolicity of the solution of PDE?

9.9 Is it reasonable to identify the k-ε models [adopted in CFD codes] as constitutive equations? Please discuss why and identify difference between k-ε models and constitutive equations discussed in this chapter.

9.10 Discuss the meaning of scaling (and scaling capability) in closure laws.

Part 2—Answers (Chapters 1–9)

Chapter 1

Answer 1.1

Thermodynamics was established since the 17th century or before. The concepts and the definitions of heat, temperature, and entropy, among the other things, belong to and characterize thermodynamics, as well as the fundamental principles.

Thermal hydraulics, looking at the semantic origin of the second word (i.e., hydraulics), focuses on water. Namely, thermal hydraulics can be associated with the evolutions (or thermodynamic transformations) of single and two phases water systems (i.e., mass or flows). Here the word "thermal" refer to broad temperature (and pressure) range, i.e., from ice formation (273 K at ambient pressure) to, typically, superheated steam temperatures consistent with the resistance of structure materials that necessarily bound the fluid (or a couple 1000 K). The term "evolutions or transformations" implies the concept of time-evolution, or transient, or time dependence. Time dependence is also part of thermodynamics, possibly not with all implications that are typical of nuclear thermal hydraulics.

Nuclear thermal hydraulics was established to address the needs coming from the design and the safety demonstration of water-cooled reactors, primarily. However, any fluid or heat transfer related issue in nuclear technology, including fusion reactors and micro-reactors, became part of the domain of interest for nuclear thermal hydraulics, as testified by the topics of papers accepted in conferences like NURETH and NUTHOS. Noticeably, fundamental principles of thermodynamics constitute the bases for balance or conservation equations (mass, momentum and energy), which are at the heart of nuclear thermal hydraulics.

Answer 1.2

Fluid dynamics primarily aims at single-phase liquid flows. River flows, propagation of pressure wave causing water hammer and deep mathematics for turbulence modeling are example of topics for fluid dynamics.

In a similar way, gas dynamics primarily aims at single-phase gas flows. Turbulence generated by aircraft, and sonic velocity are examples of topics for gas dynamics.

Thermal hydraulics is similar to fluid dynamics and, however, it is more concerned to a fluid flow with heat transfer, which results in a temperature change or a phase change. The topics of thermal hydraulics, restricted to water-cooled reactors are exemplified in the present book: two-phase flows and transient conditions are specific for thermal hydraulics (e.g., CCFL, reflood, etc.). Clearly pressure drops in single-phase flows (and not only) is of common interest to fluid dynamics, gas dynamics and thermal hydraulics.

Answer 1.3

This question has multiple answers depending upon the expertise of the reader. Furthermore, the details of any answer can be (very) different from any other answer. One possible answer follows.

Nuclear thermal hydraulics aims at improving the design of water-cooled nuclear reactors: in this connection, the methods, and the correlations for predicting the occurrence of critical heat flux constitutes one feature of nuclear thermal hydraulics. This obviously introduces the topics of two-phase flow and of excursion-increase in rod surface temperature (i.e., post CHF conditions).

Nuclear thermal hydraulics aims at characterizing the safety margins of water-cooled nuclear reactor following the occurrence of accidents: in this connection, the phenomena that occur in scaled experiments are relevant to identify the features of nuclear thermal hydraulics. The statements above introduces the discussion about facilities like LOFT (old), PKL and ATLAS (recent) and about scaling.

At the bottom end one may state, without attempting to be exhaustive and adopting acronyms widely discussed in the book, that nuclear thermal hydraulics aims at predicting CHF, TPCF, stratification, CCFL, rod surface temperature, pressure drops and heat transfer.

Answer 1.4

The modeling bases for nuclear thermal hydraulics (i.e., the theoretical tools to perform prediction addressing the issues mentioned in the answer 1.3) are the balance equations (mass, momentum, and energy). The derivation of those equations unavoidably introduces to turbulence and some models or correlations for the mass, energy and momentum transfer at a liquid-to-gas interface and a fluid-to-wall interface. These are called constitutive equation. The current knowledge of these is not sufficient to derive theoretical solutions; therefore, empirical correlations are necessary.

Empirical correlations (several dozen correlations each one depending upon dozen input parameter, are necessary to supplement-solve the balance equations) are imperfect, including range of conditions for the experiments at the basis of their derivation and terms that connect with physics.

Analysts (i.e., typically model developers) propose those terms, without demonstrating that they are sufficient or consistent with the phenomena target of modeling. Resulting modeling is imperfect and solutions from the modeling are imperfect, too. The "amount of imperfection" (or the amount of error expected when applying the solutions from the modeling to the prediction of nuclear reactor design or safety parameters) is unknown. Compensating errors may occur when performing calculations with models that depend upon hundred unknown (or not perfectly known) input parameters.

The above discussion introduces, at least, two key topics:

- Scaling analysis is necessary to design experiments to derive empirical correlations at the basis of modeling.
- Errors are expected in applications of models and codes (which are ensembles of models) to nuclear reactors: the word (and the associated procedures) uncertainty encompasses those errors.

Now the answer to the question. Validation is needed to demonstrate that the errors resulting from modeling calculations in predicting (possibly, rightly scaled) experimental data are acceptable (or small enough) compared with technological needs. Here one may note that the error in predicting experimental data is known and is referred as "accuracy"; otherwise, the error in predicting reactor parameters is unknown and is referred as "uncertainty."

Answer 1.5

DBA is an established acronym in nuclear reactor technology since the design-construction of the first reactors. DBA puts a link between the construction-operation of reactors and their acceptability from a safety (and indirectly security) viewpoint. Licensing is the process to demonstrate this link. The identification and characterization of DBA needs, triggered and fostered the developments of safety assessment or reliability analyses and thermal hydraulics, considering, among other things, probability of occurrence and consequences, respectively. Regulators from all over the world recognized the term DBA, more or less formally.

Recently (after the year 2010), the term DBC, proposed by European Regulators and introduced into IAEA documents appeared. This does not modify or affect the term DBA and is not formally recognized by all regulators. The term DBC constitutes a basis to introduce the terms DEC-A and DEC-B that substitute the previous term BDBA.

Now the answer to the question. National regulatory bodies, (or) licensing authorities, (or) regulators and licensors fix the criteria and the roadmap to select accidents within the DBA envelope and the conditions for the related (thermal hydraulic) analyses. Nuclear thermal hydraulic specialists shall be ready to address concerned questions and issues.

Answer 1.6

The scope for the question is limited to convection heat transfer: the consideration of radiation and conduction heat transfer may imply different or more specific answers.

Heat Transfer Coefficient (HTC) is the key parameter to address the question. Convection heat transfer implies the presence of a liquid and a solid (surface).

The first consideration here is that the formulation of HTC (sometimes called the Newton's law) has an empirical and complex (same quantity appears in the right-hand side and left-hand side of the equation) nature.

A wide variety of parameters affect the HTC value including, flow-system conditions (e.g., pressure, fluid velocities, fully developed flow or less, thermal flux, etc.), material properties (e.g., viscosity, density, conductivity) and geometry (e.g., shape of solid structure, e.g., flat or cylindrical, presence of spacer and mixing vane grids).

Examples of challenging parameters (i.e., parameters whose knowledge puts challenges to models and to thermal hydraulic analysts) for modeling heat transfer are the void fraction, the solid surface conditions and those parameters, part of flow-system conditions, affecting the local turbulence and the fully developed flow conditions.

Answer 1.7

Challenging (same meaning for the word "challenging" as proposed in answer 1.6) parameters for modeling pressure drops are the void fraction and the fluid velocities. The "modeling-added" two phase friction coefficient multiplies is important in this connection.

Here one may add (to the list of challenging parameters) relative velocities between steam and liquid (two-phase conditions). This opens to the consideration of interfacial drag and flow regimes. Furthermore, parameter ranges are relevant to fix the difficulties (or the challenges) in modeling and include very low fluid velocities including transition to flow reversal (or change in the mutual directions between liquid and steam), very high flow velocities (up to maximum achievable velocities, whatever is the meaning) in two-phase flows. Vaporization and condensation phenomena put additional challenges to the modeling of (not only time-) acceleration pressure drops.

Answer 1.8

The history for development framework constitutes the first difference. SYS TH codes were designed to address the licensing needs puts by regulators (primarily USNRC, or better USAEC) for safety demonstrations of nuclear reactors, during the 1960s and 1970s of the previous century. CFD codes (including two-phase modeling) developed within the framework of fluid-dynamics research having the turbulence modeling as main target as the solution of the Navier-Stokes equations. The same Navier-Stokes equation is at the origin of equations solved by SYS TH codes.

Another difference consists in the domain of integration of equations in the fluid space: this can be lower than a cubic millimeter in the case of CFD codes and (typically, but not always or in all situations) larger than one cubic decimeter in the case of SYS TH codes.

Therefore, as a key feature, CFD codes can predict parameters profiles (e.g., fluid velocities) in a cross section orthogonal to the flow direction; otherwise, SYS TH codes predict flat average values for the same parameters.

Turbulence in two-phase conditions (primarily) still represents (year 2023) an unsurmountable issue in theoretical modeling; this requires empirical equations-correlation for both CFD and SYS TH codes.

The drawback of average modeling in the case of SYS TH codes (see above in this answer), mainly in two-phase flow conditions, is partly overpassed by proper validation that does not prevent compensating errors. In the case of CFD codes, the capability of detailed modeling is overshadowed (not only) by the difficulty to measure calculated parameters and to prove correctness/acceptability of solutions.

The possibility by CFD codes to simulate an accident in the conditions of water-cooled (large) nuclear reactors is still beyond the modeling capabilities.

Answer 1.9

The key similarity between SYS TH and CFD codes is the solution of the mass momentum and energy balance equations (referred as Navier-Stokes equations in the case of CFD codes). Those equations are connectable with the same principles of thermodynamics (first principle, considering the second principle) and mechanics ($F = ma$), although they are solved in different control volume, adopting different approximations and, in some cases, using different correlations for the same parameters.

Answer 1.10

The consideration and the study of phenomena in nuclear thermal hydraulics started at NEA/CSNI within the framework of demonstrating code validation for PWR (U-tube SG and OTSG). The complexity of accident scenarios (DBA type, primarily) suggested splitting each scenario into phenomenological windows and to characterize each phenomenological window by a minimum number of phenomena. Each phenomenon is further characterized by parameters (that appear in modeling, typically) and parameter ranges (other than combination of parameters). Later on, the list of phenomena was expanded covering VVER, CANDU and selected types or accident conditions of advanced reactors.

Therefore, the main role of phenomena is to ensure that (as far as possible) SYS TH codes are validated against situations expected during accidents in water-cooled reactors (DBA envelope, based on scaled facility experiments).

More recently (starting since 2016), proposals came to expand the meaning and the role of phenomena to cover the target of developing and/or improving the code modeling.

Answer 1.11

Turbulence constitutes a challenge because (excluding DNS) empirical correlations are needed for modeling (e.g., k-ε) and their validity is a function of geometry of the modeled system and fluid boundary conditions.

Averaging (this can be space- or time-; space can be area related or volume related) constitutes a challenge because it is almost arbitrarily selected when using the codes, either SYS TH or CFD, by the analysts or code users: recommended solution convergence studies are complex and seldom finalized in nuclear thermal hydraulics investigations.

Answer 1.12

A fixed system geometry domain is involved with Eulerian-approach and a moving fluid domain is involved with Lagrangian-approach.

Fig. 1.12a provides further insight into the answer.

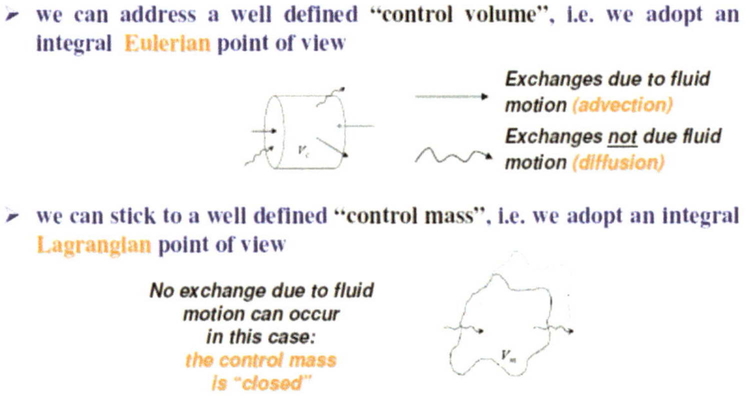

Figure 1.12a Further explanation for Lagrangian and Eulerian approaches in modeling.

Answer 1.13

Leonhard Euler (1707–1783, portrait below, Fig. 1.13a) was born in Basel (Switzerland) and was attracted by Russia and by Prussia: the most important scientist-technologist of the XVIII century. The ensemble of discoveries that he pioneered or led puts Euler as the architect of scientific revolutions not less important than those promoted by Galileo, Newton, Einstein, and Planck.

Figure 1.13a Portrait of Leonhard Euler by J.E. Handmann (1753).

The three-dimensional coordinate system to describe the fluid behavior, the differential equations for nonviscous fluid made him a precursor of Navier and Stokes. Euler was the mentor of the Turin born Lagrange, who constantly referred to him in his prolific research life. Truesdell, during the last century, identified in Euler as the predecessor of discoveries and advancements by Bernoulli and d'Alembert, other than Lagrange.

Euler is at the origin of the foundations of fluid dynamics that connects with modern nuclear thermal hydraulics.

Answer 1.14

PSA, starting from Rasmussen is at the hearth of safety demonstration for nuclear reactors. The safety demonstration implies among the other things and simplifying, (I) the triggering of accident scenarios (i.e., identification of postulated initiating events, PIE), (II) the evaluation of consequences from the analysis of each scenario, or the evaluation of the source term at the level of the reactor core, and (III) the calculation of doses to population. The path from (I) to (II) and the steps (II) and (III) require the application of nuclear thermal hydraulics, typically SYS TH codes.

Namely, SYS TH code application is necessary to address the question (among others), path from (I) to (II): has the concerned accident radiological relevance? The amount of radiation release to containment in step (II), needs nuclear thermal hydraulics as well as the diffusion of radioactive material to environment (step III). Here one may note:

- Investigations by simplified SYS TH codes (and/or adopting simplified procedures for code applications) characterize the applications of nuclear thermal hydraulics to (direct) PSA needs, as stated above. For instance, quality demonstration for nodalizations or code users (i.e., analysts performing the analyses), scaling and uncertainty studies are not requested (typically), raising concerns about the quality of those analyses.
- After the loss of geometric integrity for the core, nuclear thermal hydraulics is still necessary for the evaluation of the accident progression (DEC-B conditions, see answer 1−5), i.e., part of coupling between nuclear thermal hydraulics and severe accident phenomenology. This topic is not discussed in the present book (where consideration goes to situations before loss of geometric integrity).

Answer 1.15

Any relevant application of nuclear thermal hydraulics to water-cooled requires the calculation of thermal power produced in fuel rods which results from neutron flux and decay power. Therefore, the calculation of neutron flux is an essential boundary condition for thermal hydraulics analyses; moreover, the neutron flux is not only a boundary condition but has a feedback effect upon generated power determined by thermal hydraulic parameters like convection heat transfer coefficient (which affects Doppler effect) and void fraction (which affects stability in BWR core). Let us address the question by summarizing the connection between nuclear thermal hydraulics and neutron physics:

→ Thermal neutron fluxes are primarily concerned in water-cooled nuclear reactors (with significant exception of supercritical reactors, not a key topic for this book). This implies a moderation process with nonnegligible energy (then thermal power) generated in the moderator.
→ Fission products unavoidably generate decay heat during nominal operation and accident conditions, namely after scram or stop of fission reactions.
→ Gamma heating occurs from fission products: decay heat at previous arrow item includes a portion of gamma heating; additional portions of gamma heating are dissipated in structural materials (e.g., clad of fuel rods), in coolant and in moderator.
→ Neutron leakages from the core and gamma heating are also responsible for thermal power (typically negligible) generated out of core.
→ Radiolysis occurs because of radiation and absorbs a (typically negligible) portion of power generated into fuel rods.
→ Neutrinos power (or power associated with the massive production of neutrinos) is lost in the universe and does not contribute to energy balance in water-cooled nuclear reactors.

All above terms imply proper modeling and connection between nuclear thermal hydraulics and neutron physics. This connection is relevant from a technological viewpoint and outside the scope of the book. Moreover, one may note:

(a) A complete coupling between nuclear thermal hydraulics and neutron physics (i.e., considering the arrow items above and the three-dimensional neutron physics modeling) is within the technological capabilities of industry and research groups. However, this coupling is not mandatory for licensing analyses (we regret to note this).
(b) There is the capability to calculate local, i.e., single fuel rod, thermal and fast neutron fluxes (or energy spectra); those fluxes affect generated power and are of concern to establish CHF (or DNBR) limits, e.g., by sub channel codes. Scientific and technologic community does not exploit related capability (again, we regret to note this) in most of applications.

Chapter 2

Answer 2.1

The issue by USAEC (current US NRC) Interim Acceptance Criteria (IAC) triggered in 1971 important research in two complementary and tightly connected main directions within nuclear thermal hydraulics:

- Scaled experiments, integral type, large scale (as far as possible).
- SYS TH code development (UVUT, primarily).

There was no other way than complex SYS TH codes supported by experiments having size, or pressure, or power necessarily much lower than prototype nuclear reactors to fulfill the licensing criteria. Large investments from industry and regulators from all over the world brought to significant achievements in 2–3 decades. Ironically, once the impulse from 1971 IAC exhausted and the demonstration of safety for water cooled reactors was basically achieved (nuclear thermal hydraulics viewpoint), the interest towards the nuclear technology decreased (also due to disasters of Three Mile Island, Chernobyl and Fukushima, which have nothing or little to do, case of Three Mile Island, with nuclear thermal hydraulics).

Answer 2.2

The activity on code validation started at the Committee on the Safety of Nuclear Installations (CSNI) of the Nuclear Energy Agency (NEA) of the Organisation for Economic Co-operation and Development (OECD) at its Paris headquarters. Following the availability of rudimentary thermal hydraulic codes from US and Germany at the beginning of the 1970s, CSNI promoted the comparisons between predicted results and measured data. The series of International Standard Problems (ISP), which is still supported after 50 years, started. As an immediate result, large errors between measurements and predictions occurred as well as large scatter among different predictions. This raised the question of quality of a prediction and introduced the need for code qualification (or validation) before application to the analysis of transients in nuclear reactors.

The pioneering effort on validation in nuclear thermal hydraulics continued for a few decades at CSNI. This brought to ITF and SETF matrices for code validation, to accuracy quantification procedures, to thermal hydraulic phenomena identification and proposal for uncertainty methods, i.e., taking into account of unavoidable errors resulting from validation processes and, finally, to BEPU.

Answer 2.3

"Scaling," "scaling issue," and "addressing the scaling issue" constitute synonyms. In other words, scaling (activities) represents an issue that needs to be addressed. Glossaries in international documents (e.g., IAEA and CSNI) provide robust and consensus definitions. Here, let us consider one of the science and technology backgrounds that gives origin to the scaling synonyms in nuclear thermal hydraulics.

Let us state in advance that key scale parameters (or parameters that may not have the same values in an experimental facility and in a reactor) are:

- Pressure (up to 16 MPa in reactor conditions);
- Size, primarily flow area (up to a few square meters in reactor conditions);

- Total power (up to 5000 Mwth in reactor conditions, thus implying about 40,000 fuel rods);
- Linear power, q' (up to 55 kw/m in one individual fuel rod).

Combinations of those parameters are relevant: it is (nearly) impossible to perform experiments where all those parameters assume simultaneously the values as in the reactor.

First, one may look at the licensing process, where safety analysts (let us assume, the licensee) adopt an assigned thermal hydraulic computer codes (i.e., a SYS TH code) to demonstrate the acceptability and the safety of an assigned reactor, referring to an assigned Accident Scenario (AS). The licensor (or the regulatory body) asks to show the qualification of the code looking at the concerned AS. The licensee shows (good-acceptable) results of code calculations compared with experiments in a test facility. The licensor comments "Yes, fine, but the experiment you considered has been performed at a scale much smaller than the reactor: how do you guarantee acceptable quality of prediction at the reactor scale?" The licensee may answer pursuing (at least) two different logical paths:

(A) Declaring that the capability to predict phenomena is independent of scale: if an assigned phenomenon occurs in a properly scaled facility (see below) and the code is capable to predict that phenomenon, then the same code is capable to predict that phenomenon in nuclear reactor. The licensor may never accept this argumentation.

(B) Showing that his code is capable to predict the given phenomenon (i.e., all phenomena part of the concerned AS) measured in differently scaled experiments. Broad differences in key scaling parameters defined above (namely, size) characterize the scaled experiments. The licensor may accept this argumentation, noting that there is no better way to address the scaling issue (i.e., if a reasonable set of experiments is considered).

The declaration at item (A) reveals weak due to a number of motivations. One of those motivations deals with turbulence that is highly connected with size: a suitable model at small-scale does not imply suitability for the same model at large-scale.

Second, one may look at the design of a scaled facility. The best design for a scaled experiment is not necessarily a matter between licensee and licensor, although the designer shall make an effort to make the design suitable for code validation and licensing applications. Any designer aims at constructing the largest possible facility, i.e., adopting the largest values for the key scale parameters. Therefore, resources available (i.e., not only budget) put the key limitations. The following is suggested:

- The optimal simulation of core heat transfer including rod surface temperature (in a hypothetic experimental facility or model) tightly depends upon linear power of fuel rods: same core height, sub channel size and q' are necessary as in the reactor (or prototype).
- Simulation of gravity-governed phenomena (natural circulation, reflood, etc.) imposes same height, pressure, and time evolution during any AS for model and prototype.

Experimental facilities (in some case expensive ones) can be built and have been built without fulfilling the suggestion at dashed items. Those facilities may reveal relevant for code validation; however, the similarity of phenomena measured in the model and expected in the prototype is questionable and distortions (or differences in performance between model and prototype) are expected. Although the last statement applies to any experimental facility (model) and no experimental data can be reliably transposed from model to prototype, the amount of distortions becomes minimum if dashed item recommendations are pursued.

Third, the analysis of several counterpart tests (CT) and similar tests (ST) performed in differently scaled facilities provides additional information about scaling and a substantiation to the summary above. We note that the detailed knowledge of CT and ST findings is not spread (due to various motivations) within the scientific community and the Chapter 21 of the present book provides (only) a little hint in that direction (i.e., to spread the scaling findings).

Fourth, Fig. 2.3a shows two possible end-results from a scaling investigation. The design of experimental facilities (sketch on the left) facilities and of experiment boundary and initial conditions constitutes the end-result of scaling investigations. The demonstration of code scaling capabilities and (or) the extrapolation of accuracy of code calculations (to estimate the uncertainty, sketch on the right, curve 3) constitute a completely different end-result of scaling investigations. Connection between the two investigations do exist (not discussed here).

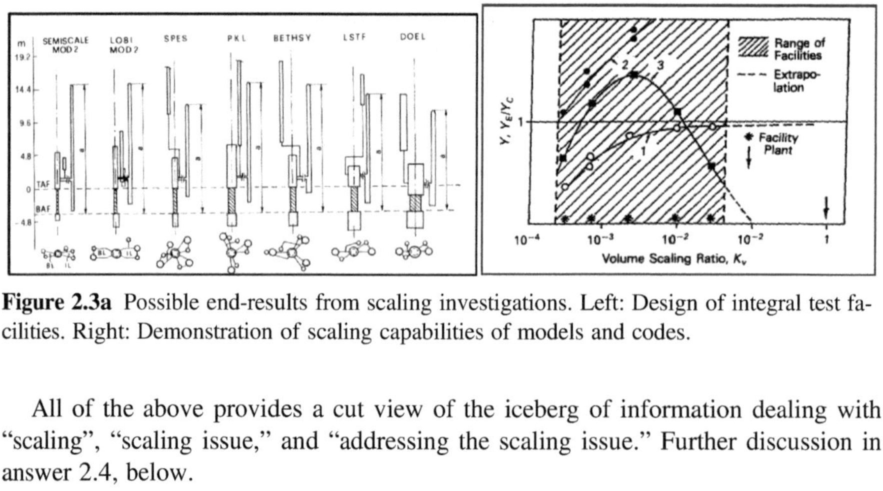

Figure 2.3a Possible end-results from scaling investigations. Left: Design of integral test facilities. Right: Demonstration of scaling capabilities of models and codes.

All of the above provides a cut view of the iceberg of information dealing with "scaling", "scaling issue," and "addressing the scaling issue." Further discussion in answer 2.4, below.

Answer 2.4

The distinction between scaling factors (derived from equations including sophisticate modeling) and design factors (derived from fundamental principles) appears essential to summarize and simplify the history of scaling in nuclear thermal hydraulics, although here we do not enter into details of that distinction. Then we distinguish very roughly the following historical phases for scaling investigations (see also answer 2.3, above):

I. Design of early experimental facilities (or reactor simulators or models, see above) by adopting design factors: this occurred starting from the 1970s of the previous century.

II. Use of scaling factors and proposal for scaling methods to determine scaling factors, to analyze experiments: this continued for the subsequent 3-4 decades and contributed to the scaling controversy or no-consensus within the scientific community, see Fig. 2.4a for more details.

III. Proposal for adopting the SYS TH codes to perform scaling analyses based on the analysis of experiments. This occurred towards the end of 1990s of the previous century; the proposal includes the concept of extrapolation of errors of code predictions of scaled experiments (or accuracy extrapolation).

IV. Attempt to establish a consensus within the scientific community in relation to the proposal at item above and introducing procedures supporting the application of SYS TH codes to scaling. This attempt occurred during the last decade and is still on going.

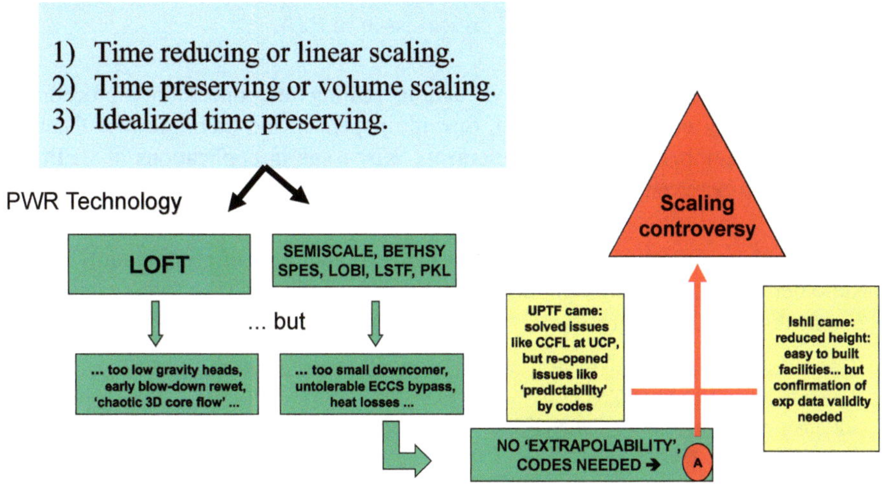

Figure 2.4a Scaling elements and background for the origins of scaling controversy.

Answer 2.5

Here the double nature for scaling should be considered, answer 2.3: scaling in the licensing process is concerned hereafter (i.e., not scaling investigations-applications aiming at the design of experiments).

When performing V&V for SYS TH codes, namely validation (e.g., validation needs properly designed scaled experiments) the scaling issue appears. When performing scaling analyses the need for validation and the need for validation procedures appear. Therefore, both statements are true: (a) validation is part of scaling; (b) scaling is part of validation. Definitely scaling and validation processes are intimately connected.

Answer 2.6

Nuclear thermal hydraulics equations (e.g., balance and constitutive equations), part of SYS TH codes, are imperfect: modeling approximations are necessary for their formulation and solution. Therefore, errors are expected in the results of their application, e.g., to demonstrate the safety of nuclear reactors. Those errors have a complex nature. Specific methods, called uncertainty methodologies exist and allow the quantification of the errors.

The application of thermal hydraulics modeling to any situation of industrial interest may reveal irrelevant without uncertainty evaluation.

Answer 2.7

Please consider answer 1.14

Any final application of PSA to accident analyses in nuclear reactors implies a nuclear thermal hydraulic calculation. The qualification level (because of the adopted simplified approaches, e.g., reduced number of control volumes) of nuclear thermal hydraulic calculation and the consideration of uncertainty in results, including bifurcations and partly working components (e.g., reduced delivery flow by a centrifugal pump, partially open valve) remain open issues. The consideration of reliability of passive systems constitutes an additional critical issue in PSA.

Answer 2.8

A suitable SYS TH code, plus validation, scaling and uncertainty procedures constitute the key elements of BEPU. Specific papers in the literature (see Chapter 14) identify a number of subsidiary elements. Key issue in applications of BEPU is the connection between those elements and the licensing process.

Answer 2.9

In general, as in other disciplines, the international scientific community is the stakeholder for nuclear thermal hydraulics. One may distinguish the following sectors that contribute to the progress:

- International Institutions like NEA and IAEA; in the area of regulators, WENRA.
- Journals like NED and STNI.
- International Conferences like NURETH and NUTHOS.
- Regulators, e.g., USNRC and KTA.
- Laboratories and Organizations (including Industries) hosting large projects (primarily aimed at performing experiments) like KAERI (Korea), JAERI (Japan), INL (USA), OSU (USA), EREC (Russia), SIET (Italy), CEA (France), FRAMATOME (Germany), LUT (Finland), BHABHA (India).
- National Institutions like ANS, ASME.

Answer 2.10

Passive systems of interest here are those systems where thermal hydraulic phenomena occur, like natural circulation. The objective for those systems is to remove thermal power from the core. The unavoidable mechanical parts or components like, piping and valves, including their working conditions are not of interest here.

The designers of water-cooled nuclear reactors adopted passive systems since the dawn of nuclear technology. Those systems became “fashion systems” after the Chernobyl accident (designers proposed new passive concepts for advanced reactors and only AP-1000 is built, so far, year 2023).

Gravity force is at the basis of passive systems. One may also state that passive systems aim at performing any given task (noticeably core cooling, as mentioned above) adopting forces that are 10—100 lower than forces (typically pump head) available for active systems performing the same tasks.

Gravity force has the highest reliability, but this is not the case for the time varying small forces generated by gravity in a thermal hydraulic environment (or gravity heads; here, one may consider the case of a chimney in a room: it works for hours or days, and sudden may fail with smoke entering the room).

The evaluation of passive system performance needs nuclear thermal hydraulics. The issues challenging the application of nuclear thermal hydraulics are:

(a) Predicting phenomena that evolve with very low fluid velocities, including zero velocities and flow reversal occurrences.
(b) Instability phenomena, not only in boiling-condensing systems.
(c) Interaction among unavoidable parallel flow paths that establish in a nuclear reactor cooling loop.
(d) Minor changes in boundary conditions not known nor under control of designers (like leakages, heat losses), therefore not part of modeling, substantially affect the performance of the system.

All of the above gave origin to a new research-technology branch, at the beginning of the present century, in-between nuclear thermal hydraulic and PSA: we refer to this as the reliability of passive systems.

Chapter 3

Answer 3.1

We first estimate the number of bubbles that are present in a cut section considering the total linear interfacial drag area and the mean diameter of the bubbles:

$$\pi dN = 89.6\ \mathrm{m} \rightarrow N = 19014$$

Then we can calculate the area occupied by the bubbles and the void fraction

$$N\pi d^2 / 4 = \alpha\pi D^2 / 4 \rightarrow \alpha = 0.12$$

Answer 3.2

Since the flow regime is inverted annular flow, a film of steam is formed at the walls of the pipe and the liquid water accumulates in the center. The linear interfacial heat transfer area can be calculated as the line that separates the liquid and the steam $2\pi r$ where r is the radius of the center water column. We know that:

$$\alpha = \frac{A_{\mathrm{steam}}}{A_{\mathrm{steam}} + A_{\mathrm{liquid}}} = 1 - \frac{A_{\mathrm{liquid}}}{A_{\mathrm{pipe}}} \rightarrow A_{\mathrm{liquid}} = (1-\alpha)A_{\mathrm{pipe}}$$

considering that:

$$A_{\mathrm{liquid}} = \pi r^2 \rightarrow \mathrm{solution} = 2\pi\sqrt{\frac{A_{\mathrm{liquid}}}{\pi}} = 1.52\ \mathrm{m}$$

Answer 3.3

Since the flow regime is annular flow, a film of water is formed at the wall of the pipe. The void fraction is expressed as:

$$\alpha = \frac{A_{\text{steam}}}{A_{\text{steam}} + A_{\text{liquid}}} = \frac{A_{\text{steam}}}{A_{\text{pipe}}}$$

To calculate A_{steam} we just need to know the outer radius of the steam, which is: $r_{\text{steam}} = D/2 - \delta = 0.345$. Then the solution can be calculated as:

$$\alpha = \frac{\pi(D/2 - \delta)^2}{\pi D^2/4} = 0.97$$

Answer 3.4

The Fauske void correlation relates the slip ratio ($S = \frac{v_g}{v_l}$) and is expressed as:

$$S = \frac{v_g}{v_l} = \sqrt{\frac{\rho_l}{\rho_g}}$$

We know that the quality can be expressed as a function of the slip ratio and the void fraction and thus we can calculate the quality as:

$$x = \frac{1}{1 + \frac{1-\alpha}{\alpha}\frac{\rho_l}{\rho_g}\frac{1}{S}} = 0.014$$

Answer 3.5

The homogeneous equilibrium model (HEM) fixes the slip ratio ($S = \frac{v_g}{v_l}$) to unity. Then, considering that the quality can be expressed as a function of the slip ratio and the void fraction, we can calculate the quality as:

$$x = \frac{1}{1 + \frac{1-\alpha}{\alpha}\frac{\rho_l}{\rho_g}\frac{1}{S}} = 0.00048$$

Answer 3.6

In normal operation with liquid forced convection heat transfer mode, one can use the Dittus-Boelter correlation to find out the Nusselt number which can be used then to calculate the heat transfer coefficient:

$$Nu = 0.023Re^{0.8}Pr^{0.4}$$

We first calculate the Reynolds and Prandtl numbers:

$$Re = \frac{\rho v D_h}{\mu} = 541163 \; Pr = \frac{\mu C_p}{k_{H_2O}} = 0.89$$

Now, using the Dittus-Boelter correlation we can calculate the Nusselt number:

$$Nu = 0.023Re^{0.8}Pr^{0.4} = 847$$

And finally, the heat transfer coefficient is:

$$Nu = \frac{h \cdot D_h}{k_{H_2O}} \rightarrow h = \frac{Nuk_{H_2O}}{D_h} = 36 \text{ kW} / \text{m}^2\text{K}$$

Answer 3.7

The following are the suggested numbers or correlations; different numbers or correlations may also be related but a justification should be provided:

- CCFL: Superficial nondimensional velocity which is then used in the Wallis correlation.
- Stratification: Froude number relates the ratio between inertial and gravitational forces.
- Convection heat transfer: Nusselt number relates the convective heat transfer with the conduction heat transfer.
- Laminar flow: Reynolds number is used to determine whether a flow is laminar or turbulent.
- Inverted annular film boiling: Nusselt number.

Answer 3.8

The following are the suggested numbers or correlations; different numbers or correlations may also be related but a justification should be provided:

- Chocking flow: Mach relates the flow velocity and the sound speed within the fluid
- offtake model: This can be related to stratification and thus we can relate it to the Froude number.
- end of natural circulation: This can be related to stratification and thus we can relate it to the Froude number.
- turbulence: clearly related to the Reynolds number
- Annular film heat transfer: Nusselt number

Answer 3.9

We can apply the definition of the hydraulic diameter

$$D_h = \frac{4A}{P_{\text{wetted}}} = \frac{4\left(78.5 \cdot 78.5 - 25\pi(10.75/2)^2\right)}{4 \cdot 78.5 + 25\pi \cdot 10.75} = 13.44 \text{ mm}$$

Considering a PWR core, the fuel assemblies do not have walls in between and the fuel rods are homogeneously distributed, therefore, we can estimate the D_h from the typical subchannel between four fuel rods. For that purpose, we can use the pitch, which is provided in the figure:

$$D_h = \frac{4A}{P_{\text{wetted}}} = \frac{4((14.3)^2 - \pi(10.75/2)^2}{\pi \cdot 10.75} = 13.47 \text{ mm}$$

Answer 3.10

We can apply the definition of the hydraulic diameter

$$Dh_{\text{wetted}} = \frac{4A}{P_{\text{wetted}}} = \frac{4\left((3.1 + 2R)^2 - \pi R^2\right)}{4 \cdot 3.1 + 2\pi R} = 8.32 \text{ mm}$$

Now for the perimeter, we only need to consider the heated regions.

$$Dh_{\text{heated}} = \frac{4A}{P_{\text{heated}}} = \frac{4\left((3.1 + 2R)^2 - \pi R^2\right)}{2\pi R} = 11.78 \text{ mm}$$

Answer 3.11

We can apply the definition of the hydraulic diameter.

$$Dh_{\text{wetted}} = \frac{4A}{P_{\text{wetted}}} = \frac{4\left((3.1 + 2R)^2 - \pi R^2\right)}{4 \cdot 3.1 + 2\pi R} = 8.32 \text{ mm}$$

Now for the perimeter, we only need to consider the heated regions.

$$Dh_{\text{heated}} = \frac{4A}{P_{\text{heated}}} = \frac{4\left((3.1 + 2R)^2 - \pi R^2\right)}{0.75\pi R} = 15.7 \text{ mm}$$

Answer 3.12

We can apply the definition of the hydraulic diameter.

$$Dh_{\text{wetted}} = \frac{4A}{P_{\text{wetted}}} = \frac{4\left((2.5 + R) \cdot 12.6 - 0.5\pi R^2\right)}{2 \cdot 2.5 + 12.6 + (12.6 - 2R) + \pi R} = 6.28 \text{ mm}$$

Now for the perimeter, we only need to consider the heated regions.

$$Dh_{\text{heated}} = \frac{4A}{P_{\text{heated}}} = \frac{4\left((2.5 + R) \cdot 12.6 - 0.5\pi R^2\right)}{\pi R} = 14.99 \text{ mm}$$

Answer 3.13

We can apply the definition of the hydraulic diameter.

$$Dh_{\text{wetted}} = \frac{4A}{P_{\text{wetted}}} = \frac{4\left((2.5 + R)^2 - \pi R^2/4\right)}{2 \cdot 2.5 + 2(2.5 + R) + \pi R/2} = 5.17 \text{ mm}$$

Now for the perimeter, we only need to consider the heated regions.

$$Dh_{\text{heated}} = \frac{4A}{P_{\text{heated}}} = \frac{4\left((2.5 + R)^2 - \pi R^2/4\right)}{\pi R/2} = 18.68 \text{ mm}$$

Answer 3.14

Since we have a total of 30 equal tubes, the hydraulic diameter will be the diameter of a single tube $D_h = 19$ mm.

For the riser part of the SG, the calculation of the hydraulic diameter becomes very complex and will also depend on the exact elevation at which we consider a cut section. Considering the figure we can follow different approaches. The first one would be to calculate the D_h precisely by calculated all wetted sections and the flow path area. A more direct approach would be to consider that the geometry is more or less uniform so we can consider a small region (as a subchannel) and calculate the D_h in that region, we can extrapolate then to the total section. Visually we can estimate that the flow path in between spaces is similar to the path inside the tubes (perhaps smaller) so the D_h will be around 15 mm.

Chapter 4

Answer 4.1

PIRT (Phenomena Identification and Ranking Table) was introduced by USNRC at the end of the 1980s of the previous century, based on early work on phenomena identification at NEA (framework of activities also attended by USNRC).

One may characterize the three parts of the questions as follows:

(a) Phenomenon: this is "a fact or situation that is observed to exist or happen, especially one whose cause or explanation is in question." Indeed, although various definitions exist, a phenomenon corresponds to something observed and measured in the nature.

(b) Phenomena Identification: this implies a picture or a framework. The accident scenarios (typically DBA or DBC) for water-cooled nuclear reactors constitute the framework. Experiments in facilities simulating those reactors constitute a need for identification.

(c) Ranking: this implies a motivation and a metric. Examples of motivation (for ranking) can be impact of phenomena in safety analysis (e.g., impact on predicting an assigned parameter), or quality level of models to calculate the phenomenon, or combinations. The metric can be qualitative or quantitative and, in general, follows the proposal of analysts who are proposing the ranking.

It is clear that the items (a) and (b) bring to a finite set of topics (i.e., the list of thermal hydraulic phenomena) around which a consensus of the international community can be established and actually has been established. Otherwise, the ranking opens to a wide variety (nearly infinite) of solutions.

Answer 4.2

The key issue is (looking at the answer 4.1) that two groups of experts asked to provide a ranking to an assigned list of phenomena definitely produce different answers. This is well established in technical literature since the introduction of the PIRT process. Allowing the interaction of those groups of experts largely decreases the differences. A simple example is the ranking of phenomena "pressure drop" (PH1) and "CCFL" (PH2), during an assigned accident scenario: one group of experts may rank high and low the two phenomena, respectively. The unavoidable concern is as follows: PH1 affects PH2 and any difference in ranking is unjustified. Furthermore,

PH1 and PH2 for assigned parameters ranges may be predicted with high or low precision by available models; however, when changing the parameter ranges (possibly during the same accident scenario) the prediction precision may largely vary.

The process of phenomena ranking brings to useful, possibly endless, discussions, which are suited for establishing a common basis for understanding by different analysts. Here, we cannot refrain to report in Fig. 4.2a, the thoughts by Novak Zuber (who was the main actor for the introduction of phenomena ranking), 20 years after the introduction of PIRT.

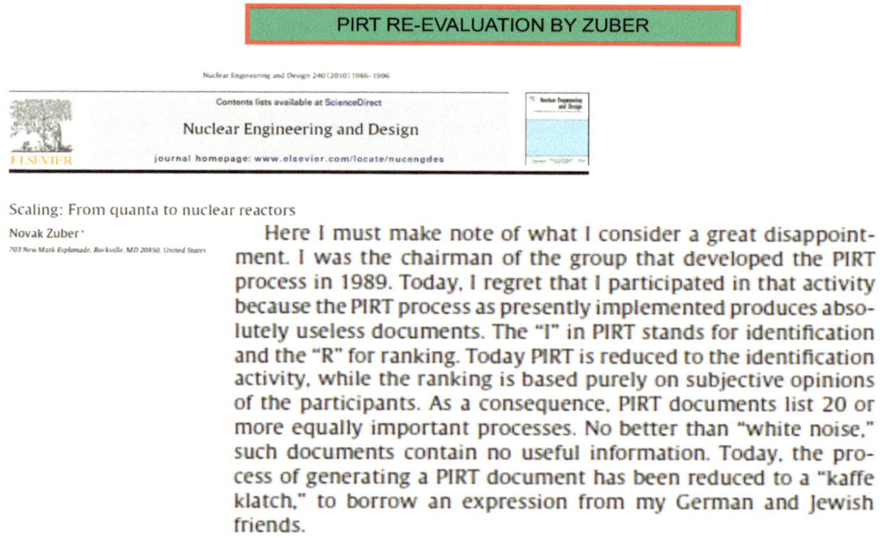
PIRT RE-EVALUATION BY ZUBER

Nuclear Engineering and Design 240 (2010) 1986–1996

Contents lists available at ScienceDirect

Nuclear Engineering and Design

journal homepage: www.elsevier.com/locate/nucengdes

ELSEVIER

Scaling: From quanta to nuclear reactors

Novak Zuber*

703 New Mark Esplanade, Rockville, MD 20850, United States

Here I must make note of what I consider a great disappointment. I was the chairman of the group that developed the PIRT process in 1989. Today, I regret that I participated in that activity because the PIRT process as presently implemented produces absolutely useless documents. The "I" in PIRT stands for identification and the "R" for ranking. Today PIRT is reduced to the identification activity, while the ranking is based purely on subjective opinions of the participants. As a consequence, PIRT documents list 20 or more equally important processes. No better than "white noise," such documents contain no useful information. Today, the process of generating a PIRT document has been reduced to a "kaffe klatch," to borrow an expression from my German and Jewish friends.

Figure 4.2a Synthesis of Novak Zuber evaluation of PIRT activities 20 years after the introduction of the process.

Answer 4.3

Key elements for proving software adequacy are verification, validation and proving scaling adequacy of individual models including scalability of closure laws. Chapters 13, 14, 21 and, mainly, 22 deal with related discussions.

Answer 4.4

The EMDAP (Evaluation Model Development and Application Process) is a logical frame to prove the applicability and the quality of software adopted for safety analyses of water-cooled (in general, not only water-cooled) reactors.

Considering selected elements or subelements of EMDAP, a not exhaustive and not systematic answer to the last statement of the question is:

- Element 1 (requirements): see Chapter 22.
- Element 1.4, just related to phenomena identification: see Chapters 6 and 15.
- Elements 2.6 and 2.7 (scaling and scaling database): see Chapter 21.
- Element 3.12 (closure laws): see Chapters 9 and 10.
- Element 4.16 (field equation and numerical solution methods): see Chapter 5, 11 and 20.
- Element 4.20 (EM uncertainties): see Chapter 14

Answer 4.5

Nuclear thermal hydraulics (investigations) allows (allow) the confirmation that safety principles and requirements, established independently of individual reactor concepts, are part of the design and the operation of water-cooled reactors, as shown by the acronym DSA (Deterministic Safety Assessment) on the left side of Fig. 2.1. Nuclear thermal hydraulics (investigations) has (have) the same role as above, specifically referring accident analysis.

Answer 4.6

The answer 4.5 puts the basis for answering the present question (question 4.6). A systematic evaluation to identify the capabilities needed for thermal hydraulics and thermal hydraulics software comes from the (thermal hydraulics) phenomena, which constitute the target of accident analysis in WCNR.

Those capabilities, together with features of numerical methods, procedure for the analyses, etc., contribute to arrive at the "thermal hydraulic analysis envelope."

Answer 4.7

We may state, adopting a simplified jargon:

- CFR (Code of Federal Regulation) is the law.
- RG (Regulatory Guides) support the interpretation of the law.
- SRP (Standard Review Plan) provides deep details streamlining the needs fixed by the law
- FSAR (Final Safety Analysis Report): is the end-result of the activities proving the fulfillment of the law referring to individual reactors.

Referring to the second (and last) statement of the question, FSAR considers all requirements involved with CFR, RG, and SRP documents. Obviously, the current answer could enter into deeper details.

Answer 4.8

The needs and the requirements of continuously updated CFR, SRP and RG are consistent with science and technology advancements (in nuclear thermal hydraulics). Those needs and requirements constitute the topic of chapter 4. The same needs and requirements are essential, a) to drive the development of thermal hydraulic codes adopted in safety and licensing and, b), to evaluate the acceptability of those codes (typically SYS TH codes).

Answer 4.9

First, let us confirm (as correctly noted in the question) that IAC document (i.e., CFR) and RG documents belong to different levels of the hierarchized set of documents, which fix the licensing process of WCNR. The main difference is as follows:

- IAC document fixes the needs and/or the targets of analyses aimed at the demonstration of the safety (and/or safety margins) of an assigned reactor or reactor-design; otherwise, RG propose roadmaps or analytical paths, including assumptions to arrive at the same demonstration.

The main similarity is:

- Both IAC and RG documents deal with same accident scenarios, phenomena and relevant (safety) parameters.

Furthermore,

- RG 1.70 proposes (and obliges to pursue) a conservative approach in the safety demonstration; otherwise, RG 1.203 opens to the possibility to adopt realistic SYS TH codes, or the codes that are considered in the present book, for the evaluation of safety margins. Both of those RG are consistent with the consolidated knowledge at the time of their issuing.

Answer 4.10

The validation (e.g., discussed in section 4.3) implies the availability of experimental data. Furthermore, the following elements of EMDAP (Fig. 4.3) refer to experimental data (not an exhaustive list):

- 1.4 (i.e., phenomena characterization [and ranking] needs experiments);
- 2.7 (i.e., ... perform IETs and SETs ...);
- 4.15 (i.e., assess scalability of models);
- 4.19 (i.e., assess scalability of integrated calculations).

Chapter 5

Answer 5.1

HEM (Homogeneous Equilibrium Model) is a synonym of EVET (Equal Velocities Equal Temperatures). HEM is the simplest modeling approach for two-phase flow, specifically in transient situations. The two-phase mixture is taken as a special (homogeneous) fluid with properties (noticeably, density) determined by void fraction. HEM is taken as reference to address the question.

EVUT (Equal Velocities Unequal Temperatures) is the same as HEM, but at least one equation is added to determine two different temperatures for liquid and steam. We can also state that EVUT modeling allows thermal nonequilibrium.

UVET (Unequal Velocities Equal Temperatures) is the same as HEM, but at least one equation is added to determine two different velocities for liquid and steam. We can also state that UVET modeling allows mechanical nonequilibrium. Drift flux model provides the added equation in majority of cases.

UVUT (Unequal Velocities Unequal Temperatures) modeling combines EVUT and UVET features and constitutes the state of art qualified modeling for SYS TH codes established towards the end of the 1970s of the previous century. UVUT modeling is synonym of six PDE (Partial Differential Equations). Six PDE imply sis unknowns.

Answer 5.2

Zuber and Findlay introduced the drift flux model in 1965. Original words from those authors are "*A general expression which can be used either for predicting the average volumetric concentration or for analyzing and interpreting experimental data is derived. The analysis takes into account both the effect of nonuniform flow and concentration profiles as well as the effect of the local relative velocity between the phases. ... are compared with experimental data ... liquid-gas mixtures in adiabatic, vertical flow over a wide pressure range. Good agreement ....*"

One may summarize as: (a) C_0 and v_{gj} are the involved empirical quantities; (b) two phase vertical flow, upward direction is involved; (c) a correlation, independent of mass, energy and momentum equations connects void fraction and (steam) velocity. Easy to conclude at this point that drift flux allows passing from HEM to UVET.

In case of UVUT, different velocities for liquid and steam depend upon the constitutive terms, namely interfacial drag; drift flux is not necessary to compute two different velocities for liquid and steam. However, drift flux model allows an optimum prediction of friction pressure drops in some of SYS TH codes.

Answer 5.3

Related to UVUT, two pressures, one for liquid and one for steam phase, are present in UVUTUP model. This implies the appearance of seven unknowns: six unknowns like in UVUT, plus one phase pressure. Among others, Bourè and Lyczkowski starting from the end of the 1970s, discuss the UVUTUP model. Without entering into details, two issues for the applicability of the model are:

- The seventh equations, necessary to make solvable the model, may not have an origin from principles of thermodynamics and mechanics as the remaining six equations.
- The measurement of two different pressures for liquid and steam in a two-phase flow system (i.e., needed for the validation of the UVUTUP model) is not within the current technological capabilities.

Furthermore, one should observe that the interfacial pressure term, which appears in the formulation of some SYS TH codes, has little or no connection with two fluid pressures; rather the interfacial pressure constitutes a numerical term allowing the solution of UVUT equations in a wide domain of (unknown-) parameter ranges.

Answer 5.4

Kocamustafaogullari and Ishii introduced the idea of Interfacial Area Transport (IAT) in 1995. The IAT equation derives from "*the statistical model of fluid particle number transport equation*," according to the words of those authors. Furthermore, "*the resulting equation includes the source and the sink terms due to the particle interactions and interfacial phase change.*"

Furthermore, interfacial area concentration (IAC), although not explicitly calculated by all codes, is necessary to take into account of, e.g., heat transfer and pressure drop between the two phases. One may state that IAT consists of a balance equation for IAC including source terms.

The idea of the IAT is to focus on a key variable-term, not always directly considered in balance equations (UVUT model), i.e., the interfacial area. Then, the IAT aims at introducing additional equations to calculate the new term and (unavoidably) at adopting the IAT to calculate (or to support the calculation of) other terms part of constitutive equations of UVUT model.

The idea is great, but following almost 30 years since the introduction of the idea, although its feasibility is demonstrated, the difficulty of a systematic validation, the tight dependency of interfacial area upon system geometry (primarily) and other thermal hydraulic system conditions, prevent its applicability and use for practical purposes (so far).

Answer 5.5

Presence of multi-fields implies the same phase (liquid or steam) belonging to fields having different velocities or temperatures. Therefore, each time one adds a field to the modeling (e.g., UVUT) at least one additional unknown appears: the temperature or the velocity of the new field, typically the velocity. This needs additional equation(s).

One may note that fields depend upon flow regimes and, in some cases, different temperatures and velocities (i.e., a spectrum of values) in a cross section or in the integration volume may characterize one phase. This suggests that whatever reasonable number of field with current modeling approaches may not reflect the reality into detail.

The modeling of droplet field with properties (temperature and velocity) different from the annular field is a typical example of multi-field modeling. The droplet field modeling appears unavoidable to predict heat transfer in post-dry-out, reflood conditions and pre-CHF modeling particularly in the presence of spacer or mixing vane grids. Typical challenges (not an exhaustive list) are: (a) fixing the average size of droplet; (b) considering the coalescence and the split of droplets; (c) calculating entrainment and de-entrainment rates.

Answer 5.6

The key difficulty (answer to the first question) is having access to suitable instrumentation that allows 3D measurements. The additional key difficulty is the dependence of 3D phenomena upon local geometric details, i.e., making impossible the consideration of the fully developed flow hypothesis. Efforts in the area of 3D modeling including current approaches may fail owing to both challenges above.

Owing to above considerations, 3D modeling approaches by SYS TH codes focus on selected components. CFD modeling of turbulence for two-phase, multi-field, geometry-dependent and transient conditions is not available or far away an acceptable validation level.

3D core models are more popular than hot leg 3D modeling. One reason is the direct connection between core models and safety parameters for the nuclear reactor. Nonetheless, 3D phenomena may appear in hot legs of typical PWR and affect system performance.

3D modeling of steam generators, both primary and secondary side, is essential to simulate some DBA scenarios. Distinguishing individual tubes of groups of tubes is a surrogate solution for the 3D modeling that may work under specific conditions in relation to which experimental data are available.

The answer to the present question is also applicable to SMR and micro-reactors where no attempt is made (or can be made) to avoid the occurrence of 3D phenomena. The only possible answer, in all situations, is to develop suitable uncertainty methods that take into account of modeling inadequacies. Again, the challenge here is the availability of experimental data.

Answer 5.7

Heat flux partitioning (the same partitioning is necessary in the case of pressure drops) derives from the availability of global or average (HEM typical) measurements and related empirical correlations: the attribution of one available value to steam, liquid, individual fields and multi-direction is necessary and raises the issue in the main question. Only one answer may address the subquestions (a)–(c): the more detailed is the modeling (e.g., 3D and multi-field) the larger is the number of (nearly or partly arbitrary) choices that model developers must introduce.

Answer 5.8

More than a unique (valuable) answer may exist to this question and any answer may not be a consensus answer within the science and technology community.

The possibility of angular momentum conservation is known since centuries. Obviously, the consideration of angular momentum conservation implies the availability of additional equations and unknowns.

During the 1950s of the last century, i.e., at the dawn of development of nuclear thermal models, fluid dynamic specialists considered the angular momentum: added complexity of modeling, low impact upon results and cylindrical pipe equivalent flow of technological interest, suggested stopping the consideration of angular momentum (together with small available computational resources).

Within the reality (current large reactors, SMR, micro-reactors), the fluid rotation generated by centrifugal pumps and the three-dimensional flow paths unavoidable for bounding the flow (e.g., in a vessel, but not only) should suggest reconsidering the assumption made in the past to neglect angular momentum.

Answer 5.9

Let us consider, (a) the UVUT set of six PDE and, (b), the blow-down from a pressurized tank or vessel into an ideal bounded (with ideal walls) thermally isolated containment. During the irreversible blow-down occurrence, the pressure decreases in the tank and increases in the containment until reaching an equilibrium.

The blow-down phenomenon is irreversible. Here, we consider the total system entropy for an isolated system, before and after an assigned (i.e., the blow-down) thermodynamic evolution. The total entropy of the fluid in the tank and in the containment plus the entropy associated with the (high) temperature of the tank walls before the blow-down start, should be less or equal than the total entropy of the fluid plus the entropy associated with the (low) temperature of the tank wall, at the end of blow-down.

The sources of entropy during and due to the thermodynamic evolution are (not an exhaustive list):

- The friction (e.g., at the exit nozzle of the blow-down tank).
- The heat transfer from the tank walls to the fluid (here we can assume, to simplify, that the walls of the tank exchange thermal power only with the fluid inside the tank, or that an adiabatic condition exists on the outer wall of the tank in contact with the containment).
- The interfacial friction and heat transfer between liquid and steam both in the tank and in the containment during the vaporization and condensation processes (to further simplify, the initial conditions in the containment may those fixed by saturated steam at a pressure much lower than the tank pressure).

The reader may perform an easy UVUT model calculation. The reader may compute the entropy before the start and at the end of the blow-down. The latter value should be equal or larger than the former value. This may not be the case, because (neglecting numerical dissipation of mass, momentum, and energy) empirical formulations needed at the three bullet items above, are "imperfect" and do not necessarily account for the second principle of thermodynamics.

SYS TH (and CFD) code calculations, having their solution depending upon empirical correlations, may contradict the entropy increase principle when an irreversible thermodynamic process is modeled.

Answer 5.10

The numerical solution in established UVUT 1D based SYS TH codes consider (owing to numerical constraints) a continuous varying flow area along the flow path axis. Therefore, code users are responsible of pressure drop (i.e., they supply pressure drop coefficients according to the formulation requested by the code manual) when discontinuities are present along the flow path. This situation is modified in some recent SYS TH codes where those pressure drop coefficients are computed by the code according to geometry data supplied by user, without proving that those coefficients allow the simulation of the actual system pressure drops[1].

The distinction requested in the question (bend, connection between downcomer and lower plenum, jet-pump, etc.) is not applicable. The challenge-issue discussed in the previous paragraph apply to any geometric discontinuity (to a different extent in terms of impact upon calculation error).

CFD code calculation may be immune from this problem, because the calculation of pressure drop at geometric discontinuities at least in single-phase flows is within the code capabilities. However, another challenge characterizes the CFD code calculations of pressure drops at geometric discontinuities in both single- and two-phase flow conditions: the amount of error is unknown and, in general, not negligible.

Accurate pressure drop measurements at geometric discontinuities (suitable ranges of Re and void fraction) are necessary to guarantee a qualified SYS TH code calculation or to confirm CFD predictions.

Answer 5.11

Definitely, UVUT 1D modeling has the capability to predict with reasonable accuracy (i.e., typically bounded by uncertainty calculations) system parameters in situations where 3D effects play a relevant role. The motivation for this is that a large number of constitutive equations, part of UVUT 1D modeling, derive from experiments where those effects unavoidably occur; the derivation of related empirical terms may as basis a reference 1D flow configuration. Thus, the constitutive terms embed the 3D effect and "supply" related information to the code calculations. This justifies proper calculation results (3D situation) outside the domain of original qualification for the 1D code.

Chapter 6

Answer 6.1

The objective and the motivation for deriving the thermal hydraulic phenomena (THP) is the validation of SYS TH codes considering (originally) selected LWR and the envelope of DBA scenarios.

NEA/CSNI specialists group derived pioneering lists of THP starting from the end of the 1980s in the previous century. Aksan, D'Auria and Glaeser proposed the 116 THP list in 2018. The NEA/CSNI phenomena, including those part of the 116 THP

[1] One may note that the calculation condition (i.e., where the code computes the pressure drop coefficients at geometric discontinuities) may reveal worse or less accurate than the condition where the user supplies his own pressure drop coefficients, because neither the code developers nor the code user have responsibility for the inaccurate value of the pressure drop coefficient.

list, are restricted to the consideration of test scenarios measured in ITF or SETF experiments. Bestion (see also D'Auria and Bestion, 2022) proposed to expand the THP list boundaries to include phenomena dealing with terms that are part of the constitutive equations of SYS TH codes.

Answer 6.2

A separate effect test facility (SETF) aims to investigate one of a few THPs or a time span part of a DBA scenario (see previous answer) considering scaling in the design of the test section and of experiments.

An Integral (Effect) Test Facility (ITF) aims at investigating the overall behavior of the primary systems (all THP expected in a DBA scenario) considering scaling in the design of the facility itself and of experiments.

A specific scaling investigation connecting ITF and SETF to the respective prototype is necessary for both SETF and ITF design and design of experiments.

Answer 6.3

Basic experiments deal with fundamental (or basic) THP. Key difference or feature of basic experiment are:

- No scaling analysis is strictly necessary for the planning, although concerned parameter ranges and combinations of those ranges "should consider" respective ranges in nuclear reactors.
- The size of experiments may not represent prototype conditions.
- Basic experiments may aim at terms of the balance equations.

Answer 6.4

Fig. 6.4a provides some details (spread of predicted data) for predictive capabilities by SYS TH codes for the TPCF phenomenon. Paper referred in the figure is by Lanfredini et al. (2020). D'Auria et al. (2021b), provide further comments on this topic. See also answer 10.7.

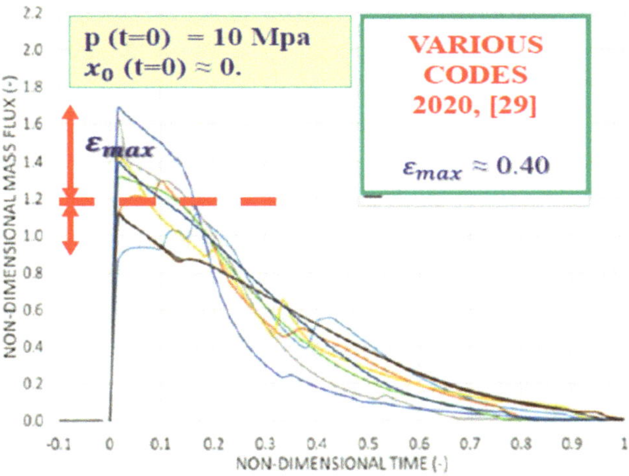

Figure 6.4a Sketch for predictive capabilities of SYS TH codes in relation to the TPCF phenomenon.

Answer 6.5

Fig. 6.5a provides some details (spread of predicted data) for predictive capabilities by SYS TH codes for the CHF phenomenon. One may observe temperature differences in rod surface before the occurrence of the CHF, spread of results in time of occurrence of CHF and temperature differences in rod surface after the occurrence of CHF. In the paper by D'Auria et al. (2021b) the reader may find further comments on this topic.

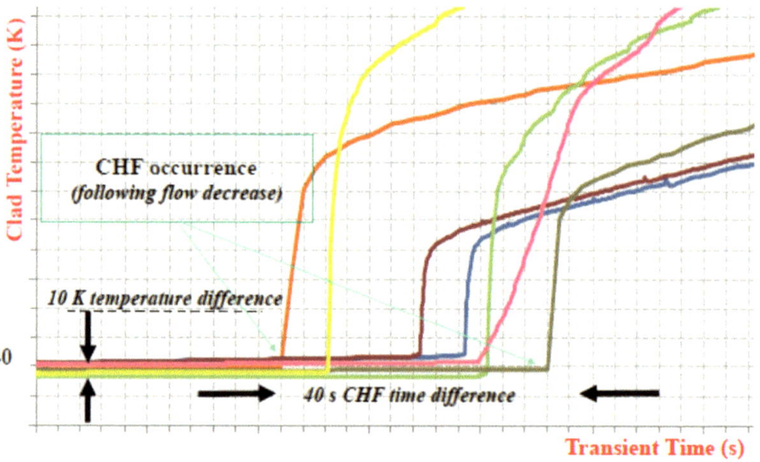

Figure 6.5a Sketch for predictive capabilities of SYS TH codes in relation to the CHF phenomenon.

Answer 6.6

Fig. 6.6a provides some details (spread of predicted and measured data) for predictive capabilities by SYS TH codes for the CCFL phenomenon, taken from Al Issa and

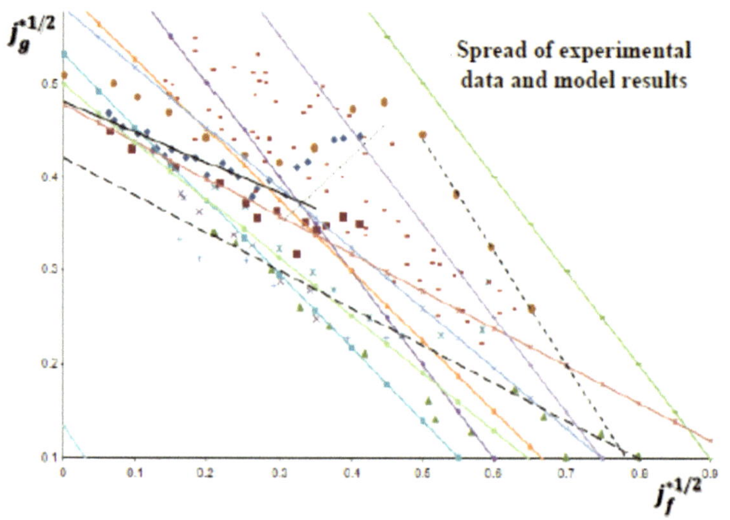

Figure 6.6a Sketch for predictive capabilities of SYS TH codes in relation to the CCFL phenomenon.

Macian (2011). In the paper by D'Auria et al. (2021b) the reader may find further comments on this topic.

As discussed by Glaeser, in chapters 10 and 13, the geometric size of the experimental apparatus where CCFL phenomenon is measured is relevant for the characterization of the same phenomenon. In particular, CCFL occurring at an assigned value of J_g in a small-scale facility may occur, with the same geometric configuration, at a higher value of J_g when the size of the facility increases. Therefore, flooding phenomena and CCFL are highly scale dependent.

Answer 6.7

SYS TH codes are capable of predicting, with reasonable (low) error, pressure and average temperature increase in containment following any interaction with primary circuit, mainly the LOCA occurrences. The average values of heat transfer coefficients, derived from experimental data, take into account of complex local phenomena and "somewhat drive" reasonable predictions.

The situation is different if one looks at local phenomena including stratification of gases, distribution of hydrogen, turbulence generated by heat transfer and condensation in the presence of noncondensable. In such conditions, accuracy of predictions is generally questionable and highly dependent upon specific scenarios.

Answer 6.8

SYS TH, CFD and devoted subchannel codes are capable to predict subchannel flow conditions in individual subchannels, a few assemblies and the entire core in single- and two-phase flow conditions.

The answer to the question must specify:

- Which type of code is concerned including its applicability for an assigned investigation.
- Whether single- or two-phase conditions are concerned.
- How many subchannels or fuel assemblies constitute the domain of integration.
- The type of subchannels (e.g., square, hexagonal, or cylindrical).
- Whether spacer or mixing vane grids are present.
- Whether the calculation focus is pressure drops, mixing or CHF prediction.
- Whether transient (e.g., including LOCA) or steady-state (e.g., focusing to maximum removable power) conditions are concerned.

Combination of those items may bring different answers.

In general, looking at CHF or pressure drop prediction in a bundle and considering unavoidable experimental uncertainties (then questionable validation of any code), inaccuracies in predictions are undistinguishable considering the code categories. In the case of CHF, predictive capabilities can be observed in Fig. 6.5a. In other words, and looking at conditions where SYS TH, CFD and subchannel codes are applicable (i.e., conditions of the test case within the boundaries of validity of those codes), spread of results from different predictions does not allow determining which code typology is preferable.

Answer 6.9

Figs. 6.9a and 6.9b depict the role of phenomena and the process for the derivation of phenomena in nuclear thermal hydraulics, respectively. The history of nuclear thermal hydraulics, Fig. 2.3 in chapter 2, is also relevant to address the question.

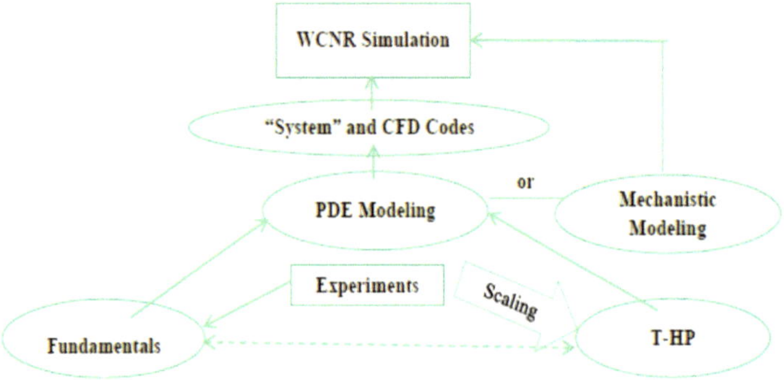

Figure 6.9a Role of phenomena (T-HP in the sketch) in nuclear reactor thermal hydraulics.

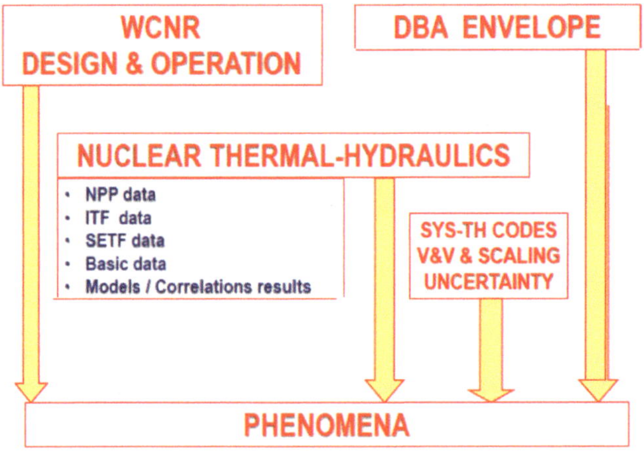

Figure 6.9b Process for derivation of phenomena in nuclear reactor thermal hydraulics.

The reader should (easily) recognize the terminology in both figures above and in Fig. 2.3 and the interconnection with different parts of the book. The answer 6.10 (below) provides further elements dealing with the connection between accident scenarios and phenomena and parameters (namely the sections in the book where one may find deeper discussion).

Answer 6.10

A discussion about PIRT (Phenomena Identification and Ranking Table) challenges can be found in answer 4.2 (i.e., PIRT is a key element of EMDAP). Here we add that Tables 15.3 and 15.4 and related text in chapter 15 of the book deal with the cross-link between phenomena and accident scenarios passing through the identification of (general) parameters[2]. The following comments apply:

[2] As reported in section 15.2.1, Table 15.4 considers the 113 phenomena list and not the updated 116 phenomena list. This does not affect current discussion.

- The cross connection between phenomena and accident scenario should be a prerequisite for PIRT (available from this book).
- Phenomena characterized by the same (general) parameters shall not have a different ranking.
- Specific parameters might be added for the identification of each phenomenon (text in chapters 6 and 15 is useful in this connection): phenomena characterized by the same specific parameters shall not have a different ranking.
- In different words, to disambiguate the description of phenomena the identification of general (done in the book) and specific (not systematically done in the book) parameters is necessary. This process provides a not sufficient hint for the ranking of phenomena (e.g., identification of parameters ranges and impact of parameters upon individual phenomena is necessary).

Chapter 7

Answer 7.1

Conjugate heat transfer refers to the connection between convection and conduction heat transfer and to the related numerical solutions in modeling: heat conduction equations in solid are solved simultaneously with equations for convection heat transfer (at the boundary between solid and fluid) and fluid balance equations. Soviet Union scientists proposed the terminology, e.g., Perelman towards the beginning of the 1960s.

All SYS TH codes have the capability of modeling the conjugate heat transfer.

Answer 7.2

The following nomenclature is not necessarily acceptable to the entire scientific community.

The Fourier equations deals with the solution of conduction heat transfer in solids (primarily), which derives from fundamental principles of thermodynamics. The Fourier law is an empirical law where the definition of heat transfer coefficient and the assumption of linearity for temperature change versus geometrical abscissa orthogonal to flow direction are relevant.

The models of SYS TH codes adopt (better, take into account in numerical solution of) both Fourier equation and Fourier law.

Answer 7.3

We consider the modeling of an individual fuel pin. We start from the Fourier equation (without boundary conditions) solved within the solid domain assumed as homogeneous (i.e., only solid UO2 considered, w/o gap and w/o clad modeling)[3]:

$$\rho c_p \frac{\partial T}{\partial t} = \nabla k \nabla T + q'''$$

[3] This assumption is drastic and un-needed in numerical solutions (e.g., adopted in SYS TH codes). It is considered here to arrive at an easy analytical solution which is indicative (and valid) for the purpose of this exercise.

where symbols have a meaning understandable. We get, assuming steady state, radial symmetry and neglecting axial effects (again without boundary conditions):

$$\frac{1}{r}\frac{d}{dr}\left[k(T)r\frac{dT}{dr}\right] = -q'''(r)$$

Solving and introducing boundary conditions, namely considering Fourier law and maximum value for centerline fuel temperature and the undue assumption (again un-needed, see the footnote called above), one gets:

$$T_{\text{max-fuel}} = T_f + \frac{q'}{4\pi k}$$

where $T_{\text{max-fuel}}$ is the centerline fuel temperature and q' (linear power of the fuel pin) is geometrically connected with q''' (volumetric generated power in the fuel directly connected with neutron flux). The last equation (please note that the radius of the pins is NOT present in the equation) contributes to confirming and justifying the core formed by thin fuel pins. It is interesting to emphasize what follows:

- Fuel surface temperature, T_f (neglecting the effects of gap and cladding) and linear power, q' (in addition to conductivity "k") affect the maximum centerline temperature.
- The equation is valid provided HTC (and heat transfer area per unit length) at the pins surface is capable of removing the power generated by fission (proportional to q''' and q', see also below).
- High melting value for fuel allows larger values for q'. This justifies the adoption of UO2 instead of metallic U as fuel.
- The equation (considering the thermal capacity of UO2) fixes the amount of energy stored in the fuel pin in nominal operating conditions.
- In case of LOCA, characterized by scram in a time of tenths of a second following vaporization and loss or moderating power, considering the jump from nucleate to film boiling, the energy stored contributes to blowdown PCT (peak cladding temperature).
- Therefore, controlling the q' value allows "the control" of DNBR and of the PCT: decreasing q', i.e., against the economic benefit for the reactor, improves safety margins in nominal operation and accident conditions[4].

Neutron flux including neutron leakages and peaking factors, among other things (e.g., spacer grids) are involved and not considered in the items above.

Answer 7.4

In a (ideal) diagram where the HTC is reported as a function of the surface temperature of the heated wall (e.g., keeping constant the inlet fluid temperature of a hypothetical test section), as soon as the wall temperature overpasses an assigned value (this is the onset of nucleate boiling) the HTC steeply increases. The same occurs in pool boiling conditions. The answer to the question is as follows: overpassing the

[4] The right design solution is: to keep the maximum value of q', consistently with capabilities to compute its value, to ensure proper safety margins. The importance of continuously improving computational capabilities (i.e., the general aim of nuclear thermal hydraulic investigations) directly appears here.

assigned value (close and connected with the saturation temperature corresponding to the fluid system pressure) causes formation of voids, typically subcooled or in a subcooled liquid domain. The formation of voids on the solid surface (which unavoidably move into the liquid) induces, among other things, local turbulence, which promotes steep HTC increase.

Answer 7.5

A multi-D HTC surface derives from the (empirical) definition of HTC, or (with all symbols established, and q'' connectable with q' and q''' in answer 7.3):

$$\text{HTC} = \frac{q''}{T_w - T_f}$$

One may write

$$\text{HTC} == > f\left(x_i, x_j, x_n \ldots x_t\right) = 0$$

where $x_i, x_j, x_n \ldots x_t$, involve sets of variables, e.g., fluid related (e.g., fluid velocity, temperature and void fraction), solid related (e.g., temperature), system conditions (e.g., pressure, vertical, etc.), material properties (e.g., fluid density, viscosity, solid density), surface conditions (roughness, presence of oxide, of microscopic fractures, oxide, CRUD, etc.). Each of i, j, n, and t may vary from 1 to n1, n2, n3 and n4, respectively. We note that only a few variables have the highest importance as far as the HTC values are concerned, i.e., pressure, void velocity, and surface conditions, whereas several dozen variables may enter in the function. Furthermore, we do not discuss here the applicability of the first formula above, or the empirical definition for HTC.

The geometrical definition of the function $f\left(x_i, x_j, x_n \ldots x_t\right) = 0$ is a multi-D surface. Noticeably, the intersection of the multi-D surface with a log-log plane identified by q'' and $(T_w - T_{\text{sat}})$ qualitatively produces the Nukiyama curve in either flow or pool boiling.

The multi-D surface for HTC has never been derived in the open literature; however, scientists postulate its existence. The HTC values calculated by codes including, in heat transfer controlled situations the jumps at CHF (from nucleate to film boiling) and at RNB or MFB (from film boiling to nucleate boiling or directly to forced convection) provide points in selected region of the multi-D surface. Those points derive from sophisticate logics and logical diagrams.

The advantage of the multi-D surface is having available a standard tool where all scientists may contribute and improve eventually introducing new variables. The disadvantage is the complexity of the surface and the eventual proprietary nature of findings that improve its applicability to nuclear reactors.

Answer 7.6

Figs. 7.77 and 7.78 (chapter 7) are relevant to address this question, together with the first equation in the answer 7.5. The answer is as follows:

- The HTC definition implies the consideration of ONLY radial heat transfer occurring, i.e., in the direction orthogonal to the heating surface; in the case of reflood, heat transfer occurs in the vertical direction, too. This makes invalid the assumption for heat transfer occurrence and opens to the need of a specific approach for heat transfer (and HTC).

Answer 7.7

The formulation of the question provides some hints for the answer below. The reader may note that the scientific community does not unanimously accept the nomenclature that we consider in the answer.

First, let us distinguish between natural convection (NCO) and natural circulation (NC). NCO is a heat transfer regime (see below) and NC is a system phenomenon or a flow mode. NC occurs in a gravity environment when a heat source and a heat sink (or equivalent heat source and sink, like in the case of a chimney in a room with the atmosphere working as heat sink) are present in belong to the same loop at different (typically) elevations. Therefore, NC implies the presence of a specific designed loop. NCO implies flow motion in a pool where heat transfer causes fluid motion, although gravity may affect the motion.

Second, at very low Re number, including laminar flow, NCO may occur in a NC loop or even in a centrifugal pump equipped loop, according to literature. Here we prefer to call this condition as forced convection (FCO).

Third, we define flow boiling and pool boiling as heat transfer modes or conditions where an external energy source (e.g., including the energy source in NC) is present or not, respectively.

Fourth, heat transfer regimes, occurring when a heated surface is in contact with a fluid in a flow boiling mode are: FCO, subcooled nucleate boiling (SNB), saturated boiling (SANB), transition boiling (TB, mostly in temperature controlled condition) and film boiling (FB). In a pool boiling mode, we have, NCO, SNB, SANB, TB and FB.

One may add, without discussion, reflood or rewet, CHF, MFB, radiation and even conduction heat transfer to the above discussion. Furthermore, condensation occurs (with specific regimes) when a cooled surface is in contact with fluid.

Answer 7.8

Radiation is always present in heat transfer. In some special situations related to containment, it may affect flow configuration or system performance even at low temperatures. In case of heat transfer in nuclear reactor core, nonnegligible radiation heat transfer (RHT, related to convection) occurs when solid temperature overpasses 600 K. Answer to the second part of the question is: although the knowledge of absorptivity and emissivity as well as of geometric-impact surface may reveal questionable, RHT should result from the application of Boltzmann formulation. The RHT amount shall receive proper consideration when determining different parameter for convection heat transfer.

Answer 7.9

Better to use the term CHF, in the question, instead of Critical Heat Transfer. Now, the answer.

Early CHF correlations (namely in the 1950s of the previous century, dealing with pool boiling) are correct. The issue is that those correlations deal with heat transfer without considering the (very rapid) change of local fluid conditions. For instance, in case of pool boiling, if liquid remains on the surface of the heating solid, the CHF can be very high. However, unavoidably (and even in a tiny fraction of a second) steam may form on the surface and the actual CHF value is much lower than the value corresponding to the situation of liquid in contact with the surface. Current CHF correlations or approaches (e.g., LUT by Groeneveld) deeply consider this issue. Liu et al. (2000) provide further details for suitable CHF calculation.

Answer 7.10

Corrosion, erosion, oxide formation, CRUD (deposit), etc., occur on a heating surface (sometimes included into the word fouling). All those processes may occur simultaneously or at different times during the lifetime of a heater and combine among each other. The impact upon heat transfer and the HTC may be large also depending on heat flux (q'', as considered in answers provided above, namely, for chapter 7).

An exhaustive or even a representative list of cases is impossible. As an example, following specific undocumented analyses, one may state that the PCT value during a LOCA depends for about 200 K upon the formation of a few dozen micron CRUD on the surface.

Answer 7.11

CRUD is the acronym for Chalk River Unidentified Deposit, where the Chalk River laboratory in Canada is recognized. CRUD is a deposit that forms on the surface of fuel rods causing feedback on (not in the order of importance) pressure drops, heat transfer and moderation-absorption processes of neutrons. A series of physical, chemical and radiation related processes occur, which contribute to the CRUD formation and in some cases disruption. Chemistry of cooling water, impurities content including boric acid are responsible for CRUD.

The thickness of CRUD may vary from a few micron to several dozen microns. CRUD affects centerline fuel temperature and energy stored in the pellet, see answers 7.3 and 7.10.

Answer 7.12

The answer to this question has been given with the example discussed in the answer 7.10. Calculating the effect of CRUD on a LOCA scenario (namely calculation of PCT) implies a multi-physics calculation accounting for (at the location of CRUD appearance, not necessarily along the entire fuel pin):

- Pressure drop increase in the core.
- Decreases of heat transfer coefficient at the contact between fuel rods and fluid.
- Local (thermal) neutron flux decrease.
- Change of thermal energy stored in fuel pin

Chapter 8

Answer 8.1

The momentum balance equations constitutes a force balance equation (i.e., according to $F = m \cdot a$) and pressure is a force per unit area.

Answer 8.2

For simplicity, we refer in the answer to the HEM (or the EVET model). The friction pressure drop can be expressed by (the reader may check whether "per unit length" applies or not):

$$\frac{\tau_w P}{A} \quad or \quad f\frac{L}{D}\rho w^2$$

(where all symbols are not ambiguous, possibly, except P = wetted perimeter). The left formula is the direct/intuitive definition of the shear stress between fluid and wall. The right formula is empirical and needs experiments (for "f").

The challenge of the left formula is the calculation of $\tau_w = \mu\left(\frac{\partial u}{\partial y}\right)_{y=0}$ (all symbols should be clear), at the wall (y = 0). This calculation may depend upon wall features other than fluid velocity profile in a cross section orthogonal to flow direction. Therefore, in nuclear thermal hydraulics (let us exclude DNS and CFD) the empirical formula on the right is adopted (SYS TH codes).

Answer 8.3

Φ^2_{LO}, or the empirical multiplier for obtaining two-phase friction pressure drops starting from single phase friction pressure drops, is defined as a nondimensional quadratic term to emphasize that it is a positive value greater than 1.

Answer 8.4

Let us answer the question in case of (ideal) adiabatic flow, first. Then, let us consider a straight horizontal pipe with constant area and steady-state flow. In these conditions, assuming constant mass flow rate, the unavoidable friction between gas and wall causes pressure at the outlet lower than inlet pressure. This implies for most of gases (due to physical properties, i.e., the link between p and ρ) lower density at the outlet. To keep constant $G = \rho w A$ at the inlet and outlet section, a decreasing ρ at the outlet implies an increasing w. Therefore, gas velocity increases in a steady state condition and supports the definition of spatial acceleration under steady state flow conditions.

In the presence of heat transfer with the wall the picture above changes and no conclusion can be derived.

Answer 8.5

Let us refer to a PWR (equipped with UTSG) during an accident scenario without pump running and primary side mass inventory decreasing: a liquid plug forms in the loop seal in a nearly/partly voided primary loop. Natural circulation flow stops and steam or two-phase mixture (high void fraction) is present in the hot side, noticeably the region including the upper plenum, the hot legs and the steam generator tubes. The loop seal is filled, left situation in left sketch of Fig. 8.5a.

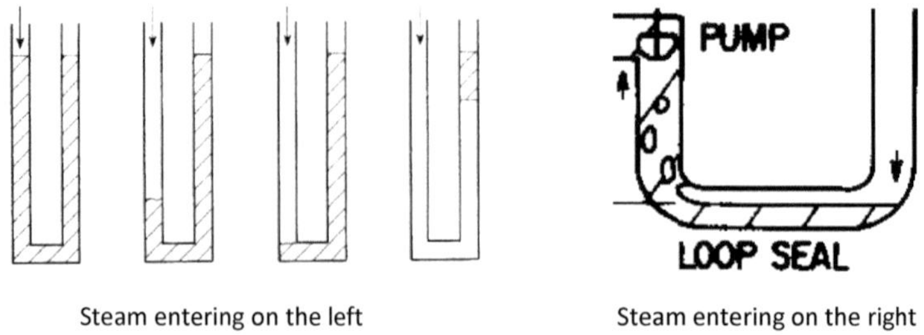

Figure 8.5a The loop seal filling and clearing process. Left: various phases from filling (left) to clearing (right). Right: Detail for flow configuration at the start of loop seal clearing.

The boil-off in the core causes pressure increase in the hot side of the primary loop and steam penetration in the loop seal starts when pressure increase is > ρgh with h being the loop seal height. A pressure increase in the core less than 0.01 MPa may induce steam in the loop seal and loop seal clearing. One may easily calculate the time needed for generating the pressure increase, assuming saturation conditions in the core, see also Kukita et al. (1990).

Answer 8.6

Here we could distinguish between support and mixing vane spacer grids (SSG and MVG, respectively). Without introducing this distinction, pressure drops in SSG and MVG range from about 100 Pa to several hundred Pa.

Answer 8.7

Fig. 8.1 is relevant for answering the question. If the pressure goes below p_{sat}, vaporization in a metastable condition may occur. This may affect thermal hydraulic evolutions downstream the geometric discontinuity.

Answer 8.8

The question concerns the modeling of pressure drops, e.g., the formulation for Φ^2_{LO}. The issue discussed hereafter applies to (almost) all empirical constitutive laws, which are part of balance equations in SYS YH codes or support the solution of balance equations. We can formulate the issue as follows. To derive qualified empirical correlations, fully developed (FD) flow and steady state (SS) constitute the prerequisite of experiments. However, the typical application domain for SYS TH codes is a nuclear reactor where time dependent thermal hydraulic phenomena occur and no fully developed flow region is possible (with possible exception of some regions of tubes in steam generators, primary side).

Therefore, we have the paradox: qualification needs FD and SS but non-FD and non-SS conditions constitute the domain of application.

Obviously, the above paradox is (unavoidably) at the origin of uncertainties (OOU), but only the UMAE-CIAU method (chapters 13 and 14) considers this OOU.

Answer 8.9

Answers 5.7 and 9.7 deal with the same or similar answer as the present one; in the former case, we adopt the word "partitioning" and in the latter case the word "repartition."

We can answer in addition to answers 5.7 and 9.7 as follows, assuming a "single 'two-phase' pressure drop measurement" (apologies for the un-necessary word repetition) and simplifying the process. We measure 300 Pa pressure drop (PD) for assigned-measured values of average void fraction, liquid and steam velocities and absolute pressure (let us call MV all Measured Values, excluding PD) in a (whatever vertical or horizontal) straight pipe over a length L.

MV allow the identification of flow regimes and a variety of correlations, e.g., Φ^2_{LO}, interfacial drag, repartition coefficient, etc., which affect the PD calculation. Each correlation (i) depends upon (j) empirical constants EC_{ij}. The measured value PD = 300 Pa should be the same as the calculated values (with some acceptable error fixed by code developers). If this is not the case and if the code developers consider the PD = 300 coming from a qualified experiment, they proceed to modify one or more EC_{ij}.

Obviously, the above process has multiple solutions. The same code developers should check that the modified set of EC_{ij} values does NOT affect (largely) previous calculations where different measured data (not only PD) supported the current stage of code development (beta code version, see chapter 22). The complexity of this process appears evident.

The answer to the second question is as follows. Code developers (typically an expert analyst or analyst's group leader) are responsible for assumptions and procedures, which may vary each time a quantity like PD = 300 Pa becomes available.

Answer 8.10

Boundary conditions or MV parameter values mentioned in answer 8.9 above largely affect typical errors in calculating PD. We expect larger (%) errors in the prediction of PD at very low (laminar flow) or very large (sonic speed) Re number. OECD/NEA/NSC (2021), and Cai et al. (2022) provide synthesis error values, which may exceed 50% (i.e., up to 100%), the measured values.

Answer 8.11

Answer 5.10 addresses both questions here.

Answer 8.12

We replace the wording "pressure drops" with "pressure drop at geometric discontinuities (PDGM)" in the question. The answer is as follows.

CFD codes have the capability to calculate PDGM as a function of Re and void fraction (and system pressure, temperature, etc.). PDGM values imply user-defined coefficients in SYS TH code applications. In case of CFD codes, the challenge is achieving a small-acceptable error in prediction (deducible when experimental data are available). Acceptable error (or precision target) is un-defined in literature, not a consensus topic in the scientific community and highly depending upon code application, region of the nuclear reactor, time progression during an accident scenario, etc. Here we can postulate that

$$\varepsilon_{PDGM} < 0.01 \quad \text{with} \quad \varepsilon_{PDGM} = \frac{|PDGM_{calc} - PDGM_{exp}|}{PDGM_{exp}}$$

constitutes an acceptable error. Definition of experimental conditions and parameter ranges are necessary for possible application of the above formula (maybe, different acceptable errors can be defined).

Chapter 9

Answer 9.1

The following terminology is necessary (at least) to answer the question: constitutive equation, closure laws (part of the question), interactions at gas liquid interface, interactions between fluid and solid (wall), state equations, numerical term, empirical term, physical term, chemical processes, radiation processes, mechanistic origin, transient and steady state.

We exclude from the consideration (for simplicity and considering current practice) the words chemical and radiation processes and transient (actually related to transient we should distinguish between development and application; here we exclude the transient in development). Therefore, typical constitutive equations do not include time as a variable (exceptions do exist).

Then, constitutive equations supplement balance equations and form two groups:

(a) Interactions between fluid and solid (noticeably, pressure drop and heat transfer).
(b) Interactions at gas liquid interface (e.g., interfacial drag).

Constitutive equations may have a mechanistic origin including physical terms, with or without empirical terms. The coefficient for convection heat transfer and the multiplier for two-phase pressure drop imply formulations that are part of the category (a); an example for category (b) is given above. Constitutive equations are derived under steady state conditions, as discussed in answer 8.8.

We prefer to state that closure laws include constitutive equations and numerical terms (necessary for the solution of balance equations).

We should refer as state equations (i.e., not constitutive equations) everything connected with water properties, e.g., Clapeyron (or Clausius and Clapeyron) equation and with thermodynamic evolutions of two phase-mixtures where Mollier diagrams are applicable.

Answer 9.2

To make feasible the solution of balance equations combined with closure laws (see answer above), all variables in constitutive equations should be the same as the unknowns of the mass, momentum and energy equations or should depend upon those unknowns.

Answer 9.3

Jacobian matrix, Jacobian determinant or simply Jacobian are established concepts or tools in mathematics to solve sets of complex equation: far away, here, to provide any insight into the topic of related formulations or applications. We recommend the reader two papers, both referring to single-phase simulations:

- Hu (2016).
- Zhang Han (H.) et al. (2017), also introducing to the method based on Jacobian-free Newton-Krylov (JFNK).

Answer 9.4

Slip ratio is a definition; drift flux is an (empirical) model. Both drift flux and slip ratio deal with the characterization of different velocities for liquid and steam. See answer 5.2 for further discussion and the paper by Ren et al. (2018) for various formulations of drift flux model.

Answer 9.5

The constitutive equations shall refer to flow regimes. In different words:

(a) The modeling of flow regime map implies a constitutive model (set of equations).
(b) Each flow regime implies different constitutive equations.

Answer 9.6

Interfacial area concentration (IAC) is given in m^2/m^3 or m^{-1}. Hibiki, Hogsett and Ishii in 1998 report for bubbly flow, values in the range 20–200 m^{-1}, approximately, Fig. 12 in their paper, under the conditions of air-water flows in a 2″ pipe at around atmospheric pressure, with superficial air and liquid velocities in the ranges 0.028–0.138 and 0.3–4.0 m/s. In different experimental conditions, Xia Wang (2010) in his PHD thesis and Ozaki and Hibiki (2021), report values up to 700 and up to 1000 m^{-1} for IAC, respectively. See also the paper by Hibiki and Jeong (2022) for recent developments in the area.

Addressing the last part of the question, if one consider a 2″ diameter pipe one gets for the contact area (IAC for 1 m^3 system volume) assuming a fully stratified condition (zero or very low velocity flow, air water, pipe length = 493 m):

(a) Horizontal flow: IAC = 25 m^2 = pipe length (corresponding to 1 m^3 volume) × pipe diameter (i.e., maximum possible value);
(b) Vertical flow: IAC = 0.00202 m^2 = cross sectional area of the pipe.

Answer 9.7

See answers 8.9 and 5.7. Solutions for the repartition issue become more and more complex (with arbitrary assumptions needed) when the number of dimensions (e.g., 3D) and the number of fields increase.

Answer 9.8

Different definitions apply and are available for hyperbolicity referring to the solution from a set of PDE equations. From an engineering viewpoint and referring to the set of UVUT 1D model equations, we may state that hyperbolicity of a solution implies the availability of the solution and values for fluid velocities, sonic speed and void fraction (primarily) close to experimental values (i.e., a qualified solution).

The consideration of appropriate constitutive equations (better, numeric closure laws according to previous answers) guarantee the hyperbolicity of a solution in the domain of interest for applications of SYS TH codes. More details can be found in the paper by Ahn et al. (2017).

Answer 9.9

Yes. k-ε model and k-ω model (for instance) in CFD codes have similarities, i.e., structure and origin (empirical), with the formulations of constitutive equations in SYS TH codes. However, k-ε and k-ω model are generally applicable (with error) for any turbulence situation; otherwise, field of applicability of each constitutive law is restricted to the domain of development or validation.

Answer 9.10

Constitutive laws are empirical (majority of cases) equations. Therefore, constitutive equations derive from the consideration of system conditions (size, pressure and heat flux or total system power) far away the reactor conditions in the majority of cases. The scaling, or the scalability of closure laws, is a process that aims at demonstrating that each constitutive law is applicable at reactor conditions.

References

Abdul-Razzak, A., 2009. Guidelines for Application of the Best Estimate Analysis and Uncertainty (BEAU) Methodology to Licensing Analysis in the Canadian CANDU Nuclear Power Industry. IAEA TM Application of Deterministic Safety Analysis, Pisa, Italy, pp. 21–29.

Abe, S., Obi, Y., Satou, A., Okagaki, Y., Sibamoto, Y., 2022. Implementation of the heat and mass transfer models for BT and post-BT regions in three-field two fluid CFD. In: 19th International Topical Meeting on Nuclear Reactor Thermal Hydraulics (NURETH-19), March 6–11, Brussels, Belgium, Paper 36197.

Abu-Zeid, M.A., 2013. Finite element analysis for prediction hydraulic performance of a rectangular flap valve. International Journal of Engineering and Technology Research 2 (1).

Abuaf, N., Jones, O.C., Zimmer, G.A., 1978. Optical probe for local void fraction and interface velocity measurements. Review of Scientific Instruments 49, 1090–1094.

Achilli, A., Congiu, C., Ferri, R., Bianchi, F., Meloni, P., Grgic, D., Dzodzo, M.B., 2012. SPES-3 facility RELAP5 sensitivity analyses on the containment system for design review. Journal of Science and Technology of Nuclear Installations. Article ID 173637.

Adamar, J., 1978. The Cauchy Problem for Linear Partial Differential Equations of Hyperbolic Type. Akademia Nauk, Moscow, Russia (in Russian).

Adamov, E.O., Vasilevskji, V.P., Ionov, A.I., et al., 1988. Analysis of the accident development first stage at the fourth unit of the Chernobyl Nuclear Power Plant. Atomnaya Energiya 64 (1), 24–28 (in Russian).

Adamov, E.O., Cherkashov, Y.M., Mironov, Y.V., Nikitin, Y.M., Bourlakov, E.V., 1995. RBMK reactors during severe accidents and related assessment techniques. In: ICHMT Seminars, Thermal Science Online, ISBN 961-91393-0-5. Issue 1.

Adamov, E.O., Cherkashov, Y.M., Ponomarev-Stepnoy, N.N., Bourlakov, E.V., 1996a. Upgrading of RBMK reactors: safety enhancement measures. In: IAEA International Conference One Decade after Chernobyl: Nuclear Safety Aspects, Vienna, Austria, April 1–3.

Adamov, E.O., Cherkashov, Y.M., Podlazov, L.N., Nikitin, Y.M., Stenbok, I.A., Novoselsky, O.Y., Ionov, A.I., Ponomarev-Stepnoy, N.N., Bouriakov, E.V., Malkin, S.D., Krayushkin, A.V., Babaitsev, M.N., Checherov, K.P., Abagyan, A.A., Vasekin, V.N., Kisil'I, M., 1996b. Chernobyl Accident Causes, overview of studies over the decade. In: IAEA International Conference One Decade after Chernobyl: Nuclear Safety Aspects, Vienna, Austria, April 1–3.

Adamsson, C., Holmkvist, C., 2015. Uncertainty analysis for lift-forces on BWR fuel assemblies. In: 16th International Topical Meeting on Nuclear Reactor Thermal Hydraulics (NURETH-16), August 30–September 4, Chicago, IL, USA.

Adamsson, C., Le Corre, J.M., 2014. Transient dryout prediction using a computationally efficient method for sub-channel film-flow analysis. Nuclear Engineering and Design 280, 316–325.

Addabbo, C., Annunziato, A., Aksan, N., D'Auria, F., Galassi, G.M., Dumont, D., Gaul, H.P., Umminger, K., Nilsson, L., Hofmann, D., Riikonen, V., Rigamonti, M., Steinhoff, F., Guba, A., Toth, I., 2002. Development and Establishment of the CERTA Network Database: European Commission Report, CERTA/SC/D8 EUR 20421 EN, Brussels, Begium, September.

Adorni, M., Bousbia-Salah, A., Hamidouche, T., Pierro, F., Di Maro, B., D'Auria, F., 2005. Analysis of partial and total flow blockage of a single fuel assembly of an MTR research reactor core. Annals of Nuclear Energy 32.

Adorni, M., Del Nevo, A., D'Auria, F., Mazzantini, O., 2011. A procedure to address the fuel rod failures during LB-LOCA transient in Atucha-2 NPP. Science and Technology of Nuclear Installations. Article ID 929358.

Adrian, R.J., 1991. Particle-imaging techniques for experimental fluid mechanics. Annual Review of Fluid Mechanics 23, 261–304.

Adrianov, V.N., 1972. The Fundamentals of Radiative and Complex Heat Transfer. Energiya, Moscow, Russia (in Russian).

Adu, S., Horvatovic, I., Darko, E.O., Emi-Reynolds, G., D'Auria, F., 2015. Application of best estimate plus uncertainty in review of research reactor safety analysis. Nuclear Technology and Radiation Protection 30 (1), 75–82.

Adu, S., Horvatovic, I., D'Auria, F., Nyarko, B.B.J.B., Emi-Reynolds, G., Emmanuel, O.D., 2015a. Analysis of channel blockage of MNSR reactor using the system thermal-hydraulic code RELAP5/MOD3.3. In: 24th International Conference Nuclear Energy for New Europe, Portoroz, Slovenia, 14–17 September.

AERB, 2010. In: The International Workshop on New Horizons in Nuclear Reactor Thermal Hydraulics and Safety, January 7–8, Niyamak Bhavan, Mumbai Region, India.

AESJ, 2015. The Fukushima-Daiichi Nuclear Accident: Final Report of AESJ Investigation Committee. Springer, Tokyo, Japan, ISBN 978-4-431-55160-7.

Agbodemegbe, V.Y., Xu, C., Akaho, E.H.K., Allotey, F.K.A., 2015. Correlation for cross-flow resistance coefficient using STAR-CCM+simulation data for flow of water through rod bundle supported by spacer grid with split-type mixing vane. Nuclear Engineering and Design 285, 134–149.

Ahn, S.H., Aksan, N., Austregesilo, H., Bestion, D., Chung, B.D., D'Auria, F., Emonot, P., Gandrille, J.-L., Hanninen, M., Horvatovic, I., Kim, K.D., Kovtonyuk, A., Petruzzi, A., 2015. FONESYS: the FOrum & NEtwork of SYStem thermal-hydraulic codes in nuclear reactor thermal-hydraulics. Nuclear Engineering and Design 281, 103–113.

Ahn, S.H., Aksan, N., Austregesilo, H., Bestion, D., Chung, B.D., Coscarelli, E., D'Auria, F., Emonot, P., Gandrille, J.-L., Sauvage, J.-Y., Hanninen, M., Horvatovic, I., Kim, K.D., Kovtonyuk, A., Lutsanych, S., Petruzzi, A., 2017. Hyperbolicity and numerics in SYS-TH codes: the FONESYS point of view. Journal of Nuclear Engineering and Design 322, 227–239.

Ahnesjö, M., Andersson, P., Le Corre, J.-M., Andersson, S., 2015. Tomographic reconstructions and predictions of radial void distribution in BWR fuel bundle with Part Length rods. In: 16th International Topical Meeting on Nuclear Reactor Thermal Hydraulics (NURETH-16), August 30–September 4, Chicago, IL, USA.

AIAA, 1998. Guide for the Verification and Validation of Computational Fluid Dynamics Simulations AIAA Standards. American Institute of Aeronautics and Astronautics, AIAA G-077, Reston, VA, USA, ISBN 978-1-56347-354-8.

Akimoto, H., Anoda, Y., Takase, T., Yoshida, H., Tamai, H., 2016. Nuclear Thermal-Hydraulics—An Advanced Course in Nuclear Engineering (Translation from a Japanese Version Printed in 2009). Springer, Japan, pp. 1–452.

Akiyama, N., Miwa, S., Sakashita, H., 2018. Influence of the direct contact condensation on steam injector operation. In: 12th International Topical Meeting on Nuclear Reactor Thermal-Hydraulics, Operation and Safety (NUTHOS-12), October 14–18, Qingdao, China.

Aksan, N., 2008. Overview on CSNI separate effects test facility matrices for validation of best estimate thermal-hydraulic computer codes. In: THICKET CSNI Seminar, Session III—Paper 03 Pisa, Italy.

Aksan, N., 2008a. International standard problems and small break loss-of-coolant accident (SBLOCA). Science and Technology of Nuclear Installations (J STNI) 814572, 1–22.

Aksan, N., D'Auria, F., 1996. Relevant thermal Hydraulic Aspects of Advanced Reactor Design—Status Report. OECD/NEA Report, NEA/CSNI/R (96)22, Paris, France.

Aksan, N., Bessette, D., Brittain, I., D'Auria, F., Gruber, P., Holmström, H.L., Landry, R., Naff, S., Pochard, R., Preusche, G., Reocreux, M., Sandervag, O., Stadtke, H., Wolfert, K., Zuber, N., 1987. CSNI Code Validation Matrix of Thermo-Hydraulic Codes for LWR LOCA and Transients. OECD/NEA/CSNI Report No. 132, Paris, France.

Aksan, S.N., Modro, M., Berta, V.T., Wahba, V.T., 1989. Review of LOFT Large Break Experiments. US NRC Report, NUREG/IA-0028, Washington DC, USA.

Aksan, S.N., D'Auria, F., Staedtke, H., 1993. User effects on the thermal-hydraulic transient system codes calculations. Nuclear Engineering and Design 145 (1), 159–174. Also OECD 'Status' Report NEA/CSNI R(94)35, Paris, France, January 1995.

Aksan, N., D'Auria, F., Glaeser, H., Pochard, R., Richards, C., Sjöberg, A., 1994. Separate Effects Test Matrix for Thermal-Hydraulic Code Validation. (A) Volume I: Phenomena Characterization and Selection of Facilities and Tests, (B) Volume II: Facility and experiment Characteristics. OECD/NEA Report, NEA/CSNI/R (93)14/Part 1 and Part 2, Paris, France. [also OECD/NEA/CSNI, 1993].

Aksan, S.N., D'Auria, F., Groeneveld, D., Kirillov, P., Saha, D., Badulescu, A., Cleveland, J., Lead Authors, 2001. Thermohydraulic Relationships for Advanced Water Cooled Reactors. IAEA Tecdoc 1203, Vienna (A) ISSN 1011-4289 (See Also IAEA, 2001), pp. 1–344.

Aksan, N., Schulenberg, T., Squarer, D., Cheng, X., Struwe, D., Sanchez, V., Dumaz, P., Kyrki-Rajamaki, R., Bittermann, D., Souyri, A., Oka, Y., Koshizuka, S., 2003. Potential safety features and safety analysis aspects for high performance light water reactor (HPLWR). In: International Conference on Global Environment and Advanced Nuclear Power Plants (GENES4/ANP2003), Kyoto, Japan.

Aksan, S.N., Ambrosini, W., Ammirabile, L., Anderson, M.H., Bae, Y.Y., Chen, Y., Churkin, A., Hanninen, M.J., Jackson, J.D., Kirillov, P.L., Leung, L., Novog, D., Pioro, I.L., Razumovskiy, V., Vijayan, P.K., Xiaojing, L., Yamada, K., 2014. Heat Transfer Behavior and Thermo-Hydraulics Code Testing for Supercritical Water Cooled Reactors (SCWRs). IAEA, TECDOC-1746, Vienna, Austria.

Aksan, S.N., Bestion, D., D'Auria, F., Hassan, Y.A., Moretti, F., 2018. Special issue from workshop advanced instrumentation and measurement techniques for nuclear reactor thermal-hydraulics (SWINTH). Journal of Nuclear Engineering and Design 336, 1–2.

Aksan, S.N., D'Auria, F., Glaeser, H., 2018a. Thermal-hydraulic phenomena for water cooled nuclear reactors. Journal of Nuclear Engineering and Design 330, 166–186.

Aksan, S.N., D'Auria, F., 2022. [or, 2021]+. Summary from 'aix-en-provence 1992' and 'Annapolis 1996' OECD/NEA/CSNI Specialists meetings in nuclear thermal hydraulics. In: Proceedings Of CSNI Specialists Meet. on Transient Thermal-hydraulics in Water Cooled Nuclear Reactors (SM-TH), Madrid, Spain, March 22–24, p1, pp. 1–20.

Aktas, B., 2003. Level Tracking in thermal-hydraulic Simulations of Nuclear Reactors. Ph.D. Thesis. The Pennsylvania State University, State College, PA, USA.

Al Issa, S., Macian, J.R., 2011. A review of CCFL phenomenon. Annals of Nuclear Energy 38, 1795–1819.

Al Issa, S., Macian, J.R., 2018. The effect of Hutze upon CCFL in a Konvoi type PWR hot leg pipe geometry of 1/4 scale. In: 12th International Topical Meeting on Nuclear Reactor Thermal-Hydraulics, Operation and Safety (NUTHOS-12), October 14–18, Qingdao, China.

Al Issa, S., Murase, M., Tomiyama, A., Hayashi, K., Macián, J.R., 2019. On CCFL at a PWR hot-leg pipe geometry and comparison between results in COLLIDER 1/4th- and Kobe 1/15th-scaled facilities. Nuclear Science and Engineering 193 (1–2), 147–159.

Al-Awad, A.S., Perez-Ferragut, M., Freixa, J., 2021. Application of a BEPU-based code assessment to the ATLAS upper head SB-LOCA test. Annals of Nuclear Energy 164, 108581.

Alamgir, M.D., Lienhard, J.H., 1981. Correlation of pressure undershoot during hot water depressurization. Journal of Heat and Mass Transfer 103, 522–525.

Aldemir, T., 1991. In: Apostolakis, G. (Ed.), Utilization of the Cell-To-Cell Mapping Technique to Construct Markov Failure Models for Process Control Systems. Elsevier, New York, NY, USA, pp. 1431–1436.

Alexeenko, S.V., Nakoriakov, V.E., Pokusaev, B.G., 1992. Wave Flow of Films of Liquid. Nauka, Novosibirsk, Russia (in Russian).

Alku, T., 2017. Modelling of turbulent effects in LOCA conditions with CATHARE-3. Nuclear Engineering and Design 321, 258–265.

Allen, T.B., Polmar, N., 2007. Rickover Father of the Nuclear Navy. Potomac Books Inc., Washington, DC, USA.

Allen, T.F.H., Starr, T.B., 2017. Hierarchy: Perspective for Ecological Complexity, second ed. (first ed.). University of Chicago Press, Chicago, IL, USA.

Allison, C., Balabanov, E., D'Auria, F., Jankowski, M., Misak, J., Salvatores, S., Snell, V., 2002. Accident Analysis for Nuclear Power Plants. IAEA Safety Reports Series No. 23, Vienna, Austria.

Almenas, K., Kaliatka, A., Uspuras, E., 1998. Ignalina RBMK-1500. A Source Book, Extended and Updated Version. Lithuanian Energy Institute, Kaunas, Lithuania, ISBN 9986-492-35-1, pp. 1–198.

Almenas, K., Pabarcius, R., Seporaitis, M., 2006. Design and tests of a device for the generation of controlled condensation implosion events. Heat Transfer Engineering 27 (3), 32–41. ISSN 0145-7632.

Altstadt, E., Carl, H., Weis, R., 2002. Fluid–Structure Interaction Experiments at the Water-hammer Test Facility of Forschungszentrum Rossendorf. Jahrestagung Kerntechnik, pp. 559–564.

Alves Portela, F., Sandham, N.D., 2020. A DNS/URANS approach for simulating rough-wall turbulent flows. International Journal of Heat and Fluid Flow 85 (108627), 1–9.

Alzaben, Y., Sanchez-Espinoza, V.H., Stieglitz, R., 2019. Analysis of a control rod ejection accident in a boron-free small modular reactor with coupled neutronics/thermal-hydraulics code. Annals of Nuclear Energy 134, 114–124.

Ambrosini, W., Ferreri, J.C., 1998. The effect of truncation error on numerical predictions of stability in a natural circulation single-phase loop. Nuclear Engineering and Design 183, 53–76.

Ambrosini, W., Bovalini, R., D'Auria, F., Petrini, P., 1987. Advancements in evaluating accuracy of thermal-hydraulic codes calculations. In: 9th Meet. Of CSNI Task Group on the Status and Assessment of Codes for Transients and ECCS, Paris, France, 1–3 June.

Ambrosini, W., D'Auria, F., Grassi, W., Vigni, P., 1988. Accuracy in the prediction of two-phase flow regimes. In: Eurotherm Sem. No. 3: Modeling of Nuclear and Advanced Heat Transfer Components, Bologna, Italy, June 14–15.

AmbrosinI, W., Bovalini, R., D'Auria, F., 1990. Evaluation of accuracy of thermal-hydraulics codes calculations. Energia Nucleare 7 (2).

Amendola, A., Reina, G., 1984. DYLAM-1, A Software Package for Event Sequence and Consequence Spectrum Methodology. EUR-924, CEC-JRC, Ispra, Italy.

Amin, S.A.U., Bang, Y.S., Lee, J.-S., 2017. Assessment of post-LOCA long term cooling considering the in-vessel downstream effect. Annals of Nuclear Energy 110, 63–72.

Ammirabile, L., 2010. Studies on supercritical water reactor fuel assemblies using the sub-channel code COBRA-EN. Nuclear Engineering and Design 240, 3087–3094.

Ammirabile, L., Walker, S.P., 2014. Dynamic balloning analysis of a generic PWR fuel assembly using the multi-rod coupled MATARE code. Nuclear Engineering and Design 268.

Amos, C., Schrock, V., 1983. Critical Discharge of Initially Subcooled Water through Slits. USNRC NUREG/CR, 3476, Washington, DC, USA.

Analytis, G.T., 1996. Developmental assessment of RELAP5/MOD3.1 with separate effect and integral test experiments: model changes and options. Nuclear Engineering and Design 163, 125–148.

Analytis, G.T., Coddington, P., 2002. Analysis and sensitivity studies of postulated SBLOCA in the Mühleberg (KKM) BWR/4 by Trac-BF1. Annals of Nuclear Energy 29.

Anders, D., Berry, R.A., Gaston, D., Martineau, R.C., Peterson, J.W., Zhang, H., Zhao, H., Zou, L., 2012. RELAP-7 Level 2 Milestone Report: Demonstation of a Steady State Single Phase PWR Simulation with RELAP-7, pp. 1–27. INL/EXT-12-25924, Idaho Falls, ID, USA, May.

Andersen, J.G.M., Heck, C.L., Klebanov, L.A., Shiralkar, B.S., 2009. Realistic assessment of break size in simulated BWR loss-of-coolant accidents. In: 13th International Topical Meeting on Nuclear Reactor Thermal Hydraulics (NURETH-13), September 27-October 2, Kanazawa City, Ishikawa, Japan.

Anderson, B.L., Lawton, T.J., Weaver, J.M., Somers, E.V., 1957. ATBAC: And IBM-704 Code for thermal Transients. Bettis Atomic Power Division, WAPD-TM-20, Pittsburgh, PA, USA.

Anderson, P.S., Astrup, P., Eget, L., Rathman, O., 1977. Numerical experience with the two-fluid model RISQUE. In: Proceedings from Topical Meeting on Thermal Reactor Safety, Sun Valley, ID, July 31–August 4.

Anderson, D.A., Tannehill, J.C., Pletcher, R.C., 1984. Computational Fluid Mechanics and Heat Transfer, Series in Computational Methods in Mechanics and Thermal Science. Hemisphere Publishing Corporation, Washington, DC, USA.

Andreani, M., Yadigaroglu, G., 1989. Dispersed Flow Film Boiling. Paul Scherrer Institut, Wurenlingen, Switzerland.

Andreani, M., Yadigaroglu, G., 1994. Prediction methods for dispersed flow film boiling. International Journal of Multiphase Flow 20, 1–51.

Andreoni, D., Clement, P., Deruaz, R., Porracchia, A., Regnier, P., 1980. Calculation of Single Tube Reflooding with Hexeco Code. Analysis of ERSEC Experiments. European Two Phase Flow Group Meeting, Glasgow, UK (June).

Andreussi, P., 1983. Droplet transfer in two-phase annular flow. International Journal of Multiphase Flow 697–713.

Andreussi, P., Di Donfrancesco, A., Messia, M., 1988. An impedance method for the measurement of liquid hold-up in two-phase flow. International Journal of Multiphase Flow 14, 777–785.

Andrushechko, S.A., Favorov, B.Y., Vasiliev, B.Y., Kosourov, K.B., Semchenkov, Y.M., Kuchumov, A.Y., Ukraintsev, V.F., 2019. VVER-1200: Evolution of the Classics. Operation Physics, Systems and Elements, Nuclear Fuel, Safety. Rosatom Series, Logos Publisher, Moscow (Ru), pp. 1–608.

Anglart, H., 2010. Thermal Hydraulics in Nuclear Systems. © 2010 Henryk Anglart—All Rights Reserved. Royal Institute of Technology, Stockholm, Sweden, pp. 1–234.

ANL, 1999. Computational Differentiation Technical Reports. Argonne National Laboratory, Chicago. Ill, USA. http://www.mcs.anl.gov/autodiff/tech_reports.html.

Annaratone, D., 2008. Steam Generators. Springer, Berlin, Germany.

Annunziato, A., Addabbo, C., Briday, G., Deruaz, R., Juhel, D., Kumamaru, H., Kukita, Y., Medich, M., Rigamonti, M., 1992. Small break LOCA counter part test in the LSTF, BETHSY, LOBI and SPES test facilities. In: Proceedings of the 5th International Topical Meeting on Reactor Thermal Hydraulics, NURETH-5, Salt Lake City, Utah, USA, pp. 1570–1576.

Annunziato, A., Glaeser, H., Lillington, J., Marsili, P., Renault, C., Sjoberg, A., 1996. CSNI Integral Test Facility Validation Matrix for the Assessment of Thermal-Hydraulic Codes for LWR LOCA and Transients. OECD/NEA Report, NEA/CSNI/R (96)17, Paris, France.

Anoda, Y., Kukita, Y., Tasaka, K., 1990. Void fraction distribution in rod bundle under high pressure conditions, Advances in Gas-liquid Flows. In: Proceedings of ASME Winter Annual Meeting, Dallas, TX, USA, pp. 283–289.

ANS, 1987. American National Standard Guidelines for the Verification and Validation of Scientific and Engineering Computer Programs for the Nuclear Industry. ANSI/ANS-10.4-1987, Chicago, IL, USA.

ANS, 2000. In: International Meeting on Best-Estimate Methods in Nuclear Installation Safety Analysis (BE-2000), Washington, DC, USA, 10–13 November.

ANS, 2004. In: International Meeting on Best-Estimate Methods in Nuclear Installation Safety Analysis (BE-2004), Washington, DC, USA, 14–18 November.

ANSI, 1995. Revision and Addendum to Nuclear Safety Criteria for the Design of Stationary Pressurized Water Reactor Plants. ANSI-N18.2A-1995, ANSI, New York, NY, USA.

ANSYS, 2010. ANSYS CFX-Solver Theory Guide—Release 13.0. ANSYS Inc, Canonsburg, PA, USA.

ANSYS, 2011. ANSYS-Fluent Theory Guide. ANSYS Inc., Canonsburg, PA, USA.

Antoni, O., Farvacque, M., Geffraye, G., Kadri, D., Lavialle, G., Ramon, B., 2009. CATHARE 2 V2.5_2: a single version for various applications. In: NURETH 13, Kanazawa, Japan.

Apostolakis, G., Moieni, P., 1987. The foundations of models of dependence in probabilistic safety assessment. Reliability Engineering 18 (3), 177–185.

Aprile, G., D'Auria, F., Frogheri, F., Galassi, G.M., 1996. Application of a qualified WWER-1000 plant nodalization for Relap5/mod3.2 computer code. In: International Conference Nuclear Energy in Countries with Small and Medium Electricity Grid, Opatjia, Croazia, October 7–9.

Apriyanti, V., Adriansyah, W., Abdurrachim, Pasek, A.D., 2017. A review of direct contact condensation of steam on water droplets. In: International Conference on Thermal Science and Technology, 17–19 November, Bali, Indonesia, AIP Conference Proceedings, 1984, vol. 1, pp. 1–10, 020035.

Arai, T., Furuya, M., Kanai, T., Shirakawa, K., Nishi, Y., 2015. Boiled-up level and boiling two-phase flow dynamics in 5×5 heated rod bundle during boil-off process under atmospheric pressure conditions. In: 16th International Topical Meeting on Nuclear Reactor Thermal Hydraulics (NURETH-16), August 30–September 4, Chicago, IL, USA.

Araneo, D., D'Auria, F., 2012. Methodology for pressurized thermal shock analysis in nuclear power plant. In: Belov, A. (Ed.), Applied Fracture Mechanics. InTech Publisher.

Archer, A., 2002. A predictive model for cavitation erosion downstream orifices. In: ASME Joint U.S. European Fluids Engineering Division Conference, Vol 1, Montreal, Quebec, Canada, July 14–18, pp. 403–409.

Ardron, K.H., Bryce, W.M., 1990. Assessment of horizontal stratification entrainment model in RELAP5/MOD2 by comparison with separate effects experiments. Nuclear Engineering and Design 122, 263–271.

Ardron, K.H., Furness, R.H., 1976. A study of the critical flow models used in reactor analysis. Nuclear Engineering and Design 39, 257–266.

AREVA, 2014. (Date of Downloading), US EPR Final Safety Analysis Report, Section 15-3-1, Partial Loss of Forced Reactor Coolant Flow, AREVA ©.

ARIA, 2016. http://www.aria.developpement-durable.gouv.fr.

Arias, F.J., 2009. Disquisition and Hydrodynamics-Hydromagnetics Aspects of Film Boiling Heat Transfer. PhD Thesis. UPC, Barcelona, Spain.

Arias, F.J., Reventos, F., 2010. Heat transfer enhancement in film boiling due to lift forces on the Taylor-Helmholtz instability in low forced convection from a horizontal surface. Journal of Enhanced Heat Transfer 17 (2), 1–97.

Arias, F.J., Pla, P., Reventos, F., 2009. Thermocapillary on film boiling heat transfer from a non isothermal horizontal surface: a theoretical study from CANDU reactors. Progress in Nuclear Energy 51 (6/7), 672–675.

Arkadov, G., Getman, A., Rodionov, A., 2012. Probabilistic Safety Assessment of Optimum Nuclear Power Plant Life Management. Woodhead Publishing, Elsevier Ltd., Kidlington, UK.

Arkoma, A., Hänninen, M., Rantamäki, K., Kurki, J., Hämäläinen, A., 2015. Statistical analysis of fuel failures in large break loss-of-coolant accident (LBLOCA) in EPR type nuclear power plant. Nuclear Engineering and Design 285.

Arlit, M., Partmann, C., Schleicher, E., Schuster, C., Hurtado, A., Hampel, U., 2018. Instrumentation for experiments on a fuel element mock-up for the study of thermal hydraulics for loss of cooling or coolant scenarios in spent fuel pools. Nuclear Engineering and Design 336, 105–111.

Arpott, D.G., Green, C.D., 1992. Chernobyl unique safety valve for a reactor clear explosion. In: Proceedings of the British National Conference on the Legacy of Nuclear-Lessons for the UK Held at the Council House, Bristol, UK, March 11.

Asali, J.C., Hanratty, T.J., Andreussi, P., 1985. Interfacial drag and film height for vertical annular flow. AIChE Journal 31, 895–902.

Ashley, R., El-Shanawany, M., Eltawila, F., D'Auria, F., 1998. Good Practices for User Effect Reduction. OECD/CSNI Report NEA/CSNI/R(98)22, Paris, France.

ASME, 2008. Standard for verification and validation in computational fluid dynamics and heat transfer. In: ASME V&V 20, American Society of Mechanical Engineers, New York, NY, USA (See below).

ASME, 2009. Standard for Verification and Validation in Computational Fluid Dynamics and Heat Transfer. V & V 20–2009 (Reaffirmed 2016). ASME, New York, NY, USA, ISBN 9780791832097.

Asmolov, V.G., Gusev, I.N., Kazanskiy, V.R., Povarov, V.P., Statsura, D.B., 2017. New generation first-of-the kind unit—VVER-1200 design features. Nuclear Energy and Technology 3, 260—269.

ASTM, 2005. Standard Guide for Evaluating the Predictive Capability of Deterministic Fire Models. American Society for Testing and Materials, ASTM std. E1355-05a, see also American National Standards Institute (ANSI), Washington, DC, USA.

Atkinson, D.I., Berard, M., Hanssen, B.V., Segeral, G., 1999. New generation multiphase flowmeter from schlumberger and framo engineering AS. In: Proceedings 17th International North Sea Flow Meas. Workshop, Oslo, Norway, pp. 154—165.

Atkinson, D.I., Berard, M., Segeral, G., 2000. Qualification of a non-intrusive multiphase flow meter in viscous flows. SPE Pap. 63118. In: Annu. Technical Conf. and Exhibition, Dallas, TX, USA.

Atkinson, D.I., Pinguet, B., Segeral, G., Theuveny, B.C., 2002. Field implications of uncertainties in multiphase flow measurements. SPE Pap. 77403. In: An. Tech. Conf. and Exhibition, San Antonio, TX, USA.

Atomenergoproekt, 2021. VVER-TOI Design. https://www.rosatom.ru.

Aumiller, D.L., Tomlinson, E.T., Clarke, W.G., 2000. A new assessment of RELAP5-3D using a General Electric level swell problem. In: RELAP5 Users Seminar, Jackson Hole, WY, USA, 12—14 September.

Aumiller, D.L., Tomlinson, E.T., Bauer, R.C., 2001. Coupled RELAP5-3D/CFD methodology with a prof-of-principle calculation. Journal of Nuclear Engineering and Design 205 (1—2), 83—90.

Aumiller, D.L., Swartele, G.W., Meholic, M.J., Lloyd, L.J., Buschman, F.X., 2015. COBRA-IE: A New Sub-Channel Analysis Code. NURETH-16, Chicago, August 30—September 4.

Aursand, P., Gjennestad, M.A., Aursand, E., Hammer, M., Wilhelmsen, Ф., 2017. The spinodal of single- and multi-component fluids and its role in the development of modern equations of state. Fluid Phase Equilibria 436, 96—112.

Avdeev, A.A., 1982. Reynolds analogy in undeveloped surface boiling on porous-coated surface. Teploenergetika 3 (23) (in Russian).

Avdeev, A.A., 1985. Pressure drop in subcooled boiling under forced convection conditions. Teploenergetika 5 (49) (in Russian).

Avduevsky, V.S., Koshkin, V.K. (Eds.), 1992. Principles of Heat Transfer in Aircraft and Rocket-Space Technology. Mashinostroenie, Moscow, Russia (in Russian).

Avramova, M.N., 2006. CTF Input Manual. The Pennsylvania State University, State College, PA, USA.

Avramova, M.N., 2007a. CTF Theory Manual. The Pennsylvania State University, State College, PA, USA.

Avramova, M.N., 2007b. Development of an Innovative Spacer Grid Model Utilizing Computational Fluid Dynamics within a Sub-channel Analysis Tool. PHD thesis at. Pennsylvania State University (PSU), State College, Pa, USA, pp. 1—231.

Avramova, M.N., 2020. Developments in thermal-hydraulic sub-channel modeling for whole core multi-physics simulations. Nuclear Engineering and Design 358 (110387), 1—11.

Avramova, M.N., Ivanov, K.I., 2020. Verification, Validation and Uncertainty Quantification of Multi-Physics Modeling of Nuclear Reactors, first ed. Woodhead Publishing, Paperback, ISBN 9780128149546, pp. 1—300.

Avramova, M., Rubin, A., Utsuno, H., 2013. Overview and discussion of the OECD/NRC benchmark based on NUPEC PWR subchannel and bundle tests. Journal of Science and Technology of Nuclear Installations. Article ID 946173.

Avsentuk, B.P., Kutateladze, S.S., 1977. Heat transfer instability on the surfaces depleted centers of vaporization. Teplofiz. Vys. Temp. (High Temp., Engl. Transl.) 15 (1), 115–120 (in Russian).

Avvakumov, A., Malofeev, V., Sidorov, V., Scott, H.H., 2007. Spatial Effects and Uncertainty Analysis for Rod Ejection Accidents in a PWR. NUREG/IA-0215, USNRC, Washington, DC, USA.

Aydemir, N.U., 2009. Application of the relative scaling methodology to a CANDU-PHWR for the power pulse phase. In: 13th International Topical Meeting on Nuclear Reactor Thermal Hydraulics (NURETH-13), September 27–October 2, Kanazawa City, Ishikawa, Japan.

Ayegba, P.O., Lawrence, C., Edomwonyi-Otu, L.C., 2020. Turbulence statistics and flow structure in fluid flow using particle image velocimetry technique: a review. Engineering Reports 2, e12138.

Ayoub, A., Gjorgiev, B., Sansavini, G., 2018. Cooling towers performance in a changing climate: techno-economic modeling and design optimization. Energy 160, 1133–1143.

Azmy, Y., Sartori, E., 2010. Nuclear Computational Science—A century in Review. Springer, Dordrech, Germany, pp. 1–470.

Bacchiani, M., Medich, C., Rigamonti, M., Vescovi, O., Alemberti, A., Conway, L.E., 1995. SPES/2, AP-600 integral system test S01007 2" CL to core makeup tank pressure balance line break. In: 7th International Topical Meeting on Nuclear Reactor Thermal Hydraulics (NURETH-7), Saratoga Springs, NY, USA, 10–15 September.

Baccou, J., Zhang, J., Fillion, P., Damblin, G., Petruzzi, A., Mendizábal, R., Reventós, F., Skorek, T., Couplet, M., Iooss, B., Oh, D.-Y., Takeda, T., 2019. Development of good practice guidance for quantification of thermal-hydraulic code model input uncertainty. Nuclear Engineering and Design 354, 110173.

Bae, B.U., Lee, K.H., Kim, Y.S., Yun, B.J., Park, G.C., 2008. Scaling methodology for a reduced-height reduced-pressure integral test facility to investigate direct vessel injection line break SBLOCA. Journal of Nuclear Engineering and Design 238.

Bae, B.-U., Kim, S., Park, Y.-S., Kang, K.H., 2014. Integral effect test and code analysis on the cooling performance of the PAFS (passive auxiliary feed-water system) during an FLB (feed-water line break) accident. Nuclear Engineering and Design 275.

Bae, H., Ryu, S.U., Kang, J.Y., Yang, J.H., Kim, Y.S., Park, H.S., 2018. Data Analysis Report: 4-Train PSS Tests. KAERI, S-750-NV457-005, Daejeon, Korea.

Bae, B., Kim, H., Kim, K., Jeong, J.J., Yun, B.J., 2021. Force balance droplet entrainment model for the horizontal stratified flow condition. International Journal of Heat and Mass Transfer 165, 120726.

Baek, J.S., Cheng, L.-Y., Diamond, D., 2015. Analysis of LOCA scenarios in the NIST research reactor before and after fuel conversion. In: 16th International Topical Meeting on Nuclear Reactor Thermal Hydraulics (NURETH-16), August 30–September 4, Chicago, IL, USA.

Baer, M.R., Nunziato, J.W., 1986. A two-phase mixture theory for the deflagration-to-detonation transition (DDT) in reactive granular materials. International Journal of Multiphase Flows 12, 861–889.

Baglietto, E., Ninokata, H., 2004. CFD modelling of secondary flows in fuel rod bundles. In: Conf. NUTHOS-6, AESJ, Nihon Genshiryoku Gakkai, Japan.

Baglietto, E., Ninokata, H., Misawa, T., 2006. CFD and DNS methodologies development for fuel bundle simulations. Nuclear Engineering and Design 236, 1503–1510.

Bagul, R.K., Pilkhwal, D.S., Vijayan, P.K., Joshi, J.B., 2013. Entrainment phenomenon in gas-liquid two-phase flow: a review. Sadhana 38 (6), 1173–1217.

Bahmanyar, M.E., Talebi, S., 2019. A performance analysis of vertical steam generator using an entropy generation method. Annals of Nuclear Energy 125, 212–221.

Bahmanyar, M.E., Taheranpour, N., Talebi, S., 2021. Applying second law of thermodynamic for optimization of horizontal steam generator. Progress in Nuclear Energy 133 (103636), 1−12.

Bahrami, M., Culham, J.R., Yovanovich, M.M., Schneider, G.E., 2006. Review of thermal joint resistances models for non-conforming rough surfaces. Applied Mechanics Review 59, 1−12.

Bajorek, S.M., 2015. A regulator's perspective on the state of the art in nuclear thermal-hydraulics. In: NURETH-16 Conference, Chicago, IL, USA, August 30−September 4.

Bajorek, S.M., Cheung, F.B., 2009. Spacer grid rewet and drop size in RBHT reflood experiments. In: Proceedings of ANS Winter Annual Meeting, November 15−19, Washington DC, USA.

Bajorek, S.M., Bernard, M., Gingrich, C., Hoxie, C.L., Ireland, A., Kelly, J., Mahaffy, J., Murray, C., Spore, J., Staudenmeier, J., Thurgood, M., Tien, K., Whitman, J., 2015. Development, validation and assessment of the TRACE thermal-hydraulics systems code. In: 16th International Topical Meeting on Nuclear Reactor Thermal Hydraulics (NURETH-16), August 30−September 4, Chicago, IL, USA.

Baker, O., 1953. Design of pipelines for the simultaneous flow of oil and gas. Society of Petroleum Engineers (Paper Number: SPE-323-G). https://doi.org/10.2118/323-G.

Baker, O., 1954. Simultaneous flow of oil and gas. Oil and Gas Journal 53, 185−195.

Baker, C.R., 2000. Flow Measurement Handbook. Cambridge University Press, Cambridge, UK.

Baker Jr., L., Just, L.C., 1962. Studies of Meta-Water Reactions at High Temperatures—III—Experimental and Theoretical Study of Zirconium-Water Reactions, vol. 6548. ANL, Argonne, IL, USA.

Balakrishnan, A.R., 1998. Pool boiling of saturated pure liquids and binary mixtures: effect of surface characteristics. In: Proceedings 11th IHTC Heat Transfer, vol. 1, pp. 71−87.

Balaras, E., Benocci, C., Piomelli, U., 1996. Two layer approximate boundary condition for large eddy simulation. Journal of AIAA 34, 1111−1119.

Bandini, G., De Rosa, F., 2014. ASTEC validation on TMI-2 and LOFT LP-FP-2. Nuclear Engineering and Design 272, 163−172.

Banerjee, S., 2003. Industrial Hazard and Plant Safety. Taylor & Francis, New York, NY, USA.

Banerjee, S., 2013. What can we expect from DNS and LES?. In: Invited at NURETH-15 Conference, Pisa, Italy.

Banerjee, S., Hancox, W.T., 1978. Transient thermohydraulics analysis for nuclear reactors. In: 6th International Heat Transfer Conference, August 7−11, Toronto, Canada, pp. 311−337 (Authors' reply during discussion session).

Banerjee, S., Lahey Jr., R.T., 1980. Advances in two-phase flow instrumentation. In: Lewins, J., Becker, M. (Eds.), Advances in Nuclear Science and Technology, vol. 13. Plenum Co., New York, NY, USA, pp. 227−414.

Banerjee, S., Ortiz, M.G., Larson, T.K., Reeder, D.L., 1998. Scaling in the safety of next generation reactors. Nuclear Engineering and Design 186, 111−133.

Bang, Y.S., Oh, D.Y., Lee, J., Kim, J., Mon, S.-K., 2022. Prediction of reflood behavior in a bundle with deformed and un-deformed rods in ATHER 5 × 5 experiment using MARS-KS code. Nuclear Engineering and Design 392 (111770), 1−10.

Bankoff, S.G., Lee, S.C., 1985. A brief review of countercurrent flooding models applicable to PWR geometries. Nuclear Safety 26 (2), 139−152.

Bankoff, S.G., Lee, S.C., 1986. A critical review of the flooding literature. In: Hewitt, G.F., Delhaye, J.M., Zuber, N. (Eds.), Chapter 2, Multiphase Science and Technology. Hemisphere Publishing Corporation, New York, NY, USA.

Bankoff, S.G., Tankin, R.S., Yuen, M.C., Hsieh, C.L., 1981. Countercurrent flow of air/water and steam/water through a horizontal perforated plate. International Journal of Heat and Mass Transfer 24 (8), 1381–1395.

Banowskia, M., Hampel Krepper, U.E., Beyer, M., Lucas, D., 2018. Experimental investigation of two-phase pipe flow with ultrafast X-ray tomography and comparison with state-of-the-art CFD simulations. Nuclear Engineering and Design 336, 90–104.

Bao, H., Feng, J., Nam, D., Zhang, H., 2020. Computationally efficient CFD prediction of bubbly flow using physics-guided deep learning. International Journal of Multiphase Flow 131 (103378), 1–20.

Barna, I.F., Imre, A.R., Baranyai, G., Ezsöl, G., 2010. Experimental and theoretical study of steam condensation induced water hammer phenomena. Nuclear Engineering and Design 240, 146–150.

Baroczy, C.J., 1966. A systematic correlation for two-phase pressure drop. Chemical Engineering Progress Symposium Series 62, 232–249.

Bartak, J., 1990. A study of the rapid depressurization of hot water and the dynamics of vapour bubble generation in superheated water. International Journal of Multiphase Flow 16, 789–798.

Bartak, J., Bestion, D., Haapalehto, T., 1994. The top-down reflooding model in the CATHARE code. Nuclear Engineering and Design 149, 141–152.

Barten, W., Jasiulevicius, A., Manera, A., Macian, R., Zerkak, O., 2008. Analysis of the capability of system codes to model cavitation water hammers: simulation of UMSICHT water hammer experiments with TRACE and RELAP5. Nuclear Engineering and Design 238, 1129–1145.

Bartosiewicz, Y., Seynhaeve, J.-M., 2014. Delayed equilibrium model (DEM) of flashing choked flows relevant to LOCA and implementation in system codes. In: Conference ICONE 22, Praha, Czech Republic, Paper 30957.

Bartosiewicz, Y., Giot, M., Seynhaeve, J.-M., 2010. Revisiting modeling techniques and validation experiments for two-phase choked flows relevant to LOCA. In: International Topical Meeting on Nuclear Thermal-Hydraulics, Operation and Safety (NUTHOS-8) N8P0317, Shanghai, PRC, 10–14 October.

Bartosiewicz, Y., Seynhaeve, J.-M., Serre, G., 2011. Delayed equilibrium model and validation experiments for two-phase choked flows relevant to LOCA. In: The 14th International Topical Meeting on Nuclear Reactor Thermalhydraulics, NURETH-14, Toronto, Ontario, Canada, September 25–30.

Basile, D., Beghi, M., Chierici, R., Salina, E., Brega, E., 1999. COBRA-EN: An Upgraded Version of the COBRA-3C/MIT Code for thermal Hydraulic Transient Analysis of Light Water Reactor Fuel Assemblies and Cores. ENEL CRTN, Report 1010/1, Milano, Italy.

Bassenghi, F., 2013. Validation of the CFD Code NEPTUNE for a Full-Scale Simulator for Decay Heat Removal Systems with In-Pool Heat Exchangers (Ph.D. thesis). University of Bologna, Bologna (I).

Batet, L., 2004. Contribution to the Development of Passive Safety Systems for Advanced Light Water Reactors. PhD Thesis. UPC, Barcelona.

Batt, D.L., Berta, V.T., 1978. ECC delivery and distribution in scaled PWR experiments. In: Consideration for the American Nuclear Society, Winter Meeting at Washington, DC, USA, 12–17 November.

Batta, A., Class, A.G., 2017. CFD analysis of pressure drop across grid spacers in rod bundles compared to correlations and heavy liquid metal experimental data. Nuclear Engineering and Design 312, 121–127.

Baudin, N., Colin, C., Ruyer, P., Sebilleau, J., 2016. Turbulent flow and transient convection in a semi-annular duct. International Journal of Thermal Science 108, 40–51.

Baumeister, K.J., Simon, F.F., 1973. Leidenfrost temperature—its correlation for liquid metals, cryogens, hydrocarbons and water. ASME Journal of Heat Transfer 95, 166–173.

Bazin, P., 1995. ROSCO programme: experimental effects of oscillations and of fuel simulator structure on reflooding. In: Proceedings of the 3rd International Conference on Nuclear Engineering (ICONE), Kyoto, Japan.

Bazin, P., Clement, P., Deruaz, R., Dumont, D., Gully, P., Noel, B., 1990. Investigation of PWR accident situations at BETHSY facility. Nuclear Engineering and Design 124, 285–297.

Bazin, P., Deruaz, R., Yonomoto, T., Kukita, Y., 1992. BETHSY-LSTF counterpart test on natural circulation in a pressurized water reactor. In: ANS National Heat Transfer Conference, San Diego, Ca, USA.

Bazin, P., Dor, I., Morel, C., 2005. Assessment of the CATHARE 3-D module for LBLOCA simulation. In: International Conference on Nuclear Reactor Thermal Hydraulics (NURETH-11), Avignon, France.

Beale, S.B., 1997. Tube banks, cross-flow over. In: International Encyclopedia of Heat and Mass Transfer. (Canadian) NRC Publications Archive, pp. 1–21.

Beale, S.B., 1997a. Tube banks, single-phase heat transfer in. In: International Encyclopedia of Heat and Mass Transfer. (Canadian) NRC Publications Archive, pp. 1–22.

Beckmeyer, R.R., Gregory, M.V., Buckner, M.R., 1979. Complete BWR-EM LOCA analysis using the WRAP-EM system. In: ANS Meeting, Atlanta, GA, USA.

Beckner, W.D., Reyes Jr., J.N., 1981. Lower Plenum Refill Research Results. USNRC Research Information Letter No 128, Washington, DC, USA. www.nrc.gov/docs/ ML17254A308.pdf.

Beckner, W.D., Reyes Jr., J.N., Anderson, R., 1979. Analysis of ECC Bypass Data. USNRC, NUREG-0573, Washington, DC, USA.

Beeny, B., Vaghetto, R., Vierow, K.M., Hassan, Y., 2016. MELCOR and GOTHIC analyses of a large dry pressurized water reactor containment to support resolution of GSI-191. Nuclear Technology 196 (2), 292–302.

Behzadi, M., Zolfaghari, A., Abbassi, M.R., Norouzi, A., Mirzaeegoudarzi, M.M., 2022. Subchannels thermal-hydraulic analysis of rod bundle nuclear fuel elements under rolling motions. Annals of Nuclear Energy 168 (108874), 1–13.

Bejan, A., 2021. Nationalism and forgetfulness in the spreading of thermal sciences. International Journal of Thermal Sciences 163, 1–9, 106802.

Bejan, A., Tsatsaronis, G., Moran, M., 1996. Thermal Design & Optimization. John Wiley & Sons, New York, NY, USA.

Belliard, M., Grandotto, M., 2000. Computation of two-phase flow in steam generator using domain decomposition and local zoom methods. In: International Conference on Nuclear Engineering, ICONE-8, Baltimore, Md, USA.

Belsito, S., D'Auria, F., 1997. Comparison of advanced computer codes in the simulation of CHF occurrence in PMK facility. In: 4th World Conf. Experimental Heat Transfer, Fluid Mechanics and Thermodynamics—Brussels, Belgium, 2–6 June.

Belsito, S., D'Auria, F., Galassi, G.M., 1994. Application of a Statistical model to the evaluation of Counterpart Test data base. Kerntechnik 59 (3).

Benedict, M., Pigford, T.H., Levy, H.W., 1981. Nuclear Chemical Engineering. McGraw-Hill, New York, NY, USA.

Benhamadouce, S., 2015. Pressure drop prediction using Code-SATURNE in NESTOR CFD benchmark. In: NURETH-16 Conference, Chicago, IL, USA, August 30–September 4.

Benhamadouche, S., 2015a. On the use of (U)RANS and LES approaches for turbulent incompressible single phase flows in nuclear engineering applications. In: The 16th International Topical Meeting on Nuclear Reactor Thermal Hydraulics (NURETH-16), Chicago, IL, USA.

Benhamadouche, S., Malouf, W.J., Arenas, M., 2015. Effects of spatial discretisation and RANS turbulence modelling on the numerical simulation of a flow through a square-edged orifice in a round pipe. In: E-Proceedings of the 36th IAHR World Congress, The Hague, Netherlands.

Benteboula, S., Dabbene, F., 2020. Modeling of wall condensation in the presence of noncondensable light gas. International Journal of Heat and Mass Transfer 151, 119313.

Berenson, P.J., 1961. Film boiling heat transfer from a horizontal surface. International Journal of Heat and Mass Transfer 83, 351–358.

Bergagio, M., Hedberg, S., Rydström, S., Anglart, H., 2015. Instrumentation for temperature and heat flux measurement on a solid surface under BWR operating conditions. In: 16th International Topical Meeting on Nuclear Reactor Thermal Hydraulics (NURETH-16), August 30–September 4, Chicago, IL, USA.

Bergles, A.E., 1977. Burnout in boiling heat transfer. Part II: sub-cooled and low quality forced-convection systems. Nuclear Safety 18 (2), 154–167.

Berman, B.R., 1976. Thermal Conduction in Solids. Oxford University Press, New York, NY, USA.

Berman, L.D., Gordon, B.G., Bogdan, S.H., 1983. Experimental relationships for prediction of heat transfer at vapour condensation on the water dispersed jet. Teploenergetika 4, 19–22 (in Russian).

Berna, C., Escrivá, A., Muñoz-Cobo, J.L., Herranz, L.E., 2014. Review of droplet entrainment in annular flow: interfacial wave sand onset of entrainment. Progress in Nuclear Energy 74, 14–43.

Bernardin, J.D., Mudawar, J., 1999. The Leidenfrost point: experimental study and assessment of existing models. ASME Journal of Heat Transfer 121, 894–903.

Berne, P., 1983. Contribution à la modelisation du taux de production de vapeur par auto-vaporisation dans les ecoulements diphasiques en conduites. These de Doctorat, Ecole des Arts et Manufactures, Paris, France).

Bernier, R.N., Brennen, C.E., 1983. Use of the electromagnetic flowmeter in a two-phase flow. International Journal of Multiphase Flow 9 (3), 251–257.

Berry, R.A., 2003. Ensemble Averaged Conservation Equations for Multiphase, Multi-Component, and Multi-Material Flows. Idaho National Engineering and Environmental Laboratory Report, INEEL/EXT-03-01011, Idaho Falls, ID, USA.

Berry, R.A., 2005. Notes on Well-Posed, Ensemble Averaged Conservation Equations for Multiphase, Multi-Component, and Multimaterial Flows. Idaho National Laboratory Report INL/EXT-05-00516, Idaho Falls, ID, USA.

Berry, R.A., 2005a. Notes on Well-Posed, Ensemble Averaged Conservation Equations for Multiphase, Multi-Component, and Multi-Material Flows. Idaho National Laboratory Report INL/EXT-05-00516-modified, Idaho Falls, ID, USA.

Berry, R.A., Saurel, R., Petitpas, F., Daniel, E., Le Métayer, O., Gavrilyuk, S., Dovetta, N., 2008. Progress in the Development of Compressible, Multiphase Flow Modeling Capability for Nuclear Reactor Flow Applications. Idaho National Laboratory Report INL/EXT-08-15002, Idaho Falls, Id, USA.

Berry, R.A., Saurel, R., LeMetayer, O., 2010. The discrete equation method (DEM) for fully compressible, two-phase flows in ducts of spatially varying cross-section. Nuclear Engineering and Design 240, 3797–3818.

Berry, R.A., Zou, L., Zhao, H., Andrs, D., Peterson, J.W., Zhang, H., Martineau, R.C., 2013. RELAP-7: Demonstrating Seven-Equation, Two-phase Flow Simulation in a Single Pipe, Two-phase Reactor Core and Steam Separator/Dryer, pp. 1–41. INL/EXT-13-28750, Idaho Falls, ID, USA.

Berry, R.A., Peterson, J.W., Zhang, H., Martineau, R.C., Zhao, H., Zou, L., Andrs, D., 2015. RELAP-7 Theory Manual. INL/EXT-14-31366 (rev. 1, rev. 3 issued 2018), Idaho Falls, ID, USA.

Bersano, A., Agnello, G., D'Auria, F., Zio, E., Mascari, F., 2021. Methodology and Application to Characterize the Sump Clogging Issue in Case of Long Term Core Cooling, pp. 1–10. RCCS-2021-OECD/NEA Specialist Workshop on Reactor Core and Containment Cooling Systems—Long Term Management and Reliability, Paper 1330, October 18–21, Levice, Slovakia.

Berselli, L.C., 2021. Three-Dimensional Navier-Stokes Equations for Turbulence. Elsevier, Academic Press Publisher, Duxford (UK), pp. 1–328.

Bertola, V., 2003. Two-Phase flow measurement techniques: modelling and experimentation in two-phase flow. International Centre for Mechanical Sciences, CISM, Springer, 450, pp. 281–323.

Bestion, D., 1985. Interfacial friction determination for the 1-D six equation two-fluid mode used in the CATHARE code. In: European Two-phase Flow Group Meeting, June 3–7, Southampton, UK.

Bestion, D., 1989. Fluid dynamic effects in the fuel element top nozzle area and two-phase flow in piping junctions. In: Krishner, W. (Ed.), Proceedings Of the Seminar on the Commission Contribution to Reactor Safety Research, 20–24 November. Elsevier Applied Science, Varese, Italy.

Bestion, D., 1990. The physical closure laws in the CATHARE code. Nuclear Engineering and Design 124 (3), 229–245.

Bestion, D., 1990a. Recent developments on interfacial friction models. In: European Two-phase Flow Group Meeting, Varese, Italy, 21–24 May, pp. 229–245.

Bestion, D., 2008. Extension of CFD codes to two-phase flow safety problems. In: Workshop CFD4NRS-4 Grenoble, France, September 10–12.

Bestion, D., 2012. Applicability of two-phase CFD to nuclear reactor thermal hydraulics and elaboration of Best Practice Guidelines. Journal of Nuclear Engineering and Design 253, 311–321.

Bestion, D., 2014a. The difficult challenge of a two-phase CFD model for all flow regimes. Nuclear Engineering and Design 279, 116–125 ([see also).

Bestion, D., 2014b. Validation data needs for CFD simulation of two-phase flow in a LWR core. In: CFD4NRS-5 Conf., Zurich, Switzerland, September].

Bestion, D., 2017. System thermalhydraulics for design basis accident analysis and simulation: status of tools and methods and direction for future R&D. Nuclear Engineering and Design 312, 12–29.

Bestion, D., 2017a. Multiscale simulation of Light Water reactor thermalhydraulics. (M&C 2017). In: International Conference On Mathematics & Computational Methods Applied to Nuclear Science & Engineering, Jeju, Korea, April 16–20.

Bestion, D., 2018. Specific requirements for BEPU methods using system thermal hydraulics codes with pressure vessel modeling. In: Best Estimate Plus Uncertainty International Conference (BEPU-2018), Paper 194, Lucca, Italy, May 13–19.

Bestion, D., 2018a. The concept of multiscale validation. In: 12th International Topical Meeting on Nuclear Reactor Thermal-Hydraulics, Operation and Safety (NUTHOS-12), Qingdao, China, October 14–18.

Bestion, D., 2019. About phenomena identification in a PIRT. In: 18th International Topical Meeting on Nuclear Reactor Thermalhydraulics NURETH-18, August 19–23, Portland, OR, USA.

Bestion, D., D'Auria, F., 2022. Identification of thermal hydraulic phenomena for PWR transient analysis and simulation. In: 19th International Topical Meeting on Nuclear Reactor Thermal Hydraulics (NURETH-19), Brussels, Belgium, March 6–11.

Bestion, D., Gros d'Aillon, I., 1989. Condensation tests analysis and correlation for the CATHARE code. In: Proceedings of the Fourth International Topical Meeting on Nuclear Reactor Thermal-Hydraulics (NURETH-4), Karlsruhe, Germany.

Bestion, D., Matteo, L., 2015. Scaling considerations about LWR core thermal hydraulics. In: Proceedings of the 16th International Topical Meeting on Nuclear Reactor Thermal Hydraulics (NURETH-16), Chicago, IL, USA, Aug. 30-Sept., p. 13757.

Bestion, D., Micaelli, J.C., 1986. A two-fluid stratified model suitable for pressurized water reactor safety code. In: 4th Miami Internationall Symp. On Multiphase Transport and Particulate Phenomena, Miami, Fl, USA, 15–17 December.

Bestion, D., Serre, G., 2012. On the modelling of two-phase Flow in horizontal legs of a PWR. Nuclear Engineering and Technology 44 (8), 871–888.

Bestion, D., Serre, G., 2013. Transport of interfacial area in system codes: status and perspectives. In: 15th International Topical Meeting on Nuclear Reactor Thermalhydraulics, NURETH-15, Pisa, Italy, May 12–15.

Bestion, D., Coste, P., Niceno, B., Mimouni, S., Lakehal, D., Bartosiewicz, Y., 2011. Two-phase CFD: the various approaches and their applicability to each flow regime. Multiphase Science and Technology 23 (2–4), 101–128.

Bestion, D., de Crecy, A., Camy, R., Barth, A., Bellet, S., Badillo, A., Niceno, B., Hedberg, P., Muñoz-Cobo, J.L., Moretti, F., Scheuerer, M., Nickolaeva, A., 2016. Review of uncertainty methods for computational fluid dynamics application to nuclear reactor thermal hydraulics. In: NEA/CSNI/R(2016)4, Paris, France.

Bestion, D., D'Auria, F., Lien, P., Nakamura, H., Austregesilo, H., Skorek, T., Bae, B.U., Choi, K.-Y., Kim, K.D., Mon, S.-K., Martinez-Quiroga, V., Reventos, F., Mascari, F., Schollenberger, S., Umminger, K., Reyes, J.N., Rohatgi, U.S., Wang, W., Zaki, T., 2017. The OECD /NEA/CSNI Scaling State of the Art Report—the S-SOAR. In: NEA/CSNI/R(2016)14, Paris, France, pp. 1–393.

Bestion, D., Valette, M., Fillion, P., Gaillard, P., 2017a. 3D Core thermal hydraulic phenomena in PWR SBLOCAs and IBLOCAs. In: 17th International Topical Meeting on Nuclear Reactor Thermal Hydraulics, NURETH-17, September 2017, Xi'an, China.

Bestion, D., Fillion, P., Préa, R., Bernard-Michel, G., 2018. Improved PWR LOCA simulations through refined core 3D simulations: an advanced 3D modelling and the associated METERO validation program. In: 12th International Topical Meeting on Nuclear Reactor Thermal-Hydraulics, Operation and Safety (NUTHOS-12), Paper 329, Qingdao, China, October 14–18.

Bestion, D., Fillion, P., Rabe, C., Préa, R., Mimouni, S., Merigoux, N., 2020. Some Progress Made in Multiscale Analysis, Multiscale Validation in Reactor Thermalhydraulics. ANS ATH-2020, Saclay, France. October 20–23.

Bestion, D., Fillion, P., Préa, R., 2020a. Some Progress in System Code Modelling Validation and Application Including 3D Modelling. ANS ATH-2020, Saclay, France. October 20–23.

Bestion, D., D'Auria, F., Lanfredini, M., 2022. Compensation of errors in reactor numerical and experimental simulations. In: Proceedings of the 19th International Topical Meeting on Nuclear Reactor Thermal Hydraulics (NURETH-19), Brussels Belgium, March 6–11.

Beus, S.G., 1972. Two Phase Turbulent Mixing Model for Flow in Rod Bundle. Bettis Atomic Power Laboratory, WAPD-T-2438, Pittsburgh, Pa, USA.

Bevir, M.K., 1970. Theory of induced voltage electromagnetic flow measurement. Journal of Fluid Mechanics 43, 577–590.

Bieder, U., Barré, F., 1997. Adaptation of CATHARE 2 to the containment—qualification of the models. In: 8th International Topical Meeting on Nuclear Reactor Thermalhydraulics (NURETH-8), Kyoto, 30 September–4 October.

Bieder, U., Herbette, A., 2022. CFD analysis of water vapor condensation in large containments: numerical model, verification and validation. Nuclear Engineering and Design 395 (111861), 1–18.

Bieder, U., Falk, F., Fauchet, G., 2014. LES analysis of the flow in a simplified PWR assembly with mixing grid. Journal of Progress in Nuclear Energy 75, 15–24.

IAEA, Baglietto, E., Bieder, U., Brewster, R., Cartier, O., Golibrodo, L., Hoehne, T., Jevremovic, T., Krause, M., Maheshwari, N.K., Niceno, B., Nickolaeva, A., Parshikov, I., Podila, K., Rao, Y., Simoneau, J.P., Skibin, A., Smith, B.L., Soloviev, S., Starodubtsev, M., Volkov, V., Xiong, J., 2022a. Summary Review on the Application of Computational Fluid Dynamics in Nuclear Power Plant Design, pp. 1–94. Nuclear Energy Series, NR-T-1.20, IAEA, Vienna, Austria.

Billa, C., D'Auria, F., Debrecin, N., Galassi, G.M., 1991. Application of Relap5/mod2 to PWR international standard problems. In: ANS Winter Meeting, San Francisco, CA, USA, 10–15 November.

Billone, M., Yan, Y., Burtseva, T., Daum, R., 2008. Cladding Embrittlement during Postulated Loss-Of-Coolant Accidents. USNRC NUREG/CR-6967, Washington, DC, USA.

Bird, R.B., Stewart, W.E., Lightfoot, E.N., 1960. Transport Phenomena, first ed. John Wiley & Sons Inc., New York, NY, USA.

Bischof, C.H., Carle, A., Khademi, P., Mauer, A., 1996. ADIFOR 2.0 Automatic differentiation of Fortran 77 programs. IEEE Computational Science and Eng 3 (3), 18–32.

Bischof, C.H., Bücker, H.M., Rasch, A., Slusanschi, E., Lang, B., 2007. Automatic differentiation of the general-purpose computational fluid dynamics package FLUENT. ASME Journal of of Fluids Engineering 129 (5), 1–17.

Bissen, E., Medale, M., Alpy, N., 2018. Asymptotic Numerical Method applied to stability and bifurcation analysis of a boiling flow at low pressure in a pipe. In: 12th International Topical Meeting on Nuclear Reactor Thermal-Hydraulics, Operation and Safety (NUTHOS-12), October 14–18, Qingdao, China.

Bittan, J., 2015. Simulation of thermal-hydraulics accidental transients: evaluation of MAAP5.02 versus CATHAREv2.5. In: 7th European Review Meeting on Severe Accident Research (ERMSAR-2015), Marseilles, 24–26 March.

Bittermann, D., Squarer, D., Schulenberg, T., Oka, Y., 2003. Economic prospects of the HPLWR. In: International Conference GENES4/ANP2003, Paper 1003, Kyoto, Japan.

Bittermann, D., Starflinger, J., Schulenberg, T., 2004. Turbine technologies for high performance light water reactors. In: Proceedings of International Congress on Advances in Nuclear Power Plants (ICAPP), Paper 4195, Pittsburgh, PA, USA.

Bjorge, R.W., Griffith, P., 1984. Initiation of waterhammer in horizontal and nearly horizontal pipes containing steam and subcooled water. Journal of Heat Transfer 106 (4), 835–840.

Blanco, D., Córdova, Y., Berna, C., Muñoz-Cobo, J.L., Escrivá, A., 2022. Experimental investigation of submerged gas jet discharges in stagnant water by spargers. In: 19th International Topical Meeting on Nuclear Reactor Thermal Hydraulics (NURETH-19), March 6–11, Brussels, Belgium, Paper 35543.

Blasius, P.R.H., 1913. Das ähnlichkeitsgesetz bei Reibungsvorgängen in Flüssigkeiten. Mitteilungen über Forschungsarbeiten auf dem Gebiete des Ingenieurwesens 131, 1–41 (in German).

Blinkov, V.N., Jones, O.C., Nigmatulin, B., 1993. Nucleation and flashing in nozzles—2. Comparison with experiments using a five-equation model for vapor void development. International Journal of Multiphase Flow 19 (6), 965–986.

Blinkov, V.N., Melikhov, O.I., Kapustin, A.V., Lipatov, I.A., Dremin, G.I., Nikonov, S.M., Rovnov, A.A., Elkin, I.V., 2005. PSB-VVER counterpart experiment simulating a small cold leg break LOCA. In: Proceedings Of the International Topical Meeting on Nuclear Reactor Thermal Hydraulics (NURETH-11), Avignon, France.

Blinkov, V.N., Melikhov, O.I., Melikhov, V.I., Davydov, M.V., Wolff, H., Arndt, S., 2012. Experimental studies for the VVER-440/213 bubble condenser system for Kola NPP at the integral test facility BC V-213. Science and Technology of Nuclear Installations. Article ID 275693.

Blinkov, V.N., Melikhov, O.I., Melikhov, V.I., Volkov, G.Y., 2022. Experimental study of the features of the occurrence and development of condensation-induced water hammer in a horizontal pipe. Nuclear Engineering and Design 396 (111890), 1–25.

Bloch, H.P., Singh, M.P., 2009. Steam Turbines, Design Applications and Re-rating. McGraw-Hill, New York, NY, USA.

Blokh, A.G., 1962. Basic Principles of Thermal Radiation Heat Transfer. Gosenergoizdat, Moscow, Russia (in Russian).

Blokh, A.G., 1991. In: Blokh, A.G., Zhuravlev, Y.A., Ryzhkov, L.N. (Eds.), Thermal Radiation Heat Transfer Handbook. Energoatomizdat, Moscow, Russia (in Russian).

Blomstrand, J., Eklund, R., Nylund, O., 1990. Flow Stability Investigation of SVEA BWR Fuel: Workshop on Boiling Water Reactor Stability, Brookhaven, NY, USA (see also OECD/ NEA/ CSNI, 1990).

Blowe, T.N., Kang, S.K., Hassan, Y.A., 2015. CFD methodologies for a PWR fuel rod assembly. In: 16th International Topical Meeting on Nuclear Reactor Thermal Hydraulics (NURETH-16), Chicago, IL, USA.

BMU, 2015. Safety Requirements for Nuclear Power Plants, Annex 5: Requirements for Safety Demonstration and Documentation, Germany. http://regelwerk.grs.de/en/node/123.

Bobkov, V.P., Vinogradov, V.N., Zyatnina, O.A., Kozina, N.V., 1995. A method of evaluating the critical heat flux in channels and cells of arbitrary geometries. Thermal Engineering 42 (3), 221–231.

Bobkov, V.P., Vinogradov, V.N., Zyatnina, O.A., Kozina, N.V., 1997. Considerations in describing burnout in rod bundles. Thermal Engineering 44 (3), 2–7.

Boeschofen, F., 1955. On the possibility to improve the heat transfer of uranium and aluminum surfaces in contact. In: Proceedings Of the First International Conference on the Peaceful Uses of Atomic Energy, Geneva, Switzerland, vol. 9, pp. 208–209.

Bois, G., Jamet, D., Lebaigue, O., 2010. Towards large eddy simulation of two-phase flow with phase-change: direct numerical simulation of a pseudo-turbulent two-phase condensing flow. In: 7th International Conference On Multiphase Flow, ICMF 2010, Tampa, FL, USA, May 30–June 4.

Bonn, D., Ross, D., 2001. Wetting transitions. Reports on Progress in Physics 64, 1085–1163.

Bonuccelli, M., D'Auria, F., Debrecin, N., Galassi, G.M., 1993. A methodology for the qualification of thermal-hydraulic codes nodalizations. In: International Top. Meet. On Nuclear Reactor Thermal-Hydraulics (NURETH-6), Grenoble, France.

Boothroyd, R.G., 1967. An anemometric isokinetic sampling probe for aerosols. Journal of Scientific Instruments 44, 249–253.

Borishanskii, B.M., 1961. The effect of pressure on heat transfer and critical loads. In: Problems of Two-phase Heat Transfer. GosEnergoizdat, Moscow, Russia (In Russian).
Borishanskii, B.M., Kozyrev, A.P., Svetlova, L.S., 1964. Researches of heat transfer in nucleate boiling. In: Convective Heat Transfer in Two- and Single Phase Flows. Energiya, Moscow, Russia, p. 71 (in Russian).
Borishanskii, B.M., Andrievsky, A.A., Fokin, B.S., 1973. The generalized dependence for calculation of heat transfer at the two-phase flow in pipes and channels. In: Advances in the Researches of Two-phase Heat Transfer and Hydraulics in Power Elements. Nauka, Leningrad, Russia (in Russian).
Borishansky, V.M., Kutateladze, S.S., Novikov, I.I., Fedynsky, O.S. (Eds.), 1976. Liquid Metal Coolants, third ed. Atomizdat, Moscow (in Russian).
Borisov, E., Grigorov, D., Mancheva, K., 2013. Study of long-term loss of all AC power supply sources for VVER-1000/V320 in connection with application of new engineering safety features for SAMG. Annals of Nuclear Energy 59.
Boure, J.A., 1975. In: Lahey, R.T., Wallis, G.B. (Eds.), On a Unified Presentation of the Non-equilibrium Two-phase Flow Model. ASME Annual Winter Meeting, Houston, TX, USA.
Boure, J.A., Delhaye, J.M., 1982. General equations and two-phase flow modeling. In: Hetsroni, G. (Ed.), Handbook of Multiphase Systems. Hemisphere, Washington, DC, USA.
Bousbia Salah, A., 2022. Assessment of CATHARE capabilities in simulating the PKL natural circulation interruption tests. In: 19th International Topical Meeting on Nuclear Reactor Thermal Hydraulics (NURETH-19), March 6–11 (Via Web), Brussels, Belgium.
Bousbia-Salah, A., D'Auria, F., 2002. Analysis of the peach bottom 2 BWR turbine trip experiment by Relap5/3.2 code. Nuclear Technology and Radiation Protection 17 (1–2).
Bousbia-Salah, A., Vedovi, J., D'Auria, F., Ivanov, K., Galassi, G.M., 2004. Analysis of the peach bottom turbine trip 2 experiment by coupled Relap5-Parcs three dimensional codes. Nuclear Science and Engineering 148, 337–353.
Bousbia-Salah, A., D'Auria, F., Hamidouche, T., 2010. Natural Circulation Cooling and Safety Features in Engineering and Industrial Applications. Insights into Natural Circulation, Phenomena, Models and Issues. LAP Lambert Academic Publishing, Saarbrücken, Germany.
Bovalini, R., D'Auria, F., 1993. Scaling of the accuracy of the Relap5/mod2 code. Nuclear Engineering and Design 139 (2), 187–204.
Bovalini, R., D'Auria, F., De Varti, A., Maugeri, P., Mazzini, M., 1992. Analysis of counterpart tests performed in BWR experimental simulators. [XV annual meet. Of Spanish nuclear society, Alicante (E), October 23–25 1989]. J. Nuclear Technology 97 (1), 113–130.
Bovalini, R., D'Auria, F., Galassi, G.M., 1993. Scaling of complex phenomena in system thermal-hydraulics. [International Workshop on boiling heat transfer and two-phase flow—Bariloche (Ar), October 5–9 1992]. J. Nuclear Science and Engineering 115, 89–111.
Boyack, B.E., 1996. Methodology, status and plans for development and assessment of TUF and TRAC code. In: USNRC and OECD/CSNI Workshop on Transient Thermal-Hydraulic and Neutronic Codes Requirements, Annapolis, MD, USA, 5–8 November.
Boyack, B.E., Catton, I., Duffey, R.B., Griffith, P., Katsma, K.R., Lellouche, G.S., Levy, S., Rohatgi, U.S., Wilson, G., Wulff, W., Zuber, N., 1989. Quantifying Reactor Safety Margins—Application of Code Scaling, Applicability and Uncertainty Evaluation Methodology to a Large Break, Loss-Of-Coolant Accident, USNRC NUREG/CR-5249, Washington, DC, USA, December (See Also USNRC, 1989a).

Boyack, B.E., Catton, I., Duffey, R.B., Griffith, P., Katsma, K.R., Lellouche, G.S., Levy, S., May, R., Rohatgi, U.S., Shaw, R.A., Wilson, G.E., Wulff, W., Zuber, N., 1990. Quantifying reactor safety margins. Nuclear Engineering and Design 119, 1–117 [identifying 7 papers with co-author Boyack in this J NED issue, see also next].

Boyack, B.E., Catton, I., Duffey, R.B., Griffith, P., Katsma, K.R., Lellouche, G.S., Levy, S., Rohatgi, U.S., Wilson, G.E., Wulff, W., Zuber, N., 1990a. Quantifying reactor safety margins. Part I: an overview of the code scaling, applicability and uncertainty evaluation methodology. Journal of Nuclear Engineering and Design 119 (1), 1–15.

Boyd, C., 2016. Perspectives on CFD analysis in nuclear reactor regulation. Journal of Nuclear Engineering and Design 299, 12–17.

Breber, G., Palen, G.W., Taborek, J., 1980. Prediction of horizontal tubeside condensation of pure components using flow regime criteria. ASME Journal of Heat Transfer 102 (3), 471–476.

Breden, C.R., 1963. Boiling Water Reactor Technology, Status of the Art Report. Argonne National Laboratory, Chicago, IL, USA, pp. 1–254. ANL-6562, Vol II (Water Chemistry and Corrosion).

Breuer, M., Jaffrezic, B., Arora, K., 2008a. Hybrid LES–RANS technique based on a one-equation near-wall model. Journal of Theoretical and Computational Fluid Dynamics 22 (3–4), 157–187.

Breuer, M., Peller, N., Rapp, C., Manhart, M., 2008b. Flow over periodic hills—numerical and experimental study in a wide range of Reynolds numbers. International Journal of Computers and Fluids 38, 433–457.

British Nuclear Energy Society, 2000. Water chemistry of nuclear reactor systems. In: Proceedings of the Conference Organized by BNES, No 8, Bournemouth, UK.

Brittain, I., Karwat, H., D'Auria, F., Vigni, P., Hall, D.G., Reocreux, M., 1982. CSNI Critical Flow Modelling in Nuclear Safety. NEA/OECD Report (CSNI Report No. 4), ISBN 92-64-12366-0 (Paris, France).

Bromley, L.A., 1950. Heat transfer in stable film boiling. Chemical Engineering Progress 46, 221–227.

Brooks, G.L., 2002. A Short History of the CANDU Nuclear Power System accessible at. https://canteach.candu.org/Info/Documents/. (Accessed 27 December 2021).

Brooks, C.S., Hibiki, T., Ishii, M., 2012. Interfacial drag force in one-dimensional two-fluid model. Progress in Nuclear Energy 61, 57–68.

Brooks, C.S., Ozar, B., Hibiki, T., Ishii, M., 2014. Interfacial area transport of subcooled boiling flow in a vertical annulus. Nuclear Engineering and Design 268, 152–163.

Broughton, J.M., Kuan, P., Petti, D.A., Tolman, E.L., 1989. A scenario of the Three Mile Island unit 2 accident. Nuclear Technology 87, 34–53.

Brown, G., 1951. Heat transmission by condensation of steam on a spray of water drops. In: Proceedings of General Discussion on Heat Transfer, Institute of Mechanical Engineers, USA, pp. 49–52.

Brown, R.C., Andreussi, P., Zanelli, S., 1978. The use of wire probes for the measurement of liquid thickness in annular gas-liquid flows. The Canadian Journal of Chemical Engineering 56, 754–757.

Brown, L.D., Cai, T.T., Gupta, A., 2001. Interval estimation for a binomial proportion. Statistical Sciences 16 (2), 101–133.

Bruun, H.H., 1995. Hot-wire Anemometry. Oxford University Press, Oxford, UK.

Bucalossi, A., 2008. Current Use of Best Estimate Plus Uncertainty Methods on Operational Procedures Addressing Normal and Emergency Conditions. European Commission, JRC EUR 23717 EN, Brussels, Belgium.

Bucalossi, A., Petruzzi, A., 2010. Role of best estimate plus uncertainty methods in major nuclear power plant modifications. Journal of Nuclear Science and Technology 47, 8.

Bucalossi, A., Petruzzi, A., Kristof, M., D'Auria, F., 2010. Comparison between best-estimate plus-uncertainty methods and conservative tools for nuclear power plant licensing. Nuclear Technology 172, 29–47.

Bucalossi, A., Del Nevo, A., Moretti, F., D'Auria, F., Elkin, I.V., Melikhov, O.I., 2012. Investigation of accident management procedures related to loss of feed-water and station blackout in PSB-VVER integral test facility. Nuclear Engineering and Design 250, 633–645.

Bucci, M., 2022. Towards a better understanding of surface effects on boiling heat transfer and the boiling crisis. In: 19th International Topical Meeting on Nuclear Reactor Thermal Hydraulics (NURETH-19), March 6–11 (Via Web), Brussels, Belgium.

Bungartz, H.-J., Schäfer, M., 2006. Fluid Structure Interaction. Springer, Berlin, Germany.

Buschman, F.X., Aumiller, D.L., 2017. Use of fundamental condensation heat transfer experiments for the development of a sub-grid liquid jet condensation model. Nuclear Engineering and Design 312, 147–160.

Busco, G., Hassan, Y.A., 2018. Space and energy-based turbulent scale-resolving simulations of flow in a 5×5 nuclear reactor core fuel assembly with a spacer grid. Journal of Heat and Fluid Flow 71, 420–441.

Butterworth, D., Hewitt, G.F. (Eds.), 1977. Two-Phase Flow and Heat Transfer. Oxford University Press, Oxford, UK.

Cabot, W., 1996. Near wall model in large eddy simulation of flow behind a backward facing step. In: Annual Research Briefs, Center for Turbulence Research, NASA. Stanford University, CA, USA.

Cabot, W., Moin, P., 2000. Approximate wall boundary conditions in the large-eddy simulation of high Reynolds number flow. Journal of Flow, Turbulence and Combustion 63 (1), 269–291.

Cacuci, D.G. (Ed.), 2010. Handbook of Nuclear Engineering. Springer, Berlin, Germany.

Cacuci, D., Arslan, E., 2014. Reducing uncertainties via predictive modeling: FLICA4 calibration using BFBT benchmarks. Nuclear Science and Engineering 176 (3), 339–349.

Cadinu, F., Kozlowski, T., Dinh, N., Nam, D., 2007. Relating system-to-CFD coupled code analyses to theoretical framework of a multiscale method. In: Proceedings of International Congress on Advances in Nuclear Power Plants (ICAPP), Nice, France, May 13–18, Paper 7539.

Caha, V., Krejčí, J., Čížek, J., 2018. SUBCAL-SCW: subchannel analysis for supercritical parameters. In: 12th International Topical Meeting on Nuclear Reactor Thermal-Hydraulics, Operation and Safety (NUTHOS-12), October 14–18, Qingdao, China.

Cai, Q., D'Auria, F., Umminger, K., Bestion, D., Shan, J., 2022. Prioritizing pressure drop research in nuclear thermal hydraulics. Progress in Nuclear Energy 153 (104358), 1–13.

Camargo, C., Galetti, M.R., D'Auria, F., Mazzantini, O., 2008. Exploitation of BEPU approach for licensing purposes. In: International Topical Meeting on Safety of Nuclear Installations (ENS TOPSAFE), Dubrovnik, Croatia, 30 September–3 October.

Camiciola, P., D'Auria, F., Marsili, P., Petrini, P., 1987. OECD/CSNI ISP 20: post-test calculations and sensitivity analyses by RELAP5/MOD2 performed with reference to the SGTR accident occurred at Doel-2 NPP. In: University of Pisa Report, DCMN—RL 290 (87), Pisa (I), Dec. 1987, OECD CSNI Workshop on ISP 20, Brussels Belgium, December 8–10.

Camp, A.L., Cummings, J.C., Sherman, M.P., Kupiec, C.F., Healy, R.J., Caplan, J.S., Sandhop, J.R., Saunders, J.H., 1983. Light Water Reactor Hydrogen Manual. Sandia Report NUREG/CR-2726, SAND82-1137, Albuquerque, NM, USA.

CANTEACH, 2022. The CANTEACH Project. https://www.nuceng.ca/canteach/index.htm.

Capps, N., Sweet, R., 2022. Characterization of modeling and experimental data inconsistencies from burst testing for high-burnup commercial fuel rod applications. Journal of Nuclear Materials 563 (153621), 1–12.

Capps, N., Jensen, C.B., Cappia, F., Harp, J., Terrani, K., Woolstenhulme, N.E., Wachs, D., 2021. A critical review of high burnup fuel fragmentation, relocation, and dispersal under loss-of-coolant accident conditions. Journal of Nuclear Materials 546 (152750), 1–27.

Carbajo, J.J., 1985. A study on the rewetting temperature. Nuclear Engineering and Design 85, 21–52.

Carbajo, J.J., Siegel, A.D., 1980. Review and comparison among the different models for rewetting in LWRs. Nuclear Engineering and Design 58 (1), 33–44.

Carbone, C., D'Auria, F., Mazzini, M., Oriolo, F., 1980. Determination of Electrical Power to Be Supplied to LOBI Heaters. University of Pisa Report, DCMN—NT 011(80), Pisa, Italy, June.

Carcassi, M., Mazzini, M., Vigni, P., 1983. Research on the behavior of Mark II pressure suppression containment system during LOCA, carried out at the University of Pisa. In: 2nd International Topical Meeting on Nuclear Reactor Thermalhydraulics, NURETH-2, Santa Barbara, CA, USA, January 11–14.

Carlucci, L.N., Hammouda, N., Rowe, D.S., 2004. Two-phase turbulent mixing and buoyancy drift in rod bundles. Nuclear Engineering and Design 227 (1), 65–84.

Carr, A.B., 2021. Thirty minutes before the dawn. Journal of Nuclear Technology 207 (1), 1–24.

Carra, S., Morbidelli, M., 1986. Transient mass transfer onto small particlea and drops. In: Cheremisinoff, N.P. (Ed.), Handbook of Heat and Mass Transfer, vol. 2, pp. 59–109. Houston, Tx, USA.

Carslaw, H.S., Jaeger, J.C., 1959. Conduction of Heat in Solids, second ed. Clarendon Press, Oxford, UK.

Carstens, N., George, T., 2010. Gothic Overview. (Ppt Presentation), NuScale Pre-application Meeting, USNRC, Rockville Md, USA, December 13.

Cartellier, A., 1990. Optical probes for local void fraction measurements: characterization of performance. Review of Scientific Instruments 61, 874–886.

Cartellier, A., 1992. Simultaneous void fraction measurement, bubble velocity, and size estimate using a single optical probe in gas–liquid two-phase flows. Review of Scientific Instruments 63, 5442–5453.

Cartellier, A., Achard, J.L., 1991. Local phase detection probes in fluid/fluid two-phase flows. Review of Scientific Instruments 62, 279–303.

Cartellier, A., Barrau, E., 1998. Monofiber optical probes for gas detection and gas velocity measurements: conical probes. International Journal of Multiphase Flow 24 (8), 1265–1294.

Carver, M.B., Kiteley, J.C., Zhou, R.Q.N., Junop, S.V., 1993. Validation of ASSERT subchannel code for standard and non-standard geometries. In: ARD-TD-454P, 2nd International Sem. On Sub-channel Analysis, EPRI, Palo Alto, Ca, USA.

Casamor, M., Martínez-Quiroga, V., Reventós, F., Mendizabal, R., Freixa, J., 2020. On the scaling of uncertainties in thermal hydraulic system codes. Annals of Nuclear Energy 136 (107026), 1–12.

Castellanos-Gonzalez, D.A., Maiorino, J.R., Moreira, J.M.L., 2022. Validation of the thermal-hydraulic THUNDER code and application for PWR core in transient conditions. Nuclear Engineering and Design 390 (111711), 1–21.

Catton, I., Wulff, W., Zuber, N., Rohatgi, U.S., 2005. Application of fractional scaling analysis (FSA) to loss of coolant accidents (LOCA). Part 3. Component level scaling for peak clad temperature. In: International Topical Meeting on Nuclear Reactor Thermal Hydraulics (NURETH-11), Avignon, France.

Catton, I., Zuber, N., Wulff, W., Rohatgi, U.S., 2009. Application of fractional scaling analysis (FSA) to loss of coolant accidents (LOCA); component level scaling for peak clad temperature. Journal of Fluids Engineering 131 (121401), 1–8. December.

Cavlina, N., Grgic, D., 2021. Support of safe operation of nuclear power plant Krško. In: Invited Seminar—Lecture at University of Pisa, DESTEC, (Slide Presentations, pp 1–80) Pisa, Italy, October 25.

CD-adapco, 2006. In: STAR-CCM+ Provider: STAR-CAT5 at COE; Also CD-Adapco—the Leading Provider of CAD-Embedded and Advanced Multidisciplinary Turbomachinery Simulations Exhibit at the Annual ASME Turbo Expo in Barcelona, Spain.

Ceccio, S.L., George, D.L., 1996. A review of electrical impedance techniques for the measurement of multiphase flow. ASME Journal of of Fluids Engineering 118, 391–399.

Celata, G.P., Cumo, M., Farello, G.E., Focardi, G., 1989. A comprehensive analysis of direct contact condensation of saturated steam on subcooled liquid jets. Journal of Heat Mass Transfer (Pergamon Press) 32 (4), 639–654.

Celata, G.P., Cumo, M., D'Annibale, F., Farello, G.E., 1991. Direct contact condensation of steam on droplets. International Journal of Multiphase Flow 17, 191–211.

Cess, R.D., 1964. The interaction of thermal radiation with conduction and convection heat transfer. In: Irvine, J.F., Hartnett, J.P. (Eds.), Advances in Heat Transfer, vol. 1. Academic Press Inc., New York, NY, USA, pp. 1–51.

CFR, 1989. 10 CFR 50.46, Acceptance Criteria for Emergency Core Cooling Systems for Light Water Nuclear Power Reactors, to 10 CFR Part 50, Code of Federal Regulations. US Office of the Federal Register, Washington, DC, USA.

Cha, J.E., Ahn, Y.C., Kim, M.H., 2001. Flow measurement with an electromagnetic flowmeter in two-phase bubbly and slug flow regimes. Flow Measurement and Instrumentation 12 (5–6), 329–340.

Chai, L.H., Shoji, M., 2001. Boiling curves-bifurcation and catastrophe. International Journal of Heat Mass Transfer 44, 4175–4179.

Chaikin, P.M., Lubensky, T.C., 1995. Principles of Condensed Matter Physics. Cambridge University Press, Cambridge, UK.

Chan, C.K., Lee, C.K.B., 1982. A regime map for direct contact condensation. International Journal of Multiphase Flow 8 (1).

Chan, A.M.C., Kawaji, M., Nakamura, H., Kukita, Y., 1999. Experimental study of two-phase pump performance using a full size nuclear reactor pump. Nuclear Engineering and Design 193 (1–2), 159–172.

Chandesris, M., Serre, G., Sagaut, P., 2006. A macroscopic turbulence model for flow in porous media suited for channel, pipe and rod bundle flows. International Journal of Heat Mass Transfer 49, 2739–2750.

Chandesris, M., Mazoyer, M., Serre, G., Valette, M., 2013. Rod bundle thermalhydraulics mixing phenomena: 3-D analysis with CATHARE 3 of various experiments. In: 15th International Topical Meeting on Nuclear Reactor Thermalhydraulics, NURETH-15, Pisa, Italy, May 12–15.

Chang, C.-J., Lahey Jr., R.T., 1997. Analysis of chaotic instabilities in boiling systems. Nuclear Engineering and Design 167, 307–334.

Chang, K.A., Lim, H.J., Su, C.B., 2003. Fiber optic reflectometer for velocity and fraction ratio measurements in multiphase flows. Review of Scientific Instruments 74, 3559–3565.

Chang, H., Li, Y., Zhao, R., Hu, T., Wang, H., Dai, Q., 2011. Integral thermal-hydraulic test facilities to support the development of the large passive nuclear power plant in China. Nuclear Safety Analysis and Simulation 2 (4), 317–325.

Chang, S., Kim, S., Song, C.-H., 2014. Turbulent mixing in a rod bundle with vaned spacer grids: OECD/NEA–KAERI CFD benchmark exercise test. Nuclear Engineering and Design 279, 19–36.

Chao, J.J., 2022. A Complete Perspective on Nuclear Energy. Nova Science Publishers, Inc., Hauppauge, NY, USA, ISBN 978-1-68507-821-8 https://doi.org/10.52305/PMAX6288.

Chaouki, J., Larachi, F., Dudukovic, M.P. (Eds.), 1997. Non-Invasive Monitoring of Multiphase Flows. Elsevier Sience, New York, NY, USA, pp. 1–585.

Chapman, D.R., 1979. Computational aerodynamics development and outlook. AIAA Journal 17 (12), 1293–1313.

Chaskalovic, J., 2014. Mathematical and Numerical Methods for Partial Differential Equations—Applications for Engineering Sciences. Springer, Berlin, Germany, ISBN 978-3-319-03562-8, pp. 1–358.

Chatzikyriakou, D., Walker, S.P., Hewitt, S.P., 2010. The contribution of non-wetting droplets to direct cooling of the fuel during PWR post-LOCA reflood. Nuclear Engineering and Design 240, 3108–3114.

Chaudri, K.S., Kim, J., Kim, Y., 2019. Development and validation of a fast sub-channel code for LWR multi-physics analyses. Nuclear Engineering and Technology 51, 1218–1230.

Chaxton, K.T., Collier, J.G., Ward, J.A., 1972. H.T.F.S. Correlation for Two-phase Pressure Drop and Void Fraction in Tubes, AERE-R7162, UK.

Chen, J.C., 1966. A correlation for boiling heat transfer to saturated fluids in convective flow. Industrial and Engineering Chemistry Process Design and Development 5 (3), 322–327.

Chen, R.C., Fan, L.S., 1992. Particle image velocimetry for characterizing the flow structure in three-dimensional gas–liquid–solid fluidized beds. Chemical Engineering Science 47, 3615–3622.

Chen, Y.-S., Yuann, Y.-R., 2015. Negative pressure difference evaluation of Lungmen ABWR containment by using Gothic. Annals of Nuclear Energy 75.

Chen, H., Zhang, X., 2015. Development of a steady-state sub-channel code for small reactor on the basis of combined cross momentum and non-linear conduction. Nuclear Engineering and Design 289, 186–194.

Chen, J.C., Sundaram, R.K., Ozkaynak, F.T., 1977. A Phenomenological Correlation for post-CHF Heat Transfer. NUREG-0237, Washington, DC, USA.

Chen, Y.S., Fischer, S.R., Sullivan, L.H., 1979. Calculations of Flow Oscillations during Reflood Using RELAP4/MOD6 (Idaho National Engineering Lab., Idaho Falls, Id, USA).

Chen, B.C.J., Sha, W.T., Doria, M.L., Schmitt, R., Thompson, J.F., 1980. BODYFIT-1FE: a computer code for three-dimensional steady-state/transient single-phase rod-bundle thermal-hydraulic analysis. In: ANL-80-127, USNRC NUREG/CR-1874, Washington, DC, USA.

Chen, Y., Yang, C., Zhang, C., Zhao, M., Du, K., 2009. An experimental study of critical flow of water under near-critical and supercritical pressures. In: The 13th International Topical Meeting on Nuclear Reactor Thermal Hydraulics (NURETH-13), N13P1048, Kanazawa, Japan, September 27–October 2.

Chen, C.-H., Wang, J.-R., Lin, H.-T., Shih, C., 2014. ATWS analysis for Maanshan PWR using TRACE/SNAP code. Annals of Nuclear Energy 72.

Chen, S.K., Todreas, N.E., Nguyen, N.T., 2014. Evaluation of existing correlations for the prediction of pressure drop in wire-wrapped hexagonal array pin bundles. Nuclear Engineering and Design 267, 109–131.

Chen, S.K., Chen, Y.M., Todreas, N.E., 2018. The upgraded Cheng and Todreas correlation for pressure drop in hexagonal wire-wrapped rod bundles. Nuclear Engineering and Design 335, 356–373.

Chen, S.K., Chen, Y.M., Todreas, N.E., 2019. A note on the accuracy of the upgraded Cheng and Todreas correlation for predicting pressure drop in hexagonal bare rod bundles. Applied Thermal Engineering 161 (114193), 1–7.

Cheng, S.K., Todreas, N.E., 1986. Hydrodynamic models and correlations for bare and wire-wrapped hexagonal rod bundles—bundle friction factors, subchannel friction factors and mixing parameters. Nuclear Engineering and Design 92 (2), 227–251.

Cheng, L.-Y., Baek, J.-S., Cuadra, A., Aronson, A., Diamond, D., 2014. TRACE simulation of a BWR anticipated transient without scram leading to emergency depressurization. In: ANS Annual Meeting, Reno, NV, USA.

Cheng, L.-Y., Baek, J., Cuadra, A., Aronson, A., Diamond, D., Yarsky, P., 2015. TRACE/PARCS analysis of ATWS with instability for a MELLLA+ BWR/5. In: 16th International Topical Meeting on Nuclear Reactor Thermal Hydraulics (NURETH-16), August 30–September 4, Chicago, IL, USA.

Cheng, Z., Xu, T., Vasic, A., Rao, Y.F., Aydemir, N.U., 2018. Assessment of CATHENA 4 subchannel capability against RD-14M tests. In: 12th International Topical Meeting on Nuclear Reactor Thermal-Hydraulics, Operation and Safety (NUTHOS-12), October 14–18, Qingdao, China.

Cherubini, M., Giannotti, W., Araneo, D., D'Auria, F., 2008a. Use of the natural circulation flow map for natural circulation systems evaluation. Science and Technology of Nuclear Installations. Article ID 479673.

Cherubini, M., Muellner, N., D'Auria, F., Petrangeli, G., 2008b. Application of an optimized AM procedure following a SBO in a VVER-1000. Nuclear Engineering and Design 238, 74–80.

Cherubini, M., Moretti, F., D'Auria, F., 2009. Independent analysis for the assessment of multidimensional features of the MARS code. In: KINS/RR-669, RETAS-MARS-CA-03, Daejeon, Korea, [initially Issued as University of Pisa, DIMNP/GRNSPG NT 637(08)-rev.0, Pisa, Italy, 2008].

Cherubini, M., Moretti, F., D'Auria, F., Ahn, S.H., Cho, Y.J., Hoehne, T., 2010. Independent assessment of MARS 3D features: use of experimental data and CFD support. In: 8th International Conference On Nuclear Option in Countries with Small and Medium Electricity Grids, Dubrovnik, Croatia, Article ID S6-67, pp 1–23, May 16–20.

Cheung, F.-B., 2021. In: NED Special Issue for 40th NURETH Celebration, Journal of Nuclear Engineering and Design (VSI Covering 2020–2021, 13 Papers Published).

Cheung, Y.K., Parameswaran, V., Shaug, J.C., 1983. BWR Refill-Reflood Program, Task 4.7, Model Development, TRAC-BWR Component Models. USNRC NUREG/CR-2574 (also EPRI report NP-2376, GEAP-22052, April 1983), Washington, DC, USA.

Chexal, B., Lellouche, G., 1986. A Full-Range Drift-Flux Correlation for Vertical Flows (Rev. 1). EPRI NP-3989-SR, Palo Alto, CA, USA.

Chexal, B., Lellouche, G., Horowitz, J., Healtzer, J., 1992. A void fraction correlation for general applications. Progress in Nuclear Energy 27 (4), 255–295.

Chikhi, N., Fichot, F., 2010. Reflooding model for quasi-intact rod configuration: quench front tracking and heat transfer closure laws. Nuclear Engineering and Design 240, 3387–3396.

Chikhi, N., Fleurot, J., 2012. Revisiting the QUENCH-11 integral reflood test with a new thermal–hydraulic model: existence of a minimum injection rate. Annals of Nuclear Energy 49, 12–22.

Chisholm, D., 1967. A theoretical basis for the Lockhart-Martinelli correlation for two-phase flow. International Journal of Heat Mass Transfer 10, 1767–1778.

Chisholm, D., 1968. The influence of mass velocity on friction pressure gradient during steam-water flow. In: Proceedings Thermodynamics and Fluid Mechanics Conv., Proceedings of the Institution of Mech Engrs., 182 (Pt 3H), Paper 35, pp 336–341 (First Published September 1, 1967), UK.

Chisholm, D., 1973. Pressure gradients due to friction during the flow of evaporating two-phase mixtures in smooth tubes and channels. International Journal of Heat Mass Transfer 16, 347–358.

Chisholm, D., Sutherland, L.A., 1969. Prediction of pressure gradient in systems during two-phase flow. In: Proceedings Inst. Of Mechanical Engineers, Symposium on Two-phase Flow Systems, University of Leeds, Leeds, UK.

Cho, Y.J., Jeun, G.D., 2004. Modeling of liquid entrainment and vapor pull-through in header-feeder pipes of CANDU. Journal of Korean Nuclear Society 36, 142–152.

Cho, S., Song, C.-H., Park, C.K., Yang, S.K., Chung, M.K., 1998. Experimental study on dynamic Pressure Pulse in direct contact condensation of steam discharging into subcooled water. In: 1st Korea-Japan Symp. On Nuclear Thermal Hydraulics and Safety (NTHASl), Pushan, Korea.

Cho, S., Park, H.-S., Choi, K.-Y., Kang, K.-H., Baek, W.-P., Kim, Y.-S., 2009. Core thermal-hydraulic behavior during the reflood phase of cold leg LBLOCA experiments using the ATLAS test facility. Nuclear Engineering and Technology 41 (10).

Cho, S., Kim, S., Bae, B.-U., Cho, Y.-J., Kim, Y.-S., Yun, Y.-J., Jeon, W.-J., 2018. Qualification test of Korean localized RCP's seal under extended SBO condition of APR-1400. In: 12th International Topical Meeting on Nuclear Reactor Thermal-Hydraulics, Operation and Safety (NUTHOS-12), October 14–18, Qingdao, China.

Choe, W.G., Weinberg, L., Weisman, J., 1978. Observation and correlation of flow pattern transition in horizontal, Co-current gas-liquid flow. In: Veziroglu, T.N., Kakac, S. (Eds.), Two-Phase Transport and Reactor Safety. Hemisphere Publishing Corporation, Washington, DC, USA.

Choi, Y.J., 2017. Critical heat flux (CHF) update to RELAP5-3D. RELAP5-3D Bi-annual Newsletter April-September 1–5.

Choi, T.S., No, H.C., 2012. An improved RELAP5/MOD3.3 reflood model considering the effect of spacer grids. Nuclear Engineering and Design 250, 613–625.

Choi, J.S., McLaughlin, D.K., Thompson, D.E., 2003. Experiments on the unsteady flow field and noise generation in a centrifugal pump impeller. Journal of Sound Vibration 263 (3), 493–514.

Choi, K.-Y., Kang, K.H., Kwon, T.S., Kim, Y.S., Kim, J.-R., Mon, S.K., Park, Y.-S., Park, H.S., Bae, B.U., Song, C.-H., Yi, S.J., Cho, S., Yun, B.J., 2014. Scaling Analysis Report of the ATLAS Facility. KAERI Report, KAERI/TR-5465/2014, Daejeon, Korea.

Choi, K.-Y., Kang, K.H., Song, C.H., 2019. Recent achievement and future prospects of the ATLAS program. Nuclear Engineering and Design 354.

Christensen, H., 1961. MS, Royal Technical University of Denmark (1954). Doctoral Dissertation, Massachusetts Institute of Technology, Boston, Ma, USA.

Chu, I.-C., Euh, D.-J., Song, C.-H., 2015. Observation of CHF phenomena based on a visualization of near wall boiling structure in a vertical narrow channel submerged in a pool. In: 16th International Topical Meeting on Nuclear Reactor Thermal Hydraulics (NURETH-16), August 30–September 4, Chicago, IL, USA.

Chudnovskaya, I.I., Shtern, Z.Y., Zarichniak, Y.P., Muratov, B.L., 1978. Method and results of researches of in-tube formations. In: Proceedings Of the Temperature Regime and Hydraulics in Steam Generators. Nauka, Leningrad, Russia, pp. 21–32 (in Russian).

Chudnovsky, A.F., 1954. Heat and Mass Transfer in Disperse Mediums. GITTL, Moscow, Russia (in Russian).

Chugh, D., Roy, S., Shao, J., Al-Dahhan, M.H., 2017. Experimental investigation of gas–liquid flow in monolith channels using monofiber optical probes. AIChE Journal 63, 327–336.

Chun, M.H., Kang, M.G., 1998. Effect of heat exchanger tube parameters on nucleate pool boiling heat transfer. Journal of Heat Transfer 120 (2), 468–476.

Chun, T.H., Oh, D.S., 1998. A pressure drop model for spacer grids with and without flow mixing vanes. Journal of Nuclear Science and Technology 35 (7), 508–510.

Chung, M.S., Chang, K.S., Lee, S.J., 2000. Wave propagation in two-phase flow based on a new hyperbolic two-fluid model. Journal of Numerical Heat Transfer, Part A 38, 169–191.

Churchill, S.W., 1977. Friction factor equations spans all fluid-flow regimes. Chemical Engineering 84, 91–92.

Churchill, S.W., Chu, H.H.S., 1975a. Correlating equations for laminar and turbulent free convection from a horizontal cylinder. International Journal of Heat Mass Transfer 18, 1049.

Churchill, S.W., Chu, H.H.S., 1975b. Correlating equations for laminar and turbulent free convection from a vertical plate. International Journal of Heat Mass Transfer 18, 1323–1329.

Cicchitti, A., Lombardi, C., Silvestri, M., Soldaini, G., Zavattarelli, R., 1960. Two-phase cooling experiments: pressure drop, heat transfer and burnout experiments. Journal of Energia Nucleare 7, 407–425.

Ciofalo, M., 2022. Lezioni di Termoidraulica. Società Editrice Esculapio, Bologna, Italy, ISBN 978-88-9385-295-1, pp. 1–202 (in Italian).

Cioni, L., D'Auria, F., Grassi, W., 1986. Critical analysis of reflood models. In: 4th Conf. Of Italian Society of Heat Transport, Genova, Italy, June 5–7.

Clark, W.W., 2002. Liquid film thickness measurement. Multiphase Science and Technology 14, 1–74.

Clausing, A.M., Chao, B.T., 1965. Thermal contact resistance in vacuum. Journal of Heat Transfer Transactions, ASME 87 (2), 243–250.

Clement, P., Regnier, P., 1978. Heat Transfer Modelling at the Quench Front during Reflood Phase in a LOCA -interpretation of Experimental Results. European Two-Phase Flow Group Meeting, Stockholm, Sweden.

Clement, P., Deruaz, R., Veteau, J.M., 1982. Reflooding of a PWR rod bundle. Effect of inlet flow oscillations and spacer grids. In: Proceedings International Meeting on Thermal Reactor Safety, Chicago, IL, 29 August–2 September, NUREG/CP-0027, vol. 3. USNRC, Washington, DC, USA, pp. 1763–1770.

Clifford, I., Ferroukhi, H., 2021. Assessing the effects of switching from multi-channel to three-dimensional nodalisations of the core in TRACE. Nuclear Engineering and Design 383 (111446), 1–15.

Cocci, R., Ghione, A., Sargentini, L., Damblin, G., Lucor, D., 2022. Model assessment for direct contact condensation induced by a sub-cooled water jet in a circular pipe. International Journal of Heat Mass Transfer 195 (123162), 1–17.

Coddington, P., Macian, J.R., 2002. A study of the performance of void fraction correlations used in the context of drift-flux two-phase flow models. Nuclear Engineering and Design 215, 199–216.

Colburn, A.P., Hougen, O.A., 1934. Design of cooler condensers for mixtures of vapors with non-condensing gases. Industrial and Engineering Chemistry Research 26, 1178–1182.

Colebrook, C.F., 1939. Turbulent flow in pipes, with particular reference to the transition region between the smooth and rough pipe laws. Journal of the Institution of Civil Engineers 11 (4), 133–156.

Colebrook, C.F., White, C.M., 1937. Experiments with fluid friction in roughened pipes. Proceedings of the Royal Society of London, Series A 161 (906), 367–381.

Collier, J.G., 1975. Two Phase Flow and Heat Transfer in Water Cooled Nuclear Reactors, Lecture Series. Dartmouth College, Hanover, NH, USA.

Collier, J.G., Thome, J.R., 1980. Convective Boiling and Condensation. Oxford University Press, Oxford, UK (also McGraw—Hill, London, UK, 1981).

Collier, J.G., Thome, J.R., 1994. Convective Boiling and Condensation, third ed. Clarendon Press, Oxford, UK.

Colombo, M., Colombo, L., Cammi, A., Ricotti, M.E., 2015. A scheme of correlation for frictional pressure drop in steam-water two-phase flow in helicoidal tubes. Chemical Engineering Sciences 123, 460–473.

Colombo, M., Fairweather, M., Walker, S.P., Kumar, M., Moharana, A., Nayak, A.K., Joshi, J.B., Dasgupta, A., Chandraker, D.K., Tanskanen, V., Giteshkumar, P., CFD model development for two-phase flows, Joshi, J.B., Nayak, A.K., 2018. Advances of Computational Fluid Dynamics in Nuclear Reactor Design and Safety Assessment. Elsevier, Woodhead Publishing, Duxford, UK, pp. 239–262 (Chapter 3) of.

Comley, G.C.W., 1978. Some aspects of system chemistry of BWRs. Journal of Progress Nuclear Energy 2, 153–181.

Comolet, R., 1976. Mecanique Experimentale des Fluides, Bd. 2: Dynamique des Fluides Reels. Masson, Paris, France.

Coney, M.W.E., 1973. The theory and application of conductance probes for the measurement of liquid film thickness in two phase flow. Journal of Physics E: Scientific Instruments 6, 903–910.

Congiu, C., D'Auria, F., Frogheri, M., Galassi, G.M., Madeira, A., Zhao, H., 1996. Relevant thermal-hydraulic aspects in four 1% Small break LOCA experiments in LOBI facility. Kerntechnik 61 (4).

Cooper, C.A., Pearson, K.G., Jowitt, D., 1984. The THETIS 80% blocked cluster experiment. In: Part 3: Forced Reflood Experiments. AEEW—R 1765, Winfrith, UK.

Cornwell, K., Kew, P.A., 1993. Boiling in small parallel channels. In: Pilavachi, P.A. (Ed.), Energy Efficiency in Process Technology. Elsevier, New York, NY, USA, pp. 624–638.

COST, 2007. Best practice guideline for the CFD simulation of flows in the urban environment. In: Franke, J., Hellsten, A., Schlünzen, H., Carissimo, B. (Eds.), COST Action 732: Quality Assurance and Improvement of Microscale Meteorological Models. European Cooperation in Science and Technology, pp. 16–18.

Coste, P., 2004. Computational simulation of multi-D liquid-vapor thermal shock with condensation. In: Proceedings of ICMF'04, Yokohama, Japan, May 30–June 4.

Coste, P., Laviéville, J., 2009. A wall function-like approach for two-phase CFD condensation modelling of the pressurized thermal shock. In: The 13th International Topical Meeting on Nuclear Reactor Thermal Hydraulics (NURETH-13), Kanazawa, Japan, September 27–October 2.

Courant, R., Friedrichs, K., Lewy, H., 1928. Über die partiellen Differenzengleichungen der mathematischen physik, (in German). Mathematische Annalen 100, 32–74.

Courtin, S., 2013. High cycle thermal fatigue damage prediction inmixing zones of nuclear power plants: engineering issues illustrated on the FATHER case. Procedia Engineering 66, 240–249.

Cowen, H., Pfeffer, S., Vedovi, J., Ibarra, L., Vreeland, D., Miranda, D., 2015. TRACG analysis and results of phase 1 of the OECD/NEA Oskarshamn-2 BWR stability benchmark. In: 16th International Topical Meeting on Nuclear Reactor Thermal Hydraulics (NURETH-16), August 30–September 4, Chicago, IL, USA.

Craft, T.J., Gerasimo, A.V., Launder, B.E., Robinson, C.M.E., 2006. A computational study of the near-field generation and decay of wingtip vortices. International Journal of Heat Fluid Flow 27, 684–695.

Creer, J.M., Rowe, D.S., Bates, J.M., Sutey, A.M., 1976. Effects of Sleeve Blockages on Axial Velocity and Intensity of Turbulence in an Unheated 7x7 Rod Bundle. Battelle Pacific Northwest Laboratories, BNWL-1965, Richland, WA, USA.

Cremer, I., Tap, R., Hupp, M., 2022. Updated modelling and simulation of hydrogen recombination in full-scale containment analysis. In: The 19th International Topical Meeting on Nuclear Reactor Thermal Hydraulics (NURETH-19), 35508, Brussels, Belgium, March 6–11.

Creveling, H.F., De Paz, J.F., Baladi, J.Y., Schoenals, R.J., 1975. Stability characteristics of a single-phase free convection loop. Journal of Fluid Mechanics 67 (1), 65–84.

Crouzet, F., Daude, F., Galon, P., Hérard, J.-M., Hurisse, O., Liu, Y., 2015. Validation of a two-fluid model on unsteady liquid–vapor water flows. Computers and Fluids 119, 131–142.

Crowley, C.J., Block, J.A., Cary, C.N., 1977. Down-comer Effects in a 1/15-scale PWR Geometry, Experimental Data Report. USNRC, NUREG-0281, Washington, DC, USA.

Crowley, C.J., Rothe, P.H., Sam, R.G., 1981. 1/5-scale Countercurrent Flow Data Presentation and Discussion. CREARE Inc., NUREG/CR-2106, November, Washington, DC, USA.

CS & A, 2001. RETRAN-3D—A Program for Transient Thermal-Hydraulic Analysis of Complex Fluid Flow Systems, Volume 1: Theory and Numerics, EPRI-NP-7500(A), Rev. 5, Palo Alto, CA, USA.

CSNI, Aksan, N., D'Auria, F., Glaeser, H., Pochard, R., Richards, C., Sjöberg, A., 1994. Separate Effects Test Matrix for thermal-hydraulic Code Validation, (A) Volume I: Phenomena Characterization and Selection of Facilities and Tests, (B) Volume II: Facility and experiment Characteristics. NEA/CSNI/R(93)14/Part 1 and Part 2, Paris, France [also OECD/NEA/CSNI, 1993].

CSNI, Aksan, S.N., D'Auria, F., Städtke, H., 1995. User Effects on the Transient System Code Calculations. NEA/CSNI/R(94)35, Paris, France [also Aksan et al., 1993].

CSNI, Annunziato, A., Glaeser, H., Lillington, J., Marsili, P., Renault, C., Sjöberg, A., 1996. CSNI Integral Test Facility Validation Matrix for the Assessment of thermal-hydraulic Codes for LWR LOCA and Transients. NEA/CSNI/R(96)17, Paris, France [also OECD/NEA/CSNI, 1996].

CSNI, Ashley, R., El-Shanawany, M., Eltawila, F., D'Auria, F., 1998. Good Practices for User Effect Reduction. NEA/CSNI/R(98)22, Paris, France [also Ashley et al., 1998].

CSNI, 2000. CSNI International Standard Problems (ISP), Brief Descriptions (1975–1999). NEA/CSNI/R(2000)5, Paris France.

CSNI, 2001. Validation Matrix for the Assessment of thermal-hydraulic Codes for VVER LOCA and Transients. NEA/CSNI/R(2001)4, Paris, France [also OECD/NEA/CSNI, 2001].

CSNI, 2004. International Standard Problem Procedure. CSNI Report 17 Revision 4, NEA/ CSNI/ R(2004)8, Paris, France.

CSNI, Glaeser, C.H., 2011. BEMUSE Phase VI Report, Status Report on the Area, Classification of the Methods, Conclusions and Recommendations. GRS NEA/CSNI/R(2011)4, Paris, France (also OECD/NEA/CSNI, 2011).

CSNI, 2016a. Post-BEMUSE Reflood Model Input Uncertainty Methods (PREMIUM) Benchmark. Final Report, NEA/CSNI/R(2016)18, Paris, France.

CSNI, 2016b. Scaling in System thermal-hydraulics Applications to Nuclear Reactor Safety and Design: A State-Of-The-Art Report. NEA/CSNI/R(2016)14, Paris, France [also Bestion et al., 2017].

CSNI, 2016c. Final Integration Report of the Rig-Of-Safety Assessment (ROSA-2) Project (2009–2012) (Authors: Nakamura et al.). NEA/CSNI/R(2016)10 [distributed 2017], Paris, France.

CSNI, Petruzzi, C.A., D'Auria, F., 2006. BEMUSE Phase II Report, Re-analysis of the ISP-13 Exercise, post-test Analysis of the LOFT L2-5 Test Calculation. University of Pisa NEA/CSNI/R (2006)2, Paris, France [also OECD/NEA/CSNI, 2006].

CSNI, De Crecy, C.A., Bazin, P., 2007. BEMUSE Phase III Report, Uncertainty and Sensitivity Analysis of the LOFT L2-5 Test. CEA NEA/CSNI/R(2007)4, Paris, France.

CSNI, Reventos, C.F., Pérez, M., Batet, L., Pericas, R., 2008. BEMUSE Phase IV Report, Simulation of a LB-LOCA in Zion Nuclear Power Plant. UPC NEA/CSNI/R(2008)6, Paris, France (also OECD/NEA/CSNI, 2007a).

CSNI, Reventos, C.F., Batet, L., Pérez, M., 2009. BEMUSE Phase V Report, Uncertainty and Sensitivity Analysis of a LB-LOCA in Zion Nuclear Power Plant. UPC NEA/CSNI/R(2009)13, Paris, France (also OECD/NEA/CSNI, 2009a).

CSNI, Choi, K.-Y., Baek, W.-P., Cho, S., Park, H.S., Kang, K.H., Kim, Y.S., Kim, H.T., 2012. International Standard Problem No. 50, Volume 1, Analysis of Blind/open, Calculations. NEA/CSNI/R(2012)6, Paris, France.

Cuamatzi-Meléndez, R., Flores-Cuamatzi, E., 2020. Modelling fluid-structure interaction of water recirculating flow to predict damage and/or failure in a jet-pump assembly of a nuclear boiling water reactor. Engineering Structures 206 (110155), 1–15.

Cumo, M., 1982. Experimental Thermo-Technics (In Italian). ENEA Publisher, Roma, Italy.

Cumo, M., Naviglio, A., 1990. Thermal Hydraulic Design of Components for Steam Generation Plants. CRC Press, Boca Raton, FL, USA.

Cumo, M., Farello, G.E., Ferrari, G., 1977. Heat transfer in condensing jets of steam in water (pressure-suppression systems). In: Research Report CNEN RT/ING(77)8, Casaccia, Rome, Italy.

Cuta, J.M., Kontz, A.S., Stewart, C.W., Montgomery, S.D., Nomura, K.K., 1985. VIPRE-01: A thermal-hydraulic Code for Reactor Cores. Volume 2: User's Manual (Rev. 2). Battelle Pacific Northwest Laboratories, EPRI-NP-2511-CCM-Vol 2, Richland, Wa, USA.

D'Auria, F., Faluomi, V., 1996. Qualitative analysis of scaling effects in isolation condenser behaviour with Relap5/mod3 thermal-hydraulic code. In: ASME-JSME International Conference On Nuclear Engineering (ICONE-4)—New Orleans, Louisiana, USA, March 10–14.

D'Auria, F., Galassi, G.M., 1990c. Outline of user effect on codes predictions. In: University of Pisa Report, DCMN—NT 158(90), Pisa (I), June 1990, 7th Meet. Of CSNI THSB Task Group, Paris, France, June 26–28.

D'Auria, F., Pieve, L., 1983. Density measurements for two-phase critical flows (in Italian). In: 38th Conf. Italian Thermo-Technic Association (ATI), Bari, Italy, September 28–30 [only Abstract in Proceedings].

D'Auria, F., Dini, R., Maugeri, P., Oriolo, F., 1989. Application of RELAP5/MOD2 to design calculations of more passive safety reactors. In: ANS Winter Meet., Session on Thermal-Hydraulic Aspects of Passive Safety and New Generation Reactors, San Francisco (CA), USA, November 26–December 1, vol. 60. ANS Trans., pp. 1–792

D'Auria, F., Mazzini, M., Oriolo, F., Galassi, G.M., 1989a. Assessment of RELAP5/MOD2 and RELAP5/MOD1-EUR Codes on the Basis of LOBI-MOD2 Test Results. CEC Report EUR 12403 EN, Luxembourg (L).

D'Auria, F., Ferri, R., Galassi, G.M., Sugaroni, F., 1992. Evaluation of the data base from the small break LOCA Counterpart tests performed in LOBI, SPES, BETHSY and LSTF facilities. University of Pisa Report, DCMN—NT 193(92), Pisa, Italy, pp. 1–190.

D'Auria, F., Galassi, G.M., Annunziato, A., Addabbo, C., 1992a. Comparative analysis of accident management procedures in LOBI and SPES facilities. In: 5th International Topical Meeting on Nuclear Reactor Thermal-hydraulics, Salt Lake City, UT, USA, September 20–24.

D'Auria, F., Modro, M., Oriolo, F., Tasaka, K., 1993a. Relevant thermal hydraulic aspects of new generation LWR's. CSNI special meeting on transient two-phase flow—system thermal-hydraulics, Aix-en-Provience, France, April 6–8. Journal of Nuclear Engineering and Design 145 (1, 2).

D'Auria, F., Galassi, G.M., Leonardi, M., Galetti, R., 1997d. Application of the Fast Fourier Transform Method to Evaluate the Accuracy of SBLOCA Data Base. International Topical Meeting on Nuclear Reactor Thermal Hydraulics (NURETH-8), Kyoto, Japan.

D'Auria, F., Frogheri, M., Giannotti, W., 1999. Relap5/mod3.2 post-test Analysis and Accuracy Quantification of LOBI Test BL-44. USNRC NUREG/IA-0153, Washington, DC, USA.

D'Auria, F., Fontani, E., Frogheri, M., Misale, M., Garcia, A., 1999a. Analysis of single phase natural circulation experiment with Cathare V1.3U and Relap5/Mod3.2 codes. International Journal of Thermal Science 38, 977–983.

D'Auria, F., Hassan, Y.A., 2019. Special issue 'trends and perspectives in nuclear thermal-hydraulics'—to the memory of B. R. Sehgal, G. Yadigaroglu and G. F. Hewitt. Journal of Nuclear Engineering and Design 354, 1–450. December.

Da Silva, M.J., Schleicher, E., Hampel, U., 2007. Capacitance wire-mesh sensor for fast measurement of phase fraction distributions. Measurement Science and Technology 18, 2245–2251.

Damerell, P.S., Simons, J.W. (Eds.), 1993a. Reactor safety issues resolved by the 2D/3D program. International Agreement Report, USNRC NUREG/IA 0127, Washington, DC, USA.

Damerell, P.S., Simons, J.W., 1992. 2D/3D Program Work Summary Report. GRS Report No. 100, ISBN 3-923875-50-9 (Koln, Germany).

Damerell, P.S., Simons, J.W. (Eds.), 1993. 2D/3D Program work summary report. International Agreement Report, USNRC NUREG/IA 0126, Washington, DC, USA. [also Damerell & Simons, 1992].

Daw, J., Rempe, J.L., Knudson, D.L., Crepeau, J., Wilkins, S.C., 2009. High temperature irradiation-resistant thermocouple performance improvements. In: Conference on NPIC-HMIT 2009, INL/CON-09-15267 (TRN: US1000631), Knoxville, TN, USA.

de Cachard, F., 1986. Development Implementation and Assessment of Specific Closure Laws for Inverted-Annular Film Boiling in a Two-Fluid Model. USNRC NUREG/IA-0133, Washington, DC, USA.

de Crecy, F., 1986. Modeling of stratified two-phase flow in pipes, pumps and other devices. International Journal of Multiphase Flow 12, 307–323.

Glaeser, H., Skorek, T., Joucla, J., Probst, P., Chung, B., Oh, D.-Y., Kynck, M., Pernica, R., Macek, J., Meca, R., Macian, J.R., D'Auria, F., Petruzzi, A., Perez, M., Reventos, F.,

Fujioka, K., 2007. BEMUSE phase III report, uncertainty and sensitivity analysis of the LOFT L2-5 test. In: de Crecy, A., Bazin, P. (Eds.), OECD/NEA/CSNI Report NEA/CSNI/R (2007)4, Paris, France, October.

de Crécy, A., Bazin, P., 2004. Quantification of the Uncertainties of the Physical Models of CATHARE 2. Best Estimate-2004, ANS Winter Meeting, Washington, DC, USA, November 14–18.

de Crecy, A., Bazin, P., Glaeser, H., Skorek, T., Joucla, J., Probst, P., Fujioka, K., Chung, B.D., Oh, D.Y., Kyncl, M., Pernica, R., Macek, J., Meca, R., Macian, J.R., D'Auria, F., Petruzzi, A., Batet, L., Perez, M., Reventos, F., 2008. Uncertainty and sensitivity analysis of the LOFT L2-5 test: results of the BEMUSE programme. Nuclear Engineering and Design 238 (12), 3561–3578.

De Geer, L.-E., Persson, C., Rodhe, H., 2017. A nuclear jet at chernobyl around 21:23:45 UTC on April 25, 1986. Nuclear Technology 1, 1–12.

De Kruijf, W.J.M., Van Der Hagen, T.H.J.J., Zboray, R., Manera, A., Mudde, R.F., 2002. CIRCUS and DESIRE: Experimental Facilities for Research on Natural-Circulation-Cooled Boiling Water Reactors. IAEA-TECDOC-1281, Vienna, Austria.

De Kruijf, W.J.M., Ketelaar, K.C.J., Avakian, G., Gubernatis, P., Caruge, D., Manera, A., Van Der Hagen, T.H.J.J., Yadigaroglu, G., Dominicus, G., Rohde, U., Prasser, H.-M., Castrillo, F., Huggenberger, M., Hennig, D., Munoz-Cobo, J.L., Aguirre, C., 2003. Planned experimental studies on natural-circulation and stability performance of boiling water reactors in four experimental facilities and first results (NACUSP). Journal of Nuclear Engineering and Design 221.

De Lorenzo, M., Lafon, P., Di Matteo, M., Pelanti, M., Seynhaeve, J.-M., Bartosiewicz, Y., 2017. Homogeneous two-phase flow models and accurate steam-water table look-up method for fast transient simulations. International Journal of Multiphase Flow 95, 199–219.

De Lorenzo, M., Lafon, P., Pelanti, M., 2019. A hyperbolic phase-transition model with non-instantaneous EoS-independent relaxation procedures. Journal of Computational Physics 379, 279–308.

De Lorenzo, M., Lafon, P., Pelanti, M., Pantano, A., Di Matteo, M., Bartosiewicz, Y., Seynhaeve, J.-M., 2021. A hyperbolic phase-transition model coupled to tabulated EoS for two-phase flows in fast depressurizations. Nuclear Engineering and Design 371 (110954), 1–16.

De Moura, C.A., Kubrusly, C. (Eds.), 2013. The Courant-Friedrichs-Lewy (CFL) Condition—80 Years after its Discovery, Birkhauser Basel. Springer Science, Germany, ISBN 978-0-8176-8393-1.

De Salve, M., Panella, B., Scorta, G., 1986. Heat and mass transfer by direct condensation of steam on a subcooled turbulent water jet. In: Proceedings 8th International Heat Transfer Conference, Volume 4, August, San Francisco, CA, USA, pp. 1653–1658.

De Santi, G., Piplies, L., Sanders, J., 1986. Mass flow instabilities in LOBI steam generators U-tubes under natural circulation conditions. In: 2nd International Meeting on NPP Thermal-Hydraulics and Operation, Tokyo, Japan.

De Santis, D., Kottapalli, S., Shams, A., 2018. Numerical simulations of rod assembly vibration induced by turbulent axial flows. Nuclear Engineering and Design 335, 94–105.

De With, P.A., Calay, R.K., de With, G., 2007. Three-dimensional condensation regime diagram for direct contact condensation of steam injected into water. International Journal of Heat Mass Transfer 50, 1762–1770.

Deendarlianto, T., Lucas, D., Vierow, K.M., 2012. Gas–liquid countercurrent two-phase flow in a PWR hot leg: a comprehensive research review. Nuclear Engineering and Design 243, 214–233.

Dehbi, A.A., 1991. The Effects of Noncondensable Gases on Steam Condensation under Turbulent Natural Convection Conditions. Ph.D. dissertation. Massachusets Institute of Technology, Department of Nuclear Eng., Boston, Ma, USA.

Dehbi, A.A., 2015. A generalized correlation for steam condensation rates in the presence of air-helium mixtures. In: 16th International Topical Meeting on Nuclear Reactor Thermal Hydraulics (NURETH-16), August 30–September 4, Chicago, IL, USA.

Dehbi, A.A., 2021. A correlation spanning the entire laminar convection regime for vapor condensation over a horizontal tube in the presence of a non-condensable gas. International Communications in Heat and Mass Transfer 128 (105619), 1–12.

Deimel, C., Günther, M., Skoda, R., 2014. Application of a pressure based CFD code with mass transfer model based on the Rayleigh equation for the numerical simulation of the cavitating flow around a hydrofoil with circular leading edge. In: EPJ Web of Conferences, vol. 67. EDP Sciences, pp. 1–8.

Del Frate, L., 2016. Assessment of CFD Methods for Single and Two-phase Flows in Nuclear Reactors (Ph.D. Thesis). University of Pisa, Pisa, Italy.

Del Frate, L., Moretti, F., Galassi, G., D'Auria, F., 2016. Limitations in the use of hydraulic diameter. Journal of Nuclear Science and Technology 6 (1), 53–62.

Del Nevo, A., D'Auria, F., Mazzini, M., Bykov, M., Elkin, I.V., Suslov, A., 2008. The design of PSB-VVER experiments relevant to accident management. Power and Energy Systems 2 (1), 371–385.

Del Nevo, A., Rozzia, D., Moretti, F., D'Auria, F., 2012a. OECD/NRC PSBT benchmark: investigating the CATHARE2 capability to predict void fraction in PWR fuel bundle. Science and Technology of Nuclear Installations. Article ID 757356.

Del Nevo, A., Adorni, M., D'Auria, F., Melikhov, O.I., Elkin, I.V., Schekoldin, V.I., Zakutaev, M.O., Zaitsev, S.I., Bencık, M., 2012b. Validation of advanced computer codes for VVER technology: LB-LOCA transient in PSB-VVER facility. Science and Technology of Nuclear Installations. Article ID 480948.

Delchini, M.O., Ragusa, J.C., Berry, R.A., 2017. Simulations of single- and two-phase shock tubes and gravity-driven wave problems with the RELAP-7 nuclear reactor system analysis code. Nuclear Engineering and Design 319, 106–116.

Delhaye, J.M., Achard, J.L., 1977. Thermal and hydraulic aspects of nuclear reactor safety. In: Jones, O.C., Bankoff, S.G. (Eds.), Light Water Reactors, vol. I. ASME, New York, NY, USA, pp. 289–332.

Delhaye, J.M., Giot, M., Riethmuller, M.L. (Eds.), 1981. Thermohydraulics of the Two-phase Systems for Industrial Design and Nuclear Engineering. Hemisphere Publishing Co., McGraw-Hill Book Company, New York, NY, USA, pp. 1–525.

Demick, N.T., Galvin, M.R., Groome, J.T., Woods, B.G., 2007. OSU MASLWR Test Facility Description Report. OSU-MASLWR-07001, Corvallis, Or, USA.

Deng, J., Deng, -J.), Lu, Q., Liu, Y., Wang, X., Yang, X., Zhang, Y., Zhang, J., 2020. Review of sub-channel code development for pressurized water reactor and introduction of CORTH-V2.0 sub-channel code. Progress in Nuclear Energy 125, 103373.

Deng, K., Deng, -K.), Wang, C., Qiu, S., Su, G.H., Tian, W., Wang, Y., He, X., 2020. The characteristics and correlation of saturated pool boiling critical heat flux for downward-facing plain heating surface. Applied Thermal Engineering 166 (114724), 1–10.

Denisov, V.P., Dragunov, E.G., 2002. VVER Reactor Facilities for Nuclear Power Plants (In Russian). Iz-dAT Publ., Moscow, Russia, pp. 1–480.

Dennis, R., Samples, W.R., Anderson, D.M., Silverman, L., 1957. Isokinetic sampling probes. Industrial and Engineering Chemistry 49 (2), 294−302.

Deruaz, R., Clement, P., Veteau, J., 1985. Study of Two-Dimensional Effects in the Core of a Light Water Reactor during the ECCS Phase Following a Loss of Coolant Accident, pp. 1−200. CEA/CENG, SR–002-F, EUR–10076 Grenoble, France. http://inis.iaea.org/Search/search.aspx?origq=RN:17022937.

Deruaz, R., Clement, P., Veteau, J.M., 1985a. 2D effects in the core during the reflooding phase of a LOCA. In: Skupinski, E., Tolley, B., Vilain, J. (Eds.), Safety of Thermal Water Reactors. Graham and Trotman, London, UK, pp. 103−113.

Deshpande, S.S., Anumolu, L., Trujillo, M.F., 2012. Evaluating the performance of the two-phase flow solver interFoam. Computational Science and Discovery 5 (1), 014016.

Deych, M.Y., Filipov, G.A., 1969. The Gas Dynamics of Two Phase Media. Gazodinamika Dvukhfaznykh Sred, Moscow, Russia.

Dhotre, M.T., Smith, B., Niceno, B., 2007. CFD simulation of bubbly flows: random dispersion model. Chemical Engineering Sciences 62, 7140−7150.

Dhotre, M.T., Niceno, B., Smith, B., 2008. Large eddy simulation of a bubble column using dynamic sub-grid scale model. Chemical Engineering Journal 136, 337−348.

Diaz, J., Liu, Q., Petrov, V., Manera, A., Sun, X., 2022. High-Resolution X-ray radiography methods developed for post-CHF experiment. In: 19th International Topical Meeting on Nuclear Reactor Thermal Hydraulics (NURETH-19), March 6−11 (Via Web), Brussels, Belgium.

Diessler, G., Taylor, M.F., 1956. Analysis of axial turbulent flow heat transfer through banks of rods or tube. In: Reactor Heat Transfer Conference, AEC TID-7529, Vol. 2, Washington, DC, USA.

Dinh, N., (Nam, D.), Nourgaliev, R., Bui, A., Lee, H., 2013. Perspectives on nuclear reactor thermal hydraulics. In: Proceedings of the International Top. Meet. on Nuclear Reactor Thermal Hydraulics (NURETH-15), Pisa Italy, May 12−17.

Dittus, F.W., Boelter, L.M.K., 1930. Heat transfer in automobile radiators of the tubular type. In: Publications in Engineering, vol. 2. University of California, Berkeley, CA, USA, pp. 443−461.

Divo, E., Kassab, A.J., Kapat, J.S., Tapley, J., 2000. Determination of heat transfer coefficient maps using an inverse BEM algorithm. In: Tanaka, M., Dulikravich, G.S. (Eds.), Inverse Problems in Engineering Mechanics. Elsevier, Kidlington, Oxford, UK.

Diwan, S.S., Chetan, S.J., Ramesh, 0.N., 2006. On the bursting criterion for laminar separation bubbles. In: Proceedings of the IUTAM Symposium on Laminar-Turbulent Transition. Springer Academic Publisher, Dordrecht, The Netherlands, pp. 401−407.

Dix, G.E., 1983. BWR loss of coolant technology review. In: 2nd International Topical Meeting on Nuclear Reactor Thermalhydraulics, Santa Barbara, CA, USA, January 11−14.

Dixon, S.L., Hall, C.A., 2014. Fluid Mechanics and Thermodynamics of Turbomachinery. Imprint Butterworth-Heinemann, © Elsevier Inc., Amsterdam, Netherlands, ISBN 978-0-12-415954-9, pp. 1−537.

Doi, T., Futatsugi, T., Murase, M., Hayashi, K., Hosokawa, S., Tomiyama, A., 2012. Countercurrent flow limitation at the junction between the Surge Line and the pressurizer of a PWR. Science and Technology of Nuclear Installations. Article ID 754724.

Dokhane, A., Judd, J., Gajev, I., Zerkak, O., Ferroukhi, H., Kozlowski, T., 2017. Analysis of Oskarshamn-2 stability event using TRACE/SIMULATE-3K and comparison to TRACE/PARCS and SIMULATE-3K stand-alone. Annals of Nuclear Energy 102, 190−199.

Dollezhal, N.A., Emeljanov, I.Y., 1980. The Channel Nuclear Power Reactor. Atomizdat, Moscow (in Russian).

Dominguez-Ontiveros, E.E., Hassan, Y.A., 2009. Non-intrusive experimental investigation of flow behavior inside a 5×5 rod bundle with spacer grids using PIV and MIR. Nuclear Engineering and Design 239 (5), 888–898.

Dominick, J., Durst, F., Raszillier, H., Zeisel, H., 1987. A method to measure mass and volume flow rates of two-phase flows. International Journal of Multiphase Flow 13, 685–698.

Dong, F., Xu, Y.B., Xu, L.J., Hua, L., Qiao, X.T., 2005. Application of dual-plane ERT system and cross-correlation technique to measure gas–liquid flows in vertical upward pipe. Flow Measurement and Instrumentation 16, 191–197.

Dong, S., Yang, J., He, C., 2018. Thermal-hydraulic analysis of filtered containment venting system for long-term operation under station blackout. In: 12th International Topical Meeting on Nuclear Reactor Thermal-Hydraulics, Operation and Safety (NUTHOS-12), October 14–18, Qingdao, China.

Dopchie, H., 1971. Letter from Association Vincotte, to C. K. Beck, AEC, subject. In: Possible Inadequacy of ECCS Protection against Break of Vapor Line on Pressurizer. Association Vincotte, Brussels, Belgium, April 27.

Dor, I., Germain, P., 2011. Core radial profile effect during reflooding, validation of CATHARE-2 3D module using SCTF tests. In: International Topical Meeting on Nuclear Reactor Thermalhydraulics, NURETH-14, Toronto, Ontario, Canada.

Dor, I., Bazin, P., Boudier, P., 2005. Assessment of the CATHARE 3-D module against LOFT test L2-5 and L2-6. In: International Conference On Nuclear Eng. (ICONE-13), May, 2005, Beijing, China.

Dor, I., Lavialle, G., Germain, P., Mieusset, T., 2007. Investigation of downcomer level evolution during late reflooding with CATHARE 3D module. In: International Topical Meeting on Nuclear Reactor Thermalhydraulics (NURETH-12), Pittsburgh, PA, USA.

Doroshchuk, V.E., Levitan, L.L., Lantzman, F.P., 1975. Investigation into Burnout in Uniformly Heated Tubes. ASME Publication, Chicago, Il, USA. 75-WA/HT-22.

Dougherty, T.J., Fighetti, C., Reddy, G., Yang, B., Jafri, T., McAssey, E., Qureshi, Z., 1990. Flow boiling in vertical down flow. In: Proceedings 9th International Heat Transfer Conference, Jerusalem, Israel, vol. 2, pp. 9–14.

Douglas, J., Rachford Jr., H.H., 1956. On the numerical solution of heat conduction problems in two and three space variables. Transactions of the American Mathematical Society 82 (2), 421.

Dreyer, W., Müller, W.H., Weiss, W., 2000. Tales of thermodynamics and obscure applications of the second law. Continuum Mechanics and Thermodynamics 12, 151–184.

Drikakis, D., 2002. Embedded turbulence model in numerical methods for hyperbolic conservation laws. International Journal for Numerical Methods in Fluids 39 (9), 763–781.

Du, Y., Yan, X., Zang, J., Zhang, Y., 2020. Review of regime maps of steam-submerged direct contact condensation. Frontiers in Energy Research 8, 1–9. https://doi.org/10.3389/fenrg.2020.00119 article 119.

Dudlik, A., 2000. Vergleichende Untersuchungen zur Beschreibung von transienten Stromungsvorgangen in Rohrleitungen. Univ. Diss. Dortmund 1999, volume 20 of UMSICHT-Schriftenreihe, Fraunhofer IRB Verlag, Stuttgart, Germany.

Duenas, A.M., Wachs, D.M., Yoshiura, R.K., Mignot, G.P., Wu, Q., Marcum, W.R., Reyes Jr., J.N., 2022. Application of dynamical system scaling for accelerated fuel qualification efforts. In: Proceedings Of the 19th International Topical Meeting on Nuclear Reactor Thermal Hydraulics, NURETH-19, Brussels, Belgium, March 6–11.

Duffey, R.B., 1985. Comment during the panel session management of SBLOCA sequences. In: Specialists Meet. On Small Break LOCA Analyses in LWRs, Vol 3, Pisa, Italy, June 23–27.

Duffey, R.B., 2021. Personal Communication to F. D'Auria, April 13.

Duffey, R.B., D'Auria, F., 2020. Nuclear energy and its history: past consequences, present inadequacies and a perspective for success. Energy and Power Engineering 12, 193−236. ISSN Online: 1947-3818. https://www.scirp.org/journal/epe.

Duffey, R.B., Porthouse, D.T.C., 1973. The physics of rewetting in water reactor emergency core cooling. Nuclear Engineering and Design 25, 379−394.

Dul'nev, G.N., 1963. Heat Transfer in Radioelectronic Devices. Gosenergoizdat, Moscow, Russia (in Russian).

Dul'nev, G.N., Novikov, V.V., 1991. Processes of Transfer in Nonuniform Mediums. Energoatomizdat, Moscow, Russia (in Russian).

Dul'nev, G.N., Zarichniak, Y.P., 1974. Heat Conduction of Mixture and Composite Materials. Energiya, Moscow, Russia (in Russian).

Dunbar, W.R., Moody, S.D., Lior, N., 1995. Exergy analysis of an operating Boiling Water Reactor nuclear power station. Energy Conversion and Management 36 (3), 149−159.

Dupleac, D., Prisecaru, I., Ghitescu, P., Negut, G., 2007. Effect of CANDU 6 core modeling on large break LOCA analysis. In: 12th International Topical Meeting on Nuclear Reactor Thermal Hydraulics (NURETH-12), September 30−October 4, Pittsburgh, PA, USA.

Duponcheel, M., Seynhaeve, J.M., Bartosiewic, Y., 2015. Implementation and assessment of the delayed equilibrium model for computing flashing choked flows in a multifield CFD code. In: NURETH-16, 16th International Topical Meeting on Nuclear Reactor Thermal Hydraulics, Chicago, IL, USA, August 30−September 4.

Dupont, T.F., Liu, Y., 2003. Back and forth error compensation and correction methods for removing errors induced by uneven gradients of the level set function. Journal of Computational Physics 190 (1), 311−324.

Durbin, P.A., 1991. Near-wall turbulence closure modeling without "damping functions". Theoretical and Computational Fluid Dynamics 3, 1−13.

Durmayaz, A., Yavuz, H., 2001. Exergy analysis of a pressurizes-water reactor nuclear-power plant. Applied Energy 69, 39−61.

Durran, D.R., 2010. Numerical Methods for Fluid Dynamics. Springer, New York, NY, USA, ISBN 978-1-4419-6411-3, pp. 1−516.

Dusic, M., Dutton, M., Glaeser, H., Herb, J., Hortal, J., Mendizabal, R., Pelayo, F., 2014. Combining insights from probabilistic and deterministic safety analyses in option 4 from the IAEA Specific Safety Guide SSG-2. Nuclear Technology 188, 63−77.

Dyomkin, N.B., 1962. The Actual Contact Area of Solid Surfaces. Academy of Sciences USSR, Moscow, Russia (in Russian).

Dzodzo, M.B., 2009. EMDAP Method of Scaling Analysis. 3-D S.UN.COP 2009 Seminar, KTH, Stockholm, Sweden, October 12−30.

Dzubenko, B.V., Dreitser, G.A., Ashmantas, L.A., 1988. Transient Heat Transfer in Bundles of Twisted Tubes. Mashinostroenie, Moscow, Russia.

D'Auria, F., 1992. Scaling and counterpart tests. In: OECD/NEA/CSNI Specialist Meeting on Transient Two-phase Flow—Current Issues in System Thermal Hydraulics. Extended Abstract in Proceedings Of CSNI Spec. Meet. On Transient Two-phase Flow—System Thermal-hydraulics, Aix-En-Provence, France, April 6−8, pp. 391−392.

D'Auria, F., 1998. Proposal for training of thermal-hydraulic system codes users. In: IAEA Specialists Meeting on User Qualification and User Effects on Accident Analysis for Nuclear Power Plants. Specialists Meetings, Vienna, Austria, 31 August−4 September.

D'Auria, F., 2001. The role of the experimental database in the validation of thermal-hydraulic system codes. In: Invited at FLUIDOS 2001—VII International Seminors on Recent Advances in Fluid Mechanics, Physics of Fluid and Associated Complex Systems, Buenos Aires, Argentina, 17–19 October.

D'Auria, F., 2009. Overview of uncertainty methods in nuclear reactor thermal-hydraulics. In: Invited at International Topical Meeting on Nuclear Reactor Thermal Hydraulics (NURETH-13), Kanazawa (J), September 27–October 2.

D'Auria, F., 2012. Perspectives in system thermal-hydraulics. Nuclear Engineering and Technology 44 (8), 855–871.

D'Auria, F., 2012a. Natural circulation situations relevant to NPP. In: Joint ICTP-IAEA Course on Natural Circulation Phenomena and Passive Safety Systems in Advanced Water Cooled Reactors, Trieste, Italy.

D'Auria, F., 2013. Nuclear thermal-hydraulics: from fundamentals to applications. In: Conferencia pronunciada por el Dr. F. D'Auria el 18 de marzo (2013) en la Academia Nacional de Ciencias de Buenos Aires—Anticipo de Anales de la Academia Nacional de Ciencias de Buenos Aires, Buenos Aires, Argentina.

D'Auria, F., 2014. The BWR stability issue. IAEA training course on natural circulation phenomena and passive safety systems in advanced water-cooled reactors. In: School of Advanced Nuclear Energy System Studies (SANESS), Global Centre for Nuclear Energy Partnership (GCNEP), Mumbai, India, (29 September–3 October).

D'Auria, F., 2015. Presentation at panel 'data, data science and data-driven thermal-hydraulics. In: NURETH-16 Conference, Chicago, IL, USA, August 30–September 4.

D'Auria, F., 2018. Prioritization of nuclear thermal-hydraulics researches. Journal of Nuclear Engineering and Design 340, 105–111.

D'Auria, F., 2019. Nuclear Fission: from E. Fermi to Adm. Rickover, to industrial exploitation, to nowadays challenges. Journal of Advancement in Scientific and Engineering Research 4 (1), 17–30. https://doi.org/10.33495/aser_v7i1.19.102. ISSN 2315-9821 @2019 Scienceweb Publishing. www.sciencewebpublishing.net.

D'Auria, F., 2019a. Best estimate plus uncertainty (BEPU): status and perspective. Journal of Nuclear Engineering and Design 352, 110190.

D'Auria, F., 2021. An old issue and a new challenge for nuclear reactor safety. Frontiers in Energy, (Short Communication) 1–6. https://doi.org/10.1007/s11708-021-0729-0.

D'Auria, F., 2021a. Passive systems and nuclear thermal hydraulics. Journal of Nuclear Engineering and Design 385 (111513), 1–7.

D'Auria, F., 2021b. Nuclear Energy: Since 1900 until Enrico Fermi and Admiral Rickover and Today Situation. Invited VEBLEO Webinar. May 24 (ppt, slides 1–19) Vebleo—Science, Engineering and Technology. https://vebleo.com.

D'Auria, F., Bestion, D., 2022. Nuclear thermal hydraulic phenomena: bases and challenges. Journal of Nuclear Technology 208, 990–1011.

D'Auria, F., Bousbia Salah, A., 2004. The BWR Stability Issue. Invited at OECD/CSNI & IRS [N] 'Seminar on Transfer of Competence, Knowledge and Experience Gained through CSNI Activities in the Field of Thermal-hydraulics'—THICKET 2004 Session X—Paper 26, INSTN, Saclay (F), June 7–11.

D'Auria, F., Debrecin, N., 2014. Perspectives in licensing and nuclear reactor safety technology. In: Third International Scientific and Technical Conference "Innovative Designs and Technologies of Nuclear Power" (ISTC NIKIET-2014), Moscow, Russia, 7–10 October.

D'Auria, F., Duffey, R.B., 2022. Innovations needs in nuclear reactor safety and risk. Journal of Nuclear Energy and Technology (NUCET) 8 (2), 77–90. https://doi.org/10.3897/nucet.8.82296.

D'Auria, F., Frogheri, M., 2002. Use of natural circulation map for assessing PWR performance. Nuclear Engineering and Design 215 (1–2), 111–126.

D'Auria, F., Galassi, G.M., 1990. Characterization of instabilities during two-phase natural circulation in PWR typical conditions. Experimental Thermal and Fluid Science 3 (90).

D'Auria, F., Galassi, G.M., 1990a. Assessment of RELAP5/MOD2 code on the basis of experiments performed in LOBI facility. Nuclear Technology 90 (3).

D'Auria, F., Galassi, G.M., 1990b. Flowrate and density oscillations during two-phase natural circulation in PWR typical conditions. Nuclear Engineering and Design 122.

D'Auria, F., Galassi, G.M., 1998. Code validation and uncertainties in system thermal-hydraulics. Progress in Nuclear Energy 33 (1/2), 175–216.

D'Auria, F., Galassi, G.M., 2000. Methodology for the Evaluation of the Reliability of Passive Systems. University of Pisa Report, DIMNP-NT 420(00)-rev. 1, Pisa, Italy. October).

D'Auria, F., Galassi, G.M., 2001. Best Estimate Analysis and Uncertainty Evaluation of the Angra-2 LBLOCA DBA, University of Pisa Report, DIMNP—NT 433(01), Pisa, Italy—rev. 1, July, (Endorsed by CNEN, Rio de Janeiro, Brazil).

D'Auria, F., Galassi, G.M., 2004. Best-Estimate Analysis of the Angra-2 ATWS Event Category. University of Pisa Report, DIMNP—NT 526(04)-rev.4, Pisa, Italy.

D'Auria, F., Galassi, G.M., 2010. Scaling in nuclear reactor system thermal-hydraulics. Nuclear Engineering and Design 240 (10), 3267–3293.

D'Auria, F., Galassi, G.M., 2014. Natural circulation situations relevant to nuclear power plants. In: IAEA Training Course on Natural Circulation Phenomena and Passive Safety Systems in Advanced Water Cooled Reactors SANESS-GCNEP, Mumbai, India, 29 September–3 October, Paper ID L19.

D'Auria, F., Galassi, G.M., 2019. Best Estimate Plus Uncertainty Approach in nuclear reactor safety and licensing: brief history and elements after licensing. Journal of Nuclear Technology and Radiation Protection 34 (3), 299–312. https://doi.org/10.2298/NTRP190413022D.

D'Auria, F., Galassi, G.M., 2022. Modeling, Validation, Scaling, Uncertainty and Application in Nuclear Thermal Hydraulics. Invited at ATH'22, June 12–16, Anaheim, CA, USA.

D'Auria, F., Giannotti, W., 2000. Development of code with capability of internal assessment of uncertainty. Nuclear Technology 131 (1), 159–196.

D'Auria, F., Hassan, Y.A., 2021. Challenges and concerns for development of nuclear thermal hydraulics. Solicited paper for VSI "40 years of NURETH". Nuclear Engineering and Design 375 (111074), 1–12.

D'Auria, F., Ingegneri, M., 1997. Qualitative and quantitative analysis of Cathare-2 code results of counterpart test calculations in LOBI, SPES, BETHSY and LSTF facilities. In: 8th International Topical Meeting on Nuclear Reactor Thermalhydraulics (NURETH-8), Kyoto, 30 September–4 October.

D'Auria, F., Ingegneri, M., 1998. Use and qualification of the UMAE uncertainty Methodology having as basis the Cathare2 code. Journal of Nuklearna Tehnologija—Biblid 0351-689X (13 2), 3–16.

D'Auria, F., Lanfredini, M., 2018. Introducing V&V&C in nuclear thermal-hydraulics. In: ASME 2018 Verification and Validation Symposium. Minneapolis (MN, USA), May 16–18, Paper 9321.

D'Auria, F., Mazzantini, O., 2011. The best estimate plus uncertainty challenge in the current licensing process of present reactors. Science and Technology of Nuclear Installations. Article ID 958218.

D'Auria, F., Pellicoro, V., 1996. Local instability in BWR reactor simulator. In: International Conference on Nuclear Engineering (ICONE-4), New Orleans, LA, USA, vol. 3.

D'Auria, F., Vigni, P., 1980. Two-phase Critical Flow Models. OECD/NEA/CSNI Report No. 49, Paris, France.

D'Auria, F., Vigni, P., 1981. Fluid-dynamic analysis of steam-water flows from a pressure vessel. In: Workshop on Jet Impingement and Pipe Whip, Genova, Italy, June 29–July 1.

D'Auria, F., Vigni, P., 1983. BLOWAES: a method for evaluating blowdown two phase flowrate from pressure and thrust measurement. In: 2nd International Topical Meeting on Nuclear Reactor Thermalhydraulics, NURETH-2, Santa Barbara, CA, USA, January 11–14.

D'Auria, F., Vigni, P., 1985. Proposed set of criteria in designing nuclear power plants experimental simulators. In: 3rd International Topical Meeting on Reactor Thermal-hydraulics (NURETH-3), Newport, RI, USA, October 15–18.

D'Auria, F., Galassi, G.M., Oriolo, F., Vigni, P., 1985. Assessment of scaling principles for the simulation of small break LOCA experiments in PWRs. In: Spec. Meet. On Small Break LOCA Analyses in LWR, Pisa, Italy, 23–27 June. (Also Nuclear Engineering and Design, Vol 102, 1987).

D'Auria, F., De Sanctis, N., Di Marco, P., Lahey Jr., R.T., Podowski, M.Z., 1987. A linear model to study fluid dynamic instabilities in boiling channels due to density oscillations. Heat Technology 5 (3–4), 55–72.

D'Auria, F., Karwat, H., Mazzini, M., 1988. Planning of counterpart tests in LWR experimental simulators. In: Proceedings of the ANS National Heat Transfer Conference, Houston, TX, USA.

D'Auria, F., Galassi, G.M., Vigni, P., Calastri, A., 1991. Scaling of natural circulation in PWR systems. Nuclear Engineering and Design 2, 187–206.

D'Auria, F., Debrecin, N., Galassi, G.M., Galeazzi, S., 1993. Application of RELAP5/MOD3 to the evaluation of loss of feedwater in test facilities and in nuclear plant. Nuclear Engineering and Design 141 (3).

D'Auria, F., Galassi, G.M., Ingegneri, M., 1994. Evaluation of the data base from high power and low power small break LOCA counterpart tests performed in LOBI, SPES, BETHSY and LSTF facilities. University of Pisa Report, DCMN NT 237 (94) (Pisa, Italy).

D'Auria, F., Frogheri, M., Leonardi, M., 1994a. Natural circulation performance in western type and eastern type PWR. In: XI Annual Simulator Conference, San Diego, CA, USA, 10–14 April.

D'Auria, F., Debrecin, N., Galassi, G.M., 1995. Outline of the uncertainty methodology based on accuracy extrapolation (UMAE). Nuclear Technology 109 (1), 21–38.

D'Auria, F., Leonardi, M., Glaeser, H., Pochard, R., 1995a. Current status of methodologies evaluating the Uncertainties in the prediction of thermal-hydraulic phenomena in nuclear reactors. In: International Symposium on Two Phase Flow Modelling and Experimentation, Rome, Italy, October 9–11. Edizioni ETS, Pisa, Italy, pp. 501–509.

D'Auria, F., Galassi, G.M., Gatta, P., Mastrantonio, L., Marsili, P., 1996. Comparison between small LOCA scenarios in Eastern and Western type PWRs. In: ASME-JSME International Conference On Nuclear Engineering (ICONE-4), New Orleans, LO, USA, 10–14 March.

D'Auria, F., Clement, P., Faluomi, V., Staedtke, H., Vanhoenacker, L., 1996a. Lesson learned from OECD/CSNI ISP on small break LOCA. OECD-CSNI Report OECD/GD 97, 10 (Paris, France).

D'Auria, F., Frogheri, M., Monasterolo, U., 1997. Removable power by natural circulation in PWR systems. In: ASME-JSME Conference (ICONE-5), Nice, France, May 26–30.

D'Auria, F., Ambrosini, W., Anegawa, T., Blomstrand, J., De Betou, J., Langenbuch, S., Lefvert, T., Valtonen, K., 1997a. State-of-the-Art Report on Boiling Water Reactor Stability (SOAR on BWRS). NEA/CSNI/R(96)21, OCDE/GD(97)13. OECD/NEA, Paris, France.

D'Auria, F., Frogheri, M., Misale, M., 1997b. System codes capabilities in predicting instabilities in single phase natural circulation. In: 4th Regional Meet. Nuclear Energy in Central Europe, Bled, Slovenia, September 7–10.

D'Auria, F., Galassi, G.M., Mastrantonio, L., 1997c. Accident management studies in the WWER-1000 plant. In: ASME-JSME International Conference On Nuclear Engineering (ICONE-5, 2458), Nice, France, 26–30 May.

D'Auria, F., Chojnacki, E., Glaeser, H., Lage, C., Wickett, T., 1998. Overview of uncertainty issues and methodologies. In: Invited at OECD/CSNI Seminar on Best Estimate Methods in Thermal-Hydraulic Analyses, Ankara, Turkey, June 29–July 1.

D'Auria, F., Galassi, G.M., Spadoni, A., 1998a. Optimisation of the Elevation of a Heat Exchanger in Dual Natural Circulation (In Italian), University of Pisa Report, DCMN—NT 363(98), Pisa, Italy.

D'Auria, F., Giannotti, W., Piagentini, A., 2000. Consideration of bifurcations within the internal assessment of uncertainty. In: ASME-JSME International Conference On Nuclear Engineering (ICONE), Baltimore, MD, USA, April 2–6.

D'Auria, F., Galassi, G.M., Giannotti, W., 2001. Confirmatory Safety Analyses Carried Out by Relap5 and Cathare Codes Related to the Kozloduy VVER 440/230 Unit No. 3, University of Pisa Report, DIMNP—NT 440(01)-rev.1, Pisa, Italy, November.

D'Auria, F., Galassi, G.M., Giannotti, W., 2003. Post-Test Analysis by Relap5/mod3.3β of VVER-440 Bubble Condenser Tests 01 to 05 Performed in BC V-213 Facility. University of Pisa Report, DIMNP-NT 509(03), Pisa, Italy, July.

D'Auria, F., Gago, J.L., Galassi, G.M., Grgic, D., Spadoni, A., 2003a. Three mile Island unit 1 main steam line break three dimensional neutronics/thermalhydraulics analysis: application of different coupled codes. Nuclear Technology 142 (1).

D'Auria, F., Galassi, G.M., Giannotti, W., 2003b. Transient analyses of the BETHSY 4.2b test carried out by the code cathare 2. University of Pisa, DIMNP NT 516 (03).

D'Auria, F., Galassi, G.M., Giannotti, W., Petruzzi, A., Moussavian, K., Pierro, F., Lo Nigro, A., Muellner, N., 2003c. Qualification and application of coupled reactor cooling systems and containment nodalizations. In: IAEA TM Progress in Development and Use of Coupled Codes for Accident Analysis, Vienna, Austria, 26–28 November.

D'Auria, F., (Project Coordinator), P., Bousbia Salah, A., Galassi, G.M., Vedovi, J., Reventos, F., Cuadra, A., Gago, J.L., Sjoberg, A., Yitbarek, M., Sandervag, O., Garis, N., Anhert, C., Aragones, J.M., Verdu, G., Miro`, R., Hadek, J., Macek, J., Ivanov, K., Rizwan-Uddin, Sartori, E., Rindelhardt, U., Rohde, U., Frid, V., Panayotov, D., 2004. Neutronics/thermal-hydraulics Coupling in LWR Technology—CRISSUE-S. Vol. I, OECD/NEA Report No. 4452, Paris, France; Vol II, OECD/NEA Report No. 5436, ISBN 92-64-02084-5 (Paris, France).

D'Auria, F., Galassi, G.M., Giannotti, W., 2004a. Accident with Guillotine Rupture of Main Coolant Tube Dn 500 for (VVER 440) Units 3 and 4 of Kozloduy NPP: Best Estimate and Uncertainty Evaluation of LBLOCA 500 Mm for Kozloduy Unit 3 NPP. University of Pisa Report, DIMNP—NT 544(04)-rev.0, Pisa, Italy, September.

D'Auria, F., Galassi, G.M., Giannotti, W., Araneo, D., Cherubini, M., Del Nevo, A., 2005. The boron Issue in PWR and VVER-1000, OECD/NEA/CSNI PKL Project. PKL Analytical Workshop. University of Pisa, Pisa, Italy.

D'Auria, F., Soloviev, S., Novoselsky, O., Moskalev, A.M., Radkevitch, V., Malofeev, V., Parisi, C., Cherubini, M., Pierro, F., Moretti, F., Lead Authors, 2005a. Deterministic safety technology in RBMK—EC TACIS project R2.03/97 <Software development for accident analysis of VVER and RBMK reactors in Russia>. Final Technical Report, Part B, University of Pisa—DIMNP NT 580 (05), 1−838. Pisa, Italy.

D'Auria, F., Cherubini, M., Galassi, G.M., Muellner, N., 2005b. Analysis of measured and calculated counterpart test data in PWR and VVER1000 simulators. Journal of Nuclear Technology and Radiation Protection XX, 16−22.

D'Auria, F., Bousbia Salah, A., Petruzzi, A., Del Nevo, A., 2006. State of the art in using best estimate calculation tools in nuclear technology. Nuclear Engineering and Technology 38, 11−32.

D'Auria, F., Melikhov, O., Suslov, A., Bykov, M., Elkin, I.V., Araneo, D., Cherubini, M., Del Nevo, A., Giannotti, W., Muellner, N., Lead Authors, 2006a. Accident management technology in VVER-1000. University of Pisa—DIMNP NT 581 (06), 1−1250. Pisa, Italy.

D'Auria, F., Glaeser, H., Lee, S., Misak, J., Modro, M., Schultz, R.R., 2008. Best Estimate Safety Analysis for Nuclear Power Plants: Uncertainty Evaluation. IAEA Safety Report Series (SRS) No. 52, Vienna, Austria.

D'Auria, F., Gabaraev, B., Novoselsky, O., Radkevich, V., Filinov, N., Mazzini, D., Moretti, F., Pierro, F., Vigni, A., Parafilo, L., Kryuchkov, D., 2008a. The multiple pressure tube rupture (MPTR) issue in RBMK safety technology. Nuclear Engineering and Design 238.

D'Auria, F., Cherubini, M., Pierro, F., Giannotti, W., 2008b. The Individual Channel Monitoring (ICM) proposal to improve the safety performance of RBMK. Nuclear Engineering and Design 238.

D'Auria, F., Gabaraev, B., Radkevitch, V., Moskalev, A.M., Uspuras, E., Kaliatka, A., Parisi, C., Cherubini, M., Pierro, F., 2008c. Thermal-hydraulic performance of primary system of RBMK in case of accidents. Nuclear Engineering and Design 238.

D'Auria, F., Mazzantini, O., Cherubini, M., Giannotti, W., Parisi, C., Moretti, F., Melideo, D., Del Nevo, A., Galassi, G.M., Araneo, D., Terzuoli, F., Adorni, M., Muellner, N., Petruzzi, A., Lazzerini, D., Santoro, R., Bousbia-Salah, A., 2008d. DEGB LBLOCA (2 X 100% Break in CL) in Atucha-2 NPP, University of Pisa Report, DIMNP NT 628(08)-rev. 1, Pisa, Italy.

D'Auria, F., Gabaraev, B., Soloviev, S., Novoselsky, O., Moskalev, A.M., Uspuras, E., Galassi, G.M., Parisi, C., Petrov, A., Radkevich, V., Parafilo, L., Kryuchkov, D., 2008e. Deterministic accident analysis for RBMK. Nuclear Engineering and Design 238, 975−1001.

D'Auria, F., Novoselsky, O., Safonov, V., Uspuras, E., Galassi, G.M., Cherubini, M., Giannotti, W., 2008f. Thermal−hydraulic performance of confinement system of RBMK in case of accidents. Journal of Nuclear Engineering and Design 238, 925−939.

D'Auria, F., Lombardi-Costa, A., Bousbia Salah, A., 2009. The boiling water reactor stability—updated. In: IAEA/ICTP Course on Natural Circulation in Water-Cooled Nuclear Power Plants, International Center for Theoretical Physics (ICTP)—Lecture ID T12 & T13, Trieste, Italy, June 22−26.

D'Auria, F., Muellner, N., Parisi, C., Petruzzi, A., 2010. BEPU approach in licensing framework, including 3D NK applications. In: Er, M.J. (Ed.), New Trends in Technologies: Devices, Computer, Communication, and Industrial Systems, ISBN 978-953-307-212-8, pp. 197−216. © 2010 (Chapter 11).

D'Auria, F., Bestion, D., Jeong, J.J., Kim, M., 2012. V & V in system thermal-hydraulics. In: ASME Verification and Validation Symposium (V & V 2012) Meeting, Las Vegas, NV, USA, May 2−4.

D'Auria, F., Camargo, C., Mazzantini, O., 2012a. The Best Estimate Plus Uncertainty (BEPU) approach in licensing of current nuclear reactors. Nuclear Engineering and Design 248, 317–328.

D'Auria, F., Camargo, C., Muellner, N., Lanfredini, M., Mazzantini, O., 2012b. The simulation of I & C in accident analyses of nuclear power plants. Nuclear Engineering and Design 250, 656–663.

D'Auria, F., Galassi, G., Pla, P., Adorni, M., 2012c. The Fukushima Event: The Outline and the Technological Background. Science and Technology of Nuclear Installations. J. STNI, Hindawi Publisher. ID 507921.

D'Auria, F., Aksan, S.N., Petruzzi, A., 2013. General view of the NURETH-15 conference. ANS Transactions 109 (1), 1735–1738.

D'Auria, F., Pierro, F., Araneo, D., Galassi, G.M., 2014. Reliability of TH passive safety systems. In: IAEA Training Course on Natural Circulation Phenomena and Passive Safety Systems in Advanced Water Cooled Reactors SANESS-GCNEP, Mumbai, India, 29 September–3 October, Paper ID L20.

D'Auria, F., Del Nevo, A., Muellner, N., 2014a. Insights into natural circulation stability. In: IAEA Training Course on Natural Circulation Phenomena and Passive Safety Systems in Advanced Water Cooled Reactors, SANESS-GCNEP, Mumbai, September 29–October 4, Paper ID L08.

D'Auria, F., Glaeser, H., Kim, M., 2015. A vision for nuclear reactor safety. In: 46th Jahrestagung Kerntechnik Annual Meeting on Nuclear Technology, Berlin, Germany, May 5–9.

D'Auria, F., Lombardi-Costa, A., Bousbia Salah, A., 2016. The boiling water reactor stability. In: OECD/NEA/CSNI, THICKET Seminar, Paper 21, Budapest, Hungary, June 26–July 1.

D'Auria, F., Bestion, D., Lien, P., Nakamura, H., 2016a. The OECD/NEA/CSNI SOAR (state of art report) on scaling (the S- SOAR). In: Invited at 11th International Topical Meeting on Nuclear Thermal Hydraulics Operation and Safety (NUTHOS-11), Gyeongju, Korea, October 9–13, pp. 1–17.

D'Auria, F., Glaeser, H., Debrecin, N., 2017. Independent assessment for new nuclear reactor safety. EPJ Nuclear Sciences and Technologies 3, 31.

D'Auria, F., Debrecin, N., Glaeser, H., 2017a. Strengthening nuclear reactor safety and analysis. Journal of Nuclear Engineering and Design 324, 209–219.

D'Auria, F., Debrecin, N., Glaeser, H., 2018. New safety barrier for current and future nuclear reactors. In: Invited at 5th International Scientific and Technical Conf. Innovative Designs and Technologies of Nuclear Power (ISTC NIKIET—2018). Moscow (Ru), Oct. 2–5.

D'Auria, F., Lanfredini, M., 2019a. V&V&C in nuclear reactor thermal hydraulics. In: Sehgal, R., Yadigaroglu, G., Hewitt, G.F. (Eds.), Special Issue Trends and Perspectives in Nuclear Thermal-hydraulics—To the Memory of B, Journal of Nuclear Engineering and Design, vol. 354, pp. 1–10 paper 110162 (paper PIII-19).

D'Auria, F., Aksan, S.N., Hassan, Y.A., 2019b. Nuclear thermal-hydraulics: what is it? In: Sehgal, R., Yadigaroglu, G., Hewitt, G.F. (Eds.), Special Issue Trends and Perspectives in Nuclear Thermal-hydraulics—To the Memory of B, Journal of Nuclear Engineering and Design, vol. 354 paper 110246. pp. 1–5. (paper PI–2). https://doi.org/10.1016/j.nucengdes.2019.110246.

D'Auria, F., Debrecin, N., Glaeser, H., 2019c. The technological challenge for current generation nuclear reactors. Nuclear Energy and Technology (NUCET) 5 (3), 183–199.

D'Auria, F., 2020. Passive systems and nuclear thermal hydraulics. Invited at ANS-SFEN International Topical Meeting on advances in thermal hydraulics. In: Crouzet, F., Wald, J.-F., Hamel-Bloch, P., Le Goff, M., Baron-Alves, L. (Eds.), ATH'20, Proceedings, pp. 1–12. Paris (F), October 20–23 [see also D'Auria, 2021a].

D'Auria, F., Aksan, N., Hassan, Y.A., 2021. What is nuclear thermal-hydraulics?. In: OECD/NEA/CSNI Specialists Meeting on Transient Thermal-hydraulics in Water Cooled Nuclear Reactors (SM-TH), December 13–17, CIEMAT, Madrid (Spain), p. 08.

D'Auria, F., Mazzantini, O., Galassi, G.M., 2021a. Atucha II Nuclear Reactor: Design Safety and Licensing. Libreriauniversitaria edizioni Webster srl, Padova, Italy, ISBN 978-88-3359-351-7, pp. 1–673.

D'Auria, F., Aksan, N., Glaeser, H., 2021b. Physical phenomena in nuclear thermal hydraulics and current status. Italian Journal of Engineering Science 65 (1), 1–11. March. http://iieta.org/journals/ti-ijes.

D'iachenko, P.E., Tolkacheva, N.N., Andreev, G.A., Karpov, T.M., 1963. The Actual Contact Area of Mating Surfaces. Academy of Sciences USSR, Moscow, Russia (in Russian).

D'Onorio, M., Giampaolo, A., Caruso, G., Giannetti, F., 2021. Preliminary uncertainty quantification of the core degradation models in predicting the Fukushima Daiichi unit 3 severe accidents. Nuclear Engineering and Design 382 (111383), 1–22.

Ebadian, M.A., Dong, Z.F., 1998. Forced convection, internal flow in ducts. In: Rohsenow, W.M., Hartnett, J.P., Cho, Y.I. (Eds.), Handbook of Heat Transfer, third ed. McGraw Hill, New York, NY, USA (Chapter 5).

Ebert, D. (Ed.), 1995. Proceedings of the OECD/CSNI Specialists Meeting on Boron Dilution Reactivity Transients Held in State College, Pennsylvania, PA, 18–20 October, NUREG/CP-0158, Washington, DC; USA.

EC, 2016a. https://clearinghouse-oef.jrc.ec.europa.eu/.

EC, 2016b. http://stresa.jrc.ec.europa.eu/.

EC-EURATOM, 2013. ASAMPSA2 Best Practices Guidelines for L2 PSA Development and Application, Vol. 2—Best Practices for the Gen II PWR, Gen II BWR. Extension to Gen III Reactors. Technical Report ASAMPSA2/WP2-3/D3.3/2013-35, Brussels, Belgium.

Eckert, E.R.G., Drake, R.M., 1972. Analysis of Heat and Mass Transfer, first ed. McGraw Hill, New York, NY, USA.

Edwards, A.R., O'Brien, T.P., 1970. Studies of phenomena connected with the depressurization of water reactors. Journal of the British Nuclear Energy Society 9, 125–135.

Eklund, M., Alamaniotis, M., Hernandez, H., Jevremovic, T., 2015. Method of characteristics: a review with applications to science and nuclear engineering computation. Progress in Nuclear Energy 85, 548–567.

Ekman, V.W., 1905. On the influence of the Earth's rotation on ocean currents. Archive for Mathematics, Astronomy and Physics (translated from German) 2, 1–52.

El-Wakil, M.M., 1962. Nuclear Power Engineering. McGraw-Hill Book Company, Inc., New York, NY, USA.

El-Wakil, M.M., 1971a. Nuclear Energy Conversion. Intext Educational Publisher, Scranton, PA, USA.

El-Wakil, M.M., 1971b. Nuclear Heat Transport. Intext Educational Publisher, Scranton, PA, USA.

Elias, E., Lellouche, J., 1994. Two-phase critical flow. International Journal of Multiphase Flow 20, 91–168.

Elkow, K.J., Rezkallah, K.S., 1996. Void fraction measurements in gas–liquid flows using capacitance sensors. Measurement Science and Technology 7, 1153–1163.

EM-DAT, 2016. http://www.emdat.be.

Emonot, P., Souyri, A., Gandrille, J.-L., Barre, F., 2009. CATHARE-3: a new system code for thermal-hydraulics in the context of the NEPTUNE project. In: The 13th International Top. Meet. On Nuclear Reactor Thermal Hydraulics (NURETH-13), Kanazawa City, Ishikawa Prefecture, Japan, September 27–October 2.

Eom, S., 2016. Overview of Aging Management Program for CANDU Feeders—Canadian Regulatory Perspectives. ASME PVP 2016-63545, July.

Eom, S., Riznic, J., Nitheanandan, T., Cho, D.H., Taylor, H., Li, M., 2021. Design and aging management for feeder pipe and feeder supports, chapter. In: Riznic, J. (Ed.), Pressurized Heavy Water Reactors: CANDU. Elsevier Publisher, Amsterdam, Netherlands, pp. 171−227.

EPRI, 1980. Analysis of Three Mile Island Unit 2 Accident. NSAC-80-1 (NSAC-1 Revised), Palo Alto, CA, USA.

EPRI, 2014. Computational Fluid Dynamics Benchmark of High Fidelity Rod Bundle Experiments: Industry Round Robin Phase 1−Rod Bundle with Simple Support Grids. EPRI, 3002000504, Palo Alto, CA, USA.

EPRI, 2015. Computational Fluid Dynamics Benchmark of High Fidelity Rod Bundle Experiments: Industry Round Robin Phase 2—Rod Bundle with Mixing Vane Grids. EPRI, 3002005401, Palo Alto, Ca, USA.

Erbacher, F.J., Neitzel, H.J., Wiehr, K., 1990. Cladding Deformation and Emergency Core Cooling of a Pressurized Water Reactor in a LOCA. Summary Description of the REBEKA Program. Scientific Report KfK 4781, Karlsruhe, Germany.

Ericson, L., Hall, D.G., 1978. The Marviken critical flow tests—a description and early results. In: ENS/ANS International Topical Meeting on Nuclear, Power Reactor Safety, Brussels, Belgium, October 16−19.

Espinosa-Paredes, G., Batet, L., Nunez-Carrera, A., Sugimoto, J., 2012. Severe accident analysis in nuclear power plants. Science and Technology of Nuclear Installations. Article ID 430471.

Estevez-Albuja, S., Vazquez-Rodriguez, C., 2021. AP1000 IRWST numerical analysis with Gothic. Nuclear Engineering and Design 372 (110991), 1−16.

Estevez-Albuja, S., Fernandez-Cosials, K., Vazquez-Rodríguez, C., Goni-Velilla, Z., Jimenez, G., 2021. AP1000 Passive Containment Cooling System study under LBLOCA conditions using the Gothic code. Nuclear Engineering and Design 384, 111442.

Estrada-Perez, C.E., Hassan, Y., 2010. PTV experiments of subcooled boiling flow through a vertical rectangular channel. International Journal of Multiphase Flow 36, 691−706.

Estrada-Perez, C.E., Yoo, J., Hassan, Y.A., 2015. Feasibility investigation of experimental visualization techniques to study subcooled boiling flow. International Journal of Multiphase Flow 73, 17−33.

Etherington, H. (Ed.), 1958. Nuclear Engineering Handbook. McGraw-Hill Book Company, Inc., New York, NY, USA.

Ezsol, G., Perneczky, L., Szabados, L., Toth, I., 2012. PMK-2: the first integral thermal-hydraulics tests for the safety evaluation of VVER-440/213 nuclear power plants. Journal of Science and Technology of Nuclear Installations 780472.

Fair, J.R., 1960. What you need to design thermosiphon reboilers. Petrol Refinery 39 (2), 105−123.

Faragher, J., 2004. Probabilistic methods for the quantification of uncertainty and error in computational fluid dynamics simulations. In: DSTO-TR-1633, Sydney, Australia. http://dspace.dsto.defence.gov.au/dspace/handle/1947/4214.

Farmer, F.R., 1967. Siting criteria—a new approach. In: Atom, vol. 128, pp. 152−170.

Farrar, B., Bruun, H.H., 1989. Interaction effects between a cylindrical hot-film anemometer probe and bubbles in air/water and oil/water flow. Journal of Physics E: Scientific Instrumentation 22, 114−123.

Fasel, H.F., Seidel, J., Wernz, S., 2002. A methodology for simulations of complex turbulent flows. Journal of Fluids Engineering 124 (4), 933−942.

Fauske, H.K., 1961. Contribution to the Theory of Two-phase One-Component Critical Flow. ANL-6633. Argonne National Laboratory, Chicago, IL, USA.

Fauske, H.K., 1965. The discharge of saturated water through tubes. Chemical Engineering Programme Symposium, Series 61, 210–216.

Faya, A.J.G., Wolf, L., Todreas, N.E., 1979. Development of a Method for BWR Subchannel Analysis. MIT Energy Laboratory, Boston, Ma, USA. MIT-EL-79-027 (PHD thesis).

Feburie, V., Giot, M., Granger, S., Seynhaeve, J.M., 1993. A model for choked flow through cracks with inlet subcooling. International Journal of Multiphase Flow 19.

Fedotovsky, V.C., 1994. Effective thermal conductivity of heterogeneous materials. In: Proceedings of the 1st Russian National Conference on Heat Transfer, vol. 10. Izdatel'stvo MEI, Moscow, Russia, pp. 116–120 (in Russian).

Fenech, H. (Ed.), 1981. Heat Transfer and Fluid Flow in Nuclear Systems. Pergamon Press Inc., Elmford, NY, USA.

Fenech, H., Rohsenow, W.M., 1963. Thermal Conductance of Metallic Surfaces in Contact, pp. 15–24. ASME Paper 62-HT-32. See also Transactions ASME 85C.

Fennern, L., 2018. Design evolution of BWRs: dresden to generation III+. Journal of Progress in Nuclear Energy 102, 38–57.

Ferguson, J.A., Kophazi, J., Eaton, M.D., 2021. Virtual element methods for the spatial discretization of the multi-group neutron diffusion equation on polygonal meshes with application to nuclear reactor physics. Annals of Nuclear Energy 151 (107884), 1–14.

Fernández-Cosials, K., Goñi, Z., Jiménez, G., Queral, C., Montero, J., 2017. Three-dimensional simulation of a LBLOCA in an AP1000 containment building. Energy Procedia 127, 234–241.

Fernandez-Moguel, L., Birchley, J., 2015. Analysis of the accident in the Fukushima Daiichi nuclear power station Unit 3 with MELCOR-2.1. Annals of Nuclear Energy 83, 193–215.

Ferng, Y.-M., 1996. Evaluation of inherent distortions in the IIST facility using the RELAP5/MOD3 code. Nuclear Science and Engineering 123, 190–205.

Ferng, Y.-M., Liu, Y.-T., Shih, C., 2010. Investigating the execution of EOPs in LOTDFP+ATWS+LOCA scenarios for an ABWR using MAAP code. Nuclear Engineering and Design 240.

Ferrell, J.K., Bylund, D.M., 1966. A Study of Convection Boiling inside Channels, Fina Report, vol. II. North Carolina State University, Dept. of Chemical Eng., NC, USA.

Ferreri, J.C., 1985. A note on the advection-diffusion equation. International Journal of Numerical Methods in Fluids 5, 593–596.

Ferreri, J.C., Ambrosini, W., 1998. User qualification and NPP safety analysis: a case dependent issue or a discipline of learning?. In: IAEA Spec. Meet. On User Qualification for and User Effect on Accident Analysis for Nuclear Power Plants, 31 August–2 September, Vienna, Austria.

Ferreri, J.C., Ambrosini, W., 2001. Calculation of sensitivity to parameters in single-phase natural circulation using ADIFOR. International Journal of Computational Fluid Dynamics 16 (4), 277–281.

Ferreri, J.C., Ambrosini, W., 2002. On the analysis of thermal-fluid-dynamic instabilities via numerical discretization of conservation equations. Nuclear Engineering and Design 215, 153–170.

Ferri, R., Achilli, A., Cattadori, G., Bianchi, F., Meloni, P., 2005. Design, experiments and Relap5 code calculations for the PERSEO facility. Nuclear Engineering and Design 235 (10–12), 1201–1214.

Ferroni, L., Natale, A., 2018. Exergy analysis of a PWR nuclear steam supply system—Part I, general theoretical model. Energy Procedia 148, 1230–1237.

Ferziger, J.H., 1996. Large eddy simulation, simulation and modeling of turbulent flow. In: Gatski, T.B., Hussaini, M.Y. (Eds.), ICASE/LArc Series in Comp. Science and Eng. Oxford Univ. Press, NY, USA.

Ferziger, J.H., Peric, M., 1999. Computational Methods for Fluid Dynamics. Springer, Berlin, Germany.

Ferziger, J.H., Peric, M., Street, R.L., 2019. Computational Methods for Fluid Dynamics. Springer, Berlin, Germany, pp. 1−596. ISBN 10 3319996916.

Fillion, P., Chanoine, A., Dellacherie, S., Kumbaro, A., 2011. FLICA-OVAP: a new platform for core thermal−hydraulic studies. Nuclear Engineering and Design 241 (11), 4348−4358.

Filonenko, G.K., 1948. On Friction Factor for a Smooth Tube, vol. 10. All Union Thermotechnical Institute, Izvestija VTI, Russia.

Fineschi, F., Carcassi, M., Carnasciali, F., 1990. Inerting and venting of Mark II containment in BWR nuclear plants. Nuclear Engineering and Design 120, 57−65.

Fischer, M., 1969. The dynamics of waves including shocks in two-phase flow. Nuclear Engineering and Design 11, 103−131.

Fischer, K., Ambrosini, W., Forgione, N., Oriolo, F., Bartonicek, B., Caruso, G., Cheng, X., Krieg, R., Coudert, L., Touret, J.-P., Danisch, R., Jackson, D., Kleinhessling, W., Weber, G., Malet, J., Orden, A., Tkac, A., Wolf, G., Yadigaroglu, G., 2003. Scaling of containment experiments. In: 10th International Topical Meeting on Nuclear Reactor Thermal Hydraulics (NURETH-10), Seoul, Korea, October 5−9.

Fischer, K., 2018. Personal communication to N. Aksan, F. D'Auria and H. Glaeser. In: Relation to "Relevance of Ventilation Blowers for Containment Performance in Case of Accident".

Fitzsimmons, D.E., 1964. Two-Phase Pressure Drop in Piping Components. Hanford Laboratory Rep., HW-80970, Hanford, WA, USA.

Fletcher, C.A.J., 1991. Computational Techniques for Fluid Dynamics, (Vol 1: Fundamental and General Techniques; Vol 2: Specific Techniques for Different Flow Categories), second ed. Springer Verlag, Berlin Germany.

Fletcher, C.D., Schultz, R.R., 1992. RELAP5/MOD3 Code Manual: User's Guidelines. EGG-2596, USNRC NUREG/CR-5535, Vol. 5, Washington, DC, USA.

Florschuetz, L.W., Chao, B.T., 1965. On the mechanics of vapor bubble collapse. Journal of Heat Transfer 87 (2), 209−220.

Fluent, 2006. Fluent 6.3 User Guide. Fluent Inc., Centerra Resource Park. Lebanon, NH USA).

Fokin, B.S., Goldberg, Y.N., 1980. Simulation of free convection boiling crisis in vapor blanketing of a horizontal tube bundle. Heat Transfer/Soviet Research 12 (3), 77−81.

Fokken, J., Krohn, B., Kapulla, R., Niceno, B., Prasser, H.-M., Badillo, A., 2019. NEA Benchmark Exercise: Computational Fluid Dynamic Prediction and Uncertainty Quantification of a GEMIX Mixing Layer Test, pp. 1−98. NEA/CSNI/R(2017)19, February 6, Paris, France.

Ford, J.D., Lekic, A., 1973. Rate of growth of drops during condensation. International Journal of Heat and Mass Transfer 16, 61−64.

Fore, L.B., Dukler, A.E., 1995. Droplet deposition and momentum transfer in annular flow. AIChE Journal 41 (9), 2040−2046.

Forge, A., Pochard, R., Porracchia, A., Miro, J., Sonnenburg, H.G., Steinhoff, F., Teschendorff, V., 1988. Comparison of thermal-hydraulic safety codes for PWR systems. Gesellschaft für Reaktorsicherheit GmbH (GRS), Federal Republic of Germany, CEC Report EUR 11522, Graham & Trotman.

Forslund, R.P., Rohsenow, W.M., 1968. Dispersed flow film boiling. Journal of Heat and Mass Transfer 90 (6), 399−407.

Forster, H.K., Zuber, N., 1954. Growth of a vapor bubble in a superheated liquid. Journal of Applied Physics 25, 474–478.

Fort, C., Bardet, P.M., 2022. High-Speed molecular tagging velocimetry in water to measure wall shear stress. In: 19th International Topical Meeting on Nuclear Reactor Thermal Hydraulics (NURETH-19), March 6–11 (Via Web), Brussels, Belgium.

Fossa, M., 1998. Design and performance of a conductance probe for measuring the liquid fraction in two-phase gas–liquid flows. Flow Measurement and Instrumentation 9, 103–109.

Fournier, Y., Bonelle, J., Moulinec, C., Shang, Z., Sunderland, A.G., Uribe, J.C., 2011. Optimizing code SATURNE computations on petascale systems. Computational Fluids 45, 103–108.

Foussat, A.J.M., Hulin, J.H., 1984. Vertical liquid-liquid and liquid-gas two-phase flow measurements with a vortex flowmeter. In: Delhaye, J.M., et al. (Eds.), Measuring Techniques in Gas-Liquid Two-phase Flows. Sptinger-Verlag, Berlin, Germany, pp. 651–676.

Freitas, R., Bestion, D., 1993. On the prediction of the flooding phenomenon with the Cathare code. In: 6th International Topical Meeting on Nuclear Reactor Thermal Hydraulics, NURETH-6, Grenoble, France, October 5–8.

Freitas, C.J., Ghia, U., Celik, I., Roache, P.J., Raad, P., 2003. ASME's quest to quantify numerical uncertainty. In: 41st Aerospace, Sciences Meeting, Reno, NV, USA, 6–9 January.

Freixa, J., 2007. SBLOCA with boron Dilution in Pressurized Water Reactors. Impact on Operation and Safety. PhD Thesis. UPC, Barcelona, Spain.

Freixa, J., Manera, A., 2010. Analysis of an RPV upper head SBLOCA at the ROSA facility using TRACE. Nuclear Engineering and Design 240, 1779–1788.

Freixa, J., Manera, A., 2011. Verification of a TRACE EPR ™ model on the basis of a scaling calculation of an SBLOCA ROSA test. Nuclear Engineering and Design 241, 888–896.

Freixa, J., Batet, L., Pretel, C., Reventos, F., 2007. Boron transport model with physical diffusion for RELAP5. Nuclear Technology 160, 205–215.

Freixa, J., Reventos, F., Pretel, C., Batet, L., Sol, I., 2009a. SBLOCA with boron dilution in pressurized water reactors. Impact on operation and safety. Nuclear Engineering and Design 239 (4), 749–760.

Freixa, J., Reventos, F., Manera, A., 2009b. TRACE and RELAP5 thermal-hydraulic analysis on boron dilution tests at the PKL facility. In: NURETH-13, Kanazawa, Japan, September 27–October 1.

Freixa, J., Zerkak, O., Belaid, S., 2013. Thermal-hydraulic system code performance for SBLOCA phenomenology under different geometries and scales. In: 15th International Topical Meeting on Nuclear Reactor Thermalhydraulics, NURETH-15, Pisa, Italy, May 12–15.

Freixa, J., Pretel, C., Batet, L., Reventós, F., Tien, K., 2014. Post-test Thermal-Hydraulic Analysis of PKL Tests F1.1 and F1.2. USNRC, NUREG/IA-0417, Washington, DC, USA, pp. 1–73.

Freixa, J., Perez, M., Reventos, F., Allison, C.M., Hohorst, J.K., 2015a. Revisiting ISP-13 with RELAP/SCDAPSIM/MOD3.5 using core SCDAP components. In: NURETH-16, Chicago, IL, USA, September.

Freixa, J., Martınez-Quiroga, V., Zerkak, O., Reventos, F., 2015b. Modelling guidelines for core exit temperature simulations with system codes. Nuclear Engineering and Design 286, 116–129.

Freixa, J., Martinez-Quiroga, V., Reventos, F., 2015c. Core exit temperature response during an SBLOCA event in the Asco' NPP. In: The 16th International Topical Meeting on Nuclear Reactor Thermal-Hydraulics, NURETH-16, Chicago, IL, USA, September.

Freixa, J., de Alfonso, E., Reventos, F., 2016. Testing methodologies for quantifying physical models uncertainties. A comparative exercise using CIRCE and IPREM (FFTBM). Nuclear Engineering and Design 305, 116–129.

Freixa, J., Martínez-Quiroga, V., Reventós, F., 2016a. Qualification of a full plant nodalization for the prediction of the core exit temperature through a scaling methodology. Nuclear Engineering and Design 308, 115–132.

Frepoli, C., 2008. An overview of westinghouse realistic large break LOCA evaluation model. Journal of Science and Technology of Nuclear Installations 1–15. Article ID 498737.

Frepoli, C., Mahaffy, J.H., Ohkawa, K., 2003. Notes on the implementation of a fully implicit numerical scheme for a two-phase three-field flow model. Nuclear Engineering and Design 225, 191–217.

Frepoli, C., Mahaffy, J.H., Hochreiter, L.E., 2003a. Calculation of dry out and post-dry out heat transfer in rod bundles using a three field model. Nuclear Engineering and Design 224, 131–148.

Frepoli, C., Yurko, J., Reyes Jr., J.N., 2015. Experimental Test Facility Data Synthesis with the Dynamical System Scaling Methodology. International Topical Meeting on Nuclear Reactor Thermal Hydraulics (NURETH-16), Chicago, IL, USA.

Frepoli, C., Fricano, J.W., Buschman, F., Aumiller, D., 2015a. Simulation of Faraday waves with the RELAP-3D thermal hydraulic code. In: 16th International Topical Meeting on Nuclear Reactor Thermal Hydraulics (NURETH-16), August 30–September 4, Chicago, IL, USA.

Fried, E.A., Costello, F.A., 1962. Interface thermal contact resistance problem in space vehicles. ARS Journal 32 (2), 237–243.

Friedel, L., 1979. Improved friction pressure drop correlations for horizontal and vertical two-phase flow. In: European Two-phase Flow Group Meeting, Ispra (I).

Friedel, L., 1980. Pressure drop during gas/vapor-liquid flow in pipes. International Chemical Engineering 20, 352–367.

Fröhlich, J., von Terzi, V., 2008. Hybrid LES/RANS methods for the simulation of turbulent flows. Progress in Aerospace Sciences 44, 349–377.

Fu, X., 2001. Interfacial Area Measurement and Transport Modeling in Air-Water Two-phase Flow (Ph.D. thesis). School of Nuclear Engineering, Purdue University, West Lafayette, IN, USA.

Fu, S., Xiao, Z.X., Huang, J.B., Zhang, Y.F., 2010. Investigation of practical flow control methodologies with RANS/LES hybrid methods. In: Progress in Hybrid RANS-LES Modelling, Series Notes on Numerical Fluid Mechanics and Multidisciplinary Design, vol. 111, pp. 41–58.

Fujita, K., 1990. Flow induced vibration and fluid structure interaction in nuclear power plant components. Journal of Wind Engineering and Industrial Aerodynamics 33, 405–418.

Fukuda, K., Kobori, T., 1979. Classification of two-phase flow stability by density-wave oscillation model. Nuclear Science and Technology 16.

Fulwood, R.R., 2000. Probabilistic Risk Assessment in Chemical and Nuclear Industries. Butterworth-Heinemann Publisher, Oxford, UK.

Funahashi, H., Hayashi, K., Hosokawa, S., Tomiyama, A., 2017. Improvement of separator performance with modified pick-off ring and swirler. Journal of Nuclear Engineering and Design 322, 360–367.

Fureby, C., 2009. Large eddy simulation modelling of combustion for propulsion applications. Philosophical Transactions of the Royal Society A: Mathematical, Physical and Engineering Sciences 367, 2957–2969. https://doi.org/10.1098/rsta.2008.0271.

Fureby, C., Henriksson, M., Parmhed, O., Sjökvist, L., Tegner, J., 2008. CFD predictions of jet exhaust plumes. In: 38th AIAA Fluid Dynamics Conference and Exhibit, June 23–26, Seattle, WA, USA, vol. 1, pp. 136–152.

Furuya, O., 1985. An analytical model for prediction of two-phase (non-condensable) flow pump performance. ASME Journal of of Fluids Engineering 107 (1), 139–147.

Gaillard, P., Bestion, D., Dor, I., Germain, P., Moutin, F., 2015. The CATHARE code condensation modelling confronted to the TOPFLOW-PTS steady state experiments. In: International Conference on Nuclear Reactor Thermal Hydraulics (NURETH-16), August 30–September 4, Chicago, IL, USA.

Galassi, M.C., Coste, P., Morel, C., Moretti, F., 2009. Two-Phase flow simulations for PTS investigation by means of neptune CFD code. Journal of Science and Technology of Nuclear Installations (J STNI) 950536. https://doi.org/10.1155/2009/950536.

Galetti, M.R., D'Auria, F., 2000. Questions arising from the application of best-estimate methods to the Angra-2 NPP licensing process in Brasil. In: International Meeting on Best-Estimate Methods in Nuclear Installation Safety Analysis (BE-2000) IX, Washington, DC, USA, 10–13 November.

Galetti, M.R., D'Auria, F., 2004. Technical and regulatory concerns in the use of best estimate methodologies in LBLOCA analysis licensing process. In: International Meeting on Best-Estimate Methods in Nuclear Installation Safety Analysis (BE-2004) IX, Washington, DC, USA, 14–18 November.

Galetti, M.R., D'Auria, F., 2005. Licensing experience with BE in Brazil: regulatory audit analysis. In: IAEA TM Use of a Best-Estimate Approach in Licensing with Evaluations of Uncertainties, Pisa, Italy, 12–16 September.

Gallagher, P., Woodburn, P., 2003. An overview of European research in CFD-based fluid loading and fluid structure interaction. In: MARNET-CFD Offshore Technology Conference, March 20–21.

Gallego-Marcos, I., Grishchenko, D., Kudinov, P., 2019. Thermal stratification and mixing in a Nordic BWR pressure suppression pool. Annals of Nuclear Energy 132, 442–450.

Gamble, R.E., Fanning, A.F., Vaishnavi, B., 2006. ESBWR Scaling Report. GE NEDO-33082, Revision 1, Wilmington, NC, USA.

Ganic, E.N., Rohsenov, W.M., 1977. Dispersed flow heat transfer. International Journal of Heat Mass Transfer 20 (8), 855–866.

Garis, N., 2016. Personal Communication to F. D'Auria in Relation to Oskarshamn-2 Event, Stockholm, Sweden.

Garland, W.J. (Ed.), 2014. The Essential CANDU, A Textbook on the CANDU Nuclear Power Plant Technology. Copyright by UNENE (Canada), Hamilton, Ontario, Canada, pp. 1–1614. ISBN 0-9730040.

Garland, W.J., 2022. Nuclear Engineering Website Maintained by Professor William (Bill) J. Garland. https://www.nuceng.ca/.

Garnier, G., Manon, E., Cubizolles, G., 2001. Local measurements of flow boiling of refrigerant 12 in a vertical tube. Multiphase Science and Technology 13, 1–11.

Garrett, G.R., Lowery, B., Hanson, M., Miller, D.J., Almudhhi, T., Cheung, F.-B., Bajorek, S.M., Tien, K., Hoxie, C.L., 2022. Measurements of droplet size and velocity distributions during reflood. In: 19th International Topical Meeting on Nuclear Reactor Thermal Hydraulics (NURETH-19), March 6–11 (Via Web), Brussels, Belgium.

Gatski, T.B., Jongen, T., 2000. Nonlinear eddy viscosity and algebraic stress models for solving complex turbulent flows. Progress in Aerospace Sciences 36, 655–682.

Gavezzotti, A., 2020. Carnot, at the Origins of Thermodynamics (In Italian), RCS MediaGroup S.p.A. (Grafica Veneta), Trebaseleghe (PD), Italy.

Gavrilov, P.M., 1999. Onset of boiling in highly overheated liquid. Teplofiz. Vys. Temp. 37 (4), 602–605 (in Russian).

GE, 2014. Advanced Technology Manual Chapter 6.2 Primary Containments. General Electrics, GE-0522-R504b and USNRC, HRDT, ML14140A438, Washington DC, USA, pp. 1–39.

GE-Hitachi, 2007. The ESBWR General Description Book. GE Hitachi Nuclear Energy, Wilmington, NC, USA, pp. 1–201.

GE-Hitachi, 2011. Status Report 100—Economic Simplified Boiling Water Reactor (ESBWR). https://aris.iaea.org/pdf/ESBWR.pdf.

Geffraye, G., Laroche, S., 1999. Analysis of the MHYRESA hot-leg entrainment tests. In: 9th International Topical Meeting on Nuclear Reactor Thermal Hydraulics (NURETH-9), San Francisco, CA, USA, October 3–8.

Geffraye, G., Bazin, P., Pichon, P., Bengaouer, A., 1995. CCFL in hot legs and steam generators and its prediction with the CATHARE code. In: Proceedings Of the 7th International Topical Meeting on Nuclear Reactor Thermal Hydraulics (NURETH-7), Saratoga Springs, NY, USA, Sept. 10–15, NUREG/CP-0142.

Geffraye, G., Faydide, B., Laroche, S., Ratel, G., 2000. Analysis of the MHYRESA hot-leg entrainment tests. In: 8th International Conference On Nuclear Engineering (ICONE-8), Baltimore, MD, USA, April 2–6.

George, D.L., Torczynski, J.R., Shollenberger, K.A., O'Hern, T.J., Ceccio, S.L., 2000. Validation of electrical-impedance tomography for measurements of material distribution in two-phase flows. International Journal of Multiphase Flow 26 (4), 549–581.

Gertner, R.F., 1965. A photographic study of bubble-wise boiling in a large volume. Teploperedacha 87 (1), 20–35 (in Russian).

Ghafari, M., Ghofrani, M.B., D'Auria, F., 2018. New turbulence modeling for simulation of Direct Contact Condensation in two-phase pressurized thermal shock. Progress in Nuclear Energy 108, 358–371.

Gharari, R., Kazeminejad, H., Mataji-Kojouri, N., Hedayat, A., 2018. A review on hydrogen generation, explosion, and mitigation during severe accidents in light water nuclear reactors. International Journal of Hydrogen Energy 43, 1939–1965.

Ghione, A., Noe, B., Vinai, P., Demazière, C., 2018. Assessment of criteria for Onset of Flow Instability in vertical narrow rectangular channels with downward flow. In: 12th International Topical Meeting on Nuclear Reactor Thermal-Hydraulics, Operation and Safety (NUTHOS-12), October 14–18, Qingdao, China.

Gibeling, H., Mahaffy, J., 2002. Benchmarking simulations with CFD to 1-D coupling. In: Joint IAEA/OECD Technical Meeting on Use of CFD Codes for Safety Analysis of Reactor Systems, Including Containment, Pisa, Italy.

Gilles, J., Bourouga, B., Bardon, J.-P., 2002. Estimation of the part-bath heat transfer coefficient during a quenching. In: Proceedings 12th International Heat Transfer Conference Grenoble, France, vol. 4, pp. 767–772.

Giustini, G., Murallidharan, J., Sato, Y., Ničeno, B., Badalassi, V., Walker, S.P., 2015. Numerical study of heat diffusion controlled bubble growth in a pressurized liquid. In: 16th International Topical Meeting on Nuclear Reactor Thermal Hydraulics (NURETH-16), August 30–September 4, Chicago, IL, USA.

Glaeser, H., 1989. Analysis of down-comer and tie plate countercurrent flow in the Upper Plenum Test Facility (UPTF). In: NURETH-4 Conference, Karlsruhe, Germany.

Glaeser, H., 1990a. Post Test Calculations of UPTF Test 12 with the Advanced Computer Code TRAC-Pf1/mod1. GRS-A-1 727, Garching, Germany.

Glaeser, H., 1990b. Post Test Calculation of UPTF Test 13 with the Advanced Computer Code TRAC-Pf1/mod1. GRS-A-1 728, Garching, Germany.

Glaeser, H., 1992. Downcomer and tie plate countercurrent flow in the upper plenum test facility (UPTF). Nuclear Engineering and Design 113, 259–283.

Glaeser, H., 2000. Uncertainty evaluation of thermal-hydraulic code results. In: Proceedings of Best Estimate Methods in Nuclear Installations Safety Analysis, Washington, DC, USA.

Glaeser, H., 2008. GRS method for uncertainty and sensitivity evaluation of code results and applications. Science and Technology of Nuclear Installations 1–7. Article ID 798901.

Glaeser, H., 2010. Evaluation of licensing margins of reactors using best estimate methods including uncertainty analysis. In: IAEA Regional Workshop on Application of Deterministic Best Estimate (BE) Safety Analysis for Advanced NPP, AERB, Mumbai, India, 13–17 December.

Glaeser, H., 2011. BEMUSE Phase 6 Report—Status Report on the Area, Classification of the Methods, Conclusions and Recommendations, NEA/CSNI (2011)4, Paris, France (See Also CSNI, 2011).

Glaeser, H., 2019. Requirements for statistical safety demonstration for loss of coolant analyses. Nuclear Engineering and Design 354, 110143.

Glaeser, H., Karwat, H., 1993. The contribution of UPTF experiments to resolve some scale-up uncertainties in countercurrent two phase flow. Nuclear Engineering and Design 145, 1–2.

Glaeser, H., Rohatgi, U.S., 2019. Scaling ability of the counter-current flow limitation (CCFL) correlations for application to reactor thermal hydraulics. Nuclear Engineering and Design 354, 110226.

Glaeser, H., Wolfert, K., 2012. Scaling of thermal-hydraulic Phenomena and System Code Assessment, Seminar on the Transfer of Competence, Knowledge and Experience Gained through CSNI Activities in the Field of thermal-hydraulics. OECD/NEA Head-quarters, Issy-les-Moulineaux, France, 25–29 June.

Glaeser, H., Hofer, E., Kloos, M., Skorek, T., 1994. Uncertainty and sensitivity analysis of a post-experiment calculation in thermal-hydraulics. Reliability Engineering and System Safety 45, 19–33.

Glaeser, H., Hora, A., Krzykacz-Hausmann, B., Skorek, T., 2005. Aussagesicherheit von ATHLET-Rechenprogrammergebnissen für eine deutsche Referenz-DWR-Anlage mit 200%-Bruch im kalten Strang. GRS-A-3279, Garching, Germany.

Glaeser, H., Krzykacz-Hausmann, B., Luther, W., Schwarz, S., Skorek, T., 2008. Methodenentwicklung und exemplarische Anwendungen zur Bestimmung der Aussagesicherheit von Rechenprogrammergebnissen. GRS-A-3443, Garching, Germany.

Glaeser, H., D'Auria, F., De Crecy, A., Reventos, F., 2011. Main results of the OECD best estimate methods. In: Uncertainty and Sensitivity Evaluation (BEMUSE) Programme, International Topical Meeting on Nuclear Reactor Thermal-Hydraulics (NURETH-14), Article N. 71, Toronto (Ontario, Canada), September 25–30.

Glasstone, S., Edlund, M.C., 1952. The Elements of Nuclear Reactor Theory. Van Nostrand Company, Inc., Princeton, NJ, USA.

Glasstone, S., Sesonske, A., 1955. Principles of Nuclear Reactor Engineering. Van Nostrand Company, Inc., Princeton, NJ, USA.

Glebov, V.P., Eskin, V.P., Trubachev, V.M., Taratuta, V.A., Kaar, H.A., 1983. Deposits in Tubes in Supercritical Pressure Boilers (In Russian). Energoatomizdat Publisher, Moscow, Russia, pp. 1–240.

Glück, M., 2007. Sub-channel analysis with F-COBRA-TF—code validation and approaches to CHF prediction. Nuclear Engineering and Design 237 (6), 655–667.

Glück, M., 2008a. Validation of the sub-channel code F-COBRA-TF part I: recalculation of single-phase and two-phase pressure loss measurements. Nuclear Engineering and Design 238 (9), 2308–2316.

Glück, M., 2008b. Validation of the sub-channel code F-COBRA-TF part II: recalculation of void measurements. Nuclear Engineering and Design 238 (9), 2317–2327.

Gnielinski, G., 1976. New equations flow regime heat and mass transfer in turbulent pipe and channel flow. International Chemical Engineering 16, 359–368.

Gogonin, N.I., 2006. Velocity dependence of heat transfer at vapor condensation inside vertical tubes. In: Proceedings of the 4th Russian National Conference on Heat Transfer, vol. 5. Izdatel'stvo MEI, Moscow, Russia, pp. 93–96 (in Russian).

Goldbrunner, M., Karl, J., Hein, D., 2002. Experimental Investigation of Heat Transfer Phenomena during Direct Contact Condensation in the Presence of Non Condensable Gas by Means of the Linear Raman Spectroscopy. Laser Techniques for Fluid Mechanics. Springer, pp. 113–131.

Golovnyov, I.F., Zamuraev, V.P., Katznel'son, S.S., 1984. In: Soloukhin, R.I. (Ed.), Radiative Heat Transfer in High Temperature Gases. Energoatomizdat, Moscow, Russia (in Russian).

Gonzales-Santalo, J.M., Griffith, P., 1972. Two Phase Flow Mixing in Rod Bundle Sub-Channels. ASME Winter Annual Meeting, New York, NY, USA.

Goodwin, I.R., Hunt, C.H., Maxwell, D.C., Taylor, R.B., 1960. SPY (D1G) and IBM-704 Digital Computer Program for the Calculation of Nuclear Reactor thermal Transients. Knolls Atomic Power Lab., KAPL-M-D1G-TD-13, Schenectady, NY, USA.

Gorbachev, B.I., 2002. The Causes and the Scenario of the Chernobyl Accident and Radionuclide Release on the CHNPP Unit 4 Site (Article in the Report by T. Imanaka [Ed.]).

Govan, A.H., Hewitt, G.F., Owen, D.G., Burnett, G., 1989. Wall shear stress measurements in vertical air-water annular two-phase flow. International Journal of Multiphase Flow 15 (3), 307–325.

Govier, G.W., Aziz, K., 1976. The Flow of Complex Mixtures in Pipes. Van Nostrand Reinhold Company, New York, NY, USA.

Graf, U., 1998. Implicit coupling of fluid-dynamic systems: application to multidimensional countercurrent two-phase flow of water and steam. Nuclear Science and Engineering 129 (3), 305–310.

Graffard, E., Goux, F., 2006. CFX Code Application to the French Reactor for Inherent boron Dilution Safety Issue, CFD4NRS, Garching, Germany, 5–7 September.

Gregu, G., Takahashi, M., Pellegrini, M., Mereu, R., 2017. Experimental study on steam chugging phenomenon in a vertical sparger. International Journal of Multiphase Flow 88, 87–98.

Gresho, P., Lee, R., 1981. Don't suppress the wiggles, they are telling you something. Computers and Fluids 12, 223–231.

Grgic, D., Bajs, T., Oriani, L., Conway, L.E., 2003. Coupled RELAP5/Gothic model for accident analysis of the IRIS reactor. In: IAEA Technical Meeting on Use of Computation Fluid Dynamics (CFD) Codes for Safety Analysis of Reactor Systems, Pisa, Italy, Nov. 11–15.

Griffith, P., 1997. Screening Reactor Steam/water Piping Systems for Water Hammer. NUREG/CR-6519. Division of Systems Technology, Office of Nuclear Regulatory Research. U.S. Nuclear Regulatory Commission.

Gröber, H., Erk, S., 1933. Die Grundgesetze Wärmeübetragung. Springer, Berlin, Germany.

Groeneveld, D.C., 1973. The thermal Behavior of a Heated Surface at and beyond Dry-Out. PhD Thesis. University of Western Ontario, London, Canada.

Groeneveld, D.C., 1973a. Post-Dry Heat Transfer at Reactor Operating Conditions. AECL-4513. Chalk River, Ontario, Canada.

Groeneveld, D.C., 1992. A review of inverted annular and low quality film boiling. In: Hewitt, G.F., Delhaye, J.M., Zuber, N. (Eds.), Post-dryout Heat Transfer. CRC Press, Cleveland, OH, USA, pp. 327–366.

Groeneveld, D.C., 1996. On the definition of critical heat flux margin. Nuclear Engineering and Design 163, 245–257.

Groeneveld, D.C., 2019. Critical Heat Flux Data Used to Generate the 2006 Groeneveld Lookup Tables, pp. 1–129. USNRC NUREG/KM-0011, Washington DC, USA.

Groeneveld, D.C., Snoek, C.W., 1986. A comprehensive examination of heat transfer correlations suitable for reactor safety analysis. Multiphase Science and Technology 2 (1–4), 181–274.

Groeneveld, D.C., Cheng, S.C., Doan, T., 1986. 1986 AECL-UO critical heat flux lookup table. Heat Transfer Engineering 7 (1–2), 46–62.

Groeneveld, D.C., Leung, L.K.H., Kirillov, P.L., Bobkov, V.P., Smogalev, I.P., Vinogradov, V.N., Huang, X.C., Royer, E., 1996. The 1995 look-up table for critical heat flux in tubes. Nuclear Engineering and Design 163, 1–23.

Groeneveld, D.C., Leung, L.K.H., Aksan, N., Badulescu, A., Cleveland, J., Baek, W.-P., Chung, M.K., D'Auria, F., Cheng, X., Kirillov, P.L., Macek, J., 1999. A general method of predicting critical heat flux in Advanced Water Cooled Reactors. In: 9th International Topical Meeting on Nuclear Reactor Thermal-hydraulics, San Francisco (Ca), October 3–8.

Groeneveld, D.C., Shan, J.Q., Vasìc, A.Z., Leung, L.K.H., Durmayaz, A., Yang, J., Cheng, S.C., Tanase, A., 2007. The 2006 CHF look-up table. Nuclear Engineering and Design 237, 1900–1922.

Groetzbach, G., 1987. Direct numerical and large eddy simulation of turbulent channel flows. In: Cheremisinoff, N.P. (Ed.), Encyclopedia of Fluid Mechanics, vol. 6. Gulf Publishing, Houston, TX, USA, pp. 1337–1391.

Grossetete, C., 1995. Caracterisation Experimetnale et Simulations de l'evolution d'un Ecoulement Diphasique a Bulles Ascendant Dans Une Conduite Vertical (Ph.D. thesis). Ecole Centrale Paris, Paris, France (in French).

Groudev, P., Stefanova, A., 2006. Validation of Relap5/Mod3.2 model on trip of one main coolant pump for VVER 440/V320. Nuclear Engineering and Design 236.

Groß (or Gross), E.M., Cammiade, L.M.F., Kelm, S., Kampili, M., Allelein, H.-J., Rietz, M., Rohlfs, W., Kneer, R., Dues, M., 2022. Improved near wall velocity measurements in the momentum boundary layer using laser Doppler velocimetry profile sensor. In: 19th International Topical Meeting on Nuclear Reactor Thermal Hydraulics (NURETH-19), March 6–11 (Via Web), Brussels, Belgium.

GRS, 1992. 2D/3D program work summary report. In: Gesellschaft für Anlagen-und-Reaktorsicherheit (GRS) mbH, GRS-100, Garching, Germany (also Damerell and Simons, 1992).

GRS, 1993. Reactor safety issues resolved by the 2D/3D Program. In: Gesellschaft für Anlagenund Reaktorsicherheit (GRS) mbH GRS-101, Garching, Germany (see also Damerell and Simons, 1993b).

Gu, Z., 2018. History review of nuclear reactor safety. Annals of Nuclear Energy 120, 682–690.

Gu, Z., Yu, H., Dai, J., Huang, D., Yan, M., Shen, Y., Guo, C., 2022. Code development of Single-phase 2-D pressure wave propagation. Annals of Nuclear Energy 175 (109211), 1–11.

Gudoshnikov, A.N., Migrov, Y.A., 2008. Verification of the computer code KORSAR taking into account the effect of non-condensable on thermal-hydraulic processes. Thermal Engineering 55 (11), 971–977.

Guelfi, A., Bestion, D., Boucker, M., Boudier, P., Fillion, P., Grandotto, M., Hérard, J.-M., Hervieu, E., Péturaud, P., 2007. Neptune—a new software platform for advanced nuclear thermal hydraulics. Nuclear Science and Engineering 156, 281–324.

Guenther, R., 1990. Modern Optics. Wiley, New York, NY, USA.

Gui, M., Bi, Q., Wang, T., Shan, J., 2022. Experimental investigation on pressure drop characteristics of two-phase flow in a rod bundle geometry under high pressure conditions. Annals of Nuclear Energy 165 (108787), 1–14.

Guillen, D.P., Aksenova, A.E., Canabal, F., Chen, Y.-S., Cheng, G., Chudanov, V.V., Elkawi, O.S.A., Fischer, P.F., Goloviznin, V.M., Hoeld, A., Huang, W., Ito, Y., Karabasov, S.A., Li, K., Obabko, A.V., Pervichko, V.A., Tautges, T.J., Tinoco, H., Wang, T.-S., Zaytsev, M.A., 2013. Nuclear Reactor Thermal Hydraulics and Other Applications, IntechOpen, pp. 1–193. www.intechOpen.com.

Guimaraes, L., (Leonam, made available), 2020. Marine Applications of Nuclear Power, Aruvian's R'search (From Web, Downloaded 2020), pp. 1–576.

Guimaraes, L.N.F., Lamartine-Nogueira-Frutuoso, 1992. Non-Linear Dynamics of a Once-Through Steam Generator (PhD Dissertation). University of Tennessee, Knoxville, TN, USA.

Gülich, J.F., 2010. Centrifugal Pumps, second ed. Springer, Berlin and Heidelberg, Germany.

Gungor, A., Menon, S., 1996. Direct Simulation of subgrid turbulence in high Re, wall-bounded flows. In: 36th AIAA Fluid Dynamics Conference and Exhibit, Fluid Dynamics and Co-located Conferences, San Francisco, CA, USA.

Gunther, A.Y., Shaw, W.A., 1945. A general correlation of friction factors for various types of surfaces in crossflow. Transactions of the American Society of Mechanical Engineers 57, 643–660.

Guo, Y., Hammouda, N., 2015. On the aspect of evaluation of critical channel power and associated uncertainty in CANDU slow loss of regulation event analysis. In: 16th International Topical Meeting on Nuclear Reactor Thermal Hydraulics (NURETH-16), August 30-September 4, Chicago, IL, USA.

Guo, S., Dsouza, B., Xie, Y., Leong, A., Leavitt, J.J., Zhang, J., 2019. Aluminum corrosion in reactor containment environment following a loss of coolant accident (LOCA): high-temperature flow loop tests. Journal of Corrosion Science 151, 122–131.

Gupta, S., Langer, G., Colombet, M., 2014. Hydrogen Combustion during Spray Operation: Tests HD-30, HD-31 and HD-32. OECD-NEA THAI-2 Project, Technical Report 1501420-TR-HD-30-32, Eschborn, Germany.

Gupta, S., Schmidt, E., Von Laufenberg, B., Freitag, M., Poss, G., Funke, F., Weber, G., 2015. Thai test facility for experimental research on hydrogen and fission product behaviour in light water reactor containments. Journal of Nuclear Engineering and Design 294.

Gustafsson, B., Kjellen, B., 1971. Two-phase Flow in a Nine-Rod Bundle with Tilted Power Distribution and Flowsplit in the Outlet (In Swedish). Studsvik, Aktiebolaget Atomenergi, Nykoping, Sweden.

Gustafsson, B., Harju, R., Imset, O., 1974. Flow and enthalpy distribution in a 9-rod bundle. In: European Two-phase Flow Meeting, June 3–7, Harwell, UK.

Guyer, B.L., 2013. Nuclear Power Plant Performance Assessment Pertaining to Plant Aging in France and the United States. PhD Thesis at. MIT, Boston, MA.

Guyot, M.K., Ormiston, S.J., Soliman, H.M., 2016. Numerical analysis of two-phase flow from a stratified region through a small circular side branch. International Journal of Multiphase Flow 87, 175–183.

Ha, K.S., Jeong, J.J., Sim, S.K., 1998. Improvement of liquid droplet entrainment model in the COBRA-TF code. Journal of Korean Nuclear Society 30 (3), 181–193.

Ha, S.J., Park, C.E., Kim, K.D., Ban, C.H., 2011. Development of the SPACE code for nuclear power plants. Nuclear Engineering and Technology 43, 45–62.

Ha, K.S., Lee, S.W., Kim, K.D., 2019. Consideration of H2 generation in fluid governing equations of SPACE code. Annals of Nuclear Energy 131, 291–296.

Haar, L., Gallagher, J.S., Kell, G.S., 1984. NBS/NRC Steam Tables: Thermodynamic and Transport Properties and Computer Programs for Vapor and Liquid States of Water in SI Units. Hemisphere, New York, NY, USA.

Habensky, V.B., Gerliga, V.A., 1994. Flow Instability in Elements of Equipment. Nauka, S. Petersburg, Russia (in Russian).

Hackbush, W., Wittum, G., 1995. Numerical treatment of coupled systems. In: Eleventh-GAMMA Seminar Kiel (G), Vol 51, January 20–22.

Hackford, H.L., 1960. Infrared Radiation. McGraw-Hill, New York, NY, USA.

Hadamard, J., 1902. Sur les problemes aux derivees partielles et leur signification physique. Princeton University Bulletin, pp. 49–52.

Haddad, K.H., Cheung, F.-B., 1998. Steady-state subcooled nucleate boiling on a downward-facing hemispherical surface. Journal of Heat Transfer 120, 365–370.

Hafner, W., Fischer, K., 1990. Contact condensation effects in the main coolant pipe. In: Krischer, W. (Ed.), Reactor Safety Research—The CEC Contribution. Elsevier Applied Science, London, UK, pp. 197–206.

Hahne, E., Spindler, K., Shen, N., 1990. Incipience of flow boiling in subcooled well wetting fluids. In: Proceedings 9th International Heat Transfer Conference, Jerusalem, Israel, vol. 2, pp. 69–74.

Hamad, F.A., Imberton, F., Bruun, H.H., 1997. An optical probe for measurements in liquid-liquid two-phase flow. Measurement Science and Technology 8, 1122–1132.

Hamba, F., 2003. A hybrid RANS/LES simulation of turbulent channel flow. Theoretical and Computational Fluid Dynamics 16, 387–403.

Hamba, F., 2006. A hybrid RANS/LES simulation of high-Reynolds-number channel flow using additional filtering at the interface. Theoretical and Computational Fluid Dynamics 20 (2), 89–101.

Hamidouche, T., Bousbia Salah, A., Adorni, M., D'Auria, F., 2005. Comparative Study between Relap5/3.2 system code and channel codes for the safety analysis of research reactors. In: International Conference on Nuclear Engineering (ICONE-13), May 16–20, Beijing, China.

Hammersley, R.J., Elicson, G.T., Henry, R.E., 2000. Two phase flow and water hammer transient assessment with the TREMOLO computer code. In: Conf. Best-Estimate Methods in Nuclear Installation Safety Analysis (BE-2000), Washington, DC, USA, November.

Han, B., Yang, B.-W., 2019. Effect of indirect heating of rod bundle in fuel assembly thermal hydraulic experiment on local heat flux measurement. In: 18th International Topical Meeting on Nuclear Reactor Thermal Hydraulics (NURETH-18), August 18–23, Portland, OR, USA.

Han, B.S., Bang, Y.S., Kim, K., Seul, K.W., Woo, S.W., 2012. Prediction of number of breached rods following a LBLOCA of CANDU plants using a BEPU approach. In: International Conference ICAPP, Chicago, IL, USA.

Han, B., Yang, B.-W., Zhang, H., Zha, Y., 2017. Numerical analysis of dimple structure effect on grids performance under subcooled boiling condition. Nuclear Power Engineering 38 (3), 158–163.

Han, K.H., Yoo, J.M., Yun, B.J., Jeong, J.J., 2019. Development of a droplet entrainment rate model for a vertical adiabatic two-phase flow. Annals of Nuclear Energy 132, 181–190.

Handbook, 1974. Methods and Correlations for Calculation of Heat Transfer and Pressure Drop in Heat Exchange Equipment of NPP. PTM 24.031.05-72 (in Russian).

Handbook, 1983. Nuclear Steam Generator with Tubes in the Form of Helical Coils. NPO CKTI. PTM. 1983.108.300.01-81 (in Russian).

Handbook, 1986. Heat-Exchange Equipment of NPP. Thermal and Hydraulic Calculation. NPO CKTI. PTM. 1986.108.031.05-84 (in Russian).

Hänsch, S., Walker, S., Narayanan, C., 2016. Mechanistic studies of single bubble growth using interface-tracking methods. Nuclear Engineering and Design 321, 230–243.

Hanson, R.G., Wilson, G.E., Ortiz, M.G., Griggs, D.P., 1992. Development of a phenomena identification and ranking table (PIRT) for a postulated double-ended guillotine break in a production reactor. Nuclear Engineering and Design 136 (3), 335–346.

Haratyk, G., Gourmel, V., 2015. Preliminary accident analysis of Flexblue® underwater reactor. Nuclear Science and Technology 1 (6).

Harding, J.H., Martin, D.G., Potter, P.E., 1989. Thermophysical and Thermochemical Properties of Fast Reactor Materials: Report EUR 12402 EN, Commission of the European Communities, Nuclear Science and Technology. ECSC-EEC-EAEC, Brussels, Belgium.

Hari, S., Hassan, Y.A., 2002. Improvement of the subcooled boiling model for low-pressure conditions in thermal-hydraulics code. Nuclear Engineering and Design 216, 139–152.

Harlow, F.H., Amsden, A.A., 1968. Numerical calculation of almost incompressible flow. Journal of Computational Physics 3 (80).

Harlow, F.H., Amsden, A.A., 1971. Fluid Dynamics, second ed. Los Alamos Scientific Laboratory, Los Alamos, NM, USA. Monograph, LA-4700.

Harlow, F.H., Amsden, A.A., 1975. Flow of interpenetrating material phases. Journal of Computational Physics 18.

Harris, D.R., Reupke, W.A., Wilson, W.B., 1975. Consistency Among Differential Nuclear Data and Integral Observations: The ALVIN Code for Data Adjustment for Sensitivity Calculations and for Identification of Inconsistent Data. Los Alamos scientific laboratory, Los Alamos, NM, USA. Report No LA-5987.

Harvey, A.H., Peskin, A.P., Klein, S.A., 2000. NIST/ASME Steam Properties, Version 2.2, Users' Guide, vol. 10. National Institute of Standards and Technology. U.S. Department of Commerce, NIST Standard Reference Database, USA (March).

Hassan, Y.A., 2002. Multiphase bubbly flow visualization using particle image velocimetry. Annals of the New York Academy of Sciences 972, 223–228.

Hassan, Y.A., 2005. In: Buchlin, J.M. (Ed.), Two-Phase Flow CFD in Nuclear Reactors (I). Institute for Fluid Dynamics, Berlin. Published in the Industrial Two-Phase Flow CFD, Lecture Series 2005-04.

Hassan, Y.A., 2019. High fidelity experimental measurements for modeling and simulation of nuclear engineering applications. In: Sehgal, B.R., Yadigaroglu, G., Hewitt, G.F., D'Auria, [F. (Eds.), Special Issue Trends and Perspectives in Nuclear Thermal-hydraulics—To the Memory of, Journal of Nuclear Engineering and Design, vol. 354 paper 110181 pp. 1–10. (paper PIII-09). https://doi.org/10.1016/j.nucengdes.2019.110181.

Topical issue on CFD4NRS. In: Hassan, Y.A., Smith, B.L. (Eds.), Nuclear Engineering and Design 238, 771–785.

Hassan, Y.A., Estrada-Perez, C.E., Yoo, J.S., 2014. Measurement of subcooled flow boiling using Particle TrackingVelocimetry and infrared thermographic technique. Nuclear Engineering and Design 268, 185–190.

Haste, T.J., Birchley, J., Richner, M., 2010. Accident management following loss of coolant accident during cooldown in a Westinghouse two-loop PWR. Nuclear Engineering and Design 240.

Havet, M., Haedens, M., Schneidesch, C., Zhang, J., 2021. A commensurate industry approach to DEC-A safety analysis. In: OECD/NEA/CSNI Specialists Meeting on Transient Thermal-hydraulics in Water Cooled Nuclear Reactors (SM-TH), December 13–17, CIEMAT, Madrid (Spain), P65.

Hays, W.L., 1981. Statistics, third ed. Holt Rinehrt, Winston, Germany. ISBN-13: 978-0030744679.

He, M., Villarreal, E., Ban, H., Chen, M., 2021. Boiling heat transfer experiments under power transients for LWRs: a review study. Progress in Nuclear Energy 141 (103952), 1–22.

Heagy, S.A., Frepoli, C., Reyes Jr., J.N., 2019. Investigation of scaling distortions for NRE-LAP5 model of NIST facility realized using the dynamical systems scaling methodology. In: Proceedings of the 18th International Topical Meeting on Nuclear Reactor Thermal Hydraulics, NURETH-18, August 18–22, Portland, OR, USA, pp. 6083–6095.

Heerens, W.C., 1986. Application of capacitance techniques in sensor design. Journal of Physics E: Scientific Instrumentation 19, 897–906.

Heinze, D., Schulenberg, T., Behnke, L., 2017. A physically based, one-dimensional three-fluid model for direct contact condensation of steam jets in flowing water. International Journal of Heat and Mass Transfer 106, 1041–1050.

Hellsten, A., 2004. New Two Equation Turbulence Model for Aerodynamics Application. Dissertation for the Degree of Doctor of Science in Technology. Helsinki University of Technology, Espoo, Finland.

Hellsten, A., 2005. New advanced k-omega turbulence model for high-lift aerodynamics. AIAA Journal 43 (9), 1857–1869.

Hellsten, A., Bezard, H., 2005. Behaviour of nonlinear two-equation turbulence models at the free-stream edges of turbulent flows. In: Rodi, W., Mulas, M. (Eds.), Engineering Turbulence Modelling and Experiments. Elsevier, Amsterdam, NL.

Hellsten, A., Wallin, S., 2009. Explicit algebraic Reynolds stress and non-linear eddy viscosity models. International Journal of Computational Fluid Dynamics 23, 349–361.

Henry, R.E., 1968. A Study of One and Two-Component Two-phase Critical Flows at Low Qualities. ANL 7430, Chicago, IL, USA, March.

Henry, R.E., 1970. The two-phase critical discharge of initially saturated or subcooled liquid. Nuclear Science and Engineering 41, 336–342.

Henry, R.E., 2011. TMI-2: An Event in Accident Management for Light-Water-Moderated Reactors. ANS Monograph, LaGrange Park, IL, USA.

Henry, R.E., Fauske, H.K., 1971. The two-phase critical flow of one-component mixtures in nozzles, orifices, and short tubes. ASME Transactions C, Journal of Heat and Mass Transfer 93 (2), 179–187.

Henry, R.E., Fauske, H.K., McComas, S.T., 1970. Two-phase critical flow at low qualities, part II. Nuclear Science and Engineering 41, 92–98.

Henry, R.E., Epstein, M., Fauske, H.K., 2015. An explanation of BORAX-1, SPERT-1 and the SL-1 accident. In: 16th International Topical Meeting on Nuclear Reactor Thermal Hydraulics (NURETH-16), August 30–September 4, Chicago, IL, USA.

Heo, J., Kim, K.D., 2015. PAPIRUS, a parallel computing framework for sensitivity analysis, uncertainty propagation, and estimation of parameter distribution. Nuclear Engineering and Design 292, 237–247.

Heo, J., Yon, S.H., Kim, K.D., Ha, K.S., Hibiki, T., 2020. Validation of interfacial area concentration model for simulating bubbly and cap/slug flow behaviors in large diameter pipes using LSTF, ATLAS, and horizontal pipe experiment data. Progress in Nuclear Energy 128, 103499.

Heo, J., Kim, K.D., Ha, K.S., 2021. A numerical derivation of internal energy equation under a staggered mesh system for 1-D thermal hydraulic system analysis codes. Nuclear Engineering and Design 371, 110935.

Herer, C., 2015. Margins to critical heat flux in pressurized water reactors using modern thermal-hydraulic methods. In: 16th International Topical Meeting on Nuclear Reactor Thermal Hydraulics (NURETH-16), August 30−September 4, Chicago, IL, USA.

Herer, C., Bestion, D., Fillion, P., Prea, R., Parrinello, V., Bousbia Salah, A., Kim, K., Jeong, J.J., 2019. 3-DSYSTH: an OECD/NEA activity on multi-dimensional capabilities of thermalhydraulic system. In: International Congress on Advances in Nuclear Power Plants (ICAPP), Juan-les-Pins, France, May 12−15.

Herkenrath, H., Hufschmidt, W., 1979. Experimental investigation of the enthalpy and mass flow distribution between subchannels in a BWR cluster geometry (PELCO-S). In: Joint Research Centre of the European Community, EUR 6585, Ispra, Italy, pp. 1−78.

Herkenrath, H., Hufschmidt, W., Jung, U., Weckermann, F., 1981. Experimental Investigation of the Enthalpy and Mass Flow Distribution in 16-rod Clusters with BWR-PWR Geometries and Conditions. Joint Research Centre of the European Community, Ispra, Italy. EUR 7575.

Hernandez, R., Folsom, C.P., Woolstenhulme, N.E., Jensen, C.B., Bess, J.D., Gorton, J.P., Brown, N.R., 2020. Review of pool boiling critical heat flux (CHF) and heater rod design for CHF experiments in TREAT. Progress in Nuclear Energy 123 (103303), 1−14.

Herranz, L.E., López, C., 2020. Challenges and sensitivities in the modelling of Fukushima Daiichi unit 1 unfolding with MELCOR 2.2. Annals of Nuclear Energy 141 (107348), 1−7.

Herranz, L.E., Anderson, M.H., Corradini, M.L., 1998. A diffusion layer model for steam condensation within the AP600 containment. Nuclear Engineering and Design 183, 133−150.

Herviou, K., 2005. Development of a Methodology and of a Computer Tool for Source Term Estimation in Case of Nuclear Emergency in a Light Water Reactor (ASTRID). EC Report ASTRID/04.39, Brussels, Belgium.

Hetsroni, G., 1982. Handbook of Multiphase Systems. Hemisphere—McGraw-Hill, New York, NY, USA.

Hewitt, G.F., 1973. Role of experiments in two-phase systems with particular reference to measurement techniques. Journal of the British Nuclear Energy Society 12, 213−240.

Hewitt, G.F., 1978. Measurement of Two Phase Flow Parameters. Academic Press Inc., London, UK.

Hewitt, G.F., 1989. Countercurrent two-phase flow. In: Hewitt, G.F. (Ed.), Proceedings 4th International Topical Meeting on Nuclear Reactor Thermal-Hydraulics (NURETH-4), vol. 2 (Karlsruhe, Germany).

Hewitt, G.F., Govan, A.H., 1990. Phenomenological modelling of non-equilibrium flows with phase change. International Journal of Heat Mass Transfer 33 (2), 229−242.

Hewitt, G.F., Hall-Taylor, N.S., 1970. Annular Two-phase Flow. Pergamon Press, Oxford, UK (see also below).

Hewitt, G.F., Hall-Taylor, N.S., 1972. Annular Two-phase Flow. Pergamon Press, Oxford, UK (see also above).

Hewitt, G.F., Roberts, D.N., 1969. Studies of Two-phase Flow Patterns by Simultaneous X-ray and Flash Photography. AERE-M-2159, HMSO, Harwell, UK.

Hewitt, G.F., Kearsey, H.A., Lacey, P.M.C., Pulling, D.J., 1963. Burn-Out and Nucleation in Climbing Film Flow. UKAEA Rep. AERE-R437, Harwell, UK.

Hewitt, G.F., Jayanti, S., Hope, C.B., 1990. Structure of thin liquid films in gas−liquid horizontal flow. International Journal of Multiphase Flow 16, 951−957.

Hewitt, G.F., Delhaye, J.M., Zuber, N. [E.s], 1992. Post-dryout Heat Transfer. CRC Press Inc., Boca Raton, FL, USA.

Heywood, N.I., Mehta, K.B., 1999. Performance evaluation of eight electromagnetic flowmeters with fine sand slurries. Hydrotransport 14, 413–431. BHRG.

Heywood, N.I., Mehta, K.B., Poplar, D., 1993. Evaluation of seven commercially available electromagnetic flowmeters with Newtonian and non-Newtonian China clay slurry in pipe flow. In: 12th International Conference Slurry Handling and Pipeline Transport. Hydrotransport, vol. 12. BHRG, pp. 353–376.

Hibiki, T., Ishii, M., 2000. Two-group interfacial area transport equations at bubbly-to-slug flow transition. Nuclear Engineering and Design 202, 39–76.

Hibiki, T., Ishii, M., 2002. Development of one-group interfacial area transport equation in bubbly flow systems. International Journal of Heat Mass Transfer 45, 2351–2372.

Hibiki, T., Jeong, J.J., 2022. Partition method of wall friction and interfacial drag force model for horizontal two phase flows. Nuclear Engineering and Technology 54, 1495–1507.

Hibiki, T., Hogsett, S., Ishii, M., 1998. Local measurement of interfacial area interfacial velocity and liquid turbulence in two-phase flow. Nuclear Engineering and Design 184, 287–304.

Hibiki, T., Ishii, M., Xiao, Z., 2001. Axial interfacial area transport of vertical bubbly flows. International Journal of Heat Mass Transfer 44, 1869–1888.

Hibiki, T., Mi, Y., Situ, R., Ishii, M., 2003. Interfacial area transport of vertical upward bubbly two-phase flow in an annulus. International Journal of Heat Mass Transfer 46, 4949–4962.

Hinata, S., 1972. A study on the measurement of the local void fraction by the optical fiber glass probe. Bulletin of JSME 15 (88), 1228–1235.

Hirano, M., Watanabe, T., 1992. Analysis of the Mihama-2 SGTR event and ROSA-IV experiment. In: Proceedings of International Topical Meeting on Nuclear Reactor Thermal Hydraulics (NURETH-5), Salt Lake City, UT USA, September 21–24, pp. 165–173.

Hirt, C.W., 1968. Heuristic stability theory for finite-difference equations. Journal of Computational Physics 2, 339–355.

Hochreiter, L.E., 1985. Flecht Seaset Program, Final Report, pp. 1–364. NRC/EPRI/Westinghouse Report No. 16, USNRC NUREG/CR-4167, Washington, DC, USA.

Hochreiter, L.E., Cheung, F.-B., Lin, T.F., Baratta, A.J., Frepoli, C., Sridharan, A., Todd, D.R., Rosal, E.L., 2000. Dispersed Flow Heat Transfer under Reflood Conditions in a 49 Rod Bundle: Test Plan and Design—Results from Tasks 1–10, pp. 1–138. PSU ME/NE -NRC-04-98-041 Report 1, Revision 1, College Park, PA, USA.

Hochreiter, L.E., Cheung, F.-B., Lin, T.F., Frepoli, C., Sridharan, A., Todd, D.R., Rosal, E.R., 2010. Rod Bundle Heat Transfer Test Facility, Test Plan and Design. USNRC, NUREG/CR-6975, Washington, DC, USA.

Hofer, E., 1990. The GRS programme package for uncertainty and sensitivity analysis. In: Seminar on Methods and Codes for Assessing the Off-Site Consequences of Nuclear Accidents. EUR 13013. CEC, Brussels, Belgium.

Hofer, E., 1993. Probabilistische Unsicherheitsanalyse von Ergebnissen umfangreicher Rechenmodelle. GRS-A-2002, Garching, Germany.

Hofer, E., 2003. Statistik und Probabilistik in der Schadensumfangsanalyse. Note, GRS Garching, Germany, 11 February.

Hofmeister, J., Schulenberg, T., Starflinger, J., 2005. Optimization of a fuel assembly for a HPLWR. In: Proceedings of ICAPP'05, Paper No. 5077, Seoul, Korea.

Höhne, T., Hänsch, S., 2015. A droplet entrainment model for horizontal segregated flows. Nuclear Engineering and Design 286, 18–26.

Holian, B.L., Hoover, W.G., Moran, B., Straub, G.K., 1980. Shock-wave structure via non-equilihrium molecular dynamics and Navier-Stokes continuum mechanics. Phys. Rev. A 22 (6), 1−11.

Holm, R., 1967. Electric Contacts Handbook. Springer-Verlag, Berlin, Germany.

Holmes, B.J., Allen, E.J., 1990. A Review of Critical Flow Data for Pressurized Water Reactor Safety Studies. AEEW-R 2592, Winfrith, UK.

Holowach, M.J., Hochreiter, L.E., Mahaffy, J.H., Cheung, F.-B., 2003. Modeling of droplet entrainment phenomena at a quench front. International Journal of Heat and Fluid Flow 24, 902−918.

Holpanov, L.P., Shkadov, V.Y., 1990. Hydrodynamics and Heat and Mass Transfer with the Interface. Nauka, Moscow, Russia (in Russian).

Hong, S.J., Choo, Y.J., Hwang, S.H., Lee, B.C., Ha, S.J., 2015. Development of CAP code for nuclear power plant containment: lumped model. Nuclear Engineering and Design 291, 47−63.

Horhoianu, G., Ionescu, D.V., Sefan, I., Olteanu, G., 1998. CANDU type fuel behavior during rapid overpower transients. Nuclear Engineering and Design 179.

Hou, Y., Wang, L., Wang, M., Zhang, K., Zhang, X., Hu, W., Wu, Y., Tian, W., Qiu, S., Su, G.H., 2019. Experimental study of liquid sodium flow and heat transfer characteristics along a hexagonal 7-rod bundle. Applied Thermal Engineering 149, 578−587.

Housiadas, C., 1987. Contribution à l'analyse des phénomènes bidimensionnels intervenant lors du renoyage d'un coeur de Réacteur à Eau sous Pression. Propagation du front de trempe (in French). Thèse de Docteur de l'Institut National Polytechnique de Grenoble, Grenoble, France.

Housiadas, C., Veteau, J.M., 1989. Two-dimensional quench front progression in a multi-assembly rod bundle. Nuclear Engineering and Design 13 (1), 87−98.

Houser, R., Young, E., Rasmussen, A., 2013. Overview of NuScale testing programs. Transactions of the American Nuclear Society 109, 1585−1586.

Howard, R.J.A., Pourquie, M., 2002. Large eddy simulation of an Ahmed reference model. Journal of Turbulence 3, N12. https://doi.org/10.1088/1468-5248/3/1/012.

Hsiang, L.P., Faeth, G.M., 1992. Near-limit drop deformation and secondary breakup. International Journal of Multiphase Flow 18, 635−652.

Hsu, Y.Y., Graham, R.W., 1986. Transport Processes in Boiling and Two-phase Systems. ANS Monograph, Hinsdale, IL, USA (also McGraw-Hill, 1976).

Hsu, L.C., Chen, I.Y., Chyu, C.M., Wang, C.C., 2015. Two-phase pressure drops and flow pattern observations in 90° bends subject to upward, downward and horizontal arrangements. Experimental Thermal and Fluid Science 68, 484−492.

Hu, R., 2015. A high-order system thermal-hydraulics model for advanced reactor safety analyses. In: 16th International Topical Meeting on Nuclear Reactor Thermal Hydraulics (NURETH-16), August 30-September 4, Chicago, IL, USA.

Hu, R., 2016. A fully-implicit high-order system thermal hydraulics model for advanced non LWR safety analyses. Annals of Nuclear Energy 101.

Hu, R., Fanning, T.H., 2013. A momentum source model for wire-wrapped rod bundles—concept, validation, and application. Nuclear Engineering and Design 262, 371−389.

Huang, S.M., Stott, A.L., Green, R.G., Beck, M.S., 1988. Electronic transducers for industrial measurement of low value capacitances. Journal of Physics E: Scientific Instrumentation 21, 242−250.

Huang, H.-W., Shih, C., Yih, S., Chen, M.-H., Lin, J.-M., 2007. Model extension and improvement for simulator-based software safety analysis. Nuclear Engineering and Design 237.

Huang, J., Zhang, J., Wang, L., 2015. Review of vapor condensation heat and mass transfer in the presence of non-condensable gas. Applied Thermal Engineering 89, 469–484.

Huang, J., Gou, J., Ding, W., Zhanf, B., Shan, J., 2018. Development of a sub-channel analysis code based on the two-fluid model and its preliminary assessment. Nuclear Engineering and Design 340, 17–30.

Huang, S., Huang, W., Zhu, Z., 2018. Experimental study on the effect of the tracer particles concentration in ultrasonic Doppler velocimetry measurement. In: 12th International Topical Meeting on Nuclear Reactor Thermal-Hydraulics, Operation and Safety (NUTHOS-12), October 14–18, Qingdao, China.

Huck, P.D., Chaussonet, G., Fort, C., Bardet, P.M., 2022. Direct measurement of wall shear stress with volumetric microscopic molecular tagging velocimetry. In: 19th International Topical Meeting on Nuclear Reactor Thermal Hydraulics (NURETH-19), March 6–11 (Via Web), Brussels, Belgium.

Hughes, E.D., 2020. Ramifications of Two-Pressure Two-phase Flow Models and the RELAP-7 Characteristics. Personal communication to one author of the (Chapter 5) (DB) and to one editor of the book (FDA), November.

Hughes, E.D., Duffey, R.B., 1991. Direct contact condensation and momentum transfer in turbulent separated flows. International Journal of Multiphase Flow 17, 599–619.

Hughes, E.D., Ka-lam, I., Baker, A.N., Carbon, M.W., 1974. A compilation of rod array critical heat flux data sources and information. Nuclear Engineering and Design 30, 20–35.

Hughes, E.D., Lyczkowski, R.W., McFadden, J.H., Niederauer, G.F., 1976. An evaluation of state-of-the-art two-velocity two-phase flow models and their applicability to nuclear reactor transient analysis. In: Technical Reports Electrical Power Research Institute (EPRI) NP-143, Vol. 1, 2, 3. Research Project TSA SOA 75-317. Energy Inc., Palo Alto, CA, USA.

Hung, Z.-Y., Ferng, Y.-M., Hsu, W.-S., Pei, B.-S., Chen, Y.S., 2015. Analysis of AP-1000 containment passive cooling system during a loss-of-coolant accident. Annals of Nuclear Energy 85, 717–724.

Huning, A.J., Chandrasekaran, S., Garimella, S., 2021. A review of recent advances in HTGR CFD and thermal fluid analysis. Nuclear Engineering and Design 373 (111013), 1–23.

Hussaini, M.Y., Thangam, S., Woodruff, S.L., Zhou, Y., 2006. Development of a continuous model for simulation of turbulent flows. Journal of Applied Mechanics 73 (3). https://doi.org/10.1115/1.2173006.

Husted, B.P., Peterson, P., Lund, I., Holmstedt, G., 2009. Comparison of PIV and PDA droplet velocity measurement techniques on two high-pressure water mist nozzles. Fire Safety Journal 44 (8), 1030–1045.

Hutton, A.G., 2009. The merging role of large eddy simulation in industrial practice: challenges and opportunities. Philosophical Transactions of the Royal Society A 367, 2819–2826.

Hwang, D., Jeong, J.J., Chung, B.D., 2009. NUPEC BFBT subchannel void distribution analysis using the MATRA and MARS codes. Nuclear Engineering and Technology 41 (3), 295–306.

Hwang, M., Bae, S.W., Chung, B.D., 2012. APR1400 LBLOCA uncertainty quantification using Monte Carlo method and comparison with Wilks' formula. In: International Conference ICAPP, Chicago, IL, USA.

Hyvarinen, J., 1992. An inherent boron dilution mechanism in pressurized water reactors. In: CSNI Specialist Meeting on Transient Two-phase Flow, OECD/NEA/CSNI, Aix-en-Provence, France.

Hyvarinen, J., 1996. Heat transfer characteristics of horizontal steam generators under natural circulation conditions. Nuclear Engineering and Design 66.

Hyvärinen, J., Telkkä, J., Kauppinen, O., Purhonen, H., 2017. Motel—modular design and physical scaling principles of the next generation thermal hydraulics test facility at LUT. In: 17th International Topical Meeting on Nuclear Reactor Thermal Hydraulics, NURETH-17, Sept 2017, Xi'an, China.

IAEA, 1991. Safety Related Terms for Advanced Nuclear Plants. IAEA-TECDOC-626, Vienna, Austria.

IAEA, 1993. Off-gas and Air Cleaning Systems for Accident Conditions in Nuclear Power Plants, Technical Report Series, No 358, STI/DOC/010/358, October, Vienna (A), pp. 1–132.

IAEA, 1994. Status of Advanced Containment Systems for Next Generation Water Reactors, TECDOC-752, Vienna, Austria.

IAEA, 1995. Experimental Design Verification of VVER-440 Model 213 Nuclear Power Plants, Reference Plant: Bohunice V2 (Sk). Technical Co-operation Project RER/9/004 on Evaluation of Safety Aspects of VVER-440 Model 213 Nuclear Power Plants, TECDOC-810, Vienna, Austria.

IAEA, 1995a. Progress in design, research and development and testing of safety systems for advanced water cooled reactors. In: Proceedings of Technical Committee Meeting, Piacenza (I), TECDOC-872, Vienna, Austria.

IAEA, 1996. Defense in Depth in Nuclear Safety, INSAG-10, Vienna, Austria.

IAEA, 1997. Guidelines on the Pressurized Thermal Shock Analysis for WWER Nuclear Power Plants, EBP-WWER-08, Vienna, Austria.

IAEA, 1999. Final Report of the Programme on the Safety of WWER and RBMK Nuclear Power Plants. IAEA-EBP-WWER-15, Vienna, Austria.

IAEA, 2000. Safety of Nuclear Power Plants: Design, Requirements. NS-R-1, Vienna, Austria.

IAEA, 2000a. [Aksan N. Chairman of the Technical Meeting]. Experimental Tests and Qualification of Analytical Methods to Address Thermo-Hydraulic Phenomena in Advanced Water Cooled Reactors. IAEA-TECDOC-1149, Vienna, Austria.

IAEA, 2000b. Irradiation Damage in Graphite Due to Fast Neutrons in Fission and Fusion Systems, pp. 1–176. TECDOC-1154, Vienna, Austria.

IAEA, 2001. [Aksan N. Chairman of CRP]. Thermo-Hydraulic Relationships for Advanced Water Cooled Reactors. IAEA, Vienna, Austria IAEA-TECDOC-1203 (see Aksan et al., 2001).

IAEA, 2002. Accident Analysis for Nuclear Power Plants. IAEA Safety Reports Series (SRS) No. 23, ISSN 1020-6450; ISBN 92-0-115602-2, Vienna, Austria.

IAEA, 2002a. Natural Circulation Data and Methods for Advanced Water Cooled Nuclear Power Plants Designs, TECDOC 1281, Vienna, Austria.

IAEA, 2002b. Heavy Water Reactors, Status and Projected Development, pp. 1–702. IAEA Technical Report Series No 407, Vienna, Austria.

IAEA, 2003. Accident Analysis for Nuclear Power Plants with Pressurized Heavy Water Reactors, IAEA Safety Reports Series (SRS) No 29, Vienna, Austria.

IAEA, D'Auria, F., Durin, M., Mahaffy, J.H., Mišák, J.-C., Micaelli, J., Réocreux, M., Royen, J., Sandervåg, O., Scheuerer, M., Smith, B., Téchy, Z., Vitale, E., 2003a. Use of computational fluid dynamics codes for safety analysis of nuclear reactor systems. In: Summary Report of a Technical Meeting Jointly Organized by the IAEA and NEA/OECD, Held in Pisa, Italy, 11–14 November 2002, Tecdoc, vol. 1379. IAEA, Vienna, Austria, pp. 1–53.

IAEA, 2005. Aksan N. Chairman of CRP]. Natural Circulation in Water Cooled Nuclear Power Plants: Phenomena, Models, and Methodology for System Reliability Assessments. IAEA-TECDOC-1474, Vienna, Austria.

IAEA, 2005a. Assessment of Defense in Depth for Nuclear Power Plants, IAEA Safety Reports Series No 46, Vienna, Austria.
IAEA, 2006. Fundamental Safety Principles, Safety Fundamentals. SF-1, Vienna, Austria.
IAEA, 2006a. The Management System for Facilities and Activities, Safety Requirement No. GS-R-3, IAEA, Vienna, (A).
IAEA, 2006b. Application of the Management System for Facilities and Activities, Safety Guide No. GS-G-3.1, IAEA, Vienna (A).
IAEA, 2007. Nuclear Power Plant Design Characteristics, IAEA TECDOC 1544, ISBN 92-0-102507-6, pp. 1−40.
IAEA, 2008a. Best Estimate Safety Analysis for Nuclear Power Plants: Uncertainty Evaluation. Safety Report Series, No. 52, Vienna, Austria. (see also D'Auria et al., 2008).
IAEA, 2008b. Approaches and Tools for Severe Accident Analysis for Nuclear Power Plants. Safety Report Series, No. 56, Vienna, Austria.
IAEA, 2009a. Safety Assessment for Facilities and Activities, General Safety Requirements. GSR Part 4, Vienna, Austria.
IAEA, 2009b. Passive Safety Systems and Natural Circulation in Water Cooled Nuclear Power Plants. TECDOC 1624, Vienna, Austria.
IAEA, 2009c. Deterministic Safety Analysis for Nuclear Power Plants. IAEA SSG-2, Vienna, Austria.
IAEA, 2010a. Development and Application of Level 1 Probabilistic Safety Assessment for Nuclear Power Plants. IAEA, Safety Standard Series, SSG-3. Vienna, Austria.
IAEA, 2010b. Development and Application of Level 2 Probabilistic Safety Assessment for Nuclear Power Plants. Safety Standard Series, SSG-4, IAEA, Vienna, Austria.
IAEA, 2010c. Pressurized Thermal Shock in Nuclear Power Plants: Good Practices for Assessment. TECDOC 1627, IAEA, Vienna, Austria.
IAEA, 2011. A Framework for an Integrated Risk Informed Decision Making Process. INSAG-25, Vienna, Austria.
IAEA, 2012. Natural Circulation Phenomena and Modeling for Advanced Water Cooled Reactors, TECDOC 1677, Vienna, Austria.
IAEA, 2012a. Efficient Water Management in Water Cooled Reactors, Nuclear Energy Series, No. NP-T-2.6, STI/PUB/1569, ISBN 978-92-0-132610-2, pp. 1−131. Vienna, Austria.
IAEA, 2014. Evaluation of Advanced Thermohydraulic System Codes for Design and Safety Analysis of Integral Type Reactors. TECDOC 1733, Vienna, Austria.
IAEA, 2014a. [Aksan N. Chairman of CRP]. Heat Transfer Behavior and Thermohydraulics Code Testing for Supercritical Water Cooled Reactors (SCWRs). Vienna, Austria, IAEA-TECDOC-1746. (see also Aksan et al., 2014).
IAEA, 2014b. The Fukushima Daiichi Accident: Report by the Director General. GC(59)14, Vienna, Austria.
IAEA, 2014c. Safety Classification of Structures, Systems and Components in Nuclear Power Plants. IAEA SSG-30, Vienna, Austria.
IAEA, 2015. Verification and Validation of System Codes for Nuclear Safety Analyses. Unpublished draft report (at the basis of (Chapter 22)), Vienna, Austria.
IAEA, 2015a. Consultancy Activity for Connecting DSA and PSA (Un-published Material), Vienna, Austria.
IAEA, 2015b. Technical Challenges in the Application and Licensing of Digital Instrumentation and Control Systems in Nuclear Power Plants, Nuclear Energy Series, NP-T-1.13, Vienna, Austria.
IAEA, 2016a. http://www-ns.iaea.org/home/ni/databases.asp#irs.
IAEA, 2016b. http://www-ns.iaea.org/home/ni/databases.asp#finas.

IAEA, 2016c. http://www-ns.iaea.org/home/ni/databases.asp#irsrr.
IAEA, 2019. Deterministic Safety Analysis for Nuclear Power Plants, Specific Safety Guide, SSG-2 (Rev. 1), Vienna, Austria.
IAEA, 2019a. Design of the Reactor Containment and Associated Systems for Nuclear Power Plants, SSG-53, Vienna, Austria, pp. 1−110.
IAEA, 2020. Advances in Small Modular Reactor Technology Developments; A Supplement to: Advanced Reactors Information System (ARIS). IAEA, Vienna, Austria, pp. 1−354.
IAEA, 2020a. Advances in Small Modular Reactor Technology Developments. A Supplement to: IAEA Advanced Reactors Information System (ARIS)—2020 Edition. IAEA, Vienna (A), pp. 1−354.
IAEA, 2020b. Nuclear power reactors in the world. Reference Data Series No 2, 1−86. Vienna, Austria.
IAEA, 2022. Advanced Reactors Information System (ARIS). http://aris.iaea.org.
IAEA, Bukin, N., D'Auria, F., Elter, J., Gudoshnikov, A., Kantee, H., Kostka, P., Kristof, M., Kvidza, B., Macek, J., Makihara, Y., Matejovic, P., Misak, J., Sartmadjiev, A., Spalj, S., 2009d. Safety Analysis of WWER-440 Nuclear Power Plants: Potential Consequences of a Large Primary to Secondary System Leakage Accident, IAEA TECDOC 1610, Vienna, Austria, February.
IAEA-TM, 2005. TM on Use of a Best Estimate Approach in Licensing with Evaluation of Uncertainties, Pisa, Italy, 12−16 September.
IAEA-TM, 2006. TM on Effective Integration of Deterministic and Probabilistic Safety Analysis in Plant Safety Management, Barcelona, Spain, 4−8 September.
IAEA-TM, 2007. TM on Advanced Safety Assessment Methods for Nuclear Reactors. KAERI, Daejeon, Korea, 30 October−2 November.
IAEA-TM, 2008. TM on the Use of Safety Margins and Advanced Safety Assessment Methods in Plant Modifications. HAEA, Budapest, Hungary, 15−19 September.
IAEA-TM, 2009. TM on Application of Deterministic Best Estimate Safety Analysis, Pisa, Italy, 21−25 September.
IAEA-TM, 2010. TM (In Cooperation with OECD/NEA/CSNI) on Application of Advanced Safety Assessment Methods. SNSA, Bled, Slovenia, 7−11 June.
IAEA-TM, 2012. TM (In Cooperation with OECD/NEA/CSNI) on Combining Insights from Deterministic and Probabilistic Safety Analyses, Pisa, Italy, 11−15 June.
IAEA-TM, 2013. TM on Best Estimate Plus Uncertainty (BEPU) Methods in Safety Analyses, Pisa, Italy, 10−14 June.
Ibrahimbegovic, A., Markovic, D., 2003. Strong coupling methods in multi-phase and multi-scale modeling of inelastic behavior of heterogeneous structures. Computer Methods in Applied Mechanics and Engineering 192, 3089−3107.
Idaho National Lab, 2001. RELAP5-3D Code Manual, Volume I: Code Structure, System Models and Solution Methods, INEEL-EXT-98-00834 Revision 1.3a, Idaho Falls, ID, USA, February.
Idaho National Lab, 2012a. RELAP5-3D Code Manual, Volume I: Code Structure, System Models and Solution Methods. INEEL-EXT-98-00834, Idaho Falls, ID, USA.
Idaho National Lab, 2012b. RELAP5-3D Code Manual, Volume IV: Models and Correlations. INEEL-EXT-98-00834, Idaho Falls, ID, USA.
Idel'chik, I.E., 1960. Handbook of Hydraulic Resistance—Coefficients of Local Resistance and of Friction. Gosudarstvennoe Energeticheskoe Izdatel'stvo, Moskva-Leningrad. (Russia, USSR at the Time)—Translated from Russian by Israel Program for Scientific Translations, Jerusalem (Israel) 1966.

Idel'chik, I.E., 1979. Hand Book of Hydraulic Resistances. Hemisphere Publishing Company, New York, NY.

Iguchi, T., 1998. Study of Thermal-Hydraulics during a PWR Reflood Phase. JAERI-Research-98-054, Tokai-Mura, Japan.

Iguchi, T., Iwamura, T., Akimoto, H., Onuki, A., Abe, Y., Hojo, T., Sakaki, I., Minato, A., Adachi, H., Murao, Y., 1988. SCTF-III test plan and recent SCTF-III test results. Nuclear Engineering and Design 108, 241−247.

Ihle, P., Rust, K., 1984. FEBA Flooding Experiments with Blocked Arrays. Evaluation Report. Scientific Report KfK 3657, Karlsruhe, Germany.

Ihle, P., Rust, K., 1986. SEFLEX Fuel Rod Simulator Effects in Flooding Experiments. Part 1: Evaluation Report. Scientific Report KfK 4024, Karlsruhe, Germany.

Ihle, P., Rust, K., 1987. PWR reflood experiments using full length bundles of rods with zircaloy claddings and alumina pellets: results of the SEFLEX program. Journal of Nuclear Engineering and Design 99, 223−237.

Ikeda, K., 2014. CFD application to advanced design for high efficiency spacer grid. Nuclear Engineering and Design 279, 73−82.

Imanaka, T. (Ed.), 2002. Recent Research Activities about the Chernobyl NPP Accident in Belarus, Ukraine and Russia. Research Reactor Institute, Kyoto University, Kyoto, Japan.

Imke, U., Sanchez, V.H., 2012. Validation of the subchannel code SUBCHANFLOW using the NUPEC PWR tests (PSBT). Science and Technology of Nuclear Installations (J STNI) 1−12.

Imke, U., Sanchez, V., Gomez, R., 2010. Subchanflow: a new empirical knowledge based subchannel code. In: Annual Meeting on Nuclear Technology, Berlin, Germany.

Inayatov, A.Y., 1975. Correlation of data on heat transfer flow parallel to tube bundles at relative tube pitches of $1.1 < s/d < 1.6$. Heat Transfer/Soviet Research 7 (3), 84−88.

Incropera, F.P., DeWitt, D.P., 1981. Fundamentals of Heat Transfer. John Wiley & Sons, New York, NY, USA.

Incropera, F.P., DeWitt, D.P., 1990. Introduction to Heat Transfer. John Wiley & Sons, New York, NY, USA, pp. 1−501.

Ingersoll, L.R., Zobel, O.J., Ingersoll, A.C., 1954. Heat Conduction with Engineering, Geological and Other Applications. University of Wisconsin Press, Madison, WI, USA.

IRSN, 2013. ASAMPSA2 Best Practice Guidelines for Level 2 PSA Development and Applications. PSN-RES/SAG/2013-0177, Paris France.

Isachenko, V.I., 1977. Heat Transfer at Condensation. Energiya, Moscow, Russia (in Russian).

Isachenko, V.P., Osipova, V.A., Sukomel, A.S., 1981. Heat Transfer. Energoizdat, Moscow, Russia (in Russian).

Ishida, I.T.K., 1990. Thermal hydraulic behavior of a marine reactor under oscillations. Nuclear Engineering and Design 120, 213−225.

Ishii, M., 1975. Thermo-fluid dynamic theory of two-phase flow. Collection de la Direction des Etudes et Researches d'Electricite de France, Eyrolles, Paris, France.

Ishii, M., 1977. One-Dimensional Drift-Flux Model and Constitutive Equations for Relative Motion between Phases in Various Flow Regimes. Argonne National Laboratory Report ANL-77-47, USA.

Ishii, M., Chawla, T.C., 1979. Local Drag Laws in Dispersed Two-phase Flow. ANL-79-105. Argonne National Laboratory (ANL), Lemont, IL, USA.

Ishii, M., De Jarlais, G., 1986. Flow regime transitions and interfacial characteristics of inverted annular flow. Nuclear Engineering and Design 95, 171−184.

Ishii, M., De Jarlais, G., 1987. Flow visualizations of inverted annular flow of post-dryout heat transfer region. Nuclear Engineering and Design 99, 187−199.

Ishii, M., Hibiki, T., 2006. Thermo-Fluid Dynamics of Two-phase Flow. Springer, New York, NY, USA.

Ishii, M., Hibiki, T., 2011. Thermo-Fluid Dynamics of Two-phase Flow, second ed. Springer, New York, NY, USA.

Ishii, M., Jarlais, G., 1982. Inverted annular flow modeling. In: Advanced Code Review Group Meeting, Idaho Falls, ID, USA, July 27.

Ishii, M., Kataoka, I., 1982. Scaling Criteria for LWR's under Single-phase and Two-phase Natural Circulation. Joint NRC/ANS Meeting on Basic Thermal Hydraulic Mechanisms in LWR Analysis, Bethesda, MD, USA.

Ishii, M., Kataoka, I., 1983. Scaling Criteria for LWR's under Single-phase and Two-phase Natural Circulation. ANL-83-32, NUREG/CR-3267. Argonne National Laboratory, Chicago, IL, USA.

Ishii, M., Kataoka, I., 1984. Mechanistic modeling of pool entrainment phenomenon. International Journal of Heat Mass Transfer 27 (11), 1999–2014.

Ishii, M., Kataoka, I., 1984a. Scaling laws for thermal-hydraulic system under single phase and two-phase natural circulation. Journal of Nuclear Engineering and Design 81.

Ishii, M., Kim, S., 2004. Development of one-group and two-group interfacial area transport equation. Nuclear Science and Engineering 146, 257–273.

Ishii, M., Kojasoy, G., 1993. Interfacial Area Transport Equation and Preliminary Considerations for Closure Relations, PU/NE-93-6. School of Nuclear Engineering, Purdue University, IN, USA.

Ishii, M., Mishima, K., 1980. Study of Two-Fluid Model and Interfacial Area: NUREG/CR-1873, ANL-80-111. Argonne National Laboratory (ANL), Lemont, IL, USA.

Ishii, M., Mishima, K., 1984. Two-fluid model and hydrodynamic constitutive relations. Nuclear Engineering and Design 82, 107–126.

Ishii, M., Mishima, K., 1989. Droplet entrainment correlation in annular two-phase flow. International Journal of Heat Mass Transfer 32 (10), 1835–1845.

Ishii, M., Zuber, N., 1970. Thermally induced flow instabilities in two phase mixtures. In: 4th International Heat Transfer Conference, Paris, France.

Ishii, M., Wu, Q., Assad, A., Uhle, J., 1998. Interfacial area transport equation for two-fluid model formulation. In: Proceedings of IMuST Meeting, USA.

Ishii, M., Revankar, S.T., Leonardi, T., Dowlati, R., Bertodano, M.L., Babelli, I., Wang, W., Pokharna, H., Ransom, V.H., Viskanta, R., Han, J.T., 1998a. The three-level scaling approach with application to the Purdue University multi-dimensional integral test assembly (PUMA). Journal of Nuclear Engineering and Design 186, 177–211.

Ishii, M., Kim, S., Uhle, J., 2000. Interfacial area transport: data and models. In: Proceedings of OECD/CSNI Workshop on Advanced Thermal-hydraulics and Neutronics Codes: Current and Future Applications, Barcelona, Spain, April 10–13.

Ishii, M., Kim, S., Kelly, J.M., 2005. Development of interfacial area transport equation. Nuclear Engineering and Technology 37, 525–536.

Ismail, I., Gamio, J.C., Bukhari, S.F.A., Yang, W.Q., 2005. Tomography for multi-phase flow measurement in the oil industry. Flow Measurement and Instrumentation 16, 145–155.

Ismail, I., Shafquet, A., Karsiti, M.N., 2011. Void fraction estimation by using electrical capacitance tomography and differential pressure in an air-water co-current bubble column. Australian Journal of Basic and Applied Sciences 5 (11), 1533–1541.

ITDB, 2016. http://www-ns.iaea.org/security/itdb.asp.

Ito, D., Saito, Y., 2015. Natural convection heat transfer characteristics of KUR fuel assembly during loss of coolant accident. In: 16th International Topical Meeting on Nuclear Reactor Thermal Hydraulics (NURETH-16), August 30–September 4, Chicago, IL, USA.

Ito, D., Papadopoulos, P., Prasser, H.-M., 2016. Liquid film dynamics of two-phase annular flow in square and tight lattice sub-channels. Nuclear Engineering and Design 300, pp467–474.

Ivanov, K.N., Baratta, A., 1999a. Coupling methodologies for best estimate safety analysis. In: International Conference On Mathematics, Computation, Reactor Physics and Environmental Analysis (M & C), Madrid, Spain.

Ivanov, K.N., Beam, T.M., Baratta, A.J., Irani, A., Trikouros, N., 1999b. Pressurized Water Reactor Main Steam Line Break (MSLB) Benchmark—Final Specifications, NEA/NSC/ DOC (99)8, Paris, France.

Ivanovsky, M.N., Milovanov, Y.V., Subbotin, V.I., 1966. About the mechanism of dropwise condensation of metal vapours. Atomnaya Energiya 21 (1), 17–21 (in Russian).

Ivanovsky, M.N., Sorokin, V.P., Subbotin, V.I., 1976. Evaporation and Condensation of Metals. Atomizdat, Moscow (in Russian).

Iwamura, T., Iguchi, T., Adachi, H., Murao, Y., 1989. Quantitative evaluation of the heat transfer enhancement due to radial power distribution during reflood phase of a PWR-LOCA. Using SCTF. Journal of Nuclear Science and Technology 26, 428–440.

Izenson, M.G., Rothe, P.H., Wallis, G.B., 1988. Diagnosis of Condensation-Induced Waterhammer—Methods and Background, pp. 1–162. USNRC, NUREG/CR—5220-Vol.1, Washington, DC, USA.

Jacobson, S., 1992. Risk Evaluation of Local Dilution Transients in a Pressurized Water Reactor. Division of Energy Systems, Department of Mechanical Engineering, Dissertation No. 275, Linkoping, Sweden.

Jaeger, W., Perez-Manes, J., Imke, U., Jimenez-Escalante, J., Sanchez-Espinoza, V.H., 2013. Validation and comparison of two-phase flow modeling capabilities of CFD, sub channel and system codes by means of post-test calculations of BFBT transient tests. Nuclear Engineering and Design 263, 313–326.

Jafari, J., D'Auria, F., Kazeminejad, H., Davilu, H., 2003. Reliability evaluation of a natural circulation system. Nuclear Engineering and Design 224, 79–104.

Jagota, A.K., Rhodes, E., Scott, D.S., 1973. Tracer measurements in two-phase annular flow to obtain interchange and entrainment. The Canadian Journal of Chemical Engineering 51 (2), 139–148.

Jakob, M., 1949. Heat Transfer. John Wiley & Sons Inc., New York, NY, USA.

Jang, C., 2007. Treatment of the thermal-hydraulic uncertainties in the pressurized thermal shock analysis. Nuclear Engineering and Design 237.

Janicot, A., Bestion, D., 1993. Condensation modelling for ECC injection. Nuclear Engineering and Design 145, 37–45.

Janssen, E., 1971. Two-phase Flow and Heat Transfer in Multi-Rod Geometries-Final Report. General Electric, GEAP-10347, San Jose, Ca, USA.

Jayanti, S., Reddy, K.R., 2013. Effect of spacer grids on CHF in nuclear rod bundles. Nuclear Engineering and Design 261, 66–75.

Jayanti, S., Valette, M., 2005. Calculation of dry out and post-dry out heat transfer in rod bundles using a three field model. International Journal of Heat Mass Transfer 48, 1825–1839.

Jens, W.H., Lottes, P.A., 1951. Analysis of Heat Transfer, Burnout, Pressure Drop and Density Data for High-Pressure Water. Argonne National Lab., Chicago, IL, USA. ANL-4627.

Jeong, J.J., Sim, S.K., Ban, C.H., Park, C.E., 1997. Assessment of the COBRA/RELAP5 code using the LOFT L2-3 large-break loss-of-coolant experiment. Annals of Nuclear Energy 24 (14), 1171–1182.

Jeong, J.J., Ha, K.S., Chung, B.D., Lee, W.J., 1999. Development of a multi-dimensional thermal-hydraulic system code, MARS 1.3.1. Annals of Nuclear Energy 26 (18), 1611–1642.

Jeong, J.J., Hwang, M., Lee, Y.J., Chung, B.D., 2004. Non-uniform flow distribution in the steam generator U-tubes of A pressurized water reactor plant during single- and two-phase natural circulations. Nuclear Engineering and Design 231 (3), 303–314.

Jeong, J.J., Bae, S.W., Hwang, D.H., Lee, W.J., Chung, B.D., 2005. Hot Channel analysis capability of the best-estimate multi-dimensional system code, MARS 3.0. Nuclear Engineering and Technology 37 (5), 469–478.

Jeong, J.J., Lee, W.J., Chung, B.D., 2006. Simulation of a main steam line break accident using a coupled system thermal-hydraulics three dimensional reactor kinetics and hot channel analysis code. Annals of Nuclear Energy 33, 820–828.

Jeong, J.J., Ozar, B., Dixit, A., Julia, J.E., Hibiki, T., Ishii, M., 2008. Interfacial area transport of vertical upward air-water two-phase flow in an annulus channel. International Journal of Heat Fluid Flow 29, 178–193.

Jeong, J.J., Yon, H.Y., Park, I.K., Cho, H.K., 2010. The CUPID code development and assessment strategy. Nuclear Engineering and Technology 42 (6), 636–655.

Jepson, D.M., Azzopardi, B.J., Whalley, P.B., 1989. The effect of gas properties on drops in annular flow. International Journal of Multiphase Flow 15 (3), 327–339.

Jimenez, J., Sanchez, V., 2015. Final Report on BWR thermal-hydraulics at System Scale and Sub-channel Scale. NURESAFE, D33.12.5, EC Project, Brussels, Belgium.

Jimenez, G., Queral, C., Rebollo-Mena, M.J., Martınez-Murillo, J.C., Lopez-Alonso, E., 2013. Analysis of the operator action and the single failure criteria in a SGTR sequence using best estimate assumptions with TRACE 5.0. Annals of Nuclear Energy 58.

Jimenez, G., Herrero, J.J., Gommlich, A., Kliem, S., Cuervo, D., Jimenez, J., 2015. Boron dilution transient simulation analyses in a PWR with neutronics/ thermal-hydraulics coupled codes in the NURISP project. Annals of Nuclear Energy 84.

Jimenez, G., Serrano, C., Lopez-Alonso, E., del Carmen Molina, M., Calvo, D., García, J., Queral, C., Vicente Zuriaga, J., González, M., 2015a. BWR Mark III containment analyses using a Gothic 8.0 3D model. Annals of Nuclear Energy 85, 687–703.

Jin, H., Lian, Y., Yang, S., He, G., Guo, Z., 2013. The parameters measurement of air–water two phase flow using the electrical resistance tomography (ERT) technique in a bubble column. Flow Measurement and Instrumentation 31, 55–60.

Jin, Y., Cheung, F.-B., Bajorek, S.M., Tien, K., Hoxie, C.L., 2019. Development of a two-phase flow mass quality correlation in the post-dryout DFFB regime during reflood transients. International Journal of of Heat and Mass Transfer 137, 1076–1087.

Jin, Y., Cheung, F.-B., Shirvan, K., Bajorek, S.M., Tien, K., Hoxie, C.L., 2020. Development of a new spacer grid pressure drop model in rod bundle for the post-dryout two-phase flow regime during reflood transients. Nuclear Engineering and Design 368 (110815), 1–12.

Jin, Y., Garrett, G.R., Cheung, F.-B., Bajorek, S.M., Tien, K., Hoxie, C.L., 2022. Validation of the sub-channel analysis code COBRA-TF based on the NRC/PSU RBHT benchmark data. In: 19th International Topical Meeting on Nuclear Reactor Thermal Hydraulics (NURETH-19), March 6–11 (Via Web), Brussels, Belgium.

Jo, Y., Song, Y.J., Jo, D., 2022. Gas accumulation induced by temperature rise and sudden pressure drop in an inverse U-bend. Annals of Nuclear Energy 169 (108928), 1–19.

Johnson, R.W., 2008. Modeling strategies for unsteady turbulent flows in the lower plenum of the VHTR. Nuclear Engineering and Design 238, 482–491.

Johnson, M.K., Farroll, S., 1995. Development of a turbine meter for two-phase flow measurement in vertical pipes. Flow Measurement and Instrumentation 6 (4), 279–282.

Johnson, S.O., Curlee, N.J., Reihing, J.V., 1958. Simulation of hot channel boiling in water-cooled reactors. IRE Transactions on Nuclear Science 5 (1), 1–9.

Johnson, R.W., Guillen, D.P., Gallaway, T., 2006. Processes and procedures for application of CFD to nuclear reactor safety analysis. In: INL/EXT-06-11789, Idaho Falls, ID, USA, September.

Jones, O.C., 1980. Flashing Inception in Flowing Liquids. Brookhaven National Laboratory Report BNL-NUREG-51221, USNRC, Washington, DC, USA (Also NUREG/CR-1515, USA).

Jones, O.C., Delhaye, J.M., 1976. Transient and statistical measurement techniques for two-phase flows: a critical review. International Journal of Multiphase Flow 3, 89−116.

Jones, W., Launder, B., 1972. The prediction of laminarization with a two-equation model of turbulence. International Journal of Heat Mass Transfer 15, 301−314.

Joshi, J.B., Nayak, A.K. [E.s], 2018. Advances of Computational Fluid Dynamics in Nuclear Reactor Design and Safety Assessment. Elsevier, Woodhead Publishing, Duxford, UK, pp. 1−850.

Joyce, M., 2018. Nuclear Engineering, A Conceptual Introduction to Nuclear Power. Chapter 6, pp. 111−127, Copyright 2018 @ Elsevier Ltd,. https://doi.org/10.1016/B978-0-08-100962-8.00006-8.

Ju, S.H., No, H.C., Mayinger, F., 2000. Measurement of heat transfer coefficients for direct contact condensation in core makeup tanks using holographic interferometer. Nuclear Engineering and Design 199, 75−83.

Judd, R.A., Tahir, A., Carver, M.B., Kiteley, J.C., Rowe, D.S., Stewart, D.G., Thibeault, P.R., 1984. ASSERT-4 User's Manual. Atomic Energy of Canada, AECL-8573, Chalk River, Ontario, Canada, pp. 1−93.

Juhel, D., 1984. A Study of Interfacial and wall Heat Transfer Downstream of a Quench Front. European Two-Phase Flow Group Meeting, Rome, Italy.

Julia, J.E., Hibiki, T., Ishii, M., Yun, B.J., Park, G.C., 2009. Drift-flux model in a sub-channel of rod bundle geometry. International Journal of Heat Mass Transfer 52 (13−14), 3032−3041.

Jung, S., Kim, H., 2013. Simultaneous measurements of liquid-vapour phase and temperature distributions on boiling surface with synchronized infrared thermometry and total internal reflection techniques. In: 15th International Topical Meeting on Nuclear Reactor Thermal Hydraulics (NURETH-15), Pisa, Italy.

Kaganer, M.G., Zukova, R.I., 1966. Determination of thermal resistance in vacuum for contacts between metallic surfaces with various extent of roughness. Inzhenerno-Phyzichesky Zhurnal 11 (3), 329−337 (in Russian).

Kaiser, A., 1979. Two-phase pressure drop of boiling sodium in rod-bundle geometries; comparison of experimental results with Lockhart-Martinelli correlation. Process Energy Energy Systems 2, 583−596.

Kaizer, J.S., Heller, A.K., Oberkampf, W.L., 2015. Scientific computer simulation review. Journal of Reliability Engineering and System Safety 138, 210−218.

Kaizer, J.S., Anzalone, R., Brown, E., Panicker, M., Haider, S., Gilmer, J., Drzewiecki, T., Attard, A., 2019. Credibility Assessment Framework for Critical Boiling Transition Models. A Generic Safety Case to Determine the Credibility of Critical Heat Flux and Critical Power Models, pp. 1−121. USNRC NUREG/KM-0013, Washington, DC, USA.

Kaliatka, A., Vaisnoras, M., Valincius, M., Kaliatka, T., 2022. RELAP5 application for water hammer analysis in different thermal-hydraulic systems. International Journal of of Thermofluids 14 (100154), 1−14.

Kalinin, E.K., Dreitser, G.A., Yarcho, S.A., 1986. Heat Transfer and Hydrodynamics in the Channels of Complex Configuration. Mashinostroenie, Moscow, Russia (in Russian).

Kalinin, E.K., Dreitser, G.A., Yarcho, S.A., 1990. Heat Transfer Intensification in the Channels, third ed. Mashinostroenie, Moscow, Russia.

Kalli, H., Miettinen, A., Purhonen, H., D'Auria, F., Frogheri, M., Leonardi, M., 1995. Quantitative code accuracy evaluation of ISP 33. In: 7th International Topical Meeting on Nuclear Reactor Thermal Hydraulics (NURETH-7).

Kaltenbach, C., Laurien, E., 2018. CFD simulation of spray cooling in the model containment Thai. Nuclear Engineering and Design 328, 359−371.

Kamenchshikov, F.T., Reshetov, B.A., Ryabov, A.N., et al., 1984. Mechanical Problems of Rotating Flows and Heat Transfer Intensification in Nuclear Power. Energoatomizdat, Moscow, Russia (in Russian).

Kandlikar, S.G., 1990. A general correlation of saturation two-phase flow boiling heat transfer inside horizontal and vertical tubes. Journal of Heat Transfer 112, 219−238.

Kandlikar, S.G., 1991. Development of a flow boiling map for subcooled and saturated flow boiling of different fluids inside circular tubes. Journal of Heat Transfer 113, 190−200.

Kandlikar, S.G., 1998. Heat transfer characteristics in partial boiling, fully development boiling and significant void flow regions of subcooled flow boiling. Journal of Heat Transfer 120, 395−401.

Kang, S.K., Hassan, Y.A., 2016. Computational fluid dynamics (CFD) round robin benchmark for a pressurized water reactor (PWR) rod bundle. Nuclear Engineering and Design 301, 204−231.

Kanno, H., Shikazono, N., 2015. Experimental and modeling study on adiabatic two-phase expansion in a cylinder. International Journal of of Heat and Mass Transfer 86, 755−763.

Kano, K., Sadatomi, M., Kawahara, A., Kuno, T., 2003. Flow distribution and slug flow characteristics for hydraulically equilibrium air-water two-phase flows in a vertical 2 × 3 rod channel. In: 4th ASME-JSME Joint Fluids Engineering Conference, July 6−10, Honolulu, Hawaii, USA.

Kanzleiter, T.F., 1980. Investigation of the Phenomena Occurring within a Multi-Compartment Containment after Rupture of the Primary Cooling Circuit in Water-Cooled Reactors. AbschluBbericht BF- RS-50-01, Battelle Frankfurt, Germany.

Kapila, A.K., Menikoff, R., Bdzil, J.B., Son, S.F., Stewart, D.S., 2001. Two-phase modeling of deflagration-to-detonation transition in granular materials: reduced equations. Physics of Fluids 13, 3002−3024.

Karapantios, T.D., Paras, S.V., Karabelas, A.J., 1989. Statistical characteristics of free falling films at high Reynolds numbers. International Journal of Multiphase Flow 15, 1−21.

Karoutas, Z.E., Sung, Y., Chang, Y.R., Kogan, G.A., Joffre, P., 2002. Subcooled Boiling Data from Rod Bundles. EPRI, 1003383, Palo Alto, CA, USA, pp. 1−130.

Karplus, H.B., 1961. Propagation of Pressure Waves in a Mixture of Water and Steam. ARF of Illinois Institute of Technology, 4132-12, (United States Atomic Energy Commission Contract No. AT(ll-l)-528), Chicago, Ill, USA, pp. 1−92.

Karwat, H., 1985. Principal Characteristics of Experimental Simulators Suitable for SBLOCA Events of LWRs and Scaling Principles Adopted in Their Design. Invited at Specialist Meeting on Small Break LOCA Analyses in LWRs, Pisa, Italy).

Karwat, H., 1985a. Containments—design principles and background. Journal of Nuclear Engineering and Design 90.

Karwat, H., 1986. Problems of scaling and extrapolation of experimental results in the area of fluid dynamics and associated heat transfer related to reactor safety. European Applied Research Reports, Nuclear Science and Technology 7, 2.

Karwat, H., 1992. The Assessment of Containment Codes by Experiments Simulating Severe Accident Scenarios. Final Report, Nuclear Science and Technology, Commission of the European Communities, EUR 14159 EN Bruxelles, Belgium.

Karwat, H., Austregesilo, F.H., 1985. Scaling Effects Related to the Analysis and Interpretation of Small Break Tests. Specialist Meeting on Small Break LOCA Analyses in LWRs, Pisa, Italy.

Kasagi, N., Nishino, K., 1991. Probing turbulence with three-dimensional particle-tracking velocimetry. Experimental Thermal and Fluid Science 4 (5), 601–612.

Kasahara, F., Ninokata, H., 2000. The multi-fluid multi-phase subchannel analysis code KAMUI for subassembly accident analysis of an LMFR. Journal of Nuclear Science and Technology 37 (8), 654–669.

Kastanya, D., Boyle, S., Hopwood, J., Park, J.-H., 2013. The impact of power coefficient of reactivity on CANDU-6 reactors. Journal of Nuclear Energy and Technology 45 (5), 573–581.

Kataoka, I., Ishii, M., 1982. Mechanism and Correlation of the Droplet Entrainment and Deposition in Annular Two-phase Flow. NUREG:CR-2885, ANL-82-44, Washington DC, USA.

Kataoka, I., Ishii, M., 1987. Drift flux model for large diameter pipe and new correlation for pool void fraction. International Journal of Heat Mass Transfer 30, 1927–1939.

Kataoka, I., Ishii, M., Serizawa, A., 1986. Local formulation and measurements of interfacial area concentration in two-phase flow. International Journal of Multiphase Flow 12 (4), 505–529.

Katono, K., Ishida, N., Sumikawa, T., Yasuda, K., 2014. Air–water downscaled experiments and three-dimensional two-phase flow simulations of improved steam separator for boiling water reactor. Journal of Nuclear Engineering and Design 278, 465–471.

Katto, Y., 1994. Critical heat flux. International J Multiphase Flow 20, 53–90.

Kaviany, M., 1999. Principles of Heat Transfer in Porous Media, second ed. Springer, New York, NY, USA.

Kaviany, M., 2002. Principles of Heat Transfer. John Willey & Sons, Inc., New York, NY, USA.

Kawaji, M., Anoda, A., Nakamura, H., Tasaka, K., 1987. Phase and velocity distributions and holdup in high-pressure steam/water stratified flow in a large diameter horizontal pipe. International Journal of Multiphase Flow 13 (2), 145–159.

Kays, W.M., 1950. Loss Coefficients for Abrupt Changes in Flow Cross Section with Low Reynolds Number Flow in Single and Multiple Tube Systems, vol. 72. Transactions of the ASME, Chicago, Ill, USA, pp. 1067–1074.

Kays, W.M., 1955. Numerical solution for laminar flow heat transfer in circular tubes. Transactions of the American Society of Mechanical Engineers 77, 1265–1274.

Kays, W.M., 1975. Convective Heat and Mass Transfer. Tata-McGraw Hill Publishing Company Ltd., New Delhi, India.

Keating, A., Piomelli, U., 2006. A dynamic stochastic forcing method as a wall-layer model for large-eddy simulation. Journal of Turbulence 7 (12).

Keim, E., Huber, N., Kastner, W., 1999. Crack opening and leakage rates of real cracks. In: 7th International Topical Meeting on Nuclear Reactor Thermalhydraulics, NURETH-7, Tokyo, Japan.

Kelly, J.E., Kao, S.P., Kazimi, M.S., 1981. THERMIT-2: A Two-Fluid Model for Light Water Reactor Sub-channel Transient Analysis. Massachusetts Institute of Technology. MIT-EL-81-014, Boston, Ma, USA, pp. 1–317.

Kelly, J.M., Stewart, C.W., Cuta, J.M., 1992. VIPRE-02: two-fluid thermal-hydraulics code for reactor core and vessel analysis: mathematical modeling and solution methods. Nuclear Technology 100 (2), 246–259.

Kennedy, M.D., Peters, F.E., Carelli, M.D., Pieczynski, A.T., 1994. Advanced PWR passive containment cooling system testing. In: Proceedings ARS'94 International Topical Meeting on Advanced Reactor Safety, vol. 1 (Pittsburgh, Pa, USA).

Kerney, P.J., Faeth, G.M., Olson, D.R., 1972. Penetration characteristics of a submerged steam jet. AIChE Journal 18 (3), 548–553.

Kestin, J., 1991. Spurious solutions in critical two-phase flow numerical codes. In: International Conference on Multiphase Flow, Tsukuba, Japan.

Khadamakar, H.P., Patwardhan, A.W., Padmakumar, G., Vaidyanathan, G., 2011. Flow distribution in the inlet plenum of steam generator. Nuclear Engineering and Design 241.

Khan, S.A., 1990. The Chernobyl source term: a critical review. Nuclear Safety 31 (3), 353–374.

Khan, I., Wang, M., Zhang, Y., Tian, W., Su, G., Qiu, S., 2020. Two-phase bubbly flow simulation using CFD method: a review of models for interfacial forces. Progress in Nuclear Energy 125 (103360), 1–17.

Kharitonov, V.V., Kokorev, L.S., Del'vin, N.N., 1973. About the role accommodation coefficient at contact heat transfer. Atomnaya Energiya 35 (5), 360–361 (in Russian).

Kieffer, S.W., 1977. Sound speed in liquid-gas mixture: water-air and water-steam. Journal of Geophysical Research 82 (20), 2895–2904.

Kim, J., 2021. Thermal hydraulic modelling of grating effect for application to 3-dimensional analysis of hydrogen behavior in NPP containment. Nuclear Engineering and Design 380 (111291), 1–11.

Kim, J., Bang, Y.-S., 2017. Development of RCS-containment coupled analysis model and evaluation of LBLOCA for APR-1400. In: 26th International Conference New Energy for New Europe (NENE), Bled, Slovenia, September.

Kim, Y.-S., Choi, K.-Y., 2014. An analytical investigation of loop seal clearings for the SBLOCA tests. Annals of Nuclear Energy 68.

Kim, B.J., Kim, K.-D., 2015. Rayleigh-taylor instability of viscous flow with heat and mass transfer. In: 16th International Topical Meeting on Nuclear Reactor Thermal Hydraulics (NURETH-16), August 30–September 4, Chicago, IL, USA.

Kim, S., Kong, R., 2020. Separate-effect experiments and modeling for two-phase flow under geometric restrictions. Nuclear Engineering and Design 367 (110786), 1–14.

Kim, S., Fu, X.Y., Wang, X., Ishii, M., 2000. Development of the miniaturized four-sensor probe and the signal processing scheme. International Journal of Heat Mass Transfer 43, 4101–4118.

Kim, W.S., Kim, Y.G., Kim, J.K., 2002. A subchannel analysis code MATRA-LMR for wire wrapped LMR subassembly. Annals of Nuclear Energy 29 (3), 303–321.

Kim, S., Park, J.H., Kojasoy, G., Kelly, J.M., Marshall, S.O., 2007. Geometric effects of 90-degree Elbow in the development of interfacial structures in horizontal bubbly flow. Nuclear Engineering and Design 237, 2105–2113.

Kim, K.D., Lee, S.W., Bae, S.W., Mon, S.-K., Kim, S.H., 2009. Development of physical models and correlation packages for the SPACE code. In: The 13th International Topical Meeting on Nuclear Reactor Thermal Hydraulics (NURETH-13), Kanazawa City, Ishikawa Prefecture, Japan, September 27–October 2009.

Kim, S., Kojasoy, G., Guo, T., 2010. Two-phase minor loss in horizontal bubbly flow with elbows: 45° and 90° elbows. Nuclear Engineering and Design 240, 284–289.

Kim, B.J., Kim, J., Kim, K.D., 2014. On the wall drag term in the averaged momentum equation for dispersed flows. Nuclear Science and Engineering 178, 225–239.

Kim, K.K., Lee, W.L., Choi, S., Kim, H.R., Ha, J., 2014. SMART: the first licensed advanced integral reactor. Journal of Energy and Power Engineering 8, 94–102.

Kim, B.J., Lee, S.W., Kim, K.-D., 2015. New wall drag and form loss models for one-dimensional dispersed two-phase flow. Nuclear Engineering and Technology 47, 416–423.

Kim, J., Jung, E., Kang, S., 2015. Large eddy simulation of hydrogen dispersion from leakage in a nuclear containment model. International Journal of of Hydrogen Energy 40, 11762–11770.

Kim, K., Kihwan), Kim, B.-J., Choi, H.-S., Mon, S.-K., Song, C.-H., 2015. Effect of a blockage length on the coolability during reflood in a 2x2 rod bundle with a 90% partially blocked region. In: 16th International Topical Meeting on Nuclear Reactor Thermal Hydraulics (NURETH-16), August 30–September 4, Chicago, IL, USA.

Kim, S., Jeon, B.-G., Youn, Y.-J., Choi, H.-S., Euh, D.-J., 2016. Test facility construction for flow visualization on mixing flow inside subchannels of PWR rod bundle. In: Proceedings of the KNS 2016 Autumn Meeting, 26–28 October, Kyungju, Korea.

Kim, S.I., Park, J.H., Ha, K.S., Cho, S.-W., Song, J.H., 2016. Analysis of Fukushima unit 2 accident considering the operating conditions of RCIC system. Nuclear Engineering and Design 298, 183–191.

Kim, D.Y., No, H.C., Yon, H.J., Lim, S.J., 2018. Thermal hydraulic study of air cooled passive decay heat removal system for APR+ under extended station blackout. In: 12th International Topical Meeting on Nuclear Reactor Thermal-Hydraulics, Operation and Safety (NUTHOS-12), October 14–18, Qingdao, China.

Kim, J., Cho, S., Park, J.-K., Youn, Y.-J., Mon, S.-K., 2018. Experimental study to assess effects of balloning and fuel relocation on the coolability of fuel rod bundle. Nuclear Engineering and Design 332, 1–10.

Kim, S., (Kim, -S.), Jeon, B.G., Choi, H.S., Euh, D.J., Mon, S.K., 2020. Experimental visualization of flow structure inside subchannels of a 4 X 6 rod-bundle. Annals of Nuclear Energy 140, 107097.

Kim, K., Lee, J.B., Kim, W.-S., Choi, H.-S., Kim, J.I., 2021. Development of a prediction model relating the two-phase pressure drop in a moisture separator using an air/water test facility. Nuclear Engineering and Technology 53, 3892–3901.

Kim, J., (Jongrok), Lee, J.B., Kim, K., (Kihwan), Byeon, H., Park, J.-K., Youn, Y.-J., Choi, H.S., Seol, H., Mon, S.-K., 2022. Comparison of quenching characteristics under different reflood conditions using a thermo-mechanics and thermal hydraulics coupled experimental facility. In: 19th International Topical Meeting on Nuclear Reactor Thermal Hydraulics (NURETH-19), March 6–11 (Via Web), Brussels, Belgium.

Kim, M.G., Wisudhaputra, A., Lee, J.-H., Kim, K.-D., Park, H.-S., Jeong, J.J., 2022. Development of a special thermal-hydraulic component model for the core makeup tank. Nuclear Engineering and Technology 54, 1890–1901.

Kimber, G., Leveque, C., Lavialle, G., 1998. ISP-38 on BETHSY Test 6.9C Final Report. NEA/CSNI/R(97)38, Vol. 1 and 2, Paris, France.

King, L.V., 1914. On the convection of heat from small cylinders in a stream of fluid: determination of the convection constants of small platinum wires with applications to hot wire anemometry. Phil. Trans. R. Soc., A 214 (509–522), 373–433.

Kinoshita, I., Murase, M., Tomiyama, A., 2012. Numerical simulation of size effects on countercurrent flow limitation in PWR hot leg models. Journal of Science and Technology of Nuclear Installations. Article ID 907364.

Kirillov, P.L. (Ed.), 2008. Thermophysical Properties of Materials for Nuclear Engineering: A Tutorial and Collection of Data. IAEA, Vienna, Austria.

Kirillov, P.L., Bogoslovskaya, G.P., 2000. Heat and Mass Transfer in Nuclear Power Plants, 2nd rev. ed. Energoatomizdat, Moscow, Russia (in Russian).

Kirillov, P.L., Yushenko, S.S., 1996. Diameter effect on CHF. In: 2nd Research Coordination Meeting. IAEA CRP on Thermal Hydraulic Relationships for Advanced Water-Cooled Reactors, Vienna, Austria.

Kirillov, P.L., Bobkov, V.P., Smogalev, I.P., Vinogradov, V.N., 1996. Prediction of Critical Heat Flux in Channels Relevant to Water Cooled Reactors. IPPE (Obninsk) Contract 8219R1. IAEA, Vienna, Austria.

Handbook on Thermohydraulic Calculations in Nuclear Power Engineering. Hydrodynamic Calculations and Heat Transfer. In: Kirillov, P.L., Bobkov, V.P., Zhukov, A.V., Yur'ev, Y.S. (Eds.), 2010, vol. 1. IzdAt Publishing House, Moscow, Russia (In Russian).

Handbook on Thermohydraulic Calculations in Nuclear Power Engineering. Nuclear Reactors, Structures and Core (Calculations), Heat Exchangers, Steam Generators, Heat Pipes. In: Kirillov, P.L., Zhukov, A.V., Loginov, N.I., Makhin, V.M., Pioro, I.L., Yur'ev, Y.S. (Eds.), 2013, vol. 2. IzdAt Publishing House, Moscow, Russia (In Russian).

Kirillov, P.L., Yur'ev, Y.S., D'Auria, F., Migrov, Y.A., Soloviev, S., 2014. Handbook on Thermohydraulic Calculations in Nuclear Power Engineering. Thermohydraulic Processes under Transient and Irregular Regimes, vol. 3. IzdAt Publishing House, Moscow, Russia, pp. 1−688 (In Russian).

Kiselev, A.N., Checherov, K.P., 2001. The model of destruction process of the unit 4 the Chernobyl power plant. Atomnaya Energiya 91 (6), 425−434 (in Russian).

Kishor, G.G., Prasad, G.V.D., Pandey, M., 2007. Parametric studies of primary heat transport loop of natural circulation boiling water reactor. In: 12th International Topical Meeting on Nuclear Reactor Thermal Hydraulics (NURETH-12), September 30−October 4, Pittsburgh, Pa, USA.

Kitzel', V.A., 1973. Reflection of Light. Nauka, Moscow, Russia (in Russian).

Klausner, J.F., Mei, R., Bernhard, D.M., Zeng, L.Z., 1993. Vapor bubble departure in forced convection boiling. International Journal of Heat Mass Transfer 36, 651−662.

Klein-Heßling, W., Arndt, S., Zemitis, A., Iliev, O., Steiner, K., Freitag, M., Colombet, M., 2018. Validation and application of a coupled CFD/LP code system for containment simulations. In: 12th International Topical Meeting on Nuclear Reactor Thermal-Hydraulics, Operation and Safety (NUTHOS-12), October 14−18, Qingdao, China.

Kliem, S., Höhne, T., Rohde, U., Weiss, F.-P., 1999. Main steam line break analysis of a VVER-440 reactor using the coupled thermohydraulics system/3D-neutron kinetics code DYN3D/ATHLET in combination with the CFD code CFX-4. In: Proceedings 9th International Topical Meeting on Nuclear Reactor Thermal Hydraulics (NURETH-9), San Francisco, CA, USA, October 3−8.

Kliem, S., Hemström, B., Bezrukov, Y., Höhne, T., Rohde, U., 2007. Comparative evaluation of coolant mixing experiments at the ROCOM, vattenfall, and gidropress test facilities. Science and Technology of Nuclear Installations, (J STNI).

Kliem, S., Mittag, S., Rohde, U., Weiss, F.-P., 2009. ATWS analysis for PWR using the coupled code system DYN3D/ATHLET. Annals of Nuclear Energy 36.

Klimenko, V.V., 1984. Processes of Two-phase Heat Transfer in Cryogenic Liquids (Pool Boiling, Forced Flow, Film Pool Boiling) and the Development of Optimal Methods of Their Calculation. Abstract of Doctor thesis, Moscow, Russia (in Russian).

Klimenko, V.V., Snytin, S.Y., 1988. Film boiling crisis on a submerged heating surface. In: Proceedings of the 1st World Conf. Exp. Heat Transfer, Fluid Mech., Thermod. Dubrovnik, Croatia, September 4–9, pp. 551–563.

3rd rev.Klimenko, A.V., Zorin, V.M. (Eds.), 2001, Theoretical Fundamentals of Heat Engineering, Thermothechnical Experiment Izdatel'stvo MEI, Moscow, Russia. 2nd Book (in Russian).

Kloos, M., 2014. Statistischer Nachweis bei einem LOCA, (in German) GRS-Handout, AST-AG LOCA, BMUB, Bonn, Germany, 23 July.

Kloos, M., Hofer, E., 1999. SUSA—PC, A Personal Computer version of the program system for uncertainty and sensitivity analysis of results from computer models, Version 3.2, User's Guide and Tutorial, Gesellschaft für Anlagen- und Reaktorsicherheit mbH (GRS), Garching, Germany.

Knief, R.A., 2008. Nuclear Engineering Theory and Technology of Commercial Nuclear Power (Reprint of 1992, second ed). ANS, Hinsdale, IL, USA.

Kniesner, B., Jester-Zürker, R., Jakirlic, S., Hanjalic, K., 2007. RANS–SMC and hybrid LES/RANS modelling of a backward-facing step flow subjected to increasingly enhanced wall heating. In: 5th International Symposium on Turbulence and Shear Flow Phenomena, Garching, Germany, 27–29 August.

Knoll, G.F., 2010. Radiation Detection and Measurement. John Wiley & Sons Inc., New York, NY, USA.

Knoll, D.A., Keyes, D.E., 2004. Jacobian-free Newton-Krylov methods: a survey of approaches and applications. Journal of Computational Physics 193, 357–397.

Knoll, D., Morel, J., Margolin, L., Shashkov, M., 2005. Physically motivated discretization methods A strategy for increased predictiveness. Los Alamos Science 29, 188–212 (Los Alamos, NM, USA).

Knudsen, M., 1915. Maximum rate of vaporization of mercury. Annals of Physics (Leipzig) 47, 697.

Kocamustafaogullari, G., Ishii, M., 1995. Foundation of the interfacial area transport equation and its closure relations. Journal of Heat Mass Transfer 38 (3), 481–993.

Kock, F., Herwig, H., 2005. Entropy production calculation for turbulent shear flows and their implementation in CFD codes. International Journal of of Heat and Fluid Flow 26 (4), 672–680.

Koizumi, Y., Nakamura, H., Tasaka, K., Findlay, J.A., Lee, L.S., Sutherland, W.A., 1987. BWR small break LOCA counterpart tests at ROSA-III and FIST test facilities. Journal of Nuclear Engineering and Design 102.

Kok, J.C., 2000. Resolving the dependence on freestream values of the k-ω turbulence model. AIAA Journal 38, 1292–1295.

Kok, H.V., Van Der Hagen, T.H.J.J., Mudde, R.F., 1997. Measurements of the void-fraction distribution in a simulated fuel assembly and the role of the void-fraction on the dynamics of a natural circulation loop. Annals of Nuclear Energy 24 (16).

Kokorev, B.V., Farafonov, V.A., 1990. Steam Generators of Liquid Metal Cooled Nuclear Power Plants. Energoatomizdat, Moscow, Russia (in Russian).

Kokorev, L.C., Kharitonov, V.V., 1986. Thermohydraulic Design and Optimization of Nuclear Power Plants. Energoatomizdat, Moscow, Russia (in Russian).

Kolev, N.I., 2006. Uniqueness of the elementary physics driving heterogeneous nucleate boiling and flashing. Nuclear Engineering and Technology 38 (2).

Kolev, N.I., 2011–2015. Multiphase Flow Dynamics, Vols. I to V, (various Editions) Springer, Berlin, Germany.

Komen, E.M.J., Mathur, A., Roelofs, F., Tiselj, I., 2022. Status, perspectives, and added value of high fidelity simulations for safety and design. In: 19th International Topical Meeting on Nuclear Reactor Thermal Hydraulics (NURETH-19), March 6–11 (Via Web), Brussels, Belgium.

Koncar, B., Morel, C., Mimouni, S., Vyskocil, L., Galassi, M.C., 2011. CFD modeling of boiling bubbly flow for DNB investigations. Multiphase Science and Technology 23 (2–4).

Kondrat'ev, G.M., 1954. Regular Heat Regime. Gostekhizdat, Moscow, Russia (in Russian).

Kong, R., Kim, S., Bajorek, S.M., Tien, K., Hoxie, C., 2018. Effects of pipe size on horizontal two-phase flow: flow regimes, pressure drop, two-phase flow parameters, and drift-flux analysis. Experimental Thermal and Fluid Science 96, 75–89.

Korhonen, R., Hietanen, O., 1994. Erosion corrosion on the parallel feed water discharge lines at the Loviisa VVER 440. In: Specialists Meeting on Erosion and Corrosion of NPP Materials, Kiev, Ukraine.

Koskie, J.E., Mudawar, I., Tiederman, W.G., 1989. Parallel wire probes for measurements of thick liquid films. International Journal of Multiphase Flow 15, 521–530.

Kossolapov, A., Phillips, B., Bucci, M., 2022. Direct visualization and quantification of dry area formation leading to the boiling crisis in subcooled flow boiling of water at prototypical pressures of light water reactors. In: 19th International Topical Meeting on Nuclear Reactor Thermal Hydraulics (NURETH-19), March 6–11 (Via Web), Brussels, Belgium.

Kothe, D.B., 2011. CASL: consortium for advanced simulation of light water reactors. In: 2nd DOE/NRC/NEI Workshop on U.S. Nuclear Power Plant Life Extension R & D, Washington, DC, USA, February 22–24.

Kothe, D.B., 2015. CASL: Consortium for Advanced Simulation of Light Water Reactors Various Presentations. ORNL, Oak Ridge, TN, USA.

Kothe, D.B., Szilard, R., Turinsky, P., 2010. CASL: The Consortium for Advanced Simulation of Light Water Reactors. US Department of Energy, Oak Ridge, TN, USA.

Kottowski-Dümenil, H.M., 1994. Liquid Metal thermal Hydraulics. Inforum Verlag, Bonn, Germany, ISBN 3926956224, pp. 1–397.

Kouhia, V., Purhonen, H., Riikonen, V., Puustinen, M., Kyrki-Rajamaki, R., Vihavainen, J., 2012. PACTEL and PWR PACTEL test facilities for versatile LWR applications. Journal of Science and Technology of Nuclear Installations 548513.

Kovtonyuk, A., D'Auria, F., Petruzzi, A., 2008. Proposal of integration between CIAU and GRS method for ATHLET code. In: Presentation at a Meeting Held at GRS Headquarter, Garching, Germany, April 3.

Kovtonyuk, A., Fiori, F., Petruzzi, A., Kovtonyuk, M., D'Auria, F., 2013. Use of qualified experimental database for a system code assessment. In: International Topical Meeting on Nuclear Reactor Thermal-Hydraulics (NURETH-13), Pisa, Italy, 13–17 May.

Kozlowski, T., Wysocki, A., Gajev, I., Xu, Y., Downar, T., Ivanov, K., Magedanz, J., Hardgrove, M., March-Leuba, J., Hudson, N., Ma, W., 2014. Analysis of the OECD/NRC Oskarshamn-2 BWR stability benchmark. Annals of Nuclear Energy 67, 4–12.

Kral, P., Hyvärinen, J., Prošek, A., Guba, A., 2015. Sources and effect of non-condensable gases in reactor coolant system of LWR. In: 16th International Topical Meeting on Nuclear Reactor Thermal Hydraulics (NURETH-16), August 30–September 4, Chicago, IL, USA.

Kramer, A.W., 1958. Boiling Water Reactors. Addison-Wesley Publisher, Reading, MA, USA, pp. 1–563.

Krane, K.S., 1988. Introductory Nuclear Physics. Wiley, New York, NY, USA.

Kremlevsky, P.P., 2000. Flow Rate Measurement in Multiphase FLows. Begell House, New York, USA.

Krepper, E., Morel, C., Niceno, B., Ruyer, P., 2011. CFD modelling of adiabatic bubbly flow. Multiphase Science and Technology 23, 2–4.

Krištof, M., Kliment, T., Petruzzi, A., Lipka, J., 2009. RELAP5 simulation of surge line break accident using combined and best estimate plus uncertainty approaches. Nuclear Engineering and Design 239, 2500–2513.

Kronig, R., Brink, J., 1950. On the theory of extraction from falling droplets. Applied Science Research A2, 142–154.

KTA, 2012. Design of Reactor Cores of Pressurized Water and Boiling Water Reactors; Part 1: Principles of Thermal-hydraulic Design (in German), KTA 3101.1, Kerntechnischer Ausschuss (KTA), c/o Bundesamt für Strahlenschutz (BfS), Salzgitter, Germany.

KTA, 2017. Ventilation Systems in Nuclear Power Plants, Safety Standards of the Nuclear Safety Standards Commission (KTA), KTA 3601 (11/2017), Salzgitter, Germany.

Kucukboyaci, V., Sung, Y., Xu, Y., Cao, L., Salko, R.K., 2016. VERA-CS modeling and simulation of PWR main steam line break core response. In: Proceedings of the 24th International Conference on Nuclear Engineering (ICONE-24), Charlotte, NC, USA.

Kudirka, A.A., Gluntz, D.M., 1974. Development of jet pump for boiling water reactor recirculation system. ASME Journal of Engineering for Power 96 (1), 7–12.

Kudryashev, L.I., Zhemkov, L.I., 1963. Generalization of the theory of regular thermal regime for the case of variable coefficients of heat conductivity and specific heat. In: Collection of Scientific Papers, Heat Power Engineering (Selected Articles), FTD-TT- 62-1200/1+2+4.

Kuhn, S.Z., Schrock, V.E., Peterson, P.F., 1994. Final Report on U.C. Berkeley Single Tube Condensation Studies, UCB-NE-4201. U.C. Berkeley, CA, USA.

Kukita, Y., Namatame, K., Shiba, M., 1984. The LOCA air-injection loads in BWR Mark II pressure suppression containment systems. Nuclear Engineering and Design 77, 117–129.

Kukita, Y., Katayama, T., Nakamura, H., Tasaka, K., 1990. Loop seal clearing during a PWR small break LOCA. Journal of Nuclear Engineering and Design 121, 431–440.

Kukita, Y., Yonomoto, T., Asaka, H., Nakamura, H., Kumamaru, H., Anoda, Y., 1996. ROSA/AP-600 Testing: facility modifications and initial test results. Journal of Nuclear Science and Technology 33 (3), 259.

Kumamaru, H., Tasaka, K., 1985. ROSA-III test facility for BWR integral simulation test. In: Proceedings Specialists Meeting on Small Break LOCA Analyses in LWRs, Pisa, Italy.

Kumar, S.B., Moslemian, D., Duduković, M.P., 1995. A γ-ray tomographic scanner for imaging voidage distribution in two-phase flow systems. Flow Measurement and Instrumentation 6, 61–73.

Kunz, R.F., Kasmala, G.F., Mahaffy, J.H., Murray, C.J., 2002. On the automated assessment of nuclear reactor systems code accuracy. Nuclear Engineering and Design 211 (2–3), 245–272.

Kurul, N., Podowski, M.Z., 1990. Multi-dimensional effects in sub-cooled boiling. In: 9th International Heat Transfer Conference, Jerusalem, Israel, 1990.

Kutateladze, S.S., 1952. Heat Transfer in Boiling and Condensation, USAEC Rep. AEC-tr-3770, Washington, DC, USA, pp. 1–207.

Kutateladze, S.S., 1963. Fundamentals of Heat Transfer. Edward Arnold Publishers Ltd., London, UK.

Kutateladze, S.S., 1979. Fundamentals of Heat Transfer Theory, fifth ed. Atomizdat, Moscow, Russia (in Russian).

Kutateladze, S.S., Nakoriakov, V.E., 1982. Heat and Mass Transfer and Waves in Gas-Liquid Flows. Nauka, Novosibirsk, Russia (in Russian).

Kutateladze, S.S., Gogonin, N.I., Dorokhov, A.P., Sosunov, V.I., 1979. Dropwise condensation of moving vapour at the horizontal tube bundle. Teploenergetika 5, 12–15 (in Russian).

KWU-Siemens, 1997. FSAR: Final Safety Analysis Report Central Nuclear Almirante Alvaro Alberto, Unit 2—Rev. 0, September, Erlangen, Germany.

Labois, M., Lakehal, D., 2011. Very-large eddy simulation (V-LES) of the flow across a tube bundle. Nuclear Engineering and Design 241, 2075−2085.

Labuntsov, D.A., 1957. Heat transfer at dropwise condensation of pure vapours on vertical surfaces and horizontal tubes. Teploenergetika 7, 72−80 (in Russian).

Labuntsov, D.A., 1960. Generalized dependence to at nucleate boiling heat transfer fluids. Teploenergetika 5, 576−581.

Labuntsov, D.A., 1960a. Heat transfer at vapour condensation on vertical surface under conditions of turbulent flow of condensate film. Inzhenerno-Phyzichesky Zhurnal 3 (8) (in Russian).

Labuntsov, D.A., 1972. Issues in nucleate boiling heat transfer. Teploenergetika 9C, C14−C19.

Lahey Jr., R.T., 1978. A mechanistic subcooled boiling model. In: Proceedings 6th International Heat Transfer Conference, Toronto, Canada, pp. 293−297.

Lahey Jr., R.T., 1986. Subchannel measurements of the equilibrium quality and mass flux distribution in a rod bundle. In: Proceedings of International Heat Transfer Conf., August 17−22, San Francisco, CA, USA, pp. 2399−2404.

Lahey Jr., R.T., Drew, D.A., 2001. The analysis of two-phase flow and heat transfer using a multidimensional, four field, two-fluid model. Nuclear Engineering and Design 204, 29−44.

Lahey Jr., R.T., Moody, F.J., 1993. The Thermal-hydraulics of a Boiling Water Reactor. ANS Monograph, Hinsdale, IL, USA, pp. 1−631 (originally published, 1977).

Lakehal, D., 2002. On the modelling of multiphase turbulent flows for environmental and hydrodynamic applications. International Journal of Multiphase Flow 28, 823−863.

Lakehal, D., 2008. LEIS for the prediction of turbulent multifluid flows applied to thermal hydraulics applications. In: International Meet CFD4NRS (Proceedings Available by NEA/CSNI, Paris), Grenoble, France, 10−12 September.

Lakehal, D., 2010. LEIS for the prediction of turbulent multifluid flows applied to thermal-hydraulics applications. Nuclear Engineering and Design 240, 2096−2106.

Lakehal, D., 2019. A 20-year journey with George Yadigaroglu through computational thermal hydraulics. In: Sehgal, R., Yadigaroglu, G., Hewitt, G.F., D'Auria, F. (Eds.), Special Issue Trends and Perspectives in Nuclear Thermal-hydraulics—To the Memory of B, Journal of Nuclear Engineering and Design, vol. 354 paper 110186 pp. 1−20. (paper PIII-02). https://doi.org/10.1016/j.nucengdes.2019.110186.

Lakehal, D., Labois, M., 2011. A new modelling strategy for phase-change heat transfer in turbulent. International Journal of Multiphase Flow 37, 627−639.

Lakehal, D., Fulgosi, M., Banerjee, S., Yadigaroglu, G., 2008. Turbulence and heat transfer in condensing vapor-liquid flow. Physics of Fluids 20, 065101.

Lakehal, D., Fulgosi, M., Yadigaroglu, G., 2008a. DNS of condensing stratified steam water flow. ASME Journal of Heat Transfer 130, 021501−021510.

Lamarsh, J.R., 1966. Introduction to Nuclear Reactor Theory. Addison-Wesley, Reading, MA, USA.

Lamarsh, J.R., Baratta, A.J., 2011. Introduction to Nuclear Engineering, International Edition. Prentice Hall, Upper Saddle River, NJ, USA.

Lamb, H., 1932. Hydrodynamics. Cambridge University Press, Cambridge, UK, pp. 1−632.

Landau, L., Lifchitz, E., 1954. Mécanique des Fluides (original edition in Russian; translated in French in 1971). MIR, Moscow, USSR.

Laney, C.B., 1998. Computational Gas Dynamics. Cambridge University Press, Cambridge, UK.

Lanfredini, M., Bestion, D., D'Auria, F., Aksan, N., Fillion, P., Gaillard, P., Heo, J., Karppinen, J., Kim, K.D., Kurki, J., Liu, L., Shen, A., Vacher, J.-L., Wang, D., 2020. Critical flow prediction by system codes—recent analyses made within the FONESYS network. Nuclear Engineering and Design 366 (110731), 1−8.

Lanfredini, M., Bestion, D., D'Auria, F., Aydemir, N., Carnevali, S., Fillion, P., Gaillard, P., Jeong, J.J., Karppinen, I., Kim, K.-D., Kurki, J., Lee, J.H., Schoeffel, P., Sha, H., Skorek, T., Vacher, J.L., Waddington, G., 2022. Horizontal stratification criteria of 8 system codes and direct confrontation to TPTF and MANTILLA data. In: The 19th International Topical Meeting on Nuclear Reactor Thermal Hydraulics (NURETH-19), 35791, Brussels, Belgium, March 6−11.

Lanfredini, M., Moretti, F., D'Auria, F., Bestion, D., Umminger, K., Kim, K.D., Aksan, N., 2022a. Key Achievements from the FONESYS and SILENCE Nuclear thermal-hydraulic Networks. NURETH-19, Brussels, Belgium, March 6−11.

Lanfredini, M., Bestion, D., D'Auria, F., Aydemir, N., Carnevali, S., Fillion, P., Gaillard, P., Jeong, J.J., Lee, J.H., Lee, J., Karppinen, I., Kim, K.-D., Kurki, J., Schoeffel, P., Sha, H., Vacher, J.-L., Waddington, G., 2022b. Droplet Entrainment Prediction in Horizontal Flow by SYS-TH Codes—Recent Analyses Made within the FONESYS Network. NURETH-19, Brussels, Belgium, March 6−11.

Lanfredini, M., Bestion, D., D'Auria, F., Aydemir, N., Carnevali, S., Fillion, P., Gaillard, P., Jeong, J.J., Karppinen, L., Kim, K.-D., Kurki, J., Lee, J.H., Junk, M., Schoeffel, P., Sha, H., Vacher, J.-L., Waddington, G., 2022c. TPTF Horizontal Flow Prediction in Horizontal Flow by SYS-TH Codes—Recent Analyses Made within the FONESYS Network. NURETH-19, Brussels, Belgium, March 6−11.

LANSCE, 1981. Los Alamos Science, 2, Summer/fall, Los Alamos Nat. Lab., NM, USA.

LaPresti, M., Lee, N.E., Espinosa, R.J., Volodzko, M., Azuma, M., 2015. A dynamic model of hydrostatic mechanical face seals for RELAP5/MOD3.3. In: 16th International Topical Meeting on Nuclear Reactor Thermal Hydraulics (NURETH-16), August 30-September 4, Chicago, IL, USA.

Latyev, L.N., Petrov, V.A., Cekhovsky, V.Y., Shestakov, E.N., 1974. In: Sheindlin, A.A. (Ed.), Radiative Properties of Solid Materials. Energiya, Moscow, Russia (in Russian).

Laufer, J., 1953. Structure of Fully Developed Turbulent Pipe Flow. NACA Tech. Note, 2954, Washington, DC, USA, pp. 1−54.

Launder, B.E., Reece, G.J., Rodi, W., 1975. Progress in the development of a Reynolds stress turbulence model. Journal of Fluid Mechanics 68, 537−566.

Lavalle, A.G., Carrica, P., Garea, V., Jaime, M., 1992. A boiling heat transfer paradox. American Journal of Physics 60 (7), 593−597.

Lawrence, J.D., 1995. An Overview of Software Safety Standards. IFAC Workshop on Safety and Reliability in Emerging Control Technology, Daytona Beach, Fl, USA, October.

Le Corre, J.-M., Robers, L., Prasser, H.-M., 2022. Multi-Field transport model and local precipitation of dissolved reactor coolant impurities in BWR fuel bundle. In: 19th International Topical Meeting on Nuclear Reactor Thermal Hydraulics (NURETH-19), March 6−11 (Via Web), Brussels, Belgium.

Le, T.T., Melikhov, V.I., Melikhov, O.I., 2021. Recommended set of interfacial drag correlations for the two-phase flow under thermal−hydraulic conditions of a horizontal steam generator. J Nuclear Engineering and Design 379 (111249), 1−20.

Ledinegg, M., 1938. Instability during Natural and Forced Circulation, vol. 16. Die Wärme.

Lee, W.H., 1980. A pressure iteration scheme for two-phase flow modeling. In: Multi-phase Transport: Fundamentals and Reactor Safety Applications. University of California, Los Alamos, NM, USA, pp. 407−432.

Lee, S.Y., Garrett, A.J., 2015. Thermal hydraulics analysis for cooling tower performance. In: 16th International Topical Meeting on Nuclear Reactor Thermal Hydraulics (NURETH-16), August 30–September 4, Chicago, IL, USA.

Lee, K., Ryley, D.J., 1968. The evaporation of water droplets in superheated steam. Transactions of ASME Journal of Heat and Mass Transfer 90, 445–456.

Lee, S.Y., Tankin, R.S., 1984. Study of liquid spray (water) in a condensable environment (steam). International Journal of of Heat and Mass Transfer 27, 363–374.

Lee, H.M., McCarthy, G.E., Tien, C.L., 1982. Liquid Carryover and Entrainment in Air-Water Countercurrent Flooding. EPRI report NP-2344, Palo Alto (CA), USA.

Lee, D.H., Swinnerton, D., 1983. Evaluation of Critical Flow for Supercritical Steam-Water (UKAEA, Winfrith, UK), EPRI report NP-3086, Palo Alto, CA, USA.

Lee, J.-I., Lee, J.-Y., Hong, S.J., Part, J.Y., Ryu, Y.-H., Kim, M., 2011. Evaluation of recirculation sump performance for OPR1000 plant: Part I debris transport during the blowdown phase of LOCA. Annals of Nuclear Energy 38, 681–693.

Lee, D.Y., Liu, Y., Hibiki, T., Ishii, M., Buchanan, J.R., 2013. A study of adiabatic two-phase flows using the two-group interfacial area transport equations with a modified two-fluid model. International Journal of Multiphase Flow 57, 115–130.

Lee, J.R., Kim, J., Song, C.-H., 2014. Synthesis of the turbulent mixing in a rod bundle with vaned spacer grids based on the OECD-KAERI CFD benchmark exercise. Nuclear Engineering and Design 279, 3–18.

Lee, J.H., Kim, B.J., Kim, K.D., 2019. Improvement of horizontal stratification criteria in the SPACE code. Annals of Nuclear Energy 133, 122–128.

Lee, J.H., Kim, B.J., Lee, S.W., Kim, K.D., 2019a. Two-fluid equation for two-phase flow considering change in volume porosity. Nuclear Engineering and Design 347, 1–9.

Lee, J.Y., Jeong, J.J., Yun, B.J., 2020. Modifications of the turbulent diffusion layer model for the condensation heat transfer under the presence of noncondensable gases. Annals of Nuclear Energy 137, 107060.

Lee, D.H., Yoo, J.M., Kim, H.Y., Hong, D.J., Yun, B.J., Jeong, J.J., 2022. Application of the Machine Learning Technique for the Development of a Condensation Heat Transfer Model for a Passive Containment Cooling System. To Be Published in Nuclear Engineering and Technology. https://doi.org/10.1016/j.net.2021.12.023.

Lee, H., Hyunwong), Kim, H., Hyungdae), 2022. Experimental measurement of dynamic leidenfrost point temperature for water droplets. In: 19th International Topical Meeting on Nuclear Reactor Thermal Hydraulics (NURETH-19), March 6–11 (Via Web), Brussels, Belgium.

Lee, J., Skorek, T., Junk, M., Schöffel, P.J., 2022. Improved correlation of liquid entrainment fraction in horizontal pipes. Nuclear Engineering and Design 388, 111615.

Lee, J., Jeon, S.-S., Choo, Y.-J., Bae, J.-H., Hong, S.-J., Park, J.-Y., 2022. Analysis of experimental facilities for the assessment of natural circulation flow predicting capability of thermal hydraulic code. In: 19th International Topical Meeting on Nuclear Reactor Thermal Hydraulics (NURETH-19), March 6–11 (Via Web), Brussels, Belgium.

Leidenfrost, J.G., 1966. On the fixation of water in diverse fire. International Journal of Heat and Mass Transfer 9, 1153–1166.

Lel'chuk, B.L., Elfimov, G.I., 1968. Convective heat transfer in turbulent flow in tubes. In: Convective Heat Transfer in Single-phase Flow. Generalized Data to Regulatory Calculation. CKTI, Leningrad, Russia. Issue 82, pp. 35–47 (in Russian).

Lemaitre, P., Nuboer, A., Porcheron, E., 2005. TOSQAN Experimental Programme, Spray Test No 101. Technical report, DSU/SERAC/LECEV/05-11, IRSN 91192, Gif-sur-Yvette, France.

Leman, G.W., 1985. Atomization and Deposition in Two-phase Annular Flow: Measurement and Modelling. Ph.D. Thesis. University of Illinois at Urbana-Champaign, Urbana, Illinois, USA.

Lemmon, E.W., Huber, M.L., McLinden, M.O., 2010. NIST Standard Reference Database 23: Reference Fluid Thermodynamic and Transport Properties-REFPROP, Version 9.0. National Institute of Standards and Technology, Standard Reference Data Program, Gaithersburg, Va, USA.

Lerchl, G., Austregesilo, H., 2000. ATHLET Mod 1.2 User's Manual. GRS-P-1 / Vol.1, Rev. 2°, GRS, Garching, Germany.

Lerchl, G., Austregesilo, H., 2013. ATHLET Mod 3.0 cycle A—Models and methods. In: GRS-P-1/Vol. 4, Garching, (G).

Leung, L.K.H., 1997. Personal Communication to Lead Authors of the Reference. Aksan et al., (2001).

Leung, L.K.H., Novog, D.R., 2018. Single bubble fragmentation phenomena near flow obstructions and the impact on interfacial area concentration. In: 12th International Topical Meeting on Nuclear Reactor Thermal-Hydraulics, Operation and Safety (NUTHOS-12), October 14−18, Qingdao, China.

Leveque, R.J., 2002. Finite Volume Methods for Hyperbolic Problems. Cambridge University Press, Cambridge, UK.

Levy, S., 1999. Two-Phase Flow in Complex Systems. John Wiley & Sons, New York, USA.

Levy, S., Abdollahian, D., Healzer, J., Janssen, E., Amos, C., 1982. Critical-flow Data Review and Analysis. EPRI report NP-2192, Palo Alto, CA, USA.

Lewis, E.E., 1978. Nuclear Power Reactor Safety. John Wiley & Sons, Inc., New York, NY, USA.

Lewis, H.W., Budnitz, R.J., Castleman, A.W., Dorfan, D.E., Finlayson, F.C., Garwin, R.L., Hebel, L.C., Keeny Jr., S.M., Muller, R.A., Taylor, T.B., Smoot, G.F., von Hippel, F., Bethe, H., Panofsky, W.K.H., Weisskopf, W.F., 1975. Report to the American Physical Society by the study group on light-water reactor safety. Reviews of Modern Physics 47, S1.

Lewis, M.J., Pochard, R., D'Auria, F., Karwat, H., Wolfert, K., Yadigaroglu, G., Holmstrom, H.L.O., 1989. Thermohydraulics of Emergency Core Cooling in Light Water Reactors—A State-Of-The-Art Report (Lead Authors) OECD/NEA/CSNI Report No. 161, Paris, France.

Lewis, B.J., Onder, E.N., Prudil, A.A., 2017. Fundamentals of Nuclear Engineering. Wiley, New York, NY, USA.

Leyer, S., Wich, M., 2012. The integral test facility Karlstein. Journal of Science and Technology of Nuclear Installations. Article ID 439374.

Li, M., Bolotnov, I.A., 2020. The evaporation and condensation model with interface tracking. International Journal of of Heat and Mass Transfer 150 (119256), 1−15.

Li, Q., Luo, K.H., Kang, Q.J., He, Y.L., Chen, Q., Liu, Q., 2016. Lattice Boltzmann methods for multiphase flow and phase-change heat transfer. Progress in Energy and Combustion Science 52, 62−105.

Li, M., Zeng, K., Wonnell, L., Bolotnov, I.A., 2019. Development of a new contact angle control algorithm for level-set method. Journal of Fluids Engineering 141 (6), 061301.

Li, W., Meng, Z., Wang, J., Liu, J., Sun, Z., 2019. Experimental investigations on pressure oscillation induced by steam-air mixture gas sonic jets in sub-cooled water. International Journal of of Heat and Mass Transfer 128, 450−458.

Li, X., Prasser, H.-M., Niceno, B., 2020. URANS simulations of the single-phase mixing in a 4*4 LWR fuel bundle with mixing vanes, validated against a sub-channel experiment. Nuclear Engineering and Design 363 (110612), 1–13.

Li, Y., Zhanh, D.L., Sun, D.C., Zan, Y.F., Xi, Z., Zhuo, W.B., Li, P.Z., 2021. Experimental investigation on steam contact condensation in emergency makeup tank. Nuclear Engineering and Design 384 (111470), 1–12.

Liang, G., Mudawar, I., 2018. Pool boiling critical heat flux (CHF)—Part 1: review of mechanisms, models, and correlations. International Journal of of Heat and Mass Transfer 117, 1352–1367.

Liao, L.-Y., 1992. Study and application of boiling water reactor jet pump characteristic. Journal of Nuclear Engineering and Design 132, 339–350.

Liao, J., Frepoli, C., 2007. Development of a condensation heat transfer model for the analysis of ECC water injection in cold leg during small and intermediate break LOCA. In: 12th International Topical Meeting on Nuclear Reactor Thermal Hydraulics (NURETH-12), September 30–October 4, Pittsburgh, Pa, USA.

Liao, Y., Lucas, D., 2017. Computational modelling of flash boiling flows: a literature survey. International Journal of of Heat and Mass Transfer 111, 246–265.

Liao, Y., Lucas, D., 2021. A review on numerical modelling of flashing flow with application to nuclear safety analysis. Applied Thermal Engineering 182 (116002), 1–20.

Liao, Y., Vierow, K.M., 2007. A generalized diffusion layer model for condensation of vapor with non-condensable gases. Journal of Heat Transfer 129, 988–994.

Liaw, S.P., Dhir, V.K., 1989. Void fraction measuring during saturated pool boiling of water on partially wetted vertical surfaces. Journal of Heat Transfer 111, 731–738.

Liebert, J., Emmerling, R., 1998. UPTF experiment—flow phenomena during full-scale loop seal clearing of a PWR. Nuclear Engineering and Design 179, 51–64.

Lien, P., 2016. Personal Communication to F. D'Auria in Relation to La Salle-2 Event, Washington, DC, USA.

Liepmann, H.W., Roshko, A., 1957. Elements of Gas Dynamics. John Wiley and Sons, New York, NY, USA.

Liesch, K., Reocreux, M., 1995. Concerted Actions on Safety Research for WWER Reactors: Verification Matrix for Thermal-hydraulic System Codes Applied for WWER Analysis. Common Report IPSN/GRS No. 25, Paris-Garching (France and Germany).

Liles, D.R., Spore, J.W., Knight, T.D., Nelson, R.A., Cappiello, M.W., Pasamehmetoglu, K.O., Mahaffy, J.H., Guffee, L.A., Stumpf, H.J., Dotson, P.J., 1988. TRAC-PF1/MOD1 Correlations and Models. Technical Report NUREG/CR-5069. Los Alamos National Laboratory, Los Alamos, NM, USA.

Lilly, D.K., 1991. A proposed modification of the germano subgrid-scale closure method. Physics of Fluids A 4 (3), 633–635.

Lind, T., Pellegrini, M., Herranz, L.E., Sonnenkalb, M., Nishi, Y., Tamaki, H., Kim, S.I., Cousin, F., Fernandez-Moguel, L., Andrews, N., Sevon, T., 2020. Overview and outcomes of the OECD/NEA benchmark study of the accident at the Fukushima Daiichi NPS (BSAF), phase 2—results of severe accident analyses for unit 3. Nuclear Engineering and Design 376 (111138), 1–12.

Lioce, D., Asztalos, M., Alemberti, A., Barucca, L., Frogheri, M., Saiu, G., AP1000 passive core cooling system pre-operational tests procedure definition and simulation by means of Relap5 Mod. 3.3 computer code. Nuclear Engineering and Design, 250, pp 538– 547.

Lisovvy, O., Cherubini, M., Lazzerini, D., D'Auria, F., 2015. Base irradiation simulation and its effect on fuel behavior prediction by TRANSURANUS code: application to reactivity initiated accident condition. Nuclear Engineering and Design 283, 162–167.

Liu, H., 2000. Droplets: Fundamentals and Applications. Noyes Publ., Park Ridge NJ. Distributed in USA by William Andrew Publications, Norwich NY, USA.

Liu, W., Nariai, H., Inasaka, F., 2000. Prediction of critical heat flux for subcooled flow boiling. International Journal of Heat and Mass Transfer 43, 3371–3390.

Liu, L., Fu, X., Du, Z., Yu, N., Yang, Y., 2013. In: COSINE Validation Experiment Plan: 1st Phase, 21st International Conference on Nuclear Engineering, vol. 4. Thermal Hydraulics, Chengdu, China. July 29–August 2.

Liu, A., (Liu, -A.), Yang, B.-W., 2015. Heat loss analysis in different size rod bundle CHF experiments. In: International Seminar on Subchannel Analysis, CFD, and CHF (ISACC-2015). Shenzhen, China.

Liu, W., (Liu, -W.), Shan, J., 2015. Research of the bundle CHF prediction based on the minimum DNBR point and the BO point methods. In: 16th International Topical Meeting on Nuclear Reactor Thermal Hydraulics (NURETH-16), August 30–September 4, Chicago, IL, USA.

Liu, Y., (Liu, -Y.), Talley, J.D., Hogan, K.J., Buchanan Jr., J.R., 2015. Derivation of a generic multi-field two-fluid model. In: 16th International Topical Meeting on Nuclear Reactor Thermal Hydraulics (NURETH-16), August 30–September 4, Chicago, IL, USA.

Liu, A., (Liu, -A.), Yang, B.-W., 2015a. Heat loss simulation and uncertainty analysis in fuel bundle CHF and other large scale thermal-hydraulic experiments. In: Proceedings 16th International Topical Meeting on Nuclear Reactor Thermal Hydraulics (NURETH-16), Chicago, Illinois, USA.

Liu, Y., (Liu, -Y.), Pan, J., Wang, X., Xu, L., Deng, J., Tan, C., 2017. Development of self-reliant subchannel analysis code CORTH (in Chinese). Nuclear Power Engineering 38 (6), 157–162.

Liu, A., (Liu, -A.), Yang, B.-W., Zhu, X., 2018. Different turbulent mixing coefficient determination methods in rod bundle mixing analysis. In: 12th International Topical Meeting on Nuclear Reactor Thermal-Hydraulics, Operation and Safety (NUTHOS-12), October 14–18, Qingdao, China.

Liu, A., (Liu, -A.), Yang, B.-W., Han, B., Wang, S., 2019. Measurement uncertainty and quenching phenomena in uniform heating rod bundle CHF tests. Nuclear Engineering and Design 348, 107–120.

Liu, B., (Liu, -B.), He, S., Moulinec, C., Uribe, J., 2019. Sub-channel CFD for nuclear fuel bundles. Nuclear Engineering and Design 355 (110318), 1–16.

Liu, X.J., Song, M.Q., Cheng, X., 2019. Current status of research of heat transfer in forced convection of fluids at supercritical pressures. In: Sehgal, R., Yadigaroglu, G., Hewitt, G.F., D'Auria, [F. (Eds.), Special Issue Trends and Perspectives in Nuclear Thermal-hydraulics—To the Memory of B, Journal of Nuclear Engineering and Design, vol. 354, pp. 1–9. https://doi.org/10.1016/j.nucengdes.2019.110176 paper 110176 (paper PIII-06).

Liu, A., (Liu, -A.), Yang, B.-W., Han, B., Zhu, X., 2020. Turbulent mixing models and other mixing coefficients in subchannel codes-A review Part A: single phase. Nuclear Technology 206 (9), 1253–1295.

Liu, W., Liu-W, Peng, S., Jiang, G., Liu, Y., Du, S., Zhang, Y., Deng, J., 2020. Development and assessment of a new rod-bundle CHF correlation for China fuel assemblies. Annals of Nuclear Energy 138, 107175.

Liu, B., (Liu, -B.), He, S., Moulinec, C., Uribe, J., 2021. Coupled porous media approaches in sub-channel CFD. Nuclear Engineering and Design 377 (111159), 1–14.

Liu, Q., (Liu, -Q.), Kelly, J., Sun, X., 2021. Flow regime transition in the post-CHF flow regimes under sub-cooled and low quality conditions. International Journal of Multiphase Flow 136, 103543.

Liu, X., Qj, X., Zhang, N., Meng, Z., Sun, Z., 2021. Scaling analysis of pressure suppression containment test facility for the small, pressurized water reactor. Nuclear Engineering and Technology 53, 793–803.

Llopis, C., 2006. Modelos avanzados de sistemas de control y proteccion para reactores de agua a presion. Contribucion a la seguridad y la disponibilidad (in Spanish). PhD Thesis. UPC, Barcelona, Spain.

Llopis, C., Reventos, F., Batet, L., Pretel, C., Sol, I., 2007. Analysis of low load transients for the Vandellos-II NPP: application to operation and control support. Nuclear Engineering and Design 237 (18), 2014–2023.

Lochon, H., Daude, F., Galon, P., Hérard, J.-M., 2017. Computation of fast depressurization of water using a two-fluid model: revisiting Bilicki modelling of mass transfer. Computers and Fluids 156, 162–174.

Lockhart, R.W., Martinelli, R.C., 1949. Proposed correlation of data for isothermal two-phase, two-component flow in pipes. Chemical Engineering Progress 45 (1), 39–48.

Lodi, F., Grasso, G., 2017. Extension of the sub-channel code ANTEO+ to the mixed convection regime. Nuclear Engineering and Design 322, 368–378.

Lodi, F., Grasso, G., Mattioli, D., Sumini, M., 2016. ANTEO+: a subchannel code for thermal-hydraulic analysis of liquid metal cooled systems. Nuclear Engineering and Design 301, 128–152.

Lombardi, C., Pedrocchi, E., 1972. A pressure drop correlation in two-phase flow. Journal of Energia Nucleare 19 (2), 91–99.

Lombardi-Costa, A., Pereira, C., Ambrosini, W., D'Auria, F., 2008. Simulation of a hypothetical out-of-phase instability case in boiling water reactor by Relap5/Parcs coupled codes. Annals of Nuclear Energy 35, 947–957.

Long, J., Zhang, B., Yang, B.-W., Wang, S., 2021. Review of researches on coupled system and CFD codes. Nuclear Engineering and Technology 53, 2775–2787.

Lopes, S., Koloszar, L., Planquart, P., Visser, D.C., Van Tichelen, K., 2021. Hunting for the correct pressure drop in a scaled reactor pool: effect of geometry, mesh resolution, turbulence model and mass flow. Nuclear Engineering and Design 384 (111452), 1–10.

Lopez de Bertodano, M., Fullmer, W.D., 2015. Effect of viscosity on a well-posed one-dimensional two-fluid model for wavy two-phase flow beyond the Kelvin-Helmholtz instability: limit cycles and chaos. In: 16th International Topical Meeting on Nuclear Reactor Thermal Hydraulics (NURETH-16), August 30–September 4, Chicago, IL, USA.

Lopez de Bertodano, M.A., Jan, C.S., Beus, S.G., 1997. Annular flow entrainment rate experiment in a small vertical pipe. Nuclear Engineering and Design 178 (1), 61–70.

Lopez de Bertodano, M.A., Fullmer, W., Clausse, A., Ransom, V.H., 2017. Two-Fluid Model Stability Simulation and Chaos. Springer International Publishing, Switzerland, pp. 1–351.

López, R., Lecuona, A., Nogueira, J., Vereda, C., 2017. Numerical solution of one-dimensional transient, two-phase flows with temporal fully implicit high order schemes: subcooled boiling in pipes. Nuclear Engineering and Design 313, 319–329.

Lopez, T., Matteo, L., Bestion, D., 2020. Validation of a drift flux model used in the CATHARE Code for rod bundle geometry at low pressure and low liquid flow conditions. In: International Topical Meeting on Nuclear Reactor Thermal Hydraulics (NURETH-18), Portland, or, USA, August 19–23.

Lorduy-Alós, M., Gallardo, S., Verdú, G., 2020. Scaling analysis of an IBLOCA counterpart test between the ATLAS and LSTF facilities. Progress in Nuclear Energy 127.

Lottes, P.A., Anderson, R.P., Hoglund, B.M., Marchaterre, J.F., Petrick, M., Popper, G.F., Weatherhead, R.J., 1963. Boiling Water Reactor Technology, Status of the Art Report.

Argonne National Laboratory, Chicago, IL, USA, pp. 1–194. ANL-6562, Vol I (Heat Transfer and Hydraulics).

Lu, P., Semião, V., 2003. A new second-moment closure approach for turbulent swirling confined flows. International Journal for Numerical Methods in Fluids 41, 133–150.

Luangdilok, W., 2020. The explosions at Fukushima Daiichi Unit 3 and Unit 4 and implications on the evaluation of 1F3 accident. Nuclear Engineering and Design 362 (110536), 1–10.

Lübbesmeyer, D., Aksan, S.N., 2003. ISP 42 (PANDA Tests) Blind Phase Comparison Report. NEA/CSNI/R(2003), Paris, France.

Lubin, B., Horwitz, M., 1966. Vapor pull through at a tank drain with and without dielectrophoretic buffing. In: Conference on Long Term Cryo-Propellant Storage in Space. NASA Spatial Space Center, Huntsville, AL, USA, p. 173.

Lucas, S., 2014. Millora i adaptació del model simplificat d'Ascó per l'ùs de càlculs escalats. Aplicació a l'anàlisi de distorsions en escenaris LOBI. Universitat Politècnica de Catalunya (UPC), Barcelona, Catalunya, Spain, pp. 1–85 (in Catalan) Mster Thesis.

Lucas, P., Mishra, R., 2005. Measurement of bubble velocity components in a swirling gas–liquid pipe flow using a local four-sensor conductance probe. Measurement Science and Technology 16, 749–758.

Lucas, D., Bestion, D., Bodèle, E., Coste, P., Scheuerer, M., D'Auria, F., Mazzini, D., Smith, B., Tiselj, I., Martin, A., Lakehal, D., Seynhaeve, J.-M., Kyrki-Rajamaki, R., Ilvonen, M., Macek, J., 2009. An Overview of the Pressurized Thermal Shock Issue in the Context of the NURESIM Project. Science and Technology of Nuclear Installations, pp. 1–13. Article ID 583259.

Lucas, D., Coste, P., Höhne, T., Lakehal, D., Bartosiewicz, Y., Bestion, D., Scheuerer, M., Galassi, M.C., 2011. CFD modeling of free surface flow with and without condensation. Multiphase Science and Technology 23 (2–4).

Lutsanych, S., Moretti, F., D'Auria, F., 2015. Assessment of NEPTUNE_CFD code capabilities to simulate two-phase flow in the OECD/NRC PSBT subchannel experiments. In: International Conference Nuclear Energy in Central Europe (NENE), September 14–17, Portoroz, Slovenia.

Lutsanych, S., Moretti, F., D'Auria, F., 2017. Validation of the CATHARE 1-D and 3-D reflood models against FEBA and ACHILLES experimental tests. Nuclear Engineering and Design 321, 266–277.

Lutsanych, S., Moretti, F., D'Auria, F., 2017a. Validation of NEPTUNE-CFD two-phase flow models against OECD/NRC PSBT subchannel experiment. Journal of Nuclear Engineering and Design 321, 82–91.

Luxat, J.C., Richards, D.J., Liu, W.S., Hanna, B., Leung, R.K., Krishnam, V.S., 1996. Methodology, status and plans for development and assessment of TUF and CATHENA codes. In: USNRC and OECD/NEA/CSNI Workshop on Transient Thermal-Hydraulic and Neutronic Codes Requirements, Annapolis, MD, USA, 5–8 November.

Lyczkowski, R.W., 2018. The History of Multiphase Science and Computational Fluid Dynamics: A Personal Memoir. Mechanical Engineering Series. In: Kulacki, F.A. (Ed.). Springer International Publishing AG, Cham, Switzerland, pp. 1–334.

Lyczkowski, R.W., Gidaspow, D., Solbrig, C.W., Hughes, E.D., 1978. Characteristics and stability analyses of transient onedimensional two-phase flow equations and their finite difference approximations. Nuclear Science and Engineering 66 (3), 378–396.

Lykov, A.V., 1954. Phenomena of Transfer in Capillary-Porous Bodies. Gostekhizdat, Moscow, Russia (in Russian).

Lykov, A.V., 1958. Transporterscheinungen in Kapillarporösen Körpen. Springer Verlag, Germany.

Lykov, A.V., 1961. The Theoretical Principles of Building Thermophysics. Izdatel'stvo Academy of Sciences BSSR, Minsk, USSR (in Russian).
Lykov, A.V., 1967. Theory of Heat Conduction. Vysshaya Shkola, Moscow, Russia (in Russian).
Lykov, A.V., 1978. In: Heat and Mass Transfer/Handbook, 2nd rev. Energiya, Moscow, Russia (in Russian).
Lykov, A.V., Mikhailov, Y.A., 1961. Theory of Energy and Mass Transfer. Prentice Hall, New York, NY, USA.
Lyu, H.M., Yang, B.-W., Han, B., 2017. Rod-To-Wall Gap calculation through CFD modeling. In: The 17th International Topical Meeting on Nuclear Reactor Thermal Hydraulics, 3–8 September, Xi'an, China.
Macek, J., 1998. Personal communication to Lead Authors of the paper Aksan et al, 2001.
Macken, N.A., 1990. Momentum Effects on Countercurrent Flow in a Horizontal Pipe with Saturated Conditions. Siemens Work Report 4200 90/09, Erlangen, Germany.
Madeira, A., D'Auria, F., Alvim, A.C.M., 2003. A PWR recovery option for a total loss of feedwater beyond design basis scenario. In: 10th International Topical Meeting on Nuclear Reactor Thermalhydraulics (NURETH-10), Seoul, Korea, 5–9 October.
Madokoro, H., Sato, I., 2021. Estimation of the core degradation and relocation at the Fukushima Daiichi nuclear power station unit 2 based on RELAP/SCDAPSIM analysis. Nuclear Engineering and Design 376 (111123), 1–15.
Magdeleine, S., Mathieu, B., Lebaigue, O., Morel, C., 2010. DNS up-scaling applied to volumetric interfacial area transport equation. In: 7th International Conference On Multiphase Flow (ICMF 2010), Tampa, FL, USA, May 30–June 4.
Mahaffy, J.H., 1982. A stability enhancing two-step method for fluid flow calculations. Journal of Computational Physics 46, 329–341.
Mahaffy, J.H., 1993. Numerics of codes: stability, diffusion, and convergence. Nuclear Engineering and Design 145, 131–145.
Mahaffy, J.H. (Ed.), 2007. Best Practice Guidelines for the Use of CFD in Nuclear Reactor Safety Applications. OECD, Nuclear Energy Agency. Technical Report, NEA/CSNI/R(2007)5, Paris, France. (see also Mahaffy et al., 2007a).
Mahaffy, J.H., Chung, B., Dubois, F., Ducros, F., Graffard, E., Heitsch, M., Henriksson, M., Komen, E., Moretti, F., Morii, T., Mühlbauer, P., Rohde, U., Scheuerer, M., Smith, B.L., Song, C.-H., Watanabe, T., Zigh, G., 2007a. Best practice guidelines for the use of CFD in nuclear reactor safety applications. In: OECD Report, NEA/CSNI/R(2007)5, Paris, France.
Mahaffy, J.H., Chung, B., Dubois, F., Ducros, F., Graffard, E., Heitsch, M., Henriksson, M., Komen, E., Moretti, F., Morii, T., Mühlbauer, P., Rohde, U., Scheuerer, M., Smith, B.L., Song, C.-H., Watanabe, T., Zigh, G., 2015. Best practice guidelines for the use of CFD in nuclear reactor safety applications Revision. In: OECD Report, NEA/CSNI/R(2014)11, Paris, France, pp. 1–176.
Mahmoodi, R., Zolfaghari, A., Minuchehr, A., 2019. Simulation of pressure waves propagation following LOCA in piping systems using Laplace Transform Finite Volume. Annals of Nuclear Energy 124, 164–171.
Maiani, L., 2018. Enrico Fermi Nobel: His Heritage. Presentation at the Meet. "Commemoration of E. Fermi, 80 Years since Nobel Prize". Department of Physics, University of Pisa, Pisa (I). Dec. 10.
Maity, S., 2000. Effect of Velocity and Gravity on Bubble Dynamics. MS thesis. UCLA, Los Angeles, CA, USA.

Majumdar, P., Lele, H.G., Gupta, S.K., 1999. Modeling the effects of vapor pull through and entrainment in the simulation of stratified header in pressurized heavy water reactor. In: 7th International Conference in Nuclear Engineering (ICONE-7365), Tokyo, Japan, 19–23 April.

Makihara, Y., Sartmandjev, A., Spalj, S., Macek, J., Kantee, H., Elter, J., Horvath, G.L., D'Auria, F., Bukin, N., Gudosnikov, A., Kristof, M., Kvidza, B., Matejovic, P., 2005. Assessment of the potential consequences of large primary to secondary leakage accident for a VVER-440. In: International Conference On Nuclear Engineering (ICONE-13, 50556), Beijing China, May 16–20.

Mali, C.R., Patwardhan, A.W., Pandey, G.K., Banerjee, I., Vinod, V., 2020. CFD study on the effect of various geometrical parameters of honeycomb type orifices on pressure drop and cavitation characteristics. Nuclear Engineering and Design 370 (110880), 1–27.

Malko, M.V., 2002. The Chernobyl Reactor: Design Features and Reasons for Accident (Article in the Report by T. Imanaka [Ed.]).

Mancusi, D., Zoia, A., Chaos in eigenvalue search methods. Annals of Nuclear Energy, 112, pp 354–363.

Mandhane, J.M., Gregary, G.A., Aziz, K., 1974. A flow pattern map for gas-liquid flow in horizontal pipes. International Journal of Multiphase Flow 1, 537–553 (Cited in "Two-Phase Flow and Heat Transfer", Butterworth, D., Hewitt, G.F., Eds., 1977).

Manera, A., Macian, R., 2007. Assessment of TRACE film condensation models with and without non-condensable gases. In: 12th International Topical Meeting on Nuclear Reactor Thermal Hydraulics (NURETH-12), September 30–October 4, Pittsburgh, PA, USA.

Mansour, M., Kopparthy, S., Thevenin, D., 2022. Investigations on the effect of rotational speed on the transport of air-water two-phase flows by centrifugal pumps. International Journal of of Heat and Fluid Flow 94 (108939), 1–16.

Man'kina, N.N., 1977. Physical and Chemical Processes in the Steam-Cycle of Power Plants. Energiya, Moscow, Russia (in Russian).

Mao, H., 2018. Research on Separation and Modeling of Mixing Vane Grids' Mixing Effects in PWR Rod Bundles. Jiaotong University. Doctor of Engineering Science, Xi'an, China.

Mao, H., Yang, B.W., Han, B., 2016. Study on effects of mixing vane grids on coolant temperature distribution by subchannel analysis. Kerntechnik 81 (3), 244–250.

Mao, H., Yang, B.-W., Han, B., Liu, A., 2017. Modeling of spacer grid mixing effects through mixing vane crossflow model in subchannel analysis. Nuclear Engineering and Design 320, 41–152.

Mao, H., Yang, B.-W., Wang, S., 2019. A review of void drift models in sub-channel analysis. Nuclear Science and Engineering 193 (1–2), 33–45.

Marcel, C., Lazo, P., Cantero, M., 2022. Experimental assessment of thermal mixing in a 37-rod nuclear fuel element. In: 19th International Topical Meeting on Nuclear Reactor Thermal Hydraulics (NURETH-19), March 6–11 (Via Web), Brussels, Belgium.

March-Leuba, J., Rey, J.M., 1993. Coupled thermo-hydraulic-neutronic instabilities in boiling water nuclear reactors: a review of the state of the art. Nuclear Engineering and Design 145, 97–111.

Marchuk, G.I., 1975. Methods of Numerical Mathematics. Springer Verlag, Berlin, Germany.

Marcucci, L.E., 2018. Nuclear Physics from 1909 to 1934 and More. Presentation at the Meet. "Commemoration of E. Fermi, 80 Years since Nobel Prize". Department of Physics, University of Pisa, Pisa (I). Dec. 10.

Marcus, G.H., Mirsky, S.M., 2021. The History and Future of Nuclear Power Afloat. Nuclear News, pp. 31–39. December.

Margoulova, T., 1969. Le Centrales Nucleaires. MIR, Moscow, USSR (translated in French in 1977) (first Russian edition).
Mark, P.A., Johnson, M.W., Sproston, J.L., Millington, B.C., 1990. The turbine meter applied to void fraction determination in two-phase flow. Flow Measurement and Instrumentation 1, 246–252.
Marques, M., Pignatel, J.F., Saignes, P., D'Auria, F., Burgazzi, L., Muller, C., Bolado-Lavin, L., Kirchsteiger, C., La Lumia, V., Ivanov, I., 2005. Methodology for the reliability evaluation of a passive system and its integration into a Probabilistic Safety Assessment. Nuclear Engineering and Design 235, 2612–2632.
Martin, R.P., Dunn, B.M., 2004. Application and licensing requirements of the Framatome ANP RLBLOCA methodology. In: International Meeting On Updates in Best Estimate Methods in Nuclear Installation Safety Analysis (BE-2004), Washington, DC, USA.
Martin, R.P., Nutt, W.T., 2011. Perspectives on the application of order-statistics in best-estimate plus uncertainty nuclear safety analysis. Nuclear Engineering and Design 241, 274–284.
Martin, R.P., O'Dell, L.D., 2005. AREVA's realistic large break LOCA analysis methodology. Nuclear Engineering and Design 235, 1713–1725.
Martin, R.P., O'Dell, L.D., 2008. Development considerations of AREVA NP Inc.'s realistic LBLOCA analysis methodology. Science and Technology of Nuclear Installations 2008, 13. https://doi.org/10.1155/2008/239718. Article ID 239718.
Martinelli, R.C., Nelson, D.B., 1948. Prediction of pressure drop during forced-circulation boiling of water. Transactions of the American Society of Mechanical Engineers 70, 695–702.
Martinez Val, J.M., Aragones, J.M., Mingues, E., Perlado, J.M., Velarade, G., 1990. An analysis of the physical causes of the Chernobyl accident. Nuclear Technology 90 (3), 371–388.
Martinez-Quiroga, V., 2014. Scaling-up Methodology, a Systematic Procedure for Qualifying NPP Nodalizations. Application to the OECD/NEA ROSA-2 and PKL-2 Counterpart Test. PhD Thesis. UPC, Barcelona, Spain.
Martinez-Quiroga, V., Reventos, F., 2014. The use of system codes in scaling studies: relevant techniques for qualifying NPP nodalizations for particular scenarios. Science and Technology of Nuclear Installations 13. Article ID 138745.
Martinez-Quiroga, V., Reventos, F., Pretel, C., Sol, I., 2008. Code validation and scaling of the ROSA/LSTF test 3-1 experiment. ENS, TopSafe Conf. Dubrovnik, Proceedings. Croatia, September 30–October 3.
Martinez-Quiroga, V., Reventos, F., Freixa, J., 2014. Applying UPC scaling-up methodology to the LSTF-PKL counterpart test. Science and Technology of Nuclear Installations 18. Article ID 292916.
Martínez-Quiroga, V., Freixa, J., Reventós, F., 2018. PVST, a tool to assess the Power to Volume scaling distortions associated to code simulations. Nuclear Engineering and Design 332.
Martinson, Z.R., Gasparini, M., Hobbins, R.R., Petti, D.A., Allison, C.M., Hohorst, J.K., Hagrman, D.L., Vinjamuri, K., 1989. PBF Severe Fuel Damage Test 1–3 Test Results Report. EG&G Idaho Inc., Washington, DC, USA. EGG-2565, USNRC, NUREG/CR-5354.
Martorell, S., Lazaro, A., Marton, I., Sanchez, F., Villanueva, J.F., Carlos, S., 2015. A procedure to develop EBEPU methodology merging PSA based assumptions and BEPU methods. In: Safety and Reliability of Complex Engineered Systems. Taylor & Francis, London, UK.

Maruyama, S., 2021. Validation of unit 1 of the Fukushima Daiichi nuclear power plant during its accident. Global Journal of Research in Engineering: Electrical and Electronics Engineering 21 (1), 1−37 online ISSN 2249-4596.

Maruyama, Y., Satou, A., Asaka, H., Nakamura, H., 2009. Experimental study on void behavior in fast power transients simulating hot stand-by reactivity initiated accidents of BWR. In: 13th International Topical Meeting on Nuclear Reactor Thermal Hydraulics (NURETH-13), September 27-October 2, Kanazawa City, Ishikawa, Japan.

Mascari, F., De Rosa, F., 2015. Use of the MELCOR code at ENEA for the analysis of current and advanced reactor designs. In: 7th Meeting of the European MELCOR User Group, March 17−18, Bruxelles, Belgium.

Mascari, F., Vella, G., Woods, B.G., Welter, K., Pottorf, J., Young, E., Adorni, M., D'Auria, F., 2011. Sensitivity analysis of the MASLWR helical coil steam generator using TRACE. Journal of Nuclear Engineering and Design 241.

Mascari, F., Vella, G., Woods, B.G., D'Auria, F., 2012. Analyses of the OSU-MASLWR experimental test facility. Science and Technology of Nuclear Installations. Article ID 528241.

Mascari, F., Vella, G., Woods, B.G., Welter, K., D'Auria, F., 2012a. Analysis of Primary/Containment coupling phenomena characterizing the MASLWR design during a SBLOCA scenario, Nuclear Power Plants. In: Chang, S.H. (Ed.), Chapter 8 of the INTECH Book 'Nuclear Power Plants', ISBN 978-953-51-0408-7, pp. 203−233.

Mascari, F., Nakamura, H., De Rosa, F., D'Auria, F., 2014. Scaling rationale design and extrapolation problem for ITF and SETF. In: Proceedings International Workshop on Nuclear Safety and Severe Accident (NUSSA-2014), Kashiwa, Chiba, Japan, September 3−5.

Mascari, F., Nakamura, H., Umminger, K., De Rosa, F., D'Auria, F., 2015. Scaling issues for the experimental characterization of Reactor Coolant System in Integral Test Facilities and role of system codes as extrapolation tool. In: International Conference NURETH-16, Chicago, IL, USA, Aug. 30−Sept, vol. 4, pp. 4921−4934.

Mascari, F., De Rosa, F., Woods, B.G., Welter, K., Vella, G., D'Auria, F., Tien, K., 2016. Analysis of the OSU-MASLWR 001 and 002 Tests by Using the TRACE Code. USNRC NUREG-IA 0466, Washington, DC, USA, pp. 1−117.

Mascari, F., Nakamura, H., Umminger, K., Mon, S.-K., Lien, P., Bestion, D., D'Auria, F., 2017. Scaling approaches and system code as extrapolation tool. In: Spring CAMP Meeting, May 22−24, Warsaw, Poland.

Mascari, F., Woods, B.G., Welter, K.B., D'Auria, F., 2019. Validation of the TRACE code against small modular integral reactor natural circulation phenomena. In: 18th International Topical Meeting on Nuclear Reactor Thermal-hydraulics (NURETH-18), Portland (Or, US), August 18−22, © 2019 American Nuclear Society. Greyden Press paper 804-28325. https://nureth2019conference.wordpress.co.

Mascari, F., Bersano, A., Alcaro, F., Stempniewicz, M., Albright, L., Jevremovic, T., Andrews, N., Gauntt, R., Austregesilo, H., Buchholz, S., Bellomo, A., D'Auria, F., Di Palma, G., Lanfredini, M., Spina, G., Bertani, C., De Salve, M., Falcone, N., Caruso, G., Giannetti, F., Narcisi, V., Choi, C., Ha, K., Jeon, B.G., Kang, K.H., Kim, K., Park, H.S., Karppinen, I., Lahovský, L.F., Meca, R., Parduba, Z., Lien, P., Tomashchik, D.Y., Burgazzi, L., Lombardo, C., Meloni, P., Ferri, R., 2022. OECD/NEA/CSNI/WGAMA PERSEO benchmark: main outcomes and conclusions. In: OECD/NEA/CSNI Specialists Meeting on Transient Thermal-hydraulics in Water Cooled Nuclear Reactors (SM-TH), March 22−24, CIEMAT, Madrid, Spain, P26.

Mascari, F., D'Auria, F., Bestion, D., Lien, P.H., Nakamura, H., Austregesilo, H., Mon, S.-K., Reventos, F., Umminger, K., Reyes Jr., J.N., Rohatgi, U.S., 2022a. OECD/NEA/CSNI State-of-the-art report on scaling in system thermal-hydraulics applications to nuclear reactor safety and design (the S-SOAR). In: OECD/NEA/CSNI Specialists Meeting on Transient Thermal-hydraulics in Water Cooled Nuclear Reactors (SM-TH), March 22–24 (Via Web), CIEMAT, Madrid (Spain), P22.

Mascari, F., Woods, B.G., Welter, K., D'Auria, F., Bersano, A., Maccari, P., 2022b. Small Modular Reactors and insights on passive mitigation strategy modeling. In: OECD/NEA/ CSNI Specialists Meeting on Transient Thermal-hydraulics in Water Cooled Nuclear Reactors (SM-TH), P55, March 22–24, CIEMAT, Madrid, Spain, pp. 1–20.

Mascari, F., Bersano, A., Woods, B.G., Reyes Jr., J.N., Welter, K., Nakamura, H., D'Auria, F., 2022c. Scaling-up capabilities of TRACE integral reactor nodalization against natural circulation phenomena in Small Modular Reactors. In: 19th International Topical Meeting on Nuclear Reactor Thermal Hydraulics (NURETH-19), March 6–11, Brussels, Belgium.

Mass, H.G., Gruen, A., Papantoniou, D., 1993. Particle tracking velocimetry in three-dimensional flows: Part 1—photogrammetric determination of particle coordinates. Experiments in Fluids 15, 133–146.

Masterson, R.E., 2019. Nuclear Reactor Thermal Hydraulics—an Introduction to Nuclear Heat Transfer and Fluid Flow. Taylor & Francis Group, Boca Ratom, Fl, USA, pp. 1–1388.

Masuda, Y., Nishikawa, M., Ichijo, B., 1980. New methods of measuring capacitance and resistance of very high loss materials at high frequencies. IEEE Trans. Instrum. Meas. 29, 28–36.

Mavko, B., Prosek, A., D'Auria, F., 1997. Determination of code accuracy in predicting small break LOCA experiment. Nuclear Technology 120, 1–18.

Mayinger, F. (Ed.), 1994. Optical Measurements. Springer-Verlag GmbH, Berlin, Heidelberger, Germany.

Mayinger, F., Chavez, A., 1992. Measurement of direct-contact condensation of pure saturated vapour on an injection spray by applying pulsed laser holography. International Journal of of Heat and Mass Transfer 35 (3), 691–702.

Mayinger, F., Langer, R., 1978. PDO heat transfer. In: Proceedings 6th International Heat Transfer Conference, Toronto, Canada, vol. 6, p. 181.

Mayinger, F., Weiss, P., Wolfert, K., 1993. Two-phase flow phenomena in full-scale reactor geometry. Nuclear Engineering and Design 145 (1–2), 47–61.

Mazzantini, O., Galassi, G.M., D'Auria, F., 2019. The role of nuclear thermal hydraulics in the licensing of Atucha-II: the LBLOCA. In: Sehgal, R., Yadigaroglu, G., Hewitt, G.F., D'Auria, [F. (Eds.), Special Issue Trends and Perspectives in Nuclear Thermal-hydraulics—To the Memory of B, Journal of Nuclear Engineering and Design, vol. 354, pp. 1–27 paper 110292 (paper PIII-27).

McAdams, W.H., Woods, W.K., Heroman, R.H., 1942. Vaporization inside horizontal tubes. II. Benzene-oil mixtures. Transactions of ASME 64, 193.

McCoy, D.D., Hanratty, T.J., 1977. Rate of deposition of the droplets in annular two-phase flow. International Journal of Multiphase Flow 3 (4), 319–331.

McWald, T.H., Marschall, E.A., 1993. Comparison of elastic and plastic contact models for the prediction of thermal contact conductance. Wärme- und Stoffübertragung 28, 441–448.

Mehedinteanu, S., 2009. The evaluation of RCS depressurization to prevent core melting in pressure tube reactors (CANDU-type). Annals of Nuclear Energy 36 (4), 537–545.

Melikhov, V.I., Melikhov, O.I., Parfenov, Y., Nerovnov, A., 2011. Simulation of the thermal hydraulic processes in the horizontal steam generator with the use of the different interfacial friction correlations. Science and Technology of Nuclear Installations. Article ID 181393.

Melnikov, V.I., Kontelev, V.V., 1999. Two-phase flow diagnostic acoustic system based on ultrasound waveguides. In: Celata, G.P., Di Marco, P., Shah, K. (Eds.), Two-Phase Flow Modelling and Experimentation 1999. Edizioni ETS, Pisa, Italy, pp. 1515–1519.

Melnikov, V.I., Nigmatulin, B.I., 1994. The newest two-phase control devices in LWR equipment based on ultrasonic and WAT technology. Nuclear Engineering and Design 149, 349–355.

Mendizabal, R., Pelayo, F., 2010. BEPU methodologies and plant technical specifications. In: ASME 2010 3rd Joint US-European Fluids Engineering Summer Meeting, Montreal, Quebec, Canada, 1–5 August.

Menezes, C., Vaghetto, R., Hassan, Y.A., 2022. Subchannel pressure drop measurements and friction factor calculations in a wire-wrapped fuel bundle. In: 19th International Topical Meeting on Nuclear Reactor Thermal Hydraulics (NURETH-19), March 6–11 (Via Web), Brussels, Belgium.

Mengyao, W., Bo, W., Xin, G., Jiayi, Z., Zhiyang, C., Yang, W., Wang, Chuan, L., Yang, W., Wu, Ruifeng, T., 2022. Review on the steam-liquid separation in the steam generator of nuclear power plants. Annals of Nuclear Energy 175 (109207), 1–27.

Menter, F.R., 1992. Improved Two Equation K-ω Turbulence Model for Aerodynamics Flows, NASA Technical Memorandum 103975, Houston, TX, USA.

Menter, F.R., 1994. Two-equation eddy-viscosity turbulence models for engineering applications. AIAA Journal 32 (8).

Menter, F.R., Egorov, Y., 2005. A scale-adaptive simulation model using two-equation models. In: AIAA Paper 2005-1095, Reno, NV, USA.

Menter, F.R., Esch, T., 2001. Advanced turbulence modeling in CFX. CFX Update 20, 4–5.

Menter, F.R., Kuntz, M., Langtry, R., 2003a. Ten years of industrial experience with the SST turbulence model. In: 4th International Symposium on Turbulence, Heat and Mass Transfer, ICHMT 4, Antalya, Turkey.

Menter, F.R., Kuntz, M., Bender, R., 2003b. A scale-adaptive simulation model for turbulent flow prediction. In: AIAA Paper 2003-0767, Reno, NV, USA.

Menzel, F., Sabundjan, G., D'Auria, F., Madeira, A., 2015. Using of BEPU methodology in a final safety analysis report. In: International Conference INAC-ENFIR, Sao Paulo, Brazil, 5–9 October.

Menzel, F., Sabundjan, G., D'Auria, F., Madeira, A., 2016. Application of best estimate plus uncertainty (BEPU) methodology in a final safety analysis report (FSAR) of a generic plant. In: 11th International Conference of the Croatian Nuclear Society, Zadar, Croatia, 5–8 June.

Menzel, F., Sabundijan, G., D'Auria, F., Madeira, A., 2016a. Proposal for systematic application of BEPU in the licensing process of nuclear power plants. Nuclear Energy Science and Technology 10 (4), 323–338.

Menzel, F., Sabundijan, G., D'Auria, F., Madeira, A., 2018. BEPU-FSAR: discussion of V&V principles to non thermal-hydraulic codes. In: Best Estimate Plus Uncertainty International Conference (BEPU 2018), Lucca, Italy, May 13–19.

Meredith, R.E., Tobias, C.W., 1961. Conductivity of emulsions. Journal of the Electrochemical Society 103, 286–290.

Meri, A., Wengle, H., 2002. DNS and LES of turbulent backward-facing step flow using 2nd and 4th-order discretization. In: Advances in LES of Complex Flows. Springer, Netherlands, pp. 99–114.

Meri, A., Wengle, H., Raddaoui, M., Chauve, M.-P., Schiestel, R., 1999. Large-eddy simulation of non-equilibrium inflow conditions and of the spatial development of confined plane jet with co-flowing streams. In: Rodi, W., Laurence, D. (Eds.), Proceedings Of the 4th International Symp. on Engineering Turbulence Modelling and Measurements, Porticcio-Ajaccio, Corse, France, 24–26 May. Elsevier, Amsterdam, NL, pp. 1–197.

Merilo, M., Dechene, R.L., Cicowlas, W.M., 1977. Void fraction measurement with a rotating electric field conductance gauge. Journal of Heat and Mass Transfer Transactions, ASME 99, 330–331.

Merzari, E., Fischer, P., Yuan, H., Van Tichelen, K., Keijers, S., De Ridder, J., Degroote, J., Vierendeels, J., Doolaard, H., Gopala, V.R., Roelofs, F., 2016. Benchmark exercise for fluid flow simulations in a liquid metal fast reactor fuel assembly. Nuclear Engineering and Design 298, 218–228.

Meyer, C.A., McClintock, R.G., Silvestri, G.J., Spencer Jr., R.C., 1967. ASME Steam Tables—Thermodynamic and Transport Properties of Steam. The American Society of Mechanical Engineers, New York, NY, USA.

MHI, 2017. The Advanced Accumulator. Non-proprietary Mitsubishi Heavy Industries, MUAP-07001-NP (R6), Minato-Ku, Tokyo, Japan, pp. 1–121.

Michal, R., 2001. Nuclear News. ANS, IL, USA (November).

Michelotti, L., Rozzia, D., Del Nevo, A., D'Auria, F., 2011. Prediction of void fraction in PWR subchannel by CATHARE2 code. In: International Congress on Advances in Nuclear Power Plants (ICAPP 2011), Paper 11396, Pp 1921-1927, Nice (F), May 2–6, ISBN 978-1-61839-809-3. https://www.sfen.fr/ICAPP-11.

Miller, B.S., 1962. The features of contact heat transfer in heat-generating units of reactors. Izvestiya Vuzov. Energetika 24, 133–139 (in Russian).

Miller, R.W., 1996. Flow Measurement Engineering Handbook. McGraw Hill, New York NY, USA.

Miller, J., 2010. Fundamental of Nuclear Engineering-Module 13. http://www.nrc.gov/docs/ML1214/ ML12142A162.pdf.

Miller, A.I., Luxat, J.C., Price, E.G., Duffey, R.B., Fehrenbach, P.J., 2009. Heavy water reactors. In: Kok, K. (Ed.), Nuclear Engineering Handbook. CRC Press, Boca Raton, Fl, USA, pp. 141–196.

Milne-Thompson, L.M., 1968. Theoretical Hydrodynamics, fifth ed. Macmillan, London, UK, pp. 1–743.

Mimouni, S., Archambeau, F., Boucker, M., Lavieville, J., Morel, C., 2008. A second-order turbulence model based on a Reynolds stress approach for two-phase flow—part I: adiabatic cases. Science and Technology of Nuclear Installations 1–14.

Mimouni, S., Boucker, M., Laviéville, J., Guelfi, A., Bestion, D., 2008a. Modeling and computation of cavitation and boiling bubbly flows with the NEPTUNE CFD code. Nuclear Engineering and Design 238, 680–692.

Mimouni, S., Lamy, J.-S., Lavieville, J., Guieu, S., Martin, M., 2010. Modelling of sprays in containment applications with A CMFD code. Nuclear Engineering and Design 240, 2260–2270.

Minchenko, F.P., 1960. On the issue of heat transfer in nucleate boiling. Power Machines and Buildings 6, S17–S21.

Miro, J., Sonnenburg, H.G., Steinhoff, F., Teschendorff, V., Forge, A., Pochard, R., Porracchia, A., 1990. Comparison of Thermal-Hydraulic Safety Codes for PWR Systems. Graham & Trotman, London, UK.

Miropol'sky, Z.L., 1962. Heat transfer at condensation of high pressure vapour inside tubes. Teploenergetika 3, 79–82 (in Russian).

Miropol'sky, Z.L., 1975. Heat transfer to the superheated steam with heat supply and dissipation. Teploenergetika 3, 75–78 (in Russian).

Mitchell, A., Griffiths, D., 1980. The Finite-Difference Method in Partial Differential Equation. John Wiley & Sons, New York, NY, USA.

Mitin, I., Kalinkin, A., Laevsky, Y., 2012. A parallel iterative solver for positive-definite systems with hybrid MPI–OpenMP parallelization for multi-core clusters. Journal of Computational Sciences 3 (6), 463–468.

Mitrophanova, O.V., 2002. Methods of Mathematical Modeling of Fluid Flow and Heat Transfer of Swirling Flows in the Channels with Promotions. Diss. Doctor Techn. Sciences, MEPHI, Moscow, Russia (in Russian).

Miwa, S., Hibiki, T., 2022. Inverted annular two-phase flow in multiphase flow systems. International Journal of of Heat and Mass Transfer 186, 122340.

Mizutani, Y., Tomiyama, A., Hosokawa, S., Sou, A., Kudo, Y., Mishima, K., 2007. Two-phase flow patterns in a four by four rod bundle. Journal of Nuclear Science and Technology 44 (6), 894–901.

Modro, M., Goodrich, L.D., McPherson, G.D., 1985. Scaling and instrumentation of the LOFT facility. In: Proceedings Of Specialist Meet. on Small Break LOCA Analysis in LWRs, Pisa, Italy.

Modro, S.M., Fisher, J.E., Weaver, K.D., Reyes Jr., J.N., Groome, J.T., Babka, P., Carlson, T.M., 2003. Multi-Application Small Light Water Reactor Final Report. INEEL/EXT-04-01626, Idaho Falls, ID, USA.

Mokizuki, H., 2007. Analysis of the Chernobyl accident from 1:19:00 to the first power excursion. Nuclear Engineering and Design 237, 300–307.

Moody, F.J., 1965. Maximum flow rate of a single component, two-phase mixture. ASME Transactions Series C, Journal of Heat and Mass Transfer 87 (1), 134–142.

Moody, F.J., 1966. Maximum two-phase vessel blowdown from pipes. Journal of Heat and Mass Transfer 87, 285–295.

Moore, K.V., Rettig, W.H., 1968. RELAP2: A Digital Program for Reactor Blow-Down and Power Excursion Analysis. US DOE, IDO 17263, Idaho Falls, ID, USA.

Moorthi, A., Sharma, A.K., Velusamy, K., 2018. A review of sub-channel thermal hydraulic codes for nuclear reactor core and future directions. Nuclear Engineering and Design 332, 329–344.

Morel, C., 2015. Mathematical Modelling of Disperse Two-phase Flow. Springer-Verlag, Cham, Switzerland.

Morel, C., Bestion, D., 1999. Validation of the Cathare code against PERICLES 2D boil-up tests. In: Proceedings of 9th International Topical Meeting on Nuclear Reactor Thermal Hydraulics (NURETH-9), San Francisco, CA, USA, October 3–8.

Morel, C., Boudier, P., 1999. Validation of the CATHARE code against PERICLES 2D reflooding tests. In: Proceedings of 9th International Topical Meeting on Nuclear Reactor Thermal Hydraulics (NURETH-9), San Francisco, CA, USA, October 3–8.

Morel, C., Goreaud, N., Delhaye, J.M., 1999. The local volumetric interfacial area transport equation: derivation and physical significance. International Journal of Multiphase Flow 25, 1099–1128.

Morel, C., Bestion, D., Serre, G., Dor, I., 2000. Application of the 6-equation Cathare 3-D module to complex flows in a reactor core under core uncovery and reflooding conditions. In: AMIF-ESF Workshop on Computing Methods for Two-phase Flow, Aussois, Frances, January 12–14.

Moretti, F., 2016. Information about SWINTH and proposal for cooperation and support from WGAMA. In: 19th WGAMA Meeting, OECD Conf. Centre, Sept. 23–26, Paris, France.

Moretti, F., Del Nevo, A., Galassi, M.C., D'Auria, F., 2007. Study of a Boiling Channel with System and CFD Codes (SP2)—EC-EURATOM NURESIM Project, Report–D2.2.10.1, October, Pisa, Italy, (Submitted to EC, Brussels, Belgium).

Moretti, F., Terzuoli, F., D'Auria, F., Mazzantini, O., 2018. Instrumentation for full-scale boron injection test facility to support Atucha-2 NPP licensing. Journal of Nuclear Engineering and Design 336, 154–162.

Morozov, A.V., Sakhipgareev, A.R., 2017. Experimental estimation of the effect of contact condensation of steam–gas mixture on VVER passive safety systems operation. Nuclear Energy and Technology 3, 98–104.

MPR Associates, 1990. Summary of Results from the UPTF Downcomer Separate Effects Tests, Comparison to Previous Scaled Tests and Application to U.S. Pressurized Water Reactors. MPR-1163, USA, July.

Mueller-Dietsche, W., Katzenmeier, G., 1985. Reactor safety research at the large scale facility HDR. Journal of Nuclear Engineering and Design 88 (3), 241–251.

Muellner, N., Cherubini, M., Kromp, W., D'Auria, F., Petrangeli, G., 2007. Aprocedure to optimize the timing of operator actions of accident management procedures. Nuclear Engineering and Design 237.

Muellner, N., Camusso, P., Mazzantini, O., D'Auria, F., 2011. RELAP5 based subchannel analysis for licensing of Atucha2 NPP. In: International Topical Meeting on Nuclear Reactor Thermal-Hydraulics (NURETH-14), Article No 514, Toronto, Ontario, Canada, September 25–30.

Muftuoglu, K., Ohkawa, K., Frepoli, C., Nissley, M., 2004. Comparison of realistic large break LOCA analyses of a 3-Loop Westinghouse plant using response surface and statistical sampling techniques. In: Proceedings of ICONE-12, 25–29 April, Arlington, VA, USA.

Mull, T., Perst, M., Umminger, K., 1997. Effectiveness of accident management procedures under small break LOCA conditions. In: 5th International Conference On Nuclear Engineering (ICONE-5), Nice, France, May 26–30, pp. 1–8. Paper 2192.

Mull, T., Umminger, K., Bucalossi, A., D'Auria, F., Monnier, P., Toth, I., Schwarz, W., 2007. Final Report of the OECD-PKL Project. NTCTP-G/2007/en/0009, Areva, Erlangen.

Mullen, E., Stumpf, H.J., Siebe, D.A., 1992. Summary of Downcomer Injection Phenomena for UPTF and TRAC post-test Analyses. Los Alamos National Laboratory document LA-CP-92-188, Los Alamos, NM, USA.

Muller, T., Mourgues, A., Pedler, F., Lassauce, A., Guillodo, M., Dolleans, P., Caron-Charles, M., 2015. Experimental and theoretical study of iron concentration on clogging phenomenon in secondary circuit of pressurized nuclear power plant. In: 16th International Topical Meeting on Nuclear Reactor Thermal Hydraulics (NURETH-16), August 30–September 4, Chicago, IL, USA.

Muñoz-Cobo, J.L., Escrivá, A., Mendizabal, R., Pelayo, F., Melara, J., 2014. CSAU methodology and results for an ATWS event in a BWR using information theory methods. Nuclear Engineering and Design 278.

Muñoz-Cobo, J.L., Berna, C., Escrivá, A., 2018. Top-down scaling methodology from the LSTF facility to a three loop PWR plant applied to a SBLOCA event—the ROSA 1.2 test. Nuclear Engineering and Design 327, 248–273.

Murai, Y., Ohta, S., Shigetomi, A., Tasaka, Y., Takeda, Y., 2009. Development of an ultrasonic void fraction profiler. Measurement Science and Technology 20 (11).

Murakami, M., Maruo, K., Yoshiki, T., 1990. Development of an electromagnetic flowmeter for studying gas liquid two phase flow. International Chemical Engineering 30 (4), 699–702.

Murallidharan, J., Giustini, G., Sato, Y., Ničeno, B., Badalassi, V., Walker, S.P., 2015. Interface tracking based evaluation of bubble growth rates in high pressure pool boiling conditions. In: 16th International Topical Meeting on Nuclear Reactor Thermal Hydraulics (NURETH-16), August 30–September 4, Chicago, IL, USA.

Murao, Y., Akimoto, H., Sudoh, T., Okubo, T., 1982. Experimental study of system behavior during reflood phase of PWR LOCA using CCTF. Nuclear Science and Technology 19, 705–719.

Murao, Y., Higuchi, T., Akimoto, H., Iwamura, T., 1993. Large scale multidimensional phenomena found in CCTF and SCTF experiments. Journal of Nuclear Engineering and Design 145, 85–95.

Murase, M., Takaki, T., Nishida, K., Tomiyama, A., 2022. Counter-Current flow limitation and void fraction in a rod bundle under flooding conditions. In: 19th International Topical Meeting on Nuclear Reactor Thermal Hydraulics (NURETH-19), March 6–11 (Via Web), Brussels, Belgium.

Murata, K.K., Williams, D.C., Tills, J., Griffith, R.O., Gido, R.G., Tadios, E.L., Davis, F.J., Martinez, G.M., Washington, K.E., 1997. Code Manual for CONTAIN 2.0: A Computer Code for Nuclear Reactor Containment Analysis. SAND97-1735, USNRC NUREG/CR-6533, Washington, DC, USA.

Murray, R., Holbert, K.E., 2014. An introduction to the concepts, systems and applications of nuclear processes. In: Nuclear Energy, seventh ed. Elsevier, Amsterdam, Netherlands.

Murrone, A., Guillard, H., 2005. A five equation reduced model for compressible two phase flow problems. Journal of Computational Physics 202, 664–698.

Nagy, I., Hózer, Z., Novotny, T., Windberg, P., Vimi, A., 2015. Extended study of coolability of VVER bundle with balloned region. In: 16th International Topical Meeting on Nuclear Reactor Thermal Hydraulics (NURETH-16), August 30–September 4, Chicago, IL, USA.

Nahavandi, A.N., Castellana, F.S., Moradkhanian, E.N., 1979. Scaling laws for modeling nuclear reactors systems. Nuclear Science and Engineering 72, 75.

Nainer, O., Pun-Quach, D., Sermer, P., Phan, B., Hoppe, F.M., 2013. A BEPU analysis separating epistemic and aleatory errors to compute accurate dry-out power uncertainties. Nuclear Technology 181 (1), 170–183.

Nakamura, H., Watanabe, T., Takeda, T., Maruyama, Y., Suzuki, M., 2009. Overview of recent efforts through ROSA/LSTF experiments. Nuclear Engineering and Technology 41 (6).

Nakamura, H., Toth, I., Sandervag, O., Umminger, K., Dreier, J., Prior, R., Alonso, J.R., Muellner, N., D'Auria, F., Amri, A., Muhleisen, A., 2011. Effectiveness of core exit thermocouple (CET) temperature indication in accident management of light water reactors. In: International Topical Meeting on Nuclear Reactor Thermal-Hydraulics (NURETH-14), Article No 275, Toronto, Ontario, Canada, September 25–30.

Narayanan, C., 2021. Numerical simulation of flashing using a pressure-based compressible multiphase approach and a thermodynamic cavitation model. International Journal of Multiphase Flow 135 (103511), 1–14.

Nariai, H., Aya, I., 1986. Fluid and pressure oscillations occurring at direct contact condensation of steam flow with cold water. Nuclear Engineering and Design 95, 35–45.

Navarro, M.A., 2005. Study of contercurrent limitation in a horizontal pipe connected to an inclined one. Nuclear Engineering and Design 235, 1139–1148.

Navarro, M.A., Santos, A.A.C., 2009. Numerical evaluation of flow through a 5 X 5 PWR rod bundle: effect of the vane arrangement in a spacer grid. In: Proceedings Of International Nuclear Atlantic Conf. INAC, Rio de Janeiro, Brazil.

Neal, L.G., Bankoff, S.G., 1963. A high resolution resistivity probe for determination of local void properties in gas-liquid flow. AIChE Journal 9, 490–494.

NED, 1990. Special Issue of Journal Nuclear Engineering and Design, vol. 119. CSAU.
NED, 1998. Special Issue of Journal Nuclear Engineering and Design, vol. 186. Scaling.
NED, 2019. Special issue of Journal nuclear engineering and design, trends and perspectives in nuclear thermal hydraulics, to the memory of B. In: Sehgal, R., Yadigaroglu, G., Hewitt, G.F. (Eds.), (same as D'Auria and Hassan [Ed.S], 2019), vol. 354.
Nelson, L., 2008. Boiling Water Reactor Basics. GE Global Research, the materials ageing institute, Wilmington, NC, USA).
NEWS, 2016. http://www-ns.iaea.org/tech-areas/emergency/news.asp.
Neykov, D., Aydogan, F., Hochreiter, L., Utsuno, H., Kasahara, F., Sartori, E., Martin, M., 2006. NUPEC BWR Full-Size Fine-mesh Bundle Test (BFBT) Benchmark, pp. 1−132. OECD/NEA Papers, Vol 6/7, Paris, France.
NGDC, 2016. http://www.ngdc.noaa.gov/hazard/hazards.shtml.
Niceno, B., Dhotre, M.T., Deen, N.G., 2008. One equation sub-grid scale (SGS) modeling for Euler-Euler large eddy simulation (EELES) of dispersed bubbly flow. Chemical Engineering Sciences 63, 3923−3931.
Niceno, B., Boucker, M., Smith, B.L., 2009. Euler-Euler large eddy simulation of a square cross-sectional bubble column using the NEPTUNE CFD code. Science and Technology of Nuclear Installations. Article ID 410272.
Nijsing, R., Gargantini, I., Eifler, W., 1964. Fundamental studies of fluid flow and heat transfer in fuel element geometries, part I. In: Joint Nuclear Research Centre, EUR 2193.e, Ispra, Italy.
Nikitin, K., Clifford, I., Dokhane, A., Ferroukhi, H., 2022. Methodology for CPR estimations of BWR cycle specific transient reload analyses using CTF sub-channel code. Nuclear Engineering and Design 389 (111649), 1−14.
Fourier transforms—approach to scientific principles. In: Nikolic, G.S. (Ed.), 2011. Tech Publisher, Rijeka, Croatia, ISBN 978-953-307-231-9, pp. 1−468.
Nikuradse, J., 1932. Laws of turbulent flow in smooth pipes (English translation). Mitt. Forsch. Geb. Ing.-Wesen. 356 (1).
Nilsen, T., Kudrik, I., Nikitin, A., 1997. Bellona Report 2.96. The Russian Northern Fleet Nuclear Submarine Accidents. © Bellona. http://spb.org.ru/bellona/ehome/russia/nfl/nfl8.htm.
Ninokata, H., 1986. Analysis of low-heat-flux sodium boiling test in a 37-pin bundle by the two-fluid model computer code SABENA. Nuclear Engineering and Design 97 (2), 233−246.
Ninokata, H., Okano, T., 1985. SABENA: an advanced sub-channel code for sodium boiling analysis. In: Proceedings of 3rd International Topical Meeting on Reactor Thermal Hydraulics (NURETH-3), Vol 2, Paper 16.K-1−7, Newport, RI, USA.
Ninokata, H., Deguchi, A., Baba, K., 1985. ASFRE-III: A Computer Program for Triangular Rod Array Thermo Hydraulic Analysis of Fast Breeder Reactors, pp. 85−106. Power Reactor and Nuclear Fuel Development Corporation, PNC N941 report, O-arai, Japan.
Ninokata, H., Efthimiadis, A., Todreas, N.E., 1987. Distributed resistance modeling of wire wrapped rod bundles. Nuclear Engineering and Design 104 (1), 93−102.
Ninokata, N., Aritomi, A., 1992. 1st international seminar on sub-channel analysis in nuclear reactors (ISSCA-I), Tokyo. Institute of Applied Energy (IAE) and the THD of the Atomic Energy Society of Japan, Seminar Proceedings, Tokyo, Japan, ISBN 4-9900222-0-3, pp. 1−301 [also, Ninokata, H., Aritomi, M., 1992. Sub-Channel Analysis in Nuclear Reactors. Inst. of Applied Energy, Nihon Genshiryoku, Gakkai, Japan.
Ninokata, H., Kusuno, S., Hirusawa, S. (Eds.), 1997. 4th International Seminar on Subchannel Analysis (ISSCA-IV). Institute of Applied Energy (IAE) and the THD of the Atomic Energy Society of Japan, Seminar Proceedings, Tokyo, ISBN 4-9900222-1-1, pp. 1−354. Tokyo, Japan.

Ninokata, H., Mishima, K., Sadatomi, M., So, A., Yamamoto, Y., Nishida, K., Yabushita, Y., Hara, T., Kusuno, S., 2000. Development and improvement of transient subchannel analysis code NASCA. Developmental Status. Nihon Kikai Gakkai Nenji Taikai Koen Ronbunshu, Japan, 1, pp. 655–656.

Ninokata, H., Aritomi, M., Anegawa, T., Hara, T., Kamo, H., Kusuno, S., Mishima, K., Morooka, S., Yamamoto, Y., Nishida, K., Sadatomi, M., Sou, A., Yabushita, Y., 2000a. Development of the NASCA code for prediction of transient BT and post BT phenomena in BWR rod bundles. In: Proceedings Of Workshop Advanced Thermal-hydraulic and Neutronic Codes: Current and Future Applications Workshop, Barcelona, Spain, April 10–13, NEA/CSNI/R(2001)9, Paris France.

Ninokata, H., Merzari, E., Khakim, A., 2009. Analysis of low Reynolds number turbulent flow phenomena in nuclear fuel pin subassemblies of tight lattice configuration. Nuclear Engineering and Design 239 (5), 855–866.

Ninokata, H., Yang, B.-W., Baglietto, E. [E.s], 2018. International seminar on nuclear reactor core thermal hydraulic analysis (IS-ReCTHA). In: Abstracts from the Seminar Held in Lecco, Italy, August 29–31, 2018. www.eko.polimi.it/index.php/rectha.

Ninokata, H., Yang, B.-W., Baglietto, B., Merzari, E. (Eds.), 2021. Special Issue on IS-ReCTHA-2018. Nuclear Engineering and Design. https://www.sciencedirect.com/journal/nuclear-engineering-and-design/special-issue/10BZ1RQ3CJ2.

Nishikawa, T., Inoue, H., Motohashi, S., Ebisawa, K., 2014. Lessons learned from Kashiwazaki-Kariwa NPP after Niigataken Chuetsu-Oki earthquake (2007) in view of SSI effect. In: Hsu, T.T.C., Wu, C.-L., Li, J.-L. (Eds.), Infrastructure Systems for Nuclear Energy. John Wiley & Sons, Ltd, Chichester, UK.

Nissen, W.H.M., Van Der Voet, J., Karuza, J., 1994. The start-up of the Dodewaard natural circulation boiling water reactor—experiences. Journal of Nuclear Technology 107.

NMED, 2011. https://ire.org/nicar/database-library/databases/nuclear-material-events-databasenmed/.

No, H.C., Lee, K.-W., Song, C.-H., 2005. An experimental study on air/water countercurrent flow limitation in upper plenum with multi-hole plate. Nuclear Engineering and Technology 37 (6), 557–564.

No, H.C., Ha, S.J., Kim, K.D., Lim, H.S., Lee, E.H., Jin, H.G., 2013. Korean development of advanced thermal-hydraulic codes for water reactors and HTGRs. Nuclear Technology 181 (1), 24–43.

Noori-Kalkhoran, O., Shirani, A.S., Ahangari, R., 2016. Simulation of containment pressurization in a large break-loss of coolant accident using single-cell and multicell models and CONTAIN code. Nuclear Engineering and Technology 48 (5), 1140–1153.

Nourbakhsh, H.P., Banerjee, S., 2013. Assessing the adequacy of thermal–hydraulic analyses for safety decisions. In: NURETH-15 Conference, Pisa, Italy, May.

Nourbakhsh, H.P., Cheng, Z., 1995. Mixing phenomena of interest to boron dilution during small break LOCAs in PWRs. In: International Conference NURETH-7, Saratoga Springs, NY, USA.

Novikov, I.I., 1961. Conditions for the Similarity of Heat Transfer Processes with Variable Fluid Properties, Problems of Two-phase Heat Transfer. Moscow, GosEnergoizdat, pp. 7–14 (in Russian).

Novozhilov, I.V., 1997. Fractional Analysis—Methods of Motion Decomposition. Springer, Science & Business Media, Birkhauser Publisher, Boston, Ma, USA.

NT, 2021. Special Issue of Journal Nuclear Technology 207 (S1-S23). Supplement 1.

Nukiyama, S., 1934. Maximum and minimum values of heat q transmitted from metal to boiling water under atmospheric pressure. Journal of the Society of Mechanical Engineers 37 (53–54), 367–374.

Nukiyama, S., 1966. The maximum and minimum values of heat transmitted from metal surface to boiling water under atmosphere pressure. International Journal of Heat Mass Transfer 9, 1419–1433 (also Journal of the Japan Society of Mechanical Engineers, 37, 206).

NuScale, 2014. Plant Design Overview. NuScale, LLC, Corvallis, OR, USA, pp. 3–39. RP-1114-9375 rev.0.

Nusselt, W., 1916. Die Oberflachenkondensation des Wasserdampfes. Ver. Dtsch. Ing. 60, 541.

Nusselt, W., 1943. Technische Thermodynamik. Walter de Gruyter & Co, Berlin, Germany.

Oberkampf, W.L., Roy, C.J., 2010. Verification and Validation in Scientific Computing. Cambridge University Press, USA, ISBN 978-0-521-11360-1, pp. 1–790.

Oberkampf, W.L., Trucano, T.G., 2008. Verification and validation benchmarks. Nuclear Engineering and Design 238, 716–743.

Oberkampf, W.L., Balch, M.S., 2020. Closure on the discussion of "models, uncertainty, and the Sandia V&V challenge problem" (Hazelrigg, G.A., and Klutke G.A., ASME Journal Verif. Valid. Uncert., 5(1), paper 015501). ASME Journal Verif., Valid. Uncert. 5 (3) paper 035501 (VVUQ-20-1021).

Obry, P., Cheissoux, J.L., Grandotto, M., Gaillard, J.P., De Langre, E., Bernard, M., 1990. An advanced steam generator design 3-D code. In: Thermal Hydraulics of Advanced Heat Exchangers, 1990 ASME Winter Annual Meeting, Dallas, Tx, USA, 25–30 November.

Odar, F., 2001. Assessment of the TRAC-M Codes Using FLECHT-SEASET Reflood and Steam Cooling Data. NUREG-1744. USNRC, Washington, DC, USA.

Oddie, G., Pearson, J.R.A., 2004. Flow rate measurement in two-phase flow. Annual Review of Fluid Mechanics 36, 149–172.

Odelevsky, V.I., 1951. Calculation of generalized conductivity of heterogeneous systems. Zhurnal Tekhnicheskoy Phiziky 21 (6), 667–685 (in Russian).

OECD/NEA, 2004. Neutronics/thermal-hydraulics Coupling in LWR Technology—OECD/NEA Report No. 4452, ISBN 92-64-02083-7 (Vol. I, 103 pages), Paris, France. (see D'Auria et al., 2004).

OECD/NEA, 2016. SETF data; and. https://www.oecd-nea.org/dbprog/ccvm/indexset.html (ITF data).

OECD/NEA/CNRA, 2014. Report on Fukushima Daiichi NPP Precursor Events. NEA/CNRA (2014)1, Paris, France.

OECD/NEA/CSNI, D'Auria, F., 1997b. In: Ambrosini, W., Anegawa, T., Blomstrand, J., De Betou, J., Langenbuch, S., Lefvert, T., Valtonen, K. (Eds.), State-of-the-Art Report on Boiling Water Reactor Stability (SOAR on BWRS). NEA/CSNI/R(96)21, OCDE/GD(97) 13, OECD/NEA Paris, France.

OECD/NEA/CSNI, Wickett, T., 1998. In: D'Auria, F., Glaeser, H., Chojnacki, E., Lage, C., Sweet, D., Neil, A., Galassi, G.M., Belsito, S., Ingegneri, M., Gatta, P., Skorek, T., Hofer, E., Kloos, M., Ounsy, M., Sanchez, J.I. (Eds.), Report of the Uncertainty Methods Study for advanced best estimate thermal-hydraulic code applications—Vols. I and II. OECD/CSNI Report NEA/CSNI/R(97)35, Paris, France. [see also Wickett et al., 1998].

OECD/NEA/CSNI, 2001. A Report by OECD Support Group, 2001. Validation Matrix for the Assessment of Thermal-Hydraulic Codes for VVER LOCA and Transients. NEA/CSNI/R (2001) 4, JT00108841, Paris, France.

OECD/NEA/CSNI, Allelein, H.-J., Auvinen, A., Ball, J., Güntay, S., Herranz, L.E., Hidaka, A., Jones, A., Kissane, M., Powers, D., Weber, G., 2009. State-Of-the-Art Report (SOAR) on Nuclear Aerosols (NARSOAR), CSNI /R(2009)5, Paris, France.

OECD/NEA/CSNI, D'Auria, F., 2020. In: Herer, C., Bousbia Salah, A., Di Maio, F., Zio, E., Kang, K.H., Chabod, A., Do, K., Morin, F., Toth, B., Mascari, F., Lanfredini, M. (Eds.), Status Report on Reliability of thermal Hydraulic Passive Systems, CSNI Report (Approved) to Be Issued, Paris, France.

OECD/NEA/CSNI, 2022. In: Proceedings of Specialists Meeting on Transient Thermal-hydraulics in Water Cooled Nuclear Reactors (SM-TH)—March 22–24, Madrid (Spain) [to be issued]).

OECD/NEA/CSNI, Karwat, H., Bardelay, J., Hashimoto, T., Koroll, G.W., Krause, M., L'Heriteau, J.-P., Lundstrom, P., Notafranesco, A., Royl, P., Schwinges, B., Tezuka, H., Tills, J., Royen, J., 1999. State-of-the-Art Report (SOAR) on Containment Thermal-hydraulics and Hydrogen Distribution, CSNI/R(1999)16, Paris, France.

OECD/NEA/CSNI, Bestion, D., D'Auria, F., 2017. In: Lien, P., Nakamura, H., Austregesilo, H., Skorek, T., Bae, B.U., Choi, K.Y., Kim, K.D., Mon, S.-K., Martinez-Quiroga, V., Reventos, F., Mascari, F., Schollenberger, S., Umminger, K., Reyes, J.N., Rohatgi, U.S., Wang, W., Zaki, T. (Eds.), Scaling in System Thermal-Hydraulics Applications to Nuclear Reactor Safety and Design: A State-Of-The-Art-Report, NEA/ CSNI/R(2016)14, JT03411050, Paris, March, pp. 1–401.

OECD/NEA/CSNI, USNRC, 1997. Proceedings of OECD/NEA/CSNI workshop on transient thermal-hydraulic and neutronics codes requirements. In: Eltawila, s F., Ybarrondo, L. (Eds.), Held in Annapolis (Md, US), November 5–8, 1996, OECD/NEA/CSNI R(97)4, Paris (F)—NUREG/CP-0159, Washington, DC, USA.

OECD/NEA/CSNI, 1980. Two Phase Critical Flow Models, CSNI Report 49, Paris (see also D'Auria and Vigni, 1980).

OECD/NEA/CSNI, 1982. Critical flow modelling in nuclear safety. In: A State of the Art Report by a Group of Experts of the NEA Committee on the Safety of Nuclear Installations. Nuclear Energy Agency, Paris, France.

OECD/NEA/CSNI, 1983. In: Proceedings of the OECD/CSNI Specialist Meeting on Water Reactor Containment Safety, October 17–21, Toronto, Ontario, Canada, pp. 1–502.

OECD/NEA/CSNI, 1986. Pressure Suppression System Containments, a State of Art Report. CSNI Report 126, Paris, France.

OECD/NEA/CSNI, 1987. CSNI code validation matrix of thermo-hydraulic codes for LWR LOCA and transients. In: CSNI Report 132, Paris, France. (see also Aksan et al., 1987).

OECD/NEA/CSNI, 1988. ISP-20—Steam Generator Tube Rupture Behavior, Nuclear Power Plant Doel 2, Belgium, CSNI Report No. 154, Paris, France.

OECD/NEA/CSNI, 1989. Thermohydraulics of emergency core cooling in light water reactors—a state of the art report. In: CSNI Report 161, October, Paris, France. (see also Lewis et al., 1989).

OECD/NEA/CSNI, 1989a. Containments for Pressurized Water Reactors: A State-Of-The-Art-Report (SOAR), CSNI Report 166, Paris.

OECD/NEA/CSNI, 1990. USNRC and CSNI International Workshop on Boiling Water Reactor Stability. Brookhaven, New York, NY, USA. October 17–19 CSNI Report 178, Paris, France.

OECD/NEA/CSNI, 1992a. International Standard Problem 27 (ISP 27)—BETHSY experiment 9.1b, 2" Cold Leg Break without HPSI and with Delayed Ultimate Procedure, Comparison Report, pp. 1–157. NEA/CSNI/R(92)20, Vol I, Paris, France.

OECD/NEA/CSNI, 1993. Separate Effects Test Matrix for thermal-hydraulic Code Validation: Phenomena Characterization and Selection of Facilities and Tests. OCDE/GD(94)82, NEA/ CSNI/R(93)14, Vols. I and II, Paris, France [see also Aksan et al., 1994, and CSNI, 1994].

OECD/NEA/CSNI, 1994. Survey of Containment Designs for New/advanced Water Reactors. NEA/CSNI/R(94)5, Paris, France.
OECD/NEA/CSNI, 1996. CSNI Code Validation Matrix of Thermo-Hydraulic Codes for LWR LOCA and Transients. CSNI Report, 132-Rev. 1 (Rev. 0 published in 1987), Paris, France.
OECD/NEA/CSNI, 1996a. Evaluation of the Separate Effects Tests (SET) Validation Matrix, NEA/CSNI/R(96)16, Paris, France.
OECD/NEA/CSNI, 1996b. Relevant Thermal-Hydraulic Aspects of Advanced Reactor Design—Status Report. CSNI/R(1996)22, OECD/GD(97)8, Paris, France.
OECD/NEA/CSNI, 1996c. Lessons Learned from OECD/CSNI ISP on Small Break LOCA, Final Report. CSNI R(1996)20, also referenced as: OCDE/GD(97)10, Paris, France.
OECD/NEA/CSNI, 1996d. CSNI Integral Test Facility Validation Matrix for the Assessment of Thermal-Hydraulic Codes for LWR LOCA and Transients. NEA/CSNI/R(96)17, Paris, France.
OECD/NEA/CSNI, 1997. In: Proceedings of OECD/CSNI Specialists Meeting on Advanced Instrumentation and Measurement Techniques, Santa Barbara, CA, USA, March 17–20, 1997. NEA/CSNI/R(97)33, Paris, France.
OECD/NEA/CSNI, 1997a. ISP-brief Descriptions 1975–1997. Report NEA/CSNI/R(97)3, Paris, France.
OECD/NEA/CSNI, 1999a. In: Proceedings of the Seminar on Best Estimate Methods in Thermal-hydraulics Safety Analysis, Held in Ankara, Turkey June 29–July 1, 1998. NEA/CSNI/R(99)/22, Paris (F), pp. 1–35.
OECD/NEA/CSNI, 2000. In: OECD/CSNI Workshop on Advanced Thermal-hydraulics and Neutronics Codes: Current and Future Applications, Barcelona, Spain, April 10–13.
OECD/NEA/CSNI, 2000a. CSNI International Standard Problems (ISP) Brief Descriptions (1975-1999), CSNI/R(2000)5, Paris, France.
OECD/NEA/CSNI, 2004. OECD SETH Project—Final Report of the PKL Experimental Programme Part (2000–2004). NEA/CSNI/R(2004)24, Paris, France.
OECD/NEA/CSNI, 2006. BEMUSE Phase II Report, Re-analysis of the ISP-13 Exercise, post-test Analysis of the LOFT L2-5 Test Calculation. NEA/CSNI/R (2006)2. Paris, France.
OECD/NEA/CSNI, 2007. State of Art Report on Iodine Chemistry. NEA/CSNI/R(2007)1, Paris, France.
OECD/NEA/CSNI, 2007a. BEMUSE Phase III Report, Uncertainty and Sensitivity Analysis of the LOFT L2-5 Test. NEA/CSNI/R(2007)4, Paris, France.
OECD/NEA/CSNI, 2007b. Best Practice Guidelines for the Use of CFD in Nuclear Reactor Safety Applications, NEA/CSNI/R(2007)5, Paris, France.
OECD/NEA/CSNI, 2007c. International Standard Problem 47 (ISP-47) on Containment Thermal-hydraulics—Final Report, NEA/CSNI/R(2007)10, Paris, France.
OECD/NEA/CSNI, 2008. BEMUSE Phase IV Report, Simulation of a LB-LOCA in Zion Nuclear Power Plant. NEA/CSNI/R(2008)6, Paris, France.
OECD/NEA/CSNI, 2009a. BEMUSE Phase V Report, Uncertainty and Sensitivity Analysis of a LB-LOCA in Zion Nuclear Power Plant. NEA/CSNI/R(2009)13, Paris, France.
OECD/NEA/CSNI, 2010. Nuclear Fuel Behavior under Reactivity Initiated Accident (RIA) Conditions, NEA/CSNI/R(2010)1, Paris, France.
OECD/NEA/CSNI, 2010a. Core Exit Temperature Effectiveness in Accident Management of Nuclear Power Reactors. NEA/CSNI/R(2010)9, Paris, France.
OECD/NEA/CSNI, 2011. BEMUSE Phase VI Report, Status Report on the Area, Classification of the Methods, Conclusions and Recommendations. NEA/CSNI/R(2011)4, Paris, France (see also Glaeser, 2011).
OECD/NEA/CSNI, 2011a. Safety Margins Evaluation, SMAP Framework Assessment and Application, NEA/CSNI/R(2011)3, Paris, France.

OECD/NEA/CSNI, 2012. International Standard Problem No. 50 (ISP 50), ATLAS Test, SB-DVI-09: 50% (6-inch) Break of DVI Line of the APR-1400. Final Report, Volume I, Analysis of Blind/open Calculations, CSNI/R(2012)6, Paris, France.

OECD/NEA/CSNI, 2013. Workshop on best estimate methods and uncertainty evaluations. In: Proceedings [Workshop Held in Barcelona, Spain, 16–18 November 2011], NEA/CSNI (2013)8, Paris, France.

OECD/NEA/CSNI, 2013a. Final Integration Report of the OECD / NEA ROSA Project (2005-2009), pp. 1–101. NEA/CSNI/R(2013)1, Paris, France.

OECD/NEA/CSNI, 2015. Benchmark Study of the Accident at the Fukushima Daiichi Nuclear Power Plant (BSAF Project). NEA/CSNI/R(2015)18, Paris, France.

OECD/NEA/CSNI, 2016. The scaling state of the art report—the S-SOAR. In: NEA/CSNI/R(2016)14, Paris, France [replaced by OECD/NEA/CSNI, 2017, see also Bestion et al., (D'Auria Ed.), 2017].

OECD/NEA/CSNI, 2016a. PREMIUM a Benchmark on the Quantification of the Uncertainty of the Physical Models in the System thermal-hydraulic Codes: Methodologies and Data Review. NEA/CSNI/R(2016)9, Paris, France.

OECD/NEA/CSNI, 2016b. Post-BEMUSE Reflood Model Input Uncertainty Methods (PRE-MIUM) Benchmark. Final report, NEA/CSNI, R(2016)18 [issued August 2017], Paris, France.

OECD/NEA/CSNI, 2017a. Final Integration Report of the Rig-Of-Safety Assessment (ROSA-2) Project (2009-2012), pp. 1–217. NEA/CSNI/R(2016)10, Paris, France.

OECD/NEA/CSNI, 2022a. Final Report by Senior Expert Group on Safety Research / Capabilities and Facilities, OECD 2021, NEA No. 7565, Paris, France, pp. 1–104.

OECD/NEA/CSNI, 1992. In: Reocreux, M., Rubinstein], M.C. (Eds.), Proceedings of CSNI Spec. Meet. On Transient Two-phase Flow—System Thermal-Hydraulics, Aix-En-Provence, France, April 6–8. NEA, NEA/CSNI R(92)12, Paris, France. ISBN 2-7272-0157.

OECD/NEA/CSNI-TM, 2005. In: Proceedings of Workshop on Evaluation of Uncertainties in Relation to Severe Accidents and Level 2 PSA. Aix-En-Provence, France, 7–9 November.

OECD/NEA/CSNI, Albiol, T., Amri, A., Andreani, M., Arndt, S., Barber, D., Bentaib, A., Chin, Y.-S., Clement, B., Dickson, R., Dreier, G., Fontanet, J., Garcia de la Rua, C., Gelain, T., Glowa, G., Herranz, L., Journeau, C., Klein-Hessling, W., Kotchourko, A., Kuznetosov, M., Lee, R.J., Leitch, B., Liang, Z., Malet, J., Mathew, P.M., Maynet, N., Michielsen, N., Paladino, D., Peillon, S., Piluso, P., Porcheron, E., Sangiorgi, M., Santiago, A.E., Sonnenkalb, M., Studer, E., Vasic, A., Weber, G., Yanez, G., 2014. Containment Code Validation Matrix, NEA/CSNI/R (2014)3, Paris, France, pp. 1–202.

OECD/NEA/CSNI, Jacquemain, D., Guentay, S., Basu, S., Sonnenkalb, M., Lebel, L., Allelein, H.J., Liebana Martinez, B., Eckardt, B., Ammirabile, L., 2014a. Status Report on Filtered Containment Venting, pp. 1–175. NEA/CSNI/R (2014)7, Paris, France, July.

OECD/NEA/CSNI-TM, 2011. In: Proceedings of Workshop on Best Estimate Methods and Uncertainty Evaluations. UPC and CSN, Barcelona, Spain, 16–18 November.

OECD/NEA/NSC, 2015. State-of-the-Art Report on Multi-Scale Modelling of Nuclear Fuels. NSC/R/(2015)5, Paris, October.

OECD/NEA/NSC, Avramova, M., Neykov, B., Ivanov, K., Aydogan, F., Hochreiter, L., 2021. NUPEC BWR Full-Size Fine-mesh Bundle Test BFBT Benchmark, Volume III: Results of Phase I on Void Distribution, pp. 1–179. NEA/NSC/R(2020)6, Paris, France (available Sept. 3, 2021).

OECD/NEA/NSC, Avramova, M., Neykov, B., Ivanov, K., Aydogan, F., Hochreiter, L., 2021a. NUPEC BWR Full-Size Fine-mesh Bundle Test BFBT Benchmark, Volume IV: Results of Phase II on Pressure Drop and Critical Power, pp. 1–185. NEA/NSC/R(2020)7, Paris, France (available Sept. 15, 2021).

Ofner, B., 2001. Phase Doppler anemometry (PDA). In: Mayinger, F., Feldmann, O. (Eds.), Optical Measurements. Heat and Mass Transfer. Springer, Berlin, Heidelberg, Germany.
Oh, D.S., Chun, T.H., 2003. Simulation of turbulent flow in rod bundles using eddy viscosity models and the Reynolds stress model. In: Proceedings 10th International Topical Meeting on Nuclear Reactor Thermal Hydraulics (NURETH-10), Seoul, South Korea, October 5–9.
Oh, S., Hizoum, B., Saha, P., Dooies, B., Miranda, D., Andersen, J.G.M., 2015. Film-Droplet split correlation at the onset of annular-mist flow. In: 16th International Topical Meeting on Nuclear Reactor Thermal Hydraulics (NURETH-16), August 30–September 4, Chicago, IL, USA.
Ohba, K., Kitada, H., Nishiguchi, A., 1982. Direct contact condensation on a high speed spray jet of subcooled water. In: Bankoff, s S.G., Afgan, N.H. (Eds.), Heat Transfer in Nuclear Reactor Safety. Hemisphere, Washington, DC, USA.
Oikawa, H., Sato, T., Sato, K., Matsuura, M., 2003. ABWR design and its evolution—safety system design of ABWR and ABWR-II. In: International Conference GENEA4/ANP2003, Kyoto, Japan, September 15–19, pp. 1–8 paper 1211.
Oka, Y. (Ed.), 2014. Nuclear Reactor Design. Springer, Tokyo, Japan.
Okawa, T., Kitahara, T., Yoshida, K., Matsumoto, T., Kataoka, I., 2002. New entrainment rate correlation in annular two-phase flow applicable to wide range of flow condition. International Journal of Heat Mass Transfer 45 (1), 87–98.
Okawa, T., Kotani, A., Kataoka, I., Naito, M., 2003. Prediction of critical heat flux in annular flow using a film flow model. Journal of Nuclear Science and Technology 40 (6), 388–396.
Okawa, T., Goto, T., Minamitani, J., Yamagoe, Y., 2009. Liquid film dryout in a boiling channel under flow oscillation conditions. International Journal of of Heat and Mass Transfer 52, 3665–3675.
Okrent, D., 1981. Nuclear Reactor Safety: On the History of the Regulatory Process. University of Wisconsin Press, USA.
Ono, A., Yamashita, S., Suzuki, T., Yoshida, H., 2022. Numerical simulation of two-phase flow in the fuel assemblies with the spacer grid. In: 19th International Topical Meeting on Nuclear Reactor Thermal Hydraulics (NURETH-19), March 6–11 (Via Web), Brussels, Belgium.
Ortiz, M.G., Ghan, L.S., 1992. Uncertainty Analysis of Minimum Vessel Liquid Inventory during a Small Break LOCA in a Babcock and Wilcox Plant: An Application of the CSAU Methodology Using the RELAP5/MOD3 Computer Code. Rep. NUREG/CR-5818, EGG-2665. Idaho Natl. Engineering Lab., Idaho Falls, ID, USA.
Ose, Y., Kunugi, T., 2015. Improvement of wall surface temperature evaluation procedure during subcooled nucleate boiling in non empirical boiling and condensation model. In: 16th International Topical Meeting on Nuclear Reactor Thermal Hydraulics (NURETH-16), August 30–September 4, Chicago, IL, USA.
Osif, B.A., Baratta, A.J., Conkling, T.W., 2004. TMI 25 Years Later: The Three Mile Island Nuclear Power Plant Accident and its Impact. The Pennsylvania State University, College Park, PA, USA.
Owen, D.G., Hewitt, G.F., 1987. A Proposed Entrainment Correlation. AERE-R122279, UK.
Owoeye, E.J., Schubring, D., 2015. Computational analysis of bubble micro-layer in sub-cooled boiling. In: 16th International Topical Meeting on Nuclear Reactor Thermal Hydraulics (NURETH-16), August 30–September 4, Chicago, IL, USA.
Ozaki, T., Hibiki, T., 2021. Effect of interfacial area concentration on one-dimensional code simulation of adiabatic two-phase flows in vertical large size channels. Nuclear Engineering and Design 385 (111502), 1–15.
Ozar, B., Brooks, C.S., Euh, D.J., Hibiki, T., Ishii, M., 2013a. Investigation of one-dimensional interfacial area transport for vertical upward air-water two-phase flow in an annular channel at elevated pressures. Nuclear Engineering and Design 263, 362–379.

Ozar, B., Brooks, C.S., Hibiki, T., Ishii, M., 2013b. Interfacial area transport of vertical upward steam-water two-phase flow in an annular channel at elevated pressures. International Journal of Heat Mass Transfer 57 (2), 504–518.

Ozdemir, O.E., George, T.L., Marshall, M.D., 2015. Fukushima Daiichi Unit 1 power plant containment analysis using Gothic. Annals of Nuclear Energy 85, 621–632.

Ozdemir, O.E., Kondo, M., Erkan, N., Okamoto, K., 2019. CFD based wall condensation model for evaluating PCV conditions in Fukushima Daiichi Unit-1. Nuclear Engineering and Design 352 (110170), 1–10.

Ozisik, M.N., 1973. Radiative Transfer and Interactions with Conduction and Convection. John Wiley & Sons, New York, NY, USA.

Païdoussis, M.P., Price, S.J., de Langre, E., 2014. Fluid Structure Interactions. Cambridge University Press, Cambridge, UK.

Paik, C.Y., Hochreiter, L.E., Kelly, J.M., Kohrt, R.J., 1985. Analysis of FLECHT-SEASET 163 Rod Blocked Bundle Data Using COBRA-TF. NUREG/CR4166, WCAP 103 75, EPRI N-4111, USA.

Pakhomov, S.A., Krivokhvatsty, K.S., Sokolov, I.A., 1991. Assessment of the prompt energy release by the accident at the reactor of the Chernobyl NPP based on estimation of the ratio of activities of Xenon-133 and Xenon-133m in the air. Radiokhimiya 33 (6), 125–132 (in Russian).

Pakonen, A., Buzhinsky, I., Björkman, K., 2021. Model checking reveals design issues leading to spurious actuation of nuclear instrumentation and control systems. Reliability Engineering and System Safety 205 (107237), 1–15.

Paladino, D., Dreier, J., 2012. PANDA: a multipurpose integral test facility for LWR safety investigations. Journal of Science and Technology of Nuclear Installations. ID 239319.

Paladino, D., Kapulla, R., Paranjape, S., Suter, S., Andreani, M., 2022. PANDA experiments within the OECD/NEA HYMERES-2 project on containment hydrogen distribution, thermal radiation and suppression pool phenomena. Nuclear Engineering and Design 392 (111777), 1–18.

Pan, L.-M., He, H., Wu, Y., Ju, P., Hibiki, T., Ishii, M., 2015. Experimental study of pressure drop and modeling of interfacial shear for vertical annular flow. In: 16th International Topical Meeting on Nuclear Reactor Thermal Hydraulics (NURETH-16), August 30–September 4, Chicago, IL, USA.

Pang, B., 2014. Numerical Study of Void Drift in Rod Bundle with Sub-channel and CFD Codes. KIT Scientific Reports 7669, Karlsruhe, Germany, pp. 1–142.

Papini, D., Adamsson, C., Andreani, M., Prasser, H.-M., 2014. Assessment of GOTHIC and TRACE codes against selected PANDA experiments on a passive containment condenser. Nuclear Engineering and Design 278, 542–557.

Parisi, C., 2008. PhD Thesis in Nuclear Engineering. University of Pisa, Pisa, Italy.

Park, H., Bang, I.C., 2018. Mixing performance and pressure drop evaluation for flow driven rotational mixing vane in subchannel. In: 12th International Topical Meeting on Nuclear Reactor Thermal-Hydraulics, Operation and Safety (NUTHOS-12), October 14–18, Qingdao, China.

Park, H.S., No, H.C., Bang, Y.S., 2003. Analysis of experiments for in-tube steam condensation in the presence of non-condensable gases at a low pressure using the RELAP5/MOD3.2 code modified with a non-iterative condensation model. Nuclear Engineering and Design 225.

Park, H.S., Euh, D.J., Choi, K.Y., Kwon, T.S., Baek, W.P., 2007. An assessment of a LBLOCA similarity for a reduced-height integral effect test loop design for PWRs. Annals of Nuclear Energy 34, 931–937.

Park, H.S., Yi, S.J., Song, C.-H., 2013. SMR accident simulation in experimental test loop. Journal of Nuclear Engineering International (November), 12–15.

Park, H.S., Min, B.Y., Jung, Y.G., Shin, Y.C., Ko, Y.J., Yi, S.J., 2014. Design of the VISTA-ITL test facility for an integral type reactor of SMART and a post-test simulation of a SBLOCA Test. Journal of Science and Technology of Nuclear Installations. Article ID 840109.

Park, H.S., Kwon, T.S., Mon, S.K., Cho, S., Euh, D.J., Yi, S.J., 2017. Contribution of thermal-hydraulic validation tests to the standard design approval of SMART. Nuclear Engineering and Technology 49, 1537–1546.

Park, Y., Bae, B.U., Kim, J.R., Cho, S., Kang, K.H., Choi, K.-Y., 2020. Counterpart test for LSTF 1% cold-leg break LOCA (SB-CL-32) utilizing ATLAS test facility. Nuclear Engineering and Design 370 (110912), 1–12.

Parlatan, Y., Rohatgi, U.S., 1997. Simple model of boiling heat transfer on tubes in large pools. Journal of Heat Transfer 119, 376–379.

Pasamehmetoglu, K.O., Nelson, R.A., 1987. Transient direct contact condensation on liquid droplet. In: ASME-ANS-AIChE Nat. Heat Transfer Conference, Pittsburg, PA, USA.

Patankar, S.V., 1980. Numerical Heat Transfer and Fluid Flow. Hemisphere Publishing Corporation, McGraw-Hill Inc., Washington, DC, USA, pp. 1–197.

Patankar, S.V., Spalding, D.B., 1967. Heat and Mass Transfer in Boundary Layers. Morgan-Grampian, London, UK.

Patel, G., Tanskanen, V., Kyrki-Rajamäki, R., 2014. Numerical modelling of low-Reynolds number direct contact condensation in a suppression pool test facility. Annals of Nuclear Energy 71, 376–387.

Patel, G., Tanskanen, V., Hujala, E., Hyvärinen, J., 2017. Direct contact condensation modeling in pressure suppression poolsystem. Nuclear Engineering and Design 321, 328–342. See also: Corrigendum to Direct contact condensation modeling in pressure suppression pool system [Nuclear Engineering and Design 321 (2017) 328–342], G. Patel, V. Tanskanen, E. Hujala, J. Hyvärinen, Nuclear Engineering and Design, February, 2018.

Patel, G., Puustinen, M., Hyvarinen, J., Hujala, E., 2022. Modeling of horizontal steam injection in a water pool. In: 19th International Topical Meeting on Nuclear Reactor Thermal Hydraulics (NURETH-19), March 6–11, Brussels, Belgium, Paper 35461.

Patterson, G., Novog, D.R., 2018. Design of a fast-neutron computed tomography system for measurement of subchannel void in CANDU thermalhydraulic experiments. In: 12th International Topical Meeting on Nuclear Reactor Thermal-Hydraulics, Operation and Safety (NUTHOS-12), October 14–18, Qingdao, China.

Pättikangas, T., Niemi, J., Laine, J., Puustinen, M., Purhonen, H., 2010. CFD modelling of condensation of vapour in the pressurized PPOOLEX facility. In: CFD for Nuclear Reactor Safety Applications (CFD4NRS-3) Workshop, Bethesda, MD, USA, 14–16 September, p. 12.

Pavliuk, A.O., Kotlyarevsky, S.G., Bespala, E.V., Novoselov, I.Y., 2019. Dynamics of temperature fields during Wigner energy release in bulk graphite irradiated at low temperature. Journal of Nuclear Materials 515, 303–311.

Peaceman, D.W., Rachford Jr., H.H., 1955. The numerical solution of parabolic and elliptic differential equations. Journal of the Society for Industrial and Applied Mathematics 3, 28–41.

Pearson, K.G., 1984. Rewetting during Bottom Flooding. United Kingdom Atomic Energy Authority (UKAEA) report EEWM2100, Winfrith, UK.

Pearson, K.G., Denham, M.K., 1989. Achilles Un-balloned Cluster Experiments Part 1: Description of the Achilles Rig, Test Section and Experimental Procedure. Technical Report, AEEW-R 2336, United Kingdom Atomic Energy Authority (UKAEA). Winfrith Technology Center, Winfrith, UK.

Pearson, K.G., Piggott, B.D.G., Duffey, R.B., 1977. The effect of thermal diffusion from fuel pellets on rewetting of overheated water reactor pins. Nuclear Engineering and Design 41, 165–173.

Pecchia, M., Parisi, C., D'Auria, F., Mazzantini, O., 2011. Development and Application of MCNP5 and KENO-VI Monte Carlo Models for the Atucha-2 PHWR Analysis. Science and Technology for Nuclear Installations, pp. 1–8. Article ID 683147.

Pecchia, M., Parisi, C., D'Auria, F., Mazzantini, O., 2015. Application of MCNP for predicting power excursion during LOCA in Atucha-2 PHWR. J. Annals of Nuclear Energy 85, 271–278.

Pedras, M.H.J., De Lemos, M.J.S., 2001. Macroscopic turbulence modeling for incompressible flow through undeformable media. International Journal of Heat Mass Transfer 44, 1081–1093.

Pelanti, M., 2022. Arbitrary-rate relaxation techniques for the numerical modeling of compressible two-phase flows with heat and mass transfer. International Journal of Multiphase Flow 153 (104097), 1–23.

Pellegrini, M., Naitoh, M., Josey, C., Baglietto, E., 2015. Modeling of Rayleigh-taylor instability for steam direct contact condensation. In: 16th International Topical Meeting on Nuclear Reactor Thermal Hydraulics (NURETH-16), August 30–September 4, Chicago, IL, USA.

Pellegrini, M., Naitoh, M., Kudo, Y., Mizokami, S., 2019. Confirmation of severe accident code modeling in light of the findings at Fukushima Daiichi NPPs. Nuclear Engineering and Design 354 (110217), 1–9.

Perelman, T.L., 1961. On conjugated problems of heat transfer. International Journal of of Heat and Mass Transfer 3 (4), 293–303.

Perepelitsa, N.I., Sapankevich, A.P., 1976. Ways to Improve Parameters at Heat Transfer Crisis. Analytical Review. IPPE, Obninsk, Russia. AR-31 (in Russian).

Perez, M., 2011. Integration of a Quantitative-Based Selection Procedure in an Uncertainty Analysis Methodology for NPP Safety Analysis. PhD Thesis. UPC, Barcelona, Spain.

Perez, M., Reventos, F., Wagner, R., Allison, C., 2009. Integrated Uncertainty Analysis Using RELAP/SCDAPSIM/mod4.0. NURETH-13 Conference, Kanazawa, Japan, September 27–October 2.

Perez, M., Reventos, F., Batet, L., Pericas, R., Toth, I., Bazin, P., de Crecy, A., Germain, P., Borisov, S., Glaeser, H., Skorek, T., Joucla, J., Probst, P., Ui, A., Chung, B.D., Oh, D.Y., Kyncl, M., Pernica, R., Manera, A., D'Auria, F., Petruzzi, A., Del Nevo, A., 2010. Main results of phase IV BEMUSE project: simulation of LBLOCA in an NPP. Science and Technology of Nuclear Installations. Article ID 219294.

Perez, or Perez-Ferragut), M., Reventos, F., Batet, L., Guba, A., Toth, I., Mieusset, T., Bazin, P., de Crecy, A., Borisov, S., Skorek, T., Glaeser, H., Joucla, J., Probst, P., Ui, A., Chung, B.D., Oh, D.Y., Pernica, R., Kyncl, M., Macek, J., Manera, A., Freixa, J., Petruzzi, A., D'Auria, F., Del Nevo, A., 2011. Uncertainty and sensitivity analysis of a LBLOCA in a PWR nuclear power plant: results of the phase V of the BEMUSE programme. Nuclear Engineering and Design 241, 4206–4222.

Perez-Ferragut, M., Martinez-Quiroga, V., Casamor, M., Freixa, J., 2022. Methodology for phenomenological code assessment with integral test data. Nuclear Engineering and Design 387 (111608), 1–10.

Pericas, R., 2015. Contribution to the Validation of Best Estimate Plus Uncertainties Coupled Codes for the Analysis of NK-TH Nuclear Transients. PhD Thesis. UPC, Barcelona, Spain.

Pericas, R., Ivanov, K., Reventos, F., 2013. Comparison of best estimate plus uncertainties and conservative methodologies for a PWR MSLB analysis using coupled 3D neutron kinetics/ thermal-hydraulic code. In: NURETH-15 Pisa, Italy, May 12–15.

Pericas, R., Ivanov, K., Reventos, F., Batet, L., 2016. Code improvement and model validation for Asco-II Nuclear Power Plant model using a coupled 3D neutron kinetics/thermal-hydraulic code. Annals of Nuclear Energy 87 (2), 653–665.

Petelin, S., Gunel, I., 1995. Ideal scaling of BETHSY 9.1.B test results to NPP. In: Proceedings 2nd Regional Meeting: Nuclear Energy in Central Europe, Portoroz, Slovenia, 11–14 September, pp. 567–574.

Petelin, S., Ravnikar, I., 1997. Scaling-Up of BETHSY 9.1b experiment to real NPP size transient. In: Proceedings 4th Regional Meeting: Nuclear Energy in Central Europe, Bled, Slovenia, 7–10 September.

Petelin, S., Mavko, B., Gortnar, O., 1994. Scaling for the analysis of ISP-27 accident scenario realized on Krsko NPP. In: Proceedings Annual Meeting of the Nuclear Society of Slovenia (The First Conference), Rogaska Slatina, Slovenia, September 18–21.

Petelin, S., Mavko, B., Koncar, B., Hassan, Y.A., 2007. Scaling of the small-scale thermal hydraulic transient to the real nuclear power plant. Nuclear Technology 158 (1), 56–68.

Peters, S., Tschaepe, L., Zhang, B., Ruggles, A., 2011. V&V methodology comparisons: AIAA G-77, (1998), ASME V&V 20, (2009), ASTM E1355-05a, (2005), NEA/CSNI/ R(2007), and NRC CSAU, (1988). In: Proceedings 14th International Topical Meeting on Nuclear Reactor Thermal Hydraulics, NURETH-14, Toronto, Ontario, Canada, September 25–30.

Peterson, P.F., 1996. Theoretical basis for the Uchida correlation for condensation in reactor containments. Journal of Nuclear Engineering and Design 162, 301–306.

Petrangeli, G., Tononi, R., D'Auria, F., Mazzini, M., 1993. The SSN: an emergency system based on intentional depressurization for PWRs. In: Nuclear Engineering and Design, 143, (Also, CSNI Spec. Meet. On Intentional Coolant System Depressurization, Garching, Germany, 12–14 June 1989).

Petruzzi, A., 2008. Development and Application of Methodologies for Sensitivity Analysis and Uncertainty Evaluation of the Results of the Best Estimate System Codes Applied in Nuclear Technology. PHD Thesis. Universitá di Pisa, Pisa, Italy, pp. 1–364. URNetd-11082008-094257.

Petruzzi, A., D'Auria, F., 2008. Approaches, relevant topics, and internal method for uncertainty evaluation in predictions of thermal-hydraulics system codes. Journal of Science and Technology of Nuclear Installations. Article ID 325071.

Petruzzi, A., D'Auria, F., 2008a. Evaluation of uncertainties in system thermal-hydraulic calculations and key applications of the CIAU method. International Journal of Nuclear Energy Science and Technology (IJNEST) 4 (2), 111–131.

Petruzzi, A., D'Auria, F., 2014. Supporting qualified database for V&V and uncertainty evaluation of best-estimate system codes. In: ASME-JSME International Conference On Nuclear Engineering (ICONE-22), Prague, Czech, 7–11 July.

Petruzzi, A., D'Auria, F., 2016. Standardized consolidated calculated and reference experimental database (SCCRED): a supporting tool for V&V and uncertainty evaluation of best-estimate system codes for licensing applications. Nuclear Science and Engineering 182 (1), 13–53.

Petruzzi, A., Giannotti, W., 2020. Kv-scaling analysis to support the validation of Atucha-II best estimate evaluation model. Nuclear Engineering and Design 363, 110609.

Petruzzi, A., Bousbia Salah, A., D'Auria, F., 2004. Sub-channel analysis by Relap5 system code of Boil-off experiment (test 5002) with Neptun facility. In: Canadian Nuclear Society 6th International Conference On Simulation Methods in Nuclear Engineering, Montreal (Que, Canada), October 12–15. ISBN 0-919781-80-1 © 2004 CNS-SNC.

Petruzzi, A., D'Auria, F., Bajs, T., Reventos, F., 2005. In: International Training Program: 3D S.UN.COP-Scaling, Uncertainty and 3D Thermal-Hydraulics/Neutron-Kinetics Coupled Codes Seminar. ENC, Versailles, France.

Petruzzi, A., D'Auria, F., 2006. BEMUSE Programme. Phase 2 Report. Re-analysis of the ISP-13 Exercise, Post Test Analysis of the LOFT L2-5 Experiment, OECD/CSNI Report NEA/CSNI/R(2006)2, © OECD 2006, Paris, France. In: De Crecy, A., Bazin, P., Borisov, S., Skorek, T., Glaeser, H., Benoit, J.P., Chojnacki, E., Fujioka, K., Inoue, S., Chung, B.D., Trosztel, I., Toth, I., Oh, D.Y., Pernica, R., Kyncl, M., Macek, J., Macian, J.R., Tanker, E., Soyer, A.E., Ozdere, O., Perez, M., Reventos, F. (Eds.), pp. 1–625.

Petruzzi, A., D'Auria, F., Bajs, T., Reventos, F., Hassan, Y.A., 2008. International course to support nuclear licensing by user training in the areas of scaling, uncertainty and 3D thermal-hydraulics/neutron-kinetics coupled codes: 3D S.UN.COP seminars. Science and Technology of Nuclear Installations 874023, 1–16.

Petruzzi, A., Cacuci, D.G., D'Auria, F., 2010. Best estimate model calibration and prediction through experimental data assimilation-II: application to a blow-down benchmark experiment. Nuclear Science and Engineering 165 (1), 45–101.

Petruzzi, A., D'Auria, F., Bajs, T., Reventos, F., 2011. International training program in support of safety analysis: 3C S.UN.COP—scaling, uncertainty and 3D thermal-hydraulics/neutron-kinetics coupled codes seminars. In: International Congress on Advances in Nuclear Power Plants (ICAPP), Nice, France, 2–6 May. Paper 11451.

Petruzzi, A., Fiori, F., Kovtonyuk, A., D'Auria, F., 2012. Supporting qualified database for uncertainty evaluation. In: International Congress on Advances in Nuclear Power Plants (ICAPP), Chicago, IL, USA, 24–28 June. Paper 12085.

Petruzzi, A., Cherubini, M., Lanfredini, M., D'Auria, F., Mazzantini, O., 2016. The BEPU evaluation model with RELAP5-3D for the licensing of the Atucha-II NPP. Nuclear Technology 193, 113–160.

Petukhov, B.S., 1970. Heat Transfer and Friction in Turbulent Pipe Flow with Variable Physical Properties, Advances in Heat Transfer. Academic Press, New York, pp. 503–565.

Petukhov, B.S., 1987. Problems of Heat Transfer. Nauka, Moscow, Russia, pp. 68–86.

Petukhov, B.S., Kurganov, V.A., 1974. Analysis and generalization of experimental data on heat transfer in tubes at turbulent gas flow with variable physical properties. Teplofiz. Vys. Temp. 12 (2), 304 (in Russian).

Petukhov, B.S., Shilov, V.K. (Eds.), 1987, vol. 1. Energoatomizdat, Moscow, Russia, pp. 1–383 (in Russian).

Péturaud, P., Hampel, U., Barbier, A., Dreier, J., Dubois, F., Hervieu, E., Martin, A., Prasser, H.-M., 2011. General overview of the TOPFLOW-PTS experimental program. In: Proceedings of the 14th International Topical Meeting on Nuclear Reactor Thermal-Hydraulics (NURETH-14), Toronto, Canada.

Pham, T.Q.D., Choi, S., 2021. Numerical analysis of direct contact condensation-induced water hammering effect using OpenFOAM in realistic steam pipes. International Journal of of Heat and Mass Transfer 171 (121099), 1–14.

Piggott, B.D.G., Duffey, R.B., 1975. The quenching of irradiated fuel pins. Nuclear Engineering and Design 32, 182–190.

Pineda, H., Biazussi, J., Lopez, F., Oliveira, B., Carvalho, R.D.M., Bannwart, A.C., Ratkovich, N., 2016. Phase distribution analysis in an Electrical Submersible Pump (ESP) inlet handling water–air two-phase flow using Computational Fluid Dynamics (CFD). Journal of Petroleum Science and Engineering 139, 49–61.

Piomelli, U., 1999. Large-eddy simulation, achievements and challenges. Progress in Aerospace Sciences 35, 335.

Piomelli, U., 2008. Wall-layer models for large-eddy simulations. Progress in Aerospace Sciences 44, 437–446.

Piomelli, U., Balaras, E., 2002. Wall-layer models for large-eddy simulations. Annual Review of Fluid Mechanics 34, 349–374.

Piomelli, U., Ferziger, J., Moin, P., Kim, J., 1989. New approximate boundary conditions for large eddy simulations of wall-bounded flows. Physics of Fluids A1, 1061.

Piomelli, U., Balaras, E., Pasinato, H., Squires, K.D., Spalart, P.R., 2003. The inner-outer layer interface in large-eddy simulations with wall-layer models. International Journal of Heat Fluid Flow 24, 538–550.

Pioro, J.L., 1999. Experimental evaluation of constants for the Rohsenow pool boiling correlation. International Journal of Heat Mass Transfer 42 (11), 2003–2013.

Pioro, J.L., 2019. Current status of research of heat transfer in forced convection of fluids at superctitical pressures. In: Sehgal, R., Yadigaroglu, G., Hewitt, G.F., D'Auria, [F. (Eds.), Special Issue Trends and Perspectives in Nuclear Thermal-hydraulics—To the Memory of B, Journal of Nuclear Engineering and Design, vol. 354, pp. 1–14. https://doi.org/10.1016/j.nucengdes paper 110207 (paper PIII-05), 110207.

Pla, P., 2004. Assessment of Size Aspects in Modelling Molten Fuel Coolant Interaction. PhD Thesis. UPC, Barcelona, Spain.

Pla, P., Reventós, F., Pretel, C., Giannotti, W., D'Auria, F., Annunziato, A., Sol, I., 2007. Code validation and scaling of the LOBI BL-30 experiment. In: Proceedings of International Congress on Advances in Nuclear Power Plants (ICAPP), Nice, France, May 13–18.

Pla, P., Parisi, C., D'Auria, F., 2010. Boron dilution and boron transport after SBLOCA in PWR and VVER-1000 nuclear reactors. In: Boron Dilution in Pressurized Water Reactors. Lambert Academic Publishing, LAP GmbH & Co., Saarbrücken, Germany, ISBN 978-3-8383-8373-6.

Pla, P., Annunziato, A., Addabbo, C., Galassi, G., D'Auria, F., 2012. Preservation and use of integral system test facilities data: the experience of the LOBI data and the STRESA database. Progress in Nuclear Energy 56, 79–90.

Pla, P., Ammirabile, L., Annunziato, A., 2013. The Stresa database: a token for the future. Annals of Nuclear Energy 62, 8–16.

Pla, P., Pascal, G., Tanarro, J., Annunziato, A., 2015. The new STRESA tool for preservation of thermal-hydraulic experimental data produced in the European Commission. Annals of Nuclear Energy 85, 893–902.

Pla, P., Reventos, F., Martin, R.M., Strucic, M., Sol, I., 2016a. Simulation of steam generator plugging tubes in a PWR to analyze the operating impact. Nuclear Engineering and Design 305, 132–145.

Pla, P., Tanarro, J., Ammirabile, L., Strucic, M., Wastin, F., 2016b. Long term preservation of knowledge in the area of nuclear thermal-hydraulics—the new STRESA database. In: 11th International Conference Of Croatian Nuclear Society, Zadar, Croatia, June 5–8.

Planck, M., 1923. Vorlesungen über die Theorie der Wärmestrahlung, 4 Aufl. Leipzig, Germany.

Plas, R., 1975. Flica III: a digital computer program for thermal-hydraulic analysis of reactors and experimental loops. In: France. Centre d'Etudes Nucleaires de Saclay, CEA-CONF, 3096, Gif-sur-Yvette, France, pp. 1−14.

Plesset, M.S., Zwick, S.A., 1954. Growth of vapor bubbles in superheated liquids. Journal of Applied Physics 25, 493−500.

Podowski, M.Z., Sehgal, R., Yadigaroglu, G., Hewitt, G.F., D'Auria, s F., Hassan, Y., 2019. Understanding Two-phase Flow and Boiling Heat Transfer: Challenges and Paradoxes, Nuclear Engineering and Design [Special Issue Devoted to B, vol. 354, pp. 1−10.

Poling, B.E., Prausnitz, J.M., O'Connell, J.P., 2001. The Properties of Gases and Liquids. McGraw-Hill, New York, NY, USA.

Polley, G.T., Ralston, T., Grant, I.D.R., 1981. Forced Crossflow Boiling in an Ideal In-Line Tube Bundle, ASME 80-HT-46.

Pon, G.A., 1978. Evolution of CANDU Reactor Design. Atomic Energy of Canada Limited, Toronto, Ontario, Canada. AECL-6351.

Ponomarenko, G.L., Ryzhov, S.B., Bykov, M.A., Moskalev, A.M., 2009. Use of BEPU technique for analyses of BDBAs with cooling in WWER-1000. In: International Conference ICONE-17, Brussels, Belgium.

Pope, C., Lum, E., 2020. Nuclear reactor thermal expansion reactivity effect determination using finite element analysis coupled with Monte Carlo neutron transport analysis. In: Baccouch, M. (Ed.), Chapter or the Book "Finite Element Methods and Their Application. Intech Open, London, UK.

Popov, V.M., 1977. Heat Transfer in a Contact Zone of Permanent Connections. Energiya, Moscow, Russia (in Russian).

Popov, N.K., 2014. Thermal hydraulics design (of CANDU). Chapter 6 of the book. In: Garland, W.J. (Ed.), The Essential CANDU, A Textbook on the CANDU Nuclear Power Plant Technology, @ UNENE (Canada), pp. 1−274. ISBN 0-9730040.

Porcheron, E., Lemaitre, P., Nuboer, A., Rochas, V., Vendel, J., 2007. Experimental investigation in the TOSQAN facility of heat and mass transfers in a spray for containment application. Journal of Nuclear Engineering and Design 237.

Potter(Principal Investigator), W.C., 1990. Soviet Decision Making for Chernobyl: An Analysis of System Performance and Policy Change. Contract No. 802-12. University of California, Los Angeles, CA, USA.

Pradhan, S.K., Bang, Y.S., 2011. Best estimate prediction of NFR with uncertainty following a LBLOCA of CANDU plants. In: KNS Spring Meeting, Taebaek, Korea, 26−27 May.

Prakash, B., Parmar, H., Shah, M.T., Pareek, V.K., Anthony, L., Utikar, R.P., 2019. Simultaneous measurements of two phases using an optical probe. Experimental and Computational Multiphase Flow 1 (4), 233−241.

Prandtl, L., 1956. Führer durch die Strömungslehre, 4 Aufl. Vieweg, Braunschweig, Germany.

Prasser, H.M., Böttger, A., Zschau, J., 1998. A new electrode-mesh tomograph for gas−liquid flows. Flow Measurement and Instrumentation 9, 111−119.

Prasser, H.-M., Ezsol, G., Baranyai, G., 2004. PMK-2 Water Hammer Tests, Condensation Caused by Cold Water Injection into Main Steam-Line of VVER-440-type PWR—Data Evaluation Report (DER). WAHALoads project deliverable D51, EURATOM, Brussels, Belgium.

Prasser, H.-M., Stucki, S., Betschart, T., Eisenberg, J., 2015. Interfacial area density measurement using a three layer wire-mesh sensor. In: 16th International Topical Meeting on Nuclear Reactor Thermal Hydraulics (NURETH-16), August 30-September 4, Chicago, IL, USA.

Pretel, C., 1997. Analisis de transitorios con perdida de refrigerante primario en reactores de agua a presion y optimizacion de la gestion de accidentes (in Spanish). PhD Thesis. UPC, Barcelona, Spain.

Prosek, A., Mavko, B., 2003. Review of best estimate plus uncertainty methods of thermal-hydraulic safety analysis. In: International Conference Nuclear Energy for New Europe, Portoroz, Slovenia.

Prosek, A., D'Auria, F., Mavko, B., 2002. Review of quantitative accuracy assessment with fast Fourier transform based method (FFTBM). Journal of Nuclear Engineering and Design 217 (1–2), 179–206.

Prosek, A., D'Auria, F., Richards, D.J., Mavko, B., 2005. Quantitative assessment of thermal-hydraulics codes used for heavy water reactor calculations. Nuclear Engineering and Design 236, 295–308.

Prosperetti, A., 1982. A generalization of the Rayleigh–Plesset equation of bubble dynamics. The Physics of Fluids 25, 409.

Punetha, M., Yadav, M.K., Khandekar, S., Sharma, P.K., Ganju, S., 2020. Intrinsic transport and combustion issues of steam-air-hydrogen mixtures in nuclear containments. International Journal of of Hydrogen Energy 45, 3340–3371.

Punetha, M., Yadav, M.K., Jain, S., Khandekar, S., Sharma, P.K., 2021. Thermal-hydraulic test facility for nuclear reactor containment: engineering design methodology and benchmarking. Progress in Nuclear Energy 138 (103837), 1–20.

Pushkina, O.L., Sorokin, Y.L., 1969. Breakdown of liquid film motion in vertical tubes. Heat Transfer-Soviet Research 1 (5), 56–64.

Putney, J.M., 1988a. Proposals for Improving Interphase Drag Modeling for the Bubbly and Slug Regimes in RELAP5. CERL Report RD/L/3306/R88, PWR/HTWG/P(88)622, UK, June.

Putney, J.M., 1988b. Implementation of a New Bubbly-Slug Interphase Drag Model in RELAP5/MOD2. CERL Report RD/L/3369/R88, PWR/HTWG/P(88)597, UK, November.

Putney, J.M., 1988c. Equations for Calculating Interfacial Drag and Shear from Void Fraction Correlations. CERL Report RD/L/3429/R88, PWR/HTWG/P(88)630, UK, December.

Putney, J.M., 1989. Development of a New Bubbly-Slug Interfacial Friction Model for RELAP5—Final Report. National Power Report ESTD/L/0075/R89, PWR/HTWG/ P(89) 722, UK, October.

Putney, J.M., 1991. Development of a new bubbly-slug interfacial friction model for RELAP5. Nuclear Engineering and Design 131, 223–240.

Puustinen, M., Kyrki-Rajamäki, R., Tanskanen, V., Räsänen, A., Purhonen, H., Riikonen, V., Laine, J., Hujala, E., 2013. BWR suppression pool studies with POOLEX and PPOOLEX test facilities at LUT. In: The 15th International Topical Meeting on Nuclear Thermal Hydraulics (NURETH-15), 12–17 May, Pisa, Italy.

Qin, H., Zhou, S., Anjum, M.N., 2018. A mesh sensors system for two phase flow measurement between fuel assemblies. In: 12th International Topical Meeting on Nuclear Reactor Thermal-Hydraulics, Operation and Safety (NUTHOS-12), October 14–18, Qingdao, China.

Qiu, S.Z., Sun, D.C., Tian, W.X., Xiang, Y., Su, G.H., Zhang, P., 2015. Experimental and theoretical research on liquid entrainment in AP1000 ADS blow-down phase of SBLOCA. In: NURETH-16 Conf., Chicago, IL, USA, August 30–September 4.

Qu, W., Xiong, J., Chen, S., Qiu, Z., Deng, J., Cheng, X., 2019. PIV measurement of turbulent flow downstream of mixing vane spacer grid in 5x5 rod bundle. Annals of Nuclear Energy 132, 277–287.

Quddus, A., Shah, A., Qureshi, K.R., Tahir, A., Iqbal, M., Ayub, M.K., 2022. Numerical study on pressure oscillations amplitude distribution induced by water flow into pipe filled with steam. Annals of Nuclear Energy 172 (109095), 1–15.

Queral, C., Montero-Mayorga, J., Gonzales-Cadelo, J., Jimenez, G., 2015. AP1000 large break LOCA BEPU analysis with TRACE code. Annals of Nuclear Energy 85.

Querol, A., Gallardo, S., Verdú, G., 2015. Simulation of a SBLOCA in a hot leg. Scaling considerations and application to a nuclear power plant. Nuclear Engineering and Design 283, 81–99.

Quigley, M.S., McMonagle, C.A., Bates, J.M., 1977. Investigation of combined free and forced convection in a 2 x 6 rod bundle during controlled flow transients. In: Pacific Northwest Laboratories BNWL-2216, Richland, Wa, USA.

Rachkov, V.I., Kalyakin, S.G., Kukharchuk, O.F., Orlov, Y.I., Sorokin, A.P., 2014. From the first nuclear power plant to fourth-generation nuclear power installations [on the 60th Anniversary of the World's First nuclear power plant]. Thermal Engineering 61 (5), 327–336.

Radhakrishnan, S., Piomelli, U., 2008. Large eddy simulation of oscillating boundary layers: model comparison and validation. Journal of Geophysical Research 113C02022. https://doi.org/10.1029/2007JC004518.

Radhakrishnan, S., Piomelli, U., Keating, A., Lopes, A.S., 2006. Reynolds-averaged and large-eddy simulations of turbulent non-equilibrium flows. Journal of Turbulence 7 (63), 1–30.

Radhakrishnan, S., Piomelli, U., Keating, A., 2008. Wall-modeled large-eddy simulations of flows with curvature and mild separation. ASME Journal of of Fluids Engineering 130, 101203.

Rahim, R.A., Rahiman, M.H.F., Chan, K.S., Nawawi, S.W., 2007. Non-invasive imaging of liquid/gas flow using ultrasonic transmission-mode tomography. Sensors and Actuators A: Physical 135, 337–345.

Rahimi, P., Ward, C.A., 2005. Kinetics of evaporation: statistical rate theory approach. International Journal of of Thermodynamics 8 (1), 1–14.

Rajita, G., Mandal, N., 2016. Review on transit time ultrasonic flowmeter. In: 2nd International Conference On Control. Instrumentation, Energy & Communication (CIEC), pp. 88–92.

Rakhimov, A.C., Visser, D.C., Komen, E.M.J., 2022. Direct evaluation of the turbulence model uncertainty. In: 19th International Topical Meeting on Nuclear Reactor Thermal Hydraulics (NURETH-19), March 6–11 (Via Web), Brussels, Belgium.

Rakhmatulin, K.A., 1956. Fundamentals of gas-dynamics of interpenetrating flows of compressible media. (in Russian). Prikl. Mat. Mekh. 20, 184–195.

Rakhmatulin, K.A., 1969. Propagation of waves in multicomponent media. PMM 33 (4), 598–601.

Ramesh, O.N., Manisekaran, S., Shivsai, A., Dixit, S., 2006. The structure of the sink flow turbulent boundary layer. In: American Physical Society, 59th Annual Meeting of the APS Division of Fluid Dynamics, November 19–21, 2006APS.DFD.KL008R.

Ramilison, I.M., Lienhard, J.H., 1988. Transition boiling heat transfer and the film transition regime. ASME Journal of Heat Transfer 110 (3), 746–752.

Ramos, R.T., Holmes, A., Wu, X., Dussan, E., 2001. A local optical probe using fluorescence and reflectance for measurement of volume fractions in multi-phase flows. Measurement Science and Technology 12, 871–876.

Ransom, V.H., Wang, W., Ishii, M., 1998. Use of an ideal scaled model for scaling evaluation. Nuclear Engineering and Design 186, 135–148.

Ranz, E., Marshall, W.R., 1952. Evaporation from droplets: part I and part II. Chemical Engineering Progress 48, 141–146, pp. 173–180.

Rao, T.S., Bera, S., 2021. Protective layer dissolution by chlorine and corrosion of aluminum brass condenser tubes of a nuclear power plant. Engineering Failure Analysis 123 (105307), 1−12.

Rao, R.S., Kumar, A., Gupta, S.K., Lele, H.G., 2012. Uncertainty and sensitivity analysis of TMI-2 accident scenario using simulation based techniques. Nuclear Engineering and Technology 44 (7), 807−816.

Rao, Y.F., Cheng, Z., Waddington, G.M., Nava-Dominguez, A., 2014. ASSERT-PV 3.2: advanced subchannel thermal hydraulics code for CANDU fuel bundles. Nuclear Engineering and Design 275, 69−79.

Rapier, A.C., Jones, T.M., Macintosch, J.E., 1963. The Thermal conductance of uranium dioxide/stainless steel interfaces. International Journal of Heat Mass Transfer 6 (5), 397−416.

Rassokhin, N.G., 1980. Steam Generators of Nuclear Power Plants, second ed. Atomizdat, Moscow, Russia, pp. 1−360 (in Russian).

Ratel, G., Geffraye, G., 2000. Analysis with CATHARE code of startified flow regime in the MHYRESA hot leg entrainment test. In: 38th European Two-phase Flow Group Meeting, Karlsruhe, Germany, May 29−31.

Ravva, S.R., Iyer, K.R., Gupta, S.K., Gaikwad, A.J., 2014. Implementation and validation of the condensation model for containment hydrogen distribution studies. Nuclear Engineering and Design 270, 34−47.

Reeder, D.L., 1978. LOFT System and Test Description (5.5 Foot Nuclear Core 1 Loss-Of-Coolant Experiments). TREE-NUREG-1208, EG&G Idaho, Idaho Falls, ID, USA.

Rehme, K., 1972. Pressure drop performance of rod bundles in hexagonal arrangements. International Journal of Heat Mass Transfer 15 (12), 2499−2517.

Rehme, K., 1973. Pressure drop correlations for fuel element spacers. Nuclear Technology 17 (1), 15−23.

Rehme, K., 1992. The structure of turbulence in rod bundles and the implications on natural mixing between the subchannels. International Journal of Heat Mass Transfer 35 (2), 567−581.

Reichardt, H., 1941. Über eine Theorie der freiren Turbulenz. Angewandte Mathematics and Mechanics (Journal of Applied Mathematics and Mechanics) 21 (5), 257−264.

Reimann, J., Khan, M., 1984. Flow through a small break at the bottom of large pipe with stratified flow. Journal of Nuclear Science and Engineering 88, 297−310.

Reimann, J., Kusterer, H., John, H., 1984. Two-phase Mass flow rate measurements with Pitot tubes and density measurements. In: Delhaye, J.M., Cognet, G. (Eds.), Measuring Techniques in Gas-Liquid Two-phase Flows, International Union of Theoretical and Appl. Mechanics (IUTAM). Springer, Berlin, Heidelberg, Germany, pp. 625−650.

Reinhardt, H.J., 1989. A Study of 2-Dimensional Effects in the Core of a PWR during the Reflooding Phase of a LOCA. Analysis of Data of PERICLES Experiments with the COBRA-NC Code. Final Report, Commission of the European Communities, Nuclear Science and Technology, Reactor Safety Programme 1985-1987, EUR 12408 EN, Bruxelles, Belgium.

RELAP5-3D, 2001. Code Development Team, RELAP5-3-D Code Manual Volume I: Code Structure, System Models and Solution Methods. Idaho National Engineering and Environmental Laboratory, Idaho Falls, ID, USA.

RELAP5/MOD3, 1995. Code Manual, Models and Correlations. NUREG/CR-5535, vol. 4. USNRC, Washington, DC, USA.

RELAP5/MOD3, 1998. Code manual volume I: code structure, system models and solution procedure. In: NUREG/CR-5535. USNRC, Washington, DC, USA (see also RELAP5/MOD3, 1995).

Rempe, J.L., Knudson, D.L., 2013. TMI-2, A Case Study for PWR Instrumentation Performance during a Severe Accident, pp. 1−118. INL/EXT-13-28043, Idaho Falls, ID, USA.

Ren, Q.-Y., Pan, L.-M., Zhou, W.-X., Liu, H., Ye, T.-P., 2018. Drift-flux model of sub-channel in vertical rod bundles with spacer grids. International Journal of of Heat and Mass Transfer 126, 946−956.

Ren, Q.-Y., Zhou, W.-X., Du, S.-J., Li, Z.-C., Pan, L.-M., 2018a. Sub-channel flow regime maps in vertical rod bundles with spacer grids. International Journal of of Heat and Mass Transfer 122, 1138−1152.

Reocreux, M., 1974. Contribution à l'etude des debits critiques en ecoulement diphasique eau-vapeur. (PhD thesis, in French). Universite Scientifique et Medicale de Grenoble, Grenoble, France.

Reocreux, M., 1994. Current and future trends in nuclear reactor thermal-hydraulics. In: International Conference New Trends in Nuclear System Thermal-Hydraulics, Pisa, Italy.

Reocreux, M., 2008. A few examples of ISPs addressing reactor specific problems. In: OECD/NEA/CSNI, THICKET Seminar, Pisa, Italy, 5−9 May.

Reocreux, M., 2012. Historical perspective on the role and the achievements of CSNI in the field of thermal-hydraulics. In: THICKET-3: Seminar on the Transfer of Competence, Knowledge and Experience Gained through CSNI in the Field of thermal-hydraulics, OECD/NEA Headquarters, Issy-Les-Moulineaux, France, 25−29 June.

Reocreux, M., Aksan, S.N., 1998. Contribution from Twenty-Two Years of CSNI International Standard Problems. OECD/NEA report, NEA/CSNI/R(97) 29, Paris, France.

Reocreux, M., Rubinstein, M.C., 1992. In: Proceedings Of Spec. Meet. on Transient Two-phase Flow "Current Issues in System Thermal-Hydraulics", Aix-En-Provence, April 6−8, NEA/CSNI R(92)12, Paris (F). ISBN 2-7272-0157 (see also OECD/NEA/CSNI, 1992).

Repetto, G., Dominguez, C., Durville, B., Carnemolla, S., Campello, D., Tardif, N., Gradeck, M., 2015. The R&D PERFROI project on thermal mechanical and thermal hydraulics behaviors of a fuel rod assembly during a loss of coolant accident. In: 16th International Topical Meeting on Nuclear Reactor Thermal Hydraulics (NURETH-16), August 30-September 4, Chicago, IL, USA.

Repetto, G., Grando, Q., Bruyere, B., Eymery, S., 2022. COAL experiments investigating the reflooding of a 7 X 7 rods bundle during a loss of coolant accident thermalhydraulics results. In: 19th International Topical Meeting on Nuclear Reactor Thermal Hydraulics (NURETH-19), March 6−11 (Via Web), Brussels, Belgium, Paper 34148.

Revankar, S.T., Ishii, M., 1992. Local interfacial area measurement in bubbly flow. International Journal of Heat Mass Transfer 35, 913−925.

Reventos, F., Batet, L., Llopis, C., Pretel, C., Salvat, M., Sol, I., 2007a. Advanced qualification process of ANAV NPP integral dynamic models for supporting plant operation and control. Nuclear Engineering and Design 237 (1), 54−63.

Reventos, F., Batet, L., Pretel, C., Rìos, M., Sol, I., 2007b. Analysis of the feed and bleed procedure for the Asco' NPP: first approach study for operation support. Nuclear Engineering and Design 237 (18), 2006−2013.

Reventos, F., Batet, L., Llopis, C., Pretel, C., Sol, I., 2008a. Thermal-hydraulic analysis tasks for ANAV NPPs in support of plant operation and control. Science and Technology of Nuclear Installations. Article ID 153858.

Reventos, F., Freixa, J., Batet, L., Pretel, C., Luebbesmeyer, D., Spaziani, D., Macek, J., Kasahara, F., Umminger, K., 2008b. An analytical comparative exercise on the OECD-SETH PKL E2.2 experiment. Nuclear Engineering and Design 238 (4), 1146−1154.

Reventos, F., Llopis, C., Batet, L., Pretel, C., Sol, I., 2010. Analysis of an actual reactor trip operating event due to a high variation of neutron flux occurring in the Vandellos-II nuclear power plant. Nuclear Engineering and Design 240 (10), 2999–3008.

Reventos, F., Glaeser, H., D'Auria, F., De Crecy, A., 2011. Uncertainty and sensitivity analysis for a nuclear power plant large break loss of coolant accident (LB-LOCA) in the context of OECD BEMUSE programme. In: NURETH-14 Conf., Toronto (Ca), September 25–30.

Reventos, F., Pla, P., Matteoli, C., Nacci, G., Cherubini, M., Del Nevo, A., D'Auria, F., 2012. Consistent post-test calculations for LOCA scenarios in LOBI integral facility. Journal of Science and Technology for Nuclear Installations. Article ID 474162.

Reventos, F., Martínez-Quiroga, V., Freixa, J., 2019. Perfecting the use of hybrid models in scaling analysis. Nuclear Engineering and Design 354, 110187.

Reyes Jr., J.N., 2001. Scaling Analysis for the OSU APEX-CE Integral System Test Facility, USNRC NUREG/CR-6731, Washington, DC, USA.

Reyes Jr., J.N., 2005. Integral System Experiments Scaling Methodology, pp. 321–355. IAEA-TECDOC-1474, Vienna, Austria.

Reyes Jr., J.N., 2007. Integral System Experiment Scaling Methodology. (Lecture T14). IAEA-ICTP Natural Circulation Training Course, Trieste, Italy.

Reyes Jr., J.N., 2015. The dynamical system scaling methodology. In: International Topical Meeting on Nuclear Reactor Thermal Hydraulics (NURETH-16), Chicago, IL, USA.

Reyes Jr., J.N., 2022. Application of dynamical system scaling to bubble dynamics. In: OECD/NEA/CSNI Specialists Meeting on Transient Thermal-hydraulics in Water Cooled Nuclear Reactors (SM-TH), March 22–24, CIEMAT, Madrid, Spain, P13.

Reyes Jr., J.N., Frepoli, C., 2019. Similarity and scaling. Chapter 6 of book "design-basis accident analysis methods for light-water nuclear power plants". In: Martin, R.P., Frepoli, C. (Eds.), Vol 3 of Modern Nuclear Energy Analysis Methods. World Scientific Publishing Co. Pte. Ltd., Singapore, ISBN 978-981-3275-65-2.

Reyes Jr., J.N., Hochreiter, L., 1998. Scaling analysis for the OSU AP-600 test facility (APEX). Journal of Nuclear Engineering and Design 186.

Reyes Jr., J.N., King, J., 2003. Scaling Analysis for the OSU System Integral System Test Facility. OSU-NE-2003-01. Oregon State University, Corvallis, Or, USA.

Reyes Jr., J.N., Young, E., 2011. The NuScale advanced passive system design. In: ASME 2011 Small Modular Reactors Symposium, Washington, DC, USA, September 28–30, pp. 193–198.

Reyes Jr., J.N., Groome, J.T., Woods, B.W., Young, E., Abel, K., Yao, Y., Yoo, Y.J., 2007. Testing of the multi-application small light water reactor (MASLWR) passive safety systems. Journal of Nuclear Engineering and Design 237.

Reyes Jr., J.N., Groome, J.T., Woods, B.W., Jackson, B., Marshall, T.D., 2010. Scaling analysis for the high temperature gas reactor test section (GRTS). Journal of Nuclear Engineering and Design 240 (2).

Reyes Jr., J.N., Frepoli, C., Yurko, J., 2015. The dynamical system scaling methodology: comparing dimensionless governing equations with the H2TS and FSA methodologies. In: International Topical Meeting on Nuclear Reactor Thermal Hydraulics (NURETH-16), Chicago, IL, USA.

Reynolds, O., 1896. On the dynamical theory of incompressible viscous fluid and the determination of the criterion. Philosophical Transactions of the Royal Society 186, 123–164 (London, UK).

Reynolds, W.C., 1990. In: Lumley, J.L. (Ed.), Whither Turbulence? Turbulence at the Crossroads. Springer-Verlag, Heidelberg, Germany, p. 313.

Reynolds, W.C., Thompson, D.W., Fischer, C.R., 1961. HECTIC, an IBM 704 Computer Program for Heat Transfer Analysis of Gas Cooled Reactors. Aerojet-General Nucleonics, AGN-TM-381, San Ramon, CA, USA.

Rhee, B.W., Kim, H.T., Atkins, M.D., Kim, T., 2015. Validation test plan of a CANDU-6 moderator tank scaled-down test facility. In: 16th International Topical Meeting on Nuclear Reactor Thermal Hydraulics (NURETH-16), August 30–September 4, Chicago, IL, USA.

Richenderfer, A., Kossolapov, A., Seong, J.H., Saccone, G., Demarly, E., Kommajosyula, R., Baglietto, E., Buongiorno, J., Bucci, M., 2018. Investigation of subcooled flow boiling and CHF using high-resolution diagnostics. Experimental Thermal and Fluid Science 99, 35–58.

Richter, H.J., 1983. Separated two phase flow model: application to critical two phase flow. International Journal of Multiphase Flow 9 (5), 511–530.

Ricotti, M.E., Fomin, R.V., 2020. Small modular reactors. In: Salah Ud-Din Khan, Alexander Nakhabov, Publishing Series in Energy, Nuclear Reactor Technology Development and Utilization. Woodhead Publishing, Elsevier Ltd., Kidlington, UK, ISBN 9780128184837, pp. 187–211.

Riebold, W.L., Addabbo, C., Piplies, L., Staedtke, H., 1984. LOBI Project: Influence of Primary Loops on Blowdown, Part 1. EU Joint Research Centre, Ispra, Italy.

Riegel, B., 1978. Contribution a l'etude de la d́ecompression d'une capacité en régime diphasique (in French). PhD thesis. Institut National Polytechnique de Grenoble, Grenoble, France.

Riegel, B., 1990a. Post Test Calculation of UPTF Test 8 with the Advanced Computer Code TRAC-PF1/MOD 1. GRS-A-1666, Garching, Germany.

Riegel, B., 1990b. Post Test Calculation of UPTF Test 9 with the Advanced Computer Code TRAC-PFI/Mod 1. GRS-A-1738, Garching, Germany.

Riemke, R.A., 1993. Modification to Unal's subcooled flow boiling bubble model. Nuclear Technology 102, 416–417.

Riga, C., Dourneau, D., Diakodimitris, C., Douxchamps, P.-A., Planquart, P.P., 2022. CFD analysis of complex mixing phenomena in dead-end lines. In: 19th International Topical Meeting on Nuclear Reactor Thermal Hydraulics (NURETH-19), March 6–11 (Via Web), Brussels, Belgium.

Riznic, J.R. (Ed.), 2021a. Pressurized Heavy Water Reactors: CANDU. Publisher Elsevier, Amsterdam, Netherlands.

Riznic, J.R., 2022. From NURETH-2013 to NURETH-2019: non-critical summary and brief contribution to the history of nuclear reactor thermal hydraulics. In: OECD/NEA/CSNI Specialists Meeting on Transient Thermal-hydraulics in Water Cooled Nuclear Reactors (SM-TH), March 22–24, CIEMAT, Madrid (Spain), via Web, P02.

Riznic, J.R., D'Auria, F., Galassi, G.M., Mazzantini, O., Riznic, J., 2021. In: Koizumi, Y. (Ed.), Pressurized Heavy Water Reactors—Atucha II, vol. 8. Publisher Elsevier, Amsterdam, Netherlands, ISBN 978-0-323-85382-8, pp. 1–587.

Roache, P.J., 1972. Computational Fluid Dynamics. Hermosa Publishers, Albuquerque-Socorro, NM, USA, pp. 1–434.

Roache, P.J., 1998. Verification and Validation in Computational Science and Engineering. Hermosa Publishers, Albuquerque, NM, USA, pp. 1–446.

Roache, P.J., 2009. Fundamentals of Verification and Validation. Hermosa Publishers, Albuquerque-Socorro, NM, USA.

Robert, M., Farvacque, M., Parent, M., Faydide, B., 2003. Cathare 2 V25: a fully validated CATHARE version for various applications. In: 10th International Topical Meeting on Nuclear Reactor Thermal Hydraulics, (NURETH-10), October 5–11, Seoul, Korea.

Roelofs, F., Gopala, V.R., Jayaraju, S., Shams, A., Komen, E., 2013. Review of fuel assembly and pool thermal hydraulics for fast reactors. Nuclear Engineering and Design 265, 1205–1222.

Rogers, J.T., Rosehart, R.G., 1972. Mixing by Turbulent Interchange in Fuel Bundles—Correlations and Influences. ASME, paper 72-HT-53, New York, NY, USA.

Rogers, J.T., Tahir, A.E.E., 1975. Turbulent Interchange Mixing in Rod Bundles and the Role of Secondary Flows. ASME paper 75-HT-31, New York, NY, USA.

Rogovin, M., Frampton Jr., G.T., 1980. Three Mile Island: A Report to the Commissioners and to the Public, Vol. II, Part I. USNRC GPO Sales Program, Washington, DC, USA.

Rohatgi, U.S., 1980. Independent Assessment of TRAC-P1A with Marviken Critical Flow Test, vol. 35. ANS Transaction, pp. 304–306.

Rohatgi, U.S., 1981. Assessment of TRAC-PD2 with university of Houston counter current flow tests. American Nuclear Society (ANS) Transactions 39.

Rohatgi, U.S., 1985. Assessment of TRAC Codes with Dartmouth College Counter-current Flow Tests. Nuclear Technology, April, pp. 100–106.

Rohatgi, U.S., 2015. Role of scaling in validation. In: Keynote at ASME Verification and Validation Symp., Las Vegas, NV, USA.

Rohatgi, U.S., 2022. Verification, validation, and uncertainty quantification in thermal hydraulics, freeman scholar lecture (2019). ASME Journal of Fluids Engineering 144 (4), 1–8, 040801.

Rohatgi, U.S., Kaizer, J., 2020. Historical perspectives of BEPU research in US. Nuclear Engineering and Design 358, 110430.

Rohatgi, U.S., Neymotin, L., 1993. Assessment of scaling of pressurized water reactor emergency core coolant bypass flow tests. In: Ashpis, D.E., Gatski, T.B., Hirsh, R. (Eds.), Instabilities and Turbulence in Engineering Flows. Fluid Mechanics and its Applications, vol. 16. Springer, Dordrecht, Germany.

Rohatgi, U.S., Reshotko, E., 1975. Non-equilibrium one-dimensional two-phase flow in variable area channels. In: Lahey, R.T., Wallis, G.B. (Eds.), Non-Equilibrium Two Phase Flow. ASME Conference Paper. ASME, Houston, TX, USA.

Rohatgi, U.S., Neymotin, L., Saha, P., 1981. Independent assessment of TRAC code with various blowdown experiments. In: Proceedings of the 3rd CSNI Specialist Meeting on Transient Two-phase Flow, OECD-NEA. Hemisphere Publishing Co., Pasadena, California, pp. 573–584.

Rohatgi, U.S., Neymotin, B., Wulff, W., 1990. Bias in Peak Clad Temperature Predictions Due to Uncertainties in Modeling of ECC Bypass and Dissolved Non-condensable Gas Phenomena. USNRC, Washington, DC, USA. NUREG/CR-5354.

Rohatgi, U.S., Cheng, H., Khan, H., Neymotin, L., 1998. RAMONA-4B, A computer code with three dimensional neutron kinetics for BWR and SBWR system transients. In: USNRC NUREG/CR-6359, Vol 1 and 2, March, Washington, DC, USA.

Rohde, U., Kliem, S., Höhne, T., Karlsson, R., Hemström, B., Lillington, J., Toppila, T., Elter, J., Bezrukov, Y., 2005. Fluid mixing and flow distribution in the reactor circuit, measurement data base. Nuclear Engineering and Design 235 (2–4).

Rohde, M., Marcel, C.P., Manera, A., Van Der Hagen, T.H.J.J., Shiralkar, B., 2010. Investigating the ESBWR stability with experimental and numerical tools: a comparative study. Nuclear Engineering and Design 240, 375–384.

Rohsenow, W.M., 1952. A method of correlating heat transfer data for boiling of liquids. Journal of Heat Transfer 74, 969–976.

ROSA, -V., Suzuki, M., Nakamura, H., Ohtsu, I., Kondo, M., Shibamoto, Y., Takeda, T., Yonomoto, T., Asaka, H., Matsumoto, T., Ohwada, T., Ogawa, M., Ohwada, A., Ito, H., 2003. [Authors. In: ROSA-V Large Scale Test Facility (LSTF) System Description for the Third and Fourth Simulated Fuel Assemblies. JAERI-Tech 2003-037, Tokai-Mura, Naka-Gun, Ibaraki-Ken, Japan, pp. 1−491.

Rosal, E.R., Lin, T.F., McClellan, I.S., Brewer, R.C., Tien, K., 2010. Rod Bundle Heat Transfer Test Facility Description. Pennsylvania State University, USNRC NUREG/CR-6976, Washington, DC, USA, pp. 1−207.

Rose, J.W., 1997. On interphase matter transfer, the condensation coefficient and dropwise condensation. Proceedings of the Royal Society of London, United Kingdom, Series A: Mathematical and Physical Sciences 411, 305−311.

Rose, J.W., 2002. Dropwise condensation theory and experiment: a review. Proceedings of the Institution of Mechanical Engineers, Part A: Journal of Power and Energy 216, 115−128.

Rossi, L., de Fayard, R., Kassab, S., 2015. Measurements of the vertical distribution of the void fraction using X-ray attenuation for different flow regimes in a horizontal pipe. In: 16th International Topical Meeting on Nuclear Reactor Thermal Hydraulics (NURETH-16), August 30-September 4, Chicago, IL, USA.

Roth, G.A., Aydogan, F., 2014a. Theory and implementation of nuclear safety system codes. Part I: conservation equations, flow regimes, numerics and significant assumptions. Progress in Nuclear Energy 76, 160−182.

Roth, G.A., Aydogan, F., 2014b. Theory and implementation of nuclear safety system codes. Part II: system code closure relations, validation, and limitations. Progress in Nuclear Energy 76, 55−72.

Roth, G., Aydogan, F., 2015. Six-Field governing equation development for advanced system codes. In: 16th International Topical Meeting on Nuclear Reactor Thermal Hydraulics (NURETH-16), August 30-September 4, Chicago, IL, USA.

Rotta, J.C., 1972. Turbulente Strömumgen. BG Teubner, Stuttgart, Germany.

Rouben, B., 1997. Coupled neutronics-thermal-hydraulics LOCA analysis. In: Seminar at Chulalongkorn University, Bangkok, Thailand.

Rousseau, J.C., 1987. Flashing flow. Multiphase Science and Technology 3, 378−389.

Rousseau, J.C., Ferch, R.L., 1979. A note on two-phase separated flow models. International Journal of Multiphase Flow 5, 489−494.

Rousselet, Y.L., 2014. Interacting Effects of Inertia and Gravity on Bubble Dynamics, pp. 1−330. PHD Thesis at UCLA (Abstract co-author and Thesis Director Prof. V.K. Dhir), Los Angeles, Ca, USA.

Rowe, D.S., 1967. Cross-flow Mixing between Parallel Flow Channels during Boiling. Part I, COBRA: Computer Program for Coolant Boiling in Rod Arrays. Pacific Battelle-Northwest, BNWL-371, Richland, Wa, USA.

Rowe, D.S., 1970. COBRA-II: A Digital Computer Program for thermal-hydraulic Subchannel Analysis of Rod Bundle Nuclear Fuel Elements, pp. 1−80. Pacific Battelle-Northwest, BNWL-1229 (UC-80, reactor technology), Richland, Wa, USA.

Rowe, D.S., 1971. COBRA III, A Digital Computer Program for Steady State and Transient Thermal-Hydraulics Analysis of Rod-Bundle Nuclear Fuel Elements. Battelle-Northwest Report BNWL-B-82, Washington State, USA.

Rowe, D.S., 1973. COBRA-IIIC: Digital Computer Program for Steady State and Transient thermal-hydraulic Analysis of Rod Bundle Nuclear Fuel Elements. Pacific Battelle-Northwest, BNWL-1695, Richland, Wa, USA.

Royl, P., Travis, J.R., Breitung, W., Kim, J., Kim, S.B., 2009. GASFLOW validation with PANDA tests from the OECD SETH benchmark covering steam/air and steam/helium/ air mixtures. Science and Technology of Nuclear Installations. Article ID 759878.

Rozzia, D., Adorni, M., Del Nevo, A., D'Auria, F., 2011. Capabilities of TRANSURANUS code in simulating power ramp tests from the IFPE database. Nuclear Engineering and Design 241.

RSK, 2005. RSK-Recommendation « Anforderungen an die Nachweisführung bei Kühlmittelverluststörfall—Analysen », (in German) Anlage 1 zum Ergebnisprotokoll der 385. Sitzung der Reaktor-Sicherheitskommission (RSK) am July 20/21, Germany.

RSK, 2015. RSK-Recommendation « Anforderungen an die Nachweisführung bei Kühlmittelverluststörfall—Analysen », (in German) Anlage 1 zum Ergebnisprotokoll der 475. Sitzung der Reaktor-Sicherheitskommission (RSK) am April 15, Germany.

Rubin, A., Schoedel, A., Avramova, M.N., Utsuno, H., Bajorek, S., Velazquez-Lozada, A., 2010. OECD/NRC Benchmark Based on NUPEC PWR Subchannel and Bundle Tests (PSBT), Volume I: Experimental Database and Final Problem Specifications. US NRC and OECD Nuclear Energy Agency, NEA/NSC/DOC(2010)1, Paris, France.

Ruder, Z., Hanratty, T.J., 1990. A definition of gas-liquid plug flow in horizontal pipes. International Journal of Multiphase Flow 16, 233−242.

Ruspini, C.L., Marcel, C.P., Clausse, A., 2014. Two-phase flow instabilities: a review. International Journal of of Heat and Mass Transfer 71, 521−548.

Ryskin, G., 1987. Heat and mass transfer from a moving drop—some approximate relations for the Nusselt number. International Communications in Heat and Mass Transfer 14, 741−749.

Sadatomi, M., Kawahara, A., Sato, Y., 1994. Flow redistribution due to void drift in two phase flow in a multiple channel consisting of two subchannels. Nuclear Engineering and Design 148 (2−3), 463−474.

Sadatomi, M., Kawahara, A., Sato, Y., 1997. Treatment of two-phase turbulent mixing, void drift and diversion cross-flow in a hydraulically non-equilibrium subchannel flow. In: Proceedings Of 4th International Seminar on Subchannel Analysis (ISSCA-4), September 25−26, Tokyo, Japan, pp. 87−104.

Sadatomi, M., Kawahara, A., Kuno, T., Kano, K., 2005. Two-Phase void drift phenomena in a 2×3 rod bundle: flow redistribution data and their analysis. Nuclear Technology 152 (1), 23−37.

Sadatomi, M., Kano, K., Kawahara, A., Mori, N., 2006. Void fraction and pressure drop in two-phase equilibrium flows in a vertical 2×3 rod bundle channel: assessment of correlations against the present subchannel data. JSME International Journal, Series B, Fluids and Thermal Engineering 279−286.

Šadek, S., Grgić, D., Benčik, V., Vlahović, S., 2020. Analysis of the up-flow conversion modification and influence on selected LOCA accidents in a PWR plant. Nuclear Engineering and Design 369 (110854), 1−14.

Sagaut, P., Deck, S., 2009. Large eddy simulation for aerodynamics: status and perspectives. Philosophical Transactions of the Royal Society, Series A 367, 2849−2860.

Saha, P., 1978. A Review of Two-phase Steam-Water Critical Flow Models with Emphasis on Thermal Non-equilibrium. NUREG/CR-0417, USNRC, Washington, DC, USA.

Saha, P., Zuber, N., 1974. Point of net vapor generation and vapor void fraction in subcooled boiling. In: Proceedings Fifth International Heat Transfer Conference, vol. 4, pp. 175−179.

Saha, P., Rohatgi, U.S., Jo, J.H., Neymotin, L., Slovik, G.C., Yuelys-Miksis, C., 1983. Independent code assessment at BNL in FY 1982. In: Proceedings of the U.S. Nuclear Regulatory Commission 10th Water Reactor Safety Research Information Meeting, NUREG/CP-0041, vol. 2, pp. 145–168.

Saha, P., Rohatgi, U.S., Jo, J.H., Neymotin, L., Slovik, G.C., Yuelys-Miksis, C., Pu, J., 1983a. Independent assessment of TRAC and RELAP5 codes through separate effects tests. In: Proceedings of the 11th Water Reactor Safety Research Information Meeting, October 24–28, USNRC, Washington, DC, USA.

Saha, P., Jo, J.H., Neymotin, L., Rohatgi, U.S., Slovik, G.C., Yuelys-Miksis, C., 1985. Independent Assessment of TRAC -PF1 (Version 7.0), RELAP5 /MOD1 (Cycle 14), and TRAC -BD1 (Version 12.0) Codes Using Separate- Effects Experiments, pp. 1–163. Brookhaven National Lab, BNL-NUREG_51919, USNRC, NUREG/CR 4359, Washington, DC, USA.

Saha, P., Saha, P., 2017. Nuclear Reactor Thermal-Hydraulics: Past, Present and Future. (Monograph Series Ed.S J. Riznic, R. Schultz). ASME, New York, NY 10016, USA, pp. 1–143.

Sakashita, H., 2011. Bubble growth rates and nucleation site densities in saturated pool boiling of water at high pressures. Journal of Nuclear Science and Technology 48, 734–743.

Sakurai, A., 1990. Film boiling heat transfer. In: Proceedings 9th International Heat Transfer Conference, Jerusalem, Israel, vol. 1, pp. 157–186.

Sakurai, A., Shiotsu, M., Hata, K., 1990. A general correlation for pool film boiling heat transfer from a horizontal cylinder to sub-cooled liquid: Part 1—a theoretical pool film boiling heat transfer model including radiation contributions and its analytical solution. Journal of Heat and Mass Transfer 112.

Salay, M.A., 2001. Evaluation of Interfacial Drag Models for Use in TRAC-M Rod Bundle Components with Validation against Steady-State and Transient Boil-Off Experiments (Ph.D. thesis). Univ. of Maryland, College Park, MD, USA.

Salko, R.K., Avramova, M.N., 2015. COBRA-TF Subchannel Thermal-Hydraulics Code (CTF) Theory Manual–Revision 0. CASL-U-2015-0054, USA.

Salko, R.K., Avramova, M.N., 2016. CTF Theory Manual. PSU, USDOE, CASL-U-2016-1110-000, pp. 1–312.

Salko, R.K., Schmidt, R.C., Avramova, M.N., 2015. Optimization and parallelization of the thermal–hydraulic sub-channel code CTF for high-fidelity multi-physics applications. Annals of Nuclear Energy 84, 122–130.

Salvetti, C., 2001. The birth of nuclear energy: the Fermi Pile (in Italian). Book "Conoscere Fermi nel centenario della nascita 29 settembre 1901-2001". In: Bernardini, C., Bonolis, L. (Eds.), Società Italiana di Fisica, Bologna, Casa Editrice Compositori, ISBN 88-7794-284-3, pp. 178–204.

Sanderson, P.D., 1961. Heat Transfer from uranium fuel to the Magnox Can in a gas-cooled reactor. In: Proceedings Of International Development in Heat Transfer, vol. 1. ASME, New York, NY, USA, pp. 53–64.

Sankovich, M.F., McDonald, B.N., 1971. Once through steam generation. In: XVI Nuclear Congress, Rome, Italy, 25–26 March.

Sarkisov, A.A., Vysotskii, V.L., 2021. The largest nuclear accident in the history of the nuclear fleet. Reconstruction of events and analysis of the accident consequences to assess the risks and hazards of small nuclear power facilities. Nuclear Engineering and Design 384 (111440), 1–25.

Sarkodie, K., Fergusson-Rees, A., Diaz, P., 2018. A review of the application of non-intrusive infrared sensing for gas–liquid flow characterization. Journal of Computational Multiphase Flows 10 (1), 43–56.

Sato, Y., 1992. Cross flow of gas-liquid mixture between subchannels. In: Ninokata, N., Aritomi, A. (Eds.), Proceedings Of Sub-channel Analysis in Nuclear Reactors, Institute of Applied Energy (IAE) and the THD of the Atomic Energy Society of Japan, ISSCA-I, ISBN 4-9900222-0-3. Tokyo, Japan.

Sato, I., 2021. Analysis of Fukushima-Daiichi Nuclear Power Plant Unit 3 pressure data and obtained insights on accident progression behavior. Nuclear Engineering and Design 383 (111426), 1–19.

Saurel, R., Abgrall, R., 1999. On a multiphase Godunov method for compressible multifluid and multiphase flows. Journal of Computational Physics 150, 425–467.

Saurel, R., Abgrall, R., 1999a. A simple method for compressible multifluid flows. SIAM Journal of Scientific Computing 21, 1115–1145.

Saurel, R., Lemetayer, O., 2001. A multiphase model for compressible flows with interfaces, shocks, detonation waves and cavitation. Journal of Fluid Mechanics 431, 239–271.

Sawant, P., Marking, J., Delfino, C., 2015. NRELAP5 predictions of KAIST high pressure condensation data using existing and extended Shah condensation correlation. In: 16th International Topical Meeting on Nuclear Reactor Thermal Hydraulics (NURETH-16), August 30-September 4, Chicago, IL, USA.

Saxena, A., Prasser, H.-M., 2018. Flow structure of liquid films in annular flow. In: 12th International Topical Meeting on Nuclear Reactor Thermal-Hydraulics, Operation and Safety (NUTHOS-12), October 14–18, Qingdao, China.

Sayyaadi, H., Sabzaligol, T., 2009. Various approaches in optimization of a typical pressurized water reactor power plant. Applied Energy 86, 1301–1310.

Scannapieco, E., Harlow, F.H., 1995. Introduction to Finite-Difference Methods for Numerical Fluid Dynamics, LA 12984, Los Alamos, NM, USA.

Schäfer, T., Bieberle, A., Hampel, U., 2015. Investigations on centrifugal pumps under air entrainment conditions. In: 16th International Topical Meeting on Nuclear Reactor Thermal Hydraulics (NURETH-16), August 30–September 4, Chicago, IL, USA.

Schaffrath, A., Krussenberg, A.-K., Fjodorow, A., Gocht, U., Lischke, W., 2001. Modeling of condensation in horizontal tubes. Nuclear Engineering and Design 204.

Schenk, J.R., Groeneveld, D.C., 1990. Measurement of thermal hydraulic parameters inside multi-element bundles. In: Proceedings International Symp. On Multi-phase Flow, Miami, Fl, USA, December 9–12.

Schikorr, M., Bubelis, E., Mansani, L., Litfin, K., 2010. Proposal for pressure drop prediction for a fuel bundle with grid spacers using Rehme pressure drop correlations. Nuclear Engineering and Design 240, 1830–1842.

Schlaberg, H.I., Yang, M., Hoyle, B.S., 1996. Real time ultrasonic process tomography for two-component flows. Electronic Letters 32 (17), 1571–1572.

Schlegel, J.P., Hibiki, T., Shen, X., Appathurai, S., Subramani, H., 2015. Evaluation of interfacial area transport equation in coupled two-fluid model computation. In: 16th International Topical Meeting on Nuclear Reactor Thermal Hydraulics (NURETH-16), August 30–September 4, Chicago, IL, USA.

Schlünder, E.U., Spalding, D.B., Taborek, J., 1983. Heat Exchanger Design Handbook. VDI-Verlag GmbH, Düsseldorf Hemisphere Public Corp., Washington, DC, USA.

Schmid, S.D., 2015. Producing Power—The Pre-chernobyl History of Soviet Nuclear Industry. MIT Press, Cambridge, MA, USA.

Schmitt, R.C., Bingham, G.E., Norberg, J.A., 1970. Simulated Design Basis Accident Test of Carolina Virginia Tube Reactor Containment, Idaho Nuclear Corporation, IN—1403, Idaho Falls, (ID), USA, [OSTI Identifier 4015074].

Schneider, P.J., 1973. Conduction. In: Rohsenow, W.H., Hartnett, J.P. (Eds.), Handbook of Heat Transfer. McGraw-Hill Book Co., New York, NY, USA.

Schneider, K., Winkler, F.J., 1988. Physical model for reactor coolant pumps. Journal of Nuclear Engineering and Design 108, 99–105.

Schollenberger, S.P., Umminger, K., Schoen, B., Dennhardt, L., 2017. PKL-ROSA / LSTF counterpart testing on beyond-design-basis SB-LOCA. In: 17th International Topical Meeting on Nuclear Reactor Thermal Hydraulics, NURETH-17, Sept 2017, Xi'an, China.

Schraub, F.A., Simpson, R.L., Janssen, E., 1969. Two-Phase Flow and Heat Transfer in Multirod Geometries: Air-Water Flow Structure Data for a Round Tube, Concentric and Eccentric Annulus and Nine-Rod Bundle. General Electric Company, San Jose, Ca, USA. GEAP-5739.

Schrock, V.E., Revankar, S.T., Mannheiner, R., Wang, C.H., 1986. Small Break Critical Discharge—The Role of Valor and Liquid Entrainment in a Stratified Two-phase Region Upstream of Break. Lawrence Berkeley Laboratory LBL-22024, USNRC NUREG/CR-4761, Washington, DC, USA.

Schubring, D., Ashwood, A.C., Shedd, T.A., Hurlburt, E.T., 2010. Planar laser-induced fluorescence (PLIF) measurements of liquid film thickness in annular flow. Part I: methods and data. International Journal of Multiphase Flow 36 (10), 815–824.

Schultz, M.A., 1955. Control of Nuclear Reactors and Power Plants. McGraw-Hill Book Company, Inc., New York, NY, USA.

Schultz, R.R., 1993. Next generation nuclear plant—high-level functions and requirements: Addendum 1, requirements for system design models. In: INEEL/EXT-03-01163, Addendum 1, Idaho Falls, ID, USA, December.

Schultz, R.R., Weaver, W.L., 2003. Using the RELAP5-3D advanced systems code with commercial and advanced CFD software. In: Proceedings 11th International Conference On Nuclear Engineering, Tokyo, Japan, April 20–23.

Schultz, R.R., Kondo, M., Anoda, Y., 2001. Baseline study to model a typical condensation-induced water hammer event measured at the two-phase flow test facility (TPTF) in Japan. In: Pressure Vessel & Piping Conference, Atlanta, GA, USA, July 22–26.

Schumann, U., 1975. Subgrid scale model for finite difference simulations of turbulent flows in plane channels and annuli. Journal of Computational Physics 18, 376–404.

Schwer, L.E., 2007. An Overview of the PTC 60 / V&V 10—Guide for Verification and Validation in Computational Solid Mechanics. The American Society of Mechanical Engineers (ASME), New York, NY, USA.

Scriven, L., 1959. On the dynamics of phase growth. Chemical Engineering Sci. 10 (1–2), 1–13.

Scuro, N.L., Angelo, E., Angelo, G., Andrade, D.A., 2018. A CFD analysis of the flow dynamics of a directly-operated safety relief valve. Nuclear Engineering and Design 328, 321–332.

Sehgal, B.R., 2012. Nuclear Safety in Light Water Reactors—Severe Accident Phenomenology. Elsevier Academic Press, Oxford, UK.

Sehgal, B.R., Bechta, S., 2016. Severe accident progression in the BWR lower plenum and the modes of vessel failure. Annals of Nuclear Energy 93, 28–34.

Semenov, N.I., Kosterin, S.I., 1964. Results of studying the speed of sound in moving gas-liquid systems. Teploenergetika 11 (6), 46–51.

Semeria, R., Martinet, B., 1965. Calefaction spots on a heating wall, temperature distribution and resorption. Proceedings of the Institutions of Mechanical Engineers 180, 192.

Sene, K.J., 1984. Aspects of Bubbly Two-phase Flow. PhD Thesis. Trinity College, Cambridge, UK.

Seo, K.W., Kwon, H., Kim, S.J., Hwang, D.H., 2016. Comparison of MATRA-S and COBRA-SFS for Low Flow Subchannel Analysis. Korean Nuclear Society (KNS, Spring Meeting), May, 11−13, Jeju, Korea.

Sergeev, V.V., 1978. Post Dryout Heat Transfer in Annulus and Rod Bundles, Analytical Review, FEI OB-67. Institute of Physics and Power Engineering, Obninsk, Russia (in Russian).

Sergeev, V.V., 1987. Calculation of Heat Transfer in post Dryout Heat Transfer in Vertical Cylindrical Channels. Preprint IPPE-779, Obninsk, Russia (in Russian).

Sergeev, V.V., 2002. Verification of the model of post-CHF heat transfer in dispersed flow. In: Proceedings 3rd Russian National Heat Transfer Conference, vol. 4. MPEI Publishing House, Moscow, Russia, pp. 165−168 (in Russian).

Sergeev, V.V., 2007. Transfer in Post-CHF Zone of Steam Generating Channels and Subassembly Thermohydraulics in Transient Regimes and Accidents. PhD Thesis. IPPE, Obninsk, Russia.

Sergeev, V.V., Gal'chenko, E.F., Remizov, O.V., 1985. Engineering Method of post Dryout Heat Transfer in Annular Channels. Preprint IPPE-1649, Obninsk, Russia (in Russian).

Sergeev, V.V., Gonin, A.I., Remizov, O.V., 1990. Post critical heat transfer in the channels with spacer elements. Atomnaya Energiya 68 (6), 445−447 (in Russian).

Serre, G., Bestion, D., Franco, M., Bottin, M., Marchand, M., 2011. On the stratification criterion in the CATHARE code. In: The 14th International Topical Meeting on Nuclear Reactor Thermalhydraulics, NURETH-14, Toronto, Ontario, Canada, September 25−30.

Seul, K.W., Bang, Y.S., Kim, I.G., Yonomoto, T., Anoda, Y., 2003. Simulation of multiple steam Generator Tube Rupture (SGTR) event scenario. Journal of the Korean Nuclear Society 35 (3).

Sevon, T., 2015. A melcor model of Fukushima Daiichi unit 1 accident. Annals of Nuclear Energy 85, 1−11.

Seyyedi, S.M., Hashemi-Tilehnoee, M., Rosen, M.A., 2018. Exergy and exergoeconomic analyses of a novel integration of a 1000 MW pressurized water reactor power plant and a gas turbine cycle through a superheater. Annals of Nuclear Energy 115, 161−172.

Sha, W., Chao, B.T., 1981. Local Volume-Averaged Transport Equations for Single-phase Flow in Regions Containing Fixed, Dispersed Heat-Generating (Or Absorbing) Solids. Argonne National Laboratory, ANL-80-124, Chicago, Illinois, USA, pp. 1−125.

Sha, W.T., Domanus, H.M., Szweczyk, A.A., 1976. Development of a Computer Program for Reactor Core thermal Hydraulics: Finite Differencing in a 3-D Staggered Mesh, Suitable for Irregular Boundary Applications. Tech. Memorandum, ANL-CT-76-54, USNRC-7 (Contract A 2045), Washington, DC, USA, pp. 1−45.

Shah, M.M., 1979. A general correlation for heat transfer during film condensation inside pipes. International Journal of Heat Mass Transfer 22, 547−556.

Shapiro, A.H., 1953. The Dynamics and Thermodynamics of Compressible Fluid Flow, vol. 1. Ronald Press. www.openlibrary.org/publishers/Ronald_Press.

Shapiro, A.H., Erickson, A.J., 1957. On the changing size spectrum of particle clouds undergoing evaporation, combustion or acceleration. Transactions of ASME 79, 775.

Sharma, P.K., Gera, B., Singh, R.K., Vaze, K.K., 2012. Computational Fluid Dynamics Modeling of Steam Condensation on Nuclear Containment Wall Surfaces Based on Semiempirical Generalized Correlations. Science and Technology of Nuclear Installations, Hindawi, pp. 1−7, 106759.

Sharma, S.L., Hibiki, T., Ishii, M., Brooks, C.S., Schlegel, J.P., Liu, Y., Buchanan Jr., J.R., 2015. Turbulence-Induced bubble collision force model development and assessment for adiabatic dispersed air-water two-phase flow with the two-fluid model. In: 16th International Topical Meeting on Nuclear Reactor Thermal Hydraulics (NURETH-16), August 30-September 4, Chicago, IL, USA.

Shchukin, V.K., 1980. Heat Transfer and Hydrodynamics of Internal Flows in the Fields of Mass Forces, second ed. Mashinostroenie, Moscow, Russia.

Shchukin, V.K., Halatov, A.A., 1982. Heat Transfer, Mass Transfer and Hydrodynamics of Swirled Flows in Axisymmetrical Channels. Mashinostroenie, Moscow, Russia (in Russian).

Shen, X., Hibiki, T., 2015. Interfacial area concentration in gas-liquid bubbly to churn flow regimes in large diameter pipes. International Journal of Heat and Fluid Flow 87, 136−155.

Shen, X., Hibiki, T., 2018. An evaluation of one-group interfacial area transport equation with its constitutive bubble coalescence and breakup models in bubbly two-phase flows. In: 12th International Topical Meeting on Nuclear Reactor Thermal-Hydraulics, Operation and Safety (NUTHOS-12), October 14−18, Qingdao, China.

Shen, W., Rouben, B., 2021. Fundamentals of CANDU Reactor Physics. ASME Press, New York, NY, USA.

Shen, Y.F., Cao, Z.D., Ganglu, Q., 1991. An investigation of crossflow mixing effect caused by grid spacer with mixing blades in a rod bundle. Nuclear Engineering and Design 125, 111−119.

Shen, X., Hibiki, T., Nakamura, H., 2012. Developing structure of two-phase flow in a large diameter pipe at low liquid flow rate. International Journal of Heat Fluid Flow 34, 70−84.

Sheng, Y.Y., Irons, G.A., 1991. A combined laser Doppler anemometry and electrical probe diagnostic for bubbly two-phase flow. International Journal of Multiphase Flow 17 (5), 585−598.

Sheng, D.-Y., Seidl, M., 2015. Determination of the pressure loss coefficient at control-rod guide tube flow-hole for a PWR nuclear fuel assembly by using CFD and Bernoulli solutions. In: 16th International Topical Meeting on Nuclear Reactor Thermal Hydraulics (NURETH-16), August 30-September 4, Chicago, IL, USA.

Shercliff, J.A., 1962. The Theory of Electromagnetic Flow Measurement. Cambridge Univ. Press, Cambridge, UK.

Shi, S., Lin, Y.-C., Yang, W.S., Ishii, M., 2015. Experimental stability maps for a BWR-type small modular reactor. In: 16th International Topical Meeting on Nuclear Reactor Thermal Hydraulics (NURETH-16), August 30-September 4, Chicago, IL, USA.

Shih, C., Yang, J.H., Wang, J.R., Lin, H.-T., Chiang, S.C., Liu, C.C., Tien, K., 2015. BEPU Analysis and Benchmark with IIST 2% SBLOCA Experiment Using TRACE/Dakota. USNRC NUREG/IA 0456, Washington, DC, USA.

Shin, T.S., Jones, O.C., 1986. A active cavity model for flashing. Nuclear Engineering and Design 95, 185−196.

Shin, T.S., Jones, O.C., 1993. Nucleation and flashing in nozzles—1. A distributed nucleation model. International Journal of Multiphase Flow 19 (6), 943−964.

Shih, T.H., Zhu, J., Lumley, J.L., 1993. A realizable Reynolds stress algebraic equation Model. In: Symposium on Turbulence Shear Flow, August 10−18, 1993, Kyoto, Japan. NASA-TM-105993.

Shiraishi, T., 2011. Design of the advanced accumulator for the pressurized water reactor. Nuclear Engineering and Design 241 (9), 3910–3924.

Shiralkar, B.S., Lahey, R.T., 1972. Diabatic local void fraction measurements in freon-114 with a hot-wire anemometer. Transactions of the American Nuclear Society 15 (2), 880.

Shlykov, Y.P., 1965. Prediction of contact thermal resistance of processed metallic surfaces. Teploenergetika 10, 79–83 (in Russian).

Shlykov, Y.P., Ganin, E.A., Zarevskii, C.N., 1977. Contact thermal Resistance. Moscow, Russia Energia (in Russian).

Shoji, M., 1998. Boiling chaos and modeling. In: Proceedings 11th International Heat Transfer Conference, vol. 1, pp. 3–21.

Shotkin, L.M., 1997. How the code analyst can influence the calculation and use good practices to minimize pitfalls. Nuclear Technology 117, 40–48.

Shvetz, I.T., Dyban, E.P., 1964. Heat transfer at contact of plain metallic surfaces. Inzhenerno-Phyzichesky Zhurnal 7 (3), 3–9 (in Russian).

Shwageraus, E., Volaski, D., Fridman, E., 2009. Investigation of fuel assembly design options for high conversion thorium fuel cycle in PWRs. In: Advances in Nuclear Fuel Management IV (ANFM), Hilton Head Island, South Carolina, USA, April 12–15 [CD-ROM, American Nuclear Society, LaGrange Park, IL (2009) ANFM Conference, South Carolina, USA].

Siebe, D.A., Stumpf, H.J., 1990. Posttest Analysis of the Upper Plenum Test Facility Downcomer Separate Effects Tests with TRAC-Pf1/mod2. Los Alamos National Laboratory, Los Alamos, NM, USA. LA-CP-90-299.

Siegel, R., Howell, J.R., 1981. Thermal Radiation Heat Transfer, third ed. Hemisphere Public Corp., Washington, DC, USA.

Silverman, I., 2022. Topology based multi-phase flow model for nuclear safety applications: a 2D version. In: 19th International Topical Meeting on Nuclear Reactor Thermal Hydraulics (NURETH-19), March 6–11 (Via Web), Brussels, Belgium.

Simon, H.A., 1962. The architecture of complexity. In: Proceedings Of the American Philosophical Society (Publisher), vol. 106, pp. 467–482 (Dec. 12).

Simon, U., Paul, J., Schorsch, R., Werner, K.D., Liebert, J., 1986. 2D/3D Project, Calibration of the UPTF Tieplate Flow Module with 'Advanced Instrumentation' and Investiogation of the Function from the CORE Simulator Feedbnack Control System with Break-Through Detectors. Final Report. DE88752023/GAR KraftwerkUnion, pp. 1–183, Erlangen, Germany.

Sjoden, G.E., 2015. Foundations in Applied Nuclear Engineering Analysis. Imperial College Press, London, UK.

SKI, 1995. The 3rd International Sem. On subchannel analysis (ISSCA-III), Stockholm. In: Seminar Proceedings in SKI Report 95:16, ISRN SKI-R-95/16-SE, ISSN: 1104-1374, Sastaholm, Taby, Sweden.

Skillman, G.R., Rempe, J.L., 2021. The Three Mile Island Unit 2 Accident. Encyclopedia of Nuclear Energy, pp. 17–29.

Skipper, R.G.S., 1958. Heat transfer within reactor fuel elements. Atomic Energy Reviews 1 (3), 141–146.

Skipper, R.G.S., Wootton, K.J., 1958. Thermal resistance between uranium and can. In: Proceedings of the Second United Nation International Conference On the Peaceful Uses of Atomic Energy, Geneva, Switzerland, vol. 7, pp. 687–690.

Skorek, T., 2018. Verification & validation and uncertainty quantification of code models, ANS best estimate plus uncertainty. In: International Conference BEPU-2018, Paper 110, Real Collegio, Lucca, Italy, May 13–19.

Skorek, T., de Crécy, A., Kovtonyuk, A., Petruzzi, A., Mendizábal, R., de Alfonso, E., Reventós, F., Freixa, J., Sarrette, C., Kyncl, M., Pernica, R., Baccou, J., Fouet, F., Probst, P., Chung, B.-D., Tram, T.T., Oh, D.-Y., Gusev, A., Falkov, A., Shvestov, Y., Li, D., Liu, X., Zhang, J., Alku, T., Kurki, J., Jäger, W., Sánchez, V., Wicaksono, D., Zerkak, O., Pautz, A., 2019. Quantification of the uncertainty of the physical models in the system thermal-hydraulic codes—PREMIUM benchmark. Nuclear Engineering and Design 354, 110199.

Skorek, T., von der Cron, D., Guo, L., Schöffel, P., 2022. The three-field model in the code ATHLET. In: 19th International Topical Meeting on Nuclear Reactor Thermal Hydraulics (NURETH-19), March 6–11 (Via Web), Brussels, Belgium.

Skripova, V.P., 1972. Metastable Liquid. Nauka, Moscow, Russia (in Russian).

Slobodchuk, V., Leskin, S., Shelegov, A., Sidorov, A., 2009. Method and algorithms for analyzing the condition of emergency system pumps based on their testing results. In: 13th International Topical Meeting on Nuclear Reactor Thermal Hydraulics (NURETH-13), September 27-October 2, Kanazawa City, Ishikawa, Japan.

Slugen, V., 2014. Safety of VVER-440 Reactors: Barriers against Fission Products Release. Springer Publisher, pp. 1–178. ISBN-10 1447161149.

Smagorinsky, J., 1963. General circulation experiments with the primitive equations. Mon. Weather Rev. 91, 99–110.

Smith, J., 2017. Nuclear steam generators design. In: Riznic, J. (Ed.), Steam Generators for Nuclear Power Plants. Woodhead Publishing, Duxford, UK, pp. 35–53.

Smith, B.L., Bieder, U., Graffard, E., Heitsch, M., Henriksson, M., Höhne, T., Komen, E., Mahaffy, J., Moretti, F., Morii, T., Mühlbauer, P., Rohde, U., Scheuerer, M., Song, C.-H., Zigh, G., Andreani, M., Bestion, D., Ducros, F., Houkema, M., Lucas, D., Menter, F., Watanabe, T., 2008. Assessment of Computational Fluid Dynamics (CFD) for Nuclear Reactor Safety Problems. NEA/CSNI/R(2007)13, Paris, France.

Smith, T.R., Schlegel, J.P., Hibiki, T., Ishii, M., 2012a. Two-phase flow structure in large diameter pipes. International Journal of Heat Fluid Flow 33, 156–167.

Smith, T.R., Schlegel, J.P., Hibiki, T., Ishii, M., 2012b. Mechanistic modeling of inter-facial area transport in large diameter pipes. International Journal of Multiphase Flow 47, 1–16.

Smith, B.L., Song, C.-H., Chang, S.-K., Lee, J.R., Kim, J.W., 2013. Report of the OECD/NEA KAERI Rod Bundle CFD Benchmark Exercise. NEA/CSNI R(2013)5, Paris, France.

Smith, B.L., Andreani, M., Bieder, U., Ducros, F., Graffard, E., Heitsch, M., Henriksson, M., Höhne, T., Houkema, M., Komen, E., Mahaffy, J.H., Menter, F., Moretti, F., Morii, T., Mühlbauer, P., Rohde, U., Scheuerer, M., Song, C.-H., Watanabe, T., Zigh, G., 2015. Assessment of CFD Codes for Nuclear Reactor Safety Problems—Revision 2, pp. 1–226. NEA/CSNI/R(2014)12 Nuclear Energy Agency, Paris, France.

Smoglie, C., 1984. Two-Phase Flow through Small Branches in a Horizontal Pipe with Stratified Flow, KFK 3861, Karlsruhe, Germany.

Sobajima, M., 1981. Spray cooling test, A separate effect test and ROSA-III, A BWR-LOCA simulation facility. Journal of Nuclear Science and Technology 18 (8), 629–639.

Sokolowski, L., Koszela, Z., 2012. RELAP5 capability to predict pressure wave propagation phenomena in single- and two-phase flow conditions. Journal of Power and Technology 92 (3).

Sokolowski, L., Kozlowski, T., Calvo, A., 2012. An Assessment of Two-phase Critical Flow Models Performance in RELAP5 and TRACE against Marviken Critical Flow Tests. USNRC, Washington, DC, USA. NUREG/IA-0401.

Soloviev, S.L., 1998. Boiling and evaporation of liquid on porous surface. Diss. Doctor. tehn. Sciences. MEI, Moscow. Heat transfer in developed nucleate boiling on the capillary-porous surfaces. In: Proceedings of the Heat Transfer in Modern Technology. Institute of High Temperatures, RAS, Moscow, Russia, pp. 173–195 (in Russian).

Soma, S., Abe, S., Sibamoto, Y., Ishigaki, M., 2022. Measurement of velocity and temperature profiles in boundary layer with steam condensation. In: 19th International Topical Meeting on Nuclear Reactor Thermal Hydraulics (NURETH-19), March 6–11 (Via Web), Brussels, Belgium.

Son, Y.-S., Shin, J.-Y., 2007. Thermal-hydraulic analysis on the effect of operator action in the loss of shutdown cooling system accident in a PWR. Nuclear Engineering and Design 237.

Son, S.M., Choo, J.H., Hong, M., Kim, H.S., Choi, C.G., 2016. Preliminary CFD analysis for HVAC system design of a containment building. In: Transactions of the Korean Nuclear Society Autumn Meeting, Gyeongju, Korea, October 27–28, pp. 1–2.

Song, C.-H., 2006. Scaling of multi-dimensional thermal-hydraulic phenomena in advanced nuclear reactor systems. In: Fifth Korea-Japan Symposium on Nuclear Thermal Hydraulics and Safety (NTHAS5) Conference, Jeju, Korea.

Song, C.-H., 2015. Issues and challenges on advanced thermal-hydraulic researches. In: NURETH-16 Conference, Chicago, IL, USA, August 30–September 4.

Song, J., Bae, K., 2000. Evaluation of analytically scaled models of a pressurized water reactor using the RELAP5/MOD3 computer code. Journal of Nuclear Engineering and Design 199, 215–225.

Song, C.-H., Kim, Y.S., Cho, Y.I., 2011. Direct contact condensation of steam jet in a pool. Advances in Heat Transfer 43, 227–288.

Song, C.-H., Choi, K.-Y., Kang, K.H., 2015. ATLAS program for advanced thermal-hydraulic safety research. Nuclear Engineering and Design 294, 242–261.

Song, J.Y., Lee, J., Chang, S.H., Jeong, Y.H., 2020. Correction factor development for the 2006 Groeneveld CHF look-up table for rectangular channels under low pressure. Nuclear Engineering and Design 370 (110869), 1–10.

Sonnenburg, H.G., 1989. Full-range drift flux model based on drift-flux theory with envelope theory. In: 4th International Topical Meeting on Nuclear Reactor Thermal Hydraulics, NURETH-4, Karlsruhe, Germany, October 10–13.

Sonnenburg, H.G., 1990. Analysis of UPTF-Test 26 Run 230 by ATHLET-Code with Full-Range Drift Flux Model. GRS-A-1 723, Garching, Germany.

Sonnenkalb, M., Pellegrini, M., Herranz, L.E., Lind, T., Morreale, A.C., Kanda, K., Tamaki, H., Kim, S.I., Cousin, F., Fernandez-Moguel, L., Andrews, N., Sevon, T., 2020. Overview and outcomes of the OECD/NEA benchmark study of the accident at the Fukushima Daiichi NPS (BSAF), phase 2—results of severe accident analyses for unit 2. Nuclear Engineering and Design 369 (110840), 1–10.

Sorensen, J.N., Apostolakis, G.E., Kress, T.S., Powers, D.A., 1999. On the role of defense in depth in risk-informed regulation. In: Probability and Safety Analysis Conference, Washington, DC, USA, Aug. 22–25.

Souza Dutre, A., Sbragio, R., D'Auria, F., 2002. Analysis of depressurization in an experimental containment installation using Relap5 code. In: International Nuclear Atlantic Conference (INAC 2002 and XIII ENFIR), Rio de Janeiro, Brasil, Aug. 11–16.

Sozzi, G.L., Sutherland, W.A., 1975. Critical flow of saturated and subcooled water at high pressure. In: Non-Equilibruim Two Phase Flows. ASME Winter Annual Meeting, Houston, TX, USA, 30 November–5 December, pp. 19–25.

Spalart, P.R., 2000. Strategies for turbulence modelling and simulations. International Journal of Heat and Fluid Flow 21, 252–263.

Spalart, P.R., 2009. Detached eddy simulation. Annual Review of Fluid Mechanics 41, 181–202.

Spalart, P.R., Allmaras, S.R., 1992. A one-equation turbulence model for aerodynamic flows. In: 30th Aerospace Sciences Meeting and Exhibit—American Institute of Aeronautics and Astronautics, Reno, NV, USA, January 6–9.

Spalding, D.B., 1963. Convective Mass Transfer: An Introduction. Edward Arnold Publishers, Ltd., London, UK.

Spalding, D.B., Taborek, J. (Eds.), 1983. Heat Exchanger Design Book. VDI-Verlag GmbH, Dusseldorf, Germany. Hemisphere Public Corp., Washington DC, USA.

Sparrow, E.M., Cess, R.D., 1966. Radiation Heat Transfer. Brooks/Cole Publishing Corp., Belmout, CA, USA.

Sparrow, E.M., Cess, R.D., 1978. Radiative Heat Transfer. Hemisphere Public Corp., New York, NY, USA.

Speziale, C.G., Sarkar, S., Gatski, T.B., 1991. Modelling the pressure-strain correlation of turbulence: an invariant dynamical systems approach. Journal of Fluid Mechanics 277, 245–272.

Spindler, K., 1994. Flow boiling. In: Proceedings 10th International Heat Transfer Conf., vol. 1, pp. 349–368.

Spiridonov, Y.G., Tzykanov, V.A., Samsonov, B.V., Fomin, N.A., 1974. Experimental Determination of UO2 Thermal Conductivity Coefficient and of Fuel-Cladding Contact Conduction in Fuel Elements under Reactor Irradiation/Preprint NIIAR P-250, Dimitrovgrad, Russia (In Russian).

Spore, J.W., Guffee, L.A., Woodruff, S.B., Steinke, R.G., 1993. TRAC-PF1/MOD2: Volume I. Theory Manual, NUREG/CR-5673. USNRC, Washington, DC, USA.

Squarer, D., Schulenberg, T., Struwe, D., Oka, Y., Bittermann, D., Aksan, S.N., Maraczy, C., Kyrki-Rajamäki, R., Souyri, A., Dumaz, P., 2003. High performance light water reactor. Nuclear Engineering and Design 222, 167–180.

Srikantiah, G., Agee, L., 1993. The 2nd International Seminar on Subchannel Analysis (ISSCA-II), Palo Alto (Eds.) EPRI, Seminar Proceedings, EPRI TR-103188, Palo Alto, Ca, USA.

Staedtke, H., 2006. Gas Dynamic Aspects of Two-phase Flow. Wiley-VCH Verlag GmbH & Co. KGaA, Weinheim, Germany.

Stanev, I., D'Auria, F., 2001. Analysis of LOCA D=200 mm for Kozloduy—units 3&4 with Relap5/mod3.2 and Cathare codes. Evaluation of the results uncertainty. In: IAEA TCM on Safety Margins of Operating Reactors and Implications for Decisions Makers Including Consideration of Uncertainties of Analysis, Vienna, 15–19 October, ISBN 92-0-118102-7, pp. 107–125 (also IAEA TECDOC No 1332, January 2003.

Steinbrück, M., Birchley, J., Boldyrev, A.V., Goryachev, A.V., Grosse, M., Haste, T.J., Hozer, Z., Kisselev, A.E., Nalivaev, V.I., Semishkin, V.P., Sepold, L., Stuckert, J., Ver, N., Veshchunov, V.S., 2010. High-temperature oxidation and quench behaviour of Zircaloy-4 and E110 cladding alloys. Progress in Nuclear Energy 52, 19–36.

Stephenson, R., 1958. Introduction to Nuclear Engineering. McGraw-Hill Book Company, Inc., New York, NY, USA.

Stevanovic, V., Studovic, M., 1995. A simple model for vertical annular and horizontal stratified two-phase flows with liquid entrainment and phase transitions: one-dimensional steady state conditions. Nuclear Engineering and Design 154, 357–379.

Stewart, C.W., 1985. VIPRE-01: A Thermal-Hydraulic Code for Reactor Cores, EPRI NP-2511-CCM, Palo Alto, CA, USA.

Stewart, J.C., Groeneveld, D.C., 1981. Low quality subcooled film boiling of water at elevated pressures. Journal of Nuclear Engineering and Design 67, 259–272.

Stewering, J., Schramm, B., Sonnenkalb, M., 2015. Validation of CFD-codes for natural convection and condensation phenomena in containments with German Thai-experiments. In: 16th International Topical Meeting on Nuclear Reactor Thermal Hydraulics (NURETH-16), August 30-September 4, Chicago, IL, USA.

STNI, 2008. Journal Science and Technology of Nuclear Installations, Special Issue, Volume "Scaling, Uncertainty and 3D Coupled Code Calculations in Nuclear Technology". Hindawi Publisher.

Strogatz, S.H., Non Linear Dynamics and Chaos. Perseus Books, Westview Press, Cambridge, Ma, USA.

Strömgren, T., Zhezherun, A., Le Moigne, Y., Le Corre, J.-M., Sheng, D.Y., Perzon, S., 2015. A CFD approach to spacer grid optimization for improved dryout performance in BWR fuel bundle. In: 16th International Topical Meeting on Nuclear Reactor Thermal Hydraulics (NURETH-16), August 30–September 4, Chicago, IL, USA.

Strouse, G.F., 2008. Standard Platinum Resistance Thermometer Calibrations from the Ar TP to the Ag FP. Gaithersburg, MD. Nat. Inst. Standards Technol. NIST Special Publication, USA, pp. 250–281.

Štrubelj, L., Tiselj, I., 2007. Numerical modelling of condensation of saturated steam on subcooled water surface in horizontally stratified flow. In: 12th International Topical Meeting on Nuclear Reactor Thermal Hydraulics (NURETH-12), September 30–October 4, Pittsburgh, Pa, USA.

Strubelj, L., Ezsol, G., Tiselj, I., 2010. Direct contanct condensation induced transition from stratified to slug flow. Nuclear Engineering and Design 240, 266–274.

Stuckert, J., Veshchunov, M.S., 2008. Behaviour of Oxide Layer of Zirconium-Based Fuel Rod Cladding under Steam Starvation Conditions. Forschungszentrum Karlsruhe, Karlsruhe, Germany. Report FZKA-7373.

Stuckert, J., Steinbrück, M., Stegmaier, U., 2001. Single-Rod Quench Tests with Zr–1Nb Cladding. Comparison with Zircaloy-4 Cladding Tests and Modelling. Forschungszentrum Karlsruhe, Karlsruhe, Germany. Report FZKA-6604.

Stuckert, J., Goryachev, A., Große, M., Heck, M., Ivanova, I., Schanz, G., Sepold, L., Stegmaier, U., Steinbrück, M., 2008. Results of the QUENCH-12 Experiment on Reflood of a VVER-type Bundle. Forschungszentrum Karlsruhe, Karlsruhe, Germany. Report FZKA-7307.

Stuckert, J., Birchley, J., Große, M., Jaeckel, B., Steinbrück, M., 2011. Experimental and calculation results of the integral reflood test QUENCH-15 with ZIRLO-TM cladding tubes in comparison with results of previous QUENCH tests. Nuclear Engineering and Design 241, 3224–3233.

Studer, E., Magnaud, J.P., Dabbene, F., Tkatschenko, I., 2007. International Standard Problem on containment thermal–hydraulics ISP 47 step 1—results from the MISTRA exercise. Journal of Nuclear Engineering and Design 237.

Studer, E., Abdo, D., Benteboula, S., Bernard-Michel, G., Cariteau, B., Coulon, N., Dabbene, F., Debesse, P., Koudriakov, S., Ledier, C., Magnaud, J.-P., Norvez, O., Widloecher, J.-L., Beccantini, A., Gounand, S., Brinster, J., 2020. Challenges in containment thermal hydraulics. Nuclear Technology 206 (9), 1361–1373.

Stuhmiller, J.H., 1977. The influence of interfacial pressure forces on the character of two-phase flow model equations. International Journal of Multiphase Flow 3, 551–560.

Stuhmiller, J.H., 1986. A Dynamic Flow Regime Model of Two-phase Flow. JAYCOR Final Report. EPRI RP-888-1, Palo Alto Ca, USA.

Su, G.-Y., Phillips, B., Buongiorno, J., D'Aleo, F.P., Prasser, H.-M., Al-Safran, E., 2018. Suppression of nucleate boiling in upward two-phase annular flow: a close-up study using advanced diagnostics. In: 12th International Topical Meeting on Nuclear Reactor Thermal-Hydraulics, Operation and Safety (NUTHOS-12), October 14–18, Qingdao, China.

Suárez-Ántola, R., 2019. An analytical approach to bifurcations and stability in simplified mathematical models of nuclear reactors. Progress in Nuclear Energy 114, 171–190.

Subbotin, V.I., Ibragimov, M.K., Ushakov, P.A., Bobkov, V.P., Zhukov, A.V., Yur'ev, Y.S., 1975. Hydrodynamics and Heat Transfer in Nuclear Power Facilities. Atomizdat, Moscow, Russia (in Russian).

Sudo, Y., 1980. Film boiling heat transfer during reflood phase in postulated PWR loss-of-coolant accident. Journal of Nuclear Science and Technology 17, 516–530.

Sudo, Y., 2012. Film boiling heat transfer during reflood phase in postulated PWR loss of coolant accident. Journal of Science and Technology 17, 516–530.

Sugawara, S., Sakai, T., Watanabe, K., Rummens, H.E.C., 1991. Subchannel analysis by the FIDAS code based on three-fluid model. Nuclear Engineering and Design 132, 253–264.

Suh, K.Y., Todreas, N.E., Rohsenow, W.M., 1989. Mixed convective low flow pressure drop in vertical rod assemblies: II—experimental validation. Journal of Heat Transfer 111 (4), 966–973.

Sukhatme, S.P., 1990. Condensation on enhanced surface horizontal tubes. In: Proceedings of the 9th International Heat Transfer Conference, August 19–24, Jerusalem, Israel, vol. 1, pp. 305–327.

Sun, H., 1979. Flooding correlations for BWR bundle upper tie plate and side-entry orifices. In: 2nd Multi-phase Flow and Heat Transfer Symposium Workshop, Miami Beach, FL, USA.

Sun, X., Ishii, M., Kelly, J.M., 2003. Modified two-fluid model for the two-group interfacial area transport equation. Annals of Nuclear Energy 30, 1601–1622.

Sun, J., Ran, X., Zhang, Z., Cao, X., Fan, G., Ding, M., 2020. Effects of direct contact condensation on flow characteristics of natural circulation system at low pressure. Frontiers in Energy Research 8 (173), 1–15.

Sun, J., Wang, H., Li, J., 2022. Analysis on containment thermal hydraulic behaviour under passive containment heat removal system operation condition for HPR1000. International Journal of of Advanced Nuclear Reactor Design and Technology 4, 36–42.

Sung, Y., Cao, L., Kucukboyaci, V., Gordon, N., Salko, R.K., Tatli, E., 2016. CASL COBRA-TF improvements for PWR DNB analysis. In: Proceedings Of International Conference Top Fuel, September 11–15, Boise, ID, USA.

Sursock, J.-P., Kiang, R.L., 1985. The EPRI facility at SRI: description and connection with IST program. In: Specialist Mtg. On Small Break LOCA Analysis in LWRs, Pisa, Italy, 23–27 June.

Suzuki, M., Nakamura, H., 2009. Reliability of core exit thermocouple (CET) for accident management action during SBLOCA and abnormal transient tests at ROSA/LSTF. In: 13th International Topical Meeting on Nuclear Reactor Thermal Hydraulics (NURETH-13), September 27-October 2, Kanazawa City, Ishikawa, Japan.

Suzuki, M., Findlay, J.A., Tasaka, K., Sutherland, W.A., Nakamura, H., 1986. Similarity study of large steam line break LOCA's in ROSA-III, FIST and BWR/6. Nuclear Engineering and Design 98, 39–55.

Svanholm, K., Breghi, M.P., D'Auria, F., Ianiri, R., 1995. Halden reactor IFA-511.2 and IFA-54X: experimental series under adverse core cooling conditions. J. Experimental Thermal and Fluid Science 11 (1), 80–100.

Szabó, T., Kretzschmar, F., Schulenberg, T., 2014. Obtaining a more realistic hydrogen distribution in the containment by coupling MELCOR with GASFLOW. Nuclear Engineering and Design 269, 330–339.

Taitel, Y., Dukler, A.E., 1976. A model of predicting flow regime transitions in horizontal and near horizontal gas-liquid flow. AIChE Journal 22 (1), 47–66.

Taitel, Y., Bornea, D., Dukler, A.E., 1980. Modeling flow pattern transitions for steady upward gas-liquid flow in vertical tubes. AIChE Journal 26, 345–354.

Takahashi, M., Nayak, A.K., Kitagawa, S., Murakoso, H., 2001. Heat transfer in direct contact condensation of steam to subcooled water spray. Journal of Heat Transfer 123 (4), 703–710.

Takeda, T., Ohwada, A., Nakamura, H., 2013. Measurement of non-condensable gas in a PWR small-break LOCA simulation test with LSTF for OECD/NEA ROSA Project and RELAP5 post-test analysis. Experimental Thermal and Fluid Science 51.

Tambling, T.N., Harpster, T.L., 1977. Inspection Report No. 50-346/77-32: Davis-Besse Nuclear Power Station, Unit 1, Docket 50-346, OH, USA (November).

Tanaka, M., Dulikravich, G.S. (Eds.), 2000. Inverse Problems in Engineering Mechanics. Elsevier, Kidlington, Oxford, UK.

Tanasawa, I., 1991. Advances in Condensation Heat Transfer. Adv. Heat Transfer. Elsevier, pp. 55–139.

Tanasawa, I., 1994. Recent advances in condensation heat transfer. In: Proceedings of the 10th International Heat Transfer Conference, August 14–18, Brighton, UK, vol. 1, pp. 297–312.

Tanskanen, V., 2012. CFD Modelling of Direct Contact Condensation in Suppression Pools by Applying Condensation Models of Separated Flow. Acta Universitatis Lappeenrantaensis, vol. 472. Lappeenranta University of Technology, ISBN 978-952-265-222-5 (PDF), ISSN 1456-4491.

Tarasova, N.V., Orlov, V.M., 1964. Heat transfer and hydraulic resistance in surface boiling of water in the annular channels. In: Proceedings Of the Konvektivnaya Heat Transfer in Two-phase and Single-phase Flows. Energy, Moscow, Russia, pp. 162–186 (in Russian).

Tasaka, K., Tamaki, M., Imai, S., Kohketsu, H., Anoda, Y., Murata, H., Kukita, Y., 1992. Atmospheric-pressure small-scale thermal-hydraulic experiment of a PIUS-Type reactor. Nuclear Science and Technology 29 (12).

TEC, 1977. Toledo Edison Corporation—Licensee Event Report, NP-32-77-16, Docket 50-346, OH, USA. October).

TEPCO, 2012. Fukushima Nuclear Accident Analysis Report, Tokyo, Japan, June 20.

Teruel, F.E., Uddin, R., 2009. A new turbulence model for porous media flows. Part II: analysis and validation using microscopic simulations. International Journal of Heat Mass Transfer 52, 5193–5203.

Terzi, R., Tükenmez, I., Kurt, E., 2016. Energy and exergy analyses of a VVER type nuclear power plant. International Journal of Hydrogen Energy 41, 12465–12476.

Teschendorff, V., Austregesilo, H., Lerchl, G., 1996. Methodology, status and plans for development and assessment of ATHLET code. In: USNRC and OECD/NEA/CSNI Workshop on Transient Thermal-Hydraulic and Neutronic Codes Requirements, Annapolis, MD, USA, 5–8 November.

Teschendorff, V., Glaeser, H., Kliem, S., 2009. Bedeutung von Experimenten für die Reaktorsicherheit; Jahrestagung Kerntechnik, Dresden, Germany, 2–14 May.

Teyssedou, A., Dipama, J., Hounkonnou, W., Aubé, F., 2010. Modeling and optimization of a nuclear power plant secondary loop. Nuclear Engineering and Design 240, 1403–1416.

Theofanous, T.G., 1979. Modeling of basic condensation processes. In: The Water Reactor Safety Research Workshop on Condensation, Silver Springs, MD, USA, May 24–25.

Theofanous, T.G., Shabana, E.A., 1991. Boron mixing in the lower plenum of a BWR. Nuclear Engineering and Design 126 (2).

Theofanous, T.G., Nourbakhsh, H.P., Gherson, P., Iyer, K., 1984. Decay of Buoyancy-Driven Stratified Layers with Applications to Pressurized Thermal Shock. NUREG/CR-3700, USNRC, Washington, DC, USA.

Thielen, H., Brücher, W., Martens, R., Sogalla, M., 2007. Advanced Atmospheric Dispersion Modelling and Probabilistic Consequence Analysis for Radiation Protection Purposes in Germany. Springer, Berlin, Germany.

Thom, J.R.S., Walker, W.M., Fallon, T.A., Reising, G.R.S., 1966. Boiling in sub-cooled water during flow up hasted tubes or annuli. Proceedings Inst. Mech. Eng. 180 (Part 3C), 226–246.

Thome, J.R., 1990. Enhanced Boiling Heat Transfer. Hemisphere, New York, NY, USA.

Thulin, E., 2000. Report from the Oskarshamn-2 Stability Event of 99-02-25. ABB Atom AB, SPD-99-237-rev-0, Vasteras, Sweden (in Swedish).

Thurgood, M.J., 1983. COBRA/TRAC A thermal Hydraulic Code for Transient Analysis of Nuclear Reactor Vessels and Primary Coolant Systems. NUREG/CR-3046, Washington DC, USA.

Thurgood, M.J., Crowell, K., Kelly, J.M., 1981. COBRA-TF equations and constitutive models. FATE-81-106. Pacific Northwest Laboratory, Richland, WA, USA.

Thurgood, M.J., Kelly, J.M., Guidotti, T.E., Kohrt, R.J., Crowell, K.R., 1983. COBRA/TRAC, A Thermal-Hydraulics Code for Transient Analysis of Nuclear Reactor Vessels and Primary Coolant Systems Equations and Constitutive Models. Pacific Northwest Laboratory, Washington, DC, USA, pp. 1–196. PNL-4385, USNRC, NUREG/CR-3046.

Tiep, N.H., Kim, K.D., Heo, J., 2021. Improvement in the accuracy of SPACE prediction for the unblocked FLECHT SEASET reflood tests by data assimilation. Annals of Nuclear Energy 161, 2021.

Tiep, N.H., Kim, K.D., Heo, J., Choi, C.W., Jeong, H.Y., 2022. A Newly Proposed Data Assimilation Framework to Enhance Predictions for Reflood Tests. Nuclear Engineering and Design, accepted for publication).

Tietsch, W., 1999. RELAP5/MOD3.2 Assessment Using GERDA Small Break Test, 1605AA. USNRC NUREG/IA-0166, Washington, DC, USA.

Tikadar, A., Najeeb, U., Paul, T.C., Oudah, S.K., Salman, A.S., Abir, A.M., Carrilho, L.A., Khan, J.A., 2018. Numerical investigation of heat transfer and pressure drop in nuclear fuel rod with three-dimensional surface roughness. International Journal of of Heat and Mass Transfer 126, 493–507.

Tiselj, I., Kljenak, I., Prosek, A., 1997. Simulation of containment response during a large break loss of coolant accident scenario with the CONTAIN computer code. In: Regional Meeting: Nuclear Energy in Central Europe, Bled (Slovenia), September 7–10.

Todorova, N.K., Ivanov, K.N., 2003. Investigation of spatial coupling aspects for coupled code application in PWR safety analysis. Annals of Nuclear Energy 30.

Todreas, N.E., Kazimi, M., 1990. Nuclear Systems (Vols. I and II). Hemisphere Pub. Corp., Washington, DC, USA.

Todreas, N.E., Kazimi, M., 2012. Nuclear Systems (Vols. I and II). Hemisphere Pub. Corp., Washington, DC, USA (1st time published in 1990, see above).

Todreas, N., Chen, S.-K., Pacio, J., 2020. Adventures in study of the pressure losses in wire wrapped rod bundle arrays. Nuclear Engineering and Design 370, 110910.

Todreas, N.E., Kazimi, M.S., Massoud, M., 2021/2022. Nuclear Systems Volume II—Elements of Thermal Hydraulic Design. CRC Press, Boca Raton (Fl), US, pp. 1−657 (eBook) 9780429157608 (Also).

Todreas, N.E., Kazimi, M.S., Massoud, M., 2022. Nuclear Systems—Elements of thermal Hydraulic Design, second ed. CRC Press (an inprint of Taylor & Francis Group LLC), Boca Raton, Fl, USA, ISBN 978-1-4822-3958-4, pp. 1−635 (hbk).

Tomaz, G.C.Q., Chen, Y.M., Vaghetto, R., Hassan, Y., 2022. Recalibration of the UCTG correlation for friction factor in wire-wrapped rod bundles using multi-objective genetic algorithms. In: 19th International Topical Meeting on Nuclear Reactor Thermal Hydraulics (NURETH-19), March 6−11 (Via Web), Brussels, Belgium.

Tompkins, C., Prasser, H.M., Corradini, M., 2018. Wire-mesh sensors: a review of methods and uncertainty in multiphase flows relative to other measurement techniques. Nuclear Engineering and Design 337, 205−220.

Tong, L.S., 1965. Boiling Heat Transfer and Two-phase Flow. John Wiley and Sons, New York, NY, USA.

Tong, L.S., 1972. Boiling Crisis and Critical Heat Flux. USAEC Rep. TID-25887, Washington, DC, USA.

Tong, L.S., 1978. Heat transfer in reactor safety. Journal of IHTC 285−309.

Tong, L.S., Tang, Y.S., 1997. Boiling Heat Transfer and Two-phase Flow, second ed. CRC Press, Cleveland, OH, USA.

Tong, L.S., Weisman, J., 1970. Thermal Analysis of Pressurized Water Reactors. ANS Monograph, Hinsdale, IL, USA (2nd Edition, 1979; 3rd Edition, 1996).

Tong, L.S., Weisman, J., 1979. See Tong and Weisman, 1970.

Tong, L.S., Weisman, J., 1996. See Tong and Weisman, 1970.

Tong, L.L., Chen, J.B., Cao, X.W., Deng, J., 2014. Thermal-hydraulic behavior under station blackout for CANDU-6. Progress in Nuclear Energy 74.

Torchigin, V.P., 2021. Ball lightning and the Chernobyl accident. Optik—International Journal of for Light and Electron Optics 241 (167224), 1−7.

Tortora, G., 2009. Application of Fractional Scaling Analysis for IRIS and SPES3 Containment Vessels Pressure Responses during DVI-SBLOCA Event [Dissertation]. Politecnico di Milano, Milano, Italy.

Toth, I., Prior, R., Sandervag, O., Umminger, K., Nakamura, H., Muellner, N., Cherubini, M., Del Nevo, A., D'Auria, F., Dreier, J., Alonso, J.R., Amri, A., 2010. Core Exit Temperature Effectiveness in Accident Management of Nuclear Power Reactor. NEA/CSNI/R(2010)9, Paris, France.

Toumi, I., Bergeron, A., Gallo, D., Royer, E., Caruge, D., 2000. FLICA-4: a 3-D two-phase flow computer code with advanced numerical methods for nuclear applications. Nuclear Engineering and Design 200 (1−2), 139−155.

Toutant, A., Chandesris, M., Jamet, D., Lebaigue, O., 2009a. Jump conditions for filtered quantities at an under resolved interface. Part 1: theoretical development. International Journal of Multiphase Flow 35 (12), 1100−1118.

Toutant, A., Chandesris, M., Jamet, D., Lebaigue, O., 2009b. Jump conditions for filtered quantities at an under resolved interface. Part 2: a priori tests. International Journal of Multiphase Flow 35 (12), 1119−1129.

TRACE, 2007a. V5.0, Theory Manual. USNRC, Washington, DC, USA.

TRACE, 2007b. V5.0 User Manual. USNRC, Washington, DC, USA.

TRACE, 2007c. V5.0 Assessment Manual. USNRC, Washington, DC, USA.

TRACE V5.0, 2007. Theory Manual—Field Equations, Solution Methods and Physical Models. USNRC, Washington, DC.

Tran, T.N., Chyu, M.C., Wambsganss, M.W., France, D.M., 2000. Two-phase pressure drop of refrigerants during flow boiling in small channels: an experimental investigation and correlation development. International Journal of Multiphase Flow 26, 1739–1754.

Trapp, J.A., Ransom, V.H., 1982. A choked-flow calculation criterion for non-homogeneous, non-equilibrium two-phase flows. International Journal of Multiphase Flow 8 (6), 669–681.

Trapp, J.A., Riemke, R.A., 1986. A nearly-implicit hydrodynamics numerical scheme for two-phase flow. Journal of Computational Physics 66, 66–82.

Travin, A., Shur, M., Strelets, M., Spalart, P., 1999. In: Detached-eddy Simulations Past a Circular cylinder. Flow Turbul. Combust., vol. 63. Kluwer Academic Publishers, The Netherlands, pp. 293–313.

Trefethen, L.N., 1982. Group velocity in finite difference schemes. Society for Industrial and Applied Mathematics (SIAM) Review 24, 113–135.

Trunov, N.B., Logvinov, S.A., Dragunov, Y.G., 2001. Hydrodynamic and Thermochemical Processes in the Steam Generators of Nuclear Power Plants with VVER. Energoatomizdat, Moscow, Russia (in Russian).

Tsochatzidis, N.A., Karapantios, T.D., Kostoglou, M.V., Karabelas, A.J., 1992. A conductance method for measuring liquid fraction in pipes and packed beds. International Journal of Multiphase Flow 5, 653–667.

Tuomisto, H., Kajanto, P., 1988. Two phase Flow in full scale loops seal facility. Journal of Nuclear Engineering and Design 107, 295–305.

Tuunanen, J., Vihavainen, J., D'Auria, F., Kimber, G., 1999. Assessment of Passive Safety Injection Systems of ALWRs. Final Report of the European Commission 4th Framework Programme Project FI4I-CT95-0004 (APSI), pp. 1–79. VTT report 1957, Espoo, Finland.

Tuunanen, J., Vihavainen, J., D'Auria, F., Frogheri, M., Galassi, G.M., Kimber, G., Lillington, J., Allen, E.J., Williams, T.G., 2000. Analyses of PACTEL passive safety injection experiments with APROS, CATHARE and RELAP5 codes. J. Nuclear Engineering & Design 198, 261–286.

Tuzson, J., 2000. Centrifugal Pump Design. John Wiley & Sons Inc., New York, NY, USA.

Tyagi, P., Buwa, V.V., 2017. Experimental characterization of dense gas–liquid flow in a bubble column using voidage probes. Chemical Engineering J. 308, 912–928.

Uchida, H., Oyama, A., Togo, Y., 1965. Evaluation of post-incident cooling systems of light-water power reactors. In: Proceedings Of the 3rd International Conference on the Peaceful Uses of Atomic Energy, Geneva, August 31–September 9, 1964, vol. 13. United Nations, New York, NY, USA, pp. 93–104.

Uhle, J., Wu, Q., Ishii, M., 1998. Dynamic flow regime modeling. In: Proceedings of the 6th International Conference On Nuclear Engineering, Paper No. ICONE-6509, San Diego, CA, USA, May 10–14.

Umehara, Y., Okawa, T., 2022. Phenomenological interpretation of heat transfer coefficient distribution near the rewetting region. In: 19th International Topical Meeting on Nuclear Reactor Thermal Hydraulics (NURETH-19), March 6–11, Brussels, Belgium, Paper 34055.

Umminger, K., Del Nevo, A., 2012. Integral test facilities and thermal-hydraulic system codes in nuclear safety analysis. Science and Technology of Nuclear Installations (Preface to Special Issue) 1–3. Article ID 826732.

Umminger, K., Dennhardt, L., Schollenberger, S., Schoen, B., 2012. Integral test facility PKL: experimental PWR accident investigation. Science and Technology for Nuclear Installations. Article ID 891056.

Unal, H.C., 1976. Maximum bubble diameter maximum bubble-growth time and bubble growth rate during the subcooled nucleate flow boiling of water up to 17.7 MN/m^2. International Journal of Heat Mass Transfer 19, 643−649.

Unal, C., Williams, B., Hemez, F., Atamturktur, S.H., McClure, P., 2011. Improved best estimate plus uncertainty methodology, including advanced validation concepts, to license evolving nuclear reactors. Nuclear Engineering and Design 241 (5), 1813−1833.

UNIPI-GRNSPG, 2008. A Proposal for Performing the Atucha II Accident Analyses for Licensing Purposes. The BEPU Report—Rev. 3, Pisa, Italy [endorsed by Regulatory Authority in Buenos Aires, Argentina].

USAEC, 1971. Interim Acceptance Criteria (IAC) for ECCS. USAEC, Washington, DC, USA.

USNRC, 1974. Title 10, Code of Federal Regulations, Part 50, Appendix K. Licensing of Production and Utilization Facilities, Washington, DC, USA.

USNRC, 1975. Reactor Safety Study—An Assessment of Accident Risk in U.S. Commercial Nuclear Power Plants. WASH-1400, NUREG-75/014, Washington, DC, USA.

USNRC, 1978. Regulatory Guide 1.70, Standard Format and Content of Safety Analysis Reports for Nuclear Power Plants, LWR Edition. U.S. Nuclear Regulatory Commission, Office of Standards Development, Washington, DC, USA.

USNRC, 1983. Criteria for Accident Monitoring Instrumentation for Nuclear Power Plant. NRC RG 1.97, Washington, DC, USA [rev. 5 issued in 2019].

USNRC, 1987. Compendium of ECCS Research for Realistic LOCA Analysis. US NRC Report NUREG-1230, Washington, USA (see USNRC, 1988).

USNRC, 1988. Compendium of ECCS Research for Realistic LOCA Analysis. USNRC NUREG-1230 R4, USNRC, Washington, DC, USA (see also USNRC, 1987).

USNRC, 1988a. Inspection on March 16 through 24, 1988, to La Salle County Station at La Salle Site Marseilles, IL, USA: Augmented Inspection Team Report No 50-373/880008; 50-374/880008.

USNRC, 1988b. Information Notice No. 88-39: LaSalle Unit 2 Loss of Recirculation Pumps with Power Oscillation Event, Washington, DC, USA, June 15.

USNRC, Baranowsky, P.W., 1988c. Evaluation of Station Blackout Accidents at Nuclear Power Plants Technical Findings Related to Unresolved Safety Issue A-44 Final Report. NUREG, 1032, Washington, DC, USA, pp. 1−175.

USNRC, 1989a. Quantifying Reactor Safety Margins: Application of CSAU to a LBLOCA. NUREG/CR-5249, Washington, DC, USA.

USNRC, 1989b. Best-estimate Calculations of Emergency Core Cooling System Performance. USNRC Regulatory Guide 1.157, Washington, DC, USA.

USNRC, 1990. Severe Accident Risk: An Assessment for Five US Nuclear Power Plants: NUREG-1150, Washington, DC, USA.

USNRC, 1995. Accident Source Terms for Light Water Nuclear Power Plants. NUREG-1465, Washington, DC, USA.

USNRC, 1995a. Domestic Licensing of Production and Utilization Facilities, Code of Federal Regulations 10, Part 50. US Government Printing Office, Washington, DC, USA.

USNRC, 1997. Regulatory Guide 1.206—Combined License Applications for Nuclear Power Plants. Washington, DC, USA).

USNRC, 1998a. RELAP5/MOD3 Code Manual Volume I: Code Structure, System Models, and Solution Methods, NUREG/CR-5535, Washington, DC, USA.

USNRC, 1998b. RELAP5/MOD3 Code Manual Volume IV: Models and Correlations, NUREG/CR-5535, Washington, DC, USA.

USNRC, 1998c. Risk Assessment of Severe Accident-Induced Steam Generator Tube Rupture, NUREG-1570, Washington, DC, USA.

USNRC, 2000. Draft Section 15.0.2, Review of Analytical Computer Codes. NUREG-0800 [Standard Review Plan for the Review of Safety Analysis Reports for Nuclear Power Plants], Washington, DC, USA.

USNRC, 2005. Transient and Accident Analysis Methods. RG 1.203. USNRC, Washington, DC, USA.

USNRC, 2005a. APEX-AP1000 Confirmatory Testing to Support AP-1000 Design Certification (Non-proprietary). Office of Nuclear Regulatory Research, U.S. Nuclear Regulatory Commission, Washington, DC, USA.

USNRC, 2005b. Transient and Accident Analysis Methods. U.S. Nuclear Regulatory Commission, Office of Nuclear Regulatory Research, Regulatory Guide 1.203,Washington, DC, USA.

USNRC, 2007. Combined License Application for Nuclear Power Plants, LWR Edition. USNRC, Washington, DC, USA. RG 1.206.

USNRC, 2013. TRACE V5.840 Theory Manual: Field Equations. Solution Methods and Physical Models, Washington, DC, USA.

USNRC, 2013a. Standard Review Plan for the Review of Safety Analysis Reports for Nuclear Power Plants. NUREG 0800, Washington, DC, USA (continuously updated).

USNRC, 2014. US Government, Federal Register Notice, 79 FR 16106, March 24, Washington, DC, USA.

USNRC, 2015. Code of Federal Regulation (CFR), Washington, USA [continuously Updated—See the Website of USNRC].

USNRC, 2018. Measuring Breakaway Oxidation Behavior. Regulatory Guide (RG) 1.222, Washington, DC, USA.

USNRC, 2018a. Determining post Quench Ductility, Regulatory Guide (RG) 1.223, Washington, DC, USA.

USNRC, 2018b. Preliminary Draft (Draft Was Issued as DG-1263, Dated March 2014), Establishing Analytical Limits for Zirconium-Alloy Cladding Material, pp. 1−32. Regulatory Guide (RG) 1.224, Washington, DC, USA.

USNRC, 2022. Appendix A to Part 50—General Design Criteria for Nuclear Power Plants. USNRC website, Washington, DC, USA. https://www.nrc.gov/reading-rm/doc-collections/cfr/part050/part050.

Uspuras, E., Kaliatka, A., 2006. Accident and Transient Processes at NPPs with Channel Type Reactors. Monograph. Lithuanian Energy Institute, Kaunas, Lithuania, ISBN 9986-492-87-4, pp. 1−299.

Utanohara, Y., Kinoshita, I., Murase, M., Minami, N., Nariai, T., Tomiyama, A., 2011. Numerical simulation using CFD software of countercurrent gas−liquid flow in a PWR hot leg under reflux condensation. Nuclear Engineering and Design 241, 1643−1655.

Vaghetto, R., Hassan, Y., 2013. Study of debris-generated core blockage scenarios during loss of coolant accidents using RELAP5-3D. Nuclear Engineering and Design 261, 144−155.

Vaghetto, R., Hassan, Y., 2014. Modeling the thermal-hydraulics behavior of reactor cavity cooling system using RELAP5-3D. Annals of Nuclear Energy 73, 75−83.

Vaghetto, R., Yang, S., Hodge, D., Hassan, Y., 2021. Two-phase flow measurements and observations in a cooling panel of the reactor cavity cooling system. Progress in Nuclear Energy 131, 103578.

Valencia, J.E.L., Glantz, T., Oliveira, A.V.S., Labergue, A., Gradeck, M., 2022. Simulation of dispersed flow film boiling in LOCA conditions considering different fuel rod blockage ratios. In: 19th International Topical Meeting on Nuclear Reactor Thermal Hydraulics (NURETH-19), March 6−11 (Via Web), Brussels, Belgium.

Valette, M., 2007. Analysis of boiling two phase flow in rod bundle for NUPEC BFBT benchmark with 3-fields NEPTUNE system code. In: Proceedings of the 12th International Topical Meeting on Nuclear Reactor Thermal Hydraulics (NURETH-12) Pittsburgh, PA, USA, September 30—October 4.

Valette, M., 2011. PSBT simulations with Cathare 3. In: Proceedings of the 14th International Topical Meeting on Nuclear Reactor Thermal Hydraulics, NURETH-14, Toronto, Ontario, Canada, September 25—30.

Valette, M., Pouvreau, J., Bestion, D., Emonot, P., 2009. Revisiting large break LOCA with the CATHARE 3 three field model. In: NURETH 13, Kanazawa, Japan, 27 September—2 October (see also Valette et al., 2011).

Valette, M., Pouvreau, J., Bestion, D., Emonot, P., 2011. Revisiting large break LOCA with the CATHARE 3 three field model. Nuclear Engineering and Design 241 (11), 4487—4496.

Van De Graaf, R., Van Der Hagen, T.H.J.J., Muddle, R.F., 1994. Two-phase flow scaling laws for a simulated BWR assembly. Journal of Nuclear Engineering and Design 148, 455—462.

Van Der Hagen, T.H.J.J., Van Bragt, D.D.B., Van Der Kaa, F.J., Karuza, J., Killian, D., Nissen, W.H.M., Stekelenburg, A.J.C., Touters, J.A.A., 1997. Exploring the Dodewaard Type-I and Type-II stability: from start-up to shut-down, from stable to unstable. Annals of Nuclear Energy 24 (8), 659—669.

Van Der Hagen, T.H.J.J., Stekelenburg, A.J.C., Van Bragt, D.D.B., 2000. Reactor experiments on type-I and type-II BWR stability. Nuclear Engineering and Design 200, 177—185.

Van Leer, B., 1977. Towards the ultimate conservative difference scheme. A new approach to numerical convection. Journal of Computational Physics 23 (3), 189—196.

Van Rooyen, E., Thome, J.R., 2013. Pressure drop data and prediction method for enhanced external boiling tube bundles with R-134a and R-236fa. International Journal of Refrigeration 36, 1669—1680.

Vasilchenko, I., Kushmanov, S., Vjalitsyn, V., Vasilchenko, R., 2014. Core Designs of Modern VVER Projects. Semi-centennial Anniversary of VVER Safe Operation. Novovoronezh NPP (NvNPP), Russia.

Vazquez, M., Tsige-Tamiratb, H., Ammirabile, L., Martin-Fuertes, F., 2012. Coupled neutronics thermal-hydraulics analysis using Monte Carlo and sub-channel codes. Nuclear Engineering and Design 250, 403—411.

VDI Heat Atlas, 2010, second ed. Springer-Verlag, Berlin, Heidelberg, Germany.

Veloso, M.A., de Mattos, J.R.L., 2015. Parametric representation of centrifugal pump homologous curves. In: International Nuclear Atlantic Conference—INAC 2015, São Paulo, SP, Brazil, October 4—9.

Venn, J., 1880. On the diagrammatic and mechanical representation of propositions and reasonings. The London, Edinburgh and Dublin Philosophical Magazine and Journal of Science 9, 1—18.

Venn, J., 1881. Symbolic Logic. MacMillan, London, UK.

Verma, D., Kalra, M.S., Wahi, P., 2017. Effect of nonlinear void reactivity on bifurcation characteristics of a lumped-parameter model of a BWR: a study relevant to RBMK. Nuclear Engineering and Design 315, 179—193.

Veteau, J.M., Clement, P., Deruaz, R., 1983. Reflooding of a PWR Bundle. PERICLES Programme and First Results. European Two Phase Flow Group Meeting, Zurich, Switzerland. June 1983.

Veteau, J.M., Clément, P., Deruaz, R., 1984. 2D Effects during Reflooding of a PWR: Forced Reflood Tests of PERICLES Programme. European Two-Phase Flow Group Meeting, Rome, Italy, 19—21 June.

Veteau, J.M., Clement, P., Deruaz, R., 1985. PERICLES Programme: Main Conclusions on 2D Effects during Re-Flooding of a PWR. European Two Phase Flow Group Meeting, Marchwood, UK (June).

Vierow, K.M., Schrock, V.E., 1992. Condensation in a natural circulation loop with non-condensable gas present. In: Japan-U.S. Seminar on Two-phase Flow Dynamics, Berkeley, CA, USA.

Vigni, P., D'Auria, F., 1979. Unsteady two-phase jet forces (in Italian). Ingegneria Nucleare 10, 23–32.

Vigni, P., D'Auria, F., Rosa, U., 1978. Blowdown Experiments from a Pressure Vessel with Internal Structures (in Italian), University of Pisa Report, IIN—RL 317(78), Pisa, Italy.

Vijayan, P.K., 2002. Experimental observation on the general trends of the steady state and stability behavior of single-phase natural circulation loops. Nuclear Engineering and Design 215, 139–152.

Vijayan, P.K., 2014. Local phenomena associated with natural circulation. In: IAEA Training Course on Natural Circulation Phenomena and Passive Safety Systems in Advanced Water Cooled Reactors SANESS-GCNEP, Mumbai, India, 29 September–3 October, Paper ID L04/05.

Vijayan, P.K., Nayak, A.K., 2012. Introduction to instabilities in natural circulation systems. In: Joint ICTP-IAEA Course on Natural Circulation Phenomena and Passive Safety Systems in Advanced Water Cooled Reactors. Trieste, Italy, 3–7 December, Paper 2349-15.

Visentini, R., Colin, C., Ruyer, P., 2014. Experimental investigation of heat transfer in transient boiling. Experimental Thermal and Fluid Science 55.

Vivaldi, D., 2022. Assessment of the 3D pressure drop tensor formulation in porous media simulations for tube bundle configurations. In: 19th International Topical Meeting on Nuclear Reactor Thermal Hydraulics (NURETH-19), March 6–11 (Via Web), Brussels, Belgium.

Von Neumann, J., Richtmyer, R.D., 1950. A method for the numerical calculation of hydrodynamic shocks. Journal of Physics 21, 232–237.

Voronov, V.N., Larin, B.M., Senina, V.A., 2006. Chemical and Technological Modes of NPP with VVER. MEI, Moscow, Russia (in Russian).

Vyskocil, L., Macek, J., 2015. CFD simulation of the departure from nucleate boiling. In: 16th International Topical Meeting on Nuclear Reactor Thermal Hydraulics (NURETH-16), August 30–September 4, Chicago, IL, USA.

Wadekar, V., 1990. Flow boiling—a simple correlation for convective heat transfer coefficient. In: Proceedings 9th International Heat Transfer Conference, Jerusalem, Israel, vol. 2, pp. 87–91.

Wadekar, V., 1993. Onset of boiling in vertical up flow. In: ASME Symposium Series, vol. 89. Heat Transfer—Atlanta, GA, USA, pp. 293–299.

Wagner, W., Kruse, A., 1997. The Industrial Standard IAPWS-IF97 for the Thermodynamic Properties and Supplementary Equations for Other Properties—Properties of Water and Steam. Springer-Verlag, Heidelberg, Germany.

Wald, A., 1943. An extension of Wilk's method for setting tolerance limits. Annals of Mathematical Statistics 14, 45–55.

Walker, J.D., 1977. The Flying Circus of Physics. Journal Wiley Publishers, New York, NY, USA.

Walker, J.D., 2022. Boiling and the Leidenfrost. https://www.reed.edu. (Accessed 15 October 2022).

Wallin, S., 2000. Engineering Turbulence Modelling for CFD with a Focus on Explicit Algebraic Reynolds Stress Models. Technical Reports from Royal Institute of Technology, Department of Mechanics, Stockholm, Sweden.

Wallin, S., Johansson, A.V., 2000. An explicit algebraic Reynolds stress model for incompressible and compressible turbulent flows. Journal of Fluid Mechanics 403, 89–132.

Wallis, G.B., 1969. One Dimensional Two-phase Flow. McGraw-Hill, New York, NY, USA.

Wallis, G.B., 1970. One Dimensional Two-phase Flow. McGraw-Hill Book Company, New York, NY, USA.

Wallis, G.B., 1972. One-Dimensional Two-phase Flows. McGraw-Hill, New York, NY, USA.

Waltar, A.E., Reynolds, A.B., 1981. Fast Breeder Reactors. Pergamon Press, New York, NY, USA.

Walunj, A., Sathyabhama, A., 2018. Transient CHF enhancement in high pressure pool boiling on rough surface. Chemical Engineering and Processing: Process Intensification 127, 145–158.

Wan, P.T., 2007. A derivation of Fractional Scaling Analyses (FSA) method for time-dependent complex problems. In: Proceedings of the 12th International Topical Meeting on Nuclear Reactor Thermal Hydraulics (NURETH-12), Pittsburgh, Pa, USA, Sept. 30–Oct. 4.

Wang, X., 2010. Simulations of Two-phase Flows Using Interfacial Area Transport Equations. PHD Thesis dissertation. Ohio State University, pp. 1–225.

Wang, X., Cheng, X., 2018. Analysis of inter-channel sweeping flow in wire wrapped 19-rod bundle. Nuclear Engineering and Design 333, 115–121.

Wang, M., Moin, P., 2002. Dynamic wall modeling for LES of complex turbulent flows. Physics of Fluids 14, 2043–2051.

Wang, S., Yang, B.-W., 2019. Effects of ocean motions on density wave oscillations under natural circulation. Annals of Nuclear Energy 131, 185–195.

Wang, F., Marashdeh, Q., Fan, L.S., Williams, R.A., 2009. Electrical capacitance, electrical resistance, and positron emission tomography techniques and their applications in multiphase flow systems. Advances in Chemical Engineering 37, 179–222.

Wang, X., Sun, X., Doup, B., Zhao, H., 2011. Dynamic modeling strategy for flow regime transition in gas-liquid two-phase flows. In: The 14th International Topical Meeting on Nuclear Reactor Thermalhydraulics, NURETH-14, Toronto, Ontario, Canada, September 25–30.

Wang, S., Yang, B.-W., Liu, A., Liu, X., 2016. RELAP5 investigation on subchannel flow instability. Kerntechnik 81 (3), 337–345.

Wang, S., Yang, B.-W., Zhou, Z., Long, J., 2018a. Flow characteristic of single-phase natural circulation under ocean motions. Journal of Nuclear Science and Technology 55 (12), 1366–1380.

Wang, S., Yang, B.-W., Zhou, Z., Long, J., 2018b. Review on thermal-hydraulic characteristics of nuclear reactors under ocean conditions. Nuclear Science and Engineering 193 (1–2), 14–32.

Wang, H., Leung, L.K.H., Wang, W., Bi, Q., 2018c. A review on recent heat transfer studies to supercritical pressure water in channels. Applied Thermal Engineering 142, 573–596.

Wang, D., Niu, F., Liang, R., Zhuo, W., Wang, S., 2020. Assessment of head loss through LOCA-generated debris deposited in PWR fuel assemblies. Annals of Nuclear Energy 137 (107037), 1–11.

Wang, J.R., Huang, P.C., Ferng, Y.M., 2020. Experimentally investigating phase separation and pressure drop as bubbly flow in a vertical branching T-junction. Annals of Nuclear Energy 149 (107772), 1–9.

Wang, J., Chen, L., Cai, Q., Qi, Chen, H., Wang, C., 2021. Direct contact condensation of steam jet in subcooled water: a review. Nuclear Engineering and Design 377 (111142), 1–28.

Wang, N., Bhushan, S., Cho, H., Li, L., 2022. Modeling of vapor-liquid interactions in condensing ejectors. Applied Thermal Engineering 206 (118111), 1–11.

Wang, X., Grishchenko, D., Kudinov, P., 2022. Modeling of large pool response to steam injection though spargers using a 'unit cell' model. In: 19th International Topical Meeting on Nuclear Reactor Thermal Hydraulics (NURETH-19), March 6–11, Brussels, Belgium, Paper 35839.

An, W., Cao, X., Chen, G., Chen, P., Chen, Z., Cong, T., Ding, M., Dong, X., Fan, G., Fang, H., Gu, H., Huang, W., Li, W., Liu, D., Liu, X., Lv, X., Meng, Z., Miao, Y., Peng, C., Qiu, S., Revankar, S.T., She, L., Shi, K., Song, J., Su, G., Sun, B., Sun, H., Sun, Q., Sun, Z., Tang, H.L.C., Tian, W., Tian, Y., Tian, Z., Wang, C., Wang, J., Wang, J., Wang, L., Wen, J., Wu, Y., Yang, D., Yin, S., Zhang, R., Zhang, X., Zhang, Y., Zhang, Z., Zhou, J., Zhou, Y., 2018. Nuclear Thermal-hydraulic and Two-phase Flow. In: Wang, J., Shi, K., Meng, Z., Revankar, S.T. (Eds.). Frontiers Media, Lausanne (CH), pp. 1–120.

Wantland, J.L., Fontana, M.H., Gnadt, P.A., Kress, T.S., McPherson, R.E., Parsly, L.F., 1974. Duct-wall temperatures due to a heated-zone edge blockage in a sodium-cooled 19-rod bundle. In: Oak Ridge National Lab., CONF-741017-15 (ANS Meeting, Washington, DC), Oak Ridge, TN, USA, pp. 1–4.

Wantland, J.L., Fontana, M.H., Gnadt, P.A., Kress, T.S., McPherson, R.E., Parsly, L.F., 1974a. Thermal effects of half-size edge gaps in sodium-cooled 19-rod bundles. In: Oak Ridge National Laboratory, CONF-741017-17 (ANS Meeting, Washington, DC), Oak Ridge, TN, USA, pp. 1–4.

Watanabe, T., 2007. Simulation of temperature stratification during ECCS water injection using FLUENT: preparatory analysis for OECD/NEA ROSA project. In: 12th International Topical Meeting on Nuclear Reactor Thermal Hydraulics (NURETH-12), September 30–October 4, Pittsburgh, PA, USA.

Watanabe, K., Dooley, R.B., 1996. IAPWS Release on the IAPWS Formulation 1995 for the Thermodynamic Properties of Ordinary Water Substance for General and Scientific Use. Fredericia, Denmark, September.

Watanabe, T., Takata, T., Yamaguchi, A., 2015. Numerical study on modeling of liquid film flow under countercurrent flow limitation in volume of fluid method. In: 16th International Topical Meeting on Nuclear Reactor Thermal Hydraulics (NURETH-16), August 30–September 4, Chicago, IL, USA.

Webb, R., 1993. Principles of Enhanced Heat Transfer. Wiley & Son, New York, NY, USA.

Wedin, R., Davoust, L., Cartellier, A., Dahlkild, A., 2000. A mono-modal fiber-optics velocimeter for electrochemically generated bubbles. In: Proceedings 10th International Symposium Applications of Laser Techniques to Fluid Mechanics, Lisbon, Portugal, pp. 10–13.

Weimer, J.C., Faeth, G.M., Olson, D.R., 1973. Penetration of vapor jets submerged in subcooled liquids. AIChE Journal 19 (3), 552–558.

Weinberg, S., 1952. Heat transfer to low pressure sprays of water in a steam atmosphere. Proceedings of the Institution of Mechanical Engineers 06 (1B), 240–253.

Weinberg, G.M., Weinberg, D., 1988. General Principles of Systems Design. Dorset House Publishing, New York, NY, USA. ISBN-13: 978-0932633071, ISBN-10: 0932633072.

Weisman, J., 1992. The current status of theoretically based approaches to the prediction of the critical heat flux in flow boiling. Journal of Nuclear Technology 99, 1–21.

Weisman, J., Bowring, R.W., 1975. Methods for detailed thermal and hydraulic analysis of water-cooled reactors. Nuclear Science and Engineering 57, 255–276.

Weisman, J., Pei, B.S., 1983. Prediction of critical heat flux in flow boiling at low quality conditions. International Journal of Heat Mass Transfer 26, 1463–1478.

Weiss, F.-P., 1989. UPTF experiments: principal full-scale test results for enhanced knowledge of large break LOCA scenarios in PWR's. In: Proceedings of the 4th International Topical Meeting on Nuclear Reactor Thermal-Hydraulics (NURETH-4), Karlsruhe, Germany.

Weiss, F.-P., Sawitzki, M., Winkler, F.J., 1986. UPTF, a full-scale PWR loss-of-coolant accident experiment program. Atomkernenergie Kerntechnik 49 (1/2), 61–67.

Weiss, F.-P., Watzinger, H., Hertlein, R., 1990. UPTF experiment: a synopsis of full-scale test results. Nuclear Engineering and Design 121, 219–234.

Welander, P., 1967. On the oscillatory instability of a differentially heated fluid loop. Journal of Fluid Mechanics 29 (1), 17–30.

Wells, D., 2015. Computational Fluid Dynamics Benchmark of High Fidelity Rod Bundle Experiments. Industry Round Robin Phase 2—Rod Bundle with Mixing Vane Grids. Electric Power Research Institute, Final Report 3002005401, Palo Alto, California, USA, pp. 1–158.

Welter, K.B., Bajorek, S.M., Reyes Jr., J.N., Woods, B.G., Groome, J., Hopson, J., Young, E., DeNoma, J., Abel, K., 2005. APEX-AP1000 Confirmatory Testing to Support AP1000 Design Certification, U.S. Nuclear Regulatory Commission, NUREG-1826, Washington, DC, USA (USNRC, 2005a, in Current List).

Werner, H., Wengle, H., 1991. Large eddy simulation of turbulent flow over and round a tube in a plate channel. In: 8th Symposium of Turbulent Shear Flow, Munich, Germany.

Wert, K.L., 1995. A rationally-based correlation of mean fragment size for drop secondary breakup. International Journal of Multiphase Flow 21, 1063–1071.

Westinghouse, 2014. Gothic Code, NS-ES-0002, ©2014. Westinghouse Electric Company LLC, Pittsburg, PA, USA.

Wheeler, C.L., Stewart, C.W., Cena, R., Rowe, D.S., Sutey, A.M., 1976. COBRA-IV-I: An Interim Version of COBRA for thermal-hydraulic Analysis of Rod Bundle Nuclear Fuel Elements and Cores. Battelle Pacific Northwest Labs., Richland, WA, USA https://doi.org/10.2172/7359576.

Wickett, A.J., Yadigaroglu, G., 1994. Report of a CSNI workshop on uncertainty analysis methods. In: Proceedings Of the Workshop Held in London on March 1–4, NEA/CSNI/R(94)20/Part L, Paris, France, pp. 1–247.

Wickett, T., Sweet, D., Neil, A., D'Auria, F., Galassi, G., Belsito, S., Ingegneri, M., Gatta, P., Glaeser, H., Skorek, T., Hofer, E., Kloos, M., Chojnacki, E., Ounsy, M., Lage Perez, C., Sanchez-Sanchis, J.I., 1998. Report of the Uncertainty Methods Study for Advanced Best Estimate thermal Hydraulic Code Applications, Vol. 1 (Comparison) and Volume 2 (Report by the Participating Institutions). NEA/CSNI/R(97)35, Paris, France.

Wieckhorst, O., Gabriel, H., Harne, R., Anghelescu, M., Martinie, O., 2015. ORFEO™—a CHF correlation form applied to GAIA, AREVA's advanced PWR fuel assembly design. In: 16th International Topical Meeting on Nuclear Reactor Thermal Hydraulics (NURETH-16), August 30-September 4, Chicago, IL, USA.

Wikipedia, 2015. List of 20th-Century Earthquakes & List of 21st-Century Earthquakes.

Wilcox, D.C., 1998. Turbulence Modeling for CFD. DCW Industries, La Canada, CA, USA.

Wilcox, D.C., 2007. Basic Fluid Mechanics. DCW Industries, San Diego, CA, USA.

Wilks, S.S., 1941. Determination of sample sizes for setting tolerance limits. Annals of Mathematical Statistics 12, 91–96.

Wilks, S.S., 1942. Statistical prediction with special reference to the problem of tolerance limits. Annals of Mathematical Statistics 13, 400–409.

Wilson, G.E., Boyack, B.E., 1997. The role of the PIRT process in identifying code improvements and executing code development. In: USNRC and OECD/NEA/CSNI Workshop on Transient Thermal-Hydraulic and Neutronic Codes Requirements, Annapolis, MD, USA, 5−8 November (1996).

Wilson, G.E., Boyak, B.E., 1998. The role of the PIRT process in experiments, code development and code applications associated with reactor safety analysis. Journal of Nuclear Engineering and Design 186, 23−37.

Wilson, D.G., Chan, T.-C., Goldfinch, A.L., Zegley, R.D., 1977. Analytical Models and Experimental Studies of Centrifugal Pump Performance in Two-phase Flow. MIT Research Project, EPRI NP-170, Cambridge, MA, USA, pp. 1−183.

Wilson, G.E., Boyack, B.E., Catton, I., Duffey, R.B., Griffith, P., Katsma, K.R., Lellouche, G.S., Levy, S., Rohatgi, U.S., Wulff, W., Zuber, N., 1990. Quantifying reactor safety margins part 2: characterization of important contributors to uncertainty. Nuclear Engineering and Design 119 (1), 17−31.

WNA, 2014. Outline of History of Nuclear Energy. www.world-nuclear.org/info/Current-and-Future-Generation/.

Wolf, L., 1982. Experimental results of coupled fluid-structure interactions during blowdown of the HDR-vessel and comparisons with pre- and post-test predictions. Nuclear Engineering and Design 70, 269−308.

Wolfert, K., 2008. Scaling of thermal-hydraulic phenomena and system code assessment. In: Seminar on the Transfer of Knowledge Gained through CSNI Activities in the Field of Thermal-Hydraulics, THICKET, Pisa, Italy, 5−9 May.

Wolfert, K., Frisch, W., 1983. Proposal for the Formulation of a Validation Matrix. OECD/NEA-CSNI-SINDOC (83) 117, Paris, France.

Wong, Y.L., Groeneveld, D.C., Cheng, S.C., 1990. CHF predictions in horizontal tubes. International Journal of Multiphase Flow 16, 123−138.

Woo, K., Ishii, M., Kuran, S., Yoo, Y.J., Yamamoto, Y., Kawamura, S., 2007. Experimental study on stability of natural circulation boiling water reactor during start-up transient. In: 12th International Topical Meeting on Nuclear Reactor Thermal Hydraulics (NURETH-12), September 30−October 4, Pittsburgh, PA, USA.

Woodcock, J., Spencer, D.R., 1999. Overview of observations of water coverage on the AP-600 tests for passive containment cooling. In: Proceedings 7th International Conference In Nuclear Engineering ICONE-7, Paper 7366, Tokyo, Japan, April 19−23.

Wu, T., Yang, B.-W., 2019. Subchannel analysis of mixed convection in rod bundle. In: 18th International Topical Meeting on Nuclear Reactor Thermal Hydraulics (NURETH-18), August 18−23, Portland, OR, USA.

Wu, Q., Kim, S., Ishii, M., Beus, S.G., 1998. One-group interfacial area transport in vertical bubbly flow. International Journal of Heat Mass Transfer 41, 1103−1112.

Wu, X., Fordham, E.J., Mullins, O.C., Ramos, R.T., 2000. Single point Optical Probe for Measuring Three-phase Characteristics of Fluid Flow in a Hydrocarbon Well. USA Patent 6023340.

Wu, X.Z., Yan, J.J., Shao, S.F., Cao, Y., Liu, J.P., 2007. Experimental study on the condensation of supersonic steam jet submerged in quiescent subcooled water: steam plume shape and heat transfer. International Journal of Multiphase Flow 33, 1296−1307.

Wu, X.Z., Yan, J.J., Pan, D.D., Liu, G.Y., Li, W.J., 2009. Condensation regime diagram for supersonic/sonic steam jet in subcooled water. Nuclear Engineering and Design 239 (12), 3142−3150.

Wu, D., Yu, H., Yu (Junchong), J., Yu (Jiyang), J., 2012. Reflooding characteristics investigation and predicted models improvement for the pressurized water reactor with the tight lattice core design. Nuclear Engineering and Design 249, 335−347.

Wu, Q., Lee, D.-Y., Mullin, E., Wuth, B., Bristol, B., Corradini, M.L., Yon, D., 2015. Integral Reactor Containment Condensation Model and Experimental Validation—Reactor Concepts Research Development and Demonstration, pp. 1−225. US DOE Project, NEUP 12-3630, Washington, DC, USA.

Wu, P., Shan, J., Xiang, X., Zhang (Bo), B., Gou, J., Zhang (Bin), B., 2016. The development and application of a sub-channel code in ocean environment. Annals of Nuclear Energy 95, 12−22.

Wu, Z., Lu, C., Morgan, S., Bilbao y Leon, S., Bucknor, M., 2020. A status review on the thermal stratification modeling methods for Sodium-cooled Fast Reactors. Progress in Nuclear Energy 125 (103369), 1−15.

Wulff, W., 1996. Scaling of thermohydraulic systems. Journal of Nuclear Engineering and Design 163, 359−395.

Wulff, W., Rohatgi, U.S., 1998. System Scaling for the Westinghouse AP600 Pressurized Water Reactor and Related Test Facilities; Analyses and Results. USNRC NUREG/CR-5541 (BNL-NUREG-52550), Washington DC, USA, pp. 1−232.

Wulff, W., Cheng, H.S., Mallen, A.N., 1990. Causes of Instability at La Salle and Consequences Form Postulated Scram Failure. BNL-NUREG 45273, Brookhaven, NY, USA.

Wulff, W., Cheng, H.S., Mallen, A.N., Rohatgi, U.S., 1992. BWR Stability Analysis with the BNL Engineering Plant Analyzer. USNRC NUREG/CR-5816 (also BNL-NUREG-52312), Washington, DC, USA.

Wulff, W., Cheng, H.S., Mallen, A.N., Rohatgi, U.S., 1992a. BWR Stability Analysis at Brookhaven Engineering Laboratory, USNRC/NUREG 46531, Brookhaven NY, USA, pp. 1−20.

Wulff, W., Zuber, N., Rohatgi, U.S., Catton, I., 2005. Application of fractional scaling analysis (FSA) to loss of coolant accidents (LOCA)—Part 2: system level scaling for system depressurization. In: International Topical Meeting on Nuclear Reactor Thermal Hydraulics (NURETH-11), Avignon, France.

Wulff, W., Zuber, N., Rohatgi, U.S., Catton, I., 2009. Application of fractional scaling analysis to loss of coolant accidents, system level scaling for system depressurization. Journal of Fluids Engineering 131 (8).

Wyatt, D.G., 1986. Electromagnetic flowmeter sensitivity with two phase flow. International Journal of Multiphase Flow 12 (6), 1009−1017.

Wysocki, A., Manera, A., Downar, T., March-Leuba, J., 2015. A physical mechanism for rotating lines of symmetry in BWR out-of-phase limit cycle oscillations. In: 16th International Topical Meeting on Nuclear Reactor Thermal Hydraulics (NURETH-16), August 30−September 4, Chicago, IL, USA.

Xiao, H., Cinnella, P., 2019. Quantification of model uncertainty in RANS simulations: a review. Progress in Aerospace Sciences 208, 1−31.

Xiao, J., Travis, J.R., 2013. How critical is turbulence modeling in gas distribution simulations of large-scale complex nuclear reactor containment? Annals of Nuclear Energy. 56, 227−242.

Xiao, X., Edwards, J.R., Hassan, H.A., Baurle, R.A., 2003. Inflow boundary conditions for LES/RANS simulations with applications to shock wave boundary layer interactions. In: AIAA Paper 2003−0079, Reno, NV, USA.

Xiao, X., Edwards, J.R., Hassan, H.A., 2004. Blending functions in hybrid large-eddy/Reynolds-averaged Navier-Stokes simulations. AIAA Journal 42 (12), 2508−2515.

Xiao, J., Travis, J., Royl, P., Necker, G., Svishchev, A., Jordan, T., 2016. GASFLOW-MPI: A Scalable Computational Fluid Dynamics Code for Gases, Aerosols and Combustion Volume 2: Users' Manual (Revision 1.0). KIT Scientific Report 7711, Karlsruhe, Germany.

Xu, Q., Guo, L., 2016. Direct contact condensation of steam jet in crossflow of water in a vertical pipe. Part I: experimental investigation on condensation regime diagram and jet penetration length. International Journal of Heat Mass Transfer 94, 528−538.

Xu, J., Ma, H., 2009. Applications of URANS on predicting unsteady turbulent separated flows. Acta Mechanica Solida Sinica 25 (3), 319−324.

Xu, Q., Guo, L., Zou, S., Chen, J., Zhang, X., 2013. Experimental study on direct contact condensation of stable steam jet in water flow in a vertical pipe. International Journal of Heat Mass Transfer 66, 808−817.

Xu, C., Liu, X., Liu, Y., Pan, Z., An, J., 2018. Integral test on performance of passive safety system of CAP-1400 in SBO scenario. In: 12th International Topical Meeting on Nuclear Reactor Thermal-Hydraulics, Operation and Safety (NUTHOS-12), October 14−18, Qingdao, China.

Xu, H., Scobel, J.H., Wright, R.F., 2022. Thermal stratification in the CMT balance line during borated water makeup injection. In: 19th International Topical Meeting on Nuclear Reactor Thermal Hydraulics (NURETH-19), March 6−11 (Via Web), Brussels, Belgium.

Yadav, M.K., Khandekar, S., Sharma, P.K., 2016. An integrated approach to steam condensation studies inside reactor containments: a review. Nuclear Engineering and Design 300, 181−209.

Yadigaroglu, G., 1975. Heat and Mass Transfer between Droplets and the Atmosphere—State of the Art. Department of Nuclear Engineering, University of California, Berkeley, CA, USA.

Yadigaroglu, G., 2014. CMFD and the critical-heat-flux grand challenge in nuclear thermal−hydraulics—a letter to the Editor of this special issue. International Journal of Multiphase Flow 67, 3−12.

Yadigaroglu, G., Andreani, M., 1989. Two-fluid modeling of thermal-hydraulic phenomena for best-estimate LWR safety analysis. In: Proceedings NURETH-4, Karlsruhe, Germany.

Yadigaroglu, G., Chan, K.C., 1979. Analysis of flow instabilities. In: Proceedings of the Japan−US Seminar on Two-phase Flow Dynamics July 31−August 3, Kobe, Japan.

Yadigaroglu, G., Lahey Jr., R.T., 1976. On the various forms of the conservation equations in two phase flows. International Journal of Multiphase Flow 2, 477−494.

Yadigaroglu, G., Lakehal, D., 2015. Nuclear safety and thermal-hydraulics: personal thoughts and some recent progress. In: NURETH-16 Conf., Chicago, IL, USA, August 30−September 4.

Yadigaroglu, G., Andreani, M., Aksan, S.N., Analytis, G.T., Lubbesmeyer, D., Olek, S., 1990. Modelling of Thermohydraulic Emergency Core Cooling Phenomena. Labor für Thermohydraulik, Paul Scherrer Institut, PSI Report No. 27, Wurenlingen, Switzerland.

Yadigaroglu, G., Nelson, R.A., Teschendorff, V., Murao, Y., Kelly, J., Bestion, D., 1993. Modeling of reflooding. Nuclear Engineering and Design 145, 1−35.

Yadigaroglu, G., Andreani, M., Dreier, J., Coddington, P., 2003. Trends and needs in experimentation and numerical simulation for LWR safety. Nuclear Engineering and Design 221, 205−223.

Yagnik, S., Wells, D., 2014. Computational Fluid Dynamics Benchmark of High Fidelity Rod Bundle Experiments. Industry Round Robin Phase 1—Rod Bundle with Simple Support Grids. Electric Power Research Institute, EPRI 3002000504, Palo Alto, Ca, USA.

Yagov, V.V., Puzin, V.F., Sukomel, L.A., 1998. Heat transfer during developed boiling of halocarbons and high velocity of forced motion. Teploenergetika 3, 11−19 (in Russian).

Yamada, E., Ota, H., 1980. Effective thermal conductivity of dispersed materials. Wärme-und Stoffübertragung 13, 27–37.

Yamada, K., Tajima, S., Tsubaki, M., Soneda, H., 2003. ABWR design and its evolution—primary system design of ABWR and ABWR-II. In: International Conference GENEA4/ANP2003, Kyoto, Japan, September 15–19, Paper 1161, pp. 1–8.

Yamaguchi, A., Mizokami, S., Kudo, Y., Hotta, A., 2009. Uncertainty and conservatism in safety evaluations based on a BEPU approach. In: NURETH-13 Conference, Kanazawa, Japan, September 27–October 2.

Yamamoto, Y., Akiba, M., Morooka, S., Shirakawa, K., Abe, N., 2006. Thermal hydraulic performance of tight lattice bundle. JSME International Journal, Series B 49 (2).

Yamanouchi, A., 1968. Effect of core spray cooling in transient state after loss of coolant accident. Nuclear Science and Technology 5 (11), 547–558.

Yan, B.H., 2017. Review of the nuclear reactor thermal hydraulic research in ocean motions. Nuclear Engineering and Design 313, 370–385.

Yan, J., Bolger, F., 2009. Acoustic impact on steam dryer induced by main steam line break in boiling water reactor. Nuclear Engineering and Design 239.

Yan, Y., Brown, E., Marcum, W., 2022. Experimental study of the transient boiling phenomenon on a vertically placed ribbon. Nuclear Engineering and Design 390 (111705), 1–13.

Yang, B.-W., 2013. CFD to rod bundle CHF and some issues concerning large-scale thermal hydraulic experiments and data analysis. In: International Seminar on Sub-channel Analysis, CFD Model and Verification, CHF experiment and Benchmarking, ISACC-2013, August, Xi'an, China.

Yang, B.-W., 2017. CHF mechanisms in rod bundle subchannel systems. In: 17th International Topical Meeting on Nuclear Reactor Thermal Hydraulics (NURETH-17), September 3–8, Xi'an, China.

Yang, S.K., Chung, M.K., 1998. Turbulent flow through spacer grids in rod bundles. Journal of Fluids Engineering 120, 786–791.

Yang, B., Dougherty, T.J., 1997. CHF and flow instability in rod bundles. In: 4th International Sem. On Subchannel Analysis (ISSCA-4), Tokyo, Japan.

Yang, C., Nakayama, A., 2010. A synthesis of tortuosity and dispersion in effective thermal conductivity of porous media. International Journal of Heat Mass Transfer 53, 3222–3230.

Yang, X., Schlegel, J.P., Liu, Y., Paranjape, S., Hibiki, T., Ishii, M., 2012. Measurement and modeling of two-phase flow parameters in scaled 8 x 8 BWR rod bundle. International Journal of Heat Fluid Flow 34.

Yang, J., Wang, W.W., Qiu, S.Z., Tian, W.X., Su, G.H., Wu, Y.W., 2012a. Simulation and analysis on 10-in. cold leg small break LOCA for AP1000. Annals of Nuclear Energy 46.

Yang, J., Choi, S.-W., Lim, J., Lee, D.-Y., Rassame, S., Hibiki, T., Ishii, M., 2013. Counterpart experimental study of ISP-42 PANDA tests on PUMA facility. Journal of Nuclear Engineering and Design 258.

Yang, X., Schlegel, J.P., Liu, Y., Paranjape, S., Hibiki, T., Ishii, M., 2013. Experimental study of interfacial area transport in air-water two phase flow in a scaled 8 X 8 BWR rod bundle. International Journal of Multiphase Flow 50, 16–32.

Yang, B.-W., Hassan, Y.A., Shan, J., Zhang, B., Gou, J., Cao, L., 2014. Subchannel Analysis, CFD modeling and verification, CHF experiments and benchmarking (ISACC). Editorial of Journal Special issue devoted to ISACC-2013 Seminar held in Xi'an. Science and Technology of Nuclear Installations 209747, 1–2 volume 2014.

Yang, B.-W., Shan, J., Gou, J., Zhang, H., Liu, A., Mao, H., 2014a. Uniform versus non uniform axial power distribution in rod bundle CHF experiments. Science and Technology of Nuclear Installations 462460, 1–9 volume 2014.

Yang, B.-W., Zhang, H., Zhang, B., 2015. Heat loss evaluation in large scale rod bundle CHF experiments. In: Proceedings 16th International Topical Meeting on Nuclear Reactor Thermal Hydraulics (NURETH-16). ANS Publisher. American Nuclear Society, Chicago, Illinois, USA.

Yang, L.X., Zhou, M.J., Zheng, Y., 2016. A comparison of the CFD simulation results in 5 x 5 sub-channel with mixing grids using different turbulence models. Kerntechnik 81 (3), 321–330.

Yang, B.-W., Hassan, Y.A., Ninokata, H., Shan, J., Zhang, Y., Zhang, B., 2016. Subchannel Analysis, CFD modeling and verification, CHF experiments and benchmarking (ISACC)—challenges in reactor core thermal-hydraulics: sub-channel analysis, CFD modeling and rod bundle CHF. Editorial of Journal Special Issue Devoted to ISACC-2015 Seminar Held in Shenzhen. Kerntechnik 81 (3), 213–214.

Yang, B.-W., Zhang, H., Han, B., Zha, Y.D., Shan, J.Q., 2016a. CFD analysis on mixing effects of spacer grids with different dimples and sizes for advanced fuel assemblies. Kerntechnik 81 (3), 221–232.

Yang, B.-W., Han, B., Zhang, H., Shan, J., 2016b. CFD analysis on key components of mixing vane grids vane angle/dimple shape (in Chinese). Nuclear Power Engineering 37 (2), 165–170.

Yang, B.-W., Han, B., Liu, A., Wang, S., 2019. Recent challenges in subchannel thermal-hydraulics-CFD modeling, subchannel analysis, CHF experiments, and CHF prediction. Nuclear Engineering and Design 354 (110236), 1–19.

Yang, M., Yang, -M.), Deng, Z.A., Cao, Z.B., Li, J., Wen, Z.Y., 2019. Study on distribution characteristics of gas-liquid two phase flow in Pitot tube sampler. In: IOP Conf. Series: Earth and Environmental Science, vol. 252, p. 032113.

Yang, S.R., Seo, J., Hassan, Y.A., 2019. Thermal hydraulic characteristics of unstable bubbling of direct contact condensation of steam in subcooled water. International Journal of of Heat and Mass Transfer 138, 580–596.

Yang, B.-W., Ninokata, H., Long, J., Liu, A., Han, B., 2021. Subchannel analysis—current practice and development for the future. Nuclear Engineering and Design 385 (111477), 1–28.

Yang, B.-W., Anglart, H., Han, B., Liu, A., 2021a. Progress in rod bundle CHF in the past 40 years. Nuclear Engineering and Design 376 (111076), 1–37.

Yang, B.-W., Bajorek, S.M., Jin, Y., Cheung, F.-B., 2021b. Progress in reflood thermal hydraulics studies in the past 40 years. Nuclear Engineering and Design 376 (111073), 1–24.

Yao, W., Morel, C., 2004. Volumetric interfacial area prediction in upward bubbly two-phase flow. International Journal of Heat Mass Transfer 47, 307–328.

Yao, J.C., Hochreiter, L.E., Leech, W.J., 1982. Heat transfer augmentation in rod bundles near grid spacers. Transactions of ASME Journal Heat Transfer 104 (1), 76–81.

Ybarrondo, L.J., Fabic, S., Griffith, P., McPherson, G.D., 1974. In: Examination of LOFT Scaling, Winter Annual Meeting of the ASME, Paper 74-WA/HT-53, New York, NY, USA.

Yeoh, G.H., Tu, J.Y., 2002. Improvements in reactor code RELAP5/MOD2 to predict distribution. In: Proceedings 12th International Heat Transfer Conf., Grenoble, France, vol. 4, pp. 621–626.

Yi, S.-J., Chung, M.-S., Toth, I., 2016. Pressure wave propagation phenomena in the pipe system of a nuclear power plant. Annals of Nuclear Energy 87, 157–161.

Yin, S.T., Groeneveld, D.C., Wakayama, M., 1991. The effect of radial flux distribution on critical heat flux in 37-rod bundles. In: Proceedings of the 12th Nat. Heat Transfer Conf., Minneapolis, Mn, USA, July 28–31, vol. 5, pp. 277–283.

Yin, S., Wang, N., Wang, H., 2020. Nucleation and flashing inception in flashing flows: a review and model comparison. International Journal of of Heat and Mass Transfer 146 (118898), 1−17.

Ying, S.H., Weisman, J., 1986. Prediction of critical heat flux in flow boiling at intermediate qualities. International Journal of Heat Mass Transfer 29, 1639−1648.

Ylijoki, J., Normann, S., Silde, A., 2008. Validation of APROS Version 5.08. Research Report, VTT-R-03692-08, Espoo, Finland.

Ylonen, A., 2008. Large Break Blowdown Test Facility Study. Lappeenranta University of Technology, Lappeenranta, Finland.

Ylonen, A., Bissels, W.-M., Prasser, H.-M., 2011. Single-phase cross-mixing measurements in a 4 × 4 rod bundle. Nuclear Engineering and Design 241, 2484−2493.

Yokobori, S., Nakamura, H., 2008. Optimized apportion between large-scale nuclear thermal-hydraulic tests and simulation. Japan Atomic Energy 27, 461−468.

Yonomoto, T., 1996. A Study of Entrainment at a Break and in the Core during SBLOCA in PWR. JAERI-RESEARCH-96-024. Japan Atomic Energy Research Institute, Tokai-Mura, Japan.

Yonomoto, T., Kondo, M., Kukita, Y., Scott Ghan, L., Schultz, R.R., 1997. Core makeup tank behavior observed during the ROSA-AP-600 experiments. J. Nuclear Technology 119 (2), 112−122.

Yoo, Y., Hwang, D., 1999. Development of a subchannel analysis code MATRA applicable to PWRs and ALWRs. Nuclear Engineering and Technology 31 (3), 314−327.

Yoo, J., Estrada-Perez, C.E., Hassan, Y.A., 2015. An accurate wall temperature measurement using infrared thermometry with enhanced two-phase flow visualization in a convective boiling system. International Journal of Thermal Science 90, 248−266.

Yoo, J.M., Cho, Y.J., Yon, H.Y., Jeong, J.J., 2019. Analysis of the effect of liquid droplet models on the reflood heat transfer using the CUPID code. Nuclear Engineering and Design 354, 110148.

Yoo, J.M., Yun, B.J., Yon, H.Y., Jeong, J.J., 2020. Modeling of the droplet entrainment rate in the post-dryout regime for the analysis of a reflood phase. Annals of Nuclear Energy 148, 107757.

Yoo, J.M., Yun, B.J., Yon, H.Y., Jeong, J.J., 2021. Improvement of the heat transfer enhancement model considering the droplet-wall heat transfer downstream of the flow blockage in the reflood phase. Nuclear Engineering and Design 380, 111277.

Yoshiura, R.K., Wu, Q., Wachs, D.M., Reyes Jr., J.N., Duenas, A.M., 2022. Application of dynamical system scaling to accelerated fuel testing. In: Proceedings Of the 19th International Topical Meeting on Nuclear Reactor Thermal Hydraulics, NURETH-19, Brussels, Belgium, March 6−11.

Young, A., 1990. Coriolis flowmeters for accurate measurement of liquid properties. Advances in Instrumentations and Control: ISA Proceedings 45 (4), 1891−1898.

Young, M.Y., Bajorek, S.M., Nissley, M.E., Hochreiter, L.E., 1998. Application of code scaling applicability and uncertainty methodology to the large break loss of coolant. Nuclear Engineering and Design 186, 39−52.

Yovanovich, M.M., 1986. Recent development in thermal contact, gap and joint conductance theories and experiment. In: Proceedings of the 8th International Heat Transfer Conference, August 17−22, San Francisco, CA, USA, vol. 1, pp. 35−45.

Yovanovich, M.M., 1998. Conduction and thermal contact resistances (conductances). In: Rohsenow, W.M., Hartnett, J.P., Cho, Y. (Eds.), Handbook of Heat Transfer. McGraw-Hill, New York, NY, USA. Chapt. 3).

Yovanovich, M.M., Marotta, M.E., 2003. Thermal spreading and contact resistances. In: Bejan, A., Kraus, D. (Eds.), Handbook of Heat Transfer. J. Wiley & Sons, New York, NY, USA (Chapter 4).

Yu, Y.Q., Gu, H.Y., Cheng, X., Yang, Y.H., 2012. CFD simulation on turbulent flow in tight lattice fuel assemblies. Atomic Energy Science and Technology 46 (4), 396–403.

Yu, X.-G., Park, H.S., Kim, Y.S., Kang, K.H., Cho, S., Choi, K.-Y., 2013. Systematic analysis of a station blackout scenario for APR1400 with test facility ATLAS and MARS code from scaling viewpoInternational. Nuclear Engineering and Design 259, 205–220.

Yu, A., Tang, Q., Zhou, D., Liu, J., 2020. Entropy production analysis in two-phase cavitation flows with thermodynamic cavitation model. Applied Thermal Engineering 171 (115099), 1–13.

Yu, H., Ju, H., Wang, M., Zhang (Jing), J., Qiu, S., Tian, W., Su, G.H., 2020. Study of boron diffusion models and dilution accidents in nuclear reactor: a comprehensive review. Annals of Nuclear Energy 148 (107659), 1–12.

Yuan, W., Sauer, J., Schnerr, G.H., 2001. Modeling and computation of unsteady cavitation flows in injection nozzles. Mécanique and Industries 2 (5), 383–394.

Yuan, Y., Lu, D., Zhang, Y., Ouyang, B., 2018. Experimental investigation on characteristics of direct contact condensation of steam jet under depressurization condition. In: 12th International Topical Meeting on Nuclear Reactor Thermal-Hydraulics, Operation and Safety (NUTHOS-12), October 14–18, Qingdao, China.

Yun, B.J., Cho, H.K., Euh, D.J., Song, C.-H., Park, G.C., 2004. Scaling for ECC bypass phenomena during LBLOCA reflood phase. Journal of Nuclear Engineering and Design 231, 315–325.

Yun, Y., Khvostov, G., Freixa, J., Zimmermann, M.A., 2010. Analysis of fuel rod behaviour in a EPRTM during the base irradiation and LOCA-transient utilizing the FALCON code. In: LWR Fuel Performance Meeting, Orlando, Fl, USA, September 25–29, pp. 345–353.

Yurko, J., Frepoli, C., Reyes Jr., J.N., 2015. Demonstration of test facility design optimization with the dynamical system scaling methodology. In: International Topical Meeting on Nuclear Reactor Thermal Hydraulics (NURETH-16), Chicago, IL, USA.

Zaepffel, D., Morel, C., Lhuillier, D., 2012. A multi-size model for boiling bubbly flows. Multiphase Science and Technology 24 (2), 105–179.

Zeng, Y.J., Hale, C.P., Hewitt, G.F., Walker, S.P., 2010. Flow and heat transfer in pressurized water reactor reflood. Multiphase Science and Technology 22 (4), 279–370.

Zeng, Q., Aydemir, N.U., Lien, F.-S., Xu, T., 2017. Extension of staggered-grid-based AUSM-family schemes for use in nuclear safety analysis codes. International Journal of Multiphase Flow 93, 17–32.

Zhai, L.S., Bian, P., Gao, Z.K., Jin, N.D., 2016. The measurement of local flow parameters for gas–liquid two-phase bubbly flows using a dual-sensor probe array. Chemical Engineering Sciences 144, 346–363.

Zhang, Y., 2014. Research on Pressure Drop Model for Spacer Grid and Sub-channel and Code Development. Key Laboratory on Nuclear Reactor Thermal Hydraulics Technology (Annual Meeting of CNNC), Chengdu, Sichuan, China.

Zhang, G.J., Ishii, M., 1995. Isokinetic sampling probe and image processing systems for droplet size measurement in two-phase flow. International Journal of Heat Mass Transfer 38 (11), 2019–2027.

Zhang, K.L., Sanchez-Espinoza, V.H., 2020. Coupling of TRACE and SuChanFlow based on the exterior communication interface. Progress in Nuclear Energy 119 (103040), 1–13.

Zhang, J., Seynhaeve, J.M., Giot, M., 1992. Experiments on the hydrodynamics of air-water countercurrent flow through vertical short multitube geometries. Experimental Thermal and Fluid Science 5, 499–514.

Zhang, H.L., Bachmann, C.R., Fasel, H.F., 2000. Application of a new methodology for simulations of complex turbulent flows. In: Fluids Conference and Exhibit, Fluid Dynamics and Co-located Conferences, Denver, CO, USA.

Zhang, J., Schneidesch, C., Segurado, J., 2011. Towards an industrial application of statistical uncertainty analysis methods to multi-physical modelling and safety analyses. In: Proceedings OECD/CSNI Workshop on Best Estimate Methods and Uncertainty Evaluations, Barcelona, Spain, NEA/CSNI/R(2013)8/Part 2, Paris, France.

Zhang, H., Guo, J., Lu, J., Niu, J., Li, F., 2017. A comparison of coupling algorithms for N/TH transient problems in HTR. In: International Conference On M&C Methods Applied to Nuclear Science and Engineering, Jeju, Korea, April 16–20.

Zhang, S., Shen, F., Cheng, X., 2018. Condensation of vapor with non-condensable outside a vertical tube based on HMTA. In: 12th International Topical Meeting on Nuclear Reactor Thermal-Hydraulics, Operation and Safety (NUTHOS-12), October 14–18, Qingdao, China.

Zhang, L., Kumbaro, A., Ghidaglia, J.-M., 2019. A conservative pressure based solver with collocated variables on unstructured grids for two-fluid flows with phase change. Journal of Computational Physics 390, 265–289.

Zhang, X., Zhang, K., Sanchez-Espinoza, V.H., Chen, H., 2020. Multi-scale coupling of CFD code and sub-channel code based on a generic coupling architecture. Annals of Nuclear Energy 141 (107353), 1–11.

Zhang, J., Schneidesch, C., 2021. Application of BEPU safety analysis methods to quantify margins in Belgian nuclear power plants. In: OECD/NEA/CSNI Specialists Meeting on Transient Thermal-hydraulics in Water Cooled Nuclear Reactors (SM-TH), December 13–17, CIEMAT, Madrid (Spain), P32.

Zhang, J., Jinming), Ding, W., Hampel, U., 2022. Mathematical model of the inertia-controlled microlayer profile in nucleate boiling. In: 19th International Topical Meeting on Nuclear Reactor Thermal Hydraulics (NURETH-19), March 6–11 (Via Web), Brussels, Belgium.

Zhao, Q., Hibiki, T., 2018. Review: condensation regime maps of steam submerged jet condensation. Progress in Nuclear Energy 107, 31–47.

Zhao, Q., Zhang, D.C., Zhu, X.B., Xu, D.Q., Lin, Z.Q., Lin, J.F., 1990. Industrial application of dropwise condensation. In: Proceedings of the 9th International Heat Transfer Conference, August 19–24, Jerusalem, Israel, vol. 4, pp. 391–394.

Zhao, X., Shirvan, K., Salko, R.K., Wysocki, A.J., 2017. Validation and benchmarking of CTF for single- and two-phase flow using VIPRE-01. In: 17th International Topical Meeting on Reactor Thermal Hydraulics, (NURETH-17), Xi'an, China.

Zhao, C.-Y., Qi, D., Ji, W.-T., Jin, P.-H., Tao, W.-Q., 2022. A comprehensive review on computational studies of falling film hydrodynamics and heat transfer on the horizontal tube and tube bundle. Applied Thermal Engineering 202 (117869), 1–46.

Zhu, M., Chang, H., Wang, H., Zhu, Q., Wang, J., 2021. Steam condensation separate effect tests in core make-up tank. Progress in Nuclear Energy 142 (104031), 1–11.

Zienkiewicz, O., 1984. Methode der Finiten Elemente. Carl Hanser Verlag, Leipzig, Germany.

Zigrang, D.J., Sylvester, N.D., 1985. A review of explicit friction factor equations. Transactions of ASME Journal Energy Resources Technology 107, 280–283.

Zimmermann, M.A., 2011. Safety margin assessment (SM2A): stimulation for further development of BEPU approaches. In: OECD/NEA/CSNI Workshop on Best Estimate Methods and Uncertainty Evaluations, Barcelona, Spain, November.

Zohuri, B., Fathi, N., 2015. Thermal-hydraulic Analysis of Nuclear Reactors. Springer International Publishing, Cham, Switzerland, pp. 1−646.

Zong, X., Liu, J.P., Yang, X.P., Yan, J.J., 2015. Experimental study on the direct contact condensation of steam jet in subcooled water flow in a rectangular mix chamber. International Journal of Heat Mass Transfer 80, 448−457.

Zou, L., Zhao, H., Zhang, H., 2015. On the analytical solutions and numerical verifications of the two-phase water faucet problem. In: 16th International Topical Meeting on Nuclear Reactor Thermal Hydraulics (NURETH-16), August 30−September 4, Chicago, IL, USA.

Zuber, N., 1958. Stability of boiling heat transfer. ASME Trans 80, 711−720.

Zuber, N., 1959. Hydrodynamic Aspects of Boiling Heat Transfer. Atomic Energy Commission Report AECU-4439, Washington, DC, USA.

Zuber, N., 1961. The dynamics of vapor bubbles in nonuniform temperature fields. International Journal of Heat Mass Transfer 2, 83−98.

Zuber, N., 1980. Problems in Modelling of Small Break LOCA. NUREG-0724, U.S. Nuclear Regulatory Commission, Washington, DC, USA.

Zuber, N., 1991. A Hierarchical, Two-Tiered Scaling Analysis, an Integrated Structure and Scaling Methodology for Severe Accident Technical Issue Resolution. USNRC, NUREG/CR-5809 R4, Washington, DC, USA.

Zuber, N., 2010. Scaling: from quanta to nuclear reactors. Nuclear Engineering and Design 240 (8), 1986−1996.

Zuber, N., Findlay, J.A., 1965. Average volumetric concentration in two-phase flow systems. Journal of Heat and Mass Transfer 87, 453−468.

Zuber, N., Wilson, G.E., Boyack, B.E., Catton, I., Duffey, R.B., Griffith, P., Katsma, K.R., Lellouche, G.S., Levy, S., Rohatgi, U.S., Wulff, W., 1990. Quantifying reactor safety margins, part 5: evaluation of scale-up capabilities of best estimate codes. Journal of Nuclear Engineering and Design 119, 97−107.

Zuber, N., Wilson, G.E., Ishii, M., Wulff, W., Boyack, B.E., Dukler, A.E., Griffith, P., Healzer, J.M., Henry, R.E., Lehner, J.R., Levy, S., Moody, F.J., Pilch, M., Sehgal, B.R., Spencer, B.W., Theofanous, T.G., Valente, J., 1998. An integrated structure and scaling methodology for severe accident technical issue resolution: development of methodology. Journal of Nuclear Engineering and Design 186.

Zuber, N., Wulff, W., Rohatgi, U.S., Catton, I., 2005. Application of fractional scaling analysis (FSA) to loss of coolant accidents (LOCA), Part 1: methodology development. In: International Topical Meeting on Nuclear Reactor Thermal Hydraulics (NURETH-11), Avignon, France.

Zuber, N., Rohatgi, U.S., Wulff, W., Catton, I., 2007. Application of fractional scaling analysis (FSA) to loss of coolant accidents (LOCA) methodology development. Nuclear Engineering and Design 237, 1593−1607.

Zugazagoitia, E., Queral, C., Fernández-Cosials, K., Gómez, J., Durán, L.F., Sánchez-Torrijos, J., Posada, J.M., 2020. Uncertainty and sensitivity analysis of a PWR LOCA sequence using parametric and non-parametric methods. Reliability Engineering and System Safety 193, 1−12 paper 106607.

Zvetkov, F.F., Grigor'ev, B.A., 2005. Heat and mass transfer. In: Izdatel'stvo MEI, Moscow, Russia, 2nd rev. (in Russian).

Zysin, L.V., Dorfman, E.A., 1986. Heat Transfer in the Channels during the Transition from Convection to Developed Boiling. Heat Transfer in NPP Power Equipment. Nauka, Leningrad, Russia, pp. 143−149 (in Russian).

Index

Note: 'Page numbers followed by *f* indicate figures and *t* indicate tables.'

J

K